QA 9.5 BAR

D1353271

WITHDRAWN
FROM STOCK
QMUL LIBRARY

Lambda Calculus with Types

This handbook with exercises reveals in formalisms, hitherto mainly used for hardware and software design and verification, unexpected mathematical beauty.

The lambda calculus forms a prototype universal programming language, which in its untyped version is related to Lisp, and was treated in the first author's classic The Lambda Calculus (1984). The formalism has since been extended with types and used in functional programming (Haskell, Clean) and proof assistants (Coq, Isabelle, HOL), used in designing and verifying IT products and mathematical proofs.

In this book, the authors focus on three classes of typing for lambda terms: simple types, recursive types and intersection types. It is in these three formalisms of terms and types that the unexpected mathematical beauty is revealed. The treatment is authoritative and comprehensive, complemented by an exhaustive bibliography, and numerous exercises are provided to deepen the readers' understanding and increase their confidence using types.

HENK BARENDREGT holds the chair on the Foundations of Mathematics and Computer Science at Radboud University, Nijmegen, The Netherlands.

WIL DEKKERS is an Associate Professor in the Institute of Information and Computing Sciences at Radboud University, Nijmegen, The Netherlands.

RICHARD STATMAN is a Professor of Mathematics at Carnegie Mellon University, Pittsburgh, USA.

PERSPECTIVES IN LOGIC

The *Perspectives in Logic* series publishes substantial, high-quality books whose central theme lies in any area or aspect of logic. Books that present new material not now available in book form are particularly welcome. The series ranges from introductory texts suitable for beginning graduate courses to specialized monographs at the frontiers of research. Each book offers an illuminating perspective for its intended audience.

The series has its origins in the old *Perspectives in Mathematical Logic* series edited by the Ω-Group for "Mathematische Logik" of the Heidelberger Akademie der Wissenschaften, whose beginnings date back to the 1960s. The Association for Symbolic Logic has assumed editorial responsibility for the series and changed its name to reflect its interest in books that span the full range of disciplines in which logic plays an important role.

Thomas Scanlon, Managing Editor
Department of Mathematics, University of California Berkeley

Editorial Board:

Michael Benedikt
Department of Computing Science, University of Oxford

Steven A. Cook
Computer Science Department, University of Toronto

Michael Glanzberg
Department of Philosophy, University of California Davis

Antonio Montalban
Department of Mathematics, University of Chicago

Michael Rathjen
School of Mathematics, University of Leeds

Simon Thomas
Department of Mathematics, Rutgers University

ASL Publisher
Richard A. Shore
Department of Mathematics, Cornell University

For more information, see www.aslonline.org/books_perspectives.html

PERSPECTIVES IN LOGIC

Lambda Calculus with Types

HENK BARENDREGT
Radboud University, Nijmegen

WIL DEKKERS
Radboud University Nijmegen

RICHARD STATMAN
Carnegie Mellon University

With contributions from
FABIO ALESSI, MARC BEZEM, FELICE CARDONE, MARIO COPPO, MARIANGIOLA DEZANI-CIANCAGLINI, GILLES DOWEK, SILVIA GHILEZAN, FURIO HONSELL, MICHAEL MOORTGAT, PAULA SEVERI, PAWEŁ URZYCZYN.

ASSOCIATION FOR SYMBOLIC LOGIC

CAMBRIDGE UNIVERSITY PRESS
Cambridge, New York, Melbourne, Madrid, Cape Town,
Singapore, São Paulo, Delhi, Mexico City

Cambridge University Press
The Edinburgh Building, Cambridge CB2 8RU, UK

Published in the United States of America by Cambridge University Press, New York

www.cambridge.org
Information on this title: www.cambridge.org/9780521766142

Association for Symbolic Logic
Richard Shore, Publisher
Department of Mathematics, Cornell University, Ithaca, NY 14853
http://www.aslonline.org

© Association for Symbolic Logic 2013

This publication is in copyright. Subject to statutory exception
and to the provisions of relevant collective licensing agreements,
no reproduction of any part may take place without the written
permission of Cambridge University Press.

First published 2013

Printed and bound in the United Kingdom by the CPI Group Ltd, Croydon CR0 4YY

A catalogue record for this publication is available from the British Library

ISBN 978-0-521-766-142 Hardback

Cambridge University Press has no responsibility for the persistence or
accuracy of URLs for external or third-party internet websites referred to
in this publication, and does not guarantee that any content on such
websites is, or will remain, accurate or appropriate.

Contents

Preface

This book is about lambda terms typed using *simple*, *recursive* and *intersection* types. In some sense it is a sequel to Barendregt (1984). That book is about untyped lambda calculus. Types give the untyped terms more structure: function applications are allowed only in some cases. In this way one can single out untyped terms having special properties. But there is more to it. The extra structure makes the theory of typed terms quite different from the untyped ones.

The emphasis of the book is on syntax. Models are introduced only insofar as they give useful information about terms and types or if the theory can be applied to them.

The writing of this book has been different from the one on untyped lambda calculus. First of all, since many researchers are working on typed lambda calculus, we were aiming at a moving target. Moreover there has been a wealth of material to work with. For these reasons the book was written by several authors. Several long-term open problems have been solved during the period the book was written, notably the undecidability of lambda definability in finite models, the undecidability of second-order typability, the decidability of the unique maximal theory extending $\beta\eta$-conversion and the fact that the collection of closed terms of not every simple type is finitely generated, and the decidability of matching at arbitrary types of order higher than 4. The book has not been written as an encyclopedic monograph: many topics are only partially treated; for example, reducibility among types is analyzed only for simple types built up from only one atom.

One of the recurring distinctions made in the book is the difference between the implicit typing due to Curry versus the explicit typing due to Church. In the latter case the terms are an enhanced version of the untyped terms, whereas in the Curry theory to some of the untyped terms a collection of types is being assigned. The book is mainly about Curry typing, although some chapters treat the equivalent Church variant.

The applications of the theory are within the theory itself, or in the theory of programming languages, or in proof theory, including the technology of fully formalized proofs used for mechanical verification, or in linguistics. Often the applications are given in an exercise with hints.

We hope that the book will attract readers and inspire them to pursue the topic.

Acknowledgments

Many thanks are due to many people and institutions. The first author obtained substantial support in the form of a generous personal research grant by the Board of Directors of Radboud University, the Spinoza Prize by The Netherlands Organisation for Scientific Research (NWO), and the Distinguished Lorentz Fellowship by the Lorentz Institute at Leiden University and The Netherlands Institue of Advanced Studies (NIAS) at Wassenaar. Not all of these means were used to produce this book, but they have been important. The Mathematical Forschungsinstitut at Oberwolfach, Germany, provided generous hospitality through their 'Research in Pairs' program. The Residential Centre at Bertinoro of the University of Bologna hosted us in their stunning castle. The principal regular sites where the work was done are the Institute for Computing and Information Sciences of Radboud University at Nijmegen, The Netherlands, the Department of Mathematics of Carnegie–Mellon University at Pittsburgh, USA, the Departments of Informatics at the Universities of Torino and Udine, both Italy.

The three main authors wrote the larger part of Part I and thoroughly edited Part II, drafted by Mario Coppo and Felice Cardone, and Part III, drafted by Mariangiola Dezani-Ciancaglini, Fabio Alessi, Furio Honsell, and Paula Severi. Various chapters and sections were drafted by other authors as follows: Chapter 4 by Gilles Dowek, Sections 5.3–5.5 by Marc Bezem, Section 6.4 by Michael Moortgat, and Section 17.5 by Pawel Urzyczyn, while Section 6.3 was co-authored by Silvia Ghilezan. This 'thorough editing' consisted of rewriting the material to bring it all into one style, but in many cases also in adding results and making corrections. It was agreed upon beforehand with all co-authors that this would happen.

Since 1974 Jan Willem Klop has been a close colleague and friend and we have been engaged with him in many inspiring discussions on λ-calculus and types.

Several people helped during the later phases of writing the book. The reviewer Roger Hindley gave invaluable advice. Vincent Padovani carefully read Section 4.3. Other help came from Jörg Endrullis, Clemens Grabmeyer, Tanmay Inamdar, Thierry Joly, Jan Willem Klop, Pieter Koopman, Dexter Kozen, Giulio Manzonetto, James McKinna, Vincent van Oostrom, Andrew Polonsky, Rinus Plasmeijer, Arnoud van Rooij, Jan Rutten, Sylvain Salvati, Christian Urban, Bas Westerbaan, and Bram Westerbaan.

Use has been made of the following macro packages: 'prooftree' of Paul Taylor, 'xypic' of Kristoffer Rose, 'robustindex' of Wilbert van der Kallen, and several lay-out commands of Erik Barendsen.

At the end producing this book turned out a time-consuming enterprise. But that seems to be the way: while the production of the content of Barendregt (1984) was expected to take two months, it took fifty; for this book our initial estimate was four years, while it turned out to be twenty.

Our partners were usually patiently understanding when we spent yet another period of writing and rewriting. We cordially thank them for their continuous and continuing support and love.

Nijmegen and Pittsburgh April 23, 2013

Henk Barendregt[1,2]
Wil Dekkers[1]
Rick Statman[2]

[1] Faculty of Science
Radboud University, Nijmegen, The Netherlands
[2] Departments of Mathematics and Computer Science
Carnegie Mellon University, Pittsburgh, USA

Contributors

Fabio Alessi, *Department of Mathematics and Computer Science, Udine University*
Marc Bezem, *Department of Informatics, Bergen University*
Felice Cardone, *Department of Informatics, Torino University*
Mario Coppo, *Department of Informatics, Torino University*
Mariangiola Dezani-Ciancaglini, *Department of Informatics, Torino University*
Gilles Dowek, *Department of Informatics, École Polytechnique;* and *INRIA*
Silvia Ghilezan, *Center for Mathematics & Statistics, University of Novi Sad*
Furio Honsell, *Department of Mathematics and Computer Science, Udine University*
Michael Moortgat, *Department of Modern Languages, Utrecht University*
Paula Severi, *Department of Computer Science, University of Leicester*
Paweł Urzyczyn, *Institute of Informatics, Warsaw University*

Our Founders

The founders of the topic of this book are Alonzo Church (1903–1995), who invented the lambda calculus (Church (1932), Church (1933)), and Haskell Curry (1900–1982), who invented 'notions of functionality' (Curry (1934)) that later got transformed into types for the hitherto untyped lambda terms. As a tribute to Church and Curry below are shown pictures of them at an early stage of their careers. Church and Curry were honored jointly for their timeless invention by the Association for Computing Machinery in 1982.

Alonzo Church (1903–1995), Studying mathematics at Princeton University (1922 or 1924). Courtesy of Alonzo Church and Mrs. Addison-Church.

Haskell B. Curry (1900–1982), BA in mathematics at Harvard (1920). Courtesy of Town & Gown, Penn State.

Introduction

The rise of lambda calculus

Lambda calculus is a formalism introduced by Church in 1932 that was intended to be used as a foundation for mathematics, including its computational aspects. Supported by his students Kleene and Rosser – who showed that the prototype system was inconsistent – Church distilled a consistent computational part and ventured in 1936 the Thesis that exactly the intuitively computable functions could be captured by it. He also presented a function that could not be captured by the λ-calculus. In that same year Turing introduced another formalism, describing what are now called Turing Machines, and formulated the related Thesis that exactly the mechanically computable functions are able to be captured by these machines. Turing also showed in the same paper that the question of whether a given statement could be proved (from a given set of axioms) using the rules of any reasonable system of logic is not computable in this mechanical way. Finally Turing showed that the formalism of λ-calculus and Turing Machines define the same class of functions.

Together Church's Thesis, concerning computability by *homo sapiens*, and Turing's Thesis, concerning computability by mechanical devices, using formalisms that are equally powerful and that have their computational limitations, made a deep impact on the 20th century philosophy of the power and limitations of the human mind. So far, cognitive neuropsychology has not been able to refute the combined Church–Turing Thesis. On the contrary, that discipline also shows the limitation of human capacities. On the other hand, the analyses of Church and Turing indicate an element of reflection (universality) in both Lambda Calculus and Turing Machines, that according to their combined thesis is also present in humans.

Turing Machine computations are relatively easy to implement on electronic devices, as started to happen early in the 1940s. The above-mentioned universality was employed by von Neumann[1] enabling the construction not only of ad hoc computers but even a universal one, capable of performing different tasks depending on a program. This resulted in what is now called *imperative programming*, with the C language presently the most widely used for programming in this paradigm. As with Turing Machines a computation consists of repeated modifications of some data stored in memory. The essential difference between a modern computer and a Turing Machine is that the former has random access memory[2].

Functional Programming

The computational model of Lambda Calculus, on the other hand, has given rise to *functional programming*. The input M becomes part of an expression FM to be evaluated, where F represents the intended function to be computed on M. This expression is reduced (rewritten) according to some rules (indicating the possible computation steps) and some strategy (indicating precisely which steps should be taken).

To show the elegance of functional programming, here is a short functional program generating primes using Eratosthenes sieve (Miranda program by D. Turner):

```
primes = sieve [2..]
         where
         sieve (p:x) = p : sieve [n | n<-x ; n mod p > 0]
primes_upto n = [p | p<- primes ; p<n]
```

while a similar program expressed in an imperative language looks like (Java program from `<rosettacode.org>`)

```
public class Sieve{
  public static LinkedList<Integer> sieve(int n){
    LinkedList<Integer> primes = new LinkedList<Integer>();
    BitSet nonPrimes = new BitSet(n+1);

    for (int p = 2; p <= n; p = nonPrimes.nextClearBit(p+1)){
        for (int i = p * p; i <= n; i += p)
            nonPrimes.set(i);
```

[1] It was von Neumann who visited Cambridge UK in 1935 and invited Turing to Princeton during 1936–1937, so he probably knew Turing's work.

[2] Also, the memory on a TM is infinite: Turing wanted to be technology-independent, but was restricting a computation with a given input to one using finite memory and time.

```
                primes.add(p);
            }
            return primes;
        }
}
```

Of course the algorithm is extremely simple, one of the first ever invented. However, the gain for more complex algorithms remains, as functional programs do scale up.

The power of functional programming languages derives from several facts.

(1) All expressions of a functional programming language have a constant meaning (i.e. independent of a hidden state). This is called 'referential transparency' and makes it easier to reason about functional programs and to make versions for parallel computing, important for quality and efficiency.
(2) Functions may be arguments of other functions, usually called 'functionals' in mathematics and higher-order functions in programming. There are functions acting on functionals, etc; in this way one obtains functions of arbitrary order. Both in mathematics and in programming, higher-order functions are natural and powerful phenomena. In functional programming this enables the flexible composition of algorithms.
(3) Algorithms can be expressed in a clear goal-directed mathematical way, using various forms of recursion and flexible data structures. The bookkeeping needed for the storage of these values is handled by the language compiler instead of the user of the functional language[3].

Types

The formalism as defined by Church is untyped. The early functional languages, of which Lisp (McCarthy et al. (1962)) and Scheme (Abelson et al. (1991)) are best known, are also untyped: arbitrary expressions may be applied to each other. Types first appeared in *Principia Mathematica*, Whitehead and Russell (1910-1913). In Curry (1934) types are introduced and assigned to expressions in 'combinatory logic', a formalism closely related to lambda calculus. In Curry and Feys (1958) this type assignment mechanism was adapted to λ-terms, while in Church (1940) λ-terms were ornamented

[3] In modern functional languages there is a palette of techniques (such as overloading, type classes and generic programming) to make algorithms less dependent of specific data types and hence more reusable. If desired the user of the functional language can help the compiler to achieve a better allocation of values.

by fixed types. This resulted in the closely related systems $\boldsymbol{\lambda}_{\rightarrow}^{\mathrm{Cu}}$ and $\boldsymbol{\lambda}_{\rightarrow}^{\mathrm{Ch}}$ treated in Part I.

Types are being used in many, if not most, programming languages. These are of the form

$$\mathtt{bool}, \mathtt{nat}, \mathtt{real}, \ldots$$

and occur in compounds like

$$\mathtt{nat} \to \mathtt{bool}, \mathtt{array}(\mathtt{real}), \ldots$$

Using the formalism of types in programming, many errors can be prevented if terms are required to be typable: arguments and functions should match. For example M of type A can be an argument only of a function of type $A \to B$. Types act in a way similar to the use of dimensional analysis in physics. Physical constants and data obtain a 'dimension'. Pressure p, for example, has a dimension expressed as

$$M/L^2$$

giving the constant R in Boyle's law,

$$\frac{pV}{T} = R,$$

that has a dimension which prevents one from writing an equation like $E = TR^2$. By contrast Einstein's famous equation

$$E = mc^2$$

is already meaningful from the viewpoint of dimensional analysis.

In most programming languages the formation of function space types is usually not allowed to be iterated as in

$(\mathtt{real} \to \mathtt{real}) \to (\mathtt{real} \to \mathtt{real})$ for indefinite integrals $\int f(x)dx$;

$(\mathtt{real} \to \mathtt{real}) \times \mathtt{real} \times \mathtt{real} \to \mathtt{real}$ for definite integrals $\int_a^b f(x)dx$;

$([0,1] \to \mathtt{real}) \to (([0,1] \to \mathtt{real}) \to \mathtt{real}) \to (([0,1] \to \mathtt{real}) \to \mathtt{real})$,

where the latter is the type of a map occuring in functional analysis, see Lax (2002). Here we have written "$[0,1] \to \mathtt{real}$" for what should be more accurately the set $C[0,1]$ of continuous functions on $[0,1]$.

Because there is the Hindley–Milner algorithm (see Theorem 2.3.14 in Chapter 2) that decides whether an untyped term does have a type and computes the most general type, types have found their way to functional programming languages. The first such language to incorporate the types of the simply typed λ-calculus is ML (Milner et al. (1997)). An important aspect of typed expressions is that if a term M is correctly typed by type A, then also during the computation of M the type remains the same (see Theorem 1.2.6, the 'subject reduction theorem'). This is expressed as a feature in functional programming: one only needs to check types during compile time.

In functional programming languages, however, types come of age and are allowed in their full potential by giving a precise notation for the type of data, functions, functionals, higher-order functionals, ... up to arbitrary degree of complexity. Interestingly, the use of higher-order types given in the mathematical examples is modest compared to higher-order types occurring in a natural way in programming situations:

$$\begin{aligned}
&[(a \to ([([b],c)] \to [([b],c)]) \to [([b],c)] \to [b] \to [([b],c)]) \to \\
&([([b],c)] \to [([b],c)]) \to [([b],c)] \to [b] \to [([b],c)]] \to \\
&a \to (d \to ([([b],c)] \to [([b],c)]) \to [([b],c)] \to [b] \to [([b],c)]) \\
&\to ([([b],c)] \to [([b],c)]) \to [([b],c)] \to [b] \to [([b],c)]] \to \\
&d \to ([([b],c)] \to [([b],c)]) \to [([b],c)] \to [b] \to [([b],c)]] \to \\
&([([b],c)] \to [([b],c)]) \to [([b],c)] \to [b] \to [([b],c)].
\end{aligned}$$

This type (it does not actually occur in this form in the program, but is notated using memorable names for the concepts being used) is used in a functional program for efficient parser generators, see Koopman and Plasmeijer (1999). The type $[a]$ denotes that of lists of type a and (a, b) denotes the 'product' $a \times b$. Product types can be simulated by simple types, while for list types one can use the recursive types developed in Part 2 of this book. Although in the pure typed λ-calculus only a rather restricted class of terms and types is represented, relatively simple extensions of this formalism have universal computational power. Since the 1970s the following programming languages have appeared: ML (not yet purely functional); Miranda (Thompson (1995), `<www.cs.kent.ac.uk/people/staff/dat/miranda/>`) the first purely functional typed programming language, well-designed, but slowly interpreted; Clean (van Eekelen and Plasmeijer (1993), Plasmeijer and van Eekelen (2002), `<wiki.clean.cs.ru.nl/Clean>`); and Haskell (Hutton (2007), Peyton Jones (2003), `<www.haskell.org>`). Both Clean and Haskell are state of the art pure functional languages with fast compiler generating fast code). They show that functional programming based on λ-calculus can be efficient and apt for industrial software. Functional programming languages are also being used for the design (Sheeran (2005)) and testing (Koopman and Plasmeijer (2006)) of hardware. In each case it is the compact mathematical expressivity of the functional languages that makes them fit for the description of complex functionality.

Semantics of natural languages

Typed λ-calculus has also been employed in the semantics of natural languages (Montague (1973), van Benthem (1995)). An early indication of this possibility can already be found in Curry and Feys (1958), Section 8S2.

Certifying proofs

In addition to its use in design, the λ-calculus has also been used for verification, not just for the correctness of IT products, but also of mathematical proofs. The underlying idea is the following. Ever since Aristotle's formulation of the axiomatic method and Frege's formulation of predicate logic, one could write down mathematical proofs in full detail. Frege, who captured reasoning by his introduction of the predicate logic, started to formalize mathematics, but unfortunately began from an axiom system that turned out to be inconsistent, as shown by the Russell paradox. In *Principia Mathematica* Whitehead and Russell used types to prevent the paradox. They had the same formalization goal in mind and developed some elementary arithmetic. Based on their work, Gödel was able to state and prove his fundamental incompleteness result. In spite of the intention behind *Principia Mathematica*, proofs in the underlying formal system were not fully formalized. Substitution was left as an informal operation and in fact the way *Principia Mathematica* treated free and bound variables was implicit and incomplete. Here begins the role of the λ-calculus. As a formal system dealing with manipulating formulas, distinguishing carefully between free and bound variables and their interaction, it was the missing link towards a full formalization. Now, if an axiomatic mathematical theory is fully formalized, a computer can verify the correctness of the definitions and proofs. The reliability of computer-verified theories relies on the fact that logic has only about a dozen rules and their implementation poses relatively few problems. This idea was pioneered in the late 1960s by N. G. de Bruijn in the proof-checking language and system Automath (Nederpelt et al. (1994), `<www.win.tue.nl/automath>`).

The methodology has given rise to proof-assistants. These are computer programs that help the human user to develop mathematical theories. The initiative comes from the human who formulates notions, axioms, definitions, proofs and computational tasks. The computer verifies the well-definedness of the notions, the correctness of the proofs, and performs the computational tasks. In this way arbitrary mathematical notions can represented and manipulated on a computer. Many of the mathematical assistants are based on extensions of typed λ-calculus. See Section 6.2 for more information.

What this book is and is not about

None of the fascinating applications, mentioned above, of lambda calculus with types are treated in this book. We will study the formalism for its mathematical beauty. In particular this monograph focuses on mathematical properties of three classes of typing for lambda terms.

Simple types, constructed freely from type atoms, cause strong normalization, subject reduction, decidability of typability and inhabitation, undecidability of lambda-definability. There turn out to be five canonical term models based on closed terms. Powerful extension with respectively a discriminator, surjective pairing, operators for primitive recursion, bar recursion, and a fixed point operator are being studied. Some of these extensions remain constructive, others are utterly non-constructive, and some will be at the boundary of these two methods.

Recursive types allow functions to fit as input for themselves, losing strong normalization (restored by allowing only positive recursive types). Typability remains decidable. Unexpectedly, α-conversion, dealing with the hygienic treatment of free and bound variables among recursive types, has interesting mathematical properties.

Intersection types allow functions to take arguments of different types simultaneously. Under certain mild conditions this leads to subject conversion, turning the filters of types of a given term into a lambda model. Classical lattice models can be described as intersection type theories. Typability and inhabitation now becomes undecidable, the latter being equivalent to undecidability of lambda-definability for models of simple types.

A flavor of some of the applications of typed lambda calculus is given: functional programming (Section 6.1), proof-checking (Section 6.2), and formal semantics of natural languages (Section 6.4).

What this book could have been about

This book could have been also about dependent types, higher-order types and inductive types, all used in some of the mathematical assistants. Originally we had planned a second volume to do so. But given the effort needed to write this one, we will probably not do so. Higher-order types are treated in a mathematically oriented style in Girard et al. (1989), and Sørensen and Urzyczyn (2006). Research monographs on dependent and inductive types are lacking. This is an invitation to the community of the next generation of researchers!

Some notational conventions

A *partial function* from a set X to a set Y is a collection of ordered pairs $f \subseteq X \times Y$ such that $\forall x \in X, y, y' \in Y.[\langle x, y\rangle \in f \ \&\ \langle x, y'\rangle \in f \ \Rightarrow\ y = y']$.

The set of partial functions from a set X to a set Y is denoted by $X \rightharpoonup Y$. If $f \in (X \rightharpoonup Y)$ and $x \in X$, then $f(x)$ is *defined*, notation $f(x)\downarrow$ or $x \in \mathrm{dom}(f)$, if for some y one has $\langle x, y\rangle \in f$. In that case one writes $f(x) = y$. On the other hand $f(x)$ is *undefined*, notation $f(x)\uparrow$, means that for no $y \in Y$ one has $\langle x, y\rangle \in f$. An expression E in which partial functions are involved, may be defined or not. If two such expressions are compared, then, following Kleene (1952), we write $E_1 \simeq E_2$ for

$$\text{if } E_1\downarrow, \text{ then } E_2\downarrow \text{ and } E_1 = E_2, \text{ and vice versa.}$$

The set of natural numbers is denoted by ω. The notation $\triangleq$ is used for "equality by definition". Similarly '$\stackrel{\Delta}{\Longleftrightarrow}$'. is used for the definition of a concept. By contrast ::= stands for the more specific introduction of a syntactic category defined by the Backus–Naur form. The notation $\equiv$ stands for syntactic equality (for example to remind the reader that the left hand side was defined previously as the right hand side In a definition we do not write 'M is *closed* iff $\mathrm{FV}(M) = \emptyset$' but '$M$ is *closed* if $\mathrm{FV}(M) = \emptyset$'. The end of a proof is indicated by '$\blacksquare$'.

PART I

SIMPLE TYPES $\boldsymbol{\lambda}_{\rightarrow}^{\mathbb{A}}$

The systems of *simple types* considered in Part I are built up from atomic types $\mathbb{A}$ using, as the only operator, the constructor $\rightarrow$ for forming function spaces. For example, from the atoms $\mathbb{A} = \{\alpha, \beta\}$ one can form types $\alpha\rightarrow\beta$, $(\alpha\rightarrow\beta)\rightarrow\alpha$, $\alpha\rightarrow(\alpha\rightarrow\beta)$ and so on. Two choices of the set of atoms that will be made most often are $\mathbb{A} = \{\alpha_0, \alpha_1, \alpha_2, \dots\}$, an infinite set of type variables giving $\boldsymbol{\lambda}_{\rightarrow}^{\infty}$, and $\mathbb{A} = \{0\}$, consisting of only one atomic type giving $\boldsymbol{\lambda}_{\rightarrow}^{0}$. Particular atomic types that occur in applications are e.g. Bool, Nat, Real. Even for these simple type systems, the ordering effect is quite powerful.

Requiring terms to have simple types implies that they are strongly normalizing. For an untyped lambda term one can find the collection of its possible types. Similarly, given a simple type, one can find the collection of its possible inhabitants (in normal form). Equality of terms of a certain type can be reduced to equality of terms in a fixed type. Insights coming from this reducibility provide five canonical term models of $\boldsymbol{\lambda}_{\rightarrow}^{0}$. See pages 2 and 3 for types and terms involved in this analysis.

The problem of unification

$$\exists X{:}A.MX =_{\beta\eta} NX$$

is for complex enough A undecidable. That of pattern matching

$$\exists X{:}A.MX =_{\beta\eta} N$$

will be shown to be decidable for A up to 'rank 3'. The recent proof by Stirling of general decidability of matching is not included. The terms of finite type are extended by δ-functions, functionals for primitive recursion (Gödel) and bar recursion (Spector). Applications of the theory in computing, proof-checking and semantics of natural languages will be presented.

Other expositions of the simply typed lambda calculus are Church (1941), Lambek and Scott (1981), Girard et al. (1989), Hindley (1997), and Nerode et al. (In preparation). Part of the history of the topic, including the untyped lambda calculus, can be found in Crossley (1975), Rosser (1984), Kamareddine et al. (2004) and Cardone and Hindley (2009).

Sneak preview of $\boldsymbol{\lambda}_{\rightarrow}$ (Chapters 1, 2, 3)

Terms

Term *variables* $V \triangleq \{c, c', c'', \dots\}$

$$\textit{Terms } \Lambda \quad \begin{cases} x \in V & \Rightarrow \quad x \in \Lambda \\ M, N \in \Lambda & \Rightarrow \quad (MN) \in \Lambda \\ M \in \Lambda, x \in V & \Rightarrow \quad (\lambda x M) \in \Lambda \end{cases}$$

Notations for terms
$x, y, z, \dots, F, G, \dots, \Phi, \Psi, \dots$ range over V
$M, N, L, \dots$ range over Λ
Abbreviations

$$N_1 \dots N_n \triangleq (\cdot\cdot(MN_1)\dots N_n)$$
$$\lambda x_1 \dots x_n.M \triangleq (\lambda x_1(\dots(\lambda x_n.M)\cdot\cdot))$$

Standard terms: combinators

$$\mathsf{I} \triangleq \lambda x.x$$
$$\mathsf{K} \triangleq \lambda xy.x$$
$$\mathsf{S} \triangleq \lambda xyz.xz(yz)$$

Types

Type *atoms* $\mathbb{A}_\infty \triangleq \{\mathsf{c}, \mathsf{c}', \mathsf{c}'', \dots\}$

$$\textit{Types } \mathbb{T} \quad \begin{cases} \alpha \in \mathbb{A} & \Rightarrow \quad \alpha \in \mathbb{T} \\ A, B \in \mathbb{T} & \Rightarrow \quad (A \to B) \in \mathbb{T} \end{cases}$$

Notations for types
$\alpha, \beta, \gamma, \dots$ range over $\mathbb{A}_\infty$
$A, B, C, \dots$ range over $\mathbb{T}$
Abbreviation
$A_1 \to A_2 \to \dots \to A_n \triangleq (A_1 \to (A_2 \to \dots (A_{n-1} \to A_n)\cdot\cdot))$

Standard types: each $n \in \omega$ is interpreted as type $n \in \mathbb{T}$

$$0 \triangleq \mathsf{c}$$
$$n+1 \triangleq n \to 0$$
$$(n+1)_2 \triangleq n \to n \to 0$$

Assignment of types to terms $\vdash M : A$ $(M \in \Lambda, A \in \mathbb{T})$

Basis: a set $\Gamma = \{x_1{:}A_1, \dots, x_n{:}A_n\}$, with $x_i \in V$ distinct
Type *assignment* (relative to a basis Γ) axiomatized by

$$\begin{cases} (x{:}A) \in \Gamma & \Rightarrow \quad \Gamma \vdash x : A \\ \Gamma \vdash M : (A{\to}B), \Gamma \vdash N : A & \Rightarrow \quad \Gamma \vdash (MN) : B \\ \Gamma, x{:}A \vdash M : B & \Rightarrow \quad \Gamma \vdash (\lambda x.M) : (A{\to}B) \end{cases}$$

Notations for assignment
'$x{:}A \vdash M : B$' stands for '$\{x{:}A\} \vdash M : B$'
'$\Gamma, x{:}A$' for '$\Gamma \cup \{x{:}A\}$' and '$\vdash M : A$' for '$\emptyset \vdash M : A$'

Standard assignments: for all $A, B, C \in \mathbb{T}$ one has

$\vdash \mathsf{I}$	$: A{\to}A$	as $x{:}A \vdash x : A$
$\vdash \mathsf{K}$	$: A{\to}B{\to}A$	as $x{:}A, y{:}B \vdash x : A$
$\vdash \mathsf{S}$	$: (A{\to}B{\to}C){\to}(A{\to}B){\to}A{\to}C$	similarly

Canonical term-models built up from constants

The following types A play an important role in Sections 3.4, 3.5. Their normal inhabitants (i.e. terms M in normal form such that $\vdash M : A$) can be enumerated by the following schemes.

Type	Inhabitants (all possible $\boldsymbol{\beta\eta}^{-1}$-normal forms are listed)
1_2	$\boldsymbol{\lambda xy.x, \lambda xy.y}$.
$1{\to}0{\to}0$	$\lambda fx.x, \lambda fx.fx, \lambda fx.f(fx), \lambda fx.f^3x, \ldots$; general pattern: $\boldsymbol{\lambda fx.f^n x}$.
3	$\lambda F.F(\lambda x.x), \lambda F.F(\lambda x.F(\lambda y.x)), \ldots; \boldsymbol{\lambda F.F(\lambda x_1.F(\lambda x_2.\ldots F(\lambda x_n.x_i)\cdot\cdot))}$.
$1{\to}1{\to}0{\to}0$	$\lambda fgx.x, \lambda fgx.fx, \lambda fgx.gx,$ $\lambda fgx.f(gx), \lambda fgx.g(fx), \lambda fgx.f^2x, \lambda fgx.g^2x,$ $\lambda fgx.f(g^2x), \lambda fgx.f^2(gx), \lambda fgx.g(f^2x), \lambda fgx.g^2(fx), \lambda fgx.f(g(fx)), \ldots;$ $\boldsymbol{\lambda fgx.w_{\{f,g\}}x}$, where $w_{\{f,g\}}$ is a 'word over $\Sigma = \{f, g\}$' which is 'applied' to x by interpreting juxtaposition 'fg' as function composition '$f \circ g = \lambda x.f(gx)$'.
$3{\to}0{\to}0$	$\lambda\Phi x.x, \lambda\Phi x.\Phi(\lambda f.x), \lambda\Phi x.\Phi(\lambda f.fx), \lambda\Phi x.\Phi(\lambda f.f(\Phi(\lambda g.g(fx)))), \ldots$ $\lambda\Phi x.\Phi(\lambda f_1.w_{\{f_1\}}x), \lambda\Phi x.\Phi(\lambda f_1.w_{\{f_1\}}\Phi(\lambda f_2.w_{\{f_1,f_2\}}x)), \ldots;$ $\boldsymbol{\lambda\Phi x.\Phi(\lambda f_1.w_{\{f_1\}}\Phi(\lambda f_2.w_{\{f_1,f_2\}}\cdots\Phi(\lambda f_n.w_{\{f_1,\ldots,f_n\}}x)\cdot\cdot))}$.
$1_2{\to}0{\to}0$	$\lambda bx.x, \lambda bx.bxx, \lambda bx.bx(bxx), \lambda bx.b(bxx)x, \lambda bx.b(bxx)(bxx), \ldots; \boldsymbol{\lambda bx.t}$, where t is an element of the context-free language generated by the grammar `tree ::= x | (b tree tree)`.

This follows by considering the inhabitation machine, see Section 1.3, for each mentioned type.

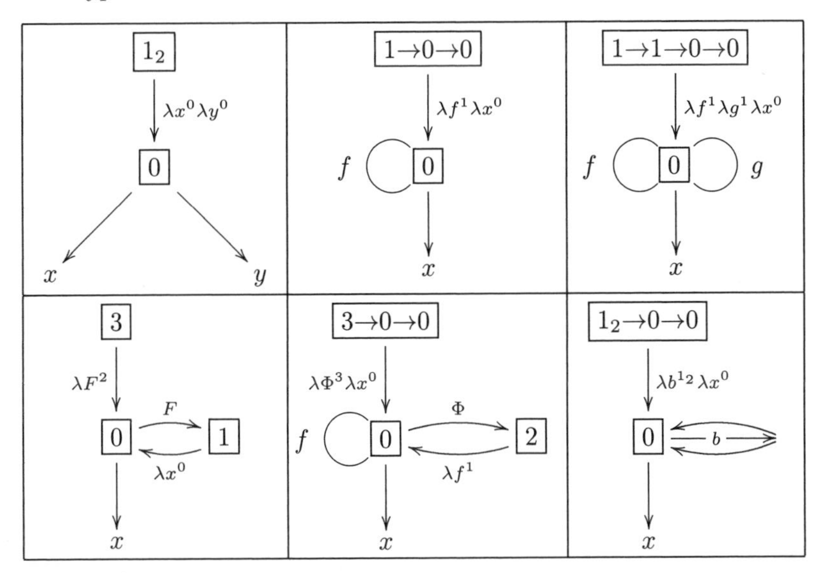

We have juxtaposed the machines for types $1{\to}0{\to}0$ and $1{\to}1{\to}0{\to}0$, as they are similar, and also those for 3 and $3{\to}0{\to}0$. According to the type reducibility theory of Section 3.4 the types $1{\to}0{\to}0$ and 3 are equivalent and therefore they are presented together in the statement.

From the types 1_2, $1{\to}0{\to}0$, $1{\to}1{\to}0{\to}0$, $3{\to}0{\to}0$, and $1_2{\to}0{\to}0$ five canonical λ-theories and term-models will be constructed, that are strictly increasing (decreasing). The smallest theory is the good old simply typed $\lambda\boldsymbol{\beta\eta}$-calculus, and the largest theory corresponds to the minimal model, Definition 3.5.47, of the simply typed λ-calculus.

1

The Simply Typed Lambda Calculus

1.1 The systems $\boldsymbol{\lambda}_{\to}^{\mathbb{A}}$

Untyped lambda calculus

Recall the untyped lambda calculus denoted by $\boldsymbol{\lambda}$, see e.g. B[1984][1].

1.1.1 Definition The set of untyped λ-terms Λ is defined by the so-called '*simplified syntax*' in Fig. 1.1. This basically means that parentheses are left implicit.

$$\begin{array}{lcl} \mathsf{V} & ::= & x \mid \mathsf{V}' \\ \Lambda & ::= & \mathsf{V} \mid \lambda \mathsf{V}\, \Lambda \mid \Lambda\, \Lambda \end{array}$$

Figure 1.1 Untyped lambda terms

This makes $\mathsf{V} = \{x, x', x'', \dots\}$.

1.1.2 Notation

(i) $x, y, z, \dots, x_0, y_0, z_0, \dots, x_1, y_1, z_1, \dots$ denote arbitrary variables.
(ii) $M, N, L, \dots$ denote arbitrary lambda terms.
(iii) $MN_1 \cdots N_k \equiv (..(MN_1) \cdots N_k)$, *association to the left.*
(iv) $\lambda x_1 \cdots x_n.M \equiv (\lambda x_1(..(\lambda x_n(M))..))$, *association to the right.*

1.1.3 Definition Let $M \in \Lambda$.

(i) The set of *free variables* of M, written $\mathrm{FV}(M)$, is defined as in Fig. 1.2. The variables in M that are not free are called *bound.*
(ii) If $\mathrm{FV}(M) = \emptyset$, then we say that M is *closed* or that it is a *combinator*:

$$\Lambda^{\emptyset} = \{M \in \Lambda \mid M \text{ is closed}\}.$$

[1] This is an abbreviation for the reference Barendregt (1984).

M	$\mathrm{FV}(M)$
x	$\{x\}$
PQ	$\mathrm{FV}(P) \cup \mathrm{FV}(Q)$
$\lambda x.P$	$\mathrm{FV}(P) - \{x\}$

Figure 1.2

Well-known combinators are $\mathsf{I} \equiv \lambda x.x, \mathsf{K} \equiv \lambda xy.y, \mathsf{S} \equiv \lambda xyz.xz(yz)$, $\mathbf{\Omega} \equiv (\lambda x.xx)(\lambda x.xx)$, and $\mathsf{Y} \equiv \lambda f.(\lambda x.f(xx))(\lambda x.f(xx))$. Officially one has $\mathsf{S} \equiv (\lambda x(\lambda x'(\lambda x''((xx'')(x'x'')))))$, according to Definition 1.1.1, so we see that the effort of learning the Notation 1.1.2 pays off.

1.1.4 Definition On Λ the equational theory $\boldsymbol{\lambda\beta\eta}$ is defined by the usual equality axiom and rules (reflexivity, symmetry, transitivity, congruence), including congruence with respect to abstraction:

$$M = N \;\Rightarrow\; \lambda x.M = \lambda x.N,$$

and the special axiom (schemes) in Fig. 1.3.

$$\begin{array}{rcll}
(\lambda x.M)N & = & M[x := N] & \qquad\qquad\qquad (\boldsymbol{\beta}\text{-rule}) \\
\lambda x.Mx & = & M, \qquad\qquad \text{if } x \notin \mathrm{FV}(M) & (\boldsymbol{\eta}\text{-rule})
\end{array}$$

Figure 1.3 The theory $\boldsymbol{\lambda\beta\eta}$

This theory can be analyzed by an appropriate notion of reduction.

1.1.5 Definition On Λ we define in Figure 1.4 the notions of *$\boldsymbol{\beta}$-reduction* and *$\boldsymbol{\eta}$-reduction*.

$$\begin{array}{rcll}
(\lambda x.M)N & \to & M[x{:}= N] & \qquad\qquad\qquad (\boldsymbol{\beta}) \\
\lambda x.Mx & \to & M, \qquad\qquad \text{if } x \notin \mathrm{FV}(M) & (\boldsymbol{\eta})
\end{array}$$

Figure 1.4 $\boldsymbol{\beta\eta}$-contraction rules

As usual, see B[1984], these notions of reduction generate the corresponding reduction relations $\to_{\boldsymbol{\beta}}, \twoheadrightarrow_{\boldsymbol{\beta}}, \to_{\boldsymbol{\eta}}, \twoheadrightarrow_{\boldsymbol{\eta}}, \to_{\boldsymbol{\beta\eta}}$ and $\twoheadrightarrow_{\boldsymbol{\beta\eta}}$. Also there are the corresponding conversion relations $=_{\boldsymbol{\beta}}, =_{\boldsymbol{\eta}}$ and $=_{\boldsymbol{\beta\eta}}$. Terms in Λ will often be considered modulo $=_{\boldsymbol{\beta}}$ or $=_{\boldsymbol{\beta\eta}}$.

1.1.6 Notation If we write $M = N$, then by default we mean $M =_{\boldsymbol{\beta\eta}} N$. This is the extensional version of the theory. See B[1984], where the default was $=_{\boldsymbol{\beta}}$.]

1.1.7 Proposition *For all $M, N \in \Lambda$ one has*

$$\vdash_{\boldsymbol{\lambda}\boldsymbol{\beta}\boldsymbol{\eta}} M = N \;\Leftrightarrow\; M =_{\boldsymbol{\beta}\boldsymbol{\eta}} N.$$

Proof See B[1984], Proposition 3.3.2. ∎

1.1.8 Remark As in B[1984], Convention 2.1.12, we will not be concerned with α-conversion, renaming bound variables in order to avoid confusion between free and bound occurrences of variables. So we write $\lambda x.x \equiv \lambda y.y$. We do this by officially working on the α-equivalence classes; when dealing with a concrete term as representative of such a class, the bound variables will be chosen maximally fresh: different from the free variables and from each other. See, however, Section 7.4, in which we introduce λ-conversion on recursive types and show how it can be avoided in a way that is more effective than for terms.

One reason why the analysis in terms of the notion of reduction $\boldsymbol{\beta}\boldsymbol{\eta}$ is useful is that the following holds.

1.1.9 Proposition (Church–Rosser theorem for $\boldsymbol{\lambda}\boldsymbol{\beta}\boldsymbol{\eta}$) *For the notions of reduction $\twoheadrightarrow_{\boldsymbol{\beta}}$, $\twoheadrightarrow_{\boldsymbol{\eta}}$, and $\twoheadrightarrow_{\boldsymbol{\beta}\boldsymbol{\eta}}$ one has the following.*

(i) *Let $M, N_1, N_2 \in \Lambda$. Then (the* diamond property*)*

$$M \twoheadrightarrow_{\boldsymbol{\beta}(\boldsymbol{\eta})} N_1 \;\&\; M \twoheadrightarrow_{\boldsymbol{\beta}(\boldsymbol{\eta})} N_2 \;\Rightarrow\; \exists Z \in \Lambda . N_1 \twoheadrightarrow_{\boldsymbol{\beta}(\boldsymbol{\eta})} Z \;\&\; N_2 \twoheadrightarrow_{\boldsymbol{\beta}(\boldsymbol{\eta})} Z.$$

One also says that the reduction relations $\twoheadrightarrow_R$, for $R \in \{\boldsymbol{\beta}, \boldsymbol{\eta}, \boldsymbol{\beta}\boldsymbol{\eta}\}$ are confluent.

(ii) *Let $M, N \in \Lambda$. Then*

$$M =_{\boldsymbol{\beta}(\boldsymbol{\eta})} N \;\Rightarrow\; \exists Z \in \Lambda . M \twoheadrightarrow_{\boldsymbol{\beta}(\boldsymbol{\eta})} Z \;\&\; N \twoheadrightarrow_{\boldsymbol{\beta}(\boldsymbol{\eta})} Z.$$

Proof See Theorems 3.2.8 and 3.3.9 in B[1984]. ∎

1.1.10 Definition

(i) Let T be a set of equations between λ-terms. Write

$$T \vdash_{\boldsymbol{\lambda}\boldsymbol{\beta}\boldsymbol{\eta}} M = N, \text{ or simply } T \vdash M = N$$

if $M = N$ is provable in $\boldsymbol{\lambda}\boldsymbol{\beta}\boldsymbol{\eta} + T$ with the equations in T added as axioms.

(ii) We say T is *inconsistent* if T proves every equation; otherwise consistent.

(iii) The equation $P = Q$, with $P, Q \in \Lambda$, is called *inconsistent*, written $P \# Q$, if $\{P = Q\}$ is inconsistent. Otherwise $P = Q$ is *consistent*.

The set $T = \emptyset$, i.e. the $\boldsymbol{\lambda}\beta\eta$-calculus itself, is consistent, as follows from the Church–Rosser theorem. Examples of inconsistent equations are $\mathsf{K}\#\mathsf{I}$ and $\mathsf{I}\#\mathsf{S}$. On the other hand $\boldsymbol{\Omega} = \mathsf{I}$ is consistent.

Simple types

Types in this part, also called *simple types*, are syntactic objects built from atomic types using the operator $\rightarrow$. In order to classify untyped lambda terms, such types will be assigned to a subset of these terms. The main idea is that if M gets type $A \rightarrow B$ and N gets type A, then the application MN is 'legal' (as M is considered as a function from terms of type A to those of type B) and gets type B. In this way types help determine what terms fit together.

1.1.11 Definition

(i) Let $\mathbb{A}$ be a non-empty set. An element of $\mathbb{A}$ is called a *type atom*. The set of *simple types* over $\mathbb{A}$, written $\mathbb{T} = \mathbb{T}^{\mathbb{A}}$, is recursively defined as follows:

$$\begin{array}{rcll} \alpha \in \mathbb{A} & \Rightarrow & \alpha \in \mathbb{T} & \textit{type atoms;} \\ A, B \in \mathbb{T} & \Rightarrow & (A \rightarrow B) \in \mathbb{T} & \textit{function space types.} \end{array}$$

We assume that no relations like $\alpha \rightarrow \beta = \gamma$ hold between type atoms: $\mathbb{T}^{\mathbb{A}}$ is freely generated. Often one finds $\mathbb{T} = \mathbb{T}^{\mathbb{A}}$ given by a simplified syntax, see Fig. 1.5.

$$\boxed{\mathbb{T} \quad ::= \quad \mathbb{A} \mid \mathbb{T} \rightarrow \mathbb{T}}$$

Figure 1.5 Simple Types

(ii) Let $\mathbb{A}_0 = \{0\}$. Then we write $\mathbb{T}^0 = \mathbb{T}^{\mathbb{A}_0}$.

(iii) Let $\mathbb{A}_\infty = \{\mathsf{c}_0, \mathsf{c}_1, \mathsf{c}_2, \ldots\}$. Then we write $\mathbb{T}^\infty = \mathbb{T}^{\mathbb{A}_\infty}$

We usually take $0 = \mathsf{c}_0$, then $\mathbb{T}^0 \subseteq \mathbb{T}^\infty$. If we write simply $\mathbb{T}$, then this refers to $\mathbb{T}^{\mathbb{A}}$ for an unspecified $\mathbb{A}$.

1.1.12 Notation

(i) If $A_1, \ldots, A_n \in \mathbb{T}$, then

$$A_1 \rightarrow \cdots \rightarrow A_n \triangleq (A_1 \rightarrow (A_2 \rightarrow \cdots \rightarrow (A_{n-1} \rightarrow A_n)..)).$$

That is, we use association to the right.

(ii) $\alpha, \beta, \gamma, \ldots, \alpha_0, \beta_0, \gamma_0, \ldots \alpha', \beta'\gamma', \ldots$ denote arbitrary elements of $\mathbb{A}$.

(iii) $A, B, C, \ldots$ denote arbitrary elements of $\mathbb{T}$.

1.1.13 Definition (Type substitution) Let $A, C \in \mathbb{T}^{\mathbb{A}}$ and $\alpha \in \mathbb{A}$. The result of substituting C for the occurrences of α in A, written $A[\alpha{:}= C]$, is defined as follows:

$$
\begin{array}{rcll}
\alpha[\alpha{:}= C] & \triangleq & C; & \\
\beta[\alpha{:}= C] & \triangleq & \beta, & \text{if } \alpha \not\equiv \beta; \\
(A \to B)[\alpha{:}= C] & \triangleq & (A[\alpha{:}= C]) \to (B[\alpha{:}= C]). &
\end{array}
$$

Assigning simple types

1.1.14 Definition ($\boldsymbol{\lambda}_{\to}^{\mathrm{Cu}}$)

(i) A (type assignment) *statement* is of the form

$$M : A,$$

with $M \in \Lambda$ and $A \in \mathbb{T}$. This statement is read as 'M in A'. The type A is the *predicate* and the term M is the *subject* of the statement.

(ii) A *declaration* is a statement with as subject a term variable.

(iii) A *basis* is a set of declarations with distinct variables as subjects.

(iv) A statement $M{:}A$ is *derivable from a basis* Γ, written

$$\Gamma \vdash_{\boldsymbol{\lambda}_{\to}}^{\mathrm{Cu}} M{:}A$$

(or $\Gamma \vdash_{\boldsymbol{\lambda}_{\to}} M : A$, or even $\Gamma \vdash M{:}A$ if there is little danger of confusion) if $\Gamma \vdash M{:}A$ can be produced by the following rules:

$$
\begin{array}{rcl}
(x{:}A) \in \Gamma & \Rightarrow & \Gamma \vdash x : A; \\[1em]
\Gamma \vdash M : (A \to B),\ \ \Gamma \vdash N : A & \Rightarrow & \Gamma \vdash (MN) : B; \\[1em]
\Gamma,\ x{:}A \vdash M : B & \Rightarrow & \Gamma \vdash (\lambda x.M) : (A \to B).
\end{array}
$$

In the last rule $\Gamma, x{:}A$ is required to be a basis.

These rules are usually written as in Fig. 1.6.

This is the modification to the lambda calculus of the system in Curry [1934], as developed in Curry et al. [1958].

1.1.15 Definition Let $\Gamma = \{x_1{:}A_1, \dots, x_n{:}A_n\}$. Then

(i) $\mathrm{dom}(\Gamma) = \{x_1, \dots, x_n\}$, the *domain* of Γ.

(ii) $x_1{:}A_1, \dots, x_n{:}A_n \vdash_{\boldsymbol{\lambda}_{\to}} M : A$ denotes $\Gamma \vdash_{\boldsymbol{\lambda}_{\to}} M : A$.

(iii) In particular $\vdash_{\boldsymbol{\lambda}_{\to}} M : A$ stands for $\emptyset \vdash_{\boldsymbol{\lambda}_{\to}} M : A$.

(iv) $x_1, \dots, x_n{:}A \vdash_{\boldsymbol{\lambda}_{\to}} M : B$ stands for $x_1{:}A, \dots, x_n{:}A \vdash_{\boldsymbol{\lambda}_{\to}} M : B$.

(axiom)	$\Gamma \vdash x : A,$	if $(x{:}A) \in \Gamma$;
($\to$-elimination)	$\dfrac{\Gamma \vdash M : (A \to B) \quad \Gamma \vdash N : A}{\Gamma \vdash (MN) : B};$	
($\to$-introduction)	$\dfrac{\Gamma, x{:}A \vdash M : B}{\Gamma \vdash (\lambda x.M) : (A \to B)}.$	

Figure 1.6 The system $\boldsymbol{\lambda}_{\to}^{\mathrm{Cu}}$ à la *Curry*

1.1.16 Example

(i) $\vdash_{\boldsymbol{\lambda}\to} \mathsf{I} : A{\to}A;$
$\vdash_{\boldsymbol{\lambda}\to} \mathsf{K} : A{\to}B{\to}A;$
$\vdash_{\boldsymbol{\lambda}\to} \mathsf{S} : (A{\to}B{\to}C){\to}(A{\to}B){\to}A{\to}C.$

(ii) Also one has

$$\begin{aligned} x{:}A &\vdash_{\boldsymbol{\lambda}\to} \mathsf{I}x : A; \\ x{:}A, y{:}B &\vdash_{\boldsymbol{\lambda}\to} \mathsf{K}xy : A; \\ x{:}(A{\to}B{\to}C), y{:}(A{\to}B), z{:}A &\vdash_{\boldsymbol{\lambda}\to} \mathsf{S}xyz : C. \end{aligned}$$

(iii) The terms $\mathsf{Y}, \boldsymbol{\Omega}$ do not have a type. This is obvious after some effort. As we will see later, a systematic reason is that all typable terms have a normal form, but these two do not.

(iv) The term $\omega \equiv \lambda x.xx$ is in normal form but does not have a type either.

1.1.17 Notation Another way of writing these rules is sometimes found in the literature, see Fig. 1.7. In this version the basis is considered as implicit

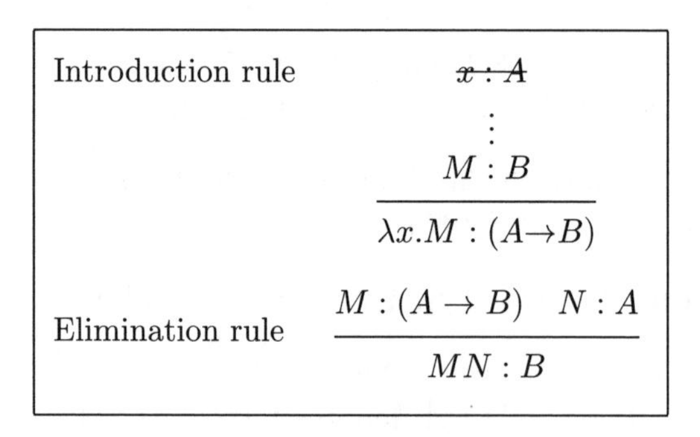

Figure 1.7 $\boldsymbol{\lambda}_{\to}^{\mathrm{Cu}}$ alternative version

and is not notated. The notation

$$\begin{array}{c} x : A \\ \vdots \\ M : B \end{array}$$

denotes that $M : B$ can be derived from $x{:}A$ and the 'axioms' in the basis. Striking through $x{:}A$ means that for the conclusion $\lambda x.M : A{\to}B$ the assumption $x{:}A$ is no longer needed; it is *discharged.*

1.1.18 Example

(i) $\vdash (\lambda xy.x) : (A \to B \to A)$ for all $A, B \in \mathbb{T}$.

We will use the notation of version 1 of $\boldsymbol{\lambda}_{\to}^{\mathbb{A}}$ for a derivation of this statement:

$$\dfrac{\dfrac{a{:}A, y{:}B \vdash x : A}{x{:}A \vdash (\lambda y.x) : B{\to}A}}{\vdash (\lambda x\lambda y.x) : A{\to}B{\to}A\ .}$$

" Note that $\lambda xy.x \equiv \lambda x\lambda y.x$ by definition.

(ii) A *natural deduction* derivation (for the alternative version of the system) of the same type assignment is

$$\dfrac{\dfrac{\dfrac{\cancel{x{:}A}\ 2 \qquad \cancel{y{:}B}\ 1}{x{:}A}}{(\lambda y.x)\ :\ (B \to A)}\ 1}{(\lambda xy.x)\ :\ (A \to B \to A)}\ 2.$$

The indices 1 and 2 are bookkeeping devices that indicate at which application of a rule a particular assumption is being discharged.

(iii) A more explicit way of dealing with cancellations of statements is the 'flag-notation' used by Fitch (1952) and in the languages Automath of de Bruijn (1980). In this notation the above derivation becomes

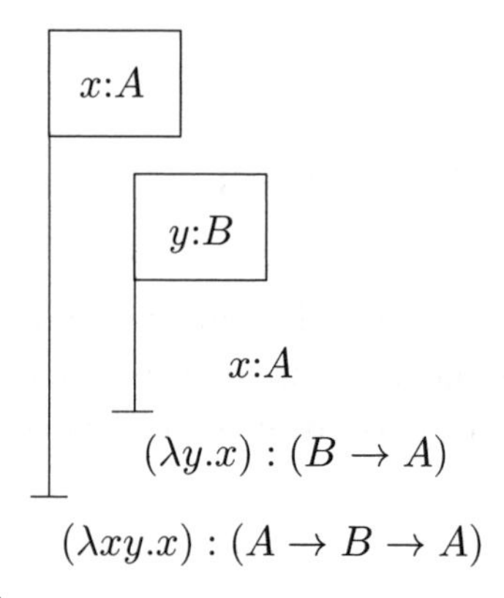

As one sees, the bookkeeping of cancellations is very explicit; on the other hand it is less obvious how a statement is derived from previous statements whenever applications are used.

(iv) Similarly one can show that, for all $A \in \mathbb{T}$,

$$\vdash (\lambda x.x) : (A \to A).$$

(v) An example with a non-empty basis is $y{:}A \vdash (\lambda x.x)y : A$.

In the rest of this chapter and in fact in the rest of this book we usually will introduce systems of typed lambda calculi in the style of the first variant of $\boldsymbol{\lambda}^{\mathbb{A}}_{\to}$.

1.1.19 Definition Let Γ be a basis and $A \in \mathbb{T} = \mathbb{T}^{\mathbb{A}}$. Then:

(i) $\Lambda^{\Gamma}_{\to}(A) = \{M \in \Lambda \mid \Gamma \vdash_{\boldsymbol{\lambda}^{\mathbb{A}}_{\to}} M : A\}$;
(ii) $\Lambda^{\Gamma}_{\to} = \bigcup_{A \in \mathbb{T}} \Lambda^{\Gamma}_{\to}(A)$;
(iii) $\Lambda_{\to}(A) = \Lambda^{\emptyset}_{\to}(A)$;
(iv) $\Lambda_{\to} = \Lambda^{\emptyset}_{\to}$.
(v) Emphasizing the dependency on $\mathbb{A}$, we write $\Lambda^{\mathbb{A}}_{\to}(A)$ or $\Lambda^{\mathbb{A},\Gamma}_{\to}(A)$, etc.

1.1.20 Definition Let Γ be a basis, $A \in \mathbb{T}$ and $M \in \Lambda$. Then:

(i) If $M \in \Lambda_{\to}(A)$, then we say that
M has type A or *A is inhabited by M*;
(ii) If $M \in \Lambda_{\to}$, then M is called *typable*.
(iii) If $M \in \Lambda^{\Gamma}_{\to}(A)$, then *$M$ has type A relative to* Γ;
(iv) If $M \in \Lambda^{\Gamma}_{\to}$, then M is called *typable relative to* Γ;
(v) If $\Lambda^{\Gamma}_{\to}(A) \neq \emptyset$, then A is *inhabited relative to* Γ.

1.1.21 Example We have

$$\begin{aligned}\mathsf{K} &\in \Lambda_{\rightarrow}^{\emptyset}(A{\rightarrow}B{\rightarrow}A);\\ \mathsf{K}x &\in \Lambda_{\rightarrow}^{\{x:A\}}(B{\rightarrow}A).\end{aligned}$$

1.1.22 Definition Let $A \in \mathbb{T}$.

(i) The *depth* of A, written $\mathrm{dpt}(A)$, is defined as follows:

$$\begin{aligned}\mathrm{dpt}(\alpha) &= 1\\ \mathrm{dpt}(A{\rightarrow}B) &= \max\{\mathrm{dpt}(A), \mathrm{dpt}(B)\} + 1.\end{aligned}$$

(ii) The *rank* of A, written $\mathrm{rk}(A)$, is defined as follows:

$$\begin{aligned}\mathrm{rk}(\alpha) &= 0\\ \mathrm{rk}(A{\rightarrow}B) &= \max\{\mathrm{rk}(A) + 1, \mathrm{rk}(B)\}.\end{aligned}$$

(iii) The *order* of A, written $\mathrm{ord}(A)$, is defined as follows:

$$\begin{aligned}\mathrm{ord}(\alpha) &= 1\\ \mathrm{ord}(A{\rightarrow}B) &= \max\{\mathrm{ord}(A) + 1, \mathrm{ord}(B)\}.\end{aligned}$$

(iv) The depth of a basis Γ is

$$\mathrm{dpt}(\Gamma) = \max_i\{\mathrm{dpt}(A_i) \mid (x_i{:}A_i) \in \Gamma\}.$$

Similarly we define $\mathrm{rk}(\Gamma)$ and $\mathrm{ord}(\Gamma)$. Note that $\mathrm{ord}(A) = \mathrm{rk}(A) + 1$.

The notion of 'order' comes from logic, where dealing with elements of type 0 is done in 'first order' predicate logic. The reason is that in first-order logic one deals with domains and their elements. In second-order logic one deals with functions between first-order objects. In this terminology 0th-order logic can be identified with propositional logic. The notion of 'rank' comes from computer science.

1.1.23 Definition For $A \in \mathbb{T}$ we define $A^k{\rightarrow}B$ by recursion on k:

$$\begin{aligned}A^0{\rightarrow}B &= B;\\ A^{k+1}{\rightarrow}B &= A{\rightarrow}A^k{\rightarrow}B.\end{aligned}$$

Note that $\mathrm{rk}(A^k{\rightarrow}B) = \mathrm{rk}(A{\rightarrow}B)$, for all $k > 0$.

Several properties can be proved by induction on the depth of a type. This holds for example for Lemma 1.1.26(i).

The asymmetry in the definition of rank is intended because the meaning of a type like $(0{\rightarrow}0){\rightarrow}0$ is more complex than that of $0{\rightarrow}0{\rightarrow}0$, as can be seen by looking to the inhabitants of these types: functionals with functions as arguments versus binary functions. Some authors use the name *type level* instead of 'rank'.

The minimal and maximal systems $\boldsymbol{\lambda}_{\to}^{0}$ and $\boldsymbol{\lambda}_{\to}^{\infty}$

The collection $\mathbb{A}$ of type variables serves as set of base types from which other types are constructed. We have $\mathbb{A}_0 = \{0\}$ with just one type atom and $\mathbb{A}_\infty = \{\alpha_0, \alpha_1, \alpha_2, \dots\}$ with infinitely many of them. These two sets of atoms and their resulting type systems play a major role in this Part I of the book.

1.1.24 Definition We define the following systems of type assignment:

(i) $\boldsymbol{\lambda}_{\to}^{0} = \boldsymbol{\lambda}_{\to}^{\mathbb{A}_0}$;
(ii) $\boldsymbol{\lambda}_{\to}^{\infty} = \boldsymbol{\lambda}_{\to}^{\mathbb{A}_\infty}$.

Focusing on $\mathbb{A}_0$ or $\mathbb{A}_\infty$ we write $\Lambda_{\to}^{0}(A) = \Lambda_{\to}^{\mathbb{A}_0}(A)$ or $\Lambda_{\to}^{\infty}(A) = \Lambda_{\to}^{\mathbb{A}_\infty}(A)$ respectively.

Many of the interesting features of the 'larger' $\boldsymbol{\lambda}_{\to}^{\infty}$ are already present in the minimal version $\boldsymbol{\lambda}_{\to}^{0}$.

1.1.25 Definition

(i) The following types of $\mathbb{T}^0 \subseteq \mathbb{T}^{\mathbb{A}}$ are often used.

$$0 = 0,\ 1 = 0{\to}0,\ 2 = (0{\to}0){\to}0,\ \dots .$$

In general

$$0 = 0 \text{ and } k+1 = k{\to}0.$$

Note that $\mathrm{rk}(n) = n$.

(ii) Define n_k by cases on n:

$$\begin{aligned} 0_k &= 0; \\ (n+1)_k &= n^k{\to}0. \end{aligned}$$

For example,

$$\begin{aligned} 1_0 &= 0; \\ 1_2 &= 0{\to}0{\to}0; \\ 2_3 &= 1{\to}1{\to}1{\to}0; \\ 1^2{\to}2{\to}0 &= (0{\to}0){\to}(0{\to}0){\to}((0{\to}0){\to}0){\to}0. \end{aligned}$$

Notice that $\mathrm{rk}(n_k) = \mathrm{rk}(n)$, for $k > 0$.

The notation n_k is used only for $n \in \omega$. In the following lemma the notation $A_1, \dots, A_a$ with subscripts denotes as usual a sequence of types.

1.1.26 Lemma

(i) *Every type* A *of* $\lambda_{\to}^{\infty}$ *is of the form*

$$A = A_1 \to A_2 \to \cdots \to A_a \to \alpha.$$

(ii) *Every type* A *of* $\boldsymbol{\lambda}_{\to}^{0}$ *is of the form*

$$A = A_1 \to A_2 \to \cdots \to A_a \to 0.$$

(iii) *We have* $\mathrm{rk}(A_1 \to A_2 \to \cdots \to A_a \to \alpha) = \max\{\mathrm{rk}(A_i) + 1 \mid 1 \le i \le n\}$.

Proof (i) By induction on the structure (depth) of A. If $A = \alpha$, then this holds for $n = 0$. If $A = B \to C$, then by the induction hypothesis one has $C = C_1 \to \cdots \to C_c \to \gamma$. Hence $A = B \to C_1 \to \cdots \to C_c \to \gamma$.

(ii) Similar to (i).

(iii) By induction on n. ■

1.1.27 Notation Let $A \in \mathbb{T}^{\mathbb{A}}$ and suppose $A = A_1 \to A_2 \to \cdots \to A_a \to \alpha$. Then the A_i are called the *components* of A. We write

$$\begin{array}{rcll} \mathtt{arity}(A) & = & a; & \\ A(i) & = & A_i, & \text{for } 1 \le i \le a; \\ \mathtt{target}(A) & = & \alpha. & \end{array}$$

Iterated components are denoted by

$$A(i,j) = A(i)(j).$$

1.1.28 Remark We usually work with $\boldsymbol{\lambda}_{\to}^{\mathbb{A}}$ for an unspecified $\mathbb{A}$, but will be more specific in some cases.

Different versions of $\boldsymbol{\lambda}_{\to}^{\mathbb{A}}$

We will introduce several variants of $\boldsymbol{\lambda}_{\to}^{\mathbb{A}}$.

The Curry version of $\boldsymbol{\lambda}_{\to}^{\mathbb{A}}$

1.1.29 Definition The system $\boldsymbol{\lambda}_{\to}^{\mathbb{A}}$ that was introduced in Definition 1.1.14 assigns types to untyped lambda terms. To be explicit it will be referred to as the *Curry version* and be denoted by $\boldsymbol{\lambda}_{\to}^{\mathbb{A},\mathrm{Cu}}$ or $\boldsymbol{\lambda}_{\to}^{\mathrm{Cu}}$, as the set $\mathbb{A}$ often does not need to be specified.

The Curry version of $\boldsymbol{\lambda}_{\to}^{\mathbb{A}}$ is called *implicitly typed* as an expression like

$$\lambda x.x\mathsf{K}$$

has a type, but this has not been done. In Section 2.3 we will see that this task is feasible. In systems more complex than $\boldsymbol{\lambda}_{\to}^{\mathbb{A}}$, finding types in the implicit version is more complicated and may even not be computable. This will be the case with second- and higher-order types, like $\boldsymbol{\lambda 2}$ (system F), see Girard et al. (1989), Barendregt (1992) or Sørensen and Urzyczyn (2006) for a description of that system and Wells (1999) for the undecidability.

The Church version $\boldsymbol{\lambda}_{\rightarrow}^{\mathrm{Ch}}$ *of* $\boldsymbol{\lambda}_{\rightarrow}^{\mathbb{A}}$

The first variant of $\boldsymbol{\lambda}_{\rightarrow}^{\mathrm{Cu}}$ is the *Church version* of $\boldsymbol{\lambda}_{\rightarrow}^{\mathbb{A}}$, denoted by $\boldsymbol{\lambda}_{\rightarrow}^{\mathbb{A},\mathrm{Ch}}$ or $\boldsymbol{\lambda}_{\rightarrow}^{\mathrm{Ch}}$. In this theory the types are assigned to embellished terms in which the variables (free and bound) come with types attached. For example the Curry style type assignments

$$\vdash_{\boldsymbol{\lambda}_{\rightarrow}}^{\mathrm{Cu}} \quad (\lambda x.x) : A{\rightarrow}A \qquad (1_{\mathrm{Cu}})$$

$$y{:}A \ \vdash_{\boldsymbol{\lambda}_{\rightarrow}}^{\mathrm{Cu}} \quad (\lambda x.xy) : (A{\rightarrow}B){\rightarrow}B \qquad (2_{\mathrm{Cu}})$$

now become

$$(\lambda x^A.x^A) \in \Lambda_{\rightarrow}^{\mathrm{Ch}}(A{\rightarrow}A) \qquad (1_{\mathrm{Ch}})$$

$$(\lambda x^{A\rightarrow B}.x^{A\rightarrow B}y^A) \in \Lambda_{\rightarrow}^{\mathrm{Ch}}((A{\rightarrow}B){\rightarrow}B) \qquad (2_{\mathrm{Ch}})$$

1.1.30 Definition Let $\mathbb{A}$ be a set of type atoms. The Church version of $\boldsymbol{\lambda}_{\rightarrow}^{\mathbb{A}}$, written $\boldsymbol{\lambda}_{\rightarrow}^{\mathbb{A},\mathrm{Ch}}$ or $\boldsymbol{\lambda}_{\rightarrow}^{\mathrm{Ch}}$ if $\mathbb{A}$ is not emphasized, is defined as follows. The system has the same set of types $\mathbb{T}^{\mathbb{A}}$ as $\boldsymbol{\lambda}_{\rightarrow}^{\mathbb{A},\mathrm{Cu}}$.

(i) *The set of term variables is different*: each such variable is coupled with a unique type. This is done in such a way that every type has infinitely many variables coupled to it. Thus we take

$$\mathsf{V}^{\mathbb{T}} = \{x^{t(x)} \mid x \in \mathsf{V}\},$$

where $t : \mathsf{V}{\rightarrow}\mathbb{T}^{\mathbb{A}}$ is a fixed map such that $t^{-1}(A)$ is infinite for all $A \in \mathbb{T}^{\mathbb{A}}$. So we have

$$\{x^A, y^A, z^A, \ldots\} \subseteq \mathsf{V}^{\mathbb{T}} \text{ is infinite for all } A \in \mathbb{T}^{\mathbb{A}};$$
$$x^A, x^B \in \mathsf{V}^{\mathbb{T}} \Rightarrow A \equiv B, \text{ for all } A, B \in \mathbb{T}^{\mathbb{A}}.$$

(ii) *Terms of type* A, written $\Lambda_{\rightarrow}^{\mathrm{Ch}}(A)$, are defined as in Fig. 1.8.

$$\begin{array}{rcl} & & x^A \in \Lambda_{\rightarrow}^{\mathrm{Ch}}(A); \\ M \in \Lambda_{\rightarrow}^{\mathrm{Ch}}(A{\rightarrow}B), N \in \Lambda_{\rightarrow}^{\mathrm{Ch}}(A) & \Rightarrow & (MN) \in \Lambda_{\rightarrow}^{\mathrm{Ch}}(B); \\ M \in \Lambda_{\rightarrow}^{\mathrm{Ch}}(B) & \Rightarrow & (\lambda x^A.M) \in \Lambda_{\rightarrow}^{\mathrm{Ch}}(A{\rightarrow}B). \end{array}$$

Figure 1.8 The system $\boldsymbol{\lambda}_{\rightarrow}^{\mathrm{Ch}}$ of typed terms *à la* Church

(iii) The set of *terms of* $\boldsymbol{\lambda}_{\rightarrow}^{\mathrm{Ch}}$, written $\Lambda_{\rightarrow}^{\mathrm{Ch}}$, is defined as

$$\Lambda_{\rightarrow}^{\mathrm{Ch}} = \bigcup_{A \in \mathbb{T}} \Lambda_{\rightarrow}^{\mathrm{Ch}}(A).$$

For example

$$\begin{aligned} y^{B \rightarrow A} x^{B} &\in \Lambda_{\rightarrow}^{\mathrm{Ch}}(A) \\ \lambda x^{A}.y^{B \rightarrow A} &\in \Lambda_{\rightarrow}^{\mathrm{Ch}}(A \rightarrow B \rightarrow A) \\ \lambda x^{A}.x^{A} &\in \Lambda_{\rightarrow}^{\mathrm{Ch}}(A \rightarrow A). \end{aligned}$$

1.1.31 Definition On $\Lambda_{\rightarrow}^{\mathrm{Ch}}$ we define the notions of reduction, given in Fig. 1.9.

$$\begin{array}{rcll} (\lambda x^{A}.M)N & \rightarrow & M[x^{A}: = N] & & (\boldsymbol{\beta}) \\ \lambda x^{A}.Mx^{A} & \rightarrow & M, & \text{if } x^{A} \notin \mathrm{FV}(M) & (\boldsymbol{\eta}) \end{array}$$

Figure 1.9 $\boldsymbol{\beta\eta}$-contraction rules for $\boldsymbol{\lambda}_{\rightarrow}^{\mathrm{Ch}}$

It will be shown in Proposition 1.2.10 that $\Lambda_{\rightarrow}^{\mathrm{Ch}}(A)$ is closed under $\boldsymbol{\beta\eta}$-reduction; i.e. this reduction preserves the type of a typed term.

As usual, see B[1984], these notions of reduction generate the corresponding reduction relations. Also there are the corresponding conversion relations $=_{\boldsymbol{\beta}}$, $=_{\boldsymbol{\eta}}$ and $=_{\boldsymbol{\beta\eta}}$. Terms in $\boldsymbol{\lambda}_{\rightarrow}^{\mathrm{Ch}}$ will often be considered modulo $=_{\boldsymbol{\beta}}$ or $=_{\boldsymbol{\beta\eta}}$. The notation $M = N$ means $M =_{\boldsymbol{\beta\eta}} N$ by default.

1.1.32 Definition (Type substitution) For $M \in \Lambda_{\rightarrow}^{\mathrm{Ch}}$, $\alpha \in \mathbb{A}$, and $B \in \mathbb{T}^{\mathbb{A}}$ we define the result of substituting B for α in M, written $M[\alpha := B]$, recursively as follows:

M	$M[\alpha := B]$
x^{A}	$x^{A[\alpha:=B]}$
PQ	$(P[\alpha := B])(Q[\alpha := B])$
$\lambda x^{A}.P$	$\lambda x^{A[\alpha:=B]}.P[\alpha := B]$

1.1.33 Notation A term like $(\lambda f^{1}x^{0}.f^{1}(f^{1}x^{0})) \in \Lambda_{\rightarrow}^{\mathrm{Ch}}(1 \rightarrow 0 \rightarrow 0)$ will also be written as

$$\lambda f^{1}x^{0}.f(fx)$$

simply indicating the types of the bound variables. This notation is analogous to that in the de Bruijn version of $\boldsymbol{\lambda}_{\rightarrow}^{\mathbb{A}}$ given below. Sometimes we will even write $\lambda fx.f(fx)$. We will return to this notational issue in Section 1.2.

The de Bruijn version $\boldsymbol{\lambda}_{\rightarrow}^{\mathrm{dB}}$ *of* $\boldsymbol{\lambda}_{\rightarrow}^{\mathbb{A}}$

There is the following disadvantage about the Church systems. Consider

$$\mathsf{I} \equiv \lambda x^A.x^A.$$

In applications one considers *dependent types* coming from the Automath language family, see Nederpelt et al. (1994), designed for formalizing arguments and proof-checking[2]. These are types that depend on a term variable (ranging over another type). An intuitive example is A^n, where n is a variable ranging over natural numbers. A more formal example is Px, where $x : A$ and $P : A \rightarrow \mathbb{T}$. In this way types may contain redexes and we may have the following reduction

$$\mathsf{I} \equiv \lambda x^A.x^A \rightarrow_{\boldsymbol{\beta}} \lambda x^{A'}.x^A,$$

in the case $A \rightarrow_{\boldsymbol{\beta}} A'$, by reducing only the first A to A'. The question now is whether $\lambda x^{A'}$ binds the x^A. If we write I as

$$\mathsf{I} \equiv \lambda x{:}A.x,$$

then this problem disappears:

$$\lambda x{:}A.x \twoheadrightarrow \lambda x{:}A'.x.$$

As the second occurrence of x is implicitly typed with the same type as the first, the intended meaning is correct. In the following system $\boldsymbol{\lambda}_{\rightarrow}^{\mathbb{A},\mathrm{dB}}$ this idea is formalized.

1.1.34 Definition The second variant of $\boldsymbol{\lambda}_{\rightarrow}^{\mathrm{Cu}}$ is the de *Bruijn version* of $\boldsymbol{\lambda}_{\rightarrow}^{\mathbb{A}}$, denoted by $\lambda_{\rightarrow}^{\mathbb{A},\mathrm{dB}}$ or $\boldsymbol{\lambda}_{\rightarrow}^{\mathrm{dB}}$. Now only bound variables get ornamented with types, but only at the binding stage. The examples (1_{Ch}), (1_{Cu}), (2_{Ch}), (2_{Cu}) now become

$$\vdash_{\boldsymbol{\lambda}_{\rightarrow}}^{\mathrm{dB}} \ (\lambda x{:}A.x) : A \rightarrow A \qquad (1_{\mathrm{dB}})$$

$$y{:}A \ \vdash_{\boldsymbol{\lambda}_{\rightarrow}}^{\mathrm{dB}} \ (\lambda x{:}(A \rightarrow B).xy) : (A \rightarrow B) \rightarrow B \qquad (2_{\mathrm{dB}})$$

[2] The proof-assistant Coq, see <coq.inria.fr> and Bertot and Castéran (2004), is a modern version of Automath in which one uses for formal proofs typed lambda terms in the de Bruijn style.

1.1.35 Definition The system $\boldsymbol{\lambda}_{\to}^{\mathrm{dB}}$ starts with a collection of *pseudo-terms*, written $\Lambda_{\to}^{\mathrm{dB}}$, defined by the following simplified syntax:

$$\boxed{\Lambda_{\to}^{\mathrm{dB}} \quad ::= \quad \mathsf{V} \mid \Lambda_{\to}^{\mathrm{dB}}\,\Lambda_{\to}^{\mathrm{dB}} \mid \lambda\mathsf{V}{:}\mathbb{T}.\Lambda_{\to}^{\mathrm{dB}}.}$$

For example $\lambda x{:}\alpha.x$ and $(\lambda x{:}\alpha.x)(\lambda y{:}\beta.y)$ are pseudo-terms. As we will see, the first one is a *legal*, i.e. actually typable, term in $\boldsymbol{\lambda}_{\to}^{\mathbb{A},\mathrm{dB}}$, whereas the second one is not.

1.1.36 Definition

(i) A basis Γ consists of a set of declarations $x{:}A$ with distinct term variables x and types $A \in \mathbb{T}^{\mathbb{A}}$. This is exactly the same as for $\boldsymbol{\lambda}_{\to}^{\mathbb{A},\mathrm{Cu}}$.
(ii) The system of type assignment obtaining statements $\Gamma \vdash M : A$ with Γ a basis, M a pseudoterm and A a type, is defined in Figure 1.10.

(axiom)	$\Gamma \vdash x : A,$	if $(x{:}A) \in \Gamma$;
($\to$-elimination)	$\dfrac{\Gamma \vdash M : (A \to B) \quad \Gamma \vdash N : A}{\Gamma \vdash (MN) : B};$	
($\to$-introduction)	$\dfrac{\Gamma, x{:}A \vdash M : B}{\Gamma \vdash (\lambda x{:}A.M) : (A \to B)}.$	

Figure 1.10 The system $\boldsymbol{\lambda}_{\to}^{\mathrm{dB}}$ à la *de Bruijn*

Provability in $\boldsymbol{\lambda}_{\to}^{\mathrm{dB}}$ is denoted by $\vdash_{\boldsymbol{\lambda}_{\to}}^{\mathrm{dB}}$. Thus the legal terms of $\boldsymbol{\lambda}_{\to}^{\mathrm{dB}}$ are defined by making a selection from the context-free language $\Lambda_{\to}^{\mathrm{dB}}$. That $\lambda x{:}\alpha.x$ is legal follows from $x{:}\alpha \vdash_{\boldsymbol{\lambda}_{\to}}^{\mathrm{dB}} x : \alpha$ using the $\to$-introduction rule. That $(\lambda x{:}\alpha.x)(\lambda y{:}\beta.y)$ is not legal follows from Proposition 1.2.12. These legal terms do not form a context-free language, see Exercise 1.7. For closed terms the Church and the de Bruijn notation are isomorphic.

1.2 First properties and comparisons

In this section we will present simple properties of the systems $\boldsymbol{\lambda}_{\to}^{\mathbb{A}}$. Deeper properties, like normalization of typable terms, will be considered in Sections 2.1, 2.2.

Properties of $\boldsymbol{\lambda}_{\to}^{\mathrm{Cu}}$

We start with properties of the system $\boldsymbol{\lambda}_{\to}^{\mathrm{Cu}}$.

1.2.1 Proposition (Weakening lemma for $\boldsymbol{\lambda}_{\to}^{\mathrm{Cu}}$) *Suppose* $\Gamma \vdash M : A$ *and* Γ' *is a basis with* $\Gamma \subseteq \Gamma'$. *Then* $\Gamma' \vdash M : A$.

Proof By induction on the derivation of $\Gamma \vdash M : A$. ■

1.2.2 Lemma (Free variable lemma for $\boldsymbol{\lambda}_{\to}^{\mathrm{Cu}}$) *For a set* $\mathcal{X}$ *of variables write*

$$\Gamma \restriction \mathcal{X} = \{x{:}A \in \Gamma \mid x \in \mathcal{X}\}.$$

(i) *Suppose* $\Gamma \vdash M : A$. *Then* $FV(M) \subseteq \mathrm{dom}(\Gamma)$.
(ii) *If* $\Gamma \vdash M : A$, *then* $\Gamma \restriction \mathrm{FV}(M) \vdash M : A$.

Proof (i), (ii) By induction on the generation of $\Gamma \vdash M : A$. ■

The following result is related to the fact that the system $\boldsymbol{\lambda}_{\to}$ is 'syntax directed', i.e. statements $\Gamma \vdash M : A$ have a unique proof.

1.2.3 Proposition (Inversion Lemma for $\boldsymbol{\lambda}_{\to}^{\mathrm{Cu}}$)

(i) $\Gamma \vdash x : A \;\Rightarrow\; (x{:}A) \in \Gamma$.
(ii) $\Gamma \vdash MN : A \;\Rightarrow\; \exists B \in \mathbb{T}\, [\Gamma \vdash M : B{\to}A \;\&\; \Gamma \vdash N : B]$.
(iii) $\Gamma \vdash \lambda x.M : A \;\Rightarrow\; \exists B, C \in \mathbb{T}\, [A \equiv B{\to}C \;\&\; \Gamma, x{:}B \vdash M : C]$.

Proof (i) Suppose $\Gamma \vdash x : A$ holds in $\boldsymbol{\lambda}_{\to}$. The last rule in a derivation of this statement cannot be an application or an abstraction, since x is not of the right form. Therefore it must be an axiom, i.e. $(x{:}A) \in \Gamma$.

(ii), (iii) These two implications are proved similarly. ■

1.2.4 Corollary *Let* $\Gamma \vdash_{\boldsymbol{\lambda}_{\to}}^{\mathrm{Cu}} xN_1 \ldots N_k : B$. *Then there exist unique* $A_1, \ldots, A_k \in \mathbb{T}$ *such that*

$$\Gamma \vdash_{\boldsymbol{\lambda}_{\to}}^{\mathrm{Cu}} N_i : A_i,\ 1 \leq i \leq k,\ \textit{and}\ x{:}(A_1 \to \cdots \to A_k \to B) \in \Gamma.$$

Proof By applying (ii) k times and then (i) of the proposition. ■

1.2.5 Proposition (Substitution lemma for $\boldsymbol{\lambda}_{\to}^{\mathrm{Cu}}$)

(i) $\Gamma, x{:}A \vdash M : B \;\&\; \Gamma \vdash N : A \;\Rightarrow\; \Gamma \vdash M[x{:} = N] : B$.
(ii) $\Gamma \vdash M : A \;\Rightarrow\; \Gamma[\alpha := B] \vdash M : A[\alpha := B]$.

Proof (i) We proceed by induction on the derivation of $\Gamma, x{:}A \vdash M : B$. Write $P^* \equiv P[x{:} = N]$.

Case 1. $\Gamma, x{:}A \vdash M : B$ is an axiom, hence $M \equiv y$ and $(y{:}B) \in \Gamma \cup \{x{:}A\}$.

Subcase 1.1. $(y{:}B) \in \Gamma$. Then $y \not\equiv x$ and $\Gamma \vdash M^* \equiv y[x{:}N] \equiv y : B$.

Subcase 1.2. $y{:}B \equiv x{:}A$. Then $y \equiv x$ and $B \equiv A$, and it follows that $\Gamma \vdash M^* \equiv N : A \equiv B$.

Case 2. $\Gamma, x{:}A \vdash M : B$ follows from $\Gamma, x{:}A \vdash F : C{\to}B$, $\Gamma, x{:}A \vdash G : C$

and $FG \equiv M$. By the induction hypothesis one has $\Gamma \vdash F^* : C{\to}B$ and $\Gamma \vdash G^* : C$. Hence $\Gamma \vdash (FG)^* \equiv F^*G^* : B$.

Case 3. $\Gamma, x{:}A \vdash M : B$ follows from $\Gamma, x{:}A, y{:}D \vdash G : E$, $B \equiv D{\to}E$ and $\lambda y.G \equiv M$. By the induction hypothesis $\Gamma, y{:}D \vdash G^* : E$, therefore $\Gamma \vdash (\lambda y.G)^* \equiv \lambda y.G^* : D{\to}E \equiv B$.

(ii) Similarly. ∎

1.2.6 Proposition (Subject reduction property for $\boldsymbol{\lambda}_{\to}^{\text{Cu}}$)

$$\Gamma \vdash M : A \;\&\; M \twoheadrightarrow_{\beta\eta} N \;\Rightarrow\; \Gamma \vdash N : A.$$

Proof It suffices to show this for a one-step $\beta\eta$-reduction, denoted by $\to$. Suppose $\Gamma \vdash M : A$ and $M \to_{\beta\eta} N$ in order to show that $\Gamma \vdash N : A$. We do this by induction on the derivation of $\Gamma \vdash M : A$.

Case 1. $\Gamma \vdash M : A$ is an axiom. Then M is a variable, contradicting $M \to N$. Hence this case cannot occur.

Case 2. $\Gamma \vdash M : A$ is $\Gamma \vdash FP : A$ directly follows from $\Gamma \vdash F : B{\to}A$ and $\Gamma \vdash P : B$. Since $FP \equiv M \to N$ we can have three subcases.

Subcase 2.1. $N \equiv F'P$ with $F \to F'$.

Subcase 2.2. $N \equiv FP'$ with $P \to P'$.

In these two subcases it follows that $\Gamma \vdash N : A$, by using the induction hypothesis twice.

Subcase 2.3. $F \equiv \lambda x.G$ and $N \equiv G[x{:}= P]$. Since

$$\Gamma \vdash \lambda x.G : B{\to}A \;\&\; \Gamma \vdash P : B$$

it follows by the Inversion Lemma 1.2.3 for $\boldsymbol{\lambda}_{\to}$ that

$$\Gamma, x \vdash G : A \;\&\; \Gamma \vdash P : B.$$

Therefore by the Substitution Lemma 1.2.5 for $\boldsymbol{\lambda}_{\to}$ it follows that

$$\Gamma \vdash G[x{:}= P] : A, \text{ i.e. } \Gamma \vdash N : A.$$

Case 3. $\Gamma \vdash M : A$ is $\Gamma \vdash \lambda x.P : B{\to}C$ and follows from $\Gamma, x \vdash P : C$.

Subcase 3.1. $N \equiv \lambda x.P'$ with $P \to P'$. By the induction hypothesis one has $\Gamma, x{:}B \vdash P' : C$, hence $\Gamma \vdash (\lambda x.P') : (B{\to}C)$, i.e. $\Gamma \vdash N : A$.

Subcase 3.2. $P \equiv Nx$ and $x \notin \text{FV}(N)$. Now $\Gamma, x{:}B \vdash Nx : C$ follows by Lemma 1.2.3(ii) from $\Gamma, x{:}B \vdash N : (B'{\to}C)$ and $\Gamma, x{:}B \vdash x : B'$, for some B'. Then $B = B'$, by Lemma 1.2.3(i), hence by Lemma 1.2.2(ii) we have $\Gamma \vdash N : (B{\to}C) = A$. ∎

The following result also holds for $\boldsymbol{\lambda}_{\to}^{\text{Ch}}$ and $\boldsymbol{\lambda}_{\to}^{\text{dB}}$; see Proposition 1.2.29 and Exercise 2.4.

1.2.7 Corollary (Church–Rosser theorem for $\boldsymbol{\lambda}_{\rightarrow}^{\mathrm{Cu}}$) *On typable terms of $\boldsymbol{\lambda}_{\rightarrow}^{\mathrm{Cu}}$ the Church–Rosser theorem holds for the notions of reduction $\twoheadrightarrow_{\beta}$, $\twoheadrightarrow_{\eta}$, and $\twoheadrightarrow_{\beta\eta}$.*

(i) *Let $M, N_1, N_2 \in \Lambda_{\rightarrow}^{\Gamma}(A)$. Then*

$$M \twoheadrightarrow_{\beta\eta} N_1 \;\&\; M \twoheadrightarrow_{\beta(\eta)} N_2 \;\Rightarrow\; \exists Z \in \Lambda_{\rightarrow}^{\Gamma}(A).N_1 \twoheadrightarrow_{\beta(\eta)} Z \;\&\; N_2 \twoheadrightarrow_{\beta(\eta)} Z.$$

(ii) *Let $M, N \in \Lambda_{\rightarrow}^{\Gamma}(A)$. Then*

$$M =_{\beta(\eta)} N \;\Rightarrow\; \exists Z \in \Lambda_{\rightarrow}^{\Gamma}(A).M \twoheadrightarrow_{\beta(\eta)} Z \;\&\; N \twoheadrightarrow_{\beta(\eta)} Z.$$

Proof By the Church–Rosser theorems for $\twoheadrightarrow_{\beta}$, $\twoheadrightarrow_{\eta}$, and $\twoheadrightarrow_{\beta\eta}$ on untyped terms, Theorem 1.1.9, and Proposition 1.2.6. ■

Properties of $\boldsymbol{\lambda}_{\rightarrow}^{\mathrm{Ch}}$

Not all the properties of $\boldsymbol{\lambda}_{\rightarrow}^{\mathrm{Cu}}$ are meaningful for $\boldsymbol{\lambda}_{\rightarrow}^{\mathrm{Ch}}$. Even those that are have to be reformulated slightly.

1.2.8 Proposition (Inversion Lemma for $\boldsymbol{\lambda}_{\rightarrow}^{\mathrm{Ch}}$)

(i) $x^B \in \Lambda_{\rightarrow}^{\mathrm{Ch}}(A) \quad\Rightarrow\quad B = A.$

(ii) $(MN) \in \Lambda_{\rightarrow}^{\mathrm{Ch}}(A) \quad\Rightarrow\quad \exists B \in \mathbb{T}.[M \in \Lambda_{\rightarrow}^{\mathrm{Ch}}(B \rightarrow A) \;\&\; N \in \Lambda_{\rightarrow}^{\mathrm{Ch}}(B)].$

(iii) $(\lambda x^B.M) \in \Lambda_{\rightarrow}^{\mathrm{Ch}}(A) \quad\Rightarrow\quad \exists C \in \mathbb{T}.[A = (B \rightarrow C) \;\&\; M \in \Lambda_{\rightarrow}^{\mathrm{Ch}}(C)].$

Proof As before. ■

Substitution of a term $N \in \Lambda_{\rightarrow}^{\mathrm{Ch}}(B)$ for a typed variable x^B is defined as usual. We show that the resulting term keeps its type.

1.2.9 Proposition (Substitution lemma for $\boldsymbol{\lambda}_{\rightarrow}^{\mathrm{Ch}}$) *Let $A, B \in \mathbb{T}$. Then:*

(i) $M \in \Lambda_{\rightarrow}^{\mathrm{Ch}}(A),\ N \in \Lambda_{\rightarrow}^{\mathrm{Ch}}(B) \;\Rightarrow\; (M[x^B := N]) \in \Lambda_{\rightarrow}^{\mathrm{Ch}}(A)$;

(ii) $M \in \Lambda_{\rightarrow}^{\mathrm{Ch}}(A) \;\Rightarrow\; M[\alpha := B] \in \Lambda_{\rightarrow}^{\mathrm{Ch}}(A[\alpha := B])$.

Proof Both (i), (ii) are proved by induction on the structure of M. ■

1.2.10 Proposition (Closure under reduction for $\boldsymbol{\lambda}_{\rightarrow}^{\mathrm{Ch}}$) *Let $A \in \mathbb{T}$. Then:*

(i) $M \in \Lambda_{\rightarrow}^{\mathrm{Ch}}(A) \;\&\; M \rightarrow_{\beta} N \;\Rightarrow\; N \in \Lambda_{\rightarrow}^{\mathrm{Ch}}(A)$;

(ii) $M \in \Lambda_{\rightarrow}^{\mathrm{Ch}}(A) \;\&\; M \rightarrow_{\eta} N \;\Rightarrow\; N \in \Lambda_{\rightarrow}^{\mathrm{Ch}}(A)$;

(iii) $M \in \Lambda_{\rightarrow}^{\mathrm{Ch}}(A)$ *and* $M \twoheadrightarrow_{\beta\eta} N$. *Then* $N \in \Lambda_{\rightarrow}^{\mathrm{Ch}}(A)$.

Proof (i) Suppose $M \equiv (\lambda x^B.P)Q \in \Lambda_{\rightarrow}^{\mathrm{Ch}}(A)$. Then by Proposition 1.2.8(ii) one has $\lambda x^B.P \in \Lambda_{\rightarrow}^{\mathrm{Ch}}(B' \rightarrow A)$ and $Q \in \Lambda_{\rightarrow}^{\mathrm{Ch}}(B')$. Therefore $B = B'$, and $P \in \Lambda_{\rightarrow}^{\mathrm{Ch}}(A)$, by Proposition 1.2.8(iii). Therefore $N \equiv P[x^B := Q] \in \Lambda_{\rightarrow}^{\mathrm{Ch}}(A)$, by Proposition 1.2.9.

(ii) Suppose $M \equiv (\lambda x^B.Nx^B) \in \Lambda_{\rightarrow}^{\mathrm{Ch}}(A)$. Therefore $A = B \rightarrow C$ and

$Nx^B \in \Lambda^{\text{Ch}}_{\to}(C)$, by Proposition 1.2.8(iii). But then $N \in \Lambda^{\text{Ch}}_{\to}(B{\to}C)$ by Proposition 1.2.8(i) and (ii).

(iii) By induction on the relation $\twoheadrightarrow_{\beta\eta}$, using (i), (ii). ■

The Church–Rosser theorem holds for $\beta\eta$-reduction on $\Lambda^{\text{Ch}}_{\to}$; see Proposition 1.2.29.

The following property called *uniqueness of types* does not hold for $\boldsymbol{\lambda}^{\text{Cu}}_{\to}$. It is instructive to find out where the proof breaks down for that system.

1.2.11 Proposition (Unicity of types for $\boldsymbol{\lambda}^{\text{Ch}}_{\to}$) *Let* $A, B \in \mathbb{T}$. *Then*

$$M \in \Lambda^{\text{Ch}}_{\to}(A) \ \&\ M \in \Lambda^{\text{Ch}}_{\to}(B) \quad \Rightarrow \quad A = B.$$

Proof By induction on the structure of M, using the Inversion Lemma 1.2.8. ■

Properties of $\boldsymbol{\lambda}^{\text{dB}}_{\to}$

The first properties of $\boldsymbol{\lambda}^{\text{dB}}_{\to}$, have proofs being similar to those for $\boldsymbol{\lambda}^{\text{Ch}}_{\to}$.

1.2.12 Proposition (Inversion Lemma for $\boldsymbol{\lambda}^{\text{dB}}_{\to}$)

(i) $\Gamma \vdash x : A \ \Rightarrow\ (x{:}A) \in \Gamma$.
(ii) $\Gamma \vdash MN : A \ \Rightarrow\ \exists B \in \mathbb{T}\ [\Gamma \vdash M : B{\to}A \ \&\ \Gamma \vdash N : B]$.
(iii) $\Gamma \vdash \lambda x{:}B.M : A \ \Rightarrow\ \exists C \in \mathbb{T}\ [A \equiv B{\to}C \ \&\ \Gamma, x{:}B \vdash M : C]$.

1.2.13 Proposition (Substitution Lemma for $\boldsymbol{\lambda}^{\text{dB}}_{\to}$)

(i) $\Gamma, x{:}A \vdash M : B \ \&\ \Gamma \vdash N : A \Rightarrow \Gamma \vdash M[x{:} = N] : B$.
(ii) $\Gamma \vdash M : A \Rightarrow \Gamma[\alpha := B] \vdash M : A[\alpha := B]$.

1.2.14 Proposition (Subject reduction property for $\boldsymbol{\lambda}^{\text{dB}}_{\to}$)

$$\Gamma \vdash M : A \ \&\ M \twoheadrightarrow_{\beta\eta} N \ \Rightarrow\ \Gamma \vdash N : A.$$

1.2.15 Proposition (Church–Rosser theorem for $\boldsymbol{\lambda}^{\text{dB}}_{\to}$) $\boldsymbol{\lambda}^{\text{dB}}_{\to}$ *satisfies* CR.

(i) *Let* $M, N_1, N_2 \in \Lambda^{\text{dB},\Gamma}_{\to}(A)$. *Then*

$$M \twoheadrightarrow_{\beta\eta} N_1 \ \&\ M \twoheadrightarrow_{\beta(\eta)} N_2 \ \Rightarrow\ \exists Z \in \Lambda^{\text{dB},\Gamma}_{\to}(A). N_1 \twoheadrightarrow_{\beta(\eta)} Z \ \&\ N_2 \twoheadrightarrow_{\beta(\eta)} Z.$$

(ii) *Let* $M, N \in \Lambda^{\text{dB},\Gamma}_{\to}(A)$. *Then*

$$M =_{\beta(\eta)} N \ \Rightarrow\ \exists Z \in \Lambda^{\text{dB},\Gamma}_{\to}(A). M \twoheadrightarrow_{\beta(\eta)} Z \ \&\ N \twoheadrightarrow_{\beta(\eta)} Z.$$

Proof See Exercise 2.4. ■

It is instructive to see why the following result fails if the two contexts are different.

1.2.16 Proposition (Unicity of types for $\boldsymbol{\lambda}_{\to}^{\mathrm{dB}}$) *Let* $A, B \in \mathbb{T}$. *Then*

$$\Gamma \vdash M : A \;\&\; \Gamma \vdash M : B \quad \Rightarrow \quad A = B.$$

Proof Similar to the proof of the Church variant. ∎

Equivalence of the systems

It may seem a bit exaggerated to have three versions of the simply typed lambda calculus, viz. $\boldsymbol{\lambda}_{\to}^{\mathrm{Cu}}$, $\boldsymbol{\lambda}_{\to}^{\mathrm{Ch}}$ and $\boldsymbol{\lambda}_{\to}^{\mathrm{dB}}$, but it proves convenient.

The Curry version inspired some implicitly typed programming languages like ML, Miranda, Haskell and Clean. Types are being derived. Since implicit typing makes programming easier, we want to consider this system.

The use of explicit typing becomes essential for extensions of $\boldsymbol{\lambda}_{\to}^{\mathrm{Cu}}$. For example in the system $\boldsymbol{\lambda}2$, also called system F, with second-order (polymorphic) types, type checking is not decidable, see Wells (1999), and hence one needs the explicit versions. The two explicitly typed systems $\boldsymbol{\lambda}_{\to}^{\mathrm{Ch}}$ and $\boldsymbol{\lambda}_{\to}^{\mathrm{dB}}$ are basically isomorphic as shown above. These systems have a very canonical semantics if the $\boldsymbol{\lambda}_{\to}^{\mathrm{Ch}}$ version is used.

We want two versions because the version $\boldsymbol{\lambda}_{\to}^{\mathrm{dB}}$ can be extended more naturally to more powerful type systems in which there is a notion of reduction on the types (those with 'dependent types' and those with higher-order types, see e.g. Barendregt (1992)) generated simultaneously. Also there are important extensions in which there is a reduction relation on types, e.g. in the system $\boldsymbol{\lambda}\omega$ with higher-order types. The classical version of $\boldsymbol{\lambda}_{\to}$ gives problems. For example, if $A \twoheadrightarrow B$, does one have that $\lambda x^A.x^A \twoheadrightarrow \lambda x^A.x^B$? Moreover, is the x^B bound by the λx^A? By denoting $\lambda x^A.x^A$ as $\lambda x{:}A.x$, as is done in $\boldsymbol{\lambda}_{\to}^{\mathrm{Ch}}$, these problems do not arise. That types are able to reduce is so important that for explicitly typed extensions of $\boldsymbol{\lambda}_{\to}$, one needs to use the dB versions.

The situation is not so bad as it may seem, since the three systems and their differences are easy to memorize. Just look at the following examples.

$$\begin{array}{rcll} \lambda x.xy & \in & \Lambda_{\to}^{\mathrm{Cu},\{y:0\}}((0{\to}0){\to}0) & \text{(Curry);} \\ \lambda x{:}(0{\to}0).xy & \in & \Lambda_{\to}^{\mathrm{dB},\{y:0\}}((0{\to}0){\to}0) & \text{(de Bruijn);} \\ \lambda x^{0\to0}.x^{0\to0}y^0 & \in & \Lambda_{\to}^{\mathrm{Ch}}((0{\to}0){\to}0) & \text{(Church).} \end{array}$$

Hence for good reasons one finds all the three versions of $\boldsymbol{\lambda}_{\to}$ in the literature.

In Part I of this book we are interested in untyped lambda terms that can be typed using simple types. We will see that up to substitution this typing is unique. For example

$$\lambda fx.f(fx)$$

can have as type $(0{\to}0){\to}0{\to}0$, but also $(A{\to}A){\to}A{\to}A$ for any type A. Also there is a simple algorithm for finding all possible types for an untyped lambda term, see Section 2.3.

We are interested in identifying, using Curry typing, typable terms M among the untyped lambda terms Λ. Since we are at the same time also interested in the types of the subterms of M, the Church typing is a convenient notation. Moreover, this information is almost uniquely determined once the type A of M is known or required. By this we mean that the Church typing is uniquely determined by A for M not containing a K-redex (of the form $(\lambda x.M)N$ with $x \notin \mathrm{FV}(M)$). If M does contain a K-redex, then the type of the $\boldsymbol{\beta}$-nf M^{nf} of M is still uniquely determined by A. For example the Church typing of $M \equiv \mathsf{KI}y$ of type $\alpha{\to}\alpha$ is $(\lambda x^{\alpha\to\alpha}y^{\beta}.x^{\alpha\to\alpha})(\lambda z^{\alpha}.z^{\alpha})y^{\beta}$. The type β is not determined. But for the $\boldsymbol{\beta}$-nf of M, the term I, the Church typing can only be $\mathsf{I}_\alpha \equiv \lambda z^{\alpha}.z^{\alpha}$. See Exercise 2.3.

If a type is not explicitly given, then possible types for M can be obtained schematically from ground types. By this we mean that. for example, the term $\mathsf{I} \equiv \lambda x.x$ has a Church version $\lambda x^{\alpha}.x^{\alpha}$ and type $\alpha{\to}\alpha$, where one can substitute any $A \in \mathbb{T}^{\mathbb{A}}$ for α. We will study this in detail in Section 2.3.

Comparing $\boldsymbol{\lambda}_{\to}^{\mathrm{Cu}}$ and $\boldsymbol{\lambda}_{\to}^{\mathrm{Ch}}$

There are canonical translations between $\boldsymbol{\lambda}_{\to}^{\mathrm{Ch}}$ and $\boldsymbol{\lambda}_{\to}^{\mathrm{Cu}}$.

1.2.17 Definition There is a *forgetful map* $|\cdot| : \Lambda_{\to}^{\mathrm{Ch}} \to \Lambda$ defined as follows:

$$|x^A| \equiv x;$$
$$|MN| \equiv |M||N|;$$
$$|\lambda x{:}A.M| \equiv \lambda x.|M|.$$

The map $|\cdot|$ just erases all type ornamentations of a term in $\Lambda_{\to}^{\mathrm{Ch}}$. The following result states that terms in the Church version 'project' to legal terms in the Curry version of $\boldsymbol{\lambda}_{\to}^{\mathbb{A}}$. Conversely, legal terms in $\boldsymbol{\lambda}_{\to}^{\mathrm{Cu}}$ can be 'lifted' to terms in $\boldsymbol{\lambda}_{\to}^{\mathrm{Ch}}$.

1.2.18 Definition Let $M \in \Lambda_{\to}^{\mathrm{Ch}}$. Then we write

$$\Gamma_M = \{x{:}A \mid x^A \in \mathrm{FV}(M)\}.$$

1.2.19 Proposition

(i) *Let $M \in \Lambda_{\to}^{\mathrm{Ch}}$. Then*

$$M \in \Lambda_{\to}^{\mathrm{Ch}}(A) \;\Rightarrow\; \Gamma_M \vdash_{\boldsymbol{\lambda}_{\to}}^{\mathrm{Cu}} |M| : A.$$

(ii) *Let* $M \in \Lambda$. *Then*

$$\Gamma \vdash^{\text{Cu}}_{\boldsymbol{\lambda}_\to} M : A \;\Leftrightarrow\; \exists M' \in \Lambda^{\text{Ch}}_\to(A).|M'| \equiv M.$$

Proof (i) By induction on the generation of $\Lambda^{\text{Ch}}_\to$. Since variables have a unique type, Γ_M is well-defined and $\Gamma_P \cup \Gamma_Q = \Gamma_{PQ}$.

(ii) ($\Rightarrow$) By induction on the proof of $\Gamma \vdash M : A$ with the induction loading that $\Gamma_{M'} = \Gamma$.

($\Leftarrow$) By (i). ∎

Notice that the converse of Proposition 1.2.19(i) is not true: one has

$$\vdash^{\text{Cu}}_{\boldsymbol{\lambda}_\to} |\lambda x^A.x^A| \equiv (\lambda x.x) : (A{\to}B){\to}(A{\to}B),$$

but $(\lambda x^A.x^A) \notin \Lambda^{\text{Ch}}((A{\to}B){\to}(A{\to}B))$.

1.2.20 Corollary *In particular, for a type* $A \in \mathbb{T}$ *one has*

$$A \textit{ is inhabited in } \boldsymbol{\lambda}^{\text{Cu}}_\to \;\Leftrightarrow\; A \textit{ is inhabited in } \boldsymbol{\lambda}^{\text{Ch}}_\to.$$

Proof Immediate. ∎

For normal terms one can do better than Proposition 1.2.19. First a structural result.

1.2.21 Proposition *Let* $M \in \Lambda$ *be in normal form. Then*

$$M \equiv \lambda x_1 \cdots x_n.yM_1 \cdots M_m,$$

with $n, m \geq 0$ *and the* $M_1, \dots, M_m$ *again in normal form.*

Proof By induction on the structure of M. See Barendregt [1984], Corollary 8.3.8 for some details if necessary. ∎

In order to prove results about the set NF of $\boldsymbol{\beta}$-nfs, it is useful to introduce the subset vNF of $\boldsymbol{\beta}$-nfs not starting with a λ, but with a free variable. These two sets can be defined by a simultaneous recursion known from context-free languages.

1.2.22 Definition The sets vNF and NF of Λ are defined by the following grammar:

$$\begin{array}{|rcl|}\hline \text{vNF} & := & \text{x} \mid \text{vNF NF} \\ \text{NF} & := & \text{vNF} \mid \lambda\text{x.NF}. \\ \hline\end{array}$$

1.2.23 Proposition *For* $M \in \Lambda$ *one has*

$$M \textit{ is in } \boldsymbol{\beta}\textit{-nf} \;\Leftrightarrow\; M \in \text{NF}.$$

Proof By simultaneous induction it follows easily that

$$M \in \mathrm{vNF} \quad \Rightarrow \quad M \equiv x\vec{N} \;\&\; M \text{ is in } \boldsymbol{\beta}\text{-nf}$$

$$M \in \mathrm{NF} \quad \Rightarrow \quad M \text{ is in } \boldsymbol{\beta}\text{-nf.}$$

Conversely, for M in $\boldsymbol{\beta}$-nf by Proposition 1.2.21 one has $M \equiv \lambda\vec{x}.yN_1 \dots N_k$, with the $\vec{N}$ all in $\boldsymbol{\beta}$-nf. It follows by induction on the structure of such M that $M \in \mathrm{NF}$. ∎

1.2.24 Proposition *Assume that $M \in \Lambda$ is in $\boldsymbol{\beta}$-nf. Then $\Gamma \vdash^{\mathrm{Cu}}_{\boldsymbol{\lambda}_{\to}} M : A$ implies that there is a unique $M^{A;\Gamma} \in \Lambda^{\mathrm{Ch}}_{\to}(A)$ such that $|M^{A;\Gamma}| \equiv M$ and $\Gamma_{M^{A;\Gamma}} \subseteq \Gamma$.*

Proof By induction on the generation of normal forms given in Definition 1.2.22.

Case $M \equiv x\vec{N}$, with N_i in $\boldsymbol{\beta}$-nf. By Proposition 1.2.4 one has

$$(x{:}A_1 \to \dots \to A_k \to A) \in \Gamma \text{ and } \Gamma \vdash^{\mathrm{Cu}}_{\boldsymbol{\lambda}_{\to}} N_i : A_i.$$

As $\Gamma_{M^{A;\Gamma}} \subseteq \Gamma$, we must have

$$x^{A_1 \to \dots \to A_k \to A} \in \mathrm{FV}(M^{A;\Gamma}).$$

By the induction hypothesis there are unique $N_i^{A_i,\Gamma}$ for the N_i. Then

$$M^{A;\Gamma} \equiv x^{A_1 \to \dots \to A_k \to A} N_1^{A_1,\Gamma} \dots N_k^{A_k,\Gamma}$$

is the unique way of typing M.

Case $M \equiv \lambda x.N$, with N in $\boldsymbol{\beta}$-nf . Then by Proposition 1.2.3 we have $\Gamma, x{:}B \vdash^{\mathrm{Cu}}_{\boldsymbol{\lambda}_{\to}} N : C$ and $A = B \to C$. By the induction hypothesis there is a unique $N^{C;\Gamma,x:B}$ for N. It is easy to verify that $M^{A;\Gamma} \equiv \lambda x^B.N^{C;\Gamma,x:B}$ is the unique way of typing M. ∎

1.2.25 Notation If M is a closed $\boldsymbol{\beta}$-nf, then we write M^A for $M^{A;\emptyset}$.

1.2.26 Corollary

(i) *Let $M \in \Lambda^{\mathrm{Ch}}_{\to}$ be a closed $\boldsymbol{\beta}$-nf. Then $|M|$ is a closed $\boldsymbol{\beta}$-nf and*

$$M \in \Lambda^{\mathrm{Ch}}_{\to}(A) \Rightarrow [\vdash^{\mathrm{Cu}}_{\boldsymbol{\lambda}_{\to}} |M| : A \;\&\; |M|^A \equiv M\,].$$

(ii) *Let $M \in \Lambda^{\emptyset}$ be a closed $\boldsymbol{\beta}$-nf and $\vdash^{\mathrm{Cu}}_{\boldsymbol{\lambda}_{\to}} M : A$. Then M^A is the unique term satisfying*

$$M^A \in \Lambda^{\mathrm{Ch}}_{\to}(A) \;\&\; |M^A| \equiv M.$$

(iii) *The following two sets are 'isomorphic'*

$$\{M \in \Lambda \mid M \text{ is closed, in } \boldsymbol{\beta}\text{-nf, and } \vdash^{\text{Cu}}_{\boldsymbol{\lambda}_\to} M : A\};$$
$$\{M \in \Lambda^{\text{Ch}}_\to(A) \mid M \text{ is closed and in } \boldsymbol{\beta}\text{-nf}\}.$$

Proof (i) By the unicity of M^A.

(ii) By the Proposition.

(iii) By (i) and (ii). ■

The applicability of this result will be enhanced once we know that every term typable in $\boldsymbol{\lambda}^{\mathbb{A}}_\to$ (whatever version) has a $\boldsymbol{\beta\eta}$-nf.

The translation $|\ |$ preserves reduction and conversion.

1.2.27 Proposition *Let $R = \boldsymbol{\beta}, \boldsymbol{\eta}$ or $\boldsymbol{\beta\eta}$. Then*

(i) *Let $M, N \in \Lambda^{\text{Ch}}_\to$. Then $M \to_R N \Rightarrow |M| \to_R |N|$. Diagrammatically,*

$$\begin{array}{ccc} M & \xrightarrow[R]{} & N \\ {\scriptstyle |\,|}\Big| & & \Big|{\scriptstyle |\,|} \\ |M| & \underset{R}{\cdots\!\!>} & |N| \end{array}$$

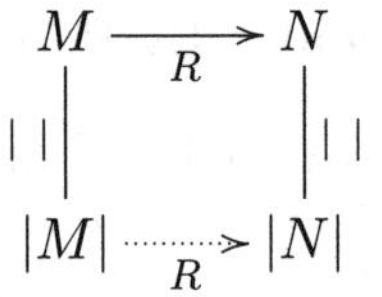

(ii) *Let $M, N \in \Lambda^{\text{Cu},\Gamma}_\to(A)$, $M = |M'|$, with $M' \in \Lambda^{\text{Ch}}_\to(A)$. Then*

$$M \to_R N \Rightarrow \exists N' \in \Lambda^{\text{Ch}}_\to(A).$$
$$|N'| \equiv N \ \&\ M' \to_R N'.$$

Diagrammatically,

$$\begin{array}{ccc} M' & \underset{R}{\cdots\!\!>} & N' \\ {\scriptstyle |\,|}\Big| & & \vdots{\scriptstyle |\,|} \\ M & \xrightarrow[R]{} & N \end{array}$$

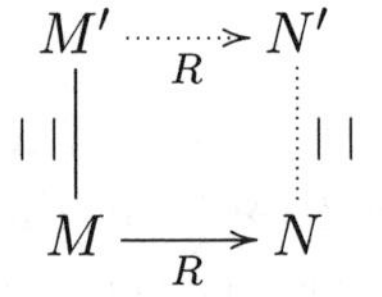

(iii) *Let $M, N \in \Lambda^{\text{Cu},\Gamma}_\to(A)$, $N = |N'|$, with $N' \in \Lambda^{\text{Ch}}_\to(A)$. Then*

$$M \to_R N \Rightarrow \exists M' \in \Lambda^{\text{Ch}}_\to(A).$$
$$|M'| \equiv M \ \&\ M' \to_R N'.$$

Diagrammatically,

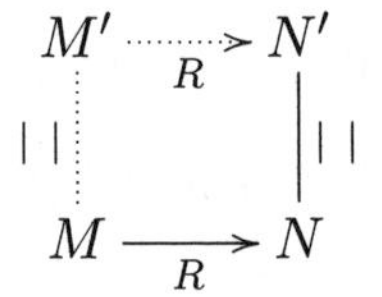

$$\begin{array}{ccc} M' & \underset{R}{\cdots\!\!>} & N' \\ {\scriptstyle |\,|}\vdots & & \Big|{\scriptstyle |\,|} \\ M & \xrightarrow[R]{} & N \end{array}$$

(iv) *The same results hold for $\twoheadrightarrow_R$ and R-conversion.*

Proof Easy. ∎

1.2.28 Corollary *Define the following two statements:*

$$\mathsf{SN}(\boldsymbol{\lambda}_{\to}^{\mathrm{Cu}}) := \forall \Gamma \forall M \in \Lambda_{\to}^{\mathrm{Cu},\Gamma}.\mathsf{SN}(M);$$

$$\mathsf{SN}(\boldsymbol{\lambda}_{\to}^{\mathrm{Ch}}) := \forall M \in \Lambda_{\to}^{\mathrm{Ch}}.\mathsf{SN}(M).$$

Then

$$\mathsf{SN}(\boldsymbol{\lambda}_{\to}^{\mathrm{Cu}}) \Leftrightarrow \mathsf{SN}(\boldsymbol{\lambda}_{\to}^{\mathrm{Ch}}).$$

In fact we will prove in Section 2.2 that both statements hold.

1.2.29 Proposition (Church–Rosser theorem for $\boldsymbol{\lambda}_{\to}^{\mathrm{Ch}}$) *On typable terms of $\boldsymbol{\lambda}_{\to}^{\mathrm{Ch}}$ the Church–Rosser theorem holds for the notions of reduction $\twoheadrightarrow_{\boldsymbol{\beta}}$, $\twoheadrightarrow_{\boldsymbol{\eta}}$, and $\twoheadrightarrow_{\boldsymbol{\beta\eta}}$.*

(i) *Let $M, N_1, N_2 \in \Lambda_{\to}^{\mathrm{Ch}}(A)$. Then*

$$M \twoheadrightarrow_{\boldsymbol{\beta\eta}} N_1 \;\&\; M \twoheadrightarrow_{\boldsymbol{\beta}(\boldsymbol{\eta})} N_2 \Rightarrow \exists Z \in \Lambda_{\to}^{\mathrm{Ch}}(A).N_1 \twoheadrightarrow_{\boldsymbol{\beta}(\boldsymbol{\eta})} Z \;\&\; N_2 \twoheadrightarrow_{\boldsymbol{\beta}(\boldsymbol{\eta})} Z.$$

(ii) *Let $M, N \in \Lambda_{\to}^{\mathrm{Ch}}(A)$. Then*

$$M =_{\boldsymbol{\beta}(\boldsymbol{\eta})} N \Rightarrow \exists Z \in \Lambda_{\to}^{\mathrm{Ch}}(A).M \twoheadrightarrow_{\boldsymbol{\beta}(\boldsymbol{\eta})} Z \;\&\; N \twoheadrightarrow_{\boldsymbol{\beta}(\boldsymbol{\eta})} Z.$$

Proof
(i) We give two proofs, both borrowing a result from Chapter 2.

First proof. We use the fact that every term of $\Lambda_{\to}^{\mathrm{Ch}}$ has a $\boldsymbol{\beta}$-nf, Theorem 2.1.13. Suppose $M \twoheadrightarrow_{\boldsymbol{\beta\eta}} N_i$, $i \in \{1,2\}$. Consider the $\boldsymbol{\beta}$-nfs N_i^{nf} of N_i. Then $|M| \twoheadrightarrow_{\boldsymbol{\beta\eta}} |N_i^{\mathrm{nf}}|$, $i \in \{1,2\}$. By the Church–Rosser theorem for untyped lambda terms one has $|N_1^{\mathrm{nf}}| \equiv |N_2^{\mathrm{nf}}|$, and is also in $\boldsymbol{\beta}$-nf. By Proposition 1.2.24 there exists unique $Z_i \in \Lambda_{\to}^{\mathrm{Ch}}$ such that $M \twoheadrightarrow_{\boldsymbol{\beta\eta}} Z_i$ and $|Z_i| \equiv |N_i^{\mathrm{nf}}|$. But then $Z_1 \equiv Z_2$ and we are done.

Second proof. This time we use that every term of $\Lambda_{\to}^{\mathrm{Ch}}$ is $\boldsymbol{\beta}$-SN, Theorem 2.2.1. It is easy to see that $\to_{\boldsymbol{\beta\eta}}$ satisfies the weak diamond property; then we are done by Newman's Lemma, Proposition 2.1.4. See for example B[1984], Definition 3.1.24 and Proposition 3.1.25.

(ii) In the usual way from (i). See for example B[1984], Theorem 3.1.12. ∎

Comparing $\boldsymbol{\lambda}_{\to}^{\mathrm{Ch}}$ and $\boldsymbol{\lambda}_{\to}^{\mathrm{dB}}$

There is a close relation between $\boldsymbol{\lambda}_{\to}^{\mathrm{Ch}}$ and $\boldsymbol{\lambda}_{\to}^{\mathrm{dB}}$. First we need the following.

1.2.30 Lemma *Let $\Gamma \subseteq \Gamma'$ be bases of $\boldsymbol{\lambda}_{\to}^{\mathrm{dB}}$. Then*

$$\Gamma \vdash_{\boldsymbol{\lambda}_{\to}}^{\mathrm{dB}} M : A \Rightarrow \Gamma' \vdash_{\boldsymbol{\lambda}_{\to}}^{\mathrm{dB}} M : A.$$

Proof By induction on the derivation of the first statement. ∎

1.2.31 Definition

(i) Let $M \in \Lambda_{\rightarrow}^{\mathrm{dB}}$ and suppose $\mathrm{FV}(M) \subseteq \mathrm{dom}(\Gamma)$.
Define M^{Γ} recursively as

$$\begin{aligned} x^{\Gamma} &= x^{\Gamma(x)}; \\ (MN)^{\Gamma} &= M^{\Gamma}N^{\Gamma}; \\ (\lambda x{:}A.M)^{\Gamma} &= \lambda x^{A}.M^{\Gamma,x:A}. \end{aligned}$$

(ii) Let $M \in \Lambda_{\rightarrow}^{\mathrm{Ch}}(A)$ in $\boldsymbol{\lambda}_{\rightarrow}^{\mathrm{Ch}}$. Define M^{-}, a pseudo-term of $\boldsymbol{\lambda}_{\rightarrow}^{\mathrm{dB}}$, as

$$\begin{aligned} (x^{A})^{-} &= x; \\ (MN)^{-} &= M^{-}N^{-}; \\ (\lambda x^{A}.M)^{-} &= \lambda x{:}A.M^{-}. \end{aligned}$$

1.2.32 Example To gain some intuition, consider

$$\begin{aligned} (\lambda x{:}A.x)^{\emptyset} &\equiv (\lambda x^{A}.x^{A}); \\ (\lambda x^{A}.x^{A})^{-} &\equiv (\lambda x{:}A.x); \\ (\lambda x{:}A{\rightarrow}B.xy)^{\{y:A\}} &\equiv \lambda x^{A\rightarrow B}.x^{A\rightarrow B}y^{A}; \\ \Gamma_{(\lambda x^{A\rightarrow B}.x^{A\rightarrow B}y^{A})} &= \{y{:}A\}, \qquad \text{cf. Definition 1.2.18.} \end{aligned}$$

1.2.33 Proposition

(i) *Let* $M \in \Lambda_{\rightarrow}^{\mathrm{Ch}}$ *and* Γ *be a basis of* $\boldsymbol{\lambda}_{\rightarrow}^{\mathrm{dB}}$. *Then*

$$M \in \Lambda_{\rightarrow}^{\mathrm{Ch}}(A) \Leftrightarrow \Gamma_{M} \vdash_{\boldsymbol{\lambda}_{\rightarrow}}^{\mathrm{dB}} M^{-} : A.$$

(ii)
$$\Gamma \vdash_{\boldsymbol{\lambda}_{\rightarrow}}^{\mathrm{dB}} M : A \Leftrightarrow M^{\Gamma} \in \Lambda_{\rightarrow}^{\mathrm{Ch}}(A).$$

Proof (i), (ii)($\Rightarrow$) By induction on the definition or the proof of the left hand side.

(i)($\Leftarrow$) By (ii)($\Rightarrow$), using $(M^{-})^{\Gamma_M} \equiv M$.

(ii)($\Leftarrow$) By (i)($\Rightarrow$), using $(M^{\Gamma})^{-} \equiv M, \Gamma_{M^{\Gamma}} \subseteq \Gamma$ and Proposition 1.2.30. ∎

1.2.34 Corollary *In particular, for a type* $A \in \mathbb{T}$ *one has*

$$A \textit{ is inhabited in } \boldsymbol{\lambda}_{\rightarrow}^{\mathrm{Ch}} \Leftrightarrow A \textit{ is inhabited in } \boldsymbol{\lambda}_{\rightarrow}^{\mathrm{dB}}.$$

Proof Immediate. ∎

Again the translation preserves reduction and conversion.

1.2.35 Proposition

(i) *Let* $M, N \in \Lambda_{\rightarrow}^{\mathrm{dB}}$. *Then*

$$M \rightarrow_{R} N \Leftrightarrow M^{\Gamma} \rightarrow_{R} N^{\Gamma},$$

where $R = \boldsymbol{\beta}, \boldsymbol{\eta}$ *or* $\boldsymbol{\beta\eta}$.

(ii) *Let* $M_1, M_2 \in \Lambda_{\rightarrow}^{\mathrm{Ch}}(A)$ *and* R *as in* (i). *Then*

$$M_1 \rightarrow_R M_2 \Leftrightarrow M_1^- \rightarrow_R M_2^-.$$

(iii) *The same results hold for conversion.*

Proof Easy. ■

Comparing $\boldsymbol{\lambda}_{\rightarrow}^{\mathrm{Cu}}$ and $\boldsymbol{\lambda}_{\rightarrow}^{\mathrm{dB}}$

1.2.36 Proposition

(i) $\Gamma \vdash_{\boldsymbol{\lambda}_{\rightarrow}}^{\mathrm{dB}} M : A \Rightarrow \Gamma \vdash_{\boldsymbol{\lambda}_{\rightarrow}}^{\mathrm{Cu}} |M| : A$;
here $|M|$ *is defined by leaving out all '*$: A$*' immediately following binding lambdas.*

(ii) *Let* $M \in \Lambda$. *Then*

$$\Gamma \vdash_{\boldsymbol{\lambda}_{\rightarrow}}^{\mathrm{Cu}} M : A \Leftrightarrow \exists M'.|M'| \equiv M \ \& \ \Gamma \vdash_{\boldsymbol{\lambda}_{\rightarrow}}^{\mathrm{dB}} M' : A.$$

Proof As for Proposition 1.2.19. ■

Again the implication in (i) cannot be reversed.

The three systems compared

Harvesting, we obtain the following for the systems $\boldsymbol{\lambda}_{\rightarrow}^{\mathrm{Ch}}$, $\boldsymbol{\lambda}_{\rightarrow}^{\mathrm{dB}}$ and $\boldsymbol{\lambda}_{\rightarrow}^{\mathrm{Cu}}$.

1.2.37 Theorem *Let* $M \in \Lambda_{\rightarrow}^{\mathrm{Ch}}$ *be in* $\boldsymbol{\beta}$-nf. *Then the following are equivalent:*

(i) $M \in \Lambda_{\rightarrow}^{\mathrm{Ch}}(A)$;

(ii) $\Gamma_M \vdash_{\boldsymbol{\lambda}_{\rightarrow}}^{\mathrm{dB}} M^- : A$;

(iii) $\Gamma_M \vdash_{\boldsymbol{\lambda}_{\rightarrow}}^{\mathrm{Cu}} |M| : A$;

(iv) $|M|^{A;\Gamma_M} \in \Lambda_{\rightarrow}^{\mathrm{Ch}}(A) \ \& \ |M|^{A;\Gamma_M} \equiv M$.

Proof By Propositions 1.2.33(i), 1.2.36, and 1.2.24, and $|M^-| = |M|$,

$$\begin{aligned} M \in \Lambda_{\rightarrow}^{\mathrm{Ch}}(A) &\Leftrightarrow \Gamma_M \vdash_{\boldsymbol{\lambda}_{\rightarrow}}^{\mathrm{dB}} M^- : A \\ &\Rightarrow \Gamma_M \vdash_{\boldsymbol{\lambda}_{\rightarrow}}^{\mathrm{Cu}} |M| : A \\ &\Rightarrow |M|^{A;\Gamma_M} \in \Lambda_{\rightarrow}^{\mathrm{Ch}}(A) \ \& \ |M|^{A;\Gamma_M} \equiv M \\ &\Rightarrow M \in \Lambda_{\rightarrow}^{\mathrm{Ch}}(A). \ \blacksquare \end{aligned}$$

1.3 Normal inhabitants

In this section we will give an algorithm that enumerates the set of closed inhabitants in $\boldsymbol{\beta}$-nf of a given type $A \in \mathbb{T}$. Since we will prove in the next chapter that all typable terms do have a normal form and that reduction preserves typing, we will thus have an enumeration of essentially all closed terms of that given type. The algorithm will be used by demonstrating that a certain type A is uninhabited or, more generally, that a certain class of terms exhausts all inhabitants of A.

Because the various versions of $\boldsymbol{\lambda}_{\rightarrow}^{\mathbb{A}}$ are equivalent as to inhabitation of closed $\boldsymbol{\beta}$-nfs, we flexibly jump between the set

$$\{M \in \Lambda_{\rightarrow}^{\mathrm{Ch}}(A) \mid M \text{ closed and in } \boldsymbol{\beta}\text{-nf}\}$$

and

$$\{M \in \Lambda \mid M \text{ closed, in } \boldsymbol{\beta}\text{-nf, and } \vdash_{\boldsymbol{\lambda}_{\rightarrow}}^{\mathrm{Cu}} M : A\};$$

thus we often write a Curry context $\{x_1{:}A_1, \ldots, x_n{:}A_n\}$ as $\{x_1^{A_1}, \ldots, x_n^{A_n}\}$ and a Church term $\lambda x^0.x^0$ as $\lambda x^0.x$, an intermediate form between the Church and the de Bruijn versions.

We do need to distinguish various kinds of normal forms.

1.3.1 Definition Let $A = A_1 \rightarrow \cdots \rightarrow A_n \rightarrow \alpha$ and suppose $M \in \Lambda_{\rightarrow}^{\mathrm{Ch}}(A)$.

(i) We say M is in *long-nf*, written lnf, if $M \equiv \lambda x_1^{A_1} \cdots x_n^{A_n}.xM_1 \cdots M_n$ and each M_i is in lnf. By induction on the depth of the type of the closure of M one sees that this definition is well founded.

(ii) We say M *has an lnf* if $M =_{\boldsymbol{\beta\eta}} N$ and N is an lnf.

In Exercise 1.14 it is proved that if M has a $\boldsymbol{\beta}$-nf, which according to Theorem 2.2.5 is always the case, then it also has a unique lnf and this will be its unique $\boldsymbol{\beta\eta}^{-1}$-nf. Here $\boldsymbol{\eta}^{-1}$ is the notion of reduction that is the converse of $\boldsymbol{\eta}$.

1.3.2 Examples

(i) $\lambda x^0.x$ is both in $\boldsymbol{\beta\eta}$-nf and lnf.
(ii) $\lambda f^1.f$ is a $\boldsymbol{\beta\eta}$-nf but not an lnf.
(iii) $\lambda f^1 x^0.fx$ is an lnf but not a $\boldsymbol{\beta\eta}$-nf; its $\boldsymbol{\beta\eta}$-nf is $\lambda f^1.f$.
(iv) The $\boldsymbol{\beta}$-nf $\lambda F_2^2 \lambda f^1.Ff(\lambda x^0.fx)$ is neither in $\boldsymbol{\beta\eta}$-nf nor lnf.
(v) A variable of atomic type α is an lnf, but of type $A \rightarrow B$ not.
(vi) A variable $f^{1\rightarrow 1}$ has as lnf $\lambda g^1 x^0.f(\lambda y^0.gy)x =_{\boldsymbol{\eta}} f^{1\rightarrow 1}$.

1.3.3 Proposition *Every $\boldsymbol{\beta}$-nf M has an lnf M^ℓ such that $M^\ell \twoheadrightarrow_{\boldsymbol{\eta}} M$.*

Proof Define M^ℓ by induction on the depth of the type of the closure of M as

$$M^\ell \equiv (\lambda\vec{x}.yM_1\cdots M_n)^\ell \equiv \lambda\vec{x}\vec{z}.yM_1^\ell\cdots M_n^\ell\vec{z}^\ell,$$

where $\vec{z}$ is the longest vector that preserves the type. Then we see that M^ℓ does the job. ■

We will define a *2-level grammar*, see van Wijngaarden (1981), for obtaining all closed inhabitants in long normal form of a given type in $\boldsymbol{\lambda}_{\rightarrow}^{\mathrm{Cu}}$.

1.3.4 Definition Let $\mathcal{L} = \{L(A;\Gamma) \mid A \in \mathbb{T}^{\mathbb{A}}; \Gamma \text{ a context of } \boldsymbol{\lambda}_{\rightarrow}^{\mathrm{Cu}}\}$. Let Σ be the alphabet of the untyped lambda terms. Define the following two-level grammar as a notion of reduction over words over $\mathcal{L} \cup \Sigma$. The elements of $\mathcal{L}$ are the non-terminals (unlike in a context-free language, there are now infinitely many of them) of the form $L(A;\Gamma)$:

$$\begin{array}{rcll} L(\alpha;\Gamma) & \Longrightarrow & xL(B_1;\Gamma)\cdots L(B_n;\Gamma), & \text{if } (x{:}\vec{B}{\rightarrow}\alpha) \in \Gamma; \\ L(A{\rightarrow}B;\Gamma) & \Longrightarrow & \lambda x^A.L(B;\Gamma,x^A). & \end{array}$$

Typical productions of this grammar are

$$\begin{aligned} L(3;\emptyset) &\Longrightarrow \lambda F^2.L(0;F^2) \\ &\Longrightarrow \lambda F^2.FL(1;F^2) \\ &\Longrightarrow \lambda F^2.F(\lambda x^0.L(0;F^2,x^0)) \\ &\Longrightarrow \lambda F^2.F(\lambda x^0.x). \end{aligned}$$

But one also has

$$\begin{aligned} L(0;F^2,x^0) &\Longrightarrow FL(1;F^2,x^0) \\ &\Longrightarrow F(\lambda x_1^0.L(0;F^2,x^0,x_1^0)) \\ &\Longrightarrow F(\lambda x_1^0.x_1). \end{aligned}$$

Hence ($\twoheadrightarrow$ denotes the transitive reflexive closure of $\Longrightarrow$)

$$L(3;\emptyset) \twoheadrightarrow \lambda F^2.F(\lambda x^0.F(\lambda x_1^0.x_1)).$$

In fact, $L(3;\emptyset)$ reduces to all possible closed long normal forms of type 3. As in simplified syntax we do not produce parentheses from the $L(A;\Gamma)$, but write them when needed.

1.3.5 Proposition *Let Γ, M, A be given. Then*

$$L(A;\Gamma) \twoheadrightarrow M \quad \Leftrightarrow \quad \Gamma \vdash M : A \;\&\; M \textit{ is in lnf.}$$

Proof ($\rightarrow$) By Definition 1.3.4.
($\Leftarrow$) By the Inversion Theorem 1.2.3. ■

Next we will modify the 2-level grammar and the inhabitation machines in order to produce all $\boldsymbol{\beta}$-nfs.

1.3.6 Definition The 2-level grammar N is defined as

$$\begin{array}{rcll} N(A;\Gamma) & \Longrightarrow & xN(B_1;\Gamma)\cdots N(B_n;\Gamma), & \text{if } (x{:}\vec{B}{\to}A)\in\Gamma; \\ N(A{\to}B;\Gamma) & \Longrightarrow & \lambda x^A.N(B;\Gamma,x^A). & \end{array}$$

Now the $\boldsymbol{\beta}$-nfs are being produced. As an example we make the following production – remember $1 = 0{\to}0$:

$$\begin{aligned} N(1{\to}0{\to}0;\emptyset) &\Longrightarrow \lambda f^1.N(0{\to}0;f^1) \\ &\Longrightarrow \lambda f^1.f. \end{aligned}$$

Similarly to the proof of Proposition 1.3.5 we obtain the following.

1.3.7 Proposition *Let Γ, M, A be given. Then*

$$N(A,\Gamma) \Longrightarrow\!\!\!\!\!\!\!\!\Longrightarrow M \quad \Leftrightarrow \quad \Gamma \vdash M : A \;\&\; M \textit{ is in } \boldsymbol{\beta}\textit{-nf.}$$

Inhabitation machines

Inspired by this proposition one can introduce for each type A a machine M_A producing the set of closed lnfs of that type. If one is interested in terms containing free variables $x_1^{A_1},\dots,x_n^{A_n}$, then one can also find these terms by considering the machine for the type $A_1{\to}\cdots{\to}A_n{\to}A$ and looking at the sub-production at node A. This means that a normal inhabitant M_A of type A can be found as a closed inhabitant $\lambda\vec{x}.M_A$ of type $A_1{\to}\cdots{\to}A_n{\to}A$.

1.3.8 Examples

(i) $A = 0{\to}0{\to}0$. Then M_A is

$$\begin{array}{ccccc} \boxed{0{\to}0{\to}0} & \xrightarrow{\lambda x^0 \lambda y^0} & \boxed{0} & \longrightarrow & x. \\ & & \downarrow & & \\ & & y & & \end{array}$$

This shows that the type 1_2 has two closed inhabitants: $\lambda xy.x$ and $\lambda xy.y$. We see that the two arrows leaving $\boxed{0}$ represent a choice.

(ii) $A = \alpha{\to}((0{\to}\beta){\to}\alpha){\to}\beta{\to}\alpha$. Then M_A is

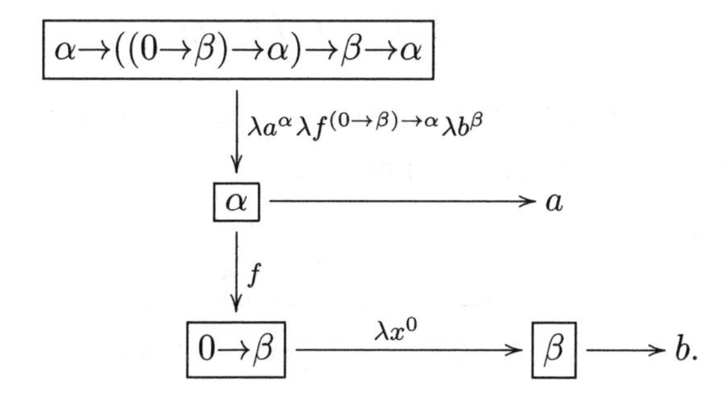

Again there are only two inhabitants, but now the production of them is rather different: $\lambda afb.a$ and $\lambda afb.f(\lambda x^0.b)$.

(iii) $A = ((\alpha{\to}\beta){\to}\alpha){\to}\alpha$. Then M_A is

This type, corresponding to Peirce's law, does not have any inhabitants.

(iv) $A = 1{\to}0{\to}0$. Then M_A is

This is the type Nat having the Church numerals $\lambda f^1 x^0.f^n x$ as inhabitants.

(v) $A = 1{\to}1{\to}0{\to}0$. Then M_A is

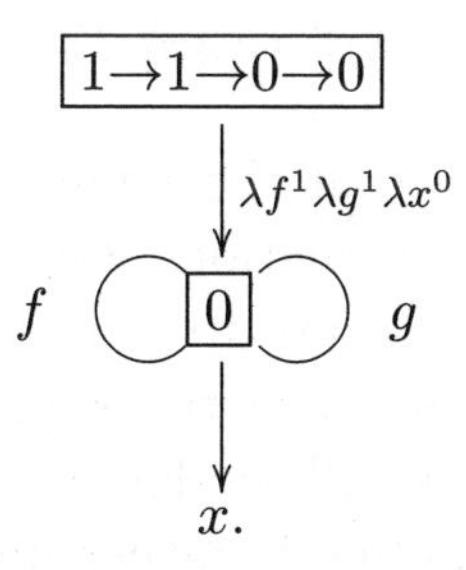

Inhabitants of this type represent words over the alphabet $\Sigma = \{f, g\}$, for example

$$\lambda f^1 g^1 x^0 . fgffgfggx,$$

where we have to insert parentheses associating to the right.

(vi) $A = (\alpha{\to}\beta{\to}\gamma){\to}\beta{\to}\alpha{\to}\gamma$. Then M_A is

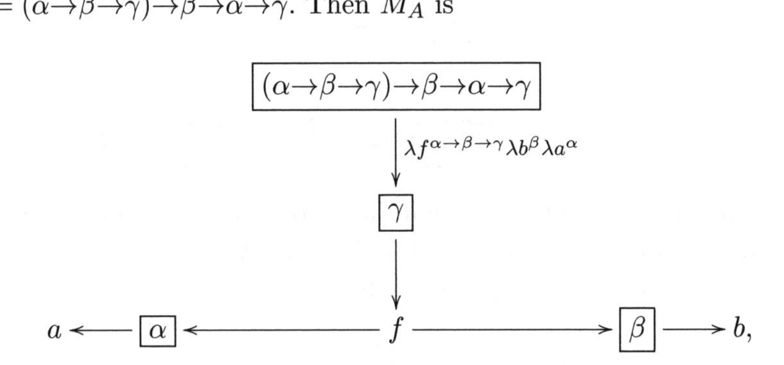

giving as term $\lambda f^{\alpha\to\beta\to\gamma}\lambda b^\beta \lambda a^\alpha . fab$. Note the way an interpretation should be given to paths going through f: the outgoing arcs (to $\boxed{\alpha}$ and $\boxed{\beta}$) should be completed both separately in order to give f its two arguments.

(vii) $A = 3$. Then M_A is

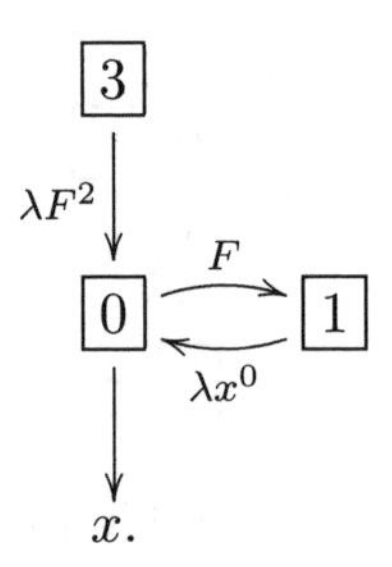

This type 3 has inhabitants having more and more binders:

$$\lambda F^2 . F(\lambda x_0^0 . F(\lambda x_1^0 . F(\cdots (\lambda x_n^0 . x_i)))).$$

The novel phenomenon that the binder λx^0 may go round and round forces us to give new incarnations λx_0^0, $\lambda x_1^0, \ldots$ each time we do this (we need a counter to ensure freshness of the bound variables). The 'terminal' variable x can take the shape of any of the produced incarnations x_k. As almost all binders are dummy, we will see that this potential infinity of binding is rather innocent and the counter is not yet really needed here.

(viii) $A = 3{\to}0{\to}0$. Then M_A is

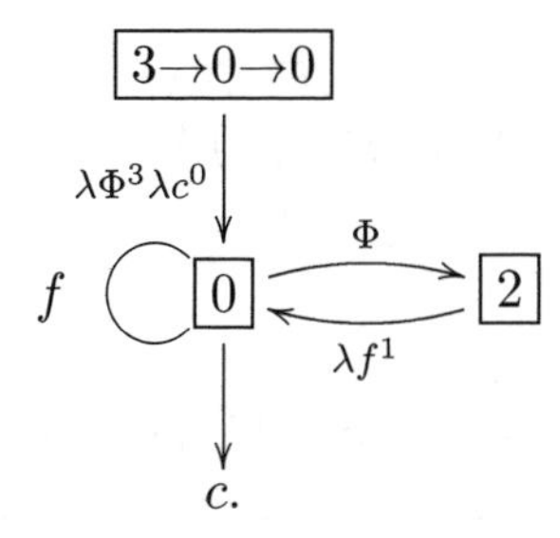

This type, called the *monster* M, does have potentially an infinite amount of binding, having as terms

$$\lambda\Phi^3 c^0.\Phi(\lambda f_1^1.f_1\Phi(\lambda f_2^1.f_2 f_1\Phi(\ldots(\lambda f_n^1.f_n\ldots f_2 f_1 c)..))),$$

for example, again with inserted parentheses associating to the right. Now a proper bookkeeping of incarnations (of f^1 in this case) becomes necessary, as the f going from $\boxed{0}$ to itself needs to be one that has already been incarnated.

(ix) $A = 1_2{\to}0{\to}0$. Then M_A is

$$\boxed{1_2{\to}0{\to}0} \xrightarrow{\lambda p^{1_2}\lambda c^0} \boxed{0} \longrightarrow c,$$

(with a double arc between $\boxed{0}$ and p)

a type of binary tree having as elements $\lambda p^{1_2}c^0.c$ and $\lambda p^{1_2}c^0.pc(pcc)$ for example. Again, as in ((vi)) the outgoing arcs from p (to $\boxed{0}$) should each be completed separately in order to give p its two arguments.

(x) $A = 1_2{\to}2{\to}0$. Then M_A is

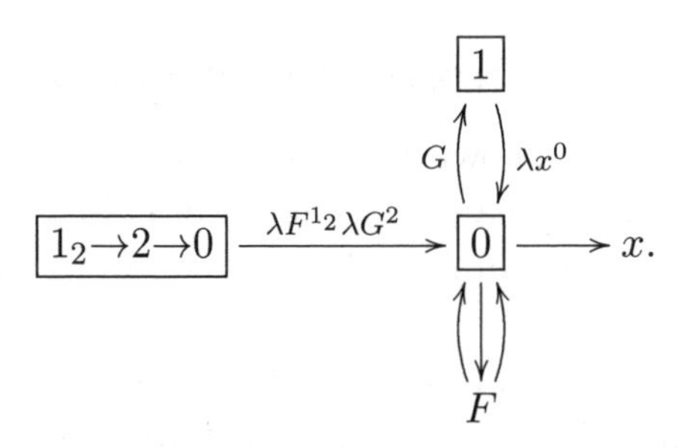

This type, which we call L, has inhabitants that can be thought of as codes for untyped lambda terms. For example the untyped terms $\omega \equiv \lambda x.xx$ and $\Omega \equiv (\lambda x.xx)(\lambda x.xx)$ can be translated to

$(\omega)^t \equiv \lambda F^{1_2} G^2.G(\lambda x^0.Fxx)$ and

$$\begin{array}{rcl}(\Omega)^t & \equiv & \lambda F^{1_2} G^2.F(G(\lambda x^0.Fxx))(G(\lambda x^0.Fxx)) \\ & =_{\beta} & \lambda FG.F((\omega)^t FG)((\omega)^t FG) \\ & =_{\beta} & (\omega)^t \cdot_L (\omega)^t,\end{array}$$

where for $M, N \in L$ one defines $M \cdot_L N = \lambda FG.F(MFG)(NFG)$. All features of producing terms inhabiting types (bookkeeping bound variables, multiple paths) are present in this example.

Following the 2-level grammar N one can make inhabitation machines for $\boldsymbol{\beta}$-nfs $M^{\boldsymbol{\beta}}_A$.

1.3.9 Example We show how the production machine for $\boldsymbol{\beta}$-nfs differs from that for lnfs. Let $A = 1\to0\to0$. Then $\lambda f^1.f$ is the (unique) $\boldsymbol{\beta}$-nf of type A that is not an lnf. This emerges from the following machine $M^{\boldsymbol{\beta}}_A$:

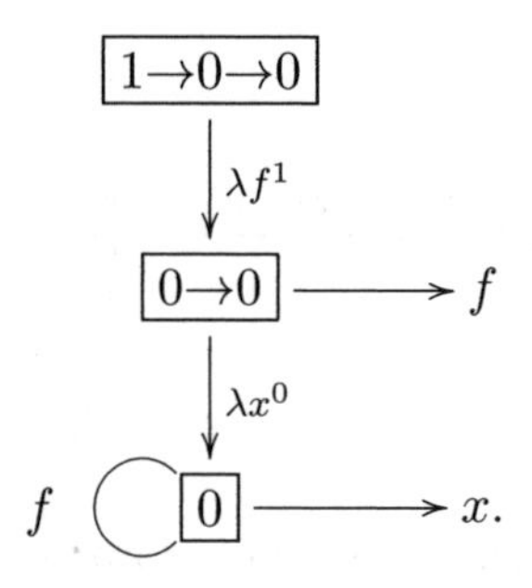

So in order to obtain the $\boldsymbol{\beta}$-nfs, one allows output at non-atomic types.

1.4 Representing data types

In this section it will be shown that first-order algebraic data types can be represented in $\boldsymbol{\lambda}^0_{\to}$. This means that an algebra $\mathcal{A}$ can be embedded into the set of closed terms in $\boldsymbol{\beta}$-nf in $\Lambda^{\mathrm{Cu}}_{\to}(A)$. That we work with the Curry version is as usual not essential.

We start with several examples: Booleans, the natural numbers, the free monoid over n generators (words over a finite alphabet with n elements) and trees with labels from a type A at the leaves. The following definitions depend on a given type A. So in fact $\mathsf{Bool} = \mathsf{Bool}_A$ etc. Often one takes $A = 0$.

Booleans

1.4.1 Definition Define

$$\mathsf{Bool} \equiv A \to A \to A;$$
$$\mathsf{true} \equiv \lambda xy.x;$$
$$\mathsf{false} \equiv \lambda xy.y.$$

Then $\mathsf{true} \in \Lambda^{\emptyset}_{\to}(\mathsf{Bool})$ and $\mathsf{false} \in \Lambda^{\emptyset}_{\to}(\mathsf{Bool})$.

1.4.2 Proposition *There are terms* not, and, or, imp, iff *with the expected behavior on Booleans. For example* $\mathsf{not} \in \Lambda^{\emptyset}_{\to}(\mathsf{Bool}\to\mathsf{Bool})$ *and*

$$\mathsf{not}\ \mathsf{true} =_{\beta} \mathsf{false},$$
$$\mathsf{not}\ \mathsf{false} =_{\beta} \mathsf{true}.$$

Proof Take $\mathsf{not} \equiv \lambda axy.ayx$ and $\mathsf{or} \equiv \lambda abxy.ax(bxy)$. From these two operations the other Boolean functions can be defined. For example, implication can be represented by

$$\mathsf{imp} \equiv \lambda ab.\mathsf{or}(\mathsf{not}\ a)b.$$

A shorter representation is $\lambda abxy.a(bxy)x$, the normal form of imp. ∎

Natural numbers

1.4.3 Definition The set of natural numbers can be represented as a type

$$\mathsf{Nat} \equiv (A\to A)\to A\to A.$$

For each natural number $n \in \omega$ we define its representation

$$\mathbf{c}_n \equiv \lambda fx.f^n x,$$

where

$$f^0 x \equiv x$$
$$f^{n+1}x \equiv f(f^n x).$$

Then $\mathbf{c}_n \in \Lambda^{\emptyset}_{\to}(\mathsf{Nat})$ for every $n \in \omega$. The representation $\mathbf{c}_n$ of $n \in \mathsf{Nat}$ is called *Church's numeral.* In B[1984] another representation of numerals was used.

1.4.4 Proposition

(i) *There exists a term* $\mathsf{S}^+ \in \Lambda^{\emptyset}_{\to}(\mathsf{Nat}\to\mathsf{Nat})$ *such that*

$$\mathsf{S}^+\mathbf{c}_n =_{\beta} \mathbf{c}_{n+1}, \text{ for all } n \in \omega.$$

(ii) *There exists a term* $\mathsf{zero_?} \in \Lambda^{\emptyset}_{\rightarrow}(\mathsf{Nat}\rightarrow\mathsf{Bool})$ *such that*

$$\mathsf{zero_?}\mathbf{c}_0 =_\beta \mathsf{true}$$
$$\mathsf{zero_?}(\mathsf{S}^+ x) =_\beta \mathsf{false}.$$

Proof (i) Take $\mathsf{S}^+ \equiv \lambda n\lambda fx.f(nfx)$. Then

$$\begin{aligned} \mathsf{S}^+\mathbf{c}_n &=_\beta \lambda fx.f(\mathbf{c}_n fx) \\ &=_\beta \lambda fx.f(f^n x) \\ &\equiv \lambda fx.f^{n+1}x \\ &\equiv \mathbf{c}_{n+1}. \end{aligned}$$

(ii) Take $\mathsf{zero_?} \equiv \lambda n\lambda ab.n(\mathsf{K}b)a$. Then

$$\begin{aligned} \mathsf{zero_?}\mathbf{c}_0 &=_\beta \lambda ab.\mathbf{c}_0(\mathsf{K}b)a \\ &=_\beta \lambda ab.a \\ &\equiv \mathsf{true}; \\ \mathsf{zero_?}(\mathsf{S}^+x) &=_\beta \lambda ab.\mathsf{S}^+x(\mathsf{K}b)a \\ &=_\beta \lambda ab.(\lambda fy.f(xfy))(\mathsf{K}b)a \\ &=_\beta \lambda ab.\mathsf{K}b(x(\mathsf{K}b)a) \\ &=_\beta \lambda ab.b \\ &\equiv \mathsf{false}. \blacksquare \end{aligned}$$

1.4.5 Definition

(i) A function $f : \omega^k \rightarrow \omega$ is called λ-*definable* with respect to Nat if there exists a term $F \in \Lambda_{\rightarrow}$ such that $F\mathbf{c}_{n_1} \cdots \mathbf{c}_{n_k} = \mathbf{c}_{f(n_1,\dots,n_k)}$ for all $\vec{n} \in \omega^k$.
(ii) For different data types represented in $\boldsymbol{\lambda}_{\rightarrow}$ one defines λ-definability similarly.

Addition and multiplication are definable in $\boldsymbol{\lambda}_{\rightarrow}$.

1.4.6 Proposition

(i) *There is a term* $\mathsf{plus} \in \Lambda^{\emptyset}_{\rightarrow}(\mathsf{Nat}\rightarrow\mathsf{Nat}\rightarrow\mathsf{Nat})$ *satisfying*

$$\mathsf{plus}\ \mathbf{c}_n\ \mathbf{c}_m =_\beta \mathbf{c}_{n+m}.$$

(ii) *There is a term* $\mathsf{times} \in \Lambda^{\emptyset}_{\rightarrow}(\mathsf{Nat}\rightarrow\mathsf{Nat}\rightarrow\mathsf{Nat})$ *such that*

$$\mathsf{times}\ \mathbf{c}_n\ \mathbf{c}_m =_\beta \mathbf{c}_{n \cdot m}.$$

Proof (i) Take $\mathsf{plus} \equiv \lambda nm\lambda fx.nf(mfx)$. Then

$$\begin{aligned}\mathsf{plus}\ \mathbf{c}_n\ \mathbf{c}_m &=_\beta \lambda fx.\mathbf{c}_n f(\mathbf{c}_m fx)\\ &=_\beta \lambda fx.f^n(f^m x)\\ &\equiv \lambda fx.f^{n+m}x\\ &\equiv \mathbf{c}_{n+m}.\end{aligned}$$

(ii) Take $\mathsf{times} \equiv \lambda nm\lambda fx.m(\lambda y.nfy)x$. Then

$$\begin{aligned}\mathsf{times}\ \mathbf{c}_n\ \mathbf{c}_m &=_\beta \lambda fx.\mathbf{c}_m(\lambda y.\mathbf{c}_n fy)x\\ &=_\beta \lambda fx.\mathbf{c}_m(\lambda y.f^n y)x\\ &=_\beta \lambda fx.\underbrace{(f^n(f^n(\dots(f^n}_{m \text{ times}}x)..)))\\ &\equiv \lambda fx.f^{n\cdot m}x\\ &\equiv \mathbf{c}_{n\cdot m}. \ \blacksquare\end{aligned}$$

The following is immediate.

1.4.7 Corollary *For every polynomial* $p \in \omega[x_1,\dots,x_k]$ *there is a closed term* $M_p \in \Lambda^{\emptyset}_{\to}(\mathsf{Nat}^k\to\mathsf{Nat})$ *such that*

$$\forall n_1,\dots,n_k \in \omega.M_p\mathbf{c}_{n_1}\cdots\mathbf{c}_{n_k} =_\beta \mathbf{c}_{p(n_1,\dots,n_k)}.$$

From the results obtained so far it follows that the polynomials extended by case distinctions (being equal or not to zero) are definable in $\boldsymbol{\lambda}^{\mathbb{A}}_{\to}$. In Schwichtenberg (1976) or Statman (1982) it is proved that exactly these so-called extended polynomials are definable in $\boldsymbol{\lambda}^{\mathbb{A}}_{\to}$. Hence primitive recursion cannot be defined in $\boldsymbol{\lambda}^{\mathbb{A}}_{\to}$; in fact not even the predecessor function, see Proposition 2.4.22.

Words over a finite alphabet

Let $\Sigma = \{a_1,\dots,a_k\}$ be a finite alphabet. Then Σ^*, the collection of words over Σ, can be represented in $\boldsymbol{\lambda}_{\to}$.

1.4.8 Definition Let Σ have k elements.

(i) The *type for words* in Σ^* of length k is

$$\mathsf{Sigma}^* \equiv (0\to0)^k\to0\to0.$$

(ii) Let $w = a_{i_1}\cdots a_{i_p}$ be a word over Σ. Define

$$\begin{aligned}\underline{w} &\equiv \lambda a_1\cdots a_k x.a_{i_1}(\cdots(a_{i_p}x)..)\\ &= \lambda a_1\cdots a_k x.\,(a_{i_1}\circ\cdots\circ a_{i_p})x.\end{aligned}$$

Note that $\underline{w} \in \Lambda^{\emptyset}_{\to}(\mathsf{Sigma}^*)$. If ϵ is the empty word ( ), then naturally

$$\begin{aligned}\underline{\epsilon} &\equiv \lambda a_1 \cdots a_k x.x \\ &= \mathsf{K}^k\mathsf{I}.\end{aligned}$$

Now we show that the operation concatenation can be defined in $\boldsymbol{\lambda}^{\mathbb{A}}_{\to}$.

1.4.9 Proposition *There exists a term*

$$\mathsf{concat} \in \Lambda^{\emptyset}_{\to}(\mathsf{Sigma}^*{\to}\mathsf{Sigma}^*{\to}\mathsf{Sigma}^*)$$

such that for all $w, v \in \Sigma^*$

$$\mathsf{concat}\ \underline{w}\ \underline{v} = \underline{wv}.$$

Proof Define

$$\mathsf{concat} \equiv \lambda wv.\vec{a}x.w\vec{a}(v\vec{a}x).$$

Then the type is correct and the definition equation holds. ■

1.4.10 Proposition

(i) *There exists a term* $\mathsf{empty}_? \in \Lambda^{\emptyset}_{\to}(\mathsf{Sigma}^*)$ *such that*

$$\begin{aligned}\mathsf{empty}_?\ \underline{\epsilon} &= \mathsf{true} \\ \mathsf{empty}_?\ \underline{w} &= \mathsf{false} \qquad , \text{ if } w \neq \epsilon.\end{aligned}$$

(ii) *Given a (represented) word* $w_0 \in \Lambda^{\emptyset}_{\to}(\mathsf{Sigma}^*)$ *and a term*

$$G \in \Lambda^{\emptyset}_{\to}(\mathsf{Sigma}^*{\to}\mathsf{Sigma}^*)$$

there exists a term $F \in \Lambda^{\emptyset}_{\to}(\mathsf{Sigma}^*{\to}\mathsf{Sigma}^*)$ *such that*

$$\begin{aligned}F\ \underline{\epsilon} &= w_0; \\ F\ \underline{w} &= G\underline{w}, \qquad \text{if } w \neq \epsilon.\end{aligned}$$

Proof (i) Take $\mathsf{empty}_? \equiv \lambda wpq.w(\mathsf{K}q)^{\sim k}p$.
(ii) Take $F \equiv \lambda w \lambda x\vec{a}.\mathsf{empty}_?\, w(w_0\vec{a}x)(Gw\vec{a}x)$. ■

There are no terms 'car' or 'cdr' such that car $\underline{aw} = \underline{a}$ and cdr $\underline{aw} = \underline{w}$.

Trees

1.4.11 Definition The set of *binary trees*, written T^2, is defined by the following simplified syntax:

$$t = \epsilon \,|\, p(t, t);$$

here ϵ is the '*empty tree*' and p is the constructor that puts two trees together. For example $p(\epsilon, p(\epsilon, \epsilon)) \in T^2$ can be depicted as

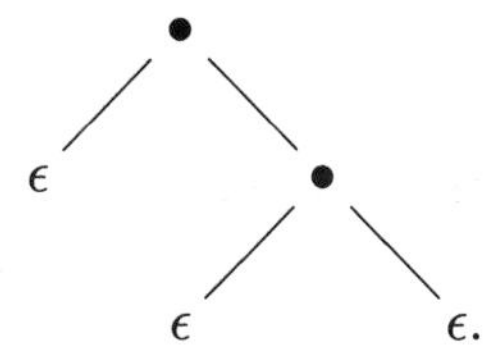

Now we will represent T^2 as a type in $\mathbb{T}^0$.

1.4.12 Definition

(i) The set T^2 will be represented by the type

$$\top^2 \equiv (0^2 \to 0) \to 0 \to 0.$$

(ii) Define for $t \in T^2$ its representation $\underline{t}$ recursively as

$$\underline{\epsilon} = \lambda pe.e;$$
$$\underline{p(t, s)} = \lambda pe.(\underline{t}pe)(\underline{s}pe).$$

(iii) Write

$$E = \lambda pe.e;$$
$$P = \lambda tspe.p(tpe)(spe).$$

Note that for $t \in T^2$ one has $\underline{t} \in \Lambda^{\emptyset}_{\to}(\top^2)$

The next result follows immediately from this definition.

1.4.13 Proposition *The map* $\underline{\ }: T^2 \to \top^2$ *can be defined recursively as*

$$\underline{\epsilon} = E;$$
$$\underline{p(t, s)} = P\underline{t}\,\underline{s}.$$

Interesting functions, like the one that selects one of the two branches of a tree, cannot be defined in $\boldsymbol{\lambda}^0_{\to}$. The type $\top^2$ will play an important role in Section 3.4.

Representing free algebras with a handicap

Now we will see that all the examples are special cases of a general construction. It turns out that first-order algebraic data types $\mathcal{A}$ can be represented in $\boldsymbol{\lambda}^0_{\rightarrow}$. The representations are said to have a *handicap* because not all primitive recursive functions on $\mathcal{A}$ are representable. Mostly it is destructors that cannot be represented. In special cases one can do better. Every finite algebra can be represented with all possible functions on them. Pairing with projections can be represented.

1.4.14 Definition

(i) An *algebra* is a set A with a specific finite set of operators of different arity:

$$\begin{array}{lll} c_1, c_2, \ldots & \in & A & \text{(constants, we may call these 0-ary operators);} \\ f_1, f_2, \ldots & \in & A{\rightarrow}A & \text{(unary operators);} \\ g_1, g_2, \ldots & \in & A^2{\rightarrow}A & \text{(binary operators);} \\ & \vdots & & \\ h_1, h_2, \ldots & \in & A^n{\rightarrow}A & \text{(n-ary operators).} \end{array}$$

(ii) An n-ary function $k : A^n{\rightarrow}A$ is called *algebraic* iff k can be defined explicitly from the given constructors by composition. For example

$$k = \lambda a_1 a_2.g_1(a_1, (g_2(f_1(a_2), c_2)))$$

is a binary algebraic function, usually specified as

$$k(a_1, a_2) = g_1(a_1, (g_2(f_1(a_2), c_2))).$$

(iii) An element a of A is called *algebraic* iff a is an algebraic 0-ary function. Algebraic elements of A can be denoted by first-order terms over the algebra.

(iv) The algebra A is called *free(ly generated)* if every element of A is algebraic and moreover if for two first-order terms t, s one has

$$t = s \Rightarrow t \equiv s.$$

In a free algebra the given operators are called *constructors*.

For example ω with constructors $0, s$ (s is the successor) is a free algebra. But $\mathbb{Z}$ with $0, s, p$ (p is the predecessor) is not free. Indeed, $0 = p(s(0))$, but $0 \not\equiv p(s(0))$ as syntactic expressions.

1.4.15 Theorem *For a free algebra $\mathcal{A}$ there is an $\underline{\mathcal{A}} \in \mathbb{T}^0$ and a map $\lambda a.\underline{a} : \mathcal{A}{\rightarrow}\Lambda^{\emptyset}_{\rightarrow}(\underline{\mathcal{A}})$ satisfying*

(i) $\underline{a}$ *is an lnf, for every* $a \in \mathcal{A}$;

(ii) $\underline{a} =_{\beta\eta} \underline{b} \Leftrightarrow a = b$;

(iii) $\Lambda^{\emptyset}_{\rightarrow}(\underline{\mathcal{A}}) = \{\underline{a} \mid a \in \mathcal{A}\}$, *up to* $\boldsymbol{\beta\eta}$*-conversion;*

(iv) *for* k*-ary algebraic functions* f *on* $\mathcal{A}$ *there is an* $\underline{f} \in \Lambda^{\emptyset}_{\rightarrow}(\underline{\mathcal{A}}^k \rightarrow \underline{\mathcal{A}})$ *such that*

$$\underline{f}\,\underline{a_1} \cdots \underline{a_k} = \underline{f(a_1, \ldots, a_k)};$$

(v) *there is a representable discriminator distinguishing between elements of the form* $c, f_1(a), f_2(a,b), \ldots, f_n(a_1, \ldots a_n)$*. More precisely, there is a term* $\mathsf{test} \in \Lambda^{\emptyset}_{\rightarrow}(\underline{\mathcal{A}} \rightarrow \underline{\omega})$ *such that for all* $a, b \in \mathcal{A}$

$$\begin{aligned} \mathsf{test}\,\underline{c} &= \mathbf{c}_0; \\ \mathsf{test}\,\underline{f_1(a)} &= \mathbf{c}_1; \\ \mathsf{test}\,\underline{f_2(a,b)} &= \mathbf{c}_2 \\ &\vdots \\ \mathsf{test}\,\underline{f_n(a_1, \ldots, a_n)} &= \mathbf{c}_n \end{aligned}$$

Proof We show this by a representative example. Let $\mathcal{A}$ be freely generated by, say, the 0-ary constructor c, the 1-ary constructor f and the 2-ary constructor g. Then an element like

$$a = g(c, f(c))$$

is represented by

$$\underline{a} = \lambda cfg.gc(fc) \in \Lambda(0 \rightarrow 1 \rightarrow 1_2 \rightarrow 0).$$

Taking $\underline{\mathcal{A}} = 0 \rightarrow 1 \rightarrow 1_2 \rightarrow 0$ we will verify the claims. First realize that $\underline{a}$ is constructed from a via $a^{\sim} = gc(fc)$ and then taking the closure $\underline{a} = \lambda cfg.a^{\sim}$.

(i) Clearly the $\underline{a}$ are in lnf.

(ii) If a and b are different, then their representations $\underline{a}, \underline{b}$ are different lnfs, hence $\underline{a} \neq_{\beta\eta} \underline{b}$.

(iii) The inhabitation machine $M_{\underline{\mathcal{A}}} = M_{0 \rightarrow 1 \rightarrow 1_2 \rightarrow 0}$ looks like

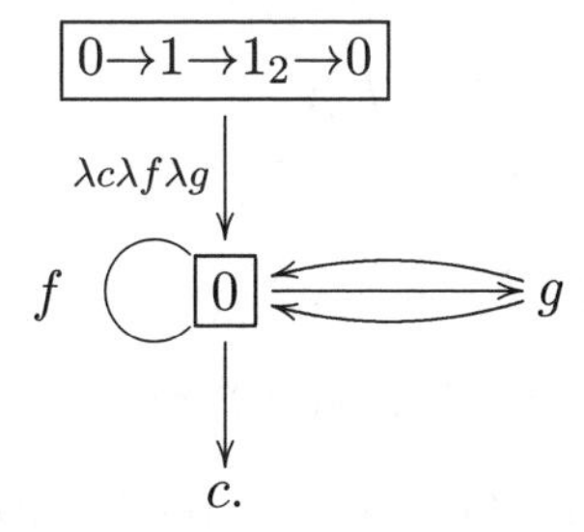

It follows that for every $M \in \Lambda^{\emptyset}_{\rightarrow}(\underline{\mathcal{A}})$ one has $M =_{\beta\eta} \lambda cfg.a^{\sim} = \underline{a}$ for some

$a \in \mathcal{A}$. This shows that $\Lambda^{\emptyset}_{\to}(\underline{\mathcal{A}}) \subseteq \{\underline{a} \mid a \in \mathcal{A}\}$. The converse inclusion is trivial. In the general case (for other data types $\mathcal{A}$) one has that $\text{rk}(\underline{\mathcal{A}}) = 2$. Hence the lnf inhabitants of $\underline{\mathcal{A}}$ have, for example, the form $\lambda c f_1 f_2 g_1 g_2.P$, where P is a typable combination of the variables $c, f_1^1, f_2^1, g_1^{1^2}, g_2^{1^2}$. This means that the corresponding inhabitation machine is similar and the argument generalizes.

(iv) An algebraic function is explicitly defined from the constructors. We first define representations for the constructors:

$$\begin{array}{rcll} \underline{c} & = & \lambda cfg.c & : \ \underline{\mathcal{A}}; \\ \underline{f} & = & \lambda acfg.f(acfg) & : \ \underline{\mathcal{A}}\to\underline{\mathcal{A}}; \\ \underline{g} & = & \lambda abcfg.g(acfg)(bcfg) & : \ \underline{\mathcal{A}}^2\to\underline{\mathcal{A}}. \end{array}$$

Then

$$\begin{array}{rcl} \underline{f}\,\underline{a} & = & \lambda cfg.f(\underline{a}cfg) \\ & = & \lambda cfg.f(a^{\sim}) \\ & \equiv & \lambda cfg.(f(a))^{\sim}, \text{ (tongue in cheek)}, \\ & \equiv & \underline{f(a)}. \end{array}$$

Similarly one has $\underline{g}\,\underline{a}\,\underline{b} = \underline{g(a,b)}$.

Now if for example $h(a,b) = g(a, f(b))$, then we can take

$$\underline{h} \equiv \lambda ab.\underline{g}a(\underline{f}b) : \underline{\mathcal{A}}^2\to\underline{\mathcal{A}}.$$

Then clearly $\underline{h}\,\underline{a}\,\underline{b} = \underline{h(a,b)}$.
Take $\mathsf{test} \equiv \lambda afc.a(\mathbf{c}_0 fc)(\lambda x.\mathbf{c}_1 fc)(\lambda xy.\mathbf{c}_2 fc)$. ∎

1.4.16 Definition The notion of free algebra can be generalized to a free *multi-sorted* algebra. We do this by giving an example. The collection of lists of natural numbers, written L_ω can be defined by the 'sorts' ω and L_ω and the constructors

$$\begin{array}{rcl} 0 & \in & \omega; \\ s & \in & \omega\to\omega; \\ \text{nil} & \in & L_\omega; \\ \text{cons} & \in & \omega\to L_\omega\to L_\omega. \end{array}$$

In this setting the list $[0,1] \in L_\omega$ is

$$\text{cons}(0,\text{cons}(s(0),\text{nil})).$$

More interesting multisorted algebras can be defined that are 'mutually recursive', see Exercise 1.13.

1.4.17 Corollary *Every freely generated multi-sorted first-order algebra can be represented in a way similar to that in Theorem* 1.4.15.

Proof Similar to that of the theorem. ■

Finite Algebras

For finite algebras one can do much better.

1.4.18 Theorem *For every finite set $X = \{a_1, \dots, a_n\}$ there exists a type $\underline{X} \in \mathbb{T}^0$ and elements $\underline{a}_1, \dots, \underline{a}_n \in \Lambda^{\emptyset}_{\to}(\underline{X})$ such that the following holds.*

(i) $\Lambda^{\emptyset}_{\to}(\underline{X}) = \{\underline{a} \mid a \in X\}$.
(ii) *For all k and $f : X^k \to X$ there exists an $\underline{f} \in \Lambda^{\emptyset}_{\to}(\underline{X}^k \to \underline{X})$ such that*

$$\underline{f}\, \underline{b}_1 \cdots \underline{b}_k = \underline{f(b_1, \dots, b_k)}.$$

Proof Take $\underline{X} = 1_n = 0^n \to 0$ and $\underline{a}_i = \lambda b_1 \cdots b_n . b_i \in \Lambda^{\emptyset}_{\to}(1_n)$.
(i) By a simple argument using the inhabitation machine M_{1_n}.
(ii) By induction on k. If $k = 0$, then f is an element of X, say $f = a_i$. Take $\underline{f} = \underline{a}_i$. Now suppose we can represent all k-ary functions. Given $f : X^{k+1} \to X$, define for $b \in X$

$$f_b(b_1, \dots, b_k) = f(b, b_1, \dots, b_k).$$

Each f_b is a k-ary function and has a representative $\underline{f_b}$. Define

$$\underline{f} = \lambda b \vec{b} . b (\underline{f_{a_1}} \vec{b}) \cdots (\underline{f_{a_n}} \vec{b}),$$

where $\vec{b} = b_2, \dots, b_{k+1}$. Then

$$\begin{aligned} \underline{f}\, \underline{b}_1 \cdots \underline{b}_{k+1} &= \underline{b}_1\, (\underline{f_{a_1}}\, \vec{\underline{b}}) \cdots (\underline{f_{a_n}}\, \vec{\underline{b}}) \\ &= \underline{f_{b_1}}\, \underline{b}_2 \cdots \underline{b}_{k+1} \\ &= \underline{f_{b_1}(b_2, \dots, b_{k+1})}, && \text{by the induction hypothesis,} \\ &= \underline{f(b_1, \dots, b_{k+1})}, && \text{by definition of } f_{b_1}. \ \blacksquare \end{aligned}$$

One even can faithfully represent the full typed structure over X as closed terms of $\boldsymbol{\lambda}^0_{\to}$; see Exercise 2.21.

Examples as free or finite algebras

The examples at the beginning of this section can all be viewed as free or finite algebras. The Booleans form a finite set and its representation is type 1_2. For this reason all Boolean functions can be represented. The natural numbers ω and the trees T are examples of free algebras with a handicapped representation. Words over a finite alphabet $\Sigma = \{a_1, \dots, a_n\}$ can be seen as an algebra with constant ϵ and further constructors $f_{a_i} = \boldsymbol{\lambda} w.a_i w$. The representations given are particular cases of the theorems about free and finite algebras.

Pairing

In the untyped lambda calculus there exists a way of storing two terms in such a manner that they can be retrieved:

$$\begin{aligned} \texttt{pair} &\equiv \lambda abz.zab; \\ \texttt{left} &\equiv \lambda z.z(\lambda xy.x); \\ \texttt{right} &\equiv \lambda z.z(\lambda xy.y). \end{aligned}$$

These terms satisfy

$$\begin{aligned} \texttt{left}(\texttt{pair}\, MN) &=_{\boldsymbol{\beta}} (\texttt{pair}\, MN)(\lambda xy.x) \\ &=_{\boldsymbol{\beta}} (\lambda z.zMN)(\lambda xy.x) \\ &=_{\boldsymbol{\beta}} M; \\ \texttt{right}(\texttt{pair}\, MN) &=_{\boldsymbol{\beta}} N. \end{aligned}$$

The triple of terms $\langle \texttt{pair}, \texttt{left}, \texttt{right} \rangle$ is called a (notion of) '$\boldsymbol{\beta}$-pairing'.

We will translate these notions to $\boldsymbol{\lambda}^0_{\to}$. We work with the Curry version.

1.4.19 Definition Let $A, B \in \mathbb{T}$ and let R be a notion of reduction on Λ.

(i) A *product* with *R-pairing* is a type $A \underline{\times} B \in \mathbb{T}$ together with terms

$$\begin{aligned} &\texttt{pair} \in \Lambda_{\to}(A \to B \to (A \underline{\times} B)); \\ &\texttt{left} \in \Lambda_{\to}((A \underline{\times} B) \to A); \\ &\texttt{right} \in \Lambda_{\to}((A \underline{\times} B) \to B), \end{aligned}$$

satisfying for variables x, y

$$\begin{aligned} \texttt{left}(\texttt{pair}\, xy) &=_R x; \\ \texttt{right}(\texttt{pair}\, xy) &=_R y. \end{aligned}$$

(ii) The type $A \underline{\times} B$ is called the *product* and the triple $\langle \texttt{pair}, \texttt{left}, \texttt{right} \rangle$ is called the *R-pairing*.

(iii) An *R-Cartesian product* is a product with R-pairing satisfying moreover for variables z

$$\texttt{pair}(\texttt{left}\, z)(\texttt{right}\, z) =_R z.$$

In that case the pairing is called a *surjective R-pairing*.

This pairing cannot be translated to a $\boldsymbol{\beta}$-pairing in $\boldsymbol{\lambda}^0_{\to}$ with a product $A \underline{\times} B$ for arbitrary types, see Barendregt (1974). But for two equal types one can form the product $A \underline{\times} A$. This makes it possible also to represent heterogeneous products using $\boldsymbol{\beta\eta}$-conversion.

1.4.20 Lemma *For every type $A \in \mathbb{T}^0$ there is a product $A \underline{\times} A \in \mathbb{T}^0$ with $\boldsymbol{\beta}$-pairing $\mathtt{pair}_0^A, \mathtt{left}_0^A$ and $\mathtt{right}_0^A$.*

Proof Take

$$\begin{aligned} A \underline{\times} A &= (A{\to}A{\to}A){\to}A; \\ \mathtt{pair}_0^A &= \lambda mnz.zmn; \\ \mathtt{left}_0^A &= \lambda p.p\mathsf{K}; \\ \mathtt{right}_0^A &= \lambda p.p\mathsf{K}_*. \blacksquare \end{aligned}$$

1.4.21 Proposition (Grzegorczyk (1964)) *Let $A, B \in \mathbb{T}^0$ be arbitrary types. There is a product $A \underline{\times} B \in \mathbb{T}^0$ with $\boldsymbol{\beta\eta}$-pairing $\langle \mathtt{pair}_0^{A,B}, \mathtt{left}_0^{A,B}, \mathtt{right}_0^{A,B} \rangle$ such that*

$$\begin{aligned} \mathtt{pair}_0^{A,B} &\in \Lambda_0; \\ \mathtt{left}_0^{A,B}, \mathtt{right}_0^{A,B} &\in \Lambda_0^{\{z:0\}}, \end{aligned}$$

and

$$\mathrm{rk}(A \underline{\times} B) = \max\{\mathrm{rk}(A), \mathrm{rk}(B), 2\}.$$

Proof Write $n = \mathtt{arity}(A), m = \mathtt{arity}(B)$. Define

$$A \underline{\times} B = A(1){\to} \dots {\to}A(n){\to}B(1){\to} \dots {\to}B(m){\to}0 \underline{\times} 0,$$

where $0 \underline{\times} 0 = (0{\to}0{\to}0){\to}0$. Then

$$\begin{aligned} \mathrm{rk}(A \underline{\times} B) &= \max_{i,j}\{\mathrm{rk}(A_i) + 1, \mathrm{rk}(B_j) + 1, \mathrm{rk}(0^2{\to}0) + 1\} \\ &= \max\{\mathrm{rk}(A), \mathrm{rk}(B), 2\}. \end{aligned}$$

Define z_A recursively: $z_0 = z; z_{A\to B} = \lambda a.z_B$. Then $z_A \in \Lambda_0^{z:0}(A)$. Write $\vec{x} = x_1, \dots, x_n, \vec{y} = y_1, \dots, y_m, \vec{z}_A = z_{A(1)}, \dots, z_{A(n)}$ and $\vec{z}_B = z_{B(1)}, \dots, z_{B(m)}$. Now define

$$\begin{aligned} \mathtt{pair}_0^{A,B} &= \lambda mn.\lambda \vec{x}\vec{y}.\mathtt{pair}_0^0(m\vec{x})(n\vec{y}); \\ \mathtt{left}_0^{A,B} &= \lambda p.\lambda \vec{x}.\mathtt{left}_0^0(p\vec{x}\vec{z}_B); \\ \mathtt{right}_0^{A,B} &= \lambda p.\lambda \vec{x}.\mathtt{right}_0^0(p\vec{z}_A\vec{y}). \end{aligned}$$

Then for example

$$\begin{aligned}\mathtt{left}_0^{A,B}(\mathtt{pair}_0^{A,B}MN) &=_\beta \lambda\vec{x}.\mathtt{left}_0^0(\mathtt{pair}_0^0 MN\vec{x}\vec{z}_B)\\ &=_\beta \lambda\vec{x}.\,\mathtt{left}_0^0[\mathtt{pair}_0^0(M\vec{x})(N\vec{z}_B)]\\ &=_\beta \lambda\vec{x}.(M\vec{x})\\ &=_\eta M. \blacksquare\end{aligned}$$

In Barendregt (1974) it is proved that $\boldsymbol{\eta}$-conversion is essential: with $\boldsymbol{\beta}$-conversion one can pair only certain combinations of types. Also it is shown that there is no *surjective pairing* in the theory with $\boldsymbol{\beta\eta}$-conversion. In Section 5.2 we will discuss systems extended with surjective pairing. By similar techniques to those in the aforementioned paper it can be shown that in $\boldsymbol{\lambda}_\rightarrow^\infty$ there is no $\boldsymbol{\beta\eta}$-pairing function $\mathtt{pair}_0^{\alpha,\beta}$ for base types.

1.4.22 Proposition *Let $A_1,\dots,A_n \in \mathbb{T}^0$. There are closed terms*

$$\begin{aligned}\mathtt{tuple}^n &: A_1\rightarrow\cdots\rightarrow A_n\rightarrow(A_1\underline{\times}\cdots\underline{\times}A_n)\\ \mathtt{proj}_k^n &: A_1\underline{\times}\cdots\underline{\times}A_n\rightarrow A_k\end{aligned}$$

such that, for $M_1,\dots,M_n$ of the right type, one has

$$\mathtt{proj}_k^n(\mathtt{tuple}^n M_1\cdots M_n) =_{\boldsymbol{\beta\eta}} M_k.$$

Proof By iterating pairing. ■

1.4.23 Notation If there is little danger of confusion and the $\vec{M}, N$ are of the right type we write

$$\begin{aligned}\langle M_1,\dots,M_n\rangle &\equiv \mathtt{tuple}^n M_1\cdots M_n;\\ N\cdot k &\equiv \mathtt{proj}_k^n N.\end{aligned}$$

Then $\langle M_1,\dots,M_n\rangle\cdot k = M_k$, for $1\le k\le n$.

1.5 Exercises

Exercise 1.1 Find types for

$$\begin{aligned}\mathsf{B} &\equiv \lambda xyz.x(yz);\\ \mathsf{C} &\equiv \lambda xyz.xzy;\\ \mathsf{C}_* &\equiv \lambda xy.yx;\\ \mathsf{K}_* &\equiv \lambda xy.y;\\ \mathsf{W} &\equiv \lambda xy.xyy.\end{aligned}$$

Exercise 1.2 Find types for SKK, $\lambda xy.y(\lambda z.zxx)x$ and $\lambda fx.f(f(fx))$.

Exercise 1.3 Show that $\text{rk}(A \rightarrow B \rightarrow C) = \max\{\text{rk}(A)+1, \text{rk}(B)+1, \text{rk}(C)\}$.

Exercise 1.4 Show that if $M \equiv P[x := Q]$ and $N \equiv (\lambda x.P)Q$, then M may have a type in $\boldsymbol{\lambda}_{\rightarrow}^{\text{Cu}}$ but N may not. A similar observation can be made for pseudo-terms of $\boldsymbol{\lambda}_{\rightarrow}^{\text{dB}}$.

Exercise 1.5 Show

(1) $\lambda xy.(xy)x \notin \Lambda_{\rightarrow}^{\text{Cu},\emptyset}$;
(2) $\lambda xy.x(yx) \in \Lambda_{\rightarrow}^{\text{Cu},\emptyset}$.

Exercise 1.6 Find inhabitants in

$$(A \rightarrow B \rightarrow C) \rightarrow B \rightarrow A \rightarrow C \text{ and } (A \rightarrow A \rightarrow B) \rightarrow A \rightarrow B.$$

Exercise 1.7 [van Benthem] Show that $\Lambda_{\rightarrow}^{\text{Ch}}(A)$ and $\Lambda_{\rightarrow}^{\text{Cu},\emptyset}(A)$ is for some $A \in \mathbb{T}^{\mathbb{A}}$ not a context-free language.

Exercise 1.8 Define in $\boldsymbol{\lambda}_{\rightarrow}^{0}$ the *pseudo-negation* $\sim A \equiv A \rightarrow 0$. Construct an inhabitant of $\sim\sim\sim A \rightarrow \sim A$.

Exercise 1.9 Prove the following (see Definition 1.2.31):

(1) Let $M \in \Lambda_{\rightarrow}^{\text{dB}}$ with $\text{FV}(M) \subseteq \text{dom}(\Gamma)$, then $(M^{\Gamma})^{-} \equiv M$ and $\Gamma_{M^{\Gamma}} \subseteq \Gamma$;
(2) Let $M \in \Lambda_{\rightarrow}^{\text{Ch}}$, then $(M^{-})^{\Gamma_M} \equiv M$.

Exercise 1.10 Construct a term F with $\vdash_{\boldsymbol{\lambda}_{\rightarrow}^{0}} F : T_2 \rightarrow T_2$ such that for trees t one has $F\underline{t} =_{\beta} \underline{t^{\text{mir}}}$, where t^{mir} is the mirror image of t, defined by

$$\epsilon^{\text{mir}} = \epsilon;$$
$$(p(t,s))^{\text{mir}} = p(s^{\text{mir}}, t^{\text{mir}}).$$

Exercise 1.11 A term M is called *proper* if all the λs appear in the prefix of M, i.e. $M \equiv \lambda\vec{x}.N$ and there is no λ occurring in N. Let A be a type such that $\Lambda_{\rightarrow}^{\emptyset}(A)$ is not empty. Show that

$$\text{Every nf of type } A \text{ is proper} \Leftrightarrow \text{rk}(A) \leq 2.$$

Exercise 1.12 Determine the class of closed inhabitants of the types 4 and 5.

Exercise 1.13 The collection of multi-ary trees can be seen as part of a multi-sorted algebra with sorts MTree and L_{MTree} as follows:

$$\text{nil} \in L_{\text{Mtree}};$$
$$\text{cons} \in \text{Mtree} \rightarrow L_{\text{Mtree}} \rightarrow L_{\text{Mtree}};$$
$$p \in L_{\text{Mtree}} \rightarrow \text{Mtree}.$$

Represent this multi-sorted free algebra in $\boldsymbol{\lambda}^0_{\to}$. Construct the lambda term representing the tree

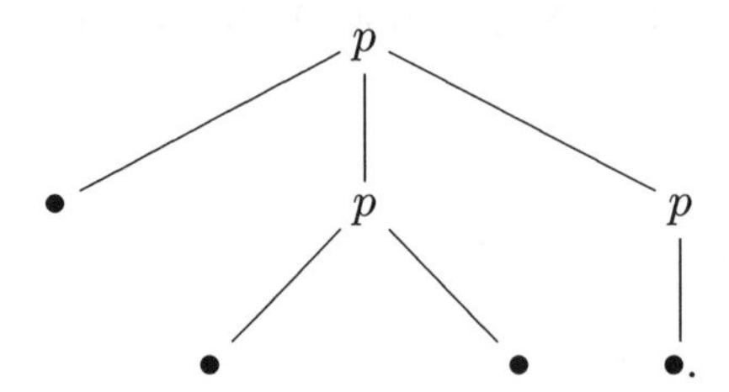

Exercise 1.14 In this exercise it will be proved that each term (having a $\boldsymbol{\beta}$-nf) has a unique lnf. A term M (typed or untyped) is always of the form $\lambda x_1 \cdots x_n.yM_1 \cdots M_m$ or $\lambda x_1 \cdots x_n.(\lambda x.M_0)M_1 \cdots M_m$. Then $yM_1 \cdots M_m$ (or $(\lambda x.M_0)M_1 \cdots M_m$) is the *matrix* of M and the $(M_0,)M_1, \ldots, M_m$ are its *components*. A typed term $M \in \Lambda^{\Gamma}(A)$ is said to be *fully eta* (f.e.) *expanded* if its matrix is of type 0 and its components are f.e. expanded. Show the following for typed terms. (For untyped terms there is no finite f.e. expanded form, but the Nakajima tree, see B[1984] Exercise 19.4.4, is the corresponding notion for the untyped terms.)

(1) M is in lnf iff M is a $\boldsymbol{\beta}$-nf and f.e. expanded.
(2) If $M =_{\boldsymbol{\beta\eta}} N_1 =_{\boldsymbol{\beta\eta}} N_2$ and N_1, N_2 are $\boldsymbol{\beta}$-nfs, then $N_1 =_{\eta} N_2$. [Hint. Use η-postponement, see B[1984] Proposition 15.1.5.]
(3) If $N_1 =_{\eta} N_2$ and N_1, N_2 are $\boldsymbol{\beta}$-nfs, then there exist $N{\downarrow}$ and $N{\uparrow}$ such that $N_i \twoheadrightarrow_{\eta} N{\downarrow}$ and $N{\uparrow} \twoheadrightarrow_{\eta} N_i$, for $i = 1, 2$. [Hint. Show that both $\to_{\eta}$ and ${}_{\eta}\!\leftarrow$ satisfy the Diamond Lemma.]
(4) If M has a $\boldsymbol{\beta}$-nf, then it has a unique lnf.
(5) If N is f.e. expanded and $N \twoheadrightarrow_{\boldsymbol{\beta}} N'$, then N' is f.e. expanded.
(6) For all M there is a f.e. expanded M^* such that $M^* \twoheadrightarrow_{\eta} M$.
(7) If M has a $\boldsymbol{\beta}$-nf, then the lnf of M is the $\boldsymbol{\beta}$-nf of M^*, its f.e. expansion.

Exercise 1.15 For which types $A \in \mathbb{T}^0$ and $M \in \Lambda_{\to}(A)$ does one have

$$M \text{ in } \boldsymbol{\beta}\text{-nf} \Rightarrow M \text{ in lnf?}$$

Exercise 1.16

(1) Let $M = \lambda x_1 \cdots x_n.x_i M_1 \cdots M_m$ be a $\boldsymbol{\beta}$-nf. Define by induction on the length of M its Φ-*normal form*, written $\Phi(M)$, as

$$\Phi(\lambda \vec{x}.x_i M_1 \ldots M_m) := \lambda \vec{x}.x_i(\Phi(\lambda \vec{x}.M_1)\vec{x}) \cdots (\Phi(\lambda \vec{x}.M_m)\vec{x}).$$

(2) Compute the Φ-nf of $\mathsf{S} = \lambda xyz.xz(yz)$.

(3) Write $\Phi^{n,m,i} := \lambda y_1 \cdots y_m \lambda x_1 \cdots x_n . x_i (y_1 \vec{x}) \cdots (y_m \vec{x})$. Then

$$\Phi(\lambda \vec{x}. x_i M_1 \cdots M_m) = \Phi^{n,m,i}(\Phi(\lambda \vec{x}. M_1)) \cdots (\Phi(\lambda \vec{x}. M_m)).$$

Show that the $\Phi^{n,m,i}$ are typable.

(4) Show that every closed normal form of type A is up to $=_{\beta\eta}$ a product of the $\Phi^{n,m,i}$.

(5) Write S in such a manner.

Exercise 1.17 (Th. Joly.) As in B[1984], the terms in this book are *abstract terms*, considered modulo α-conversion. Sometimes it is useful to be explicit about α-conversion and even to violate the variable convention that in a subterm of a term the names of free and bound variables should be distinct. For this it is useful to modify the system of type assignment.

(1) Show that $\vdash^{\text{Cu}}_{\boldsymbol{\lambda}_{\to}}$ is not closed under α-conversion. That is,

$$\Gamma \vdash M{:}A, M \equiv_\alpha M' \not\Rightarrow \Gamma \vdash M'{:}A.$$

[Hint. Consider $M' \equiv \lambda x.x(\lambda x.x)$.]

(2) Consider the following system of type assignment to untyped terms:

$$\{x{:}A\} \vdash x : A;$$

$$\frac{\Gamma_1 \vdash M : (A \to B) \quad \Gamma_2 \vdash N : A}{\Gamma_1 \cup \Gamma_2 \vdash (MN) : B}, \quad \text{provided } \Gamma_1 \cup \Gamma_2 \text{ is a basis;}$$

$$\frac{\Gamma \vdash M : B}{\Gamma - \{x{:}A\} \vdash (\lambda x.M) : (A \to B)}, \quad \text{provided } \Gamma \cup \{x{:}A\} \text{ is a basis.}$$

Provability in this system will be denoted by $\Gamma \vdash' M : A$. Show that $\vdash'$ is closed under α-conversion.

(3) Show that

$$\Gamma \vdash' M' : A \Leftrightarrow \exists M \equiv_\alpha M'. \Gamma \vdash M : A.$$

Exercise 1.18 Elements in Λ are considered in this book modulo α-conversion, by working with α-equivalence classes. If instead one works with α-conversion, as in Church (1941), then one can consider the following problems on elements M of $\Lambda^{\emptyset}$.

(1) Given M, find an α-convert of M with a smallest number of distinct variables.

(2) Given $M \equiv_\alpha N$, find a shortest α-conversion from M to N.

(3) Given $M \equiv_\alpha N$, find an α-conversion from M to N, which uses the smallest number of variables possible along the way.

Study Statman (2007) for the proofs of the following results.

(i) There is a polynomial time algorithm for solving problem (1). It is reducible to vertex coloring of chordal graphs.
(ii) Problem (2) is co-NP complete (in recognition form). The general feedback vertex set problem for digraphs is reducible to problem (2).
(iii) At most one variable besides those occurring in both M and N is necessary. This appears to be folklore but the proof is not familiar. A polynomial time algorithm for the α-conversion of M to N using at most one extra variable is given.

2

Properties

2.1 Normalization

For several applications, for example for the problem of finding all possible inhabitants of a given type, we will need the weak normalization theorem, which states that all typable terms do have a $\beta\eta$-nf (normal form). The result is valid for all versions of $\lambda_{\to}^{\mathbb{A}}$ and *a fortiori* for the subsystem $\lambda_{\to}^{0}$. The proof is due to Turing and was published posthumously in Gandy (1980). In fact all typable terms in these systems are $\beta\eta$ strongly normalizing, which means that all $\beta\eta$-reductions are terminating. This fact requires more work and will be proved in Section 2.2.

The notion of 'abstract reduction system', see Klop [1992], is useful for the understanding of the proof of the normalization theorem.

2.1.1 Definition An *abstract reduction system* (ARS) is a pair $(X, \to_R)$, where X is a set and $\to_R$ is a binary relation on X.

We usually will consider $\Lambda, \Lambda_{\to}^{\mathbb{A}}$ with reduction relations $\to_{\beta(\eta)}$ as examples of an ARS.

In the following definition WN, weak normalization, stands for having a normal form, while SN, strong normalization, stands for not having infinite reduction paths. A typical example in $(\Lambda, \to_\beta)$ is the term $\mathsf{K}\mathsf{I}\boldsymbol{\Omega}$ that is WN but not SN.

2.1.2 Definition Let (X, R) be an ARS.

(i) An element $x \in X$ is *in R-normal form* (R-*nf*) if for no $y \in X$ one has $x \to_R y$.

(ii) An element $x \in X$ is *R-weakly normalizing* (R-WN), written $x \models R$-WN (or simply $x \models$ WN), if for some $y \in X$ one has $x \twoheadrightarrow_R y$ and y is in R-nf.

(iii) (X, R) is called WN, written $(X, R) \models \mathsf{WN}$, if

$$\forall x \in X . x \models R\text{-WN}.$$

(iv) An element $x \in X$ is said to be *R-strongly normalizing* (R-SN), written $x \models R$-SN (or just $x \models \mathsf{SN}$), if each R-reduction path starting with x

$$x \to_R x_1 \to_R x_2 \to_R \cdots$$

is finite.

(v) (X, R) is said to be *strongly normalizing*, written $(X, R) \models R$-SN or simply $(X, R) \models \mathsf{SN}$, if

$$\forall x \in X . x \models \mathsf{SN}.$$

One reason why the notion of ARS is interesting is that some properties of reduction can be dealt with in ample generality.

2.1.3 Definition Let (X, R) be an ARS.

(i) We say that (X, R) is *confluent* or satisfies the *Church–Rosser property*, written $(X, R) \models \mathsf{CR}$, if

$$\forall x, y_1, y_2 \in X . [x \twoheadrightarrow_R y_1 \ \& \ x \twoheadrightarrow_R y_2 \ \Rightarrow \ \exists z \in X . y_1 \twoheadrightarrow_R z \ \& \ y_2 \twoheadrightarrow_R z].$$

(ii) We say that (X, R) is *weakly confluent* or satisfies the *weak Church–Rosser property*, written $(X, R) \models \mathsf{WCR}$, if

$$\forall x, y_1, y_2 \in X . [x \to_R y_1 \ \& \ x \to_R y_2 \ \Rightarrow \ \exists z \in X . y_1 \twoheadrightarrow_R z \ \& \ y_2 \twoheadrightarrow_R z].$$

It is not the case that $\mathsf{WCR} \Rightarrow \mathsf{CR}$, see Exercise 2.17. However, one has the following result.

2.1.4 Proposition (Newman's Lemma) *Let (X, R) be an ARS. Then for (X, R)*

$$\mathsf{WCR} \ \& \ \mathsf{SN} \ \Rightarrow \ \mathsf{CR}.$$

Proof See B[1984], Proposition 3.1.25 or Lemma 5.3.8 below, for a slightly stronger localized version. ■

In this section we will show $(\Lambda^{\mathbb{A}}_{\to}, \to_{\beta\eta}) \models \mathsf{WN}$.

2.1.5 Definition

(i) A *multiset over ω* can be thought of as a generalized set S in which each element may occur more than once. For example

$$S = \{3, 3, 1, 0\}$$

is a multiset. We say that 3 occurs in S with multiplicity 2; that 1 has multiplicity 1; etc. We also may write this multiset as

$$S = \{3^2, 1^1, 0^1\} = \{3^2, 2^0, 1^1, 0^1\}.$$

More formally, the above multiset S can be identified with a function $f \in \omega^\omega$ that is almost everywhere 0:

$$f(0) = 1, f(1) = 1, f(2) = 0, f(3) = 2, f(k) = 0,$$

for $k > 3$. Such an S is finite if f has finite *support*, where

$$\texttt{support}(f) = \{x \in \omega \mid f(x) \neq 0\}.$$

(ii) Let $\mathcal{S}(\omega)$ be the collection of all finite multisets over ω. We can identify $\mathcal{S}(\omega)$ with $\{f \in \omega^\omega \mid \texttt{support}(f) \text{ is finite}\}$. To each f in this set we let correspond the multiset, which, following our intuition, we denote by

$$S_f = \{n^{f(n)} \mid n \in \texttt{support}(f)\}.$$

2.1.6 Definition Let $S_1, S_2 \in \mathcal{S}(\omega)$. Write

$$S_1 \rightarrow_{\mathcal{S}} S_2$$

if S_2 results from S_1 by replacing some element (just one occurrence) by finitely many lower elements (in the usual order of ω). For example

$$\{3, \underline{3}, 1, 0\} \rightarrow_{\mathcal{S}} \{3, \underline{2, 2, 2, 1}, 1, 0\}.$$

The transitive closure of $\rightarrow_{\mathcal{S}}$, not required to be reflexive, is called the *multiset order*[1] and is denoted by $>$. (Another notation for this relation is $\rightarrow^+_{\mathcal{S}}$.) So for example

$$\{3, 3, 1, 0\} > \{3, 2, 2, 1, 1, 0, 1, 1, 0\}.$$

In the following result it is shown that $(\mathcal{S}(\omega), \rightarrow_{\mathcal{S}})$ is WN, using an induction up to ω^2.

[1] We consider both irreflexive, usually denoted by $<$ or its converse $>$, and reflexive, order relations, usually denoted by $\leq$ or its converse $\geq$. From $<$ we can define the reflexive version $\leq$ by

$$a \leq b \Leftrightarrow a = b \text{ or } a < b.$$

Conversely, from $\leq$ we can define the irreflexive version $<$ by

$$a < b \Leftrightarrow a \leq b \;\&\; a \neq b.$$

Also we consider partial and total (or linear) order relations for which we have, for all a, b,

$$a \leq b \text{ or } b \leq a.$$

If nothing is said the order relation is total, while partial order relations are explicitly said to be partial.

2.1.7 Lemma *We define a particular (non-deterministic) reduction strategy F on $\mathcal{S}(\omega)$. A multiset S is contracted to $F(S)$ by taking a maximal element $n \in S$ and replacing it by finitely many numbers $< n$. Then F is a normalizing reduction strategy, i.e. for every $S \in \mathcal{S}(\omega)$ the $\mathcal{S}$-reduction sequence*

$$S \to_{\mathcal{S}} F(S) \to_{\mathcal{S}} F^2(S) \to_{\mathcal{S}} \cdots$$

is terminating.

Proof By induction on the highest number n occuring in S. If $n = 0$, then we are done. If $n = k + 1$, then we can successively replace in S all occurrences of n by numbers $\leq k$ obtaining S_1 with maximal number $\leq k$. Then we are done by the induction hypothesis. ■

In fact $(\mathcal{S}(\omega), \to_{\mathcal{S}})$ is SN. Although we do not strictly need this fact until Part II, we will actually give two proofs of it. In the first place it is something one ought to know; in the second place it is instructive to see that the result does not imply that $\lambda_{\to}^{\mathbb{A}}$ satisfies SN.

2.1.8 Lemma *The reduction system $(\mathcal{S}(\omega), \to_{\mathcal{S}})$ is* SN.

Our first proof uses ordinals; the second is from first principles.

First Proof Assign to every $S \in \mathcal{S}(\omega)$ an ordinal $\#S < \omega^\omega$ as suggested by the following examples.

$$\begin{aligned}
\#\{3, 3, 1, 0, 0, 0\} &= 2\omega^3 + \omega + 3;\\
\#\{3, 2, 2, 2, 1, 1, 0\} &= \omega^3 + 3\omega^2 + 2\omega + 1.
\end{aligned}$$

More formally, if S is represented by $f \in \omega^\omega$ with finite support, then

$$\#S = \Sigma_{i \in \omega} f(i) \cdot \omega^i.$$

Notice that

$$S_1 \to_{\mathcal{S}} S_2 \Rightarrow \#S_1 > \#S_2$$

(in the example this is because $\omega^3 > 3\omega^2 + \omega$). Hence by the well-foundedness of the ordinals the result follows. ■

Second Proof Viewing multisets as functions with finite support, define

$$\begin{aligned}
\mathcal{F}_k &= \{f \in \omega^\omega \mid \forall n \geq k.\, f(n) = 0\};\\
\mathcal{F} &= \bigcup_{k \in \omega} \mathcal{F}_k.
\end{aligned}$$

The set $\mathcal{F}$ is the set of functions with finite support. Define on $\mathcal{F}$ the relation $>$ corresponding to the relation $\to_{\mathcal{S}}$ for the formal definition of $\mathcal{S}(\omega)$:

$$f > g \Leftrightarrow f(k) > g(k), \text{ where } k \in \omega \text{ is largest}$$
$$\text{such that } f(k) \neq g(k).$$

It is easy to see that $(\mathcal{F}, >)$ is a linear order. We will show that it is even a well-order, i.e. for every non-empty set $X \subseteq \mathcal{F}$ there is a least element $f_0 \in X$. This implies that there are no infinite descending chains in $\mathcal{F}$.

To show this claim, it suffices to prove that each $\mathcal{F}_k$ is well-ordered, since

$$(\mathcal{F}_{k+1} \setminus \mathcal{F}_k) > \mathcal{F}_k$$

element-wise. This will be proved by induction on k. If $k = 0$, then this is trivial, since $\mathcal{F}_0 = \{\lambda\!\!\lambda n.0\}$. Now assume (our induction hypothesis) that $\mathcal{F}_k$ is well-ordered in order to show the same for $\mathcal{F}_{k+1}$. Let $X \subseteq \mathcal{F}_{k+1}$ be non-empty. Define

$$\begin{aligned} X(k) &= \{f(k) \mid f \in X\} \subseteq \omega; \\ X_k &= \{f \in X \mid f(k) \text{ minimal in } X(k)\} \subseteq \mathcal{F}_{k+1}; \\ X_k|k &= \{g \in \mathcal{F}_k \mid \exists f \in X_k\, f|k = g\} \subseteq \mathcal{F}_k, \end{aligned}$$

where

$$\begin{aligned} f|k(i) &= f(i), && \text{if } i < k; \\ &= 0, && \text{else.} \end{aligned}$$

By the induction hypothesis $X_k|k$ has a least element g_0. Then $g_0 = f_0|k$ for some $f_0 \in X_k$. This f_0 is then the least element of X_k and hence of X. ■

2.1.9 Remark The second proof shows in fact that if $(D, >)$ is a well-ordered set, then so is $(\mathcal{S}(D), >)$, defined analogously to $(\mathcal{S}(\omega), >)$. In fact the argument can be carried out in Peano Arithmetic (PA), showing

$$\vdash_{\mathbf{PA}} \mathrm{TI}_\alpha \to \mathrm{TI}_{\alpha^\omega},$$

where TI_α is the principle of transfinite induction for the ordinal α. Since TI_ω is in fact ordinary induction we have in PA (in an iterated exponentiation, parenthesing is to the right: for example $\omega^{\omega^\omega} = \omega^{(\omega^\omega)}$)

$$\mathrm{TI}_\omega,\ \mathrm{TI}_{\omega^\omega},\ \mathrm{TI}_{\omega^{\omega^\omega}}, \ldots.$$

This implies that the proof of TI_α can be carried out in Peano Arithmetic for every $\alpha < \epsilon_0$. Gentzen [1936] shows that TI_{ϵ_0}, where

$$\epsilon_0 = \omega^{\omega^{\omega^{\cdots}}},$$

cannot be carried out in PA. See Section 5.3 for TI up to ϵ_0.

In order to prove that $\boldsymbol{\lambda}_{\to}^{\mathbb{A}}$ is WN it suffices to work with $\boldsymbol{\lambda}_{\to}^{\mathrm{Ch}}$. We will use the following notation. We write terms with extra type information, decorating each subterm with its type. For example, instead of $(\lambda x^A.M)N \in \mathtt{term}_B$ we write $(\lambda x^A.M^B)^{A\to B}N^A$.

2.1.10 Definition

(i) Let $R \equiv (\lambda x^A.M^B)^{A\to B}N^A$ be a redex. The *depth* of R, written $\mathrm{dpt}(R)$, is defined as

$$\mathrm{dpt}(R) = \mathrm{dpt}(A \to B),$$

where dpt on types is defined in Definition 1.1.22.

(ii) To each M in $\boldsymbol{\lambda}_{\to}^{\mathrm{Ch}}$ we assign the multiset

$$S_M = \{\mathrm{dpt}(R) \mid R \text{ is a redex occurrence in } M\},$$

with the understanding that the multiplicity of R in M is copied in S_M.

In the following example we study how the contraction of one redex can duplicate other redexes or create new ones.

2.1.11 Examples

(i) Let R be a redex occurrence in a typed term M. Assume

$$M \xrightarrow{R}_{\boldsymbol{\beta}} N,$$

i.e. N results from M by contracting R. This contraction can duplicate other redexes. For example (we write $M[P]$, or $M[P,Q]$ to display subterms of M)

$$(\lambda x.M[x,x])R_1 \to_{\boldsymbol{\beta}} M[R_1,R_1]$$

duplicates the other redex R_1.

(ii) (Lévy (1978)) Contraction of a $\boldsymbol{\beta}$-redex may also create new redexes. For example

$$(\lambda x^{A\to B}.M[x^{A\to B}P^A]^C)^{(A\to B)\to C}(\lambda y^A.Q^B) \to_{\boldsymbol{\beta}} M[(\lambda y^A.Q^B)^{A\to B}P^A]^C;$$
$$(\lambda x^A.(\lambda y^B.M[x^A,y^B]^C)^{B\to C})^{A\to(B\to C)}P^AQ^B \to_{\boldsymbol{\beta}} (\lambda y^B.M[P^A,y^B]^C)^{B\to C}Q^B;$$
$$(\lambda x^{A\to B}.x^{A\to B})^{(A\to B)\to(A\to B)}(\lambda y^A.P^B)^{A\to B}Q^A \to_{\boldsymbol{\beta}} (\lambda y^A.P^B)^{A\to B}Q^A.$$

In Lévy (1978), 1.8.4., Lemme 3, it is proved (for the untyped λ-calculus) that the three ways of creating redexes in Example 2.1.11(ii) are the only possibilities. It is also given as Exercise 14.5.3 in B[1984].

2.1.12 Lemma *Assume $M \xrightarrow{R}_{\beta} N$ and let R_1 be a created redex in N. Then*

$$\mathrm{dpt}(R) > \mathrm{dpt}(R_1).$$

Proof In each of three cases we can inspect that the statement holds. ∎

2.1.13 Theorem (Weak Normalization Theorem for $\boldsymbol{\lambda}_{\rightarrow}^{\mathbb{A}}$) *If $M \in \Lambda$ is typable in $\boldsymbol{\lambda}_{\rightarrow}^{\mathbb{A}}$, then M is $\boldsymbol{\beta\eta}$-WN, i.e. has a $\boldsymbol{\beta\eta}$-nf. In other words,*

$$\boldsymbol{\lambda}_{\rightarrow}^{\mathbb{A}} \models \mathsf{WN}, \; mboxormoreexplicitly, \; \boldsymbol{\lambda}_{\rightarrow}^{\mathbb{A}} \models \boldsymbol{\beta\eta}\text{-}\mathsf{WN}.$$

Proof By Proposition 1.2.27(ii) it suffices to show this for terms in $\boldsymbol{\lambda}_{\rightarrow}^{\mathrm{Ch}}$. Note that $\boldsymbol{\eta}$-reductions decrease the length of a term; moreover, for $\boldsymbol{\beta}$-normal terms $\boldsymbol{\eta}$-contractions do not create $\boldsymbol{\beta}$-redexes. Therefore in order to establish $\boldsymbol{\beta\eta}$-WN it is sufficient to prove that M has a $\boldsymbol{\beta}$-nf.

Define the following $\boldsymbol{\beta}$-reduction strategy F. If M is in normal form, then $F(M) = M$. Otherwise, let R be the *right-most redex* of maximal depth n in M. A redex occurrence $(\lambda_1 x_1.P_1)Q_1$ is called *to the right* of an other one $(\lambda_2 x_2.P_2)Q_2$, if the occurrence of its λ, viz. λ_1, is to the right of the other redex λ, viz. λ_2.

Then

$$F(M) = N$$

where $M \xrightarrow{R}_{\beta} N$. Contracting a redex can only duplicate other redexes that are to the right of that redex. Therefore by the choice of R there can only be redexes of M duplicated in $F(M)$ of depth $< n$. By Lemma 2.1.12, redexes created in $F(M)$ by the contraction $M \rightarrow_{\boldsymbol{\beta}} F(M)$ are also of depth $< n$. Therefore in case M is not in $\boldsymbol{\beta}$-nf we have

$$S_M \rightarrow_{\mathcal{S}} S_{F(M)}.$$

Since $\rightarrow_{\mathcal{S}}$ is SN, it follows that the reduction

$$M \rightarrow_{\boldsymbol{\beta}} F(M) \rightarrow_{\boldsymbol{\beta}} F^2(M) \rightarrow_{\boldsymbol{\beta}} F^3(M) \rightarrow_{\boldsymbol{\beta}} \cdots$$

must terminate in a $\boldsymbol{\beta}$-nf. ∎

2.1.14 Corollary *Let $A \in \mathbb{T}^{\mathbb{A}}$ and $M \in \Lambda_{\rightarrow}(A)$. Then M has an lnf.*

Proof Let $M \in \Lambda_{\rightarrow}(A)$. Then M has a $\boldsymbol{\beta}$-nf by Theorem 2.1.13, hence by Exercise 1.14 also an lnf. ∎

For $\boldsymbol{\beta}$-reduction this weak normalization theorem was first proved by Turing, see Gandy (1980). The proof does not really need SN for $\mathcal{S}$-reduction, requiring transfinite induction up to ω^ω. The simpler result, Lemma 2.1.7, using induction up to ω^2, suffices.

It is easy to see that a different reduction strategy does not yield an $\mathcal{S}$-reduction chain. For example the two terms

$$(\lambda x^A.y^{A\to A\to A}x^Ax^A)^{A\to A}((\lambda x^A.x^A)^{A\to A}x^A) \to_\beta$$
$$y^{A\to A\to A}((\lambda x^A.x^A)^{A\to A}x^A)((\lambda x^A.x^A)^{A\to A}x^A)$$

give the multisets $\{1,1\}$ and $\{1,1\}$. Nevertheless, SN does hold for all systems $\boldsymbol{\lambda}_{\to}^{\mathbb{A}}$, as will be proved in Section 2.2. It is an open problem whether ordinals can be assigned in a natural and simple way to terms of $\boldsymbol{\lambda}_{\to}^{\mathbb{A}}$ such that

$$M \to_\beta N \;\Rightarrow\; \mathtt{ord}(M) > \mathtt{ord}(N).$$

See Howard (1970) and de Vrijer (1987a) for steps in this direction.

Applications of normalization

We will show that $\boldsymbol{\beta}$-normal terms inhabiting the represented data types (Bool, Nat, Σ^* and T^2) all are standard, i.e. correspond to the intended elements. From WN for $\boldsymbol{\lambda}_{\to}^{\mathbb{A}}$ and the subject reduction theorem it then follows that all inhabitants of the mentioned data types are standard. The argumentation is given by a direct argument basically using the Generation Lemma. It can be streamlined, as will be done for Proposition 2.1.18, by following the inhabitation machines, see Section 1.3, for the types involved. For notational convenience we will work with $\boldsymbol{\lambda}_{\to}^{\mathrm{Cu}}$, but we could equivalently work with $\boldsymbol{\lambda}_{\to}^{\mathrm{Ch}}$ or $\boldsymbol{\lambda}_{\to}^{\mathrm{dB}}$, as is clear from Corollary 1.2.26(iii) and Proposition 1.2.33.

2.1.15 Proposition *Let* $\mathsf{Bool} \equiv \mathsf{Bool}_\alpha$*, with* α *a type atom. Then for* M *in normal form one has*

$$\vdash M : \mathsf{Bool} \;\Rightarrow\; M \in \{\mathsf{true}, \mathsf{false}\}.$$

Proof By repeated use of Proposition 1.2.21, the free variable Lemma 1.2.2 and the Generation Lemma for $\boldsymbol{\lambda}_{\to}^{\mathrm{Cu}}$, Proposition 1.2.3, one has

$$\begin{aligned}
\vdash M : \alpha{\to}\alpha{\to}\alpha \;&\Rightarrow\; M \equiv \lambda x.M_1\\
&\Rightarrow\; x{:}\alpha \vdash M_1 : \alpha{\to}\alpha\\
&\Rightarrow\; M_1 \equiv \lambda y.M_2\\
&\Rightarrow\; x{:}\alpha, y{:}\alpha \vdash M_2 : \alpha\\
&\Rightarrow\; M_2 \equiv x \text{ or } M_2 \equiv y.
\end{aligned}$$

So $M \equiv \lambda xy.x \equiv \mathsf{true}$ or $M \equiv \lambda xy.y \equiv \mathsf{false}$. ■

2.1.16 Proposition *Let* $\mathsf{Nat} \equiv \mathsf{Nat}_\alpha = (\alpha{\to}\alpha){\to}\alpha{\to}\alpha$. *Then for* M *in long normal form one has*

$$\vdash M : \mathsf{Nat} \Rightarrow M \in \{\mathbf{c}_n \mid n \in \omega\}.$$

Proof Again we have

$$\begin{aligned}\vdash M : (\alpha{\to}\alpha){\to}\alpha{\to}\alpha &\Rightarrow M \equiv \lambda f.M_1\\ &\Rightarrow f{:}\alpha{\to}\alpha \vdash M_1 : \alpha{\to}\alpha\\ &\Rightarrow M_1 \equiv \lambda x.M_2\\ &\Rightarrow f{:}\alpha{\to}\alpha, x{:}\alpha \vdash M_2 : \alpha.\end{aligned}$$

Now

$$\begin{aligned}f{:}\alpha{\to}\alpha, x{:}\alpha, \vdash M_2 : \alpha \Rightarrow\ &[M_2 \equiv x \vee\\ &[M_2 \equiv fM_3 \ \&\ f{:}\alpha{\to}\alpha,\ x{:}\alpha \vdash M_3 : \alpha]].\end{aligned}$$

Therefore by induction on the structure of M_2 it follows that

$$f{:}\alpha{\to}\alpha, x{:}\alpha \vdash M_2 : \alpha \Rightarrow M_2 \equiv f^n x,$$

with $n \geq 0$. So $M \equiv \lambda fx.f^n x \equiv \mathbf{c}_n$. ■

2.1.17 Proposition *Let* $\mathsf{Sigma}^* \equiv \mathsf{Sigma}^*_\alpha$. *Then for* M *in normal form one has*

$$\vdash M : \mathsf{Sigma}^* \Rightarrow M \in \{\underline{w} \mid w \in \Sigma^*\}.$$

Proof Again we have

$$\begin{aligned}\vdash M : \alpha{\to}(\alpha{\to}\alpha)^k{\to}\alpha &\Rightarrow M \equiv \lambda x.N\\ &\Rightarrow x{:}\alpha \vdash N : (\alpha{\to}\alpha)^k{\to}\alpha\\ &\Rightarrow N \equiv \lambda a_1.N_1 \ \&\ x{:}\alpha, a_1{:}\alpha{\to}\alpha \vdash N_1 : (\alpha{\to}\alpha)^{k-1}{\to}\alpha\\ &\ \ \vdots\\ &\Rightarrow N \equiv \lambda a_1 \cdots a_k.N \ \&\ x{:}\alpha, a_1, \ldots, a_k{:}\alpha{\to}\alpha \vdash N_k : \alpha\\ &\Rightarrow [N_k \equiv x \vee\\ &\qquad [N_k \equiv a_{i_j} N'_k \ \&\ x{:}\alpha, a_1, \ldots, a_k{:}\alpha{\to}\alpha \vdash N_k' : \alpha]]\\ &\Rightarrow N_k \equiv a_{i_1}(a_{i_2}(\cdots(a_{i_p}x)\cdot\cdot))\\ &\Rightarrow M \equiv \lambda x a_1 \cdots a_k.a_{i_1}(a_{i_2}(\cdots(a_{i_p}x)\cdot\cdot))\\ &\qquad \equiv \underline{a_{i_1}a_{i_2}\cdots a_{i_p}}.\ ■\end{aligned}$$

A more streamlined proof will be given for the data type of trees $\top^2$.

2.1.18 Proposition *Let* $\top^2 = (\alpha \to \alpha \to \alpha) \to \alpha \to \alpha$ *and* $M \in \Lambda^{\emptyset}_{\to}(\top^2)$.

(i) *If M is in long normal form, then $M \equiv \underline{t}$, for some $t \in T^2$.*
(ii) *Then $M =_{\beta\eta} \underline{t}$ for some tree $t \in T^2$.*

Proof (i) For M in long normal form use the inhabitation machine for $\top^2$ to show that $M \equiv \underline{t}$ for some $t \in T^2$.

(ii) For a general M there is, by Corollary 2.1.14, an M' in long normal form such that $M =_{\beta\eta} M'$. Then by (i) applied to M' we are done. ∎

This proof raises the question what terms in $\boldsymbol{\beta}$-nf are also in long normal form: see Exercise 1.15.

2.2 Proofs of strong normalization

We now will give two proofs showing that $\boldsymbol{\lambda}_{\rightarrow}^{\mathbb{A}} \models \mathsf{SN}$. The first one is the classical proof due to Tait (1967) that needs little technique, but uses set-theoretic comprehension. The second proof, due to Statman, is elementary, but needs results about reduction.

2.2.1 Theorem (Strong Normalization Theorem for $\boldsymbol{\lambda}_{\rightarrow}^{\mathrm{Ch}}$) *For all $A \in \mathbb{T}^\infty$ and $M \in \Lambda_{\rightarrow}^{\mathrm{Ch}}(A)$ one has $\boldsymbol{\beta\eta}$-$\mathsf{SN}(M)$.*

Proof We use an induction loading. First we add constants $d_\alpha \in \Lambda_{\rightarrow}^{\mathrm{Ch}}(\alpha)$ to $\boldsymbol{\lambda}_{\rightarrow}^{\mathbb{A}}$ for each atom α, obtaining $\boldsymbol{\lambda}_{\rightarrow}^{\mathrm{Ch}+}$. Then we prove SN for the extended system. It follows *a fortiori* that the system without the constants is SN.

One first defines for $A \in \mathbb{T}^\infty$ the following class $\mathcal{C}_A$ of *computable* terms of type A; we write SN for $\mathsf{SN}_{\beta\eta}$:

$$\begin{aligned}
\mathcal{C}_\alpha &= \{M \in \Lambda_{\rightarrow}^{\mathrm{Ch},\emptyset}(\alpha) \mid \mathsf{SN}(M)\};\\
\mathcal{C}_{A\rightarrow B} &= \{M \in \Lambda_{\rightarrow}^{\mathrm{Ch},\emptyset}(A\rightarrow B) \mid \forall Q \in \mathcal{C}_A . MQ \in \mathcal{C}_B\};\\
\mathcal{C} &= \bigcup_{A \in \mathbb{T}^\infty} \mathcal{C}_A.
\end{aligned}$$

Then one defines the classes $\mathcal{C}_A^*$ of terms that are *computable under substitution*:

$$\mathcal{C}_A^* = \{M \in \Lambda_{\rightarrow}^{\mathrm{Ch}}(A) \mid \forall \vec{P} \in \mathcal{C}.[M[\vec{x} := \vec{P}] \in \Lambda_{\rightarrow}^{\mathrm{Ch},\emptyset}(A) \Rightarrow M[\vec{x} := \vec{P}] \in \mathcal{C}_A]\}.$$

Write $\mathcal{C}^* = \bigcup\{\mathcal{C}_A^* \mid A \in \mathbb{T}^\infty\}$. For $A = A_1 \rightarrow \ldots \rightarrow A_n \rightarrow \alpha$ define

$$d_A \equiv \lambda x_1^{A_1} \ldots \lambda x_n^{A_n}.d_\alpha.$$

Then for A one has

$$M \in \mathcal{C}_A \Leftrightarrow \forall \vec{Q} \in \mathcal{C}.M\vec{Q} \in \mathsf{SN}, \tag{2.1}$$

$$M \in \mathcal{C}_A^* \Leftrightarrow \forall \vec{P}, \vec{Q} \in \mathcal{C}.M[\vec{x} := \vec{P}]\vec{Q} \in \mathsf{SN}, \tag{2.2}$$

where the $\vec{P}, \vec{Q}$ should have the correct types and $M\vec{Q}$ and $M[\vec{x}{:}= \vec{P}]\vec{Q}$ are of type α, respectively. By an easy simultaneous induction on A one can show

$$M \in \mathcal{C}_A \Rightarrow \mathsf{SN}(M); \tag{2.3}$$

$$d_A \in \mathcal{C}_A. \tag{2.4}$$

In particular, since $M[\vec{x}{:}= \vec{P}]\vec{Q} \in \mathsf{SN} \Rightarrow M \in \mathsf{SN}$, it follows that

$$M \in \mathcal{C}^* \Rightarrow M \in \mathsf{SN}. \tag{2.5}$$

Now one shows by induction on M that

$$M \in \Lambda(A) \Rightarrow M \in \mathcal{C}^*_A. \tag{2.6}$$

We distinguish cases and use (2.2).

Case $M \equiv x$. Then for $P, \vec{Q} \in \mathcal{C}$ one has $M[x{:}= P]\vec{Q} \equiv P\vec{Q} \in \mathcal{C} \subseteq \mathsf{SN}$, by the definition of $\mathcal{C}$ and (2.3).

Case $M \equiv NL$. Easy.

Case $M \equiv \lambda x.N$. Now $\lambda x.N \in \mathcal{C}^*$ iff for all $\vec{P}, Q, \vec{R} \in \mathcal{C}$ one has

$$(\lambda x.N[\vec{y}{:}= \vec{P}])Q\vec{R} \in \mathsf{SN}. \tag{2.7}$$

By the induction hypothesis one has $N \in \mathcal{C}^* \subseteq \mathsf{SN}$; therefore, if $\vec{P}, Q, \vec{R} \in \mathcal{C} \subseteq \mathsf{SN}$, then

$$N[x{:}= Q, \vec{y}{:}= \vec{P}]\vec{R} \in \mathsf{SN}. \tag{2.8}$$

Now every maximal reduction path σ starting from the term in (2.7) passes through a reduct of the term in (2.8), as reductions within $N, \vec{P}, Q, \vec{R}$ are finite, hence σ is finite. Therefore we have (2.7).

Finally by (2.6) and (2.5), every typable term of $\boldsymbol{\lambda}_{\rightarrow}^{\mathrm{Ch}+}$, hence of $\boldsymbol{\lambda}_{\rightarrow}^{\mathbb{A}}$, is SN. ∎

The idea of the proof is that one would have liked to prove by induction on M that it is SN. But this is not directly possible. One needs the *induction loading* that $M\vec{P} \in \mathsf{SN}$. For a typed system with only combinators this is sufficient and is covered by the original argument of Tait (1967). For lambda terms one needs the extra induction loading of being computable under substitution. This argument was first presented by Prawitz (1971), for natural deduction, Girard (1971) for the second-order typed lambda calculus $\boldsymbol{\lambda 2}$, and Stenlund (1972) for $\boldsymbol{\lambda}_{\rightarrow}$.

2.2.2 Corollary (SN for $\boldsymbol{\lambda}_{\rightarrow}^{\mathrm{Cu}}$) $\forall A \in \mathbb{T}^{\infty} \forall \Gamma \forall M \in \Lambda_{\rightarrow}^{\mathrm{Cu},\Gamma}(A).\mathsf{SN}_{\beta\eta}(M)$.

Proof Suppose $M \in \Lambda$ has type A with respect to Γ and has an infinite reduction path σ. By repeated use of Proposition 1.2.27(ii) lift M to $M' \in \Lambda_{\rightarrow}^{\mathrm{Ch}}$ with an infinite reduction path (that projects to σ), contradicting the Theorem. ■

An elementary proof of strong normalization

Now we present an elementary proof, due to Statman, of strong normalization of $\boldsymbol{\lambda}_{\rightarrow}^{\mathbb{A},\mathrm{Ch}}$, where $\mathbb{A} = \{0\}$. Inspiration came from Nederpelt (1973), Gandy (1980) and Klop (1980). The point of this proof is that in this reduction system, strong normalizability follows from normalizability by local structure arguments similar to and in many cases identical with those presented for the untyped lambda calculus in B[1984]. These include analysis of redex creation, permutability of head with internal reductions, and permutability of η- with $\boldsymbol{\beta}$-redexes. In particular, no special proof technique is needed to obtain strong normalization once normalization has been observed. We use some results in the untyped lambda calculus.

2.2.3 Definition

(i) Let $R \equiv (\lambda x.X)Y$ be a $\boldsymbol{\beta}$-redex. Then R is

  (1) an I*-redex* if $x \in \mathrm{FV}(X)$;
  (2) a K*-redex* if $x \notin \mathrm{FV}(X)$;
  (3) a K^o*-redex* if R is a K-redex and $x = x^0$ and $X \in \Lambda_{\rightarrow}^{\mathrm{Ch}}(0)$;
  (4) a K^+*-redex* if R is a K-redex and is not a K^o-redex.

(ii) A term M is said to have the $\lambda\mathsf{K}^o$*-property* if every abstraction $\lambda x.X$ in M with $x \notin \mathrm{FV}(X)$ satisfies $x = x^0$ and $X \in \Lambda_{\rightarrow}^{\mathrm{Ch}}(0)$.

2.2.4 Notation

(i) $\rightarrow_{\beta\mathsf{I}}$ is reduction of I-redexes.
(ii) $\rightarrow_{\beta\mathsf{I}\mathsf{K}^+}$ is reduction of I- or K^+-redexes.
(iii) $\rightarrow_{\beta\mathsf{K}^o}$ is reduction of K^o-redexes.

2.2.5 Theorem *Every $M \in \Lambda_{\rightarrow}^{\mathrm{Ch}}$ is $\boldsymbol{\beta\eta}$-SN.*

Proof The result is proved in several steps.

(i) *Every term is $\boldsymbol{\beta\eta}$-normalizable and therefore has a head normal form, hnf, a term of the form $\lambda\vec{x}.yM_1 \cdots M_n$.* This is Theorem 2.1.13.

(ii) *There are no $\boldsymbol{\beta}$-reduction cycles.* Consider a shortest term M at the beginning of a cyclic reduction. Then

$$M \rightarrow_{\boldsymbol{\beta}} M_1 \rightarrow_{\boldsymbol{\beta}} \cdots \rightarrow_{\boldsymbol{\beta}} M_n \equiv M,$$

where, by minimality of M, at least one of the contracted redexes is a head-redex. Then M has an infinite quasi-head-reduction consisting of $\twoheadrightarrow_{\beta} \circ \rightarrow_h \circ \twoheadrightarrow_{\beta}$ steps. Therefore M has an infinite head-reduction, as internal (i.e. non-head) redexes can be postponed. (This is Exercise 13.6.13 [use Lemma 11.4.5] in B[1984].) This contradicts (i), using B[1984], Corollary 11.4.8 to the Standardization Theorem.

(iii) $M \twoheadrightarrow_{\eta} N \twoheadrightarrow_{\beta}^{+} L \Rightarrow \exists P.M \twoheadrightarrow_{\beta}^{+} P \twoheadrightarrow_{\eta} N$. This is a strengthening of $\boldsymbol{\eta}$-postponement, B[1984] Corollary 15.1.6, and can be proved in the same way.

(iv) $\boldsymbol{\beta}$-$\mathsf{SN} \Rightarrow \boldsymbol{\beta\eta}$-$\mathsf{SN}$. Take an infinite $\rightarrow_{\beta\eta}$ sequence. Make a diagram with $\boldsymbol{\beta}$-steps drawn horizontally and $\boldsymbol{\eta}$-steps vertically. These vertical steps are finite, as $\boldsymbol{\eta} \models \mathsf{SN}$. Apply (iii) at each $\twoheadrightarrow_{\eta} \circ \twoheadrightarrow_{\beta}^{+}$-step. The result yields a horizontal infinite $\rightarrow_{\beta}$ sequence.

(v) We have $\boldsymbol{\lambda}_{\rightarrow}^{\mathbb{A}} \models \boldsymbol{\beta}\mathsf{I}$-$\mathsf{WN}$. By (i).

(vi) $\boldsymbol{\lambda}_{\rightarrow}^{\mathbb{A}} \models \boldsymbol{\beta}\mathsf{I}$-$\mathsf{SN}$. By Church's result in B[1984], Conservation Theorem for $\lambda\mathsf{I}$, 11.3.4.

(vii) $M \twoheadrightarrow_{\beta} N \Rightarrow \exists P.M \twoheadrightarrow_{\beta\mathsf{IK}^{+}} P \twoheadrightarrow_{\beta\mathsf{K}^{o}} N$ ($\boldsymbol{\beta}\mathsf{K}^{o}$-postponement). When contracting a K^{o} redex, no redex can be created. Realizing this, one has

$$\begin{array}{ccc} P & \xrightarrow{\beta\mathsf{IK}^{+}} & P' \\ {\scriptstyle \beta\mathsf{K}^{o}}\downarrow & & \downarrow{\scriptstyle \beta\mathsf{K}^{o}} \\ Q & \xrightarrow[\beta\mathsf{IK}^{+}]{} & R. \end{array}$$

From this the statement follows by a simple diagram chase, that, without loss of generality, looks like

$$\begin{array}{ccccccc} M & \xrightarrow{\beta\mathsf{IK}^{+}} & \cdot & \xrightarrow{\beta\mathsf{IK}^{+}} & \cdot & \xrightarrow{\beta\mathsf{IK}^{+}} & P \\ & & {\scriptstyle \beta\mathsf{K}^{o}}\downarrow & & \downarrow{\scriptstyle \beta\mathsf{K}^{o}} & & \downarrow{\scriptstyle \beta\mathsf{K}^{o}} \\ & & \cdot & \xrightarrow{\beta\mathsf{IK}^{+}} & \cdot & \xrightarrow{\beta\mathsf{IK}^{+}} & \cdot \\ & & & & {\scriptstyle \beta\mathsf{K}^{o}}\downarrow & & \downarrow{\scriptstyle \beta\mathsf{K}^{o}} \\ & & & & \cdot & \xrightarrow{\beta\mathsf{IK}^{+}} & N. \end{array}$$

(viii) *Suppose M has the $\lambda\mathsf{K}^{o}$-property. Then M $\boldsymbol{\beta}$-reduces to only finitely many N.* First observe that $M \twoheadrightarrow_{\beta\mathsf{IK}^{+}} N \Rightarrow M \twoheadrightarrow_{\beta\mathsf{I}} N$, as a contraction of an I-redex cannot create a K^{+}-redex. (But a contraction of a K redex can create a K^{+} redex.) Hence by (vi) the set $\mathcal{X} = \{P \mid M \twoheadrightarrow_{\beta\mathsf{IK}^{+}} P\}$ is finite. Since K-redexes shorten terms, also the set of K^{o}-reducts of elements of $\mathcal{X}$ form a finite set. Therefore by (vii) we are done.

(ix) *If M has the $\lambda\mathsf{K}^{o}$-property, then $M \models \boldsymbol{\beta}$-$\mathsf{SN}$.* Use (viii) and (ii).

(x) *If M has the $\lambda\mathsf{K}^o$-property, then $M \models \beta\eta$-SN.* Use (iv) and (ix).

(xi) *For each M there is an N with the $\lambda\mathsf{K}^o$-property such that $N \twoheadrightarrow_{\beta\eta} M$.* Let $R \equiv \lambda x^A.P^B$ a subterm of M, making it fail to be a term with the $\lambda\mathsf{K}^o$-property. Write $A = A_1 \to \cdots \to A_a \to 0$, $B = B_1 \to \cdots \to B_b \to 0$. Then replace the above-mentioned subterm by

$$R' \equiv \lambda x^A \lambda y_1^{B_1} \cdots y_b^{B_b}.(\lambda z^0.(P y_1^{B_1} \cdots y_b^{B_b}))(x^A u_1^{A_1} \cdots u_a^{A_a}),$$

which $\beta\eta$-reduces to R, but which does not violate the $\lambda\mathsf{K}^o$-property. That R' contains the free variables $\vec{u}$ does not matter. Treating each such subterm this way, we obtain N.

(xii) $\boldsymbol{\lambda}_{\to}^{\mathbb{A}} \models \beta\eta$-SN. Use (x) and (xi). ∎

Other proofs of Strong Normalization from Weak Normalization are in de Vrijer (1987b), Kfoury and Wells (1995), Sørensen (1997), and Xi (1997). In de Vrijer's proof, a computation is given of the longest reduction path to β-nf for a typed term M.

2.3 Checking and finding types

There are several natural problems concerning type systems.

2.3.1 Definition

(i) Given basis Γ, term M and type A, the problem of *type checking* consists of determining whether $\Gamma \vdash M : A$.

(ii) The problem of *typability* consists of determining, for a given term M, whether M has some type with respect to some Γ.

(iii) The problem of *type reconstruction* ('finding types') consists of finding all possible types A and bases Γ that type a given M.

(iv) The *inhabitation problem* consists of finding out whether a given type A is inhabited by some term M in a given basis Γ.

(v) The *enumeration problem* consists of determining, for a given type A and a given context Γ, all possible terms M such that $\Gamma \vdash M : A$.

The five problems may be summarized stylistically as:

$$\begin{array}{rlll}
\Gamma & \vdash_{\boldsymbol{\lambda}_{\to}} & M : A? & \textit{type checking}; \\
\exists A, \Gamma\ [\Gamma & \vdash_{\boldsymbol{\lambda}_{\to}} & M : A]? & \textit{typability}; \\
? & \vdash_{\boldsymbol{\lambda}_{\to}} & M : ? & \textit{type reconstruction}; \\
\exists M\ [\Gamma & \vdash_{\boldsymbol{\lambda}_{\to}} & M : A]? & \textit{inhabitation}; \\
\Gamma & \vdash_{\boldsymbol{\lambda}_{\to}} & ? : A & \textit{enumeration}.
\end{array}$$

In a different notation this is:

$$\begin{array}{rll} & M \in \Lambda^{\Gamma}_{\to}(A)? & \textit{type checking;} \\ \exists A, \Gamma & M \in \Lambda^{\Gamma}_{\to}(A)? & \textit{typability;} \\ & M \in \Lambda^{?}_{\to}(?) & \textit{type reconstruction;} \\ & \Lambda^{\Gamma}_{\to}(A) \neq \emptyset? & \textit{inhabitation;} \\ & ? \in \Lambda^{\Gamma}_{\to}(A) & \textit{enumeration.} \end{array}$$

In this section we will treat the problems of type checking, typability and type reconstruction for the three versions of $\boldsymbol{\lambda}_{\to}$. It turns out that these problems are decidable for all versions. The solutions are essentially simpler for $\boldsymbol{\lambda}^{\mathrm{Ch}}_{\to}$ and $\boldsymbol{\lambda}^{\mathrm{dB}}_{\to}$ than for $\boldsymbol{\lambda}^{\mathrm{Cu}}_{\to}$. The problems of inhabitation and enumeration will be treated in the next section.

One may wonder what is the role of the context Γ in these questions. The problem

$$\exists\Gamma\exists A\, \Gamma \vdash M : A$$

can be reduced to one without a context. Indeed, for $\Gamma = \{x_1{:}A_1, \dots, x_n{:}A_n\}$

$$\Gamma \vdash M : A \;\Leftrightarrow\; \vdash (\lambda x_1(:A_1) \cdots \lambda x_n(:A_n).M) : (A_1 \to \cdots \to A_n \to A).$$

Therefore

$$\exists\Gamma\exists A\, [\Gamma \vdash M : A] \;\Leftrightarrow\; \exists B\, [\vdash \lambda\vec{x}.M : B].$$

On the other hand the question

$$\exists\Gamma\exists M\, [\Gamma \vdash M : A]\,?$$

is trivial: take $\Gamma = \{x{:}A\}$ and $M \equiv x$. So we do not consider this question.

The solution of problems like type checking for a fixed context will have important applications for the treatment of constants.

Checking and finding types for $\boldsymbol{\lambda}^{\mathrm{dB}}_{\to}$ and $\boldsymbol{\lambda}^{\mathrm{Ch}}_{\to}$

We will see again that the systems $\boldsymbol{\lambda}^{\mathrm{dB}}_{\to}$ and $\boldsymbol{\lambda}^{\mathrm{Ch}}_{\to}$ are essentially equivalent. For these systems the solutions to the problems of type checking, typability and type reconstruction are easy. All of the solutions are computable with an algorithm of linear complexity.

2.3.2 Proposition (Type checking for $\boldsymbol{\lambda}^{\mathrm{dB}}_{\to}$) *Let Γ be a basis of $\boldsymbol{\lambda}^{\mathrm{dB}}_{\to}$. Then there is a computable function* $\mathtt{type}_\Gamma : \Lambda^{\mathrm{dB}}_{\to} \to \mathbb{T} \cup \{\mathtt{error}\}$ *such that*

$$M \in \Lambda^{\mathrm{dB},\Gamma}_{\to}(A) \;\Leftrightarrow\; \mathtt{type}_\Gamma(M) = A.$$

Proof Define

$$\begin{array}{rcll} \mathtt{type}_\Gamma(x) &=& \Gamma(x); & \\ \mathtt{type}_\Gamma(MN) &=& B, & \text{if } \mathtt{type}_\Gamma(M) = \mathtt{type}_\Gamma(N){\to}B, \\ &=& \mathtt{error}, & \text{else;} \\ \mathtt{type}_\Gamma(\lambda x{:}A.M) &=& A{\to}\mathtt{type}_{\Gamma\cup\{x:A\}}(M), & \text{if } \mathtt{type}_{\Gamma\cup\{x:A\}}(M) \neq \mathtt{error}, \\ &=& \mathtt{error}, & \text{else.} \end{array}$$

Then the statement follows by induction on the structure of M. ■

2.3.3 Corollary *Typability and type reconstruction for* $\boldsymbol{\lambda}_{\to}^{\mathrm{dB}}$ *are computable. In fact one has:*

(i) $M \in \Lambda_{\to}^{\mathrm{dB},\Gamma} \Leftrightarrow \mathtt{type}_\Gamma(M) \neq \mathtt{error}$.

(ii) *Each* $M \in \Lambda_{\to}^{\mathrm{dB},\Gamma}(\mathtt{type}_\Gamma)$ *has a unique type; in particular*

$$M \in \Lambda_{\to}^{\mathrm{dB},\Gamma}(\mathtt{type}_\Gamma(M)).$$

Proof Use the proposition. ■

For $\boldsymbol{\lambda}_{\to}^{\mathrm{Ch}}$ things are essentially the same, except that there are no bases needed since variables come with their own types.

2.3.4 Proposition (Type checking for $\boldsymbol{\lambda}_{\to}^{\mathrm{Ch}}$) *There is a computable function* $\mathtt{type} : \Lambda_{\to}^{\mathrm{Ch}} \to \mathbb{T}$ *such that*

$$M \in \Lambda_{\to}^{\mathrm{Ch}}(A) \Leftrightarrow \mathtt{type}(M) = A.$$

Proof Define

$$\begin{array}{rcll} \mathtt{type}(x^A) &=& A; & \\ \mathtt{type}(MN) &=& B, & \text{if } \mathtt{type}(M) = \mathtt{type}(N){\to}B, \\ \mathtt{type}(\lambda x^A.M) &=& A{\to}\mathtt{type}(M). & \end{array}$$

Then the statement follows again by induction on the structure of M. ■

2.3.5 Corollary *Typability and type reconstruction for* $\boldsymbol{\lambda}_{\to}^{\mathrm{Ch}}$ *are computable. In fact one has the following. Each* $M \in \Lambda_{\to}^{\mathrm{Ch}}$ *has a unique type; in particular* $M \in \Lambda_{\to}^{\mathrm{Ch}}(\mathtt{type}(M))$.

Proof Use the proposition. ■

Checking and finding types for $\boldsymbol{\lambda}_{\rightarrow}^{\mathrm{Cu}}$

We now will show the computability of the three questions for $\boldsymbol{\lambda}_{\rightarrow}^{\mathrm{Cu}}$. This occupies 2.3.6–2.3.16 and in these items $\vdash$ stands for $\vdash_{\boldsymbol{\lambda}_{\rightarrow}}^{\mathrm{Cu}}$ over a general $\mathbb{T}^{\mathbb{A}}$.

Let us first make the easy observation that in $\boldsymbol{\lambda}_{\rightarrow}^{\mathrm{Cu}}$ types are not unique. For example $\mathsf{I} \equiv \lambda x.x$ has as a possible type $\alpha \rightarrow \alpha$, but also $(\beta \rightarrow \beta) \rightarrow (\beta \rightarrow \beta)$ and in general $A \rightarrow A$. Of these types $\alpha \rightarrow \alpha$ is the 'most general' in the sense that the other ones can be obtained by a substitution in α.

2.3.6 Definition

(i) A *substitutor* is an operation $* : \mathbb{T} \rightarrow \mathbb{T}$ such that

$$*(A \rightarrow B) \equiv *(A) \rightarrow *(B).$$

(ii) We write A^* for $*(A)$.

(iii) Usually a substitution $*$ has a finite support; that is, for all but finitely many type variables α, one has $\alpha^* \equiv \alpha$, the support of $*$ being

$$\sup(*) = \{\alpha \mid \alpha^* \not\equiv \alpha\}.$$

In that case we write

$$*(A) = A[\alpha_1 := \alpha_1^*, \ldots, \alpha_n := \alpha_n^*],$$

where $\{\alpha_1, \ldots, \alpha_n\} \supseteq \sup(*)$. We also write

$$* = [\alpha_1 := \alpha_1^*, \ldots, \alpha_n := \alpha_n^*]$$

and

$$* = [\,]$$

for the identity substitution.

2.3.7 Definition

(i) Let $A, B \in \mathbb{T}$. A *unifier* for A and B is a substitutor $*$ such that $A^* \equiv B^*$.

(ii) The substitutor $*$ is a *most general unifier* for A and B if

- $A^* \equiv B^*$
- $A^{*_1} \equiv B^{*_1} \;\Rightarrow\; \exists *_2 \,.\, *_1 \equiv *_2 \circ *$.

(iii) Let $E = \{A_1 = B_1, \ldots, A_n = B_n\}$ be a finite set of equations between types. The equations do not need to be valid. A *unifier* for E is a substitutor $*$ such that $A_1^* \equiv B_1^* \;\&\; \cdots \;\&\; A_n^* \equiv B_n^*$. In that case one writes $* \models E$. Similarly one defines the notion of a most general unifier for E.

2.3.8 Examples The types $\beta \to (\alpha \to \beta)$ and $(\gamma \to \gamma) \to \delta$ have a unifier. For example $* = [\beta := \gamma \to \gamma,\ \delta := \alpha \to (\gamma \to \gamma)]$ or $*_1 = [\beta := \gamma \to \gamma,\ \alpha := \varepsilon \to \varepsilon,\ \delta := \varepsilon \to \varepsilon \to (\gamma \to \gamma)]$. The unifier $*$ is most general, $*_1$ is not.

2.3.9 Definition We say A is a *variant* of B if for some $*_1$ and $*_2$ one has

$$A = B^{*_1} \text{ and } B = A^{*_2}.$$

2.3.10 Example Now $\alpha \to \beta \to \beta$ is a variant of $\gamma \to \delta \to \delta$, but not of $\alpha \to \beta \to \alpha$.

Note that if $*_1$ and $*_2$ are both most general unifiers of, say, A and B, then A^{*_1} and A^{*_2} are variants of each other and similarly for B.

The following result due to Robinson (1965) states that (in the first-order[2] case) *unification* is decidable.

2.3.11 Theorem (Unification theorem)

(i) *There is a recursive function U having (after coding) as input a pair of types and as output either a substitutor or* `fail` *such that*

$$\begin{aligned} A \text{ and } B \text{ have a unifier} &\Rightarrow U(A,B) \text{ is a most general unifier} \\ &\qquad \text{for } A \text{ and } B; \\ A \text{ and } B \text{ have no unifier} &\Rightarrow U(A,B) = \mathsf{fail}. \end{aligned}$$

(ii) *There is (after coding) a recursive function U having as input finite sets of equations between types and as output either a substitutor or* `fail` *such that*

$$\begin{aligned} E \text{ has a unifier} &\Rightarrow U(E) \text{ is a most general unifier for } E; \\ E \text{ has no unifier} &\Rightarrow U(E) = \texttt{fail}. \end{aligned}$$

Proof Note that $A_1 \to A_2 \equiv B_1 \to B_2$ holds iff $A_1 \equiv B_1$ and $A_2 \equiv B_2$ hold.

(i) Define $U(A,B)$ by the following recursive loop, using case distinction:

$$\begin{aligned} U(\alpha, B) &= [\alpha := B], && \text{if } \alpha \notin \mathrm{FV}(B), \\ &= [\,], && \text{if } B = \alpha, \\ &= \texttt{fail}, && \text{else}; \end{aligned}$$

$$U(A_1 \to A_2, \alpha) = U(\alpha, A_1 \to A_2);$$

$$U(A_1 \to A_2, B_1 \to B_2) = U(A_1^{U(A_2,B_2)}, B_1^{U(A_2,B_2)}) \circ U(A_2, B_2),$$

[2] That is, for the algebraic signature $\langle \mathbb{T}, \to \rangle$. Higher-order unification is undecidable, see Section 4.2.

where this last expression is considered to be `fail` if one of its parts is. Let

$$\#_{\text{var}}(A, B) = \text{`the number of variables in } A \to B\text{'},$$
$$\#_{\to}(A, B) = \text{`the number of arrows in } A \to B\text{'}.$$

By induction on $(\#_{\text{var}}(A, B), \#_{\to}(A, B))$ ordered lexicographically, one can show that $U(A, B)$ is always defined. Moreover U satisfies the specification.

(ii) If $E = \{A_1 = B_1, \ldots, A_n = B_n\}$, then define $U(E) = U(A, B)$, where $A = A_1 \to \cdots \to A_n$ and $B = B_1 \to \cdots \to B_n$. ∎

See Baader and Nipkow (1998) and Baader and Snyder (2001) for more on unification. The following result due to Parikh (1973) for propositional logic (interpreted by the propositions-as-types interpretation) and Wand (1987) simplifies the proof of the decidability of type checking and typability for $\boldsymbol{\lambda}_{\to}$.

2.3.12 Proposition *For every basis Γ, term $M \in \Lambda$ and $A \in \mathbb{T}$ such that* $\text{FV}(M) \subseteq \text{dom}(\Gamma)$ *there is a finite set of equations $E = E(\Gamma, M, A)$ such that for all substitutors $*$ one has*

$$* \models E(\Gamma, M, A) \;\Rightarrow\; \Gamma^* \vdash M : A^*, \tag{2.9}$$
$$\Gamma^* \vdash M : A^* \;\Rightarrow\; *_1 \models E(\Gamma, M, A), \tag{2.10}$$

for some $*_1$ such that $*$ and $*_1$ have the same effect on the type variables in Γ and A.

Proof Define $E(\Gamma, M, A)$ by induction on the structure of M:

$$E(\Gamma, x, A) = \{A = \Gamma(x)\};$$
$$E(\Gamma, MN, A) = E(\Gamma, M, \alpha \to A) \cup E(\Gamma, N, \alpha),$$
where α is a fresh variable;
$$E(\Gamma, \lambda x.M, A) = E(\Gamma \cup \{x{:}\alpha\}, M, \beta) \cup \{\alpha \to \beta = A\},$$
where α, β are fresh.

By induction on M one can show (using the Generation Lemma (1.2.3)) that (2.9) and (2.10) hold. ∎

2.3.13 Definition

(i) Let $M \in \Lambda$. Then (Γ, A) is a *principal pair* for M, written $\text{pp}(M)$, if
  (1) $\Gamma \vdash M : A$;
  (2) $\Gamma' \vdash M : A' \;\Rightarrow\; \exists * \, [\Gamma^* \subseteq \Gamma' \;\&\; A^* \equiv A']$.
  Here $\{x_1{:}A_1, \ldots\}^* = \{x_1{:}A_1^*, \ldots\}$.

(ii) Let $M \in \Lambda$ be closed. Then A is a *principal type*, written $\text{pt}(M)$, if

(1) $\vdash M : A$;
(2) $\vdash M : A' \Rightarrow \exists * \; [A^* \equiv A']$.

Note that if (Γ, A) is a principal pair for M, then every variant (Γ', A') of (Γ, A), in the obvious sense, is also a principal pair for M. Conversely if (Γ, A) and (Γ', A') are principal pairs for M, then (Γ', A') is a variant of (Γ, A). Similarly for closed terms and principal types. Moreover, if (Γ, A) is a principal pair for M, then $\mathrm{FV}(M) = \mathrm{dom}(\Gamma)$.

The following result is due independently to Curry (1969) and Hindley (1969). It shows that for $\boldsymbol{\lambda}_\to$ the problems of type checking and typability are decidable.

2.3.14 Theorem (Principal type theorem for $\boldsymbol{\lambda}_\to^{\mathrm{Cu}}$)

(i) *There exists a computable function* pp *such that one has*

$$M \textit{ has a type} \Rightarrow \mathrm{pp}(M) = (\Gamma, A), \textit{ where } (\Gamma, A) \textit{ is a principal pair for } M;$$
$$M \textit{ has no type} \Rightarrow \mathrm{pp}(M) = \mathtt{fail}.$$

(ii) *There exists a computable function* pt *such that for closed terms M one has*

$$M \textit{ has a type} \Rightarrow \mathrm{pt}(M) = A, \textit{ where } A \textit{ is a principal type for } M;$$
$$M \textit{ has no type} \Rightarrow \mathrm{pt}(M) = \mathsf{fail}.$$

Proof (i) Let $\mathrm{FV}(M) = \{x_1, \dots, x_n\}$ and set $\Gamma_0 = \{x_1{:}\alpha_1, \dots, x_n{:}\alpha_n\}$ and $A_0 = \beta$. Note that

$$\begin{aligned} M \text{ has a type} &\Rightarrow \exists\Gamma\,\exists A\;\; \Gamma \vdash M : A \\ &\Rightarrow \exists *\;\; \Gamma_0^* \vdash M : A_0^* \\ &\Rightarrow \exists *\;\; * \models E(\Gamma_0, M, A_0). \end{aligned}$$

Define

$$\begin{aligned} \mathrm{pp}(M) &= (\Gamma_0^*, A_0^*), && \text{if } U(E(\Gamma_0, M, A_0)) = *; \\ &= \mathtt{fail}, && \text{if } U(E(\Gamma_0, M, A_0)) = \mathtt{fail}. \end{aligned}$$

Then $\mathrm{pp}(M)$ satisfies the requirements. Indeed, if M has a type, then

$$U(E(\Gamma_0, M, A_0)) = *$$

is defined and $\Gamma_0^* \vdash M : A_0^*$ by (2.9). To show that (Γ_0^*, A_0^*) is a pp, suppose that also $\Gamma' \vdash M : A'$. Let $\widetilde{\Gamma} = \Gamma' \restriction \mathrm{FV}(M)$; write $\widetilde{\Gamma} = \Gamma_0^{*_0}$ and $A' = A_0^{*_0}$. Then also $\Gamma_0^{*_0} \vdash M : A_0^{*_0}$. Hence by (2.10) for some $*_1$ (acting the same as

$*_0$ on Γ_0, A_0) one has $*_1 \models E(\Gamma_0, M, A_0)$. Since $*$ is a most general unifier (Proposition 2.3.11) one has $*_1 = *_2 \circ *$ for some $*_2$. Now indeed

$$(\Gamma_0^*)^{*_2} = \Gamma_0^{*_1} = \Gamma_0^{*_0} = \widetilde{\Gamma} \subseteq \Gamma'$$

and

$$(A_0^*)^{*_2} = A_0^{*_1} = A_0^{*_0} = A'.$$

If M has no type, then $\neg\exists * \; * \models E(\Gamma_0, M, A_0)$ hence

$$U(\Gamma_0, M, A_0) = \mathtt{fail} = \mathrm{pp}(M).$$

(ii) Let M be closed and $\mathrm{pp}(M) = (\Gamma, A)$. Then $\Gamma = \emptyset$ and we can put $\mathrm{pt}(M) = A$. ∎

2.3.15 Corollary *Type checking and typability for $\boldsymbol{\lambda}_{\rightarrow}^{\mathrm{Cu}}$ are decidable.*

Proof As to type checking, let M and A be given. Then

$$\vdash M : A \Leftrightarrow \exists * \, [A = \mathrm{pt}(M)^*].$$

This is decidable (as can be seen using an algorithm – *pattern matching* – similar to the one in Theorem 2.3.11).

As to typability, let M be given. Then M has a type iff $\mathrm{pt}(M) \neq \mathtt{fail}$. ∎

The following result is due to Hindley (1969) and Hindley (1997), Thm. 7A2.

2.3.16 Theorem (Second principal type theorem for $\boldsymbol{\lambda}_{\rightarrow}^{\mathrm{Cu}}$)

(i) *For every $A \in \mathbb{T}$ one has*

$$\vdash M : A \Rightarrow \exists M'[M' \twoheadrightarrow_{\beta} M \; \& \; \mathtt{pt}(M') = A].$$

(ii) *For every $A \in \mathbb{T}$ there exists a basis Γ and $M \in \Lambda$ such that (Γ, A) is a principal pair for M.*

Proof (i) We outline a proof by examples. We choose three situations in which we have to construct an M' that is representative for the general case. Do Exercise 2.5 for the general proof.

Case $M \equiv \lambda x.x$ and $A \equiv (\alpha \rightarrow \beta) \rightarrow \alpha \rightarrow \beta$. Then $\mathtt{pt}(M) \equiv \alpha \rightarrow \alpha$. Take $M' \equiv \lambda xy.xy$. The η-expansion of $\lambda x.x$ to $\lambda xy.xy$ makes subtypes of A correspond to unique subterms of M'.

Case $M \equiv \lambda xy.y$ and $A \equiv (\alpha \rightarrow \gamma) \rightarrow \beta \rightarrow \beta$. Then $\mathtt{pt}(M) \equiv \alpha \rightarrow \beta \rightarrow \beta$. Take $M' \equiv \lambda xy.\mathsf{K}y(\lambda z.xz)$. The $\boldsymbol{\beta}$-expansion forces x to have a functional type.

Case $M \equiv \lambda xy.x$ and $A \equiv \alpha \rightarrow \alpha \rightarrow \alpha$. Then $\mathtt{pt}(M) \equiv \alpha \rightarrow \beta \rightarrow \alpha$. Take

$M' \equiv \lambda xy.\mathsf{K}x(\lambda fz.z(fx)(fy))$. The $\boldsymbol{\beta}$-expansion forces x and y to have the same types.

(ii) Let A be given. We know that $\vdash \mathsf{I} : A{\rightarrow}A$. Therefore by (i) there exists an $\mathsf{I}' \twoheadrightarrow_{\beta\eta} \mathsf{I}$ such that $\mathtt{pt}(\mathsf{I}') = A{\rightarrow}A$. Then take $M \equiv \mathsf{I}'x$. We have $\mathsf{pp}(\mathsf{I}'x) = (\{x{:}A\}, A)$. ∎

It is an open problem whether the result also holds in the $\lambda\mathsf{I}$-calculus.

Complexity

A closer look at the proof of Theorem 2.3.14 reveals that the typability and type-checking problems (understood as yes or no decision problems) reduce to solving first-order unification, a problem known to be solvable in polynomial time, see Baader and Nipkow [1998]. Since the reduction is also polynomial, we conclude that typability and type-checking are solvable in polynomial time as well.

However, the actual type reconstruction may require exponential space (and thus also exponential time), just to write down the result. Indeed, Exercise 2.20 demonstrates that the length of a shortest type of a given term may be exponential in the length of the term. The explanation of the apparent inconsistency between the two results is this: long types can be represented by small graphs.

In order to decide whether for two typed terms $M, N \in \Lambda_{\rightarrow}(A)$ one has

$$M =_{\beta\eta} N,$$

one can normalize both terms and see whether the results are syntactically equal (up to α-conversion). In Exercise 2.19 it will be shown that the time and space costs of solving this conversion problem is hyper-exponential (in the sum of the sizes of M, N). The reason is that there are short terms having very long normal forms. For instance, the type-free application of Church numerals

$$\mathbf{c}_n\mathbf{c}_m = \mathbf{c}_{m^n}$$

can be typed, even when applied iteratively

$$\mathbf{c}_{n_1}\mathbf{c}_{n_2}\cdots\mathbf{c}_{n_k}.$$

In Exercise 2.18 it is shown that the costs of this typability problem are also at most hyper-exponential. The reason is that Turing's proof of normalization for terms in $\boldsymbol{\lambda}_{\rightarrow}$ uses a successive development of redexes of 'highest' type. Now the length of each such development depends exponentially on the length of the term, whereas the length of a term increases at most quadratically at each reduction step. The result even holds for typable terms

$M, N \in \Lambda_{\rightarrow}^{\mathrm{Cu}}(A)$, as the cost of finding types only adds a simple exponential to the cost.

One may wonder whether there is not a more efficient way of deciding $M =_{\beta\eta} N$, for example by using memory for the reduction of the terms, rather than a pure reduction strategy that only depends on the state of the term reduced so far. The sharpest question is whether there is any Turing computable method that has a better complexity class. In Statman (1979) it is proved that this is not the case, by showing that every elementary time bounded Turing machine computation can be coded as a convertibility problem for terms of some type in $\boldsymbol{\lambda}_{\rightarrow}^{0}$. A shorter proof of this result can be found in Mairson (1992).

2.4 Checking inhabitation

In this section we study for $\boldsymbol{\lambda}_{\rightarrow}^{\mathbb{A}}$ the problem of inhabitation. That is, given a type A, we study whether there is a term M such that $\vdash_{\boldsymbol{\lambda}_{\rightarrow}^{\mathbb{A}}} M : A$ holds. By Corollaries 1.2.20 and 1.2.34 it does not matter whether we work in the system *à la* Curry, Church or de Bruijn. Therefore we will focus on $\boldsymbol{\lambda}_{\rightarrow}^{\mathrm{Cu}}$. Note that by Proposition 1.2.2 the term M must be closed. From the Normalization Theorem 2.1.13 it follows that we may limit ourselves to find a term M in $\boldsymbol{\beta}$-nf.

For example, if $A = \alpha{\rightarrow}\alpha$, then we can take $M \equiv \lambda x(:\alpha).x$. In fact we will see later that this M is, modulo $\boldsymbol{\beta}$-conversion, the only choice. For $A = \alpha{\rightarrow}\alpha{\rightarrow}\alpha$ there are two inhabitants: $M_1 \equiv \lambda x_1 x_2.x_1 \equiv \mathsf{K}$ and $M_2 \equiv \lambda x_1 x_2.x_2 \equiv \mathsf{K}_*$. Again we have exhausted all inhabitants. If $A = \alpha$, then there are no inhabitants, as we will soon see.

Various interpretations will be useful to solve inhabitation problems.

The Boolean model

Type variables can be interpreted as ranging over $\mathbf{B} = \{0, 1\}$ and $\rightarrow$ as the two-ary function on $\mathbf{B}$ defined by

$$x{\rightarrow}y = 1 - x + xy$$

(classical implication). This turns every type A into a Boolean function. More formally this is done as follows.

2.4.1 Definition

(i) A *Boolean valuation* is a map $\rho : \mathbb{A}{\rightarrow}\mathbf{B}$.

(ii) Let ρ be a Boolean valuation. The *Boolean interpretation under* ρ of a type $A \in \mathbb{T}$, written $[\![A]\!]_\rho$, is defined recursively as:

$$[\![\alpha]\!]_\rho = \rho(\alpha);$$
$$[\![A_1 \to A_2]\!]_\rho = [\![A_1]\!]_\rho \to [\![A_2]\!]_\rho.$$

(iii) A Boolean valuation ρ *satisfies a type* A, written $\rho \models A$, if $[\![A]\!]_\rho = 1$. Let $\Gamma = \{x_1 : A_1, \ldots, x_n : A_n\}$, then ρ *satisfies* Γ, written $\rho \models \Gamma$, if

$$\rho \models A_1 \ \& \ \ldots \ \& \ \rho \models A_n.$$

(iv) A type A is *classically valid*, written $\models A$, iff for all Boolean valuations ρ one has $\rho \models A$.

2.4.2 Proposition *Let* $\Gamma \vdash_{\lambda_\to^{\mathbb{A}}} M{:}A$. *Then for all Boolean valuations* ρ *one has*

$$\rho \models \Gamma \ \Rightarrow \ \rho \models A.$$

Proof By induction on the derivation in $\lambda_\to^{\mathbb{A}}$. ∎

From this it follows that inhabited types are classically valid. This in turn implies that the type α is not inhabited.

2.4.3 Corollary

(i) *If* A *is inhabited, then* $\models A$.
(ii) *A type variable* α *is not inhabited.*

Proof (i) Immediate by Proposition 2.4.2, by taking $\Gamma = \emptyset$.
(ii) Immediate by (i), by taking $\rho(\alpha) = 0$. ∎

One may wonder whether the converse of 2.4.3(i), i.e.

$$\models A \ \Rightarrow \ A \text{ is inhabited} \tag{2.11}$$

holds. We will see that in $\lambda_\to^{\mathbb{A}}$ this is not the case. For $\lambda_\to^0$ (having only one base type 0), however, the implication (2.11) is valid.

2.4.4 Proposition (Statman [1982]) *Let* $A = A_1 \to \cdots \to A_n \to 0$ *be a type in* $\mathbb{T}^0$ *with* $n \geq 0$. *Then*

$$A \textit{ is inhabited} \Leftrightarrow \textit{for some } i \textit{ with } 1 \leq i \leq n \textit{ the type}$$
$$A_i \textit{ is } \text{not } \textit{inhabited.}$$

Proof ($\Rightarrow$) Assume $\vdash_{\lambda_\to^0} M : A$. Suppose towards a contradiction that all A_i are inhabited, i.e. $\vdash_{\lambda_\to^0} N_i : A_i$. Then $\vdash_{\lambda_\to^0} MN_1 \cdots N_n : 0$, contradicting 2.4.3(ii).

($\Leftarrow$) By induction on the structure of A. Assume that A_i with $1 \leq i \leq n$ is not inhabited.

Case 1. $A_i = 0$. Then

$$x_1 : A_1, \ldots, x_n : A_n \vdash x_i : 0$$

so

$$\vdash (\lambda x_1 \cdots x_n.x_i) : A_1 \to \cdots \to A_n \to 0,$$

i.e. A is inhabited.

Case 2. $A_i = B_1 \to \cdots \to B_m \to 0$. By (the contrapositive of) the induction hypothesis applied to A_i it follows that all B_j are inhabited, say $\vdash M_j : B_j$. Then

$$\begin{aligned} & x_1 : A_1, \ldots, x_n : A_n \vdash x_i : A_i = B_1 \to \cdots \to B_m \to 0 \\ \Rightarrow\ & x_1 : A_1, \ldots, x_n : A_n \vdash x_i M_1 \cdots M_m : 0 \\ \Rightarrow\ & \vdash \lambda x_1 \cdots x_n.x_i M_1 \cdots M_m : A_1 \to \cdots \to A_n \to 0 = A. \ \blacksquare \end{aligned}$$

From the proposition it easily follows that inhabitation of types in $\boldsymbol{\lambda}^0_{\to}$ is decidable with a linear time algorithm.

2.4.5 Corollary *In $\boldsymbol{\lambda}^0_{\to}$ one has, for all types A,*

$$A \text{ is inhabited} \Leftrightarrow \models A.$$

Proof ($\Rightarrow$) By Proposition 2.4.3(i).

($\Leftarrow$) Assume $\models A$ and that A is not inhabited. Then $A = A_1 \to \cdots \to A_n \to 0$ with each A_i inhabited. But then for $\rho_0(0) = 0$ one has

$$\begin{aligned} 1 &= [\![A]\!]_{\rho_0} \\ &= [\![A_1]\!]_{\rho_0} \to \cdots \to [\![A_n]\!]_{\rho_0} \to 0 \\ &= 1 \to \cdots \to 1 \to 0, \ \text{since} \models A_i \text{ for all } i, \\ &= 0, \ \text{since } 1 \to 0 = 0, \end{aligned}$$

a contradiction. $\blacksquare$

Corollary 2.4.5 does not hold for $\boldsymbol{\lambda}^{\infty}_{\to}$. In fact the type $((\alpha \to \beta) \to \alpha) \to \alpha$ (corresponding to Peirce's law) is a valid type that is not inhabited, as we will see soon.

Intuitionistic propositional logic

Although inhabited types correspond to Boolean tautologies, not all such tautologies correspond to inhabited types. Intuitionistic logic provides a precise characterization of inhabited types. The underlying idea, the *propositions-as-types* correspondence, will become clear in more detail in Sections

6.3 and 6.4. The book by Sørensen and Urzyczyn (2006) is devoted to this correspondence.

2.4.6 Definition (Implicational propositional logic)

(i) The set of formulas of the *implicational propositional logic*, written $\mathsf{form}(\mathrm{PROP})$, is defined by the following simplified syntax:

$$\begin{array}{|rcl|}\hline \mathsf{form} & ::= & \mathsf{var} \mid \mathsf{form} \supset \mathsf{form} \\ \mathsf{var} & ::= & p \mid \mathsf{var}' \\ \hline\end{array}$$

where we have written $\mathsf{form} = \mathsf{form}(\mathrm{PROP})$. For example $p', p' \supset p, p' \supset (p' \supset p)$ are formulas.

(ii) Let Γ be a set of formulas and let A be a formula. Then A is *derivable from* Γ, written $\Gamma \vdash_{\mathrm{PROP}} A$, if $\Gamma \vdash A$ can be produced by the following formal system:

$$\begin{array}{|rcl|}\hline A \in \Gamma & \Rightarrow & \Gamma \vdash A \\ \Gamma \vdash A \supset B,\ \Gamma \vdash A & \Rightarrow & \Gamma \vdash B \\ \Gamma, A \vdash B & \Rightarrow & \Gamma \vdash A \supset B. \\ \hline\end{array}$$

2.4.7 Notation

(i) We let $q, r, s, t, \ldots$ stand for arbitrary propositional variables.

(ii) As usual $\Gamma \vdash A$ stands for $\Gamma \vdash_{\mathrm{PROP}} A$ if there is little danger of confusion. Moreover, $\vdash A$ stands for $\emptyset \vdash A$.

2.4.8 Example

(i) $\vdash A \supset A$.

(ii) $A \vdash B \supset A$.

(iii) $\vdash A \supset (B \supset A)$.

(iv) $A \supset (A \supset B) \vdash A \supset B$.

2.4.9 Definition Let $A \in \mathsf{form}(\mathrm{PROP})$ and $\Gamma \subseteq \mathsf{form}(\mathrm{PROP})$.

(i) Define $[A] \in \mathbb{T}^{\infty}$ and $\Gamma_A \subseteq \mathbb{T}^{\infty}$ as follows:

A	$[A]$	Γ_A
p	p	$\emptyset$
$P \supset Q$	$[P] \to [Q]$	$\Gamma_P \cup \Gamma_Q$.

It so happens that $\Gamma_A = \emptyset$ and $[A]$ is A with the $\supset$ replaced by $\to$. But the setup will be needed for more complex logics and type theories.

(ii) Moreover, we set $[\Gamma] = \{x_A{:}A \mid A \in \Gamma\}$.

2.4.10 Proposition *Let* $A \in \mathsf{form}(\text{PROP})$ *and* $\Delta \subseteq \mathsf{form}(\text{PROP})$. *Then*

$$\Delta \vdash_{\text{PROP}} A \;\Rightarrow\; [\Delta] \vdash_{\boldsymbol{\lambda}_\to} M : [A], \textit{ for some } M.$$

Proof By induction on the generation of $\Delta \vdash A$.

Case 1. $\Delta \vdash A$ *because* $A \in \Delta$. Then $(x_A{:}[A]) \in [\Delta]$ and hence $[\Delta] \vdash x_A : [A]$. So we can take $M \equiv x_A$.

Case 2. $\Delta \vdash A$ *because* $\Delta \vdash B \supset A$ *and* $\Delta \vdash B$. Then by the induction hypothesis, $[\Delta] \vdash P : [B] \to [A]$ and $[\Delta] \vdash Q : [B]$. Therefore $[\Delta] \vdash PQ : [A]$.

Case 3. $\Delta \vdash A$ *because* $A \equiv B \supset C$ *and* $\Delta, B \vdash C$. By the induction hypothesis, $[\Delta], x_B{:}[B] \vdash M : [C]$. Hence $[\Delta] \vdash (\lambda x_B.M) : [B] \to [C] \equiv [B \supset C] \equiv [A]$. ∎

Conversely we have the following.

2.4.11 Proposition *Let* $\Delta, A \subseteq \mathsf{form}(\text{PROP})$. *Then*

$$[\Delta] \vdash_{\boldsymbol{\lambda}_\to} M : [A] \;\Rightarrow\; \Delta \vdash_{\text{PROP}} A.$$

Proof By induction on the structure of M.

Case 1. $M \equiv x$. Then by the Generation Lemma 1.2.3 one has $(x{:}[A]) \in [\Delta]$ and hence $A \in \Delta$; so $\Delta \vdash_{\text{PROP}} A$.

Case 2. $M \equiv PQ$. By the Generation Lemma for some $C \in \mathbb{T}$ one has $[\Delta] \vdash P : C \to [A]$ and $[\Delta] \vdash Q : C$. Clearly, for some $C' \in \mathsf{form}$ one has $C \equiv [C']$. Then $C \to [A] \equiv [C' \supset A]$. By the induction hypothesis, one has $\Delta \vdash C' \to A$ and $\Delta \vdash C'$. Therefore $\Delta \vdash A$.

Case 3. $M \equiv \lambda x.P$. Then $[\Delta] \vdash \lambda x.P : [A]$. By the Generation Lemma, $[A] \equiv B \to C$ and $[\Delta], x{:}B \vdash P : C$, so that $[\Delta], x{:}[B'] \vdash P : [C']$, with $[B'] \equiv B, [C'] \equiv C$ (hence $[A] \equiv [B' \supset C']$). By the induction hypothesis, it follows that $\Delta, B' \vdash C'$ and therefore $\Delta \vdash B' \to C' \equiv A$. ∎

Although intuitionistic logic gives a complete characterization of those types that are inhabited, this does not answer immediately the question of whether the type $((\alpha \to \beta) \to \alpha) \to \alpha$ corresponding to Peirce's law is inhabited.

Kripke models

Remember that a type $A \in \mathbb{T}$ is inhabited iff it is the translation of a $B \in \mathsf{form}(\text{PROP})$ that is intuitionistically provable. This explains why

$$A \text{ inhabited } \Rightarrow\; \models A,$$

but not conversely, since $\models A$ corresponds to classical validity. A tool commonly used for proving that types are not inhabited or that formulas are not intuitionistically derivable, consists of so-called *Kripke models* that we will introduce now.

2.4.12 Definition

(i) A *Kripke model* is a tuple $\mathcal{K} = \langle K, \leq, \odot, F \rangle$, such that

1 $\langle K, \leq, \odot \rangle$ is a partially ordered set with least element $\odot$;

2 $F : K \to \wp(\mathsf{var})$ is a monotonic map from K to the powerset of the set of type-variables; that is $\forall k, k' \in K\, [k \leq k' \;\Rightarrow\; F(k) \subseteq F(k')]$.

We often just write $\mathcal{K} = \langle K, F \rangle$.

(ii) Let $\mathcal{K} = \langle K, F \rangle$ be a Kripke model. For $k \in K$ define by induction on the structure of $A \in \mathbb{T}$ the notion *k forces A*, written $k \Vdash_{\mathcal{K}} A$. We often omit the subscript:

$$k \Vdash \alpha \;\Leftrightarrow\; \alpha \in F(k);$$
$$k \Vdash A_1 \to A_2 \;\Leftrightarrow\; \forall k' \geq k\, [k' \Vdash A_1 \;\Rightarrow\; k' \Vdash A_2].$$

(iii) $\mathcal{K}$ *forces* A, written $\mathcal{K} \Vdash A$, is defined as $\odot \Vdash_{\mathcal{K}} A$.

(iv) Let $\Gamma = \{x_1{:}A_1, \dots, x_n{:}A_n\}$. Then $\mathcal{K}$ *forces* Γ, written $\mathcal{K} \Vdash \Gamma$, if

$$\mathcal{K} \Vdash A_1 \;\&\; \cdots \;\&\; \mathcal{K} \Vdash A_n.$$

We say Γ *forces* A, written $\Gamma \Vdash A$, iff for all Kripke models $\mathcal{K}$ one has

$$\mathcal{K} \Vdash \Gamma \;\Rightarrow\; \mathcal{K} \Vdash A.$$

In particular *forced* A, written $\Vdash A$, if $\mathcal{K} \Vdash A$ for all Kripke models $\mathcal{K}$.

2.4.13 Lemma *Let $\mathcal{K}$ be a Kripke model. Then for all $A \in \mathbb{T}$ one has*

$$k \leq k' \;\&\; k \Vdash_{\mathcal{K}} A \;\Rightarrow\; k' \Vdash_{\mathcal{K}} A.$$

Proof By induction on the structure of A. ∎

2.4.14 Proposition $\Gamma \vdash_{\lambda_\to} M : A \;\Rightarrow\; \Gamma \Vdash A$.

Proof By induction on the derivation of $M : A$ from Γ. If $M : A$ is $x : A$ and is in Γ, then this is trivial. If $\Gamma \vdash M : A$ is $\Gamma \vdash FP : A$ and is a direct consequence of $\Gamma \vdash F : B \to A$ and $\Gamma \vdash P : B$, then the conclusion follows from the induction hypothesis and the fact that $k \Vdash B \to A \;\&\; k \Vdash B \;\Rightarrow\; k \Vdash A$. In the case that $\Gamma \vdash M : A$ is of the form $\Gamma \vdash \lambda x.N : A_1 \to A_2$ and follows directly from $\Gamma, x{:}A_1 \vdash N : A_2$, we have to do something. By the induction hypothesis we have, for all $\mathcal{K}$,

$$\mathcal{K} \Vdash \Gamma, A_1 \;\Rightarrow\; \mathcal{K} \Vdash A_2. \tag{2.12}$$

We must show $\Gamma \Vdash A_1 \to A_2$, i.e. $\mathcal{K} \Vdash \Gamma \;\Rightarrow\; \mathcal{K} \Vdash A_1 \to A_2$ for all $\mathcal{K}$.

Given $\mathcal{K}$ and $k \in K$, define

$$\mathcal{K}_k = \langle \{k' \in K \mid k \leq k'\}, \leq, k, F \rangle,$$

(where $\leq$ and F are in fact the appropriate restrictions to the subset $\{k' \in K \mid k \leq k'\}$ of K). Then it is easy to see that also $\mathcal{K}_k$ is a Kripke model and

$$k \Vdash_{\mathcal{K}} A \Leftrightarrow \mathcal{K}_k \Vdash A. \tag{2.13}$$

Now suppose $\mathcal{K} \Vdash \Gamma$ in order to show $\mathcal{K} \Vdash A_1 \to A_2$, i.e. for all $k \in \mathcal{K}$

$$k \Vdash_{\mathcal{K}} A_1 \Rightarrow k \Vdash_{\mathcal{K}} A_2.$$

Indeed,

$$\begin{array}{llll} k \Vdash_{\mathcal{K}} A_1 & \Rightarrow & \mathcal{K}_k \Vdash A_1, & \text{by (2.13)} \\ & \Rightarrow & \mathcal{K}_k \Vdash A_2, & \text{by (2.12), since } \mathcal{K}_k \Vdash \Gamma \text{ by Lemma 2.4.13} \\ & \Rightarrow & k \Vdash_{\mathcal{K}} A_2. \blacksquare & \end{array}$$

2.4.15 Corollary *Let $A \in \mathbb{T}$. Then*

$$A \textit{ is inhabited} \Rightarrow \Vdash A.$$

Proof Take $\Gamma = \emptyset$. ■

Now it can be proved, see Exercise 2.7, that (the type corresponding to) Peirce's law $P = ((\alpha \to \beta) \to \alpha) \to \alpha$ is not forced in some Kripke model. Since $\not\Vdash P$ it follows that P is not inhabited, in spite of the fact that $\models P$.

We also have a converse to Corollary 2.4.15 which theoretically answers the inhabitation question for $\boldsymbol{\lambda}_{\to}^{\mathbb{A}}$.

2.4.16 Remark [Completeness for Kripke models]

(i) The usual formulation is for provability in intuitionistic logic:

$$A \text{ is inhabited} \Leftrightarrow \Vdash A.$$

The proof is given by constructing for a type that is not inhabited a Kripke 'counter-model' $\mathcal{K}$, i.e. $\mathcal{K} \not\Vdash A$, see Kripke (1965).

(ii) In Harrop (1958) it is shown that these Kripke counter-models can be taken to be finite. This solves the decision problem for inhabitation in $\boldsymbol{\lambda}_{\to}^{\infty}$.

(iii) In Statman (1979a) the decision problem is shown to be PSPACE complete, so that further analysis of the complexity of the decision problem appears to be very difficult.

Set-theoretic models

Now we will prove using set-theoretic models that there do not exist terms satisfying certain properties. For example it is impossible to take as product $A \underline{\times} A$ just the type A itself.

2.4.17 Definition Let $A \in \mathbb{T}^{\mathbb{A}}$. An $A \times A \to A$ *pairing* is a triple $\langle \mathtt{pair}, \mathtt{left}, \mathtt{right} \rangle$ such that

$$\begin{aligned}
&\mathtt{pair} \in \Lambda^{\emptyset}_{\to}(A \to A \to A);\\
&\mathtt{left}, \mathtt{right} \in \Lambda^{\emptyset}_{\to}(A \to A);\\
&\mathtt{left}(\mathtt{pair}\, x^A y^A) =_{\beta\eta} x^A \ \&\ \mathtt{right}(\mathtt{pair}\, x^A y^A) =_{\beta\eta} y^A.
\end{aligned}$$

The definition is formulated for $\boldsymbol{\lambda}^{\mathrm{Ch}}_{\to}$. The existence of a similar $A \times A \to A$ pairing in $\boldsymbol{\lambda}^{\mathrm{Cu}}_{\to}$ (leave out the superscripts in x^A, y^A) is by Proposition 1.2.27 equivalent to that in $\boldsymbol{\lambda}^{\mathrm{Ch}}_{\to}$. We will show using a set-theoretic model that for all types $A \in \mathbb{T}$ there does not exist an $A \times A \to A$ pairing. We take $\mathbb{T} = \mathbb{T}^0$, but the argument for an arbitrary $\mathbb{T}^{\mathbb{A}}$ is the same.

2.4.18 Definition

(i) Let X be a set. The *full type structure* (for types in $\mathbb{T}^0$) over X, written $\mathcal{M}_X = \{X(A)\}_{A \in \mathbb{T}^0}$, is defined as follows. For $A \in \mathbb{T}^0$ let $X(A)$ be defined recursively as

$$\begin{aligned}
X(0) &= X;\\
X(A \to B) &= X(B)^{X(A)}, \text{ the set of functions from } X(A) \text{ into } X(B).
\end{aligned}$$

(ii) $\mathcal{M}_n = \mathcal{M}_{\{1,\dots,n\}}$.

In order to use this model, we will use the Church version $\boldsymbol{\lambda}^{\mathrm{Ch}}_{\to}$, as terms from this system are naturally interpreted in $\mathcal{M}_X$.

2.4.19 Definition

(i) A *valuation* in $\mathcal{M}_X$ is a map ρ from typed variables into $\bigcup_A X(A)$ such that $\rho(x^A) \in X(A)$ for all $A \in \mathbb{T}^0$.

(ii) Let ρ be a valuation in $\mathcal{M}_X$. The *interpretation under* ρ of a $\boldsymbol{\lambda}^{\mathrm{Ch}}_{\to}$-term into $\mathcal{M}_X$, written $[\![M]\!]_\rho$, is defined as[3]

$$\begin{aligned}
[\![x^A]\!]_\rho &= \rho(x^A);\\
[\![MN]\!]_\rho &= [\![M]\!]_\rho [\![N]\!]_\rho;\\
[\![\lambda x^A.M]\!]_\rho &= \boldsymbol{\lambda} d \in X(A).[\![M]\!]_{\rho(x^A := d)},
\end{aligned}$$

[3] Sometimes people prefer to write $[\![\lambda x^A.M]\!]_\rho$ as $\lambda d \in X(A).[\![M[x^A := \underline{d}]]\!]$, where $\underline{d}$ is a constant to be interpreted as d. Although this notation is perhaps more intuitive, we will not use it, since it also has technical drawbacks.

where $\rho(x^A:=d)=\rho'$ with $\rho'(x^A)=d$ and $\rho'(y^B)=\rho(y^B)$ if $y^B\not\equiv x^A$.

(iii) Define

$$\mathcal{M}_X \models M=N \;\Leftrightarrow\; \forall\rho\, [\![M]\!]_\rho=[\![N]\!]_\rho.$$

Before proving properties about the models it is worth doing Exercises 2.10 and 2.11.

2.4.20 Proposition

(i) $M\in\Lambda^{\mathrm{Ch}}_{\to}(A) \;\Rightarrow\; [\![M]\!]_\rho\in X(A)$.
(ii) $M=_{\beta\eta}N \;\Rightarrow\; \mathcal{M}_X\models M=N$.

Proof (i) By induction on the structure of M.

(ii) By induction on the 'proof' of $M=_{\beta\eta}N$, using

$$[\![M[x:=N]]\!]_\rho=[\![M]\!]_{\rho(x:=[\![N]\!]_\rho)}, \text{ for the } \boldsymbol{\beta}\text{-rule};$$
$$\rho\restriction\mathrm{FV}(M)=\rho'\restriction\mathrm{FV}(M) \;\Rightarrow\; [\![M]\!]_\rho=[\![M]\!]_{\rho'}, \text{ for the } \boldsymbol{\eta}\text{-rule};$$
$$[\forall d\in X(A)\; [\![M]\!]_{\rho(x:=d)}=[\![N]\!]_{\rho(x:=d)}]$$
$$\Rightarrow\; [\![\lambda x^A.M]\!]_\rho=[\![\lambda x^A.N]\!]_\rho, \text{ for the } \xi\text{-rule}. \blacksquare$$

Now we will give applications of the notion of typed structure.

2.4.21 Proposition *Let $A\in\mathbb{T}^0$. Then there does not exist an $A\times A\to A$ pairing.*

Proof Take $X=\{0,1\}$. Then for every type A the set $X(A)$ is finite. Therefore by a cardinality argument there cannot be an $A\times A\to A$ pairing, for otherwise f, defined by

$$f(x,y)=[\![\mathtt{pair}]\!]xy,$$

would be an injection from $X(A)\times X(A)$ into $X(A)$, see Exercise 2.11. $\blacksquare$

2.4.22 Proposition *There is no term* $\mathtt{pred}\in\Lambda^{\mathrm{Ch}}_{\to}(\mathsf{Nat}\to\mathsf{Nat})$ *such that*

$$\mathtt{pred}\,\mathbf{c}_0=_{\beta\eta}\mathbf{c}_0;$$
$$\mathtt{pred}\,\mathbf{c}_{n+1}=_{\beta\eta}\mathbf{c}_n.$$

Proof As before for $X=\{0,1\}$ the set $X(\mathsf{Nat})$ is finite. Therefore

$$\mathcal{M}_X\models\mathbf{c}_n=\mathbf{c}_m,$$

for some $n\neq m$. If $\mathtt{pred}$ did exist, then it would follow easily that $\mathcal{M}_X\models\mathbf{c}_0=\mathbf{c}_1$. But this implies that $X(0)$ has cardinality 1, since $\mathbf{c}_0(\mathsf{K}x)y=y$ but $\mathbf{c}_1(\mathsf{K}x)y=\mathsf{K}xy=x$, a contradiction. $\blacksquare$

Another application of semantics is that there are no fixed-point combinators in $\boldsymbol{\lambda}_{\to}^{\mathrm{Ch}}$.

2.4.23 Definition A closed term Y is a *fixed-point combinator* for type $A \in \mathbb{T}^0$ if

$$Y : \Lambda_{\to}^{\mathrm{Ch}}((A{\to}A){\to}A) \;\&\; Y =_{\beta\eta} \lambda f^{A\to A}.f(Yf).$$

2.4.24 Proposition *There exist no fixed-point combinators in* $\boldsymbol{\lambda}_{\to}^{\mathrm{Ch}}$ *for any type* A.

Proof Take $X = \{0,1\}$. Then for every A the set $X(A)$ has at least two elements, say $x, y \in X(A)$ with $x \neq y$. Then there exists an $f \in X(A{\to}A)$ without a fixed-point:

$$\begin{array}{rcll} f(z) & = & x, & \text{if } z \neq x; \\ f(z) & = & y, & \text{else.} \end{array}$$

If there is a fixed-point combinator of type A, then $[\![Y]\!]f \in \mathcal{M}_X$ is a fixed-point of f. Indeed, $Yx =_{\beta\eta} x(Yx)$ and taking $[\![\;]\!]_\rho$ with $\rho(x) = f$ the claim follows, by contradiction. ∎

Several results in this section can easily be translated to $\lambda_{\to}^{\mathbb{A}_\infty}$ with arbitrarily many type variables: see Exercise 2.12.

2.5 Exercises

Exercise 2.1 Find out which of the following terms are typable and determine for those that are the principal type:

$$\lambda xyz.xz(yz);$$
$$\lambda xyz.xy(xz);$$
$$\lambda xyz.xy(zy).$$

Exercise 2.2

(i) Let $A = (\alpha{\to}\beta){\to}((\alpha{\to}\beta){\to}\alpha){\to}\alpha$ Construct a term M such that $\vdash M : A$. What is the principal type B of M? Is there a λI-term of type B?

(ii) Find an expansion of M such that it has A as principal type.

Exercise 2.3 (Uniqueness of Type Assignments) Remember that

$$\Lambda_{\mathsf{I}} = \{M \in \Lambda \mid \text{if } \lambda x.N \text{ is a subterm of } M, \text{ then } x \in \mathrm{FV}(N)\}.$$

One has

$$M \in \Lambda_{\mathsf{I}},\ M \twoheadrightarrow_{\beta\eta} N \ \Rightarrow\ N \in \Lambda_{\mathsf{I}},$$

see e.g. B[1984], Lemma 9.1.2.

(i) Show that for all $M_1, M_2 \in \Lambda_{\rightarrow}^{\mathrm{Ch}}(A)$ one has

$$|M_1| \equiv |M_2| \equiv M \in \Lambda_{\mathsf{I}}^{\emptyset} \ \Rightarrow\ M_1 \equiv M_2.$$

[Hint. Use as induction loading towards open terms

$$|M_1| \equiv |M_2| \equiv M \in \Lambda_{\mathsf{I}}\ \&\ \mathrm{FV}(M_1) \equiv \mathrm{FV}(M_2) \ \Rightarrow\ M_1 \equiv M_2.$$

This can be proved by induction on n, the length of the shortest $\boldsymbol{\beta}$-reduction path to nf. For $n = 0$, see Propositions 1.2.19(i) and 1.2.24.]

(ii) Show that in (i) the condition $M \in \Lambda_{\mathsf{I}}^{\emptyset}$ cannot be weakened to

$$M \text{ has no } \mathsf{K}\text{-redexes.}$$

[Hint. Consider $M \equiv (\lambda x.x\mathsf{I})(\lambda z.\mathsf{I})$ and $A \equiv \alpha{\rightarrow}\alpha$.]

Exercise 2.4 Show that $\boldsymbol{\lambda}_{\rightarrow}^{\mathrm{dB}}$ satisfies the Church–Rosser Theorem. [Hint. Use Proposition 1.2.29 and translations between $\boldsymbol{\lambda}_{\rightarrow}^{\mathrm{dB}}$ and $\boldsymbol{\lambda}_{\rightarrow}^{\mathrm{Ch}}$.]

Exercise 2.5 (Hindley) Show that if $\vdash_{\boldsymbol{\lambda}_{\rightarrow}}^{\mathrm{Cu}} M : A$, then there is an M' such that

$$M' \twoheadrightarrow_{\beta\eta} M\ \&\ \mathtt{pt}(M') = A.$$

[Hints. (1) First make an $\boldsymbol{\eta}$-expansion of M in order to obtain a term with a principal type having the same tree as A. (2) Show that for any type B with a subtype B_0 there exists a context $C[\ ]$ such that

$$z{:}B \vdash C[z] : B_0.$$

(3) Use the first two hints and a term like $\lambda fz.z(fP)(fQ)$ to force identification of the types of P and Q. (For example one may want to identify α and γ in $(\alpha{\rightarrow}\beta){\rightarrow}\gamma{\rightarrow}\delta$.)]

Exercise 2.6 Each type A of $\boldsymbol{\lambda}_{\rightarrow}^{0}$ can be interpreted as an element $[\![A]\!] \in \mathbf{B}^{\mathbf{B}}$ as follows.

$$[\![A]\!](i) = [\![A]\!]_{\rho_i},$$

where $\rho_i(0) = i$. There are four elements in $\mathbf{B}^{\mathbf{B}}$

$$\{\boldsymbol{\lambda} x \in \mathbf{B}.0, \boldsymbol{\lambda} x \in \mathbf{B}.1, \boldsymbol{\lambda} x \in \mathbf{B}.x, \boldsymbol{\lambda} x \in \mathbf{B}.1 - x\}.$$

Prove that $[\![A]\!] = \boldsymbol{\lambda} x \in \mathbf{B}.1$ iff A is inhabited and $[\![A]\!] = \boldsymbol{\lambda} x \in \mathbf{B}.x$ iff A is not inhabited.

Exercise 2.7 Peirce's law is $P = ((\alpha{\to}\beta){\to}\alpha){\to}\alpha$. (i) Show that P is not is not forced in the Kripke model $\mathcal{K} = \langle K, \leq, 0, F\rangle$ with $K = \{0,1\}$, $0 \leq 1$ and $F(0) = \emptyset, F(1) = \{\alpha\}$. (ii) Show without the theory of Kripke models, only using Proposition 1.2.3 that P is not inhabited.

Exercise 2.8 Let X be a set and consider the typed λ-model $\mathcal{M}_X$. Notice that every permutation $\pi = \pi_0$ (bijection) of X can be lifted to all levels $X(A)$ by defining

$$\pi_{A\to B}(f) = \pi_B \circ f \circ \pi_A^{-1}.$$

Prove that every lambda-definable element $f \in X(A)$ in $\mathcal{M}(X)$ is invariant under all lifted permutations; i.e. $\pi_A(f) = f$. [Hint. Use the fundamental theorem for logical relations.]

Exercise 2.9 Prove in two ways that $\Lambda^{\emptyset}_{\to}(0) = \emptyset$. (i) Apply the normalization and subject reduction theorems. (ii) Apply models and the fact shown in the previous exercise that lambda-definable elements are invariant under lifted permutations.

Exercise 2.10

(i) Show that $\mathcal{M}_X \models (\lambda x^A.x^A)y^A = y^A$.
(ii) Show that $\mathcal{M}_X \models (\lambda x^{A\to A}.x^{A\to A}) = (\lambda x^{A\to A}y^A.x^{A\to A}y^A)$.
(iii) Show that $[\![\mathbf{c}_2(\mathsf{K}x^0)y^0]\!]_\rho = \rho(x)$.

Exercise 2.11 Let P, L, R be an $A \times B{\to}C$ pairing. Show that in every structure $\mathcal{M}_X$ one has

$$[\![P]\!]xy = [\![P]\!]x'y' \Rightarrow x = x' \;\&\; y = y',$$

hence $\mathrm{card}(A) \cdot \mathrm{card}(B) \leq \mathrm{card}(C)$.

Exercise 2.12 Show that Propositions 2.4.21, 2.4.22 and 2.4.24 can be generalized to $\mathbb{A} = \mathbb{A}_\infty$ and the corresponding versions of $\boldsymbol{\lambda}^{\mathrm{Cu}}_{\to}$, by modifying the notion of typed structure.

Exercise 2.13 Let ${\sim}A \equiv A{\to}0$. Show that if 0 does not occur in A, then ${\sim}{\sim}({\sim}{\sim}A{\to}A)$ is not inhabited. (One needs the *ex falso* rule to derive ${\sim}{\sim}({\sim}{\sim}A{\to}A)$.) Why is the condition on 0 necessary?

Exercise 2.14 We say that the structure of the rational numbers can be represented in $\boldsymbol{\lambda}^{\mathbb{A}}_{\to}$ if there is a type $Q \in \mathbb{T}^{\mathbb{A}}$ and closed lambda terms:

$$0, 1 : Q;$$
$$+, \cdot : Q{\to}Q{\to}Q;$$
$$-, ^{-1} : Q{\to}Q;$$

such that $(Q, +, \cdot, -, ^{-1}, 0, 1)$ modulo $=_{\beta\eta}$ satisfies the axioms of a field of characteristic 0. Show that the rationals cannot be represented in $\boldsymbol{\lambda}^{\mathbb{A}}_{\rightarrow}$. [Hint. Use a model-theoretic argument.]

Exercise 2.15 Show that there is no closed term

$$P : \mathsf{Nat}{\rightarrow}\mathsf{Nat}{\rightarrow}\mathsf{Nat}$$

such that P is a bijection in the sense that (unicity modulo $=_{\beta\eta}$)

$$\forall M{:}\mathsf{Nat}\exists! N_1, N_2{:}\mathsf{Nat}\ PN_1N_2 =_{\beta\eta} M.$$

Exercise 2.16 Show that every $M \in \Lambda^{\emptyset}((0{\rightarrow}0{\rightarrow}0){\rightarrow}0{\rightarrow}0)$ is $\beta\eta$-convertible to $\lambda f^{0\rightarrow0\rightarrow0}x^0.t$, with t given by the grammar

$$t := x \mid ftt.$$

Exercise 2.17 [Hindley] Show that there is an ARS that is WCR but not CR. [Hint. An example of cardinality 4 exists.]

The next two exercises show that the minimal length of a reduction-path of a term to normal form is in the worst case non-elementary in the length of the term[4]. See Péter (1967) for the definition of the class of (Kalmár) elementary functions. This class is the same as $\mathcal{E}_3$ in the Grzegorczyk hierarchy. To get some intuition for this class, define the family of functions $2_n{:}\omega{\rightarrow}\omega$ as

$$\begin{aligned} 2_0(x) &= x; \\ 2_{n+1}(x) &= 2^{2_n(x)}. \end{aligned}$$

Then every elementary function f is eventually bounded by some 2_n:

$$\exists n, m \forall x{>}m\ f(x) \leq 2_n(x).$$

Exercise 2.18

(i) Define the function $\mathsf{gk} : \omega{\rightarrow}\omega$ by

$$\begin{aligned} \mathsf{gk}(m) &= \#F_{\mathsf{GK}}(M), && \text{if } m = \#(M) \text{ for some untyped} \\ & && \text{lambda term } M; \\ &= 0, && \text{else.} \end{aligned}$$

Here $\#M$ denotes the *Gödel-number* of the term M and F_{GK} is the Gross–Knuth reduction strategy defined by completely developing all present redexes in M, see B[1984]. Show that gk is Kalmár elementary.

[4] In Gandy (1980) this is also proved for arbitrary reduction paths starting from typable terms. In de Vrijer (1987a) an exact calculation is given for the longest reduction paths to normal form.

(ii) For a term $M \in \Lambda_{\rightarrow}^{\text{Ch}}$ define

$$D(M) = \mathtt{max}\{\text{dpt}(A{\rightarrow}B) \mid (\lambda x^A.P)^{A\rightarrow B}Q \text{ is a redex in } M\},$$

see Definition 1.1.22(i). Show that if M is not a $\boldsymbol{\beta}$-nf, then

$$F_{\mathsf{GK}}(|M|) = |N| \Rightarrow D(M) > D(N),$$

where $|.| : \Lambda_{\rightarrow}^{\text{Ch}} {\rightarrow} \Lambda$ is the forgetful map. [Hint. Use Lévy's analysis of redex creation, see 2.1.11(ii), or Lévy (1978), 1.8.4. lemme 3.3, for the proof.]

(iii) If $M \in \Lambda$ is a term, then its *length*, written $\mathtt{lth}(M)$, is the number of symbols in M. Show that there is a constant c such that for typable lambda terms M one has, for M sufficiently long,

$$\mathtt{dpth}(\text{pt}(M)) \leq c(\mathtt{lth}(M)).$$

See the proof of Theorem 2.3.14.

(iv) Write $\sigma{:}M{\rightarrow}M^{\mathtt{nf}}$ if σ is some reduction path of M to normal form $M^{\mathtt{nf}}$. Let $\$\sigma$ be the number of reduction steps in σ. Define

$$\$(M) = \mathtt{min}\{\$\sigma \mid \sigma : M{\rightarrow}M^{\mathtt{nf}}\}.$$

Show that $\$M \leq g(\mathtt{lth}(M))$, for some function $g \in \mathcal{E}_4$. [Hint. Take $g(m) = gk^m(m)$.]

Exercise 2.19

(i) Define $\mathbf{2}_1 = \lambda f^1 x^0.f(fx)$ and $\mathbf{2}_{n+1} = (\mathbf{2}_n[0{:=}1])\mathbf{2}$. Then for all $n \in \omega$ one has $\mathbf{2}_n : 1{\rightarrow}0{\rightarrow}0$. Show that this type is the principal type of the Curry version $|\mathbf{2}_n|$ of $\mathbf{2}_n$.

(ii) (Church) Show $(\mathbf{c}_n[0{:=}1])\mathbf{c}_m =_{\boldsymbol{\beta}} \mathbf{c}_{m^n}$.

(iii) Show $\mathbf{2}_n =_{\boldsymbol{\beta}} \mathbf{c}_{2_n(1)}$, the notation is explained just before Exercise 2.18.

(iv) Let $M, N \in \Lambda$ be untyped terms. Show that if $M \twoheadrightarrow_{\boldsymbol{\beta}} N$, then

$$\mathtt{lth}(N) \leq \mathtt{lth}(M)^2.$$

(v) Conclude that $\$(M)$, see Exercise 2.18, is in the worst case non-elementary in the length of M. That is, show that there is no elementary function f such that for all $M \in \Lambda_{\rightarrow}^{\text{Ch}}$

$$\$(M) \leq f(\mathtt{lth}(M)).$$

Exercise 2.20

(i) Show that in the worst case the length of the principal type of a typable term is at least exponential in the length of the term, i.e. defining

$$f(m) = \mathtt{max}\{\mathtt{lth}(\mathrm{pt}(M)) \mid \mathtt{lth}(M) \leq m\},$$

one has $f(n) \geq c^n$, for some real number $c > 1$ and sufficiently large n. [Hint. Define

$$M_n \equiv \lambda x_n \ldots x_1.x_n(x_n x_{n-1})(x_{n-1}(x_{n-1}x_{n-2}))\ldots(x_2(x_2x_1)).$$

Show that the principal type of M_n has length $> 2^n$.]

(ii) Show that the length of the principal type of a term M is also at most exponential in the length of M. [Hint. First show that the depth of the principal type of a typable term M is linear in the length of M.]

Exercise 2.21 (Statman) We want to show that $\mathcal{M}_n \hookrightarrow \mathcal{M}_\omega$, for $n \geq 1$, by an isomorphic embedding.

(i) (Church's δ.) For $A \in \mathbb{T}^0$ define $\delta_A \in \mathcal{M}_n(A^2 \to 0^2 \to 0)$ by

$$\begin{aligned} \delta_A xyuv &= u && \text{if } x = y; \\ &= v && \text{else.} \end{aligned}$$

(ii) We add to the language $\boldsymbol{\lambda}^{\mathrm{Ch}}_{\to}$ constants $\underline{k} : 0$ for $1 \leq k \leq n$ and a constant $\underline{\delta} : 0^4 \to 0$. The intended interpretation of $\underline{\delta}$ is the map δ_0. We define the notion of reduction δ by the contraction rules

$$\begin{aligned} \underline{\delta}\,\underline{i}\,\underline{j}\,\underline{k}\,\underline{l} &\to_\delta \underline{k} && \text{if } i = j; \\ &\to_\delta \underline{l}, && \text{if } i \neq j. \end{aligned}$$

The resulting language of terms is called Λ_δ and on this we consider the notion of reduction $\to_{\beta\eta\delta}$.

(iii) Show that every $M \in \Lambda_\delta$ satisfies $\mathsf{SN}_{\beta\eta\delta}(M)$.

(iv) Show that $\to_{\beta\eta\delta}$ is Church–Rosser.

(v) Let $M \in \Lambda^{\emptyset}_\delta(0)$ be a closed term of type 0. Show that the long normal form of M is one of the constants $\underline{1}, \ldots, \underline{n}$.

(vi) (Church's theorem.) Show that every element $\Phi \in \mathcal{M}_n$ can be defined by a closed term $M_\Phi \in \Lambda_\delta$, i.e. $\Phi = [\![M_\Phi]\!]^{\mathcal{M}_n}$. [Hint. For each $A \in \mathbb{T}$ define simultaneously $\Phi \mapsto M_\Phi : \mathcal{M}_n(A) \to \Lambda_\delta(A)$ and $\underline{\delta}_A \in \Lambda_\delta(A^2 \to 0^2 \to 0)$ such that $[\![\underline{\delta}_A]\!] = \delta_A$ and $\Phi = [\![M_\Phi]\!]^{\mathcal{M}_n}$. For $A = 0$ take $M_i = \underline{i}$ and $\underline{\delta}_0 =$

$\underline{\delta}$. For $A = B \to C$, let $\mathcal{M}_n(B) = \{\Phi_1, \ldots, \Phi_t\}$ and $C = C_1 \to \ldots C_c \to 0$. Define

$$\begin{aligned}\underline{\delta}_A \equiv \lambda xyuv.\ & (\underline{\delta}_C(xM_{\Phi_1})(yM_{\Phi_1}) \\ & (\underline{\delta}_C(xM_{\Phi_2})(yM_{\Phi_2}) \\ & (\ldots \\ & (\underline{\delta}_C(xM_{\Phi_{t-1}})(yM_{\Phi_{t-1}}) \\ & (\underline{\delta}_C(xM_{\Phi_t})(yM_{\Phi_t})uv)v)v..)v)v).\end{aligned}$$

$$\begin{aligned}M_\Phi \equiv \lambda xy_1 \ldots y_c.\ & (\underline{\delta}_B x M_{\Phi_1}(M_{\Phi_1}\vec{y}) \\ & (\underline{\delta}_B x M_{\Phi_2}(M_{\Phi_2}\vec{y}) \\ & (\ldots \\ & (\underline{\delta}_B x M_{\Phi_{t-1}}(M_{\Phi_{t-1}}\vec{y}) \\ & (\underline{\delta}_B x M_{\Phi_t}(M_{\Phi_t}\vec{y})\underline{0}))..))).]\end{aligned}$$

(vii) Show that $\Phi \mapsto [\![M_\Phi]\!]^{\mathcal{M}_\omega} : \mathcal{M}_n \hookrightarrow \mathcal{M}_\omega$ is the required embedding.

(viii) (To be used later.) Let $\pi_i^n \equiv (\lambda x_1 \cdots x_n.x_i) : (0^n \to 0)$. Define

$$\begin{aligned}\Delta^n \equiv \lambda abuv\vec{x}.a\ & (b(u\vec{x})(v\vec{x}) \cdots (v\vec{x})(v\vec{x})) \\ & (b(v\vec{x})(u\vec{x}) \cdots (v\vec{x})(v\vec{x})) \\ & \ldots \\ & (b(v\vec{x})(v\vec{x}) \cdots (u\vec{x})(v\vec{x})) \\ & (b(v\vec{x})(v\vec{x}) \cdots (v\vec{x})(u\vec{x})).\end{aligned}$$

Then

$$\begin{aligned}\Delta^n \pi_i^n \pi_j^n \pi_k^n \pi_l^n &=_{\beta\eta\delta} \pi_k^n, && \text{if } i = j; \\ &=_{\beta\eta\delta} \pi_l^n, && \text{else.}\end{aligned}$$

Show that for $i \in \{1, \ldots, n\}$ one has for all $M : 0$

$$\begin{aligned}& M =_{\beta\eta\delta} \underline{i} \Rightarrow \\ & M[0 := 0^n \to 0][\delta := \Delta^n][\underline{1} := \pi_1^n] \ldots [\underline{n} := \pi_n^n] =_{\beta\eta} \pi_i^n.\end{aligned}$$

Exercise 2.22 (Th. Joly)

(i) Let $M = \langle Q, q_0, F, \delta \rangle$ be a deterministic finite automaton over the finite alphabet $\Sigma = \{a_1, \ldots, a_n\}$. That is, Q is the finite set of states, $q_0 \in Q$ is the initial state, $F \subseteq Q$ is the set of final states and $\delta : \Sigma \times Q \to Q$ is the transition function. Let $L^r(M)$ be the (regular) language consisting of words in Σ^* accepted by M by reading the words from right to left. Let $\mathcal{M} = \mathcal{M}_Q$ be the typed λ-model over Q. Show that

$$w \in L^r(M) \Leftrightarrow [\![\underline{w}]\!]^{\mathcal{M}} \delta_{a_1} \cdots \delta_{a_n} q_0 \in F,$$

where $\delta_a(q) = \delta(a, q)$ and $\underline{w}$ is defined in 1.4.8.

(ii) Similarly represent classes of trees (with elements of Σ at the nodes) accepted by a frontier-to-root tree automaton, see Thatcher (1973), by the model $\mathcal{M}$ at the type $\top_n = (0^2{\to}0)^n{\to}0{\to}0$.

3
Tools

3.1 Semantics of $\boldsymbol{\lambda}_{\rightarrow}$

So far the systems $\boldsymbol{\lambda}_{\rightarrow}^{\text{Cu}}$ and $\boldsymbol{\lambda}_{\rightarrow}^{\text{Ch}}$ (and also its variant $\boldsymbol{\lambda}_{\rightarrow}^{\text{dB}}$) had closely related properties. In this chapter we will give two rather different semantics to $\boldsymbol{\lambda}_{\rightarrow}^{\text{Ch}}$ and to $\boldsymbol{\lambda}_{\rightarrow}^{\text{Cu}}$. This will become clear from the intention one has while giving a semantics for these systems, which is as follows. For the Church systems $\boldsymbol{\lambda}_{\rightarrow}^{\text{Ch}}$, in which every λ-term comes with its unique type, there is a semantics consisting of disjoint layers, each of these corresponding with a given type. Terms of type A will be interpreted as elements of the layer corresponding to A. On the other hand, the Curry systems $\boldsymbol{\lambda}_{\rightarrow}^{\text{Cu}}$ are essentially treated as untyped λ-calculi, where one assigns to a term a set (possibly empty) of potential types. This then results in an untyped λ-model with overlapping subsets indexed by the types. This happens in such a way that if type A is assigned to term M, then the interpretation of M is an element of the subset with index A. The notion of semantics has been inspired by Henkin (1950), who dealt with the completeness in the theory of types.

Semantics for type assignment *à la* Church

In this subsection we work with the Church variant of $\boldsymbol{\lambda}_{\rightarrow}^{0}$ having one atomic type 0, rather than with $\boldsymbol{\lambda}_{\rightarrow}^{\mathbb{A}}$, having an arbitrary set of atomic types. We will write $\mathbb{T} = \mathbb{T}^0$. The reader is encouraged to investigate which results do generalize to $\mathbb{T}^{\mathbb{A}}$.

3.1.1 Definition Let $\mathcal{M} = \{\mathcal{M}(A)\}_{A \in \mathbb{T}}$ be a family of non-empty sets indexed by types $A \in \mathbb{T}$.

(i) We call $\mathcal{M}$ a *typed structure* for $\boldsymbol{\lambda}_{\rightarrow}^{0}$ if

$$\mathcal{M}(A \rightarrow B) \subseteq \mathcal{M}(B)^{\mathcal{M}(A)}.$$

Here X^Y denotes the collection of set-theoretic functions

$$\{f \mid f : Y \to X\}.$$

(ii) The typed structure $\mathcal{M}$ is called the *full typed structure* over the ground set X if $\mathcal{M}(0) = X$ and for all $A, B \in \mathbb{T}$ one has

$$\mathcal{M}(A{\to}B) = \mathcal{M}(B)^{\mathcal{M}(A)}.$$

(iii) Let $\mathcal{M}$ be provided with *application operators*

$$\begin{aligned}(\mathcal{M}, \cdot) &= (\{\mathcal{M}(A)\}_{A \in \mathbb{T}}, \{\cdot_{A,B}\}_{A,B \in \mathbb{T}}) \\ \cdot_{A,B} &: \mathcal{M}(A{\to}B) \times \mathcal{M}(A) \to \mathcal{M}(B).\end{aligned}$$

A *typed applicative structure* is such an $(\mathcal{M}, \cdot)$ satisfying *extensionality*:

$$\forall f, g \in \mathcal{M}(A{\to}B)\, [[\forall a \in \mathcal{M}(A)\ f \cdot_{A,B} a = g \cdot_{A,B} a] \Rightarrow f = g].$$

(iv) We say $\mathcal{M}$ is *trivial* if $\mathcal{M}(0)$ is a singleton. Then $\mathcal{M}(A)$ is a singleton for all $A \in \mathbb{T}$.

3.1.2 Notation For typed applicative structures we use the infix notation $f \cdot_{A,B} x$ or $f \cdot x$ for $\cdot_{A,B}(f, x)$. Often we will be even briefer, extensionality becoming

$$\forall f, g \in \mathcal{M}(A{\to}B)\, [[\forall a \in \mathcal{M}_A\ fa = ga] \Rightarrow f = g]$$

or more simply,

$$\forall f, g \in \mathcal{M}\, [[\forall a\ fa = ga] \Rightarrow f = g],$$

where f, g range over the same type $A{\to}B$ and a ranges over $\mathcal{M}_A$.

3.1.3 Proposition *The notions of typed structure and typed applicative structure are equivalent.*

Proof In a typed structure $\mathcal{M}$ define $f \cdot a = f(a)$; extensionality is obvious. Conversely, let $\langle \mathcal{M}, \cdot \rangle$ be a typed applicative structure. Define the typed structure $\mathcal{M}'$ and $\Phi_A : \mathcal{M}(A){\to}\mathcal{M}'(A)$ as

$$\begin{aligned}\mathcal{M}'(0) &= \mathcal{M}(0);\\ \Phi_0(a) &= a;\\ \mathcal{M}'(A{\to}B) &= \{\Phi_{A\to B}(f) \in \mathcal{M}'(B)^{\mathcal{M}'(A)} \mid f \in \mathcal{M}(A{\to}B)\};\\ \Phi_{A\to B}(f)(\Phi_A(a)) &= \Phi_B(f \cdot a).\end{aligned}$$

By definition Φ is surjective. By extensionality of the typed applicative structure it is also injective. Hence $\Phi_{A\to B}(f)$ is well defined. Clearly one has $\mathcal{M}'(A{\to}B) \subseteq \mathcal{M}'(B)^{\mathcal{M}'(A)}$. ∎

3.1.4 Definition Let $\mathcal{M}, \mathcal{N}$ be two typed applicative structures. A *morphism* is a type indexed family $F = \{F_A\}_{A \in \mathbb{T}}$ such that for each $A, B \in \mathbb{T}$ one has

$$F_A : \mathcal{M}(A) \to \mathcal{N}(A);$$
$$F_{A \to B}(f) \cdot F_A(a) = F_B(f \cdot a).$$

From now on we will not make a distinction between the notions 'typed structure' and 'typed applicative structure'.

3.1.5 Proposition *Let $\mathcal{M}$ be a typed structure. Then*

$$\mathcal{M} \textit{ is trivial} \Leftrightarrow \forall A \in \mathbb{T}.\mathcal{M}(A) \textit{ is a singleton.}$$

Proof ($\Leftarrow$) By definition.

($\Rightarrow$) We will show this for $A = 1 = 0 \to 0$. If $\mathcal{M}(0)$ is a singleton, then for all $f, g \in \mathcal{M}(1)$ one has $\forall x{:}\mathcal{M}(0).(fx) = (gx)$, hence $f = g$, by extensionality. Therefore $\mathcal{M}(1)$ is a singleton. ∎

3.1.6 Example The full typed structure $\mathcal{M}_X = \{X(A)\}_{A \in \mathbb{T}}$ over a non-empty set X, see Definition 2.4.18, is a typed applicative structure.

3.1.7 Definition

(i) Let $(X, \leq)$ be a non-empty partially ordered set. Let $D(0) = X$ and $D(A \to B)$ consist of the monotone elements of $D(B)^{D(A)}$, where we order this set pointwise: for $f, g \in D(A \to B)$ define

$$f \leq g \Leftrightarrow \forall a \in D(A)\, fa \leq ga.$$

The elements of the typed applicative structure $D_X = \{D(A)\}_{A \in \mathbb{T}}$ are called the *hereditarily monotonic functions*, see the chapter by Howard in Troelstra (1973).

(ii) Let $\mathcal{M}$ be a typed applicative structure. A *layered non-empty subfamily* of $\mathcal{M}$ is a family $\Delta = \{\Delta(A)\}_{A \in \mathbb{T}}$ of sets, such that

$$\forall A \in \mathbb{T}.\emptyset \neq \Delta(A) \subseteq \mathcal{M}(A).$$

We say Δ is *closed under application* if

$$f \in \Delta(A \to B), g \in \Delta(A) \Rightarrow fg \in \Delta(B).$$

We say Δ is *extensional* if

$$\forall A, B \in \mathbb{T} \forall f, g \in \Delta(A \to B).[[\forall a \in \Delta(A).fa = ga] \Rightarrow f = g].$$

If Δ satisfies all these conditions, then $\mathcal{M}{\restriction}\Delta = (\Delta, \cdot{\restriction}\Delta)$ is a typed applicative structure.

3.1.8 Definition (Environments)

(i) Let $\mathcal{D}$ be a set and V the set of variables of the untyped lambda calculus. A *(term) environment in* $\mathcal{D}$ is a total map

$$\rho : \mathsf{V} \to \mathcal{D}.$$

The set of environments in $\mathcal{D}$ is denoted by $\mathsf{Env}_{\mathcal{D}}$.

(ii) If $\rho \in \mathsf{Env}_{\mathcal{D}}$ and $d \in \mathcal{D}$, then $\rho[x := d]$ is the $\rho' \in \mathsf{Env}_{\mathcal{D}}$ defined by

$$\rho'(y) = \begin{cases} d & \text{if } y = x, \\ \rho(y) & \text{otherwise.} \end{cases}$$

3.1.9 Definition

(i) Let $\mathcal{M}$ be a typed applicative structure. Then by a (*partial*) *valuation in* $\mathcal{M}$ we mean a family of (partial) maps $\rho = \{\rho_A\}_{A \in \mathbb{T}}$ such that $\rho_A : \mathtt{Var}(A) \to \mathcal{M}(A)$.

(ii) Given a typed applicative structure $\mathcal{M}$ and a partial valuation ρ in $\mathcal{M}$, then one can define for $M \in \Lambda_{\to}(A)$ the *partial semantics*, written $[\![M]\!]_{\rho}^{\mathcal{M}} \in \mathcal{M}(A)$, as

$$\begin{aligned} [\![x^A]\!]_{\rho}^{\mathcal{M}} &= \rho_A(x); \\ [\![PQ]\!]_{\rho}^{\mathcal{M}} &= [\![P]\!]_{\rho}^{\mathcal{M}} [\![Q]\!]_{\rho}^{\mathcal{M}}; \\ [\![\lambda x^A.P]\!]_{\rho}^{\mathcal{M}} &= \boldsymbol{\lambda} d \in \mathcal{M}(A).[\![P]\!]_{\rho[x:=d]}^{\mathcal{M}}. \end{aligned}$$

We often write $[\![M]\!]_{\rho}$ for $[\![M]\!]_{\rho}^{\mathcal{M}}$, if there is little danger of confusion. The expression $[\![M]\!]_{\rho}$ may not always be defined, even if ρ is total. The problem arises with $[\![\lambda x.P]\!]_{\rho}$. Although the function

$$\boldsymbol{\lambda} d \in \mathcal{M}(A).[\![P]\!]_{\rho[x:=d]} \in \mathcal{M}(B)^{\mathcal{M}(A)}$$

is uniquely determined by $[\![\lambda x.P]\!]_{\rho} d = [\![P]\!]_{\rho[x:=d]}$, it may fail to be an element of $\mathcal{M}(A \to B)$ which is only a subset of $\mathcal{M}(B)^{\mathcal{M}(A)}$. If $[\![M]\!]_{\rho}$ *is defined*, we write $[\![M]\!]_{\rho}\downarrow$; otherwise, if $[\![M]\!]_{\rho}$ *is undefined*, we write $[\![M]\!]_{\rho}\uparrow$.

3.1.10 Definition

(i) A typed structure $\mathcal{M}$ is called a $\boldsymbol{\lambda}_{\to}^{0}$*-model* or a *typed* λ*-model* if, for every partial valuation $\rho = \{\rho_A\}_A$ and every $A \in \mathbb{T}$ and $M \in \Lambda_{\to}(A)$ such that $\mathrm{FV}(M) \subseteq \mathrm{dom}(\rho)$, one has $[\![M]\!]_{\rho}\downarrow$.

(ii) Let $\mathcal{M}$ be a typed λ-model and ρ a partial valuation. Then $\mathcal{M}, \rho$ *satisfies* $M = N$, assuming implicitly that M and N have the same type, written

$$\mathcal{M}, \rho \models M = N,$$

if $[\![M]\!]_\rho^{\mathcal{M}} = [\![N]\!]_\rho^{\mathcal{M}}$.

(iii) Let $\mathcal{M}$ be a typed λ-model. Then $\mathcal{M}$ *satisfies* $M = N$, written

$$\mathcal{M} \models M = N$$

if for all partial ρ with $\mathrm{FV}(MN) \subseteq \mathrm{dom}(\rho)$ one has $\mathcal{M}, \rho \models M = N$.

(iv) Let $\mathcal{M}$ be a typed λ-model. The *theory of* $\mathcal{M}$ is

$$\mathrm{Th}(\mathcal{M}) = \{M = N \mid M, N \in \Lambda_{\to}^{\emptyset} \ \& \ \mathcal{M} \models M = N\}.$$

3.1.11 Notation Let E_1, E_2 be partial (i.e. possibly undefined) expressions.

(i) Write $E_1 \gtrsim E_2$ for $E_1\downarrow \ \Rightarrow \ [E_2\downarrow \ \& \ E_1 = E_2]$.

(ii) Write $E_1 \simeq E_2$ for $E_1 \gtrsim E_2 \ \& \ E_2 \gtrsim E_1$.

3.1.12 Lemma

(i) *Let* $M \in \Lambda_0(A)$ *and* N *be a subterm of* M. *Then*

$$[\![M]\!]_\rho\downarrow \ \Rightarrow \ [\![N]\!]_\rho\downarrow.$$

(ii) *Let* $M \in \Lambda_0(A)$. *Then*

$$[\![M]\!]_\rho \simeq [\![M]\!]_{\rho\restriction\mathrm{FV}(M)}.$$

(iii) *Let* $M \in \Lambda_0(A)$ *and* ρ_1, ρ_2 *be such that* $\rho_1 \restriction \mathrm{FV}(M) = \rho_2 \restriction \mathrm{FV}(M)$. *Then*

$$[\![M]\!]_{\rho_1} \simeq [\![M]\!]_{\rho_2}.$$

Proof (i) By induction on the structure of M.

(ii) Similarly.

(iii) By (ii). ∎

3.1.13 Lemma *Let* $\mathcal{M}$ *be a typed applicative structure. Then*

(i) *For* $M \in \Lambda_0(A)$, $x, N \in \Lambda_0(B)$ *one has*

$$[\![M[x:=N]]\!]_\rho^{\mathcal{M}} \simeq [\![M]\!]_{\rho[x:=[\![N]\!]_\rho^{\mathcal{M}}]}^{\mathcal{M}}.$$

(ii) *For* $M, N \in \Lambda_0(A)$ *one has*

$$M \twoheadrightarrow_{\beta\eta} N \ \Rightarrow \ [\![M]\!]_\rho^{\mathcal{M}} \gtrsim [\![N]\!]_\rho^{\mathcal{M}}.$$

Proof (i) By induction on the structure of M. Write $M^\bullet \equiv M[x: = N]$. We only treat the case $M \equiv \lambda y.P$. By the variable convention we may assume that $y \notin \mathrm{FV}(N)$. We have

$$\begin{array}{lll} [\![(\lambda y.P)^\bullet]\!]_\rho & \simeq & [\![\lambda y.P^\bullet]\!]_\rho \\ & \simeq & \boldsymbol{\lambda} d.[\![P^\bullet]\!]_{\rho[y:=d]} \\ & \simeq & \boldsymbol{\lambda} d.[\![P]\!]_{\rho[y:=d][x:=[\![N]\!]_{\rho[y:=d]}]}, \quad \text{by the induction hypothesis,} \\ & \simeq & \boldsymbol{\lambda} d.[\![P]\!]_{\rho[y:=d][x:=[\![N]\!]_\rho]}, \quad \text{by Lemma 3.1.12,} \\ & \simeq & \boldsymbol{\lambda} d.[\![P]\!]_{\rho[x:=[\![N]\!]_\rho][y:=d]} \\ & \simeq & [\![\lambda y.P]\!]_{\rho[x:=[\![N]\!]_\rho]}. \end{array}$$

(ii) By induction on the generation of $M \twoheadrightarrow_{\beta\eta} N$.
Case $M \equiv (\lambda x.P)Q$ and $N \equiv P[x: = Q]$. Then

$$\begin{array}{lll} [\![(\lambda x.P)Q]\!]_\rho & \succsim & (\boldsymbol{\lambda} d.[\![P]\!]_{\rho[x:=d]})([\![Q]\!]_\rho) \\ & \succsim & [\![P]\!]_{\rho[x:=[\![Q]\!]_\rho]} \\ & \simeq & [\![P[x: = Q]]\!]_\rho, \quad \text{by (i).} \end{array}$$

Case $M \equiv \lambda x.Nx$, with $x \notin \mathrm{FV}(N)$. Then

$$\begin{array}{ll} [\![\lambda x.Nx]\!]_\rho & \succsim \boldsymbol{\lambda} d.[\![N]\!]_\rho(d) \\ & \simeq [\![N]\!]_\rho. \end{array}$$

Case $M \twoheadrightarrow_{\beta\eta} N$. This is $PZ \twoheadrightarrow_{\beta\eta} QZ$, $ZP \twoheadrightarrow_{\beta\eta} ZQ$ or $\lambda x.P \twoheadrightarrow_{\beta\eta} \lambda x.Q$, and follows directly from $P \twoheadrightarrow_{\beta\eta} Q$. Then the result follows from the induction hypothesis.

The cases where $M \twoheadrightarrow_{\beta\eta} N$ follow via reflexivity or transitivity and are easy to treat. ■

3.1.14 Definition Let $\mathcal{M}, \mathcal{N}$ be typed λ-models and let $A \in \mathbb{T}$.

(i) We say $\mathcal{M}$ and $\mathcal{N}$ are *elementary equivalent at A*, written $\mathcal{M} \equiv_A \mathcal{N}$, iff

$$\forall M, N \in \Lambda^{\emptyset}_{\to}(A).[\mathcal{M} \models M = N \Leftrightarrow \mathcal{N} \models M = N].$$

(ii) We say $\mathcal{M}$ and $\mathcal{N}$ are *elementary equivalent*, written $\mathcal{M} \equiv \mathcal{N}$, iff

$$\forall A \in \mathbb{T}.\mathcal{M} \equiv_A \mathcal{N}.$$

3.1.15 Proposition *Let $\mathcal{M}$ be a typed λ-model. Then*

$$\mathcal{M} \textit{ is non-trivial} \Leftrightarrow \forall A \in \mathbb{T}.\mathcal{M}(A) \textit{ is not a singleton.}$$

Proof ($\Leftarrow$) By definition.

($\Rightarrow$) We will show this for $A = 1 = 0\to 0$. Let c_1, c_2 be distinct elements of $\mathcal{M}(0)$. Consider $M \equiv \lambda x^0.y^0 \in \Lambda^{\emptyset}_{\to}(1)$. Let ρ_i be the partial valuation with $\rho_i(y^0) = c_i$. Then $[\![M]\!]_{\rho_i}\downarrow$ and $[\![M]\!]_{\rho_1} c_1 = c_1, [\![M]\!]_{\rho_2} c_1 = c_2$. Therefore $[\![M]\!]_{\rho_1}, [\![M]\!]_{\rho_2}$ are different elements of $\mathcal{M}(1)$. ∎

Thus with Proposition 3.1.5 one has for a typed λ-model $\mathcal{M}$

$$\begin{aligned}\mathcal{M}(0) \text{ is a singleton} &\Leftrightarrow \forall A \in \mathbb{T}.\mathcal{M}(A) \text{ is a singleton}\\ &\Leftrightarrow \exists A \in \mathbb{T}.\mathcal{M}(A) \text{ is a singleton.}\end{aligned}$$

3.1.16 Proposition *Let $\mathcal{M}, \mathcal{N}$ be typed λ-models and $F{:}\mathcal{M}\to\mathcal{N}$ a surjective morphism. Then*

(i) $F([\![M]\!]^{\mathcal{M}}_{\rho}) = [\![M]\!]^{\mathcal{N}}_{F\circ\rho}$, *for all $M \in \Lambda_{\to}(A)$;*

(ii) $F([\![M]\!]^{\mathcal{M}}) = [\![M]\!]^{\mathcal{N}}$, *for all $M \in \Lambda^{\emptyset}_{\to}(A)$.*

Proof (i) By induction on the structure of M.

Case $M \equiv x$. Then $F([\![x]\!]^{\mathcal{M}}_{\rho}) = F(\rho(x)) = [\![x]\!]^{\mathcal{N}}_{F\circ\rho}$.

Case $M = PQ$. Then

$$\begin{aligned}F([\![PQ]\!]^{\mathcal{M}}_{\rho}) &= F([\![P]\!]^{\mathcal{M}}_{\rho}) \cdot_{\mathcal{N}} F([\![Q]\!]^{\mathcal{M}}_{\rho})\\ &= [\![P]\!]^{\mathcal{N}}_{F\circ\rho} \cdot_{\mathcal{N}} [\![Q]\!]^{\mathcal{N}}_{F\circ\rho}, \qquad \text{by the induction hypothesis,}\\ &= [\![PQ]\!]^{\mathcal{N}}_{F\circ\rho}.\end{aligned}$$

Case $M = \lambda x.P$. Then we must show

$$F(\lambda\!\!\lambda d \in \mathcal{M}.[\![P]\!]^{\mathcal{M}}_{\rho[x:=d]}) = \lambda\!\!\lambda e \in \mathcal{N}.[\![P]\!]^{\mathcal{M}}_{(F\circ\rho)[x:=e]}.$$

By extensionality it suffices to show for all $e \in \mathcal{N}$

$$F(\lambda\!\!\lambda d \in \mathcal{M}.\lambda\!\!\lambda d \in \mathcal{M}.[\![P]\!]^{\mathcal{M}}_{\rho[x:=d]}) \cdot_{\mathcal{N}} e = [\![P]\!]^{\mathcal{M}}_{(F\circ\rho)[x:=e]}.$$

By surjectivity of F we can write $e = F(d)$. Now,

$$\begin{aligned}F([\![P]\!]^{\mathcal{M}}_{\rho[x:=d]}) \cdot_{\mathcal{N}} F(d) &= F([\![P]\!]^{\mathcal{N}}_{\rho[x:=d]}\\ &= [\![P]\!]^{\mathcal{N}}_{F\circ(\rho[x:=d])}, \qquad \text{by the induction hypothesis,}\\ &= [\![P]\!]^{\mathcal{N}}_{(F\circ\rho)[x:=F(d)])}.\end{aligned}$$

(ii) By (i). ∎

3.1.17 Proposition *Let $\mathcal{M}$ be a typed λ-model.*

(i) $\mathcal{M} \models (\lambda x.M)N = M[x := N]$.

(ii) $\mathcal{M} \models \lambda x.Mx = M$, *if $x \notin \mathrm{FV}(\mathcal{M})$.*

Proof (i)
$$\begin{aligned}[\![(\lambda x.M)N]\!]_\rho &= [\![\lambda x.M]\!]_\rho [\![N]\!]_\rho \\ &= [\![M]\!]_{\rho[x:=[\![N]\!]_\rho]}, \\ &= [\![M[x:=N]]\!]_\rho, \quad \text{by Lemma 3.1.13.}\end{aligned}$$

(ii)
$$\begin{aligned}[\![\lambda x.Mx]\!]_\rho d &= [\![Mx]\!]_{\rho[x:=d]} \\ &= [\![M]\!]_{\rho[x:=d]} d \\ &= [\![M]\!]_\rho d, \quad \text{as } x \notin \mathrm{FV}(M).\end{aligned}$$

Therefore by extensionality $[\![\lambda x.Mx]\!]_\rho = [\![M]\!]_\rho$. ∎

3.1.18 Lemma *Let $\mathcal{M}$ be a typed λ-model. Then*

$$\mathcal{M} \models M = N \Leftrightarrow \mathcal{M} \models \lambda x.M = \lambda x.N.$$

Proof
$$\begin{aligned}\mathcal{M} \models M = N &\Leftrightarrow \forall \rho. \quad [\![M]\!]_\rho = [\![N]\!]_\rho \\ &\Leftrightarrow \forall \rho, d. \quad [\![M]\!]_{\rho[x:=d]} = [\![N]\!]_{\rho[x:=d]} \\ &\Leftrightarrow \forall \rho, d. \quad [\![\lambda x.M]\!]_\rho d = [\![\lambda x.N]\!]_\rho d \\ &\Leftrightarrow \forall \rho. \quad [\![\lambda x.M]\!]_\rho = [\![\lambda x.N]\!]_\rho \\ &\Leftrightarrow \quad \mathcal{M} \models \lambda x.M = \lambda x.N. \; ∎\end{aligned}$$

3.1.19 Proposition

(i) *For every non-empty set X the typed structure $\mathcal{M}_X$ is a $\boldsymbol{\lambda}^0_\to$-model.*

(ii) *Let X be a poset. Then $\mathcal{D}_X$ is a $\boldsymbol{\lambda}^0_\to$-model.*

(iii) *Let $\mathcal{M}$ be a typed applicative structure. Assume that $[\![\mathsf{K}_{A,B}]\!]^{\mathcal{M}}\downarrow$ and $[\![\mathsf{S}_{A,B,C}]\!]^{\mathcal{M}}\downarrow$. Then $\mathcal{M}$ is a $\boldsymbol{\lambda}^0_\to$-model.*

(iv) *Let Δ be a layered non-empty subfamily of a typed applicative structure $\mathcal{M}$ that is extensional and closed under application. Suppose $[\![\mathsf{K}_{A,B}]\!]$, $[\![\mathsf{S}_{A,B,C}]\!]$ are defined and in Δ. Then $\mathcal{M}{\restriction}\Delta$, see Definition* 3.1.7(ii), *is a $\boldsymbol{\lambda}^0_\to$-model.*

Proof (i) Since $\mathcal{M}_X$ is the full typed structure, $[\![M]\!]_\rho$ always exists.

(ii) By induction on M one can show that $\lambda\!\!\lambda d.[\![M]\!]_{\rho(x:=d)}$ is monotonic. It then follows by induction on M that $[\![M]\!]_\rho \in \mathcal{D}_X$.

(iii) For every λ-term M there exists a typed applicative expression P consisting only of Ks and Ss such that $P \twoheadrightarrow_{\beta\eta} M$. Now apply Lemma 3.1.13.

(iv) By (iii). ∎

Operations on typed λ-models

Now we will introduce two operations on λ-models: $\mathcal{M}, \mathcal{N} \mapsto \mathcal{M} \times \mathcal{N}$, the Cartesian product, and $\mathcal{M} \mapsto \mathcal{M}^*$, the polynomial λ-model. The relationship between $\mathcal{M}$ and $\mathcal{M}^*$ is similar to that of a ring R and its ring of multivariate polynomials $R[\vec{x}]$.

Cartesian products

3.1.20 Definition If $\mathcal{M},\mathcal{N}$ are typed applicative structures, then the *Cartesian product* of $\mathcal{M},\mathcal{N}$, written $\mathcal{M}\times\mathcal{N}$, is the structure defined by

$$(\mathcal{M}\times\mathcal{N})(A) = \mathcal{M}(A)\times\mathcal{N}(A)$$
$$(M_1,N_1)\cdot(M_2,N_2) = (M_1\cdot M_2, N_1\cdot N_2).$$

3.1.21 Proposition *Let $\mathcal{M},\mathcal{N}$ be typed λ-models. For a partial valuation ρ in $\mathcal{M}\times\mathcal{N}$ write $\rho(x)=(\rho_1(x),\rho_2(x))$. Then*

(i) $[\![M]\!]^{\mathcal{M}\times\mathcal{N}}_{\rho} = ([\![M]\!]^{\mathcal{M}}_{\rho_1}, [\![M]\!]^{\mathcal{N}}_{\rho_2})$;
(ii) $\mathcal{M}\times\mathcal{N}$ *is a λ-model;*
(iii) $\mathrm{Th}(\mathcal{M}\times\mathcal{N}) = \mathrm{Th}(\mathcal{M})\cap\mathrm{Th}(\mathcal{N})$.

Proof (i) By induction on M.
(ii) By (i).
(iii)
$$\begin{aligned}\mathcal{M}\times\mathcal{N},\rho\models M=N &\Leftrightarrow [\![M]\!]_{\rho}=[\![N]\!]_{\rho}\\ &\Leftrightarrow ([\![M]\!]^{\mathcal{M}}_{\rho_1},[\![M]\!]^{\mathcal{N}}_{\rho_2}) = ([\![N]\!]^{\mathcal{M}}_{\rho_1},[\![N]\!]^{\mathcal{N}}_{\rho_2})\\ &\Leftrightarrow [\![M]\!]^{\mathcal{M}}_{\rho_1}=[\![N]\!]^{\mathcal{M}}_{\rho_1}\ \&\ [\![M]\!]^{\mathcal{M}}_{\rho_2}=[\![N]\!]^{\mathcal{M}}_{\rho_2}\\ &\Leftrightarrow \mathcal{M},\rho_1\models M=N\ \&\ \mathcal{N},\rho_2\models M=N.\end{aligned}$$

Hence for closed terms M,N

$$\mathcal{M}\times\mathcal{N}\models M=N \Leftrightarrow \mathcal{M}\models M=N\ \&\ \mathcal{N}\models M=N. \blacksquare$$

Polynomial models

3.1.22 Definition

(i) We introduce for each $m\in\mathcal{M}(A)$ a new constant $\underline{m}:A$, for each type A we choose a set of variables

$$x_0^A, x_1^A, x_2^A, \dots,$$

and we let $\underline{\mathcal{M}}$ be the set of all correctly typed applicative combinations of these typed constants and variables.

(ii) For a valuation $\rho:\mathtt{Var}\to\mathcal{M}$ define the map $(\!(-)\!)_\rho = (\!(-)\!)^{\mathcal{M}}_\rho : \underline{\mathcal{M}}\to\mathcal{M}$ by

$$\begin{aligned}(\!(x)\!)_\rho &= \rho(x);\\ (\!(\underline{m})\!)_\rho &= m;\\ (\!(PQ)\!)_\rho &= (\!(P)\!)_\rho(\!(Q)\!)_\rho.\end{aligned}$$

(iii) Write

$$P\sim_{\mathcal{M}} Q \Leftrightarrow \forall\rho\,(\!(P)\!)_\rho = (\!(Q)\!)_\rho,$$

where ρ ranges over valuations in $\mathcal{M}$.

3.1.23 Lemma

(i) $\sim_{\mathcal{M}}$ *is an equivalence relation satisfying* $\underline{de} \sim_{\mathcal{M}} \underline{d}\,\underline{e}$.
(ii) *For all* $P, Q \in \underline{\mathcal{M}}$ *one has*

$$P_1 \sim_{\mathcal{M}} P_2 \Leftrightarrow \forall Q_1, Q_2 \in \underline{\mathcal{M}}\,[Q_1 \sim_{\mathcal{M}} Q_2 \Rightarrow P_1Q_1 \sim_{\mathcal{M}} P_2Q_2].$$

Proof Note that P, Q can take all values in $\mathcal{M}(A)$ and apply extensionality. ∎

3.1.24 Definition Let $\mathcal{M}$ be a typed applicative structure. The *polynomial structure* over $\mathcal{M}$ is $\mathcal{M}^* = (|\mathcal{M}^*|, \mathsf{app})$ defined by

$$\begin{aligned} |\mathcal{M}^*| \;=\; \underline{\mathcal{M}}/{\sim_{\mathcal{M}}} &= \{[P]_{\sim_{\mathcal{M}}} \mid P \in \underline{\mathcal{M}}\}, \\ \mathsf{app}\,[P]_{\sim_{\mathcal{M}}}[Q]_{\sim_{\mathcal{M}}} &= [PQ]_{\sim_{\mathcal{M}}}. \end{aligned}$$

By Lemma 3.1.23(ii) this is well defined.

Working with $\mathcal{M}^*$ it is often convenient to use as elements those of $\underline{\mathcal{M}}$ and reason about them modulo $\sim_{\mathcal{M}}$.

3.1.25 Proposition

(i) $\mathcal{M} \subseteq \mathcal{M}^*$ *by the embedding morphism* $i = \lambda\!\!\lambda d.[\underline{d}] : \mathcal{M} \to \mathcal{M}^*$.
(ii) *The embedding* i *can be extended to an embedding* $i : \underline{\mathcal{M}} \to \underline{\mathcal{M}^*}$.
(iii) *There exists an isomorphism* $G : \mathcal{M}^* \cong \mathcal{M}^{**}$.

Proof (i) It is easy to show that i is injective and satisfies

$$i(de) = i(d) \cdot_{\mathcal{M}^*} i(e).$$

(ii) Define

$$\begin{aligned} i'(x) &= x \\ i'(\underline{m}) &= \underline{[m]} \\ i'(d_1 d_2) &= i'(d_1) i'(d_2). \end{aligned}$$

We write again i for i'.

(iii) By definition $\underline{\mathcal{M}}$ is the set of all typed applicative combinations of typed variables x^A and constants $\underline{m}^A$ and $\underline{\mathcal{M}^*}$ is the set of all typed applicative combinations of typed variables y^A and constants $(\underline{m^*})^A$. Define a map $\underline{\mathcal{M}} \to \underline{\mathcal{M}^*}$, also denoted by G, by

$$\begin{aligned} G(\underline{m}) &= \underline{[m]} \\ G(x_{2i}) &= \underline{[x_i]} \\ G(x_{2i+1}) &= y_i. \end{aligned}$$

Then we have

(1) $P \sim_{\mathcal{M}} Q \Rightarrow G(P) \sim_{\mathcal{M}^*} G(Q)$;
(2) $G(P) \sim_{\mathcal{M}^*} G(Q) \Rightarrow P \sim_{\mathcal{M}} Q$;
(3) $\forall Q \in \underline{\mathcal{M}^*} \exists P \in \underline{\mathcal{M}}[G(P) \sim Q]$.

Therefore G induces the required isomorphism on the equivalence classes. ■

3.1.26 Definition Let $P \in \underline{\mathcal{M}}$ and let x be a variable. We say that

P does not depend on x

if whenever ρ_1, ρ_2 satisfy $\rho_1(y) = \rho_2(y)$ for $y \not\equiv x$, we have $(\!(P)\!)_{\rho_1} = (\!(P)\!)_{\rho_2}$.

3.1.27 Lemma *If P does not depend on x, then $P \sim_{\mathcal{M}} P[x{:=}Q]$ for all $Q \in \underline{\mathcal{M}}$.*

Proof First show that $(\!(P[x := Q])\!)_\rho = (\!(P)\!)_{\rho[x:=(\!(Q)\!)_\rho]}$, in analogy to Lemma 3.1.13(i). Now suppose P does not depend on x. Then

$$\begin{aligned} (\!(P[x{:=}Q])\!)_\rho &= (\!(P)\!)_{\rho[x:=(\!(Q)\!)_\rho]} \\ &= (\!(P)\!)_\rho, \qquad \text{as } P \text{ does not depend on } x. \end{aligned}$$ ■

3.1.28 Proposition *Let $\mathcal{M}$ be a typed applicative structure. Then:*

(i) *$\mathcal{M}$ is a typed λ-model $\Leftrightarrow$ for each $P \in \mathcal{M}^*$ and variable x of $\underline{\mathcal{M}}$ there exists an $F \in \mathcal{M}^*$ not depending on x such that $F[x] = P$.*

(ii) *$\mathcal{M}$ is a typed λ-model $\Rightarrow$ $\mathcal{M}^*$ is a typed λ-model.*

Proof (i) Choosing representatives for $P, F \in \mathcal{M}^*$ we show
$\mathcal{M}$ is a typed λ-model $\Leftrightarrow$ for each $P \in \underline{\mathcal{M}}$ and variable x there exists an $F \in \underline{\mathcal{M}}$ not depending on x such that $Fx \sim_{\mathcal{M}} P$.

($\Rightarrow$) Let $\mathcal{M}$ be a typed λ-model and let P be given. We treat an illustrative example, namely $P \equiv \underline{f}x^0y^0$, with $f \in \mathcal{M}(1_2)$. We take $F \equiv \underline{[\![\lambda y z_f x.z_f xy]\!]}y\underline{f}$. Then

$$(\!(Fx)\!)_\rho = [\![\lambda y z_f x.z_f xy]\!]\rho(y)f\rho(x) = f\rho(x)\rho(y) = (\!(\underline{f}xy)\!)_\rho,$$

hence indeed $Fx \sim_{\mathcal{M}} \underline{f}xy$. In general for each constant $\underline{d}$ in P we take a variable z_d and define $F \equiv \underline{[\![\lambda \vec{y}\vec{z}_d\vec{x}.P]\!]}\vec{y}\vec{\underline{f}}$.

($\Leftarrow$) We show $\forall M \in \Lambda_\to(A) \exists P_M \in \underline{\mathcal{M}}(A) \forall \rho. [\![M]\!]_\rho = (\!(P_M)\!)_\rho$, by induction on $M : A$. For M a variable or application this is trivial. For $M = \lambda x.N$, we know by the induction hypothesis that $[\![N]\!]_\rho = (\!(P_N)\!)_\rho$ for all ρ. By assumption there is an F not depending on x such that $Fx \sim_{\mathcal{M}} P_N$. Then

$$(\!(F)\!)_\rho d = (\!(Fx)\!)_{\rho[x:=d]} = (\!(P_N)\!)_{\rho[x:=d]} =_{\text{IH}} [\![N]\!]_{\rho[x:=d]}.$$

Hence $[\![\lambda x.N]\!]_\rho = (\!(F)\!)_\rho$. So indeed $[\![M]\!]_\rho\!\downarrow$ for every ρ such that $\mathrm{FV}(M) \subseteq \mathrm{dom}(\rho)$. Hence $\mathcal{M}$ is a typed λ-model.

(ii) By (i) $\mathcal{M}^*$ is a λ-model if a certain property holds for $\mathcal{M}^{**}$. But $\mathcal{M}^{**} \cong \mathcal{M}^*$ and the property does hold here, since $\mathcal{M}$ is a λ-model. [To make matters concrete, one has to show for example that for all $M \in \mathcal{M}^{**}$ there is an N not depending on y such that $Ny \sim_{\mathcal{M}^*} M$. Writing $M \equiv M[x_1, x_2][y]$ one can obtain N by rewriting the y in M obtaining $M' \equiv M[x_1, x_2][x] \in \mathcal{M}^*$ and using the fact that $\mathcal{M}$ is a λ-model: $M' = Nx$, so $Ny = M$]. ∎

3.1.29 Proposition *If $\mathcal{M}$ is a typed λ-model, then* $\mathrm{Th}(\mathcal{M}^*) = \mathrm{Th}(\mathcal{M})$.

Proof This is Exercise 3.5. ∎

3.1.30 Remark In general for typed structures $\mathcal{M}^* \times \mathcal{N}^* \ncong (\mathcal{M} \times \mathcal{N})^*$, but the isomorphism holds if $\mathcal{M}, \mathcal{N}$ are typed λ-models.

Semantics for type assignment *à la* Curry

We will now employ models of untyped λ-calculus in order to give a semantics for $\boldsymbol{\lambda}^{\mathrm{Cu}}_{\to}$. The idea, due to Scott (1975a), is to interpret a type $A \in \mathbb{T}^{\mathbb{A}}$ as a subset of an untyped λ-model in such a way that it contains all the interpretations of the untyped λ-terms $M \in \Lambda(A)$. As usual one has to pay attention to $\mathrm{FV}(M)$.

3.1.31 Definition

(i) An *applicative structure* is a pair $\langle \mathcal{D}, \cdot \rangle$, consisting of a set $\mathcal{D}$ together with a binary operation $\cdot : \mathcal{D} \times \mathcal{D} \to \mathcal{D}$ on it.

(ii) An *(untyped) λ-model* for the untyped λ-calculus is of the form

$$\mathcal{D} = \langle \mathcal{D}, \cdot, [\![\ ]\!]^{\mathcal{D}} \rangle,$$

where $\langle \mathcal{D}, \cdot \rangle$ is an applicative structure and $[\![\ ]\!]^{\mathcal{D}} : \Lambda \times \mathsf{Env}_{\mathcal{D}} \to \mathcal{D}$ satisfies:

$$\begin{array}{llcll}
(1) & [\![x]\!]^{\mathcal{D}}_\rho & = & \rho(x); & \\
(2) & [\![MN]\!]^{\mathcal{D}}_\rho & = & [\![M]\!]^{\mathcal{D}}_\rho \cdot [\![N]\!]^{\mathcal{D}}_\rho; & \\
(3) & [\![\lambda x.M]\!]^{\mathcal{D}}_\rho & = & [\![\lambda y.M[x := y]]\!]^{\mathcal{D}}_\rho, & (\alpha) \\
 & & & \text{provided } y \notin \mathrm{FV}(M); & \\
(4) & \forall d \in \mathcal{D}.[\![M]\!]^{\mathcal{D}}_{\rho[x:=d]} = [\![N]\!]^{\mathcal{D}}_{\rho[x:=d]} & \Rightarrow & [\![\lambda x.M]\!]^{\mathcal{D}}_\rho = [\![\lambda x.N]\!]^{\mathcal{D}}_\rho; & (\xi) \\
(5) & \rho \restriction \mathrm{FV}(M) = \rho' \restriction \mathrm{FV}(M) & \Rightarrow & [\![M]\!]^{\mathcal{D}}_\rho = [\![M]\!]^{\mathcal{D}}_{\rho'}; & \\
(6) & [\![\lambda x.M]\!]^{\mathcal{D}}_\rho \cdot d & = & [\![M]\!]^{\mathcal{D}}_{\rho[x:=d]}. & (\boldsymbol{\beta})
\end{array}$$

We will write $[\![\ ]\!]_\rho$ for $[\![\ ]\!]^{\mathcal{D}}_\rho$ if there is little danger of confusion.

Note that by (5) for closed terms the interpretation does not depend on the ρ.

3.1.32 Definition Let a λ-model $\mathcal{D}$ be given.

(i) Let $M, N \in \Lambda$ be untyped λ-terms and let $\rho \in \mathsf{Env}_{\mathcal{D}}$. We say that $\mathcal{D}$ *with environment* ρ *satisfies the equation* $M = N$, written

$$\mathcal{D}, \rho \models M = N,$$

if $[\![M]\!]^{\mathcal{D}}_{\rho} = [\![N]\!]^{\mathcal{D}}_{\rho}$.

(ii) Let T be a set of equations between λ-terms. Then $\mathcal{D}$ satisfies T, written

$$\mathcal{D} \models T$$

if for all ρ and all $(M = N) \in T$ one has $\mathcal{D}, \rho \models M = N$. If the set T consists of equations between closed terms, then the ρ is irrelevant.

(iii) Write

$$T \models M = N$$

if for all $\mathcal{D}$ and $\rho \in \mathsf{Env}_{\mathcal{D}}$ one has

$$\mathcal{D}, \rho \models T \;\Rightarrow\; \mathcal{D}, \rho \models M = N.$$

3.1.33 Theorem (Completeness Theorem) *Let $M, N \in \Lambda$ be arbitrary and let T be a set of equations. Then*

$$T \vdash_{\lambda\beta\eta} M = N \;\Leftrightarrow\; T \models M = N.$$

Proof ($\Rightarrow$) ('Soundness') By induction on the derivation of $T \vdash M = N$.

($\Leftarrow$) ('Completeness' proper) By taking the (extensional open) term model of T, see B[1984], 4.1.17. ∎

Following Scott (1975a) a λ-model gives rise to a unified interpretation of λ-terms $M \in \Lambda$ and types $A \in \mathbb{T}^{\mathbb{A}}$. The terms will be interpreted as elements of $\mathcal{D}$ and the types as subsets of $\mathcal{D}$.

3.1.34 Definition Let $\mathcal{D}$ be a λ-model. On the powerset $\mathsf{P}D$ one can define for $X, Y \in \mathsf{P}D$ the element $(X \Rightarrow Y) \in \mathsf{P}\mathcal{D}$ as

$$X \Rightarrow Y = \{d \in D \mid d.X \subseteq Y\} = \{d \in \mathcal{D} \mid \forall x \in X.(d \cdot x) \in Y\}.$$

3.1.35 Definition Let $\mathcal{D}$ be a λ-model. Given a type environment $\xi : \mathbb{A} \to \mathsf{P}\mathcal{D}$, the *interpretation* of an $A \in \mathbb{T}^{\mathbb{A}}$ into $\mathsf{P}\mathcal{D}$, written $[\![A]\!]_{\xi}$, is defined as

$$\begin{array}{rcll} [\![\alpha]\!]_{\xi} &=& \xi(\alpha), & \text{for } \alpha \in \mathbb{A}; \\ [\![A \to B]\!]_{\xi} &=& [\![A]\!]_{\xi} \Rightarrow [\![B]\!]_{\xi}. & \end{array}$$

3.1.36 Definition Let $\mathcal{D}$ be a λ-model and let $M \in \Lambda, A \in \mathbb{T}^{\mathbb{A}}$. Let ρ, ξ range over term and type environments, respectively.

(i) We say that $\mathcal{D}$ *with* ρ, ξ *satisfies the type assignment* $M : A$, written

$$\mathcal{D}, \rho, \xi \models M : A$$

if $[\![M]\!]_\rho \in [\![A]\!]_\xi$.

(ii) Let Γ be a type assignment basis. Then

$$\mathcal{D}, \rho, \xi \models \Gamma \Leftrightarrow \text{ for all } (x{:}A) \in \Gamma \text{ one has } \mathcal{D}, \rho, \xi \models x : A.$$

(iii) $\Gamma \models M : A \Leftrightarrow \forall \mathcal{D}, \rho, \xi[\mathcal{D}, \rho, \xi \models \Gamma \Rightarrow \mathcal{D}, \rho, \xi \models M : A]$.

3.1.37 Proposition *Let* Γ, M, A *respectively range over bases, untyped terms and types in* $\mathbb{T}^{\mathbb{A}}$. *Then*

$$\Gamma \vdash^{\mathrm{Cu}}_{\boldsymbol{\lambda}^{\mathbb{A}}_{\rightarrow}} M : A \Leftrightarrow \Gamma \models M : A.$$

Proof ($\Rightarrow$) By induction on the length of proof.

($\Leftarrow$) This has been proved independently in Hindley (1983) and Barendregt et al. (1983). See Corollary 17.1.11. ∎

3.2 Lambda theories and term models

In this section we treat consistent sets of equations between terms of the same type and their term models.

3.2.1 Definition

(i) A *constant* (of type A) is a variable (of the same type) that we promise not to bind by a λ. Rather than $x, y, z, \ldots$ we write constants as $c, d, e, \ldots$, or being explicit as $c^A, d^A, e^A, \ldots$. The letters $\mathcal{C}, \mathcal{D}, \ldots$ range over sets of constants (of varying types).

(ii) Let $\mathcal{D}$ be a set of constants with types in $\mathbb{T}^0$. Write $\Lambda_{\rightarrow}[\mathcal{D}](A)$ for the set of open terms of type A, possibly containing constants in $\mathcal{D}$. Moreover $\Lambda_{\rightarrow}[\mathcal{D}] = \bigcup_{A \in \mathbb{T}} \Lambda_{\rightarrow}[\mathcal{D}](A)$.

(iii) Similarly $\Lambda^{\emptyset}_{\rightarrow}[\mathcal{D}](A)$ and $\Lambda^{\emptyset}_{\rightarrow}[\mathcal{D}]$ consist of closed terms possibly containing the constants in $\mathcal{D}$.

(iv) An *equation over* $\mathcal{D}$ (i.e. between closed λ-terms with constants from $\mathcal{D}$) is of the form $M = N$ with $M, N \in \Lambda^{\emptyset}_{\rightarrow}[\mathcal{D}]$ of the same type.

(v) A term $M \in \Lambda_{\rightarrow}[\mathcal{D}]$ is *pure* if it does not contain constants from $\mathcal{D}$, i.e. if $M \in \Lambda_{\rightarrow}$.

In this subsection we will consider sets of equations over $\mathcal{D}$. When writing $M = N$, we implicitly assume that M, N have the same type.

$(\lambda x.M)N = M[x := N]$	(β)
$\lambda x.Mx = M$, if $x \notin \mathrm{FV}(M)$	(η)
$\dfrac{}{M = N}$, if $(M = N) \in \mathcal{E}$	$(\mathcal{E})$
$M = M$	(reflexivity)
$\dfrac{M = N}{N = M}$	(symmetry)
$\dfrac{M = N \quad N = L}{M = L}$	(transitivity)
$\dfrac{M = N}{MZ = NZ}$	(R-congruence)
$\dfrac{M = N}{ZM = ZN}$	(L-congruence)
$\dfrac{M = N}{\lambda x.M = \lambda x.N}$	(ξ)

Figure 3.1

3.2.2 Definition Let $\mathcal{E}$ be a set of equations over $\mathcal{D}$.

(i) We say that $P = Q$ is *derivable from* $\mathcal{E}$, written $\mathcal{E} \vdash P = Q$ if $P = Q$ can be proved in the equational theory axiomatized in Figure 3.1. We write $M =_{\mathcal{E}} N$ for $\mathcal{E} \vdash M = N$.
(ii) We say that $\mathcal{E}$ is *consistent*, if not all equations are derivable from it.
(iii) We call $\mathcal{E}$ a *typed lambda theory* iff $\mathcal{E}$ is consistent and closed under derivability.

3.2.3 Remark A typed lambda theory always is a $\lambda\boldsymbol{\beta\eta}$-theory.

3.2.4 Notation

(i) $\mathcal{E}^+ = \{M = N \mid \mathcal{E} \vdash M = N\}$.
(ii) For $A \in \mathbb{T}^0$ write $\mathcal{E}(A) = \{M = N \mid (M = N) \in \mathcal{E} \ \&\ M, N \in \Lambda_{\rightarrow}[\mathcal{D}](A)\}$.
(iii) $\mathcal{E}_{\boldsymbol{\beta\eta}} = \emptyset^+$.

3.2.5 Proposition *If* $Mx =_{\mathcal{E}} Nx$, *with* $x \notin \mathrm{FV}(M) \cup \mathrm{FV}(N)$, *then* $M =_{\mathcal{E}} N$.

Proof Use (ξ) and (η). ∎

3.2.6 Definition Let $\mathcal{M}$ be a typed λ-model and $\mathcal{E}$ a set of equations.

(i) We say that $\mathcal{M}$ *satisfies (or is a* model *of)* $\mathcal{E}$, written $\mathcal{M} \models \mathcal{E}$, iff

$$\forall (M{=}N) \in \mathcal{E}.\mathcal{M} \models M = N.$$

(ii) We say that $\mathcal{E}$ satisfies $M = N$, written $\mathcal{E} \models M = N$, iff

$$\forall \mathcal{M}.[\mathcal{M} \models \mathcal{E} \Rightarrow \mathcal{M} \models M = N].$$

3.2.7 Proposition (Soundness) $\mathcal{E} \vdash M = N \Rightarrow \mathcal{E} \models M = N$.

Proof By induction on the derivation of $\mathcal{E} \vdash M = N$. Assume that $\mathcal{M} \models \mathcal{E}$ for a model $\mathcal{M}$ towards $\mathcal{M} \models M = N$. If $M = N \in \mathcal{E}$, then the conclusion follows from the assumption. The cases that $M = N$ falls under the axioms β or η follow from Proposition 3.1.17. The rules reflexivity, symmetry, transitivity and L,R-congruence are trivial to treat. The case falling under the rule (ξ) follows from Lemma 3.1.18. ∎

From non-trivial models one can obtain typed lambda theories.

3.2.8 Proposition *Let $\mathcal{M}$ be a non-trivial typed λ-model.*

(i) $\mathcal{M} \models \mathcal{E} \Rightarrow \mathcal{E}$ *is consistent.*
(ii) $\mathrm{Th}(\mathcal{M})$ *is a lambda theory.*

Proof (i) Suppose $\mathcal{E} \vdash \lambda xy.x = \lambda xy.y$. Then $\mathcal{M} \models \lambda xy.x = \lambda xy.y$. It follows that $d = (\lambda xy.x)de = (\lambda xy.y)de = e$ for arbitrary d, e. Hence $\mathcal{M}$ is trivial.

(ii) Clearly $\mathcal{M} \models \mathrm{Th}(\mathcal{M})$. Hence by (i) $\mathrm{Th}(\mathcal{M})$ is consistent. If $\mathrm{Th}(\mathcal{M}) \vdash M = N$, then by soundness $\mathcal{M} \models M = N$, and so $(M = N) \in \mathrm{Th}(\mathcal{M})$. ∎

The full typed λ-model over a finite set yields an interesting λ-theory.

Term models

3.2.9 Definition Let $\mathcal{D}$ be a set of constants of various types in $\mathbb{T}^0$ and let $\mathcal{E}$ be a set of equations over $\mathcal{D}$. Define the typed structure $\mathcal{M}_{\mathcal{E}}$ by

$$\mathcal{M}_{\mathcal{E}}(A) = \{[M]_{\mathcal{E}} \mid M \in \Lambda_{\to}[\mathcal{D}](A)\},$$

where $[M]_{\mathcal{E}}$ is the equivalence class modulo the congruence relation $=_{\mathcal{E}}$. Define the binary operator $\cdot$ as

$$[M]_{\mathcal{E}} \cdot [N]_{\mathcal{E}} = [MN]_{\mathcal{E}}.$$

This is well defined, because $=_{\mathcal{E}}$ is a congruence. We often will suppress the dot.

3.2.10 Proposition

(i) $(\mathcal{M}_\mathcal{E}, \cdot)$ *is a typed applicative structure.*
(ii) *The semantic interpretation of* M *in* $\mathcal{M}_\mathcal{E}$ *is determined by*

$$[\![M]\!]_\rho = [M[\vec{x}:=\vec{N}]]_\mathcal{E},$$

where $\{\vec{x}\} = \mathrm{FV}(M)$ *and the* $\vec{N}$ *are determined by* $\rho(x_i) = [N_i]_\mathcal{E}$.
(iii) $\mathcal{M}_\mathcal{E}$ *is a typed model, called the* open term model *of* $\mathcal{E}$.

Proof (i) We need to verify extensionality.

$$\begin{aligned}\forall d \in \mathcal{M}_\mathcal{E}.[M]d = [N]d \;&\Rightarrow\; [M][x] = [N][x], \text{ for a fresh } x,\\ &\Rightarrow\; [Mx] = [Nx]\\ &\Rightarrow\; Mx =_\mathcal{E} Nx\\ &\Rightarrow\; M =_\mathcal{E} N, \text{ by } (\xi), (\boldsymbol{\eta}) \text{ and (transitivity)},\\ &\Rightarrow\; [M] = [N].\end{aligned}$$

(ii) We show that $[\![M]\!]_\rho$ defined as $[M[x:=\vec{N}]]_\mathcal{E}$ satisfies the conditions in Definition 3.1.9(ii).

$$\begin{aligned}[\![x]\!]_\rho &= [x[x:=N]]_\mathcal{E}, \text{ with } \rho(x) = [N]_\mathcal{E},\\ &= [N]_\mathcal{E}\\ &= \rho(x);\\ [\![PQ]\!]_\rho &= [(PQ)[\vec{x}:=\vec{N}]]_\mathcal{E}\\ &= [P[\vec{x}:=\vec{N}]Q[\vec{x}:=\vec{N}]]_\mathcal{E}\\ &= [P[\vec{x}:=\vec{N}]]_\mathcal{E}[[Q[\vec{x}:=\vec{N}]]_\mathcal{E}\\ &= [\![P]\!]_\rho[\![Q]\!]_\rho;\\ [\![\lambda y.P]\!]_\rho[Q]_\mathcal{E} &= [(\lambda y.P)[\vec{x}:=\vec{N}]]_\mathcal{E}[Q]_\mathcal{E}\\ &= [\lambda y.P[\vec{x}:=\vec{N}]]_\mathcal{E}[Q]_\mathcal{E}\\ &= [P[\vec{x}:=\vec{N}][y:=Q]]_\mathcal{E}\\ &= [P[\vec{x},y:=\vec{N},Q]]_\mathcal{E}, \text{ because } y \notin \mathrm{FV}(\vec{N}) \text{ by the}\\ &\qquad\qquad \text{variable convention and } y \notin \{\vec{x}\},\\ &= [\![P]\!]_{\rho[y:=[Q]_\mathcal{E}]}.\end{aligned}$$

(iii) As $[\![M]\!]_\rho$ is always defined by (ii). ■

3.2.11 Corollary

(i) $\mathcal{M}_\mathcal{E} \models M = N \;\Leftrightarrow\; M =_\mathcal{E} N$.
(ii) $\mathcal{M}_\mathcal{E} \models \mathcal{E}$.

Proof (i) ($\Rightarrow$) Suppose $\mathcal{M}_\mathcal{E} \models M = N$. Then $[\![M]\!]_\rho = [\![N]\!]_\rho$ for all ρ. Choosing $\rho(x) = [x]_\mathcal{E}$ one obtains $[\![M]\!]_\rho = [M[\vec{x} := \vec{x}]]_\mathcal{E} = [M]_\mathcal{E}$, and similarly for N, hence $[M]_\mathcal{E} = [N]_\mathcal{E}$ and therefore $M =_\mathcal{E} N$.

$$\begin{array}{lll}(\Leftarrow)\ M =_\mathcal{E} N & \Rightarrow & M[\vec{x} := \vec{P}] =_\mathcal{E} N[\vec{x} := \vec{P}] \\ & \Rightarrow & [M[\vec{x} := \vec{P}]]_\mathcal{E} = [N[\vec{x} := \vec{P}]]_\mathcal{E} \\ & \Rightarrow & [\![M]\!]_\rho = [\![N]\!]_\rho \\ & \Rightarrow & \mathcal{M}_\mathcal{E} \models M = N.\end{array}$$

(ii) If $M = N \in \mathcal{E}$, then $M =_\mathcal{E} N$, hence $\mathcal{M}_\mathcal{E} \models M = N$, by (i). ■

Using this corollary we obtain completeness in a simple way.

3.2.12 Theorem (Completeness) $\mathcal{E} \vdash M = N \Leftrightarrow \mathcal{E} \models M = N$.

Proof ($\Rightarrow$) By soundness, Proposition 3.2.7.

$$\begin{array}{lll}(\Leftarrow)\ \mathcal{E} \models M = N & \Rightarrow & \mathcal{M}_\mathcal{E} \models M = N, \quad \text{as } \mathcal{M}_\mathcal{E} \models \mathcal{E}, \\ & \Rightarrow & M =_\mathcal{E} N \\ & \Rightarrow & \mathcal{E} \vdash M = N. \ ■\end{array}$$

3.2.13 Corollary *Let $\mathcal{E}$ be a set of equations. Then*

$$\mathcal{E} \textit{ has a non-trivial model} \Leftrightarrow \mathcal{E} \textit{ is consistent.}$$

Proof ($\Rightarrow$) By Proposition 3.2.8.

($\Leftarrow$) Suppose that $\mathcal{E} \nvdash x^0 = y^0$. Then by the theorem one has $\mathcal{E} \not\models x^0 = y^0$. Therefore for some model $\mathcal{M}$ one has $\mathcal{M} \models \mathcal{E}$ and $\mathcal{M} \not\models x = y$. It follows that $\mathcal{M}$ is non-trivial. ■

If $\mathcal{D}$ contains enough constants, then one can similarly define the applicative structure $\mathcal{M}^\varnothing_{\mathcal{E}[\mathcal{D}]}$ by restricting $\mathcal{M}_\mathcal{E}$ to closed terms.

Constructing Theories

The following result is due to Jacopini (1975).

3.2.14 Proposition *Let $\mathcal{E}$ be a set of equations between closed terms in $\Lambda^\varnothing_\to[\mathcal{D}]$. Then $\mathcal{E} \vdash M = N$ if for some $n \in \omega$, $F_1, \dots, F_n \in \Lambda_\to[\mathcal{D}]$ and $P_1 = Q_1, \dots, P_n = Q_n \in \mathcal{E}$ one has $\mathrm{FV}(F_i) \subseteq \mathrm{FV}(M) \cup \mathrm{FV}(N)$ and*

$$\left.\begin{array}{rcl} M & =_{\beta\eta} & F_1 P_1 Q_1 \\ F_1 Q_1 P_1 & =_{\beta\eta} & F_2 P_2 Q_2 \\ & \vdots & \\ F_{n-1} Q_{n-1} P_{n-1} & =_{\beta\eta} & F_n P_n Q_n \\ F_n Q_n P_n & =_{\beta\eta} & N. \end{array}\right\} \qquad (3.1)$$

This scheme (3.1) is called a Jacopini tableau and the sequence $F_1, \dots, F_n$ is called the list of *witnesses*.

Proof ($\Leftarrow$) Obvious, since clearly $\mathcal{E} \vdash FPQ = FQP$ if $P = Q \in \mathcal{E}$.

($\Rightarrow$) By induction on the derivation of $M = N$ from the axioms. If $M = N$ is a $\boldsymbol{\beta\eta}$-axiom or the axiom of reflexivity, then we can take as witnesses the empty list. If $M = N$ is an axiom in $\mathcal{E}$, then we can take as list of witnesses just K. If $M = N$ follows from $M = L$ and $L = N$, then we can concatenate the lists that exist by the induction hypothesis. If $M = N$ is $PZ = QZ$ (respectively $ZP = ZQ$) and follows from $P = Q$ with list $F_1, \ldots, F_n$, then the list for $M = N$ is $F_1', \ldots, F_n'$ with $F_i' \equiv \lambda ab.F_i abZ$ (respectively $F_i' \equiv \lambda ab.Z(F_i ab)$). If $M = N$ follows from $N = M$, then we have to reverse the list. If $M = N$ is $\lambda x.P = \lambda x.Q$ and follows from $P = Q$ with list $F_1, \ldots, F_n$, then the new list is $F_1', \ldots, F_n'$ with $F_i' \equiv \lambda pqx.F_i pq$. Here we use that the equations in $\mathcal{E}$ are between closed terms. ∎

Recall that $\mathsf{true} \equiv \lambda xy.x, \mathsf{false} \equiv \lambda xy.y$ both having type $1_2 = 0{\to}0{\to}0$.

3.2.15 Lemma *Let $\mathcal{E}$ be a set of equations over $\mathcal{D}$. Then*

$$\mathcal{E} \text{ is consistent } \Leftrightarrow \mathcal{E} \nvdash \mathsf{true} = \mathsf{false}.$$

Proof ($\Leftarrow$) By definition.

($\Rightarrow$) Suppose $\mathcal{E} \vdash \lambda xy.x = \lambda xy.y$. Then $\mathcal{E} \vdash P = Q$ for arbitrary $P, Q \in \Lambda_\to(0)$. But then for arbitrary terms M, N of the same type $A = A_1 {\to} \cdots {\to} A_n {\to} 0$ one has $\mathcal{E} \vdash M\vec{z} = N\vec{z}$ for fresh $\vec{z} = z_1, \ldots, z_n$ of the right type, hence $\mathcal{E} \vdash M = N$, by Proposition 3.2.5. ∎

3.2.16 Definition Let $M, N \in \Lambda^{\varnothing}_\to[\mathcal{D}](A)$ be closed terms of type A.

(i) We say M is *inconsistent with* N, written $M \# N$, if

$$\{M = N\} \vdash \mathsf{true} = \mathsf{false}.$$

(ii) We say M is *separable from* N, written $M \perp N$, iff

$$FM = \mathsf{true} \ \& \ FN = \mathsf{false}$$

for some $F \in \Lambda^{\varnothing}_\to[\mathcal{D}](A {\to} 1_2)$.

The following result, stating that inconsistency implies separability, is not true for the untyped lambda calculus: the equation $\mathsf{K} = \mathsf{YK}$ is inconsistent, but K and YK are not separable, as follows from the Genericity Lemma, see B[1984] Proposition 14.3.24.

3.2.17 Proposition *Let $M, N \in \Lambda^{\varnothing}_\to(A)$ be closed pure terms of type A. Then*

$$M \# N \Leftrightarrow M \perp N.$$

Proof ($\Leftarrow$) Separability trivially implies inconsistency.

($\Rightarrow$) Suppose $\{M = N\} \vdash \mathsf{true} = \mathsf{false}$. Then also $\{M = N\} \vdash x = y$. Hence by Proposition 3.2.14 one has

$$\begin{aligned} x &=_{\beta\eta} F_1MN \\ F_1NM &=_{\beta\eta} F_2MN \\ &\vdots \\ F_nNM &=_{\beta\eta} y. \end{aligned}$$

Let n be minimal for which this is possible. We can assume that the F_i are all pure terms with $\mathrm{FV}(F_i) \subseteq \{x, y\}$ at most. The normal form of F_1NM must be either x or y. Hence by the minimality of n it must be y, otherwise there is a shorter list of witnesses. Now consider the normal form of F_1MM. It must be either x or y.

Case 1: $F_1MM =_{\beta\eta} x$. Then put $F \equiv \lambda axy.F_1aM$ so $FM =_{\beta\eta} \mathsf{true}$ and $FN =_{\beta\eta} \mathsf{false}$.

Case 2: $F_1MM =_{\beta\eta} y$. Then put $F \equiv \lambda axy.F_1Ma$ so $FM =_{\beta\eta} \mathsf{false}$ and $FN =_{\beta\eta} \mathsf{true}$. ∎

This proposition does not hold for $M, N \in \Lambda^{\emptyset}_{\to}[\mathcal{D}]$; see Exercise 3.2.

3.2.18 Corollary *Let $\mathcal{E}$ be a set of equations over $\mathcal{D} = \emptyset$. If $\mathcal{E}$ is inconsistent, then for some equation $M{=}N \in \mathcal{E}$ the terms M and N are separable.*

Proof By the same reasoning. ∎

In the untyped theory $\boldsymbol{\lambda}$ the set

$$\mathcal{H} = \{M = N \mid M, N \text{ are closed unsolvable}\}$$

is consistent and has a unique maximal consistent extension $\mathcal{H}^*$, see B[1984]. The following result is similar for $\boldsymbol{\lambda}_{\to}$, as there are no unsolvable terms.

3.2.19 Theorem *Let*

$$\mathcal{E}_{\max} = \{M{=}N \mid M, N \in \Lambda^{\emptyset}_{\to} \textit{ and } M, N \textit{ are not separable}\}.$$

Then this is the unique maximally consistent set of equations.

Proof By the corollary this set is consistent. By Proposition 3.2.17 it contains all consistent equations. Therefore the set is maximally consistent. Moreover it is the unique such set. ∎

It will be shown in Chapter 4 that $\mathcal{E}_{\max}$ is decidable.

3.3 Syntactic and semantic logical relations

In this section we work in $\boldsymbol{\lambda}_{\to}^{0,\mathrm{Ch}}$. We introduce the well-known method of *logical relations* in two ways: one on the terms and the other on the elements of a model. Applications of the method will be given and it will be shown how the two methods are related.

Syntactic logical relations

3.3.1 Definition Let n be a fixed natural number and let $\vec{\mathcal{D}} = \mathcal{D}_1, \dots, \mathcal{D}_n$ be sets of constants of various given types.

(i) We call R an (n-ary) *family of (syntactic) relations* (or sometimes just a *(syntactic) relation*) on $\Lambda_{\to}[\vec{\mathcal{D}}]$, if $R = \{R_A\}_{A \in \mathbb{T}}$ and for $A \in \mathbb{T}$

$$R_A \subseteq \Lambda_{\to}[\mathcal{D}_1](A) \times \cdots \times \Lambda_{\to}[\mathcal{D}_n](A).$$

If we want to make the sets of constants explicit, we say that R is a relation *on terms from* $\mathcal{D}_1, \dots, \mathcal{D}_n$.

(ii) Such an R is called a *(syntactic) logical relation* if $\forall A, B \in \mathbb{T}$,

$$\forall M_1 \in \Lambda_{\to}[\mathcal{D}_1](A{\to}B), \dots, M_n \in \Lambda_{\to}[\mathcal{D}_n](A{\to}B),$$

$$\begin{aligned} R_{A\to B}(M_1, \dots, M_n) \quad \Leftrightarrow \quad & \forall N_1 \in \Lambda_{\to}[\mathcal{D}_1](A) \cdots N_n \in \Lambda_{\to}[\mathcal{D}_n](A) \\ & [R_A(N_1, \dots, N_n) \;\Rightarrow\; R_B(M_1 N_1, \dots, M_n N_n)]. \end{aligned}$$

(iii) We say R is empty if $R_0 = \emptyset$.

Given $\vec{\mathcal{D}}$, a logical family $\{R_A\}$ is completely determined by R_0. For $A \neq 0$ the R_A do depend on the choice of the $\vec{\mathcal{D}}$.

3.3.2 Lemma *If R is a non-empty logical relation, then $\forall A \in \mathbb{T}^0 . R_A \neq \emptyset$.*

Proof (For R unary.) By induction on A.

Case $A = 0$. By assumption.

Case $A = B{\to}C$. Then $R_{B\to C}(M) \Leftrightarrow \forall P \in \Lambda_{\to}(B).[R_B(P) \Rightarrow R_C(MP)]$. By the induction hypothesis one has $R_C(N)$, for some N. It follows that $M \equiv \lambda p.N \in \Lambda_{\to}(B{\to}C)$ is in R_A. ■

Even the empty logical relation is interesting.

3.3.3 Proposition *Let R be the n-ary logical relation on $\Lambda_{\to}[\vec{\mathcal{D}}]$ determined by $R_0 = \emptyset$. Then*

$$\begin{aligned} R_A &= \Lambda_{\to}[\mathcal{D}_1](A) \times \cdots \times \Lambda_{\to}[\mathcal{D}_n](A), && \text{if } \Lambda_{\to}^{\emptyset}(A) \neq \emptyset; \\ &= \emptyset, && \text{if } \Lambda_{\to}^{\emptyset}(A) = \emptyset. \end{aligned}$$

Proof For notational simplicity we take $n = 1$. By induction on A. If $A = 0$, then we are done, as $R_0 = \emptyset$ and $\Lambda^{\emptyset}_{\rightarrow}(0) = \emptyset$. If $A = A_1 \rightarrow \cdots \rightarrow A_m \rightarrow 0$, then

$$\begin{aligned} R_A(M) &\Leftrightarrow \forall P_i \in R_{A_i}.R_0(M\vec{P}) \\ &\Leftrightarrow \forall P_i \in R_{A_i}.\bot, \end{aligned}$$

seeing R as both a relation and a set, and '$\bot$' stands for the false proposition. This last statement either is always the case, which is so if

$$\begin{aligned} \exists i.R_{A_i} = \emptyset \quad &\Leftrightarrow \quad \exists i.\Lambda^{\emptyset}_{\rightarrow}(A_i) = \emptyset, && \text{by the induction hypothesis,} \\ &\Leftrightarrow \quad \Lambda^{\emptyset}_{\rightarrow}(A) \neq \emptyset, && \text{by Proposition 2.4.4;} \end{aligned}$$

or else, if $\Lambda^{\emptyset}_{\rightarrow}(A) = \emptyset$, it is never the case, by the same reasoning. ∎

3.3.4 Example Let $n = 2$ and set $R_0(M, N) \Leftrightarrow M =_{\beta\eta} N$. Let R be the logical relation determined by R_0. Then it is easily seen that for all A and $M, N \in \Lambda_{\rightarrow}[\vec{\mathcal{D}}](A)$ one has $R_A(M, N) \Leftrightarrow M =_{\beta\eta} N$.

3.3.5 Definition

(i) Let M, N be lambda terms. Then M is a *weak head expansion* of N, written $M \rightarrow_{\text{wh}} N$, if $M \equiv (\lambda x.P)Q\vec{R}$ and $N \equiv P[x := Q]\vec{R}$.

(ii) A family R on $\Lambda_{\rightarrow}[\vec{\mathcal{D}}]$ is called *expansive* if R_0 is closed under coordinate-wise weak head expansion, i.e. if $M_i' \rightarrow_{\text{wh}} M_i$ for $1 \leq i \leq n$, then

$$R_0(M_1, \ldots, M_n) \Rightarrow R_0(M_1', \ldots, M_n').$$

3.3.6 Lemma *If R is logical and expansive, then each R_A is closed under coordinate-wise weak head expansion.*

Proof Immediate by induction on the type A and the fact that

$$M' \rightarrow_{\text{wh}} M \Rightarrow M'N \rightarrow_{\text{wh}} MN. \ ∎$$

3.3.7 Example This example prepares an alternative proof of the Church–Rosser property using logical relations.

(i) Let $M \in \Lambda_{\rightarrow}$. We say that $\beta\eta$ is *confluent* from M, written $\downarrow_{\beta\eta} M$, if whenever $N_1 \;{}_{\beta\eta}\!\twoheadleftarrow M \twoheadrightarrow_{\beta\eta} N_2$, then there exists a term L such that $N_1 \twoheadrightarrow_{\beta\eta} L \;{}_{\beta\eta}\!\twoheadleftarrow N_2$. Define R_0 on $\Lambda_{\rightarrow}(0)$ by

$$R_0(M) \Leftrightarrow \downarrow_{\beta\eta} M.$$

Then R_0 determines a logical R which is expansive by the permutability of head contractions with internal ones.

(ii) Let R be the logical relation on $\Lambda_\rightarrow$ generated from

$$R_0(M) \Leftrightarrow \downarrow_{\beta\eta} M.$$

Then for an arbitrary type $A \in \mathbb{T}$ one has

$$R_A(M) \Rightarrow \downarrow_{\beta\eta} M.$$

[Hint. Write $M \downarrow_{\beta\eta} N$ if $\exists Z\, [M \twoheadrightarrow_{\beta\eta} Z \;{}_{\beta\eta}\!\twoheadleftarrow N]$. First show that for an arbitrary variable x of some type B one has $R_B(x)$. Show also that if x is fresh, then by distinguishing cases whether x gets eaten or not

$$N_1 x \downarrow_{\beta\eta} N_2 x \Rightarrow N_1 \downarrow_{\beta\eta} N_2.$$

Then use induction on A.]

3.3.8 Definition

(i) Let $R_A \subseteq \Lambda_\rightarrow[\mathcal{D}_1](A) \times \cdots \times \Lambda_\rightarrow[\mathcal{D}_n](A)$ and $*_1, \ldots, *_n$

$$*_i : \mathtt{Var}(A) \rightarrow \Lambda_\rightarrow[\mathcal{D}_i](A)$$

be substitutors, each $*$ applicable to all variables of type A. Write $R(*_1, \ldots, *_n)$ if $R_A(x^{*_1}, \ldots, x^{*_n})$ for each variable x of type A.

(ii) Define $R^*_A \subseteq \Lambda_\rightarrow[\mathcal{D}_1](A) \times \cdots \times \Lambda_\rightarrow[\mathcal{D}_n](A)$ by

$$R^*_A(M_1, \ldots, M_n) \Leftrightarrow \forall *_1 \cdots *_n\, [R(*_1, \ldots, *_n) \Rightarrow R_A(M_1^{*_1}, \ldots, M_n^{*_n})].$$

(iii) We say R is *substitutive* if $R = R^*$, i.e., for all types A,

$$R_A(M_1, \ldots, M_n) \Leftrightarrow \forall *_1 \cdots *_n\, [R(*_1, \ldots, *_n) \Rightarrow R_A(M_1^{*_1}, \ldots, M_n^{*_n})].$$

3.3.9 Lemma *Let R be logical.*

(i) *Suppose that $R_0 \neq \emptyset$. Then for closed terms*

$$M_1 \in \Lambda^\emptyset_\rightarrow[\mathcal{D}_1], \ldots, M_n \in \Lambda^\emptyset_\rightarrow[\mathcal{D}_n]$$

we have

$$R_A(M_1, \ldots, M_n) \Leftrightarrow R^*_A(M_1, \ldots, M_n).$$

(ii) *For pure closed terms $M_1 \in \Lambda^\emptyset_\rightarrow, \ldots, M_n \in \Lambda^\emptyset_\rightarrow$*

$$R_A(M_1, \ldots, M_n) \Leftrightarrow R^*_A(M_1, \ldots, M_n).$$

(iii) *For a substitutive R one has for arbitrary open $M_1, \ldots, M_n, N_1, \ldots, N_n$*

$$R_A(M_1, \ldots, M_n) \;\&\; R_B(N_1, \ldots, N_n) \Rightarrow R_A(M_1[x^B := N_1], \ldots, M_n[x^B := N_n]).$$

Proof (i) Clearly $R_A(\vec{M})$ implies $R^*_A(\vec{M})$, as the $\vec{M}$ are closed. For the converse assume $R^*_A(\vec{M})$, that is $R_A(\overrightarrow{M^*})$, for all substitutors $\vec{*}$ satisfying $R(\vec{*})$. As $R_0 \neq \emptyset$, we have $R_B \neq \emptyset$, for all $B \in \mathbb{T}^0$, by Lemma 3.3.2. So we can take $\overrightarrow{*_i}$ such that $R_B(\overrightarrow{x^{*_i}})$, for all $x = x^B$. But then $R(*)$ and hence $R(\overrightarrow{M^*})$, which is $R(\vec{M})$.

(ii) If $\Lambda^{\emptyset}_{\to}(A) = \emptyset$, then this set does not contain closed pure terms and we are done. If $\Lambda^{\emptyset}_{\to}(A) \neq \emptyset$, then by Lemma 3.3.3 we have $R_A = (\Lambda^{\emptyset}_{\to}(A))^n$ and we are also done.

(iii) Since R is substitutive we have $R^*(\vec{M})$. Let $*_i = [x{:=}N_i]$. Then $R(*_1, \ldots, *_n)$ and hence $R(M_1[x{:=}N_1], \ldots, M_n[x{:=}N_n])$. ∎

Part (i) of this lemma does not hold for $R_0 = \emptyset$ and $\mathcal{D}_1 \neq \emptyset$. Take for example $\mathcal{D}_1 = \{\boldsymbol{c}^0\}$. Then vacuously $R^*_0(\boldsymbol{c}^0)$, but not $R_0(\boldsymbol{c}^0)$.

3.3.10 Exercise (CR for $\boldsymbol{\beta\eta}$ via logical relations.) Let R be the logical relation on $\Lambda_\to$ generated by $R_0(M)$ iff $\downarrow_{\boldsymbol{\beta\eta}} M$. Show by induction on M that $R^*(M)$ for all M. [Hint. Use the fact that R is expansive.] Conclude that for closed M one has $R(M)$ and hence $\downarrow_{\boldsymbol{\beta\eta}} M$. The same holds for arbitrary open terms N: let $\{\vec{x}\} = \mathrm{FV}(M)$, then

$$\begin{array}{lll} \lambda\vec{x}.N \text{ is closed} & \Rightarrow & R(\lambda\vec{x}.N) \\ & \Rightarrow & R((\lambda\vec{x}.N)\vec{x}), \qquad \text{since } R(x_i), \\ & \Rightarrow & R(N), \qquad \text{since } R \text{ is closed under } \twoheadrightarrow_{\boldsymbol{\beta}}, \\ & \Rightarrow & \downarrow_{\boldsymbol{\beta\eta}} N. \end{array}$$

Thus the Church–Rosser property holds for $\twoheadrightarrow_{\boldsymbol{\beta\eta}}$.

3.3.11 Proposition *Let R be an arbitrary n-ary family on $\Lambda_\to[\vec{\mathcal{D}}]$. Then:*

(i) *$R^*(x, \ldots, x)$ for all variables;*
(ii) *if R is logical, then so is R^*;*
(iii) *if R is expansive, then so is R^*;*
(iv) *$R^{**} = R^*$, so R^* is substitutive;*
(v) *if R is logical and expansive, then*

$$R^*(M_1, \ldots, M_n) \Rightarrow R^*(\lambda x.M_1, \ldots, \lambda x.M_n).$$

Proof For notational simplicity we assume $n = 1$.

(i) If $R(*)$, then by Definition $R(x^*)$. Therefore $R^*(x)$.

(ii) We have to prove

$$R^*(M) \Leftrightarrow \forall N \in \Lambda_\to[\vec{\mathcal{D}}][R^*(N) \Rightarrow R^*(MN)].$$

(⇒) Assume $R^*(M)$ & $R^*(N)$ in order to show $R^*(MN)$. Let $*$ be a substitutor such that $R(*)$. Then

$$\begin{aligned} R^*(M) \,\&\, R^*(N) &\Rightarrow R(M^*) \,\&\, R(N^*) \\ &\Rightarrow R(M^*N^*) \equiv R((MN)^*) \\ &\Rightarrow R^*(MN). \end{aligned}$$

(⇐) By the assumption and (i) we have

$$R^*(Mx), \tag{3.2}$$

where we choose x to be fresh. In order to prove $R^*(M)$ we have to show $R(M^*)$, whenever $R(*)$. Because R is logical it suffices to assume $R(N)$ and show $R(M^*N)$. Choose $*' = *[x:=N]$, then also $R(*')$. Hence by (3.2) and the freshness of x we have $R((Mx)^{*'}) \equiv R(M^*N)$ and we are done.

(iii) First observe that weak head reductions permute with substitution:

$$((\lambda x.P)Q\vec{R})^* \equiv (P[x:=Q]\vec{R})^*.$$

Now let $M \to_{\text{wh}} M^w$ be a weak head reduction step. Then

$$\begin{aligned} R^*(M^w) &\Rightarrow R(M^{w*}) \equiv R(M^{*w}) \\ &\Rightarrow R(M^*) \\ &\Rightarrow R^*(M). \end{aligned}$$

(iv) For substitutors $*_1, *_2$ write $*_1*_2$ for $*_2 \circ *_1$. This is convenient since

$$M^{*_1*_2} \equiv M^{*_2\circ *_1} \equiv (M^{*_1})^{*_2}.$$

Assume $R^{**}(M)$. Let $*_1(x) = x$ for all x. Then $R^*(*_1)$, by (i), and therefore we have $R^*(M^{*_1}) \equiv R^*(M)$. Conversely, assume $R^*(M)$, i.e.

$$\forall * \, [R(*) \;\Rightarrow\; R(M^*)], \tag{3.3}$$

in order to show $\forall *_1 \, [R^*(*_1) \;\Rightarrow\; R^*(M^{*_1})]$. Now

$$\begin{aligned} R^*(*_1) &\Leftrightarrow \forall *_2 \, [R(*_2) \;\Rightarrow\; R(*_1*_2)], \\ R^*(M^{*_1}) &\Leftrightarrow \forall *_2 \, [R(*_2) \;\Rightarrow\; R(M^{*_1*_2})]. \end{aligned}$$

Therefore by (3.3) applied to $*_1*_2$ we are done.

(iv) Let R be logical and expansive. Assume $R^*(M)$. Then

$$\begin{aligned} R^*(N) &\Rightarrow R^*(M[x:=N]), && \text{since } R^* \text{ is substitutive,} \\ &\Rightarrow R^*((\lambda x.M)N), && \text{since } R^* \text{ is expansive.} \end{aligned}$$

Therefore $R^*(\lambda x.M)$ since R^* is logical. ∎

3.3.12 Theorem (Fundamental theorem for syntactic logical relations) *Let R be logical, expansive and substitutive. Then for all $A \in \mathbb{T}$ and all pure terms $M \in \Lambda_{\rightarrow}(A)$ one has*

$$R_A(M, \ldots, M).$$

Proof By induction on M we show that $R_A(M, \ldots, M)$.

Case $M \equiv x$. Then the statement follows from the assumption $R = R^*$ (substitutivity) and Proposition 3.3.11(i).

Case $M \equiv PQ$. By the hypothesis and the assumption that R is logical.

Case $M \equiv \lambda x.P$. By the hypothesis and Proposition 3.3.11(v). ∎

3.3.13 Corollary *Let R be an n-ary expansive logical relation. Then for all closed $M \in \Lambda^{\emptyset}_{\rightarrow}$ one has $R(M, \ldots, M)$.*

Proof By Proposition 3.3.11(ii), (iii), (iv) it follows that R^* is expansive, substitutive, and logical. Applying the theorem to R^* yields $R^*(M, \ldots, M)$. Then we have $R(\vec{M})$, by Lemma 3.3.9(ii). ∎

The proof in Exercise 3.3.10 was in fact an application of this corollary. In the following example we present the proof of weak normalization in Prawitz (1965).

3.3.14 Example Let R be the logical relation determined by

$$R_0(M) \Leftrightarrow M \text{ is normalizable.}$$

Then R is expansive. Note that if $R_A(M)$, then M is normalizable. [Hint. Use $R_B(x)$ for arbitrary B and x and the fact that if $M\vec{x}$ is normalizable, then so is M.] It follows from Corollary 3.3.13 that each closed term is normalizable. Hence all terms are normalizable by taking closures. For strong normalization a similar proof breaks down. The corresponding R is not expansive.

3.3.15 Example Now we 'relativize' the theory of logical relations to closed terms. A family of relations $S_A \subseteq \Lambda^{\emptyset}_{\rightarrow}[\mathcal{D}_1](A) \times \cdots \times \Lambda^{\emptyset}_{\rightarrow}[\mathcal{D}_n](A)$ which satisfies

$$\begin{aligned} S_{A\rightarrow B}(M_1, \ldots, M_n) \Leftrightarrow\ & \forall N_1 \in \Lambda^{\emptyset}_{\rightarrow}[\mathcal{D}_1](A) \cdots N_n \in \Lambda^{\emptyset}_{\rightarrow}[\mathcal{D}_n](A) \\ & [S_A(N_1, \ldots, N_n) \Rightarrow S_B(M_1N_1, \ldots, M_nN_n)] \end{aligned}$$

can be lifted to a substitutive logical relation S^* on $\Lambda_{\rightarrow}[\mathcal{D}_1] \times \cdots \times \Lambda_{\rightarrow}[\mathcal{D}_n]$ as follows. Define for substitutors $*_i : \mathtt{Var}(A) \rightarrow \Lambda^{\emptyset}_{\rightarrow}[\mathcal{D}_i](A)$

$$S_A(*_1, \ldots, *_n) \Leftrightarrow \forall x^A\ S_A(x^{*_1}, \ldots, x^{*_n}).$$

Now define S^* as follows: for $M_i \in \Lambda_{\to}[\mathcal{D}_i](A)$

$$S^*_A(M_1, \ldots, M_n) \Leftrightarrow \forall *_1 \cdots *_n\, [S_A(*_1, \ldots, *_n) \Rightarrow S_A(M_1^{*_1}, \ldots, M_n^{*_n})].$$

Show that if S is closed under coordinate-wise weak head expansions, then S^* is expansive.

The following definition is needed in order to relate the notions of logical relation and semantic logical relation, to be defined in 3.3.21.

3.3.16 Definition Let R be an $(n+1)$-ary family. The *projection* of R, written $\exists R$, is the n-ary family defined by

$$\exists R(M_1, \ldots, M_n) \Leftrightarrow \exists M_{n+1} \in \Lambda_{\to}[\mathcal{D}_{n+1}]\ R(M_1, \ldots, M_{n+1}).$$

3.3.17 Proposition

(i) *The universal n-ary relation R^U is defined by*

$$R^U_A = \Lambda_{\to}[\mathcal{D}_1](A) \times \cdots \times \Lambda_{\to}[\mathcal{D}_n](A).$$

This relation is logical, expansive and substitutive.

(ii) *Let $R = \{R_A\}_{A \in \mathbb{T}^0}, S = \{S_A\}_{A \in \mathbb{T}^0}$ with $R_A \subseteq \Lambda_{\to}[\mathcal{D}_1](A) \times \cdots \times \Lambda_{\to}[\mathcal{D}_m](A)$ and $S_A \subseteq \Lambda_{\to}[\mathcal{E}_1](A) \times \cdots \times \Lambda_{\to}[\mathcal{E}_n](A)$ be non-empty logical relations. Define*

$$(R \times S)_A \subseteq \Lambda_{\to}[\mathcal{D}_1](A) \times \cdots \times \Lambda_{\to}[\mathcal{D}_m](A) \times \Lambda_{\to}[\mathcal{E}_1](A) \times \cdots \times \Lambda_{\to}[\mathcal{E}_n](A)$$

by

$$(R \times S)_A(M_1, \ldots, M_m, N_1, \ldots, N_n) \Leftrightarrow R_A(M_1, \ldots, M_m)\ \&\ S_A(N_1, \ldots, N_n).$$

Then $R \times S$ is a non-empty logical relation. If moreover R and S are both substitutive, then so is $R \times S$.

(iii) *If R is an n-ary family and π is a permutation of $\{1, \ldots, n\}$, then R^π defined by*

$$R^\pi(M_1, \ldots, M_n) \Leftrightarrow R(M_{\pi(1)}, \ldots, M_{\pi(n)})$$

is logical if R is logical, is expansive if R is expansive, and is substitutive if R is substitutive.

(iv) *Let R be an n-ary substitutive logical relation on terms from $\mathcal{D}_1, \ldots, \mathcal{D}_n$ and let $\mathcal{D} \subseteq \bigcap_i \mathcal{D}_i$. Then the diagonal of R, written R^Δ, defined by*

$$R^\Delta(M) \Leftrightarrow R(M, \ldots, M),$$

is a substitutive logical (unary) relation on terms from $\mathcal{D}$, which is expansive if R is expansive.

(v) *If $\mathcal{R}$ is a class of n-ary substitutive logical relations, then $\bigcap\mathcal{R}$ is an n-ary substitutive logical relation, which is expansive if each member of $\mathcal{R}$ is expansive.*

(vi) *If R is an n-ary substitutive, expansive and logical relation, then $\exists R$ is a substitutive, expansive and logical relation.*

Proof (i) Trivial.

(ii) Suppose that R, S are logical. We show for $n = m = 1$ that $R \times S$ is logical;

$$\begin{aligned}(R \times S)_{A \to B}(M, N) &\Leftrightarrow R_{A \to B}(M) \ \&\ S_{A \to B}(N)\\ &\Leftrightarrow [\forall P.R_A(P) \Rightarrow R_B(MP)] \ \&\\ &\qquad [\forall Q.R_A(Q) \Rightarrow R_B(NQ)]\\ &\Leftrightarrow \forall (P,Q).(R \times S)_A(P,Q) \ \Rightarrow \ (R \times S)_B(MP, NQ).\end{aligned}$$

For the last $(\Leftarrow)$ one needs that the R, S are non-empty, and Lemma 3.3.2. If both R, S are substitutive, then trivially so is $R \times S$.

(iii) Trivial.

(iv) We show for $n = 2$ that R^Δ is logical. We have

$$\begin{aligned}R^\Delta(M) &\Leftrightarrow R(M, M)\\ &\Leftrightarrow \forall N_1, N_2.R(N_1, N_2) \ \Rightarrow \ R(MN_1, MN_2)\\ &\Leftrightarrow \forall N.R(N, N) \ \Rightarrow \ R(MN, MN),\end{aligned} \tag{3.4}$$

where the validity of the last equivalence is argued as follows. Necessity $(\Rightarrow)$ is trivial. For $(\Leftarrow)$, suppose (3.4) and $R(N_1, N_2)$, in order to show $R(MN_1, MN_2)$. By Proposition 3.3.11(i) one has $R(x, x)$, for fresh x. Hence $R(Mx, Mx)$ by (3.4). Therefore $R^*(Mx, Mx)$, as R is substitutive. Now taking $*_i = [x := N_i]$, one obtains $R(MN_1, MN_2)$.

(v) Trivial.

(vi) As in (iv) it suffices to show that

$$\forall P.[\exists R(P) \ \Rightarrow \ \exists R(MP)] \tag{3.5}$$

implies $\exists N \forall P, Q.[R(P, Q) \ \Rightarrow \ R(MP, NQ)]$. Again we have $R(x, x)$. Therefore by (3.5)

$$\exists N_1.R(Mx, N_1).$$

Choosing $N \equiv \lambda x.N_1$, we get $R^*(Mx, Nx)$, because R is substitutive. Then $R(P, Q)$ implies $R(MP, NQ)$, as in (iv). ∎

The following property R states that an M essentially does not contain the constants from $\mathcal{D}$. Remember that a term $M \in \Lambda_{\to}[\mathcal{D}]$ is called *pure* iff $M \in \Lambda_{\to}$. The property $R(M)$ states that M is convertible to a pure term.

3.3.18 Proposition *Define for $M \in \Lambda_{\to}[\mathcal{D}](A)$*

$$R_A^{\beta\eta}(M) \Leftrightarrow \exists N \in \Lambda_{\to}(A)\, M =_{\beta\eta} N.$$

Then:

(i) *$R^{\beta\eta}$ is logical;*
(ii) *$R^{\beta\eta}$ is expansive;*
(iii) *$R^{\beta\eta}$ is substitutive.*

Proof (i) If $R^{\beta\eta}(M)$ and $R^{\beta\eta}(N)$, then clearly $R^{\beta\eta}(MN)$. Conversely, suppose $\forall N\, [R^{\beta\eta}(N) \Rightarrow R^{\beta\eta}(MN)]$. Since obviously $R^{\beta\eta}(x)$ it follows that $R^{\beta\eta}(Mx)$ for fresh x. Hence there exists a pure $L =_{\beta\eta} Mx$. But then $\lambda x.L =_{\beta\eta} M$, hence $R^{\beta\eta}(M)$.

(ii) Trivial as $P \to_{\text{wh}} Q \Rightarrow P =_{\beta\eta} Q$.

(iii) We must show $R^{\beta\eta} = R^{\beta\eta *}$. Suppose $R^{\beta\eta}(M)$ and $R^{\beta\eta}(*)$. Then $M = N$, with N pure and hence $M^* = N^*$ is pure, so $R^{\beta\eta *}(M)$. Conversely, suppose $R^{\beta\eta *}(M)$. Then for $*$ with $x^* = x$ one has $R^{\beta\eta}(*)$. Hence $R^{\beta\eta}(M^*)$. But this is $R^{\beta\eta}(M)$. ■

3.3.19 Proposition *Let S be an n-ary logical, expansive and substitutive relation on terms from $\mathcal{D}_1, \ldots, \mathcal{D}_n$. Define the restriction of S to pure terms $S \restriction \Lambda$, again a relation on terms from $\mathcal{D}_1, \ldots, \mathcal{D}_n$, by*

$$(S{\restriction}\Lambda)_A(M_1, \ldots, M_n) \Leftrightarrow R(M_1)\ \&\ \cdots\ \&\ R(M_n)\ \&\ S_A(M_1, \ldots, M_n),$$

where R is as in Proposition 3.3.18. *Then $S{\restriction}\Lambda$ is logical, expansive and substitutive.*

Proof Intersection of relations preserves the notion logical, expansive and substitutive. ■

3.3.20 Proposition *Given a set of equations $\mathcal{E}$ between closed terms of the same type, define $R_{\mathcal{E}}$ by*

$$R_{\mathcal{E}}(M, N) \Leftrightarrow \mathcal{E} \vdash M = N.$$

Then:

(i) *$R_{\mathcal{E}}$ is logical;*
(ii) *$R_{\mathcal{E}}$ is expansive;*
(iii) *$R_{\mathcal{E}}$ is substitutive;*
(iv) *$R_{\mathcal{E}}$ is a congruence relation.*

Proof (i) We must show

$$\mathcal{E} \vdash M_1 = M_2 \Leftrightarrow \forall N_1, N_2[\mathcal{E} \vdash N_1 = N_2 \Rightarrow \mathcal{E} \vdash M_1N_1 = M_2N_2].$$

($\Rightarrow$) Let $\mathcal{E} \vdash M_1 = M_2$ and $\mathcal{E} \vdash N_1 = N_2$. Then $\mathcal{E} \vdash M_1N_1 = M_2N_2$ follows by (R-congruence), (L-congruence) and (transitivity).

($\Leftarrow$) For all x one has $\mathcal{E} \vdash x = x$, so $\mathcal{E} \vdash M_1x = M_2x$. Choose a fresh variable x. Then $M_1 = M_2$ follows by (ξ-rule), (η) and (transitivity).

(ii) Obvious, since provability from $\mathcal{E}$ is closed under $\boldsymbol{\beta}$-conversion, hence *a fortiori* under weak head expansion.

(iii) Assume that $R_{\mathcal{E}}(M, N)$ in order to show $R_{\mathcal{E}}{}^*(M, N)$. So suppose $R_{\mathcal{E}}(x^{*_1}, x^{*_2})$. We must show $R_{\mathcal{E}}(M^{*_1}, N^{*_2})$. Now going back to the definition of $R_{\mathcal{E}}$ this means that we have $\mathcal{E} \vdash M = N$ and $\mathcal{E} \vdash x^{*_1} = x^{*_2}$ and we must show $\mathcal{E} \vdash M^{*_1} = N^{*_2}$. Now if $\mathrm{FV}(MN) \subseteq \{\vec{x}\}$, then

$$\begin{aligned} M^{*_1} &=_{\boldsymbol{\beta}} (\lambda\vec{x}.M)\vec{x}^{*_1} \\ &=_{\mathcal{E}} (\lambda\vec{x}.N)\vec{x}^{*_2} \\ &=_{\boldsymbol{\beta}} N^{*_2}. \end{aligned}$$

(iv) Obvious. ∎

Semantic logical relations

3.3.21 Definition Let $\mathcal{M}_1, \ldots, \mathcal{M}_n$ be typed applicative structures.

(i) We say S is an n-ary *family of (semantic) relations* or just a *(semantic) relation* on $\mathcal{M}_1 \times \cdots \times \mathcal{M}_n$ iff $S = \{S_A\}_{A \in \mathbb{T}}$ and for all A

$$S_A \subseteq \mathcal{M}_1(A) \times \cdots \times \mathcal{M}_n(A).$$

(ii) We say S is a *(semantic) logical relation* iff

$$\begin{aligned} S_{A\to B}(d_1, \ldots, d_n) \quad \Leftrightarrow \quad & \forall e_1 \in \mathcal{M}_1(A), \ldots, e_n \in \mathcal{M}_n(A) \\ & [S_A(e_1, \ldots, e_n) \Rightarrow S_B(d_1e_1, \ldots, d_ne_n)]. \end{aligned}$$

for all A, B and all $d_1 \in \mathcal{M}_1(A\to B), \ldots, d_n \in \mathcal{M}_n(A\to B)$.

(iii) The relation S is called *non-empty* if S_0 is non-empty.

Note that S is an n-ary relation on $\mathcal{M}_1 \times \cdots \times \mathcal{M}_n$ iff S is a unary relation on the single structure $\mathcal{M}_1 \times \cdots \times \mathcal{M}_n$.

3.3.22 Example Define S on $\mathcal{M} \times \mathcal{M}$ by $S(d_1, d_2) \Leftrightarrow d_1 = d_2$. Then S is logical.

3.3.23 Example Let $\mathcal{M}$ be a model and let $\pi = \pi_0$ be a permutation of

$\mathcal{M}(0)$ which happens to be an element of $\mathcal{M}(0\to 0)$. Then π can be lifted to higher types by defining

$$\pi_{A\to B}(d) = \lambda e \in \mathcal{M}(A).\pi_B(d(\pi_A^{-1}(e))).$$

Now define S_π (the graph of π)

$$S_\pi(d_1, d_2) \;\Leftrightarrow\; \pi(d_1) = d_2.$$

Then S_π is logical.

3.3.24 Example (Friedman (1975)) Let $\mathcal{M}, \mathcal{N}$ be typed structures. A *partial surjective homomorphism* is a family $h = \{h_A\}_{A\in\mathbb{T}}$ of partial maps

$$h_A : \mathcal{M}(A)\to\mathcal{N}(A)$$

such that

$$h_{A\to B}(d) = e \;\Leftrightarrow\; e\in\mathcal{N}(A\to B) \text{ is the unique element (if it exists)}$$
$$\text{such that } \forall f \in \mathtt{dom}(h_A)\,[e(h_A(f)) = h_B(d\,f)].$$

This implies that, if all elements involved exist, then

$$h_{A\to B}(d)h_A(f) = h_B(d\,f).$$

Note that $h(d)$ can fail to be defined if one of the following conditions holds:

(1) for some $f\in\mathtt{dom}(h_A)$ one has $df \notin \mathtt{dom}(h_B)$;
(2) the correspondence $h_A(f) \mapsto h_B(df)$ fails to be single valued;
(3) the map $h_A(f) \mapsto h_B(df)$ fails to be in $\mathcal{N}_{A\to B}$.

Of course (3) is the basic reason for partialness, whereas (1) and (2) are derived reasons. A partial surjective homomorphism h is completely determined by its h_0. If we take $\mathcal{M} = \mathcal{M}_X$ and h_0 is any surjection $X\to\mathcal{N}_0$, then h_A is, although partial, indeed surjective for all A. Define $S_A(d,e) \Leftrightarrow h_A(d) = e$, the graph of h_A. Then S is logical. Conversely, if S_0 is the graph of a surjective partial map $h_0 : \mathcal{M}(0)\to\mathcal{N}(0)$, and the logical relation S on $\mathcal{M}\times\mathcal{N}$ induced by this S_0 satisfies

$$\forall e\in\mathcal{N}(A)\exists d\in\mathcal{M}(A)\; S_A(d,e),$$

then S is the graph of a partial surjective homomorphism from $\mathcal{M}$ to $\mathcal{N}$.

Kreisel's Hereditarily Recursive Operations are one of the first appearances of logical relations; see Bezem (1985) for a detailed account of extensionality in this context

3.3.25 Proposition *Let $R \subseteq \mathcal{M}_1 \times \cdots \times \mathcal{M}_n$ be the n-ary semantic logical relation determined by $R_0 = \emptyset$. Then*

$$\begin{array}{rcll} R_A & = & \mathcal{M}_1(A) \times \cdots \times \mathcal{M}_n(A), & \text{if } \Lambda^{\emptyset}_{\to}(A) \neq \emptyset; \\ & = & \emptyset, & \text{if } \Lambda^{\emptyset}_{\to}(A) = \emptyset. \end{array}$$

Proof Analogous to the proof of Proposition 3.3.3 for semantic logical relations, using the fact that for all $\mathcal{M}_i$ and all types A one has $\mathcal{M}_i(A) \neq \emptyset$, by Definition 3.1.1. ∎

3.3.26 Theorem (Fundamental theorem for semantic logical relations) *Let $\mathcal{M}_1, \ldots, \mathcal{M}_n$ be typed λ-models and let S be logical on $\mathcal{M}_1 \times \cdots \times \mathcal{M}_n$. Then for each term $M \in \Lambda^{\emptyset}_{\to}$ one has*

$$S([\![M]\!]^{\mathcal{M}_1}, \ldots, [\![M]\!]^{\mathcal{M}_n}).$$

Proof We treat the case $n = 1$. Let $S \subseteq \mathcal{M}$ be logical. We claim that for all $M \in \Lambda_{\to}$ and all partial valuations ρ such that $\mathrm{FV}(M) \subseteq \mathrm{dom}(\rho)$ one has

$$S(\rho) \;\Rightarrow\; S([\![M]\!]_\rho).$$

This follows by an easy induction on M. In the case $M \equiv \lambda x.N$ we need to show $S([\![\lambda x.N]\!]_\rho)$, assuming $S(\rho)$. This means that for all d of the correct type with $S(d)$ one has $S([\![\lambda x.N]\!]_\rho d)$. This is the same as $S([\![N]\!]_{\rho[x:=d]})$, which holds by the induction hypothesis.

The statement now follows immediately from the claim, by taking as ρ the empty function. ∎

We give two applications.

3.3.27 Example Let S be the graph of a partial surjective homomorphism $h : \mathcal{M} \to \mathcal{N}$. The fundamental theorem just shown implies that for closed pure terms one has $h(M) = M$, which is Lemma 15 of Friedman (1975). From this it is derived in that paper that for infinite X one has

$$\mathcal{M}_X \models M = N \;\Leftrightarrow\; M =_{\beta\eta} N.$$

We have derived this in another way.

3.3.28 Example Let $\mathcal{M}$ be a typed applicative structure. Let $\Delta \subseteq \mathcal{M}$. Write $\Delta(A) = \Delta \cap \mathcal{M}(A)$. Assume that $\Delta(A) \neq \emptyset$ for all $A \in \mathbb{T}$ and

$$d \in \Delta(A \to B), e \in \Delta(A) \;\Rightarrow\; de \in \Delta(B).$$

Then Δ may fail to be a typed applicative structure because it is not extensional. Equality as a binary relation E_0 on $\Delta(0) \times \Delta(0)$ induces a binary logical relation E on $\Delta \times \Delta$. Let $\Delta^E = \{d \in \Delta \mid E(d,d)\}$. Then the restriction of E to Δ^E is an applicative congruence and the equivalence classes form a typed applicative structure. In particular, if $\mathcal{M}$ is a typed λ-model, then write

$$\begin{aligned}\Delta^+ &= \{[\![M]\!]\,\vec{d} \mid M \in \Lambda^{\emptyset}_{\to}, \vec{d} \in \Delta\}\\ &= \{d \in \mathcal{M} \mid \exists M \in \Lambda^{\emptyset}_{\to} \exists d_1 \cdots d_n \in \Delta\ [\![M]\!]\, d_1 \cdots d_n = d\}\end{aligned}$$

for the applicative closure of Δ. The *Gandy hull* of Δ in $\mathcal{M}$ is the set Δ^{+E}. From the fundamental theorem for semantic logical relations it can be derived that

$$\mathcal{G}_{\Delta}(\mathcal{M}) = \Delta^{+E}/E$$

is a typed λ-model. This model will be also called the Gandy hull of Δ in $\mathcal{M}$. Do Exercise 3.34 to get acquainted with the notion of the Gandy hull.

3.3.29 Definition Let $\mathcal{M}_1, \ldots, \mathcal{M}_n$ be typed structures.

(i) Let S be an n-ary relation on $\mathcal{M}_1 \times \cdots \times \mathcal{M}_n$. For valuations $\rho_1, \ldots, \rho_n$ with $\rho_i : \mathtt{Var} \to \mathcal{M}_i$ we define

$$S(\rho_1, \ldots, \rho_n) \iff S(\rho_1(x), \ldots, \rho_n(x)),$$

for all variables x satisfying $\forall i.\rho_i(x)\!\downarrow$.

(ii) Let S be an n-ary relation on $\mathcal{M}_1 \times \cdots \times \mathcal{M}_n$. The *lifting of S* to $\mathcal{M}_1^* \times \cdots \times \mathcal{M}_n^*$, written S^*, is defined for $d_1 \in \mathcal{M}_1^*, \ldots, d_n \in \mathcal{M}_n^*$ as

$$\begin{aligned}S^*(d_1, \ldots, d_n) \iff{}& \forall \rho_1 : \mathsf{V} \to \mathcal{M}_1, \ldots, \rho_n : \mathsf{V} \to \mathcal{M}_n\\ & [S(\rho_1, \ldots, \rho_n) \Rightarrow S((\!(d_1)\!)^{\mathcal{M}_1}_{\rho_1}, \ldots, (\!(d_n)\!)^{\mathcal{M}_n}_{\rho_n})].\end{aligned}$$

The interpretation $(\!(-)\!)_{\rho}:\mathcal{M}^* \to \mathcal{M}$ was defined in Definition 3.1.22(ii).

(iii) For $\rho:\mathsf{V} \to \mathcal{M}^*$ define the 'substitution' $(-)^{\rho}:\mathcal{M}^* \to \mathcal{M}^*$ as

$$\begin{aligned}x^{\rho} &= \rho(x);\\ \underline{m}^{\rho} &= \underline{m};\\ (d_1 d_2)^{\rho} &= d_1^{\rho} d_2^{\rho}.\end{aligned}$$

(iv) Let now S be an n-ary relation on $\mathcal{M}_1^* \times \cdots \times \mathcal{M}_n^*$. Then S is called *substitutive* if for all $d_1 \in \mathcal{M}_1^*, \ldots, d_n \in \mathcal{M}_n^*$ one has

$$\begin{aligned}S(d_1, \ldots, d_n) \iff{}& \forall \rho_1 : \mathsf{V} \to \mathcal{M}_1^*, \ldots \rho_n : \mathsf{V} \to \mathcal{M}_n^*\\ & [S(\rho_1, \ldots, \rho_n) \Rightarrow S(d_1^{\rho_1}, \ldots, d_n^{\rho_n})].\end{aligned}$$

3.3.30 Remark If $S \subseteq \mathcal{M}_1^* \times \cdots \times \mathcal{M}_n^*$ is substitutive, then for every variable x one has $S(x, \ldots, x)$.

3.3.31 Example

(i) Let S be the equality relation on $\mathcal{M} \times \mathcal{M}$. Then S^* is the equality relation on $\mathcal{M}^* \times \mathcal{M}^*$.

(ii) If S is the graph of a surjective homomorphism, then S^* is the graph of a partial surjective homomorphism whose restriction (in the literal sense, not the analogue of 3.3.19) to $\mathcal{M}$ is S and which fixes each indeterminate x.

3.3.32 Lemma *Let $S \subseteq \mathcal{M}_1 \times \cdots \times \mathcal{M}_n$ be a semantic logical relation.*

(i) *Let $\vec{d} \in \mathcal{M}_1 \times \cdots \times \mathcal{M}_n$. Then $S(\vec{d}) \Rightarrow S^*(\underline{\vec{d}})$.*

(ii) *Suppose that S is non-empty and that the $\mathcal{M}_i$ are λ-models. Then, for $\vec{d} \in \mathcal{M}_1 \times \cdots \times \mathcal{M}_n$, one has $S^*(\underline{\vec{d}}) \Rightarrow S(\vec{d})$.*

Proof For notational simplicity, take $n = 1$. (i) Suppose that $S(d)$. Then $S^*(\underline{d})$, as $(\!(\underline{d})\!)_\rho = d$, hence $S((\!(\underline{d})\!)_\rho)$, for all ρ.

(ii) Suppose $S^*(\underline{d})$. Then for all $\rho : \mathsf{V} \to \mathcal{M}$ one has

$$\begin{aligned} S(\rho) &\Rightarrow S((\!(\underline{d})\!)_\rho^{\mathcal{M}^*}) \\ &\Rightarrow S(d). \end{aligned}$$

Since S_0 is non-empty, say $d \in S_0$, then S_A is also non-empty for all $A \in \mathbb{T}^0$, as the constant function $\lambda\vec{x}.d$ belongs to it. Hence there exists a ρ such that $S(\rho)$ and therefore $S(d)$. ∎

3.3.33 Proposition *Let $S \subseteq \mathcal{M}_1 \times \cdots \times \mathcal{M}_n$ be a semantic logical relation. Then $S^* \subseteq \mathcal{M}_1^* \times \cdots \times \mathcal{M}_n^*$ and one has:*

(i) *$S^*(x, \ldots, x)$ for all variables;*

(ii) *S^* is a semantic logical relation;*

(iii) *S^* is substitutive;*

(iv) *If S is substitutive and each $\mathcal{M}_i$ is a typed λ-model, then*

$$S^*(d_1, \ldots, d_n) \Leftrightarrow S(\lambda\vec{x}.d_1, \ldots, \lambda\vec{x}.d_n),$$

where the variables on which the $\vec{d}$ depend are included in the list $\vec{x}$.

Proof Take n=1 for notational simplicity. (i) If $S(\rho)$, then by definition one has $S((\!(x)\!)_\rho)$ for all variables x. Therefore $S^*(x)$.

(ii) We have to show

$$S^*_{A\to B}(d) \quad \Leftrightarrow \quad \forall e \in \mathcal{M}^*(A).[S^*_A(e) \Rightarrow S^*_B(de)].$$

($\Rightarrow$) Suppose $S^*_{A\to B}(d)$, $S^*_A(e)$, in order to show $S^*_B(de)$. So assume $S(\rho)$

towards $S(((de))_\rho)$. By the assumption we have $S(((d))_\rho)$, $S(((e))_\rho)$, hence indeed $S(((de))_\rho)$, as S is logical.

($\Leftarrow$) Assume the right hand side in order to show $S^*(d)$. To this end suppose $S(\rho)$ towards $S(((d))_\rho)$. Since S is logical it suffices to show $S(e) \Rightarrow S(((d))_\rho e)$ for all $e \in \mathcal{M}$. Taking $e \in \mathcal{M}$, we have

$$\begin{array}{rcll} S(e) & \Rightarrow & S^*(\underline{e}), & \text{by Lemma 3.3.32(i)}, \\ & \Rightarrow & S^*(d\underline{e}), & \text{by the right hand side}, \\ & \Rightarrow & S(((d))_\rho e), & \text{as } e = ((\underline{e}))_\rho \text{ and } S(\rho). \end{array}$$

(iii) For $d \in \mathcal{M}^*$ we show that $S^*(d) \;\Leftrightarrow\; \forall \rho{:}\mathsf{V}{\to}\mathcal{M}^*[S^*(\rho) \;\Rightarrow\; S^*(d^\rho)]$, i.e.

$$\forall \rho{:}\mathsf{V}{\to}\mathcal{M}.[S(\rho) \;\Rightarrow\; S(((d))_\rho^{\mathcal{M}})] \;\Leftrightarrow\; \forall \rho'{:}\mathsf{V}{\to}\mathcal{M}^*.[S^*(\rho') \;\Rightarrow\; S^*(d^{\rho'})].$$

As to ($\Rightarrow$), let $d \in \mathcal{M}^*$ and suppose

$$\forall \rho{:}\mathsf{V}{\to}\mathcal{M}.[S(\rho) \;\Rightarrow\; S(((d))_\rho^{\mathcal{M}})], \tag{3.6}$$

and

$$S^*(\rho'), \text{ for a given } \rho'{:}\mathsf{V}{\to}\mathcal{M}^*, \tag{3.7}$$

in order to show $S^*(d^{\rho'})$. To this end we assume

$$S(\rho'') \text{ with } \rho''{:}\mathsf{V}{\to}\mathcal{M} \tag{3.8}$$

in order to show

$$S(((d^{\rho'}))_{\rho''}^{\mathcal{M}}). \tag{3.9}$$

Now define

$$\rho'''(x) = ((\rho'(x)))_{\rho''}^{\mathcal{M}}.$$

Therefore $\rho'''{:}\mathsf{V}{\to}\mathcal{M}$ and hence by (3.6), (3.7), (3.8) one has $S(\rho'''(x))$ (being $S(((\rho'(x)))_{\rho''}^{\mathcal{M}^*}))$, so

$$S(((d))_{\rho'''}). \tag{3.10}$$

By induction on the structure of $d \in \mathcal{M}^*$ (considered as $\underline{\mathcal{M}}$ modulo $\sim_{\mathcal{M}}$) it follows that

$$((d))_{\rho'''}^{\mathcal{M}} = ((d^{\rho'}))_{\rho''}^{\mathcal{M}}.$$

Therefore (3.10) yields (3.9).

As to ($\Leftarrow$), assume the right hand side. Taking $\rho'(x) = x \in \mathcal{M}^*$ one has $S^*(\rho')$ by (i), hence $S^*(d_{\rho'}^{\mathcal{M}^*})$. Now one easily shows by induction on $d \in \underline{\mathcal{M}}$ that $d_{\rho'}^{\mathcal{M}^*} = d$, so one has $S^*(d)$.

(iv) Without loss of generality, assume that d depends only on y and that $\vec{x} = y$. As $\mathcal{M}$ is a typed λ-model, there is a unique $F \in \mathcal{M}$ such that for all $y \in \mathcal{M}$ one has $Fy = d$. This F is denoted as $\lambda y.d$:

$$\begin{aligned} S(d) &\Leftrightarrow S(Fy) \\ &\Leftrightarrow \forall \rho{:}\mathsf{V}{\to}\mathcal{M}^*[S(\rho) \Rightarrow S((\!(i(Fy))\!)_\rho)], \text{ as } S \text{ is substitutive,} \\ &\Leftrightarrow \forall \rho{:}\mathsf{V}{\to}\mathcal{M}^*[S(\rho) \Rightarrow S((\!(i(F))\!)_\rho (\!(i(y))\!)_\rho)], \\ &\Leftrightarrow \forall e \in \mathcal{M}^*.[S(e) \Rightarrow S(F\,e)], \text{ taking } \rho(x) = e, \\ &\Leftrightarrow S(F), \text{ as } S \text{ is logical,} \\ &\Leftrightarrow S(\lambda y.d). \blacksquare \end{aligned}$$

3.3.34 Proposition *Let $S \subseteq \mathcal{M}_1 \times \cdots \times \mathcal{M}_m$ and $S' \subseteq \mathcal{N}_1 \times \cdots \times \mathcal{N}_n$ be non-empty logical relations. Define $S \times S'$ on $\mathcal{M}_1 \times \cdots \times \mathcal{M}_m \times \mathcal{N}_1 \times \cdots \times \mathcal{N}_n$ by*

$$(S \times S')(d_1, \ldots, d_m, e_1, \ldots, e_n) \Leftrightarrow S(d_1, \ldots, d_m) \;\&\; S'(e_1, \ldots, e_n).$$

Then $S \times S' \subseteq \mathcal{M}_1 \times \cdots \times \mathcal{M}_m \times \mathcal{N}_1 \times \cdots \times \mathcal{N}_n$ is a non-empty logical relation. If moreover both S and S' are substitutive, then so is $S \times S'$.

Proof As for syntactic logical relations. ■

3.3.35 Proposition

(i) *The universal relation S^U defined by $S^U = \mathcal{M}_1^* \times \cdots \times \mathcal{M}_n^*$ is substitutive and logical on $\mathcal{M}_1^* \times \cdots \times \mathcal{M}_n^*$.*

(ii) *Let S be an n-ary logical relation on $\mathcal{M}^* \times \cdots \times \mathcal{M}^*$ (n copies of $\mathcal{M}^*$). Let π be a permutation of $\{1, \ldots, n\}$. Define S^π on $\mathcal{M}^* \times \cdots \times \mathcal{M}^*$ by*

$$S^\pi(d_1, \ldots, d_n) \Leftrightarrow S(d_{\pi(1)}, \ldots, d_{\pi(n)}).$$

Then S^π is a logical relation. If moreover S is substitutive, then so is S^π.

(iii) *If S is an n-ary substitutive logical relation on $\mathcal{M}^* \times \cdots \times \mathcal{M}^*$, then the diagonal S^Δ defined by*

$$S^\Delta(d) \Leftrightarrow S(d, \ldots, d)$$

is a unary substitutive logical relation on $\mathcal{M}^*$.

(iv) *If $\mathcal{S}$ is a class of n-ary substitutive logical relations on $\mathcal{M}_1^* \times \cdots \times \mathcal{M}_n^*$, then the relation $\bigcap \mathcal{S} \subseteq \mathcal{M}_1^* \times \cdots \times \mathcal{M}_n^*$ is a substitutive logical relation.*

(v) *If S is an $(n+1)$-ary substitutive logical relation on $\mathcal{M}_1^* \times \cdots \times \mathcal{M}_{n+1}^*$ and $\mathcal{M}_{n+1}^*$ is a typed λ-model, then $\exists S$ defined by*

$$\exists S(d_1, \ldots, d_n) \Leftrightarrow \exists d_{n+1}\, S(d_1, \ldots, d_{n+1})$$

is an n-ary substitutive logical relation.

Proof For convenience we take $n = 1$. We treat (v), leaving the rest to the reader.

(v) Let $S \subseteq \mathcal{M}_1^* \times \mathcal{M}_2^*$ be substitutive and logical. Define $R(d_1) \Leftrightarrow \exists d_2 \in \mathcal{M}_2^*.S(d_1.d_2)$, towards proving

$$\forall d_1 \in \mathcal{M}_1^*.[R(d_1) \Leftrightarrow \forall e_1 \in \mathcal{M}_1^*.[R(e_1) \Rightarrow R(d_1 e_1)]].$$

($\Rightarrow$) Suppose $R(d_1), R(e_1)$ in order to show $R(d_1 e_1)$. Then there are $d_2, e_2 \in \mathcal{M}_2^*$ such that $S(d_1, d_2), S(e_1, e_2)$. Then $S(d_1 e_1, d_2 e_2)$, as S is logical. Therefore $R(d_1 e_1)$ indeed.

($\Leftarrow$) Suppose $\forall e_1 \in \mathcal{M}_1^*.[R(e_1) \Rightarrow R(d_1 e_1)]$, towards proving $R(d_1)$. By the assumption

$$\forall e_1[\exists e_2.S(e_1, e_2) \Rightarrow \exists e_2'.S(d_1 e_1, e_2')].$$

Hence

$$\forall e_1, e_2 \exists e_2'.[S(e_1, e_2) \Rightarrow S(d_1 e_1, e_2')]. \tag{3.11}$$

Since S is substitutive, we have $S(x, x)$, by Remark 3.3.30. We continue as follows:

$$\begin{aligned}
S(x, x) &\Rightarrow S(d_1 x, e_2'[x]), \text{ for some } e_2' = e_2'[x] \text{ by (3.11)},\\
&\Rightarrow S(d_1 x, d_2 x), \text{ where } d_2 = \lambda x.e_2'[x] \text{ using that } \mathcal{M}_2^*\\
&\qquad\qquad \text{is a typed } \lambda\text{-model},\\
&\Rightarrow S(e_1, e_2) \Rightarrow S(d_1 e_1, d_2 e_2), \text{ by substitutivity of } S,\\
&\Rightarrow S(d_1, d_2), \text{ since } S \text{ is logical},\\
&\Rightarrow R(d_1).
\end{aligned}$$

This establishes that $\exists S = R$ is logical.

Now assume that S is substitutive, in order to show that so is R. That is, we must show

$$R(d_1) \Leftrightarrow \forall \rho_1.[[\forall x \in \mathsf{V}.R(\rho_1(x))] \Rightarrow R((d_1)^{\rho_1})]. \tag{3.12}$$

($\Rightarrow$) Assuming $R(d_1), R(\rho_1(x))$ we get $S(d_1, d_2), S(\rho_1(x), d_2^x)$, for some d_2, d_2^x. Defining ρ_2 by $\rho_2(x) = d_2^x$, for the free variables in d_2, we get $S(\rho_1(x), \rho_2(x))$, hence by the substitutivity of S we have that $S((d_1)^{\rho_1}, (d_2)^{\rho_2})$ and therefore $R((d_1)^{\rho_1})$.

($\Leftarrow$) By the substitutivity of S one has for all variables x that $S(x, x)$, by Remark 3.3.30, hence also $R(x)$. Now take, in the right hand side of (3.12), the identity valuation $\rho_1(x) = x$, for all x. Then one obtains $R((d_1)^{\rho_1})$, which is $R(d_1)$. ∎

3.3.36 Example Consider $\mathcal{M}_\omega$ and define

$$S_0(n,m) \Leftrightarrow n \leq m,$$

where $\leq$ is the usual ordering on ω. Then $\{d \in S^* \mid d =^* d\}/=^*$ is the set of hereditarily monotone functionals. Similarly $\exists(S^*)$ induces the set of hereditarily majorizable functionals, see the section by Howard in Troelstra (1973).

Relating syntactic and semantic logical relations

One may wonder whether the fundamental theorem for semantic logical relations follows from the syntactic version (but not vice versa; e.g. the usual semantic logical relations are automatically closed under $\boldsymbol{\beta\eta}$-conversion). This indeed is the case. The 'hinge' is that a logical relation $R \subseteq \Lambda_\to[\mathcal{M}^*]$ can be seen as a semantic logical relation (since $\Lambda_\to[\mathcal{M}^*]$ is a typed applicative structure) and at the same time as a syntactic one (since $\Lambda_\to[\mathcal{M}^*]$ consists of terms from some set of constants). We also need this dual vision for the notion of substitutivity. For this we have to merge the syntactic and the semantic version of these notions. Let $\mathcal{M}$ be a typed applicative structure, containing at each type A variables of type A. A valuation is a map $\rho{:}\mathsf{V} \to \mathcal{M}$ such that $\rho(x^A) \in \mathcal{M}(A)$. This ρ can be extended to a substitution $(-)^\rho{:}\mathcal{M} {\to} \mathcal{M}$. A unary relation $R \subseteq \mathcal{M}$ is substitutive if for all $M \in \mathcal{M}$ one has

$$R(M) \Leftrightarrow [\forall x{:}\mathsf{V}.[R(\rho(x)) \Rightarrow R((M)^\rho)].$$

The notion of substitutivity is analogous for relations $R \subseteq \Lambda_\to[\mathcal{D}]$, using Definition 3.3.8(iii), as for relations $R \subseteq \mathcal{M}^*$, using Definition 3.3.29(iv).

3.3.37 Notation Let $\mathcal{M}$ be a typed applicative structure. Write

$$\Lambda_\to[\mathcal{M}] = \Lambda_\to[\{\underline{d} \mid d \in \mathcal{M}\}];$$
$$\Lambda_\to(\mathcal{M}) = \Lambda_\to[\mathcal{M}]/=_{\boldsymbol{\beta\eta}}.$$

Then $\Lambda_\to[\mathcal{M}]$ is a typed applicative structure and $\Lambda_\to(\mathcal{M})$ is a typed λ-model.

3.3.38 Definition Let $\mathcal{M}$, and hence also $\mathcal{M}^*$, be a typed λ-model. For $\rho : \mathsf{V} \to \mathcal{M}^*$ extend $[\![-]\!]_\rho : \Lambda_\to \to \mathcal{M}^*$ to $[\![-]\!]_\rho^{\mathcal{M}^*} : \Lambda_\to[\mathcal{M}^*] \to \mathcal{M}^*$ as

$$\begin{aligned}
[\![x]\!]_\rho &= \rho(x) \\
[\![\underline{m}]\!]_\rho &= m, && \text{with } m \in \mathcal{M}^*, \\
[\![PQ]\!]_\rho &= [\![P]\!]_\rho [\![Q]\!]_\rho \\
[\![\lambda x.P]\!]_\rho &= d, && \text{the unique } d \in \mathcal{M}^* \text{ with } \forall e.de = [\![P]\!]_{\rho[x:=e]}.
\end{aligned}$$

Recall Definition 3.3.29 of $(-)^\rho : \mathcal{M}^* \to \mathcal{M}^*$:

$$
\begin{array}{rcll}
(x)^\rho & = & \rho(x) & \\
(\underline{m})^\rho & = & m, & \text{with } m \in \mathcal{M}^*, \\
(PQ)^\rho & = & (P)^\rho (Q)^\rho. &
\end{array}
$$

Now define the predicate $D \subseteq \Lambda_\to[\mathcal{M}^*] \times \mathcal{M}^*$ as

$$D(M, d) \Leftrightarrow \forall \rho{:}\mathsf{V}{\to}\mathcal{M}^*.[\![M]\!]_\rho^{\mathcal{M}^*} = (d)^\rho.$$

3.3.39 Lemma *The predicate D is a substitutive semantic logical relation.*

Proof First we show that D is logical. We must show for $M \in \Lambda_\to[\mathcal{M}^*]$, $d \in \mathcal{M}^*$, that

$$D(M, d) \Leftrightarrow \forall N \in \Lambda_\to[\mathcal{M}^*] \forall e \in \mathcal{M}^*.[D(N, e) \Rightarrow D(MN, de)].$$

($\Rightarrow$) Suppose $D(M, d), D(N, e)$, towards proving $D(MN, de)$. Then for all $\rho{:}\mathsf{V} \to \mathcal{M}^*$, by definition $[\![M]\!]_\rho^{\mathcal{M}^*} = (d)^\rho$ and $[\![N]\!]_\rho^{\mathcal{M}^*} = (e)^\rho$. But then $[\![MN]\!]_\rho^{\mathcal{M}^*} = (de)^\rho$, and therefore $D(MN, de)$.

($\Leftarrow$) Now suppose $\forall N \in \Lambda_\to[\mathcal{M}^*] \forall e \in \mathcal{M}^*.[D(N, e) \Rightarrow D(MN, de)]$, towards proving $D(M, d)$. Let x be a fresh variable, i.e. not in M or d. Note that $x \in \Lambda_\to[\mathcal{M}^*]$, $x \in \mathcal{M}^*$, and $D(x, x)$. Hence by assumption

$$
\begin{array}{rcll}
D(x, x) & \Rightarrow & \forall \rho [\![Mx]\!]_\rho = (dx)^\rho & \\
 & \Rightarrow & \forall \rho [\![M]\!]_\rho [\![x]\!]_\rho = (d)^\rho (x)^\rho & \\
 & \Rightarrow & \forall \rho [\![M]\!]_{\rho'} [\![x]\!]_{\rho'} = (d)^{\rho'} (x)^{\rho'}, & \text{where } \rho' = \rho[x := e], \\
 & \Rightarrow & \forall \rho \forall e \in \mathcal{M}^*.[\![M]\!]_\rho e = (d)^\rho e, & \text{by the freshness of } x, \\
 & \Rightarrow & \forall \rho [\![M]\!]_\rho = (d)^\rho, & \text{by extensionality,} \\
 & \Rightarrow & D(M, d). &
\end{array}
$$

Second, we show that D is substitutive. We must show for $M \in \Lambda_\to[\mathcal{M}^*]$, $d \in \mathcal{M}^*$, that

$$
\begin{array}{rcl}
D(M, d) & \Leftrightarrow & \forall \rho_1{:}\mathsf{V} \to \Lambda_\to[\mathcal{M}^*], \rho_2{:}\mathsf{V} \to \mathcal{M}^*. \\
 & & [\forall x \in \mathsf{V}.D(\rho_1(x), \rho_2(x)) \Rightarrow D((M)^{\rho_1}, (d)^{\rho_2})].
\end{array}
$$

($\Rightarrow$) Suppose $D(M, d)$ and $\forall x \in \mathsf{V}.D(\rho_1(x), \rho_2(x)$ towards $D((M)^{\rho_1}, (d)^{\rho_2})$. Then for all $\rho{:}\mathsf{V}{\to}\mathcal{M}^*$ one has

$$[\![M]\!]_\rho = (d)^\rho \tag{3.13}$$

$$\forall x \in \mathsf{V}.[\![\rho_1(x)]\!]_\rho = (\rho_2(x))^\rho. \tag{3.14}$$

Let $\rho_1'(x) = [\![\rho_1(x)]\!]_\rho^{\mathcal{M}^*}$ and $\rho_2'(x) = (\rho_2(x))^\rho$. By induction on M and d one can show analogous to Lemma 3.1.13(i) that

$$[\![M^{\rho_1}]\!]_\rho = [\![M]\!]_{\rho_1'} \tag{3.15}$$

$$((d)^{\rho_2})^\rho = (d)^{\rho_2'}. \tag{3.16}$$

It follows by (3.14) that $\rho_1' = \rho_2'$ and hence by (3.15), (3.16), and (3.13) that $[\![(M)^{\rho_1}]\!]_\rho = ((d)^{\rho_2})^\rho$, for all ρ. Therefore $D((M)^{\rho_1}, (d)^{\rho_2})$.

($\Leftarrow$) Assume the right hand side. Define $\rho_1(x) = x \in \Lambda_\to[\mathcal{M}^*]$, $\rho_2(x) = x \in \mathcal{M}^*$. Then we have $D(\rho_1, \rho_2)$, and so by the assumption, $D((M)^{\rho_1}, (d)^{\rho_2})$. By the choice of ρ_1, ρ_2 this is $D(M, d)$. ■

3.3.40 Lemma *Let $M \in \Lambda_\to^\varnothing$. Then $[\![M]\!]^{\mathcal{M}^*} = [\underline{[\![M]\!]^{\mathcal{M}}}] \in \mathcal{M}^*$.*

Proof Let $i{:}\mathcal{M} \to \mathcal{M}^*$ be the canonical inbedding defined by $i(d) = \underline{d}$. Then for all $M \in \Lambda_\to$ and all $\rho : \mathsf{V} \to \mathcal{M}$ one has

$$i([\![M]\!]_\rho^{\mathcal{M}}) = [\![M]\!]_{i \circ \rho}^{\mathcal{M}^*}.$$

Hence for closed terms M it follows that

$$[\![M]\!]^{\mathcal{M}^*} = [\![M]\!]_{i \circ \rho}^{\mathcal{M}^*} = i([\![M]\!]_\rho^{\mathcal{M}}) = [\underline{[\![M]\!]^{\mathcal{M}}}]. \ ■$$

3.3.41 Definition Let $R \subseteq \Lambda_\to[\mathcal{M}_1^*] \times \cdots \times \Lambda_\to[\mathcal{M}_n^*]$. Then R is called *invariant* if, for all $M_1, N_1 \in \Lambda_\to[\mathcal{M}_1^*], \ldots, M_n, N_n \in \Lambda_\to[\mathcal{M}_n^*]$, one has

$$\left.\begin{array}{l} R(M_1, \ldots, M_n) \\ \mathcal{M}_1^* \models M_1 = N_1 \ \& \ \cdots \ \& \ \mathcal{M}_n^* \models M_n = N_n \end{array}\right\} \Rightarrow R(N_1, \ldots, N_n).$$

3.3.42 Definition Let $\mathcal{M}_1, \ldots, \mathcal{M}_n$ be typed applicative structures.

(i) Let $S \subseteq \mathcal{M}_1^* \times \cdots \times \mathcal{M}_n^*$. Define the relation $S^\wedge \subseteq \Lambda_\to[\mathcal{M}_1^*] \times \cdots \times \Lambda_\to[\mathcal{M}_n^*]$ by

$$\begin{aligned} S^\wedge(M_1, \ldots, M_n) \Leftrightarrow\ & \exists d_1 \in \mathcal{M}_1^* \cdots \exists d_n \in \mathcal{M}_n^*.[S(d_1, \ldots, d_n) \ \& \\ & D(M_1, d_1) \ \& \ \cdots \ \& \ D(M_n, d_n)]. \end{aligned}$$

(ii) Let $R \subseteq \Lambda_\to[\mathcal{M}_1^*] \times \cdots \times \Lambda_\to[\mathcal{M}_n^*]$. Define $R^\vee \subseteq \mathcal{M}_1^* \times \cdots \times \mathcal{M}_n^*$ by

$$\begin{aligned} R^\vee(d_1, \ldots, d_n) \Leftrightarrow\ & \exists M_1 \in \Lambda_\to[\mathcal{M}_1^*], \ldots, M_n \in \Lambda_\to[\mathcal{M}_n^*].[R(M_1, \ldots, M_n) \ \& \\ & D(M_1, d_1) \ \& \ \cdots \ \& \ D(M_n, d_n)]. \end{aligned}$$

3.3.43 Definition Let $\iota : \mathsf{V} \to \mathcal{M}^*$ be the 'identity' valuation; that is, $\iota(x) = [x]$.

3.3.44 Lemma

(i) *Let $S \subseteq \mathcal{M}_1^* \times \cdots \times \mathcal{M}_n^*$. Then $S^\wedge$ is invariant.*

(ii) *Let* $R \subseteq \Lambda_{\to}[\mathcal{M}_1^*] \times \cdots \times \Lambda_{\to}[\mathcal{M}_n^*]$ *be invariant. Then, for all*

$$M_1 \in \Lambda_{\to}^{\emptyset}[\mathcal{M}_1^*], \ldots, M_n \in \Lambda_{\to}^{\emptyset}[\mathcal{M}_n^*],$$

one has

$$R(M_1, \ldots, M_n) \Rightarrow R^{\vee}([\![M_1]\!]_{\iota}^{\mathcal{M}_1^*}, \ldots, [\![M_n]\!]_{\iota}^{\mathcal{M}_n^*}).$$

Proof For notational convenience we take $n = 1$.

(i) $S^{\wedge}(M)$ & $\mathcal{M}^* \models M = N$
$\Rightarrow \exists d \in \mathcal{M}^*.[S(d) \;\&\; D(M,d)] \;\&\; \mathcal{M}^* \models M = N$
$\Rightarrow \exists d \in \mathcal{M}^*.[S(d) \;\&$
$\quad \forall \rho.[\,[\![M]\!]_{\rho} = (d)^{\rho} \;\&\; [\![M]\!]_{\rho} = [\![N]\!]_{\rho}]]$
$\Rightarrow \exists d.[S(d) \;\&\; D(N,d)]$
$\Rightarrow S^{\wedge}(N)$.

(ii) Suppose $R(M)$. Let $M' = \underline{[\![M]\!]_{\iota}} \in \Lambda_{\to}[\mathcal{M}^*]$. Then

$$[\![M']\!]_{\rho} = [\![M]\!]_{\iota} = [\![M]\!]_{\rho},$$

since M is closed. Hence $R(M')$ by the invariance of R and $D(M', [\![M]\!]_{\iota})$. Therefore $R^{\vee}([\![M]\!]_{\iota})$. ∎

3.3.45 Proposition *Let* $\mathcal{M}_1, \ldots, \mathcal{M}_n$ *be typed* λ*-models.*

(i) *Let* $S \subseteq \mathcal{M}_1^* \times \cdots \times \mathcal{M}_n^*$ *be a substitutive semantic logical relation. Then* $S^{\wedge}$ *is an invariant and substitutive syntactic logical relation.*
(ii) *Let* $R \subseteq \Lambda_{\to}[\mathcal{M}_1^*] \times \cdots \times \Lambda_{\to}[\mathcal{M}_n^*]$ *be a substitutive syntactic logical relation. Then* $R^{\vee}$ *is a substitutive semantic logical relation.*

Proof Again we take $n = 1$.

(i) By Lemma 3.3.44(i) $S^{\wedge}$ is invariant. Moreover, one has for $M \in \Lambda_{\to}[\mathcal{M}^*]$

$$S^{\wedge}(M) \Leftrightarrow \exists d \in \mathcal{M}^*.[S(d) \;\&\; D(M,d)].$$

By assumption, S is a substitutive logical relation, as is D, by Proposition 3.3.39. By Proposition 3.3.35(iv) and (v), so is their conjunction and its $\exists$-projection $S^{\wedge}$.

(ii) For $d \in \mathcal{M}^*$ one has

$$R^{\vee}(d) \Leftrightarrow \exists M \in \Lambda_{\to}[\mathcal{M}^*].[D(M,d) \;\&\; R(M)].$$

We conclude similarly. ∎

3.3.46 Proposition *Let* $\mathcal{M}_1, \ldots, \mathcal{M}_n$ *be typed* λ*-models. Assume that* $S \subseteq \mathcal{M}_1^* \times \cdots \times \mathcal{M}_n^*$ *is a substitutive logical relation. Then* $S^{\wedge\vee} = S$.

Proof For notational convenience take $n = 1$. Write $T = S^{\wedge}$. Then, for $d \in \mathcal{M}^*$,

$$\begin{aligned} T^{\vee}(d) &\Leftrightarrow \exists M \in \Lambda_{\to}[\mathcal{M}^*].[T(M) \ \& \ D(M,d)], \\ &\Leftrightarrow \exists M \in \Lambda_{\to}[\mathcal{M}^*]\exists d' \in \mathcal{M}^*.[S(d') \ \& \ D(M,d') \ \& \ D(M,d)], \\ &\quad \text{which implies } d' = d, \text{ as } \mathcal{M}^* = \underline{\mathcal{M}}/\sim_{\mathcal{M}}, \\ &\Leftrightarrow S(d), \end{aligned}$$

where the last $\Leftarrow$ follows by taking $M = \underline{d}, d' = d$. Therefore $S^{\wedge\vee} = S$. ∎

Using this result, the fundamental theorem for semantic logical relations can be derived from the syntactic version.

3.3.47 Proposition *The fundamental theorem for syntactic logical relations implies that for semantic logical relations. That is, letting $\mathcal{M}_1, \dots, \mathcal{M}_n$ be λ-models, for the following two statements one has* (i) $\Rightarrow$ (ii).

(i) *Let R on $\Lambda_{\to}[\vec{\mathcal{M}}]$ be an expansive and substitutive syntactic logical relation. Then for all $A \in \mathbb{T}$ and all pure terms $M \in \Lambda_{\to}(A)$ one has*

$$R_A(M, \dots, M).$$

(ii) *Let S on $\mathcal{M}_1 \times \cdots \times \mathcal{M}_n$ be a semantic logical relation. Then for each term $M \in \Lambda^{\emptyset}_{\to}(A)$ one has*

$$S_A([\![M]\!]^{\mathcal{M}_1}, \dots, [\![M]\!]^{\mathcal{M}_n}).$$

Proof We show (ii) assuming (i). For notational simplicity we take $n = 1$. Therefore let $S \subseteq \mathcal{M}$ be logical and $M \in \Lambda^{\emptyset}_{\to}$, in order to prove $S([\![M]\!])$. First we assume that S is non-empty. Then $S^* \subseteq \mathcal{M}^*$ is a substitutive semantic logical relation, by Propositions 3.3.33(iii) and (ii). Writing $R = S^{*\wedge} \subseteq \Lambda_{\to}(\mathcal{M}^*)$ we have that R is an invariant (hence expansive) and substitutive logical relation, by Proposition 3.3.45(i). For $M \in \Lambda^{\emptyset}_{\to}(A)$ we have $R_A(M)$, by (i), and proceed as follows:

$$\begin{aligned} R_A(M) &\Rightarrow R^{\vee}([\![M]\!]^{\mathcal{M}^*}), && \text{by Lemma 3.3.44(ii), as } M \text{ is closed,} \\ &\Rightarrow S_A^{*\wedge\vee}([\![M]\!]^{\mathcal{M}^*}), && \text{as } R = S^{*\wedge}, \\ &\Rightarrow S_A^{*}([\![M]\!]^{\mathcal{M}^*}), && \text{by Proposition 3.3.46(i),} \\ &\Rightarrow S_A^{*}([\underline{[\![M]\!]^{\mathcal{M}}}]), && \text{by Lemma 3.3.40,} \\ &\Rightarrow S_A([\![M]\!]^{\mathcal{M}}), && \text{by Lemma 3.3.32(ii) and the assumption.} \end{aligned}$$

In the case S is empty, then we also have $S_A([\![M]\!]^{\mathcal{M}})$, by Proposition 3.3.25. ∎

3.4 Type reducibility

In this section we study, in the context of $\boldsymbol{\lambda}_{\rightarrow}^{\text{dB}}$ over $\mathbb{T}^0$, how equality of terms of a certain type A can be reduced to equality of terms of another type B. This is the case if there is a definable injection of $\Lambda_{\rightarrow}^{\emptyset}(A)$ into $\Lambda_{\rightarrow}^{\emptyset}(B)$. The resulting poset of 'reducibility degrees' will turn out to be the ordinal $\omega+4=\{0,1,2,3,\ldots,\omega,\omega+1,\omega+2,\omega+3\}$.

3.4.1 Definition Let A, B be types of $\boldsymbol{\lambda}_{\rightarrow}^{\mathbb{A}}$.

(i) We say that there is a *type reduction* from A to B (or A is $\boldsymbol{\beta\eta}$ *reducible* to B), written $A \leq_{\boldsymbol{\beta\eta}} B$, if, for some closed term $\Phi{:}A{\rightarrow}B$, one has for all closed $M_1, M_2{:}A$

$$M_1 =_{\boldsymbol{\beta\eta}} M_2 \Leftrightarrow \Phi M_1 =_{\boldsymbol{\beta\eta}} \Phi M_2;$$

i.e., equalities between terms of type A can be uniformly translated to those of type B.

(ii) Write $A \sim_{\boldsymbol{\beta\eta}} B$ iff $A \leq_{\boldsymbol{\beta\eta}} B$ & $B \leq_{\boldsymbol{\beta\eta}} A$.

(iii) Write $A <_{\boldsymbol{\beta\eta}} B$ for $A \leq_{\boldsymbol{\beta\eta}} B$ & $B \not\leq_{\boldsymbol{\beta\eta}} A$.

The following result is easy.

3.4.2 Lemma *Let* $A = A_1{\rightarrow}\cdots{\rightarrow}A_a{\rightarrow}0$ *and* $B = A_{\pi(1)}{\rightarrow}\cdots{\rightarrow}A_{\pi(a)}{\rightarrow}0$, *where* π *is a permutation of the set* $\{1,\ldots,a\}$. *We say that* A *and* B *are equal* up to permutation of arguments. *Then:*

(i) $B \leq_{\boldsymbol{\beta\eta}} A$;

(ii) $A \sim_{\boldsymbol{\beta\eta}} B$.

Proof (i) We have $B \leq_{\boldsymbol{\beta\eta}} A$ via

$$\Phi \equiv \lambda m{:}B\lambda x_1\cdots x_a.mx_{\pi(1)}\cdots x_{\pi(a)}.$$

(ii) By (i) applied to π^{-1}. ∎

The reducibility theorem, Statman (1980a), states that there is one type to which all types of $\mathbb{T}^0$ can be reduced. At first this may seem impossible. Indeed, in a full typed structure $\mathcal{M}$ the cardinality of the sets of higher type increases arbitrarily. So one cannot always have an injection $\mathcal{M}_A{\rightarrow}\mathcal{M}_B$. But reducibility means that one restricts oneself to definable elements (modulo $=_{\boldsymbol{\beta\eta}}$) and then the injections are possible. The proof will occupy[1] 3.4.3–3.4.8. There are four main steps. In order to show that $\Phi M_1 =_{\boldsymbol{\beta\eta}} \Phi M_2 \Rightarrow M_1 =_{\boldsymbol{\beta\eta}} M_2$ in all cases a (pseudo) inverse Φ^{-1} is used. Pseudo means that

[1] A simpler alternative route discovered later by Joly is described in Exercises 3.15 and 3.16, needing also Exercise 3.19.

sometimes the inverse is not lambda-definable, but this is no problem for the implication. Sometimes Φ^{-1} is definable, but the property $\Phi^{-1}(\Phi M) = M$ only holds in an extension of the theory; because the extension will be conservative over $=_{\beta\eta}$, the reducibility will follow.

Next the type hierarchy theorem, also due to Statman (1980a), will be given. Rather unexpectedly it turns out that, under $\leq_{\beta\eta}$, types form a well-ordering of length $\omega + 4$. Finally some consequences of the reducibility theorem will be given, including the 1-section and finite completeness theorems.

In the first step towards the reducibility theorem it will be shown that every type is reducible to one of rank ≤ 3. The proof is rather syntactic. In order to show that the definable function Φ is one-to-one, a non-definable inverse is needed. A warm-up exercise for this is 3.7.

3.4.3 Proposition *Every type can be reduced to a type of rank* ≤ 3, *see Definition* 1.1.22(ii). *That is,*

$$\forall A \in \mathbb{T}^0 \exists B \in \mathbb{T}^0.[A \leq_{\beta\eta} B \ \&\ \mathrm{rk}(B) \leq 3].$$

Proof We begin by describing the intuition behind the construction of the term Φ responsible for the reducibility. If M is a term with Böhm tree (see B[1984])

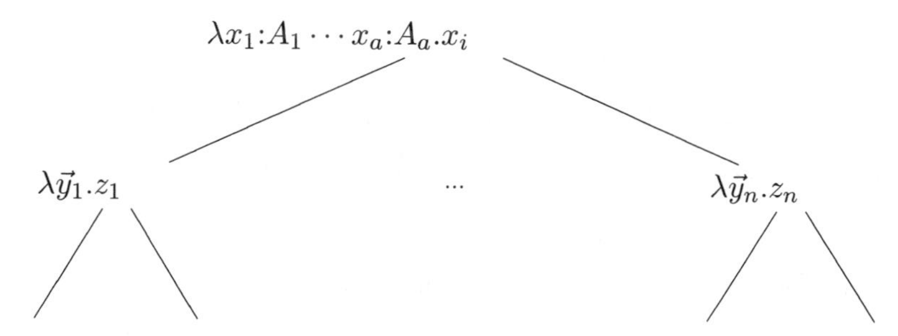

then let UM be a term with 'Böhm tree' of the form

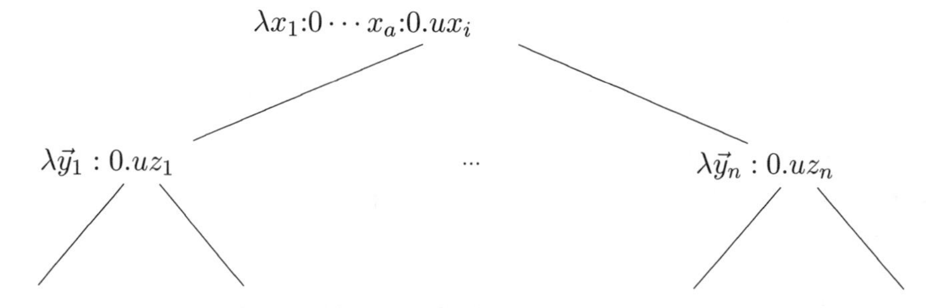

where all the typed variables are pushed down to type 0 and the variables u (each occurrence possibly different) take care that the new term remains typable. From this description it is clear that the u can be chosen in such way that the result has rank ≤ 1. It is also clear that M can be reconstructed from UM so that U is injective. The term ΦM is just UM with the auxiliary variables bound. This makes it of type with rank ≤ 3. What is less clear is that U and hence Φ are lambda-definable.

Back to the proof. Define recursively, for any type A, the types $A^\flat$ and $A^\sharp$:

$$\begin{aligned} 0^\flat &= 0; \\ 0^\sharp &= 0; \\ (A_1 \to \cdots \to A_a \to 0)^\flat &= (0^a \to 0); \\ (A_1 \to \cdots \to A_a \to 0)^\sharp &= 0 \to A_1^\flat \to \cdots \to A_a^\flat \to 0. \end{aligned}$$

Notice that $\text{rk}(A^\sharp) \leq 2$.

In the infinite context

$$\{u_A{:}A^\sharp \mid A \in \mathbb{T}\}$$

define recursively, for any type A, terms $V_A : 0 \to A, U_A : A \to A^\flat$:

$$\begin{aligned} U_0 &= \lambda x{:}0.x; \\ V_0 &= \lambda x{:}0.x; \\ U_{A_1 \to \cdots \to A_a \to 0} &= \lambda z{:}A\lambda x_1 \cdots x_a{:}0.z(V_{A_1}x_1) \cdots (V_{A_a}x_a); \\ V_{A_1 \to \cdots \to A_a \to 0} &= \lambda x{:}0\lambda y_1{:}A_1 \cdots y_a{:}A_a.u_A x(U_{A_1}y_1) \cdots (U_{A_a}y_a), \end{aligned}$$

where $A = A_1 \to \cdots \to A_a \to 0$.

Note that for $C = A_1 \to \cdots \to A_a \to B$ one has

$$U_C = \lambda z{:}C\lambda x_1 \cdots x_a{:}0.U_B(z(V_{A_1}x_1) \cdots (V_{A_a}x_a)). \tag{3.17}$$

Indeed, both sides are equal to

$$\lambda z{:}C\lambda x_1 \cdots x_a y_1 \cdots y_b{:}0.z(V_{A_1}x_1) \cdots (V_{A_a}x_a)(V_{B_1}y_1) \cdots (V_{B_b}y_b),$$

with $B = B_1 \to \cdots \to B_b \to 0$.

Notice also that for a closed term M of type $A = A_1 \to \cdots \to A_a \to 0$ one can write

$$M =_\beta \lambda y_1{:}A_1 \cdots y_a{:}A_a.y_i(M_1 y_1 \cdots y_a) \cdots (M_n y_1 \cdots y_a),$$

with the $M_1, \ldots, M_n$ closed. Write $A_i = A_{i1} \to \cdots \to A_{in} \to 0$.

Now verify that

$$\begin{aligned}
U_A M &= \lambda x_1 \cdots x_a{:}0.M(V_{A_1}x_1)\cdots(V_{A_a}x_a) \\
&= \lambda \vec{x}.(V_{A_i}x_i)(M_1(V_{A_1}x_1)\cdots(V_{A_a}x_a))\cdots(M_n(V_{A_1}x_1)\cdots(V_{A_a}x_a)) \\
&= \lambda \vec{x}.u_{A_i}x_i(U_{A_{i1}}(M_1(V_{A_1}x_1)\cdots \\
&\qquad \cdots(V_{A_a}x_a)))\cdots(U_{A_{in}}(M_n(V_{A_1}x_1)\cdots(V_{A_a}x_a))) \\
&= \lambda \vec{x}.u_{A_i}x_i(U_{B_1}M_1\vec{x})\cdots(U_{B_n}M_n\vec{x}),
\end{aligned}$$

using (3.17), where $B_j = A_1 \rightarrow \cdots \rightarrow A_a \rightarrow A_{ij}$ for $1 \le j \le n$ is the type of M_j. Hence we have that if $U_A M =_{\beta\eta} U_A N$, then, for $1 \le j \le n$,

$$U_{B_j} M_j =_{\beta\eta} U_{B_j} N_j.$$

Therefore it follows by induction on the complexity of the β-nf of M that if $U_A M =_{\beta\eta} U_A N$, then $M =_{\beta\eta} N$.

Now take as term for the reducibility $\Phi \equiv \lambda m{:}A\lambda u_{B_1}\cdots u_{B_k}.U_A m$, where the $\vec{u}$ are all the ones occurring in the construction of U_A. It follows that

$$A \le_{\beta\eta} B_1^\sharp \rightarrow \cdots \rightarrow B_k^\sharp \rightarrow A^\flat.$$

Since $\mathrm{rk}(B_1^\sharp \rightarrow \cdots \rightarrow B_k^\sharp \rightarrow A^\flat) \le 3$, we are done. ∎

For an alternative proof, see Exercise 3.15.

In the following proposition it will be proved that we can further reduce types to one particular type of rank 3. First do exercise 3.8 to get some intuition. We need the following notation.

3.4.4 Notation

(i) For $k \ge 0$ write

$$1_k = 0^k \rightarrow 0,$$

where in general $A^0 \rightarrow 0 = 0$ and $A^{k+1} \rightarrow 0 = A \rightarrow (A^k \rightarrow 0)$.

(ii) For $k_1, \ldots, k_n \ge 0$ write

$$(k_1, \ldots, k_n) = 1_{k_1} \rightarrow \cdots \rightarrow 1_{k_n} \rightarrow 0.$$

(iii) For $k_{11}, \ldots, k_{1n_1}, \ldots, k_{m1}, \ldots, k_{mn_m} \ge 0$ write

$$\begin{pmatrix} k_{11} & \cdots & k_{1n_1} \\ \vdots & \ddots & \vdots \\ k_{m1} & \cdots & k_{mn_m} \end{pmatrix} = (k_{11}, \ldots, k_{1n_1}) \rightarrow \cdots \rightarrow (k_{m1}, \ldots, k_{mn_m}) \rightarrow 0.$$

Note the 'matrix' has an uneven right side (the n_i are unequal in general).

3.4.5 Proposition *Every type A of rank ≤ 3 is reducible to*

$$1_2 \to 1 \to 1 \to 2 \to 0.$$

Proof Let A be a type of rank ≤ 3. It is not difficult to see that A is of the form

$$A = \begin{pmatrix} k_{11} & \cdots & k_{1n_1} \\ \vdots & \cdots & \vdots \\ k_{m1} & \cdots & k_{mn_m} \end{pmatrix}$$

We will first 'reduce' A to type $3 = 2 \to 0$ using an open term Ψ, containing free variables of type $1_2, 1, 1$ respectively acting as a 'pairing'. Consider the context

$$\{p{:}1_2, p_1{:}1, p_2{:}1\}.$$

Consider the notion of reduction p defined by the contraction rules

$$p_i(pM_1M_2) \to_p M_i.$$

[There now is a choice of how to proceed: if you like syntax, then carry on; if you prefer models, omit paragraphs starting with ♣ and jump to those starting with ♠.]

♣ This notion of reduction satisfies the subject reduction property. Moreover $\beta\eta p$ is Church–Rosser, see Pottinger (1981). This can be used later in the proof. [Extending the notion of reduction by adding

$$p(p_1M)(p_2M) \to_s M$$

preserves the CR property. In the untyped calculus this is not the case, see Klop (1980) or B[1984], ch. 14.] Goto ♠.

♠ Given the pairing p, p_1, p_2 one can extend it as follows. Write

$$\begin{array}{rcll} p^1 &=& \lambda x{:}0.x; & \\ p^{k+1} &=& \lambda x_1 \cdots x_k x_{k+1}{:}0.p(p^k x_1 \cdots x_k)x_{k+1}; & \\ p_1^1 &=& \lambda x{:}0.x; & \\ p_{k+1}^{k+1} &=& p_2; & \\ p_i^{k+1} &=& \lambda z{:}0.p_i^k(p_1 z), & \text{for } i \leq k; \\ P^k &=& \lambda f_1 \cdots f_k{:}1\lambda z{:}0.p^k(f_1 z)\cdots(f_k z); & \\ P_i^k &=& \lambda g{:}1\lambda z{:}0.p_i^k(gz), & \text{for } i \leq k. \end{array}$$

We have that p^k acts as a coding for k-tuples of elements of type 0 with projections p_i^k. The P^k, P_i^k do the same for type 1. In a context containing $\{f{:}1_k, g{:}1\}$ write

$$f^{k\to1} = \lambda z{:}0.f(p_1^k z)\cdots(p_k^k z);$$
$$g^{1\to k} = \lambda z_1\cdots z_k{:}0.g(p^k z_1\cdots z_k).$$

Then $f^{k\to1}$ is f moved to type 1 and $g^{1\to k}$ is g moved to type 1_k.

Using $\boldsymbol{\beta\eta p}$-convertibility one can show

$$p_i^k(p^k z_1\cdots z_k) = z_i;$$
$$P_i^k(P^k f_1\cdots f_k) = f_i;$$
$$(f^{k\to1})^{1\to k} = f.$$

For $(g^{1\to k})^{k\to1} = g$ one needs $\to_s$, the surjectivity of the pairing.

In order to define the term required for the reducibility, start with a term $\Psi{:}A\to3$ (containing p, p_1, p_2 as only free variables). We need an auxiliary term Ψ^{-1}, that acts as an inverse for Ψ in the presence of a 'true pairing':

$$\begin{aligned}\Psi \equiv\ & \lambda M{:}A\,\lambda F{:}2.M\\ & [\lambda f_{11}{:}1_{k_{11}}\cdots f_{1n_1}{:}1_{k_{1n_1}}.p_1(F(P^{n_1} f_{11}^{k_{11}\to1}\cdots f_{1n_1}^{k_{1n_1}\to1}))]\cdots\\ & [\lambda f_{m1}{:}1_{k_{m1}}\cdots f_{mn_m}{:}1_{k_{mn_m}}.p_m(F(P^{n_m} f_{m1}^{k_{m1}\to1}\cdots f_{mn_m}^{k_{mn_m}\to1}))];\\ \Psi^{-1} \equiv\ & \lambda N{:}(2\to0)\lambda K_1{:}(k_{11},\dots,k_{1n_1})\cdots\lambda K_m{:}(k_{m1},\dots,k_{mn_m}).\\ & N(\lambda f{:}1.p^m[K_1(P_1^{n_1}f)^{1\to k_{11}}\cdots(P_{n_1}^{n_1}f)^{1\to k_{1n_1}}]\cdots\\ & \qquad [K_m(P_1^{n_m}f)^{1\to k_{m1}}\cdots(P_{n_m}^{n_m}f)^{1\to k_{1n_m}}]).\end{aligned}$$

Claim. For closed terms M_1, M_2 of type A we have

$$M_1 =_{\boldsymbol{\beta\eta}} M_2 \Leftrightarrow \Psi M_1 =_{\boldsymbol{\beta\eta}} \Psi M_2.$$

It then follows that for the reduction $A \leq_{\boldsymbol{\beta\eta}} 1_2\to1\to1\to3$ we can take

$$\Phi = \lambda M{:}A.\lambda p{:}1_2\lambda p_1 p_2{:}1.\Psi M.$$

It remains to prove the claim. The only interesting direction is $(\Leftarrow)$. This follows in two ways. We first show that

$$\Psi^{-1}(\Psi M) =_{\boldsymbol{\beta\eta p}} M. \tag{3.18}$$

We will write down the computation for the 'matrix'

$$\begin{pmatrix} k_{11} & \\ k_{21} & k_{22} \end{pmatrix}$$

which is perfectly general:

$$
\begin{array}{lll}
\Psi M & =_{\beta} & \lambda F{:}2.M[\lambda f_{11}{:}1_{k_{11}}.p_1(F(P^1 f_{11}^{k_{11}\to 1}))] \\
 & & [\lambda f_{21}{:}1_{k_{21}}\lambda f_{22}{:}1_{k_{22}}.p_2(F(P^2 f_{21}^{k_{21}\to 1} f_{22}^{k_{22}\to 1}))]; \\
\Psi^{-1}(\Psi M) & =_{\beta} & \lambda K_1{:}(k_{11})\lambda K_2{:}(k_{21},k_{22}). \\
 & & \Psi M\underline{(\lambda f{:}1.p^1[K_1(P_1^1 f)^{1\to k_{11}}][K_2(P_1^2 f)^{1\to k_{21}}(P_2^2 f)^{1\to k_{22}}])} \\
 & \equiv & \lambda K_1{:}(k_{11})\lambda K_2{:}(k_{21},k_{22}).\Psi M\underline{H}, \text{ say,} \\
 & =_{\beta} & \lambda K_1 K_2.M[\lambda f_{11}.p_1(H(P^1 f_{11}^{k_{11}\to 1}))] \\
 & & [\lambda f_{21}\lambda f_{22}.p_2(H(P^2 f_{21}^{k_{21}\to 1} f_{22}^{k_{22}\to 1}))]; \\
 & =_{\beta p} & \lambda K_1 K_2.M[\lambda f_{11}.p_1(p^2[K_1 f_{11}][..\text{`junk'}..])] \\
 & & [\lambda f_{21}\lambda f_{22}.p_2(p^2[..\text{`junk'}..][K_2 f_{21} f_{22}])]; \\
 & =_{p} & \lambda K_1 K_2.M(\lambda f_{11}.K_1 f_{11})(\lambda f_{21} f_{22}.K_2 f_{21} f_{22}) \\
 & =_{\eta} & \lambda K_1 K_2.M K_1 K_2 \\
 & =_{\eta} & M,
\end{array}
$$

since

$$
\begin{array}{r}
H(P^1 f_{11}) =_{\beta p} p^2[K_1 f_{11}][..\text{`junk'}..] \\
H(P^2 f_{21}^{k_{21}\to 1} f_{22}^{k_{22}\to 1}) =_{\beta p} p^2[..\text{`junk'}..][K_2 f_{21} f_{22}].
\end{array}
$$

The argument now can be finished in a model-theoretic or syntactic way.

♣ If $\Psi M_1 =_{\beta\eta} \Psi M_2$, then $\Psi^{-1}(\Psi M_1) =_{\beta\eta} \Psi^{-1}(\Psi M_2)$. But then by (3.18) $M_1 =_{\beta\eta p} M_2$. It follows from the Church–Rosser theorem for $\beta\eta p$ that $M_1 =_{\beta\eta} M_2$, since these terms do not contain p. Goto ■.

♠ If $\Psi M_1 =_{\beta\eta} \Psi M_2$, then

$$\lambda p{:}1_2\lambda p_1 p_2{:}1.\Psi^{-1}(\Psi M_1) =_{\beta\eta} \lambda p{:}1_2\lambda p_1 p_2{:}1.\Psi^{-1}(\Psi M_2).$$

Hence

$$\mathcal{M}(\omega) \models \lambda p{:}1_2\lambda p_1 p_2{:}1.\Psi^{-1}\Psi(M_1) = \lambda p{:}1_2\lambda p_1 p_2{:}1.\Psi^{-1}(\Psi M_2).$$

Let $\mathbf{q}$ be an actual pairing on ω with projections $\mathbf{q_1}, \mathbf{q_2}$. Then in $\mathcal{M}(\omega)$

$$(\lambda p{:}1_2\lambda p_1 p_2{:}1.\Psi^{-1}(\Psi M_1))\mathbf{q}\mathbf{q_1}\mathbf{q_2} = \lambda p{:}1_2\lambda p_1 p_2{:}1.\Psi^{-1}(\Psi M_2)\mathbf{q}\mathbf{q_1}\mathbf{q_2}.$$

Since $(\mathcal{M}(\omega), \mathbf{q}, \mathbf{q_1}, \mathbf{q_2})$ is a model of $\beta\eta p$ conversion it follows from (3.18) that

$$\mathcal{M}(\omega) \models M_1 = M_2.$$

But then $M_1 =_{\beta\eta} M_2$, by a result of Friedman (1975). ■

We will see below, in Corollary 3.4.32(i), that Friedman's result will follow from the reducibility theorem. Therefore the syntactic approach is preferable.

The proof of the next proposition is again syntactic. To warm up do Exercise 3.10.

3.4.6 Proposition *Let A be a type of rank ≤ 2. Then*

$$2 \to A \leq_{\beta\eta} 1 \to 1 \to 0 \to A.$$

Proof Let $A \equiv (k_1, \ldots, k_n) = 1_{k_1} \to \cdots \to 1_{k_n} \to 0$. The term that will perform the reduction is relatively simple:

$$\Phi \equiv \lambda M{:}(2 \to A) \lambda f, g{:}1 \lambda z{:}0.M(\lambda h{:}1.f(h(g(hz)))).$$

In order to show that, for all $M_1, M_2{:}2 \to A$, one has

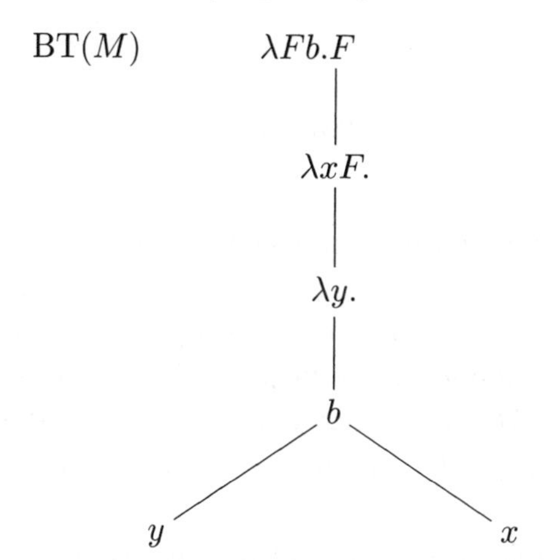

Note that its translation has the following long normal form:

$$\begin{aligned}
\Phi M &= \lambda f, g{:}1 \lambda z{:}0 \lambda b{:}1_2.f(N_x[x{:}= g(N_x[x{:}= z]])), \\
&\qquad \text{where } N_x \equiv f(b(g(bzx))x), \\
&\equiv \lambda f, g{:}1 \lambda z{:}0 \lambda b{:}1_2.f(f(b(g(bz[g(f(b(g(bzz))z))]))[g(f(b(g(bzz))z))])).
\end{aligned}$$

This term M and its translation have the trees given in Figs. 3.2 and 3.3 respectively.

Note that if we can 'read back' M from its translation ΦM, then we are done. Let $\mathtt{Cut}_{g \to z}$ be a syntactic operation on terms that replaces maximal subterms of the form gP by z. For example (omitting the abstraction prefix)

$$\mathtt{Cut}_{g \to z}(\Phi M) = f(f(bzz)).$$

Note that this gives us back the 'skeleton' of the term M, by reading $f \cdots$ as

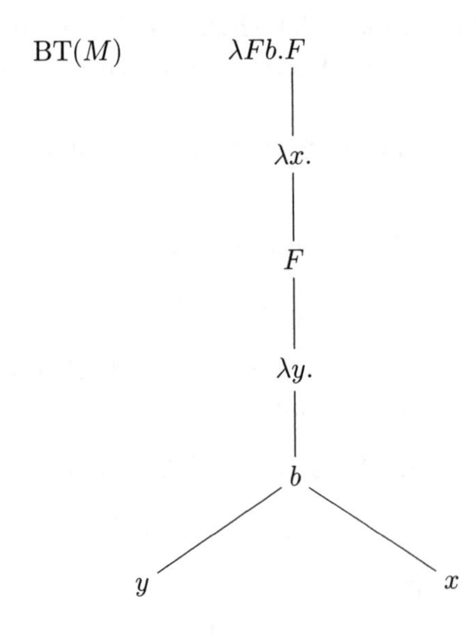

Figure 3.2

$F(\lambda\odot\cdots)$. The remaining problem is how to reconstruct the binding effect of each occurrence of the $\lambda\odot$. Using the idea of counting upwards lambdas, see de Bruijn (1972), this is done by realizing that the occurrence z coming from $g(P)$ should be bound at the position f just above where $\mathtt{Cut}_{g\to z}(P)$ matches in $\mathtt{Cut}_{g\to z}(\Phi M)$ above that z. For a precise inductive argument for this fact, see Statman (1980a), Lemma 5, or do Exercise 3.19. ∎

The following simple proposition brings almost to an end the chain of reducibility of types.

3.4.7 Proposition

$$1^4{\to}1_2{\to}0{\to}0 \leq_{\beta\eta} 1_2{\to}0{\to}0.$$

Proof As it is equally simple, let us prove instead

$$1{\to}1_2{\to}0{\to}0 \leq_{\beta\eta} 1_2{\to}0{\to}0.$$

Define $\Phi : (1{\to}1_2{\to}0{\to}0){\to}1_2{\to}0{\to}0$ by

$$\Phi \equiv \lambda M{:}(1{\to}1_2{\to}0{\to}0)\lambda b{:}1_2\lambda c{:}0.M(f^+)(b^+)c,$$

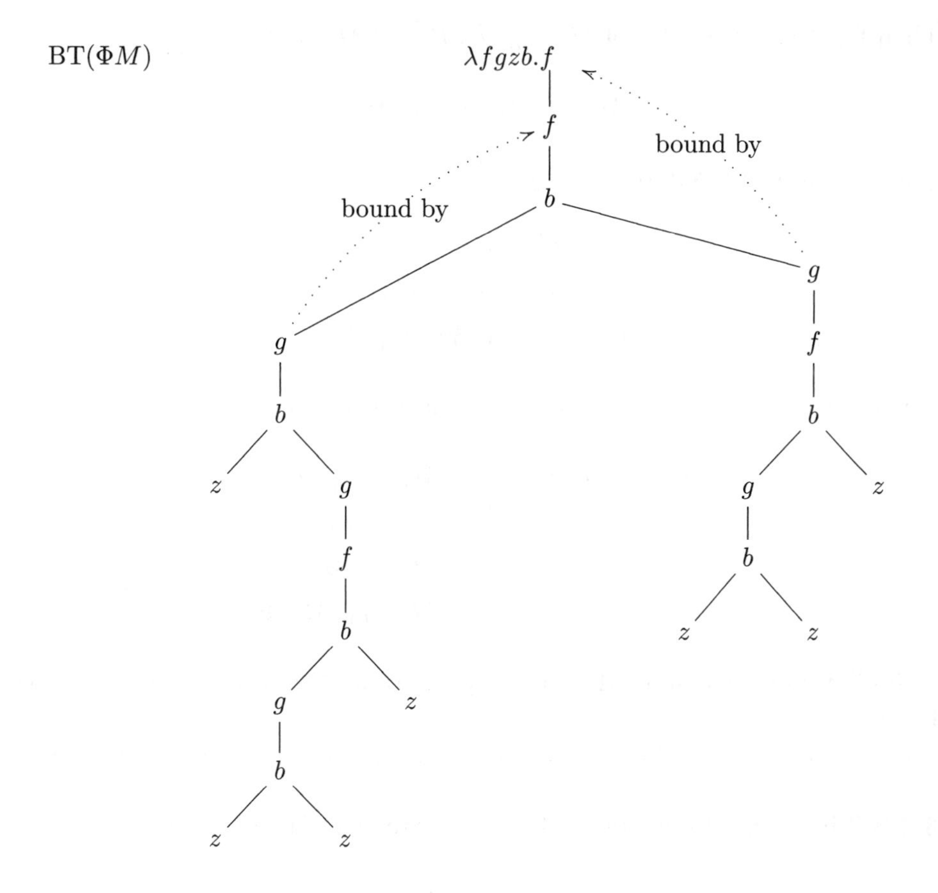

Figure 3.3

where

$$\begin{aligned} f^+ &= \lambda t{:}0.b(\#f)t; \\ b^+ &= \lambda t_1 t_2{:}0.b(\#b)(bt_1t_2); \\ \#f &= bcc; \\ \#b &= bc(bcc). \end{aligned}$$

The terms $\#f, \#b$ serve as recognizers ('Gödel numbers'). Notice that M of type $1{\to}1_2{\to}0{\to}0$ has a closed long $\boldsymbol{\beta\eta}$-nf of the form

$$M^{\mathrm{nf}} \equiv \lambda f{:}1\lambda b{:}1_2\lambda c{:}0.t,$$

with t an element of the set T generated by the grammar

$$T ::= c \mid fT \mid b\,T\,T.$$

Then for such M one has $\Phi M =_{\beta\eta} \Phi(M^{\mathrm{nf}}) \equiv M^+$ with

$$M^+ \equiv \lambda f{:}1\lambda b{:}1_2\lambda c{:}0.t^+,$$

where t^+ is recursively defined by

$$\begin{aligned} c^+ &= c; \\ (ft)^+ &= b(\#f)t^+; \\ (bt_1t_2)^+ &= b(\#b)(bt_1^+t_2^+). \end{aligned}$$

It is clear that M^{nf} can be constructed back from M^+. Therefore

$$\begin{aligned} \Phi M_1 =_{\beta\eta} \Phi M_2 &\Rightarrow M_1^+ =_{\beta\eta} M_2^+ \\ &\Rightarrow M_1^+ \equiv M_2^+ \\ &\Rightarrow M_1^{\mathrm{nf}} \equiv M_2^{\mathrm{nf}} \\ &\Rightarrow M_1 =_{\beta\eta} M_2. \blacksquare \end{aligned}$$

Similarly one can show that any type of rank ≤ 2 is reducible to $\top^2$: see Exercise 3.18.

Combining Propositions 3.4.3–3.4.7 we obtain the reducibility theorem.

3.4.8 Theorem (Reducibility Theorem, Statman (1980a)) *Let*

$$\top^2 = 1_2 \to 0 \to 0.$$

Then

$$\forall A \in \mathbb{T}^0 \ A \leq_{\beta\eta} \top^2.$$

Proof Let A be any type. Harvesting the results we obtain

$$\begin{aligned} A &\leq_{\beta\eta} B, && \text{with } \mathrm{rk}(B) \leq 3, \text{ by 3.4.3,} \\ &\leq_{\beta\eta} 1_2 \to 1^2 \to 2 \to 0, && \text{by 3.4.5,} \\ &\leq_{\beta\eta} 2 \to 1_2 \to 1^2 \to 0, && \text{by simply permuting arguments,} \\ &\leq_{\beta\eta} 1^2 \to 0 \to 1_2 \to 1^2 \to 0, && \text{by 3.4.6,} \\ &\leq_{\beta\eta} 1_2 \to 0 \to 0, && \text{by an other permutation and 3.4.7} \ \blacksquare \end{aligned}$$

Now we turn attention to the type hierarchy, Statman (1980a).

3.4.9 Definition For the ordinals $\alpha \leq \omega + 3$ define the type $A_\alpha \in \mathbb{T}^0$ as follows.

$$\begin{aligned}
A_0 &= 0;\\
A_1 &= 0{\to}0;\\
&\vdots\\
A_k &= 0^k{\to}0;\\
&\vdots\\
A_\omega &= 1{\to}0{\to}0;\\
A_{\omega+1} &= 1{\to}1{\to}0{\to}0;\\
A_{\omega+2} &= 3{\to}0{\to}0;\\
A_{\omega+3} &= 1_2{\to}0{\to}0.
\end{aligned}$$

3.4.10 Proposition *For $\alpha, \beta \leq \omega + 3$ one has*

$$\alpha \leq \beta \;\Rightarrow\; A_\alpha \leq_{\boldsymbol{\beta\eta}} A_\beta.$$

Proof For all finite k one has $A_k \leq_{\boldsymbol{\beta\eta}} A_{k+1}$ via the map

$$\Phi_{k,k+1} \equiv \lambda m{:}A_k \lambda z x_1 \cdots x_k{:}0.m x_1 \cdots x_k =_{\boldsymbol{\beta\eta}} \lambda m{:}A_k.\mathsf{K}m.$$

Moreover, $A_k \leq_{\boldsymbol{\beta\eta}} A_\omega$ via

$$\Phi_{k,\omega} \equiv \lambda m{:}A_k \lambda f{:}1 \lambda x{:}0.m(\mathbf{c}_1 f x) \cdots (\mathbf{c}_k f x).$$

Then $A_\omega \leq_{\boldsymbol{\beta\eta}} A_{\omega+1}$ via

$$\Phi_{\omega,\omega+1} \equiv \lambda m{:}A_\omega \lambda f, g{:}1 \lambda x{:}0.m f x.$$

Now $A_{\omega+1} \leq_{\boldsymbol{\beta\eta}} A_{\omega+2}$ via

$$\Phi_{\omega+1,\omega+2} \equiv \lambda m{:}A_{\omega+1} \lambda H{:}3 \lambda x{:}0.H(\lambda f{:}1.H(\lambda g{:}1.m f g x)).$$

Finally, $A_{\omega+2} \leq_{\boldsymbol{\beta\eta}} A_{\omega+3} = \top^2$ by the Reducibility Theorem 3.4.8. See Exercise 3.17 for a concrete term $\Phi_{\omega+2,\omega+3}$. ∎

3.4.11 Proposition *For $\alpha, \beta \leq \omega + 3$ one has*

$$\alpha \leq \beta \Leftarrow A_\alpha \leq_{\boldsymbol{\beta\eta}} A_\beta.$$

Proof This will be proved in 3.5.53. ∎

3.4.12 Corollary *For $\alpha, \beta \leq \omega + 3$ one has*

$$A_\alpha \leq_{\boldsymbol{\beta\eta}} A_\beta \;\Leftrightarrow\; \alpha \leq \beta.$$

For a proof that these types $\{A_\alpha\}_{\alpha \leq \omega+3}$ are a good representation of the reducibility classes we need some syntactic notions.

3.4.13 Definition A type $A \in \mathbb{T}^0$ is called *large* if it has a negative subterm occurrence, see Definition 9.3.1,of the form $B_1 \to \cdots \to B_n \to 0$, with $n \geq 2$; otherwise A is *small.*

3.4.14 Example $1_2 \to 0 \to 0$ and $((1_2 \to 0) \to 0) \to 0$ are large; $(1_2 \to 0) \to 0$ and $3 \to 0 \to 0$ are small.

Now we can partition the types $\mathbb{T} = \mathbb{T}^0$ into the following classes.

3.4.15 Definition (Type Hierarchy) Define the following sets of types:

$$\begin{array}{lcl}
\mathbb{T}_{-1} & = & \{A \mid A \text{ is not inhabited}\}; \\
\mathbb{T}_0 & = & \{A \mid A \text{ is inhabited, small, } \mathrm{rk}(A) = 1 \text{ and} \\
& & \qquad A \text{ has exactly one component of rank } 0\}; \\
\mathbb{T}_1 & = & \{A \mid A \text{ is inhabited, small, } \mathrm{rk}(A) = 1 \text{ and} \\
& & \qquad A \text{ has at least two components of rank } 0\}; \\
\mathbb{T}_2 & = & \{A \mid A \text{ is inhabited, small, } \mathrm{rk}(A) \in \{2,3\} \text{ and} \\
& & \qquad A \text{ has exactly one component of rank 1 or } 2\}; \\
\mathbb{T}_3 & = & \{A \mid A \text{ is inhabited, small, } \mathrm{rk}(A) \in \{2,3\} \text{ and} \\
& & \qquad A \text{ has at least two components of rank 1 or } 2\}; \\
\mathbb{T}_4 & = & \{A \mid A \text{ is inhabited, small and } \mathrm{rk}(A) > 3\}; \\
\mathbb{T}_5 & = & \{A \mid A \text{ is inhabited and large}\}.
\end{array}$$

Typical elements of $\mathbb{T}_{-1}$ are $0, 2, 4, \ldots$. We will not devote much attention to this class. The types in $\mathbb{T}_0, \ldots, \mathbb{T}_5$ are all inhabited. The unique element of $\mathbb{T}_0$ is $1 = 0 \to 0$ and the elements of $\mathbb{T}_1$ are 1_p, with $k \geq 2$, see the next lemma. Typical elements of $\mathbb{T}_2$ are $1 \to 0 \to 0$, $2 \to 0$ and also $0 \to 1 \to 0 \to 0$, $0 \to (1_3 \to 0) \to 0 \to 0$. The types in $\mathbb{T}_1, \ldots, \mathbb{T}_4$ are all small. Types in $\mathbb{T}_0 \cup \mathbb{T}_1$ all have rank 1; types in $\mathbb{T}_2 \cup \cdots \cup \mathbb{T}_5$ all have rank ≥ 2.

Examples of types of rank 2 not in $\mathbb{T}_2$ are

$$(1 \to 1 \to 0 \to 0) \in \mathbb{T}_3 \quad \text{and} \quad (1_2 \to 0 \to 0) \in \mathbb{T}_5.$$

Examples of types of rank 3 not in $\mathbb{T}_2$ are

$$((1_2 \to 0) \to 1 \to 0) \in \mathbb{T}_3 \quad \text{and} \quad ((1 \to 1 \to 0) \to 0 \to 0) \in \mathbb{T}_5.$$

3.4.16 Lemma *Let $A \in \mathbb{T}$. Then:*

(i) $A \in \mathbb{T}_0$ *iff* $A = (0 \to 0)$;

(ii) $A \in \mathbb{T}_1$ *iff* $A = (0^p \to 0)$, *for* $p \geq 2$;

(iii) $A \in \mathbb{T}_2$ *iff up to permutation of components*

$$A \in \{(1_p \to 0) \to 0^q \to 0 \mid p \geq 1,\ q \geq 0\} \cup \{1 \to 0^q \to 0 \mid q \geq 1\}.$$

Proof (i), (ii). If $\mathrm{rk}(A) = 1$, then $A = 0^p \to 0$, $p \geq 1$. If $A \in \mathbb{T}_0$, then $p = 1$; if $A \in \mathbb{T}_1$, then $p \geq 2$. The converse implications are obvious.

(iii) Clearly the displayed types all belong to $\mathbb{T}_2$. Conversely, let $A \in \mathbb{T}_2$. Then A is inhabited and small with rank in $\{2, 3\}$ and only one component of maximal rank.

Case $\mathrm{rk}(A) = 2$. Then $A = A_1 \to \cdots \to A_a \to 0$, with $\mathrm{rk}(A_i) \leq 1$ and exactly one A_j has rank 1. Then, up to permutation, $A = (0^p \to 0) \to 0^q \to 0$. Since A is small $p = 1$; since A is inhabited $q \geq 1$; therefore $A = 1 \to 0^q \to 0$, in this case.

Case $\mathrm{rk}(A) = 3$. Then it follows similarly that $A = A_1 \to 0^q \to 0$, with $A_1 = B \to 0$ and $\mathrm{rk}(B) = 1$. Then $B = 1_p$ with $p \geq 1$. Therefore $A = (1_p \to 0) \to 0^q \to 0$, where now $q = 0$ is possible, since $(1_p \to 0) \to 0$ is already inhabited by $\lambda m.m(\lambda x_1 \ldots x_p.x_1)$. ∎

3.4.17 Proposition *The $\mathbb{T}_i$ form a partition of $\mathbb{T}^0$.*

Proof The classes are disjoint by definition.

Any type of rank ≤ 1 belongs to $\mathbb{T}_{-1} \cup \mathbb{T}_0 \cup \mathbb{T}_1$. Any other type is either not inhabited and belongs to $\mathbb{T}_{-1}$, or belongs to $\mathbb{T}_2 \cup \mathbb{T}_3 \cup \mathbb{T}_4 \cup \mathbb{T}_5$. ∎

3.4.18 Theorem (Hierarchy Theorem, Statman (1980a))

(i) *The set of types $\mathbb{T}^0$ over the unique groundtype 0 is partitioned in the classes $\mathbb{T}_{-1}, \mathbb{T}_0, \mathbb{T}_1, \mathbb{T}_2, \mathbb{T}_3, \mathbb{T}_4, \mathbb{T}_5$.*

(ii) *Moreover,*

$$\begin{array}{lcll} A \in \mathbb{T}_5 & \Leftrightarrow & A \sim_{\beta\eta} 1_2 \to 0 \to 0; & \\ A \in \mathbb{T}_4 & \Leftrightarrow & A \sim_{\beta\eta} 3 \to 0 \to 0; & \\ A \in \mathbb{T}_3 & \Leftrightarrow & A \sim_{\beta\eta} 1 \to 1 \to 0 \to 0; & \\ A \in \mathbb{T}_2 & \Leftrightarrow & A \sim_{\beta\eta} 1 \to 0 \to 0; & \\ A \in \mathbb{T}_1 & \Leftrightarrow & A \sim_{\beta\eta} 0^k \to 0, & \textit{for some } k > 1; \\ A \in \mathbb{T}_0 & \Leftrightarrow & A \sim_{\beta\eta} 0 \to 0; & \\ A \in \mathbb{T}_{-1} & \Leftrightarrow & A \sim_{\beta\eta} 0. & \end{array}$$

(iii)

$$\begin{array}{llll} 0 & <_{\beta\eta} & 0 \to 0 & \in \mathbb{T}_0 \\ & <_{\beta\eta} & 0^2 \to 0 & \\ & <_{\beta\eta} & \cdots & \\ & <_{\beta\eta} & 0^k \to 0 & \in \mathbb{T}_1 \\ & <_{\beta\eta} & \cdots & \\ & <_{\beta\eta} & 1 \to 0 \to 0 & \in \mathbb{T}_2 \\ & <_{\beta\eta} & 1 \to 1 \to 0 \to 0 & \in \mathbb{T}_3 \\ & <_{\beta\eta} & 3 \to 0 \to 0 & \in \mathbb{T}_4 \\ & <_{\beta\eta} & 1_2 \to 0 \to 0 & \in \mathbb{T}_5. \end{array}$$

Proof (i) By Proposition 3.4.17.

(ii) By (i) and Corollary 3.4.12 it suffices to show just the implications, $\Rightarrow$.

As to $\mathbb{T}_5$, it is enough to show that $1_2{\to}0{\to}0 \leq_{\beta\eta} A$, for every inhabited large type A, since we know already the converse. For this see Statman (1980a), Lemma 7. To warm-up do Exercise 3.26.

As to $\mathbb{T}_4$, it is shown in Statman (1980a), Proposition 2, that if A is small, then $A \leq_{\beta\eta} 3{\to}0{\to}0$. It remains to show that for any small inhabited type A of rank > 3 one has $3{\to}0{\to}0 \leq_{\beta\eta} A$. Do Exercise 3.30.

As to $\mathbb{T}_3$, the implication is shown in Statman (1980a), Lemma 12. The condition about the type in that lemma is equivalent to belonging to $\mathbb{T}_3$.

As to $\mathbb{T}_2$, do Exercise 3.28(ii).

As to $\mathbb{T}_i$, with $i = 1, 0, -1$, notice that $\Lambda^{\emptyset}(0^k{\to}0)$ contains exactly k closed terms for $k \geq 0$. This is sufficient.

(iii) By Corollary 3.4.12. ∎

3.4.19 Definition Let $A \in \mathbb{T}^0$. The *class* of A, written class(A), is the unique i with $i \in \{-1, 0, 1, 2, 3, 4, 5\}$ such that $A \in \mathbb{T}_i$.

3.4.20 Remark

(i) Note that by the Hierarchy Theorem one has for all $A, B \in \mathbb{T}^0$

$$A \leq_{\beta\eta} B \Rightarrow \text{class}(A) \leq \text{class}(B).$$

(ii) Since $B \leq_{\beta\eta} A{\to}B$ via the map $\Phi = \lambda x^B y^A.x$, it follows that

$$\text{class}(B) \leq \text{class}(A \to B).$$

3.4.21 Remark Let

$$\begin{aligned} B_{-1} &= 0, \\ B_0 &= 0{\to}0, \\ B_{1,k} &= 0^k{\to}0, \qquad \text{with } k > 1, \\ B_2 &= 1{\to}0{\to}0, \\ B_3 &= 1{\to}1{\to}0{\to}0, \\ B_4 &= 3{\to}0{\to}0, \\ B_5 &= 1_2{\to}0{\to}0. \end{aligned}$$

Then for $A \in \mathbb{T}^0$ one has

(i) If $i \neq 1$, then

$$\text{class}(A) = i \Leftrightarrow A \sim_{\beta\eta} B_i.$$

(ii)
$$\begin{aligned} \text{class}(A) = 1 &\Leftrightarrow \exists k.A \sim_{\beta\eta} B_{1,k}. \\ &\Leftrightarrow \exists k.A \equiv B_{1,k}. \end{aligned}$$

This follows from the Hierarchy Theorem.

For an application in the next section we need a variant of the Hierarchy Theorem.

3.4.22 Definition Let $A \equiv A_1 \to \cdots \to A_a \to 0$, $B \equiv B_1 \to \cdots \to B_b \to 0$ be types.

(i) We say A is *head-reducible to* B, written $A \leq_h B$, iff for some term $\Phi \in \Lambda^{\emptyset}_{\to}(A \to B)$ one has

$$\forall M_1, M_2 \in \Lambda^{\emptyset}_{\to}(A)\ [M_1 =_{\beta\eta} M_2 \ \Leftrightarrow\ \Phi M_1 =_{\beta\eta} \Phi M_2],$$

and moreover Φ is of the form

$$\Phi = \lambda m{:}A \lambda x_1{:}B_0 \cdots x_b{:}B_b.m P_1 \cdots P_a, \tag{3.19}$$

with $\mathrm{FV}(P_1, \ldots, P_a) \subseteq \{x_1, \ldots, x_b\}$ and $m \notin \{x_1 \ldots x_b\}$.

(ii) We say A is *multi head-reducible to* B, written $A \leq_{h^+} B$, iff there are closed terms $\Phi_1, \ldots, \Phi_m \in \Lambda^{\emptyset}(A \to B)$ each of the form (3.19) such that

$$\forall M_1, M_2 \in \Lambda^{\emptyset}_{\to}(A)\ [M_1 =_{\beta\eta} M_2 \ \Leftrightarrow$$
$$\Phi_1 M_1 =_{\beta\eta} \Phi_1 M_2 \ \& \cdots \&\ \Phi_m M_1 =_{\beta\eta} \Phi_m M_2].$$

(iii) Write $A \sim_h B$ iff $A \leq_h B \leq_h A$ and similarly $A \sim_{h^+} B$ iff $A \leq_{h^+} B \leq_{h^+} A$.

Clearly $A \leq_h B \Rightarrow A \leq_{h^+} B$. Moreover, both $\leq_h$ and $\leq_{h^+}$ are transitive, see Exercise 3.14. We will formulate in Corollary 3.4.27 a variant of the Hierarchy Theorem.

3.4.23 Lemma $0 \leq_h 1 \leq_h 0^2 \to 0 \leq_h 1 \to 0 \to 0 \leq_h 1 \to 1 \to 0 \to 0$.

Proof By inspecting the proof of Proposition 3.4.10. ∎

3.4.24 Lemma

(i) $1 \to 0 \to 0 \not\leq_{h^+} 0^k \to 0$, *for* $k \geq 0$.
(ii) *If* $A \leq_{h^+} 1 \to 0 \to 0$, *then* $A \leq_{\beta\eta} 1 \to 0 \to 0$.
(iii) $1_2 \to 0 \to 0 \not\leq_{h^+} 1 \to 0 \to 0$, $3 \to 0 \to 0 \not\leq_{h^+} 1 \to 0 \to 0$, *and* $1 \to 1 \to 0 \to 0 \not\leq_{h^+} 1 \to 0 \to 0$.
(iv) $0^2 \to 0 \not\leq_{h^+} 0 \to 0$.
(v) *Let* $A, B \in \mathbb{T}^0$. *If* $\Lambda^{\emptyset}_{\to}(A)$ *is infinite and* $\Lambda^{\emptyset}_{\to}(B)$ *finite, then* $A \not\leq_{h^+} B$.

Proof (i) By a cardinality argument, $\Lambda^{\emptyset}_{\to}(1 \to 0 \to 0)$ contains infinitely many different elements. These cannot be mapped injectively into the finite $\Lambda^{\emptyset}_{\to}(0^k \to 0)$, not even by way of $\leq_{h^+}$.

(ii) Suppose $A \leq_{h^+} 1 \to 0 \to 0$ via $\Phi_1, \ldots, \Phi_k$. Then each element M of $\Lambda^{\emptyset}_{\to}(A)$ is mapped to a k-tuple of Church numerals $\langle \Phi_1(M), \ldots, \Phi_k(M) \rangle$.

This k-tuple can be coded as a single numeral by iterating the Cantorian pairing function on the natural numbers, which is polynomially definable and hence λ-definable.

(iii) By (ii) and the Hierarchy Theorem.

(iv) Type $0^2 \to 0$ contains two closed terms. These cannot be mapped injectively into the singleton $\Lambda^{\emptyset}_{\to}(0\to 0)$, even not by the multiple maps.

(v) Suppose $A \leq_{h^+} B$ via $\Phi_1, \dots, \Phi_k$. Then the sequences

$$\langle \Phi_1(M), \dots, \Phi_k(M) \rangle$$

are all different for $M \in \Lambda^{\emptyset}_{\to}(A)$. As B is finite (with say m elements), there are only finitely many sequences of length k (in fact m^k). This is impossible as $\Lambda^{\emptyset}_{\to}(A)$ is infinite. ■

3.4.25 Proposition *Let $A, B \in \mathbb{T}^0_i$. We have:*

(i) *If $i \notin \{1, 2\}$, then $A \sim_h B$;*
(ii) *If $i \in \{1, 2\}$, then $A \sim_{h^+} B$.*

Proof (i) Since $A, B \in \mathbb{T}_i$ and $i \neq 1$, then by Theorem 3.4.18 one has $A \sim_{\beta\eta} B$. By inspection of the proof of that theorem, in all cases except for $A \in \mathbb{T}_2$ one obtains $A \sim_h B$. Do Exercise 3.29.

(ii) *Case* $i = 1$. Write $A_k = 0^k \to 0$. It is easy to show that $A_2 \leq_h A_p$ for $p \geq 2$. One needs to check that $A_k \leq_{h^+} A_2$ for $k \geq 2$. Without loss of generality take $k = 3$. Then $M \in \Lambda^{\emptyset}_{\to}(A_3)$ is of the form $M \equiv \lambda x_1 x_2 x_3 . x_i$. Hence for $M, N \in \Lambda^{\emptyset}_{\to}(A_3)$ with $M \neq_{\beta\eta} N$ either

$$\lambda y_1 y_2 . M y_1 y_1 y_2 \neq_{\beta\eta} \lambda y_1 y_2 . N y_1 y_1 y_2 \quad \text{or} \quad \lambda y_1 y_2 . M y_1 y_2 y_2 \neq_{\beta\eta} \lambda y_1 y_2 . N y_1 y_2 y_2 .$$

Hence $A_3 \leq_{h^+} A_2$.

Case $i = 2$. Do Exercise 3.28. ■

3.4.26 Corollary *Let $A, B \in \mathbb{T}^0$, with*

$$A = A_1 \to \cdots \to A_a \to 0, \quad B = B_1 \to \cdots \to B_b \to 0.$$

Then:

(i) $A \sim_h B \;\Rightarrow\; A \sim_{\beta\eta} B$;
(ii) $A \sim_{\beta\eta} B \;\Rightarrow\; A \sim_{h^+} B$;
(iii) *Suppose $A \leq_{h^+} B$. Then for $M, N \in \Lambda^{\emptyset}(A)$*

$$M \neq_{\beta\eta} N \; (: A) \;\Rightarrow\; \lambda \vec{x} . M R_1 \dots R_a \neq_{\beta\eta} \lambda \vec{x} . N R_1 \dots R_a \; (: B),$$

for some fixed $R_1, \dots, R_a$ with $\mathrm{FV}(\vec{R}) \subseteq \{\vec{x}\} = \{x_1^{B_1}, \dots, x_b^{B_b}\}$.

Proof (i) Trivially one has $A \leq_h B \Rightarrow A \leq_{\beta\eta} B$. The result follows.
(ii) By the proposition and the Hierarchy Theorem. (iii) By the definition of $\leq_{h^+}$. ■

3.4.27 Corollary (Hierarchy Theorem Revisited, Statman (1980b))

$$\begin{array}{lcl} A \in \mathbb{T}_5 & \Leftrightarrow & A \sim_h 1_2{\to}0{\to}0; \\ A \in \mathbb{T}_4 & \Leftrightarrow & A \sim_h 3{\to}0{\to}0; \\ A \in \mathbb{T}_3 & \Leftrightarrow & A \sim_h 1{\to}1{\to}0{\to}0; \\ A \in \mathbb{T}_2 & \Leftrightarrow & A \sim_{h^+} 1{\to}0{\to}0; \\ A \in \mathbb{T}_1 & \Leftrightarrow & A \sim_{h^+} 0^2{\to}0; \\ A \in \mathbb{T}_0 & \Leftrightarrow & A \sim_h 0{\to}0; \\ A \in \mathbb{T}_{-1} & \Leftrightarrow & A \sim_h 0. \end{array}$$

Proof The Hierarchy Theorem 3.4.18 and Proposition 3.4.25 establish the implications, $\Rightarrow$. Since $\sim_h$ implies $\sim_{\beta\eta}$, we only have to prove the converses in the cases $A \sim_{h^+} 1{\to}0{\to}0$ and $A \sim_{h^+} 0^2{\to}0$. Suppose $A \sim_{h^+} 1{\to}0{\to}0$, but $A \notin \mathbb{T}_2$. Again by the Hierarchy Theorem one has $A \in \mathbb{T}_3 \cup \mathbb{T}_4 \cup \mathbb{T}_5$ or $A \in \mathbb{T}_{-1} \cup \mathbb{T}_0 \cup \mathbb{T}_1$. If $A \in \mathbb{T}_3$, then $A \sim_{\beta\eta} 1{\to}1{\to}0{\to}0$, hence $A \sim_{h^+} 1{\to}1{\to}0{\to}0$. Then $1{\to}0{\to}0 \sim_{h^+} 1{\to}1{\to}0{\to}0$, contradicting Lemma 3.4.24(ii). If $A \in \mathbb{T}_4$ or $A \in \mathbb{T}_5$, then a contradiction can be obtained similarly.

In the second case A is either empty or $A \equiv 0^k{\to}0$, for some $k > 0$; moreover $1{\to}0{\to}0 \leq_{h^+} A$. The subcase that A is empty cannot occur, since $1{\to}0{\to}0$ is inhabited. The subcase $A \equiv 0^k{\to}0$, contradicts Lemma 3.4.24(i).

Finally, suppose $A \sim_{h^+} 0^2{\to}0$ and $A \notin \mathbb{T}_1$. If $A \in \mathbb{T}_{-1} \cup \mathbb{T}_0$, then $\Lambda^{\emptyset}_{\to}(A)$ has at most one element. This contradicts $0^2{\to}0 \leq_{h^+} A$, as $0^2{\to}0$ has two distinct elements. If $A \in \mathbb{T}_2 \cup \mathbb{T}_3 \cup \mathbb{T}_4 \cup \mathbb{T}_5$, then $1{\to}0{\to}0 \leq_{\beta\eta} A \leq_{h^+} 0^2{\to}0$, giving A infinitely many closed inhabitants, contradicting Lemma 3.4.24(v). ■

Applications of the reducibility theorem

The reducibility theorem has several consequences.

3.4.28 Definition Let $\mathcal{C}$ be a class of $\boldsymbol{\lambda}^{\mathrm{Ch}}_{\to}$ models. We say $\mathcal{C}$ is *complete* iff

$$\forall M, N \in \Lambda^{\emptyset}[\mathcal{C} \models M = N \Leftrightarrow M =_{\beta\eta} N].$$

3.4.29 Definition

(i) $\mathcal{T} = \mathcal{T}_{b,c}$ is the algebraic structure of trees recursively defined as

$$\mathcal{T} = c \mid b\ \mathcal{T}\ \mathcal{T}$$

(ii) For a typed λ-model $\mathcal{M}$ we say that $\mathcal{T}$ *can be embedded into* $\mathcal{M}$, written $\mathcal{T} \hookrightarrow \mathcal{M}$, iff there exist $b_0 \in \mathcal{M}(0{\to}0{\to}0), c_0 \in \mathcal{M}(0)$ such that

$$\forall t, s \in \mathcal{T}[t \neq s \Rightarrow \mathcal{M} \models t^{\mathrm{cl}} b_0 c_0 \neq s^{\mathrm{cl}} b_0 c_0],$$

where $u^{\mathrm{cl}} = \lambda b{:}0{\to}0{\to}0 \lambda c{:}0.u$, is the closure of $u \in \mathcal{T}$.

The elements of $\mathcal{T}$ are binary trees with c on the leaves and b on the connecting nodes. Typical examples are $c, bcc, bc(bcc)$ and $b(bcc)c$. The existence of an embedding using b_0, c_0 implies for example that $b_0c_0(b_0c_0c_0), b_0c_0c_0$ and c_0 are mutually different in $\mathcal{M}$.

Note that $\mathcal{T} \not\hookrightarrow \mathcal{M}_2 (= \mathcal{M}_{\{1,2\}})$. To see this, write $gx = bxx$. One has $g^2(c) \neq g^4(c)$, but $\mathcal{M}_2 \models \forall g{:}0{\to}0 \forall c{:}0.g^2(c) = g^4(c)$: see Exercise 3.20.

Remember that $\top^2 = 1_2{\to}0{\to}0$, the type of binary trees, see Definition 1.4.12.

3.4.30 Lemma

(i) $\Pi_{i \in I} \mathcal{M}_i \models M = N \Leftrightarrow \forall i \in I. \mathcal{M}_i \models M = N$.
(ii) $M \in \Lambda^{\emptyset}(\top^2) \Leftrightarrow \exists s \in \mathcal{T}. M =_{\beta\eta} s^{\mathrm{cl}}$.

Proof (i) Since $[\![M]\!]^{\Pi_{i \in I} \mathcal{M}_i} = \boldsymbol{\lambda} i \in I. [\![M]\!]^{\mathcal{M}_i}$.

(ii) By an analysis of the possible shapes of the normal forms of terms of type $\top^2$. ∎

3.4.31 Theorem (1-section theorem, Statman (1985)) *The class $\mathcal{C}$ is complete iff there is an (at most countable) family $\{\mathcal{M}_i\}_{i \in I}$ of structures in $\mathcal{C}$ such that*

$$\mathcal{T} \hookrightarrow \Pi_{i \in I} \mathcal{M}_i.$$

Proof ($\Rightarrow$) Suppose $\mathcal{C}$ is complete. Let $t, s \in \mathcal{T}$. Then

$$\begin{aligned} t \neq s &\Rightarrow t^{\mathrm{cl}} \neq_{\beta\eta} s^{\mathrm{cl}} & \\ &\Rightarrow \mathcal{C} \not\models t^{\mathrm{cl}} = s^{\mathrm{cl}}, & \text{by completeness,} \\ &\Rightarrow \mathcal{M}_{ts} \models t^{\mathrm{cl}} \neq s^{\mathrm{cl}}, & \text{for some } \mathcal{M}_{ts} \in \mathcal{C}, \\ &\Rightarrow \mathcal{M}_{ts} \models t^{\mathrm{cl}} b_{ts} c_{ts} \neq s^{\mathrm{cl}} b_{ts} c_{ts}, & \end{aligned}$$

for some $b_{ts} \in \mathcal{M}(0{\to}0{\to}0), c_{ts} \in \mathcal{M}(0)$ by extensionality. Note that in the third implication the axiom of (countable) choice is used.

It now follows by Lemma 3.4.30(i) that we can take as countable product $\Pi_{t' \neq s'} \mathcal{M}_{t's'}$

$$\Pi_{t' \neq s'} \mathcal{M}_{t's'} \models t^{\mathrm{cl}} \neq s^{\mathrm{cl}},$$

since they differ on the pair (b_0, c_0) with $b_0(ts) = b_{ts}$ and similarly for c_0.

($\Leftarrow$) Suppose $\mathcal{T} \hookrightarrow \Pi_{i\in I}\mathcal{M}_i$ with $\mathcal{M}_i \in \mathcal{C}$. Let M, N be closed terms of some type A. By soundness one has

$$M =_{\beta\eta} N \Rightarrow \mathcal{C} \models M = N.$$

For the converse, by the Reducibility Theorem, let $F : A \to \top^2$ be such that

$$M =_{\beta\eta} N \Leftrightarrow FM =_{\beta\eta} FN,$$

for all $M, N \in \Lambda^{\emptyset}_{\to}$. Then

$$\begin{array}{llll} \mathcal{C} \models M = N & \Rightarrow & \Pi_{i\in I}\mathcal{M}_i \models M = N, & \text{by the lemma,} \\ & \Rightarrow & \Pi_{i\in I}\mathcal{M}_i \models FM = FN, & \\ & \Rightarrow & \Pi_{i\in I}\mathcal{M}_i \models t^{\mathrm{cl}} = s^{\mathrm{cl}}, & \end{array}$$

where t, s are such that

$$FM =_{\beta\eta} t^{\mathrm{cl}}, FN =_{\beta\eta} s^{\mathrm{cl}}, \tag{3.20}$$

as by Lemma 2.1.18 every closed term of type $\top^2$ is $\boldsymbol{\beta\eta}$-convertible to some u^{cl} with $u \in \mathcal{T}$. Now the chain of arguments continues as follows:

$$\begin{array}{lll} \Rightarrow & t \equiv s, & \text{by the embedding property,} \\ \Rightarrow & FM =_{\beta\eta} FN, & \text{by (3.20),} \\ \Rightarrow & M =_{\beta\eta} N, & \text{by reducibility. } \blacksquare \end{array}$$

3.4.32 Corollary

(i) [Friedman (1975)] $\{\mathcal{M}_\omega\}$ is complete.
(ii) [Plotkin (1980)] $\{\mathcal{M}_n \mid n \in \omega\}$ is complete.
(iii) $\{\mathcal{M}_{\omega_\perp}\}$ is complete.
(iv) $\{\mathcal{M}_D \mid D \text{ a finite cpo}\}$, is complete.

Proof Immediate from the theorem. ■

The completeness of the collection $\{\mathcal{M}_n\}_{n\in\omega}$ essentially states that for every pair of terms M, N of a given type A there is a number $n = n_{M,N}$ such that $\mathcal{M}_n \models M = N \Rightarrow M =_{\beta\eta} N$. Actually one can do better, by showing that n only depends on M.

3.4.33 Proposition (Finite completeness theorem, Statman (1982)) *For every type A in $\mathbb{T}^0$ and every $M \in \Lambda^{\emptyset}(A)$ there is a number $n = n_M$ such that for all $N \in \Lambda^{\emptyset}(A)$*

$$\mathcal{M}_n \models M = N \Leftrightarrow M =_{\beta\eta} N.$$

Proof By the Reduction Theorem 3.4.8 it suffices to show this for $A = \top^2$. Let M a closed term of type $\top^2$ be given. Each closed term N of type $\top^2$ has as long $\boldsymbol{\beta\eta}$-nf

$$N = \lambda b{:}1_2\lambda c{:}0.s_N,$$

where $s_N \in \mathcal{T}$. Let $\mathbf{p} : \omega{\rightarrow}\omega{\rightarrow}\omega$ be an injective pairing on the integers such that $\mathbf{p}(k_1, k_2) > k_i$. Take

$$n_M = ([\![M]\!]^{\mathcal{M}_\omega}\mathbf{p}\,0) + 1.$$

Define $\mathbf{p}'{:}X_{n+1}^2{\rightarrow}X_{n+1}$, where $X_{n+1} = \{0, \ldots, n+1\}$, by

$$\begin{aligned} \mathbf{p}'(k_1, k_2) &= \mathbf{p}(k_1, k_2), && \text{if } k_1, k_2 \le n \;\&\; \mathbf{p}(k_1, k_2) \le n; \\ &= n+1, && \text{otherwise.} \end{aligned}$$

Suppose $\mathcal{M}_n \models M = N$. Then $[\![M]\!]^{\mathcal{M}_n}\mathbf{p}'\,0 = [\![N]\!]^{\mathcal{M}_n}\mathbf{p}'\,0$. By the choice of n it follows that $[\![M]\!]^{\mathcal{M}_n}\mathbf{p}\,0 = [\![N]\!]^{\mathcal{M}_n}\mathbf{p}\,0$ and hence $s_M = s_N$. Therefore $M =_{\boldsymbol{\beta\eta}} N$. ∎

3.5 The five canonical term-models

We work with $\boldsymbol{\lambda}_{\rightarrow}^{\mathrm{Ch}}$ based on $\mathbb{T}^0$. We often will use for a term like $\lambda x^A.x^A$ its de Bruijn notation $\lambda x{:}A.x$, since it takes less space. Another advantage of this notation is that we can write $\lambda f{:}1\,x{:}0.f^2x \equiv \lambda f{:}1\,x{:}0.f(fx)$, which is $\lambda f^1\,x^0.f^1(f^1x^0)$ in Church's notation.

The open terms of $\boldsymbol{\lambda}_{\rightarrow}^{\mathrm{Ch}}$ form an extensional model, the term-model $\mathcal{M}_{\Lambda_{\rightarrow}}$. One may wonder whether there are also closed term-models, like in the untyped lambda calculus. If no constants are present, then this is not the case, since there are, for example, no closed terms of ground type 0. In the presence of constants matters change. We will first show how a set of constants $\mathcal{D}$ gives rise to an extensional equivalence relation on $\Lambda_{\rightarrow}^{\varnothing}[\mathcal{D}]$, the set of closed terms with constants from $\mathcal{D}$. Then we define canonical sets of constants and prove that for these the resulting equivalence relation is also a congruence, i.e. determines a term-model. After that it will be shown that for all sets $\mathcal{D}$ of constants with enough closed terms the extensional equivalence determines a term-model. Up to elementary equivalence (satisfying the same set of equations between closed pure terms, i.e. closed terms without any constants), all models for which the equality on type 0 coincides with $=_{\boldsymbol{\beta\eta}}$, can be obtained in this way.

3.5.1 Definition Let $\mathcal{D}$ be a set of constants, each with its own type in $\mathbb{T}^0$. Then $\mathcal{D}$ is *sufficient* if for every $A \in \mathbb{T}^0$ there is a closed term $M \in \Lambda_{\rightarrow}^{\varnothing}[D](A)$.

For example $\{x^0\}, \{F^2, f^1\}$ are sufficient. But $\{f^1\}, \{\Psi^3, f^1\}$ are not. Note that

$$\mathcal{D} \text{ is sufficient } \Leftrightarrow \Lambda^{\emptyset}_{\rightarrow}[\mathcal{D}](0) \neq \emptyset.$$

3.5.2 Definition Let $M, N \in \Lambda^{\emptyset}_{\rightarrow}[\mathcal{D}](A)$ with $A = A_1 \rightarrow \ldots \rightarrow A_a \rightarrow 0$.

(i) We say M is *$\mathcal{D}$-extensionally equivalent* with N, written $M \approx^{\mathtt{ext}}_{\mathcal{D}} N$, iff

$$\forall t_1 \in \Lambda^{\emptyset}_{\rightarrow}[\mathcal{D}](A_1) \ldots t_a \in \Lambda^{\emptyset}_{\rightarrow}[\mathcal{D}](A_a).M\vec{t} =_{\beta\eta} N\vec{t}.$$

[If $a = 0$, then $M, N \in \Lambda^{\emptyset}_{\rightarrow}[\mathcal{D}](0)$; in this case $M \approx^{\mathtt{ext}}_{\mathcal{D}} N \Leftrightarrow M =_{\beta\eta} N$.]

(ii) We say M is *$\mathcal{D}$-observationally equivalent* with N, written $M \approx^{\mathtt{obs}}_{\mathcal{D}} N$, iff

$$\forall F \in \Lambda^{\emptyset}_{\rightarrow}[\mathcal{D}](A \rightarrow 0)\, FM =_{\beta\eta} FN.$$

3.5.3 Remark

(i) Let $M, N \in \Lambda^{\emptyset}_{\rightarrow}[\mathcal{D}](A)$ and $F \in \Lambda^{\emptyset}_{\rightarrow}[\mathcal{D}](A \rightarrow B)$. Then

$$M \approx^{\mathtt{obs}}_{\mathcal{D}} N \Rightarrow FM \approx^{\mathtt{obs}}_{\mathcal{D}} FN.$$

(ii) Let $M, N \in \Lambda^{\emptyset}_{\rightarrow}[\mathcal{D}](A \rightarrow B)$. Then

$$M \approx^{\mathtt{ext}}_{\mathcal{D}} N \Leftrightarrow \forall Z \in \Lambda^{\emptyset}_{\rightarrow}[\mathcal{D}](A).MZ \approx^{\mathtt{ext}}_{\mathcal{D}} NZ.$$

(iii) Let $M, N \in \Lambda^{\emptyset}_{\rightarrow}[\mathcal{D}](A)$. Then

$$M \approx^{\mathtt{obs}}_{\mathcal{D}} N \Rightarrow M \approx^{\mathtt{ext}}_{\mathcal{D}} N,$$

by taking $F \equiv \lambda m.m\vec{t}$.

Note that in the definition of extensional equivalence the $\vec{t}$ range over closed terms (containing possibly constants). So this notion is not the same as $\beta\eta$-convertibility: M and N may act differently on different variables, even if they act the same on all those closed terms. The relation $\approx^{\mathtt{ext}}_{\mathcal{D}}$ is related to what is called in the untyped calculus the ω-rule, see B[1984], §17.3.

The intuition behind observational equivalence is that for M, N of higher type A one cannot 'see' that they are equal, unlike for terms of type 0. But one can do 'experiments' with M and N, the outcome of which is observational, i.e. of type 0, by putting these terms in a context $C[-]$ resulting in two terms of type 0. For closed terms this amounts to considering just FM and FN for all $F \in \Lambda^{\emptyset}_{\rightarrow}[\mathcal{D}](A \rightarrow 0)$.

The main result in this section is Theorem 3.5.35, it states that for all $\mathcal{D}$ and for all $M, N \in \Lambda^{\emptyset}_{\rightarrow}[\mathcal{D}]$ of the same type one has

$$M \approx^{\mathtt{ext}}_{\mathcal{D}} N \Leftrightarrow M \approx^{\mathtt{obs}}_{\mathcal{D}} N. \tag{3.21}$$

After this has been proved, we can write simply $M \approx_{\mathcal{D}} N$. The equivalence (3.21) will first be established in Corollary 3.5.18 for some 'canonical' sets of constants. The general result, Theorem 3.5.35, will follow using the theory of type reducibility.

The following obvious result is often used.

3.5.4 Remark Let $M \equiv M[\vec{\boldsymbol{d}}], N \equiv N[\vec{\boldsymbol{d}}] \in \Lambda^{\emptyset}_{\to}[\mathcal{D}](A)$, where all occurrences of $\vec{\boldsymbol{d}}$ are displayed. Then

$$M[\vec{\boldsymbol{d}}] =_{\beta\eta} N[\vec{\boldsymbol{d}}] \Leftrightarrow \lambda\vec{x}.M[\vec{x}] =_{\beta\eta} \lambda\vec{x}.N[\vec{x}].$$

The reason is that new constants and fresh variables are used in the same way and that the latter can be bound.

3.5.5 Proposition *Suppose that $\approx^{\mathtt{ext}}_{\mathcal{D}}$ is logical on $\Lambda^{\emptyset}_{\to}[\mathcal{D}]$. Then*

$$\forall M, N \in \Lambda^{\emptyset}_{\to}[\mathcal{D}]\,[M \approx^{\mathtt{ext}}_{\mathcal{D}} N \Leftrightarrow M \approx^{\mathtt{obs}}_{\mathcal{D}} N].$$

Proof By Remark 3.5.3(iii) we only have to show ($\Rightarrow$). So assume $M \approx^{\mathtt{ext}}_{\mathcal{D}} N$. Let $F \in \Lambda^{\emptyset}_{\to}[\mathcal{D}](A \to 0)$. Then trivially

$$\begin{array}{llll} & F & \approx^{\mathtt{ext}}_{\mathcal{D}} & F \\ \Rightarrow & FM & \approx^{\mathtt{ext}}_{\mathcal{D}} & FN, \quad \text{as by assumption } \approx^{\mathtt{ext}}_{\mathcal{D}} \text{ is logical,} \\ \Rightarrow & FM & =_{\beta\eta} & FN, \quad \text{because the type is } 0. \end{array}$$

Therefore $M \approx^{\mathtt{obs}}_{\mathcal{D}} N$. ∎

The converse of Proposition 3.5.5 is a good warm-up exercise. That is, if

$$\forall M, N \in \Lambda^{\emptyset}_{\to}[\mathcal{D}]\,[M \approx^{\mathtt{ext}}_{\mathcal{D}} N \Leftrightarrow M \approx^{\mathtt{obs}}_{\mathcal{D}} N],$$

then $\approx^{\mathtt{ext}}_{\mathcal{D}}$ is the logical relation on $\Lambda^{\emptyset}_{\to}[\mathcal{D}]$ determined by $\boldsymbol{\beta\eta}$-equality on $\Lambda^{\emptyset}_{\to}[\mathcal{D}](0)$.

3.5.6 Definition By $\text{BetaEta}^{\mathcal{D}} = \{\text{BetaEta}^{\mathcal{D}}_A\}_{A \in \mathbb{T}^0}$ we mean the logical relation on $\Lambda^{\emptyset}_{\to}[\mathcal{D}]$ determined by

$$\text{BetaEta}^{\mathcal{D}}_0(M, N) \Leftrightarrow M =_{\beta\eta} N,$$

for $M, N \in \Lambda^{\emptyset}_{\to}[\mathcal{D}](0)$.

3.5.7 Lemma *Let $\boldsymbol{d} = \boldsymbol{d}^{A \to 0} \in \mathcal{D}$, with $A = A_1 \to \cdots \to A_a \to 0$. Suppose*

(i) $\forall F, G \in \Lambda^{\emptyset}_{\to}[\mathcal{D}](A)[F \approx^{\mathtt{ext}}_{\mathcal{D}} G \Rightarrow F =_{\beta\eta} G]$;
(ii) $\forall t_i \in \Lambda^{\emptyset}_{\to}[\mathcal{D}](A_i)\ \text{BetaEta}^{\mathcal{D}}(t_i, t_i)$, $1 \leq i \leq a$.

Then $\text{BetaEta}^{\mathcal{D}}_{A \to 0}(\boldsymbol{d}, \boldsymbol{d})$.

Proof Write $S = \text{BetaEta}^{\mathcal{D}}$. Let $\boldsymbol{d}$ be given. Then

$$\begin{array}{llll} S(F,G) & \Rightarrow & F\vec{t} =_{\beta\eta} G\vec{t}, & \text{since } \forall \vec{t} \in \Lambda^{\emptyset}_{\rightarrow}[\mathcal{D}]\ S(t_i,t_i) \text{ by assumption (ii)}, \\ & \Rightarrow & F \approx^{\mathtt{ext}}_{\mathcal{D}} G, & \\ & \Rightarrow & F =_{\beta\eta} G, & \text{by assumption (i)}, \\ & \Rightarrow & \boldsymbol{d}F =_{\beta\eta} \boldsymbol{d}G. & \end{array}$$

Therefore we have by definition $S(\boldsymbol{d},\boldsymbol{d})$. ∎

3.5.8 Lemma *Let S be a syntactic n-ary logical relation on $\Lambda^{\emptyset}_{\rightarrow}[\mathcal{D}]$ that is closed under $=_{\beta\eta}$. Suppose $S(\boldsymbol{d},\ldots,\boldsymbol{d})$ holds for all $\boldsymbol{d} \in \mathcal{D}$. Then for all $M \in \Lambda^{\emptyset}_{\rightarrow}[\mathcal{D}]$ one has*

$$S(M,\ldots,M).$$

Proof Let $\mathcal{D} = \{\boldsymbol{d}_1^{A_1},\ldots,\boldsymbol{d}_n^{A_n}\}$. M can be written as

$$M \equiv M[\vec{\boldsymbol{d}}\,] =_{\beta\eta} (\lambda\vec{x}.M[\vec{x}])\vec{\boldsymbol{d}} \equiv M^{+}\vec{\boldsymbol{d}},$$

with M^{+} a closed and pure term (i.e. without free variables or constants). Then

$$\begin{array}{lll} & S(M^{+},\ldots,M^{+}), & \text{by the fundamental theorem} \\ & & \text{for syntactic logical relations} \\ \Rightarrow & S(M^{+}\vec{\boldsymbol{d}},\ldots,M^{+}\vec{\boldsymbol{d}}), & \text{since } S \text{ is logical and } \forall \boldsymbol{d} \in \mathcal{D}.S(\vec{\boldsymbol{d}}), \\ \Rightarrow & S(M,\ldots,M), & \text{since } S \text{ is } =_{\beta\eta} \text{ closed. } \blacksquare \end{array}$$

3.5.9 Lemma *Suppose that for all $\boldsymbol{d} \in \mathcal{D}$ one has* $\text{BetaEta}^{\mathcal{D}}(\boldsymbol{d},\boldsymbol{d})$. *Then $\approx^{\mathtt{ext}}_{\mathcal{D}}$ is* $\text{BetaEta}^{\mathcal{D}}$ *and hence logical.*

Proof Write $S = \text{BetaEta}^{\mathcal{D}}$. By the assumption and the fact that S is $=_{\beta\eta}$ closed (since S_0 is), Lemma 3.5.8 implies that

$$S(M,M) \tag{3.22}$$

for all $M \in \Lambda^{\emptyset}_{\rightarrow}[\mathcal{D}]$. It now follows that S is an equivalence relation on $\Lambda^{\emptyset}_{\rightarrow}[\mathcal{D}]$. We claim

$$S_A(F,G) \Leftrightarrow F \approx^{\mathtt{ext}}_{\mathcal{D}} G,$$

for all $F,G \in \Lambda^{\emptyset}_{\rightarrow}[\mathcal{D}](A)$. This is proved by induction on the structure of A. If $A = 0$, then this follows by definition. If $A = B \rightarrow C$, then we proceed as follows.

$$
\begin{array}{llll}
(\Rightarrow) & S_{B\to C}(F,G) & \Rightarrow & S_C(Ft,Gt), \quad \text{for all } t\in\Lambda^{\emptyset}_{\to}[\mathcal{D}](B), \text{ since } t\approx^{\mathtt{ext}}_{\mathcal{D}} t \\
& & & \text{hence, by the induction hypothesis, } S_B(t,t), \\
& & \Rightarrow & Ft\approx^{\mathtt{ext}}_{\mathcal{D}} Gt, \quad \text{for all } t\in\Lambda^{\emptyset}_{\to}[\mathcal{D}], \\
& & & \text{by the induction hypothesis,} \\
& & \Rightarrow & F\approx^{\mathtt{ext}}_{\mathcal{D}} G, \qquad \text{by definition.}
\end{array}
$$

$$
\begin{array}{llll}
(\Leftarrow) & F\approx^{\mathtt{ext}}_{\mathcal{D}} G & \Rightarrow & Ft\approx^{\mathtt{ext}}_{\mathcal{D}} Gt, \quad \text{for all } t\in\Lambda^{\emptyset}_{\to}[\mathcal{D}], \\
& & \Rightarrow & S_C(Ft,Gt)
\end{array} \tag{3.23}
$$

by the induction hypothesis. In order to prove $S_{B\to C}(F,G)$, assume $S_B(t,s)$ towards $S_C(Ft,Gs)$. Well, since also $S_{B\to C}(G,G)$, by (0), we have

$$S_C(Gt,Gs). \tag{3.24}$$

It follows from (3.23) and (3.24) and the transitivity of S (which on this type is the same as $\approx^{\mathtt{ext}}_{\mathcal{D}}$ by the induction hypothesis) that $S_C(Ft,Gs)$ indeed.

By the claim $\approx^{\mathtt{ext}}_{\mathcal{D}}$ is S and therefore $\approx^{\mathtt{ext}}_{\mathcal{D}}$ is logical. ■

3.5.10 Definition Let $\mathcal{D}=\{c_1^{A_1},\dots,c_k^{A_k}\}$ be a finite set of typed constants.

(i) The *characteristic type* of $\mathcal{D}$, written $\nabla(\mathcal{D})$, is $A_1\to\cdots\to A_k\to 0$.
(ii) We say that a type $A=A_1\to\cdots\to A_a\to 0$ is *represented in* $\mathcal{D}$ iff there are distinct constants $\boldsymbol{d}_1^{A_1},\dots,\boldsymbol{d}_a^{A_a}\in\mathcal{D}$.

In other words, $\nabla(\mathcal{D})$ is intuitively the type of $\lambda\vec{\boldsymbol{d}}_i.\boldsymbol{d}^0$, where $\mathcal{D}=\{\vec{\boldsymbol{d}}_i\}$ (the order of the abstractions is immaterial, as the resulting types are all $\sim_{\beta\eta}$ equivalent). Note that $\nabla(\mathcal{D})$ is represented in $\mathcal{D}$.

3.5.11 Definition Let $\mathcal{D}$ be a set of constants.

(i) If $\mathcal{D}$ is finite, then the *class* of $\mathcal{D}$ is the class of the type $\nabla(\mathcal{D})$, i.e. the unique i such that $\nabla(\mathcal{D})\in\mathbb{T}_i$.
(ii) In general the class of $\mathcal{D}$ is

$$\max\{\text{class}(A)\mid A \text{ represented in } \mathcal{D}\}.$$

(iii) A *characteristic type* of $\mathcal{D}$, written $\nabla(\mathcal{D})$, is any A represented in $\mathcal{D}$ such that $\text{class}(\mathcal{D})=\text{class}(A)$. That is, $\nabla(\mathcal{D})$ is any type of the highest class represented in $\mathcal{D}$.

It is not hard to see that for finite $\mathcal{D}$ the two definitions of class($\mathcal{D}$) coincide.

3.5.12 Remark Note that it follows by Remark 3.4.20 that

$$\mathcal{D}_1\subseteq\mathcal{D}_2 \Rightarrow \text{class}(\mathcal{D}_1)\le\text{class}(\mathcal{D}_2).$$

In order to show that, for arbitrary $\mathcal{D}$, extensional equivalence is the same as observational equivalence we will first prove it for the following 'canonical' sets of constants.

3.5.13 Definition The following sets of constants will play a crucial role in this section:

$$\begin{array}{lcl} \mathcal{C}_{-1} & = & \emptyset; \\ \mathcal{C}_0 & = & \{\boldsymbol{c}^0\}; \\ \mathcal{C}_1 & = & \{\boldsymbol{c}^0, \boldsymbol{d}^0\}; \\ \mathcal{C}_2 & = & \{\boldsymbol{f}^1, \boldsymbol{c}^0\}; \\ \mathcal{C}_3 & = & \{\boldsymbol{f}^1, \boldsymbol{g}^1, \boldsymbol{c}^0\}; \\ \mathcal{C}_4 & = & \{\boldsymbol{\Phi}^3, \boldsymbol{c}^0\}; \\ \mathcal{C}_5 & = & \{\boldsymbol{b}^{1_2}, \boldsymbol{c}^0\}. \end{array}$$

3.5.14 Remark The actual names of the constants are irrelevant, for example $\mathcal{C}_2$ and $\mathcal{C}_2' = \{\boldsymbol{g}^1, \boldsymbol{c}^0\}$ will give rise to isomorphic term models. Therefore we may assume that a set of constants $\mathcal{D}$ of class i is disjoint with $\mathcal{C}_i$.

From now on in this section $\mathcal{C}$ ranges over the canonical sets of constants $\{\mathcal{C}_{-1}, \dots, \mathcal{C}_5\}$ and $\mathcal{D}$ over arbitrary sets of constants.

3.5.15 Remark Let $\mathcal{C}$ be one of the canonical sets of constants. The characteristic types of these $\mathcal{C}$ are:

$$\begin{array}{lcl} \nabla(\mathcal{C}_{-1}) & = & 0; \\ \nabla(\mathcal{C}_0) & = & 0 \to 0; \\ \nabla(\mathcal{C}_1) & = & 1_2 = 0 \to 0 \to 0; \\ \nabla(\mathcal{C}_2) & = & 1 \to 0 \to 0; \\ \nabla(\mathcal{C}_3) & = & 1 \to 1 \to 0 \to 0; \\ \nabla(\mathcal{C}_4) & = & 3 \to 0 \to 0; \\ \nabla(\mathcal{C}_5) & = & 1_2 \to 0 \to 0. \end{array}$$

Also one has

$$i \leq j \;\Leftrightarrow\; \nabla(\mathcal{C}_i) \leq_{\beta\eta} \nabla(\mathcal{C}_j),$$

as follows from the theory of type reducibility.

We will need the following combinatorial lemma about $\approx^{\mathtt{ext}}_{\mathcal{C}_4}$.

3.5.16 Lemma *For every $F, G \in \Lambda[\mathcal{C}_4](2)$ one has*

$$F \approx^{\mathtt{ext}}_{\mathcal{C}_4} G \;\Rightarrow\; F =_{\beta\eta} G.$$

Proof We must show

$$[\forall h \in \Lambda[\mathcal{C}_4](1). Fh =_{\beta\eta} Gh] \;\Rightarrow\; F =_{\beta\eta} G. \tag{3.25}$$

In order to do this, we have to classify the elements of $\Lambda[\mathcal{C}_4](2)$. Define for $A \in \mathbb{T}^0$ and context Δ

$$A_\Delta = \{M \in \Lambda[\mathcal{C}_4](A) \mid \Delta \vdash M : A \ \& \ M \text{ in } \boldsymbol{\beta\eta}\text{-nf}\}.$$

It is easy to show that 0_Δ and 2_Δ are generated by the following 'two-level' grammar, see van Wijngaarden (1981):

$$\begin{aligned} 2_\Delta &= \lambda f{:}1.0_{\Delta,f:1} \\ 0_\Delta &= \boldsymbol{c} \mid \boldsymbol{\Phi}\, 2_\Delta \mid \Delta.1\, 0_\Delta, \end{aligned}$$

where $\Delta.A$ consists of $\{v \mid v^A \in \Delta\}$.

It follows that a typical element of $2_\emptyset$ is

$$\lambda f_1{:}1.\boldsymbol{\Phi}(\lambda f_2{:}1.f_1(f_2(\boldsymbol{\Phi}(\lambda f_3{:}1.f_3(f_2(f_1(f_3\, \boldsymbol{c})))))))).$$

Hence a general element can be represented by a list of words

$$\langle w_1, \ldots, w_n \rangle,$$

with $w_i \in \Sigma_i^*$ and $\Sigma_i = \{f_1, \ldots, f_i\}$, the representation of the typical element above being $\langle \epsilon, f_1 f_2, f_3 f_2 f_1 f_3 \rangle$. The inhabitation machines in Section 1.3 were inspired by this example.

Let $h_m = \lambda z{:}0.\boldsymbol{\Phi}(\lambda g{:}1.g^m(z))$; then $h_m \in 1_\emptyset$. We claim that

$$\forall F, G \in \Lambda^{\emptyset}_{\rightarrow}[\mathcal{C}_4](2)\, \exists m \in \omega.[F h_m =_{\boldsymbol{\beta\eta}} G h_m \ \Rightarrow \ F =_{\boldsymbol{\beta\eta}} G].$$

For a given $F \in \Lambda[\mathcal{C}_4](2)$ and $m \in \omega$ one can find a representation of the $\boldsymbol{\beta\eta}$-nf of $F h_m$ from the representation of the $\boldsymbol{\beta\eta}$-nf $F^{\text{nf}} \in 2_\emptyset$ of F. It will turn out that if m is large enough, then F^{nf} can be determined ('read back') from the $\boldsymbol{\beta\eta}$-nf of $F h_m$.

In order to see this, let F^{nf} be represented by the list of words $\langle w_1, \ldots, w_n \rangle$, as above. The occurrences of f_1 can be made explicit and we write

$$w_i = w_{i0} f_1 w_{i1} f_1 w_{i2} \cdots f_1 w_{ik_i}.$$

Some of the w_{ij} will be empty (certainly the w_{1j}) and $w_{ij} \in \Sigma_i^{-*}$ with $\Sigma_i^- = \{f_2, \ldots, f_i\}$. Then F^{nf} can be written as (using for application – in contrast with the usual convention – association to the right)

$$\begin{aligned} F^{\text{nf}} \equiv\ & \lambda f_1.w_{10} f_1 w_{11} \cdots f_1 w_{1k_1} \\ & \boldsymbol{\Phi}(\lambda f_2.w_{20} f_1 w_{21} \cdots f_1 w_{2k_2} \\ & \vdots \\ & \boldsymbol{\Phi}(\lambda f_n.w_{n0} f_1 w_{n1} \cdots f_1 w_{nk_n} \\ & \boldsymbol{c})..). \end{aligned}$$

Now we have

$$
\begin{aligned}
(Fh_m)^{\mathtt{nf}} \equiv\; & w_{10} \\
& \boldsymbol{\Phi}(\lambda g.g^m w_{11} \\
& \vdots \\
& \boldsymbol{\Phi}(\lambda g.g^m w_{1k_1} \\
& \boldsymbol{\Phi}(\lambda f_2.w_{20} \\
& \boldsymbol{\Phi}(\lambda g.g^m w_{21} \\
& \vdots \\
& \boldsymbol{\Phi}(\lambda g.g^m w_{2k_2} \\
& \boldsymbol{\Phi}(\lambda f_3.w_{30} \\
& \boldsymbol{\Phi}(\lambda g.g^m w_{31} \\
& \vdots \\
& \boldsymbol{\Phi}(\lambda g.g^m w_{3k_3} \\
& \vdots \\
& \boldsymbol{\Phi}(\lambda f_n.w_{n0} \\
& \boldsymbol{\Phi}(\lambda g.g^m w_{n1} \\
& \vdots \\
& \boldsymbol{\Phi}(\lambda g.g^m w_{nk_n} \\
& \mathbf{c})..))..)..)))..)))..).
\end{aligned}
$$

So if $m > \mathtt{max}_{ij}\{\mathtt{length}(w_{ij})\}$ we can read back the w_{ij} and hence $F^{\mathtt{nf}}$ from $(Fh_m)^{\mathtt{nf}}$. Therefore using an m large enough (3.25) can be derived as follows:

$$
\begin{aligned}
\forall h \in \Lambda[\mathcal{C}_4](1).Fh =_{\boldsymbol{\beta\eta}} Gh \;&\Rightarrow\; Fh_m =_{\boldsymbol{\beta\eta}} Gh_m \\
&\Rightarrow\; (Fh_m)^{\mathtt{nf}} \equiv (Gh_m)^{\mathtt{nf}} \\
&\Rightarrow\; F^{\mathtt{nf}} \equiv G^{\mathtt{nf}} \\
&\Rightarrow\; F =_{\boldsymbol{\beta\eta}} F^{\mathtt{nf}} \equiv G^{\mathtt{nf}} =_{\boldsymbol{\beta\eta}} G. \blacksquare
\end{aligned}
$$

3.5.17 Proposition *For all $i \in \{-1, 0, 1, 2, 3, 4, 5\}$ the relations $\approx^{\mathrm{ext}}_{\mathcal{C}_i}$ are logical.*

Proof Write $\mathcal{C} = \mathcal{C}_i$. For $i = -1$ the relation $\approx^{\mathtt{ext}}_{\mathcal{C}}$ is universally valid by the empty implication, as there are never terms $\vec{t}$ making $M\vec{t}, N\vec{t}$ of type 0. Therefore, the result is trivially valid.

Let S be the logical relation on $\Lambda^{\emptyset}_{\to}[\mathcal{C}]$ determined by $=_{\beta\eta}$ on the ground level $\Lambda^{\emptyset}_{\to}[\mathcal{C}](0)$. By Lemma 3.5.9 we have to check $S(\boldsymbol{c},\boldsymbol{c})$ for all constants $\boldsymbol{c}$ in $\mathcal{C}_i$. For $i \neq 4$ this is easy (trivial for constants of type 0 and almost trivial for the ones of type 1 and $1_2 = (0^2 \to 0)$; in fact for all terms $h \in \Lambda^{\emptyset}_{\to}[\mathcal{C}]$ of these types one has $S(h,h)$).

For $i = 4$ we reason as follows. Write $S = \text{BetaEta}^{\mathcal{C}_4}$. By Lemma 3.5.9 it is enough to show that $S(\Phi^3, \Phi^3)$. By Lemma 3.5.7 it suffices to prove

$$F \approx_{\mathcal{C}_4} G \Rightarrow F =_{\beta\eta} G$$

for all $F, G \in \Lambda^{\emptyset}_{\to}[\mathcal{C}_4](2)$, which has been verified in Lemma 3.5.16, and $S(t,t)$ for all $t \in \Lambda^{\emptyset}_{\to}[\mathcal{C}_4](1)$, which follows directly from the definition of S, since $=_{\beta\eta}$ is a congruence:

$$\forall M, N \in \Lambda^{\emptyset}_{\to}[0].[M =_{\beta\eta} N \Rightarrow tM =_{\beta\eta} tN]. \blacksquare$$

3.5.18 Corollary *Let $\mathcal{C}$ be one of the canonical classes of constants. Then*

$$\forall M, N \in \Lambda^{\emptyset}_{\to}[\mathcal{C}][M \approx^{\text{obs}}_{\mathcal{C}} N \Leftrightarrow M \approx^{\text{ext}}_{\mathcal{C}} N].$$

Proof By the Proposition and Proposition 3.5.5. ■

Arbitrary finite sets of constants $\mathcal{D}$

Now we pay attention to arbitrary finite sets of constants $\mathcal{D}$.

3.5.19 Remark Before starting the proof of the next results it is worth realizing the following. For $M, N \in \Lambda^{\emptyset}_{\to}[\mathcal{D} \cup \{\boldsymbol{c}^A\}] \setminus \Lambda^{\emptyset}_{\to}[\mathcal{D}]$ it makes sense to state $M \approx^{\text{ext}}_{\mathcal{D}} N$, but in general we do not have

$$M \approx^{\text{ext}}_{\mathcal{D}} N \Rightarrow M \approx^{\text{ext}}_{\mathcal{D} \cup \{\boldsymbol{c}^A\}} N. \tag{3.26}$$

Indeed, taking $\mathcal{D} = \{\boldsymbol{d}^0\}$ this is the case for $M \equiv \lambda x^0 b^{1_2}.b\boldsymbol{c}^0 x$, $N \equiv \lambda x^0 b^{1_2}.b\boldsymbol{c}^0 \boldsymbol{d}^0$. The implication (3.26) does hold if $\text{class}(\mathcal{D}) = \text{class}(\mathcal{D} \cup \{\boldsymbol{c}^A\})$, as we will see later.

We first need to show the following result.

Proposition (Lemma P_i, with $i \in \{3,4,5\}$) *Let $\mathcal{D}$ be a finite set of constants of class $i > 2$ and $\mathcal{C} = \mathcal{C}_i$. Then for $M, N \in \Lambda^{\emptyset}_{\to}[\mathcal{D}]$ of the same type we have*

$$M \approx^{\text{ext}}_{\mathcal{D}} N \Rightarrow M \approx^{\text{ext}}_{\mathcal{D} \cup \mathcal{C}} N.$$

We will assume that $\mathcal{D} \cap \mathcal{C} = \emptyset$, see Remark 3.5.14. This assumption is not yet essential since, if $\mathcal{D}, \mathcal{C}$ overlap, then the statement $M \approx^{\text{ext}}_{\mathcal{D} \cup \mathcal{C}} N$ is easier to prove. The proof occupies 3.5.20–3.5.28.

3.5.20 Notation Let $A = A_1 \to \cdots \to A_a \to 0$ and $d \in \Lambda^{\emptyset}_{\to}[\mathcal{D}](0)$. Define $\mathsf{K}^A d \in \Lambda^{\emptyset}_{\to}[\mathcal{D}](A)$ by

$$\mathsf{K}^A d = (\lambda x_1{:}A_1 \cdots \lambda x_a{:}A_a.d).$$

3.5.21 Lemma *Let $\mathcal{D}$ be a finite set of constants of class $i>1$. Then for all $A \in \mathbb{T}^0$ the set $\Lambda^{\emptyset}_{\to}[\mathcal{D}](A)$ contains infinitely many distinct long normal forms.*

Proof Because $i > -1$ there is a term in $\Lambda^{\emptyset}_{\to}[\mathcal{D}](\nabla(\mathcal{D}))$. Hence $\mathcal{D}$ is sufficient and there exists a $d^0 \in \Lambda^{\emptyset}_{\to}[\mathcal{D}](0)$ in lnf. Since $i>1$ there is a constant $\boldsymbol{d}^B \in \mathcal{D}$ with $B = B_1 \to \cdots \to B_b \to 0$, and $b > 0$. Define the sequence of elements in $\Lambda^{\emptyset}_{\to}[\mathcal{D}](0)$:

$$\begin{aligned} d_0 &= d^0; \\ d_{k+1} &= \boldsymbol{d}^{\boldsymbol{B}}(\mathsf{K}^{B_1} d_k) \cdots (\mathsf{K}^{B_b} d_k). \end{aligned}$$

As d_k is an lnf and $|d_{k+1}| > |d_k|$, the $\{\mathsf{K}^A d_0, \mathsf{K}^A d_1, \ldots\}$ are distinct long normal forms in $\Lambda^{\emptyset}_{\to}[\mathcal{D}](A)$. ∎

3.5.22 Remark We want to show that for $M, N \in \Lambda^{\emptyset}_{\to}[\mathcal{D}]$ of the same type one has

$$M \approx^{\mathsf{ext}}_{\mathcal{D}} N \Rightarrow M \approx^{\mathsf{ext}}_{\mathcal{D} \cup \{\boldsymbol{c}^0\}} N. \tag{3.27}$$

The strategy will be to show that, for all $P, Q \in \Lambda_{\to}[\mathcal{D} \cup \{\boldsymbol{c}^{\mathbf{0}}\}](0)$ in lnf, one can find a term $T_c \in \Lambda^{\emptyset}_{\to}[\mathcal{D}](A)$ such that

$$P \not\equiv Q \Rightarrow P[\boldsymbol{c}^{\mathbf{0}} := T_c] \not\equiv Q[\boldsymbol{c}^{\mathbf{0}} := T_c]. \tag{3.28}$$

Then (3.27) can be proved via the contrapositive

$$\begin{array}{rlll} M \not\approx^{\mathsf{ext}}_{\mathcal{D} \cup \{\boldsymbol{c}^0\}} N & \Rightarrow & M\vec{t} \neq_{\beta\eta} N\vec{t}\,(:0), & \text{for some } \vec{t} \in \Lambda^{\emptyset}_{\to}[\mathcal{D} \cup \{\boldsymbol{c}^0\}] \\ & \Rightarrow & P \not\equiv Q, & \text{by taking lnfs } P, Q, \\ & \Rightarrow & P[\boldsymbol{c} := T_c] \not\equiv Q[\boldsymbol{c} := T_c], & \text{by (3.28)}, \\ & \Rightarrow & M\vec{s} \neq_{\beta\eta} N\vec{s}, & \text{with } \vec{s} = \vec{t}[\boldsymbol{c} := T_c], \\ & \Rightarrow & M \not\approx_{\mathcal{D}} N. & \end{array}$$

3.5.23 Lemma *Let $\mathcal{D}$ be of class $i \geq 1$ and let $\boldsymbol{c}^0$ be an arbitrary constant of type 0. Then for $M, N \in \Lambda^{\emptyset}_{\to}[\mathcal{D}]$ of the same type*

$$M \approx^{\mathsf{ext}}_{\mathcal{D}} N \Rightarrow M \approx^{\mathsf{ext}}_{\mathcal{D} \cup \{\boldsymbol{c}^0\}} N.$$

Proof As in Remark 3.5.22 let $P, Q \in \Lambda_{\to}[\mathcal{D} \cup \{\boldsymbol{c}^0\}](0)$ be different lnfs.

Case $i > 1$. Consider the difference in the Böhm trees of P, Q at a node with smallest length. If at that node in neither trees there is a $\boldsymbol{c}$, then we can take $T_c = d^0$ for any $d^0 \in \Lambda^{\emptyset}_{\rightarrow}[\mathcal{D}]$. If at that node in exactly one of the trees there is $\boldsymbol{c}$ and in the other tree a different $s \in \Lambda^{\emptyset}_{\rightarrow}[\mathcal{D} \cup \{\boldsymbol{c}^0\}]$, then we must take d^0 sufficiently large, which is possible by Lemma 3.5.21, in order to preserve the difference; these are all cases.

Case $i = 1$. Then $\mathcal{D} = \{\boldsymbol{d}_1^0, \ldots, \boldsymbol{d}_k^0\}$, with $k \geq 2$. So one has

$$P, Q \in \{\boldsymbol{d}_1^0, \ldots, \boldsymbol{d}_k^0, \boldsymbol{c}^0\}.$$

If $\boldsymbol{c} \notin \{P, Q\}$, then take any $T_c = \boldsymbol{d}_i$. Otherwise one has $P \equiv \boldsymbol{c}$, $Q \equiv \boldsymbol{d}_i$, say. Then take $T_c \equiv \boldsymbol{d}_j$, for some $j \neq i$. ■

3.5.24 Remark Let $\mathcal{D} = \{\boldsymbol{d}^0\}$ be of class $i = 0$. Then Lemma 3.5.23 is false. Take for example $\lambda x^0.x \approx^{\mathtt{ext}}_{\mathcal{D}} \lambda x^0.\boldsymbol{d}$, as $\boldsymbol{d}$ is the only element of $\Lambda^{\emptyset}_{\rightarrow}[\mathcal{D}](0)$. But $\lambda x^0.x \not\approx^{\mathtt{ext}}_{\{\boldsymbol{d}^0,\boldsymbol{c}^0\}} \lambda x^0.\boldsymbol{d}$.

3.5.25 Lemma (P_5) *Let $\mathcal{D}$ be a finite set of class $i = 5$ and $\mathcal{C}{=}\mathcal{C}_5 = \{\boldsymbol{c}^0, \boldsymbol{b}^{1_2}\}$. Then for $M, N \in \Lambda^{\emptyset}_{\rightarrow}[\mathcal{D}]$ of the same type one has*

$$M \approx^{\mathtt{ext}}_{\mathcal{D}} N \;\Rightarrow\; M \approx^{\mathtt{ext}}_{\mathcal{D}\cup\mathcal{C}} N.$$

Proof By Lemma 3.5.23 it suffices to show for $M, N \in \Lambda^{\emptyset}_{\rightarrow}[\mathcal{D}]$ of the same type

$$M \approx^{\mathtt{ext}}_{\mathcal{D}\cup\{\boldsymbol{c}^0\}} N \;\Rightarrow\; M \approx^{\mathtt{ext}}_{\mathcal{D}\cup\{\boldsymbol{c}^0,\boldsymbol{b}^{1_2}\}} N.$$

By Remark 3.5.22, for distinct lnfs $P, Q \in \Lambda_{\rightarrow}[\mathcal{D} \cup \{\boldsymbol{c}^0, \boldsymbol{b}^{1_2}\}](0)$, it suffices to find a term $T_b \in \Lambda_{\rightarrow}[\mathcal{D} \cup \{\boldsymbol{c}^0\}](1_2)$ such that

$$P[\boldsymbol{b} := T_b] \not\equiv Q[\boldsymbol{b} := T_b]. \tag{3.29}$$

We seek such a term that is in any case injective: for all $R, R', S, S' \in \Lambda^{\emptyset}_{\rightarrow}[\mathcal{D} \cup \{\boldsymbol{c}^0\}](0)$

$$T_b RS {=_{\beta\eta}} T_b R'S' \;\Rightarrow\; R {=_{\beta\eta}} R' \;\&\; S {=_{\beta\eta}} S'.$$

Now let $\mathcal{D} = \{\boldsymbol{d}_1{:}A_1, \ldots, \boldsymbol{d}_b{:}A_b\}$. Since $\mathcal{D}$ is of class 5 the type $\nabla(\mathcal{D}) = A_1 \rightarrow \cdots \rightarrow A_b \rightarrow 0$ is inhabited and large. Let $T \in \Lambda^{\emptyset}_{\rightarrow}[\mathcal{D}](0)$.

Recall that a type $A = A_1 \rightarrow \cdots \rightarrow A_b \rightarrow 0$ is large if it has a negative occurrence of a subtype with more than one component. So we have one of the following two cases.

Case 1. For some $i \leq b$ one has $A_i = B_1 \rightarrow \cdots \rightarrow B_b \rightarrow 0$ with $b \geq 2$.

Case 2. Each $A_i = A_i' \rightarrow 0$ and some A_i' is large, $1 \leq i \leq b$.

Now we define for a type A that is large the term $T_A \in \Lambda^{\emptyset}_{\rightarrow}[\mathcal{D}](1_2)$ by induction on the structure of A, following the above-mentioned cases.

$$\begin{array}{lcll} T_A & = & \lambda x^0 y^0.\boldsymbol{d}_i(\mathsf{K}^{B_1}x)(\mathsf{K}^{B_2}y)(\mathsf{K}^{B_3}T)\cdots(\mathsf{K}^{B_b}T), & \text{if } i \le b, \text{ with } b \ge 2, \\ & & & \text{is the least such that} \\ & & & A_i = B_1 \to \cdots \to B_b \to 0, \\ & = & \lambda x^0 y^0.\boldsymbol{d}_i(\mathsf{K}^{A'_i}(T_{A'_i}xy)), & \text{if each } A_j = A'_j \to 0 \text{ and} \\ & & & i \le a \text{ is the least such} \\ & & & \text{that } A'_i \text{ is large.} \end{array}$$

By induction on the structure of the large type A we easily show using the Church–Rosser theorem that T_A is injective in the sense above.

Let $A = \nabla(\mathcal{D})$, which is large. We cannot yet take $T_b \equiv T_A$. For example the difference $\boldsymbol{bcc} \neq_{\beta\eta} T_A\boldsymbol{cc}$ gets lost. By Lemma 3.5.21 there exists a $T^+ \in \Lambda^{\emptyset}_{\to}[\mathcal{D}](0)$ with

$$|T^+| > \max\{|P|, |Q|\}.$$

Define

$$T_b = (\lambda xy.T_A(T_A x T^+)y) \in \Lambda^{\emptyset}_{\to}[\mathcal{D}](1_2).$$

Then also this T_b is injective. The T^+ acts as a 'tag' to remember where T_b is inserted. Therefore this T_b satisfies (3.29). ∎

3.5.26 Lemma (P_4) *Let $\mathcal{D}$ be a finite set of class $i = 4$ and $\mathcal{C}{=}\mathcal{C}_4 = \{\boldsymbol{c}^0, \boldsymbol{\Phi}^3\}$. Then for $M, N \in \Lambda^{\emptyset}_{\to}[\mathcal{D}]$ of the same type one has*

$$M \approx^{\mathsf{ext}}_{\mathcal{D}} N \Rightarrow M \approx^{\mathsf{ext}}_{\mathcal{D}\cup\mathcal{C}} N.$$

Proof By Remark 3.5.22 and Lemma 3.5.23 it suffices to show that for all distinct long normal forms $P, Q \in \Lambda_{\to}[\mathcal{D} \cup \{\boldsymbol{c}^0, \boldsymbol{\Phi}^3\}](0)$ there exists a term $T_\Phi \in \Lambda_{\to}[\mathcal{D} \cup \{\boldsymbol{c}^0\}](3)$ such that

$$P[\boldsymbol{\Phi} := T_\Phi] \neq Q[\boldsymbol{\Phi} := T_\Phi]. \tag{3.30}$$

Let $A = A_1 \to \cdots \to A_a \to 0$ be a small type of rank $k \ge 2$. Without loss of generality we may assume that $\mathrm{rk}(A_1) = \mathrm{rk}(A) - 1$. As A is small we have $A_1 = B \to 0$, with B small of rank $k - 2$.

Let H be a term variable of type 2. We construct a term

$$M_A \equiv M_A[H] \in \Lambda^{\{H:2\}}_{\to}(A).$$

The term M_A is defined directly if $k \in \{2, 3\}$; otherwise via M_B, with $\mathrm{rk}(M_B)$

$= \mathrm{rk}(M_A) - 2$:

$$\begin{aligned} M_A &= \lambda x_1{:}A_1 \cdots \lambda x_a{:}A_a.Hx_1, && \text{if } \mathrm{rk}(A) = 2, \\ &= \lambda x_1{:}A_1 \cdots \lambda x_a{:}A_a.H(\lambda z{:}0.x_1(\mathsf{K}^B z)), && \text{if } \mathrm{rk}(A) = 3, \\ &= \lambda x_1{:}A_1 \cdots \lambda x_a{:}A_a.x_1 M_B, && \text{if } \mathrm{rk}(A) \geq 4. \end{aligned}$$

Let $A = \nabla(\mathcal{D})$ which is small and has rank $k \geq 4$. Then without loss of generality $A_1 = B \to 0$ has rank ≥ 3. Then $B = B_1 \to \cdots \to B_b \to 0$ has rank ≥ 2. Let

$$T = (\lambda H{:}2.\boldsymbol{d}_1^{A_1}(M_B[H])) \in \Lambda^{\emptyset}_{\to}[\mathcal{D}](3).$$

Although T is injective, we cannot use it to replace Φ^3, as the difference in (3.30) may get lost in translation. Again we need a 'tag' to keep the difference between P, Q Let $n > \max\{|P|,\ |Q|\}$. Let B_i be the 'first' with $\mathrm{rk}(B_i) = k - 3$. As B_i is small, we have $B_i = C_i \to 0$. We modify the term T:

$$T_\Phi = (\lambda H{:}2.\boldsymbol{d}_1^{A_1}(\lambda y_1{:}B_1 \cdots \lambda y_b{:}B_b.(y_i \circ \mathsf{K}^{C_i})^n(M_B[H]\,\vec{y}))) \in \Lambda^{\emptyset}_{\to}[\mathcal{D}](3).$$

This term satisfies (3.30). ∎

3.5.27 Lemma (P_3) *Let $\mathcal{D}$ be a finite set of class $i = 3$ and $\mathcal{C}{=}\mathcal{C}_3 = \{\boldsymbol{c}^0, \boldsymbol{f}^1, \boldsymbol{g}^1\}$. Then for $M, N \in \Lambda^{\emptyset}_{\to}[\mathcal{D}]$ of the same type we have*

$$M \approx^{\mathtt{ext}}_{\mathcal{D}} N \;\Rightarrow\; M \approx^{\mathtt{ext}}_{\mathcal{D}\cup\mathcal{C}} N.$$

Proof Again it suffices that for all distinct long normal forms

$$P, Q \in \Lambda_{\to}[\mathcal{D} \cup \{\boldsymbol{c}^0, \boldsymbol{f}^1, \boldsymbol{g}^1\}](0)$$

there exist terms $T_f, T_g \in \Lambda_{\to}[\mathcal{D} \cup \{\boldsymbol{c}^0\}](1)$ such that

$$P[\boldsymbol{f}, \boldsymbol{g} := T_f, T_g] \not\equiv Q[\boldsymbol{f}, \boldsymbol{g} := T_f, T_g]. \tag{3.31}$$

Writing $\mathcal{D} = \{\boldsymbol{d}_1{:}A_1, \ldots, \boldsymbol{d}_a{:}A_a\}$, for all $1 \leq i \leq a$ one has $A_i = 0$ or $A_i = B_i \to 0$ with $\mathrm{rk}(B_i) \leq 1$, since $\nabla(\mathcal{D}) \in \mathbb{T}_3$. This implies that all constants in $\mathcal{D}$ can have at most one argument. Moreover there are at least two constants, say, without loss of generality, $\boldsymbol{d}_1, \boldsymbol{d}_2$, with types $B_1 \to 0, B_2 \to 0$, respectively; that is, having one argument. As $\mathcal{D}$ is sufficient there is a $d \in \Lambda^{\emptyset}_{\to}[\mathcal{D}](0)$. Define

$$\begin{aligned} T_1 &= \lambda x{:}0.\boldsymbol{d}_1(\mathsf{K}^{B_1} x) && \text{in } \Lambda^{\emptyset}_{\to}[\mathcal{D}](1), \\ T_2 &= \lambda x{:}0.\boldsymbol{d}_2(\mathsf{K}^{B_2} x) && \text{in } \Lambda^{\emptyset}_{\to}[\mathcal{D}](1). \end{aligned}$$

As P, Q are different long normal forms, we have

$$\begin{aligned} P &\equiv P_1(\lambda \vec{x}_1.P_2(\lambda \vec{x}_2.\ \cdots\ P_p(\lambda \vec{x}_p.X)..)), \\ Q &\equiv Q_1(\lambda \vec{y}_1.Q_2(\lambda \vec{y}_2.\ \cdots\ Q_q(\lambda \vec{y}_p.Y)..)), \end{aligned}$$

where the $P_i, Q_j \in (\mathcal{D} \cup \mathcal{C}_3)$, the $\vec{x}_i, \vec{y}_j$ are possibly empty strings of variables of type 0, and X, Y are variables or constants of type 0. Let (U, V) be the first pair of symbols among the (P_i, Q_i) that are different. Distinguishing cases we define T_f, T_g such that (3.31) holds. As a shorthand we write (m, n), $m, n \in \{1, 2\}$, for the choices $T_f = T_m, T_g = T_n$, respectively.

Case 1. One of U, V, say U, is a variable or in $\mathcal{D}/\{\boldsymbol{d}_1, \boldsymbol{d}_2\}$. This U will not be changed by the substitution. If V is changed, after reducing we get $U \not\equiv \boldsymbol{d}_i$. Otherwise nothing happens with U, V and the difference is preserved. Therefore we can take any pair (m, n).

Case 2. One of U, V is $\boldsymbol{d}_i$.

Subcase 2.1. The other is in $\{\boldsymbol{f}, \boldsymbol{g}\}$. Then take (j, j), where $j = 3 - i$.

Subcase 2.2. The other one is $\boldsymbol{d}_{3-j}$. Then neither is replaced; take any pair.

Case 3. We have $\{U, V\} = \{\boldsymbol{f}, \boldsymbol{g}\}$. Then both are replaced and we can take $(1, 2)$.

After deciphering what is meant, the verification that the difference is kept is trivial. ∎

3.5.28 Proposition *Let $\mathcal{D}$ be a finite set of class $i>2$ and let $\mathcal{C}=\mathcal{C}_i$. Then for all $M, N \in \Lambda^{\emptyset}_{\to}[\mathcal{D}]$ of the same type one has*

$$M \approx^{\mathtt{ext}}_{\mathcal{D}} N \;\Leftrightarrow\; M \approx^{\mathtt{ext}}_{\mathcal{D}\cup\mathcal{C}} N.$$

Proof ($\Rightarrow$) By Lemmas 3.5.25, 3.5.26, and 3.5.27.

($\Leftarrow$) Trivial. ∎

3.5.29 Remark

(i) Proposition 3.5.28 fails for $i = 0$ or $i = 2$. For $i = 0$, take $\mathcal{D} = \{\boldsymbol{d}^0\}$, $\mathcal{C} = \mathcal{C}_0 = \{\boldsymbol{c}^0\}$. Then for $P \equiv \mathsf{K}\boldsymbol{d}, Q \equiv \mathsf{I}$ one has $P\boldsymbol{c} =_{\beta\eta} \boldsymbol{d} \neq_{\beta\eta} \boldsymbol{c} =_{\beta\eta} Q\boldsymbol{c}$. But the only $u[\boldsymbol{d}] \in \Lambda^{\emptyset}_{\to}[\mathcal{D}](0)$ is $\boldsymbol{d}$, loosing the difference: $P\boldsymbol{d} =_{\beta\eta} \boldsymbol{d} =_{\beta\eta} Q\boldsymbol{d}$. For $i = 2$, take $\mathcal{D} = \{\boldsymbol{g}{:}1, \boldsymbol{d}{:}0\}$, $\mathcal{C} = \mathcal{C}_2 = \{\boldsymbol{f}{:}1, \boldsymbol{c}{:}0\}$. Then for $P \equiv \lambda h{:}1.h(h(\boldsymbol{g}\boldsymbol{d}))$, $Q \equiv \lambda h{:}1.h(\boldsymbol{g}(h\boldsymbol{d}))$ one has $P\boldsymbol{f} \neq_{\beta\eta} Q\boldsymbol{f}$, but the only $u[\boldsymbol{g}, \boldsymbol{d}] \in \Lambda^{\emptyset}_{\to}[\mathcal{D}](0)$ are $\lambda x.\boldsymbol{g}^n x$ and $\lambda x.\boldsymbol{g}^n\boldsymbol{d}$, yielding $Pu =_{\beta\eta} \boldsymbol{g}^{2n+1}\boldsymbol{d} = Qu$, respectively $Pu =_{\beta\eta} \boldsymbol{g}^n\boldsymbol{d} =_{\beta\eta} Qu$.

(ii) Proposition 3.5.28 clearly also holds for class $i = 1$.

3.5.30 Lemma *For $A = A_1 \to \cdots \to A_a \to 0$. write $\mathcal{D}_A = \{\boldsymbol{c}_1^{A_1}, \ldots, \boldsymbol{c}_a^{A_a}\}$. Let $M, N \in \Lambda^{\emptyset}_{\to}$ be pure closed terms of the same type.*

(i) *Suppose $A \leq_{h^+} B$. Then*

$$M \approx^{\mathtt{ext}}_{\mathcal{D}_B} N \;\Rightarrow\; M \approx^{\mathtt{ext}}_{\mathcal{D}_A} N.$$

(ii) *Suppose* $A \sim_{h+} B$. *Then*

$$M \approx^{\mathtt{ext}}_{\mathcal{D}_A} N \;\Leftrightarrow\; M \approx^{\mathtt{ext}}_{\mathcal{D}_B} N.$$

Proof (i) We show the contrapositive.

$$\begin{aligned}
& M \not\approx^{\mathtt{ext}}_{\mathcal{D}_A} N \\
&\quad \Rightarrow \; \exists \vec{t} \in \Lambda^{\emptyset}_{\rightarrow}[\mathcal{D}_A].M\,\vec{t}\,[\boldsymbol{a}_1,\dots,\boldsymbol{a}_a] \neq_{\boldsymbol{\beta\eta}} N\,\vec{t}\,[\boldsymbol{a}_1,\dots,\boldsymbol{a}_a]\ (:0) \\
&\quad \Rightarrow \; \exists \vec{t}\ \lambda\vec{a}.M\,\vec{t}\,[\vec{a}\,] \neq_{\boldsymbol{\beta\eta}} \lambda\vec{a}.N\,\vec{t}\,[\vec{a}\,]\ (:A), \text{ by Remark 3.5.4,} \\
&\quad \Rightarrow \; \exists \vec{t}\ \lambda\vec{b}.(\lambda\vec{a}.M\,\vec{t}\,[\vec{a}\,])\vec{R}[\vec{b}] \neq_{\boldsymbol{\beta\eta}} \lambda\vec{b}.(\lambda\vec{a}.N\,\vec{t}\,[\vec{a}\,])\vec{R}[\vec{b}]\ (:B), \\
&\qquad\quad \text{by 3.4.26(iii), as } A \leq_{h+} B, \\
&\quad \Rightarrow \; \exists \vec{t}\ \lambda\vec{b}.M\vec{t}\,[\vec{R}\,[\vec{b}\,]] \neq_{\boldsymbol{\beta\eta}} \lambda\vec{b}.N\vec{t}\,[\vec{R}\,[\vec{b}\,]]\ (:B) \\
&\quad \Rightarrow \; \exists \vec{t}\ M\vec{t}\,[\vec{R}\,[\boldsymbol{b}_1,\dots,\boldsymbol{b}_b]] \neq_{\boldsymbol{\beta\eta}} N\vec{t}\,[\vec{R}\,[\boldsymbol{b}_1,\dots,\boldsymbol{b}_b]]\ (:0), \text{ by Remark 3.5.4,} \\
&\quad \Rightarrow \; M \not\approx^{\mathtt{ext}}_{\mathcal{D}_B} N.
\end{aligned}$$

(ii) By (i). ■

3.5.31 Proposition *Let* $\mathcal{D} = \{\boldsymbol{d}_1^{B_1},\dots,\boldsymbol{d}_k^{B_k}\}$ *be of class* $i{>}2$ *and* $\mathcal{C} = \mathcal{C}_i$, *with* $\mathcal{D} \cap \mathcal{C} = \emptyset$. *Let* $A \in \mathbb{T}^0$. *Then we have the following.*

(i) *For* $P[\vec{\boldsymbol{d}}\,], Q[\vec{\boldsymbol{d}}\,] \in \Lambda^{\emptyset}_{\rightarrow}[\mathcal{D}](A)$, *such that*

$$\lambda\vec{x}.P[\vec{x}], \lambda\vec{x}.Q[\vec{x}] \in \Lambda^{\emptyset}_{\rightarrow}(B_1 \rightarrow \cdots \rightarrow B_k \rightarrow 0)$$

the following are equivalent:

(1) $P[\vec{\boldsymbol{d}}] \approx^{\mathtt{ext}}_{\mathcal{D}} Q[\vec{\boldsymbol{d}}]$;

(2) $\lambda\vec{x}.P[\vec{x}] \approx_{\mathcal{C}} \lambda\vec{x}.Q[\vec{x}]$;

(3) $\lambda\vec{x}.P[\vec{x}] \approx^{\mathtt{ext}}_{\mathcal{D}} \lambda\vec{x}.Q[\vec{x}]$.

(ii) *In particular, for pure closed terms* $P, Q \in \Lambda^{\emptyset}_{\rightarrow}(A)$, *one has*

$$P \approx^{\mathtt{ext}}_{\mathcal{D}} Q \;\Leftrightarrow\; P \approx_{\mathcal{C}} Q.$$

Proof (i) We show (1) ⇒ (2) ⇒ (3) ⇒ (1).

(1) ⇒ (2). Assume $P[\vec{\boldsymbol{d}}\,] \approx^{\mathtt{ext}}_{\mathcal{D}} Q[\vec{\boldsymbol{d}}\,]$. Then

$$\begin{aligned}
&\Rightarrow \; P[\vec{\boldsymbol{d}}\,] \approx^{\mathtt{ext}}_{\mathcal{D}\cup\mathcal{C}} Q[\vec{\boldsymbol{d}}\,], && \text{by Proposition 3.5.28,} \\
&\Rightarrow \; P[\vec{\boldsymbol{d}}\,] \approx^{\mathtt{ext}}_{\mathcal{C}} Q[\vec{\boldsymbol{d}}\,], && \\
&\Rightarrow \; P[\vec{\boldsymbol{d}}\,]\vec{t} =_{\boldsymbol{\beta\eta}} Q[\vec{\boldsymbol{d}}\,]\vec{t}, && \text{for all } \vec{t} \in \Lambda^{\emptyset}_{\rightarrow}[\mathcal{C}], \\
&\Rightarrow \; P[\vec{s}\,]\vec{t} =_{\boldsymbol{\beta\eta}} Q[\vec{s}\,]\vec{t}, && \text{for all } \vec{t}, \vec{s} \in \Lambda^{\emptyset}_{\rightarrow}[\mathcal{C}] \text{ as } \mathcal{D} \cap \mathcal{C} = \emptyset, \\
&\Rightarrow \; \lambda\vec{x}.P[\vec{x}\,] \approx^{\mathtt{ext}}_{\mathcal{C}} \lambda\vec{x}.Q[\vec{x}\,].
\end{aligned}$$

(2) $\Rightarrow$ (3). By assumption, $\nabla(\mathcal{D}) \sim_{h^+} \nabla(\mathcal{C})$. As $\mathcal{D} = \mathcal{D}_{\nabla(\mathcal{D})}$ and $\mathcal{C} = \mathcal{D}_{\nabla(\mathcal{C})}$ we have

$$\lambda\vec{x}.P[\vec{x}] \approx^{\mathtt{ext}}_{\mathcal{D}} \lambda\vec{x}.Q[\vec{x}] \Leftrightarrow \lambda\vec{x}.P[\vec{x}] \approx^{\mathtt{ext}}_{\mathcal{C}} \lambda\vec{x}.Q[\vec{x}],$$

by Lemma 3.5.30.

(3) $\Rightarrow$ (1). Assume $\lambda\vec{x}.P[\vec{x}] \approx^{\mathtt{ext}}_{\mathcal{D}} \lambda\vec{x}.Q[\vec{x}]$. Then

$$\begin{array}{rrcll} & (\lambda\vec{x}.P(\vec{x})\vec{R}\vec{S} & =_{\beta\eta} & (\lambda\vec{x}.Q(\vec{x})\vec{R}\vec{S}, & \text{for all } \vec{R},\vec{S} \in \Lambda^{\emptyset}_{\to}[\mathcal{D}], \\ \Rightarrow & P(\vec{R})\vec{S} & =_{\beta\eta} & Q(\vec{R})\vec{S}, & \text{for all } \vec{R},\vec{S} \in \Lambda^{\emptyset}_{\to}[\mathcal{D}], \\ \Rightarrow & P(\vec{\boldsymbol{d}})\vec{S} & =_{\beta\eta} & Q(\vec{\boldsymbol{d}})\vec{S}, & \text{for all } \vec{S} \in \Lambda^{\emptyset}_{\to}[\mathcal{D}], \\ \Rightarrow & P(\vec{\boldsymbol{d}}) & \approx^{\mathtt{ext}}_{\mathcal{D}} & Q(\vec{\boldsymbol{d}}). & \end{array}$$

(ii) By (i). ∎

The proposition does not hold for the class $i = 2$. To see this take $\mathcal{D} = \mathcal{C}_2 = \{\boldsymbol{f}^1, \boldsymbol{c}^0\}$ and

$$P[\boldsymbol{f},\boldsymbol{c}] \equiv \lambda h{:}0.h(h(\boldsymbol{f}\boldsymbol{c})),\ Q \equiv \lambda h{:}0.h(\boldsymbol{f}(h\boldsymbol{c})).$$

Then $P[\boldsymbol{f},\boldsymbol{c}] \approx^{\mathtt{ext}}_{\mathcal{D}} Q[\boldsymbol{f},\boldsymbol{c}]$, but $\lambda fc.P[f,c] \not\approx^{\mathtt{ext}}_{\mathcal{D}} \lambda fc.Q[f,c]$.

3.5.32 Proposition *Let $\mathcal{D}$ be a set of constants of class $i \neq 2$. Then:*

(i) *the relation $\approx^{\mathtt{ext}}_{\mathcal{D}}$ on $\Lambda^{\emptyset}_{\to}[\mathcal{D}]$ is logical;*
(ii) *the relations $\approx^{\mathtt{ext}}_{\mathcal{D}}$ and $\approx^{\mathtt{obs}}_{\mathcal{D}}$ on $\Lambda^{\emptyset}_{\to}[\mathcal{D}]$ coincide.*

Proof (i) In the case $\mathcal{D}$ is of class -1, then $M \approx^{\mathtt{ext}}_{\mathcal{D}} N$ is universally valid by the empty implication. Therefore, the result is trivially valid.

In the case $\mathcal{D}$ is of class 0 or 1, then $\nabla(\mathcal{D}) \in \mathbb{T}_0 \cup \mathbb{T}_1$. Hence $\nabla(\mathcal{D}) = 0^k \to 0$ for some $k \geq 1$. Then $\mathcal{D} = \{c^0_1, \ldots, c^0_k\}$. Now trivially $\text{BetaEta}^{\mathcal{D}}(c,c)$ for $c \in \mathcal{D}$ of type 0. Therefore $\approx^{\mathtt{ext}}_{\mathcal{D}}$ is logical, by Lemma 3.5.9.

For $\mathcal{D}$ of class $i > 2$ we reason as follows. Write $\mathcal{C} = \mathcal{C}_i$. We may assume that $\mathcal{C} \cap \mathcal{D} = \emptyset$, see Remark 3.5.14.

We must show that for all $M, N \in \Lambda^{\emptyset}_{\to}[\mathcal{D}](A \to B)$ one has

$$M \approx^{\mathtt{ext}}_{\mathcal{D}} N \Leftrightarrow \forall P, Q \in \Lambda^{\emptyset}_{\to}[\mathcal{D}](A)[P \approx^{\mathtt{ext}}_{\mathcal{D}} Q \Rightarrow MP \approx^{\mathtt{ext}}_{\mathcal{D}} NQ]. \quad (3.32)$$

($\Rightarrow$) Assume $M[\vec{\boldsymbol{d}}] \approx^{\mathtt{ext}}_{\mathcal{D}} N[\vec{\boldsymbol{d}}]$ and $P[\vec{\boldsymbol{d}}] \approx^{\mathtt{ext}}_{\mathcal{D}} Q[\vec{\boldsymbol{d}}]$, with

$$M, N \in \Lambda^{\emptyset}_{\to}[\mathcal{D}](A \to B) \text{ and } P, Q \in \Lambda^{\emptyset}_{\to}[\mathcal{D}](B),$$

in order to show $M[\vec{\boldsymbol{d}}]P[\vec{\boldsymbol{d}}] \approx^{\mathtt{ext}}_{\mathcal{D}} N[\vec{\boldsymbol{d}}]Q[\vec{\boldsymbol{d}}]$. Then $\lambda\vec{x}.M[\vec{x}] \approx_{\mathcal{C}} \lambda\vec{x}.N[\vec{x}]$ and $\lambda\vec{x}.P[\vec{x}] \approx_{\mathcal{C}} \lambda\vec{x}.Q[\vec{x}]$, by Proposition 3.5.31(i). Consider the pure closed term

$$H \equiv \lambda f{:}(\vec{E} \to A \to B)\lambda m{:}(\vec{E} \to A)\lambda\vec{x}{:}\vec{E}.f\vec{x}(m\vec{x}).$$

As $\approx_{\mathcal{C}}$ is logical, one has $H \approx_{\mathcal{C}} H$, $\lambda\vec{x}.M[\vec{x}\,] \approx_{\mathcal{C}} \lambda\vec{x}.N[\vec{x}\,]$, and $\lambda\vec{x}.P[\vec{x}\,] \approx_{\mathcal{C}} \lambda\vec{x}.Q[\vec{x}\,]$. So

$$\begin{aligned}\lambda\vec{x}.M[\vec{x}\,]P[\vec{x}\,] &=_{\boldsymbol{\beta\eta}} H(\lambda\vec{x}.M[\vec{x}\,])(\lambda\vec{x}.P[\vec{x}\,])\\ &\approx_{\mathcal{C}} H(\lambda\vec{x}.N[\vec{x}\,])(\lambda\vec{x}.Q[\vec{x}\,]),\\ &=_{\boldsymbol{\beta\eta}} \lambda\vec{x}.N[\vec{x}\,]Q[\vec{x}\,].\end{aligned}$$

But then again by the proposition

$$M[\vec{\boldsymbol{d}}\,]P[\vec{\boldsymbol{d}}\,] \approx^{\mathtt{ext}}_{\mathcal{D}} N[\vec{\boldsymbol{d}}\,]Q[\vec{\boldsymbol{d}}\,].$$

($\Leftarrow$) Assume the right hand side of (3.32) in order to show $M \approx^{\mathtt{ext}}_{\mathcal{D}} N$. That is, one has to show

$$MP_1 \ldots P_k =_{\boldsymbol{\beta\eta}} NP_1 \ldots P_k, \tag{3.33}$$

for all $\vec{P} \in \Lambda^{\emptyset}_{\to}[\mathcal{D}]$. As $P_1 \approx^{\mathtt{ext}}_{\mathcal{D}} P_1$, by assumption it follows that $MP_1 \approx^{\mathtt{ext}}_{\mathcal{D}} NP_1$. Hence one has (3.33) by definition.

(ii) That $\approx^{\mathtt{ext}}_{\mathcal{D}}$ is $\approx^{\mathtt{obs}}_{\mathcal{D}}$ on $\Lambda^{\emptyset}_{\to}[\mathcal{D}]$ follows by (i) and Proposition 3.5.5. ∎

3.5.33 Lemma *Let $\mathcal{D}$ be a finite set of constants. Then $\mathcal{D}$ is of class 2 iff one of the following cases holds:*

$$\begin{aligned}\mathcal{D} &= \{\boldsymbol{F}{:}(1_{p+1} \to 0), \boldsymbol{c}_1, \ldots, \boldsymbol{c}_q{:}0\},\ p, q \geq 0;\\ \mathcal{D} &= \{\boldsymbol{f}{:}1, \boldsymbol{c}_1, \ldots, \boldsymbol{c}_{q+1}{:}0\},\ q \geq 0.\end{aligned}$$

Proof By Lemma 3.4.16. ∎

3.5.34 Proposition *Let $\mathcal{D}$ be of class 2. Then the following hold:*

(i) *the relation $\approx^{\mathtt{ext}}_{\mathcal{D}}$ on $\Lambda^{\emptyset}_{\to}[\mathcal{D}]$ is logical;*
(ii) *the relations $\approx^{\mathtt{ext}}_{\mathcal{D}}$ and $\approx^{\mathtt{obs}}_{\mathcal{D}}$ on $\Lambda^{\emptyset}_{\to}[\mathcal{D}]$ coincide.*

Proof (i) Assume that $\mathcal{D} = \{\boldsymbol{F}, \boldsymbol{c}_1, \ldots, \boldsymbol{c}_q\}$ (the other possibility is, by Lemma 3.5.33, more easy). By Proposition 3.5.9(i) it suffices to show that for $\boldsymbol{d} \in \mathcal{D}$ one has $S(\boldsymbol{d}, \boldsymbol{d})$. This is easy for the ones of type 0. For $\boldsymbol{F} : (1_{p+1} \to 0)$ assume for notational simplicity that $k = 0$, i.e. $\boldsymbol{F} : 2$. By Lemma 3.5.7 it suffices to show $f \approx^{\mathtt{ext}}_{\mathcal{D}} g \Rightarrow f =_{\boldsymbol{\beta\eta}} g$ for $f, g \in \Lambda^{\emptyset}_{\to}[\mathcal{D}](1)$. Now elements of $\Lambda^{\emptyset}_{\to}[\mathcal{D}](1)$ are of the form

$$\lambda x_1.\boldsymbol{F}(\lambda x_2.\boldsymbol{F}(\ldots(\lambda x_{m-1}.\boldsymbol{F}(\lambda x_m.c))..)),$$

where $c \equiv x_i$ or $c \equiv \boldsymbol{c}_j$. Therefore if $f \neq_{\boldsymbol{\beta\eta}} g$, then inspecting the various possibilities (for example

$$\begin{aligned}f &\equiv \lambda x_1.\boldsymbol{F}(\lambda x_2.\boldsymbol{F}(\ldots(\lambda x_{m-1}.\boldsymbol{F}(\lambda x_m.x_n))..)) \equiv \mathsf{K}A\\ g &\equiv \lambda x_1.\boldsymbol{F}(\lambda x_2.\boldsymbol{F}(\ldots(\lambda x_{m-1}.\boldsymbol{F}(\lambda x_m.x_1))..)),\end{aligned}$$

see Exercise 3.25), we have $f(\boldsymbol{F}f) \neq_{\beta\eta} g(\boldsymbol{F}f)$ or $f(\boldsymbol{F}g) \neq_{\beta\eta} g(\boldsymbol{F}g)$, hence $f \not\approx^{\mathtt{ext}}_{\mathcal{D}} g$.

(ii) By (i) and Proposition 3.5.5. ∎

Harvesting the results we obtain the following main theorem.

3.5.35 Theorem (Statman (1980b)) *Let $\mathcal{D}$ be a finite set of typed constants of class i and $\mathcal{C} = \mathcal{C}_i$. Then:*

(i) $\approx^{\mathtt{ext}}_{\mathcal{D}}$ *is logical;*

(ii) *for closed terms $M, N \in \Lambda^{\emptyset}_{\to}[\mathcal{D}]$ of the same type one has*

$$M \approx^{\mathtt{ext}}_{\mathcal{D}} N \Leftrightarrow M \approx^{\mathtt{obs}}_{\mathcal{D}} N;$$

(iii) *for pure closed terms $M, N \in \Lambda^{\emptyset}_{\to}$ of the same type one has*

$$M \approx^{\mathtt{ext}}_{\mathcal{D}} N \Leftrightarrow M \approx^{\mathtt{ext}}_{\mathcal{C}} N.$$

Proof (i) By Propositions 3.5.32 and 3.5.34.

(ii) Similarly.

(iii) Let $\mathcal{D} = \{\boldsymbol{d}_1^{A_1}, \dots, \boldsymbol{d}_k^{A_k}\}$. Then $\nabla(\mathcal{D}) = A_1 \to \dots A_k \to 0$ and in the notation of Lemma 3.5.30 one has $\mathcal{D}_{\nabla(\mathcal{D})} = \mathcal{D}$, up to renaming constants. One has $\nabla(\mathcal{D}) \in \mathbb{T}_i$, hence by the Hierarchy Theorem revisited $\nabla(\mathcal{D}) \sim_{h^+} \mathcal{C}_i$. Thus $\approx_{\mathcal{D}_{\nabla(\mathcal{D})}}$ is equivalent to $\approx_{\mathcal{D}_{\mathcal{C}_i}}$ on pure closed terms, by Lemma 3.5.30. As $\mathcal{D}_{\nabla(\mathcal{D})} = \mathcal{D}$ and $\mathcal{D}_{\mathcal{C}_i} = \mathcal{C}_i$, we are done. ∎

From now on we can write $\approx_{\mathcal{D}}$ for $\approx^{\mathtt{ext}}_{\mathcal{D}}$ and $\approx^{\mathtt{obs}}_{\mathcal{D}}$.

Infinite sets of constants

Recall that for $\mathcal{D}$ a possibly infinite set of typed constants we defined

$$\text{class}(\mathcal{D}) = \max\{\text{class}(\mathcal{D}_f) \mid \mathcal{D}_f \subseteq \mathcal{D} \;\&\; \mathcal{D}_f \text{ is finite}\}.$$

The notion of class is well defined and one has $\text{class}(\mathcal{D}) \in \{-1, 0, 1, 2, 3, 4, 5\}$.

3.5.36 Proposition *Let $\mathcal{D}$ be a possibly infinite set of constants of class i. Let $A \in \mathbb{T}^0$ and $M \equiv M[\vec{\boldsymbol{d}}], N \equiv N[\vec{\boldsymbol{d}}] \in \Lambda^{\emptyset}_{\to}[\mathcal{D}](A)$. Then the following are equivalent:*

(i) $M \approx^{\mathtt{ext}}_{\mathcal{D}} N$;

(ii) *for all finite $\mathcal{D}_f \subseteq \mathcal{D}$ containing the $\vec{\boldsymbol{d}}$ such that* $\text{class}(\mathcal{D}_f) = \text{class}(\mathcal{D})$ *one has*

$$M \approx^{\mathtt{ext}}_{\mathcal{D}_f} N;$$

(iii) *there exists a finite* $\mathcal{D}_f \subseteq \mathcal{D}$ *containing the* $\vec{\boldsymbol{d}}$ *such that* $\text{class}(\mathcal{D}_f) = \text{class}(\mathcal{D})$ *and*

$$M \approx^{\mathtt{ext}}_{\mathcal{D}_f} N.$$

Proof (i) $\Rightarrow$ (ii). Trivial as there are fewer equations to be satisfied in $M \approx^{\mathtt{ext}}_{\mathcal{D}_f} N$.

(ii) $\Rightarrow$ (iii). Let $\mathcal{D}_f \subseteq \mathcal{D}$ be finite with $\text{class}(\mathcal{D}_f) = \text{class}(\mathcal{D})$. Let $\mathcal{D}_{f'} = \mathcal{D}_f \cup \{\vec{\boldsymbol{d}}\}$. Then $i = \text{class}(\mathcal{D}_f) \leq \text{class}(\mathcal{D}_{f'}) \leq i$, by Remark 3.5.12. Therefore $\mathcal{D}_{f'}$ satisfies the conditions of (ii) and one has $M \approx^{\mathtt{ext}}_{\mathcal{D}_{f'}} N$.

(iii) $\Rightarrow$ (i). Suppose towards a contradiction that $M \approx^{\mathtt{ext}}_{\mathcal{D}_f} N$ but $M \not\approx^{\mathtt{ext}}_{\mathcal{D}} N$. Then for some finite $\mathcal{D}_{f'} \subseteq \mathcal{D}$ of class i containing $\vec{\boldsymbol{d}}$ one has $M \not\approx^{\mathtt{ext}}_{\mathcal{D}_{f'}} N$. We distinguish cases.

Case $\text{class}(\mathcal{D}) > 2$. Since $\text{class}(\mathcal{D}_f) = \text{class}(\mathcal{D}_{f'}) = i$, Proposition 3.5.31(i) implies that

$$\lambda\vec{x}.M[\vec{x}] \approx^{\mathtt{ext}}_{\mathcal{C}_i} \lambda\vec{x}.N[\vec{x}] \ \& \ \lambda\vec{x}.M[\vec{x}] \not\approx^{\mathtt{ext}}_{\mathcal{C}_i} \lambda\vec{x}.N[\vec{x}],$$

a contradiction.

Case $\text{class}(\mathcal{D}) = 2$. Then by Lemma 3.5.33 the set $\mathcal{D}$ consists either of a constant $\boldsymbol{f}^1$ or $\boldsymbol{F}^{1_{p+1}\to 0}$ and furthermore only type 0 constants $\boldsymbol{c}^0$. So $\mathcal{D}_f \cup \mathcal{D}_{f'} = \mathcal{D}_f \cup \{\boldsymbol{c}^0_1, \ldots, \boldsymbol{c}^0_k\}$. As $M \approx^{\mathtt{ext}}_{\mathcal{D}_f} N$ by Lemma 3.5.23 one has $M \approx^{\mathtt{ext}}_{\mathcal{D}_f \cup \mathcal{D}_{f'}} N$. But then *a fortiori* $M \approx^{\mathtt{ext}}_{\mathcal{D}_{f'}} N$, a contradiction.

Case $\text{class}(\mathcal{D}) = 1$. Then $\mathcal{D}$ consists of only type 0 constants and we can reason similarly, again using Lemma 3.5.23.

Case $\text{class}(\mathcal{D}) = 0$. Then $\mathcal{D} = \{0\}$. Hence the only subset of $\mathcal{D}$ having the same class is $\mathcal{D}$ itself. Therefore $\mathcal{D}_f = \mathcal{D}_{f'}$, a contradiction.

Case $\text{class}(\mathcal{D}) = -1$. Call a type $A \in \mathbb{T}^0$ $\mathcal{D}$-inhabited if $P \in \Lambda^{\emptyset}_{\to}[\mathcal{D}](A)$ for some term P. Using Proposition 2.4.4 one can show

$$A \text{ is inhabited} \Leftrightarrow A \text{ is } \mathcal{D}\text{-inhabited}.$$

From this one can show for all $\mathcal{D}$ of class -1 that

$$A \text{ inhabited} \Rightarrow \forall M, N \in \Lambda^{\emptyset}_{\to}[\mathcal{D}](A). M \approx^{\mathtt{ext}}_{\mathcal{D}} N.$$

In fact the assumption is not necessary, since for non-inhabited types the conclusion holds vacuously. This contradicts $M \not\approx^{\mathtt{ext}}_{\mathcal{D}} N$. ∎

As a consequence of this proposition we now show that the main theorem also holds for possibly infinite sets $\mathcal{D}$ of typed constants.

3.5.37 Theorem *Let* $\mathcal{D}$ *be a set of typed constants of class* i *and* $\mathcal{C} = \mathcal{C}_i$. *Then:*

(i) $\approx^{\mathtt{ext}}_{\mathcal{D}}$ *is logical;*

(ii) *for closed terms* $M, N \in \Lambda^{\emptyset}_{\to}[\mathcal{D}]$ *of the same type one has*

$$M \approx^{\mathtt{ext}}_{\mathcal{D}} N \Leftrightarrow M \approx^{\mathtt{obs}}_{\mathcal{D}} N;$$

(iii) *for pure closed terms* $M, N \in \Lambda^{\emptyset}_{\to}$ *of the same type one has*

$$M \approx^{\mathtt{ext}}_{\mathcal{D}} N \Leftrightarrow M \approx^{\mathtt{ext}}_{\mathcal{C}} N.$$

Proof (i) Let $M, N \in \Lambda^{\emptyset}_{\to}[\mathcal{D}](A \to B)$. We must show

$$M \approx^{\mathtt{ext}}_{\mathcal{D}} N \Leftrightarrow \forall P, Q \in \Lambda^{\emptyset}_{\to}[\mathcal{D}](A).[P \approx^{\mathtt{ext}}_{\mathcal{D}} Q \Rightarrow MP \approx^{\mathtt{ext}}_{\mathcal{D}} NQ].$$

($\Rightarrow$) Suppose $M \approx^{\mathtt{ext}}_{\mathcal{D}} N$ and $P \approx^{\mathtt{ext}}_{\mathcal{D}} Q$. Let $\mathcal{D}_f \subseteq \mathcal{D}$ be a finite subset of class i containing the constants in M, N, P, Q. Then $M \approx^{\mathtt{ext}}_{\mathcal{D}_f} N$ and $P \approx^{\mathtt{ext}}_{\mathcal{D}_f} Q$. Since $\approx^{\mathtt{ext}}_{\mathcal{D}_f}$ is logical by Theorem 3.5.35 one has $MP \approx^{\mathtt{ext}}_{\mathcal{D}_f} NQ$. But then $MP \approx^{\mathtt{ext}}_{\mathcal{D}} NQ$.

($\Leftarrow$) Assume the right hand side. Let $\mathcal{D}_f$ be a finite subset of $\mathcal{D}$ of the same class containing all the constants of M, N, P, Q. We have

$$\begin{array}{rlll} P \approx^{\mathtt{ext}}_{\mathcal{D}_f} Q & \Rightarrow & P \approx^{\mathtt{ext}}_{\mathcal{D}} Q, & \text{by Proposition 3.5.36,} \\ & \Rightarrow & MP \approx^{\mathtt{ext}}_{\mathcal{D}} NQ, & \text{by assumption,} \\ & \Rightarrow & MP \approx^{\mathtt{ext}}_{\mathcal{D}_f} NQ, & \text{by Proposition 3.5.36.} \end{array}$$

Therefore $M \approx^{\mathtt{ext}}_{\mathcal{D}_f} N$. Then by Proposition 3.5.36 again we have $M \approx^{\mathtt{ext}}_{\mathcal{D}} N$.

(ii) By (i) and Proposition 3.5.5.

(iii) Let $\mathcal{D}_f$ be a finite subset of $\mathcal{D}$ of the same class. Then by Proposition 3.5.36 and Theorem 3.5.35

$$M \approx^{\mathtt{ext}}_{\mathcal{D}} N \Leftrightarrow M \approx^{\mathtt{ext}}_{\mathcal{D}_f} N \Leftrightarrow M \approx^{\mathtt{ext}}_{\mathcal{C}} N. \blacksquare$$

Term models

In this subsection we assume that $\mathcal{D}$ is a finite sufficient set of constants; that is, every type $A \in \mathbb{T}^0$ is inhabited by some $M \in \Lambda^{\emptyset}_{\to}[\mathcal{D}]$. This is the same as saying class$(\mathcal{D}) \geq 0$.

3.5.38 Definition Define

$$\mathcal{M}[\mathcal{D}] = \Lambda^{\emptyset}_{\to}[\mathcal{D}]/\approx_{\mathcal{D}},$$

with application defined by

$$[F]_{\mathcal{D}}[M]_{\mathcal{D}} = [FM]_{\mathcal{D}}.$$

Here $[-]_{\mathcal{D}}$ denotes an equivalence class modulo $\approx_{\mathcal{D}}$.

3.5.39 Theorem *Let* $\mathcal{D}$ *be sufficient. Then:*

(i) *application in* $\mathcal{M}[\mathcal{D}]$ *is well-defined;*
(ii) *For all* $M, N \in \Lambda^{\emptyset}_{\rightarrow}[\mathcal{D}]$ *on has*

$$[\![M]\!]^{\mathcal{M}[\mathcal{D}]} = [M]_{\approx_{\mathcal{D}}};$$

(iii) $\mathcal{M}[\mathcal{D}] \models M = N \Leftrightarrow M \approx_{\mathcal{D}} N;$
(iv) $\mathcal{M}[\mathcal{D}]$ *is an extensional term-model.*

Proof (i) As the relation $\approx_{\mathcal{D}}$ is logical, application is independent of the choice of representative:

$$F \approx_{\mathcal{D}} F' \;\&\; M \approx_{\mathcal{D}} M' \;\Rightarrow\; FM \approx_{\mathcal{D}} F'M'.$$

(ii) By induction on open terms $M \in \Lambda_{\rightarrow}[\mathcal{D}]$ it follows that

$$[\![M]\!]_{\rho} = [M[\vec{x}:= \rho(x_1), \ldots, \rho(x_n)]]_{\mathcal{D}}.$$

Hence (ii) follows by taking $\rho(x) = [x]_{\mathcal{D}}$.

(iii) By (ii).

(iv) Use (ii) and Remark 3.5.3(ii). ■

3.5.40 Lemma *Let* A *be represented in* $\mathcal{D}$. *Then for all* $M, N \in \Lambda^{\emptyset}_{\rightarrow}(A)$, *pure closed terms of type* A, *we have*

$$M \approx_{\mathcal{D}} N \Leftrightarrow M =_{\beta\eta} N.$$

Proof The ($\Leftarrow$) direction is trivial. As for ($\Rightarrow$),

$$\begin{aligned} M \approx_{\mathcal{D}} N \;&\Leftrightarrow\; \forall \vec{T} \in \Lambda^{\emptyset}_{\rightarrow}[\mathcal{D}].M\vec{T} =_{\beta\eta} N\vec{T} \\ &\Rightarrow\; M\vec{\boldsymbol{d}} =_{\beta\eta} N\vec{\boldsymbol{d}}, && \text{for some } \vec{\boldsymbol{d}} \in \mathcal{D} \text{ since } A \text{ is represented in } \mathcal{D}, \\ &\Rightarrow\; M\vec{x} =_{\beta\eta} N\vec{x}, && \text{by Remark 3.5.4 as } M, N \text{ are pure}, \\ &\Rightarrow\; M =_{\eta} \lambda\vec{x}.M\vec{x} =_{\beta\eta} \lambda\vec{x}.N\vec{x} =_{\eta} N. \;■ \end{aligned}$$

3.5.41 Definition

(i) If $\mathcal{M}$ is a model of $\boldsymbol{\lambda}^{\mathrm{Ch}}_{\rightarrow}[\mathcal{D}]$, then for a type A its *A-section* is simply $\mathcal{M}(A)$.
(ii) We say that $\mathcal{M}$ is *A-complete* (*A-complete for pure terms*) iff for all closed terms (pure closed terms, respectively) M, N of type A one has

$$\mathcal{M} \models M = N \Leftrightarrow M =_{\beta\eta} N.$$

(iii) We call $\mathcal{M}$ *complete* (for pure terms) if for all types $A \in \mathbb{T}^0$ it is A-complete (for pure terms).

(iv) A model $\mathcal{M}$ is called *fully abstract* if

$$\forall A \in \mathbb{T}^0 \forall x, y \in \mathcal{M}(A)[\,[\forall f \in \mathcal{M}(A{\to}0).fx = fy] \Rightarrow x = y].$$

3.5.42 Corollary *Let $\mathcal{D}$ be sufficient. Then $\mathcal{M}[\mathcal{D}]$ has the following properties:*

(i) *$\mathcal{M}[\mathcal{D}]$ is an extensional term-model;*
(ii) *$\mathcal{M}[\mathcal{D}]$ is fully abstract;*
(iii) *if A is represented in $\mathcal{D}$ then $\mathcal{M}[\mathcal{D}]$ is A-complete for pure closed terms;*
(iv) *in particular, $\mathcal{M}[\mathcal{D}]$ is $\nabla(\mathcal{D})$-complete and 0-complete for pure closed terms.*

Proof (i) By Theorem 3.5.39 the definition of application is well defined. That extensionality holds follows from the definition of $\approx_{\mathcal{D}}$. As all combinators $[\mathsf{K}_{AB}]_{\mathcal{D}}, [\mathsf{S}_{ABC}]_{\mathcal{D}}$ are in $\mathcal{M}[\mathcal{D}]$, the structure is a model.

(ii) By Theorem 3.5.39(ii). Let $x, y \in \mathcal{M}(A)$ be $[X]_{\mathcal{D}}, [Y]_{\mathcal{D}}$ respectively. Then

$$\begin{aligned}
\forall f \in \mathcal{M}(A{\to}0).fx = fy &\Rightarrow \forall F \in \Lambda^{\emptyset}_{\to}[\mathcal{D}](A{\to}0).[FX]_{\mathcal{D}} = [FY]_{\mathcal{D}} \\
&\Rightarrow \forall F \in \Lambda^{\emptyset}_{\to}[\mathcal{D}](A{\to}0).FX \approx_{\mathcal{D}} FY \ (:0) \\
&\Rightarrow \forall F \in \Lambda^{\emptyset}_{\to}[\mathcal{D}](A{\to}0).FX =_{\beta\eta} FY \\
&\Rightarrow X \approx_{\mathcal{D}} Y \\
&\Rightarrow [X]_{\mathcal{D}} = [Y]_{\mathcal{D}} \\
&\Rightarrow x = y.
\end{aligned}$$

(iii) By Lemma 3.5.40.

(iv) By (iii) and the fact that $\nabla(\mathcal{D})$ is represented in $\mathcal{D}$. For 0, the result is trivial. ∎

3.5.43 Proposition

(i) *Let $0 \leq i \leq j \leq 5$. Then for pure closed terms $M, N \in \Lambda^{\emptyset}_{\to}$*

$$\mathcal{M}[\mathcal{C}_j] \models M = N \Rightarrow \mathcal{M}[\mathcal{C}_i] \models M = N.$$

(ii) $\mathrm{Th}(\mathcal{M}[\mathcal{C}_5]) \subseteq \cdots \subseteq \mathrm{Th}(\mathcal{M}[\mathcal{C}_1])$, *see Definition* 3.1.10(iv). *All inclusions are proper.*

Proof (i) Let $M, N \in \Lambda^{\emptyset}_{\to}$ be of the same type. Then

$$\begin{array}{ll} \mathcal{M}[\mathcal{C}_i] \not\models M = N & \\ \Rightarrow & M \not\approx_{\mathcal{C}_i} N \\ \Rightarrow & M(\vec{t}\,[\vec{c}]) \neq_{\beta\eta} N(\vec{t}\,[\vec{c}]) : 0, \text{ for some } (\vec{t}\,[\vec{c}]) \in \Lambda^{\emptyset}_{\to}[\mathcal{C}], \\ \Rightarrow & \lambda\vec{c}.M(\vec{t}\,[\vec{c}]) \neq_{\beta\eta} \lambda\vec{c}.N(\vec{t}\,[\vec{c}]) : \nabla(\mathcal{C}_i), \text{ by Remark 3.5.4}, \\ \Rightarrow & \Psi(\lambda\vec{c}.M(\vec{t}\,[\vec{c}])) \neq_{\beta\eta} \Psi(\lambda\vec{c}.N(\vec{t}\,[\vec{c}])) : \nabla(\mathcal{C}_j), \\ & \quad \text{since } \nabla(\mathcal{C}_i) \leq_{\beta\eta} \nabla(\mathcal{C}_j) \text{ via some injective } \Psi, \\ \Rightarrow & \Psi(\lambda\vec{c}.M(\vec{t}\,[\vec{c}])) \not\approx_{\mathcal{C}_j} \Psi(\lambda\vec{c}.N(\vec{t}\,[\vec{c}])), \text{ since by 3.5.42(iv)} \\ & \quad \text{the model } \mathcal{M}[\mathcal{C}_j] \text{ is } \nabla(\mathcal{C}_j)\text{-complete for pure terms}, \\ \Rightarrow & \mathcal{M}[\mathcal{C}_j] \not\models \Psi(\lambda\vec{c}.M(\vec{t}\,[\vec{c}])) = \Psi(\lambda\vec{c}.N(\vec{t}\,[\vec{c}])) \\ \Rightarrow & \mathcal{M}[\mathcal{C}_j] \not\models M = N, \text{ since } \mathcal{M}[\mathcal{C}_j] \text{ is a model.} \end{array}$$

By (i) the inclusions hold; they are proper by Exercise 3.31. ■

3.5.44 Lemma *Let A, B be types such that $A \leq_{\beta\eta} B$. Suppose $\mathcal{M}[\mathcal{D}]$ is B-complete for pure terms. Then $\mathcal{M}[\mathcal{D}]$ is A-complete for pure terms.*

Proof Assume $\Phi : A \leq_{\beta\eta} B$. Then one has for $M, N \in \Lambda^{\emptyset}_{\to}(A)$

$$\begin{array}{ccc} \mathcal{M}[\mathcal{D}] \models M = N & \Leftarrow & M =_{\beta\eta} N \\ \Downarrow & & \Uparrow \\ \mathcal{M}[\mathcal{D}] \models \Phi M = \Phi N & \Rightarrow & \Phi M =_{\beta\eta} \Phi N \end{array}$$

by the definition of reducibility. ■

We immediately have

3.5.45 Corollary *Let $\approx^{\mathrm{ext}}_{\mathcal{D}}$ be logical. If $\mathcal{M}[\mathcal{D}]$ is A-complete but not B-complete for pure closed terms, then $A \not\leq_{\beta\eta} B$.*

3.5.46 Corollary *$\mathcal{M}[\mathcal{C}_5]$ is complete for pure terms, i.e. for all A and $M, N \in \Lambda^{\emptyset}_{\to}(A)$*

$$\mathcal{M}[\mathcal{C}_5] \models M = N \Leftrightarrow M =_{\beta\eta} N.$$

Proof We know $\mathcal{M}[\mathcal{C}_5]$ is $\nabla(\mathcal{C}_5)$-complete for pure terms, by Corollary 3.5.42(iii). Since for every type A one has $A \leq_{\beta\eta} \top = \nabla(\mathcal{C}_5)$, by the Reducibility Theorem 3.4.8, it follows by Lemma 3.5.44 that this model is also A-complete. ■

So $\mathrm{Th}(\mathcal{M}[\mathcal{C}_5])$, the smallest theory, is actually just $\beta\eta$-convertibility, which is decidable. At the other end of the hierarchy a dual property holds.

3.5.47 Definition We call $\mathcal{M}_{\min} = \mathcal{M}[\mathcal{C}_1]$ the *minimal model* of $\boldsymbol{\lambda}_{\to}^{\mathbb{A}}$ since it equates most terms. We call $\mathtt{Th}_{\max} = \mathrm{Th}(\mathcal{M}[\mathcal{C}_1])$ the *maximal theory.* The names will be justified below.

3.5.48 Proposition *Let $A \equiv A_1 \to \cdots \to A_a \to 0 \in \mathbb{T}^0$. Let $M, N \in \Lambda_{\to}^{\varnothing}(A)$ be pure closed terms. Then the following statements are equivalent:*

(1) *$M = N$ is inconsistent;*
(2) *for all models $\mathcal{M}$ of $\boldsymbol{\lambda}_{\to}^{\mathbb{A}}$ one has $\mathcal{M} \not\models M = N$;*
(3) *$\mathcal{M}_{\min} \not\models M = N$;*
(4) *$\exists P_1 \in \Lambda^{x,y:0}(A_1) \ldots P_a \in \Lambda^{x,y:0}(A_a).M\vec{P} = x \;\&\; N\vec{P} = y$;*
(5) *$\exists F \in \Lambda^{x,y:0}(A \to 0).FM = x \;\&\; FN = y$;*
(6) *$\exists G \in \Lambda^{\varnothing}(A \to 0^2 \to 0).GM = \lambda xy.x \;\&\; GN = \lambda xy.y$;*

Proof (1) $\Rightarrow$ (2) By soundness. (2) $\Rightarrow$ (3) Trivial. (3) $\Rightarrow$ (4) Since $\mathcal{M}_{\min}$ consists of $\Lambda^{x,y:0}/\approx_{\mathcal{C}_1}$. (4) $\Rightarrow$ (5) By taking $F \equiv \lambda m.m\vec{P}$. (5) $\Rightarrow$ (6) By taking $G \equiv \lambda mxy.Fm$. (6) $\Rightarrow$ (1) Trivial. ■

3.5.49 Corollary *The theory* $\mathrm{Th}(\mathcal{M}_{\min})$ *is the unique maximally consistent extension of* $\boldsymbol{\lambda}_{\to}^{0}$.

Proof By taking the negations in the proposition we have $M = N$ is consistent iff $\mathcal{M}_{\min} \models M = N$. Hence $\mathrm{Th}(\mathcal{M}_{\min})$ contains all consistent equations. Moreover this theory is consistent. Therefore the statement follows. ■

We already encountered $\mathrm{Th}(\mathcal{M}_{\min})$ as $\mathcal{E}_{\max}$ in Definition 3.2.19 before. In Section 4.4 it will be proved that it is decidable. The model $\mathcal{M}[\mathcal{C}_0]$ is the degenerate model consisting of one element at each type, since

$$\forall M, N \in \Lambda_{\to}^{\varnothing}[\mathcal{C}_0](0)\ M = x = N.$$

Therefore its theory is inconsistent and hence decidable.

3.5.50 Remark It is not known if the theories $\mathrm{Th}(\mathcal{M}[\mathcal{C}_2])$, $\mathrm{Th}(\mathcal{M}[\mathcal{C}_3])$ and $\mathrm{Th}(\mathcal{M}[\mathcal{C}_4])$ are decidable.

3.5.51 Theorem *Let $\mathcal{D}$ be a sufficient set of constants of class $i \geq 0$. Then:*

(i) $\forall M, N \in \Lambda_{\to}^{\varnothing}[M \approx_{\mathcal{D}} N \;\Leftrightarrow\; M \approx_{\mathcal{C}_i} N]$;
(ii) *$\mathcal{M}[\mathcal{D}]$ is $\nabla(\mathcal{C}_i)$-complete for pure terms.*

Proof (i) By Proposition 3.5.31(ii).
(ii) By (i) and Corollary 3.5.42(iv). ■

3.5.52 Remark So there are exactly five canonical term-models that are not elementary equivalent (plus the degenerate term-model equating everything).

Proof of Proposition 3.4.11

In the previous section the types A_α were introduced. The following proposition was needed to prove that these form a hierarchy.

3.5.53 Proposition *For* $\alpha, \beta \leq \omega + 3$ *one has*

$$\alpha \leq \beta \Leftarrow A_\alpha \leq_{\beta\eta} A_\beta.$$

Proof Notice that for $\alpha \leq \omega$ the cardinality of $\Lambda^{\emptyset}_{\to}(A_\alpha)$ equals α: for example $\Lambda^{\emptyset}_{\to}(A_2) = \{\lambda xy{:}0.x, \lambda xy{:}0.y\}$ and $\Lambda^{\emptyset}_{\to}(A_\omega = \{\lambda f{:}1\lambda x{:}0.f^k x \mid k \in \omega\}$. Therefore for $\alpha, \alpha' \leq \omega$ one has $A_\alpha \leq_{\beta\eta} A_{\alpha'} \Rightarrow \alpha = \alpha'$.

It remains to show that $A_{\omega+1} \not\leq_{\beta\eta} A_\omega, A_{\omega+2} \not\leq_{\beta\eta} A_{\omega+1}, A_{\omega+3} \not\leq_{\beta\eta} A_{\omega+2}$.

As for $A_{\omega+1} \not\leq_{\beta\eta} A_\omega$, consider

$$M \equiv \lambda f, g{:}1\lambda x{:}0.f(g(f(gx))),$$
$$N \equiv \lambda f, g{:}1\lambda x{:}0.f(g(g(fx)).$$

Then $M, N \in \Lambda^{\emptyset}_{\to}(A_{\omega+1})$, and $M \neq_{\beta\eta} N$. By Corollary 3.5.42(iii) we know that $\mathcal{M}[\mathcal{C}_2]$ is A_ω-complete. It is not difficult to show that $\mathcal{M}[\mathcal{C}_2] \models M = N$, by analyzing the elements of $\Lambda^{\emptyset}_{\to}[\mathcal{C}_2](1)$. Therefore, by Corollary 3.5.45, the conclusion follows.

As for $A_{\omega+2} \not\leq_{\beta\eta} A_{\omega+1}$, this is proved in Dekkers (1988) as follows. Consider

$$M \equiv \lambda F{:}3\lambda x{:}0.F(\lambda f_1{:}1.f_1(F(\lambda f_2{:}1.f_2(f_1 x))))$$
$$N \equiv \lambda F{:}3\lambda x{:}0.F(\lambda f_1{:}1.f_1(F(\lambda f_2{:}1.f_2(f_2 x)))).$$

Then $M, N \in \Lambda^{\emptyset}_{\to}(A_{\omega+2})$ and $M {\neq_{\beta\eta}} N$. In Proposition 12 of aforementioned paper it is proved that $\Phi M {=_{\beta\eta}} \Phi N$ for each $\Phi \in \Lambda^{\emptyset}_{\to}(A_{\omega+2} {\to} A_{\omega+1})$.

As for $A_{\omega+3} \not\leq_{\beta\eta} A_{\omega+2}$, consider

$$M \equiv \lambda h{:}1_2\lambda x{:}0.h(hx(hxx))(hxx),$$
$$N \equiv \lambda h{:}1_2\lambda x{:}0.h(hxx)(h(hxx)x).$$

Then $M, N \in \Lambda^{\emptyset}_{\to}(A_{\omega+3})$, and $M \neq_{\beta\eta} N$. Again $\mathcal{M}[\mathcal{C}_4]$ is $A_{\omega+2}$-complete. It is not difficult to show that $\mathcal{M}[\mathcal{C}_4] \models M = N$, by analyzing the elements of $\Lambda^{\emptyset}_{\to}[\mathcal{C}_4](1_2)$. Therefore, by Corollary 3.5.45, the conclusion follows. ∎

3.6 Exercises

Exercise 3.1 Convince yourself of the validity of Proposition 3.3.3 for $n = 2$.

Exercise 3.2 Show that there are $M, N \in \Lambda^{\emptyset}_{\rightarrow}[\{d^0\}]((1_2 \to 1_2 \to 0) \to 0)$ such that $M \# N$, but there is no term F such that $FM = \lambda xy.x$, $FN = \lambda xy.y$. [Hint. Take $M \equiv [\lambda xy.x, \lambda xy.d^0] \equiv \lambda z^{1_2 \to 1_2 \to 0}.z(\lambda xy.x)(\lambda xy.d^0)$, $N \equiv [\lambda xy.d^0, \lambda xy.y]$. The $[P, Q]$ notation for pairs is from B[1984].]

Exercise 3.3 Let $\mathcal{M}_n = \mathcal{M}_{\{1,\ldots,n\}}$. Write $\boldsymbol{c}_i = \lambda fx.f^i x$ for $i \in \omega$, the Church numerals of type $1 \to 0 \to 0$.

(i) Show that for $i, j \in \omega$ one has

$$\mathcal{M}_n \models \boldsymbol{c}_i = \boldsymbol{c}_j \iff i = j \vee [i, j \geq n-1 \;\&\; \forall k_{1 \leq k \leq n}.i \equiv j (\text{mod } k)].$$

[Hint. For $a \in \mathcal{M}_n(0), f \in \mathcal{M}_n(1)$ define the *trace of a under f* as

$$\{f^i(a) \mid i \in \omega\},$$

directed by $G_f = \{(a, b) \mid f(a) = b\}$, which by the pigeonhole principle is 'lassoo shaped'. Consider the traces of 1 under the functions f_n, g_m with $1 \leq m \leq n$, where

$$\begin{array}{rclll} f_n(k) &=& k+1, & \text{if } k < n, & \text{and } g_m(k) = k+1, \quad \text{if } k < m, \\ &=& n, & \text{if } k = n, & \qquad\qquad\quad\;\; = 1, \qquad\quad \text{if } k = m, \\ & & & & \qquad\qquad\quad\;\; = k, \qquad\quad \text{else.}] \end{array}$$

Conclude that, for example, $\mathcal{M}_5 \models \boldsymbol{c}_4 = \boldsymbol{c}_{64}$, $\mathcal{M}_6 \not\models \boldsymbol{c}_4 = \boldsymbol{c}_{64}$ and $\mathcal{M}_6 \models \boldsymbol{c}_5 = \boldsymbol{c}_{65}$.

(ii) Conclude that $\mathcal{M}_n \equiv_{1 \to 0 \to 0} \mathcal{M}_m \iff n = m$, see Definitions 3.1.14 and 3.2.4.

(iii) Show directly that $\bigcap_n \text{Th}(\mathcal{M}_n)(1) = \mathcal{E}_{\beta\eta}(1)$.

(iv) Show, using results in Section 3.4, that $\bigcap_n \text{Th}(\mathcal{M}_n) = \text{Th}(\mathcal{M}_\omega) = \mathcal{E}_{\beta\eta}$.

Exercise 3.4 The iterated exponential function 2_n is

$$2_0 = 1,$$
$$2_{n+1} = 2^{2_n}.$$

One has $2_n = 2_n(1)$, according to the definition before Exercise 2.18. Define $s(A)$ to be the number of occurrences of atoms in the type $A \in \mathbb{T}^0$, i.e.

$$s(0) = 1$$
$$s(A \to B) = s(A) + s(B).$$

Write $\#X$ for the cardinality of the set X. Show the following:

(i) $2_n \leq 2_{n+p}$.

(ii) $2_{n+2}^{2_{p+1}} \leq 2_{n+p+3}$.

(iii) $2_n^{2_p} \leq 2_{n+p}$.

(iv) If $X = \{0,1\}$, then $\forall A \in \mathbb{T}.\#(X(A)) \leq 2_{s(A)}$.

(v) For which types A do we have equality in (iv)?

Exercise 3.5 Show that if $\mathcal{M}$ is a type model, then for the corresponding polynomial type model $\mathcal{M}^*$ one has $\mathrm{Th}(\mathcal{M}^*) = \mathrm{Th}(\mathcal{M})$.

Exercise 3.6 Show that

$$A_1 \to \cdots \to A_n \to 0 \leq_{\beta\eta} A_{\pi 1} \to \cdots \to A_{\pi n} \to 0,$$

for any permutation $\pi \in S_n$.

Exercise 3.7 Let

$$A = (2 \to 2 \to 0) \to 2 \to 0 \text{ and } B = (0 \to 1^2 \to 0) \to 1_2 \to (0 \to 1 \to 0) \to 0^2 \to 0.$$

Show that

$$A \leq_{\beta\eta} B.$$

[Hint. Use the term

$$\lambda z{:}A\lambda u_1{:}(0 \to 1^2 \to 0)\lambda u_2{:}1_2\lambda u_3{:}(0 \to 2)\lambda x_1 x_2{:}0.$$
$$z[\lambda y_1, y_2{:}2.u_1 x_1(\lambda w{:}0.y_1(u_2 w))(\lambda w{:}0.y_2(u_2 w))][u_3 x_2].]$$

Exercise 3.8 Let $A = (1^2 \to 0) \to 0$. Show that

$$A \leq_{\beta\eta} 1_2 \to 2 \to 0.$$

[Hint. Use the term $\lambda M{:}A\lambda p{:}1_2\lambda F{:}2.M(\lambda f, g{:}1.F(\lambda z{:}0.p(fz)(gz)))$.]

Exercise 3.9

(i) Show that

$$\begin{pmatrix} & 2 & \\ 3 & & 4 \end{pmatrix} \leq_{\beta\eta} 1 \to 1 \to \begin{pmatrix} & 2 & \\ 3 & & 3 \end{pmatrix}.$$

(ii) Show that

$$\begin{pmatrix} & 2 & \\ 3 & & 3 \end{pmatrix} \leq_{\beta\eta} 1 \to 1 \to \begin{pmatrix} 2 \\ 3 \end{pmatrix}.$$

(iii)* Show that

$$\begin{pmatrix} 2 & 2 \\ 3 & 2 \end{pmatrix} \leq_{\beta\eta} 1_2 \to \begin{pmatrix} & 2 & \\ 3 & & 2 \end{pmatrix}.$$

[Hint. Use $\Phi = \lambda M \lambda p{:}1_2 \lambda H_1' H_2.M$
$[\lambda f_{11}, f_{12}{:}1_2.H_1'(\lambda xy{:}0.p(f_{12}xy, H_2 f_{11})]$
$[\lambda f_{21}{:}1_3 \lambda f_{22}{:}1_2.H_2 f_{21} f_{22}]$.]

Exercise 3.10 Show directly that $3{\to}0 \leq_{\beta\eta} 1{\to}1{\to}0{\to}0$. [Hint. Use

$$\Phi \equiv \lambda M{:}3\lambda f, g{:}1\lambda z{:}0.M(\lambda h{:}1.f(h(g(hz)))).$$

Typical elements of type 3 are $M_i \equiv \lambda F{:}2.F(\lambda x_1.F(\lambda x_2.x_i))$. Show that Φ acts injectively (modulo $\boldsymbol{\beta\eta}$) on these.]

Exercise 3.11 Give an example of $F, G \in \Lambda[C_4]$ such that $Fh_2 =_{\beta\eta} Gh_2$, but $F \neq_{\beta\eta} G$, where $h_2 \equiv \lambda z{:}0.\mathbf{\Phi}(\lambda g{:}1.g(gz))$.

Exercise 3.12 Suppose $(\vec{A}{\to}0), (\vec{B}{\to}0) \in \mathbb{T}_i$, with $i > 2$. Then

(i) $(\vec{A}{\to}\vec{B}{\to}0) \in \mathbb{T}_i$;
(ii) $(\vec{A}{\to}\vec{B}{\to}0) \sim_h \vec{A}{\to}0$.

Exercise 3.13

(i) Suppose that $\text{class}(A) \geq 0$. Then

$$A \leq_{\beta\eta} B \Rightarrow (C{\to}A) \leq_{\beta\eta} (C{\to}B).$$
$$A \sim_{\beta\eta} B \Rightarrow (C{\to}A) \sim_{\beta\eta} (C{\to}B).$$

[Hint. Distinguish cases for the class of A.]

(ii) Show that in (i) the condition on A cannot be dropped.
[Hint. Take $A \equiv 1_2{\to}0$, $B \equiv C \equiv 0$.]

Exercise 3.14 Show that the relations $\leq_h$ and $\leq_{h^+}$ are transitive.

Exercise 3.15 (Joly (2001a), Lemma 2, p. 981, based on an idea of Dana Scott) This exercise and the next one provide an alternative proof that all types in $\mathbb{T}^0$ can be reduced to $\top^2 = 1_2{\to}0{\to}0$. Show that any type A is reducible to

$$1_2{\to}2{\to}0 = (0{\to}(0{\to}0)){\to}((0{\to}0){\to}0){\to}0.$$

[Hint. We regard each closed term of type A as an untyped lambda term and then we retype all the variables as type 0 replacing applications XY by $fXY (=: X \bullet Y)$ and abstractions $\lambda x.X$ by $g(\lambda x.X) (=: \lambda^\bullet x.X)$ where $f : 1_2, g : 2$. Scott thinks of f and g as a retract pair satisfying $g \circ f = \mathsf{I}$

(of course in our context they are just variables which we abstract at the end). The exercise is to define terms which 'do the retyping' and insert the f and g, and to prove that they work. For $A \in \mathbb{T}$ define terms $U_A : A \to 0$ and $V_A : 0 \to A$ as follows:

$$\begin{array}{rcl} U_0 &:=& \lambda x{:}0.x; \quad V_0 := \lambda x{:}0.x; \\ U_{A\to B} &:=& \lambda u.g(\lambda x{:}0.U_B(u(V_A x))); \\ V_{A\to B} &:=& \lambda v\lambda y.V_B(fv(U_A y)). \end{array}$$

Let $A_i = A_{i_1} \to \cdots A_{ir_i} \to 0$ and write for a closed $M : A$

$$M = \lambda y_1 \cdots y_a.y_i(M_1 y_1 \ldots y_s) \cdots (M_{r_i} y_1 \ldots y_s),$$

with the M_i closed (this is the 'Φ-nf' if the M_i are written similarly). Then

$$U_A M \quad \twoheadrightarrow \quad \lambda^\bullet \vec{x}.x_i(U_{B_1}(M_1\vec{x})) \bullet \bullet \bullet (U_{B_n}(M_n\vec{x})),$$

where $B_j = A_1 \to \cdots \to A_a \to A_{ij}$, for $1 \leq j \leq n$, is the type of M_j. Show by induction on the complexity of M that, for all closed M, N,

$$U_A M =_{\beta\eta} U_A N \;\Rightarrow\; M =_{\beta\eta} N.$$

Conclude that $A \leq_{\beta\eta} 1_2 \to 2 \to 0$ via $\Phi \equiv \lambda bfg.U_A b$.]

Exercise 3.16 Show directly that

$$2 \to 1_2 \to 0 \leq_{\beta\eta} 1^2 \to 1_2 \to 0 \to 0,$$

via $\Phi \equiv \lambda M : 2 \to 1_2 \to 0 \lambda fg : 1 \lambda b : 1_2 \lambda x : 0.M(\lambda h.f(h(g(hx))))b$.

Finish the alternative proof of $\top^2 = 1_2 \to 0 \to 0$ satisfies $\forall A \in \mathbb{T}(\boldsymbol{\lambda}^0_{\to}).A \leq_{\beta\eta} \top^2$, by showing in the style of the proof of Proposition 3.4.7 the easy result that

$$1^2 \to 1_2 \to 0 \to 0 \leq_{\beta\eta} 1_2 \to 0 \to 0.$$

Exercise 3.17 Show directly that (without the Reducibility Theorem)

$$3 \to 0 \to 0 \leq_{\beta\eta} 1_2 \to 0 \to 0 = \top^2.$$

Exercise 3.18 Show directly the following:

(i) $1_3 \to 1_2 \to 0 \leq_{\beta\eta} \top^2$;
(ii) For any type A of rank ≤ 2 one has $A \leq_{\beta\eta} \top^2$.

Exercise 3.19 In this exercise the combinatorics of the argument needed in the proof of 3.4.6 is analyzed. Let $(\lambda F{:}2.M) : 3$. Define M^+ to be the lnf of $M[F{:} = H]$, where

$$H = (\lambda h{:}1.f(h(g(hz)))) \in \Lambda_{\to}^{\{f,g:1,z:0\}}(2).$$

Write $\mathtt{cut}_{g\to z}(P) = P[g{:} = \mathsf{K}z]$.

(i) Show by induction on M that if $g(P) \subseteq M^+$ is maximal (i.e. $g(P)$ is not a proper subterm of a $g(P') \subseteq M^+$), then $\mathtt{cut}_{g\to z}(P)$ is a proper subterm of $\mathtt{cut}_{g\to z}(M^+)$.

(ii) Let $M \equiv F(\lambda x{:}0.N)$. Then we know

$$M^+ =_{\beta\eta} f(N^+[x{:}= g(N^+[x{:}= z])]).$$

Show that if $g(P) \subseteq M^+$ is maximal and

$$\mathtt{length}(\mathtt{cut}_{g\to z}(P)) + 1 = \mathtt{length}(\mathtt{cut}_{g\to z}(M^+)),$$

then $g(P) \equiv g(N^+[x{:}= z])$ and is substituted for an occurrence of x in N^+.

(iii) Show that the occurrences of $g(P)$ in M^+ that are maximal and satisfy $\mathtt{length}(\mathtt{cut}_{g\to z}(P))+1 = \mathtt{length}(\mathtt{cut}_{g\to z}(M^+))$ are exactly those that were substituted for the occurrences of x in N^+.

(iv) Show that (up to $=_{\beta\eta}$) M can be reconstructed from M^+.

Exercise 3.20 Show that all elements $g \in \mathcal{M}_2(0\to0)$ satisfy $g^2 = g^4$. Conclude that $\mathcal{T} \not\hookrightarrow \mathcal{M}_2$.

Exercise 3.21 Let $\mathcal{D}$ have enough constants. Show that the class of $\mathcal{D}$ is not

$$\min\{i \mid \forall D.[D \text{ represented in } \mathcal{D} \Rightarrow D \leq_{\beta\eta} \nabla(\mathcal{C}_i)]\}.$$

[Hint. Consider $\mathcal{D} = \{\boldsymbol{c}^0, \boldsymbol{d}^0, \boldsymbol{e}^0\}$.]

Exercise 3.22 A model $\mathcal{M}$ is called finite iff $\mathcal{M}(A)$ is finite for all types A. Find out which of the five canonical term models is finite.

Exercise 3.23 Let $\mathcal{M} = \mathcal{M}_{\min}$.

(i) Determine in $\mathcal{M}(1\to0\to0)$ which of the three Church numerals $\mathbf{c}_0, \mathbf{c}_{10}$ and $\mathbf{c}_{100}$ are equal and which not.

(ii) Determine the elements in $\mathcal{M}(1_2\to0\to0)$.

Exercise 3.24 Let $\mathcal{M}$ be a model and let $|\mathcal{M}_0| \leq \kappa$. By Example 3.3.24 there exists a partial surjective homomorphism $h : \mathcal{M}_\kappa \to \mathcal{M}$.

(i) Show that $h^{-1}(\mathcal{M}) \subseteq \mathcal{M}_\kappa$ is closed under λ-definability. [Hint. Use Example 3.3.27.]

(ii) Show that as in Example 3.3.28 one has $h^{-1}(\mathcal{M})^E = h^{-1}(\mathcal{M})$.

(iii) Show that the Gandy hull $h^{-1}(\mathcal{M})/E$ is isomorphic to $\mathcal{M}$.

(iv) For the five canonical models $\mathcal{M}$ construct $h^{-1}(\mathcal{M})$ directly without reference to $\mathcal{M}$.

(v) (Plotkin) Do the same as (iii) for the free open term model.

Exercise 3.25 Let $\mathcal{D} = \{\boldsymbol{F}^2, \boldsymbol{c}_1^0, \ldots, \boldsymbol{c}_n^0\}$.

(i) Give a characterization of the elements of $\Lambda^{\emptyset}_{\rightarrow}[\mathcal{D}](1)$.

(ii) For $f, g \in \Lambda^{\emptyset}_{\rightarrow}[\mathcal{D}](1)$ show that $f \neq_{\beta\eta} g \Rightarrow f \not\approx_{\mathcal{D}} g$ by applying both f, g to $\boldsymbol{F}f$ or $\boldsymbol{F}g$.

Exercise 3.26 Prove the following.

$$
\begin{aligned}
1_2{\rightarrow}0{\rightarrow}0 \;\; \leq_{\beta\eta} \;\; & ((1_2{\rightarrow}0){\rightarrow}0){\rightarrow}0{\rightarrow}0, \text{ via} \\
& \lambda m \lambda F{:}((1_2{\rightarrow}0){\rightarrow}0)\lambda x{:}0.F(\lambda h{:}1_2.mhx) \text{ or via} \\
& \lambda m \lambda F{:}((1_2{\rightarrow}0){\rightarrow}0)\lambda x{:}0.m(\lambda pq{:}0.F(\lambda h{:}1_2.hpq))x. \\
1_2{\rightarrow}0{\rightarrow}0 \;\; \leq_{\beta\eta} \;\; & (1{\rightarrow}1{\rightarrow}0){\rightarrow}0{\rightarrow}0 \\
& \text{via } \lambda m H x.m(\lambda ab.H(\mathsf{K}a)(\mathsf{K}b))x.
\end{aligned}
$$

Exercise 3.27 Show that $\mathbb{T}_2 = \{(1_p \rightarrow 0) \rightarrow 0^q \rightarrow 0 \mid p, q > 0\}$.

Exercise 3.28 In this exercise we show that $A \sim_{\beta\eta} B \;\&\; A \sim_{h^+} B$, for all $A, B \in \mathbb{T}_2$.

(i) First establish for $p \geq 1$ that

$$1{\rightarrow}0{\rightarrow}0 \sim_{\beta\eta} 1{\rightarrow}0^p{\rightarrow}0 \;\&\; 1{\rightarrow}0{\rightarrow}0 \sim_{h^+} 1{\rightarrow}0^p{\rightarrow}0.$$

1. Show $1{\rightarrow}0{\rightarrow}0 \leq_h 1{\rightarrow}0^p{\rightarrow}0$. Therefore

$$1{\rightarrow}0{\rightarrow}0 \leq_{\beta\eta} 1{\rightarrow}0^p{\rightarrow}0 \;\&\; 1{\rightarrow}0{\rightarrow}0 \leq_{h^+} 1{\rightarrow}0^p{\rightarrow}0.$$

2. Show $1{\rightarrow}0^p{\rightarrow}0 \leq_{h^+} 1{\rightarrow}0{\rightarrow}0$. [Hint. Using inhabitation machines one sees that the long normal forms of terms in $\Lambda^{\emptyset}_{\rightarrow}(1{\rightarrow}0^p{\rightarrow}0)$ are of the form $L_i^n \equiv \lambda f{:}1\lambda x_1 \ldots x_p{:}0.f^n x_i$, with $n \geq 0$ and $1 \leq i \leq p$. Define $\Phi_i : (1{\rightarrow}0^p{\rightarrow}0){\rightarrow}(1{\rightarrow}0{\rightarrow}0)$, with $i = 1, 2$, as

$$
\begin{aligned}
\Phi_1 L &= \lambda f{:}1\lambda x{:}0.Lfx^{\sim p}; \\
\Phi_2 L &= \lambda f{:}1\lambda x{:}0.L\mathsf{I}(f^1 x) \cdots (f^p x).
\end{aligned}
$$

Then $\Phi_1 L_i^n =_{\beta\eta} \mathbf{c}_n$ and $\Phi_2 L_i^n =_{\beta\eta} \mathbf{c}_i$. So, for $M, N \in \Lambda^{\emptyset}_{\rightarrow}(1{\rightarrow}0_q{\rightarrow}0)$,

$$M \neq_{\beta\eta} N \;\Rightarrow\; \Phi_1 M \neq_{\beta\eta} \Phi_1 N \text{ or } \Phi_2 M \neq_{\beta\eta} \Phi_2 N.]$$

3. Conclude that also $1{\rightarrow}0^p{\rightarrow}0 \leq_{\beta\eta} 1{\rightarrow}0{\rightarrow}0$, by taking as reducing term

$$\Phi \equiv \lambda m f x.P_2(\Phi_1 m)(\Phi_2 m),$$

where P_2 λ-defines a polynomial injection $p_2 : \omega^2 {\rightarrow} \omega$.

(ii) Now establish for $p \geq 1, q \geq 0$ that

$$1{\to}0{\to}0 \sim_{\beta\eta} (1_p{\to}0){\to}0^q{\to}0 \ \& \ 1{\to}0{\to}0 \sim_{h^+} 1_p{\to}0^q{\to}0.$$

1. Show $1{\to}0{\to}0 \leq_h (1_p{\to}0){\to}0^q{\to}0$ using

$$\Phi \equiv \lambda m F x_1 \cdots x_q.m(\lambda z.F(\lambda y_1 \ldots y_p.z)).$$

2. Show $(1_p{\to}0){\to}0^q{\to}0 \leq_{h^+} 1{\to}0{\to}0$. [Hint. For

$$L \in \Lambda^{\emptyset}_{\to}((1_p{\to}0){\to}0^q{\to}0)$$

its lnf is of one of the following:

$$\begin{array}{lcl} L^{n,k,r} & = & \lambda F{:}(1_p{\to}0)\lambda y_1 \ldots y_q{:}0.F(\lambda \vec{z}_1. \cdots F(\lambda \vec{z}_n.z_{kr})..) \\ M^{n,s} & = & \lambda F{:}(1_p{\to}0)\lambda y_1 \ldots y_q{:}0.F(\lambda \vec{z}_1. \cdots F(\lambda \vec{z}_n.y_s)..), \end{array}$$

where $\vec{z}_k = z_{k1} \cdots z_{kp}$, $1 \leq k \leq n$, $1 \leq r \leq p$, and $1 \leq s \leq q$, in the case $q > 0$ (otherwise the $M^{n,s}$ do not exist). Define three terms $O_1, O_2, O_3 \in \Lambda^{\emptyset}_{\to}(1{\to}0{\to}1_p{\to}0)$ as

$$\begin{aligned} O_1 &\equiv \lambda f x g.g(f^1 x) \cdots (f^p x) \\ O_2 &\equiv \lambda f x g.f(g x^{\sim p}) \\ O_3 &\equiv \lambda f x g.f(g(f(g x^{\sim p}))^{\sim p}). \end{aligned}$$

Define terms $\Phi_i \in \Lambda^{\emptyset}_{\to}(((1_p{\to}0){\to}0^q{\to}0){\to}1{\to}0{\to}0)$ for $1 \leq i \leq 3$ by

$$\begin{array}{lcll} \Phi_1 L & \equiv & \lambda f x.L(O_1 f x)(f^{p+1}x) \cdots (f^{p+q}x); & \\ \Phi_i L & \equiv & \lambda f x.L(O_i f x)x^{\sim q}, & \text{for } i \in \{2,3\}. \end{array}$$

Verify that

$$\begin{array}{lcl} \Phi_1 L^{n,k,r} & = & \mathbf{c}_r \\ \Phi_1 M^{n,s} & = & \mathbf{c}_{p+s} \\ \Phi_2 L^{n,k,r} & = & \mathbf{c}_n \\ \Phi_2 M^{n,s} & = & \mathbf{c}_n \\ \Phi_3 L^{n,k,r} & = & \mathbf{c}_{2n+1-k} \\ \Phi_3 M^{n,s} & = & \mathbf{c}_n. \end{array}$$

Therefore if $M \neq_{\beta\eta} N$ are terms in $\Lambda^{\emptyset}_{\to}(1_p{\to}0^q{\to}0)$, then for at least one $i \in \{1,2,3\}$ one has $\Phi_i(M) \neq_{\beta\eta} \Phi_i(N)$.]

3. Show $1_p{\to}0^q{\to}0 \leq_{\beta\eta} 1{\to}0{\to}0$, using a polynomial injection $p_3 : \omega^3{\to}\omega$.

Exercise 3.29 Show that for all $A, B \notin \mathbb{T}_1 \cup \mathbb{T}_2$ one has $A \sim_{\beta\eta} B \Rightarrow A \sim_h B$.

Exercise 3.30 Let A be an inhabited small type of rank > 3. Show that

$$3 \to 0 \to 0 \leq_m A.$$

[Hint. For small B of rank ≥ 2 one has $B \equiv B_1 \to \ldots B_b \to 0$ with $B_i \equiv B_{i_1} \to 0$ for all i and $\mathtt{rank}(B_{i_{01}}) = \mathtt{rank}(B) - 2$ for some i_0. Define for such B the term

$$X^B \in \Lambda^{\emptyset}[F^2](B),$$

where F^2 is a variable of type 2. Thus

$$\begin{array}{rcll} X^B & \equiv & \lambda x_1 \cdots x_b.F^2 x_{i_0}, & \text{if } \mathtt{rank}(B) = 2; \\ & \equiv & \lambda x_1 \cdots x_b.F^2(\lambda y{:}0.x_{i_0}(\lambda y_1 \cdots y_k.y)), & \text{if } \mathtt{rank}(B) = 3 \text{ and} \\ & & & \text{where } B_{i_0} \text{ having} \\ & & & \text{rank 1 is } 0^k \to 0; \\ & \equiv & \lambda x_1 \cdots x_b.x_{i_0} X^{B_{i_{01}}}, & \text{if } \mathtt{rank}(B) > 3. \end{array}$$

(Here $X^{B_{i_{01}}}$ is well defined since $B_{i_{01}}$ is also small.) As A is inhabited, take $\lambda x_1 \cdots x_b.N \in \Lambda^{\emptyset}(A)$. Define $\Psi : (3 \to 0 \to 0) \to A$ by

$$\Psi(M) = \lambda x_1 \cdots x_b.M(\lambda F^2.x_i X^{A_{i1}})N,$$

where i is such that A_{i1} has rank ≥ 2. Show that Ψ works.]

Exercise 3.31 Consider the following equations.

(1) $\lambda f{:}1\lambda x{:}0.fx = \lambda f{:}1\lambda x{:}0.f(fx)$;
(2) $\lambda f, g{:}1\lambda x{:}0.f(g(g(fx))) = \lambda f, g{:}1\lambda x{:}0.f(g(f(gx)))$;
(3) $\lambda F{:}3\lambda x{:}0.F(\lambda f_1{:}1.f_1(F(\lambda f_2{:}1.f_2(f_1 x)))) =$
$\lambda F{:}3\lambda x{:}0.F(\lambda f_1{:}1.f_1(F(\lambda f_2{:}1.f_2(f_2 x))))$.
(4) $\lambda h{:}1_2\lambda x{:}0.h(hx(hxx))(hxx) = \lambda h{:}1_2\lambda x{:}0.h(hxx)(h(hxx)x)$.

(i) Show that (1) holds in $\mathcal{M}_{C_1}$, but not in $\mathcal{M}_{C_2}$.
(ii) Show that (2) holds in $\mathcal{M}_{C_2}$, but not in $\mathcal{M}_{C_3}$.
(iii) Show that (3) holds in $\mathcal{M}_{C_3}$, but not in $\mathcal{M}_{C_4}$.
[Hint. Use Lemmas 7a and 11 in Dekkers (1988).]
(iv) Show that (4) holds in $\mathcal{M}_{C_4}$, but not in $\mathcal{M}_{C_5}$.

Exercise 3.32 Construct six pure closed terms of the same type in order to show that the five canonical theories are maximally different. That is, we want terms $M_1, \ldots, M_6$ such that in $\mathtt{Th}(\mathcal{M}_{C_5})$ the $M_1, \ldots, M_6$ are mutually different; also $M_6 = M_5$ in $\mathtt{Th}(\mathcal{M}_{C_4})$, but different from $M_1, \ldots, M_4$; also $M_5 = M_4$ in $\mathtt{Th}(\mathcal{M}_{C_3})$, but different from $M_1, \ldots, M_3$; also $M_4 = M_3$ in

$\mathtt{Th}(\mathcal{M}_{\mathcal{C}_2})$, but different from M_1, M_2; also $M_3 = M_2$ in $\mathtt{Th}(\mathcal{M}_{\mathcal{C}_1})$, but different from M_1; finally $M_2 = M_1$ in $\mathtt{Th}(\mathcal{M}_{\mathcal{C}_0})$. [Hint. Use the previous exercise and a polynomially-defined pairing operator.]

Exercise 3.33 Let $\mathcal{M}$ be a typed lambda model. Let S be the logical relation determined by $S_0 = \emptyset$. Show that $S_0^* \neq \emptyset$.

Exercise 3.34 We work with $\boldsymbol{\lambda}_{\to}^{\mathrm{Ch}}$ over $\mathbb{T}^0$. Consider the full typed structure $\mathcal{M}_1 = \mathcal{M}_\omega$ over the natural numbers, the open term model $\mathcal{M}_2 = \mathcal{M}(\boldsymbol{\beta\eta})$, and the closed term model $\mathcal{M}_3 = \mathcal{M}^\emptyset[\{\boldsymbol{h}^1, \boldsymbol{c}^0\}](\boldsymbol{\beta\eta})$. For these models consider, three times, the Gandy hull:

$$\begin{aligned}\mathcal{G}_1 &= \mathcal{G}_{\{\boldsymbol{S}:1,0:0\}}(\mathcal{M}_1)\\ \mathcal{G}_2 &= \mathcal{G}_{\{[f:1],[x:0]\}}(\mathcal{M}_2)\\ \mathcal{G}_3 &= \mathcal{G}_{\{[\boldsymbol{h}:1],[\boldsymbol{c}:0]\}}(\mathcal{M}_3),\end{aligned}$$

where $\boldsymbol{S}$ is the successor function and $0 \in \omega$, f, x are variables and $\boldsymbol{h}, \boldsymbol{c}$ are constants, of type $1, 0$ respectively. Prove

$$\mathcal{G}_1 \cong \mathcal{G}_2 \cong \mathcal{G}_3.$$

[Hint. Consider the logical relation R on $\mathcal{M}_3 \times \mathcal{M}_2 \times \mathcal{M}_1$ determined by

$$R_0 = \{\langle [\boldsymbol{h}^k(\boldsymbol{c})],\, [f^k(x)],\, k\rangle \mid k \in \omega\}.$$

Apply the fundamental theorem for logical relations.]

Exercise 3.35 A function $f : \omega \to \omega$ is *slantwise λ-definable*, see also Fortune et al. (1983) and Leivant (1990) if there is a substitution operator $^+$ for types and a closed term $F \in \Lambda^\emptyset(\omega^+ \to \omega)$ such that

$$F\mathbf{c}_k{}^+ =_{\boldsymbol{\beta\eta}} \mathbf{c}_{f(k)}.$$

This can be generalized to functions of k arguments, allowing for each argument a different substitution operator.

(i) Show that $f(x, y) = x^y$ is slantwise λ-definable.
(ii) Show that the predecessor function is slantwise λ-definable.
(iii) Show that subtraction is not slantwise λ-definable. [Hint. Suppose towards a contradiction that a term $m : \mathsf{Nat}_\tau \to \mathsf{Nat}_\rho \to \mathsf{Nat}_\sigma$ defines subtraction. Use the Finite Completeness Theorem, Proposition 3.4.33, for $A = \mathsf{Nat}_\sigma$ and $M = \mathbf{c}_0$.]

Exercise 3.36 (Finite Generation, Joly (2002)) Let $A \in \mathbb{T}$. Then A is said to be *finitely generated* if there exist types $A_1, \ldots, A_t$ and terms M_1 :

$A_1, \ldots, A_t : M_t$ such that for any $M : A$, M is $\beta\eta$ convertible to an applicative combination of $M_1, \ldots, M_t$.

Example. $\mathsf{Nat} = 1{\to}0{\to}0$ is finitely generated by $\mathbf{c}_0 \equiv (\lambda fx.x) : \mathsf{Nat}$ and $S \equiv (\lambda nfx.f(fx)) : (\mathsf{Nat}{\to}\mathsf{Nat})$.

We say A *slantwise enumerates* a type B if there exists a type substitution @ and $F : @A{\to}B$ such that for each $N : B$ there exists $M : A$ such that $F@M =_{\beta\eta} N$ (F is *surjective*).

A type A is said to be *poor* if there is a finite sequence of variables $\vec{x}$, such that every $M \in \Lambda^{\emptyset}_{\to}(A)$ in $\beta\eta$-nf has $FV(M) \subseteq \vec{x}$. Otherwise A is said to be *rich*.

Example. Let $A = (1{\to}0){\to}0{\to}0$ be poor. A typical $\beta\eta$-nf of type A has the shape $\lambda F \lambda x(F(\lambda x(\cdots(F(\lambda y(F(\lambda y \cdots x \cdots)))..)))$. One allows the term to violate the variable convention (that asks different occurrences of bound variables to be different). The monster type $3{\to}1$ is rich.

The following are equivalent.

(1) A slantwise enumerates the monster type $\mathsf{M} \equiv 3{\to}0{\to}0$.
(2) The lambda-definability problem for A, , i.e. $\lambda\!\!\lambda nd.D(n, A, d)$, is undecidable.
(3) A is not finitely generated.
(4) A is rich.

However, we will not ask the reader to prove (4) $\Rightarrow$ (1) since this involves more knowledge of and practice with slantwise enumerations than available from this book. For that proof we refer the reader to Joly's paper. We have already shown that the lambda-definability problem for the monster M is undecidable. We then proceed as follows.

(i) Show A is rich iff A has rank >3 or A is large of rank 3 (for A inhabited; especially for $\Rightarrow$). Use this to show

$$(2) \Rightarrow (3) \text{ and } (3) \Rightarrow (4).$$

(ii) (Alternative way of showing (3) $\Rightarrow$ (4).) Suppose that every closed term of type A beta eta converts to a special one built up from a fixed finite set of variables. Show that it suffices to bound the length of the lambda prefix of any subterm of such a special term in order to conclude finite generation. Suppose that we consider only terms X built up only from the variables $v_1{:}A_1, \ldots, v_m{:}A_m$, both free and bound. We shall transform X using a fixed set of new variables. First we assume the set $\{A_1, \ldots, A_m\}$ is closed under subtype.

(a) Show that we can assume that X is fully expanded. For example, if X has the form

$$\lambda x_1 \cdots x_t.(\lambda x.X_0)X_1 \cdots X_s$$

then $(\lambda x.X_0)X_1 \cdots X_s$ has one of the A_i as a type (just normalize and consider the type of the head variable). Thus we can eta expand

$$\lambda x_1 \cdots x_t.(\lambda x.X_0)X_1 \cdots X_s$$

and repeat recursively. We need only double the set of variables to do this. We do this keeping the same notation.

(b) Thus given

$$X = \lambda x_1 \cdots x_t.(\lambda x.X_0)X_1 \cdots X_s$$

we have $X_0 = \lambda y_1 \cdots y_r.Y$, where $Y : 0$. Now if $r{>}m$, each multiple occurrence of v_i in the prefix $\lambda y_1 \cdots y_r$ is dummy and those that occur in the initial segment $\lambda y_1 \cdots y_s$ can be removed with the corresponding X_j. The remaining variables will be labelled $z_1, \ldots, z_k$. The remaining X_j will be labelled $Z_1, \ldots, Z_l$. Note that $r - s + t < m + 1$. Thus

$$X = \lambda x_1 \cdots x_t.(\lambda z_1 \cdots z_k Y)Z_1 \cdots Z_l,$$

where $k < 2m + 1$. We can now repeat this analysis recursively on Y, and $Z_1, \ldots, Z_l$ observing that the types of these terms must be among the A_i. We have bounded the length of a prefix.

(iii) Now for (1) $\Rightarrow$ (2). We have already shown that the lambda-definability problem for the monster M is undecidable. Suppose (1) and $\neg$(2), towards a contradiction. Fix a type B and let $B(n)$ be the cardinality of B in $P(n)$. Show that for any closed terms $M, N : C$

$$P(B(n)) \models M = N \;\Rightarrow\; P(n) \models [0 := B]M = [0 := B]N.$$

Conclude from this that lambda-definability for M is decidable, which is not the case.

Exercise 3.37 Show that for $A, B \in \mathbb{T}^0$ one has

(i) $A \leq_h B \;\Rightarrow\; A \leq_{\beta\eta} B$.
(ii) $A \leq_{\beta\eta} B \;\Rightarrow\; A \leq_{h^+} B$. [Hint. Use the hierarchy theorem. A direct proof seems to be difficult.]

4

Definability, unification and matching

4.1 Undecidability of lambda-definability

The finite standard models

Recall that the full type structure over a set X, written $\mathcal{M}_X$, is defined in Definition 2.4.18 as follows:

$$\begin{aligned} X(0) &= X, \\ X(A{\to}B) &= X(B)^{X(A)}; \\ \mathcal{M}_X &= \{X(A)\}_{A \in \mathbb{T}^0}. \end{aligned}$$

Note that if X is finite then all the $X(A)$ are finite. In that case we can represent each element of $\mathcal{M}_X$ by a finite piece of data and hence (through Gödel numbering) by a natural number. For instance for $X = \{0, 1\}$ we can represent the four elements of $X(0{\to}0)$ as follows. If 0 is followed by 0 to the right this means that 0 is mapped onto 0, etc.

0	0
1	0

0	1
1	1

0	0
1	1

0	1
1	0

Any element of the model can be expressed similarly: for instance the following table represents an element of $X((0 \to 0) \to 0)$.

<table>
<tr><td>0</td><td>0</td><td rowspan="2">0</td></tr>
<tr><td>1</td><td>0</td></tr>
<tr><td>0</td><td>1</td><td rowspan="2">0</td></tr>
<tr><td>1</td><td>1</td></tr>
<tr><td>0</td><td>0</td><td rowspan="2">0</td></tr>
<tr><td>1</td><td>1</td></tr>
<tr><td>0</td><td>1</td><td rowspan="2">1</td></tr>
<tr><td>1</td><td>0</td></tr>
</table>

We know that $\mathsf{I} \equiv \lambda x.x$ is the only closed $\boldsymbol{\beta\eta}$-nf of type $0 \to 0$. Since $[\![\mathsf{I}]\!] = \mathbf{1}_X$, the identity on X is the only function of $X(0 \to 0)$ that is denoted by a closed term.

4.1.1 Definition Let $\mathcal{M} = \mathcal{M}_X$ be a type structure over a set X and let $d \in \mathcal{M}(A)$.

(i) A closed term $F \in \Lambda^{\emptyset}(A)$ is said to λ-*define* d if $d = [\![F]\!]^{\mathcal{M}}$.
(ii) If, moreover, $\vec{e} \in \mathcal{M}$, with $e_1 \in \mathcal{M}(B_1), \ldots, e_n \in \mathcal{M}(B_n)$, then we say that $F \in \Lambda^{\emptyset}(\vec{B} \to A)$ λ-*defines* d *from the* $\vec{e}$ if one has $[\![F]\!]^{\mathcal{M}} \vec{e} = d$.
(iii) In (i) one also says that d is λ-*definable*. In (ii) one also says that d is λ-*definable from* $\vec{e}$.

The main result in this section is the undecidability of λ-definability in $\mathcal{M}_X$, for X of cardinality >6. This means that there is no algorithm for deciding whether a table describes a λ-definable element in this model. This result was published in Loader (2001b), but had already been proved by him in 1993.

4.1.2 Remark One may wonder why decidability of λ-definability doesn't hold trivially. Indeed, writing $\mathcal{M}_n = \mathcal{M}_{\{0,\ldots,n-1\}}$, then for all $A \in \mathbb{T}^0$ the set $\mathcal{M}_n(A)$ is finite, hence also the subset

$$\{f \in \mathcal{M}_n(A) | f \text{ is } \lambda\text{-definable}\}.$$

This seems to imply that λ-definability is decidable. The mistake is that λ-definability should be seen as a predicate of two arguments: the f and the A. One says loosely: decidability of λ-definability of $f \in \mathcal{M}_n(A)$ is considered uniformly in A. We return to this in Remark 4.1.26 and Theorem 4.1.28.

The method of showing that decision problems are undecidable proceeds via reducing them to well-known undecidable problems (and eventually to the Halting problem).

4.1.3 Definition

(i) A *decision problem* is a subset $P \subseteq \omega$. This P is called *decidable* if its characteristic function $K_P : \omega \to \{0, 1\}$ is computable. An *instance of a problem* is the question "$n \in P$?". Often problems are subsets of syntactic objects, like terms or descriptions of automata, that are considered as subsets of ω via some coding.
(ii) Let $P, Q \subseteq \omega$ be problems. Then P is *(many-one) reducible* to Q, written $P \leq_m Q$, if there is a computable function $f : \omega \to \omega$ such that

$$n \in P \Leftrightarrow f(n) \in Q.$$

(iii) More generally, a problem P is *Turing reducible* to a problem Q, written $P \leq_T Q$, if the characteristic function K_P is computable in K_Q, see e.g. Rogers Jr. (1967).

The following is well known.

4.1.4 Proposition *Let P, Q be problems.*

(i) *If $P \leq_m Q$, then $P \leq_T Q$.*
(ii) *If $P \leq_T Q$, then the undecidability of P implies that of Q.*

Proof (i) Suppose that $P \leq_m Q$. Then there is a computable function $f : \omega \to \omega$ such that $\forall n \in \omega.[n \in P \Leftrightarrow f(n) \in Q]$. Therefore $K_P(n) = K_Q(f(n))$. Hence $P \leq_T Q$.

(ii) Suppose that $P \leq_T Q$ and that Q is decidable, in order to show that P is decidable. Then K_Q is computable and so is K_P, as it is computable in K_Q. ∎

The next lemma follows by showing that λ-definability and relative λ-definability can be reduced to each other.

4.1.5 Lemma *Given $\mathcal{M} = \mathcal{M}_X$ for a finite X, then λ-definability is decidable iff relative λ-definability is decidable.*

Proof ($\Rightarrow$) Let $d \in \mathcal{M}(A), e \in \mathcal{M}(\vec{B})$ in order to check whether d is λ-definable from the $\vec{e}$. Consider

$$\mathcal{A} = \{f \in \mathcal{M}(\vec{B} \to A) \mid f\vec{e} = d\}.$$

This set is finite and non-empty as $\boldsymbol{\lambda}\vec{x}.d \in \mathcal{A}$. Assuming decidability of λ-definability check for each $f \in \mathcal{A}$ whether there is an $F \in \Lambda^{\emptyset}$ such that $[\![F]\!] = f$. We find such an F iff d is relatively decidable from $\vec{e}$. Therefore relative λ-definability is decidable.

($\Leftarrow$) Trivial by taking $\vec{e}$ empty, or alternatively, by taking just $e_1 = [\![\mathsf{I}]\!] \in \mathcal{M}(1)$. ∎

4.1.6 Remark In order to find an F that λ-defines $d \in \mathcal{M}(A)$ from $\vec{e} \in \mathcal{M}(\vec{B})$ it suffices to find an $M \in \Lambda^{\{\vec{b}:\vec{B}\}}(A)$ such that

$$[\![M]\!]_{[b_1:=e_1]\cdots[b_n:=e_n]} = d.$$

Indeed, then $[\![\lambda b_1 \ldots b_n.M]\!]\vec{e} = [\![M]\!]_{[b_1:=e_1]\cdots[b_n:=e_n]} = d$, so we can take $F \triangleq \lambda b_1 \ldots b_n.M$.

The proof of Loader's result proceeds by reducing the two-letter word rewriting problem, which is well known to be undecidable, to the problem of λ-definability in $\mathcal{M}_X$. By Proposition 4.1.4 the undecidability of the λ-definability follows.

4.1.7 Definition (Word rewriting problem) Let $\Sigma = \{A, B\}$, a two letter alphabet.

(i) A *word* (over Σ) is a finite sequence of letters $W = w_1 \cdots w_n$ with $w_i \in \Sigma$. The set of words over Σ is denoted by Σ^*.
(ii) If $W = w_1 \cdots w_n$, then $\mathtt{lth}(W) = n$ is called the *length* of W. If $\mathtt{lth}(W) = 0$, then W is called the *empty word* and is denoted by ϵ.
(iii) A *rewrite rule* is a pair of non-empty words C, D denoted by $C \hookrightarrow D$.
(iv) Given a word U and a finite set $\mathcal{R} = \{R_1, \ldots, R_r\}$ of rewrite rules $R_i = V_i \hookrightarrow W_i$. Then a *derivation* from U of a word S is a finite sequence of words starting with U finishing with S, and such that each word is obtained from the previous by replacing a subword C_i by D_i for some rule $C_i \hookrightarrow D_i \in \mathcal{R}$.
(v) A word S is said to be *R-derivable* from U, written $U \vdash_{\mathcal{R}} S$, if it has a derivation.

4.1.8 Example Consider the word AB and the rule $AB \hookrightarrow AABB$. Then $AB \vdash AAABBB$, but $AB \not\vdash AAB$.

We will need the following well-known result, see e.g. Post (1947).

4.1.9 Theorem *There is a word $W \in \Sigma^*$ and a finite set of rewrite rules $\mathcal{R}$ such that $\{V \in \Sigma^* \mid W \vdash_{\mathcal{R}} V\}$ is undecidable.*

4.1.10 Definition Given the alphabet $\Sigma = \{A, B\}$, define the set

$$X = X_\Sigma \triangleq \{A, B, *, L, R, Y, N\}.$$

You may read the objects L and R as *left* and *right* and the objects Y and N as *yes* and *no*. In 4.1.12-4.1.25 we write $\mathcal{M}$ for the full type structure $\mathcal{M}_X$ built over the set X.

4.1.11 Definition

(i) The following is a convenient notation. Let $A_1, \ldots, A_n$ be types. Then

$$[A_1, \ldots, A_n] \triangleq A_1 \to \cdots A_n \to 0.$$

In particular one can write $[\,] \triangleq 0$. Therefore each type is considered as a word (string) of its subtypes.

(ii) Let $W = w_1 \dots w_n \in X^*$, with $w_1, \dots, w_n \in X$, be a word over X considered as alphabet and let $v \in X$. For $1 \le i \le n$ and $1 \le j < n$ we write

$$w_1 \cdots v^{@i} \cdots w_n \triangleq w_1 \cdots w_{i-1} v w_{i+1} \cdots w_n;$$
$$w_1 \dots (LR)^{@j} \dots w_n \triangleq w_1 \cdots w_{j-1} LR w_{j+2} \dots w_n.$$

That is, in the first item at position i one replaces w_i by v, and in the second item at position j one replaces $w_j w_{j+1}$ by LR. Note that the resulting words have not changed length.

(iii) We denote by $*_1 \dots *_n$ the word $w_1 \dots w_n \in X^*$ with $w_i = *$ for $1 \le i \le n$.

For $M, N \in \Lambda$ remember that $MN^{\sim n} \equiv M \underbrace{N \cdots N}_{n}$.

4.1.12 Definition [Word encoding] Let $n \in \omega$. Let $W = w_1 \dots w_n$ be a word over $\Sigma = \{A, B\}$ of length n. We will use the types $1_n = 0^n \to 0$ to encode W.

(i) The word W is *encoded* by $h \in \mathcal{M}(1_n)$ uniquely determined as follows:

$$\begin{array}{lrcll} (W)_1 & h(*_1 \dots w_i{}^{@i} \cdots *_n) & \triangleq & Y, & \text{for } 1 \le i \le n, \\ (W)_2 & h(*_1 \dots (LR)^{@j} \cdots *_n) & \triangleq & Y & \text{for } 1 \le j < n; \\ (W)_3 & h(x_1, \dots, x_n) & \triangleq & N, & \text{otherwise.} \end{array}$$

In this case we write $\underline{W} = h$.

(ii) The word W is *weakly encoded* by $h \in \mathcal{M}(1_n)$ if only (W_1), (W_2) are valid.

For example, let $W = ABBA$. Then h weakly encodes W if

$$\begin{array}{ll} h\,L\,R * * = Y & h\,A * * * = Y; \\ h * L\,R * = Y & h * B * * = Y; \\ h * * L\,R = Y & h * * B * = Y; \\ & h * * * A = Y. \end{array}$$

If moreover $h\,a_1 \dots a_4 = N$ in all other cases, then h is uniquely determined and is denoted by $\underline{ABBA}$. Note that a word is not determined by an h that weakly encodes it: $\lambda\!\!\lambda x.Y$ weakly encodes all words. But we have the following.

4.1.13 Lemma *Let $W, V \in \Sigma^*$. Suppose $h \in \mathcal{M}(1_n)$ weakly encodes W and encodes V. Then $W = V$.*

Proof The lengths of W and V are equal, both being n. We show that $w_i = v_i$ for all $1 \le i \le n$. Suppose not, i.e. $w_i = A, v_i = B$, say, for some i. Then

$$\begin{array}{rcll} h*_1 \ldots A^{@i} \cdots *_n & = & Y, & \text{as } h \text{ weakly encodes } W, \\ h*_1 \ldots A^{@i} \cdots *_n & = & N, & \text{as } h \text{ encodes } V, \end{array}$$

which is a contradiction. ■

4.1.14 Remark Let $n > 0$. The only closed terms of type 1_n are the projections U_i^n, with $1 \le i \le n$. But their denotations do not even weakly encode any word. (If U_i^k weakly denotes $V = v_1 \ldots v_n$, then for typing reasons $k = n$ and

$$Y = \mathsf{U}_i^k v_1 *_1 \ldots v_i{}^{@i} \cdots *_k = v_1 \in \{A, B\},$$

a contradiction.) We will see, however, that encoded words are *relatively* λ-definable.

4.1.15 Definition (Encoding of a rule) —

(i) In order to define the encoding of a rule we write $]a_1 \ldots a_k[$ to denote (the "characteristic function of the singleton $\{\langle a_1, \ldots, a_k\rangle\}$", i.e.) the element $h \in \mathcal{M}(1_k)$ uniquely determined by

$$\begin{array}{rcll} hx_1 \ldots x_k & \triangleq & Y, & \text{if } \vec{x} = a_1, \ldots, a_k; \\ hx_1 \ldots x_k & \triangleq & N, & \text{otherwise.} \end{array}$$

(ii) Write $] - [= \lambda\!\!\lambda \vec{x}.N$.

(iii) Now a rule $C \hookrightarrow D$, where $\mathtt{lth}(C) = m$ and $\mathtt{lth}(D) = n$, is encoded as the object $\underline{C \hookrightarrow D} \in \mathcal{M}(1_n \to 1_m)$ uniquely determined by

$$\begin{array}{rcll} \underline{C \hookrightarrow D}\,(\underline{C}) & \triangleq & \underline{D}; & \\ \underline{C \hookrightarrow D}\,]*_1 \ldots *_n[& \triangleq &]*_1 \ldots *_m[; & \\ \underline{C \hookrightarrow D}\,]R *_2 \cdots *_n [& \triangleq &]R *_2 \cdots *_m [; & \\ \underline{C \hookrightarrow D}\,]*_1 \ldots *_{n-1} L[& \triangleq &]*_1 \ldots *_{m-1} L[; & \\ \underline{C \hookrightarrow D}\,(h) & \triangleq &] - [, & \text{otherwise.} \end{array}$$

4.1.16 Lemma *Let S, T be two words over Σ and let $C \hookrightarrow D$ be a rule. Let the lengths of the words S, T, C, D be p, q, n, m, respectively. Then $SCT \vdash SDT$ and*

$$\underline{SDT}\,\vec{s}\,\vec{d}\,\vec{t} = \underline{C \hookrightarrow D}\,(\lambda\!\!\lambda \vec{c}.\underline{SCT}\,\vec{s}\,\vec{c}\,\vec{t})\vec{d}, \tag{4.1}$$

where $\vec{s}, \vec{t}, \vec{c}, \vec{d}$ are sequences of elements in X with lengths p, q, n, m, respectively.

Proof The right hand side of (4.1) is obviously either Y or N.
It is Y iff one of the following holds:

$$\begin{array}{ll}
\boldsymbol{\lambda}\vec{c}.\underline{SCT}\,\vec{s}\,\vec{c}\,\vec{t} = \underline{V} & \text{and } \vec{d} = *_1 \dots d_i{}^{@i} \cdots *_m, \\
 & \quad \text{for some } 1 \le i \le m, \\
\boldsymbol{\lambda}\vec{c}.\underline{SCT}\,\vec{s}\,\vec{c}\,\vec{t} = \underline{V} & \text{and } \vec{d} = *_1 \dots (LR)^{@j} \cdots *_m, \\
 & \quad \text{for some } 1 \le j < m, \\
\boldsymbol{\lambda}\vec{c}.\underline{SCT}\,\vec{s}\,\vec{c}\,\vec{t} = \,]{*_1} \dots *_m[& \text{and } \vec{d} = *_1 \dots *_m \\
\boldsymbol{\lambda}\vec{c}.\underline{SCT}\,\vec{s}\,\vec{c}\,\vec{t} = \,]R *_2 \cdots *_m [& \text{and } \vec{d} = R *_2 \cdots *_m \\
\boldsymbol{\lambda}\vec{c}.\underline{SCT}\,\vec{s}\,\vec{c}\,\vec{t} = \,]{*_1} \dots *_{m-1} L[& \text{and } \vec{d} = *_1 \dots *_{m-1} L
\end{array}$$

iff one of the following holds:

$$\begin{array}{lll}
\vec{s} = *_1 \dots *_p, & \vec{d} = *_1 \dots d_i{}^{@i} \cdots *_m, & \text{and } \vec{t} = *_1 \dots *_q \\
\vec{s} = *_1 \dots *_p, & \vec{d} = *_1 \dots (LR)^{@j} \cdots *_m, & \text{and } \vec{t} = *_1 \dots *_q \\
\vec{s} = *_1 \dots s_i{}^{@i} \cdots *_p, & \vec{d} = *_1 \dots *_m, & \text{and } \vec{t} = *_1 \dots *_q \\
\vec{s} = *_1 \dots (LR)^{@j} \cdots *_p, & \vec{d} = *_1 \dots *_m, & \text{and } \vec{t} = *_1 \dots *_q \\
\vec{s} = *_1 \dots *_p, & \vec{d} = *_1 \dots *_m, & \text{and } \vec{t} = *_1 \dots t_i{}^{@i} \cdots *_q \\
\vec{s} = *_1 \dots *_p, & \vec{d} = *_1 \dots *_m, & \text{and } \vec{t} = *_1 \dots (LR)^{@j} \cdots *_q \\
\vec{s} = *_1 \dots *_{p-1} L, & \vec{d} = R *_2 \cdots *_m & \text{and } \vec{t} = *_1 \dots *_q \\
\vec{s} = *_1 \dots *_p, & \vec{d} = *_1 \dots *_{m-1} L & \text{and } \vec{t} = R *_2 \cdots *_q
\end{array}$$

iff one of the following holds:

$$\begin{array}{ll}
\vec{s}\,\vec{d}\,\vec{t} = *_1 \dots u_i{}^{@i} \cdots *_{p+m+q} & \text{and } u_i \text{ is the } i\text{th letter of } SDT \\
\vec{s}\,\vec{d}\,\vec{t} = *_1 \dots (LR)^{@j} \cdots *_{p+m+q} &
\end{array}$$

iff $\underline{SDT}\,\vec{s}\,\vec{d}\,\vec{t} = Y$. ■

4.1.17 Proposition *Let $\mathcal{R} = \{R_1, \dots, R_k\}$ be a set of rules. Then for $W, V \in \Sigma^*$*

$$W \vdash_{\mathcal{R}} V \ \Rightarrow\ \underline{V} \textit{ is } \lambda\textit{-definable from } \underline{W}, \underline{R_1}, \dots, \underline{R_k}.$$

Proof By induction on the length of the derivation of V, using the previous lemma. ■

We now want to prove the converse of this Proposition 4.1.17. That is,

$$\underline{V} \text{ is } \lambda\text{-definable from } \underline{W}, \underline{\overrightarrow{R}} \ \Rightarrow\ W \vdash_{\vec{\mathcal{R}}} V.$$

Specifying types as $\underline{W}\in\mathcal{M}(1_n), \underline{V}\in\mathcal{M}(1_m), \underline{R_i}\in\mathcal{M}(B_i)$, this is equivalent[1] with

$$\forall F\in\Lambda^{\emptyset}(1_n\to\vec{B}\to 1_m).[[\![F]\!]\underline{W}\,\underline{R}_1\ldots\underline{R}_k=\underline{V}\ \Rightarrow\ W\vdash_{\vec{\mathcal{R}}} V].$$

In the light of Remark 4.1.6 this is equivalent to

$$\forall M\in\Lambda^{\{x_W:1_n,\vec{r}:\vec{B}\}}(1_m).[[\![M]\!]_{[x_W:=\underline{W}][r_1:=\underline{R_1}]\cdots[r_k:=\underline{R_k}]}=\underline{V}\ \Rightarrow\ W\vdash_{\vec{\mathcal{R}}} V].$$

In fact something stronger will be proved: for all $M\in\Lambda^{\{x_W:1_n,\vec{r}:\vec{B}\}}(1_m)$

$$[\![M]\!]_{[x_W:=\underline{W}][r_1:=\underline{R_1}]\cdots[r_k:=\underline{R_k}]}\text{ is a weak encoding of }V\ \Rightarrow\ W\vdash_{\mathcal{R}} V.$$

The proof occupies 4.1.18–4.1.24.

4.1.18 Convention For the rest of this subsection we consider a fixed word $W\in\Sigma^*$ of length n and set of rewrite rules $\mathcal{R}=\{R_1,\ldots,R_k\}$, with $R_i=C_i\hookrightarrow D_i$ (with $\mathtt{lth}(C_i)=n_i$ and $\mathtt{lth}(D_i)=m_i$) in order to study the set $\{V\in\Sigma^*\mid W\vdash_{\mathcal{R}} V\}$. Moreover we let $x_W, r_1,\ldots,r_k$ be variables of the types of $\underline{W},\underline{R}_1,\ldots,\underline{R}_k$ (that is, of types 1_n, $1_{n_1}\to 1_{m_1},\ldots,1_{n_k}\to 1_{m_k}$) respectively. Finally ρ is a valuation such that $\rho(x_w)=\underline{W}$, $\rho(r_i)=\underline{R}_i$. A valuation φ is said to be compatible with ρ if it agrees with ρ on these variables.

4.1.19 Definition Let $\vec{z}=z_1,\ldots,z_p$ be a sequence of variables all of type 0. Write

$$0_{\vec{z}}=\{M\in\Lambda^{\emptyset}(0)\mid \mathrm{FV}(M)\subseteq\{x_w,r_1,\ldots,r_k,\vec{z}\}\}.$$

By the method of Section 1.3 the following grammar generates the long normal forms of type $0_{\vec{z}}$.

$$0_{\vec{z}}::=z_j\mid x_w 0^1_{\vec{z}}\cdots 0^n_{\vec{z}}\mid r_i(\lambda y_1\ldots y_{n_i}.0_{y_1,\ldots,y_{n_i},\vec{z}})0^1_{\vec{z}}\cdots 0^{m_i}_{\vec{z}},$$

with $1\leq j\leq p, 1\leq i\leq k$. For notational simplicity we will write

$$0_{\vec{z}}::=z_j\mid x_w 0_{\vec{z}}\cdots 0_{\vec{z}}\mid r_i(\lambda\vec{y}.0_{\vec{y},\vec{z}})0_{\vec{z}}\cdots 0_{\vec{z}},$$

but keeping in mind the numbers, or even

$$0_{\vec{z}}::=z_j\mid x_w\overrightarrow{0_{\vec{z}}}\mid r_i(\lambda\vec{y}.0_{\vec{y},\vec{z}})\overrightarrow{0_{\vec{z}}}.$$

4.1.20 Lemma *Let $M\in 0_{\vec{z}}$ not be a variable. Let φ be a valuation compatible with ρ. Then $[\![M]\!]_{\varphi}\in\{Y,N\}$.*

[1] In predicate logic one has that $[\exists x.P(x)]\ \Rightarrow\ Q$ is equivalent to $\forall x.[P(x)\ \Rightarrow\ Q]$, if $x\notin\mathrm{FV}(Q)$.

Proof By the grammar and the assumption that M is not a variable one has that the long normal form of M is $x_w\vec{Q}$ or $r_i(\lambda\vec{y}.Q)\vec{P}$. The result follows from the fact that $\underline{W}$ or $\underline{R_i}$ maximally applied yield an element of $\{Y, N\}$, as can be observed from the definition of encoding. ■

4.1.21 Lemma *Let $M \in 0_{\vec{z}}$ and φ a valuation compatible with ρ.*

(i) *Suppose $[\![M]\!]_\varphi = Y$ and $\varphi(x) \in \{A, B\}$, for some $x \in \mathrm{FV}(M)$. Then $\varphi(y) = *$ for all other $y \in \mathrm{FV}(M)$.*
(ii) *If $[\![\lambda\vec{y}.M]\!]_\varphi$ weakly encodes some $V \in \Sigma^*$, then $\varphi(x) = *$ for all $x \in \mathrm{FV}(\lambda\vec{y}.M)$.*

Proof (i) By simultaneous induction on M. That is, we show (i), assuming the induction hypothesis on M both for (i) and (ii).

Case $M \equiv z_i$. Let $x \in \mathrm{FV}(M)$. Then (i) holds trivially, as there are no free variables $y \in \mathrm{FV}(M)$ with $y \not\equiv x$.

Case $M \equiv x_w M_1 \dots M_n$. If one of the M_i is not a variable, then $[\![M_i]\!]_\varphi \in \{Y, N\}$, by Lemma 4.1.20. Then $[\![M]\!]_\varphi = \underline{W} *_1 \dots [\![M_i]\!]_\varphi{}^{@i} \cdots *_n$, by definition of $\underline{W}$. Therefore there is nothing to prove.

If all the M_i are variables, then (i) follows by definition of $\underline{W}$.

Case $M \equiv r_i(\lambda\vec{y}.Q)\vec{P}$. For every φ such that $[\![M]\!]_\varphi = Y$ one has the following.

(i) Each of the P_i is a variable, as otherwise $[\![P_i]\!]_\varphi \in \{Y, N\}$ so $[\![M]\!]_\varphi = N$.
(ii) By the definition of $\underline{R}$ the term Q cannot be a variable, so it is a preword-term.
(iii) Moreover $[\![\lambda\vec{y}.Q]\!]_\varphi \neq]-[$.

We can now verify (i) for this M.

If x is one of the variables P_j, by the definition of $\underline{R_i}$, the only way we can have $\varphi(x) \in \{A, B\}$ and $[\![M]\!]_\varphi = Y$, is if $[\![\lambda\vec{y}.Q]\!]_\varphi = \underline{C_i}$, (so that $\underline{R_i}([\![\lambda\vec{y}.Q]\!]_\varphi) = \underline{D_i}$), moreover $\varphi(x)$ is the jth letter of the word D_i, and $\varphi(P_k) = *$ for $k \neq j$. Now by the induction hypothesis (ii) for M one has $\varphi(z) = *$ for all variables $z \in \mathrm{FV}(\lambda\vec{y}.Q)$ also.

Suppose $x \in \mathrm{FV}(\lambda\vec{y}.Q)$, $\varphi(x) \in \{A, B\}$, and $[\![s]\!]_\varphi = Y$. By (3) there is a valuation φ' with $\varphi'(y) = \varphi(y)$ for all $y \in \mathrm{FV}(\lambda\vec{y}.Q)$ and $[\![Q]\!]_{\varphi'} = Y$. Applying the induction hypothesis of (i) to Q and φ', we see that $\varphi(y) = \varphi'(y) = *$ for each $y \neq x$ free in $\lambda\vec{y}.Q$, and that $[\![\lambda\vec{y}.Q]\!]_\varphi =]*_1 \dots *_n[$, so we must have $\varphi(P_k) = *$, for each P_k also.

(ii) We derive (ii) from (i). Let $x \in \mathrm{FV}(\lambda\vec{y}.M)$. Then $x \in \mathrm{FV}(M)$ and $x \neq$

y_1. Suppose $[\![\lambda\vec{y}.M]\!]_\varphi$ weakly encodes a word $V = v_1 \dots v_n$. Then

$$\begin{array}{rcll} Y & = & [\![\lambda\vec{y}.M]\!]_\varphi v_1 *_2 \cdots *_n & \\ & = & [\![M]\!]_{\varphi'}, & \text{with } \varphi' = \varphi[y_1 := v_1, y_2 := *, \dots, y_n := *]. \end{array}$$

Note that $\varphi'(y_1) = v_1 \in \{A, B\}$. Then $\varphi(x) = \varphi'(x) = *$ for any $x \in \mathrm{FV}(M)$, by (i) applied to M and φ'. As $x \not\equiv y_1$, also $\varphi'(x) = *$. ∎

4.1.22 Corollary *Let $M \in 0_{\vec{z}}$ with $\lambda\vec{z}.M \in \Lambda^\emptyset$. Suppose that $[\![\lambda\vec{z}.M]\!]$ weakly encodes some $V \in \Sigma^*$. Then each z_i occurs at most once in M.*

Proof Suppose that a variable z_i occurs more than once in M. Then we can find a term[2] $L \equiv L[x, y] \in 0_{x,y,\vec{z}}$ such that $M \equiv L[z_i, z_i]$. Now by Lemma 4.1.21(i) one has $[\![L]\!]_\varphi = N$ whenever $\varphi(x) = \varphi(y) \in \{A, B\}$. Then $[\![M]\!]_\varphi = N$ whenever $\varphi(z_i) \in \{A, B\}$. Since $[\![\lambda\vec{z}.M]\!]$ weakly encodes some $V \in \Sigma^*$, there is a valuation φ with $\varphi(z_i) \in \{A, B\}$ and $[\![M]\!]_\varphi = Y$, a contradiction. ∎

4.1.23 Proposition *Let $M \in 0_{\vec{z}}$ with $\mathrm{FV}(M) = \{\vec{z}\}$. Suppose that $[\![\lambda\vec{z}.M]\!]_\rho$ weakly encodes some $V = v_1 \dots v_q \in \Sigma^*$. Then $[\![\lambda\vec{z}.M]\!]_\rho = \underline{V}$ and $W \vdash_{\mathcal{R}} V$.*

Proof By induction on $M \in 0_{\vec{z}}$.
Case 1. $M \equiv z_i$. Then $\lambda\vec{z}.M$ is a projection, which by Remark 4.1.14 doesn't weakly encode a word. So there is nothing to prove.
Case 2. $M \equiv x_W P_1 \dots P_n$.
Claim 2.1. For every i one has $P_i = z_{p_i} \in \{\vec{z}\}$. Indeed, suppose P_i is not a variable. As $P_i \in 0_{\vec{z}}$ one has $[\![P_i]\!]_\varphi \in \{Y, N\}$, for any φ, by Lemma 4.1.20. Since $\lambda\vec{z}.M$ weakly encodes V one has by the definition of $\underline{V}$

$$\begin{array}{rcll} Y & = & \underline{V} *_1 \dots v_i{}^{@i} \cdots *_q & \\ & = & [\![\lambda\vec{z}.M]\!]_\rho *_1 \dots v_i{}^{@i} \cdots *_q & \\ & = & [\![M]\!]_{\rho[z_1 := *_1] \cdots} & \\ & = & \underline{W}\, [\![P_1]\!]_{\rho'} \cdots [\![P_n]\!]_{\rho'}, & \text{where } \rho' = \rho[z_1 := *_1] \cdots, \\ & = & N, & \text{as } [\![P_i]\!]_{\rho'} \in \{Y, N\}, \end{array}$$

a contradiction. Therefore all $\vec{P}$ are variables. It follows that

$$\lambda\vec{z}.M = \lambda z_1 \dots z_p . x_w z_{p_1} \cdots z_{p_n}.$$

Claim 2.2. All the z_i occur exactly once among the $z_{p_1}, \dots, z_{p_n}$. Indeed, if

[2] Writing $\Lambda \equiv L[x, y]$ indicates that $x, y \in \mathrm{FV}(L)$. Then $L[z_i, z_i] \triangleq L[x := z_i][y := z_i]$.

one of the $\vec{z}$, say z_i, does not occur, then

$$\begin{aligned} Y &= \underline{V} *_1 \dots v_i{}^{@i} \dots *_p \\ &= [\![\lambda z_1 \dots z_p.M]\!]_\rho *_1 \dots v_i{}^{@i} \dots *_p \\ &= [\![x_w z_{p_1} \cdots z_{p_n}]\!]_{\rho[z_1:=*_1]\dots}, && \text{by Claim 1,} \\ &= \underline{W} *_1 \dots *_n, && \text{because } z_i \notin \mathrm{FV}(M), \\ &= N, \end{aligned}$$

a contradiction. By Corollary 4.1.22 the z_i occur only once. It follows that $p = n$.

Claim 2.3. The $\vec{z}$ must occur in order: $M \equiv x_w z_1 \dots z_n$. Indeed, if for example $\lambda\vec{z}.M = \lambda z_1 \dots z_3.x_w z_2 z_3 z_1$, then

$$\begin{aligned} Y &= \underline{V} L R * \\ &= \underline{W} R * L \\ &= N, \end{aligned}$$

a contradiction.

Therefore

$$[\![\lambda\vec{z}.M]\!]_\rho = [\![\lambda\vec{z}.x_w\vec{z}]\!]_\rho = [\![x_w]\!]_\rho = \underline{W},$$

so it encodes W. Of course $W \vdash_{\mathcal{R}} W$.

Case 3. $M \equiv r_i(\lambda\vec{c}.Q)\vec{P}$. Assume $\lambda\vec{z}.M$ weakly denotes the word V. Then again, now by inspecting the definition of the encoding of rules, all P_i must be distinct variables.

Claim 3.1. One has $\vec{z} = s_1 \dots s_p, d_1 \dots d_{m_i}, t_1 \dots t_q$ and, writing $\vec{c} = c_1 \dots c_{n_i}$,

$$\lambda\vec{z}.M = \lambda\vec{s}\,\vec{d}\,\vec{t}.r_i(\lambda\vec{c}.Q[\vec{s},\vec{c},\vec{t}])\vec{d}. \tag{4.2}$$

Indeed, by the fact that the P_i are variables one has

$$\lambda\vec{z}.M = \lambda\vec{z}.r_i(\lambda\vec{c}.Q)z_{p_1} \cdots z_{p_{m_i}}.$$

We need to show that for $1 \leq j < m_i$ one has

$$p_j + 1 \leq n \;\&\; z_{p_{j+1}} = z_{p_j+1}.$$

Define environments ρ_*, ρ_1, ρ_2 as follows.

$$\begin{aligned} \rho_* &= \boldsymbol{\lambda} z.\, * \\ \rho_1 &= \rho_*[z_{p_j} := v_{p_j}] \\ \rho_2 &= \rho_*[z_{p_j} := L][z_{p_{j+1}} := R] \end{aligned}$$

By Corollary 4.1.22 one has

$$[\![\lambda\vec{v}.Q]\!]_{\rho_*} = [\![\lambda\vec{v}.Q]\!]_{\rho_1} = [\![\lambda\vec{v}.Q]\!]_{\rho_2}. \tag{4.3}$$

As before one has

$$\begin{aligned} Y &= \underline{V} * \cdots (v_{p_j})^{@p_j} \cdots n \\ &= \underline{R_i}[\![\lambda\vec{v}.Q]\!]_{\rho_1}[\![z_{p_1}]\!]_{\rho_1} \cdots [\![z_{p_{m_i}}]\!]_{\rho_1} \\ &= \underline{R_i}[\![\lambda\vec{v}.Q]\!]_{\rho_1} *_1 \cdots (v_{p_j})^{@j} \cdots *_{m_i} \end{aligned}$$

By the encoding of rules this can only happen when

$$[\![\lambda\vec{v}.Q]\!]_{\rho_*} = V_i \text{ and } \underline{R_i}[\![\lambda\vec{v}.Q]\!]_{\rho_*} = \underline{W_i}.$$

Hence by (4.3)

$$\begin{aligned} Y &= \underline{R_i}[\![\lambda\vec{v}.Q]\!]_{\rho_2} *_1 \cdots (LR)^{@j} \cdots *_{m_i} \\ &= \underline{R_i}[\![\lambda\vec{v}.Q]\!]_{\rho_2}[\![z_{p_1}]\!]_{\rho_2} \cdots [\![z_{p_{m_i}}]\!]_{\rho_2} \\ &= [\![\lambda\vec{z}.M]\!]_{\rho_2}[\![z_1]\!]_{\rho_2} \cdots [\![z_n]\!]_{\rho_2} \\ &= \underline{U} * \cdots (LR)^{@p_j} \cdots n. \end{aligned}$$

We conclude that

$$[\![z_{p_j+1}]\!]_{\rho_2} = R = [\![z_{p_{j+1}}]\!]_{\rho_2}.$$

Therefore $p_j + 1 = p_{j+1} \leq n$. This establishes Claim 3.1.

Now rewrite (4.2) as

$$M' \triangleq \lambda\vec{z}.M = \lambda\vec{s}\,\vec{d}\,\vec{t}.r_i(\lambda\vec{c}.Q'\,\vec{s}\,\vec{c}\,\vec{t})\,\vec{d}, \tag{4.4}$$

with $Q' \triangleq \lambda\vec{s}\,\vec{c}\,\vec{t}.Q \equiv \lambda s_1 \ldots s_p \lambda c_1 \ldots c_{n_i} \lambda t_1 \ldots t_q.Q$. Since $[\![M']\!]_\rho$ weakly encodes V one has $V = SDT$, for some decomposition SDT of V with respective lengths p, m_i, q. We will see later that $D = D_i$, the right hand side of the rule R_i.

By Corollary 4.1.22 the $\vec{d}$ are not among the free variables of Q. Therefore Q' is closed and the induction hypothesis holds. In order to apply this, we need to know that $[\![Q']\!]_\rho$ weakly encodes some word.

Claim 3.2. $[\![Q']\!]_\rho$ weakly encodes SC_iT, where C_i is the left hand side of R_i. Without loss of generality we assume for notational convenience that $p = 2, n_i = 1, q = 1, m_i = 2$. So we have

$$\begin{aligned} M' &= \lambda s_1 s_2 d_1 d_2 t.r_i(\lambda c.Q' s_1 s_2 c t)\, d_1 d_2, \\ Q' &= (\lambda s_1 s_2 c t.Q). \end{aligned}$$

We must show for the word $SC_iT = s_1s_2ct$ (where these letters are purposely overloaded to denote elements of $\{A, B\}$) that

$$\begin{array}{lllll} \text{(i)} & [\![Q']\!]_\rho\, s_1 * * * & = & Y, \\ \text{(ii)} & [\![Q']\!]_\rho * s_2 * * & = & Y, \\ \text{(iii)} & [\![Q']\!]_\rho * * \; c \; * & = & Y, \\ \text{(iv)} & [\![Q']\!]_\rho * * \; * \; t & = & Y, \\ \text{(v)} & [\![Q']\!]_\rho L\, R * * & = & Y, \\ \text{(vi)} & [\![Q']\!]_\rho * L\, R\, * & = & Y, \\ \text{(vii)} & [\![Q']\!]_\rho * * L\, R & = & Y. \end{array}$$

We establish this by 'probing' $[\![M]\!]_\rho$ on various sequences of elements of X.

As to (i), compute

$$\begin{array}{lll} Y & = & [\![M']\!]_\rho\, s_1 * * * * \\ & = & \underline{R_i}(\lambda\!\!\lambda c.[\![Q']\!]_\rho s_1 * c *) * *. \end{array}$$

This can only happen if the second clause in the definition of $\underline{R_i}$ applies, that is,

$$\underline{R_i}\,] * [=] * * [\text{ and } (\lambda\!\!\lambda c.[\![Q']\!]_\rho s_1 * c *) =] * [,$$

so that indeed (i) holds:

$$[\![Q']\!]_\rho s_1 * * * = (\lambda\!\!\lambda c.[\![Q']\!]_\rho s_1 * c *) * =] * [* = Y.$$

Similarly one can show (ii), (iv), and even (v).

As to (iii), the reasoning is different. In order to compute $[\![Q']\!]_\rho * * c *$, we probe $[\![M']\!]_\rho$ with $* * d_1 * *$, where d_1 is the first letter of $D = d_1 d_2$.

$$\begin{array}{lll} Y & = & [\![M']\!]_\rho * * d_1 * * \\ & = & \underline{R_i}(\lambda\!\!\lambda c.[\![Q']\!]_\rho * * c *)\, d_1 * . \end{array}$$

This can only happen if the first clause in the definition of $\underline{R_i}$ applies, that is,

$$\underline{R_i}(\underline{C_i}) = \underline{D_i} \text{ and } (\lambda\!\!\lambda c.[\![Q']\!]_\rho * * c *) = \underline{C_i},$$

so that indeed (iii) holds

$$[\![Q']\!]_\rho * * c * = (\lambda\!\!\lambda c.[\![Q']\!]_\rho * * c *)\, c = \underline{C_i}\, c = Y.$$

As to (vi), compute

$$\begin{array}{ll} Y = [\![M']\!]_\rho * LR * *, & \text{(the 'LR' crosses the border between } S \text{ and } D), \\ \quad = \underline{R_i}(\lambda\!\!\lambda c.[\![Q']\!]_\rho * Lc*)R*, & \text{('LR' is torn apart).} \end{array}$$

This can only happen if the third clause in the definition of $\underline{R_i}$ applies. Then

$$(\lambda\!\!\lambda c.[\![Q']\!]_\rho * Lc*) =]R[,$$

so that indeed (vi) holds

$$([\![Q']\!]_\rho * LR*) = (\lambda\!\!\lambda c.[\![Q']\!]_\rho * Lc*)R =]R[R = Y.$$

As to (vii), probe $[\![M']\!]_\rho$ with $* * *LR$, where now the border between D and T is crossed by 'LR'. Then $[\![Q']\!]_\rho * *LR = Y$ by the fourth clause of the definition of $\underline{R_i}$. This establishes Claim 3.2.

From Claim 3.2 it follows by the induction hypothesis for Q that $[\![Q']\!]_\rho = \underline{SC_iT}$ and $W \vdash_{\mathcal{R}} SC_iT$. Hence

$$\begin{array}{rcll} [\![M']\!]_\rho & = & [\![\lambda\vec{s}\vec{d}\vec{t}.r_i(\lambda\vec{c}.Q'\vec{s}\vec{c}\vec{t}\,)\vec{d}]\!]_\rho, & \text{by (4.4)}, \\ & = & \lambda\!\!\lambda\vec{s}\vec{d}\vec{t}.\underline{R_i}(\lambda\!\!\lambda\vec{c}.[\![Q']\!]_\rho\vec{s}\vec{c}\vec{t}\,)\vec{d} & \\ & = & \lambda\!\!\lambda\vec{s}\vec{d}\vec{t}.\underline{R_i}(\lambda\!\!\lambda\vec{c}.\underline{SC_iT}\vec{s}\vec{c}\vec{t}\,)\vec{d} & \\ & = & \underline{SD_iT}, & \text{by Proposition 4.1.16.} \end{array}$$

It follows that $[\![M']\!]_\rho$ encodes SD_iT and it weakly encodes $SDT(= V)$. Then by Lemma 4.1.13 one has $SD_iT = SDT = V$, so that $[\![\lambda\vec{z}.M]\!]_\rho$ encodes V. Moreover

$$W \vdash_{\mathcal{R}} SC_iT \vdash_{R_i} SD_iT = V,$$

showing that V is derivable. ∎

4.1.24 Corollary *Let $\mathcal{R} = \{R_1, \ldots, R_k\}$ be a set of rewrite rules. Then for $W, V \in \Sigma^*$*

$$W \vdash_{\mathcal{R}} V \Leftrightarrow \underline{V} \textit{ is definable from } \underline{W}, \vec{\underline{R}}.$$

Proof (⇒) By Proposition 4.1.17. (⇐) Let $\underline{V}$ be λ-definable from $\underline{W}$, $\vec{\underline{R}}$. Then

$$\begin{aligned} \underline{V} &= [\![F]\!]\,\underline{W}\,\vec{\underline{R}} \\ &= [\![\lambda x_w\vec{r}.Fx_w\vec{r}]\!]\,\underline{W}\,\vec{\underline{R}} \\ &= [\![Fx_w\vec{r}]\!]_\rho. \end{aligned}$$

The long normal form of $Fx_w\vec{r}$ is in $0_{\vec{z}}$ and Proposition 4.1.23 applies. ∎

Now we can harvest Loader's result, the undecidability of λ-definability.

4.1.25 Theorem (Loader) *For $d \in \mathcal{M}_X$ the notion of λ-definability is undecidable, i.e. there is no algorithm deciding whether or not a table describes a λ-definable element.*

Proof By Theorem 4.1.9 there exists a word $W \in \{A, B\}^*$, say of length n, and a set of rules $\mathcal{R} = \{R_1, \dots, R_k\}$, such that $\mathcal{W} = \{V \mid W \vdash_{\mathcal{R}} V\}$ is undecidable. By Corollary 4.1.24 one has

$$V \in \mathcal{W} \Leftrightarrow V \text{ is } \lambda\text{-definable from } \underline{W}, \vec{\underline{R}}.$$

This provides a reducibility between derivability and relative λ-definability. Hence by Proposition 4.1.4 the latter is undecidable. But then, by Lemma 4.1.5, also λ-definability (in the 'absolute sense') is undecidable. ■

4.1.26 Remark In the light of Remark 4.1.2 Thierry Joly extended Loader's result in two directions as follows. Define for $n \in \omega, A \in \mathbb{T}, d \in \mathcal{M}_n(A)$

$$D(n, A, d) \overset{\Delta}{\iff} d \in \mathcal{M}_n(A) \text{ is } \lambda\text{-definable.}$$

Since, for a fixed n_0 and A_0, the set $\mathcal{M}_{n_0}(A_0)$ is finite, it follows that $D(n_0, A_0, d)$ as predicate in d is decidable.

One has the following.

4.1.27 Proposition *Undecidability of λ-definability is monotonic in the following sense.*

$$\lambda\!\!\lambda Ad.D(n_0, A, d) \textit{ undecidable} \;\&\; n_0 \leq n_1 \;\Rightarrow\; \lambda\!\!\lambda Ad.D(n_1, A, d) \textit{ undecidable.}$$

Proof Use Exercise 3.24(i). ■

Loader's proof above shows in fact that $\lambda\!\!\lambda Ad.D(7, A, d)$ is undecidable. It was sharpened in Loader (2001a) showing that $\lambda\!\!\lambda Ad.D(3, A, d)$ is undecidable. The ultimate sharpening in this direction is proved in Joly (2005): $\lambda\!\!\lambda Ad.D(2, A, d)$ is undecidable.

Going in a different direction one has the following.

4.1.28 Theorem (Joly (2005)) *$\lambda\!\!\lambda nd.D(n, 3\to0\to0, d)$ is undecidable.*

Loosely speaking one can say that λ-definability at the monster type $\mathsf{M} = 3 \to 0 \to 0$ is undecidable. Moreover, Joly also has characterized those types A at which in this sense the λ-definability property is undecidable.

4.1.29 Definition A type A is called *finitely generated* if there are $\vec{M} = M_1, \dots, M_n$, all closed terms, not necessarily of type A, such that every element of $\Lambda^{\emptyset}(A)$ is an applicative product of the $\vec{M}$. For example $1\to0\to0$, the type of the Church numerals, is finitely generated by $\mathbf{c}_0, \mathsf{S}^+$.

4.1.30 Theorem (Joly (2002)) *Let $A \in \mathbb{T}$. Then $\lambda\!\!\lambda nd.D(n, A, d)$ is decidable iff the closed terms of type A can be finitely generated.*

For a sketch of the proof see Exercise 3.36.

4.1.31 Corollary *The monster type* $\mathsf{M} = 3 \to 0 \to 0$ *is not finitely generated.*

Proof By Theorems 4.1.30 and 4.1.28. ■

4.2 Undecidability of unification

The notions of (higher-order[3]) unification and matching problems were introduced by Huet (1975). In that paper it was proved that unification in general is undecidable. Moreover the question was asked whether matching is (un)decidable.

4.2.1 Definition

(i) Let $M, N \in \Lambda^{\emptyset}(A \to B)$. A *pure unification problem* is of the form

$$\exists X{:}A.MX = NX,$$

where one searches for an $X \in \Lambda^{\emptyset}(A)$ (and the equality is $=_{\beta\eta}$). We call A the *search-type* and B the *output-type* of the problem.

(ii) Let $M \in \Lambda^{\emptyset}(A \to B), N \in \Lambda^{\emptyset}(B)$. A *pure matching problem* is of the form

$$\exists X{:}A.MX = N,$$

where one searches for an $X \in \Lambda^{\emptyset}(A)$. Again A is the search-type and B the output type.

(iii) We often write for a unification or matching problem (when the types are known from the context or are not relevant) simply

$$MX = NX \quad \text{or} \quad MX = N$$

and speak about the unification (matching) problem with *unknown* X.

Of course matching problems are a particular case of unification problems: solving the matching problem $MX = N$ amounts to solving the unification problem

$$MX = (\lambda x.N)X.$$

4.2.2 Definition The *rank* (*order*) of a unification or matching problem is $\mathrm{rk}(A)$ ($\mathrm{ord}(A)$ respectively), where A is the search-type. Remember that $\mathrm{ord}(A) = \mathrm{rk}(A) + 1$.

[3] In contrast with the situation in 2.3.11 the present form of unification is 'higher-order', because it asks whether functions exist that satisfy certain equations.

The rank of the output-type is less relevant. Basically one may assume that it is $\top^2 = 1_2{\to}0{\to}0$. Indeed, by the Reducibility Theorem 3.4.8 one has $\Phi : B \leq_{\beta\eta} \top^2$, for some closed term Φ. Then

$$MX = NX : B \Leftrightarrow (\Phi \circ M)X = (\Phi \circ N)X : \top^2.$$

One has $\text{rk}(\top^2) = 2$. The unification and matching problems with an output type of rank <2 are decidable, see Exercise 4.6.

The main results of this section are that unification in general is undecidable from a low level onward, Goldfarb (1981), and matching up to order 4 is decidable, Padovani (2000).

In Stirling (2009) it is shown that matching in general is decidable. The paper is too complex to be included here.

As a spin-off of the study of matching problems it will be shown that the maximal theory is decidable.

4.2.3 Example The following are two examples of pure unification problems.

(i) $\exists X{:}(1{\to}0).\lambda f{:}1.f(Xf) = X$.
(ii) $\exists X{:}(1{\to}0{\to}0).\lambda fa.X(Xf)a = \lambda fa.Xf(Xfa)$.

This is not in the format of the previous definition, but we mean of course

$$(\lambda x{:}(1{\to}0)\lambda f{:}1.f(xf))X = (\lambda x{:}(1{\to}0)\lambda f{:}1.xf)X;$$
$$(\lambda x : (1{\to}0{\to}0)\lambda f{:}1\lambda a{:}0.x(xf)a)X = (\lambda x : (1{\to}0{\to}0)\lambda f{:}1\lambda a{:}0.xf(xfa))X.$$

The most understandable (provided we remember the types) form is

(i) $\lambda f.f(Xf) = X$;
(ii) $X(Xf)a = Xf(Xfa)$.

The first problem has no solution because there is no fixed-point combinator in $\boldsymbol{\lambda}^0_{\to}$. The second one does have solutions: $(\lambda fa.f(fa)$ and $\lambda fa.a)$, because $n^2 = 2n$ for $n \in \{0, 2\}$.

4.2.4 Example The following are two pure matching problems.

$$\begin{array}{rclr} X(Xf)a & = & f^{10}a & X{:}1{\to}0{\to}0; f{:}1, a{:}0; \\ f(X(Xf)a) & = & f^{10}a & X{:}1{\to}0{\to}0; f{:}1, a{:}0. \end{array}$$

The first problem is without a solution, because $\sqrt{10} \notin \omega$. The second has a solution $(X \equiv \lambda fa.f^3a)$, because $3^2 + 1 = 10$.

We will now generalize the unification and matching problems. First of all we will consider more unknowns. Then more equations. Finally, in the general versions of unification and matching problems, we do not require that the $\vec{M}, \vec{N}, \vec{X}$ are closed though they may contain a fixed finite number of constants (free variables that we will not bind). All these generalized problems will be reducible to the pure case, but (only in the transition from non-pure to pure problems) at the cost of possibly raising the rank (order) of the problem.

4.2.5 Definition

(i) Let M, N be closed terms of the same type. A pure *unification problem with unknowns* $\vec{X}$ looks like

$$M\vec{X} =_{\beta\eta} N\vec{X} \tag{4.5}$$

searches for closed terms $\vec{X}$ of the right type satisfying (4.5). The rank of a problem with unknowns $\vec{X}$ is

$$\max\{\text{rk}(A_i) \mid 1 \leq i \leq n\},$$

where the A_i are the types of the X_i. The order is defined similarly.

(ii) A *system of pure unification problems* with unknowns $\vec{X}$ starts with terms $M_1, \ldots, M_n$ and $N_1, \ldots, N_n$ such that M_i, N_i are of the same type for $1 \leq i \leq n$. We search for closed terms $\vec{X}_1, \ldots, \vec{X}_n$, all occuring among $\vec{X}$, such that

$$\begin{aligned} M_1\vec{X}_1 &=_{\beta\eta} N_1\vec{X}_1 \\ &\vdots \\ M_n\vec{X}_n &=_{\beta\eta} N_n\vec{X}_n. \end{aligned}$$

The rank (order) of such a system of problems is the maximum of the ranks (orders) of the types of the unknowns.

(iii) In the general (non-pure) case it will also be allowed to have the $M, N, \vec{X}$ range over Λ^{Γ} rather than $\Lambda^{\emptyset}$. We call this a *unification problem with constants from* Γ. The rank of a non-pure system of unknowns is defined as the maximum of the rank (orders) of the types of the unknowns.

(iv) The same generalizations are made to the matching problems.

4.2.6 Example A pure system of matching problems in the unknowns P, P_1, P_2 is the following. It states the existence of a pairing and is solvable depending on the types involved, see Barendregt (1974):

$$\begin{aligned} P_1(Pxy) &= x \\ P_2(Pxy) &= y. \end{aligned}$$

One could add a third equation (for surjectivity of the pairing),

$$P(P_1z)(P_2z) = z,$$

causing this system never to have solutions, see Barendregt (1974).

4.2.7 Example An example of a unification problem with constants from $\Gamma = \{a{:}1, b{:}1\}$ is as follows. We search for unknowns $W, X, Y, Z \in \Lambda^{\Gamma}(1)$ such that

$$\begin{aligned} X &= Y \circ W \circ Y \\ b \circ W &= W \circ b \\ W \circ W &= b \circ W \circ b \\ a \circ Y &= Y \circ a \\ X \circ X &= Z \circ b \circ b \circ a \circ a \circ b \circ b \circ Z, \end{aligned}$$

where $f \circ g = \lambda x.f(gx))$ for $f, g{:}1$, having as unique solution $W = b$, $X = a \circ b \circ b \circ a$, $Y = Z = a$. This example will be expanded in Exercise 4.5.

4.2.8 Proposition *All unification (matching) problems reduce to pure ones with just one unknown and one equation. In fact we have the following.*

(i) *A problem of rank k with several unknowns can be reduced to a problem with one unknown with rank* $\mathrm{rk}(A) = \max\{k, 2\}$.

(ii) *Systems of problems can be reduced to one problem, without altering the rank. The rank of the output type will be* $\max\{\mathrm{rk}(B_i), 2\}$, *where B_i are the output types of the respective problems in the system.*

(iii) *Non-pure problems with constants from Γ can be reduced to pure problems. In this process a problem of rank k becomes of rank*

$$\max\{\mathrm{rk}(\Gamma), k\}.$$

Proof We give the proof for unification.

(i) Following Notation 1.4.23 we have

$$\begin{aligned} &\exists \vec{X}.M\vec{X} = N\vec{X} && (4.6)\\ &\Leftrightarrow \exists X.(\lambda x.M(x \cdot 1) \cdots (x \cdot n))X = (\lambda x.N(x \cdot 1) \cdots (x \cdot n))X. && (4.7) \end{aligned}$$

Indeed, if the $\vec{X}$ work for (4.6), then $X \equiv \langle \vec{X} \rangle$ works for (4.7). Conversely, if X works for (4.7), then $\vec{X} \equiv X \cdot 1, \ldots, X \cdot n$ work for (4.6). By Proposition 1.4.22 we have $A = A_1 \times \cdots \times A_n$ is the type of X and $\mathrm{rk}(A) = \max\{\mathrm{rk}(A_1), \ldots, \mathrm{rk}(A_n), 2\}$.

(ii) Similarly for $\vec{X}_1, \ldots, \vec{X}_n$ (also are subsequences of $\vec{X}$) we have:

$$\begin{array}{llcl} \exists \vec{X} & M_1\vec{X}_1 & = & N_1\vec{X}_1 \\ & & \vdots & \\ & M_n\vec{X}_n & = & N_n\vec{X}_n \\ \Leftrightarrow \exists \vec{X} & \multicolumn{3}{l}{(\lambda\vec{x}.\langle M_1\vec{x}_1, \ldots, M_n\vec{x}_n\rangle)\vec{X} = (\lambda\vec{x}.\langle N_1\vec{x}_1, \ldots, N_n\vec{x}_n\rangle)\vec{X}.} \end{array}$$

(iii) Write a non-pure problem with $M, N \in \Lambda^{\Gamma}(A \to B)$, and $\mathrm{dom}(\Gamma) = \{\vec{y}\}$ as

$$\exists X[\vec{y}]{:}A.M[\vec{y}]X[\vec{y}] = N[\vec{y}]X[\vec{y}].$$

This is equivalent to the pure problem

$$\exists X{:}(\bigwedge\!\!\!\bigwedge \Gamma \to A).(\lambda x\vec{y}.M[\vec{y}](x\vec{y}))X = (\lambda x\vec{y}.N[\vec{y}](x\vec{y}))X. \blacksquare$$

Although the 'generalized' unification and matching problems can all be reduced to the pure case with one unknown and one equation, usually one ought not do this if one wants to get the right feel for the question.

Decidable case of unification

4.2.9 Proposition *Unification with unknowns of type* 1 *and constants of types* 0, 1 *is decidable.*

Proof The essential work to be done is the solvability of Markov's problem by Makanin. See Exercise 4.5 for the connection and a reference. ■

In Statman (1981) it is shown that the set of (bit strings encoding) decidable unification problems is itself polynomial time decidable

Undecidability of unification

The undecidability of unification was first proved by Huet. This was done before the undecidability of Hilbert's 10th problem (Is it decidable whether an arbitrary Diophantine equation over $\mathbb{Z}$ is solvable?) was established by Matiyasevič[4]. Huet reduced Post's correspondence problem to the unification problem. The theorem by Matiyasevič makes things more easy.

4.2.10 Theorem (Matiyasevič)

(i) *There are two polynomials p_1, p_2 over ω (of degree* 7 *with* 13 *variables*[5]*) such that*

$$D = \{\vec{n} \in \omega \mid \exists \vec{x} \in \omega.p_1(\vec{n}, \vec{x}) = p_2(\vec{n}, \vec{x})\}$$

[4] Also spelled in the literature as Matijasevič, Matijasevich or Matiyasevich!

[5] This can be pushed to polynomials of degree 4 and 58 variables or of degree 1.6×10^{45} and 9 variables, see Jones (1982).

is undecidable.

(ii) *There is a polynomial* $p(\vec{x},\vec{y})$ *over* $\mathbb{Z}$ *such that*

$$D = \{\vec{n}\in\omega \mid \exists\vec{x}\in\mathbb{Z}.p(\vec{n},\vec{x}) = 0\}$$

is undecidable. Therefore Hilbert's 10*th problem is undecidable.*

Proof (i) This was done by coding arbitrary recursively enumerable sets as Diophantine sets of the form D. See Matiyasevič (1972), Davis (1973) or Matiyasevič (1993).

(ii) Take $p = p_1 - p_2$ with the p_1, p_2 from (i). Using the theorem of Lagrange:

$$\forall n\in\omega\,\exists a,b,c,d\in\omega.n = a^2+b^2+c^2+d^2,$$

it follows that for $n\in\mathbb{Z}$ one has

$$n\in\omega \;\Leftrightarrow\; \exists a,b,c,d\in\omega.n = a^2+b^2+c^2+d^2.$$

Finally write $\exists x\in\omega.p(x,\dots) = 0$ as

$$\exists a,b,c,d\in\mathbb{Z}.p(a^2+b^2+c^2+d^2,\dots) = 0. \;\blacksquare$$

4.2.11 Corollary *The solvability of pure unification problems of order* 3 *(rank* 2*) is undecidable.*

Proof Take the two polynomials p_1, p_2 and D from part (i) of the theorem. Find closed terms M_{p_1}, M_{p_2} representing the polynomials, as in Corollary 1.4.7. Let $U_{\vec{n}} = \{M_{p_1}\ulcorner\vec{n}\urcorner\vec{x} = M_{p_2}\ulcorner\vec{n}\urcorner\vec{x}\}$. Using the fact that every $X\in\Lambda^{\emptyset}(\mathsf{Nat})$ is a numeral, see Proposition 2.1.16, it follows that this unification problem is solvable iff $\vec{n}\in D$. $\blacksquare$

The construction of Matiyasevič is involved. The encoding of Post's correspondence problem by Huet is a more natural way to show the undecidability of unification. It has the disadvantage that it needs to use unification at variable types. There is a way out. In Davis et al. (1961) it is proved that every recursively enumerable predicate is of the form $\exists\vec{x}\forall y_1{<}t_1\cdots\forall y_n{<}t_n.p_1 = p_2$. Using this result and higher types (Nat_A, for some non-atomic A) one can get rid of the bounded quantifiers. The analogue of Proposition 2.1.16 ($X{:}\mathsf{Nat} \Rightarrow X$ a numeral) does not hold but one can filter out the 'numerals' by a unification (with $f{:}A{\to}A$):

$$f\circ(Xf) = (Xf)\circ f.$$

This yields, without Matiyasevič's theorem, the undecidability of unification with the unknown of a fixed type.

4.2.12 Theorem *Unification of order 2 (rank 1) with constants is undecidable.*

Proof See Exercise 4.4. ∎

This implies that pure unification of order 3 is undecidable, something we already knew from Corollary 4.2.11. The interest in this result comes from the fact that unification over order 2 variables plays a role in automated deduction, and the undecidability of this problem, since it is a subcase of a more general situation, is not implied by Corollary 4.2.11.

Another proof of the undecidability unification of order 2 with constants, not using Matiyasevič's theorem, is in Schubert (1998).

4.3 Decidability of matching of rank 3

The main result will be that matching of rank 3 (which is the same as order 4) is decidable and is due to Padovani (2000). On the other hand Loader (2003) has proved that general matching modulo $=_\beta$ is undecidable. The decidability of general matching modulo $=_{\beta\eta}$, which is the intended case, was established in Stirling (2009), but will not be included here.

The structure of this section is as follows. First the notion of interpolation problem is introduced. Then by using tree automata it is shown that these problems restricted to rank 3 are decidable. Then at rank 3 the problem of matching is reduced to interpolation and hence solvable.

Let $M\vec{X} = N$, with $\vec{X} = X_1, \dots, X_k$ and $X_i : A_i$, be a matching problem with $M, N \in \Lambda^{\emptyset}[\mathcal{C}]$. The problem is of rank 3 if one has $\max_i\{\text{rk}(A_i)\} \leq 3$.

Following an idea of Statman (1982), the decidability of the matching problem can be reduced to the existence for every term N of a logical relation $\|_N \subseteq (\Lambda_\to)^2$ such that:

- $\|_N$ is an equivalence relation;
- $X_1 \|_N Y_1, \dots, X_k \|_N Y_k \Rightarrow M\vec{X} = M\vec{Y}$;
- for all types A the quotient $\mathcal{T}_A / \|_N$ is finite;
- there is an algorithm that enumerates $\mathcal{T}_A / \|_N$, i.e. that takes as argument a type A and returns a finite sequence of terms representing all the classes.

Indeed, if such a relation exists, then a simple generate and test algorithm solves the higher-order matching problem.

Similarly the decidability of the matching problem of rank n can be reduced to the existence of a relation such that $\mathcal{T}_A / \|_N$ can be enumerated up to rank n.

By the finite completeness theorem, Theorem 3.4.33, there is for a given

N a natural number n such that $\mathcal{M}_n \models M = N \Leftrightarrow M =_{\beta\eta} N$. We could try to define $X \parallel_N Y \Leftrightarrow \mathcal{M}_n \models X = Y$. Then $\parallel_N$ meets the first three requirements, but Loader's theorem 4.1.25 shows that it does not meet the fourth.

Padovani has proposed another relation – the *relative observational equivalence* – that is enumerable up to rank 3. As in the construction of the finite completeness theorem, the relative observational equivalence relation identifies terms of type 0 that are $\beta\eta$-equivalent and also all terms of type 0 that are not subterms of N. But this relation disregards the result of the application of a term to a non-definable element.

Padovani has proved that the enumerability of this relation up to rank n can be reduced to the decidability of a variant of the matching problem of rank n: the *dual interpolation problem* of rank n. Interpolation problems were introduced in Dowek (1994) as a first step towards decidability of third-order matching. The decidability of the dual interpolation problem of rank 3 has also been proved by Padovani. However, here we shall not present the original proof, but a simpler one proposed in Comon and Jurski (1998). Results will be formulated in terms of the rank.

Interpolation problems

4.3.1 Definition

(i) An *interpolation equation* is a particular matching problem

$$X\vec{M} = N,$$

where $\vec{M} = M_1, \dots, M_n$ and N are closed terms. That is, the unknown X occurs at the head. A solution of such an equation is a term P such that

$$P\vec{M} =_{\beta\eta} N.$$

(ii) An *interpolation problem* is a conjunction of such equations with the same unknown. A solution of such a problem is a term P that is a solution for all the equations simultaneously.

(iii) A *dual interpolation problem* is a conjunction of equations and negated equations. A solution of such a problem is a term that is a solution of all the equations but not a solution of any of the negated equations.

If a dual interpolation problem has a solution it has also a closed solution in long normal form. Hence, without loss of generality, we can restrict the search to such terms.

To prove the decidability of the rank 3 dual interpolation problem, we shall prove that the solutions of an interpolation equation can be recognized by a finite tree automaton. The results will then follow from the decidability of the non-emptiness of a set of terms recognized by a finite tree automaton and the closure of recognizable sets of terms by intersection and complement.

Lean solution

In fact, it is not quite correct that the solutions of a rank 3 interpolation equation can be recognized by a finite state automaton. Indeed, a solution of an interpolation equation may contain an arbitrary number of bound variables. For instance the equation

$$X\mathsf{K} = a,$$

where X is a variable of type $(0{\to}1{\to}0){\to}0$, so $\mathsf{K} = \lambda a^0 f^1.a$, has all the solutions

$$\lambda f.fa(\lambda z_1.fa(\lambda z_2.fa \cdots (\lambda z_n.fz_1(\mathsf{K}(fz_2(\mathsf{K}(fz_3 \cdots (fz_n\,(\mathsf{K}\,a))..)))))..)).$$

Moreover since each z_i has $z_1, \ldots, z_{i-1}$ in its scope it is not possible to rename these bound variables so that the variables of all these solutions are in a fixed finite set.

Thus the language of the solution cannot be *a priori* limited. In this example, it is clear however that there is another solution

$$\lambda f.(fa\Box)$$

where $\Box$ is a new constant of type $0{\to}0$. Moreover all the solutions above can be retrieved from this one by replacing the constant $\Box$ by an appropriate term (allowing captures in this replacement).

4.3.2 Definition For each simple type A, consider a constant $\Box_A : A$. Let M be a solution of an interpolation equation. A subterm occurrence of M of type A is *irrelevant* if replacing it by the constant $\Box_A$ yields a solution. A *lean* solution is a closed solution where all irrelevant subterm occurrences are the constants $\Box_A$.

Now we prove that lean solutions of an interpolation equation can be recognized by a finite tree automaton.

An example

Consider the problem

$$X\mathbf{c}_1 = ha,$$

where X is a variable of type $(1{\to}0{\to}0){\to}0$, the Church numeral $\mathbf{c}_1 \equiv \lambda fx.fx$ and a and h are constants of type 0 and 1_2. A lean solution of this equation substitutes X with the term $\lambda f.P$ where P is a lean solution of the equation $P[f := \mathbf{c}_1] = ha$.

Let $\mathcal{Q}_{ha}$ be the set of the lean solutions P of the equation $P[f := \mathbf{c}_1] = ha$. More generally, let $\mathcal{Q}_W$ be the set of lean solutions P of the equation $P[f := \mathbf{c}_1] = W$.

Notice that terms in $\mathcal{Q}_W$ can only contain the constants and the free variables that occur in W, plus the variable f and the constants $\Box_A$. We can determine membership of such a set (and in particular of $\mathcal{Q}_{ha}$) by induction over the structure of a term.

- *Analysis of membership of* $\mathcal{Q}_{ha}$
  A term is in $\mathcal{Q}_{ha}$ if it is either of the form
  $$(hP_1) \text{ and } P_1 \in \mathcal{Q}_a$$
  or
  $$(fP_1P_2) \text{ and } (P_1[f := \mathbf{c}_1]P_2[f := \mathbf{c}_1]) = ha.$$
  This means that there are terms P_1' and P_2' such that
  $P_1[f := \mathbf{c}_1] = P_1'$, $P_2[f := \mathbf{c}_1] = P_2'$, and $(P_1'P_2') = ha$;
  in other words there are terms P_1', P_2' such that
  $P_1 \in \mathcal{Q}_{P_1'}$, $P_2 \in \mathcal{Q}_{P_2'}$ and $(P_1'P_2') = ha$.
  As $(P_1'P_2') = ha$ there are three possibilities for P_1' and P_2', namely
  $P_1' = \mathsf{I}$ and $P_2' = ha$,
  $P_1' = \lambda z.hz$ and $P_2' = a$, or
  $P_1' = \lambda z.ha$ and $P_2' = \Box_0$.
  Hence $(fP_1P_2) \in \mathcal{Q}_{ha}$ if either
  $P_1 \in \mathcal{Q}_{\mathsf{I}}$ and $P_2 \in \mathcal{Q}_{ha}$,
  $P_1 \in \mathcal{Q}_{\lambda z.hz}$ and $P_2 \in \mathcal{Q}_a$, or
  $P_1 \in \mathcal{Q}_{\lambda z.ha}$ and $P_2 = \Box_0$.
  Hence, we have to analyze membership of $\mathcal{Q}_a$, $\mathcal{Q}_{\mathsf{I}}$, $\mathcal{Q}_{\lambda z.hz}$, and $\mathcal{Q}_{\lambda z.ha}$.
- *analysis of membership of* $\mathcal{Q}_a$
  A term is in $\mathcal{Q}_a$ if it has either the form
  a; or
  (fP_1P_2) with $P_1 \in \mathcal{Q}_{\mathsf{I}}$, $P_2 \in \mathcal{Q}_a$; or
  $P_1 \in \mathcal{Q}_{\lambda z.a}$, $P_2 = \Box_0$.
  Hence, we have to analyze membership of $\mathcal{Q}_{\lambda z.a}$,
- *analysis of membership of* $\mathcal{Q}_{\mathsf{I}}$ A term is in $\mathcal{Q}_{\mathsf{I}}$ if it has the form
  $$\lambda z.P_1 \text{ and } P_1 \in \mathcal{Q}_z.$$
  Hence, we have to analyze membership of $\mathcal{Q}_z$.

- *analysis of membership of* $\mathcal{Q}_{\lambda z.hz}$
  A term is in $\mathcal{Q}_{\lambda z.hz}$ if it has the form

$$\lambda z.P_1 \text{ and } P_1 \in \mathcal{Q}_{hz}.$$

  Hence, we have to analyze membership of $\mathcal{Q}_{hz}$.
- *analysis of membership of* $\mathcal{Q}_{\lambda z.ha}$
  A term is in $\mathcal{Q}_{\lambda z.ha}$ if it has the form

$$\lambda z.P_1 \text{ and } P_1 \in \mathcal{Q}_{ha}.$$

- *analysis of membership of* $\mathcal{Q}_{\lambda z.a}$
  A term is in $\mathcal{Q}_{\lambda z.a}$ if it has the form

$$\lambda z.P_1 \text{ and } P_1 \in \mathcal{Q}_{a}.$$

- *analysis of membership of* $\mathcal{Q}_{z}$
  A term is in $\mathcal{Q}_{z}$ if it has the form
  z or
  $(f P_1 P_2)$ and either
  $P_1 \in \mathcal{Q}_{\mathsf{I}}$ and $P_2 \in \mathcal{Q}_z$ or
  $P_1 \in \mathcal{Q}_{\lambda z'.z}$ and $P_2 = \Box_0$.
  Hence, we have to analyze membership of $\mathcal{Q}_{\lambda z'.z}$.
- *analysis of membership of* $\mathcal{Q}_{hz}$
  A term is in $\mathcal{Q}_{hz}$ if it has the form
  $(h P_1)$ and $P_1 \in \mathcal{Q}_z$ or
  $(f P_1 P_2)$ and either
  $P_1 \in \mathcal{Q}_{\mathsf{I}}$ and $P_2 \in \mathcal{Q}_{hz}$ or
  $P_1 \in \mathcal{Q}_{\lambda z.hz}$ and $P_2 \in \mathcal{Q}_z$ or
  $P_1 \in \mathcal{Q}_{\lambda z'.hz}$ and $P_2 = \Box_0$.
  Hence, we have to analyze membership of $\mathcal{Q}_{\lambda z'.hz}$.
- *analysis of membership of* $\mathcal{Q}_{\lambda z'.z}$
  A term is in $\mathcal{Q}_{\lambda z'.z}$ if it has the form

$$\lambda z'.P_1 \text{ and } P_1 \in \mathcal{Q}_{z}.$$

- *analysis of membership of* $\mathcal{Q}_{\lambda z'.hz}$
  A term is in $\mathcal{Q}_{\lambda z'.hz}$ if it has the form

$$\lambda z'.P_1 \text{ and } P_1 \in \mathcal{Q}_{hz}.$$

Thus we can build a tree automaton recognizing in q_W the terms of $\mathcal{Q}_W$. As long as terms contain only finitely many binders λz, we can consider these as unary function symbols.

$$
\begin{array}{rcl}
(hq_a) & \rightarrow & q_{ha} \\
(fq_{\mathsf{I}}q_{ha}) & \rightarrow & q_{ha} \\
(fq_{\lambda z.hz}q_a) & \rightarrow & q_{ha} \\
(fq_{\lambda z.ha}q_{\square_0}) & \rightarrow & q_{ha} \\
a & \rightarrow & q_a \\
(fq_{\mathsf{I}}q_a) & \rightarrow & q_a \\
(fq_{\lambda z.a}q_{\square_0}) & \rightarrow & q_a \\
\lambda z.q_z & \rightarrow & q_{\mathsf{I}} \\
\lambda z.q_{hz} & \rightarrow & q_{\lambda z.hz} \\
\lambda z.q_{ha} & \rightarrow & q_{\lambda z.ha} \\
\lambda z.q_a & \rightarrow & q_{\lambda z.a} \\
z & \rightarrow & q_z \\
(fq_{\mathsf{I}}q_z) & \rightarrow & q_z \\
(fq_{\lambda z'.z}q_{\square_0}) & \rightarrow & q_z \\
(hq_z) & \rightarrow & q_{hz} \\
(fq_{\mathsf{I}}q_{hz}) & \rightarrow & q_{hz} \\
(fq_{\lambda z.hz}q_z) & \rightarrow & q_{hz} \\
(fq_{\lambda z'.hz}q_{\square_0}) & \rightarrow & q_{hz} \\
\lambda z'.q_z & \rightarrow & q_{\lambda z'.z} \\
\lambda z'.q_{hz} & \rightarrow & q_{\lambda z'.hz}.
\end{array}
$$

Then we need a rule that recognizes $\square_0$ in the state $q_{\square_0}$

$$\square_0 \rightarrow q_{\square_0}$$

and finally a rule that recognizes in q_0 the lean solution of the equation $(X\mathbf{c}_1) = ha$

$$\lambda f.q_{ha} \rightarrow q_0.$$

Notice that as a spin off we have proved that besides f all lean solutions of this problem can be expressed with two bound variables z and z'.

The states of this automaton are labeled by the terms ha, a, I, $\lambda z.a$, $\lambda z.hz$, $\lambda z.ha$, z, hz, $\lambda z'.z$, $\lambda z'.hz$, and $\square_T$. All these terms have the form

$$N = \lambda y_1 \cdots y_p.P,$$

where P is a trunk (see Definition 4.3.3) of a subterm of ha and the free variables of P are in the set $\{z, z'\}$.

Tree automata for lean solutions

Now we show in general that solvability of dual interpolation problems is decidable by reducing the problem to the emptiness problem of tree automata, which is decidable. The proof given here is for $\boldsymbol{\lambda}^0_\to$, but can easily be generalized to the full $\boldsymbol{\lambda}^{\mathbb{A}}_\to$.

4.3.3 Definition

(i) Let M be in long normal form and $\mathcal{V}$ be a set of k variables of type 0 not occurring in M where k is the size of M. A *trunk* of M is a term P such that there exists a substitution σ mapping the variables of $\mathcal{V}$ to terms of type 0 such that $\sigma P = M$.

(ii) Consider an equation $X\vec{M} = N$, where $\vec{M} = M_1,\ldots,M_n$ and $X : B$ is a variable with $\mathrm{rk}(B) \leq 3$. Consider a finite number of constants $\Box_A : A$ for each type A subtype of B. Let k be the size of N. Consider a fixed set $\mathcal{V} = \{x^0_1,\ldots,x^0_k\}$. Let $\mathcal{N}$ be the finite set of terms of the form $\lambda y^0_1 \ldots y^0_p.P$, where the term P is a trunk of a subterm of N and the free variables of P are in $\mathcal{V}$. Also the p should be bounded as follows: if $M_i : A^i_1 \to \cdots \to A^i_{n_i} \to 0$, then $p <$ the maximal arity of all A^i_j. It is easy to check that in the special case that P is not of ground type (that is, starts with a λ which, intuitively, binds a variable in N introduced directly or hereditarily by a constant of N of higher-order type), one can take $p = 0$.

(iii) We define a tree automaton with the states q_W for W in $\mathcal{N}$ and $q_{\Box_A}$ for each constant $\Box_A$, and the transitions

$$
\begin{array}{rcll}
(f_i q_{W_1} \cdots q_{W_n}) & \to & q_W, & \text{if } (M_i\vec{W}) = W \text{ and replacing a } W_i \\
 & & & \text{different from } \Box_A \text{ by } \Box_A \text{ does not} \\
 & & & \text{yield a solution,} \\
(h q_{N_1} \cdots q_{N_n}) & \to & q_{(hN_1\cdots N_n)}, & \text{for } N_1,\ldots,N_n \text{ such that} \\
 & & & (h\,N_1 \cdots N_n) \in \mathcal{N}, \\
\Box_A & \to & q_{\Box_A} & \\
\lambda z.q_t & \to & q_{\lambda z.t} & \\
\lambda \vec{f}.q_N & \to & q_0. &
\end{array}
$$

4.3.4 Proposition *Let U and W be two elements of $\mathcal{N}$ and $X_1,\ldots,X_n$ be variables of rank at most 1. Let σ be a lean solution of the rank-1 matching problem*

$$UX_1 \ldots X_n = W.$$

Then for each i, either $\sigma X_i \in \mathcal{N}$ (modulo alpha-conversion) or $\sigma X_i = \Box_A$, with A a subtype of one of the types of the X_i.

Proof Let U' be the long normal form of $(U\sigma X_1 \cdots \sigma X_{i-1} X_i \sigma X_{i+1} \cdots \sigma X_n)$. If X_i has no occurrence in U' then as σ is lean, $\sigma X_i = \Box_A$.

Otherwise consider the higher occurrence at position l of a subterm of type 0 of U' that has the form $(X_i V_1 \cdots V_p)$. The terms $V_1, \ldots, V_p$ have type 0. Let W_0 be the subterm of W at the same position l. The term W_0 has type 0, so it is a trunk of a subterm of N.

Let V_i' be the long normal form of $V_i[X_i := \sigma X_i]$. We have $(\sigma X_i V_1' \cdots V_p') = W_0$. Consider p variables $y_1, \ldots, y_p$ of $\mathcal{V}$ that are not free in W_0. We have $\sigma X_i = \lambda y_1 \cdots y_p.P$ and

$$P[\vec{y} := \vec{V'}] = W_0.$$

Hence P is a trunk of a subterm of N and $\sigma X_i = \lambda y_1 \cdots y_p.P$ is an element of $\mathcal{N}$. ∎

4.3.5 Remark As a corollary of Proposition 4.3.4, we get an alternative proof of the decidability of rank-1 matching.

4.3.6 Proposition *Let*

$$X\vec{M} = N$$

be an equation, and $\mathcal{A}$ the associated automaton. Then a term is recognized by $\mathcal{A}$ (in q_0) iff it is a lean solution of this equation.

Proof We want to prove that a term V is recognized in q_0 if and only if it is a lean solution of the equation $V\vec{M} = N$. It is sufficient to prove that V is recognized in the state q_N iff it is a lean solution of the equation

$$V[\vec{f} := \vec{M}] = V[f_1 := M_1, \ldots, f_n := M_n] = N.$$

We prove, more generally, that for any term W of $\mathcal{N}$,

$$V \text{ is recognized in } q_W \Leftrightarrow V[\vec{f} := \vec{M}] = W.$$

($\Rightarrow$) By induction on the structure of V we show that if V is recognized in q_W, then V is a lean solution of the equation $V[\vec{f} := \vec{M}] = W$.

If $V = f_i V_1 \cdots V_p$

Case $V = f_i V_1 \cdots V_p$. Then the term V_i is recognized in a state q_{W_i}, where W_i is either a term of $\mathcal{N}$ or $\square_A$ and $M_i\vec{W} = W$. In the first case, by the induction hypothesis, V_i is a lean solution of the equation $V_i[\vec{f} := \vec{M}] = M_i$; and in the second, $V_i = \square_A$. Thus $M_i V_1[\vec{f} := \vec{M}] \cdots V_p[\vec{f} := \vec{M}] = N$, i.e. $V[\vec{f} := \vec{M}] = N$, and moreover V is lean.

Case $V = hV_1 \cdots V_p$. Then the V_i are recognized in states q_{W_i} with $W_i \in \mathcal{N}$. By induction hypothesis the V_i are lean solutions of $V_i[\vec{f} := \vec{M}] = M_i$. Hence $V[\vec{f} := \vec{M}] = N$ and moreover V is lean.

Case V is an abstraction. Similar.

($\Leftarrow$) Assume that V is a lean solution of the problem

$$V[\vec{f} := \vec{M}] = W.$$

By induction on the structure of V we show that V is recognized in q_W.

Case $V \equiv f_i V_1 \cdots V_p$. Then

$$M_i V_1[\vec{f} := \vec{M}] \cdots V_p[\vec{f} := \vec{M}] = N.$$

Let $V_i' = V_i[\vec{f} := \vec{M}]$. The V_i' are lean solutions of the rank-1 matching problem $M_i V_1' \cdots V_p' = N$. Now, by Proposition 4.3.4, one has $V_i' \in \mathcal{N}$ or $V_i' = \Box_A$. Either way V_i is a lean solution of the equation $V_i[\vec{f} := \vec{M}] = V_i'$ and by the induction hypothesis V_i is recognized in q_{W_i}. Thus V is recognized in q_W.

Case $V \equiv h V_1 \cdots V_p$. Then

$$h V_1[\vec{f} := \vec{M}] \cdots V_p[\vec{f} := \vec{M}] = W.$$

Let $W_i = V_i[\vec{f} := \vec{M}]$. We have $h\vec{W} = W$ and V_i is a lean solution of the equation $V_i[\vec{f} := \vec{M}] = W_i$. By the induction hypothesis V_i is recognized in q_{W_i}. Thus V is recognized in q_W.

Case V is an abstraction. Similar. ∎

4.3.7 Proposition *Rank* 3 *dual interpolation is decidable.*

Proof Consider a system of equations and negated equations and the automata associated with all these equations. Let L be the set of symbols occurring in these equations plus an extra constant of type 0 (to ensure all types have closed inhabitants). Obviously the system has a solution if and only if it has a solution in L. Each automaton recognizing the lean solutions can be transformed into one recognizing all the solutions in L (by adding a finite number of rules, so that the state $\Box_A$ recognizes all terms of type A in L). Then, using the fact that languages recognized by a tree automaton are closed by intersection and complement, we build an automaton recognizing all the solutions of the system in the language $\mathcal{L}$. The system has a solution if and only if the language recognized by this automaton is non-empty. Decidability follows from the decidability of the emptiness of a language recognized by a tree automaton. ∎

Decidability of rank-3 matching

A particular case

We shall start by proving the decidability of a subcase of rank-3 matching where problems are formulated in a language without any constant and the solutions also must not contain any constant.

Consider an equation $M = N$, with $M, N \in \Lambda^{\emptyset}(A)$, i.e. closed without constants. As $A \leq_{\beta\eta} \top^2 = 1_2 \to 0 \to 0$, by Theorem 3.4.8, there is a closed term $R \in \Lambda^{\emptyset}[b^{1_2}, c^0](A \to 0)$, whose constants have rank at most 1, such that for each term M of type A

$$M =_{\beta\eta} N \Leftrightarrow \forall \ell.(R_\ell M) =_{\beta\eta} (R_\ell N).$$

Indeed, let $\Phi : A \leq_{\beta\eta} \top^2 = 1_2 \to 0 \to 0$. Then take $R = \lambda z.\Phi z b^{1_2} c^0$. The long normal forms of $(R_\ell N) \in \Lambda^{\emptyset}(0)$ are closed terms whose constants have rank at most 1, thus they contain no bound variables. Let $\mathcal{U} = \mathcal{U}_N$ be the set of all subterms of type 0 of the long normal forms of $R_\ell N$. All these terms are closed. As in the relation defined by equality in the model of the finite completeness theorem, we define a congruence on closed terms of type 0 that identifies all terms that are not in $\mathcal{U}$. The number of equivalence classes of this congruence is $\text{card}(\mathcal{U}) + 1$.

4.3.8 Definition Write $M =_{\beta\eta N} M' \Leftrightarrow \forall U \in \mathcal{U}[M =_{\beta\eta} U \Leftrightarrow M' =_{\beta\eta} U]$.

Notice that if $M, M' \in \Lambda^{\emptyset}(0)$ one has the following:

$$\begin{aligned} M =_{\beta\eta N} M' &\Leftrightarrow M =_{\beta\eta} M' \text{ or } \forall U \in \mathcal{U}(M \neq_{\beta\eta} U \;\&\; M' \neq_{\beta\eta} U) \\ &\Leftrightarrow [M =_{\beta\eta} M' \\ &\qquad \text{or neither the long normal form of } M \text{ nor that of } M' \in \mathcal{U}]. \end{aligned}$$

Now we extend this to a logical relation on closed terms of arbitrary types. The following construction could be considered as an application of the Gandy hull defined in Example 3.3.28. However, we choose to do it explicitly so as to prepare for Definition 4.3.17.

4.3.9 Definition Let $\|_N$ be the logical relation lifted from $=_{\beta\eta N}$ on closed terms.

4.3.10 Lemma

(i) $\|_N$ *is head-expansive.*
(ii) *For each constant* F *of type of rank* ≤ 1 *one has* $F \|_N F$.
(iii) *For any* $X \in \Lambda(A)$ *one has* $X \|_N X$.
(iv) $\|_N$ *is an equivalence relation.*
(v) $P \|_N Q \Leftrightarrow \forall S_1, \ldots, S_k. P\vec{S} \|_N Q\vec{S}$.

Proof By the results of Section 3.3. ■

We want to prove, using the decidability of the dual interpolation problem, that the equivalence classes of this relation can be enumerated up to rank

3, i.e. that we can compute a set $\mathcal{E}_A$ of closed terms containing a term in each class.

More generally, we shall prove that if dual interpolation of rank n is decidable, then the sets $\mathcal{T}_A/\parallel_N$ can be enumerated up to rank n. We first prove the following proposition.

4.3.11 Proposition (Substitution lemma) *Let $M \in \Lambda(0)$ with $\mathrm{FV}(M) \subseteq \{x_1^{A_1}, \ldots, x_n^{A_n}\}$ be a term in long normal form. Let $V_1, \ldots, V_n, V_1', \ldots, V_n' \in \Lambda^{\emptyset}$, with $V_i, V_i' : A_i$, satisfy $V_1 \parallel_N V_1', \ldots, V_n \parallel_N V_n'$. Let $\sigma = [\vec{x} := \vec{V}]$ and $\sigma' = [\vec{x} := \vec{V'}]$*

Then

$$\sigma M =_{\beta\eta N} \sigma' M.$$

Proof By induction on the pair formed with the length of the longest reduction in σM and the size of M. The term M is in long normal form and has type 0, thus it has the form $(f W_1 \cdots W_k)$.

Case f is a constant. We can write $W_i = \overline{\lambda} S_i$, with S_i of type 0 and $\overline{\lambda} S_i$ is $\lambda \vec{y_i}.S_i$. We have $\sigma M = f \overline{\lambda} \sigma S_1 \cdots \overline{\lambda} \sigma S_k$ and $\sigma' M = f \overline{\lambda} \sigma' S_1 \cdots \overline{\lambda} \sigma' S_k$. By the induction hypothesis (as the S_i are subterms of M) we have $\sigma S_1 =_{\beta\eta N} \sigma' S_1, \ldots, \sigma S_k =_{\beta\eta N} \sigma' S_k$; thus either, for all i, $\sigma S_i =_{\beta\eta} \sigma' S_i$ and in this case $\sigma M =_{\beta\eta} \sigma' M$; or, for some i, neither the long normal forms of σS_i nor those of $\sigma' S_i$ are an element of $\mathcal{U}$. In this case neither the long normal form of σM nor that of $\sigma' M \in \mathcal{U}$ and $\sigma M =_{\beta\eta N} \sigma' M$.

Case $M \equiv x_i$. Then $M = x_i$, $\sigma M = V_i$, $\sigma' M = V_i'$, and V_i and V_i' have type 0. Thus $\sigma M =_{\beta\eta N} \sigma' M$.

Case $M \equiv x_i W_1 \ldots W_k$ and $k \neq 0$. Then $V_i \equiv \lambda z_1 \cdots \lambda z_k.S$ and $V_i' \equiv \lambda z_1 \cdots \lambda z_k.S'$. We have

$$\begin{array}{rcll} \sigma M & = & (V_i \sigma W_1 \cdots \sigma W_k) & \\ & =_{\beta\eta} & S[\vec{z} := \overrightarrow{\sigma W_1}] & \\ \sigma' M & = & (V_i' \sigma' W_1 \cdots \sigma' W_k) & \\ & =_{\beta\eta N} & S[\vec{z} := \overrightarrow{\sigma' W}], & \text{as } V_i \parallel_N V_i'. \end{array}$$

We claim $(\sigma W_i) \parallel_N (\sigma' W_i)$, for all i. Indeed, if $W_i \equiv \lambda y_1 \cdots \lambda y_p.O$, then for all closed terms $Q_1, \ldots, Q_p$, we have

$$\sigma W_i Q_1 \cdots Q_p = ([\vec{y} := \vec{Q}] \circ \sigma) O$$
$$\sigma' W_i Q_1 \cdots Q_p = ([\vec{y} := \vec{Q}] \circ \sigma') O.$$

Applying the induction hypothesis to O – that is, to a subterm of M – we get

$$(\sigma W_i) Q_1 \cdots Q_p =_{\beta\eta N} (\sigma' W_i) Q_1 \cdots Q_p$$

and thus indeed $(\sigma W_i) \parallel_N (\sigma' W_i)$. As

$$\sigma M \twoheadrightarrow S[\vec{z} := \overrightarrow{\sigma W}],$$

we can apply the induction hypothesis again and get

$$S[\vec{z} := \overrightarrow{\sigma W}] =_{\beta\eta N} S[\vec{z} := \overrightarrow{\sigma' W}].$$

Therefore $\sigma M =_{\beta\eta N} \sigma' M$. ∎

The next proposition is a direct corollary.

4.3.12 Proposition (Application lemma) *Let $V_1 \parallel_N V_1', \ldots, V_n \parallel_N V_n'$. Then for all terms $M : A_1 \to \cdots \to A_n \to 0$,*

$$MV_1 \cdots V_n =_{\beta\eta N} MV_1' \cdots V_n'.$$

Proof Apply Proposition 4.3.11 to the term $Mx_1 \cdots x_n$. ∎

Now we prove the following lemma justifying the use of the relations $=_{\beta\eta N}$ and $\parallel_N$.

4.3.13 Proposition (Discrimination lemma) *Let M be a term. Then*

$$M \parallel_N N \;\Rightarrow\; M =_{\beta\eta} N.$$

Proof As $M \parallel_N N$, by Proposition 4.3.12, we have $RM =_{\beta\eta N} RN$. Hence, as the long normal form of RN is in $\mathcal{U}$, we have $RM =_{\beta\eta} RN$. Therefore $M =_{\beta\eta} N$. ∎

Let us discuss now how we can decide the relation $\parallel_N$ and enumerate it. If M and M' are of type $A_1 \to \cdots \to A_n \to 0$, then, by definition, $M \parallel_N M'$ iff

$$\forall W_1 \in \mathcal{T}_{A_1} \cdots \forall W_n \in \mathcal{T}_{A_n}. M\vec{W} =_{\beta\eta N} M'\vec{W}.$$

The fact that $M\vec{W} =_{\beta\eta N} M'\vec{W}$ can be reformulated as

$$\forall U \in \mathcal{U}.[M\vec{W} =_{\beta\eta} U \;\Leftrightarrow\; M'\vec{W} =_{\beta\eta} U].$$

Therefore $M \parallel_N M'$ iff

$$\forall W_1 \in \mathcal{T}_{A_1} \cdots \forall W_n \in \mathcal{T}_{A_n} \forall U \in \mathcal{U}.[M\vec{W} =_{\beta\eta} U \;\Leftrightarrow\; M'\vec{W} =_{\beta\eta} U].$$

Therefore, in order to decide if $M \parallel_N M'$, we just need to list all the sequences $U, W_1, \ldots, W_n$ where U is an element of $\mathcal{U}$ and $W_1, \ldots, W_n$ are closed terms of type $A_1, \ldots, A_n$, and check that

$$\{\vec{W} \mid M\vec{W} =_{\beta\eta} U\} = \{\vec{W} \mid M'\vec{W} =_{\beta\eta} U\}.$$

Of course, the problem is that there is an infinite number of such sequences. But by Proposition 4.3.12 the fact that $M\vec{W} =_{\beta\eta N} M'\vec{W}$ is not affected if we replace the terms W_i by $\|_N$-equivalent terms. Hence, if we can enumerate the sets $\mathcal{T}_{A_1}/\|_N, \ldots, \mathcal{T}_{A_n}/\|_N$ by sets $\mathcal{E}_{A_1}, \ldots, \mathcal{E}_{A_n}$, then we can decide the relation $\|_N$ for terms of type $A_1 \rightarrow \cdots \rightarrow A_n \rightarrow 0$ by enumerating the sequences in $\mathcal{U} \times \mathcal{E}_{A_1} \times \cdots \times \mathcal{E}_{A_n}$, and checking that the set of sequences such that $M\vec{W} =_{\beta\eta} U$ is the same as the set of sequences such that $M'\vec{W} =_{\beta\eta} U$.

As the equivalence class of a term M for the relation $\|_N$ is completely determined by the set of sequences $U, W_1, \ldots, W_n$ such that $M\vec{W} =_{\beta\eta} U$, and as there are a finite number of subsets of the set $\mathcal{E} = \mathcal{U} \times \mathcal{E}_{A_1} \times \cdots \times \mathcal{E}_{A_n}$, we conclude that the set $\mathcal{T}_A/\|_N$ is finite.

To obtain an enumeration $\mathcal{E}_A$ of the set $\mathcal{T}_A/\|_N$ we need to be able to select the subsets $\mathcal{A}$ of $\mathcal{U} \times \mathcal{E}_{A_1} \times \cdots \times \mathcal{E}_{A_n}$, such that there is a term M such that $M\vec{W} =_{\beta\eta} U$ iff the sequence $U, \vec{W} \in \mathcal{A}$. This condition is exactly the decidability of the dual interpolation problem. This leads to the following proposition.

4.3.14 Proposition (Enumeration lemma) *If dual interpolation of rank n is decidable, then the sets $\mathcal{T}_A/\|_N$ can be enumerated up to rank n.*

Proof By induction on the rank of $A = A_1 \rightarrow \cdots \rightarrow A_n \rightarrow 0$. By the induction hypothesis, the sets $\mathcal{T}_{A_1}/\|_N, \ldots, \mathcal{T}_{A_n}/\|_N$ can be enumerated by sets $\mathcal{E}_{A_1}, \ldots, \mathcal{E}_{A_n}$.

Let x be a variable of type A. For each subset $\mathcal{A}$ of $\mathcal{E} = \mathcal{U} \times \mathcal{E}_{A_1} \times \cdots \times \mathcal{E}_{A_n}$ we define the dual interpolation problem containing the equation $x\vec{W} = U$ for $U, W_1, \ldots, W_p \in \mathcal{A}$ and the negated equation $x\vec{W} \neq U$ for $U, W_1, \ldots, W_p \notin \mathcal{A}$. Using the decidability of dual interpolation of rank n, we select those problems that have a solution and choose a closed solution for each of them. We obtain in this way a set $\mathcal{E}_A$.

We prove that this set is an enumeration of $\mathcal{T}_A/\|_N$, i.e. that for every term M of type A there is a term $M' \in \mathcal{E}_A$ such that $M' \|_N M$. Let $\mathcal{A}$ be the set of sequences $U, W_1, \ldots, W_p$ such that $(M\vec{W}) =_{\beta\eta} U$. The dual interpolation problem corresponding to $\mathcal{A}$ has a solution (for instance M). Thus one of its solutions $M' \in \mathcal{E}_A$. We have

$$\forall W_1 \in \mathcal{E}_{A_1} \cdots \forall W_n \in \mathcal{E}_{A_n} \forall U \in \mathcal{U}.[M\vec{W} =_{\beta\eta} U \Leftrightarrow M'\vec{W} =_{\beta\eta} U].$$

Thus

$$\forall W_1 \in \mathcal{E}_{A_1} \cdots \forall W_n \in \mathcal{E}_{A_n}.[M\vec{W} =_{\beta\eta N} M'\vec{W}],$$

hence by Proposition 4.3.12

$$\forall W_1 \in \mathcal{T}_{A_1} \cdots \forall W_n \in \mathcal{T}_{A_n}.[M\vec{W} =_{\beta\eta N} M'\vec{W}].$$

Therefore $M \parallel_N M'$. ■

Now we prove that if the sets $\mathcal{T}_A/\parallel_N$ can be enumerated up to rank n, then matching of rank n is decidable, by restricting the search of solutions to the sets $\mathcal{E}_A$.

4.3.15 Proposition (Matching lemma) *If the sets $\mathcal{T}_A/\parallel_N$ can be enumerated up to rank n, then matching problems of rank n whose right hand side is N can be decided.*

Proof Let $\vec{X} = X_1, \ldots, X_m$. Suppose the matching problem $L\vec{X} = N$ has a solution $V_1, \ldots, V_m$. Then we claim it also has a solution $\underline{V}_1, \ldots, \underline{V}_n$ with $\underline{V}_i \in \mathcal{E}_{A_i}$, for each i, where A_i is the type of X_i.

As $\vec{V}$ is a solution of the problem $M = N$, we have $M\vec{V} =_{\beta\eta} N$.

For all i, let $\underline{V}_i \in \mathcal{E}_{A_i}$ and $\underline{V}_i \parallel_N V_i$. Then

$$\begin{aligned} L\vec{\underline{V}} &=_{\beta\eta N} L\vec{V}, && \text{by Proposition 4.3.11,} \\ &=_{\beta\eta} N, && \text{as } \vec{V} \text{ is a solution,} \\ L\vec{\underline{V}} &=_{\beta\eta} N, && \text{by Proposition 4.3.13.} \end{aligned}$$

Thus for checking whether a problem has a solution it suffices to check whether it has a solution $\vec{\underline{V}}$, with each $\underline{V}_i$ in $\mathcal{E}_A$; such substitutions can be enumerated. ■

Note that the proposition can be generalized: the enumeration lets us solve every matching *inequality* of the right member N, and more generally, every dual matching problem.

4.3.16 Theorem *Rank* 3 *matching problems whose right hand side contain no constants can be decided.*

Proof Dual interpolation of rank 3 is decidable. Therefore, by Proposition 4.3.14, if N is a closed term containing no constants, then the sets $\mathcal{T}_A/\parallel_N$ can be enumerated up to rank 3. Therefore, by Proposition 4.3.15, we can decide if a problem of the form $M = N$ has a solution. ■

The general case

We consider now terms formed in a language containing an infinite number of constants of each type and we want to generalize the result. The difficulty is that we can no longer apply Statman's result to eliminate bound variables. Hence we shall define directly the set $\mathcal{U} = \mathcal{U}_N$ as the set of subterms of N of type 0. The novelty here is that the bound variables of U may now appear free in the terms of $\mathcal{U}$. It is important here to chose the names $x_1, \ldots, x_n$ of these variables, once for all.

We define the congruence $M =_{\beta\eta N} M'$ on terms of type 0 that identifies all terms that are not in $\mathcal{U}$.

4.3.17 Definition

(i) Let $M, M' \in \Lambda(0)$ (not necessarily closed). Define

$$M =_{\beta\eta N} M' \Leftrightarrow \forall U \in \mathcal{U}.[M =_{\beta\eta} U \Leftrightarrow M' =_{\beta\eta} U].$$

(ii) Define the logical relation $\|_N$ by lifting $=_{\beta\eta N}$ to all open terms at higher types.

4.3.18 Lemma

(i) *$\|_N$ is head-expansive.*
(ii) *For any variable x of arbitrary type A one has $x \|_N x$.*
(iii) *For each constant $F \in \Lambda(A)$ one has $F \|_N F$.*
(iv) *For any $X \in \Lambda(A)$ one has $X \|_N X$.*
(v) *$\|_N$ is an equivalence relation at all types.*
(vi) *$P \|_N Q \Leftrightarrow \forall S_1, \ldots, S_k.P\vec{S} \|_N Q\vec{S}$.*

Proof (i) By definition the relation is closed under arbitrary $\beta\eta$ expansion.
(ii) By induction on the generation of the type A.
(iii) Similarly.
(iv)–(vi) Easy. ∎

Now we can turn to the general case of the Enumeration Lemma, Proposition 4.3.14. Due to the presence of the free variables, the proof of this lemma introduces several novelties. Given a subset $\mathcal{A}$ of $\mathcal{E} = \mathcal{U} \times \mathcal{E}_{A_1} \times \cdots \times \mathcal{E}_{A_n}$ we cannot define the dual interpolation problem containing the equation $(x\vec{W}) = U$ for $U, W_1, \ldots, W_p \in \mathcal{A}$ and the negated equation $(x\vec{W}) \neq U$ for $U, W_1, \ldots, W_p \notin \mathcal{A}$, because the right hand side of these equations may contain free variables. Thus, we shall replace these variables by fresh constants $\vec{c} = c_1 \ldots c_n$. Define the substitution $\theta \triangleq [\vec{x} := \vec{c}]$. This substitution has an inverse $\theta^{-1} \triangleq [\vec{c} := \vec{x}]$. To each set of sequences, we associate the dual interpolation problem containing the equation $(x\vec{W}) = \theta U$ or its negation.

This introduces two difficulties: first the term θU is not a subterm of N; thus, in addition to the relation $\|_N$, we shall also need to consider the relation $\|_{\theta U}$, and one of its enumerations, for each term $U \in \mathcal{U}$. Then, the solutions of such interpolation problems could contain the constants $\vec{c}$, and we may have difficulties proving that they represent their $\|_N$-equivalence class. To solve this problem we need to duplicate the constants $\vec{c}$ with constants $\vec{d} = d_1, \ldots, d_n$. This idea goes back to Goldfarb (1981), see also Exercise 4.4.

Let us consider a fixed set of constants $\vec{c}$, $\vec{d}$ that do not occur in N. If M is a term containing constants $\vec{c}$, but not the constants $\vec{d}$, we write $\tilde{M}$ for the 'variant' of the term M, where each constant c_i is replaced by the constant d_i.

Let $A = A_1 \to \cdots \to A_n \to 0$ be a type. We assume that for any closed term U of type 0, the sets $\mathcal{T}_{A_i}/\parallel_U$ can be enumerated up to rank n by sets $\mathcal{E}^U_{A_i}$.

4.3.19 Definition Define the set of sequences $\mathcal{E}$ containing the sequence

$$\theta U, W_1, \ldots, W_n,$$

for each term $U \in \mathcal{U}$ and sequence $W_1, \ldots, W_n \in \mathcal{E}^{\theta U}_{A_1} \times \cdots \times \mathcal{E}^{\theta U}_{A_n}$. Notice that the terms in these sequences may contain the constants $\vec{c}$, but not the constants $\vec{d}$.

To each subset of $\mathcal{A}$ of $\mathcal{E}$ we associate a dual interpolation problem containing the equations $x\vec{W} = U$ and $x\tilde{W}_1 \cdots \tilde{W}_n = \tilde{U}$ for $U, W_1, \ldots, W_n \in \mathcal{A}$ and the inequalities $x\vec{W} \neq U$ and $x\tilde{W}_1 \cdots \tilde{W}_n \neq \tilde{U}$ for $U, W_1, \ldots, W_n \notin \mathcal{A}$.

The first lemma justifies the use of constants duplication.

4.3.20 Proposition *If an interpolation problem of Definition* 4.3.19 *has a solution M, then it also has a solution M' that does not contain the constants $\vec{c}, \vec{d}$.*

Proof Assume that the term M contains a constant, say c_1. Then by replacing this constant c_1 by a fresh constant e, we obtain a term M'. As the constant e is fresh, all the inequalities that M satisfies are still satisfied by M'. If M satisfies the equations $x\vec{W} = U$ and $x\tilde{W}_1 \cdots \tilde{W}_n = \tilde{U}$, then the constant e does not occur in the long normal form of $M'\vec{W}$. Otherwise the constant c_1 would occur in the long normal form of $M\tilde{W}_1 \cdots \tilde{W}_n$, i.e. in the long normal form of $\tilde{U}$ which is not the case. Thus M' also satisfies the equations $x\vec{W} = U$ and $x\tilde{W}_1 \cdots \tilde{W}_n = \tilde{U}$.

We can replace in this way all the constants $\vec{c}$, $\vec{d}$ by fresh constants, obtaining a solution where they do not occur. ∎

Now we prove that the interpolation problems of Definition 4.3.19 characterize the equivalence classes of the relation $\parallel_N$.

4.3.21 Proposition *The term M of type A not containing the constants $c_1, \ldots, c_n$, $d - 1, \ldots, d_n$ is the solution of a unique problem of Definition* 4.3.19.

Proof The term M is the solution of the interpolation problem associated to $\mathcal{A}$ and moreover $\mathcal{A}$ is the only subset $\mathcal{B}$ of E, such that M is a solution to the problem associated to $\mathcal{B}$. ■

4.3.22 Proposition *Let M and M' be two terms of type A not containing the constants $\vec{c},\vec{d}$. Then M and M' are solutions of the same unique problem of Definition* 4.3.19 *iff $M \parallel_N M'$.*

Proof By definition, if $M \parallel_N M'$, then for all $W_1,\ldots,W_n$ and for all $U \in \mathcal{U}$ we have

$$M\vec{W} =_{\beta\eta} U \Leftrightarrow M'\vec{W} =_{\beta\eta} U.$$

Thus for any $U, \vec{W} \in \mathcal{E}$, it follows that $\theta^{-1}U \in \mathcal{U}$ and

$$M(\theta^{-1}W_1)\cdots(\theta^{-1}W_n) =_{\beta\eta} \theta^{-1}U \Leftrightarrow M'(\theta^{-1}W_1)\cdots(\theta^{-1}W_n) =_{\beta\eta} \theta^{-1}U.$$

Then, as the constants $\vec{c},\vec{d}$ do not appear in M and M', we have

$$M\vec{W} =_{\beta\eta} U \Leftrightarrow M'\vec{W} =_{\beta\eta} U$$

and

$$M\tilde{W}_1\cdots\tilde{W}_n =_{\beta\eta} \tilde{U} \Leftrightarrow M'\tilde{W}_1\cdots\tilde{W}_n =_{\beta\eta} \tilde{U}.$$

Thus M and M' are the solutions of the same problem.

Conversely, assume that $M \not\parallel_N M'$. Then there exists terms $W_1,\ldots,W_n$ and a term $U \in \mathcal{U}$ such that

$$M\vec{W} =_{\beta\eta} U \text{ and } M'\vec{W} \neq_{\beta\eta} U.$$

Hence

$$M(\theta W_1)\cdots(\theta W_n) =_{\beta\eta} \theta U \text{ and } M'(\theta W_1)\cdots(\theta W_n) \neq_{\beta\eta} \theta U.$$

As the sets $\mathcal{E}_{A_i}^{\theta U}$ are an enumeration of the sets $\mathcal{T}_{A_i}/\parallel_{\theta U}$, there exist terms $\vec{S}$ such that the $S_i \parallel_{\theta U} \theta W_i$ and $\theta U, \vec{S} \in \mathcal{E}$. Using Proposition 4.3.12 we have

$$M\vec{S} =_{\beta\eta\theta U} M(\theta W_1)\cdots(\theta W_n) =_{\beta\eta} \theta U,$$

hence $M\vec{S} =_{\beta\eta\theta U} \theta U$, i.e. $M\vec{S} =_{\beta\eta} \theta U$. Similarly, we have

$$M'\vec{S} =_{\beta\eta\theta U} M'(\theta W_1)\cdots(\theta W_n) \neq_{\beta\eta} \theta U$$

hence $M'\vec{S} \neq_{\beta\eta\theta U} \theta U$, i.e. $M'\vec{S} \neq_{\beta\eta} \theta U$. Hence M and M' are not the solutions of the same problem. ■

Finally, we can prove the enumeration lemma.

4.3.23 Proposition (Enumeration lemma) *If dual interpolation of rank n is decidable, then, for any closed term N of type 0, the sets $\mathcal{T}_A/\parallel_N$ can be enumerated up to rank n.*

Proof By induction on the rank of A. Let $A = A_1 \to \cdots \to A_n \to 0$. By the induction hypothesis, for any $U \in \Lambda^{\emptyset}(0)$ the sets $\mathcal{T}_{A_i}/\parallel_U$ can be enumerated by sets $\mathcal{E}^U_{A_i}$.

We consider all the interpolation problems of Definition 4.3.19. Using the decidability of dual interpolation of rank n, we select among those problems the ones that have a solution. By Proposition 4.3.20, we can construct for each such problem a solution not containing the constants $\vec{c}, d_1, \ldots, d_n$ and by Proposition 4.3.21 and 4.3.22, these terms form an enumeration of $\mathcal{T}_A/\parallel_N$. ■

Now the general Matching Lemma (Proposition 4.3.15) follows exactly as in the particular case. Finally we prove the main result.

4.3.24 Theorem (Padovani) *Rank* 3 *matching problems can be decided.*

Proof Dual interpolation of rank 3 is decidable. Therefore, by Proposition 4.3.14, if N is a closed term, then the sets $\mathcal{T}_A/\parallel_N$ can be enumerated up to rank 3; hence, by Proposition 4.3.15, we can decide if a problem of the form $M = N$ has a solution. ■

4.4 Decidability of the maximal theory

In this section it will be proved that the maximal theory is decidable. The original proof of this result is due to Padovani (1996). This proof was later simplified by Loader (1997), based on Schmidt-Schauß (1999).

We introduce some notation for types and for terms.

4.4.1 Notation

(i) Let $A \in \mathbb{T}^0$ with $A \neq 0$. Then $A = B_1 \to \cdots \to B_n \to 0$. Remember we conveniently write $A = [B_1, \ldots, B_n]$. By allowing the notation $0 = [\,]$ all types can be written as a sequence of types. From now on in this section we will write $A = [B_1 \ldots B_n]$. Omitting the commas has as advantage that we can flexibly increase and decrease the length of a type by inserting certain strings of types.

(ii) Let $A = [B_1 \ldots B_n]$ be a type. Define

$$\begin{aligned} \#A &= n \\ A_i &= B_i, \qquad \text{for } 1 \leq i \leq \#A\,. \end{aligned}$$

In particular $\#0 = 0$. If $1 \leq i \leq \#A$ and moreover $1 \leq j \leq \#A_i$, then write

$$A_{ij} \triangleq (A_i)_j.$$

(iii) We sometimes find it convenient to write $A_\ell \triangleq A_{\#A}$. Note that this ℓ depends on its context, therefore we do this only if there is little danger of confusion. In particular $A_{i\ell} \triangleq A_{i\#A_i}$.

4.4.2 Notation

(i) Let $A = [A_1 \ldots A_n]$ be a type. Write

$$\mathsf{K}^A \triangleq \lambda z^0 x_1^{A_1} \cdots x_n^{A_n}.z \in \Lambda^\emptyset(0 \to A).$$

Then for $M \in \Lambda[\vec{c}](0)$ one has $\mathsf{K}^A M \in \Lambda[\vec{c}](A)$ and for $\vec{P} : \vec{A}$, i.e. $P_1 : A_1, \ldots, P_n : A_n$, one has $\mathsf{K}^A M\vec{P} = M$. If $A = 0$, then $\mathsf{K}^0 = \mathsf{I}$, so that $\mathsf{K}^0 M = M$.

(ii) Let $M, \vec{P} = P_1, \ldots, P_n$ be terms. Write $\vec{P}M \triangleq P_1 M, \ldots, P_n M$. If $\vec{P} = \epsilon$, the empty word, then $\vec{P}M \triangleq \epsilon$.

Remember that the maximal theory, see Definition 3.5.47, is

$$= \mathcal{T}_{\max} = \{M = N \mid M, N \in \Lambda^\emptyset(A) \ \&\ \mathcal{M}_{\min} \models M = N\},$$

where

$$\mathcal{M}_{\min} = \Lambda^\emptyset[c_1^0, c_2^0]/{\approx_{\vec{c}}^{\text{ext}}}$$

consists of all terms having the c_1, c_2 of type 0 as distinct constants, and $M \approx_{\vec{c}}^{\text{ext}} N$ on type $A = [A_1 \ldots A_n]$ is defined by

$$M \approx_{\vec{c}}^{\text{ext}} N \ \Leftrightarrow\ \forall P_1 \in \Lambda^\emptyset[\vec{c}](A_1) \cdots P_n \in \Lambda^\emptyset[\vec{c}](A_n).M\vec{P} =_{\beta\eta} N\vec{P}.$$

Theorem 3.5.35 states that $\approx_{\vec{c}}^{\text{ext}}$ is a congruence which we will denote by $\approx$.

4.4.3 Remark Let $\mathcal{M}_{\min}^{c_1,\ldots,c_n} = \Lambda^\emptyset[\vec{c}]/{\approx_{\vec{c}}^{\text{ext}}}$. Then Theorem 3.5.35 implies that also

$$\text{Th}(\mathcal{M}_{\min}^{c_1,\ldots,c_n}) = \mathcal{T}_{\max}.$$

For this reason we restrict ourselves to $n = 2$.

4.4.4 Definition Let $A \in \mathbb{T}^{\mathbb{A}}$. The *degree* of A, written $||A||$, is defined as follows:

$$\begin{array}{rcll} ||0|| & = & 2, & \\ ||A \to B|| & = & ||A||!||B||, & \text{i.e. } ||A|| \text{ factorial times } ||B||. \end{array}$$

4.4.5 Definition The following is convenient notation. Let Σ be an alphabet and $a_1 \dots a_n, b_1 \dots b_m \in \Sigma^*$. For $1 \leq i \leq n$ we write

$$a_1 \cdots (\vec{b})^{\overline{@i}} \cdots a_n \triangleq a_1 \cdots a_{i-1} b_1 \dots b_m a_{i+1} \cdots a_n.$$

Note that if $\vec{b} = b$ is a one-letter word, then $a_1 \dots b^{\overline{@i}} \dots a_n = a_1 \cdots b^{@i} \cdots a_n$, as in Definition 4.1.11, but $\mathtt{lth}(a_1 \dots (\vec{b})^{\overline{@i}} \dots a_n) = n + m - 1 > n$ for $\vec{b} = b_1 \dots b_m$ with $m > 1$. If $\vec{b} = \epsilon$, the empty word, then $\mathtt{lth}(a_1 \dots \vec{b}^{\overline{@i}} \dots a_n) = n - 1 < n$.

4.4.6 Proposition

(i) $||\,[A_1 \dots A_n]\,|| = 2||A_1||! \cdots ||A_n||!$.
(ii) $||A_i|| < ||\,[A_1 \dots A_n]\,||$.
(iii) $n < ||\,[A_1 \dots A_n]\,||$.
(iv) *Let* $A_1 \dots A_n, B_1 \dots B_m \in \mathbb{T}^0$. *Suppose* $m < ||A_i||$, *for some* $1 \leq i \leq n$, *and* $||B_j|| < ||A_i||$, *for all* $1 \leq j \leq m$. *Then*

$$||\,[A_1 \dots \vec{B}^{\overline{@i}} \dots A_n\,|| < ||\,[A_1 \dots A_n]\,||.$$

Proof Left to the reader. ∎

4.4.7 Definition

(i) Define types $\vec{C}_i^A = C_{i1}^A, \dots, C_{i\#A_i}^A \in \mathbb{T}^0$ as follows:

$$C_{ij}^A \triangleq [A_1 \dots (\vec{C}_j^{A_i})^{\overline{@i}} \dots A_\ell] = [A_1 \dots (C_{j1}^{A_i} \cdots C_{j(\#A_{i\ell})}^{A_i})^{\overline{@i}} \dots A_\ell].$$

For examples, see Example 4.4.8 and the table on page 233.

(ii) Define sequences of variables $\vec{z}_i^A : \vec{C}_i^A$, for $1 \leq i \leq \#A$, by

$$\vec{z}_i^A \triangleq z_{i1}^{C_{i1}^A} \cdots z_{i\#A_i}^{C_{i\#A_i}^A}.$$

(iii) Define by simultaneous course of value recursion, for $1 \leq i \leq \#A$ and $1 \leq j \leq \#A_i$, closed terms E_i^A, P_{ij}^A, with types

$$\begin{aligned} E_i^A &: \vec{C}_i^A \to A = [C_{i1}^A \cdots C_{i(\#A_i)}^A A_1 \cdots A_n] \\ P_{ij}^A &: A \to C_{ij}^A \end{aligned}$$

as follows (for $A = 0$ no i satisfies $1 \leq i \leq \#A$; then E_i^0 and P_{ij}^0 are not defined; $\vec{P}_i$ may be defined, but be the empty string:

$$\begin{aligned} E_i^A \;&\triangleq\; \lambda \vec{z}_i^A \lambda x_1^{A_1} \cdots x_{\#A}^{A_\ell}.x_i(\mathsf{K}^{A_{i1}}(z_{i1} x_1 \dots (\vec{P}_1^{A_i} x_i)^{\overline{@i}} \dots x_{\#A})) \\ &\qquad \cdots \\ &\qquad (\mathsf{K}^{A_{i\ell}}(z_{i\#A_i} x_1 \dots (\vec{P}_{\#A_i}^{A_i} x_i)^{\overline{@i}} \dots x_{\#A})); \\ P_{ij}^A \;&\triangleq\; \lambda f^A x_1 \dots (\vec{z}_j^{A_i})^{\overline{@i}} \dots x_{\#A}.f x_1 \dots (E_j^{A_i} \vec{z}_j^{A_i})^{\overline{@i}} \dots x_{\#A}. \end{aligned}$$

4.4.8 Example Let $A = [[B_1 B_2][C][D_1 D_2 D_3]]$. Then $\#A_1 = 2$, $\#A_2 = 1$, and $\#A_3 = 3$. Suppose $\#B_i = b_i, \#C = c, \#D_i = d_i$. Then we have, for $\vec{C}_1^A = C_{11}^A C_{12}^A$, $\vec{C}_2^A = C_{21}^A$, $\vec{C}_3^A = C_{31}^A C_{32}^A C_{33}^A$,

$$
\begin{array}{lclclclclcl}
C_{11}^A & = & ..(C_{11}^{A_1} \cdots C_{1b_1}^{A_1})^{\overline{@}1}.. & C_{21}^A & = & ..(C_{2c}^{A_2})^{\overline{@}2}.. & C_{31}^A & = & ..(C_{11}^{A_3} \cdots C_{1d_1}^{A_3})^{\overline{@}3}.. \\
C_{12} & = & ..(C_{21}^{A_1} \cdots C_{2b_1}^{A_1})^{\overline{@}1}.. & & & & C_{32}^A & = & ..(C_{21}^{A_3} \cdots C_{2d_2}^{A_3})^{\overline{@}3}.. \\
 & & & & & & C_{33}^A & = & !.(C_{31}^{A_3} \cdots C_{3d_3}^{A_3})^{\overline{@}3}..
\end{array}
$$

4.4.9 Definition Let $M \in \Lambda^{\emptyset}[\vec{c}]([\vec{A}])$ be a long normal form. Then either $M \equiv \lambda\vec{x}.c$ or $M \equiv \lambda\vec{x}.x_i\vec{N}$. In the first case, M is called *constant*, in the second it *has index* i.

Main auxiliary results. Let $A \neq 0$ be a type and $1 \leq i \leq \#A$ then

$$
\begin{array}{lrcll}
\text{(i)} & E_i^A(P_{i1}^A M) \cdots (P_{i\#A}^A M) & \approx & M, & \text{for } M \in \Lambda^{\emptyset}[\vec{c}](A) \text{ of index } i; \\
\text{(ii)} & P_{ij}^A(E_i^A L_1 \cdots L_{\#A_i}) & \approx & L_j, & \text{for } \vec{L} \in \Lambda^{\emptyset}[\vec{c}](\vec{C}_i) \text{ and } 1 \leq j \leq \#A_i.
\end{array}
$$

Before we prove this we will compute for some types the E_i^A and P_{ij}^A. It is important to realize that the C_{ij}^A are only defined if $1 \leq i \leq \#A$ and $1 \leq j \leq \#A_i$, but nevertheless the sequence of sequences of types $\vec{C}_i^A$ may be defined. It may be a sequence of empty sequences or be empty altogether.

Remember $1{\to}0{\to}0 = [[0]0]$ is the type of the Church numerals, $\top^2 = 1_2{\to}0{\to}0 = [[0,0],0]$ is the top type for reducibility, and $2 = 1{\to}0 = [1] = [[0]]$. We want to compute E_i and P_{ij} for these types and their subtypes. For $A = \top^2$ and $A = [0,0]$ we verify (i) and (ii) of the main auxiliary result.

A	aka	$\#A$	$\#A_1$	$\#A_2$	$\vec{C}_1^A$	$\vec{C}_2^A$	$\vec{z}_1^A : \vec{C}_1^A$	$\vec{z}_2^A : \vec{C}_2^A$
$[\,]$	0	0	–	–	–	–	–	–
$[0]$	1	1	0	–	ϵ	–	ϵ	–
$[00]$	1_2	2	0	0	ϵ	ϵ	ϵ	ϵ
$[[0]]$	2	1	1	–	ϵ	–	ϵ	–
$[[0]0]$	Nat	2	1	0	$[0]$	ϵ	$z_{11}^{[0]}$	ϵ
$[[0][0]0]$	Word	3	1	1	$C_{11}^A \triangleq [[0]0]$	$C_{21}^A \triangleq [[0]0]$	$z_{11}^{[[0]0]}$	$z_{21}^{[[0]0]}$
$[[00]0]$	$\top^2$	2	2	0	$C_{11}^A \triangleq [0], C_{12}^A \triangleq [0]$	ϵ	$z_{11}^{[0]}, z_{12}^{[0]}$	ϵ

A	$E_1^A : \vec{C}_1^A \to A$	$E_2^A : \vec{C}_2^A \to A$	$E_3 : \vec{C}_3^A \to A$
0	–	–	–
1	$(\mathsf{I}^1 \triangleq)\ \lambda x^0.x$	–	–
1_2	$(\mathsf{K}^{00} \triangleq)\ \lambda x_1^0 x_2^0.x_1$	$(\mathsf{K}_*^{00} \triangleq)\ \lambda x_1^0 x_2^0.x_2$	–
2	$\lambda z^0 x^1.xz$	–	–
Nat	$\lambda z_{11}^1 x_1^1 x_2^0.x_1(\mathsf{I}^1 x_2)$	–	–
Word	$\lambda z_{11}^{[1,0]} x_1^1 x_2^1 x_3^0.x_1(z_{11}x_2x_3)$	$\lambda z_{21}^{[1,0]} x_1^1 x_2^1 x_3^0.x_2(z_{21}x_2x_3)$	$\lambda x_1^1 x_2^1 x_3^0.x_3$
$\top^2$	$\lambda z_{11}^1 z_{12}^1 x_1^{1_2} x_2^0.x_1(z_{11}x_2)(z_{12}x_2)$	$\lambda x_1^{1_2} x_2^0.x_2$	–

A	$P_{1j}^A : A \to C_{1j}^A$	$P_{2j}^A : A \to C_{2j}^A$
0	–	–
1	ϵ	ϵ
1_2	ϵ	ϵ
2	$\lambda f^2.f\mathsf{I}^1$	–
Nat	$\lambda f^{\mathsf{Nat}} z_{11}^1 x_2^0.f\mathsf{I}^1 x_2$	ϵ
Word	$\lambda f^{[110]} x_2^1 x_3^0.f\mathsf{I}^1 x_2 x_3$	$\lambda f^{[110]} x_1^1 x_3^0.f x_1 \mathsf{I}^1 x_3$
$\top^2$	$P_{11}^A \triangleq \lambda f^{\top^2} x_2^0.f\mathsf{K}^{00}x_2,\ P_{12}^A \triangleq \lambda f^{\top^2} x_2^0.f\mathsf{K}_*^{1_2}x_2$	$\lambda f^{\top^2} x_1^{1_2}.f x_1$

We will verify some examples of the main auxiliary tool.

Case $A = \top^2$. (i) Assume $M \in \Lambda^{\varnothing}[\vec{c}](\top^2)$ has index 1. Then

$$\begin{aligned}
E_1^{\top^2}(P_{11}^{\top^2}M)(P_{12}^{\top^2}M)Q_1Q_2 &= Q_1[P_{11}MQ_2][P_{21}MQ_2] \\
&= Q_1[M\mathsf{K}^{00}Q_2][M\mathsf{K}_*^{00}Q_2] \\
&= \begin{cases} M\mathsf{K}^{00}Q_2 & \text{if } Q_1 \text{ has index 1, i.e. } Q_1 = \mathsf{K}^{00}, \\ M\mathsf{K}_*^{00}Q_2 & \text{if } Q_1 \text{ has index 2, i.e. } Q_1 = \mathsf{K}_*^{00}, \\ c, & \text{if } Q_1 = \lambda y_1 y_2.c \text{ is constant} \end{cases} \\
&= MQ_1Q_2, \text{in the third case because } M \text{ has index 1.}
\end{aligned}$$

Therefore $E_1^{\top^2}(P_{11}^{\top^2}M)(P_{12}^{\top^2}M) \approx M$.

If M has index 2, then $M = \lambda x_1^{1_2} x_2^0.x_2$ and

$$E_2^{\top^2} = M$$

which was to be shown, as $\vec{P}_2^{\top^2}$ is empty.

Case $A = \top^2$. (ii) Noting that $E_1 : 1 \to 1 \to \top^2$, $P_{1j} : \top^2 \to 1$, we have

$$\begin{array}{rcl} P_{1j}^{\top^2}(E_1^{\top^2} L_1 L_2) & = & P_{1j}(\lambda x_1 x_2.x_1(L_1 x_2)(L_2 x_2)) \\ & = & (\lambda f z^0.f \mathsf{K}^{1_2}_{(*)} z)(\lambda x_1 x_2.x_1(L_1 x_2)(L_2 x_2) \\ & = & \lambda z.K_{(*)}(L_1 z)(L_2 z) \\ & = & \begin{cases} \lambda z.L_1 z = L_1, & \text{if } j = 1, \\ \lambda z.L_2 z = L_2, & \text{if } j = 2. \end{cases} \end{array}$$

And for $i = 2$ one has $E_2 : \top^2$, $P_{21} : \top^2 \to \top^2$. We compute:

$$\begin{array}{rcl} P_{21}^{\top^2} E_2^{\top^2} & = & P_{21}(\lambda x_1 x_2.x_2) \\ & = & (\lambda f x_1.f x_1)(\lambda x_1 x_2.x_2) \\ & = & \lambda x_1 x_2.x_2, \end{array}$$

which is the unique element of $\Lambda^{\emptyset}[\vec{c}](\top^2)$ of index 2. Therefore

$$P_{21}^{\top^2}(E_2^{\top^2}(\epsilon L)) = P_{21}^{\top^2} E_1^{\top^2} = L.$$

Case $A = [00] = 1_2$. (i) One has

$$\begin{array}{l} E_1 = \lambda x_1^0 x_2^0.x_1 : 0 \to [00] \\ E_2 = \lambda x_1^0 x_2^0.x_2 : 0 \to [00] \end{array}$$

and the sequences $\vec{P}_1, \vec{P}_2$ are both empty. If M has index 1, then $M = \lambda x_1 x_2.x_1 = E_1$, as x_1 of type 0 cannot have any arguments. Hence

$$E_1(\epsilon M) = E_1 = M.$$

Similarly for M of index 2 the statement is correct.

Case $A = [00] = 1_2$. (ii) The general statement to be proved is

$$\vec{P}_i(E_i \vec{L}) = \vec{L}.$$

As $\mathtt{lth}(\vec{P}_i) = \mathtt{lth}(\vec{L}) = \#A_i = 0$, both strings are empty. Hence for $i = 1, 2$

$$\vec{P}_i(E_i \vec{L}) = \epsilon(E_i \epsilon) = \epsilon = \vec{L}.$$

This is a degenerate case, but valid nevertheless.

4.4.10 Proposition *Let A be a type different from 0. Suppose that the long normal form of $M \in \Lambda^{\emptyset}(A)$ has index i. Then*

$$M \approx E_i^A(P_{i1}^A M) \cdots (P_{i\#A}^A M).$$

Proof By course of value induction on $||A||$. That is, one assumes that for all A' with $||A'|| < ||A||$ the statement holds, in order to show it also holds for A. As with transfinite induction, this form of induction doesn't need a base case.

[Although it is included in the induction step, we provide the argument for the smallest possible applicable types, namely 1_n, in order to protect the reader from a possible experience of '*horror vacui*'.

Base. Let $A = 1_n$. Assume that M has index $i(\leq n)$. We compute

$$\begin{aligned} E_i^A(P_{i1}^A M)\cdots(P_{in}^A M) &= \lambda x_1 \ldots x_m.x_i, && \text{by definition of } E_i^A \text{ and } \#A_i = 0, \\ &= M, && \text{as its index is } i \text{ and } x_i \text{ is of type } 0.] \end{aligned}$$

Induction step. Let $A = [A_1 \ldots A_\ell]$. Let $M = (\lambda x_1 \ldots x_\ell.x_i N_1 \ldots N_m) : A$. Let $\vec{Q} = Q_1^{A_1}, \ldots, Q_\ell^{A_\ell} \in \Lambda^\varnothing[\vec{c}]$ in order to show

$$E_i^A(\vec{P}_i^A M)\vec{Q} =_{\beta\eta} M\vec{Q}. \tag{4.8}$$

Case 1. Q_i is not constant. Then Q_i has index j, for some $1 \leq j \leq \#A_i$, so that $\#A_i > 0$. Then

$$\begin{aligned}
& E_i^A(P_{i1}^A M)\cdots(P_{i\#A}^A M)Q_1^{A_1}\cdots Q_{\#A}^{A_\ell} \\
&= Q_i(\mathsf{K}^{A_{i1}}(P_{i1}^A M Q_1 \ldots (\vec{P}_i^{A_i} Q_i)^{\overline{@}i} \ldots Q_{\#A})) \\
&\qquad \ldots \\
&\qquad (\mathsf{K}^{A_{i\ell}}(P_{i\#A_i}^A M Q_1 \ldots (\vec{P}_i^{A_i} Q_i)^{\overline{@}i} \ldots Q_{\#A})) \\
&\equiv Q_i(\mathsf{K}^{A_{i1}}(M Q_1 \ldots (E_1^{A_i}(\vec{P}_1^{A_i} Q_i))^{\overline{@}i} \ldots Q_{\#A})) \\
&\qquad \ldots \\
&\qquad (\mathsf{K}^{A_{ij}}(M Q_1 \ldots (E_j^{A_i}(\vec{P}_j^{A_i} Q_i))^{\overline{@}i} \ldots Q_{\#A})) \\
&\qquad \ldots \\
&\qquad (\mathsf{K}^{A_{i\ell}}(M Q_1 \ldots (E_{\#A_i}^{A_i}(\vec{P}_{\#A_i}^{A_i} Q_i))^{\overline{@}i} \ldots Q_{\#A})) \\
&\approx Q_i(\mathsf{K}^{A_{i1}}(M Q_1 \ldots (E_1^{A_i}(\vec{P}_1^{A_i} Q_i))^{\overline{@}i} \ldots Q_{\#A})) \\
&\qquad \ldots \\
&\qquad (\mathsf{K}^{A_{ij}}(M Q_1 \ldots (Q_i)^{\overline{@}i} \ldots Q_{\#A})) \\
&\qquad \ldots \\
&\qquad (\mathsf{K}^{A_{i\ell}}(M Q_1 \ldots (E_{\#A_i}^{A_i}(\vec{P}_{\#A_i}^{A_i} Q_i))^{\overline{@}i} \ldots Q_{\#A})) \\
&\qquad\qquad \text{by the induction hypothesis, as } Q_i \text{ has index } j, \\
&= \mathsf{K}^{A_{ij}}(M\vec{Q})\cdots, \text{ again since } Q_i \text{ has index } j, \\
&= M\vec{Q}, \text{ as the number of arguments is correct.}
\end{aligned}$$

Case 2. $Q_i = \lambda y_1^{A_{i1}} \cdots y^{A_{i\ell}}.c$ is constant, possibly with $A_i = 0$. Then (4.8) holds in another way:

$$\begin{aligned} E_i^A(\vec{P}_i^A M)\vec{Q} &= Q_i(\mathsf{K}^{A_{i1}}\Box_1)\cdots(\mathsf{K}^{A_{i\ell}}\Box_{\#A_i}) \\ &= c \\ &= Q_i(N_1[\vec{x} := \vec{Q}])\cdots(N_{\#A_i}[\vec{x} := \vec{Q}]), \quad \text{as } Q_i \text{ is constantly } c, \\ &= M\vec{Q}, \quad \text{as } M \text{ has index } i. \end{aligned}$$

This extablishes (4.8) in all cases. Therefore $E_i^A(\vec{P}_i^A M) \approx M$. ■

The next proposition states that for every type A, the terms $M \in \Lambda^{\emptyset}[\vec{c}](A)$ with a given index can be enumerated by a term $E : \vec{C} \to A$, where the $\vec{C}$ have degrees lower than A.

4.4.11 Definition Define, by course of value recursion on $||A||$, the sets $\mathcal{R}_A \subseteq \Lambda^{\emptyset}[\vec{c}](A)$.

$$\begin{aligned} \mathcal{R}_0 &= \{c_1, c_2\} \\ \mathcal{R}_{[A_1 \ldots A_\ell]} &= \{E_i Q_1 \cdots Q_{\#A_i} \mid Q_1 \in \mathcal{R}_{C_{i1}}, \ldots, Q_{\#A_i} \in \mathcal{R}_{C_{i\#A_i}}, 1 \le i \le \#A\} \\ &\quad \cup \{\mathsf{K}^A c_1, \mathsf{K}^A c_2\}. \end{aligned}$$

4.4.12 Theorem *Let $\mathcal{M} = \Lambda^{\emptyset}[\vec{c}]/\approx$ be the minimal model. Then*

$$\forall M \in \mathcal{M}(A) \exists N \in \mathcal{R}_A . M \approx N.$$

Proof By induction on $||A||$. We say that B is smaller than C if $||B|| < ||C||$.

Case $A = 0$. Let $M \in \Lambda^{\emptyset}[\vec{c}](A)$ and N be its long normal form. Then $N \in \{c_1, c_2\} = \mathcal{R}_0$. Hence $M \approx N \in \mathcal{R}_0$.

Case $A = [A_1 \cdots A_\ell]$. By Proposition 4.4.10, for $1 \le i \le \ell$, the types $C_{i1}, \ldots, C_{i\#A_i}$ are smaller than A. Let $M \in \Lambda^{\emptyset}[\vec{c}](A)$. Suppose M has index i. Then one has

$$M \approx E_i(P_{i1}M) \cdots (P_{i\#A_i}M),$$

with $P_{ij}M \in \Lambda^{\emptyset}[\vec{c}](C_{ij}^A)$. By the induction hypothesis each $P_{ij}M \approx N_j$, for some $N_j \in \mathcal{R}_{C_{ij}}$. Hence $M \approx E_i N_1 \ldots N_{\#A_i} \in \mathcal{R}_A$.

Suppose M is constant. Then, say, $M = \mathsf{K}^A c$, so that $M \approx (\mathsf{K}^A c) \in \mathcal{R}_A$. ■

4.4.13 Corollary (Padovani) *The maximal theory is decidable.*

Proof For equivalence in the minimal model $\mathcal{M}_{\min}$ at type $A = [A_1 \cdots A_\ell]$ we have

$$M \approx N \Leftrightarrow \forall P_1 \in \Lambda^{\emptyset}[\vec{c}](A_1) \cdots P_\ell \in \Lambda^{\emptyset}[\vec{c}](A_\ell). M\vec{P} =_{\beta\eta} N\vec{P}.$$

One can restrict the $\vec{P}$ to the $\mathcal{R}_{A_j}$, hence this is decidable. ■

4.4.14 Corollary (Decidability of unification in $\mathcal{T}_{\max}$) *For terms*

$$M, N \in \Lambda^{\emptyset}[\vec{c}](A \to B),$$

of the same type, the following unification problem is decidable:

$$\exists X \in \Lambda^{\emptyset}[\vec{c}](A).MX \approx NX.$$

Proof Working in $\mathcal{M}_{\min}$, check the finitely many enumerating terms as candidates. ■

4.4.15 Corollary (Decidability of atomic higher-order matching)

(i) *For*

$$M_1 \in \Lambda^{\emptyset}[\vec{c}](A_1 \to 0), \ldots, M_n \in \Lambda^{\emptyset}[\vec{c}](A_n \to 0),$$

with $1 \leq i \leq n$, *the following problem is decidable:*

$$\begin{aligned} \exists X_1 \in \Lambda^{\emptyset}[\vec{c}](A_1), \ldots, X_n \in \Lambda^{\emptyset}[\vec{c}](A_n).[M_1 X_1 &=_{\boldsymbol{\beta\eta}} c_1 \\ &\cdots \\ M_n X_n &=_{\boldsymbol{\beta\eta}} c_n]. \end{aligned}$$

(ii) *For* $M, N \in \Lambda^{\emptyset}[\vec{c}](A \to 0)$ *the following problem is decidable.*

$$\exists X \in \Lambda^{\emptyset}[\vec{c}](A).MX =_{\boldsymbol{\beta\eta}} NX.$$

Proof (i) At type 0 the relations $\approx$ and $\boldsymbol{\beta\eta}$-convertibility coincide, so the previous Corollary applies. (ii) Similarly to (i) or by reducing this problem to the problem in (i). ■

The non-redundancy of the enumeration

We now prove that the enumeration of terms in Proposition 4.3.23 is not redundant. We follow the given construction, but actually the proof does not depend on it, see Exercise 4.2. We first prove a converse to Proposition 4.4.10.

4.4.16 Proposition *Let* $A \in \mathbb{T}^0$ *and* $1 \leq i \leq \#A$. *Suppose we are given closed terms* $\vec{L} : \vec{C}_i^A$; *that is,*

$$L_1 : C_{i1} \cdots L_{\#A_i} : C_{i\#A_i}.$$

Then for all $1 \leq j \leq \#A_i$ *one has*

$$P_{ij}^A(E_i^A L_1 \cdots L_{\#A_i}) \approx L_j.$$

Proof By induction on $||A||$. We compute $P_{ij}(E_iL_1 \ldots L_{\#A_i})$ applied to $N_1 \ldots (Q_1 \ldots Q_{\#A_{ij}})^{\overline{@i}} \ldots N_{\#A_i}$ to show that it is $L_iN_1 \ldots \vec{Q}^{\overline{@i}} \ldots N_{\#A_i}$.

$$
\begin{aligned}
& P^A_{ij}(E^A_i L_1 \cdots L_k)N_1 \ldots (\vec{Q})^{\overline{@i}} \ldots N_{\#A_i} \\
&= E^A_i \vec{L} N_1 \ldots (E^{A_i}_j \vec{Q})^{\overline{@i}} \ldots N_{\#A_i} \\
&= E^{A_i}_j \vec{Q}(\mathsf{K}^{A_{i1}}(L_1N_1 \ldots (\vec{P}^{A_i}_1(E_j\vec{Q}))^{\overline{@i}} \ldots N_{\#A_i})) \\
&\qquad\qquad \cdots \\
&\qquad\quad (\mathsf{K}^{A_{ij}}(L_1N_1 \ldots (\vec{P}^{A_i}_j(E_j\vec{Q}))^{\overline{@i}} \ldots N_{\#A_i})) \\
&\qquad\qquad \cdots \\
&\qquad\quad (\mathsf{K}^{A_{i\ell}}(L_1N_1 \ldots (\vec{P}^{A_i}_{\#A_i}(E_j\vec{Q}))^{\overline{@i}} \ldots N_{\#A_i})) \\
&= (\mathsf{K}^{A_{ij}}(L_jN_1 \ldots (\vec{P}^{A_i}_j(E_j\vec{Q}))^{\overline{@i}} \ldots N_{\#A_i})) \cdots \\
&= L_jN_1 \ldots (\vec{P}^{A_i}_j(E_j\vec{Q}))^{\overline{@i}} \ldots N_{\#A_i}, \\
&\qquad \text{as there are exactly enough arguments in } \ldots, \\
&\approx L_jN_1 \ldots (Q_1 \ldots Q_{\#A_i})^{\overline{@i}} \ldots N_{\#A_i},
\end{aligned}
$$

since

$$
\begin{aligned}
\vec{P}^{A_i}_j(E^{A_i}_j\vec{Q}) &= P_{1j}(E_j\vec{Q}), P_{2j}(E_j\vec{Q}), \ldots, P_{\#A_{ij}}(E_j\vec{Q}) \\
&= P_{1j}(E_jQ_1 \ldots Q_{\#A_i}), \ldots, P_{\#A_{ij}}(E_jQ_1 \ldots Q_{\#A_i}) \\
&\approx Q_1, Q_2, \ldots, Q_{\#A_i},
\end{aligned}
$$

by the induction hypothesis applied with i and j interchanged. ∎

4.4.17 Proposition *The enumeration in Theorem* 4.4.12 *is non-redundant, i.e.*

$$\forall A \in \mathbb{T}^0 \forall M, N \in \mathcal{R}_A . M \approx_{\mathcal{C}_2} N \ \Rightarrow \ M \equiv N.$$

Proof Consider $M, N \in \mathcal{R}_A$ with $M \approx_{\mathcal{C}_2} N$. By induction on $||A||$ we show that $M \equiv N$. Let $M = \lambda x_1 \ldots x_n . z\vec{P}$ and $N = \lambda x_1 \ldots x_n . z'\vec{Q}$, where the $z, z' \in \{\vec{x}\} \cup \{c_1, c_2\}$. If one of the z is a constant, then so is the other and one has $z = z'$ and $\vec{P} = \vec{Q} = \epsilon$, so that $M \equiv N$. If, say, $z \equiv x_i$, with $1 \le i \le n$, then $z \equiv z'$, hence both M, N have index i. By construction of $\mathcal{R}_A$ one has $M \equiv E_iM'_1 \cdots M'_{\#A_i}$ and $N \equiv E_iN'_1 \cdots N'_{\#A_i}$. But then by Proposition 4.4.16

$$M'_j \approx (P_{ij}M) \approx (P_{ij}N) \approx N'_j.$$

Hence, by the induction hypothesis $M'_j \equiv N'_j$ and therefore $M \equiv N$. ∎

4.5 Exercises

Exercise 4.1 Let $c_n = \text{card}(\mathcal{M}(1^n \to 0))$, where $\mathcal{M} = \mathcal{M}[\mathcal{C}_1]$ is the minimal model.

(i) Show that

$$c_0 = 2;$$
$$c_{n+1} = 2 + (n+1)c_n.$$

(ii) Prove that

$$c_n = 2n! \sum_{i=0}^{n} \frac{1}{i!}.$$

The $d_n = n! \sum_{i=0}^{n} \frac{1}{i!}$ "the number of arrangements of n elements" form a well-known sequence in combinatorics. See, for instance, Flajolet and Sedgewick (1993).

(iii) Can the cardinality of $\mathcal{M}(A)$ be bounded by a function of the form $k^{|A|}$ where $|A|$ is the size of $A \in \mathbb{T}^0$ and k a constant?

Exercise 4.2 Let $\mathcal{C} = \{c^0, d^0\}$. Let $\mathcal{E}$ be a computable function that assigns to each type $A \in \mathbb{T}^0$ a finite set of terms $\mathcal{X}_A$ such that

$$\forall M \in \Lambda[\mathcal{C}](A) \exists N \in \mathcal{X}_A . M \approx_{\mathcal{C}} N.$$

Show that not knowing the theory of Section 4.4 one can effectively make $\mathcal{E}$ non-redundant, i.e. such that

$$\forall A \in \mathbb{T}^0 \forall M, N \in \mathcal{E}_A . M \approx_{\mathcal{C}} N \;\Rightarrow\; M \equiv N.$$

Exercise 4.3 (Herbrand's Problem) Consider sets $\mathcal{S}$ of universally quantified equations

$$\forall x_1 \ldots x_n . [T_1 = T_2]$$

between first order terms involving constants $f, g, h, \ldots$ of various arities. Herbrand's theorem concerns the problem of whether $\mathcal{S} \models R = S$ where R, S are closed first-order terms. For example the word problem for groups can be represented this way. Now let d be a new quaternary constant i.e. $d : 1_4$ and let a, b be new 0-ary constants i.e. $a, b : 0$. We define the set $\mathcal{S}^+$ of simply typed equations by

$$\mathcal{S}^+ = \{\, (\lambda \vec{x} . T_1 = \lambda \vec{x} . T_2) \mid (\forall \vec{x} [T_1 = T_2]) \in \mathcal{S} \}.$$

Show that the following are equivalent

(i) $\mathcal{S} \not\models R = S$.

(ii) $\mathcal{S}^+ \cup \{\lambda x.dxxab = \lambda x.a,\ dRSab = b\}$ is consistent.

Conclude that the consistency problem for finite sets of equations with constants is Π^0_1-complete (in contrast to the decidability of finite sets of pure equations).

Exercise 4.4 (Undecidability of second-order unification) Consider the unification problem

$$Fx_1 \dots x_n = Gx_1 \dots x_n,$$

where each x_i has a type of rank <2. By the theory of reducibility we can assume that $Fx_1 \dots x_n$ has type $(0{\to}(0{\to}0)){\to}(0{\to}0)$ and so by introducing new constants of types 0, and $0{\to}(0{\to}0)$ we can assume $Fx_1 \dots x_n$ has type 0. Thus we arrive at the problem (with constants) in which we consider the problem of unifying 1st-order terms built up from 1st- and 2nd-order constants and variables, The aim of this exercise is to show that it is recursively unsolvable by encoding Hilbert's 10th problem, (Goldfarb, 1981). For this we shall need several constants. Begin with constants

$$\begin{aligned} a, b &: 0 \\ s &: 0{\to}0 \\ e &: 0{\to}(0{\to}(0{\to}0)) \end{aligned}$$

The nth numeral is $s^n a$.

(i) Let $F{:}0{\to}0$. F is said to be affine if $F = \lambda x.s^n x$. N is a numeral if there exists an affine F such that $Fa = N$. Show that F is affine $\Leftrightarrow$ $F(sa) = s(Fa)$.

(ii) Next show that $L = N + M$ iff there exist affine F and G such that $N = Fa$, $M = Ga$, and $L = F(Ga)$.

(iii) We can encode a computation of $n * m$ by

$$e(n * m)m(e(n * (m-1))(m-1)(\cdots(e(n*1)11)\cdots)).$$

Finally show that $L = N * M \ \Leftrightarrow\ \exists C, D, U, V$ affine and $\exists F, W$

$$\begin{aligned} Fab &= e(Ua)(Va)(Wab) \\ F(Ca)(sa)(e(Ca)(sa)b) &= e(U(Ca))(V(sa))(Fabl) \\ L &= Ua \\ N &= Ca \\ M &= Va \\ &= Da. \end{aligned}$$

Exercise 4.5 Consider $\Gamma_{n,m} = \{c_1{:}0, \ldots, c_m{:}0, f_1{:}1, \ldots, f_n{:}0\}$. Show that the unification problem with constants from Γ with several unknowns of type 1 can be reduced to the case where $m = 1$. This is equivalent to the following problem of Markov. Given a finite alphabet $\Sigma = \{a_1, \ldots, a_n\}$ consider equations between words over $\Sigma \cup \{X_1, \ldots, X_p\}$. The aim is to find for the unknowns $\vec{X}$ words $w_1, \ldots, w_p \in \Sigma^*$ such that the equations become syntactic identities. In Makanin (1977) it is proved that this problem is decidable (uniformly in n, p).

Exercise 4.6 (Decidability of unification of second-order terms) Consider the unification problem $F\vec{x} = G\vec{x}$ of type A with $\text{rk}(A) = 1$. Here we are interested in the case of pure unifiers of any types. Then $A = 1_m = 0^m \to 0$ for some natural number m. Consider for $i = 1, \ldots, m$ the systems

$$S_i = \{F\vec{x} = \lambda\vec{y}.y_i, G\vec{x} = \lambda\vec{y}.y_i\}.$$

(i) Observe that the original unification problem is solvable iff one of the systems S_i is solvable.

(ii) Show that systems whose equations have the form

$$F\vec{x} = \lambda\vec{y}.y_i$$

where $y_i : 0$ have the same solutions as single equations

$$H\vec{x} = \lambda xy.x$$

where $x, y : 0$.

(iii) Show that provided there are closed terms of the types of the x_i the solutions to a matching equation

$$H\vec{x} = \lambda xy.x$$

are exactly the same as the lambda-definable solutions to this equation in the minimal model.

(iv) Apply the method of Exercise 2.8 to the minimal model. Conclude that if there is a closed term of type A then the lambda-definable elements of the minimal model of type A are precisely those invariant under the transposition of the elements of the ground domain. Conclude that unification of terms of type of rank 1 is decidable.

Exercise 4.7 (Bram Westerbaan) Show that for the canonical term-model $\Lambda^{\emptyset}[\mathbf{f}^1, \mathbf{c}^0]/\approx$ Proposition 4.4.10 is not valid. It remains an open problem whether equality in this model is decidable. [Hint. Take $A = 2$, $M = \lambda g^1.g(g\mathbf{c})$ and $Q = \lambda x^0.\mathbf{f}x^0$.]

5
Extensions

In this chapter several extensions of $\boldsymbol{\lambda}_{\to}^{\mathrm{Ch}}$ based on $\mathbb{T}^0$ are studied. In Section 5.1 the systems are embedded into classical predicate logic by essentially adding constants δ_A (for each type A) that determine whether for $M, N \in \Lambda_{\to}^{\emptyset}(A)$ one has $M = N$ or $M \neq N$. In Section 5.2 is added a triple of terms π, π_1, π_2 that forms a surjective pairing. In both cases the resulting system becomes undecidable. In Section 5.3 the set of elements of ground type 0 is denoted by N: one thinks of it as consisting of the natural numbers. One does not work with Church numerals but with new constants $0 : \mathsf{N}, \mathsf{S}^+ : \mathsf{N} \to \mathsf{N}$, and $\mathsf{R}_A : A \to (A \to \mathsf{N} \to A) \to \mathsf{N} \to A$, for all types $A \in \mathbb{T}^0$, denoting respectively zero, successor and recursor, an operator describing primitive recursive functionals. In Section 5.4 Spector's bar recursive terms are studied. Finally in Section 5.5 fixed point combinators are added to the base system. This system is closely related to the system known as 'Edinburgh PCF'.

5.1 Lambda delta

In this section $\boldsymbol{\lambda}_{\to}^{0}$, in the form of $\boldsymbol{\lambda}_{\to}^{\mathrm{Ch}}$ based on $\mathbb{T}^0$, will be extended by constants δ $(= \delta_{A,B})$, for arbitrary A, B. Church (1940) used this extension to introduce a logical system called 'the simple theory of types', based on classical logic. The system is also refered to as 'higher-order logic', and denoted by HOL. We will introduce a variant of this system denoted by Δ. The essence of Δ will be captured in an equational way, both the formulas and the axiomatization. The intuitive idea is that $\delta = \delta_{A,B}$ satisfies for all $a, a' : A, b, b' : B$

$$\begin{aligned} \delta a a' b b' &= b \qquad \text{if } a = a'; \\ &= b' \qquad \text{if } a \neq a'. \end{aligned}$$

The type of the new constants is

$$\delta_{A,B} : A \to A \to B \to B \to B.$$

Only the Church variant of the theory, in which each term and variable carries its unique type, will be considered, but we will suppress types whenever there is little danger of confusion.

The theory Δ is a strong logical system, in fact stronger than nth-order logics for each n. It turns out that because of the presence of δs an arbitrary formula of Δ is equivalent to an equation. This fact will be an incarnation of the comprehension principle. It is because of the δs that Δ is powerful, less so because of the presence of quantification over elements of arbitrary types. Moreover, the set of equational consequences of Δ can be axiomatized by a finite subset. These are the main results in this section. It is an open question whether there is a natural (decidable) notion of reduction that is confluent and has as convertibility relation exactly these equational consequences. Since the decision problem for (higher-order) predicate logic is undecidable, such a notion of reduction will have to be non-terminating.

Higher-Order Logic

5.1.1 Definition We define a formal system called *higher-order logic*, written Δ. Terms are elements of $\Lambda_{\to}^{\mathrm{Ch},0}(\delta)$, the set of open typed terms with types from $\mathbb{T}^0$, possibly containing constants δ. Formulas are built up from equations between terms of the same type using implication ($\supset$) and typed quantification ($\forall x^A.\varphi$). Absurdity is defined by $\bot \triangleq (\mathsf{K} = \mathsf{K}_*)$, where $\mathsf{K} \equiv \lambda x^0 y^0.x$, $\mathsf{K}_* \equiv \lambda x^0 y^0.y$, and negation by $\neg\varphi \triangleq \varphi \supset \bot$. Variables always have to be given types such that the terms involved are typable and have the same type if they occur in one equation. In contrast with other sections in this book, Γ stands for a set of formulas. In Figure 5.1 the axioms and rules of Δ are given. There Γ is a set of formulas, and $\mathrm{FV}(\Gamma) = \{x \mid x \in \mathrm{FV}(\varphi),\ \varphi \in \Gamma\}$. The letters M, N, L, P, Q denote terms. Provability in this system will be denoted by $\Gamma \vdash_\Delta \varphi$, or simply by $\Gamma \vdash \varphi$.

5.1.2 Definition The other logical connectives of Δ are introduced in the usual classical manner:

$$\varphi \vee \psi \triangleq \neg\varphi \supset \psi;$$
$$\varphi \,\&\, \psi \triangleq \neg(\neg\varphi \vee \neg\psi);$$
$$\exists x^A.\varphi \triangleq \neg\forall x^A.\neg\varphi.$$

5.1.3 Lemma *For all formulas of Δ one has*

$$\bot \vdash \varphi.$$

$\Gamma \vdash (\lambda x.M)N = M[x{:}= N]$	(beta)
$\Gamma \vdash \lambda x.Mx = M,\ x \notin \mathrm{FV}(M)$	(eta)
$\Gamma \vdash M = M$	(reflexivity)
$\dfrac{\Gamma \vdash M = N}{\Gamma \vdash N = M}$	(symmetry)
$\dfrac{\Gamma \vdash M = N,\ \Gamma \vdash N = L}{\Gamma \vdash M = L}$	(trans)
$\dfrac{\Gamma \vdash M = N,\ \Gamma \vdash P = Q}{\Gamma \vdash MP = NQ}$	(cong-app)
$\dfrac{\Gamma \vdash M = N}{\Gamma \vdash \lambda x.M = \lambda x.N}\ x \notin \mathrm{FV}(\Gamma)$	(cong-abs)
$\dfrac{\varphi \in \Gamma}{\Gamma \vdash \varphi}$	(axiom)
$\dfrac{\Gamma \vdash \varphi \supset \psi \qquad \Gamma \vdash \varphi}{\Gamma \vdash \psi}$	($\supset$ -elim)
$\dfrac{\Gamma, \varphi \vdash \psi}{\Gamma \vdash \varphi \supset \psi}$	($\supset$ -intr)
$\dfrac{\Gamma \vdash \forall x^A.\varphi}{\Gamma \vdash \varphi[x{:}= M]}\ M \in \Lambda(A)$	($\forall$-elim)
$\dfrac{\Gamma \vdash \varphi}{\Gamma \vdash \forall x^A.\varphi}\ x^A \notin \mathrm{FV}(\Gamma)$	($\forall$-intr)
$\dfrac{\Gamma, M \neq N \vdash \bot}{\Gamma \vdash M = N}$	(classical)
$\Gamma \vdash M = N \supset \delta MNPQ = P$	(delta$_L$)
$\Gamma \vdash M \neq N \supset \delta MNPQ = Q$	(delta$_R$)

Figure 5.1 Δ: Higher-Order Logic

Proof By induction on the structure of φ. If $\varphi \equiv (M = N)$, then observe that by (eta)

$$
\begin{array}{lllllll}
\mathsf{K}MN & = & M & = & \lambda\vec{x}.M\vec{x} & = & \lambda\vec{x}.\mathsf{K}(M\vec{x})(N\vec{x}), \\
\mathsf{K}_*MN & = & N & = & \lambda\vec{x}.N\vec{x} & = & \lambda\vec{x}.\mathsf{K}_*(M\vec{x})(N\vec{x}),
\end{array}
$$

where the $\vec{x}$ are such that the type of $M\vec{x}$ is 0. Hence $\bot \vdash M = N$, since $\bot \equiv (\mathsf{K} = \mathsf{K}_*)$. If $\varphi \equiv (\chi \supset \psi)$ or $\varphi \equiv \forall x^A.\psi$, then the result follows immediately from the induction hypothesis for ψ. ∎

5.1.4 Proposition *We can define $\delta_{A,B}$ from $\delta_{A,0}$.*

Proof Indeed, if we only have $\delta_{A,0}$ (with their properties) and define

$$\delta_{A,B} = \lambda mnpq\vec{x}\,.\,\delta_{A,0}mn(p\vec{x})(q\vec{x}),$$

then all the $\delta_{A,B}$ satisfy the axioms. ∎

The rule (classical) is equivalent to

$$\neg\neg(M = N) \supset M = N.$$

In this rule the terms can be restricted to type 0 and the same theory Δ will be obtained.

5.1.5 Proposition *Suppose that in the formulation of Δ one requires*

$$\Gamma, \neg(M = N) \vdash_\Delta \bot \;\Rightarrow\; \Gamma \vdash_\Delta M = N \tag{5.1}$$

only for terms x, y of type 0. Then (5.1) holds for terms of all types.

Proof By (5.1) we have $\neg\neg(M = N) \supset M = N$ for terms of type 0. Assume $\neg\neg(M = N)$, with M, N of arbitrary type, towards $M = N$. Now

$$M = N \supset M\vec{x} = N\vec{x},$$

for all fresh $\vec{x}$ such that the type of $M\vec{x}$ is 0. By taking the contrapositive twice we obtain

$$\neg\neg(M = N) \supset \neg\neg(M\vec{x} = N\vec{x}).$$

Therefore by assumption and (5.1) for some terms of type 0 we get $M\vec{x} = N\vec{x}$. But then by (cong-abs) and (eta) it follows that $M = N$. ∎

5.1.6 Lemma *Let Γ, φ, ψ be formulas and let M, N and P, Q be two pairs of terms of the same type A, respectively B. Then*

(i) $\Gamma, \varphi \vdash_\Delta \psi \;\&\; \Gamma, \neg\varphi \vdash_\Delta \psi \;\Rightarrow\; \Gamma \vdash_\Delta \psi$.
(ii) $\vdash_\Delta (M = N \supset P = Q) \leftrightarrow (\delta_{A,B}MNPQ = Q)$.
(iii) $\vdash_\Delta \forall x.(M = N) \leftrightarrow (\lambda x.M = \lambda x.N)$.

Proof Left to the reader. ∎

5.1.7 Proposition *For all formulas φ one has*

$$\vdash_\Delta \neg\neg\varphi \supset \varphi.$$

Proof By induction on the structure of φ. If φ is an equation, then this is a rule of the system Δ. If $\varphi \equiv \psi \supset \chi$, then by the induction hypothesis one has $\vdash_\Delta \neg\neg\chi \supset \chi$ and we have the following derivation

$$
\begin{prooftree}
\AxiomC{$[\psi \supset \chi]^1$}
\AxiomC{$[\psi]^3$}
\BinaryInfC{χ}
\AxiomC{$[\neg\chi]^2$}
\BinaryInfC{$\bot$}
\RightLabel{1}
\UnaryInfC{$\neg(\psi \supset \chi)$}
\AxiomC{$[\neg\neg(\psi \supset \chi)]^4$}
\BinaryInfC{$\bot$}
\RightLabel{2}
\UnaryInfC{$\neg\neg\chi$}
\AxiomC{$\vdots$}
\UnaryInfC{$\neg\neg\chi \supset \chi$}
\BinaryInfC{χ}
\RightLabel{3}
\UnaryInfC{$\psi \supset \chi$}
\RightLabel{4}
\UnaryInfC{$\neg\neg(\psi \supset \chi) \supset \psi \supset \chi)$}
\end{prooftree}
$$

for $\neg\neg(\psi \supset \chi) \supset (\psi \supset \chi)$. If $\varphi \equiv \forall x.\psi$, then by the induction hypothesis $\vdash_\Delta \neg\neg\psi(x) \supset \psi(x)$. Now we have a similar derivation,

$$
\begin{prooftree}
\AxiomC{$[\forall x.\psi(x)]^1$}
\UnaryInfC{$\psi(x)$}
\AxiomC{$[\neg\psi(x)]^2$}
\BinaryInfC{$\bot$}
\RightLabel{1}
\UnaryInfC{$\neg\forall x.\psi(x)$}
\AxiomC{$[\neg\neg\forall x.\psi(x)]^3$}
\BinaryInfC{$\bot$}
\RightLabel{2}
\UnaryInfC{$\neg\neg\psi(x)$}
\AxiomC{$\vdots$}
\UnaryInfC{$\neg\neg\psi(x) \supset \psi(x)$}
\BinaryInfC{$\psi(x)$}
\UnaryInfC{$\forall x.\psi(x)$}
\RightLabel{3,}
\UnaryInfC{$\neg\neg\forall x.\psi(x) \supset \forall x.\psi(x)$}
\end{prooftree}
$$

for $\neg\neg\forall x.\psi(x) \supset \forall x.\psi(x)$. ∎

Next we will derive some equations in Δ that happen to be strong enough to provide an equational axiomatization of the equational part of Δ.

5.1.8 Proposition *The following equations hold universally (for those* *terms such that the equations make sense):*

$$
\begin{array}{rcll}
\delta MMPQ & = & P & (\delta\text{-}identity);\\
\delta MNPP & = & P & (\delta\text{-}reflexivity);\\
\delta MNMN & = & N & (\delta\text{-}hypothesis);\\
\delta MNPQ & = & \delta NMPQ & (\delta\text{-}symmetry);\\
F(\delta MNPQ) & = & \delta MN(FP)(FQ) & (\delta\text{-}monotonicity);\\
\delta MN(P(\delta MN))(Q(\delta MN)) & = & \delta MN(P\mathsf{K})(Q\mathsf{K}_*) & (\delta\text{-}transitivity).
\end{array}
$$

Proof We only show δ-reflexivity, the proof of the other assertions being similar. By the δ axioms one has

$$M = N \vdash \delta MNPP = P; \tag{5.2}$$

$$M \neq N \vdash \delta MNPP = P. \tag{5.3}$$

By the 'contrapositive' of (5.2) one has $\delta MNPP \neq P \vdash M \neq N$ and hence by (5.3) $\delta MNPP \neq P \vdash \delta MNPP = P$. Therefore $\delta MNPP \neq P \vdash \bot$, but then $\vdash \delta MNPP = P$, by the classical rule. ∎

5.1.9 Definition The *equational version* of higher-order logic, written δ, consists of equations between terms of $\Lambda_{\rightarrow}^{\mathrm{Ch},0}(\delta)$ of the same type, axiomatized as in Figure 5.2. As usual the axioms and rules are assumed to hold universally, i.e. the free variables may be replaced by arbitrary terms. We have denoted by $\mathcal{E}$ a set of equations $M = N$ with M, N terms of the same type. The system δ may be given more conventionally by leaving out all occurrences of $\mathcal{E} \vdash_\delta$ and replacing in the rule (cong-abs) the proviso '$x \notin \mathrm{FV}(\mathcal{E})$' by '$x$ not occurring in any assumption on which $M = N$ depends'.

There is a canonical map from formulas to equations, preserving provability in Δ.

5.1.10 Definition

(i) For an equation $E \equiv (M = N)$ in Δ, write $E \cdot L \triangleq M$ and $E \cdot R \triangleq N$. Here ::= denotes 'equality by definition'.

(ii) Define for a formula φ of Δ the corresponding equation φ^+ as follows:

$$
\begin{aligned}
(M = N)^+ &\triangleq M = N;\\
(\psi \supset \chi)^+ &\triangleq (\delta(\psi^+ \cdot L)(\psi^+ \cdot R)(\chi^+ \cdot L)(\chi^+ \cdot R) = \chi^+ \cdot R);\\
(\forall x.\psi)^+ &\triangleq (\lambda x.\psi^+ \cdot L = \lambda x.\psi^+ \cdot R).
\end{aligned}
$$

(iii) If Γ is a set of formulas, then $\Gamma^+ = \{\varphi^+ \mid \varphi \in \Gamma\}$.

$\mathcal{E} \vdash (\lambda x.M)N = M[x:=N]$	(β)
$\mathcal{E} \vdash \lambda x.Mx = M,\ x \notin \mathrm{FV}(M)$	(η)
$\mathcal{E} \vdash M = N$, if $(M = N) \in \mathcal{E}$	(axiom)
$\mathcal{E} \vdash M = M$	(reflexivity)
$\dfrac{\mathcal{E} \vdash M = N}{\mathcal{E} \vdash N = M}$	(symmetry)
$\dfrac{\mathcal{E} \vdash M = N,\ \mathcal{E} \vdash N = L}{\mathcal{E} \vdash M = L}$	(trans)
$\dfrac{\mathcal{E} \vdash M = N,\ \mathcal{E} \vdash P = Q}{\mathcal{E} \vdash MP = NQ}$	(cong-app)
$\dfrac{\mathcal{E} \vdash M = N}{\mathcal{E} \vdash \lambda x.M = \lambda x.N}\ x \notin \mathrm{FV}(\mathcal{E})$	(cong-abs)
$\mathcal{E} \vdash \delta MMPQ = P$	(δ-identity)
$\mathcal{E} \vdash \delta MNPP = P$	(δ-reflexivity)
$\mathcal{E} \vdash \delta MNMN = N$	(δ-hypothesis)
$\mathcal{E} \vdash \delta MNPQ = \delta NMPQ$	(δ-symmetry)
$\mathcal{E} \vdash F(\delta MNPQ) = \delta MN(FP)(FQ)$	(δ-monotonicity)
$\mathcal{E} \vdash \delta MN(P(\delta MN))(Q(\delta MN)) = \delta MN(P\mathsf{K})(Q\mathsf{K}_*)$	(δ-transitivity)

Figure 5.2 δ: Equational version of Δ

5.1.11 Remark Let $\psi^+ \equiv (M = N)$ and $\chi^+ \equiv (P = Q)$; then by definition

$$\begin{aligned}(\psi \supset \chi)^+ &\triangleq (\delta MNPQ = Q);\\ (\neg\psi)^+ &\triangleq (\delta MN\mathsf{K}\mathsf{K}_* = \mathsf{K}_*);\\ (\forall x.\psi)^+ &\triangleq (\lambda x.M = \lambda x.N).\end{aligned}$$

5.1.12 Theorem *For every formula φ one has*

$$\vdash_\Delta (\varphi \leftrightarrow \varphi^+).$$

Proof Note that $\varphi^+ = (\varphi^+)^+$, $(\psi \supset \chi)^+ = (\psi^+ \supset \chi^+)^+$, and $(\forall x.\psi)^+ = (\forall x.\psi^+)^+$. The proof of the theorem is by induction on the structure of φ. If φ is an equation, then this is trivial. If $\varphi \equiv \psi \supset \chi$, write $\psi^+ \equiv (M = N)$

and $\chi^+ \equiv (P = Q)$. Then $(\psi \supset \chi)^+ \equiv (\delta MNPQ = Q)$. We have in Δ

$$\begin{array}{rcll} \varphi & \equiv & (\psi \supset \chi) & \\ & \leftrightarrow & (M = N \supset P = Q), & \text{by the Induction Hypothesis,} \\ & \leftrightarrow & (\delta MNPQ = Q), & \text{by Lemma 5.1.6(ii),} \\ & \equiv & \varphi^+. & \end{array}$$

If $\varphi \equiv \forall x.\psi$, and ψ as before, then Lemma 5.1.6(iii) applies similarly. ∎

We will show now that Δ is conservative over δ. The proof occupies 5.1.13–5.1.19

5.1.13 Lemma

(i) $\vdash_\delta \delta MNPQz = \delta MN(Pz)(Qz)$.
(ii) $\vdash_\delta \delta MNPQ = \lambda z.\delta MN(Pz)(Qz)$, *where z is fresh.*
(iii) $\vdash_\delta \lambda z.\delta MNPQ = \delta MN(\lambda z.P)(\lambda z.Q)$, *where $z \notin \mathrm{FV}(MN)$.*

Proof (i) Use δ-monotonicity $F(\delta MNPQ) = \delta MN(FP)(FQ)$ for $F = \lambda x.xz$.
(ii) By (i) and (η).
(iii) By (ii) applied to $P' \triangleq \lambda z.P$ and $Q' \triangleq \lambda z.Q$. ∎

Note that from Lemma 5.1.13(ii) it follows that

$$\delta MNPQ\vec{R} = \delta MN(P\vec{R})(Q\vec{R}).$$

5.1.14 Lemma

(i) $\delta MNPQ = Q \vdash_\delta \delta MNQP = P$.
(ii) $\delta MNPQ = Q,\ \delta MNQR = R \vdash_\delta \delta MNPR = R$.
(iii) $\delta MNPQ = Q,\ \delta MNUV = V \vdash_\delta \delta MN(PU)(QV) = QV$.

Proof (i)

$$\begin{array}{rcll} P & = & \delta MNPP & \\ & = & \delta MN(\mathsf{K}PQ)(\mathsf{K}_*QP) & \\ & = & \delta MN(\delta MNPQ)(\delta MNQP), & \text{by } (\delta\text{-transitivity}), \\ & = & \delta MNQ(\delta MNQP), & \text{by assumption,} \\ & = & \delta MN(\delta MNQQ)(\delta MNQP), & \text{by } \delta\text{-reflexivity,} \\ & = & \delta MN(\mathsf{K}QQ)(\mathsf{K}_*QP), & \text{by } (\delta\text{-transitivity}), \\ & = & \delta MNQP. & \end{array}$$

(ii)

$$\begin{array}{rcll} R & = & \delta MNQR, & \text{by assumption,} \\ & = & \delta MN(\delta MNPQ)(\delta MNQR), & \text{by assumption,} \\ & = & \delta MN(\mathsf{K}PQ)(\mathsf{K}_*QR), & \text{by } (\delta\text{-transitivity}), \\ & = & \delta MNPR. & \end{array}$$

(iii) Suppose $\delta MNPQ = Q$ and $\delta MNUV = V$. Then we have

$$\begin{array}{rcll} \delta MN(PU)(QU) & = & \delta MN(FP)(FQ), & \text{with } F \equiv \lambda a.aU, \\ & = & F(\delta MNPQ), & \text{by } \delta\text{-monotonicity}, \\ & = & FQ, & \text{by assumption}, \\ & = & QU, & \text{by definition of } F. \\ \delta MN(QU)(QV) & = & Q(\delta MNPUV), & \text{by } \delta\text{-monotonicity}, \\ & = & QV, & \text{by assumption.} \end{array}$$

Hence the result $\delta MN(PU)(QV) = QV$ follows by (ii). ■

5.1.15 Proposition (Deduction Theorem I) *Let $\mathcal{E}$ be a set of equations. Then*

$$\mathcal{E}, M = N \vdash_\delta P = Q \;\Rightarrow\; \mathcal{E} \vdash_\delta \delta MNPQ = Q.$$

Proof By induction on the derivation of $\mathcal{E}, M = N \vdash_\delta P = Q$. If $P = Q$ is an axiom of δ, including reflexivity, or in $\mathcal{E}$, then $\mathcal{E} \vdash_\delta P = Q$ and hence $\mathcal{E} \vdash_\delta \delta MNPQ = \delta MNQQ = Q$. If $(P = Q) \equiv (M = N)$, then $\mathcal{E} \vdash_\delta \delta MNPQ \equiv \delta MNMN = N \equiv Q$. If $P = Q$ follows directly from $\mathcal{E}, M = N \vdash_\delta Q = P$, by (symmetry), then by the induction hypothesis one has $\mathcal{E} \vdash_\delta \delta MNQP = P$. But then by Lemma 5.1.14(i) one has $\mathcal{E} \vdash_\delta \delta MNPQ = Q$. If $P = Q$ follows by (transitivity), (cong-app) or (cong-abs), then the result follows from the induction hypothesis, using Lemma 5.1.14(ii), (iii) or Lemma 5.1.13(iii) respectively. ■

5.1.16 Lemma

(i) $\vdash_\delta \delta MN(\delta MNPQ)P = P$.
(ii) $\vdash_\delta \delta MNQ(\delta MNPQ) = Q$.

Proof (i) By (δ-transitivity), taking $P' \equiv \lambda a.aPQ, Q' \equiv \lambda a.P$ one has

$$\delta MN(\delta MNPQ)P = \delta MN(\mathsf{K}PQ)P = \delta MNPP = P.$$

(ii) Similarly, taking $P' \equiv \lambda a.aPQ, Q' \equiv \lambda a.aQP$. ■

5.1.17 Lemma

(i) $\vdash_\delta \; \delta\mathsf{K}\mathsf{K}_* = \mathsf{K}_*$;
(ii) $\vdash_\delta \; \delta MN\mathsf{K}\mathsf{K}_* = \delta MN$;
(iii) $\vdash_\delta \; \delta(\delta MN)\mathsf{K}_*PQ = \delta MNQP$;
(iv) $\vdash_\delta \; \delta(\delta MN\mathsf{K}\mathsf{K}_*)\mathsf{K}_*(\delta MNPQ)Q = Q$.

Proof (i)
$$\begin{aligned} \mathsf{K}_* &= \delta\mathsf{K}\mathsf{K}_*\mathsf{K}\mathsf{K}_*, && \text{by } (\delta\text{-hypothesis}),\\ &= \lambda ab.\delta\mathsf{K}\mathsf{K}_*(\mathsf{K}ab)(\mathsf{K}_*ab), && \text{by Lemma 5.1.13(ii) twice},\\ &= \lambda ab.\delta\mathsf{K}\mathsf{K}_*ab\\ &= \delta\mathsf{K}\mathsf{K}_*, && \text{by } (\eta). \end{aligned}$$

(ii)
$$\begin{aligned} \delta MN\mathsf{K}\mathsf{K}_* &= \delta MN(\delta MN)(\delta MN), && \text{by } (\delta\text{-transitivity}),\\ &= \delta MN, && \text{by } (\delta\text{-reflexivity}). \end{aligned}$$

(iii)
$$\begin{aligned} \delta MNQP &= \delta MN(\delta\mathsf{K}\mathsf{K}_*PQ)(\delta\mathsf{K}_*\mathsf{K}_*PQ), && \text{by (i), } (\delta\text{-identity}),\\ &= \delta MN(\delta(\delta MN)\mathsf{K}_*PQ)(\delta(\delta MN)\mathsf{K}_*PQ),\\ & && \text{by } (\delta\text{-transitivity}),\\ &= \delta(\delta MN)\mathsf{K}_*PQ, && \text{by } (\delta\text{-reflexivity}). \end{aligned}$$

(iv) By (ii) and (iii) we have

$$\begin{aligned} \delta(\delta MN\mathsf{K}\mathsf{K}_*)\mathsf{K}_*(\delta MNPQ)Q &= \delta(\delta MN)\mathsf{K}_*(\delta MNPQ)Q\\ &= \delta MNQ(\delta MNPQ). \end{aligned}$$

Therefore we are done by Lemma 5.1.16(ii). ■

5.1.18 Lemma

(i) $\delta MN = \mathsf{K} \vdash_\delta M = N;$
(ii) $\delta MN\mathsf{K}_*\mathsf{K} = \mathsf{K}_* \vdash_\delta M = N.$
(iii) $\delta(\delta MN\mathsf{K}\mathsf{K}_*)\mathsf{K}_*\mathsf{K}\mathsf{K}_* = \mathsf{K}_* \vdash_\delta M = N.$

Proof (i) $M = \mathsf{K}MN = \delta MNMN = N$, by assumption and ($\delta$-hypothesis).

(ii) Suppose $\delta MN\mathsf{K}_*\mathsf{K} = \mathsf{K}_*$. Then by the remark that follows Lemma 5.1.13(ii) and (δ-hypothesis)

$$M = \mathsf{K}_*NM = \delta MN\mathsf{K}_*\mathsf{K}NM = \delta MN(\mathsf{K}_*NM)(\mathsf{K}NM) = \delta MNMN = N.$$

(iii) By the assumption, Lemma 5.1.17(ii) and (iii), one has

$$\mathsf{K}_* = \delta(\delta MN\mathsf{K}\mathsf{K}_*)\mathsf{K}_*\mathsf{K}\mathsf{K}_* = \delta(\delta MN)\mathsf{K}_*\mathsf{K}\mathsf{K}_* = \delta MN\mathsf{K}_*\mathsf{K}.$$

Hence by (ii) we are done. ■

Now we are able to prove the conservativity of Δ over δ.

5.1.19 Theorem *For equations $\mathcal{E}, E$ and formulas Γ, φ of Δ we have*

(i) $\Gamma \vdash_\Delta \varphi \;\Leftrightarrow\; \Gamma^+ \vdash_\delta \varphi^+.$
(ii) $\mathcal{E} \vdash_\Delta E \;\Leftrightarrow\; \mathcal{E} \vdash_\delta E.$

Proof (i) ($\Rightarrow$) Suppose $\Gamma \vdash_\Delta \varphi$. By induction on this proof in Δ we show that $\Gamma^+ \vdash_\delta \varphi^+$.

Case 1. φ is in Γ. Then $\varphi^+ \in \Gamma^+$ and we are done.

Case 2. φ is an equational axiom. Then the result holds since δ has more equational axioms than Δ.

Case 3. φ follows from an equality rule in Δ. Then the result follows from the induction hypothesis and the fact that δ has the same equational deduction rules.

Case 4. φ follows from $\Gamma \vdash_\Delta \psi$ and $\Gamma \vdash_\Delta \psi \supset \varphi$. By the induction hypothesis $\Gamma^+ \vdash_\delta (\psi \supset \varphi)^+ \equiv (\delta MNPQ = Q)$ and $\Gamma^+ \vdash_\delta \psi^+ \equiv (M = N)$, where $\psi^+ \equiv (M = N)$ and $\varphi^+ \equiv (P = Q)$. Then $\Gamma^+ \vdash_\delta P = \delta MMPQ = \delta MNPQ = Q$, i.e. $\Gamma^+ \vdash_\delta \varphi^+$.

Case 5. $\varphi \equiv (\chi \supset \psi)$ and follows by an ($\supset$-intro) from $\Gamma, \chi \vdash_\Delta \psi$. By the induction hypothesis $\Gamma^+, \chi^+ \vdash_\delta \psi^+$ and we can apply Propostion 5.1.15.

Cases 6, 7. φ is introduced by a ($\forall$-elim) or ($\forall$-intro). Then the result follows easily from the induction hypothesis and axiom ($\boldsymbol{\beta}$) or the rule (cong-abs). One needs that $\mathrm{FV}(\Gamma) = \mathrm{FV}(\Gamma^+)$.

Case 8. $\varphi \equiv (M = N)$ and follows from $\Gamma, M \neq N \vdash_\Delta \perp$ using the rule (classical). By the induction hypothesis $\Gamma^+, (M \neq N)^+ \vdash_\delta \mathsf{K} = \mathsf{K}_*$. By the Deduction Theorem it follows that $\Gamma^+ \vdash_\delta \delta(\delta MN\mathsf{K}\mathsf{K}_*)\mathsf{K}_*\mathsf{K}\mathsf{K}_* = \mathsf{K}_*$. Hence we are done by Lemma 5.1.18(iii).

Case 9. φ is the axiom $(M = N \supset \delta MNPQ = P)$. Then φ^+ is provable in δ by Lemma 5.1.16(i).

Case 10. φ is the axiom $(M \neq N \supset \delta MNPQ = Q)$. Then φ^+ is provable in δ by Lemma 5.1.17(iv).

($\Leftarrow$) By the fact that δ is a subtheory of Δ and Theorem 5.1.12.

(ii) By (i) and the fact that $E^+ \equiv E$. ■

Logic of order n

In this subsection some results will be sketched but not (completely) proved.

5.1.20 Definition

(i) The system Δ without the two delta rules is denoted by Δ^-.
(ii) $\Delta(n)$ is Δ^- extended by the two delta rules restricted to the terms $\delta_{A,B}$ with $\mathtt{rank}(A) \leq n$.
(iii) Similarly $\delta(n)$ is the theory δ in which only terms $\delta_{A,B}$ are used with $\mathtt{rank}(\mathtt{A}) \leq n$.
(iv) The *rank of a formula* φ is $\mathtt{rank}(\varphi) = \mathtt{max}\{\,\mathtt{rank}(\delta) \mid \delta\ \mathtt{occurs\ in}\ \varphi\}$.

In the applications section we will show that $\Delta(n)$ is essentially nth-order logic.

The relation between Δ and δ that we have seen also holds level-by-level. We will only state the relevant results, the proofs being similar, but using as extra ingredient the proof-theoretic normalization theorem for Δ. This is necessary since a proof of a formula of rank n may use *a priori* formulas of arbitrarily high rank. By the normalization theorem such formulas can be eliminated.

A natural deduction is called *normal* if there is no ($\forall$-intro) immediately followed by a ($\forall$-elim), nor a ($\supset$-intro) immediately followed by a ($\supset$-elim). If a deduction is not normal, then one can subject it to reduction as follows (this idea is from Prawitz (1965)):

$$\dfrac{\dfrac{\begin{matrix}\vdots\,\Sigma\\ \varphi\end{matrix}}{\forall x.\varphi}}{\varphi[x:=M]} \quad\Rightarrow\quad \begin{matrix}\vdots\,\Sigma[x:=M]\\ \varphi[x:=M]\end{matrix}$$

$$\dfrac{\begin{matrix}\vdots\,\Sigma_2\\ \varphi\end{matrix} \qquad \dfrac{\begin{matrix}[\varphi]\\ \vdots\,\Sigma_1\\ \psi\end{matrix}}{\varphi\supset\psi}}{\psi} \quad\Rightarrow\quad \begin{matrix}\vdots\,\Sigma_2\\ [\varphi]\\ \vdots\,\Sigma_1\\ \psi\end{matrix}$$

5.1.21 Theorem *Δ-reduction on deductions is* SN. *Moreover, each deduction has a unique normal form.*

Proof This was proved, essentially, in Prawitz (1965) for first-order logic. The higher-order quantifiers pose no problems. ∎

5.1.22 Notation

(i) Denote by Γ_δ the set of universal closures of

$$\begin{aligned}\delta mmpq &= p,\\ \delta mnpp &= p,\\ \delta mnmn &= n,\\ \delta mnpq &= \delta nmpq,\\ f(\delta mnpq) &= \delta mn(fp)(fq),\\ \delta mm(p(\delta mn))(q(\delta mn)) &= \delta mn(p\mathsf{K})(q\mathsf{K}_*).\end{aligned}$$

(ii) Denote by $\Gamma_{\delta(n)}$ the set $\{\varphi\in\Gamma_\delta \mid \mathtt{rank}(\varphi)\leq n\}$.

5.1.23 Proposition (Deduction Theorem II) *Let $\mathcal{S}$ be a set of equations or negations of equations in Δ, such that for $(U=V)\in\mathcal{S}$ or $(U\neq V)\in\mathcal{S}$ one has for the type A of U,V that $\mathtt{rank}(A)\leq n$. Then:*

(i) $\mathcal{S},\Gamma_{\delta(n)},M=N\vdash_{\Delta(n)}P=Q \;\Rightarrow\; \mathcal{S},\Gamma_{\delta(n)}\vdash_{\Delta(n)}\delta MNPQ=Q;$
(ii) $\mathcal{S},\Gamma_{\delta(n)},M\neq N\vdash_{\Delta(n)}P=Q \;\Rightarrow\; \mathcal{S},\Gamma_{\delta(n)}\vdash_{\Delta(n)}\delta MNPQ=P.$

Proof In the same style as the proof of Proposition 5.1.15, but now using the Normalization Theorem 5.1.21. ∎

5.1.24 Lemma *Let $\mathcal{S}$ be a set of equations or negations of equations in Δ. Let $\mathcal{S}^*$ be $\mathcal{S}$ with each $M \neq N$ replaced by $\delta MN\mathsf{KK}_* = \mathsf{K}_*$. Then:*

(i) $\mathcal{S}, M = N \vdash_{\Delta(n)} P = Q \Rightarrow \mathcal{S}^* \vdash_{\delta(n)} \delta MNPQ = Q$;
(ii) $\mathcal{S}, M \neq N \vdash_{\Delta(n)} P = Q \Rightarrow \mathcal{S}^* \vdash_{\delta(n)} \delta MNPQ = P$.

Proof By induction on derivations. ∎

5.1.25 Theorem $\mathcal{E} \vdash_{\Delta(n)} E \Leftrightarrow \mathcal{E} \vdash_{\delta(n)} E$.

Proof ($\Rightarrow$) By taking $\mathcal{S} = \mathcal{E}$ and $M \equiv N \equiv x$ in Lemma 5.1.24(i) one obtains $\mathcal{E} \vdash_{\delta(n)} \delta xxPQ = Q$. Hence $\mathcal{E} \vdash_{\delta(n)} P = Q$, by ($\delta$-identity).
($\Leftarrow$) Trivial. ∎

The following result was established in Statman (2000).

5.1.26 Theorem

(i) *Let* $\mathtt{rank}(\mathcal{E}, M = N) \leq 1$. *Then*

$$\mathcal{E} \vdash_{\Delta} M = N \Leftrightarrow \mathcal{E} \vdash_{\delta(1)} M = N.$$

(ii) *Let Γ, A be first-order sentences. Then*

$$\Gamma \vdash_{\Delta} A \Leftrightarrow \Gamma \vdash_{\delta(1)} A^+.$$

In Statman (2000) it is also proved that $\Delta(0)$ is decidable. Since $\Delta(n)$ for $n \geq 1$ is at least first-order predicate logic, these systems are undecidable. It is observed in Gödel (1931) that the consistency of $\Delta(n)$ can be proved in $\Delta(n+1)$.

5.2 Surjective pairing

5.2.1 Definition A *pairing* on a set X consists of three maps π, π_1, π_2 such that

$$\pi : X \to X \to X$$
$$\pi_i : X \to X,$$

and for all $x_1, x_2 \in X$ one has

$$\pi_i(\pi x_1 x_2) = x_i.$$

Using a pairing one can pack two or more elements of X into one element:

$$\pi xy \in X,$$
$$\pi x(\pi yz) \in X.$$

A pairing on X is called *surjective* if one also has for all $x \in X$

$$\pi(\pi_1 x)(\pi_2 x) = x.$$

This is equivalent to saying that every element of X is a pair.

Using a (surjective) pairing one can encode data-structures.

5.2.2 Remark From a (surjective) pairing one can define $\pi^n : X^n \to X, \pi_i^n : X \to X$, $1 \leq i \leq n$ such that

$$\begin{array}{rcll} \pi_i^n(\pi^n x_1 \ldots x_n) & = & x_i, & 1 \leq i \leq n, \\ \pi^n(\pi_1^n x) \ldots (\pi_n^n x) & = & x, & \text{in the case of surjectivity.} \end{array}$$

Moreover $\pi = \pi^2$ and $\pi_i = \pi_i^2$, for $1 \leq i \leq 2$.

Proof Define

$$\begin{array}{rcll} \pi^1(x) & = & x & \\ \pi^{n+1} x_1 \ldots x_{n+1} & = & \pi(\pi^n x_1 \ldots x_n) x_{n+1} & \\ \pi_1^1(x) & = & x & \\ \pi_i^{n+1}(x) & = & \pi_i^n(\pi_1(x)), & \text{if } i \leq n, \\ & = & \pi_2(x), & \text{if } i = n+1. \blacksquare \end{array}$$

Surjective pairing is not typable in untyped λ-calculus and therefore also not in $\boldsymbol{\lambda}_{\to}$, see Barendregt (1974). In spite of this, in de Vrijer (1989), and later also in Støvring (2006) for the extensional case, it was shown that adding surjective pairing to untyped λ-calculus yields a conservative extension. Moreover normal forms remain unique, see de Vrijer (1987a) and Klop and de Vrijer (1989). By contrast the main results in this section are the following.

(1) After adding a surjective pairing to $\boldsymbol{\lambda}_{\to}^0$ the resulting system λ_{SP} becomes Hilbert–Post complete. This means that an equation between terms is either provable or inconsistent.
(2) Every recursively enumerable set $\mathcal{X}$ of terms that is closed under provable equality is Diophantine, i.e. satisfies for some terms F, G

$$M \in \mathcal{X} \Leftrightarrow \exists N\, FMN = GMN.$$

Both results will be proved by introducing Cartesian monoids and studying freely generated ones.

The system λ_{SP}

Inspired by the notion of a surjective pairing we define λ_{SP} as an extension of the simply typed lambda calculus $\boldsymbol{\lambda}^0_{\to}$.

5.2.3 Definition

(i) The set of *types of* λ_{SP} is simply $\mathbb{T}^0$.

(ii) The *terms of* λ_{SP}, written Λ_{SP} (or $\Lambda_{\text{SP}}(A)$ for terms of a certain type A or $\Lambda^{\emptyset}$, $\Lambda^{\emptyset}_{\text{SP}}(A)$ for closed terms), are obtained from $\boldsymbol{\lambda}^0_{\to}$ by adding to the formation of terms the constants $\pi : 1_2 = 0^2{\to}0$, $\pi_1 : 1$, $\pi_2 : 1$.

(iii) Equality for λ_{SP} is axiomatized by $\boldsymbol{\beta}$, η and the following scheme. For all $M, M_1, M_2 : 0$

$$\begin{aligned} \pi_i(\pi M_1 M_2) &= M_i; \\ \pi(\pi_1 M)(\pi_2 M) &= M. \end{aligned}$$

(iv) A notion of reduction SP (surjective pairing) is introduced on λ_{SP}-terms by the following contraction rules: for all $M, M_1, M_2 : 0$

$$\begin{aligned} \pi_i(\pi M_1 M_2) &\to M_i; \\ \pi(\pi_1 M)(\pi_2 M) &\to M. \end{aligned}$$

Usually we will consider SP in combination with $\boldsymbol{\beta\eta}$, obtaining $\boldsymbol{\beta\eta}$SP.

According to a well-known result in Klop (1980), reduction coming from surjective pairing in untyped lambda calculus is not confluent (i.e. does not satisfy the Church–Rosser property). This gave rise to the notion of *left-linearity* in term rewriting, see Terese (2003). We will see below, Proposition 5.2.10, that in the present typed case the situation is different.

5.2.4 Theorem *The conversion relation $=_{\boldsymbol{\beta\eta}\text{SP}}$, generated by the notion of reduction $\boldsymbol{\beta\eta}$SP, coincides with that of the theory λ_{SP}.*

Proof As usual. ∎

For objects of higher type, pairing can be defined in terms of π, π_1, π_2 as follows.

5.2.5 Definition For every type $A \in \mathbb{T}$ we define $\pi^A : A{\to}A{\to}A, \pi_i : A{\to}A$

by

$$\begin{aligned}
\pi^0 &\equiv \pi;\\
\pi_i^0 &\equiv \pi_i;\\
\pi^{A\to B} &\equiv \lambda xy{:}(A{\to}B)\lambda z{:}A.\pi^B(xz)(yz);\\
\pi_i^{A\to B} &\equiv \lambda x{:}(A{\to}B)\lambda z{:}A.\pi_i^B(xz);
\end{aligned}$$

cf. the construction in Proposition 1.4.21.

Sometimes we may suppress type annotations in π^A, π_1^A, π_2^A, but the types can always and unambiguously be reconstructed from the context.

The defined constants for higher type also form a surjective pairing, as can easily be shown.

5.2.6 Proposition *Let $\pi = \pi^A, \pi_i = \pi_i^A$. Then for $M, M_1, M_2 \in \Lambda_{\mathrm{SP}}(A)$*

$$\begin{aligned}
\pi(\pi_1 M)(\pi_2 M) &\twoheadrightarrow_{\beta\eta\mathrm{SP}} M;\\
\pi_i(\pi M_1 M_2) &\twoheadrightarrow_{\beta\eta\mathrm{SP}} M_i, \quad (i = 1, 2).
\end{aligned}$$

Proof By induction on the type A. ■

Note that the above reductions may involve more than one step, typically additional $\beta\eta$-steps.

Inspired by Remark 5.2.2 one obtains the following.

5.2.7 Proposition *Let $A \in \mathbb{T}^0$. Then there exist $\pi^{A,n} : \Lambda^{\emptyset}_{\mathrm{SP}}(A^n \to A)$, and $\pi_i^{A,n} : \Lambda^{\emptyset}_{\mathrm{SP}}(A \to A)$, $1 \le i \le n$, such that*

$$\begin{aligned}
\pi_i^{A,n}(\pi^{A,n} M_1 \dots M_n) &\twoheadrightarrow_{\beta\eta\mathrm{SP}} M_i, \qquad 1 \le i \le n,\\
\pi^{A,n}(\pi_1^{A,n} M) \cdots (\pi_n^{A,n} M) &\twoheadrightarrow_{\beta\eta\mathrm{SP}} M.
\end{aligned}$$

The original π, π_1, π_2 can be called $\pi^{0,2}, \pi_1^{0,2}, \pi_2^{0,2}$.

Now we will show that the notion of reduction $\beta\eta$SP is confluent.

5.2.8 Lemma *The notion of reduction $\beta\eta$SP satisfies* WCR.

Proof One can prove it by the critical pair lemma of Mayr and Nipkow (1998). But a simpler argument is possible, since SP reductions only reduce to terms that already did exist, and hence cannot create any redexes. ■

5.2.9 Lemma

(i) *The notion of reduction* SP *is* SN *(strongly normalizing).*
(ii) *If $M \twoheadrightarrow_{\beta\eta\mathrm{SP}} N$, then there exists P such that $M \twoheadrightarrow_{\beta\eta} P \twoheadrightarrow_{\mathrm{SP}} N$.*
(iii) *The notion of reduction $\beta\eta$SP is* SN.

Proof (i) Since SP-reductions are strictly decreasing.

(ii) Show $M \to_{\mathrm{SP}} L \to_{\beta\eta} N \Rightarrow \exists L'\, M \twoheadrightarrow_{\beta\eta} L' \twoheadrightarrow_{\beta\eta\mathrm{SP}} N$. Then (ii) follows by a staircase diagram chase.

(iii) By (i), the fact that $\beta\eta$ is SN, and a staircase diagram chase, possible by (ii). ∎

Now we show that the notion of reduction $\beta\eta$SP is confluent, in spite of being not left-linear.

5.2.10 Proposition *The notation of reduction $\beta\eta$SP is confluent.*

Proof By Lemma 5.2.9(iii) and Newman's Lemma 5.3.8. ∎

5.2.11 Definition

(i) An SP-*retraction pair from A to B* is a pair of terms $M{:}A{\to}B$ and $N{:}B{\to}A$ such that $N \circ M =_{\beta\eta\mathrm{SP}} \mathsf{I}_A$.
(ii) We say A is a SP-*retract* of B, written $A \lhd_{\mathrm{SP}} B$, if there is an SP-retraction pair from A to B.

For a proof of the following result see Exercise 5.2.

5.2.12 Proposition *Define types N_n as follows. $N_0 \equiv 0$ and $N_{n+1} \equiv N_n{\to}N_n$. Then for every type A, one has $A \lhd_{\mathrm{SP}} N_{\mathrm{rk}(A)}$.*

Cartesian monoids

We start with the definition of a Cartesian monoid, introduced in Scott (1980) and, independently, in Lambek (1980).

5.2.13 Definition

(i) A *Cartesian monoid* is a structure

$$\mathcal{C} = \langle \mathcal{M}, *, I, L, R, \langle\cdot,\cdot\rangle\rangle$$

such that $(\mathcal{M}, *, I)$ is a monoid ($*$ is associative and I is a two sided unit), $L, R \in \mathcal{M}$ and $\langle\cdot,\cdot\rangle : \mathcal{M}^2 \to \mathcal{M}$ and satisfy, for all $x, y, z \in \mathcal{M}$,

$$\begin{aligned} L * \langle x, y\rangle &= x \\ R * \langle x, y\rangle &= y \\ \langle x, y\rangle * z &= \langle x * z, y * z\rangle \\ \langle L, R\rangle &= I. \end{aligned}$$

(ii) We say $\mathcal{M}$ is *trivial* if $L = R$.

(iii) A map $f : \mathcal{M} \to \mathcal{M}'$ is a *morphism* if

$$\begin{aligned} f(m * n) &= f(m) * f(n); \\ f(\langle m, n \rangle) &= \langle f(m), f(n) \rangle, \\ f(L) &= L', \\ f(R) &= R'. \end{aligned}$$

Then automatically one has $f(I) = I'$.

Note that if $\mathcal{M}$ is trivial, then it consists of only one element: for all $x, y \in \mathcal{M}$

$$x = L * \langle x, y \rangle = R * \langle x, y \rangle = y.$$

5.2.14 Lemma *The last axiom of the Cartesian monoids can be replaced equivalently by the surjectivity of the pairing:*

$$\langle L * x, R * x \rangle = x.$$

Proof First suppose $\langle L, R \rangle = I$. Then $\langle L*x, R*x \rangle = \langle L, R \rangle * x = I * x = x$. Conversely suppose $\langle L*x, R*x \rangle = x$, for all x. Then $\langle L, R \rangle = \langle L*I, R*I \rangle = I$. ∎

5.2.15 Lemma *Let $\mathcal{M}$ be a Cartesian monoid. Then for all $x, y \in \mathcal{M}$*

$$L * x = L * y \;\&\; R * x = R * y \;\Rightarrow\; x = y.$$

Proof We have $x = \langle L * x, R * x \rangle = \langle L * y, R * y \rangle = y$. ∎

A first example of a Cartesian monoid has as carrier set the closed $\boldsymbol{\beta\eta}$SP-terms of type $1 = 0 \to 0$.

5.2.16 Definition Write for $M, N \in \Lambda^{\varnothing}_{\mathrm{SP}}(1)$:

$$\begin{aligned} \langle M, N \rangle &\equiv \pi^1 M N; \\ M \circ N &\equiv \lambda x{:}0.M(Nx); \\ \mathsf{I} &\equiv \lambda x{:}0.x; \\ \mathsf{L} &\equiv \pi^0_1; \\ \mathsf{R} &\equiv \pi^0_2. \end{aligned}$$

Define

$$\mathcal{C}^0 = \langle \Lambda^{\varnothing}_{SP}(1) / =_{\boldsymbol{\beta\eta}\mathrm{SP}}, \circ, \mathsf{I}, \mathsf{L}, \mathsf{R}, \langle \cdot, \cdot \rangle \rangle.$$

The reason for calling this structure $\mathcal{C}^0$ and not $\mathcal{C}^1$ is that we will generalize it to $\mathcal{C}^n$ based on terms of the type $1^n \to 1$.

5.2.17 Proposition *The structure $\mathcal{C}^0$ is a non-trivial Cartesian monoid.*

Proof For $x, y, z{:}1$ the following equations are valid in λ_{SP}:

$$\begin{aligned}
\mathsf{I} \circ x &= x;\\
x \circ \mathsf{I} &= x;\\
\mathsf{L} \circ \langle x, y\rangle &= x;\\
\mathsf{R} \circ \langle x, y\rangle &= y;\\
\langle x, y\rangle \circ z &= \langle x \circ z, y \circ z\rangle;\\
\langle \mathsf{L}, \mathsf{R}\rangle &= \mathsf{I}. \blacksquare
\end{aligned}$$

The third equation is intuitively right, if we remember that the pairing on type 1 is lifted pointwise from a pairing on type 0; that is, $\langle f, g\rangle = \lambda x.\pi(fx)(gx)$.

5.2.18 Example Let $[\cdot,\cdot]$ be any surjective pairing of natural numbers, with left and right projections $l, r : \omega{\to}\omega$. For example, we can take Cantor's well-known bijection[1] from ω^2 to ω. We can lift the pairing function to the level of functions by putting $\langle f, g\rangle(x) = [f(x), g(x)]$ for all $x \in \omega$. Let I be the identity function and let $\circ$ denote function composition. Then

$$\mathcal{N}^1 = \langle \omega{\to}\omega, I, \circ, l, r, \langle\cdot,\cdot\rangle\rangle.$$

is a non-trivial Cartesian monoid.

Next we will show that the equalities in the theory of Cartesian monoids are generated by a confluent rewriting system.

5.2.19 Definition

(i) Let $\mathcal{T}_{\mathrm{CM}}$ be the terms in the signature of Cartesian monoids, i.e. built up from constants $\{I, L, R\}$ and variables, using the binary constructors $\langle -, -\rangle$ and $*$.

(ii) Sometimes we need to be explicit about which variables we use and set $\mathcal{T}^n_{\mathrm{CM}}$ equal to the terms generated from $\{I, L, R\}$ and variables $x_1, \ldots, x_n$, using $\langle -, -\rangle$ and $*$. In particular $\mathcal{T}^0_{\mathrm{CM}}$ consists of the closed such terms, without variables.

(iii) Consider the notion of reduction CM on $\mathcal{T}_{\mathrm{CM}}$, giving rise to the reduction relations $\to_{\mathrm{CM}}$ and its transitive reflexive closure $\twoheadrightarrow_{\mathrm{CM}}$, introduced

[1] A variant of this function is used in Section 5.3 as a non-surjective pairing function $[x, y] + 1$, such that, deliberately, 0 does not encode a pair. This variant is specified in detail and explained in Figure 5.4.

by the contraction rules

$$\begin{aligned} L * \langle M, N\rangle &\to M \\ R * \langle M, N\rangle &\to N \\ \langle M, N\rangle * T &\to \langle M * T, N * T\rangle \\ \langle L, R\rangle &\to I \\ \langle L * M, R * M\rangle &\to M \\ I * M &\to M \\ M * I &\to M, \end{aligned}$$

modulo the associativity axioms (i.e. the terms $M * (N * L)$ and $(M * N) * L$ are considered to be the same), see Terese (2003). The following result is mentioned in Curien (1993).

5.2.20 Proposition

(i) *The notion of reduction* CM *is* WCR.
(ii) *The notion of reduction* CM *is* SN.
(iii) *The notion of reduction* CM *is* CR.

Proof (i) Examine all critical pairs. Modulo associativity there are many such pairs, but they all converge. Consider, as an example, the following reductions:

$$x * z \leftarrow (L * \langle x, y\rangle) * z = L * (\langle x, y\rangle) * z) \to L * \langle x * z, y * z\rangle \to x * z.$$

(ii) Interpret the theory of Cartesian monoids as integers by putting

$$\begin{aligned} [\![x]\!] &= 2; \\ [\![e]\!] &= 2, && \text{if } e \text{ is } L, R \text{ or } I; \\ [\![e_1 * e_2]\!] &= [\![e_1]\!].[\![e_2]\!]; \\ [\![\langle e_1, e_2\rangle]\!] &= [\![e_1]\!] + [\![e_2]\!] + 1. \end{aligned}$$

Then $[\![\cdot]\!]$ preserves associativity and

$$e \to_{\text{CM}} e' \Rightarrow [\![e]\!] > [\![e']\!].$$

Therefore the notion of reduction CM is SN.
(iii) By (i), (ii) and Newman's Lemma 5.3.8. ∎

Closed terms in CM-nf can be represented as binary trees with strings of

L, R (the empty string becomes I) at the leaves. For example

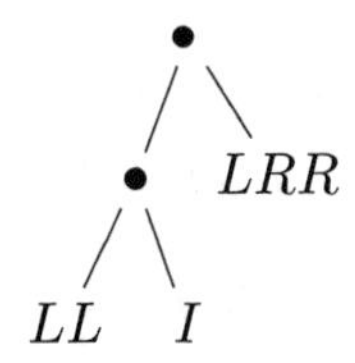

represents $\langle\langle L * L, I\rangle, L * R * R\rangle$. In such trees the subtree corresponding to $\langle L, R\rangle$ will not occur, since this term reduces to I.

The free Cartesian monoids $\mathcal{F}[x_1, \ldots, x_n]$

5.2.21 Definition

(i) The closed term model of the theory of Cartesian monoids consists of $\mathcal{T}^0_{\mathrm{CM}}$ modulo $=_{\mathrm{CM}}$ and is denoted by $\mathcal{F}$. It is the *free Cartesian monoid* with no generators.

(ii) The *free Cartesian monoid* over the generators $\vec{x} = x_1, \ldots, x_n$, written $\mathcal{F}[\vec{x}]$, is $\mathcal{T}^n_{\mathrm{CM}}$ modulo $=_{\mathcal{M}}$.

5.2.22 Proposition

(i) *For all $a, b \in \mathcal{F}$ one has*

$$a \neq b \Rightarrow \exists c, d \in \mathcal{F}\ [c * a * d = L \ \&\ c * b * d = R].$$

(ii) *The model $\mathcal{F}$ is simple: every homomorphism $g : \mathcal{F} \to \mathcal{M}$ to a non-trivial Cartesian monoid $\mathcal{M}$ is injective.*

Proof (i) We can assume that a, b are in normal form. Seen as trees (not looking at the words over $\{L, R\}$ at the leaves), the a, b can be made congruent by expansions of the form $x \leftarrow \langle L * x, R * x\rangle$. These expanded trees are distinct in some leaf, which can be reached by a string of Ls and Rs joined by $*$. Thus there is a string, say c, such that $c * a \neq c * b$ and both of these reduce to $\langle\ \rangle$-free strings of Ls and Rs joined by $*$. We can also assume that neither of these strings is a suffix of the other, since c could be replaced by $L * c$ or $R * c$ (depending on an R or an L just before the suffix). Thus there are $\langle\ \rangle$-free a', b' and integers k, l such that

$$c * a * \langle I, I\rangle^k * \langle R, L\rangle^l = a' * L \quad \text{and}$$
$$c * b * \langle I, I\rangle^k * \langle R, L\rangle^l = b' * R$$

and there exist integers n and m, which are the length of a' and of b' respectively, such that

$$a' * L * \langle\langle I, I\rangle^n * L, \langle I, I\rangle^m * R\rangle = L \quad \text{and}$$
$$b' * R * \langle\langle I, I\rangle^n * L, \langle I, I\rangle^m * R\rangle = R.$$

Therefore we can set $d = \langle I, I\rangle^k * \langle R, L\rangle^l * \langle\langle I, I\rangle^n * L, \langle I, I\rangle^m * R\rangle$.

(ii) By (i) and the fact that $\mathcal{M}$ is non-trivial. ■

Finite generation of $\mathcal{F}[x_1, \ldots, x_n]$

Next we will show that $\mathcal{F}[x_1, \ldots, x_n]$ is finitely generated as a monoid, i.e. from finitely many of its elements using the operation $*$ only.

5.2.23 Notation In a monoid $\mathcal{M}$ we define list-like left-associative and right-associative iterated $\langle\ \rangle$-expressions of length > 0 as follows. Let the elements of $\vec{x}$ range over $\mathcal{M}$. Then

$$\begin{array}{rcll} \langle\langle x\rangle & \equiv & x; & \\ \langle\langle x_1, \ldots, x_{n+1}\rangle & \equiv & \langle\langle\langle x_1, \ldots, x_n\rangle, x_{n+1}\rangle, & n > 0; \\ \langle x\rangle\rangle & \equiv & x; & \\ \langle x_1, \ldots, x_{n+1}\rangle\rangle & \equiv & \langle x_1, \langle x_2, \ldots, x_{n+1}\rangle\rangle\rangle, & n > 0. \end{array}$$

5.2.24 Definition

(i) For $\mathcal{H} \subseteq \mathcal{F}$ let $[\mathcal{H}]$ be the submonoid of $\mathcal{F}$ generated by $\mathcal{H}$ using the operation $*$.

(ii) Define the finite subset $\mathcal{G} \subseteq \mathcal{F}$ by

$$\mathcal{G} = \{\langle X * L, Y * L * R, Z * R * R\rangle\rangle \mid X, Y, Z \in \{L, R, I\}\} \cup \{\langle I, I, I\rangle\rangle\}.$$

We will show that $[\mathcal{G}] = \mathcal{F}$.

5.2.25 Lemma *Define a* string *to be an expression of the form $X_1 * \cdots * X_n$, with $X_i \in \{L, R, I\}$. Then for all strings s, s_1, s_2, s_3 one has:*

(i) $\langle s_1, s_2, s_3\rangle\rangle \in [\mathcal{G}]$;

(ii) $s \in [\mathcal{G}]$.

Proof (i) Note that

$$\langle X * L, Y * L * R, Z * R * R\rangle\rangle * \langle s_1, s_2, s_3\rangle\rangle = \langle X * s_1, Y * s_2, Z * s_3\rangle\rangle.$$

Hence, starting from $\langle I, I, I\rangle\rangle \in \mathcal{G}$ every triple of strings can be generated because the X, Y, Z range over $\{L, R, I\}$.

(ii) Notice that

$$\begin{aligned} s &= \langle L, R\rangle * s \\ &= \langle L * s, R * s\rangle \\ &= \langle L * s, \langle L, R\rangle * R * s\rangle \\ &= \langle L * s, L * R * s, R * R * s\rangle\!\rangle, \end{aligned}$$

which is in $[\mathcal{G}]$ by (i). ■

5.2.26 Lemma *Let $e_1, \dots, e_n \in \mathcal{F}$. Suppose $\langle\!\langle e_1, \dots, e_n\rangle \in [\mathcal{G}]$. Then*

(i) *$e_i \in [\mathcal{G}]$, for $1 \le i \le n$.*
(ii) *$\langle\!\langle e_1, \dots, e_n, \langle e_i, e_j\rangle\rangle \in [\mathcal{G}]$ for $0 \le i, j \le n$.*
(iii) *$\langle\!\langle e_1, \dots, e_n, X * e_i\rangle \in [\mathcal{G}]$ for $X \in \{L, R, I\}$.*

Proof (i) By Lemma 5.2.25(ii) one has $F_1 \equiv L^{(n-1)} \in [\mathcal{G}]$ and $F_i \equiv R * L^{(n-i)} \in [\mathcal{G}]$. Hence

$$\begin{aligned} e_1 &= F_1 * \langle\!\langle e_1, \dots, e_n\rangle \in [\mathcal{G}]; \\ e_i &= F_i * \langle\!\langle e_1, \dots, e_n\rangle \in [\mathcal{G}], \qquad \text{for } i = 2, \dots, n. \end{aligned}$$

(ii) By Lemma 5.2.25(i) one has $\langle I, \langle F_i, F_j\rangle\rangle = \langle I, F_i, F_j\rangle\!\rangle \in [\mathcal{G}]$. Hence

$$\langle\!\langle e_1, \dots, e_n, \langle e_i, e_j\rangle\rangle = \langle I, \langle F_i, F_j\rangle\rangle * \langle\!\langle e_1, \dots, e_n\rangle \in [\mathcal{G}].$$

(iii) Similarly $\langle\!\langle e_1, \dots, e_n, X * e_i\rangle = \langle I, X * F_i\rangle * \langle\!\langle e_1, \dots, e_n\rangle \in [\mathcal{G}]$. ■

5.2.27 Theorem *As a monoid, $\mathcal{F}$ is finitely generated. In fact $\mathcal{F} = [\mathcal{G}]$.*

Proof We have $e \in \mathcal{F}$ iff there is a sequence $e_1 \equiv L, e_2 \equiv R, e_3 \equiv I, \dots, e_n \equiv e$ such that for each $4 \le k \le n$ there are $i, j < k$ such that $e_k \equiv \langle e_i, e_j\rangle$ or $e_k \equiv X * e_i$, with $X \in \{L, R, I\}$.

By Lemma 5.2.25(i) we have $\langle\!\langle e_1, e_2, e_3\rangle \in [\mathcal{G}]$. By Lemma 5.2.26(ii), (iii) it follows that

$$\langle\!\langle e_1, e_2, e_3, \dots, e_n\rangle \in [\mathcal{G}].$$

Therefore by (i) of that lemma $e \equiv e_n \in [\mathcal{G}]$. ■

The following corollary is similar to a result of Böhm, who showed that the monoid of untyped lambda terms has two generators, see B[1984].

5.2.28 Corollary

(i) *Let $\mathcal{M}$ be a finitely generated Cartesian monoid. Then $\mathcal{M}$ is generated by two of its elements.*
(ii) *The monoid $\mathcal{F}[x_1, \dots, x_n]$ is generated by two elements.*

Proof (i) Let $\mathcal{G} = \{g_1, \ldots, g_n\}$ be the set of generators of $\mathcal{M}$. Then $\mathcal{G}$ and hence $\mathcal{M}$ is generated by R and $\langle\langle g_1, \ldots, g_n, L\rangle$.

(ii) $\mathcal{F}[\vec{x}]$ is generated by $\mathcal{G}$ and the $\vec{x}$, hence by (i) by two elements. ■

Invertibility in $\mathcal{F}$

5.2.29 Definition

(i) Denote by $\mathcal{L}$ ($\mathcal{R}$) the submonoid of the right (left) invertible elements of $\mathcal{F}$:

$$\mathcal{L} = \{a \in \mathcal{F} \mid \exists b \in \mathcal{F}\ b * a = I\};$$
$$\mathcal{R} = \{a \in \mathcal{F} \mid \exists b \in \mathcal{F}\ a * b = I\}.$$

(ii) Denote by $\mathcal{I}$ the subgroup of $\mathcal{F}$ consisting of invertible elements:

$$\mathcal{I} = \{a \in \mathcal{F} \mid \exists b \in \mathcal{F}\ a * b = b * a = I\}.$$

It is easy to see that $\mathcal{I} = \mathcal{L} \cap \mathcal{R}$. Indeed, if $a \in \mathcal{L} \cap \mathcal{R}$, then there are $b, b' \in \mathcal{F}$ such that $b * a = I = a * b'$. But then $b = b * a * b' = b'$, so $a \in \mathcal{I}$. The converse is trivial.

5.2.30 Examples

(i) $L, R \in \mathcal{R}$, since both have the right inverse $\langle I, I\rangle$.

(ii) The element $a = \langle\langle R, L\rangle, L\rangle$ having as 'tree'

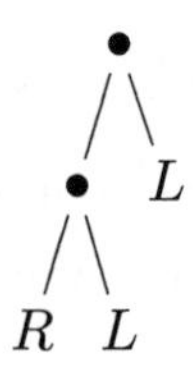

has as left inverse $b = \langle R, LL\rangle$, where we do not write the $*$ in strings.

(iii) The element 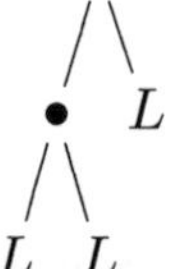 has no left inverse, since 'R cannot be obtained'.

(iv) The element $a = \langle\langle RL, LL\rangle, RR\rangle$ having the following tree:

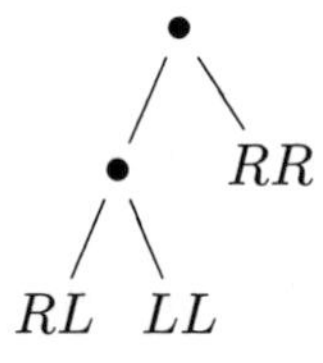

has the following right inverse: $b = \langle\langle RL, LL\rangle, \langle c, R\rangle\rangle$. Indeed

$$a * b = \langle\langle RLb, LLb\rangle, RRb\rangle = \langle\langle LL, RL\rangle, R\rangle = \langle L, R\rangle = I.$$

(v) The element 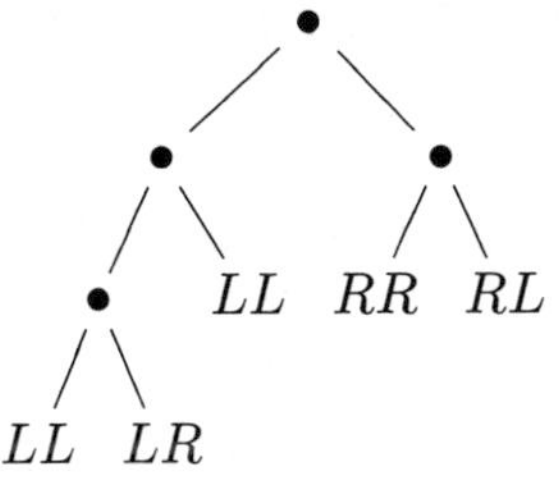 has no right inverse, as 'LL occurs twice'.

(vi) The element 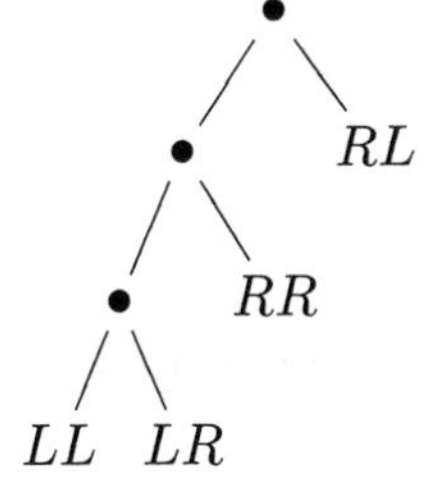 has a two-sided inverse, as 'all strings of two letters' occur exactly once, the inverse being 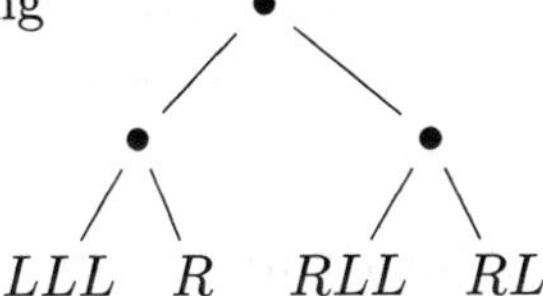 .

For normal forms $f \in \mathcal{F}$ we have the following characterizations.

5.2.31 Proposition

(i) *The normal form f has a right inverse if and only if f can be expanded (by replacing x by $\langle Lx, Rx\rangle$) so that all of its strings at the leaves have the same length and none occurs more than once.*

(ii) *The normal form f has a left inverse if and only if f can be expanded so that all of its strings at the leaves have the same length, say n, and each of the possible 2^n strings of this length actually occurs.*

(iii) *The normal form f is doubly invertible if and only if f can be expanded so that all of its strings at the leaves have the same length, say n, and each of the possible 2^n strings of this length occurs exactly once.*

Proof This is clear from the examples. ■

The following terms are instrumental for generating $\mathcal{I}$ and $\mathcal{R}$.

5.2.32 Definition Put

$$\begin{aligned} B_n &= \langle LR^0, \ldots, LR^{n-1}, LLR^n, RLR^n, RR^n \rangle\!\rangle; \\ C_0 &= \langle R, L \rangle, \\ C_{n+1} &= \langle LR^0, \ldots, LR^{n-1}, LRR^n, LR^n, RRR^n \rangle\!\rangle. \end{aligned}$$

5.2.33 Proposition

(i) *$\mathcal{I}$ is the subgroup of $\mathcal{F}$ generated (using $*$ and $^{-1}$) by*

$$\{B_n \mid n \in \omega\} \cup \{C_n \mid n \in \omega\}.$$

(ii) *$\mathcal{R} = [\{L\} \cup \mathcal{I}] = [\{R\} \cup \mathcal{I}]$, where $[\ ]$ is defined in Definition* 5.2.24.

Proof (i) In fact $\mathcal{I} = [\{B_0, B_0^{-1}, B_1, B_1^{-1}, C_0, C_1\}]$. Here $[\mathcal{H}]$ is the subset generated from $\mathcal{H}$ using only $*$. See Exercise 5.15.

(ii) By Proposition 5.2.31. ■

5.2.34 Remark

(i) The B_n alone generate the so-called *Thompson–Freyd–Heller group*, see Exercise 5.14(iv).

(ii) A related group consisting of λ-terms is $G(\boldsymbol{\lambda\eta})$ consisting of invertible closed untyped lambda terms modulo $\boldsymbol{\beta\eta}$-conversion, see B84, Section 21.3.

5.2.35 Proposition *If $f(\vec{x})$ and $g(\vec{x})$ are distinct members of $\mathcal{F}[\vec{x}]$, then there exists $\vec{h} \in \mathcal{F}$ such that $f(\vec{h}) \neq g(\vec{h})$. We say that $\mathcal{F}[\vec{x}]$ is* separable.

Proof Suppose that $f(\vec{x})$ and $g(\vec{x})$ are distinct normal members of $\mathcal{F}[\vec{x}]$. We shall find $\vec{h}$ such that $f(\vec{h}) \neq g(\vec{h})$. First transform subexpressions of the form $L \circ x_i \circ k$ and $R \circ x_j \circ k$ by substituting $\langle y, z \rangle$ for x_i, x_j and renormalizing. This process terminates, and is invertible as we can see by substituting $L * x_i$ for y and $R * x_j$ for z. Thus we can assume that $f(\vec{x})$ and $g(\vec{x})$ are distinct normal and without subexpressions of the two forms above. Indeed, such $f(\vec{x})$ and $g(\vec{x})$ can be recursively generated as a string of x_is followed by a string of Ls and Rs, or as a string of x_is followed by a single $\langle\ \rangle$ of expressions of the same form. Let m be a large number relative to $f(\vec{x}), g(\vec{x})$ ($> \#f(\vec{x}), \#g(\vec{x})$, where $\#t$ is the number of symbols in t.) For each positive integer i, with $1 \leq i \leq n$, set

$$h_i = \langle \langle R^m, \ldots, R^m, I \rangle\!\rangle, R^m \rangle$$

where the right-associative $\langle R^m, \ldots, R^m, I \rangle\!\rangle$-expression contains i occurrences of R^m. We claim that both $f(\vec{x})$ and $g(\vec{x})$ can be reconstructed from the normal forms of $f(\vec{h})$ and $g(\vec{h})$, so that $f(\vec{h}) \neq g(\vec{h})$.

Define $d_r(t)$, for a normal $t \in \mathcal{F}$, as follows:

$$\begin{aligned} d_r(\vec{w}) &= 0, && \text{if } \vec{w} \text{ is a string of } L\text{s and } R\text{s}; \\ d_r(\langle t, s\rangle) &= d_r(s) + 1. \end{aligned}$$

Note that if t is a normal member of $\mathcal{F}$ and $d_r(t) < m$, then

$$h_i * t =_{\text{CM}} \langle\langle t', \ldots, t', t\rangle\rangle, t'\rangle,$$

where $t' \equiv R^m t$ is $\langle\ \rangle$-free. Also note that if s is the CM-nf of $h_i * t$, then $d_r(s) = 1$. The normal form of, say, $f(\vec{h})$ can be computed recursively bottom up as in the computation of the normal form of $h_i * t$ above. In order to compute back $f(\vec{x})$ we consider several examples:

$$\begin{aligned} f_1(\vec{x}) &= x_3 R; \\ f_2(\vec{x}) &= \langle\langle R^2, R^2, R^2, R\rangle\rangle, R^2\rangle; \\ f_3(\vec{x}) &= x_2\langle R, R, L\rangle\rangle; \\ f_4(\vec{x}) &= x_3 x_1 x_2 R; \\ f_5(\vec{x}) &= x_3 x_1 \langle x_2 R, R\rangle. \end{aligned}$$

Then $f_1(\vec{h}), \ldots, f_5(\vec{h})$ have as trees respectively

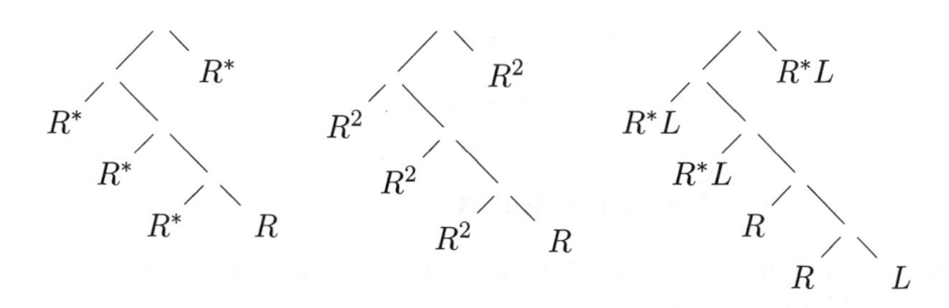

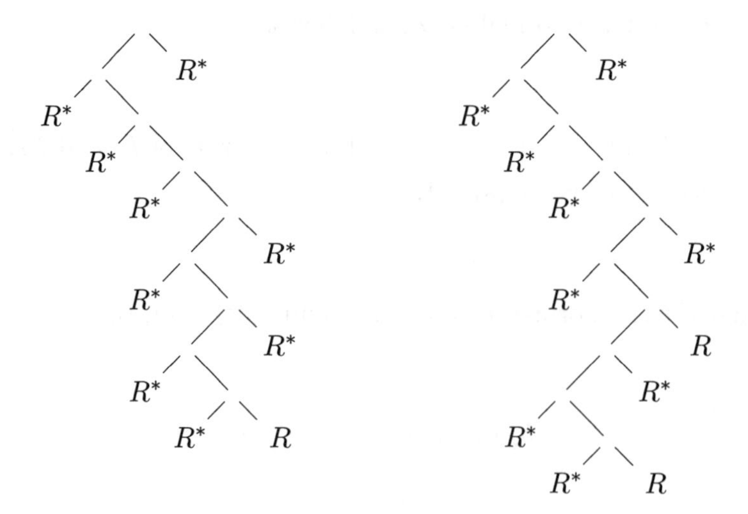

In these trees the R^* denote long sequences of Rs of possibly different lengths. ∎

Cartesian monoids inside λ_{SP}

Recall that $\mathcal{C}^0 = \langle \Lambda^{\emptyset}_{SP}(1)/{=_{\beta\eta\text{SP}}}, \circ, \mathsf{I}, \mathsf{L}, \mathsf{R}, \langle\cdot,\cdot\rangle\rangle$.

5.2.36 Proposition *There is a surjective homomorphism $h : \mathcal{F} \to \mathcal{C}^0$.*

Proof If $M : 1$ is a closed term and in long $\beta\eta$SP normal form, then M has one of the following shapes: $\lambda a.a$, $\lambda a.\pi X_1 X_2$, or $\lambda a.\pi_i X$ for $i = 1$ or $i = 2$. Then we have $M \equiv \mathsf{I}$, $M = \langle \lambda a.X_1, \lambda a.X_2\rangle$, $M = \mathsf{L} \circ (\lambda a.X)$ or $M = \mathsf{R} \circ (\lambda a.X)$, respectively. Since the terms $\lambda a.X_i$ are smaller than M, this yields an inductive definition of the set of closed terms of λ_{SP} modulo $=$ in terms of the combinators $\mathsf{I}, \mathsf{L}, \mathsf{R}, \langle\ \rangle, \circ$. Thus the elements of $\mathcal{C}^0$ are generated from $\{\mathsf{I}, \circ, \mathsf{L}, \mathsf{R}, \langle\cdot,\cdot\rangle\}$ in an algebraic way. Now define

$$\begin{aligned} h(I) &= \mathsf{I}; \\ h(L) &= \mathsf{L}; \\ h(R) &= \mathsf{R}; \\ h(\langle a, b\rangle) &= \langle h(a), h(b)\rangle; \\ h(a * b) &= h(a) \circ h(b). \end{aligned}$$

Then h is a surjective homomorphism. ∎

Now we will show in two different ways that this homomorphism is in fact injective and hence an isomorphism.

5.2.37 Theorem $\mathcal{F} \cong \mathcal{C}^0$.

First Proof We will show that the homomorphism h in Proposition 5.2.36 is injective. By a careful examination of CM-normal forms one can see the following. Each expression can be rewritten uniquely as a binary tree whose nodes correspond to applications of $\langle\cdot,\cdot\rangle$ with strings of Ls and Rs joined by $*$ at its leaves (here I counts as the empty string) and with no subexpressions of the form $\langle L * e, R * e\rangle$. Thus

$$a \neq b \Rightarrow a^{\mathtt{nf}} \not\equiv b^{\mathtt{nf}} \Rightarrow h(a^{\mathtt{nf}}) \neq h(b^{\mathtt{nf}}) \Rightarrow h(a) \neq h(b),$$

so h is injective. ■

Second Proof By Proposition 5.2.22. ■

The structure $\mathcal{C}^0$ will be generalized as follows.

5.2.38 Definition Consider the type $1^n{\to}1 = (0{\to}0)^n{\to}0{\to}0$. Define

$$\mathcal{C}^n = \langle \Lambda^{\emptyset}_{\mathrm{SP}}(1^n{\to}1)/=_{\boldsymbol{\beta\eta}\mathrm{SP}}, \mathsf{I}_n, \mathsf{L}_n, \mathsf{R}_n, \circ_n, \langle -,-\rangle_n\rangle,$$

where, writing $\vec{x} = x_1, \ldots, x_n{:}1$,

$$\begin{aligned}
\langle M, N\rangle_n &= \lambda\vec{x}.\langle M\vec{x}, N\vec{x}\rangle;\\
M \circ_n N &= \lambda\vec{x}.(M\vec{x}) \circ (N\vec{x});\\
\mathsf{I}_n &= \lambda\vec{x}.\mathsf{I};\\
\mathsf{L}_n &= \lambda\vec{x}.\mathsf{L};\\
\mathsf{R}_n &= \lambda\vec{x}.\mathsf{R}.
\end{aligned}$$

5.2.39 Proposition *The structure $\mathcal{C}^n$ is a non-trivial Cartesian monoid.*

Proof Easy. ■

5.2.40 Proposition *One has $\mathcal{C}^n \cong \mathcal{F}[x_1, \ldots, x_n]$.*

Proof As before, let $h_n : \mathcal{F}[\vec{x}]{\to}\mathcal{C}^n$ be induced by

$$\begin{array}{rclcl}
h_n(x_i) &=& \lambda\vec{x}\lambda z{:}0.x_i z &=& \lambda\vec{x}.x_i;\\
h_n(I) &=& \lambda\vec{x}\lambda z{:}0.z &=& \mathsf{I}_n;\\
h_n(L) &=& \lambda\vec{x}\lambda z{:}0.\pi_1 z &=& \mathsf{L}_n;\\
h_n(R) &=& \lambda\vec{x}\lambda z{:}0.\pi_2 z &=& \mathsf{R}_n;\\
h_n(\langle s,t\rangle) &=& \lambda\vec{x}\lambda z{:}0.\pi(s\vec{x}z)(t\vec{x}z) &=& \langle h_n(s), h_n(t)\rangle_n.
\end{array}$$

As before one can show that this is an isomorphism. ■

In what follows an important case is $n = 1$, i.e. $\mathcal{C}^{1\to1} \cong \mathcal{F}[x]$.

Hilbert–Post completeness of $\boldsymbol{\lambda}_{\rightarrow}SP$

The claim that an equation $M = N$ is either a $\boldsymbol{\beta\eta}$SP convertibility or inconsistent is proved in two steps. First it is proved for the type $1{\rightarrow}1$ by the analysis of $\mathcal{F}[x]$; then it follows for arbitrary types by reducibility of types in λ_{SP}.

Remember that $M\#_T N$ means that $T \cup \{M = N\}$ is inconsistent.

5.2.41 Proposition

(i) *Let* $M, N \in \Lambda^{\emptyset}_{\text{SP}}(1)$. *Then*

$$M \neq_{\boldsymbol{\beta\eta}\text{SP}} N \;\Rightarrow\; M\#_{\boldsymbol{\beta\eta}\text{SP}}N.$$

(ii) *The same holds for* $M, N \in \Lambda^{\emptyset}_{\text{SP}}(1{\rightarrow}1)$.

Proof (i) Since $\mathcal{F} \cong \mathcal{C}^0 = \Lambda^{\emptyset}_{\text{SP}}(1)$, by Theorem 5.2.37, this follows from Proposition 5.2.22(i).

(ii) If $M, N \in \Lambda^{\emptyset}_{\text{SP}}(1{\rightarrow}1)$, then

$$\begin{array}{rcll} M \neq N & \Rightarrow & \lambda f{:}1.Mf \neq \lambda f{:}1.Nf & \\ & \Rightarrow & Mf \neq Nf & \\ & \Rightarrow & MF \neq NF, & \text{for some } F \in \Lambda^{\emptyset}_{\text{SP}}(1), \text{ by 5.2.35,} \\ & \Rightarrow & MF\#NF, & \text{by (i) as } MF, NF \in \Lambda^{\emptyset}_{\text{SP}}(1), \\ & \Rightarrow & M\#N. \blacksquare & \end{array}$$

We now want to generalize this last result for all types by using type reducibility in the context of λ_{SP}.

5.2.42 Definition Let $A, B \in \mathbb{T}$. We say that A *is* $\boldsymbol{\beta\eta}$SP*-reducible to* B, written

$$A \leq_{\boldsymbol{\beta\eta}\text{SP}} B,$$

if there exists $\Phi : A{\rightarrow}B$ such that, for any closed $N_1, N_2 : A$,

$$N_1 = N_2 \;\Leftrightarrow\; \Phi N_1 = \Phi N_2.$$

5.2.43 Proposition *For each type A one has $A \leq_{\boldsymbol{\beta\eta}\text{SP}} 1{\rightarrow}1$.*

Proof We can copy the proof of 3.4.8 to obtain $A \leq_{\boldsymbol{\beta\eta}\text{SP}} 1_2{\rightarrow}0{\rightarrow}0$. Moreover, as $1_2{\rightarrow}0{\rightarrow}0 \leq_{\boldsymbol{\beta\eta}\text{SP}} 1{\rightarrow}1$, by $\lambda uxa.u(\lambda z_1z_2.x(\pi(xz_1)(xz_2)))a$, the conclusion follows. $\blacksquare$

5.2.44 Corollary *Let $A \in \mathbb{T}$ and $M, N \in \Lambda^{\emptyset}_{\text{SP}}$. Then*

$$M \neq_{\boldsymbol{\beta\eta}\text{SP}} N \;\Rightarrow\; M\#_{\boldsymbol{\beta\eta}\text{SP}}N.$$

Proof Let $A \leq_{\beta\eta\mathrm{SP}} 1{\to}1$ using Φ. Then

$$\begin{array}{rcl} M \neq N & \Rightarrow & \Phi M \neq \Phi N \\ & \Rightarrow & \Phi M \# \Phi N, \text{ by Corollary 5.2.41(ii),} \\ & \Rightarrow & M \# N. \blacksquare \end{array}$$

We obtain the following Hilbert–Post completeness theorem.

5.2.45 Theorem *Let $\mathcal{M}$ be a model of λ_{SP}. For any type A and closed terms $M, N \in \Lambda^{\emptyset}(A)$ the following are equivalent:*

(i) $M =_{\beta\eta\mathrm{SP}} N$;

(ii) $\mathcal{M} \models M = N$;

(iii) $\lambda_{\mathrm{SP}} \cup \{M = N\}$ *is consistent.*

Proof ((i)⇒(ii)) By soundness.
((ii)⇒(iii)) Since truth implies consistency.
((iii)⇒(i)) By Corollary 5.2.44. ■

The result also holds for equations between open terms (consider their closures). The moral is that every equation is either provable or inconsistent. In other words, that every model of λ_{SP} has the same (equational) theory.

Diophantine relations

5.2.46 Definition Let $R \subseteq \Lambda^{\emptyset}_{\mathrm{SP}}(A_1) \times \cdots \times \Lambda^{\emptyset}_{\mathrm{SP}}(A_n)$ be an n-ary relation.

(i) We call R *equational* if $\exists B \in \mathbb{T}^0 \exists M, N \in \Lambda^{\emptyset}_{\mathrm{SP}}(A_1 {\to} \cdots {\to} A_n {\to} B) \forall \vec{F}$

$$R(F_1, \ldots, F_n) \Leftrightarrow M F_1 \cdots F_n = N F_1 \cdots F_n. \tag{5.4}$$

Here equality is taken in the sense of the theory of λ_{SP}.

(ii) We say R is the *projection* of the $(n+m)$-ary relation S if

$$R(\vec{F}) \Leftrightarrow \exists \vec{G}\ S(\vec{F}, \vec{G}).$$

(iii) We say R is *Diophantine* if it is the projection of an equational relation.

Note that equational relations are closed coordinate-wise under equality and are recursive (since λ_{SP} is CR and SN). A Diophantine relation is clearly closed under equality (coordinate-wise) and recursively enumerable. Our main result will be the converse. The proof occupies 5.2.47–5.2.57.

5.2.47 Proposition

(i) *Equational relations are closed under substitution of lambda-definable functions. This means that if R is equational and R' is defined by*

$$R'(\vec{F}) \Leftrightarrow R(H_1\vec{F}, \ldots, H_n\vec{F}),$$

then R' is equational.

(ii) *Equational relations are closed under conjunction.*

(iii) *Equational relations are Diophantine.*

(iv) *Diophantine relations are closed under substitution of lambda-definable functions, conjunction and projection.*

Proof (i) Easy.

(ii) Use (simple) pairing. For example,

$$\begin{aligned} M_1\vec{F} = N_1\vec{F} \ \& \ M_2\vec{F} = N_2\vec{F} \quad &\Leftrightarrow \quad \pi(M_1\vec{F})(M_2\vec{F}) = \pi(N_1\vec{F})(N_2\vec{F}) \\ &\Leftrightarrow \quad M\vec{F} = N\vec{F}), \end{aligned}$$

with $M \equiv \lambda\vec{f}.\pi(M_1\vec{f})(M_2\vec{f})$ and N is similarly defined.

(iii) By dummy projections.

(iv) By some easy logical manipulations. For example, let

$$R_i(\vec{F}) \Leftrightarrow \exists\vec{G}_i.M_i\vec{G}_i\vec{F} = N_i\vec{G}_i\vec{F}.$$

Then

$$R_1(\vec{F}) \ \& \ R_2(\vec{F}) \Leftrightarrow \exists\vec{G}_1\vec{G}_2.[M_1\vec{G}_1\vec{F} = N_1\vec{G}_1\vec{F} \ \& \ M_2\vec{G}_2\vec{F} = N_2\vec{G}_2\vec{F}]$$

and we can use (i). ∎

5.2.48 Lemma *Let $\Phi_i : A_i \leq_{\text{SP}} (1 \to 1)$ and let $R \subseteq \prod_{i=1}^n \Lambda_{\text{SP}}^{\emptyset}(A_i)$ be equality-closed coordinate-wise. Define $R^{\Phi} \subseteq \Lambda_{\text{SP}}^{\emptyset}(1{\to}1)^n$ by*

$$\begin{aligned} &R^{\Phi}(G_1, \ldots, G_n) \Leftrightarrow \\ &\qquad \exists F_1 \cdots F_n \ [\Phi_1 F_1 = G_1 \& \cdots \Phi_n F_n = G_n \ \& \ R(F_1, \ldots, F_n)]. \end{aligned}$$

We have the following.

(i) *If R^{Φ} is Diophantine, then R is Diophantine.*

(ii) *If R^{Φ} is recursively enumerable, then so is R.*

Proof (i) By Proposition 5.2.47(iv), noting that

$$R(F_1, \ldots, F_n) \Leftrightarrow R^{\Phi}(\Phi_1 F_1, \ldots, \Phi_n F_n).$$

(ii) Similarly. ∎

From Proposition 5.2.7 we can assume without loss of generality that $n = 1$ in Diophantine equations.

5.2.49 Lemma *Let $R \subseteq (\Lambda^{\emptyset}_{\mathrm{SP}}(1{\to}1))^n$ closed under equality. Define $R^{\wedge} \subseteq \Lambda^{\emptyset}_{\mathrm{SP}}(1{\to}1)$ by*

$$R^{\wedge}(F) \Leftrightarrow R(\pi_1^{1\to1,n}(F),\ldots,\pi_n^{1\to1,n}(F)).$$

Then

(i) *R is Diophantine iff $R^{\wedge}$ is Diophantine.*
(ii) *R is recursively enumerable iff $R^{\wedge}$ is.*

Proof By Proposition 5.2.47(i) and the pairing functions $\pi^{1\to1,n}$. ∎

Note that

$$R(F_1,\ldots,F_n) \Leftrightarrow R^{\wedge}(\pi^{1\to1,n}F_1\ldots F_n).$$

5.2.50 Corollary *In order to prove that every recursively enumerable relation $R \subseteq \prod_{i=1}^n \Lambda^{\emptyset}_{\mathrm{SP}}(A_i)$ that is closed under $=_{\beta\eta\mathrm{SP}}$ is Diophantine, it suffices to do this just for such $R \subseteq \Lambda^{\emptyset}_{\mathrm{SP}}(1{\to}1)$.*

Proof By the previous two lemmas. ∎

So now we are interested in recursively enumerable subsets of $\Lambda^{\emptyset}_{\mathrm{SP}}(1{\to}1)$ closed under $=_{\beta\eta\mathrm{SP}}$. Since

$$(\mathcal{T}^1_{\mathrm{CM}}/{=_{\mathrm{CM}}}) = \mathcal{F}[x] \cong \mathcal{C}^1 = (\Lambda^{\emptyset}_{\mathrm{SP}}(1{\to}1)/ =_{\beta\eta\mathrm{SP}})$$

one can shift attention to relations on $\mathcal{T}^1_{\mathrm{CM}}$ closed under $=_{\mathrm{CM}}$. We say loosely that such relations are on $\mathcal{F}[x]$. The definition for such relations to be equational (Diophantine) is slightly different (but completely in accordance with the isomorphism $\mathcal{C}^1 \cong \mathcal{F}[x]$).

5.2.51 Definition A k-ary relation R on $\mathcal{F}[\vec{x}]$ is called *Diophantine* if there exist $s(u_1,\ldots,u_k,\vec{v}), t(u_1,\ldots,u_k,\vec{v}) \in \mathcal{F}[\vec{u},\vec{v}]$ such that

$$R(f_1[\vec{x}],\ldots,f_k[\vec{x}]) \Leftrightarrow \exists\vec{v}\in\mathcal{F}[\vec{x}].s(f_1[\vec{x}],\ldots,f_k[\vec{x}],\vec{v}) = t(f_1[\vec{x}],\ldots,f_k[\vec{x}],\vec{v}).$$

The isomorphism $h_n : \mathcal{F}[\vec{x}] \to \mathcal{C}^n$ given by Proposition 5.2.38 induces an isomorphism

$$h_n^k : (\mathcal{F}[\vec{x}])^k \to (\mathcal{C}^n)^k.$$

Diophantine relations on $\mathcal{F}$ are closed under conjunction as before.

5.2.52 Proposition (Transfer Lemma)

(i) *Let $X \subseteq (\mathcal{F}[x_1,\ldots,x_n])^k$ be equational (Diophantine). Then $h_n^k(X) \subseteq (\mathcal{C}^n)^k$ is equational (Diophantine), respectively.*

(ii) *Let* $X \subseteq (\mathcal{C}^n)^k$ *be recursively enumerable and closed under* $=_{\beta\eta\text{SP}}$. *Then* $(h_n^k)^{-1}(X) \subseteq (\mathcal{F}[x_1, \ldots, x_n])^k$ *is recursively enumerable and closed under* $=_{\text{CM}}$.

5.2.53 Corollary *In order to prove that every recursively enumerable relation on* $\mathcal{C}^1$ *closed under* $=_{\beta\eta\text{SP}}$ *is Diophantine it suffices to show that every recursively enumerable relation on* $\mathcal{F}[x]$ *closed under* $=_{\text{CM}}$ *is Diophantine.*

Before proving that every equality-closed recursively enumerable relation on $\mathcal{F}[x]$ is Diophantine, for the sake of clarity we shall give the proof first for $\mathcal{F}$. It consists of two steps: first we encode Matijasevič's solution to Hilbert's 10th problem into this setting; then we give a Diophantine coding of $\mathcal{F}$ in $\mathcal{F}$, and finish the proof for $\mathcal{F}$. Since the coding of $\mathcal{F}$ can easily be extended to $\mathcal{F}[x]$ the result then holds also for this structure and we are done.

5.2.54 Definition Write $s_0 = I$, $s_{n+1} = R^{n+1}$, elements of $\mathcal{F}$. The set of *numerals* in $\mathcal{F}$ is defined by

$$\mathcal{N} = \{s_n \mid n \in \omega\}.$$

We have the following.

5.2.55 Proposition $f \in \mathcal{N} \Leftrightarrow f * R = R * f.$

Proof This is because if f is normal and $f * R = R * f$, then the binary tree part of f must be trivial, i.e. f must be a string of Ls and Rs, therefore consists of only Rs. ∎

5.2.56 Definition A sequence of k-ary relations $R_n \subseteq \mathcal{F}$ is called *Diophantine uniformly in* n if there is a $(k+1)$-ary Diophantine relation $P \subseteq \mathcal{F}^{k+1}$ such that

$$R_n(\vec{u}) \Leftrightarrow P(s_n, \vec{u}).$$

Next we build up a toolkit of Diophantine relations on $\mathcal{F}$.

(1) $\mathcal{N}$ is equational (hence Diophantine).

Proof In 5.2.55 it was proved that

$$f \in \mathcal{N} \Leftrightarrow f * R = R * f. \blacksquare$$

(2) The sets $\mathcal{F} * L$, $\mathcal{F} * R \subseteq \mathcal{F}$ and $\{L, R\}$ are equational. In fact one has

(i) $f \in \mathcal{F} * L \;\Leftrightarrow\; f * \langle L, L\rangle = f.$
(ii) $f \in \mathcal{F} * R \;\Leftrightarrow\; f * \langle R, R\rangle = f.$
(iii) $f \in \{L, R\} \;\Leftrightarrow\; f * \langle I, I\rangle = I.$

Proof (i) Notice that if $f \in \mathcal{F}*L$, then $f = g*L$, for some $g \in \mathcal{F}$, hence $f*\langle L, L\rangle = f$. Conversely, if $f = f*\langle L, L\rangle$, then $f = f*\langle \mathsf{I}, \mathsf{I}\rangle * L \in \mathcal{F}*L$.

(ii) Similarly.

(iii) ($\Leftarrow$) By distinguishing the possibile shapes of the normal form of f. ∎

(3) **Notation**

$$\begin{aligned} [\,] &= R; \\ [f_0, \ldots, f_{n-1}] &= \langle f_0 * L, \ldots, f_{n-1} * L, R\rangle\!\rangle, \quad \text{if } n > 0. \end{aligned}$$

One easily sees that $[f_0, \ldots, f_{n-1}] * [I, f_n] = [f_0, \ldots, f_n]$. Write

$$\mathrm{Aux}_n(f) = [f, f * R, \ldots, f * R^{n-1}].$$

Then the relations $h = \mathrm{Aux}_n(f)$ are Diophantine uniformly in n.

Proof Indeed,

$$h = \mathrm{Aux}_n(f) \;\Leftrightarrow\; R^n * h = R \;\&\; h = R * h * \langle\langle L, L\rangle, \langle f * R^{n-1} * L, R\rangle\rangle.$$

To see ($\Rightarrow$), assume $h = [f, f * R, \ldots, f * R^{n-1}]$, then $h = \langle f * L, f * R * L, \ldots, f * R^{n-1} * L, R\rangle\!\rangle$, so $R^n * h = R$ and

$$\begin{aligned} R * h &= [f * R, \ldots, f * R^{n-1}] \\ R * h * \langle\langle L, L\rangle, \langle f * R^{n-1} * L, R\rangle\rangle &= [f, f * R, \ldots, f * R^{n-1}] \\ &= h. \end{aligned}$$

To see ($\Leftarrow$), note that we always can write $h = \langle h_0, \ldots, h_n\rangle\!\rangle$. By the assumptions, $h_n = R$ and $h = R * h * \langle\langle L, L\rangle, \langle f * R^{n-1} * L, R\rangle\rangle = R * h * —$, say. So by reading the following equality signs in the correct order (first the left equalities top to bottom; then the right equalities bottom to top) it follows that

$$\begin{array}{rcl} h_0 &=& h_1 \; * \; — \quad = \; f * L \\ h_1 &=& h_2 \; * \; — \quad = \; f * R * L \\ & \vdots & \\ h_{n-2} &=& h_{n-1} \; * \; — \quad = \; f * R^{n-2} * L \\ h_{n-1} &=& f * R^{n-1} * L \\ h_n &=& R. \end{array}$$

Therefore $h = \mathrm{Aux}_n(f)$. ∎

(4) Write $\mathsf{Seq}_n(f) \;\Leftrightarrow\; f = [f_0, \ldots, f_{n-1}]$, for some $f_0, \ldots, f_{n-1}$. Then Seq_n is Diophantine uniformly in n.

Proof One has $\mathtt{Seq}_n(f)$ iff

$$R^n * f = R \,\&\, \mathrm{Aux}_n(L) * \langle I, L\rangle * f = \mathrm{Aux}_n(L) * \langle I, L\rangle * f * \langle L, L\rangle,$$

as can be proved similarly (use (2)(i)). ■

(5) Define

$$\mathrm{Cp}_n(f) = [f, \ldots, f], \ (n \text{ occurrences of } f).$$

(By default $\mathrm{Cp}_0(f) = [\ ] = R$.) Then $\mathrm{Cp}_n(f) = g$ is Diophantine uniformly in n.

Proof $\mathrm{Cp}_n(f) = g$ iff

$$\mathtt{Seq}_n(g) \,\&\, g = R * g * \langle L, f * L, R\rangle\rangle. \ ■$$

(6) Let $\mathrm{Pow}_n(f) = f^n$. Then $\mathrm{Pow}_n(f) = g$ is Diophantine uniformly in n.

Proof One has $\mathrm{Pow}_n(f) = g$ iff

$$\exists h[\mathtt{Seq}_n(h) \,\&\, h = R * h * \langle f * L, f * L, R\rangle\rangle \,\&\, L * h = g].$$

This can be proved in a similar way (it helps to realize that h has to be of the form $h = [f^n, \ldots, f^1]$). ■

Now we can show that the operations $+$ and $\times$ on $\mathcal{N}$ are Diophantine.

(7) There are Diophantine ternary relations P_+, $P_\times$ such that for all n, m, k

(i) $P_+(s_n, s_m, s_k) \Leftrightarrow n + m = k$.
(ii) $P_\times(s_n, s_m, s_k) \Leftrightarrow n.m = k$.

Proof (i) Define $P_+(x, y, z) \Leftrightarrow x * y = z$. This relation is Diophantine and works: $R^n * R^m = R^k \Leftrightarrow R^{n+m} = R^k \Leftrightarrow n + m = k$.

(ii) Let $\mathrm{Pow}_n(f) = g \Leftrightarrow P(s_n, f, g)$, with P Diophantine. Then choose $P_\times = P$. ■

(8) Let $X \subseteq \omega$ be a recursively enumerable set of natural numbers. Then $\{s_n \mid n \in X\}$ is Diophantine.

Proof By (7) and the famous theorem of Matiyasevič (1972). ■

(9) Define $\mathtt{Seq}_n^{\mathcal{N}} = \{[s_{m_0}, \ldots, s_{m_{n-1}}] \mid m_0, \ldots, m_{n-1} \in \omega\}$. Then the relation $f \in \mathtt{Seq}_n^{\mathcal{N}}$ is Diophantine uniformly in n.

Proof Indeed, $f \in \mathtt{Seq}_n^{\mathcal{N}}$ iff

$$\mathtt{Seq}_n(f) \,\&\, f * \langle R * L, R\rangle = \mathrm{Aux}_n(R * L) * \langle I, R^n\rangle * f. \ ■$$

(10) Let $f = [f_0, \ldots, f_{n-1}]$ and $g = [g_0, \ldots, g_{n-1}]$. We write

$$f\#g = [f_0 * g_0, \ldots, f_{n-1} * g_{n-1}].$$

Then there exists a Diophantine relation P such that for arbitrary n and $f, g \in \mathsf{Seq}_n$ one has

$$P(f, g, h) \Leftrightarrow h = f\#g.$$

Proof Let

$$\mathsf{Cmp}_n(f) = [L * f, L * R * f * R, \ldots, L * R^{n-1} * f * R^{n-1}].$$

Then $g = \mathsf{Cmp}_n(f)$ is Diophantine uniformly in n. This requires some work. One has by the by now familiar technique

$$\begin{array}{rl}
\mathsf{Cmp}_n(f) = g & \Leftrightarrow \\
\exists h_1, h_2, h_3 & [\\
\mathsf{Seq}_n(h_1) & \& \; f = h_1 * \langle I, R^n * f \rangle \\
\mathsf{Seq}_{n^2}(h_2) & \& \; h_2 = R^n * h_2 * \langle \langle L, L \rangle, h_1 * \langle R^{n-1} * L, R \rangle \rangle \\
\mathsf{Seq}_n^{\mathcal{N}}(h_3) & \& \; h_3 = R * h_3 * \langle \langle I, I \rangle^{n+1} * L, \langle R^{n^2-1} * L, R \rangle \rangle \\
& \& \; g = \mathrm{Aux}_n(L^2) * \langle h_3, R^n \rangle * \langle h_2, R \rangle \\
&].
\end{array}$$

In order to understand this it helps to identify the h_1, h_2, h_3. Suppose $f = \langle f_0, \ldots, f_{n-1}, f_n \rangle\rangle$. Then

$$\begin{aligned}
h_1 &= [f_0, f_1, \ldots, f_{n-1}]; \\
h_2 &= [f_0, f_1, \ldots, f_{n-1}, \\
&\quad f_0 * R, f_1 * R, \ldots, f_{n-1} * R, \\
&\quad \vdots, \\
&\quad f_0 * R^{n-1}, f_1 * R^{n-1}, \ldots, f_{n-1} * R^{n-1}]; \\
h_3 &= [I, R^{n+1}, R^{2(n+1)}, \ldots, R^{(n-1)(n+1)}].
\end{aligned}$$

Now define

$$P(f, g, h) \Leftrightarrow \exists n[\mathsf{Seq}_n(f) \;\&\; \mathsf{Seq}_n(g) \;\&\; \mathsf{Cmp}_n(f * L) * \langle I, R^n \rangle * g = h].$$

Then P is Diophantine and for arbitrary n and $f, g \in \mathsf{Seq}_n$ one has

$$h = f\#g \Leftrightarrow P(f, g, h). \blacksquare$$

(11) For $f = [f_0, \ldots, f_{n-1}]$ define $\Pi(f) = f_0 * \cdots * f_{n-1}$. Then there exists a Diophantine relation P such that for all $n \in \omega$ and all $f \in \mathtt{Seq}_n$ one has

$$P(f, g) \Leftrightarrow \Pi(f) = g.$$

Proof Define $P(f, g)$ if

$$\begin{aligned} &\exists n, h\ [\\ &\mathtt{Seq}_n(f)\ \&\\ &\mathtt{Seq}_{n+1}(h)\ \&\ h = ((f * \langle I, R\rangle)\#(R * h)) * \langle L, I * L, R\rangle\rangle\\ &\quad \&\ g = L * h * \langle I, R\rangle\\ &\quad]. \end{aligned}$$

Then P works as can be seen by realizing h has to be

$$[f_0 * \cdots * f_{n-1}, f_1 * \cdots * f_{n-1}, \ldots, f_{n-2} * f_{n-1}, f_{n-1}, I]. \blacksquare$$

(12) Define $\mathrm{Byte}_n(f) \Leftrightarrow f = [b_0, \ldots, b_{n-1}]$, for some $b_i \in \{L, R\}$. Then Byte_n is Diophantine uniformly in n.

Proof Using (2) one has $\mathrm{Byte}_n(f)$ iff

$$\mathtt{Seq}_n(f)\ \&\ f * \langle\langle I, I\rangle, R\rangle = \mathrm{Cp}_n(I). \blacksquare$$

(13) Let $m \in \omega$, $[m]_2$ be its binary notation of length n and $[m]_{\mathrm{Byte}} \in \mathtt{Seq}_n^{\mathcal{N}}$ be the corresponding element, where L corresponds to a 1 and R to a 0 and the most significant bit is written last. For example $[6]_2 = 110$, hence $[6]_{\mathrm{Byte}} = [R, L, L]$. Then there exists a Diophantine relation Bin such that for all $m \in \omega$

$$\mathrm{Bin}(s_m, f) \Leftrightarrow f = [m]_{\mathrm{Byte}}.$$

Proof We need two auxiliary maps.

$$\begin{aligned} \mathrm{Pow2}(n) &= [R^{2^{n-1}}, \ldots, R^{2^0}];\\ \mathrm{Pow2}I(n) &= [\langle R^{2^{n-1}}, I\rangle, \ldots, \langle R^{2^0}, I\rangle]. \end{aligned}$$

For these the relations $\mathrm{Pow2}(n) = g$ and $\mathrm{Pow2}I(n) = g$ are Diophantine uniformly in n. Indeed, $\mathrm{Pow2}(n) = g$ iff

$$\mathtt{Seq}_n(g)\ \&\ g = ((R * g)\#(R * g)) * [I, R];$$

and $\mathrm{Pow2}I(n) = g$ iff

$$\begin{aligned} \mathtt{Seq}_n(g)\ &\&\ \mathrm{Cp}_n(L)\#g = \mathrm{Pow2}(n);\\ &\&\ \mathrm{Cp}_n(R)\#g = \mathrm{Cp}_n(I). \end{aligned}$$

It follows that Bin is Diophantine since $\mathrm{Bin}(m, f)$ iff

$$m \in \mathcal{N} \;\&\; \exists n[\mathrm{Byte}_n(f) \;\&\; \Pi(f\#\mathrm{Pow}2I(n)) = m]. \blacksquare$$

(14) We now define a surjection $\varphi : \omega \rightarrow \mathcal{F}$. Remember that $\mathcal{F}$ is generated by two elements $\{e_0, e_1\}$ using only $*$. One has $e_1 = L$. Define

$$\varphi(n) = e_{i_0} * \cdots * e_{i_{m-1}},$$

where $[n]_2 = i_{m-1} \ldots i_0$. We say that n is a *code* of $\varphi(n)$. Since every $f \in \mathcal{F}$ can be written as $L * \langle I, I\rangle * f$ the map φ is surjective indeed.

(15) $\mathrm{Code}(n, f)$ defined by $\varphi(n) = f$ is Diophantine uniformly in n.

Proof Indeed, $\mathrm{Code}(n, f)$ iff

$$\exists g\,[\mathrm{Bin}(n, g) \;\&\; \Pi(g * \langle\langle e_0, e_1\rangle, R\rangle) = f. \blacksquare$$

(16) Every equality-closed recursively enumerable subset $\mathcal{X} \subseteq \mathcal{F}$ is Diophantine.

Proof Since the word problem for $\mathcal{F}$ is decidable,

$$\#\mathcal{X} = \{m \mid \exists f \in \mathcal{X}\; \varphi(m) = f\}$$

is also recursively enumerable. By (8), $\#\mathcal{X} \subseteq \omega$ is Diophantine. Hence by (15) $\mathcal{X}$ is Diophantine via

$$g \in \mathcal{X} \;\Leftrightarrow\; \exists f\; f \in \#\mathcal{X} \;\&\; \mathrm{Code}(f, g). \blacksquare$$

(17) Every equality-closed recursively enumerable subset $\mathcal{X} \subseteq \mathcal{F}[\vec{x}]$ is Diophantine.

Proof Similarly, since also $\mathcal{F}[\vec{x}]$ is generated by two of its elements. We need to know that all the Diophantine relations $\subseteq \mathcal{F}$ are also Diophantine $\subseteq \mathcal{F}[x]$. This follows from Exercise 5.12 and the fact that such relations are closed under intersection. $\blacksquare$

5.2.57 Theorem *A relation R on closed Λ_{SP} terms is Diophantine if and only if R is closed coordinate-wise under equality and recursively enumerable.*

Proof By (17) and Corollaries 5.2.50 and 5.2.53. $\blacksquare$

5.3 Gödel's system $\mathcal{T}$: higher-order primitive recursion

In this section the extension $\boldsymbol{\lambda}_{\mathcal{T}}$ of $\boldsymbol{\lambda}_{\rightarrow}$ will be defined in which the base type 0 is considered as the set of natural numbers and denoted by N. Constants for 0, successor (S^+), and primitive recursion (R) are added. Because primitive recursion defines functionals of higher type, $\mathsf{N}\rightarrow A$, one can define on type $\mathsf{N}\rightarrow\mathsf{N}$ more than the primitive recursive functions. We will see that the Ackermann function becomes definable in $\boldsymbol{\lambda}_{\mathcal{T}}$.

5.3.1 Definition The set of number-theoretic, i.e. of type $N^k\rightarrow N$, *primitive recursive functions* is the smallest set containing zero, successor and projection functions which is closed under composition and the following schema of first-order primitive recursion:

$$
\begin{aligned}
F(0, \vec{x}) &= G(\vec{x}) \\
F(n+1, \vec{x}) &= H(F(n, \vec{x}), n, \vec{x}).
\end{aligned}
$$

This schema defines F from G and H by stating that $F(0) = G$ and by expressing $F(n+1)$ in terms of $F(n)$, H and n. The parameters $\vec{x}$ range over the natural numbers.

The primitive recursive functions were thought to consist of all computable functions. This was shown to be false in Sudan (1927) and Ackermann (1928), who independently gave examples of computable functions that are not primitive recursive. Ten years later the class of computable functions was shown to be much larger by Church and Turing. Nevertheless the primitive recursive functions include almost all functions that one encounters 'in practice', such as addition, multiplication, exponentiation, and many more.

Besides the existence of computable functions that are not primitive recursive, there is another reason to generalize the above schema, namely the existence of computable objects that are not number-theoretic functions. For example, given a number-theoretic function F and a number n, compute the maximum that F takes on arguments $<n$. Other examples of computations where inputs and/or outputs are functions are: compute the function that coincides with F on arguments less than n and is zero otherwise; compute the nth iterate of F; and so on. These computations define maps that are commonly called functionals, to emphasize that they are more general than number-theoretic functions.

Consider the full typestructure $\mathcal{M}_\omega$ over the natural numbers, see Definition 2.4.18. We allow a liberal use of currying, so the following denotations are all identified:

$$FGH \equiv (FG)H \equiv F(G,H) \equiv F(G)H \equiv F(G)(H).$$

Application is left-associative, so $F(GH)$ is notably different from the above denotations.

The aforementioned interest in higher-order computations led to the following schema of higher-order primitive recursion proposed in Gödel (1958)[2]:

$$RMN0 = M$$
$$RMN(n+1) = N(RMNn)n.$$

Here M need not be a natural number, but can have as type any $A \in \mathbb{T}^0$ (see Section 1.1). The corresponding type of N is $A \to \mathsf{N} \to A$, where N is the type of the natural numbers. We make some further observations with respect to this schema. First, the dependence of F on G and H in the first-order schema is made explicit by defining RMN, which is to be compared to F. Second, the parameters $\vec{x}$ from the first-order schema are left out above since they are no longer necessary: we can have higher-order objects as results of computations. Third, the type of R depends on the type of the result of the computation. In fact we have a family of *recursors* $R_A : A \to (A \to \mathsf{N} \to A) \to \mathsf{N} \to A$ for every type A.

5.3.2 Definition The set of *primitive recursive functionals* is the smallest set of functionals containing 0, the successor function, and functionals R of all appropriate types, that is closed under explicit $\boldsymbol{\lambda}^0_{\to}$-definition.

This definition implies that the primitive recursive functionals include projection functions and are closed under application, composition and the above schema of higher-order primitive recursion.

We shall now exhibit a number of examples of primitive recursive functionals. First, let K, K^* be defined explicitly by $K(x,y) = x$, $K^*(x,y) = y$ for all $x, y \in \omega$; that is, the first and the second projection. Obviously, K and K^* are primitive recursive functionals, as they come from $\boldsymbol{\lambda}^0_{\to}$-terms. Now consider $P \equiv R0K^*$. Then we have $P0 = 0$ and $P(n+1) = R0K^*(n+1) = K^*(R0K^*n)n = n$ for all $n \in \omega$, justifying calling P the predecessor function. Now consider $x \dot{-} y \equiv Rx(P \circ K)y$. Here $P \circ K$ is the composition of P and K; that is, $(P \circ K)xy = P(K(x,y)) = P(x)$. We have $x \dot{-} 0 = x$ and $x \dot{-} (y+1) = Rx(P \circ K)(y+1) = (P \circ K)(Rx(P \circ K)y)y = P(Rx(P \circ K)y) =$

[2] This allows one to define the so-called Dialectica interpretation, a translation of intuitionistic arithmetic into the quantifier-free theory of primitive recursive functionals of finite type, yielding a consistency proof for arithmetic.

$P(x \dot{-} y)$. Thus we have defined cut-off subtraction $\dot{-}$ as primitive recursive functional.

In the previous paragraph, we have only used $R = R_\mathsf{N}$ in order to define some functions that are, in fact, already definable with first-order primitive recursion. In this paragraph we are going to use $R_{\mathsf{N}\to\mathsf{N}}$ as well. Given functions F, F' and natural numbers x, y, define explicitly the functional G by $G(F, F', x, y) = F'(F(y))$ and abbreviate $G(F)$ by G_F. Now consider RIG_F, where R is actually $R_{\mathsf{N}\to\mathsf{N}}$ and I is the identity function on the natural numbers. We calculate $RIG_F 0 = I$ and $RIG_F(n+1) = G_F(RIG_F n)n$, which is a function assigning $G(F, RIG_F n, n, m) = RIG_F n(Fm)$ to every natural number m. In other words, $RIG_F n$ is a function which iterates F precisely n times, and we denote this function by F^n.

We finish with an example of a computable function A that is not first-order primitive recursive. The function A is a variant, due to Péter (1967) of a function by Ackermann. The essential difficulty of the function A is the nested recursion in the third equality below.

5.3.3 Definition (Ackermann function) The Ackermann function is defined recursively as:

$$
\begin{aligned}
A(0, m) &= m + 1 \\
A(n+1, 0) &= A(n, 1) \\
A(n+1, m+1) &= A(n, A(n+1, m)).
\end{aligned}
$$

Write $A(n) = \lambda\!\!\lambda m, A(n, m)$. Then $A(0)$ is the successor function and, by the last two equations, $A(n+1, m) = A(n)^{m+1}(1)$. Therefore we can define $A = RSH$, where S is the successor function and $H(F, x, y) = F^{y+1}1$. As examples we calculate $A(1, m) = H(A(0), 1, m) = A(0)^{m+1}(1) = m + 2$ and $A(2, m) = H(A(1), 1, m) = A(1)^{m+1}(1) = 2m + 3$.

Syntax of $\boldsymbol{\lambda}_\mathcal{T}$

In this subsection we formalize Gödel's $\mathcal{T}$ as an extension of the simply typed lambda calculus $\boldsymbol{\lambda}_\to^{\mathrm{Ch}}$ over $\mathbb{T}^0$, called $\boldsymbol{\lambda}_\mathcal{T}$.

5.3.4 Definition The theory *Gödel's $\mathcal{T}$*, written $\boldsymbol{\lambda}_\mathcal{T}$, is defined as follows.

(i) The set of *types of $\boldsymbol{\lambda}_\mathcal{T}$* is defined by $\mathbb{T}(\boldsymbol{\lambda}_\mathcal{T}) = \mathbb{T}^{\{\mathsf{N}\}}$, where the atomic type N is called the *natural number type*.

(ii) The *terms of $\boldsymbol{\lambda}_\mathcal{T}$* are obtained by adding to the term-formation rules of $\boldsymbol{\lambda}_\to^0$ the constants $\mathsf{0} : \mathsf{N}$, $\mathsf{S}^+ : \mathsf{N}\to\mathsf{N}$ and $\mathsf{R}_A : A\to(A\to\mathsf{N}\to A)\to\mathsf{N}\to A$ for all types A.

(iii) We denote the set of (closed) terms of type A by $\Lambda_T(A)$ (respectively $\Lambda_T^{\varnothing}(A)$) and put $\Lambda_T = \bigcup_A \Lambda_T(A)$ (respectively $\Lambda_T^{\varnothing} = \bigcup_A \Lambda_T^{\varnothing}(A)$).

(iv) Terms constructed from $\mathsf{0}$ and S^+ only are called *numerals*, with $\mathsf{S}^+(\mathsf{0})$ abbreviated by 1, $\mathsf{S}^+(\mathsf{S}^+(\mathsf{0}))$ abbreviated by 2, and so on. An arbitrary numeral will be denoted by n.

(v) We define recursively $\mathsf{n}^{A \to B} \equiv \lambda x^A.\mathsf{n}^B$, with $\mathsf{n}^{\mathsf{N}} \equiv \mathsf{n}$.

(vi) The *formulas* of $\boldsymbol{\lambda}_{\mathcal{T}}$ are equations between terms (of the same type).

(vii) The *theory* of $\boldsymbol{\lambda}_{\mathcal{T}}$ is axiomatized by equality axioms and rules, $\boldsymbol{\beta}$-conversion and the schema of higher-order primitive recursion from the previous section.

(viii) The *notion of reduction* T on $\boldsymbol{\lambda}_{\mathcal{T}}$, written $\to_T$, is defined by the following contraction rules (extending $\boldsymbol{\beta}$-reduction):

$$(\lambda x.M)N \to_T M[x := N]$$
$$\mathsf{R}_A M N \mathsf{0} \to_T M$$
$$\mathsf{R}_A M N(\mathsf{S}^+ P) \to_T N(\mathsf{R}_A M N P)P.$$

This gives rise to reduction relations $\to_T$, $\twoheadrightarrow_T$. Gödel did not consider $\boldsymbol{\eta}$-reduction.

5.3.5 Theorem *The conversion relation $=_T$ coincides with equality provable in $\boldsymbol{\lambda}_{\mathcal{T}}$.*

Proof By an easy extension of the proof of this result in untyped lambda calculus, see B[1984] Proposition 3.2.1. ∎

5.3.6 Lemma *Every closed normal form of type N is a numeral.*

Proof Consider the left-most symbol of a closed normal form of type N. This symbol cannot be a variable since the term is closed. The left-most symbol cannot be a λ, since abstraction terms are not of type N and a redex is not a normal form. If the left-most symbol is $\mathsf{0}$, then the term is the numeral $\mathsf{0}$. If the left-most symbol is S^+, then the term must be of the form S^+P, with P a closed normal form of type N. If the left-most term is R, then for typing reasons the term must be $\mathsf{R}MNP\vec{Q}$, with P a closed normal form of type N. In the latter two cases we can complete the argument by induction, since P is a smaller term. Hence P is a numeral, so also S^+P. The case $\mathsf{R}MNP$ with P a numeral can be excluded, as $\mathsf{R}MNP$ should be a normal form. ∎

We now prove SN and CR for $\boldsymbol{\lambda}_{\mathcal{T}}$, two results that could be proved independently from each other. However, the proof of CR can be simplified by using SN, which we prove first by an extension of the proof of SN for $\boldsymbol{\lambda}^0_{\rightarrow}$, Theorem 2.2.1.

5.3.7 Theorem *Every $M \in \Lambda_T$ is* SN *with respect to $\rightarrow_T$.*

Proof Recall the notion of computability from the proof of Theorem 2.2.1. We generalize it to terms of $\boldsymbol{\lambda}_T$. We shall frequently use the fact that computable terms are SN, see (2.3) on page 65. In view of the definition of computability it suffices to prove that the constants $0, \mathsf{S}^+, \mathsf{R}_A$ of $\boldsymbol{\lambda}_T$ are computable. The constant $0 : \mathsf{N}$ is computable since it is SN. Consider $\mathsf{S}^+ P$ with computable $P : \mathsf{N}$, so P is SN and hence $\mathsf{S}^+ P$. It follows that S^+ is computable. In order to prove that R_A is computable, assume that M, N, P are computable and of appropriate type such that $\mathsf{R}_A MNP$ is of type A. Since $P : \mathsf{N}$ is computable, it is SN. Since $\rightarrow_T$ is finitely branching, P has only finitely many normal forms, which are numerals by Lemma 5.3.6. Let $\#P$ be the largest of those numerals. We shall prove by induction on $\#P$ that $\mathsf{R}_A MNP$ is computable. Let $\vec{Q}$ be computable such that $\mathsf{R}_A MNP\vec{Q}$ is of type N. We have to show that $\mathsf{R}_A MNP\vec{Q}$ is SN. If $\#P = 0$, then every reduct of $\mathsf{R}_A MNP\vec{Q}$ passes through a reduct of $M\vec{Q}$, and SN follows since $M\vec{Q}$ is computable. If $\#P = \mathsf{S}^+\mathsf{n}$, then every reduct of $\mathsf{R}_A MNP\vec{Q}$ passes through a reduct of $N(\mathsf{R}_A MNP')P'\vec{Q}$, where P' is such that $\mathsf{S}^+ P'$ is a reduct of P. Then we have $\#P' = \mathsf{n}$ and by induction it follows that $\mathsf{R}_A MNP'$ is computable. Now SN follows since all terms involved are computable. We have proved that $\mathsf{R}_A MNP$ is computable whenever M, N, P are, and hence R_A is computable. ■

5.3.8 Lemma (Newman's Lemma, localized) *Let S be a set and $\rightarrow$ a binary relation on S that is* WCR. *For every $a \in S$ it follows that if $a \in$* SN, *then $a \in$* CR. (See Section 2 for the notions WCR, SN, and CR.)

Proof Call an element *ambiguous* if it reduces to two (or more) distinct normal forms. Assume $a \in$ SN, then a reduces to at least one normal form and all reducts of a are SN. It suffices for $a \in$ CR to prove that a is not ambiguous, i.e. that a reduces to exactly one normal form. Assume by contradiction that a is ambiguous, reducing to different normal forms n_1, n_2, say $a \rightarrow b \rightarrow \cdots \rightarrow n_1$ and $a \rightarrow c \rightarrow \cdots \rightarrow n_2$. Applying WCR to the diverging reduction steps yields a common reduct d such that $b \twoheadrightarrow d$ and $c \twoheadrightarrow d$. Since $d \in$ SN reduces to a normal form, say n, distinct of at least one of n_1, n_2, it follows that at least one of b, c is ambiguous. See Figure 5.3.

Hence a has a one-step reduct which is again ambiguous and SN. Iterating this argument yields an infinite reduction sequence contradicting $a \in$ SN, so a cannot be ambiguous. ■

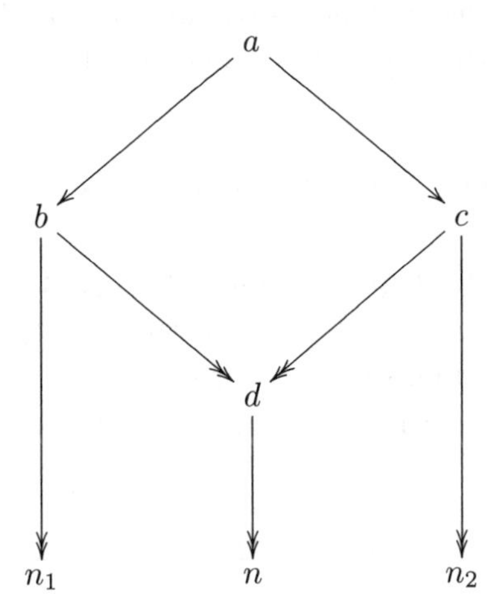

Figure 5.3 Ambiguous a has ambiguous reduct b or c.

5.3.9 Theorem *Every $M \in \Lambda_T$ is* WCR *with respect to $\rightarrow_T$.*

Proof Different redexes in the same term are either completely disjoint, or one redex is included in the other. In the first case the order of the reduction steps is irrelevant, and in the second, a common reduct can be obtained by reducing (possibly multiplied) included redexes. ∎

5.3.10 Theorem *Every $M \in \Lambda_T$ is* CR *with respect to $\rightarrow_T$.*

Proof By Newman's Localized Lemma 5.3.8, using Theorem 5.3.7. ∎

If one considers $\boldsymbol{\lambda}_{\mathcal{T}}$ also with $\boldsymbol{\eta}$-reduction, then the above results can also be obtained. For SN it simply suffices to strengthen the notion of computability for the base case to SN with also $\boldsymbol{\eta}$-reductions included. It is harder to obtain WCR, and hence CR; we require techniques like $\boldsymbol{\eta}$-postponement, see B[1984], Section 15.1.6.

Semantics of $\boldsymbol{\lambda}_{\mathcal{T}}$

In this subsection we extend the notion of model for $\boldsymbol{\lambda}^0_{\rightarrow}$ to $\boldsymbol{\lambda}_{\mathcal{T}}$.

5.3.11 Definition A *model of $\boldsymbol{\lambda}_{\mathcal{T}}$* is a typed λ-model with interpretations of the constants 0, S^+ and R_A for all A, such that the schema of higher-order primitive recursion is valid.

5.3.12 Example Recall the full type-structure over the natural numbers; that is, sets $\mathcal{M}_{\mathsf{N}} = \omega$ and $\mathcal{M}_{A\rightarrow B} = \mathcal{M}_A \rightarrow \mathcal{M}_B$, with set-theoretic application. The full type-structure becomes the canonical model of $\boldsymbol{\lambda}_{\mathcal{T}}$ by interpreting $\mathsf{0}$ as 0, S^+ as the successor function, and the constants R_A as primitive recursors of the right type. The proof that $[\![\mathsf{R}_A]\!]$ is well defined goes by induction.

Other models of Gödel's $\mathcal{T}$ can be found in Exercises 5.26–5.29.

Computational strength

As primitive recursion over higher types turns out to be equivalent with transfinite ordinal recursion, we give a brief review of the theory of ordinals.

The following are some ordinal numbers, simply called *ordinals*, in increasing order.

$$0, 1, 2, \ldots, \omega, \omega{+}1, \omega{+}2, \ldots, \omega{+}\omega = \omega{\cdot}2, \ldots, \omega{\cdot}\omega = \omega^2, \ldots, \omega^\omega, \ldots, \omega^{(\omega^\omega)}, \ldots$$

As well as ordinals themselves, some basic operations of ordinal arithmetic are also visible, namely addition, multiplication and exponentiation, denoted in the same way as in high-school algebra. The dots ... stand for many more ordinals in between, produced by iterating the previous construction process.

The most important structural property of ordinals is that $<$ is a well-order; that is, an order such that every non-empty subset contains a smallest element. This property leads to the principle of (transfinite) induction for ordinals, stating that $P(\alpha)$ holds for all ordinals α whenever P is *inductive*; that is, $P(\alpha)$ follows from $\forall\gamma < \alpha.P(\gamma)$ for all α.

In fact the arithmetical operations are defined by means of two more primitive operations on ordinals, namely the *successor* operation $+1$ and the *supremum* operation $\bigcup$. The supremum $\bigcup a$ of a set of ordinals a is the least upper bound of a, which is equal to the smallest ordinal greater than all ordinals in the set a. A typical example of the latter is the ordinal ω, the first infinite ordinal, which is the supremum of the sequence of the finite ordinals n produced by iterating the successor operation on 0.

These primitive operations divide the ordinals into three classes: the *successor* ordinals of the form $\alpha + 1$; the *limit* ordinals $\lambda = \bigcup\{\alpha \mid \alpha < \lambda\}$, i.e. ordinals which are the supremum of the set of smaller ordinals; and the *zero* ordinal 0. (In fact 0 is the supremum of the empty set, but is not considered to be a limit ordinal.) Thus we have zero, successor and limit ordinals.

Addition, multiplication and exponentiation are now defined according to Table 5.1. Ordinal arithmetic has many properties in common with ordinary arithmetic, but there are some notable exceptions. For example, addition and multiplication are associative but not commutative: $1 + \omega = \omega \neq \omega + 1$ and $2 \cdot \omega = \omega \neq \omega \cdot 2$. Furthermore, multiplication is left distributive over addition, but not right distributive: $(1 + 1) \cdot \omega = \omega \neq 1 \cdot \omega + 1 \cdot \omega$. The sum

$\alpha+\beta$ is weakly increasing in α and strictly increasing in β. Similarly for the product $\alpha\cdot\beta$ with $\alpha>0$. The only exponentiations we shall use, 2^α and ω^α, are strictly increasing in α.

Addition	Multiplication	Exponentiation ($\alpha>0$)
$\alpha+0=\alpha$	$\alpha\cdot 0=0$	$\alpha^0=1$
$\alpha+(\beta+1)=(\alpha+\beta)+1$	$\alpha\cdot(\beta+1)=\alpha\cdot\beta+\alpha$	$\alpha^{\beta+1}=\alpha^\beta\cdot\alpha$
$\alpha+\lambda=\bigcup\{\alpha+\beta \mid \beta<\lambda\}$	$\alpha\cdot\lambda=\bigcup\{\alpha\cdot\beta \mid \beta<\lambda\}$	$\alpha^\lambda=\bigcup\{\alpha^\beta \mid \beta<\lambda\}$

Table 5.1 *Ordinal arithmetic (with λ limit ordinal in the third row).*

The operations of ordinal arithmetic as defined above provide examples of a more general phenomenon called transfinite iteration, to be defined below.

5.3.13 Definition Let f be an ordinal function. Define by induction $f^0(\alpha)=\alpha$, $f^{\beta+1}(\alpha)=f(f^\beta(\alpha))$ and $f^\lambda(\alpha)=\bigcup\{f^\beta(\alpha)\mid\beta<\lambda\}$ for every limit ordinal λ. We call f^β the βth *transfinite iteration* of f.

5.3.14 Example As examples we redefine the arithmetical operations above (see Exercise 5.31):

$$\alpha+\beta=f^\beta(\alpha)$$
$$\alpha\cdot\beta=g_\alpha^\beta(0)$$
$$\alpha^\beta=h_\alpha^\beta(1),$$

with f the successor function, $g_\alpha(\gamma)=\gamma+\alpha$, and $h_\alpha(\gamma)=\gamma\cdot\alpha$.

We proceed with the canonical construction for finding, if there is one, the least fixed point of a weakly increasing ordinal function. The proof is in Exercise 5.17.

5.3.15 Lemma *Let f be a weakly increasing ordinal function. Then:*

(i) $f^{\alpha+1}(0)\ge f^\alpha(0)$ *for all* α;
(ii) $f^\alpha(0)$ *is weakly increasing in* α;
(iii) $f^\alpha(0)$ *does not surpass any fixed point of* f;
(iv) $f^\alpha(0)$ *is strictly increasing and hence* $f^\alpha(0)\ge\alpha$, *until a fixed point of* f *is reached, after which* $f^\alpha(0)$ *becomes constant.*

If a weakly increasing ordinal function f has a fixed point, then it has a smallest fixed point and Lemma 5.3.15 above guarantees that this so-called *least fixed point* is of the form $f^\alpha(0)$; that is, can be obtained by transfinite iteration of f starting at 0. This justifies the following definition.

5.3.16 Definition Let f be a weakly increasing ordinal function having a least fixed point which we denote by $\mathrm{lfp}(f)$. The *closure ordinal* of f is the smallest ordinal α such that $f^\alpha(0) = \mathrm{lfp}(f)$.

Closure ordinals can be arbitrarily large, or may not even exist. The following lemma gives a condition under which the closure ordinal exists and does not surpass ω.

5.3.17 Lemma *If f is a weakly-increasing ordinal function such that*

$$f(\lambda) = \bigcup\{f(\alpha) \mid \alpha < \lambda\}$$

for every limit ordinal λ, then the closure ordinal exists and is at most ω.

Proof Let conditions be as in the lemma. Consider the sequence of finite iterations of f: $0, f(0), f(f(0))$ and so on. If this sequence becomes constant, then the closure ordinal is finite. If the sequence is strictly increasing, then the supremum must be a limit ordinal, say λ. Then we have

$$f(\lambda) = \bigcup\{f(\alpha) \mid \alpha < \lambda\} = f^\omega(0) = \lambda,$$

so the closure ordinal is ω. ■

For example, $f(\alpha) = 1 + \alpha$ has $\mathrm{lfp}(f) = \omega$, and $f(\alpha) = (\omega + 1) \cdot \alpha$ has $\mathrm{lfp}(f) = 0$. By contrast, $f(\alpha) = \alpha + 1$ has no fixed point (note that the latter f is weakly increasing, but the condition on limit ordinals is not satisfied). Finally, $f(\alpha) = 2^\alpha$ has $\mathrm{lfp}(f) = \omega$, and the least fixed point of $f(\alpha) = \omega^\alpha$ is denoted by ϵ_0, being the supremum of the sequence

$$0, \omega^0 = 1, \omega^1 = \omega, \omega^\omega, \omega^{\omega^\omega}, \omega^{\omega^{\omega^\omega}}, \ldots$$

In the following proposition we formulate some facts about ordinals that we need later on.

5.3.18 Proposition

(i) *Every ordinal $\alpha < \epsilon_0$ can be written uniquely as*

$$\alpha = \omega^{\alpha_1} + \omega^{\alpha_2} + \cdots + \omega^{\alpha_n},$$

with $n \geq 0$ and $\alpha_1, \alpha_2, \ldots, \alpha_n$ a weakly-decreasing sequence of ordinals smaller than α.

(ii) *For all α, β we have $\omega^\alpha + \omega^\beta = \omega^\beta$ if and only if $\alpha < \beta$.*

Proof (i) This is a special case of Cantor normal forms with base ω, the generalization of the position system for numbers to ordinals, where terms of the form $\omega^\alpha \cdot n$ are written as $\omega^\alpha + \cdots + \omega^\alpha$ (n summands). The fact that the exponents in the Cantor normal form are *strictly* less than α comes from the assumption that $\alpha < \epsilon_0$.

(ii) The proof of this so-called absorption property goes by induction on β. The case $\alpha \geq \beta$ can be dealt with by using Cantor normal forms. ∎

From now on *ordinal* will mean *ordinal less than* ϵ_0, unless explicitly stated otherwise. This also applies to $\forall\alpha$, $\exists\alpha$, $f(\alpha)$ and so on.

Encoding ordinals in the natural numbers

To speak about ordinals in the language of arithmetic, we have to encode them as natural numbers. This proceeds in a number of steps. Systematic enumeration of grid points in the plane, such as shown in Figure 5.4, yields an encoding of pairs $\langle x, y\rangle$ of natural numbers x, y as given in Definition 5.3.19.

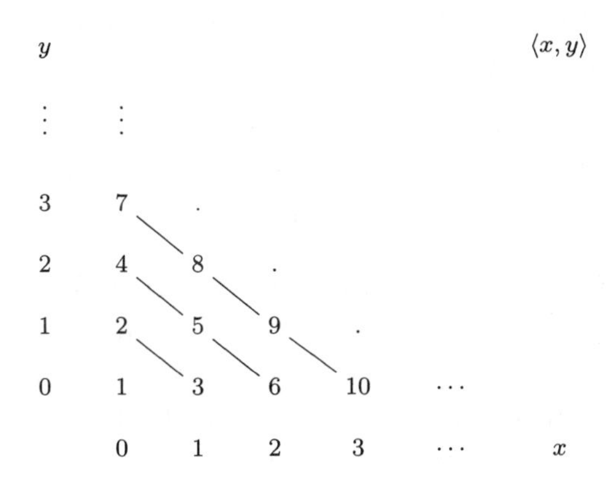

Figure 5.4 $\langle x, y\rangle$-values for $x + y \leq 3$

Finite sequences $[x_1, \ldots, x_k]$ of natural numbers, also called *lists*, can now be encoded by iterating the pairing function. The number 0 does not encode a pair and can hence be used to encode the empty list $[\,]$. All functions and relations involved, including projection functions to decompose pairs and lists, are easily seen to be primitive recursive.

5.3.19 Definition Recall that $1+2+\cdots+n = \frac{1}{2}n(n+1)$ gives the number of grid points satisfying $x+y < n$. The function $\dot{-}$ below is to be understood as cut-off subtraction, that is, $x \dot{-} y = 0$ whenever $y \geq x$. Define the following functions:

$$\begin{aligned}\langle x,y\rangle &= \tfrac{1}{2}(x+y)(x+y+1)+x+1\\ \mathrm{sum}(p) &= \min\{n \mid p \leq \tfrac{1}{2}n(n+1)\} \dot{-} 1\\ x(p) &= p \dot{-} \langle 0, \mathrm{sum}(p)\rangle\\ y(p) &= \mathrm{sum}(p) \dot{-} x(p).\end{aligned}$$

Now let $[\,] = 0$ and, for $k > 0$, $[x_1, \ldots, x_k] = \langle x_1, [x_2, \ldots, x_k]\rangle$ encode lists. Define $\mathtt{lth}(0) = 0$ and $\mathtt{lth}(p) = 1 + \mathtt{lth}(y(p))$ $(p > 0)$ to compute the length of a list.

The following lemma is a straightforward consequence of the above definition.

5.3.20 Lemma *For all $p > 0$ we have $p = \langle x(p), y(p)\rangle$. Moreover, $\langle x,y\rangle > x$, $\langle x,y\rangle > y$, $\mathtt{lth}([x_1, \ldots, x_k]) = k$ and $\langle x,y\rangle$ is strictly increasing in both arguments. Every natural number encodes a unique list of smaller natural numbers. Every natural number encodes a unique list of lists of lists and so on, ending with the empty list.*

Based on the Cantor normal form and the above encoding of lists we can represent ordinals below ϵ_0 as natural numbers in the following way. We write $\overline{\alpha}$ for the natural number representing the ordinal α.

5.3.21 Definition Let $\alpha < \epsilon_0$ have Cantor normal form $\omega^{\alpha_1} + \cdots + \omega^{\alpha_k}$. We encode α by putting $\overline{\alpha} = [\overline{\alpha_1}, \overline{\alpha_2}, \ldots, \overline{\alpha_n}]$. This representation is well defined since every α_i $(1 \leq i \leq n)$ is strictly smaller than α. Since the zero ordinal 0 has the empty sum as Cantor normal form, it is represented by the empty list $[\,]$, hence by the natural number 0.

Examples are $\overline{0} = [\,]$, $\overline{1} = [[\,]]$, $\overline{2} = [[\,],[\,]], \ldots$ and $\overline{\omega} = [[[\,]]]$, $\overline{\omega + 1} = [[[\,]],[\,]]$ and so on. Observe that $[[\,],[[\,]]]$ does not represent an ordinal as $\omega^0 + \omega^1$ is not a Cantor normal form. The following lemmas allow one to identify which natural numbers represent ordinals and to compare them.

5.3.22 Lemma *Let $\prec$ be the lexicographic ordering on lists. Then $\prec$ is primitive recursive and $\overline{\alpha} \prec \overline{\beta} \Leftrightarrow \alpha < \beta$ for all $\alpha, \beta < \epsilon_0$.*

Proof Define $\langle x,y\rangle \prec \langle x',y'\rangle \Leftrightarrow (x \prec x') \vee (x = x' \wedge y \prec y')$ and $x \not\prec 0$, $0 \prec \langle x,y\rangle$. The primitive recursive relation $\prec$ is the lexicographic ordering on pairs, and hence also on lists. Now the lemma follows using Cantor normal forms. (Note that $\prec$ is not a well-order itself, as $\cdots \prec [0,0,1] \prec [0,1], \prec [1]$ has no smallest element.) ∎

5.3.23 Lemma *For $x \in \omega$, define the following notions:*

$$\begin{array}{rl} \mathrm{Ord}(x) & \textit{if } x = \overline{\alpha} \textit{ for some ordinal } \alpha < \epsilon_0; \\ \mathrm{Succ}(x) & \textit{if } x = \overline{\alpha} \textit{ for some successor ordinal } < \epsilon_0; \\ \mathrm{Lim}(x) & \textit{if } x = \overline{\alpha} \textit{ for some limit ordinal } < \epsilon_0; \\ \mathrm{Fin}(x) & \textit{if } x = \overline{\alpha} \textit{ for some ordinal } \alpha < \omega. \end{array}$$

Then Ord, Fin, Succ *and* Lim *are primitive recursive predicates.*

Proof By course of value recursion.
(i)
Put Ord(0) and $\mathrm{Ord}(\langle x, y\rangle) \Leftrightarrow (\mathrm{Ord}(x) \wedge \mathrm{Ord}(y) \wedge (y > 0 \Rightarrow x(y) \preceq x))$.
(ii) Put $\neg\mathrm{Succ}(0)$ and $\mathrm{Succ}(\langle x, y\rangle) \Leftrightarrow (\mathrm{Ord}(\langle x, y\rangle) \wedge (x > 0 \Rightarrow \mathrm{Succ}(y)))$.
(iii) Put $\mathrm{Lim}(x) \Leftrightarrow (\mathrm{Ord}(x) \wedge \neg\mathrm{Succ}(s) \wedge x \neq [\,])$.
(iv) Put $\mathrm{Fin}(x) \Leftrightarrow (x = [\,] \vee (x = \langle 0, y\rangle \wedge \mathrm{Fin}(y)))$. ∎

5.3.24 Lemma *There exist primitive recursive functions* exp *(base ω exponentiation),* succ *(successor),* pred *(predecessor),* plus *(addition),* exp2 *(base 2 exponentiation) such that for all α, β:* $\exp(\overline{\alpha}) = \overline{\omega^\alpha}$, $\mathrm{succ}(\overline{\alpha}) = \overline{\alpha + 1}$, $\mathrm{pred}(\overline{0}) = \overline{0}$, $\mathrm{pred}(\overline{\alpha + 1}) = \overline{\alpha}$, $\mathrm{plus}(\overline{\alpha}, \overline{\beta}) = \overline{\alpha + \beta}$, $\mathrm{exp2}(\overline{\alpha}) = \overline{2^\alpha}$.

Proof Put $\exp(x) = [x]$. Put $\mathrm{succ}(0) = \langle 0, 0\rangle$ and $\mathrm{succ}(\langle x, y\rangle) = \langle x, \mathrm{succ}(y)\rangle$, then $\mathrm{succ}([x_1, \ldots, x_k]) = [x_1, \ldots, x_k, 0]$. Put $\mathrm{pred}(0) = 0$, $\mathrm{pred}(\langle x, 0\rangle) = x$ and $\mathrm{pred}(\langle x, y\rangle) = \langle x, \mathrm{pred}(y)\rangle$ for $y > 0$. For plus, use the absorption property in adding the Cantor normal forms of α and β. For exp2 we use $\omega^\beta = 2^{\omega \cdot \beta}$. Let α have Cantor normal form $\omega^{\alpha_1} + \cdots + \omega^{\alpha_k}$. Then $\omega \cdot \alpha = \omega^{1+\alpha_1} + \cdots + \omega^{1+\alpha_k}$. By absorption, $1 + \alpha_i = \alpha_i$ whenever $\alpha_i \geq \omega$. It follows that we have

$$\alpha = \omega \cdot (\omega^{\alpha_1} + \cdots + \omega^{\alpha_i} + \omega^{n_1} + \cdots + \omega^{n_p}) + n,$$

for suitable n_j, n with $\alpha_1 \geq \cdots \geq \alpha_i \geq \omega$, $n_j + 1 = \alpha_{i+j} < \omega$ for $1 \leq j \leq p$ and $n = k - i - p$ with $\alpha_{k'} = 0$ for all $i + p < k' \leq k$. Using $\omega^\beta = 2^{\omega \cdot \beta}$ we can calculate $2^\alpha = \omega^\beta \cdot 2^n$ with $\beta = \omega^{\alpha_1} + \cdots + \omega^{\alpha_i} + \omega^{n_1} + \cdots + \omega^{n_p}$ and n as above. If $\overline{\alpha} = [x_1, \ldots, x_i, \ldots, x_j, \ldots, 0, \ldots, 0]$, then $\overline{\beta} = [x_1, \ldots, x_i, \ldots, \mathrm{pred}(x_j), \ldots]$ and we can obtain $\mathrm{exp2}(\overline{\alpha}) = \overline{2^\alpha} = \overline{\omega^\beta \cdot 2^n}$ by doubling, n times, $\overline{\omega^\beta} = \exp(\overline{\beta})$ using plus. ∎

5.3.25 Lemma *There exist primitive recursive functions* num, mun *such that* $\mathrm{num}(n) = \overline{n}$ *and* $\mathrm{mun}(\overline{n}) = n$ *for all n. In particular we have*

$$\mathrm{mun}(\mathrm{num}(n)) = n \text{ and } \mathrm{num}(\mathrm{mun}(\overline{n})) = \overline{n} \textit{ for all } n.$$

In other words, num *is the order isomorphism between* $(\omega, <)$ *and* $(\{\overline{n} \mid n \in \omega\}, \prec)$ *and* mun *is the inverse order isomorphism.*

Proof Put num(0) $= 0 = [\,]$ and num$(n+1) =$ succ(num(n)) and mun(0) $=$ 0 and mun$(\langle x, y\rangle) =$ mun$(y) + 1$. ■

5.3.26 Lemma *There exists a primitive recursive function* p *such that for all limit ordinals* α *and* $\beta < \gamma + 2^\alpha$ *one has* $\alpha' < \alpha$ *and* $\beta < \gamma + 2^{\alpha'}$ *for* $\overline{\alpha'} = \mathrm{p}(\overline{\alpha}, \overline{\beta}, \overline{\gamma})$.

Proof Let the conditions be as above. The existence of α' follows directly from the definition of the operations of ordinal arithmetic on limit ordinals. The interesting point, however, is that $\overline{\alpha'}$ can be computed from $\overline{\alpha}, \overline{\beta}, \overline{\gamma}$ in a primitive recursive way, as will become clear by the following argument. If $\beta \leq \gamma$, then we can simply take $\alpha' = 0$. Otherwise, let $\beta = \omega^{\beta_1} + \cdots + \omega^{\beta_n}$ and $\gamma = \omega^{\gamma_1} + \cdots + \omega^{\gamma_m}$ be Cantor normal forms. Now $\gamma < \beta$ implies that $\gamma_i < \beta_i$ for some smallest index $i \leq m$, or no such index exists. In the latter case we have $m < n$ and $\gamma_j = \beta_j$ for all $1 \leq j \leq m$, and we put $i = m + 1$. Since α is a limit, we have $\alpha = \omega \cdot \xi$ for suitable ξ, and hence $2^\alpha = \omega^\xi$. Since $\beta < \gamma + 2^\alpha$ it follows by absorption that $\omega^{\beta_i} + \cdots + \omega^{\beta_n} < \omega^\xi$. Hence $\beta_i + 1 \leq \xi$, so $\omega^{\beta_i} + \cdots + \omega^{\beta_n} \leq \omega^{\beta_i} \cdot n < \omega^{\beta_i} \cdot 2^n = 2^{\omega \cdot \beta_i + n}$. Now take $\alpha' = \omega \cdot \beta_i + n < \omega \cdot (\beta_i + 1) \leq \omega \cdot \xi = \alpha$ and observe $\beta < \gamma + 2^{\alpha'}$. ■

From now on we will freely use ordinals in the natural numbers instead of their codes. This includes uses like α *is finite* instead of Fin$(\overline{\alpha})$, $\alpha \prec \beta$ instead of $\overline{\alpha} \prec \overline{\beta}$, and so on. Note that we avoid using $<$ for ordinals now, as it would be ambiguous. Phrases like $\forall \alpha\ P(\alpha)$ and $\exists \alpha\ P(\alpha)$ should be taken as relativized quantifications over natural numbers: that is, $\forall x\ (\mathrm{Ord}(x) \Rightarrow P(x))$, and $\exists x\ (\mathrm{Ord}(x) \wedge P(x))$, respectively. Finally, functions defined in terms of ordinals are assumed to take the value 0 for arguments that do not encode any ordinal.

Transfinite induction

Transfinite induction (TI) is a principle of proof that generalizes the usual schema of structural induction from natural numbers to ordinals. In this subsection we consider provability in Peano Arithmetic (PA), but we do this informally. Ordinals may exceed ϵ_0. In fact all ordinals below ω_1^{CK}, the first non-computable ordinal, can be represented in PA. Although TI is a valid principle, this doesn't mean that it is derivable. A similar example is the non-derivability of the ω-rule. Even if

$$P(0),\ P(1),\ P(2), \tag{5.5}$$

are all derivable in PA, the obvious consequence $\forall n.P(n)$ is valid, but not automatically derivable. From Gödel's incompleteness result it follows that there are relatively simple P (e.g. $P(n) \overset{\Delta}{\Longleftrightarrow}$ 'n is a proof of 0=1') for which this is not the case (except if PA is inconsistent). If, however, there is some uniformity in the proof of (5.5), if $\forall n.[P(n) \Rightarrow P(n+1)]$ is provable in PA, then one can conclude $\forall n.P(n)$. Transfinite Induction up to ordinal α is a different uniform way of proving a universal statement. In some cases the obvious conclusion is derivable in PA, and in other cases it is not. Gentzen has analyzed precisely when this is the case.

5.3.27 Definition Let P be a unary predicate on the ordinals.

(i) We say that P *holds below* β, written $P^{\downarrow}(\beta)$, by

$$P^{\downarrow}(\beta) \overset{\Delta}{\Longleftrightarrow} \forall \gamma{<}\beta.P(\gamma)$$

(ii) Define P to be *inductive below* β, written $\mathrm{Ind}_\beta(P)$, by

$$\mathrm{Ind}_\beta(P) \overset{\Delta}{\Longleftrightarrow} \forall \alpha{<}\beta.[P^{\downarrow}(\alpha) \;\Rightarrow\; P(\alpha)].$$

We say P is *inductive*, written $\mathrm{Ind}(P)$, if $\forall \alpha.[P^{\downarrow}(\alpha) \;\Rightarrow\; P(\alpha))]$.

(iii) The principle of *transfinite induction up to* α *applied to* P, written $\mathrm{TI}_\alpha(P)$, is

$$\mathrm{TI}_\alpha(P) \overset{\Delta}{\Longleftrightarrow} [\mathrm{Ind}(P) \;\Rightarrow\; P^{\downarrow}(\alpha)].$$

The *schema of transfinite induction up to* α, written TI_α, states

$$\mathrm{TI}_\alpha \overset{\Delta}{\Longleftrightarrow} \mathrm{TI}_\alpha(P), \text{ for all predicates } P.$$

5.3.28 Remark

(i) Suppose

$$\begin{array}{rcll} P(0) & & & \\ P(\alpha) & \Rightarrow & P(\alpha+1), & \text{for all } \alpha, \\ P^{\downarrow}(\alpha) & \Rightarrow & P(\alpha), & \text{for limit ordinals } \alpha. \end{array}$$

Then P is inductive.

(ii) If a property is inductive then TI_α implies that every ordinal up to α has this property.

(iii) Define

$$\mathrm{TI}'_\alpha(P) \overset{\Delta}{\Longleftrightarrow} [\mathrm{Ind}_\alpha(P) \;\Rightarrow\; P^{\downarrow}(\alpha)].$$

The schema TI'_α is the following:

$$\mathrm{TI}'_\alpha \overset{\Delta}{\Longleftrightarrow} \mathrm{TI}'_\alpha(P), \text{ for all predicates } P.$$

Then the schemas TI_α and TI'_a are equivalent. Indeed, TI'_α is *a priori* stronger than TI_α, as its hypothesis Ind_α is weaker than Ind. Conversely, suppose $\mathrm{Ind}_\alpha(P)$, towards $P^{\downarrow}\alpha$. Define $Q(\beta) \overset{\Delta}{\Longleftrightarrow} P(\beta)$ for $\beta \prec \alpha$ and $Q(\beta) \overset{\Delta}{\Longleftrightarrow} (0 = 0)$ else. Then $\mathrm{Ind}(Q)$, hence $Q^{\downarrow}(\alpha)$ by TI_α. Therefore $P^{\downarrow}(\alpha)$.

(iv) Do we have for all predicates P

$$\mathrm{TI}_\alpha(P) \;\Leftrightarrow\; \mathrm{TI}'_\alpha(P)?$$

As the strength of TI_α increases with α, the schema TI_α can be used to measure the proof-theoretic strength of theories: given a theory T, for which α can we prove TI_α? We shall show that TI_α is provable in Peano Arithmetic (PA) for all ordinals $\alpha < \epsilon_0$ by a famous argument due to Gentzen.

The schema TI_ω is equivalent to structural induction on the natural numbers.

5.3.29 Proposition *The schema* TI_ω *is provable in PA.*

Proof Observe that TI_ω is structural induction on an isomorphic copy of the natural numbers by Lemma 5.3.25. ∎

5.3.30 Lemma *Let P be a predicate on ordinals. Then*

$$P^{\downarrow}(\alpha + 1) \;\Leftrightarrow\; P^{\downarrow}(\alpha) \;\&\; P(\alpha).$$

Proof Since $\beta < \alpha + 1 \;\Leftrightarrow\; \beta < \alpha \vee \beta = \alpha$. ∎

5.3.31 Proposition *The schema* $\mathrm{TI}_{\omega\cdot 2}$ *is provable in PA from the schema* TI_ω.

Proof Suppose TI_ω. For given P, assume $\mathrm{Ind}(P)$ towards $\forall\alpha{<}(\omega\cdot 2).P(\alpha)$. Define

$$Q(\alpha) \triangleq P^{\downarrow}(\omega + \alpha).$$

We show $Q^{\downarrow}(\omega)$ by TI'_ω, which by Remark 5.3.28(iii) is equivalent to TI_ω. This means that we have to show $\mathrm{Ind}_\omega(Q)$. Similar to Remark 5.3.28(i) it suffices to show that one has:

(1): $Q(0)$ and;

(2): $\forall\alpha{\prec}\omega.[Q(\alpha) \;\Rightarrow\; Q(\alpha + 1)]$.

As to (1), $Q(0) \triangleq P^{\downarrow}(\omega)$ follows from $\mathrm{Ind}(P)$ and TI_ω.

As to (2), suppose $Q(\alpha)$, towards $Q(\alpha + 1)$. By definition $P^{\downarrow}(\omega + \alpha)$. By $\mathrm{Ind}(P)$ one obtains $P(\omega + \alpha)$. Therefore $P^{\downarrow}(\omega + \alpha + 1)$, by Lemma 5.3.30, which is $Q(\alpha + 1)$.

This completes the proof of $\forall\alpha{\prec}\omega.Q(\alpha)$, so $\forall\alpha{\prec}\omega\forall\beta{\prec}\omega + \alpha.P(\beta)$. Therefore $\forall\beta{\prec}\omega + \omega.P(\beta)$, as $\beta \prec \omega + \omega$ implies $\beta \prec \omega + \alpha$ for some $\alpha \prec \omega$. ∎

5.3.32 Proposition *For $\alpha \prec \epsilon_0$ the schema TI_{2^α} is provable in PA from TI_α.*

Proof Suppose TI_α. Assume $\mathrm{Ind}(P)$ for some P towards $P^{\downarrow}(2^\alpha)$. Define

$$Q(\gamma) \overset{\Delta}{\Longleftrightarrow} \forall\beta.[P^{\downarrow}(\beta) \;\Rightarrow\; P^{\downarrow}(\beta + 2^\gamma)].$$

This states that if P holds on an initial segment, then we can prolong this segment with 2^γ. The goal will be to prove $Q(\gamma)$, for all ordinals $\gamma \preceq \alpha$. Then we can prolong the empty initial segment, on which P vacuously holds, to one of length 2^α and we are done.

First we show that Q is inductive, using Remark 5.3.28(i).

Case $Q(0)$. This states $\forall\beta.[P^{\downarrow}(\beta) \Rightarrow P^{\downarrow}(\beta+1)]$, as $2^0 = 1$. Suppose $P^{\downarrow}(\beta)$. As P is inductive this implies $P(\beta)$. Therefore $P^{\downarrow}(\beta+1)$, by Lemma 5.3.30.

Case $Q(\gamma) \Rightarrow Q(\gamma+1)$. Assume $Q(\gamma)$ and $P^{\downarrow}(\beta)$ for some β. By $Q(\gamma)$ we have $P^{\downarrow}(\beta + 2^\gamma)$. Hence again by $Q(\gamma)$, but now applied to $\beta + 2^\gamma$ instead of β, we have $P^{\downarrow}(\beta + 2^\gamma + 2^\gamma)$, which is $P^{\downarrow}(\beta + 2^{\gamma+1})$, as $2^{\gamma+1} = 2^\gamma + 2^\gamma$. We conclude $Q(\gamma+1)$.

Case $\mathrm{Lim}(\gamma)$. Suppose $Q^{\downarrow}(\gamma)$ towards $Q(\gamma)$. Then $\forall\gamma'{\prec}\gamma\forall\beta.[P^{\downarrow}(\beta) \Rightarrow P^{\downarrow}(\beta + 2^{\gamma'})]$, and we must show $\forall\beta.[P^{\downarrow}(\beta) \Rightarrow P^{\downarrow}(\beta + 2^\gamma)]$. This follows from $\beta + 2^\gamma = \bigcup_{\gamma' \prec \gamma} \beta + 2^{\gamma'}$.

It follows that $\mathrm{Ind}(Q)$. By TI_α we get $Q^{\downarrow}(\alpha)$. Hence by $\mathrm{Ind}(Q)$ again we have $Q(\alpha)$. As explained above we are done. ∎

The general idea of the above proofs is that the stronger axiom schema is proved by applying the weaker schema to more complicated formulas (Q as compared to P). This procedure can be iterated as long as the more complicated formulas remain well-formed. In the case of Peano arithmetic we can iterate this procedure finitely many times.

In Exercise 5.32 one can find some laws of ordinal arithmetic needed in the proof of the following lemma.

5.3.33 Lemma *Define*

$$\begin{array}{rclcrcl} 2_0 &=& \omega \cdot 2, & \qquad & \omega_0 &=& 1, \\ 2_{n+1} &=& 2^{2_n}; & & \omega_{n+1} &=& \omega^{\omega_n}. \end{array}$$

Then $2_n = \omega_n$ for $n \geq 2$.

Proof Note that $2^{\omega\cdot\beta} = (2^\omega)^\beta = \omega^\beta$. Therefore $2_1 = 2^{\omega\cdot 2} = \omega^2$ and hence

$$2_2 = 2^{\omega^2} = 2^{\omega\cdot\omega} = \omega^\omega = \omega_2.$$

As for $n \geq 2$ one has $\omega \cdot \omega_n = \omega \cdot \omega^{\omega_{n-1}} = \omega^{1+\omega_{n-1}} = \omega^{\omega_{n-1}} = \omega_n$, it follows that

$$2_{n+1} = 2^{2_n} = 2^{\omega_n} = 2^{\omega \cdot \omega_n} = \omega^{\omega_n} = \omega_{n+1}.$$

Therefore we are done by induction. ■

5.3.34 Theorem (Gentzen (reference)) TI_α *is provable in PA for every ordinal* $\alpha < \epsilon_0$.

Proof By Proposition 5.3.29 we get TI_ω. Hence $\mathrm{TI}_{\omega \cdot 2}$ by Proposition 5.3.31. By iterating Proposition 5.3.32 a sufficient number of times, this yields TI_{2_n}. By Lemma 5.3.33 this exhausts the ordinals under $\epsilon_0 = \bigcup_{n \in \omega} \omega_n$. ■

Gentzen (1936a) showed that the schema TI_{ϵ_0} implies the consistency of Peano Arithmetic. Therefore by Gödel's second incompleteness theorem, TI_{ϵ_0} cannot be proved in PA. A direct proof was given in Gentzen (1943) that some instance of TI_{ϵ_0} is not derivable in PA. This gives another consistency proof of PA.

Transfinite recursion

The computational counterpart of transfinite induction is *transfinite recursion* TR, a principle of definition which can be used to measure computational strength. By a translation of Gentzen's argument we shall show that every function which can be defined by TR_α for some ordinal $\alpha < \epsilon_0$, is definable in Gödel's $\mathcal{T}$. Thus we will establish a lower bound to the computational strength of Gödel's $\mathcal{T}$. This subsection closely follows the development of Terlouw (1982). Quantified statements like $\forall F{:}(0 \to A)$ are to be interpreted as ranging over (open) terms of the given type, in this case over $\Lambda_{\mathcal{T}}(0 \to A)$. Alternatively this subsection can be interpreted as taking place in PA^ω, classical arithmetic of arbitrary higher type, see Troelstra (1973). The presentation is again informal.

Transfinite recursion is a generalization of course-of-value recursion, which is transfinite recursion up to ω. Let us recall this course-of-value recursion. Given a function $f : \omega \to \omega$, define $\overline{f} : \omega \to \omega$ by

$$\overline{f}(n) \triangleq \langle f(0), \ldots, f(n-1) \rangle,$$

where $\langle n_1 \ldots n_k \rangle$, with $n_1, \ldots, n_k, k \in \omega$, is some computable coding of finite lists of natural numbers. In particular $\overline{f}(0) = \langle\ \rangle$. The principle of course-of-value recursion states that if g is a computable function of two arguments, then there exists a computable function f such that[3]

$$f(n) \quad = \quad g(\overline{f}(n), n).$$

In this way $f(n)$ is computed[4] from all previous values of $f(n)$ (and n), not just from $f(n-1)$. The proof is as follows. The idea is to show first that $\overline{f}$ is computable and then f is computable as well, as $f(n) = (\overline{f}(n+1))_n$, where $(\ )_n$ takes the nth member of a list. Now $\overline{f}$ can be defined as follows

$$\begin{array}{rcl} \overline{f}(0) & = & g(\langle\ \rangle, 0); \\ \overline{f}(n+1) & = & g(\overline{f}(n) * \langle f(n) \rangle, n) \\ & = & g(\overline{f}(n) * \langle g(\overline{f}(n), n) \rangle, n) \\ & = & h(\overline{f}(n), n). \end{array}$$

Here $L * \langle k \rangle$ denotes the coded list obtained from concatenating the list (with code) L and the single element $k \in \omega$. Note that h is computable and therefore also $\overline{f}$.

This idea of obtaining f by ordinary recursion is even easier in $\boldsymbol{\lambda}_{\mathcal{T}}$, as it is simpler to code finite lists using higher types. Let $M_1, \ldots, M_n$ be a finite list of terms of type A, with $n \geq 0$. Then we can code this list by

$$[M_1, \ldots, M_n] : 0 \to A$$

defined as follows.

$$[M_1, \ldots, M_n](k) = \begin{cases} M_k, & \text{if } k < n; \\ 0^A\ (\triangleq \lambda \vec{x}.0^0), & \text{otherwise.} \end{cases}$$

5.3.35 Definition Given an $F \in \Lambda_{\mathcal{T}}(0 \to A)$ and ordinals α, β, specify $[F]^{\downarrow\alpha} : 0 \to A$ by

$$[F]^{\downarrow\alpha}(\beta) \triangleq \begin{cases} F(\beta) & \text{if } \beta \prec \alpha, \\ 0^A & \text{otherwise.} \end{cases}$$

By convention, 'otherwise' includes the cases in which α, β are not ordinals, and the cases in which $\alpha = \beta$ or $\alpha \prec \beta$. Then $[F]^{\downarrow\alpha}$ is definable as element in $\Lambda_{\mathcal{T}}(0 \to A)$.

5.3.36 Definition The class of functionals *definable* by TR_α is the smallest class of functionals which contains all primitive recursive functionals

[3] A course-of-value recursion may have parameters: in general it provides a function $f(n, \vec{x})$ such that

$$f(n, \vec{x}) \quad = \quad g(\overline{f}(n, \vec{x}), n, \vec{x}),$$

where $\overline{f}(n, \vec{x}) \triangleq \langle f(0, \vec{x}), \ldots, f(n-1, \vec{x}) \rangle$ and $\overline{f}(0, \vec{x}) \triangleq \langle\ \rangle$.

[4] Actually n can be obtained from $\overline{f}(n)$, so it is not needed. We keep the extra n to make the schema of course-of-value recursion more similar to tranfinite recursion.

and is closed under the definition schema TR_α, defining $F : 0{\to}A$ from $G : (0{\to}A){\to}0{\to}A$ as follows:

$$\mathrm{TR}_\alpha \qquad \begin{cases} F(\beta) \triangleq G([F]^{\downarrow\beta}, \beta), & \text{if } \beta \prec \alpha, \\ F(\beta) \triangleq 0^A, & \text{otherwise.} \end{cases}$$

The following lemma is to be understood as the computational counterpart of Proposition 5.3.29, with the primitive recursive functionals taking over the role of PA.

5.3.37 Proposition *Every functional definable by the schema* TR_ω *is* $\mathcal{T}$*-definable.*

Proof Given $G \in \Lambda_{\mathcal{T}}((0{\to}A){\to}0{\to}A)$, we must show $F : 0{\to}A$, determined by $F(\beta) = G([F]^{\downarrow\beta}, \beta)$ (if $\beta \prec \omega$, otherwise 0^A), is $\mathcal{T}$-definable. We first show that $[F]^{\downarrow\beta}$ is $\mathcal{T}$-definable in β. That is, there exists an $H \in \Lambda_{\mathcal{T}}(0{\to}0{\to}A)$ such that $H(\beta,\gamma) = [F]^{\downarrow\beta}(\gamma)$ (if $\beta \prec \omega$, otherwise $0^{0\to A}$), as determined by Definition 5.3.35. This H satisfies

$$\begin{aligned} H(0,\gamma) &= 0^A, \\ H(\beta+1,\gamma) &= \begin{cases} F(\beta) = H(\beta,\gamma), & \text{if } \gamma \prec \beta \\ F(\beta) = G(H(\beta),\beta), & \text{if } \gamma = \beta \\ 0^A, & \text{otherwise.} \end{cases} \end{aligned}$$

Since this is of the form

$$\begin{aligned} H(0) &= \lambda\gamma.0^A \\ H(\beta+1) &= \lambda\gamma.P(H(\beta),\beta,\gamma), \end{aligned}$$

for some P primitive recursive, it follows that H is primitive recursive. So for $\boldsymbol{\beta} \prec \omega$

$$F(\beta) = [F]^{\downarrow\beta+1}(\beta) = H(\beta+1,\beta)$$

(and otherwise by definition $F(\beta) = 0^A = H(\beta+1,\beta)$) is $\mathcal{T}$-definable. ∎

The general idea of the proofs below is that the stronger schema is obtained by applying the weaker schema to functionals of more complex types.

5.3.38 Proposition *Every functional definable by schema* $\mathrm{TR}_{\omega\cdot 2}$ *is definable by schema* TR_ω*.*

Proof Given $G \in \Lambda_{\mathcal{T}}((0{\to}A){\to}0{\to}A)$ specify $F : 0{\to}A$ by

$$\begin{aligned} F(\beta) &\triangleq G([F]^{\downarrow\beta}, \beta), & \text{if } \beta \prec \omega\cdot 2, \\ F(\beta) &\triangleq 0^A, & \text{otherwise,} \end{aligned}$$

towards its definability in TR_ω. The following $F_1 = [F]^{\downarrow\omega}$ is definable in TR_ω:

$$\begin{array}{lcll} F_1(\beta) & \triangleq & G([F_1]^{\downarrow\beta}, \beta), & \text{if } \beta \prec \omega, \\ F_1(\beta) & \triangleq & 0^A, & \text{otherwise.} \end{array}$$

Moreover, by TI_ω, one has $F(\beta) = F_1(\beta)$, for $\beta \prec \omega$. So F_1 is the 'restriction' of F to ω. Define $H(\beta) = [F]^{\downarrow\omega+\beta}$, for $\beta \prec \omega$. Then $H : 0{\to}0{\to}A$ can be specified by

$$\begin{array}{lcl} H(0,\gamma) & = & F_1(\gamma), \\ H(\beta+1,\gamma) & = & \left\{ \begin{array}{ll} H(\beta,\gamma), & \text{if } \gamma \prec \omega+\beta, \\ G(H(\beta),\gamma), & \text{if } \gamma = \omega+\beta, \\ 0^A, & \text{otherwise.} \end{array} \right. \end{array}$$

Therefore H is TR_ω-definable, by Proposition 5.3.37. Now F can be defined explicitly from F_1 and H:

$$F(\gamma) = \left\{ \begin{array}{ll} F_1(\gamma), & \text{if } \gamma \prec \omega, \\ G(H(\beta),\gamma), & \text{if } \gamma = \omega+\beta, \\ 0^A, & \text{otherwise.} \end{array} \right.$$

Therefore F is also TR_ω-definable. ■

5.3.39 Lemma *Every functional definable by the schema* TR_{2^α} *is definable by the schema* TR_α*, for all* $\alpha < \epsilon_0$.

Proof Given $G \in \Lambda_{\mathcal{T}}((0{\to}A){\to}0{\to}A)$, specify $F_0 : 0{\to}A$ by TR_{2^α}:

$$\begin{array}{lcll} F_0(\beta) & = & G([F_0]^{\downarrow\beta}, \beta), & \text{if } \beta \prec 2^\alpha, \\ F_0(\beta) & = & 0^A, & \text{otherwise,} \end{array}$$

towards its definability in TR_α.

Define for $\gamma \prec \epsilon_0$

$$\mathrm{OK}(F,\gamma) \stackrel{\Delta}{\Longleftrightarrow} \forall\beta \prec \gamma . F(\beta) = F_0(\beta).$$

This is equivalent to

$$F(\delta) = \left\{ \begin{array}{ll} G([F]^{\downarrow\delta}, \delta), & \text{if } \delta \prec \gamma, \\ 0^A, & \text{otherwise.} \end{array} \right.$$

We now look for a functional $H \in \Lambda_{\mathcal{T}}((0{\to}A){\to}0^3{\to}A)$ such that for all $\gamma \prec \epsilon_0$ we have

$$\forall F{:}(0{\to}A)\,\forall\beta.[\mathrm{OK}(F,\beta) \Rightarrow \mathrm{OK}(H(F,\gamma,\beta), \beta + 2^\gamma)]. \qquad P(\gamma)$$

Using H, take $F = H(0^{0\to A}, \alpha, 0)$. Then F is TR_α-definable and, since

$\mathrm{OK}(0^{0\to A},0)$ is vacuously true, by $P(\alpha)$ applied to $F_1 = 0^{0\to A}$ and $\beta = 0$ we get the required $\mathrm{OK}(F,2^\alpha)$. We define $H(F,\gamma,\beta)$ by TR_α on $\gamma \prec \alpha$ as follows (if $\mathrm{Lim}(\gamma)$ we use the function p from Lemma 5.3.26):

$$
\begin{aligned}
H(F,0,\beta,\delta) &\triangleq \begin{cases} F(\delta), & \text{if } \delta \prec \beta \preceq 2^\alpha, \\ G([F]^{\downarrow\beta},\delta), & \text{if } \delta = \beta \preceq 2^\alpha, \\ 0^A, & \text{otherwise;} \end{cases} \\
H(F,\gamma+1,\beta,\delta) &\triangleq \quad H(H(F,\gamma,\beta),\gamma,\beta+2^\gamma,\delta); \\
&\qquad\qquad \text{in the case } \mathrm{Lim}(\gamma) \\
H(F,\gamma,\beta,\delta) &\triangleq \begin{cases} H(F,\mathrm{p}(\gamma,\delta,\beta),\beta,\delta), & \text{if } \delta \prec \beta+2^\gamma, \\ 0^A, & \text{otherwise.} \end{cases}
\end{aligned}
$$

We continue this definition one more step to define $H(F,\gamma,\beta,\delta)$ for $\gamma = \alpha$.

Finally we show $P(\gamma)$ by transfinite induction on $\gamma \preceq \alpha$ in the metalanguage.

Case $\gamma = 0$. Suppose $\mathrm{OK}(F,\beta)$. Then for $F' = H(F,0,\beta)$ one has

$$
F'(\delta) = \begin{cases} F(\delta) = G([F]^{\downarrow\delta},\delta), & \text{if } \delta \prec \beta, \\ G([F]^{\downarrow\delta},\delta), & \text{if } \delta = \beta, \\ 0^A, & \text{otherwise.} \end{cases}
$$

As $2^0 = 1$, this entails $P(0)$, stating $\mathrm{OK}(F',\beta+1)$.

Case $\gamma+1$. Analogous to the successor case in the proof of Proposition 5.3.32, we prove $P(\gamma+1)$ by applying $P(\gamma)$ twice, once with β and once with $\beta+2^\gamma$. Indeed, given β and F we infer:

$$
\begin{aligned}
\mathrm{OK}(F,\gamma) &\Rightarrow \mathrm{OK}(H(F,\gamma,\beta),\beta+2^\gamma) \\
&\Rightarrow \mathrm{OK}(H(H(F,\gamma,\beta),\gamma,\beta+2^\gamma),\beta+2^\gamma+2^\gamma) \\
&\Leftrightarrow \mathrm{OK}(H(F,\gamma+1,\beta),\beta+2^{\gamma+1}),
\end{aligned}
$$

by definition of H and $2^\gamma + 2^\gamma = 2^{\gamma+1}$.

Case $\mathrm{Lim}(\gamma)$. Suppose that $P(\gamma')$ holds for all $\gamma' \prec \gamma$. Assume $\mathrm{OK}(F,\beta)$ towards proving $\mathrm{OK}(H(F,\gamma,\beta),\beta+2^\gamma)$. Let $\delta \prec \beta+2^\gamma$ be arbitrary. Now Lemma 5.3.26 states that $\gamma' = \mathrm{p}(\gamma,\delta,\beta)$ satisfies $\gamma' \prec \gamma$ and $\delta \prec \beta+2^{\gamma'}$. Then $\mathrm{OK}(H(F,\gamma',\beta),\beta+2^{\gamma'})$ by $P(\gamma')$. So $H(F,\gamma,\beta,\delta) = H(F,\gamma',\beta,\delta) = F_0(\delta)$. Since δ was arbitrary, it indeed follows that $\mathrm{OK}(H(F,\gamma,\beta),\beta+2^\gamma)$. ∎

5.3.40 Theorem *Every functional definable by the schema* TR_α *for some ordinal* $\alpha \prec \epsilon_0$ *is* $\mathcal{T}$*-definable.*

Proof Analogous to the proof of Theorem 5.3.34. ∎

Theorem 5.3.40 shows that ϵ_0 is a lower bound for the computational strength of Gödel's system $\mathcal{T}$. It can be shown that ϵ_0 is a sharp bound for $\mathcal{T}$, see Tait (1965), Howard (1970) and Schwichtenberg (1975).

5.4 Spector's system $\mathcal{B}$: bar recursion

Spector (1962) extends Gödel's $\mathcal{T}$ with a definition schema called bar recursion.[5] Bar recursion is a principle of definition by recursion on a well-founded tree of finite sequences of functionals of the same type. For the formulation of bar recursion we need finite sequences of functionals of type A. These can conveniently be encoded by pairs consisting of a functional of type N and one of type $\mathsf{N}{\rightarrow}A$. The intuition is that the pair $\langle x, C\rangle$ encodes the sequence of the first x values of C, that is, $C(0), \ldots, C(x-1)$. We need auxiliary functionals to extend finite sequences of any type. A convenient choice is the primitive recursive functional $\mathrm{Ext}_A : (\mathsf{N}{\rightarrow}A){\rightarrow}\mathsf{N}{\rightarrow}A{\rightarrow}\mathsf{N}{\rightarrow}A$ defined by:

$$\mathrm{Ext}_A(C, x, a, y) = \begin{cases} C(y) & \text{if } y < x, \\ a & \text{otherwise.} \end{cases}$$

We shall often omit the type subscript in Ext_A, and abbreviate $\mathrm{Ext}(C, x, a)$ by $C *_x a$ and $\mathrm{Ext}(C, x, 0^A)$ by $[C]_x$. We are now in a position to formulate the schema of *bar recursion*:[6]

$$\varphi(x, C) = \begin{cases} G(x, C) & \text{if } Y[C]_x < x, \\ H(\lambda a^A.\varphi(x+1, C *_x a), x, C) & \text{otherwise.} \end{cases}$$

The case distinction is governed by $Y[C]_x < x$, the so-called *bar condition.* The base case of bar recursion is the case in which the bar condition holds. In the other case φ is recursively called on all extensions of the (encoded) finite sequence.

A key feature of bar recursion is its proof-theoretic strength as established in Spector (1962). As a consequence, some properties of bar recursion are hard to prove, such as SN and the existence of a model. As an example of the latter phenomenon we shall show that the full set-theoretic model of Gödel's $\mathcal{T}$ is not a model of bar recursion.

Consider functionals Y, G, H defined by $G(x, C) = 0$, $H(Z, x, C) = 1 + Z(1)$ and

$$Y(F) = \begin{cases} 0 & \text{if } F(m) = 1 \text{ for all } m, \\ n & \text{otherwise, where } n = \min\{m \mid F(m) \neq 1\}. \end{cases}$$

[5] For the purpose of characterizing the provably recursive functions of analysis, yielding a consistency proof of analysis.

[6] Spector uses $[C]_x$ instead of C as last argument of G and H. Both formulations are easily seen to be equivalent since they are schematic in G, H (as well as in Y).

Let $1^{\mathsf{N}\to\mathsf{N}}$ be the constant 1 function. The crux of Y is that $Y[1^{\mathsf{N}\to\mathsf{N}}]_x = x$ for all x, so that the bar recursion is not well-founded. We calculate

$$\varphi(0, 1^{\mathsf{N}\to\mathsf{N}}) = 1 + \varphi(1, 1^{\mathsf{N}\to\mathsf{N}}) = \cdots = n + \varphi(n, 1^{\mathsf{N}\to\mathsf{N}}) = \cdots$$

which shows that φ is not well defined.

Syntax of $\boldsymbol{\lambda}_{\mathcal{B}}$

We now formalize Spector's $\mathcal{B}$ as an extension of Gödel's $\mathcal{T}$ called $\boldsymbol{\lambda}_{\mathcal{B}}$.

5.4.1 Definition The theory *Spector's* $\mathcal{B}$, written $\boldsymbol{\lambda}_{\mathcal{B}}$, is defined on the set of types $\mathbb{T}(\boldsymbol{\lambda}_{\mathcal{B}}) = \mathbb{T}(\boldsymbol{\lambda}_{\mathcal{T}})$. We use A^{N} as shorthand for the type $\mathsf{N}\to A$. The *terms* of $\boldsymbol{\lambda}_{\mathcal{B}}$ are obtained by adding constants:

$$\mathsf{B}_{(A,B)} : (A^{\mathsf{N}}\to\mathsf{N})\to(\mathsf{N}\to A^{\mathsf{N}}\to B)\to((A\to B)\to\mathsf{N}\to A^{\mathsf{N}}\to B)\to\mathsf{N}\to A^{\mathsf{N}}\to B$$
$$\mathsf{B}^c_{A,B} : (A^{\mathsf{N}}\to\mathsf{N})\to(\mathsf{N}\to A^{\mathsf{N}}\to B)\to((A\to B)\to\mathsf{N}\to A^{\mathsf{N}}\to B)\to\mathsf{N}\to A^{\mathsf{N}}\to\mathsf{N}\to B$$

for all types A, B to the constants of $\boldsymbol{\lambda}_{\mathcal{T}}$. The set of (closed) terms of $\boldsymbol{\lambda}_{\mathcal{B}}$ (of type A) is denoted by $\Lambda^{\emptyset}_{\mathcal{B}}(A)$. The *formulas* of $\boldsymbol{\lambda}_{\mathcal{B}}$ are equations between terms of $\boldsymbol{\lambda}_{\mathcal{B}}$ (of the same type). The *theory* of $\boldsymbol{\lambda}_{\mathcal{B}}$ extends the theory of $\boldsymbol{\lambda}_{\mathcal{T}}$ with the above schema of bar recursion (with $\mathsf{B}YGH$ abbreviated by φ). The *reduction relation* $\to_{\mathcal{B}}$ of $\boldsymbol{\lambda}_{\mathcal{B}}$ extends $\to_{\mathcal{T}}$ by adding the following (schematic) rules for the constants B, B^c (omitting type annotations A, B):

$$\mathsf{B}YGHXC \to_{\mathcal{B}} \mathsf{B}^cYGHXC(X \dot{-} Y[C]_X)$$
$$\mathsf{B}^cYGHXC(\mathsf{S}^+N) \to_{\mathcal{B}} GXC$$
$$\mathsf{B}^cYGHXC\mathsf{0} \to_{\mathcal{B}} H(\lambda a.\mathsf{B}YGH(\mathsf{S}^+X)(C *_X a))XC.$$

The reduction rules for B, B^c require some explanation. First note that $x \dot{-} Y[C]_x = 0$ iff $Y[C]_x \geq x$, so that testing $x \dot{-} Y[C]_x = 0$ amounts to evaluating the (negation) of the bar condition. Consider a primitive recursive functional If_0 satisfying $\mathrm{If}_0\mathsf{0}M_1M_0 = M_0$ and $\mathrm{If}_0(\mathsf{S}^+P)M_1M_0 = M_1$. A straightforward translation of the definition schema of bar recursion into a reduction rule,

$$\mathsf{B}YGHXC \to \mathrm{If}_0\ (X \dot{-} [C]_X)(GXC)(H(\lambda x.\mathsf{B}YGH(\mathsf{S}^+X)(C *_X x))XC),$$

would lead to infinite reduction sequences (the innermost B can be reduced again and again). It turns out to be necessary to evaluate the Boolean first. This has been achieved by the interplay between B and B^c.

Theorem 5.3.5, Lemma 5.3.6 and Theorem 5.3.9 carry over from $\boldsymbol{\lambda}_{\mathcal{T}}$ to $\boldsymbol{\lambda}_{\mathcal{B}}$ with proofs that are easy generalizations. We now prove SN for $\boldsymbol{\lambda}_{\mathcal{B}}$ and then obtain CR for $\boldsymbol{\lambda}_{\mathcal{B}}$ using Newman's Lemma 5.3.8. This proof is considerably more difficult than for $\boldsymbol{\lambda}_{\mathcal{T}}$, which reflects the meta-mathematical fact that

$\boldsymbol{\lambda}_{\mathcal{B}}$ corresponds to analysis (see Spector (1962)), whereas $\boldsymbol{\lambda}_{\mathcal{T}}$ corresponds to arithmetic. We start with defining hereditary finiteness for *sets* of terms, an analytical notion which plays a similar role as the arithmetical notion of computability for *terms* in the case of $\boldsymbol{\lambda}_{\mathcal{T}}$. Both are logical relations in the sense of Section 3.3, although hereditary finiteness is defined on the power set. Both computability and hereditary finiteness strengthen the notion of strong normalization: both are shown to hold by induction on terms. For meta-mathematical reasons, notably the consistency of analysis, it should not come as a surprise that we need an analytical induction loading in the case of $\boldsymbol{\lambda}_{\mathcal{B}}$.

5.4.2 Definition

(i) For every set $\mathcal{X} \subseteq \Lambda_B$, let $\mathrm{nf}(\mathcal{X})$ denote the set of B-normal forms of terms from $\mathcal{X}$. For all $\mathcal{X} \subseteq \Lambda_B(A \to B)$ and $\mathcal{Y} \subseteq \Lambda_B(A)$, let $\mathcal{X}\mathcal{Y}$ denote the set of all applications of terms in $\mathcal{X}$ to terms in $\mathcal{Y}$. Furthermore, if $M(x_1, \ldots, x_k)$ is a term with free variables $x_1, \ldots, x_k$, and $\mathcal{X}_1, \ldots, \mathcal{X}_k$ are sets of terms such that every term from $\mathcal{X}_i$ has the same type as x_i, $1 \leq i \leq k$, then we denote the set of all corresponding substitution instances by $M(\mathcal{X}_1, \ldots, \mathcal{X}_k)$.

(ii) By induction on the type A we say that a set $\mathcal{X}$ of closed terms of type A is *hereditarily finite*, written $\mathcal{X} \in \mathcal{HF}_A$:

$$\mathcal{X} \in \mathcal{HF}_{\mathsf{N}} \Leftrightarrow \mathcal{X} \subseteq \Lambda_B^{\emptyset}(\mathsf{N}) \cap \mathsf{SN} \text{ and } \mathrm{nf}(\mathcal{X}) \text{ is finite}$$
$$\mathcal{X} \in \mathcal{HF}_{A \to B} \Leftrightarrow \mathcal{X} \subseteq \Lambda_B^{\emptyset}(A \to B) \text{ and } \mathcal{X}\mathcal{Y} \in \mathcal{HF}_B \text{ whenever } \mathcal{Y} \in \mathcal{HF}_A.$$

(iii) A closed term M is *hereditarily finite*, written $M \in \mathrm{HF}^0$, if $\{M\} \in \mathcal{HF}$.

(iv) If all the free variables of a term $M(x_1, \ldots, x_k)$ occur among $x_1, \ldots, x_k$, then $M(x_1, \ldots, x_k)$ is *hereditarily finite*, written $M(x_1, \ldots, x_k) \in \mathrm{HF}$, if $M(\mathcal{X}_1, \ldots, \mathcal{X}_k)$ is hereditarily finite for all $\mathcal{X}_i \in \mathcal{HF}$, $1 \leq i \leq k$, of appropriate types.

We will show in Theorem 5.4.15 that every bar recursive term is hereditarily finite, and conclude that it is strongly normalizing.

Some basic properties of hereditary finiteness are summarized in the following lemmas. We use vector notation to abbreviate sequences of arguments of appropriate types both for terms and for sets of terms. For example, $M\vec{N}$ abbreviates $MN_1 \cdots N_k$ and $\mathcal{X}\vec{\mathcal{Y}}$ stands for $\mathcal{X}\mathcal{Y}_1 \cdots \mathcal{Y}_k$. The first two lemmas are instrumental for proving hereditary finiteness.

5.4.3 Lemma *The set $\mathcal{X} \subseteq \Lambda_B^{\emptyset}(A_1 \to \cdots \to A_n \to \mathsf{N})$ is hereditarily finite if and only if $\mathcal{X}\vec{\mathcal{Y}} \in \mathcal{HF}_{\mathsf{N}}$ for all $\mathcal{Y}_1 \in \mathcal{HF}_{A_1}, \ldots, \mathcal{Y}_n \in \mathcal{HF}_{A_n}$.*

Proof By induction on n, applying Definition 5.4.2. ∎

5.4.4 Definition Given two sets of terms $\mathcal{X}, \mathcal{X}' \subseteq \Lambda_B^{\emptyset}$, we say that $\mathcal{X}$ is *adfluent* with $\mathcal{X}'$ if every maximal reduction sequence starting in $\mathcal{X}$ passes through a reduct of a term in $\mathcal{X}'$. Let $A \equiv A_1 \to \cdots \to A_n \to \mathsf{N}$ with $n \geq 0$ and let $\mathcal{X}, \mathcal{X}' \subseteq \Lambda_B^{\emptyset}(A)$. We say that $\mathcal{X}$ is *hereditarily adfluent* with $\mathcal{X}'$ if $\mathcal{X}\vec{\mathcal{Y}}$ is adfluent with $\mathcal{X}'\vec{\mathcal{Y}}$, for all $\mathcal{Y}_1 \in \mathcal{HF}_{A_1}, \ldots, \mathcal{Y}_n \in \mathcal{HF}_{A_n}$.

5.4.5 Lemma *Let $\mathcal{X}, \mathcal{X}' \subseteq \Lambda_B^{\emptyset}(A)$ be such that $\mathcal{X}$ is hereditarily adfluent with $\mathcal{X}'$. Then $\mathcal{X} \in \mathcal{HF}_A$ whenever $\mathcal{X}' \in \mathcal{HF}_A$.*

Proof With conditions as in the lemma let $A \equiv A_1 \to \cdots \to A_n \to \mathsf{N}$. Assume $\mathcal{X}' \in \mathcal{HF}_A$. Let $\mathcal{Y}_1 \in \mathcal{HF}_{A_1}, \ldots, \mathcal{Y}_n \in \mathcal{HF}_{A_n}$; then $\mathcal{X}\vec{\mathcal{Y}}$ is adfluent with $\mathcal{X}'\vec{\mathcal{Y}}$. It follows that $\mathcal{X}\vec{\mathcal{Y}} \subseteq \mathsf{SN}$ since $\mathcal{X}'\vec{\mathcal{Y}} \subseteq \mathsf{SN}$ and $\mathrm{nf}(\mathcal{X}\vec{\mathcal{Y}}) \subseteq \mathrm{nf}(\mathcal{X}'\vec{\mathcal{Y}})$, so $\mathrm{nf}(\mathcal{X}\vec{\mathcal{Y}})$ is finite since $\mathrm{nf}(\mathcal{X}'\vec{\mathcal{Y}})$ is. Applying Lemma 5.4.3 we obtain $\mathcal{X} \in \mathcal{HF}_A$. ∎

Note that the above lemma holds in particular if $n = 0$: that is, if $A \equiv \mathsf{N}$.

5.4.6 Lemma *Let A be a type of $\boldsymbol{\lambda}_{\mathcal{B}}$. Then:*

(i) $\mathrm{HF}_A \subseteq \mathsf{SN}$;
(ii) $\mathsf{0}^A \in \mathrm{HF}_A$;
(iii) $\mathrm{HF}_A^0 \subseteq \mathsf{SN}$.

Proof We prove (ii) and (iii) by simultaneous induction on A. Then (i) follows immediately. Obviously, $\mathsf{0} \in \mathrm{HF}_{\mathsf{N}}$ and $\mathrm{HF}_{\mathsf{N}}^0 \subseteq \mathsf{SN}$. For the induction step $A \to B$, assume (ii) and (iii) hold for all smaller types. If $M \in \mathrm{HF}_{A\to B}^0$, then by the induction hypothesis (ii) $\mathsf{0}^A \in \mathrm{HF}_A^0$, and $M\mathsf{0}^A \in \mathrm{HF}_B^0$, so $M\mathsf{0}^A$ is SN by the induction hypothesis (iii), and hence M is SN. Recall that $\mathsf{0}^{A\to B} \equiv \lambda x^A.\mathsf{0}^B$. Let $\mathcal{X} \in \mathcal{HF}_A$, then $\mathcal{X} \subseteq \mathsf{SN}$ by the induction hypothesis. It follows that $\mathsf{0}^{A\to B}\mathcal{X}$ is hereditarily adfluent with $\mathsf{0}^B$. By the induction hypothesis we have $\mathsf{0}^B \in \mathrm{HF}_B$, so $\mathsf{0}^{A\to B}\mathcal{X} \in \mathcal{HF}_B$ by Lemma 5.4.5. Therefore $\mathsf{0}^{A\to B} \in \mathrm{HF}_{A\to B}$. ∎

The proofs of the following three lemmas are left to the reader.

5.4.7 Lemma *Every reduct of a hereditarily finite term is hereditarily finite.*

5.4.8 Lemma *Subsets of hereditarily finite sets of terms are hereditarily finite.*

In particular elements of a hereditarily finite set are hereditarily finite.

5.4.9 Lemma *Finite unions of hereditarily finite sets are hereditarily finite.*

In this connection of course only unions of the same type make sense.

5.4.10 Lemma *The hereditarily finite terms are closed under application.*

Proof Immediate from Definition 5.4.2. ∎

5.4.11 Lemma *The hereditarily finite terms are closed under lambda abstraction.*

Proof Let $M(x, x_1, \ldots, x_k) \in \mathrm{HF}$ be a term all of whose free variables occur among $x, x_1, \ldots, x_k$. We have to prove $\lambda x.M(x, x_1, \ldots, x_k) \in \mathrm{HF}$; that is,

$$\lambda x.M(x, \mathcal{X}_1, \ldots, \mathcal{X}_k) \in \mathcal{HF}$$

for given $\vec{\mathcal{X}} = \mathcal{X}_1, \ldots, \mathcal{X}_k \in \mathcal{HF}$ of appropriate types. Let $\mathcal{X} \in \mathcal{HF}$ be of the same type as the variable x, so $\mathcal{X} \subseteq \mathsf{SN}$ by Lemma 5.4.6. We also have $M(x, \vec{\mathcal{X}}) \subseteq \mathsf{SN}$ by the assumption on M and Lemma 5.4.6. It follows that $(\lambda x.M(x, \vec{\mathcal{X}}))\mathcal{X}$ is hereditarily adfluent with $M(\mathcal{X}, \vec{\mathcal{X}})$. Again by the assumption on M we have $M(\mathcal{X}, \vec{\mathcal{X}}) \in \mathcal{HF}$, hence $(\lambda x.M(x, \vec{\mathcal{X}}))\mathcal{X} \in \mathcal{HF}$ by Lemma 5.4.5. We conclude that $\lambda x.M(x, \vec{\mathcal{X}}) \in \mathcal{HF}$, so $\lambda x.M(x, x_1, \ldots, x_k) \in \mathrm{HF}$. ∎

5.4.12 Theorem *Every term of $\boldsymbol{\lambda}_{\mathcal{T}}$ is hereditarily finite.*

Proof By Lemma 5.4.10 and Lemma 5.4.11, the hereditarily finite terms are closed under application and lambda abstraction, so it suffices to show that the constants and the variables are hereditarily finite. Variables and the constant $\mathsf{0}$ are obviously hereditarily finite. Regarding S^+, let $\mathcal{X} \in \mathcal{HF}_{\mathsf{N}}$; then $\mathsf{S}^+\mathcal{X} \subseteq \Lambda^{\emptyset}_{\mathcal{B}}(\mathsf{N}) \cap \mathsf{SN}$ and $\mathrm{nf}(\mathsf{S}^+\mathcal{X})$ is finite since $\mathrm{nf}(\mathcal{X})$ is finite. Hence $\mathsf{S}^+\mathcal{X} \in \mathcal{HF}_{\mathsf{N}}$, so S^+ is hereditarily finite.

It remains to prove that the constants R_A are hereditarily finite. Let $\mathcal{M}, \mathcal{N}, \mathcal{X} \in \mathcal{HF}$ be of appropriate types and consider $\mathsf{R}_A\mathcal{M}\mathcal{N}\mathcal{X}$. We have in particular $\mathcal{X} \in \mathcal{HF}_{\mathsf{N}}$, so $\mathrm{nf}(\mathcal{X})$ is finite, and the proof of $\mathsf{R}_A\mathcal{M}\mathcal{N}\mathcal{X} \in \mathcal{HF}$ goes by induction on the largest numeral in $\mathrm{nf}(\mathcal{X})$. If $\mathrm{nf}(\mathcal{X}) = \{\mathsf{0}\}$, then $\mathsf{R}_A\mathcal{M}\mathcal{N}\mathcal{X}$ is hereditarily adfluent with $\mathcal{M}$. Since $\mathcal{M} \in \mathcal{HF}$ we can apply Lemma 5.4.5 to obtain $\mathsf{R}_A\mathcal{M}\mathcal{N}\mathcal{X} \in \mathcal{HF}$. For the induction step, assume $\mathsf{R}_A\mathcal{M}\mathcal{N}\mathcal{X}' \in \mathcal{HF}$ for all $\mathcal{X}' \in \mathcal{HF}$ such that the largest numeral in $\mathrm{nf}(\mathcal{X}')$ is n. Let, for some $\mathcal{X} \in \mathcal{HF}$, the largest numeral in $\mathrm{nf}(\mathcal{X})$ be $\mathsf{S}^+\mathsf{n}$. Define

$$\mathcal{X}' = \{X \mid \mathsf{S}^+X \text{ is a reduct of a term in } \mathcal{X}\}.$$

Then $\mathcal{X}' \in \mathcal{HF}$ since $\mathcal{X} \in \mathcal{HF}$, and the largest numeral in $\mathrm{nf}(\mathcal{X}')$ is n. It follows by the induction hypothesis that $\mathsf{R}_A\mathcal{M}\mathcal{N}\mathcal{X}' \in \mathcal{HF}$, so $\mathcal{N}(\mathsf{R}_A\mathcal{M}\mathcal{N}\mathcal{X}')\mathcal{X}' \in \mathcal{HF}$ and hence

$$\mathcal{N}(\mathsf{R}_A\mathcal{M}\mathcal{N}\mathcal{X}')\mathcal{X}' \cup \mathcal{M} \in \mathcal{HF},$$

by Lemmas 5.4.10, 5.4.9. We have that $\mathsf{R}_A\mathcal{MNX}$, is hereditarily adfluent with

$$\mathcal{N}(\mathsf{R}_A\mathcal{MNX}')\mathcal{X}' \cup \mathcal{M},$$

so $\mathsf{R}_A\mathcal{MNX} \in \mathcal{HF}$ by Lemma 5.4.5. This completes the induction step. ■

Before we can prove that B is hereditarily finite we need the following lemma.

5.4.13 Lemma *Let $\mathcal{Y}, \mathcal{G}, \mathcal{H}, \mathcal{X}, \mathcal{C} \in \mathcal{HF}$ be of appropriate type. Then*

$$\mathsf{B}\mathcal{YGHXC} \in \mathcal{HF},$$

whenever $\mathsf{B}\mathcal{YGH}(\mathsf{S}^+\mathcal{X})(\mathcal{C} *_{\mathcal{X}} \mathcal{A}) \in \mathcal{HF}$ for all $\mathcal{A} \in \mathcal{HF}$ of appropriate type.

Proof With the conditions as given, abbreviate $\mathsf{B}\mathcal{YGH}$ by $\mathcal{B}$ and $\mathsf{B}^c\mathcal{YGH}$ by $\mathcal{B}^c$. Assume $\mathcal{B}(\mathsf{S}^+\mathcal{X})(\mathcal{C} *_{\mathcal{X}} \mathcal{A}) \in \mathcal{HF}$ for all $\mathcal{A} \in \mathcal{HF}$. Below we will frequently and implicitly use that $\dot{-}, *, [\ ]$ are primitive recursive and hence hereditarily finite, and that hereditary finiteness is closed under application. Since hereditarily finite terms are strongly normalizable, we have that $\mathcal{BXC}$ is hereditarily adfluent with $\mathcal{B}^c\mathcal{XC}(\mathcal{X} \dot{-} \mathcal{Y}[\mathcal{C}]_{\mathcal{X}})$, and hence with $\mathcal{GCX} \cup \mathcal{H}(\lambda a.\mathcal{B}(\mathsf{S}^+\mathcal{X})(\mathcal{C} *_{\mathcal{X}} a))\mathcal{CX}$. It suffices to show that the latter set is in $\mathcal{HF}$. We have $\mathcal{GCX} \in \mathcal{HF}$, so by Lemma 5.4.9 the union is hereditarily finite if $\mathcal{H}(\lambda a.\mathcal{B}(\mathsf{S}^+\mathcal{X})(\mathcal{C} *_{\mathcal{X}} a))\mathcal{CX}$ is. It suffices that $\lambda a.\mathcal{B}(\mathsf{S}^+\mathcal{X})(\mathcal{C} *_{\mathcal{X}} a) \in \mathcal{HF}$, and this will follow by the assumption above. We first observe that $\{0^A\} \in \mathcal{HF}$ so $\mathcal{B}(\mathsf{S}^+\mathcal{X})(\mathcal{C} *_{\mathcal{X}} \{0^A\}) \in \mathcal{HF}$ and hence $\mathcal{B}(\mathsf{S}^+\mathcal{X})(\mathcal{C} *_{\mathcal{X}} a) \subseteq \mathsf{SN}$ by Lemma 5.4.6. Let $\mathcal{A} \in \mathcal{HF}$. Since $\mathcal{B}(\mathsf{S}^+\mathcal{X})(\mathcal{C} *_{\mathcal{X}} a), \mathcal{A} \subseteq \mathsf{SN}$ we have that $(\lambda a.\mathcal{B}(\mathsf{S}^+\mathcal{X})(\mathcal{C} *_{\mathcal{X}} a))\mathcal{A}$ is adfluent with $\mathcal{B}(\mathsf{S}^+\mathcal{X})(\mathcal{C} *_{\mathcal{X}} \mathcal{A}) \in \mathcal{HF}$ and hence hereditarily finite itself by Lemma 5.4.5. ■

We now arrive at the crucial step, where not only the language of analysis will be used, but also the axiom of dependent choice in combination with classical logic. We will reason by contradiction. Suppose B is not hereditarily finite. Then there are hereditarily finite $\mathcal{Y}, \mathcal{G}, \mathcal{H}, \mathcal{X}$ and $\mathcal{C}$ such that $\mathsf{B}\mathcal{YGHXC}$ is not hereditarily finite. We introduce the following abbreviations: $\mathcal{B}$ for $\mathsf{B}\mathcal{YGH}$ and $\mathcal{X}{+}\mathsf{n}$ for $\mathsf{S}^+(\cdots(\mathsf{S}^+\mathcal{X})\cdots)$ (n instances of S^+). By Lemma 5.4.13, there exists $\mathcal{U} \in \mathcal{HF}$ such that $\mathcal{B}(\mathcal{X}{+}1)(\mathcal{C} *_{\mathcal{X}} \mathcal{U})$ is not hereditarily finite. Hence again by Lemma 5.4.13, there exists $\mathcal{V} \in \mathcal{HF}$ such that

$$\mathcal{B}(\mathcal{X}{+}2)((\mathcal{C} *_{\mathcal{X}} \mathcal{U}) *_{\mathcal{X}+1} \mathcal{V})$$

is not hereditarily finite. Using dependent choice[7], let

$$\mathcal{D} = \mathcal{C} \cup (\mathcal{C} *_{\mathcal{X}} \mathcal{U}) \cup ((\mathcal{C} *_{\mathcal{X}} \mathcal{U}) *_{\mathcal{X}+1} \mathcal{V}) \cup \cdots$$

be the infinite union of the sets obtained by iterating the argument above. Note that all sets in the infinite union are hereditarily finite of type A^{N}. Since the union is infinite, it does not follow from Lemma 5.4.9 that $\mathcal{D}$ itself is hereditarily finite. However, since $\mathcal{D}$ has been built up from terms of type A^{N} having longer and longer initial segments in common, we will nevertheless be able to prove that $\mathcal{D} \in \mathcal{HF}$. Then we will arrive at a contradiction, since $\mathcal{Y}\mathcal{D} \in \mathcal{HF}$ implies that $\mathcal{Y}$ is bounded on $\mathcal{D}$, so that the bar condition is satisfied after finitely many steps, which conflicts with the construction process.

5.4.14 Lemma *The set $\mathcal{D}$ constructed above is hereditarily finite.*

Proof Let $\mathcal{N}, \vec{\mathcal{Z}} \in \mathcal{HF}$ be of appropriate type; that is, $\mathcal{N}$ of type N and $\vec{\mathcal{Z}}$ such that $\mathcal{D}\mathcal{N}\vec{\mathcal{Z}}$ is of type N. We have to show $\mathcal{D}\mathcal{N}\vec{\mathcal{Z}} \in \mathcal{HF}$. Since all elements of $\mathcal{D}$ are hereditarily finite we have $\mathcal{D}\mathcal{N}\vec{\mathcal{Z}} \subseteq \mathsf{SN}$. By an easy generalization of Theorem 5.3.9 we have WCR for $\boldsymbol{\lambda}_{\mathcal{B}}$, so by Newman's Lemma 5.3.8 we have $\mathcal{D}\mathcal{N}\vec{\mathcal{Z}} \subseteq \mathsf{CR}$. Since $\mathcal{N} \in \mathcal{HF}$ it follows that $\mathrm{nf}(\mathcal{N})$ is finite, say $\mathrm{nf}(\mathcal{N}) \subseteq \{\mathsf{0}, \ldots, \mathsf{n}\}$ for n large enough. It remains to show that $\mathrm{nf}(\mathcal{D}\mathcal{N}\vec{\mathcal{Z}})$ is finite. Since all terms in $\mathcal{D}\mathcal{N}\vec{\mathcal{Z}}$ are CR, their normal forms are unique. As a consequence we may apply a left-most innermost reduction strategy to any term $DN\vec{Z} \in \mathcal{D}\mathcal{N}\vec{\mathcal{Z}}$. At this point it might be helpful to remind the reader of the intended meaning of $*$: $C *_x A$ represents the finite sequence $C0, \ldots, C(x-1), A$. More formally,

$$(C *_x A)y = \begin{cases} C(y) & \text{if } y < x, \\ A & \text{otherwise.} \end{cases}$$

With this in mind it is easily seen that $\mathrm{nf}(\mathcal{D}\mathcal{N}\vec{\mathcal{Z}})$ is a subset of $\mathrm{nf}(\mathcal{D}_{\mathsf{n}}\mathcal{N}\vec{\mathcal{Z}})$, with

$$\mathcal{D}_{\mathsf{n}} = \mathcal{C} \cup (\mathcal{C} *_{\mathcal{X}} \mathcal{U}) \cup ((\mathcal{C} *_{\mathcal{X}} \mathcal{U}) *_{\mathcal{X}+1} \mathcal{V}) \cup \cdots \cup (\cdots (\mathcal{C} *_{\mathcal{X}} \mathcal{U}) * \cdots *_{\mathcal{X}+\mathsf{n}} \mathcal{W})$$

a finite initial part of the infinite union $\mathcal{D}$. The set $\mathrm{nf}(\mathcal{D}_{\mathsf{n}}\mathcal{N}\vec{\mathcal{Z}})$ is finite since the union is finite and all sets involved are in $\mathcal{HF}$. Hence $\mathcal{D}$ is hereditarily finite by Lemma 5.4.3. ∎

Since $\mathcal{D}$ is hereditarily finite, it follows that $\mathrm{nf}(\mathcal{Y}\mathcal{D})$ is finite. Let k be

[7] The axiom of dependent choice, DC, states the following. Let $R \subseteq X^2$ be a binary relation on a set X such that $\forall x \in X \exists y \in X. R(x, y)$. Then $\forall x \in X \exists f : \mathsf{Nat} \to X.[f(0) = x \;\&\; \forall n \in \mathsf{Nat}. R(f(n), f(n+1))]$. The axiom DC is an immediate consequence of the ordinary axiom of choice in set theory.

larger than any numeral in $\mathrm{nf}(\mathcal{Y}\mathcal{D})$. Consider

$$\mathcal{B}_{\mathsf{k}} = \mathcal{B}(\mathcal{X}{+}\mathsf{k})(\cdots(\mathcal{C} *_{\mathcal{X}} \mathcal{U}) * \cdots *_{\mathcal{X}+\mathsf{k}} \mathcal{W}')$$

as obtained in the construction above, iterating Lemma 5.4.13, hence not hereditarily finite. Since k is a strict upper bound of $\mathrm{nf}(\mathcal{Y}\mathcal{D})$ it follows that the set $\mathrm{nf}((\mathcal{X}{+}\mathsf{k}) \dot{-} \mathcal{Y}\mathcal{D})$ consists of numerals greater than $\mathsf{0}$, so that $\mathcal{B}_{\mathsf{k}}$ is hereditarily adfluent with $\mathcal{G}(\mathcal{X}{+}\mathsf{k})\mathcal{D}$. The latter set is hereditarily finite since it is an application of hereditarily finite sets (use Lemma 5.4.14). Hence $\mathcal{B}_{\mathsf{k}}$ is hereditarily finite by Lemma 5.4.5, which yields a plain contradiction.

By this contradiction, B must be hereditarily finite, and so is B^c, which follows by inspection of the reduction rules. As a consequence we obtain the main theorem of this section.

5.4.15 Theorem *Every bar recursive term is hereditarily finite.*

5.4.16 Corollary *Every bar recursive term is strongly normalizable.*

5.4.17 Remark The first normalization result for bar recursion is due to Tait (1971), who proves WN for $\boldsymbol{\lambda}_{\mathcal{B}}$. Vogel (1976) strengthens Tait's result to SN, essentially by introducing B^c and by enforcing every B-redex to reduce via B^c. Both Tait and Vogel use infinite terms. The proof above is based on Bezem (1985) and avoids infinite terms by using the notion of hereditary finiteness, which is a syntactic version of Howard's compactness of functionals of finite type, see Troelstra (1973), Section 2.8.6. All these proofs use a famous 'impredicative' argument that comes from Girard (1971).

If one considers $\boldsymbol{\lambda}_{\mathcal{B}}$ also with η-reduction, then the above results can also be obtained in a similar way to $\boldsymbol{\lambda}_{\mathcal{T}}$ with η-reduction.

Semantics of $\boldsymbol{\lambda}_{\mathcal{B}}$

In this subsection we give some interpretations of Spector's $\mathcal{B}$.

5.4.18 Definition A model of $\boldsymbol{\lambda}_{\mathcal{B}}$ is a model of $\boldsymbol{\lambda}_{\mathcal{T}}$ with interpretations of the constants $\mathsf{B}_{A,B}$ and $\mathsf{B}^c_{A,B}$ for all A, B, such that the rules for these constants can be interpreted as valid equations. In particular we have then that the schema of bar recursion is valid, with $[\![\varphi]\!] = [\![\mathsf{B}YGH]\!]$.

We saw at the beginning of this section that the full set-theoretic model of Gödel's $\mathcal{T}$ is not a model of bar recursion, due to the existence of functionals (such as Y unbounded on binary functions) for which the bar recursion is not well-founded. Designing a model of $\boldsymbol{\lambda}_{\mathcal{B}}$ amounts to ruling out such functionals, while maintaining the necessary closure properties. There are various solutions to this problem. The simplest is to take the closed terms modulo convertibility, which form a model by CR and SN. However, interpreting terms (almost) by themselves does not explain very much. For this *closed term model* the reader is asked in Exercise 5.35 to prove that it is extensional. An important model is obtained by using continuity in the form of the Kleene (1959a) and Kreisel (1959) continuous functionals. Continuity is on the one hand a structural property of bar recursive terms, since they can use only a finite amount of information about their arguments. On the other hand, continuity ensures that bar recursion is well-founded, since a continuous Y eventually gets the constant value YC on increasing initial segments $[C]_x$. In Exercise 5.34 the reader is asked to elaborate this model in detail. Refinements can be obtained by considering notions of computability on the continuous functionals, such as in Kleene (1959b), using the 'S1–S9 recursive functionals'. Computability alone, without uniform continuity on all binary functions, does not yield a model of bar recursion, see Exercise 5.30. The model of bar recursion we will elaborate in the following paragraphs is based on the same idea as the proof of strong normalization in the previous subsection. Here we consider the notion of hereditary finiteness semantically instead of syntactically. The intuition is that the set of increasing initial segments is hereditarily finite, so that any hereditarily finite functional Y is bounded on that set, and hence the bar recursion is well-founded.

5.4.19 Definition (*Hereditarily finite functionals*) Recall the full typed structure over the natural numbers: $\mathcal{M}_{\mathsf{N}} = \omega$ and $\mathcal{M}_{A\to B} = \mathcal{M}_A \to \mathcal{M}_B$. A set $\mathcal{X} \subseteq \mathcal{M}_{\mathsf{N}}$ is hereditarily finite if $\mathcal{X}$ is finite. A set $\mathcal{X} \subseteq \mathcal{M}_{A\to B}$ is hereditarily finite if $\mathcal{X}\mathcal{Y} \subseteq \mathcal{M}_B$ is hereditarily finite for every hereditarily finite $\mathcal{Y} \subseteq \mathcal{M}_A$. Here and below, $\mathcal{X}\mathcal{Y}$ denotes the set of all results that can be obtained by applying functionals from $\mathcal{X}$ to functionals from $\mathcal{Y}$. A functional F is hereditarily finite if the singleton set $\{F\}$ is hereditarily finite. Denote by $\mathcal{HF}$ the substructure of the full typed structure consisting of all hereditarily finite functionals.

The proof that $\mathcal{HF}$ is a model of $\boldsymbol{\lambda}_{\mathcal{B}}$ has much in common with the proof that $\boldsymbol{\lambda}_{\mathcal{B}}$ is SN in the previous subsection. The essential step is that the interpretation of the bar recursor is hereditarily finite. This requires the following semantic version of Lemma 5.4.13:

5.4.20 Lemma *Let $\mathcal{Y}, \mathcal{G}, \mathcal{H}, \mathcal{X}, \mathcal{C}$ be hereditarily finite sets of appropriate type. Then $[\![\mathsf{B}]\!]\mathcal{Y}\mathcal{G}\mathcal{H}\mathcal{X}\mathcal{C}$ is well defined and hereditarily finite whenever $[\![\mathsf{B}]\!]\mathcal{Y}\mathcal{G}\mathcal{H}(\mathcal{X}+1)(\mathcal{C} *_{\mathcal{X}} \mathcal{A})$ is so for all hereditarily finite $\mathcal{A}$ of appropriate type.*

The proof proceeds by iterating this lemma in the same way as in the SN proof after Lemma 5.4.13. The set of longer and longer initial sequences with elements taken from hereditarily finite sets (cf. the set $\mathcal{D}$ in Lemma 5.4.14) is hereditarily finite itself. As a consequence, the bar recursion must be well-founded when the set $\mathcal{Y}$ is also hereditarily finite. It follows that the interpretation of the bar recursor is well defined and hereditarily finite.

Following Troelstra (1973), Section 2.4.5 and 2.7.2, we define the following notion of *hereditary extensional equality*.

5.4.21 Definition We put $\approx_{\mathsf{N}}$ to be $=$; that is, convertibility of closed terms in $\Lambda^{\emptyset}_{\mathcal{B}}(\mathsf{N})$. For the type $A \equiv B{\to}B'$ we define $M \approx_A M'$ if and only if $M, M' \in \Lambda^{\emptyset}_{\mathcal{B}}(A)$ and $MN \approx_{B'} M'N'$ for all N, N' such that $N \approx_B N'$.

By (simultaneous) induction on A one shows easily that $\approx_A$ is symmetric, transitive and partially reflexive; that is, $M \approx_A M$ holds whenever $M \approx_A N$ for some N. The corresponding axiom of hereditary extensionality simply states that $\approx_A$ is (totally) reflexive: $M \approx_A M$, schematic in $M \in \Lambda^{\emptyset}_{\mathcal{B}}(A)$ and A. This is proved in Exercise 5.35.

Finally the intended model of $\boldsymbol{\lambda}_{\mathcal{B}}$ is $\mathcal{HF}/{\approx} \triangleq \langle \mathcal{HF}_A/{\approx_A}\rangle_{A\in\mathbb{T}(\boldsymbol{\lambda}_{\mathcal{B}})}$, with the obvious interpretation of T and B (being $[\mathsf{T}]_{\approx}$ respectively $[\mathsf{B}]_{\approx}$), and application satisfying $[M]_{\approx}[N]_{\approx} = [MN]_{\approx}$.

5.5 Platek's system $\mathcal{Y}$: fixed point recursion

Platek (1966) introduced a simply typed lambda calculus extended with fixed point combinators. Here we study Platek's system as an extension of Gödel's $\mathcal{T}$. An almost identical system is called PCF in Plotkin (1977).

A *fixed point combinator* is a functional Y of type $(A{\to}A){\to}A$ such that YF is a fixed point of F; that is, $YF = F(YF)$, for every F of type $A{\to}A$. Fixed point combinators can be used to compute solutions to recursion equations. The important difference with the type-free lambda calculus is that here all terms are typed, including the fixed point combinators themselves.

As an example we consider the recursion equations of the schema of higher-order primitive recursion in Gödel's system $\mathcal{T}$, see Section 5.3. We can rephrase these equations as

$$RMNn = \mathrm{If}_0\ n\ (N(RMN(n-1))(n-1))M,$$

where $\mathrm{If}_0\ nM_1M_0 = M_0$ if $n = 0$ and M_1 if $n > 0$. Hence we can write

$$\begin{aligned} RMN &= \lambda n.\ \mathrm{If}_0\ n\ (N(RMN(n-1))(n-1))M \\ &= (\lambda fn.\ \mathrm{If}_0\ n\ (N(f(n-1))(n-1))M)(RMN). \end{aligned}$$

This equation is of the form $YF = F(YF)$ with

$$F = \lambda fn.\ \mathrm{If}_0\ n\ (N(f(n-1))(n-1))M$$

and $YF = RMN$. It is easy to see that YF satisfies the recursion equation for RMN *uniformly* in M, N. This shows that, given functionals If_0 and a predecessor function (to compute $n-1$ in case $n > 0$), higher-order primitive recursion is definable by *fixed point recursion.* However, for computing purposes it is convenient to have primitive recursors at hand. By a similar argument, one can show bar recursion to be definable by fixed point recursion.

In addition to the above argument we show that every partial recursive function can be defined by fixed point recursion, by giving a fixed point recursion for minimization. Let F be a given function. Define by fixed point recursion $G_F = \lambda n.\mathrm{If}_0\ F(n)\ G_F(n+1)\ n$. Then we have

$$\begin{array}{rcl} G_F(0) & = & (\text{if } F(0) = 0 \text{ then } 0 \text{ else } G_F(1)); \\ G_F(1) & = & (\text{if } F(1) = 0 \text{ then } 1 \text{ else } G_F(2)); \\ & \vdots & \end{array}$$

By continuing this argument we see that

$$G_F(0) = \min\{n \mid F(n) = 0\};$$

that is, $G_F(0)$ computes the smallest n such that $F(n) = 0$, provided that such an n exists. If there exists no n such that $F(n) = 0$, then $G_F(0)$ as well as $G_F(1), G_F(2), \ldots$ are undefined. Given a function F of *two* arguments, minimization with respect to the second argument can now be obtained by the partial function $\lambda x.G_{F(x)}(0)$.

Already in the paragraph above we saw that fixed point recursions may be indefinite: if F does not equal zero, then $G_F(0) = G_F(1) = G_F(2) = \cdots$ does not lead to a definite value, although one could consistently assume G_F to be a constant function in this case. However, the situation is in general even worse: there is no natural number n that can consistently be assumed to be the fixed point of the successor function; that is, $n = Y(\lambda x.x+1)$, since we cannot have $n = (\lambda x.x+1)n = n+1$. This is the price to be paid for a formalism that allows one to compute all partial recursive functions.

Syntax of $\boldsymbol{\lambda}_{\mathcal{Y}}$

We now formalize Platek's $\mathcal{Y}$ as an extension of Gödel's $\mathcal{T}$ called $\boldsymbol{\lambda}_{\mathcal{Y}}$.

5.5.1 Definition The theory *Platek's* $\mathcal{Y}$, written $\boldsymbol{\lambda}_{\mathcal{Y}}$, is defined on the set of types $\mathbb{T}(\boldsymbol{\lambda}_{\mathcal{Y}}) = \mathbb{T}(\boldsymbol{\lambda}_{\mathcal{T}}) = \mathbb{T}^{\{\mathsf{N}\}}$. The *terms* of $\boldsymbol{\lambda}_{\mathcal{Y}}$ are obtained by adding constants

$$\mathsf{Y}_A : (A \to A) \to A$$

for all types A to the constants of $\boldsymbol{\lambda}_{\mathcal{T}}$. The set of (closed) terms of $\boldsymbol{\lambda}_{\mathcal{Y}}$ (of type A) is denoted by $\Lambda^{\emptyset}_{\mathsf{Y}}(A)$. The *formulas* of $\boldsymbol{\lambda}_{\mathcal{Y}}$ are equations between terms of $\boldsymbol{\lambda}_{\mathcal{Y}}$ (of the same type). The *theory* of $\boldsymbol{\lambda}_{\mathcal{Y}}$ extends the theory of $\boldsymbol{\lambda}_{\mathcal{T}}$ with the schema $\mathsf{Y}F = F(\mathsf{Y}F)$ for all appropriate types. The *reduction relation* $\rightarrow_{\mathbf{Y}}$ of $\boldsymbol{\lambda}_{\mathcal{Y}}$ extends $\rightarrow_{\mathcal{T}}$ by adding the following rule for the constants Y (omitting type annotations A):

$$\mathsf{Y} \rightarrow_{\mathbf{Y}} \lambda f.f(\mathsf{Y}f).$$

The reduction rule for Y requires some explanation, first observed by van Draanen (1995), as the rule $\mathsf{Y}F \rightarrow F(\mathsf{Y}F)$ seems simpler. However, with the latter rule we would have diverging reductions $\lambda f.\mathsf{Y}f \rightarrow_{\eta} \mathsf{Y}$ and $\lambda f.\mathsf{Y}f {\rightarrow_{\mathbf{Y}}} \lambda f.f(\mathsf{Y}f)$ that cannot be made to converge, so that we would lose CR of $\rightarrow_{\mathbf{Y}}$ in combination with $\boldsymbol{\eta}$-reduction.

The SN property does not hold for $\boldsymbol{\lambda}_{\mathcal{Y}}$: there is no Y-normal form for Y. However, the Church–Rosser property for $\boldsymbol{\lambda}_{\mathcal{Y}}$ with the notions of reduction $\boldsymbol{\beta}$, $\boldsymbol{\beta\eta}$, $\boldsymbol{\beta\eta}\mathbf{Y}$ can be proved by standard techniques from higher-order rewriting theory, for example, by using weak orthogonality; see van Raamsdonk (1996).

Although $\boldsymbol{\lambda}_{\mathcal{Y}}$ has universal computational strength in the sense that all partial recursive functions can be computed, not every computational phenomenon can be represented. For example, $\boldsymbol{\lambda}_{\mathcal{Y}}$ is inherently sequential: there is no term P such that $PMN = 0$ if and only if $M = 0$ or $N = 0$. The problem is that M and N cannot be evaluated in parallel, and if the argument that is evaluated first happens to be undefined, then the outcome is undefined even if the other argument equals 0. For a detailed account of the so-called sequentiality of $\boldsymbol{\lambda}_{\mathcal{Y}}$, see Plotkin (1977).

Semantics of $\boldsymbol{\lambda}_{\mathcal{Y}}$

In this subsection we explore the semantics of $\boldsymbol{\lambda}_{\mathcal{Y}}$ and give one model. This subject is more thoroughly studied in domain theory, see e.g. Gunter (1992) or Abramsky and Jung (1994).

5.5.2 Definition A *model of* $\boldsymbol{\lambda}_{\mathcal{Y}}$ is a model of $\boldsymbol{\lambda}_{\mathcal{T}}$ with interpretations of the constants Y_A for all A, such that the rules for these constants can be interpreted as valid equations.

Models of $\boldsymbol{\lambda}_{\mathcal{Y}}$ differ from those of $\boldsymbol{\lambda}_{\mathcal{T}}, \boldsymbol{\lambda}_{\mathcal{B}}$ in that they have to deal with partial maps. As we saw in the introduction to this section, no natural number n can consistently be assumed to be the fixed point of the successor function. Nevertheless, we have to interpret terms like YS^{+}. The canonical way of doing so is to add an element $\perp$ to the natural numbers, representing undefined objects as the fixed point of the successor function. Let $\omega^{\perp}$ denote the set of natural numbers extended with $\perp$. Now higher types are interpreted as function spaces over $\omega^{\perp}$. The basic intuition, coming from Platek (1966) and widely used and extended in model constructions by Dana Scott (see Gunter and Scott (1990)), is that $\perp$ contains less information than any natural number, and that functions and functionals give more informative output when the input becomes more informative. One way of formalizing these intuitions is by using partial orderings. We equip $\omega^{\perp}$ with the partial ordering $\sqsubseteq$ such that $\perp \sqsubset n$ for all $n \in \omega$. In order to be able to interpret Y, every function must have a fixed point. This requires some extra structure on the partial orderings, which can be formalized by the notion of complete partial ordering (cpo, see for example B[1984], Section 1.2).

The following lines bear some similarity to the introductory treatment of ordinals in Section 5.3.

We call a set *directed* if it is not empty and contains an upper bound for every two elements of it. Completeness of a partial ordering means that every directed set has a supremum. A function on a cpo is called *continuous* if it preserves suprema of directed sets. Every continuous function f of a cpo is monotone and has a least fixed point $\mathrm{lfp}(f)$, it being the supremum of the directed set enumerated by iterating f starting at $\perp$. The function lfp is itself continuous and serves as the interpretation of Y. We are now ready for the following definition.

5.5.3 Definition Define $\omega_A^{\perp}$ by induction:

$$\begin{aligned} \omega_{\mathsf{N}}^{\perp} &= \omega^{\perp}, \\ \omega_{A\to B}^{\perp} &= [\omega_A^{\perp} \to \omega_B^{\perp}], \text{ the set of all continuous maps.} \end{aligned}$$

Given the fact that cpos with continuous maps form a Cartesian closed category and that the successor, predecessor and conditional can be defined in a continuous way, the only essential step in the proof of the following lemma is to put $[\![\mathsf{Y}]\!] = \mathrm{lfp}$ for all appropriate types.

5.5.4 Lemma *The typed structure of the cpo $\omega_A^{\perp}$ is a model for $\boldsymbol{\lambda}_{\mathcal{Y}}$.*

In fact, as the essential requirement is the existence of fixed points, we could have taken monotone instead of continuous maps on cpos. This option is elaborated in detail in van Draanen (1995).

5.6 Exercises

Exercise 5.1 Prove in δ the following equations.

(i) $\delta MN\mathsf{K}_*\mathsf{K} = \delta(\delta MN)\mathsf{K}_*$.
(ii) $\delta(\lambda z.\delta(Mz)(Nz))(\lambda z.\mathsf{K}) = \delta MN$.
[Hint. Start by observing that $\delta(Mz)(Nz)(Mz)(Nz) = Nz$.]

Exercise 5.2 Prove Proposition 5.2.12.

Exercise 5.3 Let λ_P be $\boldsymbol{\lambda}^0_\rightarrow$ extended with a simple (not surjective) pairing. Show that Theorem 5.2.45 does not hold for this theory. [Hint show that in this theory the equation $\lambda x{:}0.\langle \pi_1 x, \pi_2 x\rangle = \lambda x{:}0.x$ does not hold by constructing a non-trivial counter model.]

Exercise 5.4 Does every model of λ_{SP} have the same first-order theory?

Exercise 5.5

(i) Show that if a pairing function $\langle\ ,\ \rangle : 0{\rightarrow}(0{\rightarrow}0)$ and projections $L, R : 0{\rightarrow}0$ satisfying $L\langle x,y\rangle = x$ and $R\langle x,y\rangle = y$ are added to $\boldsymbol{\lambda}^0_\rightarrow$, then for a non-trivial model $\mathcal{M}$ one has

$$\forall A \in \mathbb{T}\, \forall M, N \in \Lambda^{\emptyset}(A)\, [\mathcal{M} \models M = N \;\Rightarrow\; M =_{\beta\eta} N].$$

(ii) (Schwichtenberg and Berger (1991)) Show that for $\mathcal{M}$ a model of λ_T one has

$$\forall A \in \mathbb{T}\, \forall M, N \in \Lambda^{\emptyset}(A)\, [\mathcal{M} \models M = N \;\Rightarrow\; M =_{\beta\eta} N].$$

Exercise 5.6 Show that $\mathcal{F}[x_1, \ldots, x_n]$ for $n \geq 0$ does not have one generator. [Hint. Otherwise this monoid would be commutative, which is not the case.]

Exercise 5.7 Show that $R \subseteq \Lambda^{\emptyset}(A) \times \Lambda^{\emptyset}(B)$ is equational (cf. Definition 5.2.46(i)) iff

$$\exists M, N \in \Lambda^{\emptyset}(A{\rightarrow}B{\rightarrow}1{\rightarrow}1)\, \forall F\, [R(F) \;\Leftrightarrow\; MF = NF].$$

Exercise 5.8 Show that there is a Diophantine equation $\mathtt{lt} \subseteq \mathcal{F}^2$ such that for all $n, m \in \omega$

$$\mathtt{lt}(\mathsf{R}^n, \mathsf{R}^m) \;\Leftrightarrow\; n < m.$$

Exercise 5.9 Define $\mathtt{Seq}^{\mathcal{N}_k}_n(h)$ iff $h = [\mathsf{R}^{m_0}, \ldots, \mathsf{R}^{m_{n-1}}]$, for some $m_0, \ldots, m_{n-1} < k$. Show that $\mathtt{Seq}^{\mathcal{N}_k}_n$ is Diophantine uniformly in n.

Exercise 5.10 Let $\mathcal{B}$ be some finite subset of $\mathcal{F}$, defined in 5.2.21(i). Define $\mathtt{Seq}^{\mathcal{B}}_n(h)$ iff $h = [g_0, \ldots, g_{n-1}]$, with each $g_i \in \mathcal{B}$. Show $\mathtt{Seq}^{\mathcal{B}}_n$ is Diophantine uniformly in n.

Exercise 5.11 For $\mathcal{B} \subseteq \mathcal{F}$ define $\mathcal{B}^+$ to be the submonoid generated by $\mathcal{B}$. Show that if $\mathcal{B}$ is finite, then $\mathcal{B}^+$ is Diophantine.

Exercise 5.12 Show that $\mathcal{F} \subseteq \mathcal{F}[x]$ is Diophantine.

Exercise 5.13 Construct two concrete terms $t(a,b), s(a,b) \in \mathcal{F}[a,b]$ such that for all $f \in \mathcal{F}$ one has

$$f \in \{\mathsf{R}^n \mid n \in \omega\} \cup \{\mathsf{L}\} \;\Leftrightarrow\; \exists g \in \mathcal{F}\,[t(f,g) = s(f,g)].$$

[Remark. It is not sufficient that Diophantine sets are closed under union. But the solution is not hard and the terms are short.]

Exercise 5.14 Let $2 = \{0,1\}$ be the discrete topological space with two elements. Let Cantor space be $\mathbf{C} = 2^\omega$ endowed with the product topology. Define $Z, O : \mathbf{C}{\to}\mathbf{C}$ 'shift operators' on Cantor space as follows:

$$\begin{aligned} Z(f)(0) &= 0; \\ Z(f)(n+1) &= f(n); \\ O(f)(0) &= 1; \\ O(f)(n+1) &= f(n). \end{aligned}$$

Write $0f = Z(f)$ and $1f = O(f)$. If $\mathcal{X} \subseteq \mathbf{C}{\to}\mathbf{C}$ is a set of maps, let $\mathcal{X}^+$ be the closure of $\mathcal{X}$ under the rule

$$A_0, A_1 \in \mathcal{X} \;\Rightarrow\; A \in \mathcal{X},$$

where A is defined by

$$\begin{aligned} A(0f) &= A_0(f); \\ A(1f) &= A_1(f). \end{aligned}$$

(i) Show that if $\mathcal{X}$ consists of continuous maps, then so does $\mathcal{X}^+$.
(ii) Show that $A \in \{Z, O\}^+$ iff

$$A(f) = g \;\Rightarrow\; \exists r, s \in \omega\, \forall t > s. g(t) = f(t - s + r).$$

(iii) Define on $\{Z, O\}^+$ the following:

$$\begin{array}{rcll} I &=& \lambda x \in \{Z, O\}^+ . z; & \\ L &=& Z; & \\ R &=& O; & \\ x * y &=& y \circ x; & \\ \langle x, y \rangle &=& x(f), & \text{if } f(0) = 0; \\ &=& y(f), & \text{if } f(0) = 1. \end{array}$$

Then $\langle\{Z,O\}^+, *, I, L, R, \langle -,-\rangle\rangle$ is a Cartesian monoid isomorphic to $\mathcal{F}$, via $\varphi : \mathcal{F} \to \{Z,O\}^+$.

(iv) The *Thompson–Freyd–Heller group* can be defined by

$$\{f \in \mathcal{I} \mid \varphi(f) \text{ preserves the lexicographical ordering on } \mathbf{C}\}.$$

Show that the B_n introduced in Definition 5.2.32 generate this group.

Exercise 5.15 Let

$$\begin{array}{rclrcl} B_0 & = & \langle LL, RL, R\rangle\rangle, & B_0^{-1} & = & \langle\langle L, LR\rangle, LRR, RRR\rangle\rangle \\ B_1 & = & \langle L, LLR, RLR, RR\rangle\rangle, & B_1^{-1} & = & \langle L, \langle LR, LRR\rangle, RRR\rangle\rangle \\ C_0 & = & \langle R, L\rangle, & C_1 & = & \langle LR, L, RR\rangle\rangle. \end{array}$$

Show that for the invertible elements of the free Cartesian monoid $\mathcal{F}$ one has

$$\mathcal{I} = [\{B_0, B_0^{-1}, B_1, B_1^{-1}, C_0, C_1\}].$$

[Hint. Show that

$$\begin{aligned} B_0\langle\langle A, B, C\rangle &= \langle A, B, C\rangle\rangle \\ B_1\langle A, \langle B, C\rangle, D\rangle\rangle &= \langle A, B, C, D\rangle\rangle \\ C_0\langle A, B\rangle &= \langle B, A\rangle \\ C_1\langle A, B, C\rangle\rangle &= \langle B, A, C\rangle\rangle. \end{aligned}$$

Use this to transform any element $M \in \mathcal{I}$ into I. By the inverse transformation we get M as the required product.]

Exercise 5.16 Show that the B_n in Definition 5.2.32 satisfy

$$B_{n+2} = B_n B_{n+1} B_n^{-1}.$$

Exercise 5.17 Prove Lemma 5.3.15. [Hint. Show for all α, β:

(i) $\alpha \le \beta \Rightarrow f^\alpha(0) \le f^\beta(0)$;
(ii) $f(\beta) = \beta \Rightarrow f^\alpha(0) \le \beta$;
(iii) $\alpha < \beta \Rightarrow f^\alpha(0) < f^\beta(0)$, if $f^\alpha(0)$ is below any fixed point of f.]

Exercise 5.18 Justify the equation $f(\lambda) = \lambda$ in the proof of 5.3.17.

Exercise 5.19 Let A be the modified Ackermann function. Evaluate $A(3, m)$ and verify that $A(4, 0) = 13$ and $A(4, 1) = 65533$.

Exercise 5.20 This exercise refers to Definition 5.3.3. With one occurrence hidden in H, the term RSH contains $R_{\mathsf{N}\to\mathsf{N}}$ twice. Define A using R_{N} and $R_{\mathsf{N}\to\mathsf{N}}$ only once. Is it possible to define A with R_{N} only, possibly with multiple occurrences?

Exercise 5.21 Show that the first-order schema of primitive recursion is subsumed by the higher-order schema, expressing F in terms of R, G and H.

Exercise 5.22 Which function is computed if we replace P in $Rx(P \circ K)y$ by the successor function? Define multiplication, exponentiation and division with remainder as primitive recursive functionals.

Exercise 5.23 [Simultaneous primitive recursion] Assume G_i, H_i $(i = 1, 2)$ have been given and define F_i $(i = 1, 2)$ as follows:

$$\begin{aligned} F_i(0, \vec{x}) &= G_i(\vec{x}); \\ F_i(n+1, \vec{x}) &= H_i(F_1(n, \vec{x}), (F_2(n, \vec{x}), n, \vec{x}). \end{aligned}$$

Show that F_i $(i = 1, 2)$ can be defined by first-order primitive recursion. [Hint. Use a pairing function such as in Figure 5.4.]

Exercise 5.24 [Nested recursion, Péter (1967)] Define

$$\begin{aligned} F(n, m) &= 0 \qquad \text{if } m \cdot n = 0 \\ F(n+1, m+1) &= G(m, n, F(m, H(m, n, F(m+1, n))), F(m+1, n)) \end{aligned}$$

Show that F can be defined from G, H using higher-order primitive recursion.

Exercise 5.25 [Dialectica translation] We closely follow Troelstra (1973), Section 3.5; the solution can be found there. Let HA^ω be the theory of higher-order primitive recursive functionals equipped with many-sorted intuitionistic predicate logic with equality for natural numbers and axioms for arithmetic, in particular the schema of arithmetical induction:

$$(\varphi(0) \wedge \forall x\ (\varphi(x) \Rightarrow \varphi(x+1))) \Rightarrow \forall x\ \varphi(x).$$

The *Dialectica interpretation* of Gödel (1958), D-interpretation for short, assigns to each formula φ in the language of HA^ω a formula $\varphi^{\mathrm{D}} \equiv \exists \vec{x}\, \forall \vec{y}\, \varphi_{\mathrm{D}}(\vec{x}, \vec{y})$ in the same language. The types of $\vec{x}, \vec{y}$ depend on the logical structure of φ only. We define φ_{D} and φ^{D} by induction on φ:

(i) If φ is prime – that is, an equation of lowest type – then $\varphi^{\mathrm{D}} \equiv \varphi_{\mathrm{D}} \equiv \varphi$.

For the binary connectives, assume

$$\varphi^D \equiv \exists \vec{x}_1\ \forall \vec{y}_1\ \varphi_D(\vec{x}_1, \vec{y}_1) \text{ and } \psi^D \equiv \exists \vec{x}_2\ \forall \vec{y}_2\ \psi_D(\vec{x}_2, \vec{y}_2).$$

Then

(ii) $(\varphi \wedge \psi)^D \triangleq \exists \vec{x}_1, \vec{x}_2 \, \forall \vec{y}_1, \vec{y}_2 \, (\varphi \wedge \psi)_D$, with
$(\varphi \wedge \psi)_D \triangleq (\varphi_D(\vec{x}_1, \vec{y}_1) \wedge \psi_D(\vec{x}_2, \vec{y}_2))$.

(iii) $(\varphi \vee \psi)^D \triangleq \exists z, \vec{x}_1, \vec{x}_2 \, \forall \vec{y}_1, \vec{y}_2 \, (\varphi \vee \psi)_D$, with
$(\varphi \vee \psi)_D \triangleq ((z = 0 \Rightarrow \varphi_D(\vec{x}_1, \vec{y}_1)) \wedge (z \neq 0 \Rightarrow \psi_D(\vec{x}_2, \vec{y}_2)))$.

(iv) $(\varphi \Rightarrow \psi)^D \triangleq \exists \vec{x}'_2, \vec{y}'_1 \, \forall \vec{x}_1, \vec{y}_2 \, (\varphi \Rightarrow \psi)_D$, with
$(\varphi \Rightarrow \psi)_D \triangleq (\varphi_D(\vec{x}_1, \vec{y}'_1 \vec{x}_1 \vec{y}_2) \Rightarrow \psi_D(\vec{x}'_2 \vec{x}_1, \vec{y}_2))$.

Note that the clause for $\varphi \Rightarrow \psi$ introduces quantifications over higher types than those used for the formulas φ, ψ. This is also the case for formulas of the form $\forall z \, \varphi(z)$, see the sixth case below. For both quantifier clauses below, assume $\varphi^D(z) \equiv \exists \vec{x} \, \forall \vec{y} \, \varphi_D(\vec{x}, \vec{y}, z)$.

(v) $(\exists z \, \varphi(z))^D \triangleq \exists z, \vec{x} \, \forall \vec{y} \, (\exists z \, \varphi(z))_D$, with $(\exists z \, \varphi(z))_D \triangleq \varphi_D(\vec{x}, \vec{y}, z)$.

(vi) $(\forall z \, \varphi(z))^D \triangleq \exists \vec{x}' \, \forall z, \vec{y} \, (\forall z \, \varphi(z))_D$, with $(\forall z \, \varphi(z))_D \triangleq \varphi_D(\vec{x}' z, \vec{y}, z)$.

With φ, ψ as in the case of a binary connective, determine $(\varphi \Rightarrow (\varphi \vee \psi))^D$ and give a sequence $\vec{P} \in \Lambda^{\emptyset}_{\mathcal{T}}$ of higher-order primitive recursive functionals such that $\forall \vec{y} \, (\varphi \Rightarrow (\varphi \vee \psi))_D(\vec{P}, \vec{y})$. We say that in this way the D-interpretation of $(\varphi \Rightarrow (\varphi \vee \psi))^{\mathrm{D}}$ is *validated* by higher-order primitive recursive functionals. Validate the D-interpretation of $(\varphi \Rightarrow (\varphi \wedge \varphi))^{\mathrm{D}}$. Validate the D-interpretation of induction. The result of Gödel (1958) can now be rendered as: the D-interpretation of every theorem of HA^ω can be validated by higher-order primitive recursive functionals. This yields a consistency proof for HA^ω, since $0 = 1$ cannot be validated. Note that the D-interpretation and the successive validation translates arbitrarily quantified formulas into universally quantified propositional combinations of equations.

Exercise 5.26 Consider for any type B the set of closed terms of type B modulo convertibility. Prove that this yields a model for Gödel's $\mathcal{T}$. This model is called the *closed term model* of Gödel's $\mathcal{T}$.

Exercise 5.27 Let $*$ be *Kleene application*; that is, $i*n$ stands for applying the ith partial recursive function to the input n. If this yields a result, then we flag $i * n\!\downarrow$, otherwise $i * n\!\uparrow$. Equality between expressions with Kleene application is taken to be strict; that is, equality only holds if the left and right hand sides yield a result and the results are equal. Similarly, $i * n \in S$ should be taken in the strict sense of $i * n$ actually yielding a result in S.

By induction we define a family of sets, the *hereditarily recursive operators* $\mathrm{HRO}_B \subseteq \omega$ for every type B, as follows:

$$\mathrm{HRO}_{\mathsf{N}} = \omega$$

$$\mathrm{HRO}_{AB \to B} = \{x \in \omega \mid x * y \in \mathrm{HRO}_B \text{ for all } y \in \mathrm{HRO}_{BA}\}.$$

Prove that HRO with Kleene application constitutes a model for Gödel's $\mathcal{T}$.

Exercise 5.28 By simultaneous induction we define a family of sets, the *hereditarily extensional operators* $\text{HEO}_B \subseteq \omega$ for every type B, equipped with an equivalence relation $=_B$, as follows:

$$\begin{aligned}
\text{HEO}_{\mathsf{N}} &\triangleq \omega \\
x =_{\mathsf{N}} y &\overset{\Delta}{\iff} x = y \\
\text{HEO}_{A\to B} &\triangleq \{x \in \omega \mid x * y \in \text{HEO}_B \text{ for all } y \in \text{HEO}_A \text{ and} \\
&\qquad x * y =_B x * y' \text{ for all } y, y' \in \text{HEO}_A \text{ with } y =_A y'\} \\
x =_{A\to B} x' &\overset{\Delta}{\iff} x, x' \in \text{HEO}_{A\to B} \text{ and } x * y =_B x' * y \text{ for all } y \in \text{HEO}_A.
\end{aligned}$$

Prove that HEO with Kleene application constitutes a model for Gödel's $\mathcal{T}$.

Exercise 5.29 Recall that extensionality essentially means that objects having the same applicative behaviour can be identified. Which of the above models of $\boldsymbol{\lambda}_{\mathcal{T}}$, the full typed structure, the closed term model, HRO and HEO, is extensional?

Exercise 5.30 This exercise shows that HEO is *not* a model for bar recursion. Recall that $*$ stands for partial recursive function application. Consider functionals Y, G, H defined by $G(x, C) = 0$, $H(Z, x, C) = 1 + Z(0) + Z(1)$ and $Y(F)$ is the smallest number n such that $i * i$ converges in less than n steps for some $i < n$ and, moreover, $i * i = 0$ if and only if $F(i) = 0$ does *not* hold. The crux of the definition of Y is that no *total* recursive function F can distinguish between $i * i = 0$ and $i * i > 0$ for *all* i with $i * i\!\downarrow$. But for any finite number of such i's we do have a total recursive function making the correct distinctions. This implies that Y, although continuous and well-defined on all total recursive functions, is not uniformly continuous and not bounded on total recursive binary functions. Show that all functionals involved can be represented in HEO and that the latter model of $\boldsymbol{\lambda}_{\mathcal{T}}$ is not a model of $\boldsymbol{\lambda}_{\mathcal{B}}$.

Exercise 5.31 Verify that the redefinition of the ordinal arithmetic in Example 5.3.14 is correct.

Exercise 5.32

(i) Show for all ordinals α, β, γ that $(\alpha^\beta)^\gamma = \alpha^{\beta\cdot\gamma}$.
(ii) Show that $2^\omega = \omega$.
(iii) Show for $\alpha \geq \omega$ that $1 + \alpha = \alpha$.

Exercise 5.33 Justify the equation $f(\lambda) = \lambda$ in the proof of Lemma 5.3.17.

Exercise 5.34 This exercise introduces the *continuous functionals*, Kleene (1959a). Define for $f, g \in \omega \to \omega$ the (partial) application of f to g by $f(g) = f(\overline{g n}) - 1$, where n is the smallest number such that $f(\overline{g n}) > 0$, provided there is such n. If there is no such n, then $f * g$ is undefined. The idea is that f uses only a finite amount of information about g for determining the value of $f * g$ (if any). Define recursively for every type A a set $\mathcal{C}_A$ together with an association relation between elements of $\omega \to \omega$ and elements of $\mathcal{C}_A$. For the base type we put $\mathcal{C}_{\mathsf{N}} = \omega$ and let the constant functions be the associates of the corresponding natural numbers. For higher types we say that $f \in \omega \to \omega$ is an associate of $F \in \mathcal{C}_A \to \mathcal{C}_B$ if, for any associate g of $G \in \mathcal{C}_A$, the function h, defined by $h(n) = f(n{:}g)$, is an associate of $F(G) \in \mathcal{C}_B$. Here $n{:}g$ is shorthand for the function taking value n at 0 and value $g(k-1)$ for all $k > 0$. (Note that we have implicitly required that h is total.) Now $\mathcal{C}_{A \to B}$ is defined as the subset of those $F \in \mathcal{C}_A \to \mathcal{C}_B$ that have an associate. Show that $\mathcal{C}$ is a model for bar recursion.

Exercise 5.35 Show that for any closed term $M \in \Lambda_B^{\emptyset}$ one has $M \approx M$, see Definition 5.4.21. [Hint. Type subscripts are omitted. Define a predicate $\mathrm{Ext}(M(\vec{x}))$ for any open term M with free variables among $\vec{x} = x_1, \ldots, x_n$ by

$$M(X_1, \ldots, X_n) \approx M(X'_1, \ldots, X'_n)$$

for all $X_1, \ldots, X_n, X'_1, \ldots, X'_n \in \Lambda_B^{\emptyset}$ with $X_1 \approx X'_1, \ldots, X_n \approx X'_n$. Then prove by induction on terms that Ext holds for any open term, so in particular for closed terms. For B, prove first the following. Suppose

$$Y \approx Y', G \approx G', H \approx H', X \approx X', C \approx C',$$

and for all $A \approx A'$

$$\mathsf{B}YGH(\mathsf{S}^+X)(C *_X A) \approx \mathsf{B}Y'G'H'(\mathsf{S}^+X')(C' *_{X'} A').$$

Then

$$\mathsf{B}YGHXC \approx \mathsf{B}Y'G'H'X'C'.]$$

6
Applications

6.1 Functional programming

Lambda calculi are prototype programming languages. Just as with imperative programming languages, some of which are untyped (machine code, assembler, Basic) and some typed (Algol-68, Pascal), systems of lambda calculi exist in untyped and typed versions. There are also other differences in the various lambda calculi. The one introduced in Church (1936) is the untyped λI-calculus in which an abstraction $\lambda x.M$ is only allowed if x occurs among the free variables of M. Nowadays, 'lambda calculus' refers to the λK-calculus developed under the influence of Curry, in which $\lambda x.M$ is allowed even if x does not occur in M. This book treats the typed versions of the λK-calculus. Of these, the most elementary are the versions of the simply typed lambda calculus $\boldsymbol{\lambda}_{\rightarrow}^{\mathbb{A}}$ introduced in Chapter 1.

Computing on data types

In this subsection we explain how it is possible to represent data types in a very direct manner in the various lambda calculi.

Lambda-definability was introduced for functions on the set of natural numbers ω. In the resulting mathematical theory of computation (recursion theory, now called computability theory), other domains of input or output have been treated as second-class citizens by coding them as natural numbers. In more practical computer science, algorithms are also directly defined on other data types like trees or lists.

Instead of coding such data types as numbers one can treat them as first-class citizens by coding them directly as lambda terms *while preserving their structure*. Indeed, lambda calculus is strong enough to do this, as was emphasized in Böhm (1966) and Böhm and Gross (1966). As a result, a much more efficient representation of algorithms on these data types can be given, than if these types were represented via numbers. This methodology

Figure 6.1

was perfected in two different ways: in Böhm and Berarducci (1985) and in Böhm et al. (1994) or Berarducci and Böhm (1993). The first paper does the representation in a way that can be typed; the other papers in an essentially stronger way, but one that cannot be typed. We present the methods of these papers by treating labeled trees as an example.

Let the (inductive) data-type of labeled trees be defined by the following simplified syntax:

$$\begin{aligned} \mathtt{tree} &= \bullet \mid \mathtt{leaf\ nat} \mid \mathtt{tree} + \mathtt{tree} \\ \mathtt{nat} &= 0 \mid \mathtt{succ\ nat}. \end{aligned}$$

We see that a label can be either a bud ($\bullet$) or a leaf with a number written on it. A typical such tree is $(\mathtt{leaf}\ 3) + ((\mathtt{leaf}\ 5) + \bullet)$. This tree, together with its mirror image, looks as in Figure 6.1 ('leaf 3' is essentially 3, but we officially need to write the constructor to warrant unicity of types; in the examples below we do not write it).

Operations on such trees can be defined by recursion. For example the action of mirroring can be defined by

$$\begin{aligned} f_{\mathtt{mir}}(\bullet) &= \bullet; \\ f_{\mathtt{mir}}(\mathtt{leaf}\ n) &= \mathtt{leaf}\ n; \\ f_{\mathtt{mir}}(t_1 + t_2) &= f_{\mathtt{mir}}(t_2) + f_{\mathtt{mir}}(t_1). \end{aligned}$$

Then one has for example that

$$f_{\mathtt{mir}}((\mathtt{leaf}\ 3) + ((\mathtt{leaf}\ 5) + \bullet)) = ((\bullet + \mathtt{leaf}\ 5) + \mathtt{leaf}\ 3).$$

We will now show in two different ways how trees can be represented as lambda terms and how operations like $f_{\mathtt{mir}}$ on these objects become lambda-definable. The first method is from Böhm and Berarducci (1985). The resulting data objects and functions can be represented by lambda terms typable in the second-order lambda calculus $\boldsymbol{\lambda}2$: see Girard et al. (1989) or Barendregt (1992).

6.1.1 Definition

(i) Let b, l, p be variables (used as mnemonics for `bud`, `leaf` and `plus`). Define $\varphi = \varphi^{b,l,p} : \texttt{tree} \to \texttt{term}$, where `term` is the collection of untyped lambda terms, as

$$\begin{aligned}\varphi(\bullet) &= b;\\ \varphi(\texttt{leaf}\ n) &= l\ulcorner n\urcorner;\\ \varphi(t_1 + t_2) &= p\,\varphi(t_1)\varphi(t_2).\end{aligned}$$

Here $\ulcorner n\urcorner \equiv \lambda fx.f^n x$ is Church's numeral representing n as lambda term.

(ii) Define $\psi_1 : \texttt{tree} \to \texttt{term}$ as

$$\psi_1(t) = \lambda blp.\varphi(t).$$

6.1.2 Proposition *Define*

$$\begin{aligned}B_1 &\equiv \lambda blp.b;\\ L_1 &\equiv \lambda nblp.ln;\\ P_1 &\equiv \lambda t_1 t_2 blp.p\,(t_1 blp)(t_2 blp).\end{aligned}$$

Then one has

(i) $\psi_1(\bullet) = B_1$;
(ii) $\psi_1(\texttt{leaf}\ n) = L_1\ulcorner n\urcorner$;
(iii) $\psi_1(t_1 + t_2) = P_1\,\psi_1(t_1)\psi_1(t_2)$.

Proof (i) Trivial.
(ii) We have

$$\begin{aligned}\psi_1(\texttt{leaf}\ n) &= \lambda blp.\varphi(\texttt{leaf}\ n)\\ &= \lambda blp.l\ulcorner n\urcorner\\ &= (\lambda nblp.ln)\ulcorner n\urcorner\\ &= L_1\ulcorner n\urcorner.\end{aligned}$$

(iii) Similarly, using the fact that $\psi_1(t)blp = \varphi(t)$. ∎

This proposition states that the trees we considered are representable as lambda terms in such a way that the constructors ($\bullet$, `leaf` and $+$) are lambda-definable. In fact, the lambda terms involved can be typed in $\boldsymbol{\lambda}2$. A nice connection between these terms and proofs in second-order logic is given in Leivant (1983b).

Now we will show that iterative functions over these trees, such as $f_{\texttt{mir}}$, are also lambda-definable.

6.1.3 Proposition (Iteration) *Given lambda terms A_0, A_1, A_2 there exists a lambda term F such that, for variables n, t_1, t_2,*

$$\begin{aligned} FB_1 &= A_0; \\ F(L_1\, n) &= A_1\, n; \\ F(P_1 t_1 t_2) &= A_2(Ft_1)(Ft_2). \end{aligned}$$

Proof Take $F \equiv \lambda w.wA_0A_1A_2$. ∎

As is well known, primitive recursive functions can be obtained from iterative functions.

There is a way of coding a finite sequence of lambda terms $M_1, \ldots, M_k$ as one lambda term

$$\langle M_1, \ldots, M_k \rangle \equiv \lambda z.zM_1 \cdots M_k$$

such that the components can be recovered. Indeed, take

$$U_k^i \equiv \lambda x_1 \cdots x_k.x_i;$$

then

$$\langle M_1, \ldots, M_k \rangle U_k^i = M_i.$$

6.1.4 Corollary (Primitive recursion) *Given lambda terms C_0, C_1, C_2 there exists a lambda term H such that*

$$\begin{aligned} HB_1 &= C_0; \\ H(L_1\, n) &= C_1\, n; \\ H(P_1 t_1 t_2) &= C_2 t_1 t_2 (Ht_1)(Ht_2). \end{aligned}$$

Proof Define the auxiliary function $F \equiv \lambda t.\langle t, Ht \rangle$. Then, by the proposition, F can be defined using iteration. Indeed,

$$F(P_1 t_1 t_2) = \langle Pt_1t_2, H(Pt_1t_2) \rangle = A_2(Ft_1)(Ft_2),$$

with

$$A_2 \equiv \lambda t_1 t_2.\langle P(t_1U_2^1)(t_2U_2^1), C_2(t_1U_2^1)(t_2U_2^1)(t_1U_2^2)(t_2U_2^2) \rangle.$$

Now take $H = \lambda t.FtU_2^2$. [This was a trick Kleene thought up while being treated with laughing gas at the dentist: see Kleene (1975).] ∎

Now we will present the method of Böhm et al. (1994) and Berarducci and Böhm (1993) for representing data types. Again we consider the example of labeled trees.

6.1.5 Definition Define $\psi_2 : \texttt{tree} \to \texttt{term}$ as

$$\begin{aligned}\psi_2(\bullet) &= \lambda e.eU_3^1 e;\\ \psi_2(\texttt{leaf}\ n) &= \lambda e.eU_3^2 \ulcorner n \urcorner e;\\ \psi_2(t_1 + t_2) &= \lambda e.eU_3^3 \psi_2(t_1)\psi_2(t_2)e.\end{aligned}$$

Then the basic constructors for labeled trees are definable by

$$\begin{aligned}B_2 &\equiv \lambda e.eU_3^1 e;\\ L_2 &\equiv \lambda n \lambda e.eU_3^2 ne;\\ P_2 &\equiv \lambda t_1 t_2 \lambda e.eU_3^3 t_1 t_2 e.\end{aligned}$$

6.1.6 Proposition *Given lambda terms A_0, A_1, A_2 there exists a term F such that*

$$\begin{aligned}FB_2 &= A_0 F;\\ F(L_2 n) &= A_1 n F;\\ F(P_2 xy) &= A_2 xyF.\end{aligned}$$

Proof Try $F \equiv \langle\langle X_0, X_1, X_2\rangle\rangle$, the 1-tuple of a triple. Then we must have

$$\begin{aligned}FB_2 &= B_2\langle X_0, X_1, X_2\rangle\\ &= U_3^1 X_0 X_1 X_2 \langle X_0, X_1, X_2\rangle\\ &= X_0\langle X_0, X_1, X_2\rangle\\ &= A_0\langle\langle X_0, X_1, X_2\rangle\rangle\\ &= A_0 F,\end{aligned}$$

provided $X_0 = \lambda x.A_0\langle x\rangle$. Similarly one can find X_1, X_2. ∎

This second representation is essentially untypable, at least in typed lambda calculi in which all typable terms are normalizing. This follows from the following consequence of a result similar to Proposition 6.1.6. Let $\mathsf{K} = \lambda xy.x$, $\mathsf{K}_* = \lambda xy.y$ represent true and false respectively. Then writing

$$\underline{\texttt{if}}\ \texttt{bool}\ \underline{\texttt{then}}\ X\ \underline{\texttt{else}}\ Y\ \underline{\texttt{fi}}$$

for

$$\texttt{bool}\ X\,Y,$$

the usual behavior of the conditional is obtained. Now if we represent the natural numbers as a data type in the style of the second representation, we immediately get that the lambda-definable functions are closed under minimization. Indeed, let

$$\chi(x) = \mu y[g(x, y) = 0],$$

and suppose that g is lambda-defined by G. Then there exists a lambda term H such that

$$Hxy = \underline{\texttt{if}}\ \underline{\texttt{zero}_?}\,(Gxy)\ \underline{\texttt{then}}\ y\ \underline{\texttt{else}}\ (Hx(\underline{\texttt{succ}}\ \ y))\ \underline{\texttt{fi}}.$$

Indeed, we can write this as $Hx = AxH$ and apply Proposition 6.1.6, but now formulated for the recursively defined type num. Then $F \equiv \lambda x.Hx\ulcorner 0\urcorner$ does represent χ. Here $\underline{\texttt{succ}}$ represents the successor function and $\underline{\texttt{zero}_?}$ a test for zero; both are lambda-definable, again by the analog to Proposition 6.1.6. Since minimization enables us to define all partial recursive functions, the terms involved cannot be typed in a normalizing system.

Self-interpretation

A lambda term M can be represented internally as a lambda term $\ulcorner M\urcorner$. This representation should be such that, for example, one has lambda terms P_1, P_2 satisfying $P_i\ulcorner X_1X_2\urcorner = X_i$. Kleene (1936) already showed that there is a ('meta-circular') self-interpreter E such that, for closed terms M, one has $\mathsf{E}\ulcorner M\urcorner = M$. The fact that data types can be represented directly in the lambda calculus was exploited by Mogensen (1992) to find a simpler representation for $\ulcorner M\urcorner$ and E.

The difficulty of representing lambda terms internally is that they do not form a first-order algebraic data type because of the binding effect of the lambda. Mogensen (1992) solved this problem as follows. Consider the data type with signature

const, app, abs

where const and abs are unary, and app is a binary constructor. Let $\underline{\texttt{const}}$, $\underline{\texttt{app}}$ and $\underline{\texttt{abs}}$ be a representation of these in lambda calculus (in the style of Definition 6.1.5).

6.1.7 Proposition (Mogensen (1992)) *Define*

$$\begin{aligned}\ulcorner x\urcorner &\equiv \underline{\texttt{const}}\ x;\\ \ulcorner PQ\urcorner &\equiv \underline{\texttt{app}}\ \ulcorner P\urcorner\ulcorner Q\urcorner;\\ \ulcorner \lambda x.P\urcorner &\equiv \underline{\texttt{abs}}(\lambda x.\ulcorner P\urcorner).\end{aligned}$$

Then there exists a self-interpreter E *such that for all lambda terms* M

(possibly containing variables) one has

$$\mathsf{E}\ulcorner M \urcorner = M.$$

Proof By an analog to Proposition 6.1.6 there exists a lambda term E such that

$$\begin{aligned}\mathsf{E}(\underline{\texttt{const}}\ x) &= x;\\ \mathsf{E}(\underline{\texttt{app}}\ p\,q) &= (\mathsf{E}p)(\mathsf{E}q);\\ \mathsf{E}(\underline{\texttt{abs}}\ z) &= \lambda x.\mathsf{E}(zx).\end{aligned}$$

Then by an easy induction one can show that $\mathsf{E}\ulcorner M \urcorner = M$ for all terms M. ■

Following the construction of Proposition 6.1.6 by Böhm et al. (1994), this term E is given the following very simple form:

$$\mathsf{E} \equiv \langle\langle \mathsf{K}, \mathsf{S}, \mathsf{C} \rangle\rangle,$$

where $\mathsf{S} \equiv \lambda xyz.xz(yz)$ and $\mathsf{C} \equiv \lambda xyz.x(zy)$. This happens to be a tribute to Kleene, Stephen Cole, and is a great improvement over Kleene (1936) or B[1984]. See also Barendregt (1991, 1994, 1995) for more about self-interpreters.

A short history of functional programming

We now describe in brief a history of how lambda calculi (untyped and typed) inspired (either consciously or unconsciously) the creation of functional programming. See also Barendregt et al. (2013).

Imperative versus functional programming

While Church had captured the notion of computability via the lambda calculus, Turing had done the same via his model of computation based on Turing machines. When in the Second World War computational power was needed for military purposes, the first electronic devices were built basically as Turing machines with random access memory. Statements in the instruction set for these machines, like $x{:} = x + 1$, are directly related to the instructions of a Turing machine. Such statements are much more easily interpreted by hardware than the act of substitution fundamental to the lambda calculus. Early on, the hardware of the early computers was modified each time a different computational job had to be done. Then von Neumann, who must have known[1] Turing's concept of a universal Turing machine, suggested building one machine that could be programmed to do

[1] Church had invited Turing to the United States in the mid-1930s. After his first year it was von Neumann who invited Turing to stay for a second year. See Hodges (1983).

all possible computational jobs using *software*. In the resulting computer revolution, almost all machines are based on this so-called von Neumann computer, consisting of a programmable universal machine. It would have been more appropriate to call it the Turing computer.

The model of computability introduced by Church (lambda-definability) – although equivalent to that of Turing – was harder to interpret in hardware. Therefore the emergence of the paradigm of functional programming, which is based essentially on lambda-definability, took much more time. Because functional programs are closer to the specification of computational problems than imperative ones, this paradigm is more convenient than the traditional imperative one. Another important feature of functional programs is that parallelism is much more naturally expressed in them than in imperative programs. See Turner (1981) and Hughes (1989) for some evidence for the elegance of the functional paradigm. The implementation difficulties for functional programming have to do with memory usage, compilation time and actual run time of functional programs. In the contemporary state of the art of implementing functional languages, these problems have been solved satisfactorily.[2]

Classes of functional languages

Let us describe some languages that have been – and in some cases still are – influential in the expansion of functional programming. These languages come in several classes.

Lambda calculus by itself is not yet a complete model of computation, since an expression M may be evaluated by different so-called reduction strategies that indicate which sub-term of M is evaluated first (see B[1984], Chapter 12). By the Church–Rosser theorem this order of evaluation is not important for the final result: the normal form of a lambda term is unique if it exists. But the order of evaluation makes a difference for efficiency (both time and space) and also for the question whether or not a normal form is obtained at all.

So-called 'eager' functional languages have a reduction strategy that evaluates an expression like FA by first evaluating F and A (in no particular order) to, say, $F' \equiv \lambda a. \cdots a \cdots a \cdots$ and A' and then contracting $F'A'$ to $\cdots A' \cdots A' \cdots$. This evaluation strategy has definite advantages for the efficiency of the implementation. The main reason for wanting this is that if A is large, but its normal form A' is small, then it is advantageous both

[2] Logical programming languages also have the same advantages. But so far pure logical languages of industrial quality have not been developed. (Prolog is not pure and λ-Prolog, see Nadathur and Miller (1988), although pure, is presently a prototype.)

for time and space efficiency to perform the reduction in this order. Indeed, evaluating FA directly to

$$\cdots A \cdots A \cdots$$

takes more space and if A is now evaluated twice, it also takes more time.

Eager evaluation, however, is not a normalizing reduction strategy in the sense of B[1984], Chapter 12. For example, if $F \equiv \lambda x.\mathsf{I}$ and A does not have a normal form, then evaluating FA eagerly diverges, while

$$FA \equiv (\lambda x.\mathsf{I})A = \mathsf{I},$$

if it is evaluated left-most outermost (roughly 'from left to right'). This kind of reduction is called 'lazy evaluation'.

It turns out that eager languages are, nevertheless, computationally complete, as we will soon see. The implementation of these languages was the first milestone in the development of functional programming. The second milestone consisted of the efficient implementation of lazy languages.

In addition to the distinction between eager and lazy functional languages there is another one of equal importance. This is the difference between untyped and typed languages. The difference comes directly from the difference between the untyped lambda calculus and the various typed lambda calculi; see B[1984]. Typing is useful, because many programming bugs (errors) result in a typing error that can be detected automatically prior to one running a program. On the other hand, typing is not too cumbersome, since in many cases the types need not be given explicitly. The reason for this is that, by the type reconstruction algorithm of Curry (1969) and Hindley (1969) (later rediscovered by Milner (1978)), one can automatically find the type (in a certain context) of an untyped but typable expression. Therefore, the typed versions of functional programming languages are often based on the implicitly typed lambda calculi *à la Curry*. Types also play an important role in making implementations of lazy languages more efficient, see below.

As well as the functional languages that will be treated below, the languages APL and FP are important historically. The former, introduced in Iverson (1962), was, and still is, relatively widespread. The language FP was designed by Backus, who gave, in the lecture on the occasion of receiving his Turing award (for his work on imperative languages) a strong and influential plea for the use of functional languages (Backus (1978)). Both APL and FP programs consist of a set of basic functions that can be combined to define operations on data structures. The language APL has, for example, many functions for matrix operations. In both languages composition is the only way to obtain new functions and, therefore, they are less complete than a full functional language in which user-defined functions can be created. As a consequence, these two languages are essentially limited in their ease of expressing algorithms.

Eager functional languages

Let us first give the promised argument that eager functional languages are computationally complete. Every computable (recursive) function is lambda-definable in the λI-calculus (see Church (1941) or B[1984], Theorem 9.2.16). In the λI-calculus a term having a normal form is strongly normalizing (see Church and Rosser (1936) or B[1984], Theorem 9.1.5). Therefore an eager evaluation strategy will find the required normal form.

The first functional language, LISP, was designed and implemented by McCarthy et al. (1962). The evaluation of expressions in this language is eager. LISP had (and still has) considerable impact on the art of programming. Since it has a good programming environment, many skillful programmers were attracted to it and produced interesting programs (so-called 'artificial intelligence'). LISP is not a pure functional language for several reasons. Assignment is possible in it; there is a confusion between local and global variables[3] ('dynamic binding'; some LISP users even like it); LISP uses the '`Quote`', where (`Quote` M) is like $\ulcorner M \urcorner$. In later versions of LISP, such as Common LISP (see Steele Jr. (1984)) and Scheme (see Abelson et al. (1991)), dynamic binding is no longer present. The '`Quote`' operator, however, is still present in these languages. Since $\mathsf{I}a = a$ but $\ulcorner \mathsf{I}a \urcorner \neq \ulcorner a \urcorner$ adding '`Quote`' to the lambda calculus is inconsistent. As one may not reduce in LISP within the scope of a '`Quote`', however, having a '`Quote`' in LISP is not inconsistent. '`Quote`' is not an available function but only a constructor. That is, if M is a well-formed expression, then so is (`Quote` M)[4]. Also, LISP has a primitive fixed-point operator 'LABEL' (implemented as a cycle) that is also found in later functional languages.

[3] This means substitution of an expression with a free variable into a context in which that variable becomes bound. The originators of LISP were in good company: in Hilbert and Ackermann (1928) the same was done, as was noticed by von Neumann in his review of that book. Church may have known von Neumann's review and avoided confusing local and global variables by introducing α-conversion.

[4] Using '`Quote`' as a function would violate the Church–Rosser property. An example is

$$(\lambda x.x(\mathsf{I}a))\ \mathtt{Quote}$$

that then would reduce to both

$$\mathtt{Quote}\ (\mathsf{I}a) \to \ulcorner \mathsf{I}a \urcorner$$

and to

$$(\lambda x.xa)\ \mathtt{Quote} \to \mathtt{Quote}\ a \to \ulcorner a \urcorner$$

and there is no common reduct for these two expressions $\ulcorner \mathsf{I}a \urcorner$ and $\ulcorner a \urcorner$.

In the meantime, Landin (1964) developed an abstract machine – the SECD machine – for the implementation of reduction. Many implementations of eager functional languages, including some versions of LISP, have used, or are still using, this computational model. (The SECD machine also can be modeled for lazy functional languages; see Henderson (1980).) Another way of implementing functional languages is based on the so-called CPS-translation. This was introduced in Reynolds (1972) and used in compilers by Steele Jr. (1978) and Appel (1992). See also Plotkin (1975) and Reynolds (1993).

The first important typed functional language with an eager evaluation strategy was Standard ML, see Milner (1978). This language was based on the Curry variant $\boldsymbol{\lambda}_{\rightarrow}^{\mathrm{Ch}}$, the simply typed lambda calculus with implicit typing. Expressions are type-free, but are only legal if a type can be derived for them. By the algorithm of Curry and Hindley cited above, it is decidable whether an expression does have a type and, moreover, its most general type can be computed. Milner added two features to $\boldsymbol{\lambda}_{\rightarrow}^{\mathbb{A}}$. The first was the addition of new primitives. One has the fixed-point combinator $\mathcal{Y}$ as primitive, with essentially all types of the form $(A{\rightarrow}A){\rightarrow}A$, with $A \equiv (B{\rightarrow}C)$, assigned to it. Indeed, if $f : A{\rightarrow}A$, then $\mathcal{Y}f$ is of type A so that both sides of

$$f(\mathcal{Y}f) = \mathcal{Y}f$$

have type A. Primitives for basic arithmetic operations are also added. With these additions, ML becomes a universal programming language, while $\boldsymbol{\lambda}_{\rightarrow}^{\mathbb{A}}$ is not (since all its terms are normalizing). The second addition to ML is the 'let' construction

$$\underline{\mathtt{let}}\ x\ \underline{\mathtt{be}}\ N\ \underline{\mathtt{in}}\ M\ \underline{\mathtt{end}}. \tag{6.1}$$

This language construct has as its intended interpretation

$$M[x{:}= N], \tag{6.2}$$

so that one may think that $\underline{\mathtt{let}}$ is not necessary. If, however, N is large, then this translation of (6.1) becomes space inefficient. Another interpretation of (6.1) is

$$(\lambda x.M)N, \tag{6.3}$$

but this interpretation has its limitations, as N has to be given one fixed type, whereas in (6.2) the various occurrences of N may have different types. The expression (6.1) is a way to make use of both the space reduction ('sharing') of the expression (6.3) and the 'implicit polymorphism' in which N can have more than one type of (6.2). An example of the let expression is

$$\underline{\texttt{let}}\ \texttt{id}\ \underline{\texttt{be}}\ \lambda x.x\ \underline{\texttt{in}}\ \lambda fx.(\texttt{id}\,f)(\texttt{id}\,x)\ \underline{\texttt{end}}.$$

This is typable by

$$(A{\to}A){\to}(A{\to}A),$$

if the second occurrence of `id` gets type $(A{\to}A){\to}(A{\to}A)$ and the third $(A{\to}A)$.

Because of its relatively efficient implementation and the possibility of type checking at compile time (for finding errors), the language ML has evolved into important industrial variants (like Standard ML of New Jersey).

A language inspired by ML and the 'Categorical Abstract Machine' (CAM, Cousineau et al. (1987)) is CaML, see `<caml.inria.fr>`. A version with an object-oriented layer is OCaML, see `<ocaml.org>`. A major application written in this language is the proof-assistant Coq. CaML was inspired by the categorical foundations of the lambda calculus, see Smyth and Plotkin (1982), Koymans (1982) and Curien (1993). All of these papers have been inspired by the work on denotational semantics of Scott, see Scott (1972) and Gunter and Scott (1990).

Lazy functional languages

Although all computable functions can be represented in an eager functional programming language, not all reductions in the full $\lambda\mathsf{K}$-calculus can be performed using eager evaluation. We already saw that if $F \equiv \lambda x.\mathsf{I}$ and A does not have a normal form, then eager evaluation of FA does not terminate, while this term does have a normal form. In 'lazy' functional programming languages the reduction of FA to I is possible, because the reduction strategy for these languages is essentially left-most outermost reduction which is normalizing.

One of the advantages of having lazy evaluation is that one can work with 'infinite' objects. For example there is a legal expression for the potentially infinite list of primes

$$[2, 3, 5, 7, 11, 13, 17\ \ldots],$$

from which one can take the nth projection in order to get the nth prime. See Turner (1981) and Hughes (1989) for interesting uses of the lazy programming style.

Earlier we explained why eager evaluation can be implemented more efficiently than lazy evaluation: copying large expressions is expensive because of space and time costs. In Wadsworth (1971) the idea of *graph reduction* was introduced in order to also do lazy evaluation efficiently. In this model of computation, an expression like $(\lambda x. \cdots x \cdots x \cdots)A$ does not reduce to $\cdots A \cdots A \cdots$ but to $\cdots @ \cdots @ \cdots$; $@ : A$, where the first two occurrences of @ are pointers referring to the A behind the third occurrence. In this way lambda expressions become dags (directed acyclic graphs).[5]

Based on the idea of graph reduction, using carefully chosen combinators as primitives, the experimental language SASL, see Turner (1976, 1979), was one of the first implemented lazy functional languages. The notion of graph reduction was extended by Turner by implementing the fixed-point combinator (one of the primitives) as a cyclic graph. (Cyclic graphs were already described in Wadsworth (1971) but were not used there.) Like LISP, the language SASL is untyped. It is fair to say that – unlike programs written in the eager languages such as LISP and Standard ML – the execution of SASL programs was orders of magnitude slower than that of imperative programs in spite of the use of graph reduction.

In the 1980s typed versions of lazy functional languages did emerge, accompanied by a considerable speed-up of their performance. A lazy version of ML, called Lazy ML (LML), was implemented efficiently by a group at Chalmers University; see Johnsson (1984). As underlying computational model they used the so-called G-machine, which avoids building graphs whenever it is efficient not to. For example, if an expression is purely arithmetical (this can be seen from type information), then the evaluation can be done more efficiently than by using graphs. Another implementation feature of the LML is the compilation into super-combinators, see Hughes (1984), that do not form a fixed set, but are created on demand depending on the expression to be evaluated. Emerging from SASL, the first fully developed typed lazy functional language, called Miranda^{TM}, was developed by Turner (1985). Special mention should be made of its elegance and its functional I/O interface (see below).

Notably, the ideas in the G-machine made lazy functional programming much more efficient. In the late 1980s very efficient implementations of two typed lazy functional languages appeared that we will discuss below: Clean, see van Eekelen and Plasmeijer (1993), and Haskell, see Peyton Jones and Wadler (1993), Hudak et al. (1992). These languages, with their implementations, execute functional programs in a way that is comparable to the speed of contemporary imperative languages such as C.

[5] Robin Gandy mentioned at a meeting for the celebration of his seventieth birthday that already in the early 1950s Turing had told him that he wanted to evaluate lambda terms using graphs. In Turing's description of the evaluation mechanism he made the common mistake of confusing free and bound variables. Gandy pointed this out to Turing, who then said: "Ah, this remark is worth 100 pounds a week!"

Interactive functional languages

The versions of functional programming that we have considered so far could be called 'autistic'. A program consists of an expression M, its execution of the reduction of M and its output of the normal form M^{nf} (if it exists). Although this is quite useful for many purposes, no interaction with the outside world is made. Even just dealing with input and output (I/O) requires interaction.

For I/O and other forms of interaction we need the concept of a 'process' as opposed to a function. Intuitively a process is something that (in general) is geared towards continuation while a function is geared towards termination. Processes have an input channel on which an input stream (a potentially infinite sequence of tokens) is coming in and an output channel on which an output stream is coming out. A typical process is the control of a traffic light system: it is geared towards continuation, there is an input stream (coming from the push-buttons for pedestrians) and an output stream (regulating the traffic lights). Text editing is also a process. In fact, even the most simple form of I/O is already a process.

A primitive way of dealing with I/O in a functional language is used in some versions of ML. There is an input stream and an output stream. Suppose one wants to perform the following process P:

read the first two numbers x, y of the input stream;
put their difference $x - y$ onto the output stream.

Then one can write in ML the following program

$$\mathtt{write}\,(\mathtt{read} - \mathtt{read}).$$

This is not very satisfactory, since it relies on a fixed order of evaluation of the expression '`read` − `read`'.

A more satisfactory way consists of so-called continuations, see Gordon (1994). To the lambda calculus one adds primitives `Read`, `Write` and `Stop`. The operational semantics of an expression is now as follows:

$$\begin{array}{rcll} M & \Rightarrow & M^{\mathrm{hnf}}, & \text{where } M^{\mathrm{hnf}} \text{ is the head normal form}^6 \text{ of } M;\\ \mathtt{Read}\, M & \Rightarrow & M\,a, & \text{where } a \text{ is taken off the input stream;}\\ \mathtt{Write}\, b\, M & \Rightarrow & M, & \text{and } b \text{ is put into the output stream;}\\ \mathtt{Stop} & \Rightarrow & & \text{i.e., do nothing.} \end{array}$$

[6] A head normal form in lambda calculus is of the form $\lambda\vec{x}.yM_1\cdots M_n$, with the $M_1\cdots M_n$ possibly not in normal form.

Now the process P above can be written as

$$P = \texttt{Read}\ (\lambda x.\texttt{Read}\ (\lambda y.\texttt{Write}\ (x - y)\ \texttt{Stop})).$$

If, instead, one wants a process Q that continuously takes two elements of the input stream and puts the difference on the output stream, then one can write as a program the extended lambda term

$$Q = \texttt{Read}\ (\lambda x.\texttt{Read}\ (\lambda y.\texttt{Write}\ (x - y)\ Q)),$$

which can be found using the fixed-point combinator.

Now, every interactive program can be written in this way, provided that special commands written on the output stream are interpreted. For example one can imagine that writing

`'echo'` 7 or `'print'` 7

on the output channel will put 7 on the screen or print it out respectively. The use of continuations is equivalent to that of monads in programming languages like Haskell, as shown in Gordon (1994). (The present version of Haskell I/O is more refined than this; we will not consider this issue.)

If $A_0, A_1, A_2, \ldots$ is an effective sequence of terms (i.e., $A_n = F \ulcorner n \urcorner$ for some F), then this infinite list can be represented as a lambda term

$$\begin{aligned} [A_0, A_1, A_2, \ldots\,] &\equiv [A_0, [A_1, [A_2, \ldots\,]]] \\ &= H \ulcorner 0 \urcorner, \end{aligned}$$

where $[M, N] \equiv \lambda z.zMN$ and

$$H \ulcorner n \urcorner \ = \ [F \ulcorner n \urcorner, H \ulcorner n+1 \urcorner].$$

This H can be defined using the fixed-point combinator.

Now the operations `Read`, `Write` and `Stop` can be made explicitly lambda-definable if we use

$$\begin{aligned} \texttt{In} &= [A_0, A_1, A_2, \ldots], \\ \texttt{Out} &= [\ldots, B_2, B_1, B_0\,], \end{aligned}$$

where `In` is a representation of the potentially infinite input stream given by 'the world' (i.e., the user and the external operating system) and `Out` one of the potentially infinite output stream given by the machine running the interactive functional language. Every interactive program M should be acting on $[\texttt{In}, \texttt{Out}]$ as argument. So M in the continuation language becomes

$$M\ [\texttt{In}, \texttt{Out}].$$

The following definition then matches the operational semantics:

$$\left.\begin{array}{rcl} \texttt{Read}\; F\; [[A, \texttt{In}'], \texttt{Out}] & = & F\; A\; [\texttt{In}', \texttt{Out}]; \\ \texttt{Write}\; F\; B\; [\texttt{In}, \texttt{Out}] & = & F\; [\texttt{In}, [B, \texttt{Out}]] \\ \texttt{Stop}\; [\texttt{In}, \texttt{Out}] & = & [\texttt{In}, \texttt{Out}]. \end{array}\right\} \qquad (6.4)$$

In this way $[\texttt{In}, \texttt{Out}]$ acts as a dynamic state. An operating system should take care that the actions on [In,Out] are actually performed to the I/O channels. Also we have to take care that statements like `'echo'` 7 are being interpreted. It is easy to find pure lambda terms `Read`, `Write` and `Stop` satisfying (6.4). This seems to be a good implementation of the continuations and therefore a good way of dealing with interactive programs.

There is, however, a serious problem. Define

$$M \equiv \lambda p.[\texttt{Write}\; b_1\; \texttt{Stop}\; p, \texttt{Write}\; b_2\; \texttt{Stop}\; p].$$

Now consider the evaluation

$$\begin{aligned} M\; [\texttt{In}, \texttt{Out}] &= [\texttt{Write}\; b_1\; \texttt{Stop}\; [\texttt{In}, \texttt{Out}], \texttt{Write}\; b_2\; \texttt{Stop}\; [\texttt{In}, \texttt{Out}]] \\ &= [[\texttt{In}, [b_1, \texttt{Out}]], [\texttt{In}, [b_2, \texttt{Out}]]. \end{aligned}$$

What will happen to the actual output channel: should b_1, or perhaps b_2, be added to it?

The dilemma is caused by the duplication of the I/O channels [In,Out]. One solution is not to explicitly mention the I/O channels, as in the lambda calculus with continuations. This is essentially what happens in the method of monads in the interactive functional programming language Haskell. If one writes something like

$$\texttt{Main}\; f_1 \circ \cdots \circ f_n$$

the intended interpretation is $(f_1 \circ \cdots \circ f_n)[\texttt{In}, \texttt{Out}]$.

The solution put forward in the functional language Clean is to use a typing system that guarantees that the I/O channels are never duplicated. For this purpose a so-called 'uniqueness' typing system is designed, see Barendsen and Smetsers (1993, 1996), that is related to linear logic (see Girard (1995)). Once this is done, one can improve the way in which parts of the world are used explicitly. A representation of all aspects of the world can be incorporated in lambda calculus. Instead of having just [In,Out], the world can now be extended to include (a representation of) the screen, the printer, the mouse, the keyboard and whatever gadgets one would like to add to the computer periphery (e.g., other computers to form a network). So interpreting

`'print'` 7

now simply becomes something like

$$\texttt{put}\ 7\ \texttt{printer}.$$

This has the advantage that if one wants to echo a 7 and to print a 3, but the order in which this happens is immaterial, then one is not forced to make an over-specification, like sending first `'print'` 3 and then `'echo'` 7 to the output channel:

$$[\ldots, \texttt{'echo'}\ 7, \texttt{'print'}\ 3]$$

By representing inside the lambda calculus with uniqueness types as many gadgets of the world as one would like, one can write something like

$$\begin{aligned} &F\,[\texttt{keyboard, mouse, screen, printer}] \\ &\quad = [\texttt{keyboard, mouse, put}\ 3\ \texttt{screen, put}\ 7\ \texttt{printer}]. \end{aligned}$$

What happens first depends on the operating system and parameters, which we do not know (for example on how long the printing queue is). But we are not interested in this. The system satisfies the Church–Rosser theorem and the eventual result (7 is printed and 3 is echoed) is unambiguous. This makes Clean somewhat more natural than Haskell (also in its present version) and definitely more appropriate for an implementation on parallel hardware.

Both Clean and Haskell are state-of-the-art functional programming languages producing efficient code; as to compiling time, Clean belongs to the class of fast compilers (including those for imperative languages). Many serious applications are written in these languages. The interactive aspect of both languages is made possible by lazy evaluation and the use of higher-type[7] functions, two themes that are at the core of the lambda calculus ($\lambda\mathsf{K}$, that is). It is to be expected that they will have a significant impact on the production of modern (interactive window-based) software.

Other aspects of functional programming

In several of the following viable applications there is a price to pay. Types can no longer be derived by the Hindley–Milner algorithm, but need to be deduced by an assignment system more complex than that of the simply typed λ-calculus $\boldsymbol{\lambda}_{\rightarrow}$.

Type classes

Certain types come with standard functions or relations. For example on the natural numbers and integers one has the successor function, and the

[7] In the functional programming community these are called 'higher-*order* functions'. We prefer to use the more logically correct expression 'higher-*type*', since 'higher-order' refers to quantification over types (like in the system $\boldsymbol{\lambda}2$).

equality and order relations. A type class is like a signature in computer science or a similarity type in logic: it states to which operations, constants, and relations the data type is coupled. In this way one can write programms not for one type but for a class of types.

If the operators on classes are not only first order but higher order, one obtains 'type constructor classes', that are much more powerful. See Jones (1993), where the idea was introduced and Voigtländer (2009) for recent results.

Generic programming

The idea of type classes can be pushed further. Even if data types are different, in the sense that they have different constructors, one can share code. For

$$[\mathtt{a}_0, \mathtt{a}_1, \mathtt{a}_2, \ldots]$$

a stream, there is the higher type function '$\mathtt{map}_s$' that acts like

$$\mathtt{map}_\mathtt{s}\mathtt{f}[\mathtt{a}_0, \mathtt{a}_1, \mathtt{a}_2, \ldots] \twoheadrightarrow [\mathtt{fa}_0, \mathtt{fa}_1, \mathtt{fa}_2, \ldots].$$

But there is also a '$\mathtt{map}_t$' that distributes a function over all data present at nodes of the tree.

Generic programming makes it possible to write one program 'map' that acts both for streams and trees. What happens here is that this 'map' works on the code for data types and recognizes its structure. Then 'map' transforms itself, when requested, into the right version for doing the intended work. See Hinze et al. (2007) for an elaboration of this idea. In Plasmeijer et al. (2007) generic programming is exploited for efficient programming of web-interfaces for work flow systems.

Dependent types

These types come from the language Automath, see next section, intended for expressing mathematical properties as a type depending on a term. This breaks the independence of types from terms, but is quite useful in proof-checking. A typical dependent type is an n-dimensional vector space F^n that depends on the element n of another type. In functional programming, dependent types have been used to enable more functions to be typed. See Augustson (1999).

Dynamic types

The underlying computational model for functional programming consists of reducing λ-terms. From the λ-calculus point of view, one can pause a reduction of a term towards some kind of normal form in order to continue work later with the intermediate expression. In many efficient compilers of functional programming languages one does not reduce any term, but translates it into some machine code and works on it until there is (the code of) the normal form. There are no intermediate expressions: in particular the type information is lost during (partial) execution. The mechanism of 'dynamic types' makes it possible to store the intermediate values in such a way that a reducing computer can be switched off and work is continued the next day. Even more exciting applications of this idea to distributed or even parallel computing is to exchange partially evaluated expressions and continue the computation process elsewhere.

In applications like web-browsers one may want to ask for 'plug-ins', that employ functions involving types that are not yet known to the designer of the application. This becomes possible using dynamic types. See Pil (1999).

Generalized algebraic data types

These form another powerful extension of the simple types for functional languages. See Peyton Jones et al. (2006) and Schrijvers et al. (2009).

Major applications of functional programming

Among the many functional programs for an impressive range of applications, two major ones stand out. The first consists of the proof-assistants, to be discussed in the next section. The second consists of design languages for hardware, see Sheeran (2005) and Nikhil (2008).

6.2 Logic and proof-checking

The Curry–de Bruijn–Howard correspondence

One of the main applications of type theory is its connection with logic. For several logical systems L there is a type theory λ_L and a map translating formulas A of L into types $[A]$ of λ_L such that

$$\vdash_L A \Leftrightarrow \Gamma_A \vdash_{\lambda_L} M : [A], \text{ for some } M,$$

where Γ_A is some context 'explaining' A. The term M can be constructed canonically from a natural deduction proof D of A. So in fact one has

$$\vdash_L A, \text{ with proof } D \Leftrightarrow \Gamma_A \vdash_{\lambda_L} [D] : [A], \tag{6.5}$$

where the map $[\,]$ is extended to cover also derivations. For deductions from a set of assumptions one has

$$\Delta \vdash_L A, \text{ with proof } D \Leftrightarrow \Gamma_A, [\Delta] \vdash_{\lambda_L} [D] : [A].$$

Curry did not observe the correspondence in this precise form. He noted that inhabited types in $\boldsymbol{\lambda}_{\rightarrow}$, like $A{\rightarrow}A$ or $A{\rightarrow}B{\rightarrow}A$, all had the form of a tautology of (the implication fragment of) propositional logic.

Howard (1980) (the work was done in 1968 and written down in the unpublished but widely circulated notes from 1969), inspired by the observation of Curry and by Tait (1963), gave the more precise interpretation (6.5). He coined the term *propositions-as-types* and *proofs-as-terms*.

On the other hand, de Bruijn independently of Curry and Howard developed type systems satisfying (6.5). The work was started also in 1968 and the first publication was de Bruijn (1968), see also de Bruijn (1970, 1994a), and de Bruijn (1994b). The motivation of de Bruijn was his visionary view that proof checking by machine would one day be feasible and important. The collection of systems he designed was called the Automath family, derived from 'AUTOmatic MATHematics verification'. The type systems were such that the right hand side of (6.5) was efficiently verifiable by machine, so that one had machine verification of provability. Also de Bruijn and his students were engaged in developing, using and implementing these systems.

Initially the Automath project received little attention from mathematicians. They did not understand the technique and worse, they did not see the need for machine verification of provability. Also the verification process was rather painful. After five 'monastic' years of work, van Benthem Jutting (1977) came up with a machine verification of a result from Landau (1900) fully rewritten in the terse 'machine code' of one of the Automath languages. Since then modern versions of the family of proof-assistants, such as Mizar (`<www.mizar.org>`, COQ (Bertot and Castéran (2004)), HOL (Gordon and Melham (1993)), and Isabelle (Nipkow et al. (2002)) have been developed, in which considerable help from the computer environment is obtained for the formalization of proofs. With these systems the task of verifying Landau's result took something like five months. An important contribution to these second-generation systems came from Scott and Martin-Löf, who added inductive data-types to the systems in order to make formalizations more natural.[8] In Kahn (1995) methods are developed that enable proof objects to be translated automatically into natural language. It is hoped that

[8] For example, proving Gödel's incompleteness theorem contains the following technical point. The main step in the proof essentially consists of constructing a compiler from a universal programming language into arithmetic. For this one needs to describe strings over an alphabet in the structure of numbers with 'plus' and 'times'. This is involved and Gödel used the Chinese remainder theorem to do it. Having available the data-type of strings, together with the corresponding operators, makes the translation much more natural. The incompleteness of this stronger theory is stronger than that of arithmetic. But then the customary *essential* incompleteness result states incompleteness for all extensions of an arithmetical theory with inductive types, which is weaker than the essential incompleteness of just arithmetic.

in the near future new proof-checkers will emerge in which formalizing is not that much more difficult than, say, writing an article in TEX.

Computer Mathematics

Systems for computer algebra (CA) are able to represent mathematical notions on a machine and compute with them. These objects can be integers, real or complex numbers, polynomials, integrals and the like. The computations are usually symbolic, but can also be numerical to a virtually arbitrary degree of precision. It is fair to say – as is sometimes done – that "computer algebra systems can represent $\sqrt{2}$ exactly". In spite of the fact that this number has an infinite decimal expansion, this is not a miracle. The number $\sqrt{2}$ is represented in a computer simply as a symbol (as we do on paper or in our mind), and the machine knows how to manipulate it. The common feature of this kind of notion represented in computer algebra systems is that in one way or other they are all computable. Computer algebra systems have reached a high level of sophistication and efficiency and are commercially available. Scientists and both pure and applied mathematicians have made good use of them in their research.

There is now emerging a new technology, namely that of systems for Computer Mathematics (CM). In these, virtually all mathematical notions can be represented exactly, including those that do not have a computational nature. How is this possible? Suppose, for example, that we want to represent a non-computable object like the co-Diophantine set

$$X = \{n \in \omega \mid \neg\exists\vec{x}\, D(\vec{x}, n) = 0\}.$$

Then we can do as before and represent it by a special symbol. But now the computer in general cannot operate on it because the object may be of a non-computational nature.

Before answering the question in the previous paragraph, it is good to realize that non-computability comes from the quantifiers $\forall$ (for all) and $\exists$ (exists). Indeed, these quantifiers usually range over an infinite set and therefore one loses decidability.

Nevertheless, for ages mathematicians have been able to obtain interesting information about these non-computable objects. This is because there is a notion of *proof*. Using proofs one can state with confidence that, e.g.,

$$3 \in X, \text{ i.e., } \neg\exists\vec{x}\, D(\vec{x}, 3) = 0.$$

The statement "it is in general hard to find proofs, but the verification of a putative one can be done in a relatively easy way" is attrbuted to Aristotle. Another contribution of Aristotle was his quest for the formalization of logic. About 2300 years later, when Frege had found the right formulation of predicate logic and Gödel had proved that it is complete, this quest was fulfilled. Mathematical proofs can now be completely formalized and verified by computers. This is the underlying basis for CM systems.

Present day CM systems are able to help a user to develop from primitive notions and axioms many theories that consist of defined concepts, theorems and proofs.[9] All CM systems have been inspired by the Automath project of de Bruijn (see de Bruijn (1970, 1994b) and Nederpelt et al. (1994)) for the automated verification of mathematical proofs.

Representing proofs as lambda terms

Now that mathematical proofs can be fully formalized, the question arises how this can best be done (for efficiency reasons, as far as machines are concerned, and pragmatic reasons for humans). Hilbert represented a proof of statement A from a set of axioms Γ as a finite sequence $A_0, A_1, \ldots, A_n$ such that $A = A_n$ and each A_i, for $0 \leq i \leq n$, is either in Γ or follows from previous statements using the rules of logic.

A more efficient way of representing proofs employs typed lambda terms and is called the *propositions-as-types* interpretation discovered by Curry, Howard and de Bruijn. This interpretation maps propositions into types and proofs into the corresponding inhabitants. The method is as follows. A statement A is transformed into the type (i.e., collection)

$$[A] = \text{ the set of proofs of } A.$$

So A is provable if and only if $[A]$ is 'inhabited' by a proof p. Now a proof of $A \Rightarrow B$ consists (according to the Brouwer–Heyting interpretation of implication) of a function having as argument a proof of A and as value a proof of B. In symbols

$$[A \Rightarrow B] = [A] \rightarrow [B].$$

Similarly

$$[\forall x \in X.Px] = \Pi x{:}X.[Px],$$

where $\Pi x{:}A.[Px]$ is the Cartesian product of the $[Px]$, because a proof of $\forall x \in A.Px$ consists of a function that assigns to each element $x \in A$ a proof of Px. In this way proof objects become isomorphic to the intuitionistic natural deduction proofs of Gentzen (1969). Using this interpretation, a proof of

[9] This way of doing mathematics, the axiomatic method, was also described by Aristotle. It was Euclid of Alexandria (-300) who first used this method very successfully in his *Elements*.

$\forall y \in A.Py \Rightarrow Py$ is $\lambda y{:}A\lambda x{:}Py.x$. Here $\lambda x{:}A.B(x)$ denotes the function that assigns to input $x \in A$ the output $B(x)$. A proof of

$$(A \Rightarrow A \Rightarrow B) \Rightarrow A \Rightarrow B$$

is

$$\lambda p{:}(A \Rightarrow A \Rightarrow B)\lambda q{:}A.pqq.$$

A description of the typed lambda calculi in which these types and inhabitants can be formulated is given in Barendregt (1992), which also gives an example of a large proof object. Verifying whether p is a proof of A boils down to verifying whether, in the given context, the type of p is equal (convertible) to $[A]$. This is concerned with the $\{\to, \forall\}$ fragment of predicate logic. In Martin-Löf (1984) the other logical connectives, such as & and $\vee$, are represented in type systems. The idea hinges on an extension of typed lambda calculus with the addition of 'inductive data types' and the corresponding recursion operators (introduced in Scott (1970)). Translating propositions as types has as default intuitionistic logic. Classical logic can be dealt with by adding the excluded middle as an axiom. An interesting alternative is the $\lambda\mu$-calculus in Parigot (1992) that can deal with classical logic.

If a complicated computer system claims that a certain mathematical statement is correct, then one may wonder whether this is indeed the case. For example, there may be software errors in the system. A satisfactory methodological answer was given by de Bruijn. Proof objects should be public and written in such a formalism that a reasonably simple proof-checker can verify them. One should be able to verify the program for this proof-checker 'by hand'. We call this the *de Bruijn criterion*. The proof-development systems Isabelle/HOL, (see Nipkow et al. (2002)), HOL-light and Coq (see Bertot and Castéran (2004)), all satisfy this criterion.

A way of keeping proof objects from growing too large is to employ the so-called Poincaré principle. Poincaré (1902), p. 12, stated that an argument showing that $2+2=4$ "is not a proof in the strict sense, it is a verification" (moreover he claimed that an arbitrary mathematician will agree with this remark). In the Automath project of de Bruijn, the following interpretation of the Poincaré principle was given. If p is a proof of $A(t)$ and $t =_R t'$, then the same p is also a proof of $A(t')$. Here, R is a notion of reduction consisting of ordinary $\boldsymbol{\beta}$-reduction and $\boldsymbol{\delta}$-reduction in order to deal with the unfolding of definitions. Since $\boldsymbol{\beta\delta}$-reduction is not too complicated to be programmed, the type systems enjoying this interpretation of the Poincaré principle still satisfy the de Bruijn criterion[10].

In spite of their compact representation in typed lambda calculi and the use of the Poincaré principle, proof objects become large, something like 4 to 10 times the length of a complete informal proof. Large proof objects are tiresome to generate by hand. With the necessary persistence van Benthem Jutting (1977) has written lambda after lambda to obtain the proof objects showing that all proofs (but one) in Landau (1960) are correct. Using a modern CM system one can do better. The user introduces the context consisting of the primitive notions and axioms. The definitions necessary for formulating a theorem to be proved (the goal) are then given. The proof is developed in an interactive session with the machine. Thereby the user only needs to give certain 'tactics' to the machine. (The interpretation of these tactics by the machine does nothing mathematically sophisticated, only the necessary book-keeping. The sophistication comes from giving the correct tactics.) The final goal of this research is that the necessary effort for interactively generating formal proofs is no more complicated than producing some text in, say, LaTeX. This goal has not been reached yet: see Barendregt and Wiedijk (2005) for references, including those about other approaches to computer mathematics. These include NuPrl, HOL, Otter, Mizar and the Boyer–Moore theorem prover. These systems do not satisfy the de Bruijn criterion, but some of them probably can be modified easily so that they do.

Computations in proofs

The following is taken from Barendregt and Barendsen (1997). There are several computations that are needed in proofs. This happens, for example, if we want to prove formal versions of the following intuitive statements.

$$
\begin{array}{lrcl}
\text{(i)} & [\sqrt{45}] & = & 6, \text{where } [r] \text{ is the integer part of a real;} \\
\text{(ii)} & \mathtt{Prime}(61); & & \\
\text{(iii)} & (x+1)(x+1) & = & x^2 + 2x + 1.
\end{array}
$$

A way of handling (i) is to use the Poincaré principle extended to the reduction relation $\twoheadrightarrow_\iota$ for primitive recursion on the natural numbers. Operations like $f(n) = [\sqrt{n}\,]$ are primitive recursive and hence are lambda-definable (using $\twoheadrightarrow_{\beta\iota}$) by a term, say F, in the lambda calculus extended by an operation

[10] The reductions may sometimes cause the proof-checking to be of an unacceptable time complexity. We have that p is a proof of A iff $\mathtt{type}(p) =_{\beta\delta} A$. Because the proof is coming from a human, the necessary conversion path is feasible, but to find it automatically may be hard. The problem probably can be avoided by enhancing proof objects with hints for a reduction strategy.

for primitive recursion R satisfying

$$R\,A\,B\,\texttt{zero} \to_\iota A$$
$$R\,A\,B\,(\texttt{succ}\,x) \to_\iota B\,x\,(R\,A\,B\,x).$$

Then, writing $\ulcorner 0 \urcorner = \texttt{zero}$, $\ulcorner 1 \urcorner = \texttt{succ zero}, \ldots$, since

$$\ulcorner 6 \urcorner = \ulcorner 6 \urcorner$$

is formally derivable, it follows from the Poincaré principle that the same is true for

$$F \ulcorner 45 \urcorner = \ulcorner 6 \urcorner$$

(with the same proof object), since $F \ulcorner 45 \urcorner \twoheadrightarrow_{\beta\iota} \ulcorner 6 \urcorner$. Usually, a proof obligation arises that F is adequately constructed. For example, in this case it could be

$$\forall n\, (F\,n)^2 \;\le\; n\; < ((F\,n) + 1)^2.$$

Such a proof obligation needs to be formally proved, but only once; after that, reductions such as

$$F \ulcorner n \urcorner \twoheadrightarrow_{\beta\iota} \ulcorner f(n) \urcorner$$

can be freely used many times.

In a similar way, a statement like (ii) can be formulated and proved by constructing a lambda-defining term $K_{\texttt{Prime}}$ for the characteristic function of the predicate Prime. This term should satisfy

$$\forall n\, [(\texttt{Prime}\, n \;\leftrightarrow\; K_{\texttt{Prime}}\, n = \ulcorner 1 \urcorner)\; \&$$
$$(K_{\texttt{Prime}}\, n = \ulcorner 0 \urcorner \;\vee\; K_{\texttt{Prime}}\, n = \ulcorner 1 \urcorner)],$$

which is the proof obligation.

Statement (iii) corresponds to a symbolic computation. This computation takes place on the syntactic level of formal terms. There is a function g acting on syntactic expressions satisfying

$$g((x+1)(x+1)\,) = x^2 + 2x + 1,$$

that we want to lambda-define. While $(x+1)$:Nat (in context x:Nat), the expression on a syntactic level represented internally satisfies 'x+1' : $\texttt{term}(\mathsf{Nat})$, for the suitably defined inductive type $\texttt{term}(\mathsf{Nat})$. After introducing a reduction relation $\twoheadrightarrow_\iota$ for primitive recursion over this data type, one can use techniques similar to those of Section 6.1 to lambda-define g, say by G, so that

$$G\,\text{`}(x+1)(x+1)\,\text{'} \twoheadrightarrow_{\beta\iota} \text{`}x^2 + 2x + 1\text{'}.$$

Now in order to finish the proof of (iii), one needs to construct a self-interpreter E, such that for all expressions $p : \mathsf{Nat}$ one has

$$\mathsf{E}`p\text{'} \twoheadrightarrow_{\beta\iota} p$$

and prove the proof obligation for G which is

$$\forall t{:}\mathtt{term}(\mathsf{Nat})\ \mathsf{E}(G\,t) \;=\; \mathsf{E}\,t.$$

It follows that

$$\mathsf{E}(G`(x+1)(x+1)\text{'}) \;=\; \mathsf{E}`(x+1)(x+1)\text{'};$$

now since

$$\begin{aligned}\mathsf{E}(G`(x+1)(x+1)\text{'}) &\twoheadrightarrow_{\beta\iota} \mathsf{E}`x^2+2x+1\text{'}\\ &\twoheadrightarrow_{\beta\iota} x^2+2x+1\\ \mathsf{E}`(x+1)(x+1)\text{'} &\twoheadrightarrow_{\beta\iota} (x+1)(x+1),\end{aligned}$$

we have by the Poincaré principle

$$(x+1)(x+1) \;=\; x^2+2x+1.$$

The use of inductive types like Nat and term(Nat) and the corresponding reduction relations for primitive reduction was suggested by Scott (1970) and the extension of the Poincaré principle for the corresponding reduction relations of primitive recursion by Martin-Löf (1984). Since such reductions are not too hard to program, the resulting proof-checking still satisfies the de Bruijn criterion.

In Oostdijk (1996) a program is presented that, for every primitive recursive predicate P, constructs both the lambda term K_P defining its characteristic function and the proof of the adequacy of K_P. The resulting computations for $P = \mathtt{Prime}$ are not efficient, because a straightforward (non-optimized) translation of primitive recursion is given and the numerals (the represented numbers) used are in a unary (rather than n-ary) representation; but the method is promising. In Elbers (1996), a more efficient ad hoc lambda-definition of the characteristic function of Prime is given, using Fermat's small theorem about primality; the required proof obligation was also given. Using these ideas Caprotti and Oostdijk (2001) could generate proofs of primality of numbers having up to hundred digits. The method was refined in Grégoire et al. (2006), providing proof generators for the primality of numbers of several hundred digits.

Foundations for existing proof-assistants

Early indications of the possibility of relating logic and types were Church (1940) and a remark in Curry and Feys (1958). The former is worked out in Andrews (2002). The latter has led to the Curry–Howard correspondence between formulas and types (Howard (1980) written in 1969, Martin-Löf (1984), Barendregt (1992), de Groote (1995), and Sørensen and Urzyczyn (2006)).

Higher-order logic as foundations yielded the mathematical assistants HOL (Gordon and Melham (1993), `</hol.sourceforge.net>`), HOL Light (Harrison (2009b), `<www.cl.cam.ac.uk/~jrh13/hol-light/>`), and Isabelle[11], (Nipkow et al. (2002), `<www.cl.cam.ac.uk/research/hvg/isabelle>`). Type theory as foundations gave rise to the systems Coq (based on constructive logic, but with the possibility of impredicativity; Bertot and Castéran (2004), `<coq.inria.fr>`) and Agda (based on Martin-Löf's type theory: intuitionistic and predicative; Bove et al. (2009)). We also mention the proof assistant Mizar (Muzalewski (1993), `<mizar.org>`) which is based on an extension of ZFC set theory. On the other end of the spectrum there is ACL_2 (Kaufmann et al. (2000)), which is based on primitive recursive arithmetic.

All these systems give (usually interactive) support for the fully formal proof of a mathematical theorem, derived from user-specified axioms. For an insightful comparison of these and many more existing proof assistants see Wiedijk (2006), in which the irrationality of $\sqrt{2}$ has been formalized using seventeen different assistants.

Highlights

By the end of the twentieth century the technology for formalizing mathematical proofs was in place, but impressive examples were missing. The situation changed dramatically during the first decade of the twenty-first century. The full formalization and computer verification of the Four Color Theorem in was achieved in Coq by Gonthier (2008) (formalizing the proof in Robertson et al. (1997)); the Prime Number Theorem in Isabelle by Avigad et al. (2007) (elementary proof by Selberg) and in HOL Light by Harrison (2009a) (the classical proof by Hadamard and de la Vallée Poussin using complex function theory). Building upon the formalization of the Four Color Theorem the Jordan Curve Theorem has been formalized by Tom Hales, who did this as one of the ingredients needed for the full formalization of his proof of the Kepler Conjecture, Hales (2005).

[11] Isabelle is actually a '*logical framework*' in which a proof assistant proper can be defined. The main version is Isabelle/HOL, which representing higher-order logic.

Certifying software and hardware

This development of high quality mathematical proof assistants was accelerated by the industrial need for reliable software and hardware. The method for certifying industrial products is to fully formalize both their specification and their design and then to provide a proof that the design meets the specification[12]. This reliance on so-called 'Formal Methods' had been advocated since the 1970s, but remained unconvincing for many years. Proofs of correctness were much more complex than the mere correctness itself. So if a human had to judge the long proofs of certification, then nothing was gained. The situation changed dramatically after proof assistants came of age. The ARM6 processor – predecessor of the ARM7 embedded in the great majority of mobile phones, personal organizers and MP3 players – was certified, Fox (2003), by the above-mentioned method. The seL4 operating system has been fully specified and certified, Klein et al. (2009). The same holds for a realistic kernel of an optimizing compiler for the C programming language, Leroy (2009).

Illative lambda calculus

Work on type systems for the representation of logic has resulted in an unexpected side effect. By making a modification inspired by the TSs, it became possible, after all, to give an extension of the untyped lambda calculus, called *Illative Lambda Calculi* (ILC; the expression 'illative' comes from '*illatum*' past participle of the Latin word *inferre* which means to infer), such that first-order logic can be faithfully and completely embedded into it. The method can be extended for an arbitrary PTS[13], so that higher-order logic can be represented too.

The resulting ILCs are in fact simpler than the TSs. But doing computer mathematics via ILC is probably not very practical, as it is not clear how to do proof-checking for these systems.

One nice thing about the ILC is that the old dream of Church and Curry came true; namely, there is a system based on untyped lambda calculus (or combinators) on which logic, hence mathematics, can be based. More importantly there is a 'combinatory transformation' between the ordinary interpretation of logic and its propositions-as-types interpretation. Basically, the situation is as follows. The interpretation of predicate logic in ILC is such

[12] This presupposes that the distance between the desired behavior and the specification on the one hand, and that of the design and realization on the other is small enough to be bridged properly.

[13] For first-order logic, the embedding is natural, but for second-order logic, for example, this is less so. It is an open question whether there exists a natural representation of second- and higher-order logic in ILC.

that

$$\begin{aligned} \vdash_{\text{logic}} A \text{ with proof } p &\Leftrightarrow \forall r \vdash_{\text{ILC}} [A]_r[p] \\ &\Leftrightarrow \vdash_{\text{ILC}} [A]_{\mathsf{I}}[p] \\ &\Leftrightarrow \vdash_{\text{ILC}} [A]_{\mathsf{K}}[p] = \mathsf{K}[A]'_{\mathsf{I}}[p] = [A]'_{\mathsf{I}}, \end{aligned}$$

where r ranges over untyped lambda terms. Now if $r = \mathsf{I}$, then this translation is the propositions-as-types interpretation; if, on the other hand, one has $r = \mathsf{K}$, then the interpretation becomes an isomorphic version of first-order logic denoted by $[A]'_{\mathsf{I}}$. See Barendregt et al. (1993) and Dekkers et al. (1998) for these results. A short introduction to ILC (in its combinatory version) can be found in B[1984], Appendix B.

6.3 Proof theory

Lambda terms for natural deduction, sequent calculus and cut elimination

It is well known that there is a good correspondence between natural deduction derivations and typed lambda terms. Moreover normalizing these terms is equivalent to eliminating cuts in the corresponding sequent calculus derivations. Several papers have been written on this topic. The correspondence between sequent calculus derivations and natural deduction derivations is, however, not one-to-one. This causes some syntactic technicalities. The correspondence is best explained by two extensionally equivalent type assignment systems for untyped lambda terms, one corresponding to natural deduction (λN) and the other to sequent calculus (λL). These two systems constitute different grammars for generating the same (type assignment relation for untyped) lambda terms. The second grammar is ambiguous, but the first one is not. This fact explains the many-one correspondence mentioned above. Moreover, the second type assignment system has a 'cut–free' fragment (λL^{cf}). This fragment generates exactly the typable lambda terms in normal form. The cut elimination theorem becomes a simple consequence of the fact that typed lambda terms posses a normal form.

Introduction

The relation between lambda terms and derivations in sequent calculus, between normal lambda terms and cut-free derivations in sequent calculus and finally between normalization of terms and cut elimination of derivations has been observed by several authors (Prawitz (1965), Zucker (1974) and Pottinger (1977)). This relation is less than perfect because several cut-free sequent derivations correspond to one lambda term. In Herbelin (1995) a

`form`	$=$	`atom` $\mid$ `form`$\to$`form`
`atom`	$=$	`p` $\mid$ `atom`$'$

Figure 6.2

lambda calculus with explicit substitution operators is used in order to establish a perfect match between terms of that calculus and sequent derivations. We will not avoid the mismatch, but get a satisfactory view of it, by seeing the sequent calculus as a more intensional way to do the same as natural deduction: assigning lambda terms to provable formulas.

Next to the well-known system $\boldsymbol{\lambda}_{\to}$ of Curry type assignment to type free terms, which here will be denoted by λN, there are two other systems of type assignment: λL and its cut-free fragment λL^{cf}. The three systems λN, λL and λL^{cf} correspond exactly to the natural deduction calculus NJ, the sequent calculus LJ and the cut-free fragment of LJ, here denoted by N, L and L^{cf} respectively. Moreover, λN and λL generate the same type assignment relation. The system λL^{cf} generates the same type assignment relation as λN restricted to normal terms, and cut elimination corresponds exactly to normalization. The mismatch between the logical systems that was observed above, is due to the fact that λN is a syntax-directed system, whereas both λL and λL^{cf} are not. (A syntax-directed version of λL is possible if rules with arbitrarily many assumptions are allowed, see Capretta and Valentini (1998).)

The type assignment system of this section, from Barendregt and Ghilezan (2000), is a subsystem of one in Barbanera et al. (1995) and also implicitly present in Mints (1996).

For simplicity the results are presented only for the essential kernel of intuitionistic propositional logic, i.e. for the minimal implicational fragment. The method probably can be extended to the full first-order intuitionistic logic, using the terms that are in Mints (1996). For a classical version based on the already mentioned $\lambda\mu$-calculus of Parigot, see Ghilezan (2007).

The logical systems N, L and L^{cf}

6.3.1 Definition The set `form` of formulas (of minimal implicational propositional logic) is defined by the simplified syntax given in Fig. 6.2.

Note that the set of formulas is $\mathbb{T}^{\mathbb{A}}$ with $\mathbb{A} = \{p, p', p'', \dots\}$, i.e. a notational variant of $\mathbb{T}^{\infty}$. The intention is *a priori* different: the formulas are intended to denote propositions, with the $\to$-operation denoting implica-

N	
$\dfrac{A \in \Gamma}{\Gamma \vdash A}$	axiom
$\dfrac{\Gamma \vdash A \to B \qquad \Gamma \vdash A}{\Gamma \vdash B}$	$\to$ elim
$\dfrac{\Gamma, A \vdash B}{\Gamma \vdash A \to B}$	$\to$ intr

Figure 6.3

tion; the types denote collections of lambda terms, with the $\to$ denoting the functionality of these.

We write $p, q, r, \ldots$ for arbitrary atoms and $A, B, C, \ldots$ for arbitrary formulas. Sets of formulas are denoted by $\Gamma, \Delta, \ldots$ The set Γ, A stands for $\Gamma \cup \{A\}$.

6.3.2 Definition

(i) A formula A is *derivable* in the system N from the set[14] Γ, written $\Gamma \vdash_N A$, if $\Gamma \vdash A$ can be generated by the axiom and rules in Fig. 6.3.
(ii) A formula A is *derivable* from a set of assumptions Γ in the system L, written $\Gamma \vdash_L A$, if $\Gamma \vdash A$ can be generated by the axiom and rules in Fig. 6.4.
(iii) The system L^{cf} is obtained from the system L by omitting the rule (cut); see Fig. 6.5.

6.3.3 Lemma *Suppose $\Gamma \subseteq \Gamma'$. Then, in all systems,*

$$\Gamma \vdash A \ \Rightarrow\ \Gamma' \vdash A.$$

Proof By a trivial induction on derivations. ∎

6.3.4 Proposition *For all Γ and A we have*

$$\Gamma \vdash_N A \ \Leftrightarrow\ \Gamma \vdash_L A.$$

Proof ($\Rightarrow$) By induction on derivations in N. For the rule ($\to$ elim) we

[14] In contrast with the situation for bases, Definition 1.1.14(iii), the set Γ is arbitrary.

L	
$\dfrac{A \in \Gamma}{\Gamma \vdash A}$	axiom
$\dfrac{\Gamma \vdash A \qquad \Gamma, B \vdash C}{\Gamma, A{\to}B \vdash C}$	$\to$ left
$\dfrac{\Gamma, A \vdash B}{\Gamma \vdash A{\to}B}$	$\to$ right
$\dfrac{\Gamma \vdash A \qquad \Gamma, A \vdash B}{\Gamma \vdash B}$	cut

Figure 6.4

L^{cf}	
$\dfrac{A \in \Gamma}{\Gamma \vdash A}$	axiom
$\dfrac{\Gamma \vdash A \qquad \Gamma, B \vdash C}{\Gamma, A{\to}B \vdash C}$	$\to$ left
$\dfrac{\Gamma, A \vdash B}{\Gamma \vdash A{\to}B}$	$\to$ right

Figure 6.5

need the rule (cut):

$$\dfrac{\Gamma \vdash_L A{\to}B \qquad \dfrac{\Gamma \vdash_L A \qquad \dfrac{}{\Gamma, B \vdash_L B}\,(\text{axiom})}{\Gamma, A{\to}B \vdash_L B}\,(\to \text{left})}{\Gamma \vdash_L B}\,(\text{cut}).$$

($\Leftarrow$) By induction on derivations in L. The rule ($\to$ left) is treated as follows:

$$\dfrac{\dfrac{\dfrac{\Gamma \vdash_N A}{\Gamma, A{\to}B \vdash_N A}\,(6.3.3) \quad \dfrac{}{\Gamma, A{\to}B \vdash_N A{\to}B}\,(\text{axiom})}{\Gamma, A{\to}B \vdash_N B}\,(\to \text{elim}) \quad \dfrac{\Gamma, B \vdash_N C}{\Gamma \vdash_N B{\to}C}\,(\to \text{intr})}{\Gamma, A{\to}B \vdash_N C}\,(\to \text{elim}).$$

The rule (cut) is treated as follows:

$$\dfrac{\Gamma \vdash_N A \quad \dfrac{\Gamma, A \vdash_N B}{\Gamma \vdash_N A{\to}B}\,(\to \text{intr})}{\Gamma \vdash_N B}\,(\to \text{elim}).\ \blacksquare$$

6.3.5 Definition Consider the following rule as alternative to the rule (cut):

$$\dfrac{\Gamma, A{\to}A \vdash B}{\Gamma \vdash B}\,(\text{cut}')$$

The system L' is defined by replacing the rule (cut) by (cut$'$).

6.3.6 Proposition *For all* Γ *and* A

$$\Gamma \vdash_L A \Leftrightarrow \Gamma \vdash_{L'} A.$$

Proof ($\Rightarrow$) The rule (cut) is treated as follows.

$$\dfrac{\dfrac{\Gamma \vdash_{L'} A \qquad \Gamma, A \vdash_{L'} B}{\Gamma, A{\to}A \vdash_{L'} B}\,(\to \text{left})}{\Gamma \vdash_{L'} B}\,(\text{cut}').$$

($\Leftarrow$) The rule (cut$'$) is treated as follows:

$$\dfrac{\Gamma, A{\to}A \vdash_L B \quad \dfrac{\dfrac{}{\Gamma, A \vdash_L A}\,(\text{axiom})}{\Gamma \vdash_L A{\to}A}\,(\to \text{right})}{\Gamma \vdash_L B}\,(\text{cut}).\ \blacksquare$$

Note that we have not yet investigated the role of L^{cf}.

The type assignment systems λN, λL and λL^{cf}

6.3.7 Definition

(i) A *type assignment* is an expression of the form

$$P : A,$$

where $P \in L$ is an untyped lambda term and A is a formula.

λN	
$\dfrac{(x{:}A) \in \Gamma}{\Gamma \vdash x : A}$	axiom
$\dfrac{\Gamma \vdash P : (A{\to}B) \qquad \Gamma \vdash Q : A}{\Gamma \vdash (PQ) : B}$	$\to$ elim
$\dfrac{\Gamma, x{:}A \vdash P : B}{\Gamma \vdash (\lambda x.P) : (A{\to}B)}$	$\to$ intr

Figure 6.6

(ii) A *declaration* is a type assignment of the form

$$x : A.$$

(iii) A *context* Γ is a set of declarations such that for every variable x there is at most one declaration $x{:}A$ in Γ.

In the following definition, the system $\boldsymbol{\lambda}_{\to}$ over $\mathbb{T}^{\infty}$ is called λN. The formulas of N are isomorphic to types in $\mathbb{T}^{\infty}$ and the derivations in N of a formula A are isomorphic to the closed terms M of A considered as a type. If the derivation is from a set of assumptions $\Gamma = \{A_1, \ldots, A_n\}$, then the derivation corresponds to an open term M under the basis $\{x_1{:}A_1, \ldots, x_n{:}A_n\}$. This correspondence is called the *Curry–Howard isomorphism.* One can consider a proposition as the type of its proofs. Under this correspondence proofs of $A{\to}B$ are functions mapping the collection of proofs of A into that of B. See Howard (1980), de Groote (1995), and Sørensen and Urzyczyn (2006) and the references therein for more on this topic.

6.3.8 Definition

(i) A type assignment $P : A$ is *derivable* from the context Γ in the system λN, written

$$\Gamma \vdash_{\lambda N} P : A,$$

if $\Gamma \vdash P : A$ can be generated by the axiom and rules given in Fig. 6.6.

(ii) A type assignment $P : A$ is *derivable* form the context Γ in the system λL, written

$$\Gamma \vdash_{\lambda L} P : A,$$

if $\Gamma \vdash P : A$ can be generated by the axiom and rules given in Fig. 6.7.

$$\lambda L$$

$$\begin{array}{ll}
\dfrac{(x{:}A) \in \Gamma}{\Gamma \vdash x{:}A} & \text{axiom} \\[2ex]
\dfrac{\Gamma \vdash Q : A \qquad \Gamma, x{:}B \vdash P : C}{\Gamma, y : A{\to}B \vdash P[x{:=}yQ] : C} & \to \text{left} \\[2ex]
\dfrac{\Gamma, x{:}A \vdash P : B}{\Gamma \vdash (\lambda x.P) : (A{\to}B)} & \to \text{right} \\[2ex]
\dfrac{\Gamma \vdash Q : A \qquad \Gamma, x{:}A \vdash P : B}{\Gamma \vdash P[x{:=}Q] : B} & \text{cut}
\end{array}$$

Figure 6.7

$$\lambda L^{\text{cf}}$$

$$\begin{array}{ll}
\dfrac{(x{:}A) \in \Gamma}{\Gamma \vdash x{:}A} & \text{axiom} \\[2ex]
\dfrac{\Gamma \vdash Q : A \qquad \Gamma, x{:}B \vdash P : C}{\Gamma, y : A{\to}B \vdash P[x{:=}yQ] : C} & \to \text{left} \\[2ex]
\dfrac{\Gamma, x{:}A \vdash P : B}{\Gamma \vdash (\lambda x.P) : (A{\to}B)} & \to \text{right}
\end{array}$$

Figure 6.8

In the rule ($\to$ left) it is required that $\Gamma, y{:}A{\to}B$ is a context. This is the case if y is fresh or if $\Gamma = \Gamma, y{:}A{\to}B$, i.e. $y{:}A{\to}B$ already occurs in Γ.

(iii) The system λL^{cf} is obtained from the system λL by omitting the rule (cut): see Fig. 6.8.

6.3.9 Remark The alternative rule (cut′) could also have been used to define the variant $\lambda L'$. The right version for the rule (cut′) with term assignment is as in Fig. 6.9.

6.3.10 Notation Let $\Gamma = \{A_1, \dots, A_n\}$ and $\vec{x} = \{x_1, \dots, x_n\}$. Write

$$\Gamma_{\vec{x}} = \{x_1{:}A_1, \dots, x_n{:}A_n\}$$

Rule cut′ for $\lambda L'$
$\dfrac{\Gamma,\ x{:}A{\rightarrow}A \vdash P : B}{\Gamma \vdash P[x{:=}\mathsf{I}] : B}$ cut′

Figure 6.9

and

$$\Lambda^{\circ}(\vec{x}) = \{P \in \mathtt{term} \mid FV(P) \subseteq \vec{x}\},$$

where $FV(P)$ is the set of free variables of P.

The following result has been observed for N and λN by Curry, Howard and de Bruijn. (See Troelstra and Schwichtenberg (1996) 2.1.5. and Hindley (1997) 6B3, for some fine points about the correspondence between deductions in N and corresponding terms in λN.)

6.3.11 Proposition (Propositions-as-types interpretation) *Let S be one of the logical systems N, L or L^{cf} and let λS be the corresponding type assignment system. Then*

$$\Gamma \vdash_S A \Leftrightarrow \exists \vec{x}\ \exists P \in \Lambda^{\circ}(\vec{x})\ \Gamma_{\vec{x}} \vdash_{\lambda S} P : A.$$

Proof ($\Rightarrow$) By an easy induction on derivations, just observing that the correct lambda term can be constructed.

($\Leftarrow$) By omitting the terms. ∎

Since λN is exactly $\boldsymbol{\lambda}_{\rightarrow}$, the simply typed lambda calculus, we know the following results from previous chapters: Theorem 2.2.1 and Propositions 1.2.6 and 1.2.3. From Corollary 6.3.15 it follows that the results also hold for λL.

6.3.12 Proposition

(i) (Normalization Theorem for λN).

$$\Gamma \vdash_{\lambda N} P : A \ \Rightarrow\ P \text{ is strongly normalizing.}$$

(ii) (Subject Reduction Theorem for λN).

$$\Gamma \vdash_{\lambda N} P : A \ \&\ P \twoheadrightarrow_{\beta} P' \ \Rightarrow\ \Gamma \vdash_{\lambda N} P' : A.$$

(iii) (Inversion Lemma for λN). *Type assignment for terms of a certain syntactic form can only be caused in the obvious way:*

$$\begin{array}{llcl}
(1) & \Gamma \vdash_{\lambda N} \quad x : A & \Rightarrow & (x{:}A) \in \Gamma. \\
(2) & \Gamma \vdash_{\lambda N} \quad PQ : B & \Rightarrow & \Gamma \vdash_{\lambda N} P : (A{\to}B) \;\&\; \Gamma \vdash_{\lambda N} Q : A, \\
 & & & \text{for some type } A. \\
(3) & \Gamma \vdash_{\lambda N} \quad \lambda x.P : C & \Rightarrow & \Gamma, x{:}A \vdash_{\lambda N} P : B \;\&\; C \equiv A{\to}B, \\
 & & & \text{for some types } A, B.
\end{array}$$

Relating λN, λL and λL^{cf}

Now the proof of the equivalence between systems N and L will be 'lifted' to that of λN and λL.

6.3.13 Proposition $\Gamma \vdash_{\lambda N} P : A \Rightarrow \Gamma \vdash_{\lambda L} P : A.$

Proof By inductions on derivations in λN. Modus ponens ($\to$ elim) is treated as follows:

$$\dfrac{\Gamma \vdash_{\lambda L} P : A{\to}B \qquad \dfrac{\Gamma \vdash_{\lambda L} Q : A \quad \Gamma, x{:}B \vdash_{\lambda L} x{:}B}{\Gamma, y{:}A{\to}B \vdash_{\lambda L} yQ : B}\,(\to \text{left})}{\Gamma \vdash_{\lambda L} PQ : B}\,(\text{cut}). \;\blacksquare$$

6.3.14 Proposition

(i) $\Gamma \vdash_{\lambda L} P : A \Rightarrow \Gamma \vdash_{\lambda N} P' : A$, *for some* $P' \twoheadrightarrow_\beta P$.

(ii) $\Gamma \vdash_{\lambda L} P : A \Rightarrow \Gamma \vdash_{\lambda N} P : A.$

Proof (i) By induction on derivations in λL. The rule ($\to$ left) is treated as follows (the justifications are left out, but they are as in the proof of 6.3.4):

$$\dfrac{\dfrac{\dfrac{\Gamma \vdash_{\lambda N} Q : A}{\Gamma, y{:}A{\to}B \vdash_{\lambda N} Q : A} \qquad \dfrac{}{\Gamma, y{:}A{\to}B \vdash_{\lambda N} y{:}A{\to}B}}{\Gamma, y{:}A{\to}B \vdash_{\lambda N} yQ : B} \qquad \dfrac{\Gamma, x{:}B \vdash_{\lambda N} P : C}{\Gamma \vdash_{\lambda N} (\lambda x.P) : B{\to}C}}{\Gamma, y{:}A{\to}B \vdash_{\lambda N} (\lambda x.P)(yQ) : C}$$

Now $(\lambda x.P)(yQ) \to_\beta P[x{:=}yQ]$ as required. The rule (cut) is treated as follows:

$$\dfrac{\Gamma \vdash_{\lambda N} Q : A \quad \dfrac{\Gamma, x{:}A \vdash_{\lambda N} P : B}{\Gamma \vdash_{\lambda N} (\lambda x.P) : A{\to}B}\,(\to \text{intr})}{\Gamma \vdash_{\lambda N} (\lambda x.P)Q : B}\,(\to \text{elim}).$$

Now $(\lambda x.P)Q \to_\beta P[x{:=}Q]$ as required.

(ii) By (i) and the Subject Reduction Theorem for λN (6.3.12(ii)). $\blacksquare$

6.3.15 Corollary $\Gamma \vdash_{\lambda L} P : A \Leftrightarrow \Gamma \vdash_{\lambda N} P : A$.

Proof By Propositions 6.3.13 and 6.3.14(ii). ∎

Now we will investigate the role of the cut-free system.

6.3.16 Proposition

$$\Gamma \vdash_{\lambda L^{\mathrm{cf}}} P : A \Rightarrow P \textit{ is in } \boldsymbol{\beta}\text{-nf}.$$

Proof By an easy induction on derivations. ∎

6.3.17 Lemma *Suppose*

$$\Gamma \vdash_{\lambda L^{\mathrm{cf}}} P_1 : A_1, \ldots, \Gamma \vdash_{\lambda L^{\mathrm{cf}}} P_n : A_n.$$

Then

$$\Gamma, x{:}A_1 \to \cdots \to A_n \to B \vdash_{\lambda L^{\mathrm{cf}}} x P_1 \cdots P_n : B$$

for those variables x such that $\Gamma, x{:}A_1 \to \cdots \to A_n \to B$ is a context.

Proof We treat the case $n = 2$, which is perfectly general. We abbreviate $\vdash_{\lambda L^{\mathrm{cf}}}$ as $\vdash$.

$$\dfrac{\Gamma \vdash P_1 : A_1 \qquad \dfrac{\Gamma \vdash P_2 : A_2 \qquad \dfrac{}{\Gamma, z{:}B \vdash z : B}\,(\text{axiom})}{\Gamma, y{:}A_2 \to B \vdash yP_2 \equiv z[z{:=}yP_2] : B}\,(\to \text{left})}{\Gamma, x{:}A_1 \to A_2 \to B \vdash xP_1P_2 \equiv (yP_2)[y{:=}xP_1] : B}\,(\to \text{left}).$$

Note that x may occur in some of the P_i. ∎

6.3.18 Proposition *Suppose that P is a* $\boldsymbol{\beta}$-nf. *Then*

$$\Gamma \vdash_{\lambda N} P : A \Rightarrow \Gamma \vdash_{\lambda L^{\mathrm{cf}}} P : A.$$

Proof By induction on the following generation of normal forms:

$$\mathrm{nf} = \mathtt{var}\ \mathrm{nf}^* \mid \lambda\mathtt{var}.\mathrm{nf}.$$

Here `var nf`* stands for `var` followed by 0 or by more occurrences of `nf`. The case $P \equiv \lambda x.P_1$ is easy. The case $P \equiv xP_1 \cdots P_n$ follows from the previous lemma, using the Inversion Lemma for λN, Proposition 6.3.12(iii). ∎

Now we get as bonus the Hauptsatz of Gentzen (1936b) for minimal implicational sequent calculus.

6.3.19 Theorem (Cut-Elimination)

$$\Gamma \vdash_L A \Rightarrow \Gamma \vdash_{L^{\mathrm{cf}}} A.$$

Proof

$$\begin{array}{rll}\Gamma \vdash_L A & \Rightarrow \Gamma_{\vec{x}} \vdash_{\lambda L} P : A, & \text{for some } P \in \Lambda^{\circ}(\vec{x}), \text{ by 6.3.11,}\\ & \Rightarrow \Gamma_{\vec{x}} \vdash_{\lambda N} P : A, & \text{by 6.3.14(ii),}\\ & \Rightarrow \Gamma_{\vec{x}} \vdash_{\lambda N} P^{\text{nf}} : A, & \text{by 6.3.12(i),(ii),}\\ & \Rightarrow \Gamma_{\vec{x}} \vdash_{\lambda L^{\text{cf}}} P^{\text{nf}} : A, & \text{by 6.3.18,}\\ & \Rightarrow \Gamma \vdash_{L^{\text{cf}}} A, & \text{by 6.3.11.} \blacksquare\end{array}$$

As it is clear that the proof implies that cut-elimination can be used to normalize terms typable in $\lambda N = \lambda\!\rightarrow$, Statman (1979) implies that the expense of cut-elimination is beyond elementary time (Grzegorczyk class 4). Moreover, as the cut-free deduction is of the same order of complexity as the corresponding normal lambda term, the size of the cut-free version of a derivation is non-elementary in the size of the original derivation.

Discussion

The main technical tool is the type assignment system λL corresponding exactly to sequent calculus (for minimal propositional logic). This type assignment system is a subsystem of one studied in Barbanera et al. (1995). The terms involved in λL are also in Mints (1996). The difference between the present approach and that of Mints is that in that paper, derivations in L are first-class citizens, whereas in λL the provable formulas and the lambda terms are.

In λN typable terms are built up as usual (following the grammar of lambda terms). In λL^{cf} only normal terms are typable. They are built up from variables by transitions like

$$P \longmapsto \lambda x.P$$

and

$$P \longmapsto P[x{:=}yQ].$$

This is an ambiguous way of building terms, in the sense that one term can be built up in several ways. For example, one can assign to the term $\lambda x.yz$ the type $C{\rightarrow}B$ (in the context $z{:}A, y{:}A{\rightarrow}B$) via two different cut-free derivations:

$$\dfrac{\dfrac{x{:}C, z{:}A \vdash z : A \qquad x{:}C, z{:}A, u{:}B \vdash u : B}{x{:}C, z{:}A, y{:}A{\rightarrow}B \vdash yz : B}\ (\rightarrow \text{left})}{z{:}A, y{:}A{\rightarrow}B \vdash \lambda x.yz : C{\rightarrow}B}\ (\rightarrow \text{right})$$

and

$$\dfrac{z{:}A \vdash z{:}A \qquad \dfrac{x{:}C, z{:}A, u{:}B \vdash u : B}{z{:}A, u{:}B \vdash \lambda x.u : C{\rightarrow}B}\ (\rightarrow \text{right})}{z{:}A, y{:}A{\rightarrow}B \vdash \lambda x.yz : C{\rightarrow}B}\ (\rightarrow \text{left}).$$

These correspond, respectively, to the following two formations of terms:

$$\begin{array}{lllll} u & \longmapsto & yz & \longmapsto & \lambda x.yz, \\ u & \longmapsto & \lambda x.u & \longmapsto & \lambda x.yz. \end{array}$$

Therefore there are more sequent calculus derivations giving rise to the same lambda term. This is the cause of the mismatch between sequent calculus and natural deduction as described in Zucker (1974), Pottinger (1977) and Mints (1996). See also Dyckhoff and Pinto (1999), Schwichtenberg (1999) and Troelstra (1999).

In Herbelin (1995) the mismatch between L-derivations and lambda terms is repaired by translating these into terms with explicit substitution:

$$\lambda x.(u\langle u{:=}yz\rangle),$$
$$(\lambda x.u)\langle u{:=}yz\rangle.$$

In this section lambda terms are also considered as first-class citizens for sequent calculus. This gives an insight into the aforementioned mismatch by understanding it as an intensional view of how the sequent calculus generates these terms.

It is interesting to note how, in the full system λL, the rule (cut) generates terms not in β-normal form. The extra transition now is

$$P \longmapsto P[x{:=}F].$$

This will introduce a redex if x occurs actively (in a context xQ) and F is an abstraction ($F \equiv \lambda x.R$), the other applications of the rule (cut) being superfluous. Also, the alternative rule (cut$'$) can be understood better. Using this rule the extra transition becomes

$$P \longmapsto P[x{:=}\mathsf{I}].$$

This will have the same effect (modulo one β-reduction) as the previous transition, if x occurs in a context xFQ. So with the original rule (cut) the argument Q (in the context xQ) is waiting for a function F to act on it. With the alternative rule (cut$'$) the function F comes close (in context xFQ), but the 'couple' FQ has to wait for the 'green light' provided by I.

Also, it can be observed that if one wants to manipulate derivations in order to obtain a cut-free proof, then the term involved gets reduced. By the Strong Normalization Theorem for λN ($= \boldsymbol{\lambda}_{\rightarrow}$) it follows that eventually a cut-free proof will be reached.

We have not studied in detail whether, along the lines of this section, cut-elimination can be done for the full system of intuitionistic predicate logic, but there seems to be no problem.

6.4 Grammars, terms and types

Typed lambda calculus is widely used in the study of the semantics of natural language, in combination with a variety of rule-based syntactic engines. In this section, we focus on categorial type logics. The type discipline in these systems is responsible both for the construction of grammatical form (syntax) and for meaning assembly. We address two central questions. First, what are the *invariants* of grammatical composition, and how do they capture the uniformities of the form–meaning correspondence across languages? Second, how can we reconcile grammatical invariants with structural diversity, i.e. variation in the realization of the form–meaning correspondence in the 6000 or so languages of the world?

The grammatical architecture to be unfolded below has two components. Invariants are characterized in terms of a minimal *base system*: the pure logic of residuation for composition and structural incompleteness. Viewing the types of the base system as formulas, we model the syntax–semantics interface along the lines of the Curry–Howard interpretation of derivations. Variation arises from the combination of the base logic with a *structural module*. This component characterizes the structural deformations under which the basic form–meaning associations are preserved. Its rules allow reordering and/or restructuring of grammatical material. These rules are not globally available, but keyed to unary type-forming operations, and thus anchored in the lexical type declarations.

It will be clear from this description that the type–logical approach has its roots in the type calculi developed by Jim Lambek in the late 1950s. The technique of controlled structural options is a more recent development, inspired by the modalities of linear logic.

Grammatical invariants: the base logic

Compared with the systems used elsewhere in this book, the type system of categorial type logics can be seen as a specialization designed to take linear order and phrase structure information into account; the set $\mathcal{F}$ of formulas (types) is defined as

$$\mathcal{F} ::= \mathcal{A} \mid \mathcal{F}/\mathcal{F} \mid \mathcal{F} \bullet \mathcal{F} \mid \mathcal{F}\backslash\mathcal{F}.$$

The set of type atoms $\mathcal{A}$ represents the basic ontology of phrases that one can think of as grammatically ‘complete’. Examples, for English, could be np for noun phrases, s for sentences, n for common nouns. There is no claim of universality here: languages can differ as to which ontological choices they make. Formulas A/B, $B\backslash A$ are directional versions of the implicational type $B \to A$. They express incompleteness in the sense that expressions with slash types produce a phrase of type A in composition with a phrase of type

B to the right or to the left. Product types $A \bullet B$ explicitly express this composition.

Frame semantics provides the tools for making precise the informal description of the interpretation of the type language in the structural dimension. In this setting, frames $F = (W, R_\bullet)$ consist of a set W of linguistic resources (expressions, 'signs'), structured in terms of a ternary relation $R_\bullet$, the relation of grammatical composition or 'Merge' as it is known in the generative tradition. A valuation $V : \mathcal{F} \mapsto \mathcal{P}(W)$ interprets types as sets of expressions. For complex types, the valuation respects the clauses below, i.e. expressions x with type $A \bullet B$ can be disassembled into an A part y and a B part z. The interpretation for the directional implications is dual with respect to the y and z arguments of the Merge relation, thus expressing incompleteness with respect to composition:

$$\begin{array}{rcl} x \in V(A \bullet B) & \Leftrightarrow & \exists yz.R_\bullet xyz \text{ and } y \in V(A) \text{ and } z \in V(B) \\ y \in V(C/B) & \Leftrightarrow & \forall xz.(R_\bullet xyz \text{ and } z \in V(B)) \text{ implies } x \in V(C) \\ z \in V(A\backslash C) & \Leftrightarrow & \forall xy.(R_\bullet xyz \text{ and } y \in V(A)) \text{ implies } x \in V(C). \end{array}$$

Algebraically, this interpretation turns the product and the left and right implications into a residuated triple in the sense of the following biconditionals:

$$A \longrightarrow C/B \Leftrightarrow A \bullet B \longrightarrow C \Leftrightarrow B \longrightarrow A\backslash C \qquad \text{(Res)}.$$

In fact, we have the *pure* logic of residuation here: (Res), together with Reflexivity ($A \longrightarrow A$) and Transitivity (from $A \longrightarrow B$ and $B \longrightarrow C$, conclude $A \longrightarrow C$), fully characterizes the derivability relation, as the following completeness result shows.

Completeness (Došen (1992), Kurtonina (1995)) The relation $A \longrightarrow B$ is provable in the grammatical base logic iff for every valuation V on every frame F we have $V(A) \subseteq V(B)$.

Notice that we do not impose any restrictions on the interpretation of the Merge relation. In this sense, the laws of the base logic capture grammatical *invariants*: properties of type combination that hold no matter what the structural particularities of individual languages may be. And indeed, at the level of the base logic, important grammatical notions, rather than being postulated, can be seen to emerge from the type structure:

Valency Selectional requirements distinguishing verbs that are intransitive $np\backslash s$, transitive $(np\backslash s)/np$, ditransitive $((np\backslash s)/np)/np$, etc., are expressed in terms of the directional implications. In a context-free grammar, these would require the postulation of new non-terminals.

Case The distinction between phrases that can fulfill any noun phrase selectional requirement versus phrases that insist on playing the role of subject $s/(np\backslash s)$, or direct object $((np\backslash s)/np)\backslash(np\backslash s)$, or prepositional object $(pp/np)\backslash pp$, and so on, is expressed through higher-order type assignment.

Complements versus modifiers Compare exocentric types (A/B with $A \neq B$) versus endocentric types A/A. The latter express modification; optionality of A/A type phrases follows.

Filler-gap dependencies Nested implications $A/(C/B)$, $A/(B\backslash C)$, etc., signal the withdrawal of a gap hypothesis of type B in a domain of type C.

Parsing-as-deduction

For automated proof search, one turns the algebraic presentation in terms of (Res) into a sequent presentation enjoying cut-elimination. Sequents for the grammatical base logic are statements $\Gamma \Rightarrow A$ with Γ a structure, A a type formula. Structures are binary branching trees with formulas at the leaves: $\mathcal{S} ::= \mathcal{F} \mid (\mathcal{S}, \mathcal{S})$. In the rules, we write $\Gamma[\Delta]$ for a structure Γ containing a substructure Δ. Lambek (1958, 1961) proved that Cut is a redundant rule in this presentation. Top-down backward-chaining proof search in the cut-free system respects the subformula property and yields a decision procedure:

$$\frac{}{A \Rightarrow A}\ \text{Ax} \qquad \frac{\Delta \Rightarrow A \quad \Gamma[A] \Rightarrow B}{\Gamma[\Delta] \Rightarrow B}\ \text{Cut}$$

$$\frac{\Gamma \Rightarrow A \quad \Delta \Rightarrow B}{(\Gamma, \Delta) \Rightarrow A \bullet B}\ (\bullet R) \qquad \frac{\Gamma[(A, B)] \Rightarrow C}{\Gamma[A \bullet B] \Rightarrow C}\ (\bullet L)$$

$$\frac{\Delta \Rightarrow B \quad \Gamma[A] \Rightarrow C}{\Gamma[(\Delta, B\backslash A)] \Rightarrow C}\ (\backslash L) \qquad \frac{(B, \Gamma) \Rightarrow A}{\Gamma \Rightarrow B\backslash A}\ (\backslash R)$$

$$\frac{\Delta \Rightarrow B \quad \Gamma[A] \Rightarrow C}{\Gamma[(A/B, \Delta)] \Rightarrow C}\ (/L) \qquad \frac{(\Gamma, B) \Rightarrow A}{\Gamma \Rightarrow A/B}\ (/R).$$

To specify a grammar for a particular language it suffices now to give its lexicon. The relation $Lex \subseteq \Sigma \times \mathcal{F}$ associates each word with a finite number of types. A string belongs to the language for lexicon Lex and goal type B, $w_1 \cdots w_n \in L(Lex, B)$ iff for $1 \leq i \leq n$, $(w_i, A_i) \in Lex$, and $\Gamma \Rightarrow B$, where Γ is a tree with 'yield' at its endpoints $A_1, \ldots, A_n$. Buszkowski and Penn (1990) model the acquisition of lexical type assignments as a process of solving type equations. Their unification-based algorithms take function–argument structures as input (binary trees with a distinguished daughter); one obtains variations depending on whether the solution should assign a unique type to every vocabulary item, or whether one accepts multiple assignments. Kanazawa (1998) studied learnable classes of grammars from this perspective, in the sense of Gold's notion of identifiability 'in the limit'; the formal theory of learnability for type–logical grammars has recently developed into a quite active field of research.

Meaning assembly

Lambek's original work looked at categorial grammar from a purely syntactic point of view, which probably explains why this work was not taken into account by Richard Montague when he developed his theory of model-theoretic semantics for natural languages. In the 1980s, van Benthem played a key role in bringing the two traditions together, by introducing the Curry–Howard perspective, with its dynamic, derivational view on meaning assembly rather than the static, structure-based view of rule-based approaches.

For semantic interpretation, we want to associate every type A with a semantic domain D_A, the domain where expressions of type A find their denotations. It is convenient to set up semantic domains via a map from the directional syntactic types used so far to the undirected type system of the typed lambda calculus. This indirect approach is attractive for a number of reasons. On the level of atomic types, one may want to make different basic distinctions depending on whether one uses syntactic or semantic criteria. For complex types, a map from syntactic to semantic types makes it possible to forget information that is relevant only for the way expressions are to be configured in the form dimension. For simplicity, we focus on implicational types here – accommodation of product types is straightforward.

For a simple extensional interpretation, the set of atomic semantic types could consist of types e and t, with D_e the domain of discourse (a non-empty set of entities, objects), and $D_t = \{0, 1\}$, the set of truth values. By $D_{A \to B}$, the semantic domain for a functional type $A \to B$, we mean the set of functions from D_A to D_B. The map from syntactic to semantic types $(\cdot)'$ could now stipulate for basic syntactic types that $np' = e$, $s' = t$, and $n' = e \to t$. Sentences, in this way, denote truth values; (proper) noun phrases, individuals; common nouns, functions from individuals to truth values. For complex syntactic types, we set $(A/B)' = (B\backslash A)' = B' \to A'$. On the level of semantic types, the directionality of the slash connective is no longer taken into account. Of course, the distinction between numerator and denominator – domain and range of the interpreting functions – is kept. Some common parts of speech with their corresponding syntactic and semantic types are

determiner	$(s/(np\backslash s))/n$	$(e \to t) \to (e \to t) \to t$
intransitive verb	$np\backslash s$	$e \to t$
transitive verb	$(np\backslash s)/np$	$e \to e \to t$
reflexive pronoun	$((np\backslash s)/np)\backslash(np\backslash s)$	$(e \to e \to t) \to e \to t$
relative pronoun	$(n\backslash n)/(np\backslash s)$	$(e \to t) \to (e \to t) \to e \to t.$

Formulas-as-types, proofs as programs

Curry's basic insight was that one can see the functional types of type theory as logical implications, giving rise to a one-to-one correspondence between typed lambda terms and natural deduction proofs in positive intuitionistic logic. Translating Curry's 'formulas-as-types' idea to the categorial type logics we are discussing, we have to take the differences between intuitionistic logic and the grammatical resource logic into account. Below we give the slash rules of the base logic in natural deduction format, now taking term-decorated formulas as basic declarative units. Judgements take the form of sequents $\Gamma \vdash M : A$. The antecedent Γ is a structure with leaves $x_1 : A_1, \ldots, x_n : A_n$. The x_i are unique variables of type A'_i. The succedent is a term M of type A' with exactly the free variables $x_1, \ldots, x_n$, representing a program which, given inputs $k_1 \in D_{A'_1}, \ldots, k_n \in D_{A'_n}$, produces a value of type A' under the assignment that maps the variables x_i to the objects k_i. In other words the x_i are the parameters of the meaning assembly procedure; for these parameters we will substitute the actual lexical meaning recipes when we rewrite the leaves of the antecedent tree to terminal symbols (words). A derivation starts from axioms $x : A \vdash x : A$. The Elimination and Introduction rules have a version for the right and the left implication. On the meaning assembly level, this syntactic difference is ironed out, since we already saw that $(A/B)' = (B\backslash A)'$. As a consequence, we don't have the *isomorphic* (one-to-one) correspondence between the terms and proofs of Curry's original program. But we do read off meaning assembly from the categorial derivation:

$$\frac{(\Gamma, x : B) \vdash M : A}{\Gamma \vdash \lambda x.M : A/B}\, I/ \qquad \frac{(x : B, \Gamma) \vdash M : A}{\Gamma \vdash \lambda x.M : B\backslash A}\, I\backslash$$

$$\frac{\Gamma \vdash M : A/B \quad \Delta \vdash N : B}{(\Gamma, \Delta) \vdash MN : A}\, E/ \qquad \frac{\Gamma \vdash N : B \quad \Delta \vdash M : B\backslash A}{(\Gamma, \Delta) \vdash MN : A}\, E\backslash.$$

A second difference between the programs/computations that can be obtained in intuitionistic implicational logic, and the recipes for meaning assembly associated with categorial derivations has to do with the resource management of assumptions in a derivation. In Curry's original program, the number of occurrences of assumptions (the 'multiplicity' of the logical resources) is not critical. One can make this style of resource management explicit in the form of structural rules of Contraction and Weakening, allowing for the duplication and waste of resources:

$$\frac{\Gamma, A, A \vdash B}{\Gamma, A \vdash B}\ C \qquad \frac{\Gamma \vdash B}{\Gamma, A \vdash B}\ W.$$

By contrast, the categorial type logics are resource sensitive systems where each assumption has to be used exactly once. We have the following correspondence between resource constraints and restrictions on the lambda terms that code derivations:

(1) no empty antecedents: each subterm contains a free variable;
(2) no Weakening: each λ operator binds a variable free in its scope;
(3) no Contraction: each λ operator binds at most one occurrence of a variable in its scope.

Taking into account also word order and phrase structure (in the absence of Associativity and Commutativity), the slash introduction rules responsible for the λ operator can only reach the immediate daughters of a structural domain.

These constraints imposed by resource sensitivity put severe limitations on the expressivity of the derivational semantics. There is an interesting division of labor here in natural language grammars between derivational and lexical semantics. The proof term associated with a derivation is a *uniform* instruction for meaning assembly that fully abstracts from the contribution of the particular lexical items on which it is built. At the level of the lexical meaning recipes, we do not impose linearity constraints. Below are some examples of non-linearity; syntactic type assignment for these words was given above. The lexical term for the reflexive pronoun is a pure combinator: it identifies the first and second coordinate of a binary relation. The terms for relative pronouns or determiners have a double bind λ for computing the intersection of their two $(e \to t)$ arguments (noun and verb phrase), and for testing the intersection for non-emptiness in the case of 'some':

a, some (determiner)	$(e \to t) \to (e \to t) \to t$	$\lambda P \lambda Q.(\exists\ \lambda x.((P\ x) \wedge (Q\ x)))$
himself (reflexive pronoun)	$(e \to e \to t) \to e \to t$	$\lambda R \lambda x.((R\ x)\ x)$
that (relative pronoun)	$(e \to t) \to (e \to t) \to e \to t$	$\lambda P \lambda Q \lambda x.((P\ x) \wedge (Q\ x)))$.

The interplay between lexical and derivational aspects of meaning assembly is illustrated with the natural deduction below. Using variables $x_1, \ldots, x_n$ for the leaves in left to right order, the proof term for this derivation is $((x_1\ x_2)\ (x_4\ x_3))$. Substituting the above lexical recipes for 'a' and 'himself' and non-logical constants $\mathbf{boy}^{e \to t}$ and $\mathbf{hurt}^{e \to e \to t}$, we obtain, after β-conversion, $(\exists\ \lambda y.((\mathbf{boy}\ y) \wedge ((\mathbf{hurt}\ y)\ y)))$. Notice that the proof term reflects the derivational history (modulo directionality); after lexical substitution this transparency is lost. The full encapsulation of lexical semantics is one of the strong attractions of the categorial approach:

$$\dfrac{\dfrac{\dfrac{\text{a}}{(s/(np\backslash s))/n} \quad \dfrac{\text{boy}}{n}}{(\text{a, boy}) \vdash s/(np\backslash s)}\ (/E) \qquad \dfrac{\dfrac{\text{hurt}}{(np\backslash s)/np} \quad \dfrac{\text{himself}}{((np\backslash s)/np)\backslash(np\backslash s)}}{(\text{hurt, himself}) \vdash np\backslash s}\ (\backslash E)}{((\text{a, boy}), (\text{hurt, himself})) \vdash s}\ (/E).$$

Structural variation

A second source of expressive limitations of the grammatical base logic is of a more structural nature. Consider situations where a word or phrase makes a uniform semantic contribution, but appears in contexts which the base logic cannot relate derivationally. In generative grammar, such situations are studied under the heading of 'displacement', a suggestive metaphor from our type-logical perspective. Displacement can be *overt* (as in the case of question words, relative pronouns and the like: elements that enter into a dependency with a 'gap' following at a potentially unbounded distance, e.g. 'Who do you think that Mary likes (gap)?'), or *covert* (as in the case of quantifying expressions with the ability for non-local scope construal, e.g. 'Alice thinks someone is cheating', which can be construed as 'there is a particular x such that Alice thinks x is cheating'). We have seen already that such expressions have higher-order types of the form $(A \to B) \to C$. The Curry–Howard interpretation then effectively dictates the uniformity of their contribution to the meaning assembly process as expressed by a term of the form $(M^{(A \to B) \to C}\ \lambda x^A.N^B)^C$, where the 'gap' is the λ-bound hypothesis. What remains to do is to provide the fine-structure for this abstraction process, specifying which subterms of N^B are in fact 'visible' for the λ-binder. To work out this notion of visibility or structural accessibility, we introduce structural rules, in addition to the logical rules of the base logic studied so far. From the pure residuation logic, one obtains a hierarchy of categorial calculi by adding the structural rules of Associativity, Commutativity or both. For reasons of historical precedence, the system of Lambek (1958), with an associative composition operation, is known as **L**; the more fundamental system of Lambek (1961) as **NL**, i.e. the non-associative version of

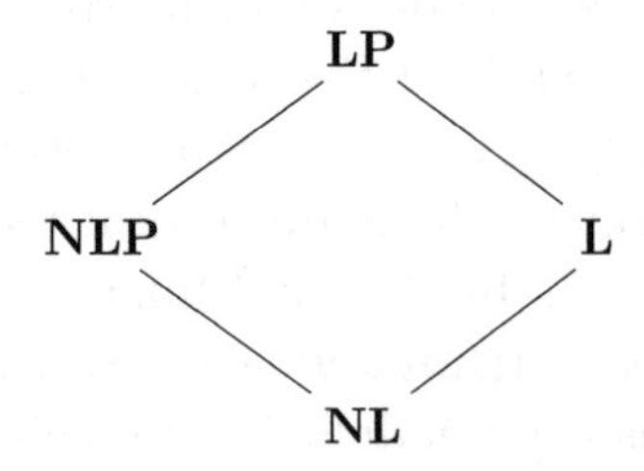

Figure 6.10 Various Lambek calculi

L. Addition of commutativity turns these into **LP** and **NLP**, respectively. See Fig. 6.10. For linguistic applications, it is clear that *global* options of associativity and/or commutativity are too crude: they would entail that arbitrary changes in constituent structure and/or word order cannot affect well-formedness of an expression. What is needed, is a *controlled* form of structural reasoning, anchored in lexical type assignment.

Control operators

The strategy is familiar from linear logic: the type language is extended with a pair of unary operators ('modalities'). They are constants in their own right, with logical rules of *use* and of *proof.* In addition, they can provide controlled access to structural rules:

$$\mathcal{F} ::= \mathcal{A} \mid \Diamond\mathcal{F} \mid \Box\mathcal{F} \mid \mathcal{F}\backslash\mathcal{F} \mid \mathcal{F} \bullet \mathcal{F} \mid \mathcal{F}/\mathcal{F}.$$

Consider the logical properties first. The truth conditions below characterize the control operators $\Diamond$ and $\Box$ as inverse duals with respect to a binary accessibility relation $R_\Diamond$. This interpretation turns them into a residuated pair, just like composition and the left and right slash operations, i.e. we have $\Diamond A \longrightarrow B$ iff $A \longrightarrow \Box B$ (Res):

$$x \in V(\Diamond A) \text{ iff } \exists y. R_\Diamond xy \text{ and } y \in V(A)$$
$$x \in V(\Box A) \text{ iff } \forall y. R_\Diamond yx \text{ implies } y \in V(A).$$

We saw that for composition and its residuals, completeness with respect to the frame semantics doesn't impose restrictions on the interpretation of the merge relation $R_\bullet$. Similarly, for $R_\Diamond$ in the pure residuation logic of $\Diamond, \Box$. This means that consequences of (Res) characterize grammatical invariants, in the sense indicated above. From (Res) one easily derives the fact that the control operators are monotonic ($A \longrightarrow B$ implies $\Diamond A \longrightarrow \Diamond B$ and $\Box A \longrightarrow \Box B$), and that their compositions satisfy $\Diamond\Box A \longrightarrow A \longrightarrow \Box\Diamond A$. These properties can be put to good use in refining lexical type assignment so that selectional dependencies are taken into account. Compare the effect of an assignment A/B versus $A/\Diamond\Box B$. The former will produce an expression of type A in composition both with expressions of type B and $\Diamond\Box B$, the latter only with the more specific of these two, $\Diamond\Box B$. An expression typed as $\Box\Diamond B$ will *resist* composition with either A/B or $A/\Diamond\Box B$.

For sequent presentation, the antecedent tree structures now have unary in addition to binary branching: $\mathcal{S} ::= \mathcal{F} \mid (\mathcal{S}) \mid (\mathcal{S}, \mathcal{S})$. The residuation pattern then gives rise to the following rules of use and proof – cut-elimination carries over straightforwardly to the extended system, and with it decidability and the subformula property:

$$\frac{\Gamma[(A)] \Rightarrow B}{\Gamma[\Diamond A] \Rightarrow B}\ \Diamond L \qquad \frac{\Gamma \Rightarrow A}{(\Gamma) \Rightarrow \Diamond A}\ \Diamond R$$

$$\frac{\Gamma[A] \Rightarrow B}{\Gamma[(\Box A)] \Rightarrow B}\ \Box L \qquad \frac{(\Gamma) \Rightarrow A}{\Gamma \Rightarrow \Box A}\ \Box R.$$

Controlled structural rules

Let us turn then to the use of $\Diamond, \Box$ as control devices, providing restricted access to structural options that would be destructive in a global sense. Consider the role of the relative pronoun 'that' in the phrases below:

(a)	the paper that appeared today	$(n\backslash n)/(np\backslash s)$
(b)	the paper that John wrote	$(n\backslash n)/(s/np)$ + Ass
(c)	the paper that John wrote today	$(n\backslash n)/(s/np)$ + Ass,Com.

The first, (a), where the gap hypothesis is in subject position, is derivable in the structurally-free base logic with the type-assignment given. The second might suggest that the gap in object position is accessible via re-bracketing of $(np, ((np\backslash s)/np, np))$ under associativity. The third shows that as well as re-bracketing, reordering is also required to access a non-peripheral gap. The controlled structural rules $(P1)$ and $(P2)$ below allow the required restructuring and reordering only for $\Diamond$ marked resources. In combination with a type assignment $(n\backslash n)/(s/\Diamond\Box np)$ to the relative pronoun, they make the right branches of structural configurations accessible for gap introduction. As long as the gap subformula $\Diamond\Box np$ carries the licensing $\Diamond$, the structural rules are applicable; as soon as it has found the appropriate structural position where it is selected by the transitive verb, it can be used as a regular np, given $\Diamond\Box np \longrightarrow np$:

$$\begin{array}{ll} (P1) & (A \bullet B) \bullet \Diamond C \longrightarrow A \bullet (B \bullet \Diamond C) \\ (P2) & (A \bullet B) \bullet \Diamond C \longrightarrow (A \bullet \Diamond C) \bullet B. \end{array}$$

Frame constraints, term assignment

Whereas the structural interpretation of the pure residuation logic does not impose restrictions on the $R_\Diamond$ and $R_\bullet$ relations, completeness for structurally extended versions requires a frame constraint for each structural postulate. In the case of $(P2)$ above, the constraint guarantees that whenever we can connect root r to leaves x, y, z via internal nodes s, t, one can rewire root and leaves via internal nodes s', t':

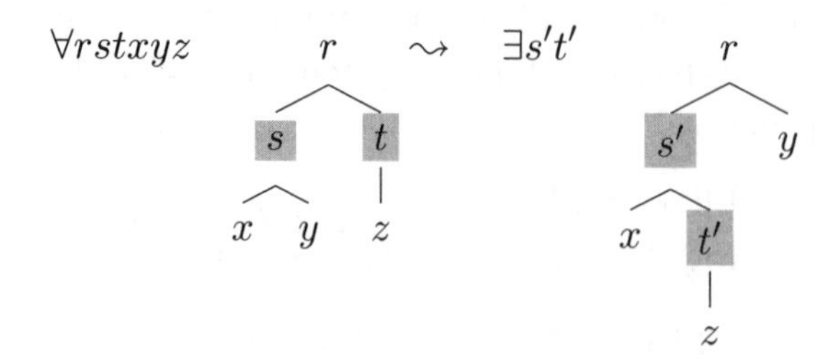

As for term assignment and meaning assembly, we have two options. The first is to treat $\Diamond, \Box$ purely as syntactic control devices. One then sets $(\Diamond A)' = (\Box A)' = A'$, and the inference rules affecting the modalities leave no trace in the term associated with a derivation. The second is to actually provide denotation domains $D_{\Diamond A}$, $D_{\Box A}$ for the new types, and to extend the term language accordingly. This is done by Wansing (2002), who developed a set-theoretic interpretation of minimal temporal intuitionistic logic. The temporal modalities of future possibility and past necessity are indistinguishable from the control operators $\Diamond, \Box$, both proof-theoretically and as far as their relational interpretation is concerned, which in principle would make Wansing's approach a candidate for linguistic application.

Embedding translations

A general theory of sub-structural communication in terms of $\Diamond, \Box$ was given in Kurtonina and Moortgat (1997). Let $\mathcal{L}$ and $\mathcal{L}'$ be neighbors in the landscape of Fig. 6.10. We have translations $\cdot^\natural$ from $\mathcal{F}(/, \bullet, \backslash)$ of $\mathcal{L}$ to $\mathcal{F}(\Diamond, \Box, /, \bullet, \backslash)$ of $\mathcal{L}'$ such that

$$\mathcal{L} \vdash A \longrightarrow B \quad \text{iff} \quad \mathcal{L}' \vdash A^\natural \longrightarrow B^\natural.$$

The $\cdot^\natural$ translation decorates formulas of the source logic $\mathcal{L}$ with the control operators $\Diamond, \Box$. The modal decoration has two functions. In the case where the target logic $\mathcal{L}'$ is more discriminating than $\mathcal{L}$, it provides access to controlled versions of structural rules that are globally available in the source logic. This form of communication is familiar from the embedding theorems of linear logic, showing that no expressivity is lost by removing free duplication and deletion (Contraction/Weakening). The other direction of communication is obtained when the target logic $\mathcal{L}'$ is less discriminating than $\mathcal{L}$. The modal decoration in this case blocks the applicability of structural rules that by default are freely available in the more liberal $\mathcal{L}$.

As an example, consider the grammatical base logic **NL** and its associative neighbor **L**. For $\mathcal{L} = \mathbf{NL}$ and $\mathcal{L}' = \mathbf{L}$, the $\cdot^\natural$ translation below effectively removes the conditions for applicability of the associativity postulate $A \bullet (B \bullet C) \longleftrightarrow (A \bullet B) \bullet C$ (Ass), restricting the set of theorems to those of **NL**. For $\mathcal{L} = \mathbf{L}$ and $\mathcal{L}' = \mathbf{NL}$, the $\cdot^\natural$ translation provides access to a controlled form of associativity (Ass$_\diamond$) $\diamond(A \bullet \diamond(B \bullet C)) \longleftrightarrow \diamond(\diamond(A \bullet B) \bullet C)$, the image of (Ass) under $\cdot^\natural$:

$$\begin{array}{rcl} p^\natural & = & p \quad (p \in \mathcal{A}) \\ (A \bullet B)^\natural & = & \diamond(A^\natural \bullet B^\natural) \\ (A/B)^\natural & = & \Box A^\natural / B^\natural \\ (B \backslash A)^\natural & = & B^\natural \backslash \Box A^\natural. \end{array}$$

Generative capacity, computational complexity

The embedding results discussed above allow one to determine the Cartesian coordinates of a language in the logical space for diversity. Which regions of that space are actually populated by natural language grammars? In terms of the Chomsky hierarchy, recent work in a variety of frameworks has converged on the so-called mildly context-sensitive grammars: formalisms more expressive than context-free, but strictly weaker than context-sensitive, and allowing polynomial parsing algorithms. The minimal system in the categorial hierarchy **NL** is strictly context-free and has a polynomial recognition problem, but, as we have seen, needs structural extensions. Such extensions are not innocent, as shown in Pentus (1993, 2006): whereas **L** remains strictly context-free, the addition of global associativity makes the derivability problem NP-complete. Also for **LP**, coinciding with the multiplicative fragment of linear logic, we have NP-completeness. Moreover, van Benthem (1995) shows that **LP** recognizes the full permutation closure of context-free languages, a lack of structural discrimination making this system unsuited for actual grammar development. The situation with $\diamond$-controlled structural rules is studied in Moot (2002), who establishes a PSPACE complexity ceiling for linear (for $\bullet$), non-expanding (for $\diamond$) structural rules via simulation of lexicalized context-sensitive grammars. By imposing tighter restrictions on allowable structure rules one can obtain interesting fragments with mildly context-sensitive expressivity and polynomial parsing complexity, as shown in Moot (2008).

For a grammatical framework assigning equal importance to syntax and semantics, strong generative capacity is more interesting than weak capacity.

Tiede (2001, 2002) studied the natural deduction proof trees that form the skeleton for meaning assembly from a tree-automata perspective, arriving at a strong generative capacity hierarchy. The base logic **NL**, though strictly context-free at the string level, can assign *non-local* derivation trees, making it more expressive than context-free grammars in this respect. Normal form **NL** proof trees remain regular; those of the associative neighbor **L** can be non-regular, but do not extend beyond the expressivity of indexed grammars, generally considered to be an upper bound for the complexity of natural language grammars.

Variants, further reading

The material discussed in this section is covered in greater depth in the chapters of Moortgat and Buszkowski in van Benthem and ter Meulen (1997). The van Benthem (1995) monograph is indispensible for the relations between categorial derivations, type theory and lambda calculus and for discussion of the place of type-logical grammars within the general landscape of resource-sensitive logics. Morrill (1994) provides a detailed type-logical analysis of syntax and semantics for a rich fragment of English grammar, and situates the type-logical approach within Richard Montague's Universal Grammar framework. The textbook Moot and Retoré (2012) covers standard categorial grammars and recent developments, and discusses the grammar development environment Grail, a general type-logical theorem prover based on proof nets and structural graph rewriting. Bernardi (2002) and Vermaat (2006) are PhD theses that study syntactic and semantic aspects of cross-linguistic variation for a wide variety of languages.

This section has concentrated on the Lambek-style approach to type-logical deduction. The framework of Combinatory Categorial Grammar, studied by Steedman and his co-workers, takes its inspiration more from the Curry–Feys tradition of combinatory logic. The particular combinators used in CCG are not so much selected for completeness with respect to some structural model for the type-forming operations (such as the frame semantics introduced above) but for their computational efficiency, which places CCG among the mildly context-sensitive formalisms. Steedman (2000) is a good introduction to this line of work, whereas Baldridge (2002) shows how one can fruitfully import the technique of lexically-anchored modal control into the CCG framework.

Another variation elaborating on Curry's distinction between an abstract level of *tectogrammatical* organization and its concrete *phenogrammatical* realizations is the framework of Abstract Categorial Grammar (ACG, De Groote, Muskens). An ACG is a structure $(\Sigma_1, \Sigma_2, \mathcal{L}, s)$, where the Σ_i are higher-order linear signatures, the abstract vocabulary Σ_1 versus the object vocabulary Σ_2, the map $\mathcal{L}$ is from the abstract to the object vocabulary, and s the distinguished type of the grammar. In this setting, one can model the syntax–semantics interface in terms of the abstract versus object vocabulary distinction. But one can also study the composition of natural language *syntax* from the perspective of non-directional linear implicational types, using the canonical λ-term encodings of strings and trees and operations on them discussed elsewhere in this book. Expressive power for this framework can be measured in terms of the maximal order of the constants in the abstract vocabulary and of the object types interpreting the atomic abstract types. A survey of results for the ensuing complexity hierarchy can be found in de Groote and Pogodalla (2004). Whether one approaches natural language grammars from the top (non-directional linear implications at the **LP** level) or from the bottom (the structurally-free base logic **NL**) of the categorial hierarchy is to a certain extent a matter of taste, reflecting the choice, for the structural regime, between allowing everything except what is explicitly forbidden, or forbidding everything except what is explicitly allowed. The theory of structural control, see Kurtonina and Moortgat (1997), shows that the two viewpoints are feasible.

PART II

RECURSIVE TYPES $\boldsymbol{\lambda}_{=}^{\mathcal{A}}$

The simple types $\boldsymbol{\lambda}_{\rightarrow}$ of Part I were *freely* generated from the type atoms $\mathbb{A}$. This means that there are no identifications like $\alpha = \alpha{\rightarrow}\beta$ or $0{\rightarrow}0 = (0{\rightarrow}0){\rightarrow}0$.

With the recursive types of this part the situation changes. Now one allows extra identifications between types; for this purpose one considers types modulo a congruence determined by some set $\mathcal{E}$ of equations between types. Another way of obtaining type identifications is to add the 'fixed-point operator' μ for types as a syntactic type constructor, together with a canonical congruence $\sim$ on the resulting terms. Given a type $A[\alpha]$ in which α may occur, the type $\mu\alpha.A[\alpha]$ has as intended meaning a solution X of the equation $X = A[X]$. Following a suggestion of Dana Scott (1975b), both approaches (types modulo a set of equations $\mathcal{E}$ or using the operator μ) can be described by considering *type algebras*, consisting of a set $\mathcal{T}$ on which a binary operation $\rightarrow$ is defined (one then can have in such structures e.g. $a = a{\rightarrow}b$). For example for $A \equiv \mu\alpha.\alpha{\rightarrow}B$ one has $A \sim A{\rightarrow}B$, which will become an equality in the type algebra.

We mainly study systems with only $\rightarrow$ as type constructor, since this restriction focuses on the most interesting phenomena. For applications, sometimes other constructors, like $+$ and $\times$ are needed; these can be added easily. Recursive type specifications are used in programming languages. One can, for example, define the type of lists of elements of type A by the equation

$$\texttt{list} = 1 + (A \times \texttt{list}).$$

For this we need a type constant 1 for the one element type (intended to contain `nil`), and type constructors $+$ for disjoint union of types and $\times$ for Cartesian product. Recursive types have been used in several programming languages since ALGOL-68, see van Wijngaarden (1981) and Pierce (2002).

Using type algebras one can define a notion of type assignment to lambda terms, that is stronger than the one using simple types. In a type algebra in which one has a type $C = C \rightarrow A$ one can give the term $\lambda x.xx$ the type C

as follows:

$$\dfrac{\dfrac{\dfrac{x{:}C \vdash x : C}{x{:}C \vdash x{:}C{\to}A}\,{\scriptstyle C=C\to A} \qquad x{:}C \vdash x : C}{x{:}C \vdash xx : A}}{\dfrac{\vdash \lambda x.xx : C \to A}{\vdash \lambda x.xx : C}\,{\scriptstyle C\to A=C.}}$$

Another example is the fixed-point operator $\mathsf{Y} \equiv \lambda f.(\lambda x.f(xx))(\lambda x.f(xx))$ that now will have as type $(A \to A) \to A$ for all types A such that there exists C satisfying $C = C \to A$.

Several properties of the simple type systems are valid for the recursive type systems. For example Subject Reduction and the Decidability of Type Assignment. Some other properties are lost, for example Strong Normalization of typable terms and the canonical connection with logic in the form of the formulas-as-type interpretation. By making some natural assumption on the type algebras the Strong Normalization property is regained.

Finally, we also consider type structures in which type algebras are enriched with a partial order, so that now one can have $a \leq a \to b$. Subtyping could be pursued much further, looking at systems of inequalities as generalized simultaneous recursions. Here we limit our treatment to a few basic properties: type systems featuring subtyping will be dealt with thoroughly in Part III.

7

The Systems $\boldsymbol{\lambda}_{=}^{\mathcal{A}}$

In Part II of this book we will again consider the set of types $\mathbb{T} = \mathbb{T}^{\mathbb{A}}$ freely generated from atomic types $\mathbb{A}$ and the type constructor $\to$. (Sometimes other type constructors, including constants, will be allowed.) But now the freely generated types will be 'bent together' by making identifications like $A = A{\to}B$. This is done by considering types modulo a congruence relation $\approx$ (an equivalence relation preserved by $\to$). Then one can define the operation $\to$ on the equivalence classes. As suggested by Scott (1975b) this can be described by considering type algebras consisting of a set with a binary operation $\to$ on it. In such structures one can have for example $a = a \to b$. The notion of type algebra was anticipated in Breazu-Tannen and Meyer (1985) expanding on a remark of Scott (1975b); it was taken up in Statman (1994) as an alternative to the presentation of recursive types via the μ-operator. It will be used as a unifying theme throughout this Part.

7.1 Type algebras and type assignment

Type algebras

7.1.1 Definition

(i) A *type algebra* is a structure

$$\mathcal{A} = \langle X, \to_{\mathcal{A}} \rangle,$$

where $\to_{\mathcal{A}}$ is a binary operation on X. For X we also write $|\mathcal{A}|$.

(ii) The type algebra $\langle \mathbb{T}^{\mathbb{A}}, \to \rangle$, consisting of the simple types under the operation $\to$, is called the *free type algebra* over $\mathbb{A}$. This terminology will be justified in 7.2.1 below.

7.1.2 Notation

(i) If $\mathcal{A}$ is a type algebra we write $a \in \mathcal{A}$ for $a \in |\mathcal{A}|$. In the same style, if

(axiom)	$\Gamma \vdash x : a \qquad \text{if } (x{:}a) \in \Gamma$
$(\to\text{E})$	$\dfrac{\Gamma \vdash M : a \to b \qquad \Gamma \vdash N : a}{\Gamma \vdash (MN) : b}$
$(\to\text{I})$	$\dfrac{\Gamma, x{:}a \vdash M : b}{\Gamma \vdash (\lambda x.M) : (a \to b)}$

Figure 7.1 The systems $\boldsymbol{\lambda}_{=}^{\mathcal{A}}$.

there is little danger of confusion, we often write $\mathcal{A}$ for $|\mathcal{A}|$ and $\to$ for $\to_{\mathcal{A}}$.

(ii) We will use $\alpha, \beta, \ldots$ for arbitrary elements of $\mathbb{A}$ and $A, B, C, \ldots$ to range over $\mathbb{T}^{\mathbb{A}}$. On the other hand $a, b, c, \ldots$ range over a type algebra $\mathcal{A}$.

Type assignment à la Curry

We now introduce formal systems for assigning elements of a type algebra to λ-terms. We will focus our presentation mainly on type inference systems à la Curry, but for any of them a corresponding typed calculus à la Church can be defined.

The formal rules for assigning types to λ-terms are defined as in Section 1.1, but here the types are elements in an arbitrary type algebra $\mathcal{A}$. This means that the judgments of the systems are of the following shape:

$$\Gamma \vdash M : a,$$

where $a \in \mathcal{A}$, and Γ, called a *basis* over $\mathcal{A}$, is a set of statements of the shape $x{:}a$, where x is a term variable and $a \in \mathcal{A}$. As before, the subjects in $\Gamma = \{x_1{:}a_1, \ldots, x_n{:}a_n\}$ should be distinct, i.e. $x_i = x_j \;\Rightarrow\; i = j$.

7.1.3 Definition Let $\mathcal{A}$ be a type algebra, $a, b \in \mathcal{A}$, and let $M \in \Lambda$. Then the system of type assignment $\boldsymbol{\lambda}_{=}^{\mathcal{A},\mathrm{Cu}}$, or simply $\boldsymbol{\lambda}_{=}^{\mathcal{A}}$, is defined by the rules in Fig. 7.1. In rule $(\to\text{I})$ it is assumed that $\Gamma, x{:}a$ is a basis.

We write $\Gamma \vdash_{\boldsymbol{\lambda}_{=}^{\mathcal{A}}} M : a$, or simply $\Gamma \vdash_{\mathcal{A}} M : a$, in the case $\Gamma \vdash M : a$ can be derived in $\boldsymbol{\lambda}_{=}^{\mathcal{A}}$.

We could have denoted this system by $\boldsymbol{\lambda}_{\to}^{\mathcal{A}}$, but instead we write $\boldsymbol{\lambda}_{=}^{\mathcal{A}}$ to emphasize the difference with the system $\boldsymbol{\lambda}_{\to}^{\mathbb{A}}$, which is $\boldsymbol{\lambda}_{=}^{\mathcal{A}}$ over the free type algebra $\mathcal{A} = \mathbb{T}^{\mathbb{A}}$. In a general $\mathcal{A}$ we can have identifications, for example $b = b \to a$ and then of course we have

$$\Gamma \vdash_{\mathcal{A}} M : b \;\Rightarrow\; \Gamma \vdash_{\mathcal{A}} M : (b \to a).$$

This makes a dramatic difference. There are examples of type assignment in $\boldsymbol{\lambda}_{=}^{\mathcal{A}}$ to terms which have no type in the simple type assignment system $\boldsymbol{\lambda}_{\rightarrow}^{\mathbb{A}}$.

7.1.4 Example Let $\mathcal{A}$ be a type algebra with $b = (b \to a)$ for $a, b \in \mathcal{A}$. Then

(i) $\vdash_{\mathcal{A}} (\lambda x.xx) : b$.
(ii) $\vdash_{\mathcal{A}} \Omega : a$, where $\Omega = (\lambda x.xx)(\lambda x.xx)$.
(iii) $\vdash_{\mathcal{A}} \mathsf{Y} : (a{\to}a) \to a$,

where $\mathsf{Y} \equiv \lambda f.(\lambda x.f(xx))(\lambda x.f(xx))$ is the fixed point combinator.

Proof (i) The following is a deduction of $\vdash_{\mathcal{A}} (\lambda x.xx) : b$.

$$\cfrac{\cfrac{x{:}b \vdash x : b \quad x{:}b \vdash x : b}{x{:}b \vdash xx : a}\,({\to}\mathrm{E}), \quad b = (b \to a).}{\vdash (\lambda x.xx) : (b \to a) = b}$$

(ii) As $\vdash_{\mathcal{A}} (\lambda x.xx) : b$, we also have $\vdash_{\mathcal{A}} (\lambda x.xx) : (b \to a)$, since $b = b \to a$. Therefore $\vdash_{\mathcal{A}} (\lambda x.xx)(\lambda x.xx) : a$.

(iii) We can prove $\vdash_{\mathcal{A}} \mathsf{Y} : (a \to a) \to a$ in $\boldsymbol{\lambda}_{=}^{\mathcal{A}}$ in the following way. First modify the deduction constructed in (i) to obtain $f{:}a \to a \vdash_{\mathcal{A}} \lambda x.f(xx) : b$. Since $b = b \to a$ we have as in (ii) by rule $({\to}\mathrm{E})$

$$f : a \to a \vdash_{\mathcal{A}} (\lambda x.f(xx))(\lambda x.f(xx)) : a$$

from which we get

$$\vdash_{\mathcal{A}} \lambda f.(\lambda x.f(xx))(\lambda x.f(xx)) : (a \to a) \to a. \blacksquare$$

7.1.5 Proposition *Suppose that* $\Gamma \subseteq \Gamma'$. *Then*

$$\Gamma \vdash_{\mathcal{A}} M : a \;\Rightarrow\; \Gamma' \vdash_{\mathcal{A}} M : A.$$

We say that the rule 'weakening' is admissible.

Proof By induction on derivations. ∎

Quotients and syntactic type algebras and morphisms

A 'recursive type' b satisfying $b = (b \to a)$ can be easily obtained by working modulo the correct equivalence relations.

7.1.6 Definition

(i) A *congruence* on a type algebra $\mathcal{A} = \langle \mathcal{A}, \to \rangle$ is an equivalence relation $\approx$ on $\mathcal{A}$ such that for all $a, b, a', b' \in \mathcal{A}$ one has

$$a \approx a' \;\&\; b \approx b' \;\Rightarrow\; (a \to b) \approx (a' \to b').$$

(ii) In this situation define for $a \in \mathcal{A}$ its *equivalence class*, written $[a]_{\approx}$, by

$$[a]_{\approx} = \{b \in \mathcal{A} \mid a \approx b\}.$$

(iii) The *quotient type algebra* of $\mathcal{A}$ under $\approx$, written $\mathcal{A}/{\approx}$, is defined by

$$\langle \mathcal{A}/{\approx}, \rightarrow_{\approx} \rangle,$$

where

$$\begin{aligned} \mathcal{A}/{\approx} &= \{[a]_{\approx} \mid a \in \mathcal{A}\} \\ [a]_{\approx} \rightarrow_{\approx} [b]_{\approx} &= [a \rightarrow b]_{\approx}. \end{aligned}$$

Since $\approx$ is a congruence, the operation $\rightarrow_{\approx}$ is well defined.

A special place among type algebras is taken by quotients of the free type algebras modulo some congruence. In fact, in Proposition 7.1.18 we shall see that every type algebra has this form, up to isomorphism.

7.1.7 Definition Let $\mathbb{T} = \mathbb{T}^{\mathbb{A}}$.

(i) A *syntactic* type algebra over $\mathbb{A}$ is of the form

$$\mathcal{A} = \langle \mathbb{T}/{\approx}, \rightarrow_{\approx} \rangle,$$

where $\approx$ is a congruence on $\langle \mathbb{T}, \rightarrow \rangle$.

(ii) We usually write $\mathbb{T}/{\approx}$ for the syntactic type algebra $\langle \mathbb{T}/{\approx}, \rightarrow_{\approx} \rangle$, as no confusion can arise about $\rightarrow_{\approx}$ since it is determined by $\approx$.

7.1.8 Remark

(i) We often simply write A for $[A]_{\approx}$, for example in "$A \in \mathbb{T}/{\approx}$", thereby identifying $\mathbb{T}/{\approx}$ with $\mathbb{T}$ and $\rightarrow_{\approx}$ with $\rightarrow$.

(ii) The free type algebra over $\mathbb{A}$ is also syntactic, in fact it is the same as $\mathbb{T}^{\mathbb{A}}/{=}$, where $=$ is the ordinary equality relation on $\mathbb{T}^{\mathbb{A}}$. This algebra will henceforth be denoted simply by $\mathbb{T}^{\mathbb{A}}$.

7.1.9 Definition Let $\mathcal{A}$ and $\mathcal{B}$ be type algebras.

(i) A map $h : \mathcal{A} \rightarrow \mathcal{B}$ is called a *morphism* between $\mathcal{A}$ and $\mathcal{B}$, written[1] $h : \mathcal{A} \rightarrow \mathcal{B}$, iff for all $a, b \in \mathcal{A}$ one has

$$h(a \rightarrow_{\mathcal{A}} b) = h(a) \rightarrow_{\mathcal{B}} h(b).$$

(ii) An *isomorphism* is a morphism $h : \mathcal{A} \rightarrow \mathcal{B}$ that is injective and surjective. Note that in this case the inverse map h^{-1} is also a morphism. We call $\mathcal{A}$ and $\mathcal{B}$ *isomorphic*, written $\mathcal{A} \cong \mathcal{B}$, if there is an isomorphism $h : \mathcal{A} \rightarrow \mathcal{B}$.

[1] This is an overloading of the symbol "$\rightarrow$" with little danger of confusion.

(iii) We say that $\mathcal{A}$ is *embeddable* in $\mathcal{B}$, written $\mathcal{A} \hookrightarrow \mathcal{B}$, if there is an injective morphism $i : \mathcal{A} \to \mathcal{B}$. In this case we also write $i : \mathcal{A} \hookrightarrow \mathcal{B}$.

Constructing type algebras by equating elements

The following construction makes extra identifications in a given type algebra. It will serve in the next subsection as a tool for building a type algebra satisfying a given set of equations. What we do here is just bending together elements (like considering numbers modulo p). In the next subsection we also extend type algebras in order to get new elements that will be cast with a special role (like extending the real numbers with an element X, obtaining the ring $\mathbb{R}[X]$ and then bending $X^2 = -1$ to create the imaginary number i).

7.1.10 Definition Let $\mathcal{A}$ be a type algebra.

(i) An *equation over* $\mathcal{A}$ is of the form $(\underline{a} \doteq \underline{b})$ with $a, b \in \mathcal{A}$.
(ii) We say $\mathcal{A}$ *satisfies* such an equation $\underline{a} \doteq \underline{b}$ (or $\underline{a} \doteq \underline{b}$ *holds* in $\mathcal{A}$), written

$$\mathcal{A} \models \underline{a} \doteq \underline{b},$$

if $a = b$.
(iii) We say $\mathcal{A}$ *satisfies a set* $\mathcal{E}$ *of equations over* $\mathcal{A}$, written

$$\mathcal{A} \models \mathcal{E},$$

if every equation $\underline{a} \doteq \underline{b} \in \mathcal{E}$ holds in $\mathcal{A}$.

Here $\underline{a}$ is the corresponding constant for an element $a \in \mathcal{A}$. But usually we will simply write $a = b$ for $\underline{a} \doteq \underline{b}$.

7.1.11 Definition Let $\mathcal{A}$ be a type algebra and let $\mathcal{E}$ be a set of equations over $\mathcal{A}$.

(i) The *smallest congruence relation on* $\mathcal{A}$ *extending* $\mathcal{E}$ is introduced via an equality defined by the axioms and rules in Fig. 7.2, where a, a', b, b', c range over $\mathcal{A}$. The system of *equational logic* extended by the statements in $\mathcal{E}$, written $(\mathcal{E})$, is also defined in Fig. 7.2. If $\mathcal{E}'$ is another set of equations over $\mathcal{A}$ we write

$$\mathcal{E} \vdash \mathcal{E}'$$

if $\mathcal{E} \vdash a = b$ for all $a = b \in \mathcal{E}'$.
(ii) Write $=_{\mathcal{E}} \; = \{(a, b) \mid a, b \in \mathcal{A} \;\&\; \mathcal{E} \vdash a = b\}$. This is the smallest congruence relation extending $\mathcal{E}$.
(iii) The *quotient type algebra* $\mathcal{A}$ *modulo* $\mathcal{E}$, written $\mathcal{A}/\mathcal{E}$ is defined as

$$\mathcal{A}/\mathcal{E} = (\mathcal{A}/=_{\mathcal{E}}).$$

(axiom)	$\mathcal{E} \vdash a = b$	if $(a = b) \in \mathcal{E}$
(refl)	$\mathcal{E} \vdash a = a$	
(symm)	$\dfrac{\mathcal{E} \vdash a = b}{\mathcal{E} \vdash b = a}$	
(trans)	$\dfrac{\mathcal{E} \vdash a = b \quad \mathcal{E} \vdash b = c}{\mathcal{E} \vdash a = c}$	
($\to$-cong)	$\dfrac{\mathcal{E} \vdash a = a' \quad \mathcal{E} \vdash b = b'}{\mathcal{E} \vdash a \to b = a' \to b'}$	

Figure 7.2 The system of equational logic ($\mathcal{E}$).

If for example we want to construct recursive types a, b such that $b = b \to a$, then we simply work modulo $=_{\mathcal{E}}$, with $\mathcal{E} = \{b = b \to a\}$.

7.1.12 Definition Let $h : \mathcal{A} \to \mathcal{B}$ be a morphism between type algebras.

(i) For $a_1, a_2 \in \mathcal{A}$ define $h(a_1 = a_2) = (h(a_1) = h(a_2))$.
(ii) Put $h(\mathcal{E}) = \{h(a_1 = a_2) \mid a_1 = a_2 \in \mathcal{E}\}$.

7.1.13 Lemma *Let $\mathcal{E}$ be a set of equations over $\mathcal{A}$ and let $a, b \in \mathcal{A}$. Then*

(i) $\mathcal{A} \models \mathcal{E} \;\&\; \mathcal{E} \vdash a = b \;\Rightarrow\; \mathcal{A} \models a = b$.

Let moreover $h{:}\mathcal{A} \to \mathcal{B}$ be a morphism. Then:

(ii) $\mathcal{A} \models a_1 = a_2 \;\Rightarrow\; \mathcal{B} \models h(a_1 = a_2)$;
(iii) $\mathcal{A} \models \mathcal{E} \;\Rightarrow\; \mathcal{B} \models h(\mathcal{E})$.

Proof (i) By induction on the proof of $\mathcal{E} \vdash a = b$.
(ii) Since $h(a_1 = a_2) = (h(a_1) = h(a_2))$.
(iii) By (ii). ∎

7.1.14 Remark

(i) Often we simply state that $a = b$, instead of $[a] = [b]$, holds in $\mathcal{A}/\mathcal{E}$. This is comparable to saying that 1+2=0 holds in $\mathbb{Z}/(3)$, rather than saying that $[1]_{(3)} + [2]_{(3)} = [0]_{(3)}$ holds. This explains the interest of Lemma 7.1.13(ii), which actually is trivial.
(ii) Similarly, we write $h(a) = b$ instead of $h([a]) = [b]$.

7.1.15 Lemma *Let $\mathcal{E}$ be a set of equations over $\mathcal{A}$ and let $a, b \in \mathcal{A}$. Then*

(i) $\mathcal{A}/\mathcal{E} \models a = b \;\Leftrightarrow\; \mathcal{E} \vdash a = b$;
(ii) $\mathcal{A}/\mathcal{E} \models \mathcal{E}$.

Proof (i) By the definition of $\mathcal{A}/\mathcal{E}$.
(ii) By (i). ∎

7.1.16 Remark

(i) The system $\mathcal{E}$ is a congruence relation on $\mathcal{A}$ iff $=_{\mathcal{E}}$ coincides with $\mathcal{E}$.
(ii) The definition of a quotient type algebra $\mathcal{A}/\!\approx$ is a particular case of the construction 7.1.11(iii), since by (i) one has $\approx\, = (=_{\approx})$. In most cases a syntactic type algebra is given by $\mathbb{T}/\mathcal{E}$ where $\mathcal{E}$ is a set of equations between elements of the free type algebra $\mathbb{T}$.

7.1.17 Example

(i) Let $\mathbb{T}^0 = \mathbb{T}^{\{0\}}$, $\mathcal{E}_1 = \{0 = 0{\to}0\}$. Then all elements of $\mathbb{T}^0$ are equated in $\mathbb{T}^0/\mathcal{E}_1$. As a type algebra, $\mathbb{T}^0/\mathcal{E}_1$ contains therefore only one element $[0]_{\mathcal{E}_1}$ (which will be identified with 0 itself by Remark 7.1.8(i)). For instance we have

$$\mathbb{T}^0/\mathcal{E}_1 \models 0 = 0 \to 0 \to 0.$$

Moreover we have that 0 is a solution for $X = X \to 0$ in $\mathbb{T}^0/\mathcal{E}_1$.

At the semantic level an equation like $0 = 0 \to 0$ is satisfied by many models of the type-free λ-calculus. Indeed using such a type it is possible to assign type X to all pure type-free terms (see Exercise 7.12).

(ii) Let $\mathbb{T}^\infty = \mathbb{T}^{\mathbb{A}_\infty}$ be a set of types with $\infty \in \mathbb{A}_\infty$. Define $\mathcal{E}_\infty$ as the set of equations

$$\infty = T \to \infty, \quad \infty = \infty \to T,$$

where T ranges over $\mathbb{T}^\infty$. Then in $\mathbb{T}^\infty/\mathcal{E}_\infty$ the element ∞ is a solution of all equations of the form $X = A(X)$ over $\mathbb{T}^\infty$, where $A(X)$ is any type expression over $\mathbb{T}^\infty$ with at least one free occurrence of X. Note that in $\mathbb{T}^\infty/\mathcal{E}_\infty$ one does not have that $a \to b = a' \to b' \;\Rightarrow\; a = a' \;\&\; b = b'$.

We now show that every type algebra can be considered as a syntactic one.

7.1.18 Proposition *Every type algebra is isomorphic to a syntactic one.*

Proof Given $\mathcal{A} = \langle \mathcal{A}, \to \rangle$, take $\mathbb{A} = \{\underline{a} \mid a \in \mathcal{A}\}$ and

$$\mathcal{E} = \{\underline{a}{\to}\underline{b} = \underline{a \to b} \mid a, b \in \mathbb{A}\}.$$

Then $\mathcal{A}$ is isomorphic to $\mathbb{T}^{\mathbb{A}}/\mathcal{E}$ via the isomorphism $a \mapsto [\underline{a}]_{\mathcal{E}}$. ∎

7.1.19 Definition Let $\mathcal{E}$ be a set of equations over $\mathcal{A}$ and let $\mathcal{B}$ be a type algebra.

(axiom)	$\Gamma \vdash x : A \qquad \text{if } (x{:}A) \in \Gamma$
$(\to \mathrm{E})$	$\dfrac{\Gamma \vdash M : A \to B \qquad \Gamma \vdash N : A}{\Gamma \vdash (MN) : B}$
$(\to \mathrm{I})$	$\dfrac{\Gamma, x{:}A \vdash M : B}{\Gamma \vdash (\lambda x.M) : (A \to B)}$
(equal)	$\dfrac{\Gamma \vdash M : A \qquad A \approx B}{\Gamma \vdash M : B}$

Figure 7.3 The system $\boldsymbol{\lambda}_=^{\mathbb{T}/\approx}$.

(i) We say $\mathcal{B}$ *justifies* $\mathcal{E}$ if for some $h{:}\mathcal{A} \to \mathcal{B}$

$$\mathcal{B} \models h(\mathcal{E}).$$

(ii) We say $\mathcal{E}'$ over $\mathcal{B}$ *justifies* $\mathcal{E}$ if $\mathcal{B}/\mathcal{E}'$ justifies $\mathcal{E}$.

The intention is that h interprets the constants of $\mathcal{E}$ in $\mathcal{B}$ in such a way that the equations as seen in $\mathcal{B}$ become valid. We will see in Proposition 7.2.7 that

$$\mathcal{B} \text{ justifies } \mathcal{E} \;\Leftrightarrow\; \text{there exists a morphism } h : \mathcal{A}/\mathcal{E} \to \mathcal{B}.$$

Type assignment in a syntactic type algebra

7.1.20 Notation If $\mathcal{A} = \mathbb{T}/\approx$ is a syntactic type algebra, then we write

$$x_1{:}A_1, \ldots, x_n{:}A_n \vdash_{\mathbb{T}/\approx} M : A$$

for

$$x_1{:}[A_1]_\approx, \ldots, x_n{:}[A_n]_\approx \vdash_{\mathbb{T}/\approx} M : [A]_\approx.$$

We will present systems often in the following form.

7.1.21 Proposition *The system of type assignment $\boldsymbol{\lambda}_=^{\mathbb{T}/\approx}$ can be axiomatized by the axioms and rules given in Fig. 7.3, where now A, B range over $\mathbb{T}$ and Γ is of the form $\{x_1{:}A_1, \ldots, x_n{:}A_n\}$, $\vec{A} \in \mathbb{T}$.*

Proof Easy. ∎

Systems of type assignment can be related via the notion of type algebra morphism. The following property can easily be proved by induction on derivations.

7.1.22 Lemma *Let $h : \mathcal{A} \to \mathcal{B}$ be a type algebra morphism. Then for* $\Gamma = \{x_1{:}A_1, \ldots, x_n{:}A_n\}$

$$\Gamma \vdash_{\mathcal{A}} M : A \;\Rightarrow\; h(\Gamma) \vdash_{\mathcal{B}} M : h(A),$$

where $h(\Gamma) = \{x_1{:}h(A_1), \ldots, x_n{:}h(A_n)\}$.

In Chapter 9 we will prove the following properties of type assignment.

(1) A type assignment system $\boldsymbol{\lambda}_{=}^{\mathcal{A}}$ has the subject reduction property for β-reduction iff $\mathcal{A}$ is invertible: $a \to b = a' \to b' \;\Rightarrow\; a = a' \;\&\; b = b'$, for all $a, a', b, b' \in \mathcal{A}$.
(2) For the type assignment introduced in this section there is a notion of 'principal type scheme' with properties similar to those of the basic system $\boldsymbol{\lambda}_{\to}$. As a consequence of this, most questions about typing λ-terms in given type algebras are decidable.
(3) There is a simple characterization of the collection of type algebras for which a Strong Normalization theorem holds. It is decidable whether a given λ-term can be typed in them.

Explicitly typed systems

Explicitly typed versions of λ-calculus with recursive types can also be defined as for the simply typed lambda calculus in Part I, where now, as in the previous section, the types are from a (syntactic) type algebra.

In the explicitly typed systems each term is defined as a member of a specific type that is uniquely determined by the term itself. In particular, as in Section 1.4, we assume now that each variable is coupled with a unique type which is part of it. We also assume without loss of generality that all terms are well named: see Definition 1.3.4.

The Church version

7.1.23 Definition Let $\mathcal{A} = \mathbb{T}^{\mathbb{A}}/{\approx}$ be a syntactic type algebra and $A, B \in \mathcal{A}$. We introduce a *Church version of* $\boldsymbol{\lambda}_{=}^{\mathcal{A}}$, written $\boldsymbol{\lambda}_{=}^{\mathcal{A},\mathrm{Ch}}$. The set of *typed terms of the system* $\boldsymbol{\lambda}_{=}^{\mathcal{A},\mathrm{Ch}}$, written $\Lambda_{=}^{\mathcal{A},\mathrm{Ch}}(A)$ for each type A, is defined by the term formation rules given in Fig. 7.4.

This is not a type assignment system but a disjoint family of typed terms.

The de Bruijn version

A formulation of the system in the 'de Bruijn' style is possible as well. The 'de Bruijn' formulation is indeed the most widely used to denote explicitly typed systems in the literature, especially in the field of computer science. The 'Church' style, on the other hand, emphasizes the distinction between

$$
\begin{array}{rcl}
& & x^A \in \Lambda_{=}^{\mathcal{A},\mathrm{Ch}}(A); \\
M \in \Lambda_{=}^{\mathcal{A},\mathrm{Ch}}(A \to B), N \in \Lambda_{=}^{\mathcal{A},\mathrm{Ch}}(A) & \Rightarrow & (MN) \in \Lambda_{=}^{\mathcal{A},\mathrm{Ch}}(B); \\
M \in \Lambda_{=}^{\mathcal{A},\mathrm{Ch}}(B) & \Rightarrow & (\lambda x^A.M) \in \Lambda_{=}^{\mathcal{A},\mathrm{Ch}}(A \to B); \\
M \in \Lambda_{=}^{\mathcal{A},\mathrm{Ch}}(A) \text{ and } A \approx B & \Rightarrow & M \in \Lambda_{=}^{\mathcal{A},\mathrm{Ch}}(B).
\end{array}
$$

Figure 7.4 The family $\Lambda_{=}^{\mathcal{A},\mathrm{Ch}}$ of typed terms.

$$
\begin{array}{ll}
\text{(axiom)} & \Gamma \vdash x : A \qquad \text{if } (x{:}A) \in \Gamma \\[2ex]
(\to\mathrm{E}) & \dfrac{\Gamma \vdash M : A \to B \qquad \Gamma \vdash N : A}{\Gamma \vdash MN : B} \\[2ex]
(\to\mathrm{I}) & \dfrac{\Gamma, x{:}A \vdash M : B}{\Gamma \vdash (\lambda x{:}A.M) : A \to B} \\[2ex]
\text{(equiv)} & \dfrac{\Gamma \vdash M : A \qquad A \approx B}{\Gamma \vdash M : B}
\end{array}
$$

Figure 7.5 The system $\boldsymbol{\lambda}_{=}^{\mathcal{A},\mathrm{dB}}$.

explicitly and implicitly typed systems, and is more suitable for the study of models in Chapter 10. Given a syntactic type algebra $\mathcal{A} = \mathbb{T}/\approx$ the formulation of the system $\boldsymbol{\lambda}_{=}^{\mathcal{A},\mathrm{dB}}$ in the de Bruijn style is given by the rules in Fig. 7.5.

Theorems 1.2.19, 1.2.33, 1.2.36, and 1.2.37, relating the systems $\boldsymbol{\lambda}_{\to}^{\mathrm{Cu}}, \boldsymbol{\lambda}_{\to}^{\mathrm{Ch}}$, and $\boldsymbol{\lambda}_{\to}^{\mathrm{dB}}$, also hold after a change of notation, for example $\boldsymbol{\lambda}_{\to}^{\mathrm{Ch}}$ must be changed into $\boldsymbol{\lambda}_{=}^{\mathcal{A},\mathrm{Ch}}$, for the systems of recursive types $\boldsymbol{\lambda}_{=}^{\mathcal{A},\mathrm{Cu}}, \boldsymbol{\lambda}_{=}^{\mathcal{A},\mathrm{Ch}}$, and $\boldsymbol{\lambda}_{=}^{\mathcal{A},\mathrm{dB}}$. The proofs are equally simple.

The Church version with coercions

In an explicitly typed calculus we expect that a term completely codes the deduction of its type. Now any type algebra introduced in the previous sections is defined via a notion of equivalence on types which is used, in general, to prove that a term is well typed. But in the systems $\boldsymbol{\lambda}_{=}^{\mathcal{A},\mathrm{Ch}}$ the way in which type equivalences are proved is not coded in the term. To do this we must introduce new terms representing equivalence proofs. With this in mind we introduce new constants representing, in a syntactic type algebra, the equality axioms between types. The most interesting case is when these equalities are of the form $\alpha = A$ with α an atomic type. Equations of this form will be extensively studied and motivated in Section 7.3).

$$\begin{array}{rcl} & & x^A \in \Lambda_{=}^{\mathcal{A},\mathrm{Cho}}(A); \\ \alpha = A \in \mathcal{E} & \Rightarrow & \mathit{fold}_\alpha \in \Lambda_{=}^{\mathcal{A},\mathrm{Cho}}(A \to \alpha); \\ \alpha = A \in \mathcal{E} & \Rightarrow & \mathit{unfold}_\alpha \in \Lambda_{=}^{\mathcal{A},\mathrm{Cho}}(\alpha \to A); \\ M \in \Lambda_{=}^{\mathcal{A},\mathrm{Cho}}(A{\to}B), N \in \Lambda_{=}^{\mathcal{A},\mathrm{Cho}}(A) & \Rightarrow & (MN) \in \Lambda_{=}^{\mathcal{A},\mathrm{Cho}}(B); \\ M \in \Lambda_{=}^{\mathcal{A},\mathrm{Cho}}(B) & \Rightarrow & (\lambda x^A.M) \in \Lambda_{=}^{\mathcal{A},\mathrm{Cho}}(A{\to}B). \end{array}$$

Figure 7.6 The family $\Lambda_{=}^{\mathcal{A},\mathrm{Cho}}$ of typed terms.

$$\begin{array}{llll} (R_{\mathcal{E}}^{\mathrm{uf}}) & \mathit{unfold}_\alpha(\mathit{fold}_\alpha\ M^A) & \to & M^A, \quad \text{if } \alpha = A \in \mathcal{E}; \\ (R_{\mathcal{E}}^{\mathrm{fu}}) & \mathit{fold}_\alpha(\mathit{unfold}_\alpha\ M^\alpha) & \to & M^\alpha, \quad \text{if } \alpha = A \in \mathcal{E}. \end{array}$$

Figure 7.7 The reduction rules on typed terms in $\Lambda_{=}^{\mathcal{A},\mathrm{Cho}}$.

7.1.24 Definition Let $\mathcal{A} = \mathbb{T}/{=_\mathcal{E}}$, where $\mathcal{E}$ is a set of type equations of the form $\alpha = A$ with α an atomic type. We introduce a system $\boldsymbol{\lambda}_{=}^{\mathcal{A},\mathrm{Cho}}$.

(i) The set of *typed terms of the system* $\boldsymbol{\lambda}_{=}^{\mathcal{A},\mathrm{Cho}}$, written $\Lambda_{=}^{\mathcal{A},\mathrm{Cho}}(A)$ for each type A, is defined as in Fig. 7.6. The terms fold_α, unfold_α are called *coercions* and represent the two ways in which the equation $\alpha = A$ can be applied. This will be exploited in Section 7.3.
(ii) Add for each equation $\alpha = A \in \mathcal{E}$ the reduction rules given in Fig. 7.7. The rules $(R_{\mathcal{E}}^{\mathrm{uf}})$ and $(R_{\mathcal{E}}^{\mathrm{fu}})$ represent the isomorphism between α and A expressed by the equation $\alpha = A$.

7.1.25 Example Let $\mathcal{E} = \{\alpha = \alpha \to \beta\}$. The following term is the version of $\lambda x.xx$ in the system $\boldsymbol{\lambda}_{=}^{\mathcal{A},\mathrm{Cho}}$ above:

$$\mathit{fold}_\alpha(\lambda x^\alpha.(\mathit{unfold}_\alpha\, x^\alpha)\, x^\alpha) \in \Lambda_{\mathcal{A}}^{\mathrm{Cho}}(\alpha).$$

The system $\boldsymbol{\lambda}_{=}^{\mathcal{A},\mathrm{Cho}}$ in which all type equivalences are expressed via coercions is equivalent to the system $\boldsymbol{\lambda}_{=}^{\mathcal{A},\mathrm{Ch}}$, in the sense that for each term $M \in \Lambda_{=}^{\mathcal{A},\mathrm{Ch}}(A)$ there is a term $M' \in \Lambda_{=}^{\mathcal{A},\mathrm{Cho}}(A)$ obtained from an η-expansion of M by adding some coercions. Conversely for each term $M' \in \Lambda_{=}^{\mathcal{A},\mathrm{Cho}}(A)$ there is a term $M \in \Lambda_{=}^{\mathcal{A},\mathrm{Ch}}(A)$ which is η-equivalent to a term $M'' \in \Lambda_{=}^{\mathcal{A},\mathrm{Ch}}(A)$ obtained from M by erasing all its coercions.

For instance working with $\mathcal{E} = \{\alpha = \alpha \to \beta\}$ of Example 7.1.25 and the term $x^{\alpha\to\gamma}$ one has $\lambda y^{\alpha\to\beta}.x^{\alpha\to\gamma}(\mathit{fold}_\alpha y^{\alpha\to\beta}) \in \Lambda_{=}^{\mathcal{A},\mathrm{Ch}}((\alpha \to \beta) \to \gamma)$, as $\alpha \to \gamma =_\mathcal{E} (\alpha \to \beta) \to \gamma$. See also Exercise 7.16.

For many interesting terms of $\boldsymbol{\lambda}_{=}^{\mathcal{A},\mathrm{Cho}}$, however, η-conversion is not needed to obtain the equivalent term in $\boldsymbol{\lambda}_{=}^{\mathcal{A},\mathrm{Ch}}$, as in the case of Example 7.1.25.

Definition 7.1.23 identifies equivalent types, and therefore one term can have infinitely many types (though all equivalent to each other). Such presentations have been called *equi-recursive* in the recent literature, see Gapeyev et al. (2002), and are more interesting both from the practical and the theoretical point of view, especially when designing corresponding type checking algorithms. The formulation with explicit coercions is classified as *iso-recursive*, due to the presence of explicit coercions from a recursive type to its unfolding and conversely. We shall not pursue this matter, but refer the reader to Abadi and Fiore (1996) which is, to the best of our knowledge, the only study of this issue, in the context of a call-by-value formulation of the system FPC, see Plotkin (1985).

7.2 More on type algebras

Free algebras

7.2.1 Definition Let $\mathbb{A}$ be a set of atoms, and let $\mathcal{A}$ be a type algebra such that $\mathbb{A} \subseteq \mathcal{A}$. We say that $\mathcal{A}$ is *the free type algebra over* $\mathbb{A}$ if, for any type algebra $\mathcal{B}$ and any function $f : \mathbb{A} \to \mathcal{B}$, there is a unique morphism $f^+ : \mathcal{A} \to \mathcal{B}$ such that, for any $\alpha \in \mathbb{A}$, one has $f^+(\alpha) = f(\alpha)$, diagrammatically:

$$\begin{array}{ccc} \mathbb{A} & \xrightarrow{f} & \mathcal{B} \\ {\scriptstyle i}\downarrow & \nearrow_{f^+} & \\ \mathcal{A}, & & \end{array} \tag{7.1}$$

where $i : \mathbb{A} \to \mathcal{A}$ is the embedding map.

The following result, see, e.g. Goguen et al. (1977), Proposition 2.3, characterizes the free type algebra over a set of atoms $\mathbb{A}$:

7.2.2 Proposition *$\langle \mathbb{T}^{\mathbb{A}}, \to \rangle$ is the free type algebra over $\mathbb{A}$.*

Proof Given a map $f : \mathbb{A} \to \mathcal{B}$, define a morphism $f^+ : \mathbb{T}^{\mathbb{A}} \to \mathcal{B}$ as follows:

$$\begin{aligned} f^+(\alpha) &= f(\alpha) \\ f^+(A \to B) &= f^+(A) \to_{\mathcal{B}} f^+(B). \end{aligned}$$

This is clearly the unique morphism that makes diagram (7.1) commute. ∎

Subalgebras, quotients and morphisms

7.2.3 Definition Let $\mathcal{A} = \langle \mathcal{A}, \to_{\mathcal{A}} \rangle$, $\langle \mathcal{B} = \mathcal{B}, \to_{\mathcal{B}} \rangle$ be two type algebras. Then $\mathcal{A}$ is a *subtype algebra* of $\mathcal{B}$, written $\mathcal{A} \subseteq \mathcal{B}$, if $|\mathcal{A}| \subseteq |\mathcal{B}|$ and

$$\to_{\mathcal{A}} = \to_{\mathcal{B}} \upharpoonright \mathcal{A},$$

i.e. for all $a_1, a_2 \in \mathcal{A}$ one has $a_1 \to_{\mathcal{A}} a_2 = a_1 \to_{\mathcal{B}} a_2$. That is, the identity $i : \mathcal{A} \to \mathcal{B}$ is a morphism.

Clearly any subset of $\mathcal{B}$ closed under $\to_{\mathcal{B}}$ induces a subtype algebra of $\mathcal{B}$.

7.2.4 Proposition *Let $\mathcal{A}, \mathcal{B}$ be type algebras and $\approx$ be a congruence on $\mathcal{A}$.*

(i) *Given a morphism $f : \mathcal{A} \to \mathcal{B}$ such that $\mathcal{B} \models f(\approx)$, i.e. $\mathcal{B} \models \{f(a) = f(a') \mid a \approx a'\}$, then there is a unique morphism $f^\sharp : \mathcal{A}/{\approx} \to \mathcal{B}$ such that $f^\sharp([a]_\approx) = f(a)$:*

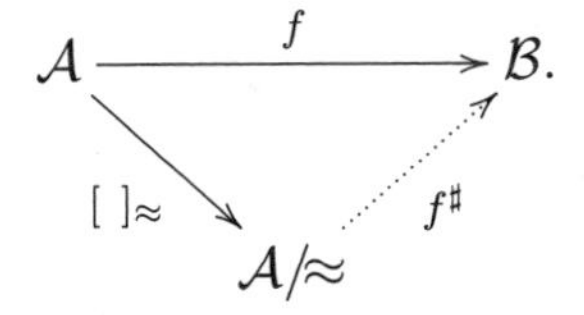

Moreover, $[\ ]_\approx$ is surjective.

(ii) *If $\forall a, a' \in \mathcal{A}.[f(a) = f(a') \Rightarrow a \approx a']$, then $f^\sharp$ is injective.*

(iii) *Given a morphism $f : \mathcal{A}/{\approx} \to \mathcal{B}$, write $f^\natural = f \circ [\ ]_\approx$:*

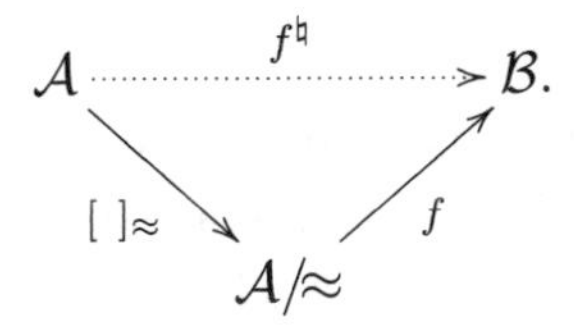

Then $f^\natural : \mathcal{A} \to \mathcal{B}$ is a morphism such that $\mathcal{B} \models f^\natural(\approx)$.

(iv) *Given a morphism $f : \mathcal{A} \to \mathcal{B}$ as in* (i), *then one has $f^{\sharp\natural} = f$.*

(v) *Given a morphism $f : \mathcal{A}/{\approx} \to \mathcal{B}$ as in* (iii), *then one has $f^{\natural\sharp} = f$.*

Proof (i) The map $f^\sharp([a]_\approx) = f(a)$ is uniquely determined by f and well defined:

$$\begin{array}{rcll} [a] = [a'] & \Rightarrow & a \approx a' & \\ & \Rightarrow & f(a) = f(a'), & \text{as } \mathcal{B} \models f(\approx), \\ & \Rightarrow & f^\sharp([a]) = f^\sharp([b]). & \end{array}$$

The map $[\ ]_\approx$ is surjective by the definition of $\mathcal{A}/{\approx}$; it is a morphism by the definition of $\to_\approx$.

(ii)–(v) Equally simple. ∎

7.2.5 Corollary *Let $\mathcal{A}, \mathcal{B}$ be two type algebras and $f{:}\mathcal{A} \to \mathcal{B}$ a morphism. Define*

(1) $f(\mathcal{A}) = \{b \mid \exists a \in \mathcal{A}.f(a) = b\} \subseteq \mathcal{B}$;

(2) $a \approx_f a' \;\Leftrightarrow\; f(a) = f(a')$, *for* $a, a' \in \mathcal{A}$.

Then

(i) $f(\mathcal{A})$ *is a sub-type algebra of* $\mathcal{B}$;

(ii) *the morphisms* $[\ ]_{\approx_f} : \mathcal{A} \to (\mathcal{A}/\!\approx_f)$ *and* $f^\sharp : (\mathcal{A}/\!\approx_f) \to \mathcal{B}$ *are an 'epi-mono' factorization of* f: $f = f^\sharp \circ [\ ]_f$, *with* $[\ ]_f$ *surjective and* $f^\sharp$ *injective.*

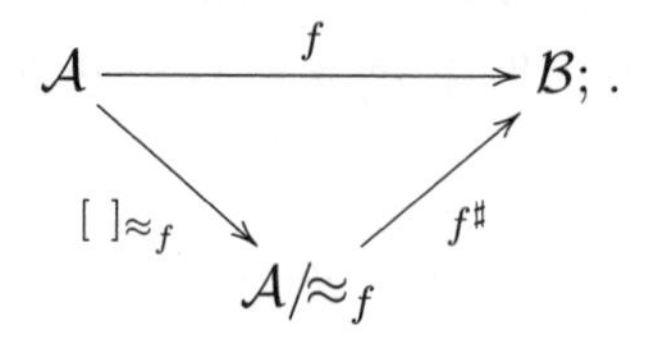

(iii) $(\mathcal{A}/\!\approx_f) \cong f(\mathcal{A}) \subseteq \mathcal{B}$.

Proof (i) $f(\mathcal{A})$ is closed under $\to_{\mathcal{B}}$. Indeed, $f(a) \to_{\mathcal{B}} f(a') = f(a \to_{\mathcal{A}} a')$.

(ii) By the definition of $\approx_f$ one has $\mathcal{B} \models \approx_f$, hence Proposition 7.2.4(i) applies.

(iii) Easy. ∎

7.2.6 Remark

(i) In the case $\mathcal{A} = \mathbb{T}/\!\approx$ is a syntactic type algebra and $\mathcal{B} = \langle \mathcal{B}, \to \rangle$, the morphisms $h : \mathbb{T}/\!\approx \to \mathcal{B}$ correspond exactly to the morphisms $h^\natural : \mathbb{T} \to \mathcal{B}$ such that for all $A, B \in \mathbb{T}$

$$A \approx B \Rightarrow h^\natural(A) = h^\natural(B).$$

The correspondence is given by $h^\natural(A) = h([A])$.

(ii) If $\mathbb{T} = \mathbb{T}^{\mathbb{A}}$ for some set $\mathbb{A}$ of atomic types then $h^\natural$ is uniquely determined by its restriction $h^\natural \upharpoonright \mathbb{A}$.

(iii) If moreover $\mathcal{B} = \mathbb{T}'/\!\approx'$ then $h^\natural(A) = [B]_{\approx'}$ for some $B \in \mathbb{T}'$. Identifying B with its equivalence class in $\approx'$, we can write simply $h^\natural(A) = B$. The first condition in (i) then becomes $A \approx B \;\Rightarrow\; h^\natural(A) \approx' h^\natural(B)$.

7.2.7 Proposition *Let* $\mathcal{E}$ *be a set of equations over* $\mathcal{A}$. *Then*

(i) $\mathcal{B}$ *justifies* $\mathcal{E} \;\Leftrightarrow\;$ *there is a morphism* $g{:}\mathcal{A}/\mathcal{E} \to \mathcal{B}$;

(ii) $\mathcal{E}'$ *over* $\mathcal{B}$ *justifies* $\mathcal{E} \;\Leftrightarrow\;$ *there is a morphism* $g{:}\mathcal{A}/\mathcal{E} \to \mathcal{B}/\mathcal{E}'$.

Proof (i) ($\Rightarrow$) Suppose $\mathcal{B}$ justifies $\mathcal{E}$. Then there is a morphism $h{:}\mathcal{A} \to \mathcal{B}$ such that $\mathcal{B} \models h(\mathcal{E})$. By Proposition 7.2.4(i) there is a morphism $h^\sharp : \mathcal{A}/\mathcal{E} \to \mathcal{B}$. So take $g = h^\sharp$.

($\Leftarrow$) Given a morphism $g{:}\mathcal{A}/\mathcal{E} \to \mathcal{B}$. Then $h = g^\natural$ is such that $\mathcal{B} \models h(\mathcal{E})$, according to Proposition 7.2.4(iii).

(ii) By (i). ∎

Invertible type algebras and prime elements

7.2.8 Definition

(i) A relation $\sim$ on a type algebra $\langle\mathcal{A},\to\rangle$ is called *invertible* if for all $a,b,a',b'\in\mathcal{A}$

$$(a\to b)\sim(a'\to b')\ \Rightarrow\ a\sim a'\ \&\ b\sim b'.$$

(ii) A type algebra $\mathcal{A}$ is *invertible* if the equality relation $=$ on $\mathcal{A}$ is invertible.

Invertibility has a simple characterization for syntactic type algebras.

7.2.9 Remark A syntactic type algebra $\mathbb{T}/{\approx}$ is invertible if one has

$$(A\to B)\ \approx\ (A'\to B')\ \Rightarrow\ A\ \approx\ A'\ \&\ B\ \approx\ B',$$

i.e. if the congruence $\approx$ on the free type algebra $\mathbb{T}$ is invertible.

The free syntactic type algebra $\mathbb{T}$ is invertible. See Example 7.1.17(ii) for an instance of a non-invertible type algebra. Another useful notion concerning type algebras is that of prime element.

7.2.10 Definition Let $\mathcal{A}$ be a type algebra.

(i) An element $a\in\mathcal{A}$ is *prime* if $a\neq(b\to c)$ for all $b,c\in\mathcal{A}$.
(ii) We write $||\mathcal{A}||=\{a\in\mathcal{A}\mid a \text{ is a prime element}\}$.

7.2.11 Remark If $\mathcal{A}=\mathbb{T}/{\approx}$ is a syntactic type algebra, then an element $A\in\mathbb{T}$ is prime if $A\ \not\approx\ (B\to C)$ for all $B,C\in\mathbb{T}$. In this case we also say that A is prime with respect to $\approx$.

In Exercise 7.17(i) it is shown that a type algebra is not always generated by its prime elements. Moreover in item (iii) of that exercise it is shown that a morphism $h{:}\mathcal{A}\to\mathcal{B}$ is not uniquely determined by $h\restriction||\mathcal{A}||$.

Well-founded type algebras

7.2.12 Definition A type algebra $\mathcal{A}$ is *well founded* iff $\mathcal{A}$ is generated by $||\mathcal{A}||$. That is, if $\mathcal{A}$ is the smallest subset of $\mathcal{A}$ containing $||\mathcal{A}||$ and closed under $\to$.

The free type algebra $\mathbb{T}^{\mathbb{A}}$ is well founded, while, for example, $\mathbb{T}^{\{\alpha,\beta\}}[\alpha=\alpha\to\beta]$ is not. A well-founded invertible type algebra is isomorphic to a free type algebra.

7.2.13 Proposition *Let $\mathcal{A}$ be an invertible type algebra. Then:*

(i) $\mathbb{T}^{||\mathcal{A}||} \hookrightarrow \mathcal{A}$;

(ii) *if moreover* $\mathcal{A}$ *is well founded, then* $\mathbb{T}^{||\mathcal{A}||} \cong \mathcal{A}$.

Proof (i) Let i be the morphism determined by $i(a) = a$ for $a \in ||\mathcal{A}||$. Then $i : \mathbb{T}^{||\mathcal{A}||} \hookrightarrow \mathcal{A}$. Indeed, note that the type algebra $\mathbb{T}^{||\mathcal{A}||}$ is free and deduce the injectivity of i by induction on the structure of the types, using the invertibility of $\mathcal{A}$.

(ii) By (i) and well-foundedness. ■

In Exercise 7.17(ii) it will be shown that this embedding is not necessarily surjective: some elements may not be generated by prime elements.

7.2.14 Proposition *Let* $\mathcal{A}$, $\mathcal{B}$ *be type algebras and let* $\sim$, $\approx$ *be congruence relations on* $\mathcal{A}$, $\mathcal{B}$, *respectively.*

(i) *Let* $h_0 : \mathcal{A} \to \mathcal{B}$ *be a morphism such that*

$$\forall x, y \in \mathcal{A}. x \sim y \;\Rightarrow\; h_0(x) \approx h_0(y). \tag{7.2}$$

Then there exists a morphism $h : \mathcal{A}/\sim \to \mathcal{B}/\approx$ *such that*

$$\begin{array}{ccc} \mathcal{A} & \xrightarrow{h_0} & \mathcal{B} \\ {\scriptstyle [\,]_\sim}\downarrow & & \downarrow{\scriptstyle [\,]_\approx} \\ \mathcal{A}/\sim & \xrightarrow[h]{} & \mathcal{B}/\approx \end{array} \qquad \forall x \in \mathcal{A}. h([x]_\sim) = [h_0(x)]_\approx. \tag{7.3}$$

(ii) *Suppose moreover that* $\mathcal{A}$ *is well founded and invertible. Let* $h : \mathcal{A}/\sim \to \mathcal{B}/\approx$ *be a map. Then* h *is a morphism iff there exists a morphism* $h_0 : \mathcal{A} \to \mathcal{B}$ *such that* (7.2) *holds.*

Proof (i) By (7.2) the expression (7.3) is a proper definition of h. One easily verifies that h is a morphism.

(ii) ($\Rightarrow$) Define for $x, y \in \mathcal{A}$

$$h_0(x) = b, \quad \text{if } x \in ||\mathcal{A}||, \text{ for some chosen } b \in h([x]_\approx);$$
$$h_0(x \to_{\mathcal{A}} y) = h_0(x) \to_{\mathcal{B}} h_0(y).$$

Then by well-founded induction one has that $h_0(x)$ is defined for all $x \in \mathcal{A}$ and $h([x]_\sim) = [h_0(x)]_\approx$, using also the fact that $\mathcal{A}$ is invertible. The map h_0 is by definition a morphism.

($\Leftarrow$) By (i). ■

Enriched type algebras

The notions can be generalized in a straightforward way to type algebras having more constructors, including constants (0-ary constructors). This will happen only in exercises and applications.

7.2.15 Definition

(i) A type algebra $\mathcal{A}$ is called *enriched* if there are, in addition to $\rightarrow$, other type constructors (of arity ≥ 0) present in the signature of $\mathcal{A}$ that denote operations over $\mathcal{A}$.

(ii) An *enriched set of types* over the atoms $\mathbb{A}$, written $\mathbb{T} = \mathbb{T}^{\mathbb{A}}_{C_1,\ldots,C_k}$ is the collection of types freely generated from $\mathbb{A}$ by $\rightarrow$ and some other constructors $C_1, \ldots, C_k$.

For enriched type algebras (of the same signature), the definitions of morphisms and congruences are extended by also taking into account the new constructors. A *congruence* over an enriched set of types $\mathbb{T}$ is an equivalence relation $\approx$ that is preserved by all constructors. For example, if C is a constructor of arity 2, we must have $a \approx b,\ a' \approx b' \ \Rightarrow\ C(a,b) \approx C(a',b')$.

In particular, an enriched set of types $\mathbb{T}$ together with a congruence $\approx$ yields in a natural way an *enriched syntactic type algebra* $\mathbb{T}/{\approx}$. For example, if $+$, $\times$ are two new binary type constructors and 1 is a (0-ary) type constant, we have an enriched type algebra $\langle \mathbb{T}^{\mathbb{A}}_{1,+,\times}, \rightarrow, +, \times, 1\rangle$ which is useful for applications (think of it as the set of types for a small meta-language for denotational semantics).

Sets of equations over type algebras

7.2.16 Proposition (Ackermann (1928)) *If $\mathcal{E}$ is a finite set of equations over $\mathbb{T}^{\mathbb{A}}$, then $=_{\mathcal{E}}$ is decidable.*

Proof Write $A =_n B$ if there is a derivation of $A =_{\mathcal{E}} B$ using a derivation of length at most n. It can be shown by a routine induction on the length of derivations that

$$\begin{aligned} A =_n B \ \Rightarrow\ & A \equiv B \ \vee \\ & [A \equiv A_1 {\rightarrow} A_2 \ \&\ B \equiv B_1 {\rightarrow} B_2 \ \& \\ & A_1 =_{m_1} B_1 \ \&\ A_2 =_{m_2} B_2, \text{ with } m_1, m_2 < n] \vee \\ & [A =_{m_1} A' \ \&\ B =_{m_2} B' \ \& \\ & ((A' = B') \in \mathcal{E} \vee (B' = A') \in \mathcal{E}) \text{ with } m_1, m_2 < n]; \end{aligned}$$

the most difficult case is when $A =_{\mathcal{E}} B$ has been obtained using rule (trans).

This implies that if $A =_{\mathcal{E}} B$, then every type occurring in a derivation is a subtype of a type in $\mathcal{E}$ or of A or of B. From this we can conclude that for finite $\mathcal{E}$ the relation $=_{\mathcal{E}}$ is decidable: trying to decide that $A = B$ leads to a list of finitely many such equations with types in a finite set; eventually one will hit an equation that is immediately provable. For the details see Exercise 7.19. ∎

In the following lemma, (i) states that working modulo some systems of equations is compositional and (ii) states that a quotient of a syntactic type algebra $\mathcal{A} = \mathbb{T}/\approx$ is just a syntactic type algebra $\mathbb{T}/\mathcal{E}$ with $\approx \subseteq \mathcal{E}$. Point (i) implies that type equations can be solved incrementally.

7.2.17 Lemma

(i) *Let $\mathcal{E}_1$, $\mathcal{E}_2$ be sets of equations over $\mathcal{A}$. Then*

$$\mathcal{A}/(\mathcal{E}_1 \cup \mathcal{E}_2) \cong (\mathcal{A}/\mathcal{E}_1)/\mathcal{E}_{12},$$

where $\mathcal{E}_{12}$ is defined by

$$([A]_{\mathcal{E}_1} = [B]_{\mathcal{E}_1}) \in \mathcal{E}_{12} \Leftrightarrow (A = B) \in \mathcal{E}_2.$$

(ii) *Let $\mathcal{A} = \mathbb{T}/\approx$ and let $\mathcal{E}$ be a set of equations over $\mathcal{A}$. Then*

$$\mathcal{A}/\mathcal{E} \cong \mathbb{T}/\mathcal{E}',$$

where

$$\mathcal{E}' = \{A = B \mid A \approx B\} \cup \{A = B \mid ([A]_{\approx} = [B]_{\approx}) \in \mathcal{E}\}.$$

Proof (i) By induction on derivations it follows that for $A, B \in \mathcal{A}$ one has

$$\vdash_{\mathcal{E}_1 \cup \mathcal{E}_2} A = B \Leftrightarrow \vdash_{\mathcal{E}_{12}} [A]_{\mathcal{E}_1} = [B]_{\mathcal{E}_1}.$$

It follows that the map $h{:}\mathbb{T}/(\mathcal{E}_1 \cup \mathcal{E}_2) \to (\mathbb{T}/\mathcal{E}_1)/\mathcal{E}_{12}$, given by

$$h([A]_{\mathcal{E}_1 \cup \mathcal{E}_2}) = [[A]_{\mathcal{E}_1}]_{\mathcal{E}_{12}},$$

is well defined and an isomorphism.

(ii) Define

$$\begin{aligned} \mathcal{E}_1 &= \{A = B \mid A \approx B\}, \\ \mathcal{E}_2 &= \{A = B \mid ([A]_{\approx} = [B]_{\approx}) \in \mathcal{E}\}. \end{aligned}$$

Then $\mathcal{E}_{12}$ in the notation of (i) is $\mathcal{E}$. Now we can apply (i):

$$\begin{aligned} \mathcal{A}/\mathcal{E} &= (\mathbb{T}/\approx)/\mathcal{E} \\ &= (\mathbb{T}/\mathcal{E}_1)/\mathcal{E}_{12} \\ &= \mathbb{T}/(\mathcal{E}_1 \cup \mathcal{E}_2). \ \blacksquare \end{aligned}$$

7.2.18 Notation In general to ease notation we often identify the level of types with that of equivalence classes of types. We do this whenever the exact nature of the denoted objects can be recovered unambiguously from the context. For example, if $\mathcal{A} = \mathbb{T}^{\mathbb{A}}/\approx$ is a syntactic type algebra and A denotes as usual an element of $\mathbb{T}^{\mathbb{A}}$, then in the formula $A \in \mathcal{A}$ the A stands for $[A]_{\approx}$. If we consider this $\mathcal{A}$ modulo $\mathcal{E}$, then $A =_{\mathcal{E}} B$ is equivalent to $A =_{\mathcal{E}'} B$, with $\mathcal{E}'$ as in Lemma 7.2.17(ii).

7.3 Recursive types via simultaneous recursion

In this section we construct type algebras containing elements satisfying recursive equations, like $a = a \to b$ or $c = d \to c$. There are essentially two ways to do this: defining the recursive types as the solutions of a given system of recursive type equations; or via a general fixed point operator μ in the type syntax. Recursive type equations let us define explicitly only a finite number of recursive types, while the introduction of a fixed-point operator in the syntax makes all recursive types expressible without an explicit separate definition.

For both ways one considers types modulo a congruence relation. Some of these congruence relations will be defined proof-theoretically (inductively), as in Definition 7.1.11. Other congruence relations will be defined semantically, using possibly infinite trees (co-inductively), as is done in Section 7.5.

Adding indeterminates

In algebra one constructs, for a given ring R and set of indeterminates $\vec{X}$, a new object $R[\vec{X}]$, the ring of polynomials over $\vec{X}$ with coefficients in R. A similar construction will be made for type algebras. Intuitively $\mathcal{A}(\vec{X})$ is the type algebra obtained by 'adding' to $\mathcal{A}$ one new object for each indeterminate in $\vec{X}$ and taking the closure under $\to$. Since this definition of $\mathcal{A}(\vec{X})$ is somewhat syntactic we assume, using Proposition 7.1.18, that $\mathcal{A}$ is a syntactic type algebra. Often we will take for $\mathcal{A}$ the free syntactic type algebra $\mathbb{T}^{\mathbb{A}}$ over an arbitrary non-empty set of atomic types $\mathbb{A}$.

7.3.1 Definition Let $\mathcal{A} = \mathbb{T}^{\mathbb{A}}/\approx$ be a syntactic type algebra. Let $\vec{X} = X_1, \ldots, X_n$ $(n \geq 0)$ be a set of *indeterminates*, i.e. a set of type symbols such that $\vec{X} \cap \mathbb{A} = \emptyset$. The extension of $\mathcal{A}$ with $\vec{X}$ is defined as

$$\mathcal{A}(\vec{X}) = \mathbb{T}^{\mathbb{A} \cup \{\vec{X}\}}/\approx.$$

Note that $\mathbb{T}/\approx$ stands for for $\mathbb{T}/=_{\approx}$. So in $\mathcal{A}(\vec{X}) = \mathbb{T}^{\mathbb{A} \cup \{\vec{X}\}}/\approx$ the relation $\approx$ is extended with the identity on the $\vec{X}$. Note also that in $\mathcal{A}(\vec{X})$ the indeterminates are not related to any other element, since $\approx$ is not defined for elements of $\vec{X}$. By Proposition 7.1.18 this construction can be applied to arbitrary type algebras as well.

7.3.2 Notation The type $A(\vec{X})$ ranges over arbitrary elements of $\mathcal{A}(\vec{X})$.

7.3.3 Proposition $\mathcal{A} \hookrightarrow \mathcal{A}(\vec{X})$.

Proof Immediate. ∎

We consider extensions of a type algebra $\mathcal{A}$ with indeterminates in order to build solutions to $\mathcal{E}(\vec{a}, \vec{X})$, where $\mathcal{E}(\vec{a}, \vec{X})$ (or simply $\mathcal{E}(\vec{X})$ with $\vec{a}$ understood) is a set of equations over $\mathcal{A}$ with indeterminates $\vec{X}$. This solution may not exist in $\mathcal{A}$, but via the indeterminates we can build an extension $\mathcal{A}'$ of $\mathcal{A}$ containing elements $\vec{c}$ solving $\mathcal{E}(\vec{X})$.

For simplicity consider the free type algebra $\mathbb{T} = \mathbb{T}^{\mathbb{A}}$. Our first way of extending $\mathbb{T}$ with elements satisfying a given set of equations $\mathcal{E}(\vec{X})$ is to consider the type algebra $\mathbb{T}(\vec{X})/\mathcal{E}$ whose elements are the equivalence classes of $\mathbb{T}(\vec{X})$ under $=_{\mathcal{E}}$.

7.3.4 Definition Let $\mathcal{A}$ be a type algebra and $\mathcal{E} = \mathcal{E}(\vec{X})$ be a set of equations over $\mathcal{A}(\vec{X})$. Denote $\mathcal{A}(\vec{X})/\mathcal{E}$ by $\mathcal{A}[\mathcal{E}]$.

Satisfying existential equations

We now want to state when existential statements like $\exists X.a = b \to X$, with $a, b \in \mathcal{A}$, hold in a type structure. We say that $\exists X.a = b \to X$ holds in $\mathcal{A}$, written

$$\mathcal{A} \models \exists X.a = b \to X,$$

if for some $c \in \mathcal{A}$ one has $a = b \to c$.

The following definitions are stated for sets of equations $\mathcal{E}$ but apply to a single equation $a = b$ as well, by considering it as a singleton $\{a = b\}$.

7.3.5 Definition Let $\mathcal{A}$ be a type algebra and $\mathcal{E}{=}\mathcal{E}(\vec{X})$ a set of equations over $\mathcal{A}(\vec{X})$.

(i) We say $\mathcal{A}$ *solves* $\mathcal{E}$ (or $\mathcal{A}$ *satisfies* $\exists\vec{X}.\mathcal{E}$ or $\exists\vec{X}.\mathcal{E}$ *holds* in $\mathcal{A}$), written $\mathcal{A} \models \exists\vec{X}.\mathcal{E}$, if there is a morphism $h{:}\mathcal{A}(\vec{X}) \to \mathcal{A}$ such that $h(a) = a$, for all $a \in \mathcal{A}$ and $\mathcal{A} \models h(\mathcal{E}(\vec{X}))$.

(ii) For any h satisfying (i), the sequence $\langle h(X_1), \ldots, h(X_n) \in \mathcal{A}\rangle$ is called a *solution* in $\mathcal{A}$ of $\mathcal{E}(\vec{X})$.

7.3.6 Remark

(i) Note that $\mathcal{A} \models \exists\vec{X}.\mathcal{E}$ iff $\mathcal{A} \models \mathcal{E}[\vec{X}{:} = \vec{a}]$ for some $\vec{a} \in \mathcal{A}$. Indeed, choose $a_i = h(X_i)$ as the definition of the $\vec{a}$ or of the morphism h.

(ii) If $\mathcal{A}$ solves $\mathcal{E}(\vec{X})$, then $\mathcal{A}(\vec{X})$ justifies $\mathcal{E}(\vec{X})$, but not conversely. During justification one may reinterpret the constants, via a morphism.

7.3.7 Remark

(i) The set of equations $\mathcal{E}(\vec{X})$ over $\mathcal{A}(\vec{X})$ is interpreted as the problem of finding appropriate $\vec{X}$ in $\mathcal{A}$. This is similar to stating that the polynomial $x^2 - 3 \in \mathbb{R}[x]$ has $\sqrt{3} \in \mathbb{R}$ as (non-unique) root.

(ii) In the previous definition we tacitly changed the indeterminates $\vec{X}$ in a bound variable: by $\exists\vec{X}.\mathcal{E}$ or $\exists\vec{X}.\mathcal{E}(\vec{X})$ we intend $\exists\vec{x}.\mathcal{E}(\vec{x})$. We will allow this abuse of language regarding $\vec{X}$ as bound variables, since it is clear what we mean.

(iii) If $\vec{X} = \emptyset$, then

$$\mathcal{A} \models \exists\vec{X}.\mathcal{E} \Leftrightarrow \mathcal{A} \models \mathcal{E}.$$

7.3.8 Example There exists a type algebra $\mathcal{A}$ such that

$$\mathcal{A} \models \exists X.(X \to X) = (X \to X \to X). \tag{7.4}$$

Take $\mathcal{A} = \mathbb{T}[\mathcal{E}]$, with $\mathcal{E} = \{X \to X = X \to X \to X\}$, with solution

$$\mathbf{X} = [X]_{\{X \to X = X \to X \to X\}}.$$

7.3.9 Remark Let $\mathcal{R} = \{X = a \to X,\ Y = a \to a \to Y\}$ over $\mathbb{T}^{\{a\}}(X, Y)$. Then $[X]_{\mathcal{R}}, [Y]_{\mathcal{R}} \in \mathbb{T}[\mathcal{R}]$ is a solution of $\exists X Y.\mathcal{R}$. Note that also $[X]_{\mathcal{R}}, [X]_{\mathcal{R}}$ is such a solution and intuitively $[X]_{\mathcal{R}} \neq [Y]_{\mathcal{R}}$, as we will see later more precisely. Hence solutions are not unique.

Simultaneous recursions

In general $\mathbb{T}/\mathcal{E}$ is not invertible. For instance, in Example 7.1.17(ii) take $\mathbb{A}_\infty = \{\alpha, \infty\}$. Then in $\mathbb{T}^{\mathbb{A}_\infty}/\mathcal{E}_\infty$ one has $\alpha \to \infty = \infty \to \infty$, but $\alpha \neq \infty$.

Note also that in a system of equations $\mathcal{E}$ the same type can be the left hand side of more than one equation of $\mathcal{E}$. For instance, this is the case for ∞ in Example 7.1.17 ((ii)).

The following notion will specialize to a particular $\mathcal{E}$, such that $\mathcal{A}[\mathcal{E}]$ is invertible. A simultaneous recursion (sometimes abbreviated to 'sr' even for the plural) is represented by a set $\mathcal{R}(\vec{X})$ of type equations of a particular shape over $\mathcal{A}$, in which the indeterminates $\vec{X}$ represent the recursive types to be added to $\mathcal{A}$. Such types occur in programming languages, for the first time in Algol-68; see van Wijngaarden (1981).

7.3.10 Definition Let $\mathcal{A}$ be a type algebra.

(i) A *simultaneous recursion* over $\mathcal{A}$ with *indeterminates* $\vec{X} = \{X_1, \dots, X_n\}$ is a finite set $\mathcal{R} = \mathcal{R}(\vec{X})$ of equations over $\mathcal{A}(\vec{X})$ of the form

$$\left.\begin{array}{rcl} X_1 & = & A_1(\vec{X}) \\ & \vdots & \\ X_n & = & A_n(\vec{X}) \end{array}\right\} \quad \mathcal{R}$$

where all indeterminates $X_1, \dots, X_n$ are different.

(ii) The *domain* of $\mathcal{R}$, written $\mathrm{Dom}(\mathcal{R})$, consists of the set $\{\vec{X}\}$.

(iii) If $\mathrm{Dom}(\mathcal{R}) = \vec{X}$, then $\mathcal{R}$ is said to be a simultaneous recursion over $\mathcal{A}(\vec{X})$.

(iv) The equational theory on $\mathcal{A}(\vec{X})$ axiomatized by $\mathcal{R}$ is denoted by $(\mathcal{R})$.

It is useful to consider restricted forms of simultaneous recursion.

7.3.11 Definition (Proper simultaneous recursion)

(i) A simultaneous recursion $\mathcal{R}(\vec{X})$ is *proper* if

$$(X_i = X_j) \in \mathcal{R} \Rightarrow i < j.$$

(ii) A simultaneous recursion $\mathcal{R}(\vec{X})$ is *simple* if no equation $X_i = X_j$ occurs in $\mathcal{R}$.

Note that a simple simultaneous recursion is proper. The definition of proper is intended to rule out circular definitions like $X = X$ or $X = Y, Y = X$. Proper simultaneous recursions are convenient from the Term Rewriting System (TRS) point of view introduced in Section 8.3: the reduction relation will be SN. We always can make a simultaneous recursion proper, as will be shown in Proposition 7.3.25

7.3.12 Example For example let $\alpha, \beta \in \mathbb{A}$. Then

$$X_1 = \alpha \to X_2$$
$$X_2 = \beta \to X_1$$

is a simultaneous recursion with indeterminates $\{X_1, X_2\}$ over $\mathbb{T}^{\mathbb{A}}$.

Intuitively it is clear that in this example one has $X_1 =_{\mathcal{R}} \alpha \to \beta \to X_1$, but $X_1 \neq_{\mathcal{R}} X_2$. To show this the following is convenient.

A simultaneous recursion can be considered as a TRS, see Klop (1992) or Terese (2003). The reduction relation is denoted by $\Rrightarrow_{\mathcal{R}}$; we will later encounter its converse $\Rrightarrow_{\mathcal{R}}^{-1}$ as another useful reduction relation.

7.3.13 Definition Let $\mathcal{R}$ on $\mathcal{A}$ be given.

(i) Define on $\mathcal{A}(\vec{X})$ the *$\mathcal{R}$-reduction relation*, written $\twoheadrightarrow_{\mathcal{R}}$, induced by the notion of reduction

$$\left.\begin{array}{lcl} X_1 & \Rightarrow_{\mathcal{R}} & A_1(\vec{X}) \\ & \vdots & \\ X_n & \Rightarrow_{\mathcal{R}} & A_n(\vec{X}) \end{array}\right\} \qquad (\Rightarrow_{\mathcal{R}}).$$

So $\twoheadrightarrow_{\mathcal{R}}$ is the smallest reflexive, transitive, and compatible relation on $\mathcal{A}(\vec{X})$ extending $\Rightarrow_{\mathcal{R}}$.

(ii) The relation $=_{\mathcal{R}}$ is the smallest compatible equivalence relation extending $\twoheadrightarrow_{\mathcal{R}}$

(iii) We denote the resulting term rewriting system by

$$\mathrm{TRS}(\mathcal{R}) = (\mathcal{A}(\vec{X}), \Rightarrow_{\mathcal{R}}).$$

It is important to note that the $\vec{X}$ are not variables in the TRS sense: if $a(X) \twoheadrightarrow_{\mathcal{R}} b(X)$, then $a(c) \twoheadrightarrow_{\mathcal{R}} b(c)$ need not hold. Rewriting in $\mathrm{TRS}(\mathcal{R})$ is between closed expressions.

In general $\Rightarrow_{\mathcal{R}}$ is not normalizing. For example for $\mathcal{R}$ as above one has

$$X_1 \Rightarrow_{\mathcal{R}} (\alpha \to X_2) \Rightarrow_{\mathcal{R}} (\alpha \to \beta \to X_1) \Rightarrow_{\mathcal{R}} \cdots .$$

Remember that a rewriting system $\langle X, \Rightarrow \rangle$ is *Church–Rosser* (CR) if

$$\forall a, b, c \in X.[a \Rightarrow^* b \;\&\; a \Rightarrow^* c \;\Rightarrow\; \exists d \in X.[b \Rightarrow^* d \;\&\; c \Rightarrow^* d]],$$

where $\Rightarrow^*$ is the transitive reflexive closure of $\Rightarrow$.

7.3.14 Proposition (Church–Rosser Theorem for $\Rightarrow_{\mathcal{R}}$) *Given a simultaneous recursion $\mathcal{R}$ over $\mathcal{A}$. Then:*

(i) *for $a, b \in \mathcal{A}$ one has $\mathcal{R} \vdash a = b \Leftrightarrow a =_{\mathcal{R}} b$;*
(ii) *$\twoheadrightarrow_{\mathcal{R}}$ on $\mathcal{A}(\vec{X})$ is CR;*
(iii) *Therefore $a =_{\mathcal{R}} b$ iff a, b have a common $\twoheadrightarrow_{\mathcal{R}}$ reduct.*

Proof (i) See for example Terese (2003), Exercise 2.4.3.
(ii) Easy, the 'redexes' are all disjoint.
(iii) By (ii). ∎

So in the example above one has $X_1 \neq_{\mathcal{R}} X_2$ and $X_1 =_{\mathcal{R}} (\alpha \to \beta \to X_1)$.

An important property of simultaneous recursions is that they do not identify elements of $\mathcal{A}$.

7.3.15 Lemma *Let $\mathcal{R}(\vec{X})$ be a simultaneous recursion over a type algebra $\mathcal{A}$. Then for all distinct $a, b \in \mathcal{A}$ we have $a \neq_{\mathcal{R}} b$.*

Proof By Proposition 7.3.14(ii). ∎

Lemma 7.3.15 is no longer true, in general, if we start from a set of equations $\mathcal{E}$ instead of a simultaneous recursion $\mathcal{R}(\vec{X})$. For example take $\mathcal{E} = \{a = a \to b, b = (a \to b) \to b\}$. In this case $a =_{\mathcal{E}} b$. In the following we will use indeterminates only in the definition of simultaneous recursion. Generic equations will be considered only between closed terms (i.e. without indeterminates).

Another application of the properties of TRS($\mathcal{R}$) is the invertibility of a simultaneous recursion.

7.3.16 Proposition *Let $\mathcal{R}$ be a simultaneous recursion over $\mathbb{T}$. Then $=_{\mathcal{R}}$ is invertible.*

Proof Suppose $A \to B =_{\mathcal{R}} A' \to B'$, in order to show $A =_{\mathcal{R}} A'$ & $B =_{\mathcal{R}} B'$. By the CR property for $\twoheadrightarrow_{\mathcal{R}}$ the types $A \to B$ and $A' \to B'$ have a common $\twoheadrightarrow_{\mathcal{R}}$-reduct which must be of the form $C \to D$. Then $A =_{\mathcal{R}} C =_{\mathcal{R}} A'$ and $B =_{\mathcal{R}} D =_{\mathcal{R}} B'$. ∎

Note that the images of $\mathcal{A}$ and the $[X_i]$ in $\mathcal{A}(\vec{X})/ =_{\mathcal{R}}$ are not necessarily disjoint. For instance if $\mathcal{R}$ contains an equation $X = a$ where $X \in \vec{X}$ and $a \in \mathcal{A}$ we have $[X] = [a]$.

7.3.17 Definition

(i) Let $\mathcal{R} = \mathcal{R}(\vec{X})$ be a simultaneous recursion in $\vec{X}$ over a type algebra $\mathcal{A}$ (i.e. a special set of equations over $\mathcal{A}(\vec{X})$). As in Definition 7.3.4 write

$$\mathcal{A}[\mathcal{R}] = \mathcal{A}(\vec{X})/\mathcal{R}.$$

(ii) For X one of the $\vec{X}$, write $\mathbf{X} = [X]_{\mathcal{R}}$.

(iii) We say that $\mathcal{A}[\mathcal{R}]$ is obtained by *adjunction* of the elements $\vec{\mathbf{X}}$ to $\mathcal{A}$.

The method of adjunction then allows us to define recursive types incrementally, according to Lemma 7.2.17(i).

7.3.18 Remark

(i) By Proposition 7.3.16 the type algebra $\mathbb{T}[\mathcal{R}]$ is invertible.

(ii) In general $\mathcal{A}[\mathcal{E}]$ is not invertible, see Example 7.1.17(ii).

(iii) Let the indeterminates of $\mathcal{R}_1$ and $\mathcal{R}_2$ be disjoint, then $\mathcal{R}_1 \cup \mathcal{R}_2$ is again a simultaneous recursion. By Lemma 7.2.17(i) $\mathcal{A}[\mathcal{R}_1 \cup \mathcal{R}_2] = \mathcal{A}[\mathcal{R}_1][\mathcal{R}_2]$. Recursive types can therefore be defined incrementally.

7.3.19 Theorem *Let $\mathcal{A}$ be a type algebra and $\mathcal{R}$ a simultaneous recursion over $\mathcal{A}$. Then:*

(i) $\varphi{:}\mathcal{A} \hookrightarrow \mathcal{A}[\mathcal{R}]$, *where* $\varphi(a) = [a]_{\mathcal{R}}$;

(ii) *$\mathcal{A}[\mathcal{R}]$ is generated from (the image under φ of) $\mathcal{A}$ and the $[\mathbf{X}_i]_{\mathcal{R}}$;*
(iii) *$\mathcal{A}[\mathcal{R}] \models \exists \vec{X}.\mathcal{R}$ and the $\mathbf{X}_1, \ldots, \mathbf{X}_n$ form a solution of $\mathcal{R}$ in $\mathcal{A}[\mathcal{R}]$.*

Proof (i) The canonical map φ is an injective morphism by Lemma 7.3.15.
(ii) Clearly $\mathcal{A}[\mathcal{R}]$ is generated by the $\mathbf{X}_i$ and the $[a]_{\mathcal{R}}$, with $a \in \mathcal{A}$.
(iii) $\mathcal{A}[\mathcal{R}] \models \exists \vec{X}.\mathcal{R}$ by Lemma 7.1.15(ii). ∎

In Theorem 7.3.19(iii) we stated that the $\mathbf{X_1}, \ldots, \mathbf{X_n}$ form a solution of $\mathcal{R}$. In fact they form a solution of $\mathcal{R}$ translated to $\mathcal{A}[\mathcal{R}](\vec{X})$. Moreover, this translation is trivial, due to the injection $\varphi{:}\mathcal{A} \hookrightarrow \mathcal{A}[\mathcal{R}]$.

Folding and unfolding

Simultaneous recursions are a natural tool for specifying types satisfying given equations. We call *unfolding (modulo $\mathcal{R}$)* the operation of replacing an occurrence of X_i by $A_i(\vec{X})$, for any equation $X_i = A_i(\vec{X}) \in \mathcal{R}$; *folding* is the reverse operation. If $a, b \in \mathcal{A}(\vec{X})$ then $a =_{\mathcal{R}} b$ if they can be transformed one into the other by a finite number of applications of the operations folding and unfolding, possibly on subexpressions of a and b.

7.3.20 Example

(i) The simultaneous recursion $\mathcal{R}_0 = \{X_0 = A \to X_0\}$, where $A \in \mathbb{T}$ is a type, specifies a type X_0 which is such that

$$X_0 =_{\mathcal{R}_0} A \to X_0 =_{\mathcal{R}_0} A \to A \to X_0 \cdots$$

i.e. $X_0 =_{\mathcal{R}_0} A^n \to X_0$ for any n. This represents the type of a function which can take an arbitrary number of arguments of type A.
(ii) The simultaneous recursion $\mathcal{R}_1 = \{X_1 = A \to A \to X_1\}$ is similar to $\mathcal{R}_0$ but not all equations modulo $\mathcal{R}_0$ hold modulo $\mathcal{R}_1$. For instance $X_1 \neq_{\mathcal{R}_1} A \to X_1$ (i.e. we cannot derive $X_1 = A \to X_1$ from the derivation rules of Definition 7.1.11(i)).

7.3.21 Remark Note that $=_{\mathcal{R}}$ is the minimal congruence with respect to $\to$ satisfying $\mathcal{R}$. Two types can be different with respect to it even if they seem to represent functions with the same behavior, like X_0 and X_1 above. As another example take $\mathcal{R} = \{X = A \to X,\ Y = A \to Y\}$. Then we have $X \neq_{\mathcal{R}} Y$ since we cannot prove $X = Y$ using only the rules of Definition 7.1.11(i). These types will, however, be identified in the tree equivalence to be introduced in Section 7.5.

We will often consider only proper simultaneous recursions. In order to do this, it is useful to transform a simultaneous recursion into an 'equivalent' one. We introduce two notions of equivalence for simultaneous recursion.

7.3.22 Definition Let $\mathcal{R} = \mathcal{R}(\vec{X})$ and $\mathcal{R}' = \mathcal{R}'(\vec{X}')$ be simultaneous recursions over $\mathcal{A}$.

(i) We say $\mathcal{R}$ and $\mathcal{R}'$ are *equivalent* if $\mathcal{A}[\mathcal{R}] \cong \mathcal{A}[\mathcal{R}']$.
(ii) Let $\vec{X} = \vec{X}'$ be the same set of indeterminates. Then $\mathcal{R}(\vec{X})$ and $\mathcal{R}'(\vec{X})$ are *logically equivalent* if

$$\forall a, b \in \mathcal{A}[\vec{X}].a =_{\mathcal{R}} b \Leftrightarrow a =_{\mathcal{R}'} b.$$

7.3.23 Remark

(i) It is easy to see that $\mathcal{R}$ and $\mathcal{R}'$ over the same $\vec{X}$ are logically equivalent iff

$$\mathcal{R} \vdash \mathcal{R}' \text{ and } \mathcal{R}' \vdash \mathcal{R}.$$

(ii) Two logically equivalent simultaneous recursions are also equivalent.
(iii) There are equivalent $\mathcal{R}, \mathcal{R}'$ that are not logically equivalent; for example

$$\mathcal{R} = \{X = \alpha\} \text{ and } \mathcal{R}' = \{X = \beta\}.$$

Note that $\mathcal{R}$ and $\mathcal{R}'$ are on the same set of indeterminates.

7.3.24 Definition Let $\mathcal{A}$ be a type algebra. Define $\mathcal{A}_\bullet := \mathcal{A}(\vec{\bullet})$, where $\vec{\bullet}$ are some indeterminates with special names different from all X_i. These $\vec{\bullet}$ are treated as new elements that are said to have been *added* to $\mathcal{A}$. Indeed, $\mathcal{A} \hookrightarrow \mathcal{A}_\bullet$.

7.3.25 Proposition

(i) *Every proper simultaneous recursion $\mathcal{R}(\vec{X})$ over $\mathcal{A}$ is equivalent to a simple $\mathcal{R}'(\vec{X}')$, where $\vec{X}'$ is a subset of $\vec{X}$.*
(ii) *Let $\mathcal{R}$ be a simultaneous recursion over $\mathcal{A}$. Then there is a proper $\mathcal{R}'$ over $\mathcal{A}_\bullet$ such that*

$$\mathcal{A}[\mathcal{R}] \cong \mathcal{A}_\bullet[\mathcal{R}'].$$

Proof (i) If $\mathcal{R}$ is not simple, then $\mathcal{R} = \mathcal{R}_1 \cup \{X_i = X_j\}$, with $i<j$. Now define

$$\mathcal{R}^-(X_1, \ldots, X_{i-1}, X_{i+1}, \ldots, X_n),$$

by $\mathcal{R}^- = \mathcal{R}_1[X_i := X_j]$. Note that $\mathcal{R}^-$ is still proper (since an equation $X_k = X_i$ in $\mathcal{R}$ becomes $X_k = X_j$ in $\mathcal{R}^-$ and $k<i<j$), equivalent to $\mathcal{R}$, and has one equation fewer. So after finitely many such steps the simple $\mathcal{R}'$ is obtained. One easily proves that

$$\mathcal{A}[\vec{X}]/\mathcal{R} \cong \mathcal{A}[X_i, \ldots, X_{i-1}, X_{i+1}, \ldots, X_n]/\mathcal{R}'$$

as follows. Note that if $\mathcal{R} = \{X_k = A_k(\vec{X}) \mid 1 \le k \le n\}$, then

$$\mathcal{R}^- = \{X_k = A_k(\vec{X})[X_i := X_j] \mid k \ne i\}.$$

Define

$$\begin{array}{rl} g^\natural : & \mathcal{A}(\vec{X}) \to \mathcal{A}[X_i, \ldots, X_{i-1}, X_{i+1}, \ldots, X_n] \\ h^\natural : & \mathcal{A}[X_i, \ldots, X_{i-1}, X_{i+1}, \ldots, X_n] \to \mathcal{A}(\vec{X}) \end{array}$$

by

$$\begin{array}{rcll} g^\natural(A) & = & A[X_i := X_j], & \text{for } A \in \mathcal{A}[\vec{X}], \\ h^\natural(A) & = & A, & \text{for } A \in \mathcal{A}[X_i, \ldots, X_{i-1}, X_{i+1}, \ldots, X_n] \end{array}$$

and show

$$\begin{array}{rcll} g^\natural(X_k) & = & g^\natural(A_k(\vec{X})), & \text{for } 1 \le k \le n, \\ h^\natural(X_k) & = & h^\natural((A_k(\vec{X}))[X_i := X_j]), & \text{for } k \ne j. \end{array}$$

Then $g^\natural, h^\natural$ induce the required isomorphism g and its inverse h.

(ii) First remove each $X_j = X_j$ from $\mathcal{R}$ and put the X_j in $\vec{\bullet}$. The equations $X_i = X_j$ with $i > j$ are treated in the same way as $X_j = X_i$ in (i). The proof that indeed $\mathcal{A}[\mathcal{R}] \cong \mathcal{A}_\bullet[\mathcal{R}']$ is very easy. Now $g^\natural$ and $h^\natural$ are in fact identities. ■

7.3.26 Lemma *Let $\mathcal{R}(\vec{X})$ be a proper simultaneous recursion over $\mathcal{A}$. Then all its indeterminates X are such that either $X =_{\mathcal{R}} a$ where $a \in \mathcal{A}$ or $X =_{\mathcal{R}} (b \to c)$ for some $b, c \in \mathcal{A}[\vec{X}]$.*

Proof Easy. ■

The prime elements of the type algebras $\mathbb{T}[\mathcal{R}]$, where $\mathcal{R}$ is proper and $\mathbb{T} = \mathbb{T}^{\mathbb{A}}$, can easily be characterized.

7.3.27 Lemma *Let $\mathcal{R}(\vec{X})$ be a proper simultaneous recursion over $\mathbb{T}^{\mathbb{A}}$. Then:*

$$||\mathbb{T}[\mathcal{R}]|| = \{[\alpha] \mid \alpha \in \mathbb{A}\};$$
$$[\alpha] \subseteq \{\alpha\} \cup \{\vec{X}\},$$

i.e. $[\alpha]$ consists of α and some of the $\vec{X}$.

Proof The elements of $\mathbb{T}[\mathcal{R}]$ are generated from $\mathbb{A}$ and the $\vec{X}$. Now note that by Lemma 7.3.26(i) an indeterminate X either is such that $X =_{\mathcal{R}} A \to B$ for some $A, B \in \mathbb{T}^{\mathbb{A} \cup \vec{X}}$ (and then $[X]$ is not prime), or $X =_{\mathcal{R}} \alpha$ for some atomic type α. Moreover, by Proposition 7.3.14 it follows that no other atomic types or arrow types can belong to $[\alpha]$. Therefore, the only prime elements in $\mathbb{T}[\mathcal{R}]$ are the equivalence classes of the $\alpha \in \mathbb{A}$. ∎

For a proper simultaneous recursion $\mathcal{R}$ we can, for instance, write $||\mathbb{T}[\mathcal{R}]|| = \mathbb{A}$, choosing α as the representative of $[\alpha]$.

Justifying sets of equations by a simultaneous recursion

Remember that $\mathcal{B}$ justifies a set of equations $\mathcal{E}$ over $\mathcal{A}$ if there is a morphism $h{:}\mathcal{A} \to \mathcal{B}$ such that $\mathcal{B} \models h(\mathcal{E})$; moreover a set $\mathcal{E}'$ over $\mathcal{B}$ justifies $\mathcal{E}$ over $\mathcal{A}$ iff $\mathcal{B}/\mathcal{E}'$ justifies $\mathcal{E}$. A particular case is that a simultaneous recursion $\mathcal{R}$ over $\mathcal{B}(\vec{X})$ justifies $\mathcal{E}$ over $\mathcal{A}$ iff $\mathcal{B}[\mathcal{R}]$ justifies $\mathcal{E}$. Proposition 7.2.7 stated that $\mathcal{B}$ justifies a set of equations $\mathcal{E}$ iff there is a morphism $h{:}\mathcal{A}/\mathcal{E} \to \mathcal{B}$. Indeed, all the equations in $\mathcal{E}$ become valid after interpreting the elements of $\mathcal{A}$ in the correct way in $\mathcal{B}$.

In Chapter 8 it will be shown that in the correct context the notion of justifying is decidable. But decidability only makes sense if $\mathcal{B}$ is given in an effective 'finitely presented' way.

7.3.28 Proposition *Let $\mathcal{A}, \mathcal{B}$ be type algebras and let $\mathcal{E}$ be a set of equations over $\mathcal{A}$.*

(i) *Let $\mathcal{E}'$ be a set of equations over $\mathcal{B}$. Then*

$$\mathcal{E}' \text{ justifies } \mathcal{E} \ \Leftrightarrow\ \exists g.\ g : \mathcal{A}/\mathcal{E} \to \mathcal{B}/\mathcal{E}'.$$

(ii) *Let $\mathcal{R}$ be a simultaneous recursion over $\mathcal{B}(\vec{X})$. Then*

$$\mathcal{R} \text{ justifies } \mathcal{E} \ \Leftrightarrow\ \exists g.\ g : \mathcal{A}/\mathcal{E} \to \mathcal{B}[\mathcal{R}].$$

Proof (i), (ii). By Proposition 7.2.7(ii). ∎

7.3.29 Example Let $\mathcal{E} = \{\alpha \to \beta = \alpha \to \alpha \to \beta\}$. Then $\mathcal{R} = \{X = \alpha \to X\}$ justifies $\mathcal{E}$ over $\mathbb{T}^{\{\alpha,\beta\}}$ as we have the homomorphism

$$h : \mathbb{T}^{\{\alpha,\beta\}}/\mathcal{E} \to \mathbb{T}^{\{\alpha\}}[\mathcal{R}]$$

determined by $h([\alpha]_{\mathcal{E}}) = [\alpha]_{\mathcal{R}}$, $h([\beta]_{\mathcal{E}}) = [X]_{\mathcal{R}}$; or, with our notational conventions, $h(\alpha) = \alpha$, $h(\beta) = X$ (where h is indeed a syntactic homomorphism).

7.3.30 Proposition *Let $\mathcal{A}, \mathcal{B}$ be type algebras. Suppose that $\mathcal{A}$ is well-founded and invertible. Let $\mathcal{E}$ be a system of equations over $\mathcal{A}$ and $\mathcal{R}(\vec{X})$ be a simultaneous recursion over $\mathcal{B}$. Then*

$$\mathcal{R} \text{ justifies } \mathcal{E} \ \Leftrightarrow\ \exists h{:}\mathcal{A} \to \mathcal{B}(\vec{X})\ \forall a, b \in \mathcal{A}.[a =_{\mathcal{E}} b \ \Rightarrow\ h(a) =_{\mathcal{R}} h(b)]. \quad (7.5)$$

Proof By Corollary 7.2.7(ii) and Proposition 7.2.14. ∎

As free type algebras are well founded and invertible, (7.5) holds for all $\mathbb{T}^{\mathbb{A}}$.

Closed type algebras

A final general notion concerning type algebras is the following.

7.3.31 Definition Let $\mathcal{A}$ be a type algebra.

(i) We say $\mathcal{A}$ is *closed* iff every simultaneous recursion $\mathcal{R}$ over $\mathcal{A}$ can be solved in $\mathcal{A}$, cf. Definition 7.3.5.
(ii) We call $\mathcal{A}$ *uniquely closed*, if every proper simultaneous recursion $\mathcal{R}$ over $\mathcal{A}$ has a unique solution in $\mathcal{A}$.

7.3.32 Remark There are type algebras that are closed but not uniquely so. For instance let $\mathcal{A} = \mathbb{T}^{\{a,b\}}/\mathcal{E}$ with $\mathcal{E} = \{a = a \to a, b = b \to b, b = a \to b, b = b \to a\}$. Then $\mathcal{A}$ is closed, but not uniquely so. A simple uniquely closed type algebra will be given in Section 7.5.

From Proposition 7.2.16 we know that $=_{\mathcal{R}}$ is decidable for any (finite) $\mathcal{R}$ over $\mathbb{T}^{\mathcal{A}}(\vec{X})$. In Chapter 8 we will prove some other properties of $\mathbb{T}[\mathcal{R}]$, in particular that it is decidable whether a simultaneous recursion $\mathcal{R}$ justifies a set $\mathcal{E}$ of equations.

7.4 Recursive types via μ-abstraction

Another way of representing recursive types is by enriching the syntax of types with a new operator μ to explicitly denote solutions of recursive type equations. The resulting (syntactic) type algebra 'solves' arbitrary type equations, i.e. is closed in the sense of Definition 7.3.31.

Some bureaucracy for renaming and substitution

The reader is advised to skip this subsection at first reading: go to 7.4.23.

7.4.1 Definition ($\dot{\mu}$-types) Let $\mathbb{A} = \mathbb{A}_\infty$ be the infinite set of type atoms considered as type variables for the purpose of binding and substitution. The set $\mathbb{T}^{\mathbb{A}}_{\dot{\mu}}$ is defined by the following 'simplified syntax', omitting parentheses. The '$\cdot$' on top of the μ indicates that we do not (yet) consider the types modulo α-conversion (renaming of bound variables).

$$\boxed{\mathbb{T}^{\mathbb{A}}_{\dot{\mu}} \;::=\; \mathbb{A} \mid \mathbb{T}^{\mathbb{A}}_{\dot{\mu}} \to \mathbb{T}^{\mathbb{A}}_{\dot{\mu}} \mid \dot{\mu}\mathbb{A}\mathbb{T}^{\mathbb{A}}_{\dot{\mu}}.}$$

Often we write $\mathbb{T}_{\dot{\mu}}$ for $\mathbb{T}^{\mathbb{A}}_{\dot{\mu}}$, leaving $\mathbb{A}$ implicit.

The subset of $\mathbb{T}^{\mathbb{A}}_{\dot{\mu}}$ containing only types without occurrences of the $\dot{\mu}$ operator coincides with the set $\mathbb{T}^{\mathbb{A}}$ of simple types.

7.4.2 Notation

(i) Just as with repeated λ-abstraction we write

$$\dot{\mu}\alpha_1 \cdots \alpha_n.A \triangleq (\dot{\mu}\alpha_1(\dot{\mu}\alpha_2 \cdots (\dot{\mu}\alpha_n(A))..)).$$

(ii) We assume that $\rightarrow$ takes precedence over $\dot{\mu}$, so that, for example, the type $\dot{\mu}\alpha.A \rightarrow B$ should be parsed as $\dot{\mu}\alpha.(A \rightarrow B)$.

According to the intuitive semantics of recursive types, a type expression of the form $\dot{\mu}\alpha.A$ should be regarded as the solution for α in the equation $\alpha = A$, and is then equivalent to the type expression $A[\alpha := \dot{\mu}\alpha.A]$.

In $\dot{\mu}\beta.A$ the operator $\dot{\mu}$ binds the variable β. We write FV(A) for the set of variables occurring free in A, and BV(A) for the set of variables occurring bound in A.

7.4.3 Notation

(i) The sets of variables occurring as *bound variables* or as *free variables* in the type $A \in \mathbb{T}^{\mathbb{A}}_{\dot{\mu}}$, written BV$(A)$, FV$(A)$, respectively, are defined recursively as

A	FV(A)	BV(A)
α	$\{\alpha\}$	$\emptyset$
$A \rightarrow B$	$\mathrm{FV}(A) \cup \mathrm{FV}(B)$	$\mathrm{BV}(A) \cup \mathrm{BV}(B)$
$\dot{\mu}\alpha.A_1$	$\mathrm{FV}(A_1) - \{\alpha\}$	$\mathrm{BV}(A_1) \cup \{\alpha\}$

(ii) If $\beta \notin \mathrm{FV}(A) \cup \mathrm{BV}(A)$ we write $\beta \notin A$.

Bound variables can be renamed by α-conversion: $\dot{\mu}\beta.A \equiv_\alpha \dot{\mu}\gamma.A[\beta := \gamma]$, provided that $\gamma \notin A$. From 7.4.23 on, we will consider types in $\mathbb{T}^{\mathbb{A}}_{\dot{\mu}}$ modulo α-convertibility, obtaining $\mathbb{T}^{\mathbb{A}}_{\mu}$. Towards this goal, items 7.4.1–7.4.22 are a preparation.

We will often assume that the names of bound and free variables in types are distinct: this can be easily obtained by a renaming of bound variables. In contrast with λ-terms we like to be explicit about this so-called α-conversion.

We will distinguish between 'naive' substitution $[\beta := A]_{\not\alpha}$ in which innocent free variables may be captured and ordinary 'smart' substitution $[\beta := A]$ that avoids this.

7.4.4 Definition Let $A, B \in \mathbb{T}_{\dot{\mu}}$.

(i) The *naive substitution* operator, written $A[\beta := B]_{\not\alpha}$, is defined in Fig. 7.8. The notation $A[\beta := B]_{\not\alpha}$ comes from Endrullis et al. (2011).
(ii) Ordinary *'smart' substitution*, written $A[\beta := B]$, which avoids the capturing of free variables ('dynamic binding') was defined by Curry as in Fig. 7.9, see B[1984], Definition C.1.

7.4.5 Lemma

(i) *If β' does not occur in A, then*

$$A[\beta := \beta'] \equiv A[\beta := \beta']_{\not\alpha}.$$

A	$A[\beta := B]_{\not\alpha}$
α	α, if $\alpha \neq \beta$,
β	B
$A_1 \to A_2$	$A_1[\beta := B]_{\not\alpha} \to A_2[\beta := B]_{\not\alpha}$
$\dot\mu\beta.A$	$\dot\mu\beta.A$
$\dot\mu\alpha.A$	$\dot\mu\alpha.(A[\beta := B]_{\not\alpha})$, if $\alpha \neq \beta$,

Figure 7.8

A	$A[\beta := B]$
α	α if $\alpha \neq \beta$
β	B
$A_1 \to A_2$	$A_1[\beta := B] \to A_2[\beta := B]$,
$\dot\mu\beta.A$	$\dot\mu\beta.A$
$\dot\mu\alpha.A_1$	$\dot\mu\alpha'.(A_1[\alpha := \alpha'][\beta := B])$, if $\alpha \neq \beta$, where $\alpha' = \alpha$ if $\beta \notin \mathrm{FV}(A_1)$ or $\alpha \notin \mathrm{FV}(B)$, else α' is the first variable in the sequence of type variables $\alpha_0, \alpha_1, \alpha_2, \ldots$ that is not in $\mathrm{FV}(A_1) \cup \mathrm{FV}(B)$.

Figure 7.9

(ii) *If $\beta \notin \mathrm{FV}(A)$, then*

$$A[\beta := B] \equiv A.$$

Proof (i) By induction on the structure of A. The interesting case is $A \equiv \dot\mu\gamma.C$, with $\gamma \not\equiv \beta$. Then

$$\begin{aligned}
(\dot\mu\gamma.C)[\beta := \beta'] &\equiv \dot\mu\gamma.C[\beta := \beta'], && \text{since } \gamma \not\equiv \beta', \text{ as } \beta' \notin (\dot\mu\gamma.C),\\
&\equiv \dot\mu\gamma.C[\beta := \beta']_{\not\alpha}, && \text{by the induction hypothesis,}\\
&\equiv (\dot\mu\gamma.C)[\beta := \beta']_{\not\alpha}, && \text{by Definition 7.4.4(i).}
\end{aligned}$$

(ii) Similarly, the interesting case being $A \equiv \dot\mu\gamma.C$, with $\gamma \not\equiv \beta$. Then

$$\begin{aligned}
(\dot\mu\gamma.C)[\beta := B] &\equiv \dot\mu\gamma.C[\beta := B], && \text{as } \beta \notin \mathrm{FV}(A) \ \&\ \beta \not\equiv \gamma, \text{ so } \beta \notin \mathrm{FV}(C),\\
&\equiv \dot\mu\gamma.C, && \text{by the induction hypothesis.} \ \blacksquare
\end{aligned}$$

7.4.6 Definition (α-conversion) On $\mathbb{T}_{\dot\mu}$ we define the notion of *α-reduction* and *α-conversion* via the contraction rule

$$\dot\mu\alpha.A \mapsto_\alpha \dot\mu\alpha'.A[\alpha := \alpha'], \text{ provided } \alpha' \notin \mathrm{FV}(A).$$

The relation $\Rightarrow_\alpha$ is the smallest compatible relation containing $\mapsto_\alpha$. The relation $\Rightarrow_\alpha^*$ is the transitive reflexive closure of $\Rightarrow_\alpha$. Finally $\equiv_\alpha$ is the smallest congruence containing $\mapsto_\alpha$.

For example $\dot{\mu}\alpha.\alpha \to \alpha \equiv_\alpha \dot{\mu}\beta.\beta \to \beta$. Also $\dot{\mu}\alpha.(\alpha \to \dot{\mu}\beta.\beta) \equiv_\alpha \dot{\mu}\beta.(\beta \to \dot{\mu}\beta.\beta)$.

7.4.7 Lemma

(i) *If* $A \Rightarrow_\alpha B$, *then* $B \Rightarrow_\alpha A$.
(ii) $A \equiv_\alpha B$ *implies* $A \Rightarrow_\alpha^* B$ & $B \Rightarrow_\alpha^* A$.

Proof (i) If $\dot{\mu}\alpha.A \Rightarrow_\alpha \dot{\mu}\alpha'.A[\alpha := \alpha']$, then $\alpha \notin \mathrm{FV}(A[\alpha := \alpha'])$, so that also

$$\dot{\mu}\alpha'.A[\alpha := \alpha'] \Rightarrow_\alpha \dot{\mu}\alpha.A[\alpha := \alpha'][\alpha' := \alpha] \equiv \dot{\mu}\alpha.A.$$

(ii) By (i). ■

7.4.8 Definition

(i) Define on $\mathbb{T}_{\dot{\mu}}$ a notion of $\dot{\mu}$-*reduction* via the contraction rule $\mapsto_{\dot{\mu}}$

$$\dot{\mu}\alpha.A \mapsto_{\dot{\mu}} A[\alpha := \dot{\mu}\alpha.A].$$

(ii) A $\dot{\mu}$-*redex* is of the form $\dot{\mu}\alpha.A$ and its *contraction* is $A[\alpha := \dot{\mu}\alpha.A]$.
(iii) The relation $\Rightarrow_{\dot{\mu}} \subseteq \mathbb{T}_{\dot{\mu}} \times \mathbb{T}_{\dot{\mu}}$ is the compatible closure of $\mapsto_{\dot{\mu}}$. That is,

$$\begin{array}{lcr} A \Rightarrow_{\dot{\mu}} A' & \Rightarrow & A \to B \Rightarrow_{\dot{\mu}} A' \to B \\ A \Rightarrow_{\dot{\mu}} A' & \Rightarrow & B \to A \Rightarrow_{\dot{\mu}} B \to A' \\ A \Rightarrow_{\dot{\mu}} A' & \Rightarrow & \dot{\mu}\alpha.A \Rightarrow_{\dot{\mu}} \dot{\mu}\alpha.A'. \end{array}$$

(iv) As usual $\Rightarrow_{\dot{\mu}}^n$ denotes reduction in n steps.
(v) The relation $\Rightarrow_{\dot{\mu}}^*$ is the reflexive and transitive closure of $\Rightarrow_{\dot{\mu}}$, i.e. 0 or more reduction steps.
(vi) The relation $\dot{\mu}$-*conversion*, written $=_{\dot{\mu}}$, is the conversion relation generated by $\dot{\mu}$-reduction, i.e. the smallest congruence relation containing $\mapsto_{\dot{\mu}}$.

7.4.9 Lemma *Let* $A, A', B \in \mathbb{T}_{\dot{\mu}}$. *Then*

(i) $A \Rightarrow_{\dot{\mu}} A' \;\Rightarrow\; A[\alpha := B] \Rightarrow_{\dot{\mu}} A'[\alpha := B]$.
(ii) $A \Rightarrow_{\dot{\mu}} A' \;\Rightarrow\; B[\alpha := A] \Rightarrow_{\dot{\mu}} B[\alpha := A']$.
(iii) *Both* (i) *and* (ii) *hold with* $\Rightarrow_{\dot{\mu}}$ *replaced by* $=_{\dot{\mu}}$.

Proof (i) By induction on the derivation of $A \Rightarrow_{\dot{\mu}} A'$.
(ii) By induction on the structure of B.
(iii) By (i) and (ii). ■

7.4.10 Lemma *Let* $A, A', B \in \mathbb{T}_{\dot{\mu}}$. *Then*

(i) $A \Rightarrow_\alpha A' \;\Rightarrow\; A[\alpha := B] \Rightarrow_\alpha^* A'[\alpha := B]$.
(ii) $A \Rightarrow_\alpha A' \;\Rightarrow\; B[\alpha := A] \Rightarrow_\alpha^* B[\alpha := A']$.
(iii) *Both* (i) *and* (ii) *hold with* $\Rightarrow_\alpha$ *and* ${\Rightarrow_\alpha}^*$ *replaced by* $\equiv_\alpha$.

Proof (i) By induction on the derivation of $A \Rightarrow_\alpha A'$.
(ii) By induction on the structure of B. (iii) By (i) and (ii). ■

7.4.11 Lemma (Substitution Lemma) *Let* $A \in \mathbb{T}_{\dot{\mu}}$. *Then for all* $B, C \in \mathbb{T}_{\dot{\mu}}$ *and type variables* β, γ *with* $\beta \not\equiv \gamma$ *and* $\beta \notin \mathrm{FV}(C)$ *one has*

$$A[\beta := B][\gamma := C] \equiv_\alpha A[\gamma := C][\beta := B[\gamma := C]]. \tag{7.6}$$

Writing $D^\gamma = D[\gamma := C]$, (7.6) *becomes*

$$(A[\beta := B])^\gamma \equiv_\alpha A^\gamma[\beta := B^\gamma].$$

Proof By induction on the number of symbols in A. The interesting cases are $A \equiv \alpha$ and $A \equiv \dot{\mu}\alpha.A_1$.

Case $A \equiv \alpha$. If $\alpha \notin \{\beta, \gamma\}$, then the result is trivial: $\alpha \equiv_\alpha \alpha$. If $\alpha \equiv \beta$, then (7.6) boils down to $B^\gamma \equiv_\alpha B^\gamma$. If $\alpha \equiv \gamma$, then, because of the assumption $\beta \notin \mathrm{FV}(C)$, (7.6) becomes $C \equiv_\alpha C$.

Case $A \equiv \dot{\mu}\alpha.A_1$. We must show

$$(\dot{\mu}\alpha.A_1)[\beta := B])^\gamma \equiv_\alpha (\dot{\mu}\alpha.A_1)^\gamma[\beta := B^\gamma]. \tag{7.7}$$

By Lemma 7.4.10(iii) it suffices to show (7.7) for an 'α-variant' $\dot{\mu}\alpha_2.A_2 \equiv_\alpha \dot{\mu}\alpha.A_1$. Take α_2 such that $\alpha_2 \notin \{\beta, \gamma\} \cup \mathrm{FV}(B) \cup \mathrm{FV}(C)$. Then, by the freshness of α_2,

$$\begin{array}{rcll}(\dot{\mu}\alpha_2.A_2)[\beta := B]^\gamma & \equiv & \dot{\mu}\alpha_2.(A_2[\beta := B])^\gamma, & \text{by Definition 7.4.4(ii)},\\ & \equiv_\alpha & \dot{\mu}\alpha_2.(A_2^\gamma)[\beta := B^\gamma], & \text{by the induction hypothesis},\\ & \equiv & (\dot{\mu}\alpha_2.A_2)^\gamma[\beta := B^\gamma]. \blacksquare & \end{array}$$

7.4.12 Lemma *Let $A, B \in \mathbb{T}_{\dot{\mu}}$ and $\alpha' \notin \mathrm{FV}(A)$. Then*

$$A[\alpha := \alpha'][\alpha' := B] \equiv_\alpha A[\alpha := B].$$

Proof By induction on the structure of A. We treat the case $A \equiv \dot{\mu}\beta.A_1$. As in the proof of the Substitution Lemma, 7.4.11, we may assume that $\beta \notin \{\alpha, \alpha'\} \cup \mathrm{FV}(B)$. Then, as before,

$$\begin{array}{rcll}(\dot{\mu}\beta.A_1)[\alpha := \alpha'][\alpha' := B] & \equiv & \dot{\mu}\beta.(A_1[\alpha := \alpha'][\alpha' := B]) & \\ & \equiv_\alpha & \dot{\mu}\beta.A_1[\alpha := B], & \text{by the induction hypothesis},\\ & \equiv & (\dot{\mu}\beta.A_1)[\alpha := B]. \blacksquare & \end{array}$$

Avoiding α-conversion on $\dot{\mu}$-types

As $\dot{\mu}$-types are built up from the variable binding $\dot{\mu}$-operator one has to take care that there will be no clash of variables during a $\dot{\mu}$-reduction. This is similar to the situation with untyped λ-terms and $\boldsymbol{\beta}$-reduction. An important difference between those λ-terms and $\dot{\mu}$-types is the possibility of choosing a correct α-variant of the $\dot{\mu}$-types that remains correct after arbitrary reduction steps. In the (un)typed lambda calculus this is not possible. The term

$$(\lambda x.xx)(\lambda yz.yz) \in \Lambda^{\emptyset}$$

has its bound variables maximally different. But after some reduction steps a clash of variables occurs:

$$\begin{array}{rl}(\lambda x.xx)(\lambda yz.yz) & \to_{\boldsymbol{\beta}} (\lambda yz.yz)(\lambda yz.yz)\\ & \to_{\boldsymbol{\beta}} \lambda z.(\lambda yz.yz)z\\ & \not\to_{\boldsymbol{\beta}} (\lambda z.(\lambda z.zz)) \equiv \lambda zz.zz,\end{array}$$

which should have been $\lambda zz'.zz'$. In order to avoid clashes in the untyped lambda calculus one therefore should be constantly alert and apply α-conversion whenever needed. This example can be modified so that a typable λ-term results: do Exercise 7.2.

In B[1984] this necessary hygienic discipline is somewhat swept under the carpet via the so-called 'variable convention': choosing the names of bound variables maximally fresh. The belief that this is sound came from the calculus with nameless binders in de Bruijn (1972), which is implemented for converting lambda terms in the proof-checker Agda. The variable convention has been cleverly implemented in the Nominal package for Isabelle/HOL, see Urban and Tasson (2005), with the goal of reasoning formally about binding. They show that the variable convention is compatible with structural inductions over terms with binders. However, Urban et al. (2007) also show that care needs to be taken in inductions over recursively defined predicates: if a bound variable occurs free in a conclusion of a rule, then the variable convention can easily lead to unsound reasoning.

7.4.13 Definition Let $A \in \mathbb{T}_{\dot{\mu}}$.

(i) The type A is called *safe* if for all subtypes $\dot{\mu}\beta.B$ of A the one step reduction

$$\dot{\mu}\beta.B \Rightarrow_{\dot{\mu}} B[\beta := \dot{\mu}\beta.B]_{\not\alpha}$$

does not result in binding a free variable of $\dot{\mu}\beta.B$ in $B[\beta := \dot{\mu}\beta.B]_{\not\alpha}$. One also could state that for all subtypes $\dot{\mu}\beta.B$ of A

$$B[\beta := \dot{\mu}\beta.B] \equiv B[\beta := \dot{\mu}\beta.B]_{\not\alpha}.$$

(ii) We say that A is *forever safe* if

$$A \Rightarrow_{\dot{\mu}}^{*} B \;\Rightarrow\; B \text{ is safe, for all } B \in \mathbb{T}_{\dot{\mu}}.$$

7.4.14 Example

(i) $\dot{\mu}\alpha.\alpha \to \beta$ is safe.
(ii) $\dot{\mu}\alpha.(\beta \to \dot{\mu}\beta.(\alpha \to \beta))$ is not safe; 'contracting $\dot{\mu}\alpha$' leads to a clash.
(iii) $\dot{\mu}\alpha.(\beta \to \dot{\mu}\gamma.(\alpha \to \gamma))$ is safe.

As with (un)typed λ-terms one has that every $A \in \mathbb{T}_{\dot{\mu}}$ has an α-variant that is safe, by renaming bound variables by fresh bound variables. We will show something better: the existence of an α-variant that is safe and remains so. In general being safe does not guarantee remaining safe after $\dot{\mu}$-reduction.

7.4.15 Definition Let $P \subseteq \mathbb{T}_{\dot{\mu}}$ be a predicate on types.

(i) We say P is *α-flexible* if

$$\forall A \in \mathbb{T}_{\dot{\mu}} \exists A' \in \mathbb{T}_{\dot{\mu}}.[A \equiv_{\alpha} A' \;\&\; P(A')].$$

(ii) We say P is *$\dot\mu$-invariant* if for all $A, B \in \mathbb{T}_{\dot\mu}$

$$A \Rightarrow_{\dot\mu} B \;\&\; P(A) \;\Rightarrow\; P(B).$$

(iii) We say P is *protective* if for all $A \in \mathbb{T}_{\dot\mu}$

$$P(A) \;\Rightarrow\; A \text{ is safe.}$$

(iv) We call P *α-avoiding* if it is α-flexible, protective, and $\dot\mu$-invariant.

Note that if a predicate $P \subseteq \mathbb{T}_{\dot\mu}$ is protective and invariant, then $P(A)$ implies that A is forever safe. It would be nice to have an α-avoiding predicate: then every type A could be replaced by an α-equivalent one that is forever safe.

7.4.16 Remark Let us try several predicates P_i on $\mathbb{T}_{\dot\mu}$.

(i) $P_1(A) \overset{\Delta}{\iff} \mathrm{FV}(A) \cap \mathrm{BV}(A) = \emptyset$.
(ii) $P_2(A) \overset{\Delta}{\iff} P_1(B)$ for all subtypes B of A.
(iii) $P_3(A) \overset{\Delta}{\iff} P_1(A)$ and different occurrences $\dot\mu$ in A bind different type variables.
(iv) $P_4(A) \overset{\Delta}{\iff} A$ is safe.
(v) $P_5(A) \overset{\Delta}{\iff} A$ is forever safe.

Unfortunately none of these predicates is α-avoiding.

7.4.17 Lemma

(i) *P_1 is α-flexible, $\dot\mu$-invariant, but not protective.*
(ii) *P_2 is α-flexible, protective, but not $\dot\mu$-invariant.*
(iii) *P_3 is α-flexible, protective, but not $\dot\mu$-invariant.*
(iv) *P_4 is α-flexible, protective, but not $\dot\mu$-invariant.*
(v) *P_5 is protective and $\dot\mu$-invariant; it is α-avoiding iff there exists an α-avoiding predicate.*

Proof For the proof we use some convenient notation of van Oostrom: the binary operation '$\to$' is denoted by an implicit 'multiplication', with association to the right. For example, $\dot\mu\alpha.\alpha\beta\alpha$ denotes $\dot\mu\alpha.\alpha \to (\beta \to \alpha)$. (i) Define $A_1 \equiv \dot\mu\alpha.B[\alpha]$, with $B[\alpha] \equiv \dot\mu\beta.\alpha\beta(\dot\mu\alpha.\beta)$. Then $P_1(A_1)$; but by contracting $B[\alpha]$ a capture of $\alpha \in \mathrm{FV}(B[\alpha])$ occurs:

$$B[\alpha] \Rightarrow_{\dot\mu} \alpha B[\alpha](\dot\mu\alpha.B[\alpha]),$$

Hence A_1 is not safe, and therefore P_1 is not protective.

(ii) Clearly P_2 is α-flexible and protective. Define $A_2 \equiv \dot{\mu}\alpha\beta.\alpha\beta$. Then $P_2(A_2)$; but

$$A_2 \Rightarrow_{\dot{\mu}} \dot{\mu}\beta.A_2\beta \equiv A_2'.$$

Now $P_2(A_2')$ fails as its subterm $A_2\beta$ contains β both as free and as bound variable. Therefore P_2 is not $\dot{\mu}$-invariant.

(iii) Again P_3 is α-flexible and protective. Define $A_3 = \dot{\mu}\alpha.\alpha\alpha$. Then $P_3(A_3)$, but

$$A_3 \Rightarrow_{\dot{\mu}} A_3A_3$$

and $P(A_3A_3)$ does not hold. Therefore P_3 is not $\dot{\mu}$-invariant.

(iv) [van Oostrom.] Once again P_4 is α-flexible and protective. In order to show it is not $\dot{\mu}$-invariant, define $A_4 \equiv \dot{\mu}\alpha\beta.\alpha(\dot{\mu}\gamma.\beta(\dot{\mu}\alpha.\gamma))$. Then $P_4(A_4)$ and

$$\begin{array}{rlr} A_4 & \Rightarrow_{\dot{\mu}} \dot{\mu}\alpha\underline{\dot{\mu}\beta.\alpha(\beta(\dot{\mu}\alpha\gamma.\beta(\dot{\mu}\alpha.\gamma)))} & \equiv \dot{\mu}\alpha.B[\alpha]. \\ & \Rightarrow_{\dot{\mu}} \dot{\mu}\alpha.\alpha(B[\alpha](\dot{\mu}\alpha\gamma.B[\alpha](\dot{\mu}\alpha.\gamma)), & \text{a variable capture.} \end{array}$$

Therefore $\dot{\mu}\alpha\beta.B[\alpha]$ is not safe, and hence P_4 not $\dot{\mu}$-invariant.

(v) The predicate P_5 is safe and $\dot{\mu}$-invariant by definition. If there is an α-avoiding predicate P, then P is protective, hence $P \subseteq P_5$; but P is also α-flexible and therefore for all $A \in \mathbb{T}_{\dot{\mu}}$ there is an α-variant $A' \in P \subseteq P_5$. The converse is trivial. ∎

After all these examples we now show that there does exist an α-avoiding predicate on $\mathbb{T}_{\dot{\mu}}$. The result is due to van Oostrom and was inspired by Melliès (1996).

7.4.18 Definition Let $A \in \mathbb{T}_{\dot{\mu}}$.

(i) An occurrence of $\dot{\mu}\alpha$ in A *binds* an occurrence of the variable α if $\dot{\mu}\alpha$ occurs as $\dot{\mu}\alpha.B$ with $\alpha \in \mathrm{FV}(B)$, and α occurs in the scope of $\dot{\mu}\alpha$.

(ii) An occurrence of $\dot{\mu}\alpha$ in A *captures* an occurrence of the variable $\beta \not\equiv \alpha$ if β freely occurs in the scope of $\dot{\mu}\alpha$.

(iii) For subterm occurrences o_1, o_2 of A of the form $\dot{\mu}\alpha$ or α, with α any type variable, define the relation

$$o_1 \prec o_2 \Leftrightarrow o_1 \text{ is captured by } o_2 \text{ or } o_1 \text{ binds } o_2.$$

(iv) An occurrence α is *self-capturing* if

$$\alpha \prec^* \dot{\mu}\alpha.$$

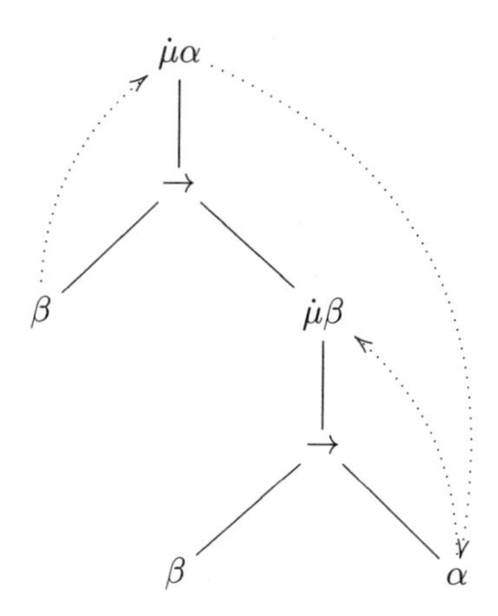

Figure 7.10

7.4.19 Example Let $A \equiv \dot{\mu}\alpha.\beta \to (\dot{\mu}\beta.\beta \to \alpha)$. In Fig. 7.10 one sees that the 'highest' occurence in A of the variable β is self-capturing, but not the other occurrence.

7.4.20 Example In Fig. 7.11, where we write $\dot{\mu}\alpha.A \to B$ as

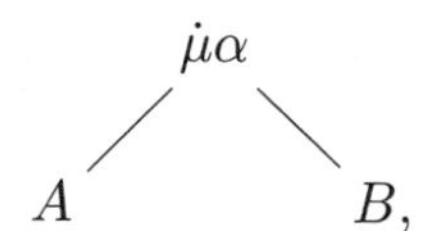

one has $\alpha_0 \prec^* \dot{\mu}\alpha_k$ for all $k \geq 1$. Convince yourself that if there is self-capturing, e.g. $\alpha_4 = \alpha_0$, then a naive $\dot{\mu}$-reduction that does not change names of bound variables may result in a variable clash. The diagram shows a typical 'topology' causing the phenomenon of self-capturing.

7.4.21 Proposition (van Oostrom (2007)) *Define $P \subseteq \mathbb{T}_{\dot{\mu}}$ by*

$$P(A) \overset{\Delta}{\iff} \textit{for no variable } \beta \textit{ there is a self-capturing occurrence in } A.$$

Then P is α-avoiding.

Proof We will show that P is: (i) α-flexible; (ii) protecting; and (iii) $\dot{\mu}$-invariant. (i) Renaming all binders in A in such a way that they are fresh and pairwise distinct we obtain $A' \equiv_\alpha A$ with $P(A')$.

(ii) We have to show that

$$P(A) \Rightarrow A \text{ is safe},$$

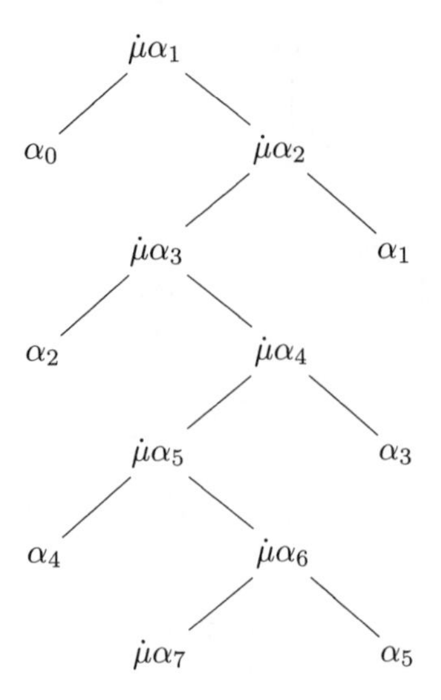

Figure 7.11

for which it suffices that for all subtypes $\dot\mu\beta.B$ of A

$$P(\dot\mu\beta.B) \;\Rightarrow\; B[\beta := \dot\mu\beta.B] = B[\beta := \dot\mu\beta.B]_{\not\alpha}. \tag{7.8}$$

It is not clear how to prove this directly. One can prove a stronger statement. Let $Q(\beta, B, C)$ be the statement

$$Q(\beta, B, C) \overset{\Delta}{\iff} [\,\beta \in \mathrm{FV}(B) \;\Rightarrow\; \forall\gamma \in \mathrm{BV}(B).[\,\beta \prec \dot\mu\gamma \;\Rightarrow\; \gamma \notin \mathrm{FV}(C)]].$$

This tells us that if β occurs in $\mathrm{FV}(B)$, an occurence which is captured by an occurrence of $\dot\mu\gamma$ in B, then the variable γ does not occur in $\mathrm{FV}(C)$. The stronger statement to be proved is

$$\forall C.[\,Q(\beta, B, C) \;\Rightarrow\; B[\beta := C] = B[\beta := C]_{\not\alpha}\,]. \tag{7.9}$$

Now we show (7.9) by induction on the structure of B. When B is a variable or of the form $B = B_1 \to B_2$ this is easy, noting that

$$B' \subseteq B \;\Rightarrow\; Q(\beta, B, C) \;\Rightarrow\; Q(\beta, B', C).$$

If $B = \dot\mu\gamma.B'$, then we must show

$$Q(\beta, \dot\mu\gamma.B', C) \;\Rightarrow\; (\dot\mu\gamma.B')[\beta := C] = (\dot\mu\gamma.B')[\beta := C]_{\not\alpha}.$$

If $\beta \notin \mathrm{FV}(B')$, then this is trivial. Otherwise $\beta \in \mathrm{FV}(B')$, hence $\beta \prec \dot\mu\gamma$. By the assumption $Q(\beta, \dot{\mu}\gamma.B', C)$ it follows that $\gamma \notin \mathrm{FV}(C)$. Then

$$\begin{array}{rcll}(\dot{\mu}\gamma.B')[\beta := C] & = & \dot{\mu}\gamma.(B'[\beta := C]), & \text{since } \dot{\mu}\gamma \text{ cannot bind a variable in } C, \\ & = & \dot{\mu}\gamma.(B'[\beta := C]_{\not\alpha}), & \text{by the induction hypothesis,} \\ & = & (\dot{\mu}\gamma.B')[\beta := C]_{\not\alpha}, & \text{by definition of } [\beta := C]_{\not\alpha}.\end{array}$$

This establishes (7.9).

Now for $(\dot{\mu}\beta.B) \subseteq A$ we have

$$P(\dot{\mu}\beta.B) \;\Rightarrow\; Q(\beta, B, \dot{\mu}\beta.B). \tag{7.10}$$

Indeed, if $\beta \in \mathrm{FV}(B)$ and γ occurs both in $\mathrm{BV}(B)$ and in $\mathrm{FV}(\dot{\mu}\beta.B)$, two different occurrences, then $\beta \prec \dot{\mu}\gamma$ is impossible since it implies $\gamma \prec \dot{\mu}\beta \prec \beta \prec \dot{\mu}\gamma$, i.e. γ is self-capturing, contradicting $P(\dot{\mu}\beta.B)$.

Now by (7.9) and (7.10) we obtain (7.8).

(iii) Suppose $A \Rightarrow_{\dot{\mu}} A'$ and $P(A)$, in order to show $P(A')$. Let

$$A = C[\dot{\mu}.\beta.B] \Rightarrow_{\dot{\mu}} C[B[\beta := \dot{\mu}\beta.B]],$$

where $C[\,]$ is some context (with possible binding effects, like $C[\,] = \dot{\mu}\epsilon.[\,]$). It suffices to show that if γ is a self-capturing occurrence of a variable in A', then also A contains a self-capturing occurrence of γ. Let σ_1', σ_2' be occurrences of an α or $\dot{\mu}\alpha$ in A'. Let σ_1, σ_2 be the unique occurrences in A such that σ_i' is the residual, see Terese (2003), of σ_i (for $1 \le i \le 2$). The symbols ($\alpha, \dot{\mu}\alpha$, or $\to$) written at occurrences σ_i and σ_i' are the same. We claim that they satisfy

$$\sigma_1' \prec \sigma_2' \;\Rightarrow\; \sigma_1 \prec^* \sigma_2. \tag{7.11}$$

Let $\{\sigma_1', \sigma_2'\} = \{\gamma, \dot{\mu}\delta\}$ and suppose $\sigma_1' \prec \sigma_2'$.

Case 1. The occurrences σ_1', σ_2' in A' both occur in the same 'component', i.e. both in $C[\,]$, in the 'body' B, or in (a copy of) $B[\beta := \dot{\mu}\beta.B]$, respectively. Then also $\sigma_1 \prec \sigma_2$.

Case 2. Now σ_1' occurs on $C[\;]$, σ_2' in B. Then $\gamma \neq \beta$, because otherwise $\dot{\mu}\beta.B$ would have been substituted for it. Hence $\sigma_1 \prec \sigma_2$.

Case 3. σ_1' occurs on $C[\;]$, σ_2' in (a copy of) $\dot{\mu}\beta.B$. Now $\gamma \neq \beta$ holds, because otherwise $\sigma_1' \not\prec \sigma_2'$. If $\gamma = \delta$, then $\sigma_1' = \dot{\mu}\gamma, \sigma_2' = \gamma$ and we also have $\sigma_1 \prec \sigma_2$. If $\gamma \neq \delta$, then $\sigma_1' = \gamma, \sigma_2' = \dot{\mu}\delta$ and we have $\sigma_1 \prec \sigma_2$.

Case 4. σ_1' occurs on B, σ_2' in (a copy of) $\dot{\mu}\beta.B$. Then again $\gamma \neq \beta$, for the same reason as in Case 3. If $\gamma = \delta$, then $\sigma_1' = \dot{\mu}\gamma, \sigma_2' = \gamma$ and we have $\sigma_1 \prec \sigma_2$. If $\gamma \neq \delta$, then $\sigma_1' = \gamma, \sigma_2' = \dot{\mu}\delta$ and we have $\sigma_1 \prec^* \sigma_2$ (this is the only place where we use $\prec^*$), because

$$\gamma \prec \dot{\mu}\beta \prec \beta \prec \dot{\mu}\delta.$$

This establishes (7.11). Now suppose $\neg P(A')$. Then A' contains a self-capturing occurrence $\gamma \prec^* \dot{\mu}\gamma$. By (7.11) and transitivity it follows that also A contains such an occurrence. Therefore indeed $\neg P(A)$. ■

At the end of the day for every $A \in \mathbb{T}_{\dot{\mu}}$ an α-variant that can be forever $\dot{\mu}$-reduced naively can be easily found.

7.4.22 Corollary *Let $A \in \mathbb{T}_{\dot{\mu}}$. Suppose that all binders in A are pairwise distinct and different from the free variables in A. Then A is forever safe.*

Proof As in (i) in the above proof the condition implies $P(A)$. Therefore A is forever safe. ■

Identifying types equal up to renaming

7.4.23 Definition

(i) The set of recursive types, written $\mathbb{T}_{\mu}$, is defined as

$$\mathbb{T}_{\mu}^{\mathbb{A}} \triangleq \mathbb{T}_{\dot{\mu}}^{\mathbb{A}}/\equiv_{\alpha},$$

i.e. $\mathbb{T}_{\mu}^{\mathbb{A}} \triangleq \{[A]_{\alpha} \mid A \in \mathbb{T}_{\dot{\mu}}^{\mathbb{A}}\}$, where $[A]_{\alpha} \triangleq \{B \in \mathbb{T}_{\dot{\mu}}^{\mathbb{A}} \mid B \equiv_{\alpha} A\}$.

(ii) On $\mathbb{T}_{\mu}^{\mathbb{A}}$ one defines

$$[A]_{\alpha} \to [B]_{\alpha} \triangleq [A \to B]_{\alpha},$$
$$\mu\beta.[A]_{\alpha} \triangleq [\dot{\mu}\beta.A]_{\alpha}.$$

(iii) For elements of $\mathbb{T}_{\mu}^{\mathbb{A}}$ instead of $[A]_{\alpha}$ we simply write A.

Note that on $\mathbb{T}_{\mu}^{\mathbb{A}}$ the operations $\to$ and $\mu\beta$ are well-defined, i.e. independent of the choice of representative in the α-equivalence classes.

Definition 7.4.4(ii) of substitution for $\mathbb{T}_{\dot{\mu}}$ can immediately be lifted to $\mathbb{T}_{\mu}$ as follows.

7.4.24 Definition For $A, B \in \mathbb{T}_{\mu}^{\mathbb{A}}$ substitution is defined as

A	$A[\beta := B]$	
α	α	if $\alpha \neq \beta$
β	B	
$A_1 \to A_2$	$A_1[\beta := B] \to A_2[\beta := B]$,	
$\mu\alpha.A_1$	$\mu\alpha.(A_1[\beta := B])$,	if $\alpha \neq \beta$

Using the variable convention we choose $\alpha \neq \beta$, $\alpha \notin B$ for the case $A \equiv \mu\alpha.A_1$.

$$\begin{array}{lll}
\text{(ident)} & \mathcal{H} \vdash A = A & \\
\text{(symm)} & \dfrac{\mathcal{H} \vdash A = B}{\mathcal{H} \vdash B = A} & \\
\text{(trans)} & \dfrac{\mathcal{H} \vdash A = B \quad \mathcal{H} \vdash B = C}{\mathcal{H} \vdash A = C} & \\
\text{(axiom)} & \mathcal{H} \vdash A = B, & \text{if } (A = B) \in \mathcal{H}, \\
(\mu\text{-eq}) & \mathcal{H} \vdash \mu\alpha.A = A[\alpha := \mu\alpha.A] & \\
(\mu\text{-cong}) & \dfrac{\mathcal{H} \vdash A = A'}{\mathcal{H} \vdash \mu\alpha.A = \mu\alpha.A'}, & \text{if } \alpha \text{ not free in } \mathcal{H}, \\
(\to\text{-cong}) & \dfrac{\mathcal{H} \vdash A = A' \quad \mathcal{H} \vdash B = B'}{\mathcal{H} \vdash A \to B = A' \to B'} &
\end{array}$$

Figure 7.12 The system of equational logic (μ)

7.4.25 Remark The above definition of substitution is correct, i.e. it is independent of the choice of representatives, and the variable convention makes sense, as follows immediately from Lemma 7.4.10(iii).

7.4.26 Lemma (Substitution Lemma) *Let $A \in \mathbb{T}_\mu$. Then for all $B, C \in \mathbb{T}_\mu$ and type variables β, γ with $\beta \not\equiv \gamma$ and $\beta \notin \mathrm{FV}(C)$ one has*

$$A[\beta := B][\gamma := C] \equiv A[\gamma := C][\beta := B[\gamma := C]]. \tag{7.12}$$

Writing $D^\gamma = D[\gamma := C]$, (7.12) becomes

$$(A[\beta := B])^\gamma \equiv A^\gamma[\beta := B^\gamma].$$

Proof Directly from Lemma 7.4.11. ■

7.4.27 Definition (Weak equivalence of recursive types)

(i) The equational theory (μ) is defined in Fig. 7.12. The variable $\mathcal{H}$ stands for a set of equations between two elements of $\mathbb{T}_\mu$. We say that $A, B \in \mathbb{T}_\mu$ are *(weakly) equivalent*, written $A =_\mu B$, if $\vdash_\mu A = B$, i.e. $\vdash A = B$ is provable in (μ) from $\mathcal{H} = \emptyset$.

(ii) We will often identify $\mathbb{T}_\mu$ with the (syntactic) type algebra $\mathbb{T}_\mu/{=_\mu}$, if there is little danger of confusion.

7.4.28 Remark Rule (μ-eq) is well-defined, as follows immediately from Lemma 7.4.10(iii).

(axiom)	$\Gamma \vdash x : A \quad \text{if } (x{:}A) \in \Gamma$
$(\to \text{E})$	$\dfrac{\Gamma \vdash M : A \to B \qquad \Gamma \vdash N : A}{\Gamma \vdash (MN) : B}$
$(\to \text{I})$	$\dfrac{\Gamma, x{:}A \vdash M : B}{\Gamma \vdash (\lambda x.M) : (A \to B)}$
(equal)	$\dfrac{\Gamma \vdash M : A \qquad A =_{\mu} B}{\Gamma \vdash M : B}$

Figure 7.13 The system $\lambda_{=}^{\mathbb{T}^!_{\mu}}$

7.4.29 Example Let $B \triangleq \mu\beta.A \to \beta$, with $\beta \notin \text{FV}(A)$. Then we have

$$B =_{\mu} A \to B =_{\mu} A \to A \to B =_{\mu} \cdots .$$

Now let $B' \triangleq \mu\beta.(A \to A \to \beta)$. Then

$$B' =_{\mu} A \to A \to B' =_{\mu} \cdots .$$

It is easy to see that $B \neq_{\mu} B'$ (see also Section 8.2). Indeed B and B' are the μ-types solving, respectively, the type equations in $\mathcal{R}_0$ and $\mathcal{R}_1$ of Example 7.3.20.

The structure $\langle \mathbb{T}_{\mu}^{\mathbb{A}}, \to \rangle$ is a type algebra and therefore by Definition 7.1.3 there is a corresponding notion of type assignment $\vdash_{\mathbb{T}_{\mu}^{\mathbb{A}}}$. In fact one of the reasons to have introduced the notion of type algebra is to cover $\vdash_{\mathbb{T}_{\mu}^{\mathbb{A}}}$ as a special case. Without introducing $\langle \mathbb{T}_{\mu}^{\mathbb{A}}, \to \rangle$ as type algebra, its essence is treated as follows.

7.4.30 Definition Let $\mathbb{T}^!_{\mu}$ be $\mathbb{T}_{\mu}$ not modulo $=_{\mu}$. For $M \in \Lambda$ one can define the system $\lambda_{=}^{\mathbb{T}^!_{\mu}}$. For $A \in \mathbb{T}^!_{\mu}$ and Γ with types from $\mathbb{T}^!_{\mu}$ the notion $\vdash_{\mathbb{T}^!_{\mu}}$ as in Fig 7.13.

7.4.31 Lemma *For $B, A_1, \ldots, A_n \in \mathbb{T}^!_{\mu}$ one has*

$$x_1{:}A_1, \ldots, x_n{:}A_n \vdash_{\mathbb{T}^!_{\mu}} M : B \Leftrightarrow x_1{:}[A_1]_{\mu}, \ldots, x_n{:}[A_n]_{\mu} \vdash_{\mathbb{T}_{\mu}} M : [B].$$

Proof By induction on derivations. For $\Leftarrow$ use the strengthening

$$x_1{:}[A_1]_{\mu}, \ldots, x_n{:}[A_n]_{\mu} \vdash_{\mathbb{T}_{\mu}} M : [A] \Rightarrow$$
$$\exists \vec{A}', B' \in \mathbb{T}^!_{\mu}.[x_1{:}A'_1, \ldots, x_n{:}A'_n \vdash_{\mathbb{T}^!_{\mu}} M : B' \ \&\ [A'_i] = [A_i], [B'] = [B]]. \blacksquare$$

7.4.32 Definition

(i) Define on $\mathbb{T}_\mu$ a notion of *μ-reduction* via the contraction rule $\mapsto_\mu$

$$\mu\alpha.A \mapsto_\mu A[\alpha := \mu\alpha.A].$$

We say that *unfolding* is the operation consisting of replacing $\mu\alpha.A$ by $A[\alpha := \mu\alpha.A]$ and call *folding* its inverse.

(ii) A *μ-redex* is of the form $\mu\alpha.A$ and its *contraction* is $A[\alpha := \mu\alpha.A]$.

(iii) The relation $\Rightarrow_\mu \subseteq \mathbb{T}_\mu \times \mathbb{T}_\mu$ is the compatible closure of $\mapsto_\mu$. That is,

$$\begin{aligned} A \Rightarrow_\mu A' &\Rightarrow A \to B \Rightarrow_\mu A' \to B; \\ A \Rightarrow_\mu A' &\Rightarrow B \to A \Rightarrow_\mu B \to A'; \\ A \Rightarrow_\mu A' &\Rightarrow \mu\alpha.A \Rightarrow_\mu \mu\alpha.A'. \end{aligned}$$

(iv) As usual $\Rightarrow_\mu^n$ denotes reduction in n steps.

(v) The relation $\Rightarrow_\mu^*$ is the reflexive and transitive closure of $\Rightarrow_\mu$, i.e. 0 or more reduction steps.

(vi) Finally $=_\mu$ *μ-conversion* is the conversion relation generated by μ-reduction, i.e. the smallest congruence relation containing $\mapsto_\mu$. We will see in Proposition 7.4.33(i) that this overloading is justified, as it is the same as the relation in Definition 7.4.27.

This turns $\mathbb{T}_\mu$ into a *Combinatory Reduction System* (CRS), written $\langle \mathbb{T}_\mu, \Rightarrow_\mu \rangle$, which generates $=_\mu$ as its convertibility relation. For an introduction to term rewriting systems (TRS) and CRS see e.g. Klop (1992), Klop et al. (1993) and Terese (2003).

7.4.33 Proposition

(i) *Convertibility corresponding to $\Rightarrow_\mu$ is $=_\mu$.*

(ii) *The reduction relation $\Rightarrow_\mu$ on $\mathbb{T}_\mu$ is Church–Rosser.*

Proof (i) By definition of $=_\mu$; do Exercise 8.2.

(ii) The notion of reduction $\Rightarrow_\mu$ induces on μ-types an orthogonal combinatory reduction system, see Terese (2003), Chapter 11. By Theorem 11.6.19 of that book it is Church–Rosser. ∎

Therefore two types in $\mathbb{T}_\mu$ are weakly equivalent iff they can be transformed one into the other by a finite number of applications of folding and unfolding, possibly on subexpressions.

Most μ-types can be unfolded to a form with main constructor (a type variable or $\to$) at top level, unless they belong to the following pathological subset of $\mathbb{T}_\mu$.

7.4.34 Definition Let $A \in \mathbb{T}_\mu$.

(i) We say A is *circular* iff $A \equiv \mu\beta_1 \cdots \beta_n.\beta_i$ for some $1 \leq i \leq n$.
(ii) The most typical circular type is $\bullet \triangleq \mu\alpha.\alpha$. Following Ariola and Klop (1996) the symbol '$\bullet$' is called 'the blackhole'.
(iii) We say A is *contractive in* α if $\mu\alpha.A$ is not circular.

Circular types correspond to non-proper simultaneous recursions.

7.4.35 Remark In Definition 7.4.34(i) we mean in fact that an element of $\mathbb{T}_\mu$ is circular if it is of the form $A = [\dot{\mu}\beta_1 \cdots \beta_n.\beta_i]_\alpha$. Suppose that also $A = [B]_\alpha$, for some $B \in \mathbb{T}_{\dot{\mu}}$. Then by Lemma 7.4.7(ii) one has $\dot{\mu}\beta_1 \cdots \beta_n.\beta_i \Rightarrow_\alpha^* B$. It follows that $B \equiv \dot{\mu}\gamma_1 \cdots \gamma_m.\gamma_j$, with $1 \leq j \leq m$. Therefore each representative of A is of the form $\dot{\mu}\beta_1 \cdots \beta_n.\beta_i$. In particular it is decidable whether or not A is circular.

7.4.36 Lemma *Let A be a circular type. Then $A =_\mu \mu\alpha.\alpha \equiv \bullet$. Therefore, the circular types form an equivalence class in $\mathbb{T}_\mu$.*

Proof Let $A \equiv \mu\beta_1 \cdots \beta_n.\beta_i$. Then

$$\begin{aligned} A &\Rightarrow_\mu^{n-i} \mu\beta_1 \cdots \beta_i.\alpha_i, \\ &\Rightarrow_\mu^{i-1} \mu\beta_i.\beta_i. \end{aligned}$$

Therefore $A \Rightarrow_\mu \mu\alpha.\alpha$. ∎

7.4.37 Lemma *Let $A \in \mathbb{T}_\mu$. One has exactly one of the following cases:*

$$\begin{aligned} A &\mapsto_\mu^* \beta; \\ A &\mapsto_\mu^* (B \to C); \\ A & \textit{ is circular.} \end{aligned}$$

Moreover, β and $B \to C$ are unique.

Proof Write $A \equiv \mu\beta_1 \cdots \beta_n.A_1$, with $A_1 \not\equiv \mu\gamma.A_2$. If $A_1 \in \{\beta_1 \cdots \beta_n\}$, then A is circular. If $A_1 \equiv \beta \notin \{\beta_1 \cdots \beta_n\}$, then $A \mapsto_\mu^n \beta$. If $A_1 \equiv B_1 \to C_1$, then there exist unique B, C such that $A \mapsto_\mu^n (B \to C)$. ∎

7.4.38 Definition Let $A \in \mathbb{T}_\mu$.

(i) [Grabmayer (2007)] The *lead symbol* of A, written $\text{ls}(A)$, is defined as

$$\begin{aligned} \text{ls}(A) &\triangleq \alpha, && \text{if } A \mapsto_\mu^* \alpha; \\ \text{ls}(A) &\triangleq \to, && \text{if } A \mapsto_\mu^* (B \to C); \\ \text{ls}(A) &\triangleq \bullet, && \text{if } A \text{ is circular.} \end{aligned}$$

(ii) The *principal reduced form* A' of A is defined as

$$\begin{array}{lcll} A' & \equiv & \alpha, & \text{if } A \mapsto_{\mu}^{*} \alpha; \\ A' & \equiv & B \to C, & \text{if } A \mapsto_{\mu}^{*} (B \to C); \\ A' & \equiv & \bullet, & \text{if } A \text{ is circular.} \end{array}$$

(iii) In untyped λ-calculus terms can have several head normal forms but only one principal head normal form. One speaks about the *principal head normal form.* Similarly one can speak about a *reduced form* of A, obtained by μ-reduction, contrasting it with the unique principal reduced form, obtained by μ-contraction when A is not circular.

The type $B \equiv \mu\beta.A \to \beta$ has $A \to B$ as reduced form.

7.4.39 Lemma

(i) *If A has $B \to C$ and $B' \to C'$ as reduced forms, then $B =_{\mu} B'$ and $C =_{\mu} C'$.*
(ii) *If A has α and α' as reduced forms, then $\alpha \equiv \alpha'$.*

Proof (i), (ii) By the Church–Rosser theorem for $\Rightarrow_{\mu}$. ■

7.4.40 Lemma

(i) *Let $\alpha, \beta \in \mathbb{A}$ be different. Then $[\alpha]_{\mu} \neq [\beta]_{\mu}$.*
(ii) *The prime elements of $\mathbb{T}_{\mu} = \mathbb{T}_{\mu}^{\mathbb{A}}$ are given by*

$$||\mathbb{T}_{\mu}|| = \{[\alpha]_{\mu} \mid \alpha \in \mathbb{A}\} \cup \{[\bullet]\}$$

(iii) $\mathbb{T}^{\mathbb{A}} \hookrightarrow \mathbb{T}_{\mu}^{\mathbb{A}}$.

Proof (i) By Proposition 7.4.33.
(ii) By Lemma 7.4.37.
(iii) One has

$$\begin{array}{lcll} \mathbb{T}^{\mathbb{A}} & \hookrightarrow & \mathbb{T}^{||\mathbb{T}_{\mu}||}, & \text{by (i) and (ii),} \\ & \hookrightarrow & \mathbb{T}_{\mu}, & \text{by Lemma 7.2.13(ii);} \end{array}$$

the necessary condition that $\mathbb{T}_{\mu}$ is invertible will be proved in Chapter 8. ■

7.4.41 Example The deductions in Proposition 7.1.4 can be obtained in $\lambda\mu$ by taking $A \in \mathbb{T}_{\mu}$ arbitrarily and B as $\mu t.t \to A$. Then one has $B = \mu t.t \to A =_{\mu} (\mu t.t \to A) \to A = B \to A$.

7.4.42 Fact In Chapter 8 we will see the following properties of $\mathbb{T}_{\mu}$.

(i) $\mathbb{T}_{\mu}$ is closed, i.e. $\mathbb{T}_{\mu} \models \exists \vec{X}.\mathcal{R}$, for all simultaneous recursions $\mathcal{R}$ over $\mathbb{T}_{\mu}$.

(ii) $=_\mu$ is invertible, and hence $\mathbb{T}_\mu$ is an invertible type algebra.
(iii) $=_\mu$ is decidable.

7.5 Recursive types as trees

We now introduce the type algebra $\mathsf{Tr}_\infty^{\mathbb{A}}$ consisting of the finite and infinite trees over a set of atoms $\mathbb{A}$. These are type algebras that solve arbitrary simultaneous recursions $\mathcal{R}$ over $\mathbb{T}^{\mathbb{A}}$ in a unique way and induce a stronger notion of equivalence between types. For example, if $A = A \to b$ and $B = (B \to b) \to b$, then in Tr_∞ one has $A = B$.

We first need to recall the basic terminology needed to present (infinite) labeled trees by means of addresses and suitable labelings for them. A node of a tree is represented by the unique path in the tree leading to it. If Σ is a set, then Σ^* denotes the set of all finite sequences of elements of Σ. The elements of Σ^* are usually called the *words* over the *alphabet* Σ. As usual concatenation is represented simply by juxtaposition; ϵ denotes the empty word.

7.5.1 Definition (Trees) A *type tree* (or just *tree*, for short) over $\mathbb{A}$ is a partial function

$$t : \{0,1\}^* \rightharpoonup \mathbb{A} \cup \{\to\} \cup \{\bullet\}$$

satisfying

(1) $\epsilon \in \mathrm{dom}(t)$ (i.e. a tree has at least the root node);
(2) if $uv \in \mathrm{dom}(t)$, then also $u \in \mathrm{dom}(t)$
  (i.e. if a node is in a tree, then all its prefixes are in that tree);
(3) if $t(u) = \to$ then $u0, u1 \in \mathrm{dom}(t)$
  (i.e. if $\to$ is given at a node, then it has exactly two successors);
(4) if $t(u) \in \mathbb{A} \cup \{\bullet\}$ then $uv \notin \mathrm{dom}(t)$ for all $v \neq \epsilon$
  (i.e. labels other than '$\to$' occur only at endpoints).

The set of trees over $\mathbb{A}$ will be denoted by $\mathsf{Tr}_\infty^{\mathbb{A}}$. We define $\mathsf{Tr}_{\mathrm{F}}^{\mathbb{A}} \subseteq \mathsf{Tr}_\infty^{\mathbb{A}}$ to be the set of *finite* trees, i.e. those with finite domain. We will often write simply Tr_∞, Tr_{F} when $\mathbb{A}$ is understood or not relevant.

7.5.2 Example The function t defined on the (finite) domain $\{\epsilon, 0, 1\}$ as

$$\begin{aligned} t(\epsilon) &= \to, \\ t(0) &= \alpha, \\ t(1) &= \bullet, \end{aligned}$$

represents the tree 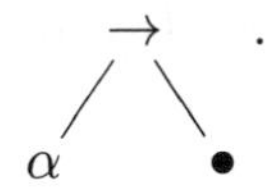 .

We say that the symbol $c \in \mathbb{A} \cup \{\to, \bullet\}$ occurs in tree t at node $u \in \Sigma^*$ if $t(u) = c$.

Since $\mathsf{Tr}_{\mathrm{F}}(\mathbb{A})$ is clearly isomorphic to $\mathbb{T}(\mathbb{A})$, we will often identify them, considering (and representing) simple types as finite trees and vice versa. The operation of substitution of types, for elements of $\mathbb{A}$, can be extended to trees. Among infinite trees, we single out trees having a certain periodic structure, which represent solutions of (systems of) equations of the form

$$\xi = A[\alpha{:=}\xi],$$

where $A \in \mathsf{Tr}_{\mathrm{F}}$. These will be called the *regular* trees.

7.5.3 Example Take the equation $\xi = \alpha \to \xi$ where A is any type. The solution of this equation is given by $\xi = t_0$ where t_0 is the following infinite tree:

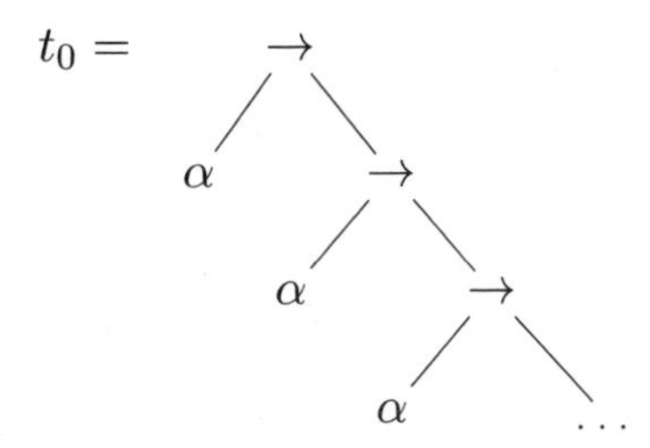

Note that t_0 has only a finite number of distinct subtrees, namely α and t_0 itself. Such a tree is called *regular.*

7.5.4 Definition (Regular trees)

(i) Let $t \in \mathsf{Tr}_\infty$ be a tree and $w \in \mathrm{dom}(t)$ be a word. The *subtree of* t *at* w, written $t|w$, is the tree defined by the following conditions:

   (1) $\mathrm{dom}(t|w) = \{u \in \{0,1\}^* \mid wu \in \mathrm{dom}(t)\}$;

   (2) $(t|w)(u) = t(wu)$, for all $u \in \mathrm{dom}(t|w)$.

   A *subtree* of t is $t|w$ for some $w \in \mathrm{dom}(t)$.

(ii) A tree is *regular* if the set of its subtrees is finite. The set of regular trees is denoted by Tr_{R}.

Note that $\mathsf{Tr}_{\mathrm{F}} \subseteq \mathsf{Tr}_{\mathrm{R}} \subseteq \mathsf{Tr}_\infty$. Moreover, we will see that regular trees are closed under substitutions, i.e. if t, s are regular trees then so is $t[\alpha{:} = s]$.

We consider $\mathsf{Tr}_\infty, \mathsf{Tr}_{\mathrm{R}}$ and Tr_{F} as type algebras. Since $\to$ is injective, all of them are invertible.

7.5.5 Lemma *The type algebras* $\mathsf{Tr}_\infty, \mathsf{Tr}_{\mathrm{R}}$ *and* Tr_{F} *are invertible.*

7.5.6 Definition Let $t, u \in \mathsf{Tr}_\infty$.

(i) Define for $k \in \omega$ the tree $(t)_k \in \mathsf{Tr}_{\mathrm{F}}$, its *truncation* at level k, as

$$\begin{aligned}(t)_0 &= \bullet; \\ (\alpha)_{k+1} &= \alpha, && \text{for } \alpha \in \mathbb{A}; \\ ((t \to u))_{k+1} &= (t)_k \to (u)_k.\end{aligned}$$

(ii) Define $t =_k u$ by $(t)_k = (u)_k$.

7.5.7 Example Let $t = \alpha{\to}\beta$. Then

$$\begin{aligned}(t)_0 &= \bullet; \\ (t)_1 &= \bullet{\to}\bullet; \\ (t)_n &= t && \text{for } n \geq 2.\end{aligned}$$

The following obvious fact will be useful.

7.5.8 Lemma *Let* $t, u, t', u' \in \mathsf{Tr}_\infty$.

(i) $t = u \;\Leftrightarrow\; \forall k \geq 0 \,.\, t =_k u$;
(ii) $t =_0 u$;
(iii) $t \to u =_{k+1} t' \to u' \;\Leftrightarrow\; t =_k u \;\&\; t' =_k u'$.

7.5.9 Definition Let $t, s \in \mathsf{Tr}_\infty$. The *substitution* of s for the occurrences of α in t, written $t[\alpha := s]$, is defined as follows (note that α may occur infinitely many times in t):

$$\begin{aligned} t[\alpha{:} = s](u) &= s(w) && \text{if } u = vw \;\&\; t(v) = \alpha, \\ & && \text{for some } v, w \in \Sigma^*; \\ &= t(u) && \text{else.}\end{aligned}$$

Bisimulation

Equalities among trees can be proved by a proof technique called *co-induction*, which allows to show that two trees are equal whenever there is a bisimulation that relates them.

7.5.10 Definition A relation $R \subseteq \mathsf{Tr}_\infty^{\mathcal{A}} \times \mathsf{Tr}_\infty^{\mathcal{A}}$ is a *bisimulation* if, for all $s, t \in \mathsf{Tr}_\infty$ sRt implies

- either $s, t \in \mathbb{A}$ and $s = t$;
- or $s = s_1 \to s_2$, $t = t_1 \to t_2$, and $s_1 R t_1$, $s_2 R t_2$.

The existence of a bisimulation between two trees s and t amounts to a proof that there is no contradiction in assuming that $s = t$. The co-induction principle for trees is as follows.

7.5.11 Proposition (Co-induction for trees) *Let R be a bisimulation over $\mathsf{Tr}_\infty \times \mathsf{Tr}_\infty$. Then*

$$\forall s, t \in \mathsf{Tr}_\infty.[sRt \;\Rightarrow\; s = t].$$

Proof Assume that $\langle s, t\rangle \in R$. We can show that $\mathrm{dom}(s) = \mathrm{dom}(t)$ and that, for all addresses $w \in \mathrm{dom}(s)$, $\langle s|w, t|w\rangle \in R$ by induction on the length of w. We have by Definition 7.5.10 either $s|w = t|w$ in $\mathbb{A}$, hence $s(w) = t(w)$, or $s|w = s_1 \to s_2$ and $t|w = t_1 \to t_2$ and then also $s(w) = t(w)$, both sides being equal to $\to$. Hence in either case $s(w) = t(w)$. So $s = t$ as partial functions. ■

In particular the equality relation is the largest bisimulation over trees. This co-induction principle may be used whenever we want to prove that two finitary presentations of trees, typically given as terms over some type algebra, have the same value.

Tree-unfolding for invertible general type algebras

Every (possibly non-well-founded) invertible type algebra can be mapped into a tree type algebra, by a morphism determined by the 'tree-unfolding' of the types in the algebra. Invertibility is needed in order to do the unfolding in a unique fashion. The construction has a flavor similar to the construction of Böhm trees for untyped lambda terms, see B[1984], in that both have a co-inductive nature.

7.5.12 Definition Let $\mathcal{A} = \langle |\mathcal{A}|, \to\rangle$ be an invertible type algebra. Write $\mathsf{Tr}_\infty^{\mathcal{A}} = \mathsf{Tr}_\infty^{||\mathcal{A}||}$. The *tree-unfolding* of a type $a \in \mathcal{A}$, written $(a)^*_{\mathcal{A}}$, is defined as

$$\begin{array}{rcll}
(a)^*_{\mathcal{A}}(\epsilon) & = & a, & \text{if } a \in ||\mathcal{A}||;\\
(a)^*_{\mathcal{A}}(\epsilon) & = & \to, & \text{if } a = b_0 \to b_1;\\
(a)^*_{\mathcal{A}}(iw) & = & \uparrow, & \text{if } a \in ||\mathcal{A}||;\\
(a)^*_{\mathcal{A}}(iw) & = & (b_i)^*_{\mathcal{A}}(w) \quad (i = 0, 1), & \text{if } a = b_0 \to b_1.
\end{array}$$

In spite of its technicality, the construction of Definition 7.5.12 is quite intuitive. The tree $(a)^*_{\mathcal{A}}$, which has the (names of the) prime elements of $\mathcal{A}$ as leaves, corresponds to the (possibly) infinite 'unfolding' of a with respect to $\to$.

7.5.13 Lemma *For invertible $\mathcal{A}$ the map $(-)^*_{\mathcal{A}} : \mathcal{A} \to \mathsf{Tr}_\infty^{||\mathcal{A}||}$ satisfies the following.*

$$
\begin{array}{rcll}
(a)^*_{\mathcal{A}} & = & a, & \text{if } a \in ||\mathcal{A}||; \\
(a)^*_{\mathcal{A}} & = & \begin{array}{c} \to \\ \diagup \quad \diagdown \\ (b_1)^*_{\mathcal{A}} \qquad (b_2)^*_{\mathcal{A}} \end{array}, & \text{if } a = b_1 \to b_2.
\end{array}
$$

Note that this is not an inductive definition: a may be as complex as $a \to b$.

Proof By Definition 7.5.12. ■

This property characterizes the map.

7.5.14 Remark

(i) The map $(-)^*_{\mathcal{A}} : \mathcal{A} \to \mathsf{Tr}_\infty^{\mathcal{A}}$ is a morphism.
(ii) In Section 7.6 we will see that the simple intuitive characterization of Lemma 7.5.13 can be considered as its official definition. See Remark 7.6.18.

7.5.15 Definition Let $\mathcal{A}$ be an invertible type algebra.

(i) *Strong equality*, written $=^*_{\mathcal{A}}$, is the relation $=^*_{\mathcal{A}} \subseteq \mathcal{A} \times \mathcal{A}$ defined by
$$a =^*_{\mathcal{A}} b \Leftrightarrow (a)^*_{\mathcal{A}} = (b)^*_{\mathcal{A}}.$$
By contrast, the relation $=$ will be called *weak equality*.
(ii) Define $\mathcal{A}^* = \mathcal{A}/{=^*_{\mathcal{A}}}$.
(iii) For $a \in \mathcal{A}$ write $a^* = [a]_{=^*_{\mathcal{A}}}$.

It is immediate from the definition that $=^*_{\mathcal{A}}$ is a congruence with respect to $\to$ and hence $\mathcal{A}^*$ is well defined.

7.5.16 Lemma *Let $\mathcal{A}$ be an invertible type algebra.*

(i) *$=^*_{\mathcal{A}}$ is the greatest invertible congruence over $\mathcal{A}$ such that for all $a \in ||\mathcal{A}||$*
$$[a]_{=^*} = \{a\}.$$
(ii) *$\mathcal{A}^*$ is an invertible type algebra and $||\mathcal{A}^*|| = \{a^* \mid a \in ||\mathcal{A}||\}$.*

Proof (i) To prove invertibility note that $a_1 \to a_2 =^* b_1 \to b_2$ implies, by definition, $(a_i)^* = (b_i)^*$ $(i = 1, 2)$ and then $a_i =^* b_i$. Note also that if $a \in ||\mathcal{A}||$, then $a =^* b$ iff $a = b$. Hence $a^* = \{a\}$ and indeed a^* is prime.

On the other hand, let now $\approx$ be any invertible congruence over $\mathcal{A}$ such that $[a]_\approx = \{a\}$ for all $a \in ||\mathcal{A}||$ and $\approx \not\subseteq =^*_{\mathcal{A}}$. Then there are two elements

a and b such that $a \approx b$ but $a \neq^* b$. In this case there must be some finite path w such that $(a)^*_{\mathcal{A}}(w) \neq (b)^*(w)$. It is easy to show by induction on w and using invertibility that this implies that for some prime element $a' \in \mathcal{A}$ either $a' \approx b' \to c'$ or $a' \approx d'$ for some other prime element d' of $\mathcal{A}$. In both cases $[a']_\approx \neq \{a'\}$.

(ii) By (i). ∎

7.5.17 Proposition *Let $\mathcal{A}$ be an invertible type algebra.*

(i) *There are canonical morphisms*

$$\begin{array}{rcll} i_{\mathcal{A}} & : & \mathbb{T}^{||\mathcal{A}||} & \hookrightarrow \mathcal{A} \\ f & : & \mathcal{A} & \to \mathcal{A}^* \\ i_* & : & \mathcal{A}^* & \hookrightarrow \mathsf{Tr}^{\mathcal{A}}_\infty \end{array}$$

with f surjective, $i_{\mathcal{A}}, i_*$ injective and $(\ )^* = i_* \circ f$; in diagrammatic form, this means

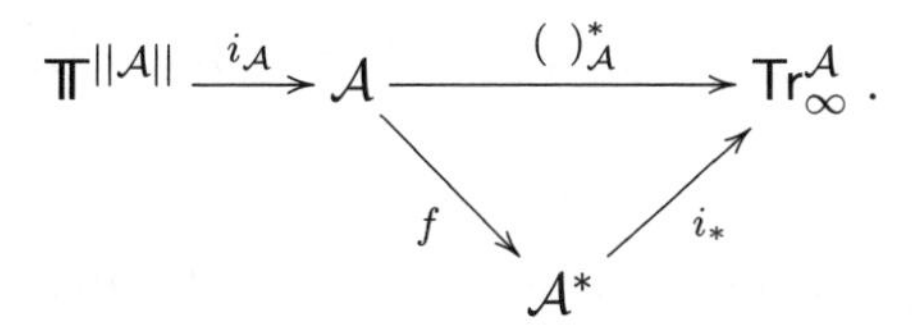

(ii) *The maps in* (i) *are uniquely determined by postulating for all $a \in ||\mathcal{A}||$*

$$\begin{aligned} i_{\mathcal{A}}(a) &= a; \\ f(a) &= (a)^*_{\mathcal{A}}; \\ i_*((a)^*_{\mathcal{A}}) &= (a)^*_{\mathcal{A}}. \end{aligned}$$

Proof (i) By Propositions 7.2.13 and 7.2.5(ii).

(ii) By the (easy) proof of these propositions. ∎

Tree-unfolding for invertible syntactic type algebras

The most interesting applications of the tree-unfolding $(-)^*_{\mathcal{A}}$, Definition 7.5.12, are for syntactic type algebras of the form $\mathcal{A} = \mathbb{T}/\approx$. In this case $\mathcal{A}^*$ can easily be described.

7.5.18 Definition Let $\mathcal{A}$ be an invertible syntactic type algebra.

(i) In the case $\mathcal{A} = \mathbb{T}^{\mathbb{A}}/\approx$, with invertible relation $\approx$, write

$$\begin{array}{rcl} (-)^*_\approx & \text{for} & (-)^*_{\mathcal{A}} : \mathcal{A} \to \mathsf{Tr}^{\mathbb{A}}_\infty, \\ =^* & \text{for} & =^*_{\mathcal{A}}, \\ \mathbb{T}^*_\approx & \text{for} & \mathcal{A}/=^*_{\mathcal{A}}. \end{array}$$

(ii) In the case $\mathcal{A} = \mathbb{T}^{\mathbb{A}}[\mathcal{R}]$, which is always invertible, write

$$\begin{aligned} (-)^*_{\mathcal{R}} &\quad \text{for} \quad (-)^*_{\mathcal{A}} : \mathcal{A} \to \mathsf{Tr}^{\mathbb{A}}_{\infty}, \\ =^*_{\mathcal{R}} &\quad \text{for} \quad =^*_{\mathcal{A}}, \\ \mathbb{T}[\mathcal{R}]^* &\quad \text{for} \quad \mathcal{A}/=^*_{\mathcal{A}}. \end{aligned}$$

(iii) In the case $\mathcal{A} = \mathbb{T}^{\mathbb{A}}_{\mu}/=_{\mu}$, which is always invertible, write

$$\begin{aligned} (-)^*_{\mu} &\quad \text{for} \quad (-)^*_{\mathcal{A}} : \mathcal{A} \to \mathsf{Tr}^{\mathbb{A}}_{\infty}, \\ =^*_{\mu} &\quad \text{for} \quad =^*_{\mathcal{A}}, \\ \mathbb{T}^*_{\mu} &\quad \text{for} \quad \mathcal{A}/=^*_{\mathcal{A}}. \end{aligned}$$

7.5.19 Remark In the case $\mathcal{A} = \mathbb{T}^{\mathbb{A}}/\approx$, Lemma 7.5.16 states that $=^*$ is the greatest invertible congruence on $\mathbb{T}$ extending $\approx$ and preserving all prime elements of $\mathcal{A}$. That is, if $\Box$ is an invertible congruence on $\mathbb{T}$ extending $\approx$, such that

$$[A]_{\approx} \in ||\mathcal{A}|| \ \& \ A \Box B \ \Rightarrow \ [A]_{\approx} = [B]_{\approx},$$

then $\Box \subseteq =^*$.

The map $(-)^*_{\mathcal{A}}$ can be considered as an *interpretation* of types in $\mathbb{T}$ as infinite trees.

7.5.20 Notation

(i) Working in a syntactic type algebra $\mathbb{T}^{\mathbb{A}}/\approx$, we view a type as if it is its equivalence class. Therefore we bend the meaning of $(-)^*_{\approx}$ so that it also applies to an element of $\mathbb{T}^{\mathbb{A}}$:

$$(-)^*_{\approx} : \mathbb{T}^{\mathbb{A}} \to \mathsf{Tr}^{\mathbb{A}}_{\infty}.$$

That is, we identify $(-)^*_{\approx} = (-)^{*\,\natural}_{\approx}$; see Lemma 7.2.4(ii) for the definition of $\natural$:

$$\begin{array}{ccc} \mathbb{T}^{\mathbb{A}} & \xrightarrow{(-)^*_{\approx} = (-)^{*\,\natural}_{\approx}} & \mathsf{Tr}^{\mathbb{A}}_{\infty}. \\ & {\scriptstyle [-]_{\approx}} \searrow \quad \nearrow {\scriptstyle (-)^*_{\approx}} & \\ & \mathbb{T}^{\mathbb{A}}/\approx & \end{array}$$

(ii) For $\mathcal{A} = \mathbb{T}^{\mathbb{A}}[\mathcal{R}]$ and $\mathcal{A} = \mathbb{T}^{\mathbb{A}}_{\mu}/=_{\mu}$ this boils down to

$$(-)^*_{\mathcal{R}} : \mathbb{T}(\vec{X}) \to \mathsf{Tr}^{\mathbb{A}}_{\infty};$$

$$(-)^*_{\mu} : \mathbb{T}^{\mathbb{A}}_{\mu} \to \mathsf{Tr}^{\mathbb{A}}_{\infty}.$$

(iii) One has by definition

$$\begin{aligned} A =^* B &\Leftrightarrow (A)^*_{\approx} = (B)^*_{\approx}, \\ A =^*_{\mathcal{R}} B &\Leftrightarrow (A)^*_{\mathcal{R}} = (B)^*_{\mathcal{R}}, \\ A =^*_{\mu} B &\Leftrightarrow (A)^*_{\mu} = (B)^*_{\mu}. \end{aligned}$$

7.5.21 Lemma *In this context we have*

(i) $\mathbb{T}^*_{\approx} \cong \mathbb{T}/=^*$.

(ii) $\mathbb{T}[\mathcal{R}]^* \cong \mathbb{T}(\vec{X})/=^*_{\mathcal{R}}$.

(iii) $\mathbb{T}^*_{\mu} \cong \mathbb{T}_{\mu}/=^*_{\mu}$.

Proof Immediate from Definition 7.5.18 and Notation 7.5.20. ■

Now we focus on the syntactic type algebras $\mathbb{T}(\vec{X})/\approx$, $\mathbb{T}[\mathcal{R}]$ and $\mathbb{T}_{\mu}/=_{\mu}$. In Chapter 9 it will be proved that these type algebras are invertible; for $\mathbb{T}[\mathcal{R}]$ this is stated below without proof.

Tree equivalence for simultaneous recursion

For syntactic type algebras coming from a simultaneous recursion $\mathcal{R}$, the tree-unfolding $(\ )^*_{\mathcal{R}} : \mathbb{T} \to \mathsf{Tr}_{\infty}$ can be characterized as follows.

7.5.22 Lemma *Let $\mathcal{R} = \mathcal{R}(\vec{X})$ be a simultaneous recursion over $\mathbb{T} = \mathbb{T}^{\mathbb{A}}$. Then we have the following.*

(i) *$\mathbb{T}[\mathcal{R}]$ is invertible.*

(ii) *Suppose moreover that $\mathcal{R}$ is proper. Then the map $(-)^*_{\mathcal{R}} : \mathbb{T}[\mathcal{R}] \to \mathsf{Tr}^{\mathbb{A}}_{\infty}$ can be characterized as*

$$\begin{aligned} (\alpha)^*_{\mathcal{R}} &= \alpha, && \text{if } \alpha \in \mathbb{A}; \\ (A \to B)^*_{\mathcal{R}} &= \begin{array}{c} \to \\ / \quad \backslash \\ (A)^*_{\mathcal{R}} \quad (B)^*_{\mathcal{R}} \end{array}; \\ (X)^*_{\mathcal{R}} &= (B)^*_{\mathcal{R}}, && \text{if } X \in \vec{X} \text{ and } (X = B) \in \mathcal{R}. \end{aligned}$$

Proof (i) By Proposition 7.3.16.

(ii) Note that by Lemma 7.3.27 one has $||\mathbb{T}[\mathcal{R}]|| = \{[\alpha] \mid \alpha \in \mathbb{A}\}$. ■

7.5.23 Remark If $\mathcal{R}$ is not proper there may be equations like $X = X$, or $X = Y$, $Y = X$. In this case the above procedure becomes circular due to the third clause in Lemma 7.5.22(ii).

7.5.24 Example

(i) Let $\mathcal{R} = \{X = A \to X, Y = A \to A \to Y\}$. We have $X \neq_{\mathcal{R}} Y$. But both types unfold to the same tree $(X)^*_{\mathcal{R}} = (Y)^*_{\mathcal{R}} = t_0$, where t_0 is the tree defined in Example 7.5.3 and one has $X =^*_{\mathcal{R}} A \to X =^*_{\mathcal{R}} Y =^*_{\mathcal{R}} Y \to A$.

(ii) Let $\mathcal{R}$ be the simultaneous recursion defined by

$$X_1 = X_2 \to X_1,$$
$$X_2 = X_1 \to X_2.$$

It is easy to see that $(X_1)^*_{\mathcal{R}} = (X_2)^*_{\mathcal{R}} = t_1$ where t_1 is the following infinite tree:

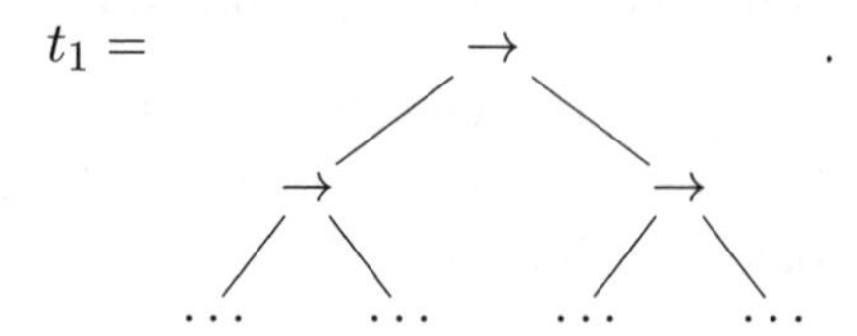

Hence $X_1 =^*_{\mathcal{R}} X_2$. Note that $X_1 \neq_{\mathcal{R}} X_2$. Also $(X)^*_{\mathcal{R}_2} = t_1$, where $\mathcal{R}_2 = \{X = X \to X\}$.

(iii) Take the simultaneous recursion $\mathcal{R} = \{X = A \to X,\ Y = A \to Y\}$. We have that $X =^*_{\mathcal{R}} Y$. In fact both $(X)^*_{\mathcal{R}}$ and $(Y)^*_{\mathcal{R}}$ are equal to the tree t_0 defined in Example 7.5.3. Note that $X \neq_{\mathcal{R}} Y$.

So we see that the relation $=^*_{\mathcal{R}}$ has a more semantic nature than $=_{\mathcal{R}}$. It turns out to be the type equivalence induced by the interpretation of types in continuous models (see Chapter 10). The relation $=^*_{\mathcal{R}}$ can be also characterized as the equational theory of a suitable simultaneous recursion $\mathcal{R}^*$. In particular in Chapter 8 we will prove constructively the following property.

7.5.25 Theorem *Let $\mathcal{R}(\vec{X})$ be a simultaneous recursion over $\mathbb{T}$. Then there is a simultaneous recursion $\mathcal{R}^*(\vec{X})$ such that for all $A, B \in \mathbb{T}(\vec{X})$ one has $A =^*_{\mathcal{R}} B \Leftrightarrow A =_{\mathcal{R}^*} B$.*

7.5.26 Corollary $\mathbb{T}[\mathcal{R}]^* \cong \mathbb{T}[\mathcal{R}^*]$.

Proof
$$\begin{aligned} \mathbb{T}[\mathcal{R}]^* &\cong \mathbb{T}(\vec{X})/=^*_{\mathcal{R}}, && \text{by Lemma 7.5.21(ii)},\\ &= \mathbb{T}(\vec{X})/=_{\mathcal{R}^*}, && \text{by the theorem},\\ &= \mathbb{T}[\mathcal{R}^*]. \blacksquare \end{aligned}$$

In Chapter 8 we will give an alternative axiomatization for the relation $=^*_{\mathcal{R}}$ using a co-induction principle.

As a consequence of Theorem 7.3.19 the simultaneous recursion $\mathcal{R}$ is solved not only by the type algebra $\mathbb{T}[\mathcal{R}]$, but also by its tree-collapse $\mathbb{T}[\mathcal{R}]^*$. But now, as bonus, the solution is unique.

7.5.27 Theorem *Let $\mathcal{R}(\vec{X}) = \{X_1 = A_1(\vec{X}), \ldots, X_n = A_n(\vec{X})\}$ be an sr over $\mathbb{T}$. Then*

(i) $\mathbb{T} \hookrightarrow \mathbb{T}[\mathcal{R}]^*$ *and* $\mathbb{T}[\mathcal{R}]^*$ *is generated by (the image of)* $\mathbb{T}$ *and the* $\vec{\mathbf{X}}$*;*
(ii) $\mathbb{T}[\mathcal{R}]^* \models \exists \vec{X} \mathcal{R}$*;*
(iii) *if moreover* $\mathcal{R}$ *is proper, then* $\vec{\mathbf{X}}$*, with* $\mathbf{X}_i = [X_i]_{=^*_{\mathcal{R}}}$ *is the unique solution of* $\mathcal{R}$ *in* $\mathbb{T}[\mathcal{R}]^*$.

Proof (i) By Theorem 7.3.19(iii) and Lemma 7.5.17(ii).
(ii) $\vec{\mathbf{X}}$ is a solution of $\mathcal{R}$: apply 7.3.19(i) and 7.5.17(ii).
Note that $\mathbb{T}[\mathcal{R}]^*$ can be seen as a subset of Tr_∞. The map $F{:}\mathsf{Tr}^n_\infty \to \mathsf{Tr}^n_\infty$ defined by $F(\vec{X}) = \langle A_1(\vec{X}), \ldots, A_n(\vec{X})\rangle$ satisfies the conditions of Proposition 7.6.8(ii), since $\mathcal{R}$ is proper. Hence by that proposition it has a unique fixed point. Clearly $\vec{\mathbf{X}}$ viewed as element of Tr^n_∞ is this fixed point. ∎

Tree equivalence for μ-types

7.5.28 Fact The type algebra $\mathbb{T}_\mu/{=_\mu}$ is invertible. See Theorem 8.2.1.

Therefore the construction of 7.5.12 can be applied to this type algebra. For example, from Lemma 7.5.21 it follows that $(\mathbb{T}_\mu/{=_\mu})^* = (\mathbb{T}_\mu/{=^*_\mu})$.

7.5.29 Lemma *The map $(-)^*_\mu : \mathbb{T}^{\mathbb{A}}_\mu \to \mathsf{Tr}^{\mathbb{A}}_\infty$ can be characterized as*

$$
\begin{array}{lll}
(A)^*_\mu = \alpha, & & \textit{if } \alpha \textit{ is a reduced form of } A,\\
(A)^*_\mu = \begin{array}{c} \to \\ / \quad \backslash \\ (B)^*_\mu \quad (C)^*_\mu \end{array}, & & \textit{if } B \to C \textit{ is a reduced form of } A,\\
(A)^*_\mu = \bullet, & & \textit{if } A \textit{ is circular.}
\end{array}
$$

Proof Similar to that of 7.5.22. ∎

In particular $=^*_\mu$ is an invertible congruence extending $=_\mu$ and there is a unique morphism $h^*{:}\mathbb{T}_\mu \to \mathbb{T}^*_\mu$ (or rather $h^*{:}\mathbb{T}_\mu/{=_\mu} \to \mathbb{T}^*_\mu$) defined by $h^*([A]_{=_\mu}) = [A]_{=^*_\mu}$.

7.5.30 Example Consider the types $B = \mu\beta.\alpha\to\beta$ and $B' = \mu\beta.\alpha\to\alpha\to\beta$. We have that $B =^*_\mu B'$. Indeed, one has $(B)^*_\mu = (B')^*_\mu = t_0$, where t_0 is the infinite tree of Example 7.5.3.

7.5.31 Lemma $(A[\alpha{:}= B])^*_\mu = (A)^*_\mu[(\alpha)^*_\mu{:}= (B)^*_\mu]$.

Proof By induction on the structure of A. ∎

7.5.32 Remark Note that by Lemma 7.5.31

$$(\mu\alpha.A)^*_\mu = (A[\alpha := \mu\alpha.A])^*_\mu = (A)^*_\mu[\alpha{:} = (\mu\alpha.A)^*_\mu].$$

So $(\mu\alpha.A)^*_\mu$ is a solution of the equation $X = (A)^*_\mu[\alpha := X]$ in Tr_R. If $A \neq \alpha$, this solution in Tr_∞ is unique, by Proposition 7.6.8 and Theorem 7.6.5.

In Chapter 8 we will give a complete axiomatization of $=^*_\mu$. As an application we obtain a constructive proof that $=^*_\mu$ is decidable.

Types as regular trees

All trees in the codomain of $(-)^*_\mathcal{R}$ are regular.

7.5.33 Lemma *Let $\mathcal{R}$ be a simultaneous recursion over $\mathbb{T}$. Then for all $A \in \mathbb{T}[\vec{X}]$ $(A)^*_\mathcal{R}$ is a regular tree, i.e. $(-)^*_\mathcal{R} : \mathbb{T}[\mathcal{R}] \to \mathsf{Tr}_\mathrm{R}$.*

Proof Let $\vec{X} = X_1, \ldots, X_n$ $(n \geq 0)$ and $\mathcal{R} = \{X_i = B_i \mid 1 \leq i \leq n\}$. If $A \in \mathbb{T}[\vec{X}]$ let $\mathcal{S}(A)$ denote the set of all subtypes of A. Obviously $\mathcal{S}(A)$ is finite for all A. Now it is easy to prove by induction on w that for all $w \in \{0,1\}^*$ for which $(A)^*_\mathcal{R}(w)$ is defined we have

$$(A)^*_\mathcal{R}|w = (C)^*_\mathcal{R} \quad \text{for some} \quad C \in \mathcal{S}(A) \cup \bigcup_{1 \leq i \leq n} \{X_i\} \cup \bigcup_{1 \leq i \leq n} \mathcal{S}(B_i).$$

Since $\mathcal{S}(A)$ and all $\mathcal{S}(B_i)$ for $1 \leq i \leq n$ are finite, $(A)^*_\mathcal{R}$ can have only a finite number of subtypes and hence is regular. ∎

The following theorem is an immediate consequence of this.

7.5.34 Theorem

(i) *Let $\mathcal{R}(\vec{X})$ be a proper simultaneous recursion over $\mathbb{T}$. Then $\mathcal{R}$ has a unique solution in Tr_R given by $\vec{t} = (X_1)^*_\mathcal{R}, \ldots, (X_n)^*_\mathcal{R}$.*
(ii) *Tr_∞ is a uniquely closed type algebra.*
(iii) *Tr_R is a uniquely closed type algebra.*

Proof (i) By Theorem 7.5.27 and Lemma 7.5.33.
(ii) As the proof of Theorem 7.5.27(iii).
(iii) Every $\mathcal{R}$ over Tr_R has a unique solution in Tr_∞. That the solution is in Tr_R follows as in the proof of Lemma 7.5.33. ∎

7.5.35 Remark

(i) On the other hand each regular tree can be obtained as a component of a solution of some simultaneous recursion $\mathcal{R}$; see Exercise 7.9 and also Courcelle (1983), Theorem 4.2.1.

(ii) It is easy to see that Lemma 7.5.33 and Theorem 7.5.34 hold also if we assume more generally that $\mathcal{R}$ is a simultaneous recursion with coefficients in Tr_{R}; see also Courcelle (1983), Theorem 4.3.1.

7.5.36 Remark The mapping $(-)^*_{\mathcal{R}}$ can be seen as a *unifier* of $\mathcal{R}$ in Tr_{R}. It is well known, see Courcelle (1983), Prop. 4.9.5(ii), that $(-)^*_{\mathcal{R}}$ is indeed the *most general* unifier of $\mathcal{R}$ in Tr_{R}. This means that each syntactic morphism $h : \mathbb{T}[\mathcal{R}] \to \mathsf{Tr}_{\mathrm{R}}$ can be written as $h = s \circ (-)^*_{\mathcal{R}}$ where s is a substitution in Tr_{R}. Note that any such substitution can be seen as a morphism $s : \mathsf{Tr}_{\mathrm{R}} \to \mathsf{Tr}_{\mathrm{R}}$.

7.5.37 Remark Simultaneous recursions of a special form can be viewed as $T_{\mathbb{A}}$-coalgebras, where $T_{\mathbb{A}}$ is the tree functor of Definition 7.6.15, as suggested by an analogy between simultaneous recursions and systems of equations defining non-well-founded sets (Aczel (1988)). See Section 4 of Cardone and Coppo (2003) for an application of this remark to the decidability of strong equivalence of simultaneous recursions.

The μ-types are just notations for regular trees, see Courcelle (1983), Theorem 4.5.7.

7.5.38 Lemma

(i) *If* $A \in \mathbb{T}_\mu$, *then* $(A)^*_\mu \in \mathsf{Tr}_{\mathrm{R}}$.
(ii) *Each regular tree* t *can be written as* $t = (A)^*_\mu$, *for some type* $A \in \mathbb{T}_\mu$.
(iii) $\mathbb{T}^*_\mu \cong \mathsf{Tr}_{\mathrm{R}}$.

Proof (i) This follows from Remark 7.5.35(i).
(ii) By the fact that $\mathbb{T}_\mu$ is a closed type-algebra (a formal proof of this will be given in Chapter 9).
(iii) Immediate from (i). ∎

Type assignment modulo tree equality

7.5.39 Definition As shorthand we denote the type assignment systems corresponding to $\mathbb{T}_\mu, \mathbb{T}^*_\mu, \mathbb{T}/\mathcal{E}, \mathbb{T}/\mathcal{E}^*, \mathbb{T}[\mathcal{R}]$, and $\mathbb{T}(\mathcal{R})^*$ by $\lambda\mu$, $\lambda\mu^*$, $\lambda\mathcal{E}$, $\lambda\mathcal{E}^*$, $\lambda\mathcal{R}$, and $\lambda\mathcal{R}^*$ respectively. The only difference in the formal presentations of these systems is in the set of types and in the definition of the corresponding equivalence.

7.5.40 Remark

(i) It is easy to see that any substitution $s : \mathbb{A} \to \mathbb{T}_\mu$ determines a type algebra homomorphism of $s : \mathbb{T}_\mu \to \mathbb{T}_\mu$. So all the typings of Lemma 7.1.4 hold in $\lambda\mu$ if we replace α by any type $A \in \mathbb{T}_\mu$ (i.e. taking the

Table 7.1

Type algebra	Type assignment system	Simplified notation
$\mathcal{A}$	$\boldsymbol{\lambda}_{=}^{\mathcal{A}}$	$\boldsymbol{\lambda}\mathcal{A}$
$\mathbb{T}/\approx$	$\boldsymbol{\lambda}_{=}^{\mathbb{T}/\approx}$	$\boldsymbol{\lambda}\approx$
$\mathbb{T}/\mathcal{E}$	$\boldsymbol{\lambda}_{=}^{\mathbb{T}/\mathcal{E}}$	$\boldsymbol{\lambda}\mathcal{E}$
$\mathbb{T}/\mathcal{E}^*$	$\boldsymbol{\lambda}_{=}^{\mathbb{T}/\mathcal{E}^*}$	$\boldsymbol{\lambda}\mathcal{E}^*$
$\mathbb{T}[\mathcal{R}]$	$\boldsymbol{\lambda}_{=}^{\mathbb{T}[\mathcal{R}]}$	$\boldsymbol{\lambda}\mathcal{R}$
$\mathbb{T}(\mathcal{R})^*$	$\boldsymbol{\lambda}_{=}^{\mathbb{T}(\mathcal{R})^*}$	$\boldsymbol{\lambda}\mathcal{R}^*$
$\mathbb{T}_\mu/=_\mu$	$\boldsymbol{\lambda}_{=}^{\mathbb{T}_\mu}$	$\boldsymbol{\lambda}\mu$
$\mathbb{T}_\mu^*$	$\boldsymbol{\lambda}_{=}^{\mathbb{T}_\mu^*}$	$\boldsymbol{\lambda}\mu^*$

substitution $[\alpha := A]$). Note in particular that $(\lambda x.xx)(\lambda x.xx)$ has all types and that the fixed point operator has all the types of the form $(A \to A) \to A$. Obviously the same property holds in the other systems, in particular in $\lambda\mu^*$.

(ii) The two systems $\lambda\mu$ and $\lambda\mu^*$ are not equivalent. Take $T_0 = \mu\beta.\alpha \to \beta$ and $T_1 = \mu\beta.\alpha \to \alpha \to \beta$ as in Example 7.4.29. Then $\{x : T_1\} \nvdash_{\lambda\mu} x : T_0$ while $\{x : T_1\} \vdash_{\lambda\mu^*} x : T_0$. Similarly $\{x : T_1\} \nvdash_{\lambda\mu} x : \alpha \to T_1$ but $\{x : T_1\} \vdash_{\lambda\mu^*} x : \alpha \to T_1$.

Notations for type assignment systems

The notion of type algebra was introduced to unify a number of type assignment systems. The general version is $\boldsymbol{\lambda}_{=}^{\mathcal{A}}$, which specializes to various cases.

We introduce the following notation.

7.5.41 Definition

(i) The type assignment system $\boldsymbol{\lambda}_{=}^{\mathcal{A}}$ is also denoted as $\boldsymbol{\lambda}\mathcal{A}$.
(ii) Similar shorthands apply to several other systems; see Table 7.1.
(iii) Using the simplified notation we can write

$$\vdash_{\boldsymbol{\lambda}\mathcal{A}}, \vdash_{\boldsymbol{\lambda}\approx}, \vdash_{\boldsymbol{\lambda}\mathcal{E}}, \vdash_{\boldsymbol{\lambda}\mathcal{E}^*}, \vdash_{\boldsymbol{\lambda}\mathcal{R}}, \vdash_{\boldsymbol{\lambda}\mathcal{R}^*}, \vdash_{\boldsymbol{\lambda}\mu}, \vdash_{\boldsymbol{\lambda}\mu^*}.$$

The approach we are taking here in Part II comes from Scott (1975b): this manifold of type assignment systems can be captured by one idea and a parameter: the type algebra.

Explicitly typed versions

The explicitly typed versions of the systems $\lambda\mathcal{R}$ and $\lambda\mu$ are usually given in 'Church' style with coercions (the *fold-unfold* constants) introduced in Definition 7.1.24. In the system $\lambda_{\mu}^{\mathrm{Ch}_0}$ in particular the fold–unfold constants are labeled with the folded version of the corresponding μ-type:

$$fold_{\mu t.A} \in \Lambda^{\mathrm{Ch}}_{\mathbb{T}_\mu}(A[t := \mu t.A] \to \mu t.A);$$
$$unfold_{\mu t.A} \in \Lambda^{\mathrm{Ch}}_{\mathbb{T}_\mu}(\mu t.A \to A[t := \mu t.A]).$$

On the terms extended by the coercion operators one postulates a notion of reduction:

$$unfold_{\mu t.A}(fold_{\mu t.A} M) \to M;$$
$$fold_{\mu t.A}(unfold_{\mu t.A} M') \to M',$$

where $M \in \Lambda^{\mathrm{Ch}}_{\mathbb{T}_\mu}(A[t := \mu t.A])$ and $M' \in \Lambda^{\mathrm{Ch}}_{\mathbb{T}_\mu}(\mu t.A)$.

7.6 Special views on trees

The set of trees, as introduced in Definition 7.5.1, can be regarded as a metric space or as a coalgebra. This view is given here for the interested reader, and may be skipped.

Trees as a metric space

We can turn Tr_∞ into a metric space in the following way Courcelle (1983).

7.6.1 Definition [Metrics over Tr_∞] Let $\alpha, \alpha' \in \mathsf{Tr}_\infty$. Define

$$\mathrm{d}(\alpha, \alpha') = \begin{cases} 0 & \text{if } \alpha = \alpha' \\ 2^{-\delta(\alpha,\alpha')} & \text{if } \alpha \neq \alpha' : \end{cases}$$

here $\delta(\alpha, \alpha')$ is the length of the minimum path w such that $w \in \mathrm{dom}(\alpha)$, $w \in \mathrm{dom}(\alpha')$ and $\alpha(w) \neq \alpha'(w)$.

7.6.2 Proposition (Courcelle (1983)) *$\langle \mathsf{Tr}_\infty, d\rangle$ is a complete metric space. With respect to the resulting topology, the set Tr_{F} is a dense subset of Tr_∞ and Tr_∞ is the topological completion of Tr_{F}.*

7.6.3 Remark In fact $\langle \mathsf{Tr}_\infty, d\rangle$ is even an *ultrametric space*, i.e. satisfies the strengthened triangle inequality

$$d(x, y) \leq \max\{d(x, z), d(z, y)\}.$$

7.6.4 Definition Let $\langle D, d\rangle$ be a metric space. A map $f : D \longrightarrow D$ is called *contractive* if there exists a real number c $(0 \leq c < 1)$ such that

$$\forall x, x' \in D \;\; d(f(x), f(x')) \;\leq\; c \cdot d(x, x')\,.$$

A basic property of complete metric spaces is the following.

7.6.5 Theorem (Banach fixed point theorem) *Let $\langle D, d\rangle$ be a complete metric space. Every contractive mapping $f : D \longrightarrow D$ has a* unique *fixed point x in D. This will be denoted by* $\mathbf{fix}(f)$.

Proof Let $x_0 \in D$. Define $x_n = f^n(x_0)$. This is a Cauchy sequence. The fixed point x can be defined as the limit of the x_n. If x, y are fixed points, then $d(x,y) = d(f^n(x), f^n(y)) \leq c^n \cdot d(x,y)$. Therefore $d(x,y) = 0$ and hence the fixed point is unique. ∎

7.6.6 Proposition *Let $D_1 = \langle D_1, d_1\rangle$, $D_2 = \langle D_2, d_2\rangle$ be metric spaces. Define on $D = D_1 \times D_2$ the map $d : D \to \mathbb{R}$ by*

$$d((x_1,x_2),(y_1,y_2)) = \max\{d_1(x_1,y_1), d_2(x_2,y_2)\}.$$

Then $D = \langle D, d\rangle$ is a metric space.

(i) *If D_1, D_2 are complete metric spaces, then so is D.*
(ii) *If D_1, D_2 are ultrametric spaces, then so is D.*

7.6.7 Definition The notion of *algebraic* map on a type algebra $\mathcal{A}$ is defined as usual. For example, if $a, b \in \mathcal{A}$, then f defined by $f(x) = a \to (b \to x) \to x$ is algebraic.

7.6.8 Proposition

(i) *Let $f : \mathsf{Tr}_\infty \to \mathsf{Tr}_\infty$ be algebraic that is not the identity. Then f is contractive.*
(ii) *Let $f_1, \ldots, f_n : \mathsf{Tr}_\infty^n \to \mathsf{Tr}_\infty$ be algebraic such that for all $a_1, \ldots, a_{i-1}, a_{i+1}, \ldots,$ $a_n \in \mathsf{Tr}_\infty$, with $1 \leq i \leq n$, one has*

$$\lambda x_i.f_i(a_1, \ldots, a_{i-1}, x_i, a_{i+1}, \ldots, a_n) : \mathsf{Tr}_\infty \to \mathsf{Tr}_\infty \text{ is not the identity.}$$

Define

$$F(\vec{x}) = \langle f_1(\vec{x}), \ldots, f_n(\vec{x})\rangle.$$

Then F is contractive on Tr_∞^n.

Proof (i) It suffices to show that the map g defined by $g(x,y) = x \to y$ is contractive in its two arguments. Indeed, $d(x \to y, x' \to y) = \frac{1}{2}d(x,x')$.

(ii) For notational simplicity we treat $n = 2$. We must show

$$d(F(x_1,x_2),F(y_1,y_2)) \leq c.d((x_1,x_2),(y_1,y_2)) \tag{7.13}$$

Now

$$\begin{aligned} d(F(x_1,x_2),F(y_1,y_2)) &= d(\langle f_1(x_1,x_2), f_2(x_1,x_2)\rangle, \langle f_1(y_1,y_2), f_2(y_1,y_2)\rangle) \\ &= \max\{d(f_1(x_1,x_2), f_1(y_1,y_2)), d(f_2(x_1,x_2), f_2(y_1,y_2))\}. \end{aligned}$$

Now for $i = 1,2$ one has

$$\begin{aligned} d((f_i(x_1,x_2), f_i(y_1,y_2))) &\leq \max\{d(f_i(x_1,x_2), f_i(y_1,x_2)), d(f_i(y_1,x_2), f_i(y_1,y_2))\} \\ &\leq \max\{c_i \cdot d(x_1,y_1), c_i \cdot d(x_2,y_2)\}, \text{ by the assumption and (i),} \\ &= c_i \cdot d((x_1,x_2),(y_1,y_2)). \end{aligned}$$

Now (7.13) easily follows. ∎

Now let $t \in \mathsf{Tr}_\infty$ and α a variable occurring in t. If $\alpha \not\equiv t$, then $\lambda\!\!\lambda x \in \mathsf{Tr}_\infty.t[\alpha := x]$ defines a contractive mapping of Tr_∞ into itself and therefore it has a fixed point.

The following property is also easy to prove, see Courcelle (1983), Theorem 4.3.1.

7.6.9 Proposition *If* $\alpha \in \mathsf{Tr}_\mathrm{R}$ *and* $\alpha \not\equiv t$ *then* $\mathbf{fix}(\lambda\!\!\lambda x \in \mathsf{Tr}_\infty.t[\alpha := x]) \in \mathsf{Tr}_\mathrm{R}$.

Trees as a coalgebra

Algebras and coalgebras

The notion of *algebra* to be introduced presently arises as a categorical generalization of the usual set-theoretical notion of algebraic structure. In order to give some background to the general definitions, we show below how natural numbers provide a simple (but fundamental) example of an algebra.

7.6.10 Example Natural numbers form a structure $\langle \omega, 0, \mathtt{suc}\rangle$ with an element $0 \in \omega$ and a unary operator $\mathtt{suc} : \omega \to \omega$, the successor function. Towards a categorical formulation of this structure, observe that two functions with a common codomain can be 'packed' into one function as follows. First of all, define the disjoint union

$$A + B = (A \times \{0\}) \cup (B \times \{1\}).$$

There are canonical maps

$$\begin{array}{rccc} \mathtt{inl}: & A & \to & A+B \\ & a & \mapsto & \langle a, 0\rangle \\ \mathtt{inr}: & B & \to & A+B \\ & b & \mapsto & \langle b, 1\rangle. \end{array}$$

Then, two functions $f : A \to C$ and $g : B \to C$ can be packed together as $[f, g] : A + B \to C$ by setting

$$\begin{aligned} [f,g](\mathtt{inl}(a)) &= f(a), \\ [f,g](\mathtt{inr}(b)) &= g(b). \end{aligned}$$

In the case of natural numbers, we have a mapping $[0, \mathtt{suc}] : 1 + \omega \to \omega$, where 1 is any singleton, say $1 = \{*\}$, and $0 : 1 \to \omega$ has as its (unique) value exactly the number 0. In categorical terms, the situation can be described as follows.

7.6.11 Definition Let a functor $T : \mathsf{Set} \to \mathsf{Set}$ be given. We call it the *signature functor*.

(i) A *T-algebra* consists of a pair $\langle X, \xi\rangle$ with $X \in \mathsf{Set}$ the *carrier* and ξ a morphism

$$\xi : T(X) \to X.$$

(ii) If $\langle X, \xi : T(X) \longrightarrow X\rangle$ and $\langle Y, \zeta : T(Y) \longrightarrow Y\rangle$ are two T-algebras, then a *morphism* is a mapping $f : X \longrightarrow Y$ such that

$$\begin{array}{ccc} T(X) & \xrightarrow{\ \xi\ } & X \\ {\scriptstyle T(f)}\downarrow & & \downarrow{\scriptstyle f} \\ T(Y) & \xrightarrow{\ \zeta\ } & Y \end{array}$$

commutes. It is easy to check that the T-algebras with these morphisms form a category.

(iii) A T-algebra $\langle A, \mathtt{in}\rangle$ is called *initial* iff it is an initial object in the category of T-algebras, i.e. for every T-algebra $\langle X, \xi\rangle$ there is a unique T-algebra morphism $\varphi{:}\langle A, \mathtt{in}\rangle \to \langle X, \xi\rangle$.

It is well known that natural numbers $\langle \omega, [0, \mathtt{suc}]\rangle$ form the initial algebra for the signature functor

$$T_\omega(X) = 1 + X = X^0 + X^1.$$

Initiality of ω is equivalent to the iteration principle: the unique morphism $h : \omega \to A$ from the T_ω-algebra $\langle \omega, [0, \mathtt{suc}]\rangle$ to any other T_ω-algebra $\langle A, [a, f]\rangle$ is precisely the unique function from ω to A satisfying the equations (see Dedekind (1901)):

$$\begin{aligned} h(0) &= a \\ h(\mathtt{suc}(n)) &= f(h(n)). \end{aligned}$$

7.6.12 Definition A *polynomial* functor $T{:}\mathsf{Set} \to \mathsf{Set}$ is of the form

$$T(X) = A_0 \times X^0 + A_1 \times X^1 + \cdots + A_n \times X^n,$$

where $\times$ denotes the Cartesian product and the $A_k \in \mathsf{Set}$ are arbitrary objects.

7.6.13 Proposition *For every polynomial functor $T{:}\mathsf{Set} \to \mathsf{Set}$ there is an initial algebra $\langle I, \mathtt{in}\rangle$ that is unique up to isomorphism in the category of T-algebras. Moreover, $\mathtt{in} : T(I) \to I$ is an isomorphism, hence $T(I) \cong I$.*

Proof Do Exercise 7.20. ∎

We now examine *coalgebras*, which are dual to algebras in the categorical sense and appear in nature as spaces of infinitely proceeding processes. A simple coalgebra is that of streams over a set A, namely infinite lists of elements of A. Another coalgebra is that of *lazy lists*, consisting of both the finite lists and the streams. Some properties of these coalgebras are left as exercises. In this section we shall be concerned only with *trees* as a coalgebra. For more examples and properties of coalgebras, and the general theory, we refer the reader to Rutten (2000).

7.6.14 Definition (Coalgebra) Let a functor $T : \mathsf{Set} \to \mathsf{Set}$ be given.

(i) A *T-coalgebra* is a pair $\langle X, \xi : X \longrightarrow T(X)\rangle$.
(ii) If $\langle X, \xi : X \longrightarrow T(X)\rangle$ and $\langle Y, \zeta : Y \longrightarrow T(Y)\rangle$ are T-coalgebras, then a *T-coalgebra morphism* is a mapping $f : X \longrightarrow Y$ such that

$$\begin{array}{ccc} X & \xrightarrow{\xi} & T(X) \\ {\scriptstyle f}\downarrow & & \downarrow{\scriptstyle T(f)} \\ Y & \xrightarrow{\zeta} & T(Y) \end{array}$$

commutes.
(iii) A T-coalgebra $\langle C, \mathtt{out}\rangle$ is called *final* iff it is a final object in the category of T-coalgebras, i.e., for any T-coalgebra $\langle X, \xi : X \longrightarrow T(X)\rangle$ there is a unique T-coalgebra morphism $\varphi : X \longrightarrow C$.

Dualizing the proof of Proposition 7.6.13, we have that every polynomial endofunctor of Set has a final coalgebra, which is unique up to a T-coalgebra isomorphism. Furthermore, final T-coalgebras are fixed points of T: if $\langle X, \xi : X \longrightarrow T(X)\rangle$ is the final T-coalgebra, then ξ is a bijection.

Trees as final coalgebras

We sketch now how it is possible – and in fact convenient – to regard (finite and infinite) trees over a set of atoms $\mathbb{A}$ as the final coalgebra of a suitable functor.

7.6.15 Definition (The tree functor) Let $\mathbb{A}$ be a set of atoms. The assignment

$$T_{\mathbb{A}}(X) = \mathbb{A} + (X \times X)$$

defines a functor over Set that will be called *the tree functor.*

We can define the $T_{\mathbb{A}}$-coalgebra

$$\omega : \mathsf{Tr}_{\infty}(\mathbb{A}) \longrightarrow \mathbb{A} + \mathsf{Tr}_{\infty}(\mathbb{A}) \times \mathsf{Tr}_{\infty}(\mathbb{A})$$

by the following clauses:

$$\begin{array}{rcll} \omega(t) & = & \mathtt{inl}(a), & \text{if } t = a \in \mathbb{A}, \\ & = & \mathtt{inr}(\langle t', t'' \rangle), & \text{if } t = t' \to t''. \end{array}$$

7.6.16 Proposition *$\langle \mathsf{Tr}_{\infty}(\mathbb{A}), \omega \rangle$ is the final $T_{\mathbb{A}}$-coalgebra.*

Proof We only give the details of the construction of the unique $T_{\mathbb{A}}$-coalgebra morphism toward $\langle \mathsf{Tr}_{\infty}(\mathbb{A}), \omega \rangle$, because this has a concrete description that shall also be exploited later. Let $\xi : X \longrightarrow \mathbb{A} + (X \times X)$ be any $T_{\mathbb{A}}$-coalgebra. First define, for any $x \in X$, an element $\ell(x) \in \mathbb{A} \cup \{\to\}$ as

$$\begin{array}{rcll} \ell(x) & = & a, & \text{if } \xi(x) = \mathtt{inl}(a) \text{ for some } a \in \mathbb{A}, \\ & = & \to, & \text{otherwise.} \end{array}$$

For any $x \in X$ we have to construct a tree $\varphi(x)$ in such a way that the resulting mapping $\varphi : X \longrightarrow \mathsf{Tr}_{\infty}(\mathbb{A})$ is a $T_{\mathbb{A}}$-coalgebra morphism. We define the corresponding partial function (see Definition 7.5.1)

$$\varphi(x) : \{0,1\}^* \rightharpoonup \mathbb{A} \cup \{\to\}$$

by induction on the length of the tree addresses:

$$\begin{array}{rcll} \varphi(x)(\epsilon) & = & \ell(x), & \\ \varphi(x)(iw) & = & \uparrow, & \text{if } \xi(x) = \mathtt{inl}(a) \text{ for some } a \in \mathbb{A}, \\ & = & \varphi(x_i)(w), & \text{if } \xi(x) = \mathtt{inr}\langle x_0, x_1 \rangle \text{ for some } x_0, x_1 \in X. \end{array}$$

We leave to the reader the verification that φ is indeed the unique $T_{\mathbb{A}}$-coalgebra morphism from X to $\mathsf{Tr}_{\infty}^{\mathcal{A}}$. ■

7.6.17 Remark A flat simultaneous recursion $\mathcal{R}(\vec{X})$ – one whose right hand sides have either the shape $\alpha \in \mathbb{A}$ or the shape $X' \to X''$ for some $X', X'' \in \vec{X}$ – may be seen directly as a $T_{\mathbb{A}}$-coalgebra. Explicitly, $\mathcal{R}(\vec{X})$ corresponds to the $T_{\mathbb{A}}$-coalgebra

$$f_{\mathcal{R}} : \vec{X} \longrightarrow \mathbb{A} + \vec{X} \times \vec{X}$$

defined by the following conditions, where $X \in \vec{X}$:

$$\begin{array}{rcll} f_{\mathcal{R}}(X) & = & \mathtt{inl}(\alpha), & \text{if } X = \alpha \in \mathbb{A}, \\ & = & \mathtt{inr}(\langle X', X'' \rangle), & \text{if } X = X' \to X''. \end{array}$$

The unique $T_{\mathbb{A}}$-morphism from $\vec{X}$ to $\mathsf{Tr}_\infty(\mathbb{A})$, which exists by finality, is easily seen to be the solution of $\mathcal{R}(\vec{X})$ in $\mathsf{Tr}_\infty(\mathbb{A})$.

Co-induction enables us to define maps in a 'coordinate-free' fashion, rather than in the way of Definition 7.5.12.

7.6.18 Remark The coalgebraic treatment of trees can also justify the format of Lemma 7.5.13 as an official definition. This is intuitively simpler than that given in Definition 7.5.12. If $\mathcal{A}$ is invertible, then there is an obvious map

$$\tau : |\mathcal{A}| \longrightarrow ||\mathcal{A}|| + (|\mathcal{A}| \times |\mathcal{A}|),$$

where the disjoint union matches the case distinction in the definition of function $(-)^*_{\mathcal{A}}$ in Lemma 7.5.13. The existence of the unique morphism $(-)^*_{\mathcal{A}} : \mathcal{A} \to \mathsf{Tr}_\infty(\mathbb{A})$ follows then by finality of $\langle \mathsf{Tr}_\infty(\mathbb{A}), \omega\rangle$, where $\mathbb{A} = ||\mathcal{A}||$.

We have seen already a direct proof of the following co-induction principle for trees, see Proposition 7.5.11. A categorical proof avoids using argumentation involving maps. This fits with the 'coordinate-free' treatment of definitions and propositions, see Section 7.5.

7.6.19 Proposition (Co-induction for trees) *Let R be a bisimulation over $\mathsf{Tr}_\infty \times \mathsf{Tr}_\infty$. Then*

$$\forall s, t \in \mathsf{Tr}_\infty.[sRt \;\Rightarrow\; s = t].$$

Proof The following categorical proof, exploiting the finality of the coalgebra of trees, is taken from Rutten (2000), Theorem 8.1. Observe that R is a bisimulation exactly when there is a function $\omega_R : R \longrightarrow T_{\mathbb{A}}(R)$ that makes the diagram

$$\begin{array}{ccccc}
\mathsf{Tr}_\infty & \xleftarrow{\pi_1} & R & \xrightarrow{\pi_2} & \mathsf{Tr}_\infty \\
\downarrow{\scriptstyle\omega} & & \downarrow{\scriptstyle\omega_R} & & \downarrow{\scriptstyle\omega} \\
T_{\mathbb{A}}(\mathsf{Tr}_\infty) & \xleftarrow[T_{\mathbb{A}}(\pi_1)]{} & T_{\mathbb{A}}(R) & \xrightarrow[T_{\mathbb{A}}(\pi_2)]{} & T_{\mathbb{A}}(\mathsf{Tr}_\infty)
\end{array}$$

commute, where π_1 (resp. π_2) extracts the first (resp. the second) component of R. Then they both are morphisms to the final coalgebra $\langle \mathsf{Tr}_\infty, \omega\rangle$, therefore $\pi_1 = \pi_2$, and all pairs in R have identical components. ■

7.7 Exercises

Exercise 7.1 Let $B = \mu\alpha.\beta \to \alpha$, with $\alpha \not\equiv \beta$. Show that $\vdash_{\lambda\mu} \mathsf{YK} : B$.

Exercise 7.2 Show that there is a term $M \in \Lambda^{\emptyset}_{\to}$ such that all occurrences of λ in M bind different variables, but M cannot be reduced naively; i.e. there is a $\boldsymbol{\beta}$-reduct N of M with a redex $(\lambda x.P)Q$ in N such that

$$(\lambda x.P)Q \not\to_{\boldsymbol{\beta}} P[x := Q]_{\not{\alpha}}.$$

[Hint. Consider a term of the form $M \equiv (\lambda x.x(Bx))\mathbf{c}_1$.]

Exercise 7.3 Let $h : \mathcal{T} \to \mathcal{S}$ be a morphism. Define the *kernel* of h, written $\mathtt{ker}(h)$, as $\mathtt{ker}(h) = \{a = b \mid h(a) = h(b)\}$.

(i) Show that $\mathtt{ker}(h)$ is always a congruence relation and hence

$$(a = b) \in \mathtt{ker}(h) \;\Leftrightarrow\; a =_{\mathtt{ker}(h)} b.$$

(ii) Let $h : \mathcal{T} \to \mathcal{S}$ be a morphism and $\mathcal{E}$ be a set of equations over $\mathcal{T}$. Show that

$$\mathcal{E} \subseteq \mathtt{ker}(h) \;\Leftrightarrow\; (=_{\mathcal{E}}) \subseteq \mathtt{ker}(h).$$

Exercise 7.4 Show that $\mathcal{R}$ over $\mathbb{T}(\vec{X})$ justifies $\mathcal{E}$ over $\mathbb{T}$ in exactly one of the following cases. See also Exercise 8.12.

(i) $\mathcal{R} = \{X = \alpha \to \alpha \to X\}$;
$\mathcal{E} = \{\beta = \alpha \to \alpha \to \alpha \to \beta,\ \beta = \alpha \to \alpha \to \alpha \to \alpha \to \beta\}$.
(ii) $\mathcal{R} = \{X = \alpha \to \alpha \to X\}$
$\mathcal{E} = \{\beta = \alpha \to \alpha \to \alpha \to \alpha \to \beta,\ \beta = \alpha \to \alpha \to \alpha \to \alpha \to \alpha \to \alpha \to \beta\}$.

Exercise 7.5 Let $\mathcal{E} = \mathcal{E}(\vec{X})$ be a set of equations over $\mathcal{A}(\vec{X})$. Show that the following are equivalent:

(i) $\mathcal{B}$ justifies $\mathcal{E}$;
(ii) $\mathcal{B} \models h^+(\mathcal{E})$ for some $h^+ : \mathcal{A}(\vec{X}) \to \mathcal{B}$;
(iii) $\mathcal{B} \models \exists \vec{X}.h(\mathcal{E})$, for some $h : \mathcal{A} \to \mathcal{B}$ (here $h : \mathcal{A} \to \mathcal{B}$ is extended naturally to $h : \mathcal{A}(\vec{X}) \to \mathcal{B}(\vec{X})$, for example, $h(a \to X) = h(a) \to X$);
(iv) There is a morphism $h^\sharp : \mathcal{A}[\mathcal{E}] \to \mathcal{B}$.

Exercise 7.6 Let $\mathcal{E}$ be a binary relation on a type algebra $\mathcal{T}$. Show the following for the category of type algebras with their morphisms.

(i) A morphism $h : \mathcal{T}/\mathcal{E} \to \mathcal{S}$ is uniquely determined by a morphism $h^\natural : \mathcal{T} \to \mathcal{S}$ such that $\mathcal{E} \subseteq \mathtt{ker}(h^\natural)$ by $h([a]_\mathcal{E}) = h^\natural(a)$.
(ii) The canonical map $[-]_\mathcal{E} : \mathcal{T} \to \mathcal{T}/\mathcal{E}$ is an epimorphism having as kernel $=_\mathcal{E}$. Conversely, if $h : \mathcal{T} \to \mathcal{S}$ is an epimorphism and $\mathcal{E}$ is its kernel, then $\mathcal{S} \cong \mathcal{T}/\mathcal{E}$.
(iii) Up to isomorphism, $\mathcal{T}/\mathcal{E}$ is determined as the initial object such that $[-]_\mathcal{E} : \mathcal{T} \to \mathcal{T}/\mathcal{E}$ has kernel $\mathcal{E}$. That is, if $\mathcal{S}$ is such that there is a $h : \mathcal{T} \to \mathcal{S}$ with kernel $\mathcal{E}$, then there is a unique arrow $k : \mathcal{T}/\mathcal{E} \to \mathcal{S}$ such that $k \circ [-]_\mathcal{E} = h$.

Exercise 7.7 Let $\mathcal{A}, \mathcal{B}$ be type algebras and let $h : \mathcal{A} \to \mathcal{B}$ be a morphism. Show:

(i) $\mathcal{B} \models h(\mathcal{E}) \;\Leftrightarrow\; \mathcal{E} \subseteq \ker(h)$;
(ii) $\mathcal{B}$ justifies $\mathcal{E} \;\Leftrightarrow\; \exists h : \mathcal{A} \to \mathcal{B}.\mathcal{E} \subseteq \ker(h)$.

Exercise 7.8 Suppose there is a morphism $h : \mathcal{B} \to \mathcal{C}$. Show that if $\mathcal{R}[\vec{X}]$ is justified by $\mathcal{B}$, then it is also justified by $\mathcal{C}$.

Exercise 7.9 Show that for every regular tree $T \in \mathsf{Tr}_{\mathrm{R}}$ there exists a proper simultaneous recursion $\mathcal{R}(\vec{X})$ over $\mathbb{T}$ such that $T = (X_1)^*_{\mathcal{R}}$. [Hint. Let $T_1, \ldots, T_n$, with $T_1 = T$ be the distinct subtrees occurring in T and take $\vec{X} = X_1, \ldots, X_n$.]

Exercise 7.10 (Klop)

(i) Draw the complete μ-reduction graph of $A = \mu\alpha\beta\gamma.\beta$, i.e. the set $\{B \mid A \Rightarrow^*_\mu B\}$ directed by $\Rightarrow_\mu$. [Warning. The set has more than 2010 elements.]
(ii) Do the same for the type $A_{n,k} = \mu\alpha_1 \cdots \alpha_n.\alpha_k$.

Exercise 7.11 Show that $||\mathbb{T}_\mu|| = \{[\alpha] \mid \alpha \in \mathbb{A}\} \cup \{[\mu t.t]\}$.

Exercise 7.12 Let U denote the type $\mu t.(t \to t)$. Show that, for all pure λ-terms M, $\Gamma_0 \vdash_{\lambda\mu} M : U$, where Γ_0 is the environment which assigns type U to all free variables of M.

Exercise 7.13 Show that the rule

$$\frac{\Gamma \vdash M : A \qquad \alpha \text{ does not occur in } \Gamma}{\Gamma \vdash M : \mu\alpha.A}$$

is admissible in $\lambda\mu$.

Exercise 7.14 Let $\mathcal{A}$ be a type algebra with elements a, b such that $b = b \to a$. Provide the de Bruijn version of the terms in Proposition 7.1.4: $\omega \equiv \lambda x.xx$, $\Omega \equiv \omega\omega$, $\mathsf{Y} \equiv \lambda f.(\lambda x.f(xx))(\lambda x.f(xx))$.

Exercise 7.15 Inserting the *fold_* and *unfold_* constants at the proper types, find a version Y_μ of the fixed point operator Y which is well typed in $\lambda_\mu^{\mathrm{Cho}}$. Assuming a reduction rule similar to $(R^{\mathrm{uf}}_{\mathcal{E}})$ of Definition 7.1.24, verify that it can be proved that this has the same reduction properties as Y.

Exercise 7.16 Let

$$\begin{aligned} M &\equiv \lambda x^{\gamma\to\mu\alpha.(\beta\to\alpha)}.x^{\gamma\to\mu\alpha.(\beta\to\alpha)} \\ &\in \Lambda_{=}^{\mathcal{A},\mathrm{Cho}}((\gamma \to \mu\alpha.(\beta \to \alpha)) \to \gamma \to \mu\alpha.(\beta \to \alpha)). \end{aligned}$$

Construct a term M' by adding, to some η-expansion of M, occurrences of *fold* or *unfold* such that

$$M' \in \Lambda_{=}^{\mathcal{A},\mathrm{Cho}}((\gamma \to \mu\alpha.(\beta \to \alpha)) \to \gamma \to \beta \to \mu\alpha.(\beta \to \alpha)).$$

In general construct for each $M \in \Lambda_{=}^{\mathcal{A},\mathrm{Ch}}(A)$ a term $M' \in \Lambda_{=}^{\mathcal{A},\mathrm{Cho}}(A)$ such that $\widehat{M'} =_\eta M$, where $\widehat{M'}$ is the term obtained from M' by erasing all occurrences of *fold* and *unfold*.

Exercise 7.17

(i) A type algebra $\mathcal{A}$ is not always generated by its prime elements. [Hint. If $\mathcal{A} = \{A\}$ with $A = A \to A$, then $||\mathcal{A}|| = \emptyset$.]
(ii) The embedding 7.2.13(i) is not always surjective. [Hint. Use (i).]
(iii) A morphism $h{:}\mathcal{A} \to \mathcal{B}$ is not uniquely determined by $h \restriction ||\mathcal{A}||$.

Exercise 7.18 Let $\mathcal{E}$ be a set of equations over the type algebra $\mathcal{A}$. We say that $\mathcal{E}$ *satisfies* $A = B$, written $\mathcal{E} \models A = B$, if for all type algebras $\mathcal{C}$ and all morphisms $h : \mathcal{A} \to \mathcal{C}$ one has $\mathcal{C} \models h(\mathcal{E}) \;\Rightarrow\; \mathcal{C} \models h(A) = h(B)$. Show that

$$\mathcal{E} \models A = B \;\Leftrightarrow\; \mathcal{E} \vdash A = B.$$

Exercise 7.19 This elaborates the proof of Proposition 7.2.16.

(i) Show that

$$\text{(refl)} \qquad \frac{}{A = A};$$

$$(\to) \qquad \frac{A_1 = B_1 \quad A_2 = B_2}{(A_1 \to A_2) = (B_1 \to B_2)};$$

$$(\mathcal{E}[P,Q]) \qquad \frac{A = P \quad B = Q}{A = B}, \qquad \text{if } (P = Q) \in \mathcal{E} \text{ or } (Q = P) \in \mathcal{E}.$$

is a 'cut-free' (no application of transitivity) axiomatization of $\vdash_\mathcal{E}$.

(ii) Given a statement $A = B$, construct a tree (see Figure 7.14) that describes all possible attempts of backwards derivations of $A = B$, following the cut-free rules. The nodes are labeled with equations $P = Q$ (expressing an equation to be proved) or a symbol '$\to$' or a pair $\mathcal{E}[P,Q]$ (expressing proof-steps in the cut-free system), with P, Q always being subtypes of a type occurring in $\mathcal{E} \cup \{A = B\}$, such that:

(1) There are exactly two 'targets' of the $\to$ or $\mathcal{E}[P,Q]$ nodes.
(2) A node with label $P = Q$ is provable if both targets of at least one of its ($\to$ or $\mathcal{E}[P',Q']$) targets (an $\to$ or an $\mathcal{E}[P',Q']$ node) are provable (in the cut-free version of $\vdash_\mathcal{E}$). [The $P = Q$ nodes are called 'or-nodes' (they have dotted lines) and the $\to$ and $\mathcal{E}[P,Q]$ are called 'and-nodes' (they have solid lines).]

(3) The tree path terminates at an or-node if the formula is $A = A$ (indicating local success) or if the formula is '$P = Q$' which is already below along the same path from the root (indicating local failure).

This tree is finite and contains all possible proofs of $A = B$ (if any exists). Check, among all possible subtrees of this tree in which at an or-node only one successor, and at an and-node both successors are chosen, whether at a terminating node one always has success. If there is at least one such subtree, then $A = B$ is provable in the cut-free system and hence in $\mathcal{E}$, otherwise it is not provable.

Exercise 7.20 Prove Proposition 7.6.13.
[*Hint.* Take I as the direct limit (union with identifications via the $T^k(f)$)

$$0 \xrightarrow{f} T(0) \xrightarrow{T(f)} T^2(0) \xrightarrow{T^2(f)} \cdots,$$

where $0 = \emptyset$ is the initial element of Set and $f{:}\,0{\to}T(0)$ the canonical map. The inverse of the morphism $\mathtt{in}$ is defined by mapping an element of I, say in $T^k(0)$, via $T^k(f)$ to the next level in I. This is an isomorphism.]

Exercise 7.21 Let $T : \mathsf{Set}{\to}\mathsf{Set}$ be a functor and let $\langle C, \mathtt{out}\rangle$ be the final T-coalgebra. Prove the following.

(i) For all $\alpha{:}X \to T(X)$ there exists a unique $f{:}X \to C$ such that the diagram in Fig. 7.15 commutes.
(ii) For all $\alpha{:}X \to T(C+X)$ there exists a unique $f{:}X \to C$ such that the diagram in Fig. 7.16 commutes.

[Hint. Verify that there exists a $g : C + X{\to}C$ such that

$$\begin{array}{ccc} C + X & \xrightarrow{T(\mathtt{nil}) \circ \mathtt{out}\, \alpha} & T(C+X) \\ {\scriptstyle g}\downarrow & & \downarrow{\scriptstyle T(g)} \\ C & \xrightarrow[\mathtt{out}]{} & T(C) \end{array}$$

commutes. Show that $g = [\mathbf{1}_C, g \circ \mathtt{nil}]$, i.e. $g \circ \mathtt{nil} = \mathbf{1}_C$.]

Exercise 7.22 The collection of *streams* over $\mathbb{A}$ is $\mathbb{S}^{\mathbb{A}}$. An element $s \in \mathbb{S}^{\mathbb{A}}$ is often written as $s = \langle s_0, s_1, s_2, \ldots\rangle$, with $s_i = s(i)$. There are two basic operations ('head' and 'tail') on $\mathbb{S}^{\mathbb{A}}$ with $\mathtt{hd} : \mathbb{S}^{\mathbb{A}} \to \mathbb{A}$; $\mathtt{tl} : \mathbb{S}^{\mathbb{A}} \to \mathbb{S}^{\mathbb{A}}$. defined by

$$\mathtt{hd}(\langle s_0, s_1, s_2, \ldots\rangle) = s_0;$$
$$\mathtt{tl}(\langle s_0, s_1, s_2, \ldots\rangle) = \langle s_1, s_2, s_3, \ldots\rangle.$$

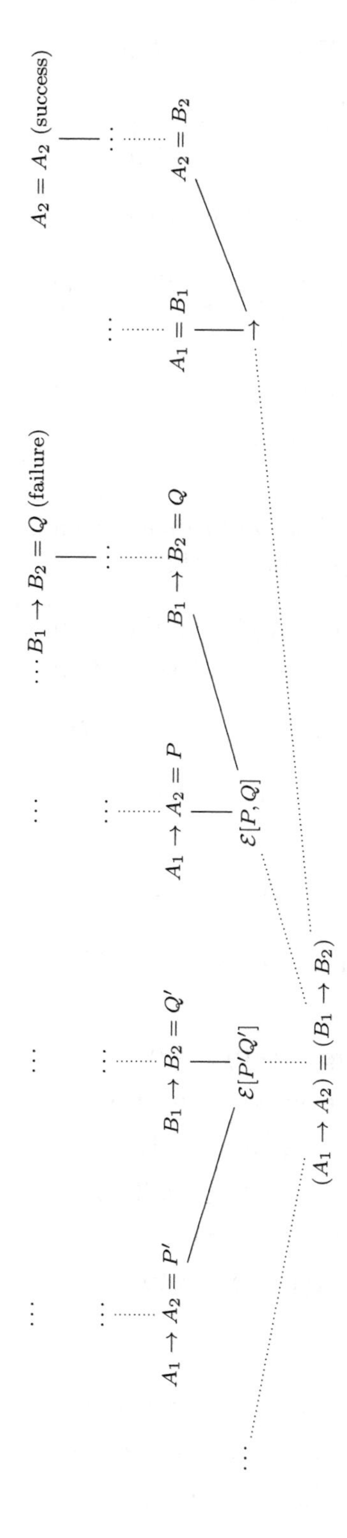

Figure 7.14 Exhaustive search for derivation in the cut-free version of $\vdash_{\mathcal{E}}$, in this case for the formula $A = B$ of the form $(A_1 \to A_2) = (B_1 \to B_2)$.

$$\begin{array}{ccc} X & \xrightarrow{\alpha} & T(X) \\ f \downarrow & & \downarrow T(f) \\ C & \xrightarrow{\mathtt{out}} & T(C) \end{array}$$

Figure 7.15 Co-iteration

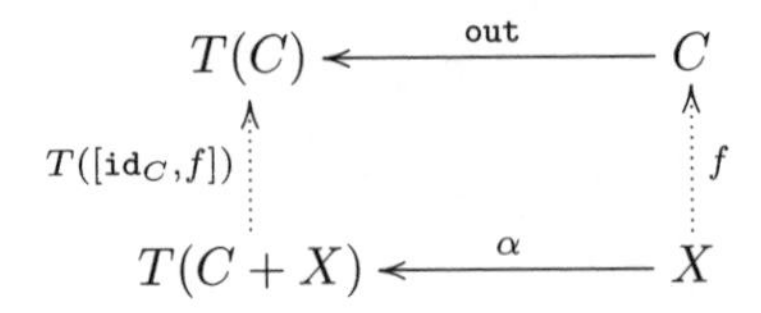

Figure 7.16 Primitive corecursion

The stream $\mathtt{tl}(s)$ is also called the *derivative* of s and is denoted by s'. One has $t = s$ iff $\forall n \in \omega . s(n) = t(n)$, for $s, t \in \mathbb{S}$.

(i) Show that $\mathbb{S}^{\mathbb{A}}$ is the final coalgebra of the functor $T_{\mathbb{S}^{\mathbb{A}}}(X) = \mathbb{A} \times X$.

(ii) A relation $R \subseteq \mathbb{S}^{\mathbb{A}} \times \mathbb{S}^{\mathbb{A}}$ is called a *bisimulation* iff for all $s, t \in \mathbb{S}^{\mathbb{A}}$

$$sRt \;\Rightarrow\; s_0 = t_0 \;\&\; s' R t'. \tag{7.14}$$

Show the principle of co-induction for streams: for R a bisimulation over $\mathbb{S}^{\mathbb{A}}$, and all $s, t \in \mathbb{S}^{\mathbb{A}}$

$$sRt \;\Rightarrow\; s = t.$$

Exercise 7.23 Define maps $\mathtt{even}, \mathtt{odd} : \mathbb{S}^{\mathbb{A}} \to \mathbb{S}^{\mathbb{A}}$ and $\mathtt{zip} : \mathbb{S}^{\mathbb{A}} \times \mathbb{S}^{\mathbb{A}} \to \mathbb{S}^{\mathbb{A}}$ as follows, *by co-recursion*:

$$\begin{aligned}
(\mathtt{even}(s))(0) &= s(0) \\
(\mathtt{even}(s))' &= \mathtt{even}(s'') \\
(\mathtt{odd}(s))(0) &= s(1) \\
(\mathtt{odd}(s))' &= \mathtt{odd}(s'') \\
(\mathtt{zip}(s,t))(0) &= s(0) \\
(\mathtt{zip}(s,t))' &= \mathtt{zip}(t, s').
\end{aligned}$$

This has the effect that

$$\begin{aligned}
\mathtt{even}(\langle a_0, a_1, \ldots \rangle) &= (\langle a_0, a_2, \ldots \rangle) \\
\mathtt{odd}(\langle a_0, a_1, \ldots \rangle) &= (\langle a_1, a_3, \ldots \rangle) \\
\mathtt{zip}(\langle a_0, a_1, \ldots \rangle, \langle b_0, b_1, \ldots \rangle) &= (\langle a_0, b_0, a_1, b_1, \ldots \rangle).
\end{aligned}$$

Show that, for all $s, t \in \mathbb{S}^{\mathbb{A}}$, one has

$$\begin{aligned} \texttt{even}(\texttt{zip}(s,t)) &= s; \\ \texttt{odd}(\texttt{zip}(s,t)) &= t. \end{aligned}$$

[Hint. Show that $R = \{\langle \texttt{even}(\texttt{zip}(s,t)), s\rangle \mid s, t \in \mathbb{S}\}$ is a bisimulation.]

Exercise 7.24 Give a definition using co-iteration of the maps

$$\texttt{even}, \texttt{odd} : \mathbb{S}^{\mathbb{A}} \to \mathbb{S}^{\mathbb{A}} mboxand\ \texttt{zip} : \mathbb{S}^{\mathbb{A}^2} \to \mathbb{S}^{\mathbb{A}}.$$

Exercise 7.25 Using (a slightly modified form of) co-induction, show that

$$\forall s \in \mathbb{S}^{\mathbb{A}}.\texttt{zip}(\texttt{even}(s), \texttt{odd}(s)) = s.$$

Exercise 7.26 Let $F, G : \mathbb{S}^{\mathbb{A}} \to \mathbb{S}^{\mathbb{A}}$. Suppose

$$\forall s \in \mathbb{S}.(F\,s)(0) = (G\,s)(0)\ \&\ (F\,s)' = F(s')\ \&\ (G\,s)' = G(s').$$

Show by co-induction that

$$\forall s \in \mathbb{S}.F\,s = G\,s.$$

Exercise 7.27 Define $\texttt{map} : (\mathbb{A} \to \mathbb{A}) \to \mathbb{S}^{\mathbb{A}} \to \mathbb{S}^{\mathbb{A}}$ by

$$\begin{aligned} (\texttt{map}\ f\ s)(0) &= f(s(0)), \\ (\texttt{map}\ f\ s)' &= \texttt{map}\ f\ s'. \end{aligned}$$

Using Exercise 7.26 show that

$$\forall f, g \in \mathbb{A} \to \mathbb{A} \forall s \in \mathbb{S}.\texttt{map}\,g\,(\texttt{map}\,f\ s) = \texttt{map}\,(g \circ f)\,s.$$

Exercise 7.28 Construct by primitive corecursion a map $g{:}\mathbb{S}^{\mathbb{A}} \to \mathbb{S}^{\mathbb{A}}$ such that

$$g(s) = \langle 0, s_0, s_1, s_2, \ldots \rangle.$$

Exercise 7.29 (Rutten (2005)) For $s, t \in \mathbb{S}$ define the sum and convolution-product corecursively as

$$\begin{aligned} (s+t)(0) &= s(0) + t(0), \\ (s+t)' &= s' + t'; \\ (s \times t)(0) &= s(0) * t(0), \\ (s \times t)' &= (s' \times t) + (s \times t'). \end{aligned}$$

(i) Show the following for all $s, t, u \in \mathbb{S}$.

$$\begin{array}{lrcl}
\text{(a)} & s+t & = & t+s; \\
\text{(b)} & (s+t)+u & = & s+(t+u); \\
\text{(c)} & (s+t)\times u & = & (s\times u)+(t\times u); \\
\text{(d)} & s\times t & = & t\times s; \\
\text{(e)} & (s\times t)\times u & = & s\times(t\times u).
\end{array}$$

[*Hint.* Use co-induction.]

(ii) Show that

$$\langle a_0, a_1, a_2, \ldots\rangle \times \langle b_0, b_1, b_2, \ldots\rangle = \langle c_0, c_1, c_2, \ldots\rangle,$$

with $c_n = \Sigma_{k=0}^{n} \binom{n}{k} a_k b_{n-k}$.

(iii) Show that

$$\langle 1, r, r^2, \ldots\rangle \times \langle 1, s, s^2, \ldots\rangle = \langle 1, r+s, (r+s)^2, \ldots\rangle.$$

[*Hint.* Use (ii). Alternatively, define $GS(r) = \langle 1, r, r^2, \ldots\rangle$ using co-recursion and show $GS(r) \times GS(s) = GS(r+s)$, using co-induction.]

Exercise 7.30 Define the coalgebra *lazy lists* consisting of finite and infinite lists. Is there a polynomial functor for which this is the final coalgebra?

Exercise 7.31 Show that $\mu\alpha.(\alpha \to \mu\beta.\beta) \equiv_\alpha \mu\beta.(\beta \to \mu\beta.\beta)$, in spite of the restriction on variables in the axioms (α-conv).

8
Properties of Recursive Types

In this chapter we study the properties of recursive types independently of their use in typing λ-terms. Most of these properties will however be useful in the study of the properties of typed terms in the next chapter.

We will discuss properties of μ-types and of systems of type equations in two separate sections, even if most of them are indeed similar. In Section 8.1 we build a solution of a simultaneous recursion using μ-types, and show the essential equivalence of the two notations for recursive types. There are, nevertheless, aspects which are treated more naturally in one approach than in the other. For instance, in Chapter 9, the notion of principal type scheme is formulated and studied in a more natural way using simultaneous recursion. In Sections 8.2, 8.3 we show the decidability of weak and strong equality for μ-types.

8.1 Simultaneous recursions vs μ-types

The μ-types and the recursive types defined via a simultaneous recursion turn out to be closely related.

From simultaneous recursions to μ-types

First we show that types in any simultaneous recursion can be simulated (in a precise sense) by μ-types.

Take any simultaneous recursion $\mathcal{R} = \mathcal{R}(\vec{X})$ over $\mathbb{T} = \mathbb{T}^{\mathbb{A}}$. We show that $\mathbb{T}_\mu \models \exists \vec{X}.\mathcal{R}(\vec{X})$, for $\mathbb{T}_\mu = \mathbb{T}_\mu^{\mathbb{A}}$. We do this in a constructive way by building an n-tuple of types $S_1, \ldots, S_n \in \mathbb{T}_\mu$ such that $\mathbb{T}_\mu \models \mathcal{R}(\vec{S})$. This means that

$$\begin{aligned} S_1 &=_\mu C_1[\vec{X} := \vec{S}] \\ &\vdots \\ S_n &=_\mu C_n[\vec{X} := \vec{S}], \end{aligned}$$

if $\mathcal{R}$ is of the form

$$\begin{aligned} X_1 &= C_1(\vec{X}) \\ &\vdots \\ X_n &= C_n(\vec{X}), \end{aligned} \tag{8.1}$$

with the $C_i(\vec{X}) \in \mathbb{T}[\vec{X}]$. The following construction is taken from Bekič (1984), p. 39, the 'Bisection Lemma', and is regarded as part of the 'folklore' of the subject.

8.1.1 Theorem *Let $\mathcal{R} = \mathcal{R}(\vec{X})$ be a simultaneous recursion over $\mathbb{T} = \mathbb{T}^{\mathbb{A}}$ as* (8.1) *above.*

(i) *There is a morphism $h : \mathbb{T}(\vec{X}) \to \mathbb{T}_\mu$, leaving $\mathbb{A}$ fixed, such that*

$$S_1 = h(X_1), \ldots, S_n = h(X_n)$$

is a solution of $\mathcal{R}(\vec{X})$ in $\mathbb{T}_\mu$. This means that for each $1 \le i \le n$

$$S_i =_\mu C_i[\vec{X} := \vec{S}].$$

(ii) *Therefore $\mathbb{T}_\mu \models \exists \vec{X}.\mathcal{R}(\vec{X})$, i.e. $\mathbb{T}_\mu$ solves $\mathcal{R}$.*

Proof (i) First define the types $\vec{P} = P_1, \ldots, P_n \in \mathbb{T}_\mu$ for $1 \le i \le n$.

$$\begin{aligned} P_1 &\triangleq \mu X_1.C_1 \\ P_2 &\triangleq \mu X_2.(C_2[X_1 := P_1]) \\ &\vdots \\ P_n &\triangleq \mu X_n.(C_n[X_1 := P_1] \cdots [X_{n-1} := P_{n-1}]). \end{aligned}$$

Then define the types $\vec{S} = S_1, \ldots, S_n \in \mathbb{T}_\mu$ by

$$\begin{aligned} S_n &\triangleq P_n \\ S_{n-1} &\triangleq P_{n-1}[X_n := S_n] \\ &\vdots \\ S_1 &\triangleq P_1[X_2 := S_2, \ldots, X_n := S_n]. \end{aligned}$$

Finally set

$$\begin{aligned} h(\alpha) &\triangleq \alpha, && \text{if } \alpha \in \mathbb{A}, \\ h(X_i) &\triangleq S_i, \\ h(A \to B) &\triangleq h(A) \to h(B). \end{aligned}$$

Clearly h is a morphism fixing $\mathbb{T}_\mu$.

For notational simplicity we write down the proof for $n = 2$. The proof in this case is similar to the first proof of the Double Fixed-Point Theorem 6.5.1 in B[1984]. Let $\mathcal{R}(X_1, X_2)$ be

$$\begin{aligned} X_1 &= C_1(X_1, X_2), \\ X_2 &= C_2(X_1, X_2). \end{aligned}$$

Write the construction of the $\vec{S}$ as

$$\begin{aligned} P_1^{X_2} &= \mu X_1.C_1(X_1, X_2), \\ P_2 &= \mu X_2.C_2(P_1^{X_2}, X_2), \\ S_2 &= P_2, \\ S_1 &= P_1^{S_2}. \end{aligned}$$

Then

$$\begin{aligned} S_2 &= P_2 \\ &= C_2(P_1^{P_2}, P_2) \\ &= C_2(S_1, S_2); \\ S_1 &= P_1^{S_2} \\ &= C_1(S_1, S_2). \end{aligned}$$

(ii) By (i) and Definition 7.3.5. ∎

8.1.2 Example Take the following simultaneous recursion $\mathcal{R}_1$:

$$\begin{array}{lcl} X_1 &=& X_2 \to X_1 \\ X_2 &=& X_1 \to X_2. \end{array}$$

Applying Definition 8.1.1 we have:

$$\begin{array}{lcl} P_1 &=& \mu X_1.(X_2 \to X_1) \\ P_2 &=& \mu X_2.((\mu X_1.(X_2 \to X_1)) \to X_2) \end{array}$$

and then

$$\begin{array}{lcl} S_2 &=& \mu X_2.((\mu X_1.(X_2 \to X_1)) \to X_2) \\ S_1 &=& \mu X_1.(\mu X_2.((\mu X_1'.(X_2 \to X_1')) \to X_2) \to X_1). \end{array}$$

It is easy to check that $S_1 =_\mu S_2 \to S_1$ and $S_2 =_\mu S_1 \to S_2$. We can prove that $S_1 \neq_\mu S_2$ using the technique developed in Section 8.2.

8.1.3 Remark The solution $\langle S_1, \dots, S_n \rangle$ is not unique in $\mathbb{T}_\mu$ (modulo $=_\mu$). In Example 8.1.2 both the pairs $\langle S_2, S_1 \rangle$ and $\langle S, S \rangle$, where $S = \mu X.X \to X$, solve $\mathcal{R}_1$. Note, however, that $(S_1)^*_\mu = (S_2)^*_\mu = (S)^*_\mu$ are the same trees.

Note also that in this proof all the needed type equivalences can be obtained by Axiom (μ-eq). No other rules of Definition 7.4.27 (for instance ($\to$-cong)) have been used. So the solution is one of a rather strong kind.

8.1.4 Corollary

(i) *For the morphism* $h : \mathbb{T}[\vec{X}] \to \mathbb{T}_\mu$ *of Theorem* 8.1.1 *one has, for all* $A, B \in \mathbb{T}(\vec{X})$,

$$A =_{\mathcal{R}} B \Rightarrow h(A) =_\mu h(B).$$

(ii) *Therefore* h *induces a morphism* $h^\sharp : \mathbb{T}[\mathcal{R}] \to \mathbb{T}_\mu$, *satisfying* $h^\sharp([\alpha]) = \alpha$ *for* $\alpha \in \mathbb{A}$ *and* $h^\sharp([A]) = h(A)$, *for* $A \in \mathbb{T}(\vec{X})$.

(iii) $\vdash_{\mathbb{T}^{\mathbb{A}}[\mathcal{R}]} M : A \Rightarrow \vdash_{\mathbb{T}_\mu} M : h^\sharp(A)$.

Proof (i) By induction on the derivation of $A =_{\mathcal{R}} B$.
(ii) By Proposition 7.2.4(i).
(iii) By Lemma 7.1.22. ∎

8.1.5 Remark In general the reverse of Corollary 8.1.4(i) is not valid. Take for example

$$\mathcal{R}_0 = \{X_1 = X_1 \to X_1,\ X_2 = X_2 \to X_2\}.$$

The construction of Theorem 8.1.1 provides $S_1 = S_2 = \mu X.(X \to X)$, but one has $X_1 \neq_{\mathcal{R}_0} X_2$.

In the case of solutions modulo strong equivalence there are other constructions in the literature which give simpler solutions, see Ariola and Klop (1996) for example.

From μ-type to simultaneous recursion

In the opposite direction, any μ-type can be represented by an equivalent simultaneous recursion. While the idea underlying this property is intuitively simple, a formal definition in the case of weak equivalence becomes complicated and not especially interesting. In the case of strong equivalence, however, this notion can be easily formalized by exploiting the interpretation of types as infinite trees.

We define for each $A \in \mathbb{T}_\mu^{\mathbb{A}}$ a simultaneous recursion $\mathcal{R}$ and a $B \in \mathbb{T}^{\mathbb{A}}$ such that $(B)^*_{\mathcal{R}}$ is the same infinite tree as $(A)^*_\mu$. In the following definition we assume, as usual, that all bound variables in A have different names, which are also different from the names of all free variables in A.

8.1.6 Definition We associate to a $A \in \mathbb{T}_\mu$ a type $\mathcal{T}(A) \in \mathbb{T}^{\mathbb{A}}$ and a simultaneous recursion $\mathcal{S}R(A)$ over $\mathbb{T}^{\mathbb{A}}$ defined in the following way.

$$\begin{array}{rclrcll} \mathcal{T}(\alpha) & \triangleq & \alpha & \mathcal{S}R(\alpha) & \triangleq & \emptyset, & \text{for } \alpha \in \mathbb{A}, \\ \mathcal{T}(A \to B) & \triangleq & \mathcal{T}(A) \to \mathcal{T}(B) & \mathcal{S}R(A \to B) & \triangleq & \mathcal{S}R(A) \cup \mathcal{S}R(B) & \\ \mathcal{T}(\mu\alpha A) & \triangleq & \alpha & \mathcal{S}R(\mu\alpha A) & \triangleq & \mathcal{S}R(A) \cup \{\alpha = \mathcal{T}(A)\}. & \end{array}$$

8.1.7 Theorem $(A)^*_\mu = (\mathcal{T}(A))^*_{\mathcal{S}R(A)}$

Proof See Exercise 8.1. ∎

8.2 Properties of μ-types

In this section we investigate properties of μ-types both under weak and strong equivalence. In particular we show that both relations are invertible and decidable.

Invertibility and decidability of weak equivalence

Invertibility and decidability will be proved using the notion of reduction $\Rightarrow_\mu$, defined in Definition 7.4.32. Note that $\Rightarrow_\mu$ is not normalizing: types can be infinitely unfolded. However, the CR property, Proposition 7.4.33(ii), suffices to prove invertibility.

8.2.1 Theorem *The relation $=_\mu$ on $\mathbb{T}_\mu$ is invertible, i.e., for all A_1, A_2, B_1, $B_2 \in \mathbb{T}_\mu$,*

$$A_1 \to B_1 =_\mu A_2 \to B_2 \quad \textit{implies} \quad A_1 =_\mu A_2 \;\&\; B_1 =_\mu B_2.$$

Proof Assume $A_1 \to B_1 =_\mu A_2 \to B_2$. By Lemma 7.4.33(ii) there is a type C such that $A_1 \to B_1 \Rightarrow^*_\mu C$ and $A_2 \to B_2 \Rightarrow^*_\mu C$. But then we must have $C \equiv C_1 \to C_2$ with $A_i \Rightarrow^*_\mu C_1$ and $B_i \Rightarrow^*_\mu C_2$ for $i = 1, 2$. By 7.4.33(i) this implies $A_1 =_\mu A_2$ and $B_1 =_\mu B_2$. ∎

Decidability of weak equivalence

The relation $=_\mu$ is also decidable. The proof occupies Sections 8.2.3–8.2.31. This result is relevant, as one of the typing rules in $\lambda\mu$, see Definition 7.4.30, is

$$\vdash_{\lambda\mu} M : A,\; A =_\mu B \Rightarrow \vdash_{\lambda\mu} M : B.$$

Hence decidability of $=_\mu$ is needed to show the system is recursively axiomatizable. But the system $\vdash_{\lambda\mu}$ can be reformulated more carefully by postulating

$$\Gamma \vdash M : A, A \Rightarrow_\mu B \text{ or } B \Rightarrow_\mu A \;\Rightarrow\; \Gamma \vdash M : B.$$

This generates the same system $\lambda\mu$ in a different way, giving obviously a recursive axiomatization.

Surprisingly, decidability is not so easy to prove, due to the transitivity of equality (a 'cut'-rule) and the presence of α-conversion. This makes proof-search potentially undecidable, as infinitely many types may have to be tried. In Cardone and Coppo (2003) a proof system without cut-rule was presented. But recently Wil Dekkers pointed out that their treatment was not completely sound. Nevertheless their result is correct and in Endrullis et al. (2011) three different proofs of the decidability of $=_\mu$ were given, two of them based on Cardone and Coppo (2003) and the third on tree-regular languages. The proof below, inspired by the second proof in Endrullis et al. (2011), is elementary and straightforward.

8.2.2 Remark One obvious way of trying to prove decidability of weak equivalence is by introducing a reduction relation $\Rightarrow_\mu^{-1}$, defined by the contraction rule

$$A[\alpha := \mu\alpha.A] \Rightarrow_\mu^{-1} \mu\alpha.A.$$

If the relation $\Rightarrow_\mu^{-1}$ were Church–Rosser, then each $=_\mu$-equivalence class would have a unique minimal representative reachable in a finite number of steps. From this, decidability of $=_\mu$ would follow. Unfortunately $\Rightarrow_\mu^{-1}$ is not Church–Rosser, see Exercise 8.7, so this proof strategy fails.

The cut-rule (trans) in the definition of system (μ) in Fig. 7.12 causes problems in the proof of decidability. In seeking a proof of $A = C$ one must find a type B such that $A = B$ and $B = C$. But in general there are many possible choices for B. In Cardone and Coppo (2003) another proof system (μ^-), without rule (trans), is defined that turns out to be equivalent with the proof-system (μ).

8.2.3 Definition (The system (μ^-))

(axiom)	$A = A$
(left μ-step)	$\dfrac{A_1[\alpha := \mu\alpha.A_1] = B}{\mu\alpha.A_1 = B}$
(right μ-step)	$\dfrac{A = B_1[\alpha := \mu\alpha.B_1]}{A = \mu\alpha.B_1}$
(μ-cong)	$\dfrac{A_1 = B_1}{\mu\alpha.A_1 = \mu\alpha.B_1}$
($\rightarrow$ -cong)	$\dfrac{A_1 = B_1 \quad A_2 = B_2}{A_1 \rightarrow A_2 = B_1 \rightarrow B_2}$

We write $\vdash_{\mu^-} A = B$ if $A = B$ is provable in the system (μ^-).

In 8.2.4–8.2.11 it will be shown that

$$A =_\mu B \Leftrightarrow \vdash_{\mu^-} A = B.$$

8.2.4 Definition (Standard reduction $\Rightarrow^*_{\text{st}}$)

(i) Consider a $\Rightarrow_\mu$-reduction sequence

$$R : A_1 \Rightarrow_\mu A_2 \Rightarrow_\mu \cdots \Rightarrow_\mu A_n.$$

In each step $A_i \Rightarrow_\mu A_{i+1}$ mark all μ occurring in A_i to the left of the contracted μ with a symbol $\diamondsuit$. Then R is called a *standard* $\Rightarrow_\mu$-*reduction* if only non-marked μ are contracted in R.

(ii) Write $A \Rightarrow^*_{\text{st}} B$ if there is a standard reduction $R : A \Rightarrow^*_\mu B$.

8.2.5 Definition Let the length $l(A)$ of a type A be defined recursively by

$$l(\alpha) = 1,\ \ l(A_1 \rightarrow A_2) = 1 + l(A_1) + l(A_2),\ \ l(\mu\alpha.A_1) = 1 + l(A_1).$$

8.2.6 Lemma *The relation* $\Rightarrow^*_{st}$ *is the smallest relation such that:*

(a) $A \Rightarrow^*_{st} A$;

(b) *If* $A_1[\alpha := \mu\alpha.A_1] \Rightarrow^*_{st} B$, *then* $\mu\alpha.A_1 \Rightarrow^*_{st} B$;

(c) *If* $A_1 \Rightarrow^*_{st} B_1$, *then* $\mu\alpha.A_1 \Rightarrow^*_{st} \mu\alpha.B_1$;

(d) *If* $A_1 \Rightarrow^*_{st} A'_1$ *and* $A_2 \Rightarrow^*_{st} A'_2$, *then* $(A_1 \rightarrow A_2) \Rightarrow^*_{st} (A'_1 \rightarrow A'_2)$.

Proof Write $A \Rightarrow'_{\text{st}} B$ if (A, B) is in the least relation satisfying (a)–(d). By induction on the generation of $\Rightarrow'_{\text{st}}$ one has

$$(A \Rightarrow'_{\text{st}} B) \;\Rightarrow\; (A \Rightarrow^*_{\text{st}} B).$$

Conversely, suppose $A \Rightarrow^*_{\text{st}} B$. Then there exists a standard reduction $A \Rightarrow^* B$. By induction on (n, l), i.e. on the ordinal $\omega n + l$, where n is the length of this reduction and $l = l(A)$, we show $A \Rightarrow'_{\text{st}} B$. The only interesting cases are the following.

Case 1. $A \equiv A_1 \to A_2$ then $B \equiv B_1 {\to} B_2$ and $A_1 \Rightarrow^*_{\text{st}} B_1$, $A_2 \Rightarrow^*_{\text{st}} B_2$. By the induction hypothesis $A_i \Rightarrow'_{\text{st}} B_i$, hence $A_1 \to A_2 \Rightarrow'_{\text{st}} B_1 {\to} B_2$.

Case 2. $A \equiv \mu\alpha.A_1$.

Subcase 2.1. $\mu\alpha.A_1 {\Rightarrow_\mu} A_1[\alpha := \mu\alpha.A_1] \Rightarrow^*_{\text{st}} B$. By the induction hypothesis $A_1[\alpha := \mu\alpha.A_1] \Rightarrow'_{\text{st}} B$ and hence $A \Rightarrow'_{\text{st}} B$, by clause (b).

Subcase 2.2. $A \Rightarrow^*_{\text{st}} B$ is $\mu\alpha.A_1 \Rightarrow_\mu \mu\alpha.A_2 \Rightarrow^*_{\text{st}} B$. Then $B \equiv \mu\alpha.A_{n+1}$ and $A_1 \Rightarrow^*_{\text{st}} A_{n+1}$. By the induction hypothesis $A_1 \Rightarrow'_{\text{st}} A_{n+1}$ hence $A \Rightarrow'_{\text{st}} B$. ∎

8.2.7 Example Let $A = \mu\alpha\mu\beta.C[\alpha, \beta]$, where $C[\alpha, \beta]$ is a type in which possibly α and β occur several times. Let $B = \mu\alpha.C[\alpha, \mu\beta.C[\alpha, \beta]]$. Consider

(i) $A \Rightarrow_\mu B \Rightarrow_\mu C[B, \mu\beta.C[B, \beta]]$.
(ii) $A \Rightarrow_\mu \mu\beta.C[A, \beta] \Rightarrow_\mu C[A, \mu\beta.C[A, \beta]] \Rightarrow_\mu C[B, \mu\beta.C[A, \beta]] \Rightarrow_\mu C[B, \mu\beta.C[B, \beta]]$.

Then (i) is not a standard reduction, but (ii) is.

The following standardization theorem for $\Rightarrow^*_\mu$ can be obtained as a particular case of the standardization theorem for Combinatory Reduction Systems; see Klop (1980), Theorem 6.2.8.8, page 193. For another proof see Endrullis et al. (2011).

8.2.8 Lemma *If* $A \Rightarrow^*_\mu B$ *then* $A \Rightarrow^*_{st} B$.

8.2.9 Corollary $A =_\mu B \;\Leftrightarrow\;$ *A and B have a common standard reduct.*

Proof Use the fact that the reduction relation $\Rightarrow^*_\mu$ is Church–Rosser. ∎

8.2.10 Proposition $\vdash_{\mu^-} A = B \;\Leftrightarrow\;$ *A and B have a common standard reduct.*

Proof ($\Rightarrow$) By induction on the derivation of $\vdash_{\mu^-} A = B$, using Lemma 8.2.6.

($\Leftarrow$) Let $A \Rightarrow^*_\mu C$, $B \Rightarrow^*_\mu C$. Referring to Lemma 8.2.6 we can distinguish the following cases.

Case (a) for both A and B, i.e. $A \equiv C$ and $B \equiv C$.

Case (b) for A (and similarly for B).

Case (c) for A and for B.

Case (d) for A and for B.

This suffices. If we have Case (c) for A and Case (a) for B, then in fact we have Case (c) for both A and B. If we have Case (d) for A and Case (a) for B, then (d) holds for both A and B. Finally, Case (c) for A and Case (d) for B (or vice versa) cannot occur at the same time. In all these four cases we easily conclude $\vdash_{\mu^-} A = B$. ■

8.2.11 Corollary $A =_\mu B \Leftrightarrow \vdash_{\mu^-} A = B$.

Proof By Corollary 8.2.9. ■

Items 8.2.3–8.2.11 all come from Cardone and Coppo (2003), but their proof of the decidability of $\vdash_{\mu^-} A = B$ incorrectly treated rule (μ-cong).

In the proof below we only focus on the decidability of $\vdash_{\mu^-} A = B$ and not on efficiency. When we give upper bounds on numbers of steps in proofs then these are not meant to be sharp and they are only given to provide a simple proof.

In order to show decidability of $A =_\mu B$, we will focus on 'proof attempts'. Some of these will succeed, other may fail. Proof attempts will be written upside down, with on top the equation $A = B$ for which the provability is being investigated and the rules of Definition 8.2.3 in reverse order. Then rule (μ-cong) has the form

$$\frac{\mu\alpha.A_1 = \mu\alpha.B_1}{A_1[\alpha := \beta] = B_1[\alpha := \beta]}\ \beta \notin \mathrm{FV}(\mu\alpha.A_1) \cup \mathrm{FV}(\mu\alpha.B_1).$$

(Of course by the variable convention also $\beta \notin \mathrm{BV}(A_1) \cup \mathrm{BV}(B_1)$.) A successful proof attempt for the equality $\mu\alpha\beta.(\alpha \to \beta) = \mu\alpha.(\alpha \to \mu\gamma\beta.(\alpha \to \beta))$ is in Example 8.2.12.

8.2.12 Example

$$\cfrac{\cfrac{\cfrac{\mu\alpha\beta.(\alpha \to \beta) = \mu\alpha.(\alpha \to \mu\gamma\beta.(\alpha \to \beta))}{\mu\beta.(\delta \to \beta) = \delta \to \mu\gamma\beta.(\delta \to \beta)}\ (\mu\text{-cong})}{\delta \to \mu\beta.(\delta \to \beta) = \delta \to \mu\gamma\beta.(\delta \to \beta)}\ (\text{left } \mu\text{-step})}{\delta = \delta \qquad \cfrac{\mu\beta.(\delta \to \beta) = \mu\gamma\beta.(\delta \to \beta)}{\mu\beta.(\delta \to \beta) = \mu\beta.(\delta \to \beta).}\ (\text{right } \mu\text{-step})}\ (\to\text{-cong})$$

We will show that each path in a proof attempt for $\vdash_{\mu^-} A = B$, with some restrictions, has finite length and conclude from this that the total number of nodes in an exhaustive proof search tree is finite. Then the decidability of $\vdash_{\mu^-} A = B$ follows.

8.2.13 Remark The example above was a successful proof attempt for $\vdash_{\mu^-} A = B$. Other such attempts may fail. For example, starting with a left μ-step and a right μ-step, one encounters again at some stage a node $A = B$, so this attempt results in an infinite tree. Of course one should stop the proof attempt at such a node.

In general if in a proof attempt for $\vdash_{\mu^-} A = B$ we arrive at a node $C = D$ encountered earlier on the path from the root to $C = D$, then we stop at such a node. In fact we also stop when we arrive at a so-called equivalent node.

8.2.14 Definition Equations $A_1 = B_1$ and $A_2 = B_2$ are called *equivalent* if the one results from the other by a change of free variables.

8.2.15 Lemma *Attempts for equivalent equations have the same structure. Let for simplicity $A_2 = B_2$ result from $A_1 = B_1$, and vice versa, by renaming the free occurrences of α_1 into α_2, and vice versa, where $\alpha_1 \neq \alpha_2$. Then a proof attempt for $A_2 = B_2$ results from one for $A_1 = B_1$.*

Proof First replace all occurrences of α_2 in the proof attempt for $A_1 = B_1$ by a fresh variable γ and then replace all α_1 by α_2. ∎

So it is not a restriction when we stop at a node that is equivalent with a node that we encountered earlier.

Now, what exactly is a proof attempt for $A = B$? It is a binary tree with $A = B$ as root. Starting with $A = B$ we apply each time one rule of Definition 8.2.3, upside down, as long as a rule is applicable, and we are not at a node that is an axiom $C = C$. In general at most one rule is applicable, but at a node $\mu\alpha.C = \mu\alpha.D$ there are typically three possibilities: (left μ-step), (right μ-step), and (μ-cong). There we choose one from the three rules that can be applied. The tree is branched in nodes $A_1 \to A_2 = B_1 \to B_2$ because there we get two nodes $A_1 = B_1$ and $A_2 = B_2$ and we continue with both these two. We stop at a node that is an axiom. We also stop at a node when no rule can be applied any longer and we stop at a node $C = D$ that is equivalent with a node $C' = D'$ that we encountered earlier on the path from the root $A = B$ to $C = D$. When all the leaves of the resulting tree are axioms then the attempt is successful. Then $\vdash_{\mu^-} A = B$ holds and we stop. Else we start a new attempt. If none of the possible proof attempts is successful then $\vdash_{\mu^-} A = B$ does not hold.

8.2.16 Remark One has $\text{FV}(A_1[\alpha := \mu\alpha.A_1]) = \text{FV}(\mu\alpha.A_1)$. Therefore

$$A =_\mu B \Rightarrow \text{FV}(A) = \text{FV}(B).$$

So if $\text{FV}(A) \neq \text{FV}(B)$ then one cannot have $\vdash_{\mu^-} A = B$. Therefore we may restrict ourselves to proof attempts such that for all occurring $C = D$ one has $\text{FV}(C) = \text{FV}(D)$.

This last equality plays an essential role in the decidability proof. We consider paths in proof attempts i.e. sequences of equations and also the corresponding sequences of left and right hand sides. We wish to put restrictions on these sequences of left and right hand sides such that they are the same for E and F occurring in an equation $E = F$. This will be reached by putting the same restriction on E and F whenever $\text{FV}(E) = \text{FV}(F)$. Of course also restrictions on the free variables in the types occurring in equations $C = D$ in a proof attempt only need to be stated for $\text{FV}(C)$.

We put some more restrictions on the proof attempts. If in a proof attempt for $A = B$ one arrives at a node $\mu\alpha.C = \mu\alpha.D$, then one may continue, applying rule (μ-cong), with $C[\alpha := \beta] = D[\alpha := \beta]$, $\beta \notin \text{FV}(C)$. In the case $\alpha \in \text{FV}(C)$ this may result in infinitely many equations. These, however, are all equivalent and hence by Lemma 8.2.15 they have proof attempts of the same structure.

8.2.17 Definition

(i) If in two different proof attempts for $A = B$ we arrive at a node $\mu\alpha.C = \mu\alpha.D$ along the same path from $A = B$ and we apply rule (μ-cong) in both cases, then we substitute the same β.

(ii) A proof attempt is *special* if for every step

$$\frac{\mu\alpha.A_1 = \mu\alpha.B_1}{A_1[\alpha := \beta_1] = B_1[\alpha := \beta_1]}\ \beta_1 \notin \text{FV}(\mu\alpha.A_1)$$

occurring in it and for every lower step

$$\frac{\mu\gamma.C = \mu\gamma.D}{C[\gamma := \beta_2] = D[\gamma := \beta_2]}\ \beta_2 \notin \text{FV}(\mu\gamma.C)$$

such that $\text{FV}(\mu\gamma.C) = \text{FV}(\mu\alpha.A_1)$, one has $\beta_2 = \beta_1$. By the variable convention we may assume that the bound variables in C, D are such that β_2 is fresh.

We may restrict ourselves to special proof attempts, as follows from the following lemma.

8.2.18 Lemma *For each proof attempt there exists a special proof attempt of the same structure and such that the one is successful iff the other is.*

Proof Change the proof attempt by steps in a successful proof attempt as follows. Consider the first application of rule (μ-cong) where an incorrect β_2 is substituted. Replace it by the correct β_1. Then first change every occurrence of β_1 below that node into a fresh variable γ and then replace each β_2 into β_1. ∎

8.2.19 Definition The binary relation $\leadsto : \mathbb{T}_\mu \to \mathbb{T}_\mu$ is defined as

$$\begin{array}{rcll} \mu\alpha.A_1 & \leadsto & A_1[\alpha := \beta], \quad \beta \notin \mathrm{FV}(\mu\alpha.A_1) \cup \mathrm{BV}(A_1); & \\ \mu\alpha.A_1 & \leadsto & A_1[\alpha := \mu\alpha.A_1]; & \\ (A_1 \to A_2) & \leadsto & A_i, & \text{for } i = 1, 2. \end{array}$$

8.2.20 Remark Note that $\leadsto$ is not meant to be a reduction relation, compatible with μ or $\to$. For example, $\mu\beta\mu\alpha.A_1 \leadsto \mu\beta.A_1[\alpha := \mu\alpha.A_1]$. is not required.

8.2.21 Example The following are examples of $\leadsto$-sequences.

(i) Let $A = \mu\alpha.\alpha \to (\alpha \to \beta)$. Then

$$A \leadsto A \to (A \to \beta) \leadsto A \leadsto A \to (A \to \beta) \leadsto A \to \beta \leadsto A \leadsto \alpha \to (\alpha \to \beta).$$

(ii) Let $A = \alpha \to \mu\beta\gamma.(\gamma \to \beta)$. Then

$$A \leadsto \mu\beta\gamma.(\gamma \to \beta) \leadsto \mu\gamma.(\gamma \to \mu\beta\gamma.(\gamma \to \beta)) = \mu\alpha.A.$$

So, in the following $\leadsto$-sequence, infinitely many different types occur

$$\mu\alpha.A \leadsto A[\alpha := \beta_1] \leadsto^* \mu\alpha.A \leadsto A[\alpha := \beta_2] \leadsto^* \mu\alpha.A \leadsto A[\alpha := \beta_3] \leadsto^* \cdots$$

In order to prevent infinitely many variables β_i in a $\leadsto$-sequence such as the one above, we put a restriction on them such that in a sequence like

$$\cdots \leadsto^* \mu\alpha.A \leadsto A[\alpha := \beta_1] \leadsto^* \mu\alpha.A \leadsto A[\alpha := \beta_2] \leadsto \cdots$$

we always have $\beta_1 = \beta_2$. In fact we define the following restriction on $\leadsto$-sequences.

8.2.22 Definition

(i) A $\leadsto$-sequence

$$\cdots \leadsto^* \mu\alpha.A \leadsto A[\alpha := \beta_1] \leadsto^* \mu\gamma.C \leadsto C[\gamma := \beta_2] \leadsto \cdots$$

is *special* if $\mathrm{FV}(\mu.\alpha.A) = \mathrm{FV}(\mu\gamma.C) \;\Rightarrow\; \beta_1 = \beta_2$.

(ii) A *quasi* $\leadsto$*-sequence* is a special $\leadsto$-sequence, with some $\leadsto$ replaced by $\equiv$.

8.2.23 Lemma *Let p be a path in a special proof attempt starting from $A = B$. Then the left hand sides of the equations in that path form a quasi $\leadsto$-sequence from A and the right hand sides a quasi $\leadsto$-sequence from B.*

Proof Immediate from Definition 8.2.17(ii) and Definition 8.2.22. ∎

Now we will show in Corollary 8.2.27 that the number of different types occurring in a quasi $\leadsto$-sequence starting from a type A is bounded by $\mathrm{dpt}(A) - 1$, where dpt is the depth, Definition 1.1.22(i). From that it follows in Lemma 8.2.29 that paths in a proof attempt for $A = B$ without repetitions have finite length implying the decidability of $A = B$, Theorem 8.2.31.

8.2.24 Definition The notion of *depth* of a type can be extended to $A \in \mathbb{T}_\mu$, and again written as $\mathrm{dpt}(A)$, as follows:

$$\begin{aligned}
\mathrm{dpt}(\alpha) &\triangleq 1; \\
\mathrm{dpt}(A \to B) &\triangleq \max\{\mathrm{dpt}(A), \mathrm{dpt}(B)\} + 1; \\
\mathrm{dpt}(\mu\alpha.A) &\triangleq \mathrm{dpt}(A) + 1.
\end{aligned}$$

Let $\leadsto'$ be defined as $\leadsto$, but without the rule $\mu\alpha.A_1 \leadsto A_1[\alpha := \mu\alpha.A_1]$. We have the following.

8.2.25 Lemma *Let $m'(A)$ be the maximal length of a special $\leadsto'$- sequence without repetitions starting with A. Then $m(A) = \mathrm{dpt}(A) - 1$.*

Proof We have $m'(A) \leq \mathrm{dpt}(A) - 1$, because at each such $\leadsto'$-step the depth of the type decreases by at least 1 and for a variable we have $m'(\alpha) = 0 = \mathrm{dpt}(\alpha) - 1$. On the other hand for each type A there exists a maximal $\leadsto'$-sequence

$$A = A_0 \leadsto' A_1 \leadsto' A_2 \leadsto' \cdots \leadsto' A_n,$$

which can be constructed as follows. If $A_i = B_1 \to B_2$, take A_{i+1} as B_j with maximal depth. If $A_i = \mu\alpha.B$, then take $A_{i+1} = B[\alpha := \beta]$ for some β, where possibly β is determined by a previous application of the rule $\mu\alpha.A_1 \leadsto A_1[\alpha := \beta]$. We have $\mathrm{dpt}(A_n) = 1$ and $\mathrm{dpt}(A_{i-1}) = \mathrm{dpt}(A_i) + 1$. Therefore $\mathrm{dpt}(A) = n + 1$ and $m'(A) = n = \mathrm{dpt}(A) - 1$. ∎

Now applications of the rule $\mu\alpha.A_1 \leadsto A_1[\alpha := \mu\alpha.A_1]$ do not add extra to the length, because we have the following.

8.2.26 Lemma *If $A = A_0 \rightsquigarrow A_1 \rightsquigarrow A_2 \rightsquigarrow \cdots \rightsquigarrow A_n$ is a special $\rightsquigarrow$-sequence without repetitions then there exists a similar $\rightsquigarrow'$-sequence, of the form $A = B_0 \rightsquigarrow' B_1 \rightsquigarrow' B_2 \rightsquigarrow' \cdots \rightsquigarrow' B_n$, such that for each i we have*

$$A_i = B_i[\gamma_1 := C_1, \ldots, \gamma_m := C_m],$$

where $C_k = \beta$ for some β or $C_k = A_j$ for some $j \leq i$.

Proof By an easy induction on n. By the variable convention choose $\mathrm{BV}(A)$ such that $\mathrm{FV}(A_i) \cap \mathrm{BV}(A) = \emptyset$ for each i; use that if $B_i = \mu\gamma.C_i$, then $\gamma \in \mathrm{BV}(A)$. ■

Now by Lemma 8.2.25 we immediately get

8.2.27 Corollary *Let $m(A)$ be the maximal length of a special $\rightsquigarrow$- sequence without repetitions starting with A. Then $m(A) = \mathrm{dpt}(A) - 1$.*

8.2.28 Lemma *Let $r : A \rightsquigarrow A_1 \rightsquigarrow A_2 \rightsquigarrow \cdots$ be a special $\rightsquigarrow$-sequence with $\#(r)$ the number of different types occurring in r. Then we have*

$$\#(r) \leq 2^{\mathrm{dpt}(A)}.$$

Proof Identifying all occurrences of the same type in the sequence r we get a directed graph with the different types of r as nodes, possibly with cycles. For example the graph corresponding with 8.2.21(i) has as set of nodes $\{A, A \to (A \to B), A \to \beta, \alpha \to (\alpha \to \beta)\}$. By the condition in Definition 8.2.22 each node of that graph has at most two outgoing arcs and by Corollary 8.2.27 it follows that each node in the graph is reachable from A via a path of length $\leq \mathrm{dpt}(A) - 1$. Now the number of nodes in a directed graph where each node has at most two outgoing arcs and each node can be reached from a fixed node A via a path of length at most k is at most $2^{k+1} - 1 \leq 2^{k+1}$, as is easily proved by induction on k. Hence $\#(r) \leq 2^{\mathrm{dpt}(A)}$. ■

8.2.29 Lemma *Let p be a path without repetitions in a proof attempt for $A = B$. Then*

$$\mathrm{length}(p) \leq 2^{\mathrm{dpt}(A) + \mathrm{dpt}(B)}.$$

Proof Immediate by the previous lemma and Lemmas 8.2.18 and 8.2.23. ■

8.2.30 Lemma $\vdash_{\mu^-} A = B$ *is decidable*

Proof Let $N = 2^{\mathrm{dpt}(A) + \mathrm{dpt}(B)}$. Then we get from the preceeding lemma that the number of nodes in a proof attempt without repetitions is at most 2^{N+1} and that the number of different proof attempts is at most $3^{(2^N)}$. Hence the number of proof steps in an exhaustive proof search for $\vdash_{\mu^-} A = B$ is finite. ■

Now the final theorem immediately follows from Corollary 8.2.11.

8.2.31 Theorem *$A =_\mu B$ is decidable.*

Invertibility of strong equivalence

Furthermore strong μ-equivalence is invertible.

8.2.32 Proposition *The type algebra $\mathbb{T}^*_\mu$ is invertible.*

Proof By the fact that

$$(A \to B)^*_\mu = \begin{array}{ccc} & \to & \\ \swarrow & & \searrow \\ (A)^*_\mu & & (B)^*_\mu \end{array}.$$

Hence $(A \to B)^*_\mu = (A' \to B')^*_\mu$ implies $(A)^*_\mu = (A')^*_\mu$ and $(B)^*_\mu = (B')^*_\mu$. ■

Clearly $A =^*_\mu B$ if the trees $(A)^*_\mu$ and $(B)^*_\mu$ have the same label at all nodes.

Decidability and axiomatization of strong equivalence

We will show that it is decidable whether or not given two types $A, B \in \mathbb{T}_\mu$ their unfolding as infinite trees are equal, i.e. $(A)^*_\mu = (B)^*_\mu$, also written as $A =^*_\mu B$. The decidability is due to Koster (1969), with an exponential algorithm. In unpublished lecture notes by Koster this was improved to an algorithm of complexity $O(n^2)$. In Kozen et al. (1995) another quadratic algorithm is given. In Moller and Smolka (1995) an $O(n \log n)$ algorithm is presented. Automata aspects of both algorithms can be found in ten Eikelder (1991).

While the notion of weak equivalence was introduced by means of a formal inference system, that of strong equivalence was introduced in Section 7.5 in a semantic way via the interpretation of recursive types as trees. We show now that strong equivalence can also be represented by a (rather simple) finite set of formal rules, exploiting implicitly the proof principle of co-induction. The formal system (BH) and the proof that

$$\vdash_{\mathrm{BH}} A = B \Leftrightarrow A =^*_\mu B$$

are taken from Brandt and Henglein (1998). Other complete formalizations of strong equivalence have been given by Ariola and Klop (1996) and Amadio and Cardelli (1993), and will be discussed below in Definition 8.2.59. See Grabmayer (2005) for a proof-theoretical analysis of these formalizations.

Here we follow Brandt and Henglein (1998), who give at the same time an axiomatization and the decidability of strong equality.

8.2.33 Definition For $A \in \mathbb{T}_\mu$ define the set $\mathcal{SC}^s(A) \subseteq \mathbb{T}_\mu$ by

$$\begin{aligned}\mathcal{SC}^s(c) &\triangleq \{c\}, \text{ if } c \text{ is a variable or a constant;}\\ \mathcal{SC}^s(A_1 \to A_2) &\triangleq \{A_1 \to A_2\} \cup \mathcal{SC}^s(A_1) \cup \mathcal{SC}^s(A_2);\\ \mathcal{SC}^s(\mu\beta A) &\triangleq \{\mu\beta A\} \cup \mathcal{SC}^s(A)[\beta := \mu\beta A].\end{aligned}$$

8.2.34 Lemma *Let $A \in \mathbb{T}_\mu$. Then the cardinality of the set $\mathcal{SC}^s(A)$ is at most the number of symbols in A and hence finite.*

Proof By induction on the generation of A. ∎

8.2.35 Lemma *For all $A \in \mathbb{T}_\mu$ one has*

$$\mathcal{SC}^s(\mathcal{SC}^s(A)) = \mathcal{SC}^s(A).$$

Proof ($\subseteq$). First, it follows by induction on A that[1]

$$\mathcal{SC}^s(A[\beta := B]) \subseteq \mathcal{SC}^s(A)[\beta := B] \cup \mathcal{SC}^s(B). \tag{8.2}$$

Next, $\mathcal{SC}^s(\mathcal{SC}^s(A)) \subseteq \mathcal{SC}^s(A)$ by induction on A. For $A = \mu\alpha.A_1$ use (8.2) and the induction hypothesis for A_1.

($\supseteq$). This immediately follows from the fact that $B \in \mathcal{SC}^s(B)$ for all B. ∎

8.2.36 Definition Following Endrullis et al. (2011) we define for $A \in \mathbb{T}_\mu$ the set $\mathcal{SC}^s_r(A) \subseteq \mathbb{T}_\mu$. Define a binary relation on $\mathbb{T}^2_\mu$, also denoted by $\leadsto$, as

$$\begin{aligned}(A_1 \to A_2) &\leadsto A_i;\\ \mu\alpha.A &\leadsto A[\alpha := \mu\alpha.A].\end{aligned}$$

Now write

$$\mathcal{SC}^s_r(A) \triangleq \{C \in \mathbb{T}_\mu \mid A \leadsto^* C\},$$

where $\leadsto^*$ is the transitive reflexive closure of $\leadsto$.

Note the difference between this relation $\leadsto$ and the one in Definition 8.2.19.

8.2.37 Lemma *For all $A \in \mathbb{T}_\mu$ one has $\mathcal{SC}^s_r(A) = \mathcal{SC}^s(A)$.*

Proof ($\subseteq$) By Lemma 8.2.35.

($\supseteq$) By induction on the structure of A, using the Substitution Lemma 7.4.26. ∎

8.2.38 Corollary *For every $A \in \mathbb{T}_\mu$ the set $\mathcal{SC}^s_r(A)$ is finite.*

[1] In Brandt and Henglein (1998) Lemma 15, under the condition $\beta \in \mathrm{FV}(A)$, even equality is proved.

(ident)	$\mathcal{H} \vdash A = A$	
(symm)	$\dfrac{\mathcal{H} \vdash A = B}{\mathcal{H} \vdash B = A}$	
(trans)	$\dfrac{\mathcal{H} \vdash A = B \quad \mathcal{H} \vdash B = C}{\mathcal{H} \vdash A = C}$	
(axiom)	$\mathcal{H} \vdash A{=}B,$	if $(A{=}B) \in \mathcal{H},$
(μ-eq)	$\mathcal{H} \vdash \mu\alpha.A = A[\alpha := \mu\alpha.A],$	
(deconstr)	$\dfrac{\mathcal{H} \vdash (A \to B) {=} (A' \to B')}{\mathcal{H} \vdash A{=}A' \quad \mathcal{H} \vdash (B'{=}B')}$	

Figure 8.1

Proof By Lemma 8.2.34. ∎

For an alternative proof of this corollary, see Exercise 8.9.

8.2.39 Definition (Ariola and Klop (1996))

(i) On $\mathbb{T}_\mu$ consider the formal system defined by the axioms and rules in Fig. 8.1. Write $\mathcal{H} \vdash_{\mu\text{-dec}} A = B$ if $\mathcal{H} \vdash A = B$ is derivable in this system.

(ii) The *deductive closure* of $\mathcal{H}$ is defined as

$$\{C = D \mid \mathcal{H} \vdash_{\mu\text{-dec}} C = D\}.$$

(iii) Define an equation $A = B$ to be *consistent* if for no $C = D$ in the deductive closure of $\{A = B\}$ one has $\mathrm{ls}(C) \not\equiv \mathrm{ls}(D)$, where ls is the lead-symbol, see Definition 7.4.38.

8.2.40 Lemma *For $A, B \in \mathbb{T}^{\mathbb{A}}$ one has*

$$A =^*_\mu B \Leftrightarrow A = B \textit{ is consistent.}$$

Proof See Exercise 8.15. ∎

8.2.41 Example Consider $A \triangleq \mu\alpha.\alpha{\to}\alpha$ and $B \triangleq \mu\alpha\beta.\alpha{\to}\beta$. Write $C \triangleq \mu\gamma.B{\to}\gamma$. The deductive closure of $\{A = B\}$ is $\{P = Q \mid P, Q \in \{A, B, C\}\}$. For all P, Q we have $\mathrm{ls}(P) = \mathrm{ls}(Q) = {\to}$. Hence $A{=}B$ is consistent: no inconsistency 'has appeared'. Therefore $A =^*_\mu B$.

8.2.42 Lemma *Let $C = D$ be in the deductive closure of $\{A = B\}$. Then*

$$C, D \in \mathcal{SC}_r^s(A) \cup \mathcal{SC}_r^s(B).$$

Therefore the cardinality of the deductive closure is at most n^2, where n is the maximum of the number of symbols in A and in B.

Proof By induction on the derivation in $\vdash_{\mu\text{-dec}}$ and Lemmas 8.2.34 and 8.2.37. ∎

8.2.43 Theorem (Koster (1969)) *The relation $=_\mu^* \subseteq (\mathbb{T}_\mu)^2$ is decidable.*

Proof Using the system $\vdash_{\mu\text{-dec}}$ one produces the elements of the deductive closure of $\{A = B\}$ as an increasing sequence of sets

$$\mathcal{C}_0 \triangleq \{A = B\} \subseteq \mathcal{C}_1 \subseteq \mathcal{C}_2 \subseteq \cdots \subseteq \mathcal{C}_n$$

such that if $\mathcal{C}_n = \mathcal{C}_{n+1}$, then $\mathcal{C}_n = \mathcal{C}_m$ for all $m > n$. This is done by adding to already produced equations only new equations produced by the formal system. At a certain moment $\mathcal{C}_n = \mathcal{C}_{n+1}$, by Lemma 8.2.42; then the deductive closure is completed and we can decide whether the result is consistent or not. This is sufficient by Lemma 8.2.40. ∎

Axiomatization of strong equality

We will introduce the deductive system (BH) for deriving strong equality. The decidability of $=_\mu^*$ will follow a second time as Corollary (8.2.56).

The proof system for strong equality can be given for both μ-types and simultaneous recursions: the approach and the proof techniques are essentially the same. Below, we will carry out the proofs for μ-types. From the completeness of this formalization we obtain a proof of the decidability of strong equivalence for μ-types. In the next section this proof strategy will be used to show the decidability of strong equivalence $=_{\mathcal{R}}^*$ for a simultaneous recursion $\mathcal{R}$.

In the system (BH) we have judgments of the form

$$\mathcal{H} \vdash A = B$$

in which $\mathcal{H}$ is a set of equations between types with lead symbols '$\rightarrow$'. The meaning of this judgment is that we can derive $A = B$ using the equations in $\mathcal{H}$. We will show in Theorem 8.2.55 that provability in (BH) from $\mathcal{H} = \emptyset$ corresponds exactly to strong equivalence.

8.2.44 Definition Let $\mathcal{H}$ denote a set of equalities between types in $\mathbb{T}_\mu$ with lead symbol '$\rightarrow$'.. The system (BH) is defined by the rules in Fig. 8.2.

We write $\mathcal{H} \vdash_{\text{BH}} A = B$ if $\mathcal{H} \vdash A = B$ can be derived by the rules of (BH).

(ident)	$\mathcal{H} \vdash A = A$	
(symm)	$\dfrac{\mathcal{H} \vdash A = B}{\mathcal{H} \vdash B = A}$	
(trans)	$\dfrac{\mathcal{H} \vdash A = B \quad \mathcal{H} \vdash B = C}{\mathcal{H} \vdash A = C}$	
(axiom)	$\mathcal{H} \vdash A = B,$	if $(A = B) \in \mathcal{H}$,
(μ-eq)	$\mathcal{H} \vdash \mu\alpha.A = A[\alpha := \mu\alpha.A]$	
(μ-cong)	$\dfrac{\mathcal{H} \vdash A = A'}{\mathcal{H} \vdash \mu\alpha.A = \mu\alpha.A'},$	if α not free in $\mathcal{H}$,
($\to$-cong)	$\dfrac{\mathcal{H} \vdash A = A' \ \& \ \mathcal{H} \vdash B = B'}{\mathcal{H} \vdash (A \to B) = (A' \to B')}$	
(coind)	$\dfrac{\begin{array}{c}\mathcal{H}, (A \to B) = (A' \to B') \vdash A = A' \\ \mathcal{H}, (A \to B) = (A' \to B') \vdash B = B'\end{array}}{\mathcal{H} \vdash (A \to B) = (A' \to B')}$	

Figure 8.2 The system (BH).

Rule (coind) is obviously the crucial one. It says that if we are able to prove $A = A'$ and $B = B'$ *assuming* $A \to B = A' \to B'$ then we can conclude that $A \to B = A' \to B'$.

By rule (coind) the system (BH) exploits its co-inductive characterization of equality of infinite trees. Given a formal derivation in system (BH) of a judgment of the form $\mathcal{H} \vdash A = B$, we can regard each application of rule (coind) as a step in the construction of a bisimulation, see Definition 7.5.10 relating the infinite trees $(A)^*_\mu$ and $(B)^*_\mu$.

8.2.45 Remark

(i) By induction on derivations one can show that weakening is an admissible rule:

$$\mathcal{H} \vdash A = B \ \Rightarrow \ \mathcal{H}, \mathcal{H}' \vdash A = B.$$

(ii) Note that the rule ($\to$-cong) can be omitted while maintaining the same provability strength. This follows from the rule (coind) and weakening. Indeed, assuming $\mathcal{H} \vdash A = A'$, $\mathcal{H} \vdash B = B'$, one has by weakening $\mathcal{H}, (A \to B) = (A' \to B') \vdash A = A'$; $\mathcal{H}, (A \to B) = (A' \to B') \vdash B = B'$ and therefore by rule (coind) $\mathcal{H} \vdash (A \to B) = (A' \to B')$.

8.2.46 Example Let $S \triangleq \mu\beta.\mu\alpha.\alpha \to \beta$, $T \triangleq \mu\alpha.\alpha \to S$, and $B \triangleq \mu\alpha.\alpha \to \alpha$.

Note that $S =_\mu T =_\mu T \to S$, and $B =_\mu B \to B$ but $S \neq_\mu B$. We will show $\vdash_{\mathrm{BH}} S = B$. Let $\mathcal{D}$ stand for the following derivation of $\mathcal{H} \vdash T = B$ with $\mathcal{H} = \{T \to S = B \to B\}$:

$$\dfrac{\dfrac{\dfrac{}{\mathcal{H}\vdash T=T\to S}(\mu\text{-eq}) \quad \dfrac{}{\mathcal{H}\vdash T\to S=B\to B}(\text{hyp})}{\mathcal{H}\vdash T=B\to B}(\text{trans}) \qquad \dfrac{\dfrac{}{\mathcal{H}\vdash B=B\to B}(\mu\text{-eq})}{\mathcal{H}\vdash B\to B=B}(\text{symm})}{\mathcal{H}\vdash T=B}(\text{trans}).$$

Using this we obtain the following derivation in (BH):

$$\dfrac{\dfrac{}{\vdash S=T}(\mu\text{-eq}) \qquad \dfrac{\dfrac{}{\vdash T=T\to S}(\mu\text{-eq}) \quad \dfrac{\dfrac{\begin{matrix}\mathcal{D}\\ \vdots\\ \mathcal{H}\vdash T=B\end{matrix} \quad \dfrac{\dfrac{}{\mathcal{H}\vdash S=T}(\mu\text{-eq}) \quad \begin{matrix}\mathcal{D}\\ \vdots\\ \mathcal{H}\vdash T=B\end{matrix}}{\mathcal{H}\vdash S=B}(\text{trans})}{\vdash T\to S=B\to B}(\text{coind}) \quad \dfrac{\dfrac{}{\vdash B=B\to B}(\mu\text{-eq})}{\vdash B\to B=B}(\text{symm})}{\vdash T\to S=B}(\text{trans})}{\vdash T=B}(\text{trans})}{\vdash S=B}(\text{trans}).$$

Now we will prove the following soundness and completeness result:

$$A =_\mu^* B \quad \Leftrightarrow \quad \vdash_{BH} A = B.$$

The proof occupies 8.2.47–8.2.55.

8.2.47 Definition

(i) Let $A, B \in \mathbb{T}_\mu$ and $k \geq 0$. Define

$$A =_k^* B \stackrel{\Delta}{\Longleftrightarrow} ((A)_\mu^*)_k = ((B)_\mu^*)_k.$$

(ii) A set $\mathcal{H}$ of formal equations of the form $A = B$ is called *k-valid* if $A =_k^* B$ for all $A = B \in \mathcal{H}$.

(iii) The set $\mathcal{H}$ is *valid* if $A =_\mu^* B$ for all $A = B \in \mathcal{H}$.

Note that $A =_\mu^* B \Leftrightarrow \forall. k \geq 0. A =_k^* B$, by Lemma 7.5.8(i).

8.2.48 Lemma *Let $\mathcal{H}$ be valid. Then*

$$\mathcal{H} \vdash A = B \Rightarrow A =_\mu^* B.$$

Proof It suffices to show for all $k \geq 0$ that if $\mathcal{H}$ is k-valid then

$$\mathcal{H} \vdash A = B \Rightarrow A =_k^* B.$$

We use induction on k. If $k = 0$ this is trivial, by Lemma 7.5.8(ii). If $k > 0$ the proof is by induction on derivations. Most rules are easy to treat. Rule (μ-cong) follows using Remark 7.5.32, showing by induction on p

$$A =^*_k B \Rightarrow \forall p \leq k.[\mu\alpha.A =^*_p \mu\alpha.B].$$

The most interesting case is when the last applied rule is (coind), i.e.

$$\text{(coind)} \quad \frac{\begin{array}{c}\mathcal{H} \cup \{A \to B = A' \to B'\} \vdash A = A' \\ \mathcal{H} \cup \{A \to B = A' \to B'\} \vdash B = B'\end{array}}{\mathcal{H} \vdash A \to B = A' \to B'.}$$

Since $\mathcal{H}$ is k-valid it is also $(k-1)$-valid. By the induction hypothesis on k we have $A \to B =^*_{k-1} A' \to B'$. But then $\mathcal{H} \cup \{A \to B = A' \to B'\}$ is also $(k-1)$-valid and hence, again by the induction hypothesis on k, we have $A =^*_{k-1} A'$ and $B =^*_{k-1} B'$. By Lemma 7.5.8(iii) we conclude $A \to B =^*_k A' \to B'$. ∎

8.2.49 Corollary (Soundness) $\vdash_{\text{BH}} A = B \Rightarrow A =^*_\mu B$.

Proof Take $\mathcal{H}$ empty. ∎

The opposite implication is the completeness of (BH). The proof of this fact is given in a constructive way. Below, in Definition 8.2.50, we define a recursive predicate $\mathcal{S}(\mathcal{H}, A, B)$, where $\mathcal{H}$ is a set of equations and A, $B \in \mathbb{T}_\mu$. The relation $\mathcal{S}$ will satisfy

$$\mathcal{S}(\mathcal{H}, A, B) \Rightarrow \mathcal{H} \vdash_{\text{BH}} A = B.$$

Note that it is trivially decidable whether a type A is non-circular or not. If A and B are circular we can easily prove $\vdash_{\text{BH}} A = B$, see Lemma 7.4.36. Otherwise A has a reduced form A' as defined in 7.4.38 such that, for all $\mathcal{H}$, $\mathcal{H} \vdash_{\text{BH}} A = A'$ by rules (μ-eq) and (trans).

8.2.50 Definition Let $A, B \in \mathbb{T}_\mu$ be two μ-types and let $\mathcal{H}$ be a set of equations of the form $A_1 \to A_2 = B_1 \to B_2$. The predicate $\mathcal{S}(\mathcal{H}, A, B)$ is defined as follows. Let A', B' be the reduced forms of A, B, respectively. Remember Definition 7.4.38(i), explaining that these can only be one of the possible forms:

(a) α, a type atom;
(b) $A \to B$, a function type; or
(c) $\bullet$ standing for a circular type.

In these cases the lead symbol of the type involved is α, $\rightarrow$, or $\bullet$, respectively. Then define

$$
\begin{array}{rcll}
\mathcal{S}(\mathcal{H}, A, B) & \triangleq & \mathsf{true}, & \text{if } A' \equiv B' \text{ or if } (A' = B') \in \mathcal{H};\ \text{else} \\
& \triangleq & \mathsf{false}, & \text{if } A' \text{ and } B' \text{ have different lead symbols; else} \\
& \triangleq & \multicolumn{2}{l}{\mathcal{S}(\mathcal{H} \cup \{A' = B'\}, A_1, B_1)\ \&\ \mathcal{S}(\mathcal{H} \cup \{A' = B'\}, A_2, B_2),} \\
& & & \text{if } A' \equiv A_1 \rightarrow A_2, B' \equiv B_1 \rightarrow B_2.
\end{array}
$$

This $\mathcal{S}(\mathcal{H}, A, B)$ will always be defined and be a truth value in $\{\mathsf{true}, \mathsf{false}\}$. At the same time $\mathcal{S}$ is considered as a partial computable function denoting this truth value, that happens to be total. Therefore it is intuitively clear also to regard $\mathcal{S}(\mathcal{H}, A, B)$ as an expression reducing to the truth value.

8.2.51 Notation

(i) Write

$$
\begin{array}{l}
\mathcal{S}(\mathcal{H}, A, B) \succ \mathcal{S}(\mathcal{H}', A_1, B_1) \\
\mathcal{S}(\mathcal{H}, A, B) \succ \mathcal{S}(\mathcal{H}', A_2, B_2),
\end{array}
$$

where $\mathcal{H}' = \mathcal{H} \cup \{A_1 \rightarrow A_2 = B_1 \rightarrow B_2\}$, if the value of $\mathcal{S}(\mathcal{H}, A, B)$ depends directly on that of $\mathcal{S}(\mathcal{H}', A_i, B_i)$ according to the last case of the previous definition.

Note that $\succ$ is a non-deterministic relation.

(ii) The notations $\succ^n$ and $\succ^*$ have the usual meaning: n-step rewriting and the reflexive transitive closure of $\succ$.

8.2.52 Lemma

(i) *If $\mathcal{S}(\mathcal{H}, A, B) \succ^* \mathcal{S}(\mathcal{H}', A', B')$, then $A' \in \mathcal{SC}^s_r(A)$, $B' \in \mathcal{SC}^s_r(B)$ and all equations in $\mathcal{H}' \backslash \mathcal{H}$ are of the form $A'' = B''$, where $A'' \in \mathcal{SC}^s_r(A)$, $B'' \in \mathcal{SC}^s_r(B)$.*

(ii) *The relation $\succ$ is well founded.*

(iii) *The predicate $\mathcal{S}$ is decidable, i.e., $\mathcal{S}$ seen as map is total and computable.*

Proof (i) Directly by Definition 8.2.36.

(ii) Suppose there exists an infinite sequence

$$
\mathcal{S}(\mathcal{H}, A, B) \succ \mathcal{S}(\mathcal{H}_1, A_1, B_1) \succ \mathcal{S}(\mathcal{H}_2, A_2, B_2) \succ \cdots .
$$

Then $\mathcal{S}(\mathcal{H}_k, A_k, B_k) \succ \mathcal{S}(\mathcal{H}_{k+1}, A_{k+1}, B_{k+1})$, with $\mathcal{S}(\mathcal{H}_k, A_k, B_k)$ being true, for otherwise, by the first clause of the definition of $\mathcal{S}(\mathcal{H}, A, B)$, the sequence stops. From (i) and Lemma 8.2.34 it follows that the sequence terminates.

(iii) By (ii) and König's lemma the evaluation of $\mathcal{S}(\mathcal{H}, A, B)$ must terminate in a finite conjunction of booleans. From this the value is uniquely determined. ■

8.2.53 Lemma *If $\mathcal{S}(\mathcal{H}, A, B) = \mathsf{true}$, then $\mathcal{H} \vdash_{\mathrm{BH}} A = B$. If moreover A, B have no circular subterms, then rule (μ-cong) is not needed.*

Proof Each step of the definition of $\mathcal{S}(\mathcal{H}, A, B)$, which does not determine a false value, corresponds to the application of one or more deduction rules in (BH). For instance the first instance corresponds to a proof, by (μ-eq), (μ-cong), and (trans), that two circular types are equal, see Lemma 7.4.36; and the last instance to an application of rule (coind). ■

See also Exercise 8.3.

8.2.54 Lemma *If $A =^*_\mu B$, then $\mathcal{S}(\mathcal{H}, A, B) = \mathsf{true}$.*

Proof Let n be the maximum number such that $\mathcal{S}(\mathcal{H}, A, B) \succ^n \mathcal{S}(\mathcal{H}', A', B')$. By Lemma 8.2.52(ii) we know that such an n must certainly exist. The proof is by induction on n. If $n = 0$, then as $A =^*_\mu B$, i.e. $(A)^*_\mu = (B)^*_\mu$, we have that $\mathcal{S}(\mathcal{H}, A, B) = \mathsf{true}$, by Lemma 7.5.29. If $n > 0$, then we are in the third instance of Definition 8.2.50. Let m_i be the maximum number of steps such that $\mathcal{S}(\mathcal{H}, A_i, B_i) \succ^{m_i} \mathcal{S}(\mathcal{H}', A_i', B_i')$, for $i = 1, 2$. We have that $m_i < n$. Now use the induction hypotheses for $A_i =^*_\mu B_i$ and the fact that $A =^*_\mu B$ implies $A_i =^*_\mu B_i$, by Lemma 8.2.32. ■

Now we can harvest.

8.2.55 Theorem (Completeness) *Let $A, B \in \mathbb{T}_\mu$. Then the following are equivalent:*

(i) $A =^*_\mu B$;
(ii) $\mathcal{S}(\emptyset, A, B) = \mathsf{true}$;
(iii) $\vdash_{\mathrm{BH}} A = B$.

Proof (i) $\Rightarrow$ (ii). By Lemma 8.2.54.
(ii) $\Rightarrow$ (iii). By Lemma 8.2.53.
(iii) $\Rightarrow$ (i). By Corollary 8.2.49. ■

Now Theorem 8.2.43, Koster (1969), follows again as a corollary.

8.2.56 Corollary *Given $A, B \in \mathbb{T}_\mu$, it is decidable whether $A =^*_\mu B$.*

Proof By Theorem 8.2.55 and Lemma 8.2.52. ■

8.2.57 Remark The connection between the present proof of this result via the system of Brandt–Henglein (BH) and the one using the deductive closure presented in Theorem 8.2.43 is elaborated in Grabmayer (2005), where it is shown that the two are a kind of mirror image of each other.

8.2.58 Corollary *If $A, B \in \mathbb{T}_\mu$ are not circular and $\vdash_{\mathrm{BH}} A = B$, then there is a derivation of $A = B$ in BH such that rule (μ-cong) is not used.*

Proof By completeness and Lemma 8.2.53. ■

The predicate $\mathcal{S}$ from Definition 8.2.50 is a computable procedure for testing equality of μ-types. In the present form $\mathcal{S}$ is $O(2^{n \times m})$ where n, m are, respectively, the number of arrow types in $\mathcal{SC}^s(A)$ and $\mathcal{SC}^s(B)$. More efficient algorithms are known, as mentioned at the beginning of this subsection.

Other systems

Other systems have been proposed in the literature for giving a complete axiomatization of $=^*_\mu$. In Amadio and Cardelli (1993) a formal system (that we will denote by (μ^*_{AC})) is mentioned for proving strong equivalence defined by adding to the rules of system (μ), see Definition 7.4.27, the following:

8.2.59 Definition

(i) The rule (AC) is defined as

$$(\mathrm{AC}) \quad \frac{A[\beta := B] = B \quad A[\beta := B'] = B'.}{B = B'}$$

In Ariola and Klop (1996) an equivalent but slightly different rule was introduced, see Exercise 8.13.

(ii) The system μ extended with the rule (AC) is denoted by (μ^*_{AC}).

The soundness of this system can be proved by a standard induction on derivations using the uniqueness of fixed points in Tr_∞, Theorem 7.6.5, when the last applied rule is (AC). In fact we have that both $(\mu\alpha.B)^*$ (by Remark 7.5.32) and $(A)^*$ (by the induction hypothesis and Lemma 7.5.31) are fixed points of $\lambda\zeta \in \mathsf{Tr}_\infty.(B)^*[\alpha := \zeta]$. In Ariola and Klop (1996) it is proved that (μ^*_{AC}) also is complete with respect to the tree semantics. So (μ^*_{AC}) is equivalent to (BH).

Rule (AC) has indeed a great expressive power; it sometimes allows more synthetic proofs than rule (coind). The system presented in this section, however, uses a more basic proof principle (co-induction) and suggests a natural algorithm (obtained by going backwards in the deduction tree) to test type equality.

8.2.60 Example In (μ^*_{AC}) we have $\vdash \mu\alpha.(\beta \to \alpha) = \mu\alpha.(\beta \to \beta \to \alpha)$. Indeed, by two applications of (μ-eq), we have $\vdash \mu\alpha.(\beta \to \alpha) = \beta \to \beta \to \mu\alpha.(\beta \to \alpha)$. Then we can apply rule (AC) with $B[\alpha] = \beta \to \beta \to \alpha$. Compare this proof with those in Exercises 8.6 and 8.5.

Some general properties of strong equality can be easily proved by rule (AC). As an example, in the following proposition we show that two consecutive applications of the μ-operator can be contracted into just one.

8.2.61 Proposition *The following are directly provable by the rule* (AC).

(i)	$\mu\alpha.A$	$=$	$A,$	*if α does not occur in A;*
(ii)	$\mu\alpha\beta.A$	$=$	$\mu\beta\alpha.A;$	
(iii)	$\mu\alpha\beta.A(\alpha,\beta)$	$=$	$\mu\alpha.A(\alpha,\alpha).$	

Proof Do Exercise 8.11. ∎

8.2.62 Remark The notion of bisimulation over infinite trees, see Definition 7.5.10, is formulated directly over their representation in $\mathbb{T}_\mu$. A relation $R_\mu \subseteq \mathbb{T}_\mu \times \mathbb{T}_\mu$ is a μ-bisimulation if the following statements hold. Again, $\mathrm{ls}(A)$ is the lead-symbol of A, defined in Definition 7.4.38:

$$\begin{aligned} A\ R_\mu\ B &\Rightarrow \mathrm{ls}(A) \equiv \mathrm{ls}(B);\\ A\ R_\mu\ \mu\alpha.B &\Rightarrow A\ R_\mu\ B[\alpha := \mu\alpha.B];\\ \mu\alpha.A\ R_\mu\ B &\Rightarrow A[\alpha := \mu\alpha.A]\ R_\mu\ B;\\ A \to B\ R_\mu\ A' \to B' &\Rightarrow A\ R_\mu\ A'\ \&\ B\ R_\mu\ B'. \end{aligned}$$

It is easy to prove that for any pair of recursive types $A, B \in \mathbb{T}_\mu$ and μ-bisimulation R_μ one has

$$A\ R_\mu\ B \Rightarrow A =^*_\mu B.$$

8.3 Properties of types defined by a simultaneous recursion

In this section we will study some fundamental properties of type algebras defined via type equations and simultaneous recursions over a set $\mathbb{T} = \mathbb{T}^{\mathbb{A}}$ of types. All results can however be easily generalized to arbitrary type algebras. Some properties of types defined in this way are essentially the same as those of μ-types, but their proofs sometimes require slightly different techniques.

Decidability of weak equivalence for a simultaneous recursion

For a simultaneous recursion $\mathcal{R}$ over $\mathbb{T}$ we have already proved invertibility of $=_\mathcal{R}$ in Proposition 7.3.16 by introducing the term rewriting system

(TRS) with, as notion of reduction, $X_i \Rightarrow_{\mathcal{R}} A_i(\vec{X})$. Decidability follows as a particular case of Proposition 7.2.16.

Decidability for recursive types defined by a simultaneous recursion $\mathcal{R}$ can also be proved via the 'inverse' TRS, see Statman (1994), Terese (2003), that generates $=_{\mathcal{R}}$ and is complete, i.e. Church–Rosser (CR) and strongly normalizing (SN). We present this proof here since the inverse TRS will also be used in the proof of Theorem 8.3.31. Moreover its properties can suggest efficient algorithms for testing type equality.

The first step consists of orienting the equations of $\mathcal{R}$.

8.3.1 Definition Let $\mathcal{R} = \{X_i = A_i \mid 1 \leq i \leq n\}$ be a proper simultaneous recursion over $\mathbb{T}$. The rewriting system $\text{TRS}^{-1}(\mathcal{R})$ is generated by the notion of reduction

$$\{A_i \Rightarrow_{\mathcal{R}}^{-1} X_i \mid X_i = A_i \in \mathcal{R}\}.$$

Then $=_{\mathcal{R}}^{-1}$ is the convertibility relation generated by $\text{TRS}^{-1}(\mathcal{R})$.

8.3.2 Proposition *If $\mathcal{R}$ is proper, then* $\text{TRS}^{-1}(\mathcal{R})$ *is SN.*

Proof Each contraction decreases the size of the type to which it is applied, except for $X_j \Rightarrow X_i$. But then we have $j > i$, since $\mathcal{R}$ is proper, Definition 7.3.11. Therefore we can make the following argument. Define for $A \in \mathbb{T}(\vec{X})$ the following numbers:

$$\begin{aligned} s(A) &\triangleq \text{number of symbols in } A; \\ n(A) &\triangleq \text{sum of the indices of variables among } \vec{X} \text{ in } A. \end{aligned}$$

Then reducing A, the pair $\langle s(A), n(A)\rangle$ decreases in the lexicographical order. ∎

However $\text{TRS}^{-1}(\mathcal{R})$ is, in general, not CR, as we show now.

8.3.3 Example Let $\mathcal{R}$ be the simultaneous recursion

$$\begin{aligned} X_0 &= X_0 \to X_2 \\ X_1 &= (X_0 \to X_2) \to X_2 \\ X_2 &= X_0 \to X_1. \end{aligned}$$

Then $\text{TRS}^{-1}(\mathcal{R})$ consists of the rules

$$\begin{aligned} X_0 \to X_2 &\Rightarrow_{\mathcal{R}} X_0 \\ (X_0 \to X_2) \to X_2 &\Rightarrow_{\mathcal{R}} X_1 \\ X_0 \to X_1 &\Rightarrow_{\mathcal{R}} X_2. \end{aligned}$$

Observe that the left hand side of the first equation is a subterm of the left hand side of the second. In particular $(X_0 \to X_2) \to X_2$ can be reduced both to X_1 and to $X_0 \to X_2$ which further reduces to X_0: it has then two distinct normal forms X_1 and X_0. Therefore $\text{TRS}^{-1}(\mathcal{R})$ is not CR.

Expressions like $X_0 \to X_2$ and $(X_0 \to X_2) \to X_2$ in this example are called *critical pairs* in the literature on term rewriting systems. In $\text{TRS}^{-1}(\mathcal{R})$ there is a critical pair whenever there are i, j such that $i \neq j$ and A_i is a subexpression of A_j. The following result is well known.

8.3.4 Theorem (Knuth–Bendix) *Let $\mathcal{T}$ be a TRS that is SN.*

(i) *If all critical pairs of $\mathcal{T}$ have a common reduct, then $\mathcal{T}$ is CR.*
(ii) *If $\mathcal{T}$ has no critical pairs, then it is CR.*

Proof (i) See Terese (2003) Theorem 2.7.16.
(ii) By (i). ∎

Now we present an algorithm for transforming any proper simultaneous recursion into a logically equivalent one, see Definition 7.3.22(ii), without critical pairs. The procedure amounts to a simple case of the Knuth–Bendix completion algorithm, see Terese (2003) Theorem 7.4.2, as the equations involved are between closed terms.

8.3.5 Proposition *Let $\mathcal{R}$ be a proper simultaneous recursion. Then there exists a proper simultaneous recursion $\mathcal{R}^\diamond$ such that:*

(i) *$\mathcal{R}^\diamond$ is logically equivalent to $\mathcal{R}$;*
(ii) *$\text{TRS}^{-1}(\mathcal{R}^\diamond)$ is complete, i.e. SN and CR.*

Proof Let $\mathcal{R}$ be a proper simultaneous recursion. We define by recursion on n a sequence of sets of equations $\mathcal{D}_n$, $\mathcal{I}_n$ $(n \geq 0)$ such that

(a) $\mathcal{D}_n$ is a proper simultaneous recursion;
(b) $\mathcal{I}_n$ is a set of equations of the form $X_i = X_j$ with $i < j$;
(c) $\mathcal{D}_n \cup \mathcal{I}_n$ is logically equivalent to $\mathcal{R}$, for all n.

Let $\mathcal{D}_0 = \mathcal{R}$ and $\mathcal{I}_0 = \emptyset$. Define $\mathcal{D}_{n+1}$, $\mathcal{I}_{n+1}$ from $\mathcal{D}_n$, $\mathcal{I}_n$ as follows.

(1) If there exists a pair of equations $X_i = A_i, X_j = A_j \in \mathcal{D}_n$ such that A_j is a proper subexpression of A_i take

$$\mathcal{D}_{n+1} = (\mathcal{D}_n - \{X_i = A_i\}) \cup \{X_i = A_i^*\}$$
$$\mathcal{I}_{n+1} = \mathcal{I}_n,$$

where A_i^* is the result of replacing all occurrences of A_j in A_i by X_j.

(2) If there exist two equations $X_i = A, X_j = A \in \mathcal{D}_n$ then, assuming $i < j$, take

$$\mathcal{D}_{n+1} = \mathcal{D}_n[X_i := X_j]$$
$$\mathcal{I}_{n+1} = \mathcal{I}_n \cup \{X_i = X_j\}.$$

(3) Otherwise take $\mathcal{D}_{n+1} = \mathcal{D}_n$ and $\mathcal{I}_{n+1} = \mathcal{I}_n$.

(4) The algorithm terminates if $\mathcal{D}_{n+1} = \mathcal{D}_n$. Let N be the least n such that $\mathcal{D}_{n+1} = \mathcal{D}_n$. Define

$$\mathcal{R}^\diamond = \mathcal{D}_N \cup \mathcal{I}_N;$$
$$\mathrm{TRS}^\diamond(\mathcal{R}) = \mathrm{TRS}^{-1}(\mathcal{R}^\diamond).$$

This algorithm terminates, as can be seen as follows. In cases (2) and (1) if $A_j \notin \mathbb{A} \cup \{\vec{X}\}$, the number of symbols in $\mathcal{D}_n$ decreases. In case (1), if $A_j \in \mathbb{A} \cup \{\vec{X}\}$, then

(i) an α is replaced by X_j; or
(ii) an X_k is replaced by X_j, with $j < k$.

After a finite number of applications of (1) or (2) we must eventually apply a rule in which the number of symbols in $\mathcal{D}_n$ decreases and so eventually the process must stop.

(i) By construction each $\mathcal{D}_n \cup \mathcal{I}_n$ is a proper simultaneous recursion that is logically equivalent to $\mathcal{R}$. In particular this holds for $\mathcal{R}^\diamond = \mathcal{D}_N \cup \mathcal{I}_N$.
(ii) Note that $\mathrm{TRS}^\diamond(\mathcal{R})$ is SN, by Proposition 8.3.2 and (i). We claim: $\mathrm{TRS}^\diamond(\mathcal{R})$ has no critical pairs. Indeed $\mathrm{TRS}^{-1}(\mathcal{D}_N)$ has no such pairs, for otherwise we could apply step 1 or 2 of the definition of $\mathcal{D}_n$ to $\mathcal{D}_N$. Moreover if $X_j = X_i \in \mathcal{I}_N$ then X_j does not occur in $\mathcal{D}_N$ and there is no other equation of the form $X_j = X_{i'}$ in $\mathcal{I}_N$. In fact, if $X_j = X_i$ has been put in $\mathcal{I}_k$ at step 2 of the definition of $\mathcal{D}_n$, for some $(0 < k \leq N)$, then X_j does not occur in $\mathcal{D}_k$, hence not in $\mathcal{D}_n$ for all $n \geq k$. Consequently no other equation containing X_j is put in any $\mathcal{I}_n$ for $n > k$. By Theorem 8.3.4(ii) it follows that $\Rightarrow$ is CR. ■

8.3.6 Example Applying the above algorithm to the simultaneous recursion $\mathcal{R}$ defined in Example 8.3.3 we obtain (assuming $X_0 < X_1 < X_2$)

$$\mathcal{D}_1 = \{X_0 = X_0 \to X_2,\ X_1 = X_0 \to X_2,\ X_2 = X_0 \to X_1\},$$
$$\mathcal{I}_1 = \emptyset;$$

$$\mathcal{D}_2 = \{X_1 = X_1 \to X_2,\ X_2 = X_1 \to X_1\},$$
$$\mathcal{I}_2 = \{X_0 = X_1\}.$$

Now no more transformations are possible, so we have $N = 2$. Note that $\mathcal{D}_2 \cup \mathcal{I}_2$ is logically equivalent to $\mathcal{R}$ and has no critical pairs. We obtain a simultaneous recursion $\mathcal{R}^\diamond$ and $\mathrm{TRS}^\diamond(\mathcal{R})$ as follows:

$$\left.\begin{array}{lcl} X_0 & = & X_1 \\ X_1 & = & X_1 \to X_2 \\ X_2 & = & X_1 \to X_1 \end{array}\right\} \mathcal{R}^\diamond; \qquad \left.\begin{array}{rcl} X_1 & \Rightarrow_{\mathcal{R}^\diamond} & X_0 \\ X_1 \to X_2 & \Rightarrow_{\mathcal{R}^\diamond} & X_1 \\ X_1 \to X_1 & \Rightarrow_{\mathcal{R}^\diamond} & X_2 \end{array}\right\} \mathrm{TRS}^\diamond(\mathcal{R}).$$

8.3.7 Corollary *Let $\mathcal{R}$ be a proper simultaneous recursion. Then $=_\mathcal{R}$ is decidable.*

Proof Since $\mathcal{R}^\diamond$ is logically equivalent to $\mathcal{R}$, it follows also that $=_\mathcal{R}$ is the convertibility relation generated by $\mathrm{TRS}^\diamond(\mathcal{R})$. For a type C, let C^{nf} be its (unique) normal form with respect to $\mathrm{TRS}^\diamond(\mathcal{R})$. This normal form exists and is computable from C, since $\mathrm{TRS}^\diamond(\mathcal{R})$ is complete. Now, given a pair of types $A, B \in \mathbb{T}[\vec{X}]$, we have

$$A =_\mathcal{R} B \Leftrightarrow A^{\mathrm{nf}} \equiv B^{\mathrm{nf}}.$$

This is decidable. ∎

Alternatively, decidability of $=_\mathcal{R}$ also follows as a particular case of Proposition 7.2.16. A more algebraic but less direct proof is given in Marz (1999).

Strong equivalence as an equational theory

In this subsection we show that for every simultaneous recursion $\mathcal{R}$ we can constructively find another $\mathcal{R}^*$ such that $=^*_\mathcal{R}$ coincides with the equational theory of $\mathcal{R}^*$. This lets us shift to tree type algebras generated by a simultaneous recursion all the results and the techniques valid for the equational theories. With this aim we need some preliminary definitions and lemmas.

8.3.8 Definition Let $\mathcal{R} = \mathcal{R}(\vec{X})$ be a simultaneous recursion over $\mathbb{T}^\mathbb{A}$.

(i) A simultaneous recursion $\mathcal{R}[\vec{X}]$ is *flat* if all equations of $\mathcal{R}$ are of the form $X = \alpha$ where $\alpha \in \mathbb{A}$ is an atomic type or $X = Y \to Z$, where $X, Y, Z \in \vec{X}$

(ii) An indeterminate $X \in \vec{X}$ is called an *orphan* in $\mathcal{R}$ if X does not occur in one of the right hand sides of $\mathcal{R}$.

8.3.9 Example Note that a flat simultaneous recursion is simplified. Any simultaneous recursion can be transformed to an equivalent flat one by adding and possibly removing indeterminates. For example $\mathcal{R}_1 = \{X = X{\to}(\alpha{\to}\beta), Y = (X{\to}X){\to}Y\}$ can be 'flattened' to

$$\mathcal{R}_1' \triangleq \{X = X{\to}Z, Z = U{\to}V, U = \alpha, V = \beta, Y = W{\to}Y, W = X{\to}X\}.$$

Note that $\mathcal{R}_2 \triangleq \{X = Y{\to}Z, Y = Z{\to}Y, Z = Z{\to}Z, W = X{\to}Y\}$ has W as orphan. Removing it by considering $\mathcal{R}_2' \triangleq \mathcal{R}_2 - \{W = X{\to}Y\}$ yields an equivalent simultaneous recursion, as $\mathbb{T}^{\mathbb{A}}[\mathcal{R}_2'] \cong \mathbb{T}^{\mathbb{A}}[\mathcal{R}_2]$ by mapping $[W]$ onto $[X{\to}Y]$. But now X has become an orphan in $\mathcal{R}_2'$. We can remove also X and obtain an $\mathcal{R}_2'' \triangleq \{Y = Z{\to}Y, Z = Z{\to}Z\}$ that is without orphans.

8.3.10 Lemma *Each proper simultaneous recursion can be transformed into an equivalent flat one without orphans.*

Proof Let $\mathcal{R}(\vec{X})$ be a proper simultaneous recursion over $\mathbb{T}$. First transform $\mathcal{R}(\vec{X})$ into an equivalent simplified simultaneous recursion $\mathcal{R}'(\vec{X}')$ applying the construction in the proof of Lemma 7.3.25. Now take any equation $X = A_1 \to A_2 \in \mathcal{R}'$ such that either A_1 or A_2 is not an indeterminate. Assume that A_1 is not an indeterminate. Then replace in $\mathcal{R}'$ this equation by the two equations $X = Y \to A_2$, $Y = A_1$, where Y is a fresh new indeterminate. Do similarly if A_2 is not an indeterminate. Repeat these steps until the resulting simultaneous recursion is flat; it is trivial to prove that this process terminates. It is also trivial to prove that at each step $\mathcal{R}'$ is transformed into an equivalent simultaneous recursion. The example given shows that orphans can be removed successively. The proof that at each step one has $\mathbb{T}^{\mathbb{A}}[\mathcal{R}] \cong \mathbb{T}^{\mathbb{A}'}[\mathcal{R}']$ can be given in a way similar to that in Proposition 7.3.25. ∎

8.3.11 Definition Let $\mathcal{R} = \mathcal{R}[\vec{X}]$ be a flat simultaneous recursion. For each equivalence class $[X]$, for $X \in \vec{X}$, with respect to the relation $=^*_{\mathcal{R}}$, choose a representative X'. We define the simultaneous recursion $\mathcal{R}^* = \mathcal{R}^*[\vec{X}]$ in two steps as follows. First let $\mathcal{R}' = \mathcal{R}'[\vec{X}']$ be the flat simultaneous recursion defined as

$$\mathcal{R}' \triangleq \{X' = \alpha \mid (X' = \alpha) \in \mathcal{R}\} \cup$$
$$\{X' = Y' \to Z' \mid \exists Y \in [Y'], Z \in [Z'].(X' = Y \to Z) \in \mathcal{R}\}.$$

Then define $\mathcal{R}^* = \mathcal{R}^*[\vec{X}]$ as

$$\mathcal{R}^* \triangleq \{X = \alpha \mid (X = \alpha) \in \mathcal{R}\} \cup$$
$$\{X = Y' \to Z' \mid X \in [X'],\ X' = Y' \to Z' \in \mathcal{R}'\}.$$

8.3.12 Example Let $\mathcal{R} \triangleq \{X = Y \to Z, Y = Z \to X, Z = X \to Y\}$. Then

$\mathcal{R}' = \{X = X \to X\}$;
$\mathcal{R}^* = \{X = X \to X, Y = X \to X, Z = X \to X\}$.

A different construction of $\mathcal{R}^*$ is given in Exercise 8.17.

We will show that weak equality with respect to $\mathcal{R}^*$ is equivalent to strong equality with respect to $\mathcal{R}$.

8.3.13 Lemma

(i) $X \in [X'] \Rightarrow X =_{\mathcal{R}^*} X'$.
(ii) $A =_{\mathcal{R}} B \Rightarrow A =_{\mathcal{R}^*} B$.

Proof (i) By the Definition of $\mathcal{R}^*$ from $\mathcal{R}'$ in Definition 8.3.11.
(ii) It suffices to show

$$(X = B) \in \mathcal{R} \Rightarrow X =_{\mathcal{R}^*} B.$$

The case $(X = \alpha) \in \mathcal{R}$ is trivial. Now let $(X = Y \to Z) \in \mathcal{R}$. Then

$$X \in [X_1],\ Y \in [Y_1],\ Z \in [Z_1],\ (X_1 = Y_1 \to Z_1) \in \mathcal{R}^*.$$

Then the result follows from (i). ∎

8.3.14 Theorem *Given a flat simultaneous recursion $\mathcal{R}(\vec{X})$, define $\mathcal{R}^*$ as in Definition* 8.3.11. *Then:*

(i) *for all $A, B \in \mathbb{T}[\vec{X}]$*

$$A =_{\mathcal{R}^*} B \Leftrightarrow A =^*_{\mathcal{R}} B;$$

(ii) $(\mathbb{T}[\mathcal{R}])^* \cong \mathbb{T}[\mathcal{R}^*]$.

Proof (i) ($\Rightarrow$) It suffices to show

$$(X = A(\vec{X})) \in \mathcal{R}^* \Rightarrow X =^*_{\mathcal{R}} A(\vec{X}).$$

Case 1. $A(\vec{X}) \equiv \alpha$. Then $(X = \alpha) \in \mathcal{R}^*$, hence also $(X = \alpha) \in \mathcal{R}$.

Case 2. $A(\vec{X}) = Y \to Z$. Then $(X = Y \to Z) \in \mathcal{R}^*$, and $(X_1 = Y_1 \to Z_1) \in \mathcal{R}$, $X =^*_{\mathcal{R}} X_1$, $Y =^*_{\mathcal{R}} Y_1$, and $Z =^*_{\mathcal{R}} Z_1$. Then clearly $X =^*_{\mathcal{R}} Y_1 \to Z_1$.

($\Leftarrow$) Assume $(A)^*_{\mathcal{R}} = (B)^*_{\mathcal{R}}$. We show $A =_{\mathcal{R}^*} B$ by induction on $s(A) + s(B)$, where $s(C)$ is the number of symbols in $C \in \mathbb{T}[\vec{X}']$. We distinguish the following cases.

(1) $A = \alpha, B = \alpha$;
(2) $A = \alpha, B = X, (X = \alpha) \in \mathcal{R}$;
(3) $A = Y, B = \alpha, (Y = \alpha) \in \mathcal{R}$;
(4) $A = Y, B = Z$;
(5) $A = Y, B = B_1 \to B_2, (Y = A_1 \to A_2) \in \mathcal{R}$;
(6) $A = A_1 \to A_2, B = Y, (Y = B_1 \to B_2) \in \mathcal{R}$;
(7) $A = A_1 \to A_2, B = B_1 \to B_2$.

The cases (1)–(3) are trivial.

Case (4). Now $Y \in [Y'], Z \in [Z']$, where $Y', Z' \in \vec{X}'$. Then $(Y)^*_{\mathcal{R}} = (Z)^*_{\mathcal{R}}$, so $[Y] = [Z]$, and $Y' = Z'$. By Lemma 8.3.13 we have $Y =_{\mathcal{R}^*} Y', Z =_{\mathcal{R}^*} Z'$. Therefore $Y =_{\mathcal{R}^*} Z$.

Case (5). $A_1, A_2 \in \vec{X}$, because $\mathcal{R}$ is flat, say $(Y = X_1 \to X_2) \in \mathcal{R}$. Now $(B_1 \to B_2)^*_{\mathcal{R}} = (X_1 \to X_2)^*_{\mathcal{R}}$, hence by invertibility, Lemma 7.5.5,

$$B_1 =^*_{\mathcal{R}} X_1, B_2 =^*_{\mathcal{R}} X_2$$

and by the induction hypothesis

$$B_1 =_{\mathcal{R}^*} X_1, B_2 =_{\mathcal{R}^*} X_2.$$

Therefore $B_1 \to B_2 =_{\mathcal{R}^*} X_1 \to X_2 =_{\mathcal{R}^*} Y$, by Lemma 8.3.13.

Case (6). Similar to Case (5).

Case (7). Similar to Case (5), but easier.

(ii) From (i) we immediately get that the identity morphism $\mathbb{T}[\vec{X}] \to \mathbb{T}[\vec{X}]$ induces an isomorphism $\mathbb{T}[\mathcal{R}^*] \to (\mathbb{T}[\mathcal{R}])^*$. ∎

8.3.15 Corollary *For every proper simultaneous recursion $\mathcal{R}$ there exists a flat simultaneous recursion $\mathcal{R}^*$ such that*

$$\mathbb{T}[\mathcal{R}^*] \cong (\mathbb{T}[\mathcal{R}])^*.$$

Proof Let $\mathcal{R}$ be proper. By Lemma 8.3.10 there exists a flat simultaneous recursion $\mathcal{R}_1$ such that $\mathbb{T}[\mathcal{R}] \cong \mathbb{T}[\mathcal{R}_1]$. Then $(\mathbb{T}[\mathcal{R}])^* \cong (\mathbb{T}[\mathcal{R}_1])^*$ as follows easily. Now apply the theorem. ∎

Axiomatization of strong equivalence for simultaneous recursion Given a proper simultaneous recursion $\mathcal{R}$ Theorem 8.3.14(ii) shows that $=^*_{\mathcal{R}}$ can be axiomatized by the rules of Definition 7.1.11 simply by taking $\mathcal{R}^*$ instead of $\mathcal{R}$. However, in a way similar to what we did in Section 8.2 for $=^*_{\mu}$, we can also define a co-inductive system $(\mathcal{R}^*)$ that directly axiomatizes $=^*_{\mathcal{R}}$. Also in this system we have judgments of the form $\mathcal{H} \vdash A = B$ in which $\mathcal{H}$ is a set of equations of the shape $A \to A' = B \to B'$. As in system (BH), Definition 8.2.44, the crucial point is the introduction of a rule (coind).

8.3.16 Definition Let $\mathcal{R}$ be a proper simultaneous recursion. The system $(\mathcal{R}^*)$ is defined by the axioms and rules in Fig. 8.3.

We write $\mathcal{H} \vdash^*_{\mathcal{R}} A = B$ to mean that $\mathcal{H} \vdash A = B$ can be derived by the above rules.

In this system rule ($\to$-cong) is missing but it is easy to prove that it is derivable. Note that in this system there are no types having properties analogous to these of the circular types in $\mathbb{T}_{\mu}$, $\mathbb{T}^*_{\mu}$.

(ident)	$\mathcal{H} \vdash A = A$	
(symm)	$\dfrac{\mathcal{H} \vdash A = B}{\mathcal{H} \vdash B = A}$	
(trans)	$\dfrac{\mathcal{H} \vdash A = B \quad \mathcal{H} \vdash B = C}{\mathcal{H} \vdash A = C}$	
(axiom)	$\mathcal{H} \vdash A{=}B,$	if $A{=}B \in \mathcal{H},$
($\mathcal{R}$-eq)	$\mathcal{H} \vdash X = A,$	if $X = A \in \mathcal{R},$
(coind)	$\dfrac{\mathcal{H}, (A \to B){=}(A' \to B') \vdash A{=}A' \quad \mathcal{H}, (A \to B){=}(A' \to B') \vdash B{=}B'}{\mathcal{H} \vdash (A \to B){=}(A' \to B')}$	

Figure 8.3 The system $(\mathcal{R}^*)$

8.3.17 Example Let $\mathcal{R}_1 \triangleq \{X = A \to A \to X\}$ where A is a any type. Then we have $\vdash^*_{\mathcal{R}_1} X = A \to X$, with the following proof, where C denotes $A \to A \to X$.

$$\dfrac{\dfrac{(\mathcal{R}\text{-eq})}{\vdash X{=}C} \quad \dfrac{\dfrac{(\text{ident})}{\{C{=}A \to X\} \vdash A{=}A} \quad \dfrac{\dfrac{(\mathcal{R}\text{-eq})}{\{C{=}A \to X\} \vdash X{=}C} \quad \dfrac{(\text{hyp})}{\{C{=}A \to X\} \vdash C{=}A \to X}}{\{C{=}A \to X\} \vdash X{=}A \to X}\,(\text{trans})}{\vdash C{=}A \to X}\,(\text{coind}),\ (\text{symm})}{\vdash X{=}A \to X}\,(\text{trans}).$$

With the same technique as used in Section 8.2 we can prove the soundness and completeness of $(\mathcal{R}^*)$ with respect to strong equivalence. The completeness proof, in particular, is based on a variant of the algorithm introduced in Definition 8.2.50 (up to some minor adjustments due to the different set of types) which, given two types A and B, builds a proof of $A = B$ iff $A =^*_{\mathcal{R}} B$. This yields the following.

8.3.18 Theorem *Let $\mathcal{R}(\vec{X})$ be a proper simultaneous recursion over $\mathbb{T}$.*

(i) $\vdash^*_{\mathcal{R}} A{=}B \Leftrightarrow A =^*_{\mathcal{R}} B$.
(ii) *Given $A, B \in \mathbb{T}[\vec{X}]$ it is decidable whether or not $A =^*_{\mathcal{R}} B$.*

Justifying type equations by a simultaneous recursion

In the study of recursive type inference it will be useful to know whether a given simultaneous recursion justifies a given set of equations, according to Definition 7.1.19(ii). We prove that this is a decidable property, a result that is needed in Section 9.2. The original proof is due to Statman (1994).

In the rest of this subsection we show the decidability of the existence of morphisms between algebras of the form $\mathbb{T}^{\mathbb{A}}[\mathcal{R}]$. By Proposition 7.3.25(ii) and Lemma 8.3.10 one has $\mathbb{T}^{\mathbb{A}}[\mathcal{R}] \cong \mathbb{T}^{\mathbb{A}'}[\mathcal{R}']$, where $\mathcal{R}'$ is flat and without orphans. Moreover this isomorphism is effective. Hence in a proof that the existence of morphisms is decidable one may assume that the $\mathcal{R}$ are flat and without orphans.

8.3.19 Definition Let $\mathcal{A}$, $\mathcal{A}'$ be type algebras.

(i) A set of *constraints* is of the form $\mathcal{C} \subseteq \mathcal{A} \times \mathcal{A}'$.
(ii) A morphism $h : \mathcal{A} \to \mathcal{A}'$ is said to *agree with* $\mathcal{C}$ if it satisfies $h(a) = a'$, for all $\langle a, a'\rangle \in \mathcal{C}$.
(iii) Two sets of constraints $\mathcal{C}, \mathcal{C}' \subseteq \mathcal{A} \times \mathcal{A}'$ are *equivalent*, if

$$\text{for all morphisms } h : \mathcal{A} \to \mathcal{A}'.[h \text{ agrees with } \mathcal{C} \Leftrightarrow h \text{ agrees with } \mathcal{C}'].$$

It will be shown that the existence of an $h : \mathbb{T}^{\mathbb{A}}/\mathcal{E} \to \mathbb{T}^{\mathbb{A}'}[\mathcal{R}']$ agreeing with a finite set of constraints $\mathcal{C}$ is decidable. The proof occupies 8.3.20–8.3.30. For the proof, substantial help was obtained from Dexter Kozen and Jan Willem Klop (personal communications).

8.3.20 Definition Let $\mathcal{A}$ be a type algebra and $a, b, c \in \mathcal{A}$.

(i) If $a = b \to c$, then b and c are called *direct descendants* of a. Direct descendants are *descendants*. And direct descendants of descendants are also descendants. Write $a \rightsquigarrow b$ if b is a direct descendant of a and $a \rightsquigarrow^+ b$ if b is a descendant of a. The relation $\rightsquigarrow^*$ is the reflexive transitive closure of $\rightsquigarrow$.
(ii) An element $a \in \mathcal{A}$ is called *cyclic* if $a \rightsquigarrow^+ a$.
(iii) Write $C(\mathcal{A}) = \{a \in \mathcal{A} \mid a \text{ is cyclic}\}$.

For example in $\mathbb{T}^{\{a\}}[X = X \to a]$ the element $[X]$ is cyclic, as $[X] = [X] \to [a]$. Notice that not every indeterminate $X \in \vec{X}$ of $\mathcal{R}$ needs to be cyclic. For example if $\mathcal{R} = \{X = X \to Y, Y = a\}$, then only $[X]$, but not $[Y]$, is cyclic in $\mathbb{T}^{\mathbb{A}}[\mathcal{R}]$.

8.3.21 Notation Let $\mathcal{R}$ be a simultaneous recursion over $\mathbb{T}^{\mathbb{A}}$.

(i) Write $I(\mathcal{R}) \triangleq \{[X] \in \mathbb{T}^{\mathbb{A}}[\mathcal{R}] \mid X \text{ an indeterminate of } \mathcal{R}\}$, for the set of indeterminates in $\mathbb{T}^{\mathbb{A}}[\mathcal{R}]$.
(ii) Write $C(\mathcal{R}) \triangleq C(\mathbb{T}^{\mathbb{A}}[\mathcal{R}])$, for the set of cyclic elements of $\mathbb{T}^{\mathbb{A}}[\mathcal{R}]$.

We will see that for a flat simultaneous recursion without orphans $\mathcal{R}$ one has $C(\mathcal{R}) \subseteq I(\mathcal{R})$.

8.3.22 Lemma *Let $h : \mathcal{A} \to \mathcal{B}$ be a morphism and $a, b \in \mathcal{A}$.*

(i) *Suppose $\mathcal{B}$ is invertible. Let $a = b \to c$. Then $h(b), h(c)$ are uniquely determined by $h(a)$.*
(ii) $a \rightsquigarrow b \Rightarrow h(a) \rightsquigarrow h(b)$.
(iii) *a is cyclic $\Rightarrow h(a)$ is cyclic.*

Proof (i) Because $\mathcal{B}$ is invertible.

(ii) Suppose $a \rightsquigarrow b$. If, say, $a = b \to c$, then $h(a) = h(b) \to h(c)$, so $h(a) \rightsquigarrow h(b)$.

(iii) By (ii). ■

8.3.23 Lemma *Let $\mathcal{R}$ be a flat simultaneous recursion over $\mathbb{T}^{\mathbb{A}}$ and let $a, b \in \mathbb{T}^{\mathbb{A}}[\mathcal{R}]$, with $a \rightsquigarrow b$ and $a = [A]_{\mathcal{R}}$.*

(i) *Then $A \equiv X$ or $A \equiv A_1 \to A_2$, for some $A_1, A_2 \in \mathbb{T}^{\mathbb{A} \cup \vec{X}}$.*
(ii) *In the case $A \equiv X$ one has $(X = X_1 \to X_2) \in \mathcal{R}$ and $b = [X_i]$, for some $i \in \{1, 2\}$.*
(iii) *In the case $A \equiv A_1 \to A_2$ one has $b = [A_i]_{\mathcal{R}}$, for some $i \in \{1, 2\}$.*

Proof Note that A has a unique pair of direct descendants, because $\mathbb{T}[\mathcal{R}]$ is invertible by Proposition 7.3.16.

(i) Suppose $A \equiv \alpha$. Then $\alpha =_{\mathcal{R}} C \to D$ for some $C, D \in \mathbb{T}[\vec{X}]$. But then $\alpha \twoheadrightarrow_{\mathcal{R}} (C' \to D')$ by the Church–Rosser Theorem for $\twoheadrightarrow_{\mathcal{R}}$, Proposition 7.3.14(ii), a contradiction.

(ii) and (iii) The case $(X = \alpha) \in \mathcal{R}$ follows the case $A \equiv \alpha$ in (i). The rest is immediate by the invertibility of $\mathbb{T}[\mathcal{R}]$. ■

8.3.24 Lemma *Let $\mathcal{R}$ be flat. Let a, b range over $\mathbb{T}^{\mathbb{A}}[\mathcal{R}]$.*

(i) $a \rightsquigarrow b \;\&\; a \in I(\mathcal{R}) \Rightarrow b \in I(\mathcal{R})$.
(ii) *If*

$$a = a_1 \rightsquigarrow a_2 \rightsquigarrow \cdots \rightsquigarrow a_n \rightsquigarrow \cdots$$

is an infinite chain of descendants, then for some k one has $a_k \in I(\mathcal{R})$.

Proof (i) Let $a \in I(\mathcal{R})$. Then $a = [X]$, hence by Lemma 8.3.23(ii) one has $(X = X_1 \to X_2) \in \mathcal{R}$ and $b = [X_i]$, for some i.

(ii) Let $a_1 = [A]$. If $A \in \vec{X}$, then we are done. Otherwise by Lemma 8.3.23(i) and (iii) one has $A \equiv A_1 \to A_2$ and $a_2 = [A_i]$, for some i. Thus continuing, by splitting the type $A_1 \to A_2 \in \mathbb{T}^{\mathbb{A} \cup \vec{X}}$ into its components, one eventually obtains $a_k = [Y]$, by Lemma 8.3.23(i). ■

8.3.25 Corollary *Let $\mathcal{R}$ be flat without orphans. Then*

(i) $\forall a \in I(\mathcal{R}) \exists b \in C(\mathcal{R}). b \rightsquigarrow^* a$;

(ii) $C(\mathcal{R}) \subseteq I(\mathcal{R})$.

Proof (i) Since $\mathcal{R}$ is without orphans, each $a \in I(\mathcal{R})$ has a 'father' $a' \in I(\mathcal{R})$. Going backwards along the ancestors, since the set $I(\mathcal{R})$ is finite, we must end up in a cyclic element $a^{(k)} \in C(\mathcal{R})$. Then $a^{(k)} \leadsto^* a$, and we are done.

(ii) Let $a \in C(\mathcal{R})$. Then

$$a \leadsto^+ a \leadsto^+ \cdots .$$

Hence for some $b \in I(\mathcal{R})$ one has $a \leadsto^+ b \leadsto^+ a$, by (ii) of the lemma. But then $a \in I(\mathcal{R})$, by (i) of the lemma. ■

8.3.26 Proposition *Let $\mathcal{R}(\vec{X}), \mathcal{R}'(\vec{X}')$ be simultaneous recursions, over $\mathbb{T} = \mathbb{T}^{\mathbb{A}}, \mathbb{T}^{\mathbb{A}'}$ respectively, and let $\mathcal{A} = \mathbb{T}^{\mathbb{A}}[\mathcal{R}]$, $\mathcal{A}' = \mathbb{T}^{\mathbb{A}'}[\mathcal{R}']$.*

(i) *It is decidable whether or not there exists a morphism $h : \mathcal{A} \to \mathcal{A}'$.*
(ii) *Let moreover $\mathcal{C} \subseteq \mathcal{A} \times \mathcal{A}'$ be finite. Then it is decidable whether or not*

there exists a morphism $h : \mathcal{A} \to \mathcal{A}'$ agreeing with $\mathcal{C}$.

Proof (i) By Proposition 7.3.25(ii) and Lemma 8.3.10 we may assume that $\mathcal{R}$ and $\mathcal{R}'$ are flat and without orphans.

Define a *proto-morphism* to be a map $h : I(\mathcal{R}) \to I(\mathcal{R}')$ such that

$$a = b \to c \;\Rightarrow\; h(a) = h(b) \to h(c).$$

Claim 1. *For a morphism $h : \mathcal{A} \to \mathcal{A}'$ the restriction $h \restriction I(\mathcal{R})$ is a proto-morphism.*

We only need to show that if $a \in I(\mathcal{R})$, then $h(a) \in I(\mathcal{R}')$. By Corollary 8.3.25(i) there exists a $b \in C(\mathcal{R})$ such that $b \leadsto^* a$. But then, as $\mathcal{R}'$ is flat, we have $h(b) \in C(\mathcal{R}') \subseteq I(\mathcal{R}')$, by Lemma 8.3.22(iii) and Corollary 8.3.25(ii). By Lemma 8.3.22(ii) we have $h(b) \leadsto^* h(a)$. Therefore $h(a) \in I(\mathcal{R}')$, by Lemma 8.3.24(i):

$$\begin{array}{ccc}
b \in C(\mathcal{R}) & \xrightarrow{\;h\;} & h(b) \in C(\mathcal{R}') \subseteq I(\mathcal{R}') \\
\big\downarrow {*} & & \big\downarrow {*} \\
a \in I(\mathcal{R}) & \overset{h}{\dashrightarrow} & h(a) \in I(\mathcal{R}').
\end{array}$$

Claim 2. *Any proto-morphism $h : I(\mathcal{R}) \to I(\mathcal{R}')$ can be extended to a morphism $h^+ : \mathcal{A} \to \mathcal{A}'$.*

For an X occurring in $\mathcal{R}$ choose $h_0(X)$ such that $h([X]) = [h_0(X)]$. Then extend h_0 to a morphism $h_0^+ : \mathbb{T}^{\mathbb{A} \cup \vec{X}} \to \mathbb{T}^{\mathbb{A}' \cup \vec{X}'}$ by recursion, with a case distinction for the base case:

$$
\begin{array}{llcll}
\text{Basis} & h_0^+(\alpha) & = & h_0(X), & \text{if } (X = \alpha) \in \mathcal{R}, \\
 & & = & \text{arbitrary}, & \text{otherwise}, \\
 & h_0^+(X) & = & h_0(X); & \\
\text{Recursion} & h_0^+(A{\to}B) & = & h_0^+(A) \to h_0^+(B). &
\end{array}
$$

Then $A =_{\mathcal{R}} B \Rightarrow h_0^+(A) =_{\mathcal{R}'} h_0^+(B)$. Hence by Proposition 7.2.14(i), h_0^+ induces the required morphism $h^+ : \mathcal{A} \to \mathcal{A}'$:

$$h^+([A]) = [h_0^+(A)].$$

Claim 3. *There exists a morphism* $h : \mathcal{A} \to \mathcal{A}'$ *iff there exists a proto-morphism* $h_0 : I(\mathcal{R}) \to I(\mathcal{R}')$.

Immediate by Claims 1 and 2.

Claim 4. *It is decidable whether or not there exists a proto-morphism* $h_0 : I(\mathcal{R}) \to I(\mathcal{R}')$.

As $I(\mathcal{R}), I(\mathcal{R}')$ are finite there are only finitely many candidates. By Proposition 8.3.7 the relation $=_{\mathcal{R}}$ is decidable, hence also the requirement to be a proto-morphism.

Now we are done by Claims 3 and 4.

(ii) Instead of considering equivalence classes, we use elements of $\mathbb{T}^{\mathbb{A}\cup\vec{X}}$, $\mathbb{T}^{\mathbb{A}'\cup\vec{X}'}$ and work modulo $=_{\mathcal{R}}, =_{\mathcal{R}'}$, respectively. Given a set of constraints $\mathcal{C}$ one can construct an equivalent set $\mathcal{C}'$ such that the elements are of the form

$$
\begin{array}{l}
\langle \alpha, A' \rangle, \\
\langle X, \alpha' \rangle, \\
\langle X, Y' \to Z' \rangle, \\
\langle A \to B, \alpha' \rangle.
\end{array}
$$

Other possible forms are $\langle X, X' \rangle, \langle A \to B, X' \rangle$ and $\langle A \to B, A' \to B' \rangle$. Using invertibility the last pair can be replaced equivalently by $\{\langle A, A' \rangle, \langle B, B' \rangle\}$. In the first pair $\langle X, X' \rangle$ the type X' can be replaced by the right hand side of X' in $\mathcal{R}'$, and similarly in the second. Let us call the resulting $\mathcal{C}'$ a *simplified set of constraints*. A proto-morphism $h : \mathcal{A} \to \mathcal{A}'$ is called *consistent with* $\mathcal{C}'$ if the following hold:

$$
\begin{array}{rcll}
\langle \alpha, A' \rangle \in \mathcal{C}' & \Rightarrow & h(X) = A', & \text{in the case } (X = \alpha) \in \mathcal{R}, \\
\langle X, \alpha' \rangle \in \mathcal{C}' & \Rightarrow & h(X) = \alpha' & \\
\langle X, Y' \to Z' \rangle \in \mathcal{C}' & \Rightarrow & h(X) = Y' \to Z' & \\
\langle A \to B, \alpha' \rangle \notin \mathcal{C}'. & & &
\end{array}
$$

The reason we require $\langle A \to B, \alpha' \rangle \notin \mathcal{C}'$ is that $h(A \to B) = \alpha$ is impossible for a morphism h, as $\alpha \neq_{\mathcal{R}'} A' \to B'$.

Now, similarly to the claims in (i), one can show the following.

Claim $3'$. *There exists a morphism $h : \mathcal{A} \to \mathcal{A}'$ agreeing with $\mathcal{C}$ iff there exists a proto-morphism $h : I(\mathcal{R}) \to I(\mathcal{R}')$ consistent with $\mathcal{C}'$, a simplified form of $\mathcal{C}$.*

Claim $4'$. *It is decidable whether or not there exists a proto-morphism $h : I(\mathcal{R}) \to I(\mathcal{R}')$ consistent with a simplified set $\mathcal{C}'$.*

From Claims $3'$, $4'$ the result follows. ∎

In Statman (1994) it is shown that the justifiability of a simultaneous recursion in another simultaneous recursion is an NP-complete problem.

This result also holds for type algebras of the form $\mathcal{A} = \mathbb{T}^{\mathbb{A}}/\mathcal{E}$. We show this by extending the axiomatization of type equality in order to force a type algebra generated by a set of equations $\mathcal{E}$ to have the invertibility property.

8.3.27 Definition Write $\mathcal{E} \vdash^{\text{inv}} a = b$ if there is a proof of $\vdash a = b$ by the axioms and rules for equality given in Definition 7.1.11 extended by the following two rules for invertibility

$$(\text{inv}) \quad \frac{\vdash a_1 \to a_2 = b_1 \to b_2}{\vdash a_i = b_i} \quad (i = 1, 2).$$

We define for $a, b \in \mathbb{T}$ the relation $a =_{\mathcal{E}}^{\text{inv}} b \Leftrightarrow \mathcal{E} \vdash^{\text{inv}} a = b$.

8.3.28 Lemma *Let $\mathcal{E}$ a system of type equations. Then there exists a $\mathcal{E}^{\text{inv}}$ which is a proper simultaneous recursion such that*

$$\mathcal{E} \vdash^{\text{inv}} \mathcal{E}^{\text{inv}} \;\&\; \mathcal{E}^{\text{inv}} \vdash \mathcal{E}.$$

In other words, $\mathcal{E}^{\text{inv}}$ is equivalent to $\mathcal{E}$ for provability with the extra rule of invertibility.

Proof Given $\mathcal{E}$ over $\mathbb{T}^{\mathbb{A}}$. The relation $=_{\mathcal{E}}^{\text{inv}}$ is the least invertible congruence containing $=_{\mathcal{E}}$. We now define a simultaneous recursion which generates $=_{\mathcal{E}}^{\text{inv}}$.

Let $\alpha_0, \alpha_1, \ldots$ denote the elements of $\mathbb{A}$. As in Definition 8.3.5(i) define by recursion on n sets of equations $\mathcal{D}_n$, $\mathcal{I}_n$ $(n \geq 0)$. Let $\mathcal{D}_0 \triangleq \mathcal{E}$ and $\mathcal{I}_0 \triangleq \emptyset$. Define $\mathcal{D}_{n+1}$, $\mathcal{I}_{n+1}$ from $\mathcal{D}_n$, $\mathcal{I}_n$, $n \geq 0$, as follows:

(1) If $(A \to B = A' \to B') \in \mathcal{D}_n$, then take

$$\mathcal{D}_{n+1} \triangleq \mathcal{D}_n - \{A \to B = A' \to B'\} \cup \{A = A', B = B'\}$$

and $\mathcal{I}_{n+1} \triangleq \mathcal{I}_n$;

(2) Replace in $\mathcal{D}_n$ all equations of the form $A \to B = \alpha$ by $\alpha = A \to B$;

(3) If $(\alpha_i = A \to B), (\alpha_i = A' \to B') \in \mathcal{D}_n$, then take

$$\mathcal{D}_{n+1} \triangleq \mathcal{D}_n - \{\alpha_i = A' \to B'\} \cup \{A = A', B = B'\},$$

assuming $\text{dpt}(A \to B) \leq \text{dpt}(A' \to B')$, see Definition 1.1.22(i) and take $\mathcal{I}_{n+1} \triangleq \mathcal{I}_n$;

(4) If $(\alpha_i = \alpha_i) \in \mathcal{D}_n$ for some i, then take $\mathcal{D}_{n+1} \triangleq \mathcal{D}_n - \{\alpha_i = \alpha_i\}$ and $\mathcal{I}_{n+1} \triangleq \mathcal{I}_n$;

(5) If $(\alpha_i = \alpha_j) \in \mathcal{D}_n$, with $i \neq j$, then take

$$\mathcal{D}_{n+1} \triangleq \mathcal{D}_n[\alpha_h := \alpha_k] - \{\alpha_i = \alpha_j\}$$
$$\mathcal{I}_{n+1} \triangleq \mathcal{I}_n \cup \{\alpha_h = \alpha_k\},$$

where $h = \min(i, j)$ and $k = \max(i, j)$;

(6) Otherwise take $\mathcal{D}_{n+1} \triangleq \mathcal{D}_n$ and $\mathcal{I}_{n+1} \triangleq \mathcal{I}_n$.

(7) Let N be the least n such that $\mathcal{D}_{n+1} = \mathcal{D}_n$ and $\mathcal{I}_{n+1} = \mathcal{I}_n$. We will show below that this number exists. Write $\mathcal{D}_\mathcal{E} \triangleq \mathcal{D}_N, \mathcal{I}_\mathcal{E} = \mathcal{I}_N$.

(8) Finally define $\mathcal{E}^{\text{inv}} \triangleq \mathcal{D}_N \cup \mathcal{I}_N$.

Claim. *The number N in Step (7) exists.* Indeed, we assign to the set $D = \{A_1 = B_1, \ldots, A_n = B_n\}$ the multiset

$$S_D = \{d(A_1), d(B_1), \ldots, d(A_n), d(B_n)\}.$$

Steps (1), (3)–(6) of the algorithm decrease the multiset order, see Definition 2.1.6. Step (2) keeps the multiset fixed. Since the multiset order is well founded, see Lemma 2.1.8, it follows that a potentially infinite path $D_1, D_2, D_3, \ldots$ ends up by only performing Step (2). But this is impossible, by counting the number of type atoms on the right hand sides.

It is easy to prove, by induction on n, that for all $n \geq 0$

$$\mathcal{D}_n \cup \mathcal{I}_n \vdash \mathcal{E} \text{ and } \mathcal{E} \vdash^{\text{inv}} \mathcal{D}_n \cup \mathcal{I}_n.$$

To show that $\mathcal{D}_N \cup \mathcal{I}_N$ is a proper simultaneous recursion, note that, by Steps (2) and (4) above, $\mathcal{D}_N$ is a proper simultaneous recursion. Now note that when an equation $\alpha_i = \alpha_j$ is put in $\mathcal{I}_n$ we must have $i < j$ and moreover α_i does not occur in $\mathcal{D}_m$ for all $m \geq n$. So no other equation of the form $\alpha_i = \alpha_h$ can be put in some $\mathcal{I}_m$ for $m \geq n$. ■

We now give some applications of this result.

8.3.29 Proposition *Let $\mathcal{A}$ be a type algebra. Then there exists an invertible type algebra $\mathcal{A}^{\text{inv}}$ initial under $\mathcal{A}$. That is, there is a morphism*

$k : \mathcal{A} \to \mathcal{A}^{\text{inv}}$ such that all $h : \mathcal{A} \to \mathcal{B}$, with $\mathcal{B}$ invertible, can be factored through k; that is $h = i \circ k$, for some $i : \mathcal{A}^{\text{inv}} \to \mathcal{B}$:

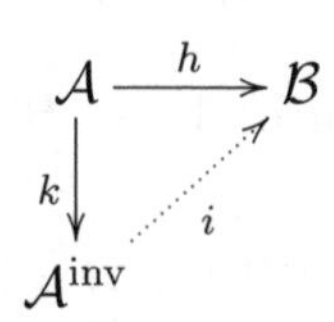

Proof By Lemma 7.1.18 we can assume that $\mathcal{A} = \mathbb{T}/\sim$ is a syntactic algebra. Then there is a morphism $h^{\natural} : \mathbb{T} \to \mathcal{B}$ corresponding to h according to Proposition 7.2.4(iii).

Now define $\mathcal{A}^{\text{inv}} = \mathbb{T}/\approx^{\text{inv}}$ where $\approx^{\text{inv}}$ is a shorthand for $=^{\text{inv}}_{\approx}$ defined in 8.3.28. Next note that $h^{\natural}$ can also be seen as a syntactic morphism from $\mathbb{T}$ to $\mathcal{B}$ preserving $\approx^{\text{inv}}$. In fact, if $A_1 \to A_2 \approx B_1 \to B_2$, we must not only have $A_i \approx^{\text{inv}} B_i$ $(i = 1, 2)$ by rule (inv), but also $h^{\natural}(A_i) = h^{\natural}(B_i)$ since $\mathcal{B}$ is invertible. Now take $k : \mathcal{A} \to \mathcal{A}^{\text{inv}}$ such that $k([A]_{\approx}) = [A]_{\approx^{\text{inv}}}$ and i such that $i([A]_{\approx^{\text{inv}}}) = h^{\sharp}(A) = h([A]_{\approx})$. ∎

8.3.30 Corollary *Let $\mathcal{A} = \mathbb{T}^{\mathbb{A}}/\mathcal{E}$, $\mathcal{A}' = \mathbb{T}^{\mathbb{A}'}[\mathcal{R}']$, with proper $\mathcal{R}'$, and $\mathcal{C} \subseteq \mathcal{A} \times \mathcal{A}'$. Then it is decidable whether or not*

there exists a morphism $h : \mathcal{A} \to \mathcal{A}'$ agreeing with $\mathcal{C}$.

Proof The existence of an $h : \mathcal{A} \to \mathcal{A}'$ is, by Proposition 8.3.29, equivalent to the existence of an $h_1 : \mathcal{A}^{\text{inv}} \to \mathcal{A}'$. Moreover, h agrees with $\mathcal{C}$ iff h_1 agrees with $\mathcal{C}_1 = \{\langle k(a), a'\rangle \mid \langle a, a'\rangle \in \mathcal{C}\}$. Therefore by Proposition 8.3.26(ii) we are done. ∎

It is possible to strengthen Proposition 8.3.26(i) as follows.

8.3.31 Theorem *Let $\mathcal{E}, \mathcal{E}'$ be finite sets of equations over $\mathbb{T}^{\mathbb{A}}$ and $\mathbb{T}^{\mathbb{A}'}$ respectively. Then it is decidable whether or not there exists a morphism $h : \mathbb{T}^{\mathbb{A}}/\mathcal{E} \to \mathbb{T}^{\mathbb{A}'}/\mathcal{E}'$.*

Proof This is proved in Kozen (1977). ∎

8.4 Exercises

Exercise 8.1 Prove Theorem 8.1.7. [Hint. Argue by induction on A. Note that the union of two simultaneous recursions, provided the indeterminates are disjoint, is still a simultaneous recursion.]

Exercise 8.2 Prove Proposition 7.4.33.

Exercise 8.3 Let (triv) be the following rule

$$\text{(triv)} \qquad M = N, \text{ for } M, N \text{ circular.}$$

Let $A_1 \equiv \mu\beta.(\mu\alpha.\alpha \to \beta)$ and $B_1 \equiv \mu\beta.(\mu\alpha.\alpha \to \beta) \to \beta$.

(i) Show that one has $\vdash_{\text{BH}} A_1 = B_1$, without using rule ($\mu$-cong).
(ii) Let ($\text{BH}^{\sim}$) be the system (BH) with (μ-cong) replaced by (triv). Show

$$\vdash_{\text{BH}} A = B \Leftrightarrow \vdash_{\text{BH}^{\sim}} A = B.$$

[Hint. Use Lemma 8.2.53 and Theorem 8.2.55(ii), (iii).]
(iii) Show that (triv) is not equivalent to (μ-cong) in system (μ). [Hint. Let system ($\mu^{\sim}$) be system (μ) with (μ-cong) replaced by (triv). Show that $A_1 = B_1$ is not derivable in ($\mu^{\sim}$).]

Exercise 8.4 Each non-circular type A has a reduced form A' which is in $\mathcal{SC}^s(A)$, see Definition 8.2.33. [Hint. If $B \in \mathcal{SC}^s(A)$ and $B \Rightarrow_\mu C$ by unfolding B at the top level, then $C \in \mathcal{SC}^s(A)$.]

Exercise 8.5 Show that $\vdash_{\text{BH}} \mu\alpha.(\beta \to \beta \to \alpha) = \mu\alpha.(\beta \to \alpha)$, where β is any type variable, see Examples 7.4.29 and 8.2.60.

Exercise 8.6

(i) Prove $\mu\alpha.((\alpha \to \alpha) \to \alpha) = \mu\alpha.(\alpha \to \alpha \to \alpha)$ in (BH).
(ii) Prove directly $\mu\alpha.((\alpha \to \alpha) \to \alpha) =^*_\mu \mu\alpha.(\alpha \to \alpha \to \alpha)$.

Compare the proofs with that in (μ^*_{AC}).

Exercise 8.7 (Grabmayer) Show that $\Rightarrow_\mu^{-1}$ is not CR. [Hint. Take the following type:

$$A = (\mu s.s \to \mu t.(s \to t)) \to \mu t'.((\mu s.s \to \mu t.(s \to t)) \to t').$$

Show that $A = B[t' := \mu t'.B] = C[s := \mu s.C]$ for two types $B \neq C$ irreducible with respect to $\Rightarrow_\mu^{-1}$.]

Exercise 8.8 This exercise, inspired by Endrullis et al. (2011), is about $\mathbb{T}_{\dot\mu}$. The relation $\leadsto$ over $\mathbb{T}_{\dot\mu}$ is defined as follows.

(1) $(A \to B) \leadsto A$;
(2) $(A \to B) \leadsto B$;
(3) $(\dot\mu\alpha.A) \leadsto A$;
(4) $(\dot\mu\alpha.A) \leadsto (A[\alpha := \dot\mu\alpha.A])$.

(i) Write $\leadsto^*$ for the transitive reflexive closure of $\leadsto$.
(ii) Write $A \leadsto_n B$ if $A \leadsto^* B$ in n steps.

(iii) Define $\mathcal{SC}^{\cdot}(A) = \{B \in \mathbb{T}_{\dot\mu} \mid A \rightsquigarrow^* B\}$. Show that $\mathcal{SC}^{\cdot}(A)$ is always finite.

[Hint. Define the set of underlined $\dot\mu$-types, written $\underline{\mathbb{T}}_{\dot\mu}$; for example $(\dot\mu\alpha.\alpha \to \dot\mu\beta\dot\mu\gamma.(\alpha \to \underline{\beta \to \gamma})$ and $(\dot\mu\alpha.\alpha \to \underline{\dot\mu\beta\dot\mu\gamma.(\alpha \to \underline{\beta \to \gamma})})$ are in $\underline{\mathbb{T}}_{\dot\mu}$. For $B \in \underline{\mathbb{T}}_{\dot\mu}$ write $|B| \in \mathbb{T}_{\dot\mu}$ for the type obtained from B by erasing all underlinings. On $\underline{\mathbb{T}}_{\dot\mu}$ we define $\underline{\rightsquigarrow}$, a variant of $\rightsquigarrow$:

(1) $(A \to B) \underline{\rightsquigarrow} A$;
(2) $(A \to B) \underline{\rightsquigarrow} B$;
(3) $(\dot\mu\alpha.A) \underline{\rightsquigarrow} A$;
(4) $(\dot\mu\alpha.A) \underline{\rightsquigarrow} (A[a := \underline{\dot\mu a.A}])$.

Show that $\{B \in \underline{\mathbb{T}}_{\dot\mu} \mid A \underline{\rightsquigarrow} B\}$ is finite, for all $A \in \underline{\mathbb{T}}_{\dot\mu}$. Finally, prove that for every $\rightsquigarrow$-reduction without repetitions $R : A \equiv B_0 \rightsquigarrow B_1 \rightsquigarrow \cdots \rightsquigarrow B_n$, there is a $\underline{\rightsquigarrow}$-reduction $R' : A \equiv B_0 \underline{\rightsquigarrow} B_1' \underline{\rightsquigarrow} \cdots \underline{\rightsquigarrow} B_n'$, such that $|B_i'| \equiv B_i$ for $1 \leq i \leq n$. This yields the conclusion.]

Exercise 8.9 Recall the definition of $\mathcal{SC}_r^s(A) = \{C \in \mathbb{T}_\mu | A \rightsquigarrow^* C\}$ in Definition 8.2.36.

(i) Show the following. If $A[\alpha := \mu\alpha.A] \rightsquigarrow^* B$ then:
   (a) either this reduction is $A[\alpha := \mu\alpha.A] \rightsquigarrow^* \mu\alpha.A \rightsquigarrow^* B$;
   (b) or this reduction is a $[\alpha := \mu\alpha.A]$ instance of $A \rightsquigarrow^* C$. (So $B = C[\alpha := \mu\alpha.A]$.)

(ii) Show that $\mathcal{SC}_r^s(\mu\alpha.A_1) \subseteq \{\mu\alpha.A_1\} \cup \mathcal{SC}_r^s(A_1)[\alpha := \mu\alpha.A_1]$.
(iii) Show that $\mathcal{SC}_r^s(A)$ is finite.

Exercise 8.10 Prove that if $A, B \in \mathbb{T}_\mu$ and $A \, R_\mu \, B$, where R_μ is a bisimulation over types defined in Remark 8.2.62 then $(A)_\mu^* = (B)_\mu^*$.

Exercise 8.11 Prove Proposition 8.2.61.

Exercise 8.12 Let $\mathcal{E} =_{\text{def}} \{X = \alpha^p \to X,\ X = \alpha^q \to X\}$ and $\mathcal{R} = \{X = \alpha^r \to X\}$. Show that $\mathcal{R}(X)$ solves $\mathcal{E}(X)$ over $\mathbb{T}^{\{\alpha\}}(X)$ iff $r | \gcd(p, q)$, where $n | m$ denotes 'n divides m'.

Exercise 8.13 In Ariola and Klop (1994) the following rule is introduced:

$$(\text{AK}) \quad \frac{A[\alpha := B] = B.}{\mu\alpha.A = B}$$

Show that (AK) is equivalent to (AC) in Definition 8.2.59.

Exercise 8.14 Give a formulation of rule (AC) for a simultaneous recursion. Use it to prove $\vdash^*_{\{c=\beta\to\beta\to c\}} c = \beta \to c$.

Exercise 8.15 Prove Lemma 8.2.40.

Exercise 8.16 Prove soundness, completeness and decidability of $(\mathcal{R}^*)$ in Definition 8.3.16. This then proves Theorem 8.3.18.

Exercise 8.17 Define, recursively on n, the equivalence relation $R_n \subseteq \vec{X} \times \vec{X}$ $(n \geq 0)$ as the minimal equivalence relation such that the following holds.

(1) XR_nX' for all $n \geq 0$ if $X = \alpha$, $X' = \alpha \in \mathcal{R}$ for the same atomic type α;
(2) XR_0X' if $X = Y \to Z$, $X' = Y' \to Z' \in \mathcal{R}$ for some Y, Y', Z, $Z' \in \vec{X}$;
(3) $XR_{n+1}X'$ if $X = Y \to Z$, $X' = Y' \to Z' \in \mathcal{R}$ and YR_nY' and ZR_nZ'.

Let N be the first such that $R_N = R_{N+1}$. Show

$$X =^*_{\mathcal{R}} Y \Leftrightarrow XR_NY.$$

Exercise 8.18 This result is due to Grabmayer (2007) and is a direct proof of both the regularity of $(A)^*_\mu$ and decidability of $=^*_\mu$.

(i) Let $A \in \mathbb{T}_\mu$. Then for all subtrees $T \subseteq (A)^*_\mu$ of the tree-unfolding of A there exists a $B \in \mathcal{SC}^s(A)$ such that $T = (B)^*_\mu$.
(ii) Show $\text{card}(\mathcal{SC}^s(A)) \leq l(A)$, where $l(A)$ is defined in Definition 8.2.5.
(iii) A number n is *large enough* for a regular tree T if all subtrees of T occur at a p of length $< n$. Show that $l(A)$ is large enough for $(A)^*_\mu$.
(iv) For $A \in \mathbb{T}_\mu$ the tree $(A)^*_\mu$ is regular with at most $l(A)$ subtrees. [Hint. Use (i).] By (i) there is a surjection of $\mathcal{SC}^s(A)$, which is of cardinality $< l(A)$ by (ii), to the set of subtrees of $(A)^*_\mu$. Therefore also the latter set is of cardinality $< l(A)$ and hence finite.
(v) Conclude that $=^*_\mu$ is decidable. [Sketch: Let $A, B \in \mathbb{T}_\mu$. Given a sequence number $p \in \{0,1\}^*$. By (iii) there are only finitely many different pairs $\langle (A)^*_\mu|p, (B)^*_\mu|p\rangle$. Therefore, in order to test $(A)^*_\mu = (B)^*_\mu$, we claim that one needs to examine these trees only up to depth $N = l(A) \cdot l(B) + 1$:

$$A =^*_\mu B \Leftrightarrow \forall p < N.(A)^*_\mu|p \cong (B)^*_\mu|p.$$

Indeed, $\Rightarrow$ is obvious. As to $\Leftarrow$, assume the right hand side and $A \neq^*_\mu B$ towards a contradiction. Then $(A)^*_\mu|p \neq (Q)^*_\mu|p$, for some p.]

9

Properties of Terms with Types

In this chapter we establish some basic properties of terms which have types in the systems introduced in earlier ones. We only consider type inference systems à la Curry. Some of the results shown in this chapter are meaningless for typed systems à la Church, while some others can be easily adapted to them (like, for instance, the Subject Reduction Theorem). In Section 9.1 we study the subject reduction property for recursive type systems, in Section 9.2 the problem of finding types for untyped terms, and in Section 9.3 we study restrictions on recursive types such that strong normalization holds for terms that inhabit these.

9.1 First properties of $\boldsymbol{\lambda}_{=}^{\mathcal{A}}$

We will show that for invertible $\mathcal{A}$ the recursive type systems $\boldsymbol{\lambda}_{=}^{\mathcal{A}}$ have the subject reduction property, i.e.

$$\left.\begin{array}{l} \Gamma \vdash_{\mathcal{A}} M : a \\ M \twoheadrightarrow_{\beta(\eta)} M' \end{array}\right\} \Rightarrow \Gamma \vdash_{\mathcal{A}} M' : a,$$

where $\vdash$ is defined in Section 7.1. This means that typings are stable with respect to ($\boldsymbol{\beta}$- or) $\boldsymbol{\beta\eta}$-reduction, which is the fundamental evaluation process for typable λ-terms. In general, however, types are not preserved under the reverse operation of expansion.

We will study the subject reduction property for a type assignment system induced by an arbitrary type algebra $\mathcal{T}$ and see that it satisfies the subject reduction property iff it is invertible. We start with some basic lemmas.

9.1.1 Lemma $\Gamma, x{:}b \vdash_{\mathcal{A}} M : a,\ \Gamma \vdash_{\mathcal{A}} N : b \ \Rightarrow\ \Gamma \vdash_{\mathcal{A}} M[x{:=}N] : a.$

Proof By induction on the derivation of $\Gamma, x{:}b \vdash_{\mathcal{A}} M : a$. ∎

9.1.2 Lemma *Suppose $x \notin \mathrm{FV}(M)$, then*

$$\Gamma, x{:}b \vdash_{\mathcal{A}} M : a \;\Rightarrow\; \Gamma \vdash M : a.$$

Proof By induction on the derivation of $\Gamma, x{:}b \vdash M : a$. ∎

9.1.3 Proposition (Inversion Lemma) *Let $\mathcal{A}$ be a type algebra.*

(i) $\Gamma \vdash_{\mathcal{A}} x : a \quad\Leftrightarrow\quad (x{:}a) \in \Gamma.$

(ii) $\Gamma \vdash_{\mathcal{A}} (MN) : a \quad\Leftrightarrow\quad \exists b \in \mathcal{T}.[\Gamma \vdash_{\mathcal{A}} M : (b \to a) \;\&\; \Gamma \vdash_{\mathcal{A}} N : b].$

(iii) $\Gamma \vdash_{\mathcal{A}} (\lambda x.M) : aa \quad\Leftrightarrow\quad \exists b, c \in \mathcal{T}.[a = (b \to c) \;\&\; \Gamma, x{:}b \vdash_{\mathcal{A}} M : c].$

Proof ($\Leftarrow$) immediate.

($\Rightarrow$) The proof is in all cases by induction on the derivation of $\Gamma \vdash P : d$.

(i) To prove $\Gamma \vdash x : A$ we can use only rule (axiom).

(ii) Similar to (i). Now we can use only rule ($\to E$).

(iii) Now we can only use rule ($\to I$). ∎

9.1.4 Remark It is good to realize that the Inversion Lemma is a consequence of the following easy observation. For a derivation of, for example, $\Gamma \vdash_{\mathcal{A}} MN : a$, there is a $b \in \mathcal{A}$ such that this statement is a direct consequence of the statements $\Gamma \vdash_{\mathcal{A}} M : (b \to a)$ and $\Gamma \vdash_{\mathcal{A}} N : b$ in that same derivation. Note that there may be several $b \in \mathcal{A}$ such that $\Gamma \vdash_{\mathcal{A}} M : (b \to a)$ and $\Gamma \vdash_{\mathcal{A}} N : b$.

For syntactic type algebras this boils down to the following.

9.1.5 Corollary (Inversion Lemma I) *Let $\mathcal{A} = \mathbb{T}/{\approx}$ be a syntactic type algebra. Then*

(i) $\Gamma \vdash_{\mathbb{T}/\approx} x : A \quad\Leftrightarrow\quad \exists A' \in \mathbb{T}.[A' \approx A \;\&\; x{:}A' \in \Gamma].$

(ii) $\Gamma \vdash_{\mathbb{T}/\approx} (MN) : A \quad\Leftrightarrow\quad \exists B \in \mathbb{T}.[\Gamma \vdash_{\mathbb{T}/\approx} M : B \to A \;\&\; \Gamma \vdash_{\mathbb{T}/\approx} N : B].$

(iii) $\Gamma \vdash_{\mathbb{T}/\approx} (\lambda x.M) : A \quad\Leftrightarrow\quad \exists B, C \in \mathbb{T}.[A \approx (B \to C) \;\&\; \Gamma, x{:}B \vdash_{\mathbb{T}/\approx} M : C].$

Proof From Proposition 9.1.3 and Notation 7.1.20. ∎

9.1.6 Corollary (Inversion Lemma II) *Let $\mathcal{A}$ be an invertible type algebra. Then*

$$\Gamma \vdash_{\boldsymbol{\lambda}\mathcal{A}} \lambda x.M : (a \to b) \;\Leftrightarrow\; \Gamma, x{:}a \vdash_{\boldsymbol{\lambda}\mathcal{A}} M : b.$$

Proof ($\Rightarrow$) If $\Gamma \vdash \lambda x.M : a \to b$, then by Proposition 9.1.3(iii) we have $a \to b = c \to d$ and $\Gamma, x{:}c \vdash M : d$. By invertibility $a = c$ and $b = d$. Hence $\Gamma, x{:}a \vdash M : b$.

($\Leftarrow$) By rule ($\to$ I). ∎

Now we can prove the subject reduction property for one-step $\boldsymbol{\beta\eta}$-reduction.

9.1.7 Lemma

(i) *If* $\Gamma \vdash_{\mathcal{A}} \lambda x.(Mx) : a$ *and* $x \notin \mathrm{FV}(M)$, *then* $\Gamma \vdash_{\mathcal{A}} M : a$.
(ii) *If moreover* $\mathcal{T}$ *is invertible, then*

$$\Gamma \vdash_{\mathcal{A}} (\lambda x.M)N : a \Rightarrow \Gamma \vdash_{\mathcal{A}} (M[x := N]) : a.$$

Proof (i) By the Inversion Lemma 9.1.3(iii) we have that $\Gamma, x{:}b \vdash (Mx) : c$ for some $b, c \in \mathcal{T}$ such that $a = (b \to c)$. Moreover, by parts (i) and (iii) of Lemma 9.1.3 we have that $\Gamma, x{:}b \vdash M : d \to c$ and $\Gamma, x{:}b \vdash x : d$ for some type d such that $d = b$. Then $(d \to c) = (b \to c) = a$. Then $\Gamma \vdash M : a$, by Lemma 9.1.2.

(ii) Suppose $\Gamma \vdash (\lambda x.M)N : a$. By the Inversion Lemma 9.1.3(iii)

$$\Gamma \vdash (\lambda x.M) : b \to a \;\&\; \Gamma \vdash N : b,$$

for some type b. Moreover, by Lemma 9.1.3(iii), there are types b' and a' such that $b \to a = b' \to a'$ and

$$\Gamma, x{:}b' \vdash M : a'.$$

Then $b' = b$ and $a' = a$, by invertibility. Hence $\Gamma \vdash N : b'$ and by Lemma 9.1.1

$$\Gamma \vdash (M[x := N]) : a'. \blacksquare$$

9.1.8 Corollary (Subject Reduction for $\boldsymbol{\lambda}_{=}^{\mathcal{A}}$) *Let $\mathcal{A}$ be an invertible type algebra. Then $\boldsymbol{\lambda}_{=}^{\mathcal{A}}$ satisfies the subject reduction property for $\boldsymbol{\beta\eta}$.*

Proof By Lemma 9.1.7. ■

Invertibility of $\mathcal{A}$ is a characterization of those type algebras $\mathcal{A}$ such that $\vdash_{\mathcal{A}}$ has the subject reduction property for $\boldsymbol{\beta}$-reduction. The result is from Statman (1994).

9.1.9 Theorem *Let $\mathcal{T}$ be a type algebra. Then*

$$\mathcal{T} \text{ is invertible} \Leftrightarrow \boldsymbol{\lambda}_{=}^{\mathcal{A}} \text{ satisfies the subject reduction property for } \boldsymbol{\beta}.$$

Proof ($\Rightarrow$) By Lemma 9.1.8.

($\Leftarrow$) Let $a, a', b, b' \in \mathcal{T}$ and assume $a \to b = a' \to b'$. We first prove that $b = b'$. Writing $\Gamma_1 = \{x{:}b,\ y{:}a'\}$ we have $\Gamma_1 \vdash ((\lambda z.x)y) : b'$. Then, by subject reduction, $\Gamma_1 \vdash x : b'$. Hence by the Inversion Lemma 9.1.3(i) $b = b'$.

Now take $\Gamma_2 = \{x{:}(a \to b),\ y{:}(a \to a),\ z{:}a'\}$. We obtain $\Gamma_2 \vdash (\lambda u.x(yu))z : b'$. Again by subject reduction we obtain $\Gamma_2 \vdash (x(yz)) : b'$. Next, by the Inversion Lemma 9.1.3(ii), there is a type $c \in \mathbb{T}$ such that $\Gamma_2 \vdash x : c \to b'$ and $\Gamma_2 \vdash (yz) : c$. Now by parts (iii) and (i) of the Inversion Lemma

9.1.3, the second statement implies $a \to a = a' \to c$. By the first part of the proof we then have $a = c$ and so $a \to a = a' \to a$. Now use this to prove $x{:}a' \vdash ((\lambda y.y)x) : a$. A final application of subject reduction and the Inversion Lemma yields $a = a'$. ∎

9.2 Finding and inhabiting types

In this section we consider the problem of finding the (possibly empty) collection of types for a given untyped λ-term $M \in \Lambda$ in a system $\boldsymbol{\lambda}_{=}^{\mathcal{A}}$ à la Curry[1]. The questions that can be asked about typing terms in recursive type systems are similar to those introduced in Section 2.3, but now we have one more parameter: the type algebra in which we want to work. For simplicity we work only with closed terms. So assume $M \in \Lambda^{\emptyset}$. The problems are the following.

(1) *Type reconstruction*

(1a) Find all type algebras $\mathcal{A}$ and all types $a \in \mathcal{A}$ such that $\vdash_{\mathcal{A}} M : a$.
(1b) Given a type algebra $\mathcal{A}$ find all types $a \in \mathcal{A}$ such that $\vdash_{\mathcal{A}} M : a$.

2) *Typability* The problems can be seen as a particular case of both 1a. and 1b. Now we ask whether the set of all possible types of a given term is empty or not. Thus for a given λ-term M the typability problems corresponding to 1a and 1b are:

(2a) Does there exist a type algebra $\mathcal{A}$ and $a \in \mathcal{A}$ such that $\vdash_{\mathcal{A}} M : a$?
(2b) Given $\mathcal{A}$, does there exist an $a \in \mathcal{A}$ such that $\vdash_{\mathcal{A}} M : a$?

(3) *Type checking* Given $\mathcal{A}$ and $a \in \mathcal{A}$, do we have $\vdash_{\boldsymbol{\lambda}\mathcal{A}} M : a$?

In the next two problems the role of terms and types are reversed: we start with a type algebra $\mathcal{A}$ and element $a \in \mathcal{A}$.

(4) *Enumeration* Find all $M \in \Lambda$ such that $\vdash_{\boldsymbol{\lambda}\mathcal{A}} M : a$.
(5) *Inhabitation* Is there a term $M \in \Lambda$ such that $\vdash_{\boldsymbol{\lambda}\mathcal{A}} M : a$?

Stylistically these problems can be rendered as follows.

[1] As in the case of simple types, this kind of problem is much simpler for explicitly typed terms. The well-formed terms of the typed calculi à la Church have a unique type and have all type information written in them. To check that they are well formed with respect to the formation rules the only non-trivial step is that of checking type equivalence in rule (equal). But we have seen in Chapter 8 that all interesting type equivalences are decidable. Type checking is even simpler in the system $\lambda_{\mu}^{\mathrm{Ch}_0}$ where also type equivalence is explicit. In the case of the typed calculi à la de Bruijn, the type information is given only partially, which sometimes leads to undecidability; see Schubert (1998).

Type reconstruction		$\vdash_? M : ?$		$\vdash_{\boldsymbol{\lambda}\mathcal{A}} M : ?.$
Typability	$\exists \mathcal{A}, a \in \mathcal{A}.$	$\vdash_{\boldsymbol{\lambda}\mathcal{A}} M : a?$	$\exists a \in \mathcal{A}.$	$\vdash_{\boldsymbol{\lambda}\mathcal{A}} M : a?$
Type checking		$\vdash_{\boldsymbol{\lambda}\mathcal{A}} M : a?$		
Enumeration		$\vdash_{\boldsymbol{\lambda}\mathcal{A}} ? : a;$		
Inhabitation	$\exists M \in \Lambda.$	$\vdash_{\boldsymbol{\lambda}\mathcal{A}} M : a?$		

Problems (1) and (2) could be called 'finding types'; problem (3) 'fitting'; and problems (4) and (5) 'finding terms'.

From Corollary 9.1.3 it follows that for invertible $\mathcal{A}$ one has

$$x_1{:}a_1, \ldots, x_n{:}a_n \vdash_{\mathcal{A}} M : a \;\Leftrightarrow \vdash_{\mathcal{A}} (\lambda x_1 \cdots x_n.M) : (a_1 \to \cdots \to a_n \to a).$$

So in this case it is not restrictive to formulate these questions for closed terms only. We will not consider the possibility of having a type environment Γ as parameter for non-invertible type algebras. The reader is invited to study this situation.

The problem of inhabitation is sometimes trivial. As pointed out in Remark 7.5.40(i), in the systems $\lambda\mu$, $\lambda\mu^*$, for instance, all types are inhabited by $\boldsymbol{\Omega} = (\lambda x.xx)(\lambda x.xx)$. We will discuss the problem of finding inhabitants in $(\boldsymbol{\lambda}\mathcal{R})$ at the end of this section.

We will present here the basic results concerning type reconstruction considering only pure λ-terms. The generalization to terms containing constants is rather straightforward and will be discussed in Section 11.3.

Next we will answer the typing questions for syntactic type algebras $\mathcal{A}$. The answer to similar questions for μ-types follows easily from this case.

Finding types in type algebras

The most natural way of describing a type algebra is via a set of equations over a set $\mathbb{T}$ of types. So in many of the above questions we will assume (without loss of generality) that $\mathcal{A}$ is of the form $\mathbb{T}/\mathcal{E}$. Since a simultaneous recursion over $\mathcal{A}$ is a particular set of equations over $\mathcal{A}(\vec{X})$ all results apply immediately to a simultaneous recursion as well. Only Theorem 9.2.7 about type checking is specific for type algebras defined via a simultaneous recursion.

Using the notion of type algebra morphism, see Definition 7.1.9, we can define, in a quite natural way, the notion of a principal type scheme for type assignments with respect to arbitrary type algebras. It is a quite natural generalization of Corollary 2.3.15, showing the existence of a principal type scheme for $\boldsymbol{\lambda}^{\mathcal{A}}_{=}$.

Principal type scheme in $\boldsymbol{\lambda}_{=}^{\mathcal{A}}$

For every $M \in \Lambda$ a *principal triple* will be constructed: a type algebra $\mathcal{A}_M = \mathbb{T}^{\mathbf{c}_M}/\mathcal{E}_M$, where $\mathbf{c}_M$ is a set of type atoms and $\mathcal{E}_M$ is a set of equations over $\mathbb{T}^{\mathbf{c}_M}$, a type $a_M = [\alpha_M]_{\mathcal{E}_M}$, and a basis Γ_M over $\mathcal{A}_M$ such that

(i) $\Gamma_M \vdash_{\mathcal{A}_M} M : a_M$.
(ii) $\Gamma \vdash_{\mathcal{A}} M : a \quad \Leftrightarrow \quad$ there is a morphism $h : \mathcal{A}_M \to \mathcal{A}$ such that $h(\Gamma_M) \subseteq \Gamma$ and $h(a_M) = a$.

Since we are now working with type algebras, we define a type reconstruction algorithm based on the use of type equations, rather than use unification as in Section 2C in Part I. This approach was first introduced, for type reconstruction in $(\boldsymbol{\lambda}_{\to})$, in Curry (1969) and has been followed by many other authors, see e.g. Wand (1987). We start with some simple examples that explain the method. To make the examples more readable we will use numbers to denote atomic types and we will add to each reconstruction rule used in the deduction the type equation which makes it valid.

9.2.1 Example In the following examples we use the notation from 7.1.20, that is elements of $\mathbb{T}^{\mathbb{A}}/\mathcal{E}$ are denoted as elements of $\mathbb{T}^{\mathbb{A}}$, to be considered modulo $\mathcal{E}$.

(i) Let $M \equiv \lambda x.xx$. In order to type M, we build from below the following derivation. (We will write α_n just as n.)

$$\dfrac{\dfrac{x{:}1 \vdash x : 1 \quad x{:}1 \vdash x : 1 \quad 1 = 1 \to 2}{x{:}1 \vdash xx : 2 \qquad 3 = 1 \to 2}}{\vdash \lambda x.xx : 3} \quad .$$

This gives the triple

$$a_M = 3, \quad \Gamma_M = \emptyset, \quad \mathcal{A}_M = \mathbb{T}^{\mathbf{c}_M}/\mathcal{E}_M,$$

where $\mathbf{c}_M = \{1, 2, 3\}$ and $\mathcal{E}_M = \{1 = 1 \to 2, 3 = 1 \to 2\}$. Simplifying this gives isomorphic $a_M = 1$, $\Gamma_M = \emptyset$, $\mathcal{A}_M = \mathbb{T}^{\{1,2\}}/\{1 = 1 \to 2\}$, i.e.

$$\mathcal{A}_M = \mathbb{T}^{\{\alpha_1,\alpha_2\}}/\{\alpha_1 = \alpha_1 \to \alpha_2\}.$$

Indeed

$$\vdash_{\mathcal{A}_M} (\lambda x.xx) : \alpha_1.$$

In order to show this assignment is initial, suppose that $\Gamma \vdash_{\boldsymbol{\lambda}\mathcal{A}} M : a_1$. Then one can reconstruct from below a derivation with the same shape, using Proposition 9.1.3 and Remark 9.1.4:

$$\dfrac{\dfrac{x{:}a_1 \vdash x : a_1 \quad x{:}a_1 \vdash x : a_1 \quad a_1 = a_1 \to a_2}{x{:}a_1 \vdash xx : a_2 \qquad a_3 = a_1 \to a_2}}{\vdash \lambda x.xx : a_3}$$

The required morphism is determined by $h(k) = a_k$ for $k \leq 3$. Indeed one has $h(a_M) = a_3$ and $h(\Gamma_M) \subseteq \Gamma$.

(ii) If we consider the (open) term $M \equiv x(xx)$ we have, using Church notation to represent deduction in a more compact way, the following:

$$\vdash_{\mathcal{A}_M} x^1\,(x^1\,x^1)^2 : 3,$$

where $a_M = 3$, $\mathcal{A}_M = \mathbb{T}^{\{1,2,3\}}/\{1 = 1 \to 2 = 2 \to 3\}$ and $\Gamma_M = \{x^1\}$.

Moreover, if we want to consider type assignment with respect to invertible type algebras, we can convert $\mathcal{A}_M$ to an invertible type algebra $\mathcal{A}_M^{\text{inv}}$. Assuming invertibility we get $1 = 2 = 3$. Then $\mathcal{A}_M^{\text{inv}}$ can be simplified to the *trivial* type algebra $\mathbb{T}^{\{1\}}/\{1 = 1 \to 1\}$.

(iii) If we consider terms having a simple type in $(\lambda \to)$, then the construction sketched above does not involve recursive definitions as in case (i). Moreover, if we assume invertibility, the resulting type algebra is isomorphic to a free one and we get the same principal type as in $(\boldsymbol{\lambda}_{\to})$. Let $M \equiv \mathbf{c}_2 \equiv \lambda fx.f(fx)$. Then

$$\dfrac{\dfrac{\dfrac{\dfrac{f{:}1, x{:}2 \vdash f{:}1 \quad f{:}1, x{:}2 \vdash x : 2 \quad 1 = 2 \to 3}{f{:}1, x{:}2 \vdash fx : 3 \qquad 1 = 3 \to 4}}{f{:}1, x{:}2 \vdash f(fx) : 4 \qquad 5 = 2 \to 4}}{f{:}1 \vdash \lambda x.f(fx) : 5 \qquad 6 = 1 \to 5.}}{\vdash \lambda fx.f(fx) : 6}$$

Simplifying, this gives the triple

$$\vdash_{\mathbb{T}/\mathcal{E}} \mathbf{c}_2 : (2 \to 3) \to (2 \to 4), \tag{9.1}$$

with $\mathcal{E} = \{2 \to 3 = 3 \to 4\}$. We can understand this by looking at

$$\vdash \lambda f^{(2\to3)=(3\to4)}\, x^2.(f^{3\to4}(f^{2\to3}x^2)^3)^4 : (2 \to 3) \to 2 \to 4.$$

Furthermore, in $\boldsymbol{\lambda}_{\to}^{\mathbb{A}}$ this term M can be typed:

$$\vdash_{\boldsymbol{\lambda}_{\to}^{\mathbb{A}}} \mathbf{c}_2 : (\alpha \to \alpha) \to \alpha \to \alpha. \tag{9.2}$$

Note that there is a morphism $h : \mathbb{T}/\mathcal{E} \to \mathbb{T}^{\alpha}$ determined by

$$h(2) = h(3) = h(4) = \alpha.$$

This h respects the equations in $\mathcal{E}$. In this way the type assignment (9.2) is seen to follow from (9.1), applying Lemma 7.1.22.

Again assuming invertibility, we get $2 = 3 = 4$ and so $\mathbb{T}/\approx$ contains only identities. Then $\mathcal{A}^{\text{inv}}_{\mathbf{c}2}$ is isomorphic to $\mathbb{T}^{\{\alpha\}}$, the free type algebra over the atomic type α.

(iv) $M \equiv \mathsf{II}$. Bottom-up we construct the following derivation-tree:

$$\dfrac{\dfrac{x{:}1 \vdash x : 1 \quad 2 = 1 \to 1}{\vdash (\lambda x.x) : 2} \qquad \dfrac{y{:}1' \vdash y : 1' \quad 2' = 1' \to 1'}{\vdash (\lambda y.y) : 2'} \qquad 2 = 2' \to 0.}{\vdash (\lambda x.x)(\lambda y.y) : 0}$$

Hence $\mathcal{E} = \{1 \to 1 = (1' \to 1') \to 0\}$ on $\mathbb{T}^{\{0,1,1'\}}$ and we have

$$\vdash_{\mathcal{E}} \mathsf{II} : 0. \tag{9.3}$$

In $\boldsymbol{\lambda}_{\to}$ this term has as principal type $\alpha \to \alpha$. This can be obtained as image of (9.3) under the morphism $h : \mathbb{T}^{\{0,1,1'\}} \to \mathbb{T}^{\{\alpha\}}$ defined by

$$\begin{aligned} h(0) &= \alpha \to \alpha; \\ h(1) &= \alpha \to \alpha; \\ h(1') &= \alpha. \end{aligned}$$

Note that it was important to keep the names of the bound variables of the two occurrences of I different.

We present now the formal algorithm for building the principal triple of a term M. To simplify its definition we make the assumption that the names of all free and bound variables of M are distinct. This can always be achieved by α-conversion.

9.2.2 Definition Let $M \in \Lambda$. Define a set of type constants $\mathbf{c}_M$, a type α_M, a basis Γ_M, a set of equations $\mathcal{E}_M$, and a type algebra $\mathcal{A}_M$ with element a_M as follows. We do this by first defining for each subterm-occurrence $L \subseteq M$ for L not a variable a distinct type atom α_L. For variables x occuring in M we choose a fixed type α_x (for different occurrences the same α_x). Then we define $\mathcal{E}_L$ for each subterm-occurrence $L \subseteq M$, obtaining this notion also for M as highest subterm of itself.

L	$\mathcal{E}_L$
x	$\emptyset$
PQ	$\mathcal{E}_P \cup \mathcal{E}_Q \cup \{\alpha_P = \alpha_Q \to \alpha_{PQ}\}$
$\lambda x.P$	$\mathcal{E}_P \cup \{\alpha_{\lambda x.P} = \alpha_x \to \alpha_P\}$

Define $\mathbf{c}_M$ as the set of all atomic types α_L and α_x occurring in M. Finally we define

$$\begin{aligned} \Gamma_M &= \{x{:}[\alpha_x] \mid x \in \mathrm{FV}(M)\}; \\ \mathcal{A}_M &= \mathbb{T}^{\mathbf{c}_M}/\mathcal{E}_M; \\ a_M &= [\alpha_M]_{\mathcal{E}_M}. \end{aligned}$$

The type a_M is called the *principal recursive type* of M and $\mathcal{A}_M$ its *principal type algebra*. We say Γ_M is the *principal recursive basis* of M, which is empty if M is closed. The triple $\langle \Gamma_M, \mathcal{A}_M, a_M \rangle$ is called the *principal recursive triple*. Note that $\mathcal{E}_M$ is a simultaneous recursion if we consider the α_L as indeterminates.

Type reconstruction

Of the following theorem, part (iii) solves both versions of the type reconstruction problem.

9.2.3 Theorem *For any $M \in \Lambda$ the principal recursive triple $\langle \Gamma_M, \mathcal{A}_M, a_M \rangle$ satisfies:*

(i) $\Gamma_M \vdash_{\mathcal{A}_M} M : a_M$.

(ii) $\Gamma \vdash_{\mathcal{A}} M : a \Leftrightarrow$ *there is a morphism* $h : \mathcal{A}_M \to \mathcal{A}$ *such that* $h(\Gamma_M) \subseteq \Gamma$ *and* $h(a_M) = a$.

(iii) *For $M \in \Lambda^{\emptyset}$ this simplifies to*

$$\vdash_{\lambda\mathcal{A}} M : a \Leftrightarrow \exists h{:}\mathcal{A}_M \to \mathcal{A}.h(a_M) = a.$$

The triple is 'principal', as it is the initial one giving M a type satisfying (i) *and* (ii).

Proof Take as triple the one defined in the previous definition.

(i) By induction on the structure of $L \subseteq M$ we show that this statement holds for M replaced by L and hence also for M itself.

Case $L \equiv x$. Then clearly $x{:}\alpha_x \vdash x : \alpha_x$.

Case $L \equiv PQ$, *and* $\Gamma_P \vdash_{\mathbb{T}/\mathcal{E}_P} P : \alpha_P$, $\Gamma_Q \vdash_{\mathbb{T}/\mathcal{E}_Q} Q : \alpha_Q$. Then $\Gamma_P \cup \Gamma_Q \vdash_{\mathbb{T}/\mathcal{E}} PQ : \alpha_{PQ}$, as $\alpha_P =_{\mathcal{E}} \alpha_Q \to \alpha_{PQ}$.

Case $L \equiv \lambda x.P$ *and* $\Gamma_P \vdash_{\mathbb{T}/\mathcal{E}_P} P : \alpha_P$. Then $\Gamma_P - \{x{:}\alpha_x\} \vdash_{\mathbb{T}/\mathcal{E}} \lambda x.P : \alpha_x \to \alpha_P = \alpha_{\lambda x.P}$.

(ii) ($\Leftarrow$) By (i), Lemma 7.1.22 and weakening, Proposition 7.1.5.

($\Rightarrow$) By Remark 7.2.6 it is enough to define a morphism $h^{\natural} : \mathbb{T}^{\mathbf{c}_M} \to \mathcal{A}$ such that $h^{\natural}(\alpha_M) = a$ and $B = C \in \mathcal{E}_M \Rightarrow h^{\natural}(B) = h^{\natural}(C)$, for all $B, C \in \mathbb{T}^{\mathbf{c}_M}$.

Take a deduction $\mathcal{D}$ of $\Gamma \vdash_{\mathcal{A}} M : a$. Note that in Definition 9.2.2 for every $L \subseteq M$ a type derivation $\mathcal{D}_L$ is constructed in which to each variable x occurring in L we assign a type α_x and to each subterm occurrence of L that is not a variable, a distinct type variable. Moreover, as in Example 9.2.1 $\mathcal{D}_M$ and $\mathcal{D}$ have the same shape corresponding to the structure of M. Now we define $h^\natural$:

$$h^\natural(\alpha_L) \;=\; b_L, \qquad \text{for all } \alpha_L \text{ assigned to the subterm occurrences } L \text{ of } M, \text{ where } b_L \text{ is assigned in } \mathcal{D} \text{ to the corresponding } L.$$

By induction on the subterm occurrences of M it easily follows, using Proposition 9.1.3 and Remark 9.1.4, that all equations in $\mathcal{E}_M$ are preserved. Moreover, by construction, $h^\natural(\alpha_M) = a$. Obviously we have $h(\Gamma_M) \subseteq \Gamma$, since Γ_M contains only assumptions for variables occurring in M.

(iii) By (ii), knowing that $\Gamma_M = \emptyset$ for $M \in \Lambda^\emptyset$. ∎

9.2.4 Remark Note that $\mathcal{E}_M, \Gamma_M$ and a_M are the generalization of the notion of principal type and basis scheme for simple types, see Hindley (1969), (1997). In typing a term in the simple type assignment system we require that $\mathcal{E}_M$ can be solved in an initial type algebra $\mathbb{T}$. In this case, $\mathcal{E}_M$ simply defines the substitution which, applied to Γ_M and a_M, gives the principal typing of M in the usual sense. Theorem 9.2.3 is then a generalization of the Principal Type Theorem.

If we want to consider only invertible type algebras, then by Proposition 8.3.29 we can take $\mathcal{B}_M = \mathbb{T}^{\mathbf{c}_M}/\mathcal{E}_M^{inv}$ as the initial type algebra for M. Let $\mathcal{R}_M = \mathcal{E}_M^{\text{inv}}$. Note that, by Lemma 8.3.28, $\mathcal{R}_M$ is a proper simultaneous recursion and we have $\mathcal{B}_M = \mathbb{T}^{\mathbf{c}_M}/\mathcal{R}_M$.

9.2.5 Corollary *For every $M \in \Lambda$ there exists a so-called* invertible principal triple *Γ_M, $\mathcal{B}_M$, a_M such that $\mathcal{B}_M$ is invertible and that*

(i) $\Gamma_M \vdash_{\mathcal{B}_M} M : b_M$.

(ii) $\Gamma \vdash_{\mathcal{B}} M : a$, *with $\mathcal{B}$ invertible* $\Leftrightarrow$ $\exists h : \mathcal{B}_M \to \mathcal{B}.[h(\Gamma_M) \subseteq \Gamma \;\&\; h(a_M) = a]$.

Proof Given M, let $\Gamma'_M, \mathcal{A}_M, a_M$ be its principal triple and let $\mathcal{B}_M = \mathcal{A}_M^{\text{inv}}$. Let $k : \mathcal{A}_M \to \mathcal{B}_M$ be the canonical morphism and take $\Gamma_M = k(\Gamma'_M), b_M = k(a_M)$. Then (i) holds by the theorem and Lemma 7.1.22. Property (ii) follows by Part (ii) of the theorem and Proposition 8.3.29. ∎

Let $\mathbf{a}_M$ be the subset of $\mathbf{c}_M$ containing all $\alpha \in \mathbf{c}_M$ such that α does not occur in the right hand of any equation in $\mathcal{R}_M$. Then $\mathcal{B}_M = \mathbb{T}^{\mathbf{a}_M}[\mathcal{R}_M]$.

The typability problems

Theorem 9.2.3 provides the abstract answer to the question of whether or not a term has a type within a given type algebra $\mathcal{A}$. The first typability question

$$\text{“Given an untyped } M \in \Lambda^{\emptyset}\text{, does there exist } \mathcal{A} \text{ and } a \in \mathcal{A} \text{ such that } \vdash_{\lambda\mathcal{A}} M : a?\text{”},$$

is trivially checked: taking take $\mathcal{A} = \mathbb{T}^{\{\alpha\}}/\{\alpha = \alpha \to \alpha\}$ and $a = [\alpha]$, one has $\vdash_{\lambda\mathcal{A}} M : a$: see Exercise 7.12. However, this problem is no longer trivial when term constants are considered, see the discussion in Section 11.3.

The decidability of the question: given $M \in \Lambda^{\emptyset}$ and type algebra $\mathcal{A}$

$$\text{“Does there exist an } a \in \mathcal{A} \text{ such that } \vdash_{\lambda\mathcal{A}} M : a?\text{”},$$

only makes sense if $\mathcal{A}$ is ‘finitely presented’, i.e. of the form $\mathbb{T}^{\mathbb{A}}/\mathcal{E}$ with $\mathbb{A}$ and $\mathcal{E}$ finite. Then typability is decidable.

9.2.6 Theorem *Let $\mathcal{E}$ be a set of equations over $\mathbb{T} = \mathbb{T}^{\mathbb{A}}$, with $\mathbb{A}$ and $\mathcal{E}$ finite. It is decidable whether or not an $M \in \Lambda^{\emptyset}$ has a type in $\mathbb{T}/\mathcal{E}$.*

Proof By Theorem 9.2.3(iii) this boils down to checking whether $\mathcal{E}$ justifies $\mathcal{E}_M$, which is decidable by Theorem 8.3.31. ∎

A related but stronger version of this problem will be discussed in the next subsection in the case that $\mathcal{E}$ is a simultaneous recursion $\mathcal{R}$.

Type checking

We work in the type algebra $\mathbb{T}[\mathcal{R}]$. When we write

$$x_1{:}A_1, \dots, x_n{:}A_n \vdash_{\lambda\mathcal{R}} M : A$$

we mean $x_1{:}[A_1], \dots, x_n{:}[A_n] \vdash_{\lambda\mathcal{R}} M : [A]_{\mathcal{R}}$, according to Notation 7.1.20, identifying $\mathbb{T}(\vec{X})/{=_{\mathcal{R}}}$ with $\mathbb{T}(\vec{X})$.

9.2.7 Theorem *Suppose we are given $M \in \Lambda$, a proper simultaneous recursion $\mathcal{R}$ over $\mathbb{T}^{\mathbb{A}}$, with $\mathbb{A}$ finite, a type basis Γ and a type $A \in \mathbb{T}$. Then it is decidable whether or not $\Gamma \vdash_{\lambda\mathcal{R}} M : A$.*

Proof Without loss of generality, we can assume $\text{dom}(\Gamma) = \text{FV}(M)$. Now it follows from Theorem 9.2.3(ii) that $\Gamma \vdash_{\lambda\mathcal{R}} M : A$ iff there is a morphism $h : \mathcal{T}_M \to \mathbb{T}[\mathcal{R}]$ such that $h(a_M) = A$ and $h(\Gamma_M) \subseteq \Gamma$. This is decidable by Corollary 8.3.30. ∎

Finding inhabitants

We will now discuss the inhabitation problem for $(\boldsymbol{\lambda}\mathcal{R})$. Following a suggestion of R. Statman we will reduce the inhabitation problem for recursive types to one for simple types discussed in Section 2.4. In the following we will consider the syntactic version of the type assignment system for $(\boldsymbol{\lambda}\mathcal{R})$ (see Proposition 7.1.21), where the type equivalence relation is given by $=_{\mathcal{R}}$.

9.2.8 Definition Let $\mathcal{R}(\vec{X}) = \{X_i = A_i(\vec{X}) \mid 1 \leq i \leq n\}$ be a simultaneous recursion over $\mathbb{T}$and $A \in \mathbb{T}[\vec{X}]$. Define the type $((\mathcal{R})) \to A \in \mathbb{T}[\vec{X}]$ and the environment $\Gamma_{\mathcal{R}}$ as:

(i) $((\mathcal{R})) \to A = (X_1 \to A_1) \to (A_1 \to X_1) \to \cdots \to A$.

(ii) $\Gamma_{\mathcal{R}} = \{c_i : X_i \to A_i \mid 1 \leq i \leq n\} \cup \{c_i' : A_i \to X_i \mid 1 \leq i \leq n\}$.

We will show that A is inhabited in $\boldsymbol{\lambda}\mathcal{R}$ if and only if $((\mathcal{R})) \to A$ is inhabited in $\boldsymbol{\lambda}_{\to}$.

9.2.9 Lemma *Let $\mathcal{R}(\vec{X})$ be a simultaneous recursion over $\mathbb{T}$ and let $A, B \in \mathbb{T}[\vec{X}]$. If $\mathcal{R} \vdash A = B$ then there are terms M and M' such that $\Gamma_{\mathcal{R}} \vdash_{\boldsymbol{\lambda}_{\to}} M : A \to B$ and $\Gamma_{\mathcal{R}} \vdash_{\boldsymbol{\lambda}_{\to}} M' : B \to A$.*

Proof By induction on the proof of $\mathcal{R} \vdash A = B$. If $A = B$ has been obtained by rule (axiom) (i.e. $A = X_i$ and $B = A_i$ where $X_i = A_i \in \mathcal{R}$) then take $M = c_i$ and $M' = c_i'$. If $A = B$ has been obtained by rule (refl) the proof is trivial.

As for the induction step the most interesting case is when $A = B$ has been obtained by rule ($\to$-cong). In this case we have $A = A_1 \to A_2$, $B = B_1 \to B_2$, $\mathcal{R} \vdash A_1 = B_1$ and $\mathcal{R} \vdash A_2 = B_2$. By the induction hypothesis there are M_1' and M_2 such that $\Gamma_{\mathcal{R}} \vdash_{\boldsymbol{\lambda}_{\to}} M_1' : B_1 \to A_1$ and $\Gamma_{\mathcal{R}} \vdash_{\boldsymbol{\lambda}_{\to}} M_2 : A_2 \to B_2$. The take $M = \lambda x.\lambda y.M_2\,(x\,(M_1'\,y))$ and note that $\Gamma_{\mathcal{R}} \vdash_{\boldsymbol{\lambda}_{\to}} M : (A_1 \to A_2) \to B_1 \to B_2$ (assume $x : A_1 \to A_2$ and $y : B_1$). The term M' can be defined in a similar way. ∎

9.2.10 Lemma *Let A be a type in $\mathbb{T}[\vec{X}]$ and Γ a type environment over $\mathbb{T}[\vec{X}]$. Then*

$$\exists M.\ \Gamma \vdash_{\boldsymbol{\lambda}\mathcal{R}} M : A \Leftrightarrow \exists M'.\ \Gamma, \Gamma_{\mathcal{R}} \vdash_{\boldsymbol{\lambda}_{\to}} M' : A.$$

Proof ($\Rightarrow$) By induction on deductions in $(\boldsymbol{\lambda}\mathcal{R})$ using Lemma 9.2.9.

($\Leftarrow$) Take a deduction $\mathcal{D}$ of $\Gamma, \Gamma_{\mathcal{R}} \vdash_{\boldsymbol{\lambda}_{\to}} M' : A$. To build a deduction of $\Gamma \vdash_{\boldsymbol{\lambda}\mathcal{R}} M : A$ for some term M we only need to get rid of the constants c_i and c_i'. The proof is by induction on the derivation of $\Gamma, \Gamma_{\mathcal{R}} \vdash_{\boldsymbol{\lambda}^{\mathbb{A}}_{\to}} M' : A$.

If $M' = x \in \mathrm{dom}(\Gamma)$, then we are done trivially. If $M' = c_i$ then, since $X_i =_{\mathcal{R}} A_i$, we have $X_i \to X_i =_{\mathcal{R}} X_i \to A_i$. Then we can take $M = \mathsf{I}$. We have $\Gamma \vdash_{\lambda\mathcal{R}} \mathsf{I} : X_i \to A_i$, using rule (equal). If $M' = c_i'$ the argument is similar.

All induction steps are trivial. ■

Lemma 9.2.10 can be generalized in the following way.

9.2.11 Proposition *Given a simultaneous recursion $\mathcal{R}(\vec{X})$ over $\mathbb{T}$ and a type A in $\mathbb{T}[\vec{X}]$ then*

$$A \text{ is inhabited in } \boldsymbol{\lambda}\mathcal{R} \text{ iff } ((\mathcal{R})) \to A \text{ is inhabited in } (\boldsymbol{\lambda}_{\to}).$$

9.2.12 Remark The reduction of the inhabitation problem for recursive types to that for simple types is immediate if we consider the system $(\boldsymbol{\lambda}\mathcal{R}^{\mathrm{Ch}_0})$ with explicit constants representing type conversion via folding and unfolding. Here the c_i play the roles of $unfold_{X_i}$ and the c_i' of $fold_{X_i}$.

Moreover, using Lemma 9.2.9 it can easily be shown that any typing in $(\boldsymbol{\lambda}\mathcal{R})$ can be represented by a term of $(\boldsymbol{\lambda}\mathcal{R}^{\mathrm{Ch}_0})$.

Finding μ-types and their inhabitants

Concerning the problems of type reconstruction, typability, type checking and inhabitation in $(\lambda\mu)$ and $(\lambda\mu^*)$, two of these are trivial. Write $\lambda\mu^{(*)}$ to mean that both the $\lambda\mu$ and the $\lambda\mu^*$ versions hold. Typability in $\lambda\mu^{(*)}$, i.e.

$$\exists A \in \mathbb{T}_\mu . \vdash_{\lambda\mu^{(*)}} M : A,$$

always holds, taking $A = \mu\alpha.\alpha \to \alpha$, see Exercise 7.12. Inhabitation in $\lambda\mu^{(*)}$,

$$\exists M \in \Lambda . \vdash_{\lambda\mu^{(*)}} M : A,$$

always holds, taking $M \equiv \Omega$, see Example 7.1.4(ii), using $\mu\alpha.\alpha \to A$.

The remaining problems, type reconstruction and type checking are first considered for $\lambda\mu^*$ using μ-terms modulo strong equality (as the situation there is somewhat simpler than the one for $\lambda\mu$). Note that both $\lambda\mu^*$ and $\lambda\mu$ are invertible, so we can consider only invertible type algebras.

The notions of principal type and principal pair, see Definitions 2.3.13 and 9.2.5, can be generalized in a straightforward way to μ-types considered modulo strong equivalence. On the one hand, in $\mathbb{T}_\mu$ all possible recursive types are present and so in considering problems about assigning types to λ-terms there is no parameter $\mathcal{R}$ describing the recursive types available in the system. On the other hand, the solutions are not unique up to weak equivalence $=_\mu$ but only up to $=_{\mu^*}$. Therefore, only for the system $(\lambda\mu^*)$ do we know how to show the Principal Type Theorem with a straightforward formulation.

9.2.13 Definition Let $\vec{X}$ denote the indeterminates of the simultaneous recursions $\mathcal{E}_M$ and $\vec{S} = \langle S_1, \ldots, S_n \rangle$ be a solution of $\mathcal{E}_M$ in $\mathbb{T}_\mu$ (typically that given by Theorem 8.1.1). We define

(i) $a^\mu_M = a_M[\vec{X} := \vec{S}]$;
(ii) $\Gamma^\mu_M = \Gamma_M[\vec{X} := \vec{S}]$.

9.2.14 Theorem

(i) $\Gamma^\mu_M \vdash_{\lambda\mu} M : a^\mu_M$.
(ii) *Suppose* $\Gamma \vdash_{\lambda\mu^*} M : A$. *Then there is a substitution* $s : \mathbf{a}_M \to \mathbb{T}_\mu$ *such that* $A =^*_\mu s(a^\mu_M)$ *and* $\Gamma \supseteq s(\Gamma^\mu_M)$ *(modulo* $=^*_\mu$*).*

Proof (i) By Theorems 9.2.3 and 8.1.1.
(ii) By Remark 7.5.36, since μ-types modulo $=^*_\mu$ are just notation for regular trees. ∎

9.2.15 Corollary $\Gamma^\mu_M \vdash_{\lambda\mu^*} M : a^\mu_M$.

Proof By Theorem 9.2.14(i). See also the remark following Lemma 7.5.29. ∎

Using a similar argument one can show the following result.

9.2.16 Theorem *For* $M \in \Lambda$, *basis* Γ *and* $A \in \mathbb{T}_\mu$ *it is decidable whether or not* $\Gamma \vdash_{\lambda\mu^*} M : A$.

Proof See Exercise 9.6. ∎

As for weak equivalence, Theorem 9.2.3 implies that the set of possible $\boldsymbol{\lambda}\mu$ types for a term M, i.e. the set of all Γ, A such that $\Gamma' \vdash_{\lambda\mu} M : A$ for all $\Gamma' \supseteq \Gamma$, is, after coding, recursively enumerable. Moreover, by Lemma 7.1.22 and Theorem 8.1.1, given a typing deduction in $\boldsymbol{\lambda}\mathcal{R}$ (for any simultaneous recursion $\mathcal{R}$) we can easily build one in $\lambda\mu$ (via the construction in Theorem 8.1.1). It is not clear, however, how a principal type can be constructed in $\boldsymbol{\lambda}\mu$, owing to the many incomparable solutions that the same simultaneous recursion can have in $\mathbb{T}_\mu$.

9.3 Strong normalization

As shown by the examples in Section 7.1, type systems with recursive types do not have, in general, the Strong Normalization property. This is easy: in the presence of a type $A = A \to A$ all untyped lambda terms can be typed. By restricting the use of the μ operator in type formation, however, it is possible to define systems in which Strong Normalization holds. The Strong Normalization theorem in this section comes from Mendler (1987),

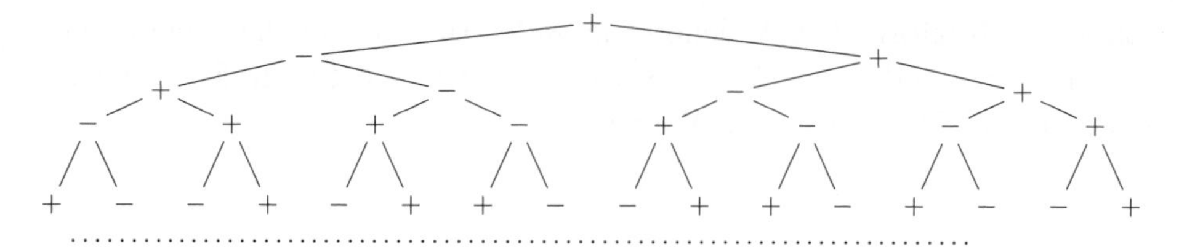

Figure 9.1 The universal binary tree and the ± nodes

(1991). We will prove it in detail for an assignment system with μ-types. Exploiting the fact, shown in Section 8.1, that any simultaneous recursion can be solved in $\mathbb{T}_\mu$, this result can be extended to assignment systems defined via a simultaneous recursion.

Strong normalization for μ-types

Using recursive types, in general, we can assign types to non-normalizing terms, such as the fixed point operator $\mathsf{Y} \equiv \lambda f.(\lambda x.f(xx))(\lambda x.f(xx))$, or even to unsolvable ones, such as $\mathbf{\Omega} \equiv (\lambda x.xx)(\lambda x.xx)$. However, there is a subset $\mathbb{T}_\mu^+ \subseteq \mathbb{T}_\mu$ such that terms that can be typed using only types in $\mathbb{T}_\mu^+$ are strongly normalizable. The set $\mathbb{T}_\mu^+$ completely characterizes the types assuring normalization in the following sense. For any type $T \in \mathbb{T}_\mu$, such that $T \notin \mathbb{T}_\mu^+$, it is possible to define a non-normalizing term which can be typed using only the type T and its subtypes.

9.3.1 Definition

(i) The notion of *positive* and *negative* occurrence of a subtype in a type of $\mathbb{T}_\mu$ (not modulo $=_\mu$) is defined (recursively) by:

(1) A is *positive* in A;

(2) if A is *positive* (*negative*) in B then A is *positive* (*negative*) in $C \to B$ and *negative* (*positive*) in $B \to C$;

(3) if A is *positive* (*negative*) in B then A is *positive* (*negative*) in $\mu t.B$.

(ii) We say that a type variable α is *positive* (*negative*) in A if all free occurrences of α in A are *positive* (*negative*). It is not required that $\alpha \in \mathrm{FV}(A)$, but we do not speak about occurrences of bound variables being positive or negative.

Note that A is positive (negative) in B if A occurs in B only on the left hand side of an even (odd) number of '$\to$'. Let this number be the *level* of the occurrence of A in B. For instance δ is positive in $((\mu\alpha.\alpha \to \delta) \to \beta) \to \gamma$, since the occurrence of δ is at level 2, but not in $((\mu\alpha.\delta \to \alpha) \to \beta) \to \gamma$, since now it is at level 3.

Finite or infinite trees can be considered as subtrees of the universal binary tree. The positive and negative subtypes in a type considered as a tree are displayed in Fig. 9.1. It is interesting to note that the sequences of the signs $+, -$ on one horizontal level, starting all the way from the left with a $+$, form longer and longer initial segments of the Morse–Thue sequence, see Allouche and Shallit (2003).

9.3.2 Definition

(i) A type $A \in \mathbb{T}_\mu$ is called *positive* if for every $\mu\alpha.B \subseteq A$ all occurrences of α in B are positive.
(ii) $\mathbb{T}_\mu^+ = \{A \in \mathbb{T}_\mu \mid A \text{ is positive}\}$.
(iii) Denote by $\lambda\mu^+$ the type assignment system defined as $\lambda\mu$, but with the the set of types restricted to $\mathbb{T}_\mu^+$.

9.3.3 Remark Note that $A, B \in \mathbb{T}_\mu^+ \Leftrightarrow (A \to B) \in \mathbb{T}_\mu^+$.

9.3.4 Examples

(i) $\mu\beta.\alpha{\to}\beta \in \mathbb{T}_\mu^+$.
(ii) $\mu\beta.\beta{\to}\alpha \notin \mathbb{T}_\mu^+$.
(iii) $\mu\beta.\beta{\to}\beta \notin \mathbb{T}_\mu^+$.
(iv) $((\mu\beta.\alpha{\to}\beta){\to}\gamma) \in \mathbb{T}_\mu^+$.
(v) $(\mu\beta.(\alpha{\to}\beta){\to}\gamma) \notin \mathbb{T}_\mu^+$.

Obviously any deduction in $\lambda\mu^+$ is also a deduction in $\lambda\mu$. So any type assignment provable in $\lambda\mu^+$ is also provable in $\lambda\mu$.

Positive recursive types are often referred to as *inductive types* in the literature, since it is possible to define for them an induction principle, see Mendler (1991). In that paper the result is proved in fact for a stronger system, defined as an extension of second-order typed lambda calculus with provable positive recursive types and induction and coinduction operators for each type.

The proof of strong normalization for terms typable in $\lambda\mu^+$ is based on the notion of saturated set introduced in Girard (1971), which is in turn based on Tait's method presented in Section 2.2 provided with an impredicative twist. We will prove it for the Curry version of $\lambda\mu$, but the proof extends to the systems with explicit typing, see Definition 7.1.23. For an arithmetical proof of the aforementioned strong normalization result, see David and Nour (2007).

Let SN be the set of strongly normalizing type-free λ-terms.

9.3.5 Definition

(i) A subset $\mathcal{X} \subseteq \mathsf{SN}$ is *saturated* if the following conditions hold.

(1) The subset $\mathcal{X}$ is *closed under reduction*, i.e. for all $M, N \in \Lambda$

$$M \in \mathcal{X} \;\&\; M \twoheadrightarrow_\beta N \;\Rightarrow\; N \in \mathcal{X}.$$

(2) For all variables x one has

$$\forall n \geq 0 \forall R_1, \ldots, R_n \in \mathsf{SN}. x R_1 \cdots R_n \in \mathcal{X}.$$

(3) For all $Q \in \mathsf{SN}$ one has

$$P[x := Q] R_1 \cdots R_n \in \mathcal{X} \implies (\lambda x.P) Q R_1 \cdots R_n \in \mathcal{X}.$$

(ii) Denote by $\mathtt{SAT}$ the set of all saturated subsets of SN.

9.3.6 Proposition $\mathtt{SAT}$ *is a complete lattice, see Definition* 10.1.4(vi), *under the operations of set inclusion, with* $\sup\{\mathcal{X}, \mathcal{Y}\} = \mathcal{X} \cup \mathcal{Y}$ *and* $\inf\{\mathcal{X}, \mathcal{Y}\} = \mathcal{X} \cap \mathcal{Y}$.

The top element of $\mathtt{SAT}$ is SN while its bottom element $\bot_{\mathtt{SAT}} = \bigcap_{\mathcal{X} \in \mathtt{SAT}} \mathcal{X}$ is the smallest set containing all terms of the shape $x R_1 \ldots R_n$ ($R_i \in \mathsf{SN}$) and that satisfies Definition 9.3.5(i)(3). We recall that a function $f : \mathtt{SAT} \to \mathtt{SAT}$ is *monotonic* if $X \subseteq Y$ (where $X, Y \in \mathtt{SAT}$) implies $f(X) \subseteq f(Y)$ and *anti-monotonic* if $X \subseteq Y$ implies $f(Y) \subseteq f(X)$. If $f : \mathtt{SAT} \to \mathtt{SAT}$ is a monotonic function, then

$$fix(f) = \bigcap \{x \mid f(x) \subseteq x\}$$

is the smallest fixed point of f.

9.3.7 Definition The operation $\Rightarrow: \mathtt{SAT} \times \mathtt{SAT} \to \mathtt{SAT}$ is defined by

$$(\mathcal{X} \Rightarrow \mathcal{Y}) = \{M \mid \forall N \in \mathcal{X}. (MN) \in \mathcal{Y}\}.$$

The proof of the following proposition is standard.

9.3.8 Proposition *The operation* $\Rightarrow$ *is well defined, i.e. for all* $\mathcal{X}, \mathcal{Y} \in \mathtt{SAT}$ $\mathcal{X} \Rightarrow \mathcal{Y} \in \mathtt{SAT}$. *Moreover* $\Rightarrow$ *is anti-monotonic in its first argument and monotonic in its second argument; that is, if* $\mathcal{X}' \subseteq \mathcal{X}$ *and* $\mathcal{Y} \subseteq \mathcal{Y}'$, *then* $(\mathcal{X} \Rightarrow \mathcal{Y}) \subseteq (\mathcal{X}' \Rightarrow \mathcal{Y}')$.

We now define an interpretation of types in $\mathbb{T}_\mu^+$ as elements of $\mathtt{SAT}$ and of terms as elements of Λ. This is inspired by standard semantical notions, see Chapter 10. Let a *type environment* be a function $\tau : \mathbb{A} \to \mathtt{SAT}$.

9.3.9 Definition The *interpretation of a type* $A \in \mathbb{T}^+_\mu$ under a type environment τ, written $[\![A]\!]_\tau$, is defined by

$$\begin{aligned} [\![\alpha]\!]_\tau &= \tau(\alpha); \\ [\![A \to B]\!]_\tau &= [\![A]\!]_\tau \Rightarrow [\![B]\!]_\tau; \\ [\![\mu\alpha.A]\!]_\tau &= fix(\lambda\!\!\lambda \mathcal{X}{:}\mathtt{SAT}.[\![A]\!]_{\tau[\alpha:=\mathcal{X}]}). \end{aligned}$$

9.3.10 Lemma

(i) *If* α *is positive (negative) in* $A \in \mathbb{T}^+_\mu$ *then*

$$\lambda\!\!\lambda \mathcal{X}{:}\mathtt{SAT}.[\![A]\!]_{\tau[\alpha:=\mathcal{X}]}$$

is a monotonic (anti-monotonic) function over the complete lattice SAT.

(ii) $[\![-]\!]_\tau$ *is well defined on* $\mathbb{T}^+_\mu$, *i.e.* $[\![A]\!]_\tau \in \mathtt{SAT}$ *for all types* $A \in \mathbb{T}^+_\mu$ *and all type environments* τ.

(iii) *For circular types* A *one has* $[\![A]\!]^\tau = \bot_{\mathtt{SAT}}$.

Proof (i), (ii) by simultaneous induction on A.

(iii) Notice that $[\![\mu\alpha.\alpha]\!]^\tau = \bot_{\mathtt{SAT}}$ and $[\![\mu\alpha.B]\!]^\tau = [\![B]\!]^\tau$, if $\alpha \notin B$. Then for a circular $A \equiv \mu\alpha_1 \cdots \alpha_n.\alpha_i$ one has $[\![A]\!]^\tau = [\![\mu\alpha_i.\alpha_i]\!]^\tau = \bot_{\mathtt{SAT}}$. ∎

9.3.11 Lemma

(i) $[\![A[\alpha := B]]\!]_\tau = [\![A]\!]_{\tau[\alpha:=[\![B]\!]_\tau]}$.

(ii) $[\![\mu\alpha.A]\!]_\tau = [\![A[\alpha := \mu\alpha.A]]\!]_\tau$.

Proof (i) By structural induction on A.

(ii) By definition, $[\![\mu\alpha.A]\!]_\tau$ is a fixed point of $\lambda\!\!\lambda \mathcal{X}{:}\mathtt{SAT}.[\![A]\!]_{\tau[t:=\mathcal{X}]}$. Therefore we have $[\![\mu\alpha.A]\!]_\tau = [\![A]\!]_{\tau[t:=[\![\mu\alpha.A]\!]_\tau]}$, which is $[\![A[\alpha := \mu\alpha.A]]\!]_\tau$, by (i). ∎

By Lemma 9.3.11, type interpretation is preserved under weak equivalence:

9.3.12 Lemma *Let* $A, B \in \mathbb{T}^+_\mu$. *If* $A =_\mu B$, *then* $[\![A]\!]_\tau = [\![B]\!]_\tau$.

We define now an interpretation of terms in Λ. This is in fact a term model interpretation. Call a *term environment* a function from the set of term variables V to elements of Λ.

9.3.13 Definition Let $\rho : \mathsf{V} \to \Lambda$ be a term environment. The *evaluation* of a term M under ρ is defined by

$$[\![M]\!]_\rho = M[x_1 := \rho(x_1), \ldots, x_n := \rho(x_n)],$$

where $x_1, \ldots, x_n$ $(n \geq 0)$ are the variables free in M.

We define now the notion of satisfiability of a statement $\Gamma \vdash_{\lambda\mu+} M : A$ with respect to the interpretations of types and terms just given.

9.3.14 Definition Let τ, ρ be a type and a term environment, respectively. Write

(i) $\tau, \rho \models M : A \quad \Leftrightarrow \quad [\![M]\!]_\rho \in [\![A]\!]_\tau$.

(ii) $\tau, \rho \models \Gamma \quad \Leftrightarrow \quad \tau, \rho \models x : A$, for all $x{:}A \in \Gamma$.

(iii) $\Gamma \models M : A \quad \Leftrightarrow \quad \forall\tau\forall\rho.[\tau, \rho \models \Gamma \Rightarrow \tau, \rho \models M : A]$.

9.3.15 Lemma $\Gamma \vdash_{\lambda\mu+} M : A \;\Rightarrow\; \Gamma \models M : A$.

Proof By induction on the derivation of $M : A$ from Γ, using Lemma 9.3.12 for rule (equal). ∎

9.3.16 Theorem (Mendler) $\Gamma \vdash_{\lambda\mu+} M : A \Rightarrow$ *M is strongly normalizing.*

Proof By Lemma 9.3.15 one has $\Gamma \models M : A$. Define ρ_0 by $\rho_0(x) = x$ for all $x \in \mathsf{V}$. Trivially, for any type environment τ, one has $\tau, \rho_0 \models \Gamma$. Hence $\tau, \rho_0 \models M : A$. Therefore $[\![M]\!]_{\rho_0} = M \in [\![A]\!]_\tau \subseteq \mathsf{SN}$. ∎

9.3.17 Remark The converse of this result does not hold. For example $\omega \equiv \lambda x.xx$ is strongly normalizing, but not typable in $\lambda\mu^+$: see Exercise 9.9.

We can reformulate Theorem 9.3.16 in terms of typability. Define the set of terms typable from a set of types as follows.

9.3.18 Definition Let $\mathcal{X} \subseteq \mathbb{T}_\mu$. Then

$$\text{Typable}(\mathcal{X}) = \{M \in \Lambda \mid M \text{ can be typed using only types in } \mathcal{X}\}.$$

By 'using only types in $\mathcal{X}$' we mean that all types (including those in the basis) in some derivation of $\Gamma \vdash M : A$ are elements of $\mathcal{X}$.

9.3.19 Definition Denote by $\mathcal{SC}^+(A)$ be the smallest set of types containing $\mathcal{SC}^s(A)$, see Definition 8.2.33, that is closed under $\to$.

9.3.20 Lemma

(i) *If $A \in \mathbb{T}_\mu^+$ and $B \in \mathcal{SC}^s(A)$, then $B \in \mathbb{T}_\mu^+$.*

(ii) *If $A \in \mathbb{T}_\mu^+$ and $B \in \mathcal{SC}^+(A)$, then $B \in \mathbb{T}_\mu^+$.*

Proof (i) By induction on the definition of $\mathcal{SC}^s(A)$.

(ii) By (i) and Remark 9.3.3. ∎

The following is a consequence of Theorem 9.3.16.

9.3.21 Corollary *If $B \in \mathbb{T}_\mu^+$, then* Typable($\mathcal{SC}^+(B)$) $\subseteq \mathsf{SN}$.

Proof Immediate from Theorem 9.3.16 and Lemma 9.3.20. ∎

Note that this corollary also implies Theorem 9.3.16. Indeed, assume $G \vdash_{\lambda\mu+} M : A$. Let $\mathcal{X} = \{A_1, \dots, A_n\} \subseteq \mathbb{T}^+_\mu$ be the set of types used in this deduction. Then we can apply the corollary to

$$B = A_1 \to A_2 \to \cdots \to A_n \in \mathbb{T}^+_\mu.$$

It is easy to extend this result and its proof to systems with other type constructors like Cartesian product and disjoint union since all these operators are monotonic in both arguments.

Conversely, all μ-types B which do not belong to $\mathbb{T}^+_\mu$ make it possible to type a non-normalizable term using only types of $\mathcal{SC}^+(B)$. We see this in the next theorem. The construction is due to Mendler (1991). As a warm-up, do Exercise 9.11.

Let $\vec{A}$ denote a sequence of types $A_1, \dots, A_n$ $(n \geq 0)$ and $\vec{A} \to B$ denote the type $A_1 \to \cdots \to A_n \to B$. Moreover if $\vec{M}$ is a sequence of terms $M_1, \dots, M_n$ then $\vec{M} : \vec{A}$ denotes the set of statements $M_1 : A_1, \dots, M_n : A_n$.

9.3.22 Theorem (Mendler) *Let $B \in \mathbb{T}_\mu \setminus \mathbb{T}^+_\mu$. Then there is a term N without normal form, typable in $\mathcal{SC}^+(B)$; or, expressed as a formula (where* N *denotes the set of normalizable terms):*

$$\forall B \in \mathbb{T}_\mu . B \notin \mathbb{T}^+_\mu \Rightarrow \text{Typable}(\mathcal{SC}^+(B)) \not\subseteq \mathsf{N}.$$

Proof By assumption there is a subtype $\mu\alpha.T$ of B such that α has a negative occurrence in T. Now there is an integer $n \geq 0$ and types Q_i, $\vec{A}_i$, B_j, with $0 \leq i \leq 2n+1$, $1 \leq j \leq 2n+1$ such that

$$\begin{aligned} \mu\alpha.T &=_\mu Q_{2n+1}; \\ Q_i &\equiv \vec{A}_i \to Q_{i-1} \to B_i, \quad \text{with } 1 \leq i \leq 2n+1; \\ Q_0 &\equiv \vec{A}_0 \to \mu\alpha.T, \end{aligned}$$

where Q_{2n+1} is the reduced form of $\mu\alpha.T$ (see Definition 7.4.38). Clearly all these types are elements of $\mathcal{SC}^+(B)$; see Example 9.3.23.

Case $n = 0$ is an exercise.

Case $n > 0$. Define terms $N_i : Q_i$ with $0 \leq i \leq 2n+1$ in the following order $N_1, N_3, \dots, N_{2n+1}, N_0, N_2, \dots, N_{2n}$:

$$\begin{aligned} N_1 &\triangleq \lambda\vec{x}_1.\lambda y_0.f^{2n+1,1}(y_0\vec{z}_0\vec{z}_{2n+1}y_{2n}); \\ N_{2i+1} &\triangleq \lambda\vec{x}_{2i+1}\lambda y_{2i}.f^{2i,2i+1}(y_{2i}\vec{z}_{2i}N_{2i-1}), && \text{for } 0 \leq i \leq n, \\ N_0 &\triangleq \lambda\vec{x}_0.N_{2n+1}; \\ N_{2i} &\triangleq \lambda\vec{x}_{2i}\lambda y_{2i-1}.f^{2i-1,2i}(y_{2i-1}\vec{z}_{2i-1}N_{2i-2}), && \text{for } 1 \leq i \leq n. \end{aligned}$$

Note that the y_i for $0 \leq i < 2n$ do not occur free in any N_j and that y_{2n} occurs free in $N_1, N_3, \dots, N_{2n-1}$, but not in $N_{2n+1}, N_0, N_2, \dots, N_{2n}$. We define these terms N_i relative to the basis $\{f^{k,h} : B_k \to B_h\}$, with $1 \leq k, h \leq 2n+1$. Moreover in these terms we have the intended types $\vec{x}_i{:}\vec{A}_i$, $\vec{z}_i{:}\vec{A}_i$, and $y_i{:}Q_i$.

Now consider the term $N = (N_{2n+1}\vec{z}_{2n+1}N_{2n})$. It is straightforward to verify that N can be typed in the following basis:

$$\Gamma_0 = \{f^{k,h} : B_k \to B_h \mid 1 \leq k, h \leq 2n+1\} \cup \{\vec{z}_i : \vec{A}_i \mid 0 \leq i \leq 2n+1\},$$

using only types of $\mathcal{SC}^+(B)$. The term N has the following (unique) reduction which is infinite:

$$\begin{array}{rl} N & = N_{2n+1}\vec{z}_{2n+1}N_{2n} \\ & \twoheadrightarrow_\beta f^{2n,2n+1}(N_{2n}\vec{z}_{2n}N_{2n-1}[y_{2n} := N_{2n}]) \\ & \twoheadrightarrow_\beta f^{2n,2n+1}(f^{2n-1,2n}(N_{2n-1}[y_{2n} := N_{2n}]\vec{z}_{2n-1}N_{2n-2})) \\ & \twoheadrightarrow_\beta f^{2n,2n+1}(\cdots(f^{1,2}(N_1[y_{2n} := N_{2n}]\vec{z}_1 N_0))\cdots) \\ & \twoheadrightarrow_\beta f^{2n,2n+1}(\cdots(f^{1,2}(f^{2n+1,1}(N_0\vec{z}_0\vec{z}_{2n+1}N_{2n})))\cdots) \\ & \to_\beta f^{2n,2n+1}(\cdots(f^{1,2}(f^{2n+1,1}(N_{2n+1}\vec{z}_{2n+1}N_{2n})))\cdots) \\ & = f^{2n,2n+1}(\cdots(f^{1,2}(f^{2n+1,1}N))\cdots). \blacksquare \end{array}$$

Theorem 9.3.22 can be immediately extended to λ_μ^{Ch} and $\lambda_\mu^{\text{Ch}_0}$.

9.3.23 Example Let $B \triangleq \mu\alpha.T$, with $T = ((\alpha \to \beta) \to \alpha) \to \beta$. We consider the left-most negative occurrence of α. [Considering the right-most occurrence of α will result in a simpler example.] Now

$$B =_\mu ((B \to \beta) \to B) \to \beta.$$

Then we have $n = 1$, $Q_3 = ((B \to \beta) \to B) \to \beta$, $Q_2 = (B \to \beta) \to B$, $Q_1 = B \to \beta$, and $Q_0 = B$. There are no $\vec{A}_i$, so one has $B_1 \equiv B_3 \equiv \beta$, and $B_2 \equiv B$. Moreover,

$$\begin{array}{rcll} \Gamma_0 & \triangleq & \{f^{12}{:}\beta \to B, f^{2,3}{:}B \to \beta, f^{3,1}{:}\beta \to \beta\}; & \\ N_1 & \triangleq & \lambda y_0.f^{3,1}(y_0\, y_2); & \\ N_0 = N_3 & \triangleq & \lambda y_2.f^{2,3}(y_2\, N_1) & = \lambda y_2.f^{2,3}(y_2\, \lambda y_0.f^{3,1}(y_0\, y_2)); \\ N_2 & \triangleq & \lambda y_1.f^{1,2}(y_1\, N_0). & \end{array}$$

Note that $\Gamma_0, y_2{:}Q_2 \vdash N_1 : Q_1$, $\Gamma_0 \vdash N_i : Q_i$, for $i \in \{0, 2, 3\}$. Then we have $\Gamma_0 \vdash_{\lambda\mu} (N_0\, N_2) : \alpha$. We obtain the infinite reduction

$$(N_0\, N_2) \to_\beta f^{2,3}(N_2\, N_1\, [y_2 := N_2]) \twoheadrightarrow_\beta f^{2,3}(f^{1,2}(f^{3,1}(N_0\, N_2))) \to \cdots .$$

Theorems 9.3.16 and 9.3.22 show that $\mathbb{T}_\mu^+$ is the largest subset $\mathcal{X} \subseteq \mathbb{T}_\mu$ such that if a term can be typed using only the types in $\mathcal{X}$, then M is strongly normalizing.

9.3.24 Corollary *Let $B \in \mathbb{T}_\mu$. Then*

(i) $\mathrm{Typable}(\mathcal{SC}^+(B)) \subseteq \mathsf{N} \Rightarrow B \in \mathbb{T}^+_\mu$.

(ii) $B \in \mathbb{T}^+_\mu \Leftrightarrow \mathrm{Typable}(\mathcal{SC}^+(B)) \subseteq \mathsf{SN}$.
$\Leftrightarrow \mathrm{Typable}(\mathcal{SC}^+(B)) \subseteq \mathsf{N}$.

Proof (i) By Theorem 9.3.22.

(ii) $B \in \mathbb{T}^+_\mu \Rightarrow \mathrm{Typable}(\mathcal{SC}^+(B)) \subseteq \mathsf{SN}$, by Corollary 9.3.21,
$\Rightarrow \mathrm{Typable}(\mathcal{SC}^+(B)) \subseteq \mathsf{N}$
$\Rightarrow B \in \mathbb{T}^+_\mu$, by (i). ∎

9.3.25 Corollary *Let $B \in \mathbb{T}_\mu$. Then*

$$\mathrm{Typable}(\mathcal{SC}^+(B)) \subseteq \mathsf{N} \Leftrightarrow \mathrm{Typable}(\mathcal{SC}^+(B)) \subseteq \mathsf{SN}.$$

This reminds us of the old question whether or not $M \in \mathsf{N} \Rightarrow M \in \mathsf{SN}$. This is not the case. $\mathsf{KI}\boldsymbol{\Omega}$ has a normal form, but is not strongly normalizing. The reason is that there is the subterm $\boldsymbol{\Omega}$ that has no normal form. There is even a term $M_0 = (\lambda z.(\lambda xy.y)(zz))(\lambda z.(\lambda xy.y)(zz))$, such that every subterm has a normal form, but the term itself is not strongly normalizing. Note that $M_0 =_\beta \mathsf{Y}(\lambda xy.y) =_\beta \mathsf{Y}(\mathsf{KI})$.

The Strong Normalization theorem holds also for the typed systems $\lambda^{\mathrm{Ch}}_\mu$ and $\lambda^{\mathrm{Ch}_0}_\mu$ (which is indeed weaker than $\lambda^{\mathrm{Ch}}_\mu$). In the case of $\lambda^{\mathrm{Ch}_0}_\mu$ it is easy to prove that the reduction rules $(R^{\mathrm{uf}}_\mathcal{E})$ and $(R^{\mathrm{fu}}_\mathcal{E})$ of Definition 7.1.24 cannot cause infinite reduction sequences. On the other hand there does not seem to be a natural definition of positive types for $\vdash_{\lambda\mu^*}$.

Strong normalization for simultaneous recursion

9.3.26 Definition We say that a simultaneous recursion $\mathcal{R}$ is *inductive* if the following holds: if $X =_\mathcal{R} C$ for an indeterminate X, then X has only positive occurrences in C.

9.3.27 Example

(i) Let $\mathcal{R}_0 = \{X_0 = X_1 \to X_0, X_1 = X_0 \to X_1\}$. Then $\mathcal{R}_0$ is inductive. Note that by unfolding we can get $X_0 =_{\mathcal{R}_0} (X_0 \to X_1) \to X_0)$ (and so on) but both occurrences of X_0 are positive.

(ii) Let $\mathcal{R}_1 = \{X_0 = X_1 \to X_1, X_1 = X_0 \to X_1\}$. Then $\mathcal{R}_1$ is not inductive. In fact $X_1 =_{\mathcal{R}_1} (X_1 \to X_1) \to X_1$.

The following is a useful property.

9.3.28 Proposition *A simultaneous recursion $\mathcal{R}$ is inductive iff the solution $(S_1, \ldots, S_n)$ of $\mathcal{R}$ in $\mathbb{T}_\mu$ found in the proof of Theorem 8.1.1 is such that $S_i \in \mathbb{T}^+_\mu$, $1 \le i \le n$.*

Proof A routine check. ■

By this proposition it is easily decidable if a given simultaneous recursion is inductive. As a consequence of Theorems 9.3.16 and 9.3.22 we can characterize those simultaneous recursions which can type only terms that are strongly normalizing.

9.3.29 Theorem *Let $\mathcal{R}$ be an inductive simultaneous recursion. Then*

$$\Gamma \vdash_{\lambda\mathcal{R}} M : A \;\Rightarrow\; M \in \mathsf{SN}.$$

Proof Let $\mathcal{R} = \mathcal{R}(X_1, \dots, X_n)$ and $h : \mathbb{T}[\mathcal{R}] \to \mathbb{T}_\mu^+$ be the type algebra morphism defined by $h(X_i) = S_i$, with $1 \leq i \leq n$, where $(S_1, \dots, S_n)$ is the solution of $\mathcal{R}$ found in Theorem 8.1.1(i), and $h(\alpha) = \alpha$, for all other atomic types α. By Proposition 9.3.28 we have $S_i \in \mathbb{T}_\mu^+$, for $1 \leq i \leq n$. Then for all $B \in \mathbb{T}[\vec{X}]$ we have $h(B) \in \mathbb{T}_\mu^+$ and by Lemma 7.1.22 we get $h(\Gamma) \vdash_{\lambda\mu+} M : h(A)$. Now Theorem 9.3.16 applies. ■

9.3.30 Theorem *Let $\mathcal{R} = \{X_i = A_i \mid 1 \leq i \leq n\}$ be a non-inductive simultaneous recursion. Then there is a term N without normal form such that for some basis Γ and some i we have $\Gamma \vdash_{\lambda\mathcal{R}} N : X_i$.*

Proof We have $X_i =_{\mathcal{R}} A_i$ for some i with X_i occurring negative in A_i. The proof is the same as for Theorem 9.3.22 where X_i replaces $\mu\alpha.T$ and $=_{\mathcal{R}}$ replaces $=_\mu$. ■

Type Inference

The typability of a term M is decidable also with respect to inductive simultaneous recursions (and with respect to $\lambda\mu^+$). This property follows easily from the following lemma.

9.3.31 Lemma

(i) *Let $\mathcal{T} = \mathbb{T}/{\approx}$ and $\mathcal{T}' = \mathbb{T}'/{\approx'}$ be syntactic type algebras and $h : \mathcal{T} \to \mathcal{T}'$ a morphism. If $A \in \mathbb{T}$ has a positive (negative) occurrence in $B \in \mathbb{T}$ then there exist $A', B' \in \mathbb{T}$ such that $h([A]_\approx) = [A']_{\approx'}$, $h([B]_\approx) = [B']_{\approx'}$, and A' has a positive (negative) occurrence in B'.*

(ii) *Let $\mathcal{R}(\vec{X})$ and $\mathcal{R}'(\vec{X}')$ be proper simultaneous recursions over $\mathbb{T}$, $\mathbb{T}'$ respectively. Assume that $\mathcal{R}$ is non-inductive and that there is a morphism $h : \mathbb{T}[\mathcal{R}] \to \mathbb{T}[\mathcal{R}']$. Then $\mathcal{R}'$ is non-inductive.*

Proof (i) By induction on B.

(ii) Take $X \in \vec{X}$ such that $X =_{\mathcal{R}} C$ for some C which contains a negative occurrence of X. Then by (i) there exist A', C' such that

$$h([X]_{\mathcal{R}}) = [A']_{\mathcal{R}'}, \quad h([C]_{\mathcal{R}}) = [C']_{\mathcal{R}'}, \text{ and}$$
$$A' \text{ has a negative occurrence in } C'.$$

As $X =_{\mathcal{R}} C$, we have $[h([X]_{\mathcal{R}}) = h([C]_{\mathcal{R}})$, hence $A' =_{\mathcal{R}'} C'$. It suffices to prove the claim: if $B' =_{\mathcal{R}'} D', B'$ has a negative occurrence in D', then there exist X', D'' such that X' has a negative occurrence in D''. The claim follows by an easy induction on B'. Note that $B' =_{\mathcal{R}'} D'$ where D' is an arrow type. Hence $B' \notin \mathbb{T}'$ by Lemma 7.3.27. So there are only the two cases $B' = X'$ or $B' = B'_1 \to B'_2$. For the second case use that $=_{\mathcal{R}'}$ is invertible. ∎

9.3.32 Definition Let $M \in \Lambda$. We say that M can be *inductively* typed if it has a type in $\lambda\mathcal{R}$, for some inductive simultaneous recursion $\mathcal{R}$.

9.3.33 Theorem *Let $M \in \Lambda$. Then*

$$M \textit{ can be inductively typed} \Leftrightarrow \mathcal{E}_M \textit{ is inductive.}$$

Proof We show

$$M \text{ cannot be inductively typed} \Leftrightarrow \mathcal{E}_M \text{ is not inductive.}$$

($\Rightarrow$) Immediate by Lemma 9.2.3.

($\Leftarrow$) Assume $\mathcal{E}_M$ is not inductive and that $\Gamma \vdash_{\lambda\mathcal{R}} M : A$. Then by Theorem 9.2.3 and Lemma 9.3.31 $\mathcal{R}$ is not inductive. ∎

9.3.34 Theorem *It is decidable whether or not an $M \in \Lambda$ can be inductively typed.*

Proof By Theorem 9.3.33 and Proposition 9.3.28. ∎

9.3.35 Example Take the term $\lambda x.xx$. We saw in Example 9.2.1((i)) that $\mathcal{E}_{\lambda x.xx}$ contains an equation $\alpha_1 = \alpha_1 \to \alpha_2$ and hence is not inductive. Therefore there is no inductive simultaneous recursion which can give a type to $\lambda x.xx$. Compare this with Theorems 9.3.29 and 9.3.30.

An inductive simultaneous recursion can be solved in $\mathbb{T}^+_\mu$, by Proposition 9.3.28. Hence if a term is typable from an inductive simultaneous recursion, then it is also typable in $\lambda\mu^+$. The converse result follows from Exercise 9.5.

9.3.36 Theorem *It is decidable whether or not a given term M can be given a type in $\lambda\mu^+$, i.e. whether or not $\Gamma \vdash_{\lambda\mu^+} M : A$ for some type A and basis Γ over $\mathbb{T}^+_\mu$.*

Proof By the above and Theorem 9.3.34. ∎

9.4 Exercises

Exercise 9.1 Let $M \equiv \lambda xy.xy(xy)$. Construct the principal recursive type and type algebra for M.

Exercise 9.2 Let $M \equiv \mathbf{c}_2\mathbf{c}_2$.

(i) Find the principal recursive type and type algebra for M.
(ii) Show that $\vdash_{\boldsymbol{\lambda}_\rightarrow} M : (\alpha \rightarrow \alpha) \rightarrow \alpha \rightarrow \alpha$. This is the principal type in $\boldsymbol{\lambda}_\rightarrow$ and in an invertible type algebra.
(iii) Show that $\vdash_{\boldsymbol{\lambda}_=^{\mathcal{A}}} M : \beta$, with $\mathcal{A} = \mathbb{T}^{\{\alpha,\beta\}}/\mathcal{E}$, where

$$\mathcal{E} = \{(\alpha \rightarrow \alpha) \rightarrow \alpha \rightarrow \alpha = ((\alpha \rightarrow \alpha) \rightarrow \alpha \rightarrow \alpha) \rightarrow \beta\}.$$

(iv) Find the morphisms mapping the recursive principal type of M onto respectively $(\alpha \rightarrow \alpha) \rightarrow \alpha \rightarrow \alpha$ and β, according to Theorem 9.2.3.

Exercise 9.3 Prove that the system $\lambda_\mu^{\mathrm{Cho}}$ defined in 7.1.24 satisfies the subject-reduction property. Note that this does not follow from Lemma 9.1.8, since type equivalence is not defined by a type algebra.

Exercise 9.4 Let $(\lambda\mu^-)$ be the system that assigns types in $\mathbb{T}_\mu$ to terms in Λ, in which the equivalence on types is not an $\rightarrow$-congruence, i.e. the rule $\rightarrow$-cong is deleted from Definition 7.4.27(i). Show that

$$x{:}(A \rightarrow \mu\alpha.B) \vdash_{\lambda\mu^-} (\lambda y.xy) : (A \rightarrow B[\alpha := \mu\alpha.B]);$$
$$x{:}(A \rightarrow \mu\alpha.B) \nvdash_{\lambda\mu^-} x : (A \rightarrow B[\alpha := \mu\alpha.B]).$$

Exercise 9.5 Show that if a term M has a type A in $\lambda\mu$, then for some simultaneous recursion $\mathcal{R}$ it also has a type B in $\lambda\mathcal{R}$, with A, B having the same tree unfolding, i.e. $(A)^* = (B)^*$.

Exercise 9.6 Prove Theorem 9.2.16. [Hint. Assume for simplicity that M is closed. By Theorem 9.2.14(ii) and Corollary 9.2.15 we have to find a substitution s such that $s(a_M^*) = A$. Use Remark 7.5.36.]

Exercise 9.7 Design an algorithm to decide directly (i.e. without passing through translation in $\mathbb{T}_\mu^+$), whether a given simultaneous recursion is inductive.

Exercise 9.8 Show that the typings for the terms in Example 7.1.4 are all principal, except the one for Y.

Exercise 9.9 Prove that $\lambda x.xx$ is not typable in $\lambda\mu^+$. [Hint. Use the Inversion Lemma 9.1.3.]

Exercise 9.10 Show that if $A \in \mathbb{T}_\mu^+$ and $A =_\mu B$, then $B \in \mathbb{T}_\mu^+$.

Exercise 9.11 For $M \in \Lambda$ write $\langle M \rangle \equiv \lambda y.yM$. Define

$$N \equiv \lambda x.x\langle x \rangle;$$
$$L \equiv N\langle N \rangle.$$

(i) Show that L is not normalizing. [Hint. Note that $\langle P \rangle Q \to_\beta QP$.] *Historical note.* This is very much like Quine's paradox (QP):

(QP)

"yields falsehood when preceded by its own quotation"
yields falsehood when preceded by its own quotation.

(QP) states that a sentence obtained by the given recipe is false; but if one follows that recipe then (QP) itself is obtained.

(ii) Show that $\vdash_{\lambda\mu} L : \alpha$, using $A = \mu\beta.((\beta \to \alpha) \to \alpha) \to \alpha$.

10

Models

Our purpose in the present chapter is to build concrete type algebras for the interpretation of recursive types. In Section 10.1 we focus on systems à la Curry, where, in general, infinitely many types can be inferred for each (type-free) λ-term. Accordingly, it is natural to regard the interpretation of a type as a collection of elements of a model of the untyped λ-calculus. This idea is due to Scott (1975a).

We shall also describe, in Sections 10.2 and 10.3, how to build models for explicitly typed systems with recursive types. Classical categories of domains yield straightforward models for these formulations. Beside these, we shall also consider models based on different constructions (like continuous closures or partial equivalence relations) that are of interest in their own.

10.1 Interpretations of type assignments in $\boldsymbol{\lambda}_{=}^{\mathcal{A}}$

Before constructing a concrete interpretation of type inference systems of the form $\boldsymbol{\lambda}_{=}^{\mathcal{A}}$ introduced in Definition 7.1.3, we need to define what structures are needed to specify such an interpretation. In what follows, we shall focus on pure λ-terms, i.e., we shall in general deal only with terms in Λ. In Chapter 11 we will indicate how the results can be extended to the situation in which there are constants, both in the types and terms.

10.1.1 Definition Let $\mathcal{D} = \langle D, \cdot, [\![\ ]\!]_{\rho}^{\mathcal{D}} \rangle$ be a λ-model: see Definition 3.1.31.

(i) We can view $\mathcal{P}(\mathcal{D})$ as a type algebra by considering $\mathcal{P}(\mathcal{D}) = \langle \mathcal{P}(D), \Rightarrow \rangle$, where $\Rightarrow$ is defined in Definition 3.1.34 as

$$X \Rightarrow Y = \{d \in D \mid d \cdot X \subseteq Y\} \triangleq \{d \in \mathcal{D} \mid \forall x \in X.(d \cdot x) \in Y\}.$$

(ii) An *interpretation* of a type algebra $\mathcal{A}$ in $\mathcal{D}$ is a morphism $h : \mathcal{A} \to \mathcal{P}(\mathcal{D})$.

10.1.2 Definition Let $\mathcal{D}$ be a λ-model, $\rho \in \mathsf{Env}_{\mathcal{D}}$, $\mathcal{A}$ be a type algebra and $h : \mathcal{A} \to \mathcal{P}(\mathcal{D})$ be a morphism. Let Γ be a basis.

(i) We say that $\mathcal{D}, \rho, h$ *satisfies* the type assignment statement $M : a$, written

$$\models_{\mathcal{D},\rho,h} M : a,$$

if $[\![M]\!]_{\rho}^{\mathcal{D}} \in h(a)$.

(ii) We say $\mathcal{D}, \rho, h$ satisfies Γ, written

$$\models_{\mathcal{D},\rho,h} \Gamma,$$

if $\models_{\mathcal{D},\rho,h} x : a$, for all $(x : a) \in \Gamma$.

(iii) We say Γ satisfies $(M : a)$ with respect to $\mathcal{D}, \rho, h$, written

$$\Gamma \models_{\mathcal{D},\rho,h} M : a,$$

if $\models_{\mathcal{D},\rho,h} \Gamma \;\Rightarrow\; \models_{\mathcal{D},\rho,h} M : A,$.

(iv) We say Γ satisfies $(M : a)$ with respect to $\mathcal{A}, \mathcal{D}$, written

$$\Gamma \models_{\mathcal{A},\mathcal{D}} M : a,$$

if $\Gamma \models_{\mathcal{D},\rho,h} M : a$, for all $\rho \in \mathsf{Env}_{\mathcal{D}}, h : \mathcal{A} \to \mathcal{P}(\mathcal{D})$.

(v) Finally we write $\Gamma \models_{\mathcal{A}} M : a$ if for all $\mathcal{D}$

$$\Gamma \models_{\mathcal{A},\mathcal{D}} M : a.$$

10.1.3 Proposition (Soundness)

$$\Gamma \vdash_{\lambda_{=}^{\mathcal{A}}} M : a \;\Rightarrow\; \Gamma \models_{\mathcal{A}} M : a.$$

Proof By induction on the length of proof of the left hand side. ∎

For well-founded type algebras $\mathcal{A}$, morphisms to $\mathcal{P}(\mathcal{D})$ can be obtained by assigning arbitrary values $\xi(a)$ for the prime elements $a \in \mathcal{A}$. In this way morphisms from $\mathbb{T}^{\mathbb{A}}$ are determined by the choice $\xi(\alpha)$ for $\alpha \in \mathbb{A}$.

Some domain theory

For non well-founded type algebras such as $\mathbb{T}_{\mu}$ or $\mathbb{T}[\mathcal{R}]$, the existence of morphisms to $\mathcal{P}(\mathcal{D})$ is less obvious. We therefore present some basic domain theory to be used in Section 10.2 in order to show that interpretations do exist, exploiting the domain structure of a well-known class of λ-models.

10.1.4 Definition

(i) A *partially ordered set*, also called a *poset*, is a structure $\mathcal{D} = \langle D, \sqsubseteq \rangle$ satisfying

$$x \sqsubseteq x;$$
$$x \sqsubseteq y \;\&\; y \sqsubseteq z \Rightarrow x \sqsubseteq z;$$
$$x \sqsubseteq y \;\&\; y \sqsubseteq x \Rightarrow x = y.$$

(ii) For $X \subseteq \mathcal{D}$ we say that $d \in \mathcal{D}$ is an *upper bound* of X, written $X \sqsubseteq d$, if

$$\forall x \in X . x \sqsubseteq d.$$

(iii) A subset X of $\mathcal{D}$ has a *supremum* (*sup*) $d \in \mathcal{D}$, written $\sqcup X = d$ if

$$X \sqsubseteq d$$
$$\forall d' \in \mathcal{D} . X \sqsubseteq d' \;\Rightarrow\; d \sqsubseteq d'$$

That is, d is the least upper bound of X.

(iv) A subset X of a poset $\mathcal{D}$ is *directed* if X is nonempty and

$$\forall x, y \in X \exists z \in X [x \sqsubseteq z \;\&\; y \sqsubseteq z].$$

(v) A *complete partial order* (CPO) is a poset $\mathcal{D}$ such that there is a smallest element $\bot$ and every directed $X \subseteq \mathcal{D}$ has a sup $\bigsqcup X \in \mathcal{D}$.

(vi) A *complete lattice* is a poset $\mathcal{D}$ such that every $X \subseteq \mathcal{D}$ has a sup $\bigsqcup X$ in $\mathcal{D}$.

10.1.5 Definition Let $\mathcal{D}$ be a complete lattice.

(i) An element $d \in \mathcal{D}$ is called *compact* (also called *finite*) if for every directed $Z \subseteq \mathcal{D}$ one has

$$d \sqsubseteq \bigsqcup Z \;\Rightarrow\; \exists z \in Z . d \sqsubseteq z.$$

(ii) Write $\mathcal{K}(\mathcal{D}) = \{d \in \mathcal{D} \mid d \text{ is compact}\}$. We say $\mathcal{D}$ is an *algebraic lattice* if for all $x \in \mathcal{D}$ the set $\{e \in \mathcal{K}(\mathcal{D}) \mid e \sqsubseteq x\}$ is directed and

$$x = \bigsqcup \{e \in \mathcal{K}(\mathcal{D}) \mid e \sqsubseteq x\}.$$

(iii) We call $\mathcal{D}$ an ω-*algebraic lattice* if in addition $\mathcal{K}(\mathcal{D})$ is countable.

10.1.6 Example

(i) For any set X, the powerset $\langle \mathsf{P}X, \subseteq \rangle$ is a complete lattice under the subset ordering. It is an ω-algebraic lattice iff X is countable.

(ii) A typical CPO is $\langle (\bot, \mathit{true}, \mathit{false}), < \rangle$, where $\bot < \mathit{true}$ and $\bot < \mathit{false}$.

(iii) Given two sets X and Y, there is the CPO of *partial functions* $X \rightharpoonup Y$ (given as graphs) ordered by subset (on the graphs).

Note that for posets $\mathcal{D} = \langle D, \sqsubseteq \rangle$ we have

$$\begin{array}{rcl} \omega\text{-algebraic lattice} & \Rightarrow & \text{algebraic lattice} \\ & \Rightarrow & \text{complete lattice} \\ & \Rightarrow & \text{complete partial order} \\ & \Rightarrow & \text{partially ordered set.} \end{array}$$

In Part III we will work mainly with ω-algebraic lattices and in this chapter mainly with CPOs, in order to promote mental flexibility.

10.1.7 Definition

(i) A CPO $\mathcal{D} = \langle D, \sqsubseteq \rangle$ induces a topology called the *Scott topology* on $\mathcal{D}$. A set $O \subseteq \mathcal{D}$ is called *Scott open* if

$$\begin{array}{rcl} x \in O \,\&\, x \sqsubseteq y & \Rightarrow & y \in O; \\ X \subseteq \mathcal{D} \text{ directed } \,\&\, \bigsqcup X \in O & \Rightarrow & \exists x \in X \cap O. \end{array}$$

(ii) The category **CPO** consists of CPOs as objects and, as morphisms, functions that are continuous with respect to the Scott topology.

10.1.8 Remark For each $x \in D$, the set $U_x = \{z \mid z \not\sqsubseteq x\}$ is open. Moreover, $\mathcal{D}$ is a T_0 space. Indeed, if $x \neq y$, then, say, $x \not\sqsubseteq y$ and U_y separates x from y. In general, $\mathcal{D}$ is not T_1 (if $x \sqsubseteq y$, then every open set containing x also contains y, so the partial order is discrete iff the Scott topology is T_1.).

10.1.9 Definition Let $f : \mathcal{D} \to \mathcal{D}'$ be a map.

(i) We say f is *strict* if $f(\bot_{\mathcal{D}}) = \bot_{\mathcal{D}'}$.
(ii) We say f is *monotonic* if

$$\forall d, d' \in \mathcal{D}.[d \sqsubseteq d' \;\Rightarrow\; f(d) \sqsubseteq f(d')].$$

(iii) If $\mathcal{D}, \mathcal{D}'$ are CPOs, then write $[\mathcal{D} \to \mathcal{D}'] = \{f : \mathcal{D} \to \mathcal{D}' \mid f \text{ is continuous}\}$.

A continuous map $f \in [\mathcal{D} \to \mathcal{D}']$ is always monotonic (consider for $d \sqsubseteq d'$ the directed set $\{d, d'\}$), but it does not need to be strict.

10.1.10 Lemma *Let $\mathcal{D}, \mathcal{D}'$ be CPOs and let $f : \mathcal{D} \to \mathcal{D}'$.*

(i) *f is continuous $\Rightarrow$ f is monotonic.*
(ii) *f is continuous with respect to the Scott topology $\Leftrightarrow$ for every directed $X \subseteq D$*

- *$\bigsqcup f(X) \in D'$ exists;*

- $f(\bigsqcup X) = \bigsqcup f(X)$.

(iii) *For* $f, g \in [\mathcal{D} \to \mathcal{D}']$ *define*

$$f \sqsubseteq g \Leftrightarrow \forall d \in \mathcal{D}. f(d) \sqsubseteq g(d).$$

Then $\langle [\mathcal{D} \to \mathcal{D}'], \sqsubseteq \rangle$ *is a CPO.*

Proof See for example B[1984], Propositions 1.2.6, 1.2.11. ∎

The following is a well-known property of CPOs for finding fixed points.

10.1.11 Theorem *Let* $\langle D, \sqsubseteq \rangle$ *be a CPO. There is a functional*

$$\mathbf{fix} \in [[D \to D] \to D]$$

such that for every $f \in [D \to D]$ *one has*

$$f(\mathbf{fix}(f)) = \mathbf{fix}(f)$$

Proof Take **fix** to be the function which assigns to $f : D \to D$ the element

$$\bigsqcup_{n \in \omega} f^{(n)}(\bot_D).$$

Then as $\{f^{(n)}(\bot) \mid n \in \omega\}$ is directed, the function **fix** is continuous; that the equation holds is left to Exercise 10.4. ∎

10.1.12 Definition A *reflexive structure* is of the form $\mathcal{D} = \langle D, F, G \rangle$, where $\mathcal{D}$ is a CPO and $F : D \to [D \to D]$ and $G : [D \to D] \to D$, are continuous and satisfy $F \circ G = 1_{[D \to D]}$, i.e. $F(G(f)) = f$, for all $f \in [\mathcal{D} \to \mathcal{D}]$.

A reflexive structure $\mathcal{D}$ can be turned into a λ-model.

10.1.13 Definition

(i) Let $\mathcal{D}$ be a reflexive structure. We turn $\mathcal{D}$ into an applicative structure by defining the binary operation $d \cdot e = F(d)(e)$.

(ii) Then we define for $M \in \Lambda$ and $\rho \in \mathsf{Env}_{\mathcal{D}}$ the interpretation $[\![M]\!]_\rho \in \mathcal{D}$ by induction on the structure of M. This will turn $\mathcal{D}$ into a λ-model in the sense of Definition 3.1.31.

$$\begin{array}{rcll} [\![x]\!]_\rho & = & \rho(x), & \text{for } x \in \mathsf{V}; \\ [\![MN]\!]_\rho & = & [\![M]\!]_\rho \cdot [\![N]\!]_\rho; & \\ [\![\lambda x.M]\!]_\rho & = & G(\lambda\!\!\lambda d.[\![M]\!]_{\rho[x \mapsto d]}). & \end{array}$$

The last equation could also have been written as $[\![\lambda x.M]\!]_\rho = G(f)$, where $f(d) = [\![M]\!]_{\rho[x \mapsto d]}$. That this is indeed a continuous function can be proved by induction on the structure of M.

10.1.14 Proposition *Let $\mathcal{D} = \langle D, F, G\rangle$ be a reflexive structure. Define the maps $\cdot : \mathcal{D} \times \mathcal{D} \to \mathcal{D}$ and $[\![\ ]\!]_\rho$ as above. Then:*

(i) $[\![M[x:=N]]\!]_\rho = [\![M]\!]_{\rho[x \mapsto [\![N]\!]_\rho]}$;

(ii) $\langle \mathcal{D}, \cdot, [\![\ ]\!]_\rho\rangle$ *is a λ-model.*

10.2 Interpreting $\mathbb{T}_\mu$ and $\mathbb{T}^*_\mu$

In this section, $\mathcal{D}$ will range over λ-models of the form $\mathcal{D} = \langle D, F, G\rangle$, with $\mathcal{D}$ a reflexive structure. If $\mathcal{A}$ is a well-founded type algebra of the form $\mathbb{T}^{\mathbb{A}}/\mathcal{E}$, then there are many morphisms $h : \mathcal{A} \to \mathsf{P}D$ determined by the images $h(\alpha)$ for $\alpha \in \mathbb{A}$.

If $\mathcal{A}$ is not well founded (examples of such are $\mathbb{T}_\mu$, $\mathbb{T}^*_\mu$, $\mathbb{T}[\mathcal{R}], \mathbb{T}[\mathcal{R}]^*$, see Definitions 7.3.17 and 7.5.18), then it is harder to construct morphisms $\mathcal{A} \to \mathsf{P}D$. We will address this in the present section for the systems $\mathbb{T}_\mu$, $\mathbb{T}^*_\mu$, which are somewhat more general since they contain solutions of all recursive type equations. The same technique can also be applied to $\mathbb{T}[\mathcal{R}], \mathbb{T}[\mathcal{R}]^*$; see Exercise 10.17.

For this construction we have to find a suitable class of subsets of a λ-model $\mathcal{D}$ closed under $\Rightarrow$, with the property that $\mathbb{T}_\mu$ and $\mathbb{T}^*_\mu$ can be mapped by a morphism to the type algebra so obtained.

Approximating λ-models

An important tool for interpreting recursive types is λ-models having a *notion of approximation.*

10.2.1 Definition (Notion of approximation) Let $\langle D, F, G\rangle$ be a reflexive structure. A family of continuous functions $\{(\cdot)_n : D \to D\}_{n \in \omega}$ is a *notion of approximation* for D if it satisfies the following conditions. Write d_n for $(d)_n$.

(i) For all $d \in D$, and all $n, m \in \omega$:

$$\bot_0 = \bot \tag{10.1}$$

$$n \leq m \Rightarrow d_n \sqsubseteq d_m \tag{10.2}$$

$$(d_n)_m = d_{\min(n,m)} \tag{10.3}$$

$$d = \bigsqcup_{n \in \omega} d_n. \tag{10.4}$$

(ii) For all $d, e \in D$ and $n \in \omega$:

$$d_0 \cdot e = d_0 = (d \cdot \bot)_0 \tag{10.5}$$

$$d_{n+1} \cdot e_n = d_{n+1} \cdot e = (d \cdot e_n)_n \tag{10.6}$$

$$d \cdot e = \bigsqcup_{n \in \omega} (d_{n+1} \cdot e_n). \tag{10.7}$$

(iii) For $X \subseteq D$ write $X_n \triangleq \{d_n \mid d \in X\}$ for $n \in \omega$.

10.2.2 Lemma *Suppose $k \leq n$. Then*

$$\begin{array}{rcl} (X)_k & \subseteq & (X)_n. \\ (X_n)_k & = & X_k. \end{array}$$

Proof Since $d_k = (d_k)_n = (d_n)_k$, by (10.3). ■

The conditions of Definition 10.2.1 are satisfied by the λ-models D_∞ built by the classical construction of Scott; see, for example, B[1984], Lemma 18.2.8 and Proposition 18.2.13. Some of these may fail for the same construction as modified by Park, see B[1984], Exercise 18.4.21. Furthermore, they also apply to λ-models not explicitly obtained by means of an inverse limit construction, such as the models D_A, see Engeler (1981), or the filter λ-model $\mathcal{F}$ introduced in Barendregt et al. (1983), see Part III. See also Exercises 16.6 and 10.5.

Complete-uniform sets

10.2.3 Notation From now on in this section we use $\mathcal{D}$ to denote a reflexive structure with a notion of approximation.

One way of interpreting elements of $\mathbb{T}_\mu$ is as the collection of *ideals* of D, i.e., the non-empty subsets of D closed with respect to the Scott topology (see MacQueen et al. (1986), Coppo (1985), Cardone and Coppo (1991)). Equivalently, these can be described as the non-empty, downward closed subsets X of D such that $\bigsqcup \Delta \in X$ whenever $\Delta \subseteq X$ is directed. Here we shall use a slightly more general semantical notion of type, by relaxing the requirement of downward closure and assuming only *completeness*: types are closed under increasing sequences and are *uniform*: if d belongs to a type, then d_n belongs to that type for all $n \in \omega$. These notions come from Abadi and Plotkin (1990).

10.2.4 Remark The closure properties that we require for our semantical notion of type are motivated by the fact that we are working essentially in *continuous* λ-models. For example, in all such models, see B[1984], Chapter 19, §3,

$$\bot_D = [\![(\lambda x.xx)(\lambda x.xx)]\!],$$

hence $\bot_D$ belongs to all types. This accounts for non-emptiness, see Example 7.1.4(ii). On the other hand, the interpretation of

$$\mathsf{Y} \equiv \lambda f.(\lambda x.f(xx))(\lambda x.f(xx)),$$

the fixed point combinator, defines the map $\mathbf{fix} \in [[D \to D] \to D]$ introduced in Theorem 10.1.11. Now, it was shown in Example 7.1.4(iii) that Y has type $(A \to A) \to A$. This motivates the definition of completeness below, which states that every type has to be closed under least upper bounds of increasing sequences. Concerning uniformity, observe that if X is an ideal of D, the set $X_n \triangleq \{d_n \mid d \in X\}$ is not, in general, an ideal of D. For example, in D_∞, the subset D_0 is not downward closed.

The construction described below can be performed, more generally, using n-ary complete and uniform relations over D as interpretations of types. This applies in particular to the interpretation of types as (complete and uniform) *partial equivalence relations* (PERs) over D, which can be exploited in the construction of an extensional model for the versions à la Church of both $\lambda\mu$ and $\lambda\mu^*$ (this is the content of Exercises 10.11, 10.12 and 10.13). The subsets that interpret types are required to respect the notion of approximation, in the following sense.

10.2.5 Definition

(i) A subset $X \subseteq D$ is *complete* if:
- $\bot_D \in X$;
- if $d^0 \sqsubseteq d^1 \sqsubseteq \cdots$ are all in X, then so is $\bigsqcup_{n\in\omega} d^n$.

(ii) Let ∇ be a complete subset of D_0. A subset $X \subseteq D$ is ∇-*uniform* if:
- $X_0 = \nabla$;
- $X_n \subseteq X$ for all $n \in \omega$.

(iii) We denote by $\mathcal{CU}_\nabla(D)$, or just $\mathcal{CU}(D)$ if ∇ is clear from the context, the set of *complete and* ∇-*uniform subsets* of D.

The following provides some properties of this set.

10.2.6 Proposition *Let $X, Y \in \mathcal{CU}_\nabla(D)$. Then, for all $n \in \omega$:*

(i) *$X_n \in \mathcal{CU}_\nabla(D)$;*
(ii) *$X = Y \Leftrightarrow \forall n \in \omega,\ X_n = Y_n$.*

Proof (i) X_n is uniform by (10.3). It also is complete: since $\bot \in X$ also $\bot = \bot_n \in X_n$; and if d^k is an increasing chain in X_n, then

$$\bigsqcup_k d^k = \bigsqcup_k d^k_n = (\bigsqcup_k d^k)_n \in X_n,$$

by continuity of the approximation mappings.

(ii) Easy, using Definition 10.2.5 and (10.4). ■

10.2.7 Proposition *Let the sequence of sets $X^{(n)} \in \mathcal{CU}_\nabla(D)$, $n \in \omega$, satisfy*

$$\forall n \in \omega \left(X^{(n)} = (X^{(n+1)})_n \right). \tag{10.8}$$

Define $X^{(\infty)} = \{d \mid d_n \in X^{(n)}\} \in \mathcal{CU}_\nabla(D)$. Then this set is the unique, complete, and ∇-uniform subset of D, such that $(X^{(\infty)})_n = X^{(n)}$, for all $n \in \omega$.

Proof We first claim that

(i) $(X^{(n)})_k = X^{(k)}$, for $k \leq n$;
(ii) $X^{(k)} \subseteq X^{(n)}$, for $k \leq n$;
(iii) $(X^{(n)})_k \subseteq X^{(k)}$, for all k, n.

Item (i) is proved by induction on $n - k$. If $n = k$, then

$$\begin{aligned}(X^{(n)})_n &= ((X^{(n+1)})_n)_n \\ &= (X^{(n+1)})_n \\ &= X^{(n)}.\end{aligned}$$

Now we show the equation for $n, k-1$, assuming it for n, k.

$$\begin{aligned}(X^{(n)})_{k-1} &= ((X^{(n)})_k)_{k-1} \\ &= (X^{(k)})_{k-1}, \qquad \text{by the induction hypothesis,} \\ &= X^{(k-1)}.\end{aligned}$$

Item (ii) follows from (i), as $(X^{(n)})_k \subseteq X^{(n)}$. Item (iii) follows for $k \leq n$ from (i). For $n \leq k$, one has, from (ii), by uniformity of of $X^{(n)}$,

$$(X^{(n)})_k \subseteq X^{(n)} \subseteq X^{(k)}.$$

To show that $X^{(\infty)} \in \mathcal{CU}_\nabla(D)$, notice that $(X^{(\infty)})_0 = \nabla$, because if $d \in \nabla$ then $d \in X^{(i)}$ for all $i \in \omega$ as each $X^{(i)}$ is uniform. But then $d \in (X^{(i)})_i$, hence $d_i = d \in X^{(i)}$ and therefore $\nabla \subseteq (X^{(\infty)})_0$. Conversely, observe that $(X^{(\infty)})_0 \subseteq X^{(0)} = \nabla$, by assumption (10.8).

To prove uniformity of $X^{(\infty)}$, let $d \in X^{(\infty)}$. Then $(d_n)_k \in (X^{(n)})_k \subseteq X^{(k)}$, by (iii). Therefore $d_n \in X^{(\infty)}$.

As for completeness, let $d^{(j)} \in X^{(\infty)}$ be a chain; then $(d^{(j)})_n \in X^{(n)}$, for all $n \in \omega$, and therefore also

$$\bigsqcup_j (d^{(j)})_n = \left(\bigsqcup_j d^{(j)}\right)_n \in X^{(n)},$$

by continuity of the approximation mappings, which entails that

$$\bigsqcup_j d^{(j)} \in X^{(\infty)}.$$

By (iii) we have $X^{(n+1)} \subseteq X^{(\infty)}$, for every n. Hence $X^{(n)} \subseteq (X^{(n+1)})_n \subseteq (X^{(\infty)})_n$. Therefore $X^{(n)} = (X^{(\infty)})_n$, because the reverse inclusion holds by definition of $X^{(\infty)}$. Finally, $X^{(\infty)}$ is unique, because if Y is another complete and uniform subset with the property that $Y_n = X^{(n)}$, then $Y_n = (X^{(\infty)})_n$ for all $n \in \omega$, and this yields $Y = X^{(\infty)}$. ■

The following result shows that each class $\mathcal{CU}_\nabla(D)$ is closed under $\Rightarrow$.

10.2.8 Proposition *If $X, Y \in \mathcal{CU}(D)$ then $(X \Rightarrow Y) \in \mathcal{CU}(D)$.*

Proof We first show that $\nabla = (X \Rightarrow Y)_0$. Let $d \in \nabla$, $a \in X$. Then

$$\begin{aligned} d \cdot a &= d_0 \cdot a, && \text{as } \nabla \subseteq D_0, \\ &= d, && \text{by (10.5)}. \end{aligned}$$

Therefore $d \cdot a \in \nabla = Y_0 \subseteq Y$, by uniformity. So $\nabla \subseteq X \Rightarrow Y$, and $\nabla \subseteq (X \Rightarrow Y)_0$ since $\nabla = \nabla_0$. Conversely, let $d \in (X \Rightarrow Y)$. Then

$$d_0 = (d \cdot \bot_D)_0, \qquad \text{by (10.5)}.$$

But $\bot_D \in \nabla = X_0 \subseteq X$, hence $d \cdot \bot_D \in Y$, because $d \in X \Rightarrow Y$, and therefore $d_0 \in (d \cdot \bot_D)_0 \in Y_0 = \nabla$.

Also we have $\bot_{\mathcal{D}} \in X \Rightarrow Y$, since $\bot_{\mathcal{D}} \in \nabla = (X \Rightarrow Y)_0 \subseteq X \Rightarrow Y$.

To prove uniformity of $X \Rightarrow Y$, assume that $d \in X \Rightarrow Y$, and consider d_n for $n > 0$ (we already know that $(X \Rightarrow Y)_0 = \nabla \subseteq X \Rightarrow Y$). For any $a \in X$

$$d_n \cdot a = (d \cdot a_{n-1})_{n-1}, \qquad \text{by (10.6)};$$

but $a_{n-1} \in X$ by uniformity of X and therefore $d \cdot a_{n-1} \in Y$. Therefore also $(d \cdot a_{n-1})_{n-1} \in Y$ by uniformity of Y, and finally $d_n \cdot a \in X \Rightarrow Y$.

To prove completeness, assume that we have an increasing chain of elements $d^{(j)} \in X \Rightarrow Y$, in order to show that the sup also is in $X \Rightarrow Y$. Let $a \in X$. Then

$$\left(\bigsqcup_{j \in \omega} d^{(j)}\right) \cdot a = \bigsqcup_{j \in \omega} (d^{(j)} \cdot a) \in Y,$$

by continuity of application and completeness of Y. Therefore

$$\bigsqcup_{j \in \omega} d^{(j)} \in (X \Rightarrow Y). \blacksquare$$

The following property of the interpretation of function types is the key to the possibility of the interpretation of recursive types.

10.2.9 Proposition *For any $X, Y \in \mathcal{CU}(D)$ and $n \in \omega$,*

$$(X \Rightarrow Y)_{n+1} = (X_n \Rightarrow Y_n)_{n+1}.$$

Proof Assume that $d_{n+1} \in (X \Rightarrow Y)_{n+1} \subseteq X \Rightarrow Y$ and that $a_n \in X_n \subseteq X$. Then $d_{n+1} \cdot a_n \in Y$. But

$$d_{n+1} \cdot a_n = (d_{n+1} \cdot a_n)_n \in Y_n$$

by (10.6), so $d_{n+1} \in X_n \Rightarrow Y_n$ and hence $d_{n+1} \in (X_n \Rightarrow Y_n)_{n+1}$. Conversely, let $d_{n+1} \in (X_n \Rightarrow Y_n)_{n+1}$, and assume that $a \in X$. Then $a_n \in X_n$. Now

$$d_{n+1} \cdot a = d_{n+1} \cdot a_n \in Y_n \subseteq Y,$$

using again (10.6) and uniformity of Y and $X_n \Rightarrow Y_n$, so $d_{n+1} \in X \Rightarrow Y$. Therefore $d_{n+1} \in (X \Rightarrow Y)_{n+1}$. $\blacksquare$

Now we define the type algebra that will be used throughout this section.

10.2.10 Definition $\mathcal{S}(D) \triangleq \langle \mathcal{CU}(D), \Rightarrow \rangle$.

As an example, let us see how the theory developed so far lets us to interpret a type $T = T \to T$ as a complete and uniform subset $\Xi = \Xi \Rightarrow \Xi$. The latter is built in denumerably many steps

$$\Xi^{(0)}, \Xi^{(1)}, \Xi^{(2)}, \ldots$$

The 0th stage is $\Xi^{(0)} = \nabla$. Then, whatever Ξ eventually will be, $\Xi_0 = \Xi^{(0)}$. Later stages are defined by the recurrence

$$\Xi^{(n+1)} = (\Xi^{(n)} \Rightarrow \Xi^{(n)})_{n+1}.$$

If we can show that for all $n \in \omega$

$$\left(\Xi^{(n+1)}\right)_n = \Xi^{(n)}, \tag{10.9}$$

then we can exploit Proposition 10.2.7 and take $\Xi = \Xi^{(\infty)}$, so that $\Xi_n = \Xi^{(n)}$ for all $n \in \omega$. The proof of (10.9) is therefore the core of the technique, and appeals in an essential way to Proposition 10.2.9. The fact that $\Xi = \Xi \Rightarrow \Xi$ is then a direct consequence of Proposition 10.2.6(ii). Of course, this process must be carried out, in parallel, for *all* type expressions, by defining a whole family of approximate interpretations of types. We shall now show how to do this in the case of μ-types. The same method will be applied to the interpretation of simultaneous recursions in Exercise 10.17.

Approximate interpretations of μ-types

In order to state the definition of the approximate interpretation of types it is convenient to introduce an auxiliary notation.

10.2.11 Notation For $X, Y \in \mathcal{CU}(D)$ and $n \in \omega$, let $X \Rightarrow^{n+1} Y$ denote $(X \Rightarrow Y)_{n+1}$.

10.2.12 Lemma

(i) *If $A \subseteq D$ is ∇-uniform, then $(A)_n = A \cap D_n$.*
(ii) *Let $X, Y \in \mathcal{CU}(D)$. Then $(X \Rightarrow^{n+1} Y) = (X \Rightarrow Y) \cap D_{n+1}$.*
(iii) $(X \Rightarrow^{n+1} Y) = X_n \Rightarrow^{n+1} Y_n$.

Proof (i) Note that $A_n \subseteq A \cap D_n$ by uniformity. Conversely, if $d \in A \cap D_n$, then $d = d_n \in A_n$, by the idempotency of $(\cdot)_n$, implied by (10.3).
(ii) By (i).
(iii) By Proposition 10.2.9. ■

10.2.13 Definition (Approximate Interpretations of Types) Given a type $A \in \mathbb{T}_\mu$, a number $n \in \omega$, and a type environment $\eta : \mathbb{A} \to \mathcal{CU}(D)$, define, by induction on n and the complexity of the type A, the nth *approximation* of the interpretation of A in the environment η, $\mathcal{I}^n[\![A]\!]_\eta$, written as:

$$\begin{aligned}
\mathcal{I}^0[\![A]\!]_\eta &= \nabla;\\
\mathcal{I}^{n+1}[\![\alpha]\!]_\eta &= (\eta(\alpha))_{n+1};\\
\mathcal{I}^{n+1}[\![A_1 \to A_2]\!]_\eta &= \mathcal{I}^n[\![A_1]\!]_\eta \Rightarrow^{n+1} \mathcal{I}^n[\![A_2]\!]_\eta;\\
\mathcal{I}^{n+1}[\![\mu\alpha.A_1]\!]_\eta &= \mathcal{I}^{n+1}[\![A_1]\!]_{\eta[\alpha \mapsto \mathcal{I}^n[\![\mu\alpha.A_1]\!]_\eta]}.
\end{aligned}$$

By a simple inductive argument (on n and then on the structure of the type) one can see that each $\mathcal{I}^n[\![A]\!]_\eta$ is a complete and uniform subset of D, in fact $\mathcal{I}^n[\![A]\!]_\eta \subseteq D_n$. We shall make frequent use below of the following properties, whose easy inductive proofs are left as an exercise.

10.2.14 Lemma *Let $A \in \mathbb{T}_\mu$, $\alpha \in \mathbb{A}$, $n \in \omega$ and η be an environment.*

(i)	$\mathcal{I}^n[\![A]\!]_\eta$	$= \mathcal{I}^n[\![A]\!]_{(\eta\restriction n)}$,	*where* $(\eta \restriction n)(\alpha) \triangleq (\eta(\alpha))_n$.
(ii)	$\mathcal{I}^{n+1}[\![A]\!]_\eta$	$= \mathcal{I}^{n+1}[\![A]\!]_{\eta[\alpha \mapsto (\eta(\alpha))_n]}$,	*if $\mu\alpha.A$ is non-circular.*
(iii)	$\mathcal{I}^n[\![\mu\alpha.A]\!]_\eta$	$= \nabla$,	*if $\mu\alpha.A$ is circular.*

10.2.15 Lemma *For any $A \in \mathbb{T}_\mu$, $n \in \omega$, and type environment η,*

$$\mathcal{I}^n[\![A]\!]_\eta = \left(\mathcal{I}^{n+1}[\![A]\!]_\eta\right)_n.$$

Proof For arbitrary $A \in \mathbb{T}_\mu$, $n \in \omega$, and η, the statement is clearly equivalent to the conjunction of the following two:

(a) $\mathcal{I}^n[\![A]\!]_\eta \subseteq \mathcal{I}^{n+1}[\![A]\!]_\eta$;
(b) $d \in \mathcal{I}^{n+1}[\![A]\!]_\eta \Rightarrow d_n \in \mathcal{I}^n[\![A]\!]_\eta$.

We show (a) and (b) by induction on n. The basis is obvious, and the induction step is proved by induction on the complexity of the type A.

Case $A \equiv \alpha$, a type variable. Then $\mathcal{I}^n[\![\alpha]\!]_\eta = (\eta(\alpha))_n$, and we can use Lemma 10.2.2.

Case $A \equiv A_1 \to A_2$. Then we have to show first that

$$\mathcal{I}^{n-1}[\![A_1]\!]_\eta \Rightarrow^n \mathcal{I}^{n-1}[\![A_2]\!]_\eta \subseteq \mathcal{I}^n[\![A_1]\!]_\eta \Rightarrow^{n+1} \mathcal{I}^n[\![A_2]\!]_\eta.$$

So assume that $d \in \mathcal{I}^{n-1}[\![A_1]\!]_\eta \Rightarrow^n \mathcal{I}^{n-1}[\![A_2]\!]_\eta$ and that $a \in \mathcal{I}^n[\![A_1]\!]_\eta$. By the induction hypothesis (b) on n, $a_{n-1} \in \mathcal{I}^{n-1}[\![A_1]\!]_\eta$, hence

$$d \cdot a_{n-1} = d \cdot a \in \mathcal{I}^{n-1}[\![A_2]\!]_\eta \subseteq \mathcal{I}^n[\![A_2]\!]_\eta$$

using the induction hypothesis (a), and equation (10.6). If

$$d \in \mathcal{I}^n[\![A_1]\!]_\eta \Rightarrow^{n+1} \mathcal{I}^n[\![A_2]\!]_\eta$$

and $a \in \mathcal{I}^{n-1}[\![A_1]\!]_\eta$, $a \in \mathcal{I}^n[\![A_1]\!]_\eta$ by induction hypothesis (a), so $d{\cdot}a \in \mathcal{I}^n[\![A_2]\!]_\eta$ and by induction hypothesis (b) we get $(d \cdot a)_{n-1} \in \mathcal{I}^{n-1}[\![A_2]\!]_\eta$. The result follows by observing again using equation (10.6), of Definition 10.2.1, $(d \cdot a)_{n-1} = d_n \cdot a$, and therefore

$$d_n \in \mathcal{I}^{n-1}[\![A_1]\!]_\eta \Rightarrow^n \mathcal{I}^{n-1}[\![A_2]\!]_\eta.$$

Case $A \equiv \mu\alpha.A_1$. If A is circular, then the property is trivially true by Lemma 10.2.14(iii). So, assume that A is non-circular. Then

$$\begin{aligned}
\mathcal{I}^n[\![\mu\alpha.A_1]\!]_\eta &= \mathcal{I}^n[\![A_1]\!]_{\eta[\alpha \mapsto \mathcal{I}^{n-1}[\![\mu\alpha.A_1]\!]_\eta]} \\
&= \mathcal{I}^n[\![A_1]\!]_{\eta[\alpha \mapsto (\mathcal{I}^n[\![\mu\alpha.A_1]\!]_\eta)_{n-1}]}, && \text{by the induction hypothesis on } n, \\
&= \mathcal{I}^n[\![A_1]\!]_{\eta[\alpha \mapsto \mathcal{I}^n[\![\mu\alpha.A_1]\!]_\eta]}, && \text{by Lemma 10.2.14(ii)}, \\
&\subseteq \mathcal{I}^{n+1}[\![A_1]\!]_{\eta[\alpha \mapsto \mathcal{I}^n[\![\mu\alpha.A_1]\!]_\eta]}, && \text{by the induction hypothesis on } A, \\
&= \mathcal{I}^{n+1}[\![\mu\alpha.A_1]\!]_\eta.
\end{aligned}$$

Now, let $d \in \mathcal{I}^{n+1}[\![\mu\alpha.A_1]\!]_\eta = \mathcal{I}^{n+1}[\![A_1]\!]_{\eta[\alpha \mapsto \mathcal{I}^n[\![\mu\alpha.A_1]\!]_\eta]}$. Then, by the induction hypothesis on the complexity of the type, one has

$$d_n \in \mathcal{I}^n[\![A_1]\!]_{\eta[\alpha \mapsto \mathcal{I}^n[\![\mu\alpha.A_1]\!]_\eta]}.$$

Hence by Lemma 10.2.14, we have $d_n \in \mathcal{I}^n[\![A_1]\!]_{\eta[\alpha \mapsto (\mathcal{I}^n[\![\mu\alpha.A_1]\!]_\eta)_{n-1}]}$, as we assumed that A_1 is contractive in α. But $(\mathcal{I}^n[\![\mu\alpha.A_1]\!]_\eta)_{n-1} = \mathcal{I}^{n-1}[\![\mu\alpha.A_1]\!]_\eta$, by the induction hypothesis on n. Therefore

$$d_n \in \mathcal{I}^n[\![A_1]\!]_{\eta[\alpha \mapsto \mathcal{I}^{n-1}[\![\mu\alpha.A_1]\!]_\eta]} = \mathcal{I}^n[\![\mu\alpha.A_1]\!]_\eta. \blacksquare$$

The interpretation, $\mathcal{I}[\![A]\!]_\eta$, of a type A can now be defined by glueing its approximate interpretations $\mathcal{I}^n[\![A]\!]_\eta$, as described in Proposition 10.2.7.

10.2.16 Definition (Type Interpretations) For every type A and type environment η, define the *type interpretation* $\mathcal{I}[\![\ ]\!]_\eta : \mathbb{T}_\mu \to \mathcal{CU}(D)$,

$$\mathcal{I}[\![A]\!]_\eta \triangleq \mathcal{I}^\infty[\![A]\!]_\eta,$$

to be $X^{(\infty)}$, where $X^{(n)} = (\mathcal{I}^n[\![A]\!]_\eta)$.

Observe that the interpretation of types depends on the choice of the base ∇. For example, let $D = D_\infty$ be defined as the inverse limit of a sequence of CPOs (see 16.3.1–16.3.24), where D_0 consists of the two points $\bot, \top$. Now if $\nabla = \{\bot\}$, then $\top \notin \mathcal{I}[\![\mu t.t \to t]\!]_\eta$, whereas $\mathcal{I}[\![\mu t.t \to t]\!]_\eta = D$ when $\nabla = \{\bot, \top\}$. Note that if $\nabla = \{\bot\}$, then we have $D_\infty \notin \mathcal{CU}(D_\infty)$. But $\mathcal{I}^\infty[\![A]\!]_\eta \in \mathcal{CU}(D_\infty)$, for all A and η.

10.2.17 Proposition *For any $A \in \mathbb{T}_\mu$, $n \in \omega$, and η, one has*

$$\left(\mathcal{I}[\![A]\!]_\eta\right)_n = \mathcal{I}^n[\![A]\!]_\eta.$$

Proof From Proposition 10.2.7. $\blacksquare$

10.2.18 Proposition

(i) *For any type $A \in \mathbb{T}_\mu$, and environment η one has*

$$\mathcal{I}[\![\mu\alpha.A]\!]_\eta = \mathcal{I}[\![A]\!]_{\eta[\alpha \mapsto \mathcal{I}[\![\mu\alpha.A]\!]_\eta]}.$$

(ii) *For $A, B \in \mathbb{T}_\mu$, any type variable α, and any η, one has*

$$\mathcal{I}[\![A[\alpha := B]]\!]_\eta = \mathcal{I}[\![A]\!]_{\eta[\alpha \mapsto \mathcal{I}[\![B]\!]_\eta]}.$$

Proof (i) If $\mu\alpha.A$ is circular, then Lemma 10.2.14(iii) applies. So, assume that $\mu\alpha.A$ is non-circular. By Proposition 10.2.6 it suffices that for all n,

$$\left(\mathcal{I}[\![\mu\alpha.A]\!]_\eta\right)_n = \left(\mathcal{I}[\![A]\!]_{\eta[\alpha\mapsto\mathcal{I}[\![\mu\alpha.A]\!]_\eta]}\right)_n.$$

By Proposition 10.2.17, for all B, η, n,

$$\left(\mathcal{I}[\![B]\!]_\eta\right)_n = \mathcal{I}^n[\![B]\!]_\eta.$$

Case $n = 0$. Then both sides equal ∇.

Case $n+1$. Then we have

$$\begin{aligned}\mathcal{I}^{n+1}[\![\mu\alpha.A]\!]_\eta &= \mathcal{I}^{n+1}[\![A]\!]_{\eta[\alpha\mapsto\mathcal{I}^n[\![\mu\alpha.A]\!]_\eta]}, && \text{by Definition 10.2.13,}\\ &= \mathcal{I}^{n+1}[\![A]\!]_{\eta[\alpha\mapsto\mathcal{I}[\![\mu\alpha.A]\!]_\eta]}, && \text{by Lemma 10.2.14(ii) applied}\\ & && \text{to } \eta' = \eta[\alpha\mapsto\mathcal{I}[\![\mu\alpha.A]\!]_\eta].\end{aligned}$$

(ii) By a double induction; see Exercise 10.7. ■

10.2.19 Theorem (Properties of Type Interpretations) *For any type environment η, and all types A, B, the following conditions are satisfied:*

(i) $\mathcal{I}[\![\alpha]\!]_\eta = \eta(\alpha)$;
(ii) $\mathcal{I}[\![A\to B]\!]_\eta = \mathcal{I}[\![A]\!]_\eta \Rightarrow \mathcal{I}[\![B]\!]_\eta$;
(iii) $\mathcal{I}[\![\mu\alpha.A]\!]_\eta = \mathcal{I}[\![A[\alpha := \mu\alpha.A]]\!]_\eta$.

Proof (i) $d\in\mathcal{I}[\![\alpha]\!]_\eta$ iff, for all $n\in\omega$, $d_n\in\mathcal{I}^n[\![\alpha]\!]_\eta = (\eta(\alpha))_n$ iff $d\in\eta(\alpha)$.

(ii) Observe that for all $n\in\omega$:

$$\begin{aligned}\left(\mathcal{I}[\![A\to B]\!]_\eta\right)_{n+1} &= \mathcal{I}^{n+1}[\![A\to B]\!]_\eta, && \text{by Proposition 10.2.17,}\\ &= \mathcal{I}^n[\![A]\!]_\eta \Rightarrow^{n+1} \mathcal{I}^n[\![B]\!]_\eta\\ &= \left(\mathcal{I}[\![A]\!]_\eta\right)_n \Rightarrow^{n+1} \left(\mathcal{I}[\![B]\!]_\eta\right)_n\\ &= \left(\mathcal{I}[\![A]\!]_\eta \Rightarrow \mathcal{I}[\![B]\!]_\eta\right)_{n+1}, && \text{by Proposition 10.2.9.}\end{aligned}$$

The result follows by induction from Proposition 10.2.6(ii), observing that

$$\left(\mathcal{I}[\![A\to B]\!]_\eta\right)_0 = \nabla = \left(\mathcal{I}[\![A]\!]_\eta \Rightarrow \mathcal{I}[\![B]\!]_\eta\right)_0.$$

(iii) By Lemma 10.2.18(i),(ii). ■

Soundness and completeness for interpreting $\mathbb{T}_\mu$ and $\mathbb{T}_\mu^*$

For any type environment η, Theorem 10.2.19 implies that $\mathcal{I}[\![-]\!]_\eta$ is a type algebra morphism from $\mathbb{T}_\mu/{=_\mu}$ to $\mathcal{D}$. We immediately have the following corollary.

10.2.20 Proposition (Soundness of $\vdash_{\lambda\mu}$) *For all* $\rho \in \mathsf{Env}_{\mathcal{D}}, \eta : \mathbb{A} \to C$ *we have*

$$\Gamma \vdash_{\lambda\mu} M : A \;\Rightarrow\; \Gamma \models_{\mathcal{D},\rho,\mathcal{I}[\![-]\!]_\eta} M : A.$$

Proof By Proposition 10.1.3. ∎

We now proceed to show that the type interpretation introduced in Definition 10.2.16 and characterized in Theorem 10.2.19 induces a type algebra morphism from $\mathbb{T}_\mu^*$ to $\mathcal{S}(D)$: see Definitions 7.5.18 and 10.2.10. To this end, we introduce a notion of approximate interpretation for the regular trees which result from unfolding types in $\mathbb{T}_\mu$ infinitely often. This interpretation is of interest in its own right, and is indeed the notion of interpretation which was taken as basic in Cardone and Coppo (1991).

10.2.21 Definition (Approximate interpretation of regular trees) Given $n \in \omega$, $t \in \mathsf{Tr}_{\mathrm{R}}$, $\alpha \in \mathbb{A}$ and $\eta : \mathbb{A} \to \mathcal{CU}(D)$ a type environment, define the nth *approximation* of the regular tree t, written $\mathcal{T}^n[\![t]\!]_\eta$, by induction on n as follows:

$$\begin{aligned}
\mathcal{T}^0[\![t]\!]_\eta &= \nabla;\\
\mathcal{T}^{n+1}[\![\alpha]\!]_\eta &= (\eta(\alpha))_{n+1};\\
\mathcal{T}^{n+1}[\![\bullet]\!]_\eta &= \nabla;\\
\mathcal{T}^{n+1}[\![t_1 \to t_2]\!]_\eta &= (\mathcal{T}^n[\![t_1]\!]_\eta \Rightarrow^{n+1} \mathcal{T}^n[\![t_2]\!]_\eta).
\end{aligned}$$

10.2.22 Lemma *For all types* $A \in \mathbb{T}_\mu$, *all type environments* η *and all* $n \in \omega$,

$$\mathcal{I}^n[\![A]\!]_\eta = \mathcal{T}^n[\![A^*]\!]_\eta,$$

where $(-)^* = (-)_\mu^* : \mathbb{T}_\mu \to \mathsf{Tr}_{\mathrm{R}}$ *is given in Notation* 7.5.20.

Proof By induction on n. The basis is obvious, while the induction step is proved by induction on $A \in \mathbb{T}_\mu$. Again the basis is clear. For the induction step we distinguish cases.

Case $A \equiv A_1 \to A_2$. Then

$$\begin{aligned}
\mathcal{T}^{n+1}[\![(A_1 \to A_2)^*]\!]_\eta &= \mathcal{T}^{n+1}[\![(A_1)^* {\to} (A_2)^*]\!]_\eta\\
&= \mathcal{T}^n[\![(A_1)^*]\!]_\eta \Rightarrow^{n+1} \mathcal{T}^n[\![(A_2)^*]\!]_\eta\\
&= \mathcal{I}^n[\![A_1]\!]_\eta \Rightarrow^{n+1} \mathcal{I}^n[\![A_2]\!]_\eta\\
&= \mathcal{I}^{n+1}[\![A_1 \to A_2]\!]_\eta.
\end{aligned}$$

Case $A \equiv \mu\alpha.A_1$. Then

$$\mathcal{T}^{n+1}[\![(\mu\alpha.A_1)^*]\!]_\eta = \mathcal{T}^{n+1}[\![(A_1)^*[\alpha := (\mu\alpha.A_1)^*]]\!]_\eta.$$

Hence we can prove by cases on the possible forms of $(A_1)^*$ that

$$\mathcal{T}^{n+1}[\![(\mu\alpha.A_1)^*]\!]_\eta = \mathcal{T}^{n+1}[\![(A_1)^*]\!]_{\eta[\alpha\mapsto\mathcal{T}^n[\![(\mu\alpha.A_1)^*]\!]_\eta]}.$$

Then we have

$$\begin{aligned}\mathcal{T}^{n+1}[\![(\mu\alpha.A_1)^*]\!]_\eta &= \mathcal{T}^{n+1}[\![(A_1)^*]\!]_{\eta[\alpha\mapsto\mathcal{T}^n[\![(\mu\alpha.A_1)^*]\!]_\eta]}\\ &= \mathcal{I}^{n+1}[\![A_1]\!]_{\eta[\alpha\mapsto\mathcal{I}^n[\![\mu\alpha.A_1]\!]_\eta]}\\ &= \mathcal{I}^{n+1}[\![\mu\alpha.A_1]\!]_\eta,\end{aligned}$$

using the induction hypotheses on n and A_1. ∎

10.2.23 Proposition *Let $A, B \in \mathbb{T}_\mu$. Then for all type environments η one has*

$$A =^*_\mu B \Rightarrow \mathcal{I}[\![A]\!]_\eta = \mathcal{I}[\![B]\!]_\eta.$$

Proof Let $A, B \in \mathbb{T}_\mu$ and assume $A =^*_\mu B$. By Lemma 10.2.22 for all n, η

$$\begin{aligned}\mathcal{I}^n[\![A]\!]_\eta &= \mathcal{T}^n[\![A^*]\!]_\eta\\ &= \mathcal{T}^n[\![B^*]\!]_\eta\\ &= \mathcal{I}^n[\![B]\!]_\eta.\end{aligned}$$

Now the statement follows from Proposition 10.2.6(ii). ∎

As an immediate consequence we have the soundness of rule (equal) in Definition 7.1.3 and therefore, by a straightforward inductive argument, also the soundness of the typing rules of the system à la Curry with strong equality of types.

10.2.24 Corollary (Soundness of $\vdash_{\lambda\mu^*}$) *For all $\rho \in \mathsf{Env}_\mathcal{D}, \eta : \mathbb{A} \to \mathcal{CU}(D)$*

$$\Gamma \vdash_{\lambda\mu^*} M : A \Rightarrow \Gamma \models_{\mathcal{D},\rho,\mathcal{I}[\![-]\!]_\eta} M : A.$$

Proof Again by Proposition 10.1.3. ∎

Applications of soundness

One application of the type interpretations described so far, Proposition 10.2.25, is the following, easy proof of a standard result on definability by *untyped* terms, see B[1984], Exercise 19.4.5. Let $\mathcal{D}$ be a λ-model. An element $d \in \mathcal{D}$ is called λ-definable if there exists an $M \in \Lambda^\emptyset$ such that $d = [\![M]\!]^\mathcal{D}$.

Now let $\mathcal{D}$ be Scott's λ-model D_∞, see 16.3.1–16.3.24, based on D_0, $\langle i_0, j_0\rangle$, where

- D_0 is the two-point lattice $\{\bot_{\mathcal{D}_0}, \top_{\mathcal{D}_0}\}$;

- $\langle i_0, j_0\rangle$ is the so-called standard embedding–projection pair,

$$i_0(d) = \lambda e.d,\ j_0(f) = \bot_{D_0};$$

- $\nabla = \{\bot_{\mathcal{D}_0}\}$.

We view the sets D_n as subsets of D_∞ by identifying them with $\Phi_{n\infty}(D_n)$. Then the maps $\lambda d.d_n : D \to D$ form a notion of approximation for D. Note that $\top = \top_D = \langle \top_{D_n}\rangle$, hence $\top_n = \top_{D_n}$.

Here $\bot \in D$ is λ-definable, as $\bot = [\![\Omega]\!]$.

10.2.25 Proposition *In the standard D_∞ model, based on $D = \{\bot, \top\}$, the element $\top = \top_D$ is not λ-definable.*

Proof Let $L = \mu\alpha.\alpha \to \alpha$, and recall that every $M \in \Lambda^{\emptyset}$ has type L. Then $[\![M]\!] \in \mathcal{I}[\![L]\!]$, by soundness, Proposition 10.2.20. Now let $d = [\![M]\!]$. Then

$$d_0 = [\![M]\!]_0 \in (\mathcal{I}[\![L]\!])_0 = \mathcal{I}^0[\![L]\!] = \nabla = \{\bot_0\}.$$

But $\top_0 = \top_{D_0} \neq \bot_{D_0} = \bot_0$. Hence $d \neq \top$. ∎

Another application of Corollary 10.2.24 shows that types of a special form are inhabited only by unsolvable terms, in the sense of B[1984], §8.3. This generalizes the easy observation that, if $\vdash_{\lambda\mu^*} M : \mu\alpha.\alpha$ for a closed term M, then M is unsolvable. In fact $\mathcal{I}[\![\mu\alpha.\alpha]\!] = \nabla$, so $[\![M]\!] = \bot$ by soundness (Corollary 10.2.24), and in D any M such that $[\![M]\!] = \bot$ is unsolvable, see B[1984], Theorem 19.2.4(i).

Two lemmas about the standard D_∞ essentially need that $i_0(\top_{D_0}) = \top_{D_1}$.

10.2.26 Lemma

(i) $i_n(\top_n) = \top_{n+1}$, *for all* $n \geq 0$;
(ii) $\Phi_{n,\infty}(\top_n) = \top$, *for all* $n \geq 0$.

Proof (i) By induction on n.
(ii) By (i). ∎

Note that $D \notin \mathcal{CU}(D)$, because $D_0 \neq \nabla$. In the next lemma we show $D - \{\top\} \in \mathcal{CU}(D)$.

10.2.27 Lemma *Let $\overline{D} = D - \{\top\}$. Then $\overline{D} \in \mathcal{CU}(D)$.*

Proof First we show that $(\overline{D})_0 = \nabla$. Let $d \in \overline{D}$. Then $\Phi_{0\infty}(d_0) = \Phi_{0\infty} \circ \Phi_{\infty 0}(d_0) \sqsubseteq d$. But $\Phi_{0\infty}(\top_0) = \top$, by Lemma 10.2.26(ii). Hence $d_0 \neq \top_0$, so $d_0 = \bot_0 \in \nabla$.

Now we show that $\overline{D}$ is complete. Note that $d_0 = \bot_0$, for $d \in \overline{D}$, because if $d_0 = \top_0$, then $\Phi_{0\infty}(d_0) = \top$, but $\Phi_{0\infty}(d_0) \sqsubseteq d \neq \top$. Now let $e = \bigsqcup_D \overline{D}$. We must show that $e \in \overline{D}$, i.e. $e \neq \top$. It suffices to show $e_0 \neq \top_0$. Indeed

$$\begin{aligned} e_0 &= \Phi_{\infty 0}(e) = \Phi_{\infty 0}(\sup\{d \mid d \in \overline{D}\}) \\ &= \sup\{\Phi_{\infty 0}(d) \mid d \in \overline{D}\} = \sup\{d_0 \mid d \in \overline{D}\} \\ &= \sup\{\bot_0\} = \bot_0 \neq \top_0. \blacksquare \end{aligned}$$

10.2.28 Theorem *Let M be a closed term. Suppose for $n, m \geq 0$*

$$\vdash_{\lambda\mu^*} M : \alpha_1 \to \cdots \to \alpha_n \to \mu\alpha.(\beta_1 \to \cdots \to \beta_m \to \alpha).$$

Then M is unsolvable.

Proof Let η be the type environment such that $\eta(\gamma) = \overline{D}$ for all atoms γ. We claim that

$$(\overline{D} \Rightarrow \cdots \Rightarrow \overline{D} \Rightarrow \overline{D}) \subsetneq \overline{D}. \tag{10.10}$$

As $\top \notin \overline{D} \Rightarrow \cdots \Rightarrow \overline{D} \Rightarrow \overline{D}$, any element of $\overline{D} \Rightarrow \cdots \Rightarrow \overline{D} \Rightarrow \overline{D}$ is also an element of $\overline{D}$. In order to show that the inclusion is strict, let the *step function* $(e \mapsto e')$, for compact elements $e, e' \in D$, be defined by

$$(e \mapsto e')(d) = \begin{cases} e', & \text{if } e \sqsubseteq d\,, \\ \bot, & \text{otherwise.} \end{cases}$$

Now consider

$$(\vec{e} \mapsto \top) \triangleq (e_1 \mapsto (e_2 \mapsto \cdots \mapsto (e_p \mapsto \top) \cdots)),$$

where $e_1, \ldots, e_p \in \overline{D}$ are compact elements, with $e_1 \neq \bot$: then $(\vec{e} \mapsto \top)(\bot) = \bot$, hence $(\vec{e} \mapsto \top) \in \overline{D}$, but $(\vec{e} \mapsto \top) \notin (\overline{D}^m \Rightarrow \overline{D})$.

Now, assume towards a contradiction that M is solvable. Then $M\vec{N} = \mathsf{I}$, for some closed $\vec{N} = N_1, \ldots, N_\ell$. Hence $[\![M]\!] \cdot [\![N_1]\!] \cdots [\![N_\ell]\!] = [\![\lambda x.x]\!]$. Note that $[\![N_i]\!] \in \overline{D}$, by Proposition 10.2.25. Let

$$T \equiv \alpha_1 \to \cdots \to \alpha_n \to \mu\alpha.(\boldsymbol{\beta}_1 \to \cdots \to \boldsymbol{\beta}_m \to \alpha).$$

Case $m = 0$. If $\ell \geq n$ then $[\![M]\!] \cdot [\![N_1]\!] \cdots [\![N_n]\!] \in \nabla$, which is impossible because then $[\![MN_1 \cdots N_n]\!] = \bot$, so $MN_1 \cdots N_n$ is unsolvable. Otherwise, $\ell < n$ and

$$[\![\lambda x.x]\!] \in \overline{D}^{n-\ell} \Rightarrow \nabla.$$

Then $[\![(\lambda x.x)\mathsf{I}^{n-\ell}]\!] \in \nabla$, hence $(\lambda x.x)\mathsf{I}^{n-\ell} = \mathsf{I}$ would be unsolvable, again a contradiction.

Case $m > 0$. We can assume that $\ell < n$, otherwise take some μ-unfoldings of T. Now

$$[\![\lambda x.x]\!] \in (\overline{D}^{n-\ell} \Rightarrow \mathcal{I}[\![\mu\alpha.(\boldsymbol{\beta}_1 \to \cdots \to \boldsymbol{\beta}_m \to \alpha)]\!]_\eta)$$

because, for $i = 1, \ldots, \ell$, we have $[\![N_i]\!] \in \overline{D} = \mathcal{I}[\![\alpha_i]\!]_\eta$. By applying $[\![\lambda x.x]\!]$ to $\mathsf{I}^{\sim(n-(\ell+1))}d$, we get

$$\forall d \in \overline{D}. d \in \mathcal{I}[\![\mu\alpha.\boldsymbol{\beta}_1 \to \cdots \to \boldsymbol{\beta}_m \to \alpha]\!]_\eta.$$

Therefore $\mathcal{I}[\![\mu\alpha.\boldsymbol{\beta}_1 \to \cdots \to \boldsymbol{\beta}_m \to \alpha]\!]_\eta = \overline{D}$. But this is impossible because then

$$\begin{aligned}\mathcal{I}[\![\mu\alpha.\boldsymbol{\beta}_1 \to \cdots \to \boldsymbol{\beta}_m \to \alpha]\!]_\eta &= \overline{D}^m \Rightarrow \mathcal{I}[\![\mu\alpha.\boldsymbol{\beta}_1 \to \cdots \to \boldsymbol{\beta}_m \to \alpha]\!]_\eta \\ &= \overline{D}^m \Rightarrow \overline{D} \\ &\subsetneq \overline{D}, \quad \text{by (10.10). } \blacksquare\end{aligned}$$

Completeness

Partial converses to the above soundness results were proved in Coppo (1985) (see also Cardone and Coppo (1991)) for an interpretation of types in a domain D of the shape $\mathbb{A} + [D \to D]$. Now, the interpretation in $\mathcal{D}$ of an unsolvable λ-term in topological models is $\perp$, see B[1984], §§8.3, 19.2. However, $\perp$ is an element of every complete and uniform subset; therefore, if M is such a term, it is true that $[\![M]\!]_\rho \in \mathcal{I}[\![A]\!]_\eta$ for any term environment ρ, any type A and any type environment η. The incompleteness of all type inference systems presented above, in particular of $\vdash_{\lambda\mu^*}$, becomes apparent from just considering the λ-term $\Delta_3\Delta_3$, where $\Delta_3 \triangleq \lambda x.xxx$, which is unsolvable of order 0, yet has principal type scheme $\mu\alpha.\alpha \to \alpha$. Therefore the system $\vdash_{\lambda\mu^*}$ is incomplete.

Clearly this cannot be remedied by adding to $\vdash_{\lambda\mu^*}$ the following rule giving the same types to $(\beta\eta)$-convertible terms:

$$(Eq) \quad \frac{\Gamma \vdash M : A \quad M =_{\beta\eta} N}{\Gamma \vdash N : A}.$$

In Definition 10.2.31, however, we extend the system $\vdash_{\lambda\mu^*}$ to a complete system as follows. The system introduced in Coppo (1985) exploits the notion of *approximant* of a λ-term in the formulation of an infinitary rule that assigns a type to a term when all its approximants can be assigned that type.

10.2.29 Definition (Approximants of λ-terms) Let $\Lambda\perp$ be the set of λ-terms with a new constant $\perp$.

$$
\begin{array}{|ll|}
\hline
(\bot) & \Gamma \vdash \bot : A \\
(C) & \dfrac{\Gamma \vdash P : A \ \text{ for all } P \in \mathcal{A}(M)}{\Gamma \vdash M : A} \\
\hline
\end{array}
$$

Figure 10.1 Extra axiom and rule for $\lambda\mu^{*\infty}$

(i) For $M \in \Lambda$, define recursively the *direct approximant* $\omega(M) \in \Lambda\bot$ of M:

(1) $\omega(x) = x$ if x is a variable;
(2) $\omega(\lambda x.M) = \lambda x.\omega(M)$ if $\omega(M) \neq \bot$, otherwise $\omega(\lambda x.M) = \bot$;
(3) $\omega(xM_1 \cdots M_k) = x\omega(M_1) \cdots \omega(M_k)$, for $k > 0$;
(4) $\omega((\lambda x.M)M_1 \cdots M_k) = \bot$, for $k > 0$.

(ii) A term $P \in \Lambda\bot$ is called an *approximant* of a λ-term M if $P = \omega(N)$ for some $N \in \Lambda$ with $M \twoheadrightarrow_\beta N$.
(iii) The set of *approximants*, written $\mathcal{A}(M)$, of the λ-term M, is defined by

$$\mathcal{A}(M) \triangleq \{P \in \Lambda\bot \mid P \text{ is approximant of } M\}.$$

A classical result is the Approximation Theorem by Wadsworth (1976), stating that the following property (10.11) holds for Scott's $\mathcal{D}_\infty$ models.

10.2.30 Definition Let $\mathcal{D}$ be a λ-model.

(i) Elements of $\Lambda\bot$ are interpreted in $\mathcal{D}$ by interpreting $\bot$ as the smallest element of D.
(ii) If for all $M \in \Lambda$ and term environments ρ one has

$$[\![M]\!]_\rho = \bigsqcup\{[\![P]\!]_\rho \mid P \in \mathcal{A}(M)\}, \tag{10.11}$$

then we say $\mathcal{D}$ satisfies the *approximation property*.

We now introduce the promised complete extension of $\vdash_{\text{BH}}$.

10.2.31 Definition The type inference system $\lambda\mu^{*\infty}$ is defined by adding to the system $\vdash_{\lambda\mu^*}$ the rules in Fig. 10.1.

For the resulting system $\lambda\mu^{*\infty}$ it is possible to prove a form of completeness.

10.2.32 Theorem *Let $\mathcal{D}$ range over λ-models satisfying* (10.11). *Then*

$$\Gamma \vdash_{\lambda\mu^{*\infty}} M : A \Leftrightarrow \forall \mathcal{D}\, [\Gamma \models_{\mathbb{T}_{\mu^*},\mathcal{D}} M : A]$$

see Definition 10.1.2(iv) *and Lemma* 7.5.21(iii).

A proof of this result and a thorough discussion of the system $\lambda\mu^{*\infty}$ is contained in Cardone and Coppo (1991), §4.

More directly relevant to applications is another completeness result, which states that $=^*_\mu$ completely describes the equivalence which identifies two recursive types whenever their interpretations are identical in all type environments: this provides a strong semantical motivation for the use of $=^*_\mu$ as the preferred notion of equality of recursive types. This can be proved following the same idea as in Coppo (1985) and Cardone and Coppo (1991), by using a domain $D \cong \mathbf{A} + [D \to D]$, where $\mathbf{A}$ is a CPO whose elements are basic values, for example the flat CPO of integers or of boolean values.

The following result on the *equivalence of recursive types* will be obtained as Theorem 11.1.39.

10.2.33 Theorem *Let $A, B \in \mathbb{T}_\mu$. Then*

$$A =^*_\mu B \Leftrightarrow \forall \eta. [\mathcal{I}[\![A]\!]_\eta = \mathcal{I}[\![B]\!]_\eta].$$

10.3 Type interpretations in systems with explicit typing

One straightforward way of interpreting the explicitly typed calculi with recursive types consists in restricting the full type structure introduced in Section 3.1 by replacing sets with domains and functions with (Scott) continuous functions. Indeed, there is a sense in which domain theory and the solution of recursive domain equations can be regarded as a purely semantical theory of recursive types. The leading idea is to define $\mathcal{M}(A \to B)$ as the domain of continuous functions from $\mathcal{M}(A)$ to $\mathcal{M}(B)$; a recursive type $\mu\alpha.A$ is interpreted as the solution of the domain equation $D \cong F(D)$, where F interprets the functor over domains defined by (the interpretation of) A: such a solution can be built, for example, as the limit of an inverse sequence of approximations, following Scott (1972).

In order to overcome some pathologies that arise in the resulting model, for example the fact that for many natural choices of categories of domains the interpretation of type $\mu\alpha.\alpha \to \alpha$ collapses to one point, alternative interpretations are available.

First, following Scott (1976), we can interpret types as closures over the λ-model $\mathcal{P}\omega$. Such a construction yields far more than a model for simple recursive types: indeed, it has been used in McCracken (1979) and Bruce et al. (1990) to interpret the polymorphic λ-calculus and in Barendregt and Rezus (1983) to give a model for calculi with dependent types and a type of all types.

On the other hand, types can be interpreted as partial equivalence relations (more precisely, as complete and uniform partial equivalence relations, PERs) over a continuous λ-model with a notion of approximation, along the lines of Section 10.2. Such models are discussed in many references, among them Amadio (1991), Cardone (1989), Abadi and Plotkin (1990). We shall sketch these model constructions later, see Exercises 10.11, 10.12 and 10.13. We also mention in passing other models that have been proposed in the literature, mostly based on operational interpretations that have not been dealt with at all in this Part: see Birkedal and Harper (1999), Appel and McAllester (2001) and Vouillon and Melliès (2004).

In this section we focus in particular on the system $\lambda_\mu^{\text{Ch}_0}$ in which all possible recursive types are present. The advantage of this system is that the terms $fold_{\mu\alpha.A}$ and $unfold_{\mu\alpha.A}$ afford an explicit notation for the isomorphisms provided by the solution of recursive domain equations and formalize the intuitive idea that a recursive type is interpreted as such a solution.

We assume below that $\mathbb{A}$ is a collection of basic types containing at least one element with a non empty interpretation.

Domain models

Solving recursive domain equations

The sketch below summarizes the results that we shall need for the model construction; complete details can be found in Smyth and Plotkin (1982) or Amadio and Curien (1998), §7.1.

10.3.1 Definition We associate to the category **CPO** a category $\mathbf{CPO}^{ep}$ whose morphisms from D to E are the embedding-projection pairs $\langle e : D \to E, p : E \to D\rangle$; that is, the pairs of continuous functions such that

$$p \circ e = 1_D \quad \text{and} \quad e \circ p \sqsubseteq 1_E.$$

Observe that, if $\langle e_1, p_1\rangle : D_1 \longrightarrow E_1$ and $\langle e_2, p_2\rangle : D_2 \longrightarrow E_2$ are embedding–projection pairs, then we have an embedding–projection pair

$$\langle [p_1 \to e_2], [e_1 \to p_2]\rangle : [D_1 \to D_2] \longrightarrow [E_1 \to E_2],$$

where $[p_1 \to e_2](h)$ is $e_2 \circ h \circ p_1$ for all $h : D_1 \to D_2$ and $[e_1 \to p_2](k)$ is $p_2 \circ k \circ e_1$ for all $k : E_1 \to E_2$.

In what follows, $F : \mathbf{CPO}^{ep} {\to} \mathbf{CPO}^{ep}$ is a functor that is continuous, i.e. preserves direct limits of chains of embeddings of the form (10.13) below. This F associates to each $D \in \mathbf{CPO}$ an element $F(D)$ and to each $\langle e, p\rangle : D {\to} E$ a pair $F(\langle e, p\rangle) : F(D) {\to} F(E)$. Note that e and p determine each other completely. Hence if $F(\langle e, p\rangle) = \langle e', p'\rangle$, then e' is completely determined by e and p' by p. Therefore we write loosely $F(e) = e'$ and $F(p) = p'$.

The solution of an equation $D \cong F(D)$ will be obtained as the inverse limit of a sequence of projections:

$$D_0 \xleftarrow{j_0} D_1 \xleftarrow{j_1} D_2 \xleftarrow{j_2} \cdots \xleftarrow{j_{n-1}} D_n \xleftarrow{j_n} D_{n+1} \xleftarrow{j_{n+2}} \cdots, \qquad (10.12)$$

or, equivalently, as the direct limit of the sequence of embeddings:

$$D_0 \xrightarrow{i_0} D_1 \xrightarrow{i_1} D_2 \xrightarrow{i_2} \cdots \qquad D_n \xrightarrow{i_n} D_{n+1} \xrightarrow{i_{n+1}} \cdots, \qquad (10.13)$$

where

$$\begin{array}{rcll} D_0 & = & \mathbf{1}, & \text{the one-element CPO,} \\ D_{n+1} & = & F(D_n) & \\ \langle i_0, j_0 \rangle & = & \langle \bot_{D_0 \to D_1}, \bot_{D_1 \to D_0} \rangle. & \end{array}$$

Compare this with Section 16.3, where the case $F(D) \cong [D \to D]$ is treated, starting from a general D_0. For the present D_0, the pair $\langle i_0, j_0 \rangle$ is the only morphism from D_0 to D_1 and the pairs $\langle i_n, j_n \rangle$, for $n > 0$, are embedding–projection pairs from D_n to D_{n+1} and are defined recursively by

$$i_n = F(i_{n-1}) : F(D_{n-1}) \longrightarrow F(D_n)$$
$$j_n = F(j_{n-1}) : F(D_n) \longrightarrow F(D_{n-1}).$$

The direct limit of the sequence (10.13) can be described explicitly as

$$D_\infty \triangleq \underrightarrow{\lim} \langle D_n, i_n \rangle = \{ \langle d_n \rangle_{n \in \omega} \mid \forall n \in \omega . d_n \in D_n \text{ and } j_{n+1}(d_{n+1}) = d_n \},$$

with embeddings $i_{n\infty} : D_n \to D_\infty$ and projections $j_{\infty n} : D_\infty \to D_n$, for all $n \in \omega$, making all the diagrams

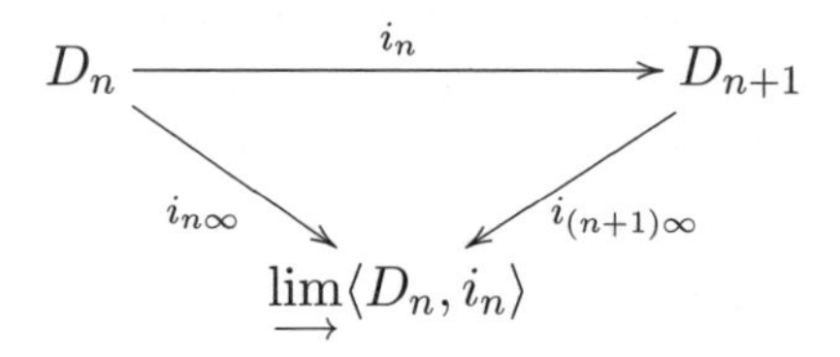

commute. From the continuity of F and universality we obtain:

10.3.2 Proposition *There is a unique isomorphism $\vartheta : F(D_\infty) \longrightarrow D_\infty$, given by*

$$\bigsqcup_{n \in \omega} \left(i_{(n+1)\infty} \circ F(j_{\infty n}) \right)$$

whose inverse ϑ^{-1} *is*

$$\bigsqcup_{n\in\omega}\left(F(i_{n\infty})\circ j_{\infty(n+1)}\right).$$

Note that the suprema do indeed exist, because $\langle i_{(n+1)\infty}\circ F(j_{\infty n})\rangle$ and $\langle F(i_{n\infty})\circ j_{\infty(n+1)}\rangle$ are chains. This follows from the fact that the $\langle F(i_n), F(j_n)\rangle$ are morphisms in $\mathbf{CPO}^{ep}$, hence $F(i_n)\circ F(j_n)\sqsubseteq \mathsf{Id}_{F(D_{n+1})}$.

For a proof, we refer the reader to Plotkin (1982), Chapter 5. The construction is based on the observation that all the following diagrams (in $\mathbf{CPO}^{ep}$) commute:

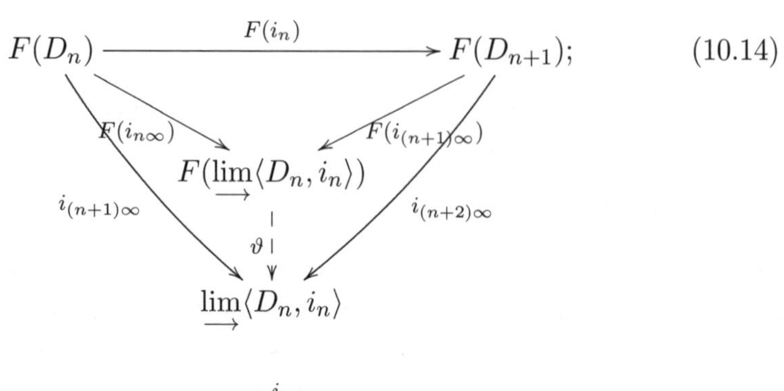

(10.14)

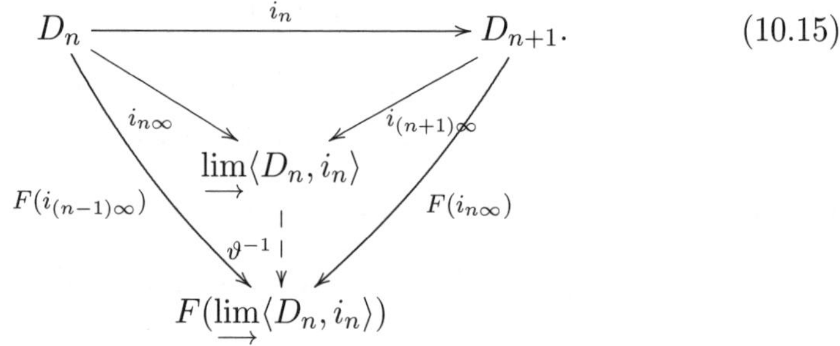

(10.15)

Interpreting μ*-types as domains*

With the machinery set up for the solution of recursive domain equations we can now easily define a typed applicative structure, see Definition 3.1.1, by interpreting each $A\in\mathbb{T}^1_\mu$, with atoms $\mathbb{A}$, as $\mathcal{M}^{\text{CPO}}(A)\in\mathbf{CPO}$, by induction on A. If η maps $\mathbb{A}$ to the objects of $\mathbf{CPO}$, then it can be extended to an interpretation of $\mathbb{T}_\mu$ as follows:

(a) $\mathcal{M}^{\text{CPO}}_\eta(\alpha)=\eta(\alpha)$;
(b) $\mathcal{M}^{\text{CPO}}_\eta(A\to B)=[\mathcal{M}^{\text{CPO}}_\eta(A)\to\mathcal{M}^{\text{CPO}}_\eta(B)]$;
(c) $\mathcal{M}^{\text{CPO}}_\eta(\mu\alpha.A)=\varinjlim\langle F^n(\mathbf{1}), i_n\rangle$, where $F(D)=\mathcal{M}^{\text{CPO}}_{\eta[\alpha\mapsto D]}(A)$.

The above clauses, especially (c), are justified by the theory developed in Lehmann and Smyth (1981), §4.

For morphisms $\langle e, p\rangle : D \to E$ we give the definition of $F(\langle e, p\rangle)$ by a few examples:

- if $A = \beta$, then $F(D) = F(E) = \eta(\beta)$ and $F(\langle e, p\rangle) = \langle \mathbf{1}, \mathbf{1}\rangle$;
- if $A = \alpha$, then $F(D) = D$, $F(E) = E$ and $F(\langle e, p\rangle) = \langle e, p\rangle$;
- if $A = \alpha{\to}\beta$, then $F(D) = [D{\to}\eta(\beta)]$, $F(E) = [E{\to}\eta(\beta)]$, where $[D{\to}E]$ is the CPO of continuous functions from D to E; see Definition 10.1.9 and $F(\langle e, p\rangle) = \langle [p{\to}\mathbf{1}], [e{\to}\mathbf{1}]\rangle$.

Summarizing we have the following result.

10.3.3 Theorem $\mathcal{M}^{CPO}$ *is an applicative type structure.*

We can straightforwardly extend Definition 3.1.9 and define an interpretation of every term $M \in \Lambda^{\mathrm{Ch}}_{\mathbb{T}_\mu}(A)$. In particular, define

$$\begin{aligned} \mathit{fold}_{\mu\alpha.A} &\triangleq \theta : F(\mathcal{M}^{\mathrm{CPO}}_\eta(\mu\alpha.A)) \longrightarrow \mathcal{M}^{\mathrm{CPO}}_\eta(\mu\alpha.A) \\ \mathit{unfold}_{\mu\alpha.A} &\triangleq \theta^{-1} : \mathcal{M}^{\mathrm{CPO}}_\eta(\mu\alpha.A) \longrightarrow F(\mathcal{M}^{\mathrm{CPO}}_\eta(\mu\alpha.A)), \end{aligned}$$

where $F(D)$ is defined, as above, as $F(D) \triangleq \mathcal{M}^{\mathrm{CPO}}_{\eta[\alpha\mapsto D]}(A)$, and θ and θ^{-1} are as in diagrams (10.14) and (10.15), respectively. It can immediately be verified that this interpretation is well defined, and not trivial.

10.3.4 Theorem $\mathcal{M}^{CPO}$ *is a typed λ-model that models the coercions and their reductions.*

The following result is a typical application of the interpretation of terms of $\lambda^{\mathrm{Cho}}_\mu$ outlined above. It relates the interpretation of the typed fixed-point combinator Y^A_μ of Exercise 7.15 to the iterative construction of fixed-points of continuous endofunctions of a CPO D (see Theorem 10.1.11). It is similar to a well-known result of Park (1976) on the interpretation of the $\mathbf{Y}$ combinator in Scott's D_∞ models, and was stated for the typed case in Cosmadakis (1989).

10.3.5 Theorem *The interpretation of Y^A_μ of type $(A \to A) \to A$ coincides, for any environment η, with the functional*

$$\mathbf{fix} \in [\mathcal{M}^{CPO}_\eta(A) \to \mathcal{M}^{CPO}_\eta(A)] \to \mathcal{M}^{CPO}_\eta(A).$$

Proof A straightforward analysis of the interpretation of recursive types in the category **CPO**, whose details are left as an exercise. ∎

Since $\mathbb{T}^{\mathbb{A}}_\mu$ is an enrichment of the set of simple types $\mathbb{T}$, the interpretation of $\mathbb{T}^{\mathbb{A}}_\mu$ yields a **CPO** model of simple types. Observe that the interpretation in $\mathcal{M}^{\rm CPO}$ of types such as, for example, $\mu\alpha.\alpha \to \alpha$ is trivial: its interpretation in $\mathcal{M}^{\rm CPO}$ has only one element. This semantical phenomenon suggests immediately that, at the operational level, terms of trivial type should be observationally equivalent. Programming languages generally possess a notion of *observable type*, for example integers or booleans. Two closed terms M, N of the same type A are *observationally equivalent* if, for any context $C[\ \ ]$ bringing a term of type A into an observable type $\alpha \in \mathbb{A}$,

$$C[M] \text{ has a value iff } C[N] \text{ has the same value.}$$

A *trivial type* is a type such that any two inhabitants are observationally equivalent. Somewhat surprisingly, we shall see that there are trivial types in $\mathbb{T}_\mu$. Indeed we can state for trivial types a result similar to the Genericity Lemma for the pure type-free λ-calculus B[1984], Proposition 14.3.24.

10.3.6 Definition (Trivial types) A type T is *trivial* if

(i) either $T =_\mu \mu\alpha_1 \cdots \mu\alpha_k.A_1 \to \cdots \to A_n \to \alpha_i$ $(k \geq 1, n \geq 0, 1 \leq i \leq k)$,
(ii) or $T =_\mu S \to T'$ where T' is trivial.

10.3.7 Lemma *let A be a trivial type. Then $\mathcal{M}^{CPO}(A) \cong \mathbf{1}$.*

Proof By induction on the structure of types, observing that $\mathcal{M}^{\rm CPO}(A) \cong \mathbf{1}$ and $A =_\mu A'$ imply that $\mathcal{M}^{\rm CPO}(A') \cong \mathbf{1}$ by a straightforward induction on the proof that $A =_\mu A'$. First, $A \equiv \mu\alpha_1 \cdots \mu\alpha_k.A_1 \to \cdots \to A_n \to \alpha_i$. Then $A =_\mu \mu\alpha_i.A_1' \to \cdots \to A_n' \to \alpha_i$, and $\mathcal{M}^{\rm CPO}(\mu\alpha_i.A_1' \to \cdots \to A_n' \to \alpha_i) \cong \mathbf{1}$ because, in **CPO**, we have $[D \to \mathbf{1}] \cong \mathbf{1}$. Therefore $\mathcal{M}^{\rm CPO}(A) \cong \mathbf{1}$. Now let $A \equiv S \to T$, with T trivial. Then $\mathcal{M}^{\rm CPO}(T) \cong \mathbf{1}$ by the induction hypothesis. Then also $\mathcal{M}^{\rm CPO}(S \to T) \cong \mathbf{1}$. ∎

Given a set $\mathcal{T}$ of equations between closed terms of the same type, we write $\mathcal{T} \vdash M = N$ if and only if the equation $M = N$ follows from (the typed version of) β-conversion using the equations in $\mathcal{T}$ as further axioms: this is called the *typed λ-theory generated by $\mathcal{T}$*. A theory with a set $\mathcal{T}$ of equations is *consistent* if there is a type A and terms M, N of type A such that $\mathcal{T} \nvdash M = N$. We can now state an easy result that shows that it is consistent to identify terms of the same trivial type.

10.3.8 Corollary *The following theory $\mathcal{T}$ is consistent:*

$$\mathcal{T} = \{M = N \mid M, N \in \Lambda^{\rm Ch}_{\mathbb{T}_\mu}(B) \textit{ and } B \textit{ is trivial}\}.$$

Proof By Lemma 10.3.7 one has $\mathcal{M}^{\rm CPO} \models \mathcal{T}$. Let $A \equiv \alpha \to \alpha \to \alpha$ and $M \equiv \lambda x^\alpha.\lambda y^\alpha.x$, $N \equiv \lambda x^\alpha.\lambda y^\alpha.y$. Let $\eta(\alpha)$ be a non-trivial CPO. Then

$[\![M]\!] \neq [\![N]\!]$ in $\eta(\alpha) \Rightarrow \eta(\alpha) \Rightarrow \eta(\alpha)$. Then $\mathcal{M}^{\text{CPO}} \not\models M = N$, hence $\mathcal{T} \not\vdash M = N$. Therefore $\mathcal{T}$ is consistent. ∎

In the following it will be often useful to decorate subterms with their types. Let us say that a subterm N^T of a term M, where T is a trivial type, is *maximally trivial* if it is not a subterm of another subterm of M of trivial type. If M is any typed term let $\overline{M}$ be the term obtained by replacing all maximal trivial subterms of M of type T by Ω^T (as shown in Example 7.1.4 there are typed versions of Ω at all types).

10.3.9 Lemma *Let $M \to_\beta M'$. Then $\overline{M} \to_\beta^= \overline{M'}$, where $\to_\beta^=$ denotes one or zero reduction steps.*

Proof A straightforward structural induction on M. The interesting case is $M^A \equiv ((\lambda x^T.P^A)^{T\to A}Q^T)^A \to_\beta P[x^T := Q^T]^A$, with T trivial and A not trivial, so that Q^T is maximally trivial in M. We immediately have $\overline{((\lambda x^T.\overline{P^A})^{T\to A}\overline{Q^T})^A \to_\beta P[x^T := \overline{Q^T}]^A}$, observing that both x and Q are of the same trivial type T. ∎

A type $S \in \mathbb{T}_\mu$ is *hereditarily non-trivial* if no trivial type is a subexpression of S. Say that an occurrence of a subterm P of a normalizable term M is *useless* if M has the same β-normal form whenever P is replaced by any other term of the same type.

10.3.10 Lemma *Let S be hereditarily non-trivial. For any closed normalizable term $M \in \Lambda^{\text{Ch}}_{\mathbb{T}_\mu}(S)$, any occurrence in M of a subterm P of trivial type T is useless.*

Proof Note that if M is in normal form (i.e. $M \equiv \lambda\vec{x}.x_iM_1\cdots M_m$, with all M_i in normal form), the types of all its subterms are hereditarily non-trivial. In fact, $S \equiv S_1 \to S_2 \to \cdots \to S_n \to B$, and $S_i \equiv B_1^i \to B_2^i \to \cdots \to B_m^i \to B$, so $B_1^i, \dots, B_m^i, B$ are hereditarily non-trivial because they are subexpressions of a hereditarily non-trivial type. Therefore, the types of $x_i, M_1, \dots, M_m$ are hereditarily non-trivial and this holds, recursively, for the subterms of $M_1, \dots, M_m$. Thus it is impossible for a normal form to have a subterm of trivial type.

If M is not in normal form let M' be a term obtained from M by replacing a subterm of a trivial type T by any other subterm of the same type. Note that $\overline{M} = \overline{M'}$ and apply Lemma 10.3.9. ∎

10.4 Exercises

Exercise 10.1 Let $\mathcal{D}$ be a CPO.

(i) Show that a subset $O \subseteq \mathcal{D}$ is closed iff it is closed downwards and closed under sups of directed sets.
(ii) Show that continuous functions are monotonic.

Exercise 10.2 Let $\mathcal{D}$ be a complete lattice.

(i) Show that $\mathcal{D}$ has a *top* element $\top$ such that $\forall d \in \mathcal{D}.d \sqsubseteq \top$.
(ii) Show that $\mathcal{D}$ has arbitrary *infima* and the *sup* and *inf* of two elements.
(iii) Show that $\bot_\mathcal{D}$ is compact and that $d, e \in \mathcal{K}(\mathcal{D}) \Rightarrow d \sqcup e \in \mathcal{K}(\mathcal{D})$.
(iv) Show that, in general, it is not true that if $d \sqsubseteq e \in \mathcal{K}(\mathcal{D})$, then $d \in \mathcal{K}(\mathcal{D})$. [Hint. Take $\omega+1$ in the ordinal $\omega+\omega = \{0, 1, 2, \ldots\ \omega, \omega+1, \omega+2, \ldots\}$. It is compact, but $\omega\ \ (\sqsubseteq \omega+1)$ is not.]

Exercise 10.3 Let $\mathcal{D}$ be a complete lattice.

(i) Show that every monotonic $f : \mathcal{D} \to \mathcal{D}$ has a unique smallest fixed point. [Hint. Take $x = \bigsqcap\{z \mid f(z) \sqsubseteq z\}$.]
(ii) In the case that f is also continuous, show that the smallest fixed point is $\mathbf{fix}(f)$.

Exercise 10.4 Fill in the details of Theorem 10.1.11.

Exercise 10.5 Let A be a non-empty set. Denote Engeler's model by D_A. Define it as follows: D_A has as underlying set $\mathcal{P}(D^*)$, where $D^* = \bigcup_{n \in \omega} D_n$ and

$$\begin{aligned} D_0 &= \emptyset \\ D_1 &= A \\ D_{n+2} &= D_{n+1} \cup \{(\beta, m) \mid m \in D_{n+1}, \beta \subseteq_{\text{fin}} D_{n+1}\}. \end{aligned}$$

Thus, D^* is the smallest set X containing A such that if β is a finite subset of X and $m \in X$, then $(\beta, m) \in X$. The complete lattice $D_A = \langle \mathcal{P}(D^*), \subseteq \rangle$ can be turned into a reflexive structure by defining, for $d, e \in D_A$

$$d \cdot e = \{m \in D^* \mid \exists \beta \subseteq_{\text{fin}} e.(\beta, m) \in d\}.$$

Show that $d_n \triangleq d \cap D_n$ defines a notion of approximation for D_A.

Exercise 10.6 Prove Lemma 10.2.14.

Exercise 10.7 Prove Lemma 10.2.18(ii). [Hint. Prove by induction that

$$\forall n \in \omega.\mathcal{I}^n[\![A[\alpha := B]]\!]_\eta = \mathcal{I}^n[\![A]\!]_{\eta[\alpha \mapsto \mathcal{I}[\![B]\!]_\eta]}.]$$

Exercise 10.8 (Bekic's Theorem for recursive types) Prove that the equation

$$\mu\beta.(A[\alpha := \mu\alpha.B]) = (\mu\beta.A)[\alpha := \mu\alpha.(B[\beta := \mu\beta.A])],$$

is valid in the interpretation described in Section 10.2. [Hint. Let $\hat{A} \triangleq A[\alpha := \mu\alpha.B]$ and $\hat{B} \triangleq B[\beta := \mu\beta.A]$. Then prove simultaneously by induction on $n \in \omega$ that the following two equations hold:

(i) $\mathcal{I}^n[\![\mu\beta.\hat{A}]\!]_\eta = \mathcal{I}^n[\![\mu\beta.A[\alpha := \mu\alpha.\hat{B}]]\!]_\eta$;
(ii) $\mathcal{I}^n[\![\mu\alpha.\hat{B}]\!]_\eta = \mathcal{I}^n[\![\mu\alpha.B[\beta := \mu\beta.\hat{A}]]\!]_\eta$.]

Exercise 10.9 (Metric interpretation of contractive recursive types) For $X, Y \in \mathcal{CU}$, let

$$\delta(X,Y) = \begin{cases} \max\{n \mid X_n = Y_n\} & \text{if } X \neq Y \\ \infty & \text{otherwise} \end{cases}$$

Define a function $d : \mathcal{CU} \times \mathcal{CU} \to \mathbf{R}^+$ by setting

$$d(X,Y) = \begin{cases} 2^{-\delta(X,Y)}, & \text{if } X \neq Y; \\ 0, & \text{otherwise.} \end{cases}$$

Show the following (see Remark 7.6.3 and Theorem 7.6.5):

(1) $\langle \mathcal{CU}, d\rangle$ is a complete ultrametric space;
(2) the functions: $\boldsymbol{\lambda} X \in \mathcal{CU}.X \Rightarrow Y$ for a fixed $Y \in \mathcal{CU}$; $\boldsymbol{\lambda} Y \in \mathcal{CU}.X \Rightarrow Y$, for a fixed $X \in \mathcal{CU}$; and $\boldsymbol{\lambda} X \in \mathcal{CU}.X \Rightarrow X$, are contractive, where the function $\Rightarrow : \mathcal{CU} \times \mathcal{CU} \to \mathcal{CU}$ is given in Definition 10.1.1;
(3) for $\mu\alpha.A \in \mathbb{T}_\mu$ define

$$[\![\mu\alpha.A]\!]_\eta = \mathbf{fix}(\boldsymbol{\lambda} X \in \mathcal{CU}.[\![A]\!]_{\eta[\alpha \mapsto X]}).$$

Use these facts to give an interpretation of all $A \in \mathbb{T}_\mu$. [Hint. See MacQueen et al. (1986), Amadio (1991).]

Exercise 10.10 Show that $\vdash_{\lambda\mu^*} \lambda y.y((\lambda x.x\,x)(\lambda x.x\,x)) : \mu\alpha.T_\alpha \to \alpha$, where $T_\alpha = \mu\beta.\beta \to (\beta \to \alpha)$. [Recall that $(\lambda x.x\,x)(\lambda x.x\,x)$ has all types.]

Exercise 10.11 A *partial equivalence relation* (PER) over D is any symmetric and transitive relation $R \subseteq D \times D$.

Let D be a λ-model with a notion of approximation and R be a PER over D. Let $R_n = \{\langle d_n, e_n\rangle \mid \langle d, e\rangle \in R\}$. We say that R is complete and uniform over D if we have:

- $R_0 = \{\langle d, d\rangle \mid d \in D_0\}$;

- (*Completeness*) if, for $\{\langle d^{(1,j)}, d^{(2,j)}\rangle \mid j\in\omega\}$ an increasing chain,

$$\left(\forall j\in\omega.\langle d^{(1,j)}, d^{(2,j)}\rangle\in R\right) \Rightarrow \langle \bigsqcup_{j\in\omega} d^{(1,j)}, \bigsqcup_{j\in\omega} d^{(2,j)}\rangle\in R;$$

- (*Uniformity*) $\langle d^1, d^2\rangle\in R \Rightarrow \forall n\in\omega\langle d^1_n, d^2_n\rangle\in R.$

Generalize Definition 10.2.13 to the case where types are interpreted as complete and uniform partial equivalence relations over a λ-model D with a notion of approximation.

Exercise 10.12 Use the interpretation of types as complete and uniform partial equivalence relations of Exercise 10.11 to build a model of $\lambda^{\mathrm{Cho}}_{\mu}$. [Hint. The main steps are:

(i) Prove the following Fundamental Lemma. If $\Gamma\vdash_{\lambda^{\mathrm{Ch}}_{\mu}} M : A$, η is a type environment and ρ_1, ρ_2 are term environments such that $\langle\rho_1(x_i), \rho_2(x_i)\rangle \in\mathcal{I}[\![A_i]\!]_\eta$ whenever $x_i : A_i\in\Gamma$, then

$$\langle[\![M]\!]_{\rho_1}, [\![M]\!]_{\rho_2}\rangle\in\mathcal{I}[\![A]\!]_\eta.$$

(ii) Define the model $\mathcal{M} = \langle T, \{\mathcal{M}(R)\}_{R\in T}, \{\Phi_{RS}\}_{R,S\in T}\rangle$ as follows. Write T for the set of all complete and uniform PERs over D. For $R, S\in T$ define:

$$\begin{aligned}(R\Rightarrow S) &= \{\langle d_1, d_2\rangle \mid \forall\langle e_1, e_2\rangle\in R.\langle d_1\cdot e_1, d_2\cdot e_2\rangle\in S\};\\ [d]_R &= \{e\in D \mid \langle d, e\rangle\in R\};\\ \mathcal{M}(R) &= \{[d]_R \mid \langle d, d\rangle\in R\};\\ \Phi_{RS}([d]_{R\Rightarrow S})([e]_R) &= [d\cdot e]_S,\end{aligned}$$

where $[d]_{R\Rightarrow S}\in\mathcal{M}(R\Rightarrow S)$ and $[e]_R\in\mathcal{M}(R)$. Show that the above construction is well defined.]

Exercise 10.13 Explore the consequences of choosing other definitions of R_0; for example, $R_0 = D_0\times D_0$ or $R_0 = \{\langle\bot_D, \bot_D\rangle\}$.

Exercise 10.14

(i) In this exercise, we sketch how to build a model $\mathcal{M}$ for the system $\lambda^{\mathrm{Cho}}_{\mu}$ where types are interpreted as *closures* over the λ-model $\mathcal{P}\omega$. The construction can be adapted easily to systems of the form $\lambda^{\mathcal{T}\text{-Ch}}$, for a type algebra $\mathcal{T}$, provided we assume that elements of $\mathcal{T}$ can be mapped homomorphically to closures. We presuppose a basic knowledge of the λ-model $\mathcal{P}\omega$, see for example Scott (1976), §5, and B[1984], §18.1, Exercises 18.4.3–18.4.9. For a related construction see Bruce et al. (1990), §7.2. Define a *closure* (over $\mathcal{P}\omega$) as a continuous function $a : \mathcal{P}\omega\to\mathcal{P}\omega$

such that $\mathbf{I} \subseteq a = a \circ a$ where $\mathbf{I} \in \mathcal{P}\omega$ is the interpretation of the identity function. Let $\mathcal{V}$ denote the collection of all closures over $\mathcal{P}\omega$. Given $a \in \mathcal{V}$, its range $\mathrm{im}(a)$ coincides with the set $\{x \in \mathcal{P}\omega \mid a(x) = x\}$. It can easily be proved that $\mathrm{im}(a)$ is an ω-algebraic lattice. Moreover, Scott (1976), Theorem 5.2, proves that, if D is an algebraic lattice with a countable basis, then there is a closure a_D over $\mathcal{P}\omega$ such that $D \cong \mathrm{im}(a_D)$. This representation of countably based algebraic lattices as closures over $\mathcal{P}\omega$ lifts nicely to continuous function spaces, since for all closures a, b there is a closure $a \leadsto b$, defined by $(a \leadsto b)(x) = b \circ x \circ a$, with a continuous bijection

$$\Phi_{ab} : \mathrm{im}(a \leadsto b) \stackrel{\sim}{\longrightarrow} [\mathrm{im}(a) \to \mathrm{im}(b)].$$

(ii) If η denotes a type environment assigning to each atom a closure, define the mapping $\mathcal{I}[\![\cdot]\!]_\eta : \mathbb{T}_\mu \to \mathcal{V}$ as follows:

(a) $\mathcal{I}[\![\alpha]\!]_\eta = \eta(\alpha)$;
(b) $\mathcal{I}[\![A \to B]\!]_\eta = \mathcal{I}[\![A]\!]_\eta \leadsto \mathcal{I}[\![B]\!]_\eta$;
(c) $\mathcal{I}[\![\mu\alpha.A]\!]_\eta = \mathbf{fix}(\lambda\!\!\lambda a \in \mathcal{V}.\mathcal{I}[\![A]\!]_{\eta[\alpha \mapsto a]})$.

Show that this mapping is well defined.

(iii) Let $A, B \in \mathbb{T}_\mu$. Show that if $A =_\mu B$ in the formal system of Definition 7.4.27(i), then $\mathcal{I}[\![A]\!]_\eta = \mathcal{I}[\![B]\!]_\eta$, for any type environment η.

(iv) For any $a \in \mathcal{V}$, define $\mathcal{M}(a)$ as $\mathrm{im}(a)$: we have isomorphisms

$$\Phi_{ab} : \mathcal{M}(a \leadsto b) \longrightarrow [\mathcal{M}(a) \to \mathcal{M}(b)]$$

for any pair $a, b \in \mathcal{V}$ and (trivial) isomorphisms

$$\Psi_{ab} : \mathcal{M}(a) \to \mathcal{M}(b)$$

whenever $a = b$. Define $\mathcal{M}$ as a structure

$$\mathcal{M} = \langle \mathcal{V}, \{\mathcal{M}(a)\}_{a \in \mathcal{V}}, \{\Phi_{ab}\}_{a,b \in \mathcal{V}}, \{\Psi_{ab}\}_{a,b \in \mathcal{V}} \rangle.$$

Show that $\mathcal{M}$ is a typed λ-model of $\lambda_\mu^{\mathrm{Cho}}$.

Exercise 10.15 Let Γ be a basis. A Γ, η-*environment* is a function ρ such that $\rho(x) \in \mathcal{M}(\mathcal{I}[\![A]\!]_\eta)$, whenever $(x{:}A) \in \Gamma$. Define the interpretation of a term (with respect to a Γ, η-environment) by induction on its construction tree in the system $\lambda_\mu^{\mathrm{Cho}}$:

$$\begin{aligned}
[\![\Gamma, x{:}A \vdash_{\lambda_\mu^{\mathrm{Ch}}} x : A]\!]_{\eta\rho} &= \rho(x),\\
[\![\Gamma \vdash_{\lambda_\mu^{\mathrm{Ch}}} MN : B]\!]_{\eta\rho} &= \Phi_{ab}\left([\![\Gamma \vdash_{\lambda_\mu^{\mathrm{Ch}}} M : A \to B]\!]_{\eta\rho}\right)[\![\Gamma \vdash_{\lambda_\mu^{\mathrm{Ch}}} N : A]\!]_{\eta\rho},\\
[\![\Gamma \vdash_{\lambda_\mu^{\mathrm{Ch}}} (\lambda x{:}A.M) : A \to B]\!]_{\eta\rho} &= \Phi_{ab}^{-1}(f),
\end{aligned}$$

where f is the function defined by the assignment:

$$f(d) = [\![\Gamma, x : A \vdash_{\lambda_\mu^{\mathrm{Ch}}} M : B]\!]_{\eta\rho[x \mapsto d]}$$

for any $d \in \mathcal{M}(\mathcal{I}[\![A]\!]_\eta)$. Now $f \in [\mathcal{M}(a) \to \mathcal{M}(b)]$, for $a = \mathcal{I}[\![A]\!]_\eta$ and $b = \mathcal{I}[\![B]\!]_\eta$.

Let $a = \mathcal{I}[\![\mu\alpha.A]\!]_\eta$ and $b = \mathcal{I}[\![A[\alpha := \mu\alpha.A]]\!]_\eta$; then we have

$$\begin{aligned} [\![\Gamma \vdash_{\lambda_\mu^{\mathrm{Ch}}} \mathit{fold}_{\mu\alpha.A}(M) : \mu\alpha.A]\!]_{\eta\rho} &= \Psi_{ab}^{-1}\left([\![\Gamma \vdash_{\lambda_\mu^{\mathrm{Ch}}} M : A[\alpha := \mu\alpha.A]]\!]_{\eta\rho}\right), \\ [\![\Gamma \vdash_{\lambda_\mu^{\mathrm{Ch}}} \mathit{unfold}_{\mu\alpha.A}(M) : A[\alpha := \mu\alpha.A]]\!]_{\eta\rho} &= \Psi_{ab}\left([\![\Gamma \vdash_{\lambda_\mu^{\mathrm{Ch}}} M : \mu\alpha.A]\!]_{\eta\rho}\right). \end{aligned}$$

Show that this yields a typed λ-model of $\lambda_\mu^{\mathrm{Cho}}$.

Exercise 10.16 Define a partial order on $\mathsf{Tr}_\infty^{\mathbb{A}}$, exploiting the "undefined" tree $\bullet$, as follows. For $s, t \in \mathsf{Tr}_\infty^{\mathbb{A}}$

$$\begin{aligned} s \preccurlyeq t \Leftrightarrow \mathrm{dom}(s) \subseteq \mathrm{dom}(t), &\quad \text{and} \\ \forall w \in \mathrm{dom}(s).s(w) \neq \bullet &\quad \Rightarrow \quad s(w) = t(w). \end{aligned}$$

(i) Prove that the partially ordered set $\langle \mathsf{Tr}_\infty^{\mathbb{A}}, \preccurlyeq \rangle$ is ω-complete and has $\bullet$ as smallest element.

(ii) Prove that $\mathsf{Tr}_\infty^{\mathbb{A}}$ is the initial ω-complete Σ-algebra, where

$$\Sigma = \mathbb{A} \cup \{\to\} \cup \{\bullet\}.$$

(iii) Conclude that for every $\eta : \mathbb{A} \to \mathcal{V}$, there is a unique continuous Σ-morphism extending η,

$$\mathcal{T}[\![\cdot]\!]_\eta : \mathsf{Tr}_\infty^{\mathbb{A}} \to \mathcal{V},$$

which satisfies the following equations:

$$\begin{aligned} \mathcal{T}[\![\alpha]\!]_\eta &= \eta(\alpha) \text{ for any } \alpha \in \mathbb{A}; \\ \mathcal{T}[\![\bullet]\!]_\eta &= \mathbf{I}; \\ \mathcal{T}[\![t_1 \to t_2]\!]_\eta &= \mathcal{T}[\![t_1]\!]_\eta \leadsto \mathcal{T}[\![t_2]\!]_\eta. \end{aligned}$$

for all $t_1, t_2 \in \mathsf{Tr}_\infty^{\mathbb{A}}$. [Hint. See for example Goguen et al. (1977), Theorem 4.8, or Courcelle (1983).]

(iv) Prove that for all types $A, B \in \mathbb{T}_\mu^{\mathbb{A}}$ and all type environments η one has:

(i) $\mathcal{T}[\![A^*]\!]_\eta = \mathcal{I}[\![A]\!]_\eta$;

(ii) $\mathcal{I}[\![A]\!]_\eta = \mathcal{I}[\![B]\!]_\eta$, whenever $A =_\mu^* B$.

[Hint. Use structural induction on $A \in \mathbb{T}_\mu^{\mathbb{A}}$. Show first that for any pair of infinite trees $s, t \in \mathsf{Tr}_\infty^{\mathbb{A}}$, we have

$$\mathcal{T}[\![s[\alpha := t]]\!]_\eta = \mathcal{T}[\![s]\!]_{\eta[\alpha \mapsto \mathcal{T}[\![t]\!]_\eta]}.$$

Then use the fact that $\mathcal{I}[\![\mu\alpha.A_1]\!]_\eta = \bigsqcup_{n \in \omega} \Theta^{(n)}(\bot)$, where $\Theta^{(0)}(\bot) = \bot$ and $\Theta^{(n+1)}(\bot) = \mathcal{I}[\![A_1]\!]_{\eta[\alpha \mapsto \Theta^{(n)}(\bot)]}$. See Cardone (2002), where also the converse of this soundness result is proved: for $A, B \in \mathbb{T}_\mu^{\mathbb{A}}$ one has

$$A =_\mu^* B \;\Leftrightarrow\; \mathcal{I}[\![A]\!]_\eta = \mathcal{I}[\![B]\!]_\eta, \text{ for all } \eta : \mathbb{A} \to \mathcal{V}.]$$

Exercise 10.17 Given a simultaneous recursion

$$\mathcal{R}(\vec{X}) = \{X_i = A_i \mid 1 \le i \le n\},$$

where $A_i \in \mathbb{T}^{\mathbb{A} \cup \{\vec{X}\}}$ for each $i = 1, \ldots, n$, show how to construct type algebra morphisms h, k: see Definitions 7.3.17(i) and 7.5.18(ii), with

$$\begin{aligned} h :\;& \mathbb{T}[\mathcal{R}] \longrightarrow \mathcal{S}(D) \\ k :\;& \mathbb{T}[\mathcal{R}]^* \longrightarrow \mathcal{S}(D). \end{aligned}$$

[Hint. Adapt the technique used in Section 10.2. Alternatively, observe that a solution $\vec{\mathbf{X}} = \mathbf{X}_1, \ldots, \mathbf{X}_n$ of $\mathcal{R}(\vec{X})$ consists of regular trees so, for each $i = 1, \ldots, n$, there is $A_{\mathbf{X}_i} \in \mathbb{T}_\mu^{\mathbb{A}}$ such that $(A_{\mathbf{X}_i})^* = \mathbf{X}_i$. Let η be the type environment defined by the mapping $X_i \mapsto \mathcal{I}[\![A_{\mathbf{X}_i}]\!]$. Consider the following equalities:

$$\begin{aligned} \mathcal{J}[\![X_i]\!]_\eta &= \eta(X_i) \\ \mathcal{J}[\![A' \to A'']\!]_\eta &= \mathcal{J}[\![A']\!]_\eta \Rightarrow \mathcal{J}[\![A'']\!]_\eta. \end{aligned}$$

They define recursively the extension of η to the required type algebra morphisms.]

Exercise 10.18 A function $f : \mathcal{CU}(D) \to \mathcal{CU}(D)$ is called *ideal* if, for all $X \in \mathcal{CU}$ and $n \in \omega$,

$$(f(X))_{n+1} = (f((X)_n))_{n+1}.$$

Show that such an ideal function f has a unique fixed-point $X \in \mathcal{CU}(D)$.

11

Applications

11.1 Subtyping

The model of recursive types discussed in the previous chapter justifies thinking of types as subsets of a model for the untyped λ-calculus (possibly extended with constants). This interpretation suggests a straightforward notion of *subtyping*: recursive types A, B are in the subtype relation, written $A \leqslant B$, if $\mathcal{I}[\![A]\!]_\eta \subseteq \mathcal{I}[\![B]\!]_\eta$ for all type environments η interpreting the free type variables occurring in them as complete and uniform subsets of D. We shall see later that a natural formal system for deriving type inequalities for recursive types is sound and complete for this interpretation. We start by introducing some basic systems of subtyping for simple types, showing by way of examples how, assuming that some types are included in others, one can achieve the same effects as having recursive types. Then, we discuss in some more detail the formal system described in Brandt and Henglein (1998) for axiomatizing the subtyping relation on recursive types introduced by Amadio and Cardelli (1993), which is by now standard.

11.1.1 Definition A *type structure* is of the form $\mathcal{S} = \langle |\mathcal{S}|, \leqslant, \rightarrow \rangle$, where $\langle |\mathcal{S}|, \leqslant \rangle$ is a poset and $\rightarrow : |\mathcal{S}|^2 \rightarrow |\mathcal{S}|$ is a binary operation such that for $a, b, a', b' \in |\mathcal{S}|$ one has

$$a' \leqslant a \;\&\; b \leqslant b' \;\Rightarrow\; (a \rightarrow b) \leqslant (a' \rightarrow b').$$

These type structures should not be confused with type structures in Part III of this book, where the name is used as an abbreviation for 'intersection type structures'. Note that a type structure $\mathcal{S} = \langle |\mathcal{S}|, \leqslant, \rightarrow \rangle$ can be considered as a type algebra $\langle |\mathcal{S}|, \rightarrow \rangle$. We call the latter the type algebra underlying the type structure $\mathcal{S}$.

(axiom)	$\Gamma \vdash x : a \quad \text{if } (x{:}a) \in \Gamma$
$(\to\text{E})$	$\dfrac{\Gamma \vdash M : a \to b \qquad \Gamma \vdash N : a}{\Gamma \vdash (MN) : b}$
$(\to\text{I})$	$\dfrac{\Gamma, x{:}a \vdash M : b}{\Gamma \vdash (\lambda x.M) : (a \to b)}$
$(\leqslant)$	$\dfrac{\Gamma \vdash M : a \quad a \leqslant b}{\Gamma \vdash M : b}$

Figure 11.1 The system $\lambda_{\leqslant}^{\mathcal{S}}$

11.1.2 Definition Let $\mathcal{S}$ be a type structure and $a, b \in \mathcal{S}$. Then the system of type assignment $\lambda_{\leqslant}^{\mathcal{S}}$ is defined by the axioms and rules in Fig. 11.1. Write $\Gamma \vdash_{\lambda_{\leqslant}^{\mathcal{S}}} M : a$ or $\Gamma \vdash_{\mathcal{S}} M : a$ if $\Gamma \vdash M : a$ can be derived in the system $\lambda_{\leqslant}^{\mathcal{S}}$.

11.1.3 Proposition *Let $\mathcal{S}$ be a type structure. Then*

$$\Gamma \vdash_{\lambda_{=}^{\mathcal{S}}} M : A \;\Rightarrow\; \Gamma \vdash_{\lambda_{\leqslant}^{\mathcal{S}}} M : A,$$

where $\Gamma \vdash_{\lambda_{=}^{\mathcal{S}}} M : A$ is type assignment using the type algebra underlying $\mathcal{S}$.

Proof Trivial as $\lambda_{=}^{\mathcal{S}}$, see Definition 7.1.3, has fewer rules than $\lambda_{\leqslant}^{\mathcal{S}}$. ∎

11.1.4 Lemma *Suppose that $\Gamma \subseteq \Gamma'$. Then for all $a \in \mathcal{S}$*

$$\Gamma \vdash_{\mathcal{S}} M : a \;\Rightarrow\; \Gamma' \vdash_{\mathcal{S}} M : a.$$

11.1.5 Example Let a, b be elements of a type structure $\mathcal{S}$.

(i) If $b \leqslant (b {\to} a)$, then $\vdash_{\mathcal{S}} (\lambda x.xx) : (b \to a)$.
(ii) If $(b {\to} a) \leqslant b$, then $\vdash_{\mathcal{S}} (\lambda x.xx) : (b \to a) \to a$.
(iii) If $((b \to a) \to a) \leqslant (b {\to} a) \leqslant b$, then $\vdash_{\mathcal{S}} (\lambda x.xx)(\lambda x.xx) : a$.
(iv) If $a \leqslant (a \to a)$, then $\vdash_{\mathcal{S}} (\lambda x.x(xx)) : (a \to a)$.

11.1.6 Definition Let $\mathcal{S}, \mathcal{S}'$ be type structures. A map $h : \mathcal{S} \to \mathcal{S}'$ is called a *morphism* of type structures if for all $a, b \in \mathcal{S}$ one has

$$\begin{aligned} a \leq_{\mathcal{S}} b &\Rightarrow h(a) \leq_{\mathcal{S}'} h(b); \\ h(a \to b) &= h(a) \to_{\mathcal{S}'} h(b). \end{aligned}$$

11.1.7 Lemma *Let $h : \mathcal{S} \to \mathcal{S}'$ be a morphism of type structures. Then for all $a \in \mathcal{S}$*

$$\Gamma \vdash_{\mathcal{S}} M : a \;\Rightarrow\; h(\Gamma) \vdash_{\mathcal{S}'} M : h(a).$$

Type structures from type theories

11.1.8 Definition

(i) Let $\mathbb{A}$ be a countable (finite or countably infinite) set of symbols, called *(type) atoms*. The set of *types* over $\mathbb{A}$, written $\mathbb{T}^{\mathbb{A}}$ (if there is little danger of confusion it is also denoted by $\mathbb{T}$), is defined by the abstract syntax

$$\mathbb{T} = \mathbb{A} \mid \mathbb{T} {\to} \mathbb{T}.$$

The set $\mathbb{A}$ varies in the applications of types. We let $\alpha, \beta, \gamma, \ldots$ range over $\mathbb{A}$ and $A, B, C, \ldots$ over $\mathbb{T}^{\mathbb{A}}$.

(ii) Some of the type atoms that will be used in different contexts are:

$\mathsf{c}_0, \mathsf{c}_1, \mathsf{c}_2, \ldots$	atoms without special role;
$\bot, \top, \mathbf{nat}, \mathbf{int}, \mathbf{bool}, \ldots,$	atoms with special properties.

(iii) The atoms $\top$ and $\bot$ are called *top* and *bottom*, respectively. The intention for the special atoms **nat**, **int**, **bool** is to have primitive natural numbers, integers and booleans. These can be defined, see Section 3.1, but having primitive ones is often more efficient.

11.1.9 Definition

(i) Let $\mathcal{H}$ be a set of type inequalities. Define for $A, B \in \mathbb{T}^{\mathbb{A}}$ *derivability* of $A \leqslant B$ from basis $\mathcal{H}$, written $\mathcal{H} \vdash A \leqslant B$, by the axioms and rules in Fig. 11.2.

(ii) A set of type inequalities $\mathcal{T}$ is a *type theory* if, for all $A, B \in \mathbb{T}$,

$$\mathcal{T} \vdash A \leqslant B \;\Rightarrow\; (A \leqslant B) \in \mathcal{T}.$$

(iii) If $\mathcal{T} = \{A \leq B \mid \mathcal{H} \vdash A \leq B\}$, then $\mathcal{T}$ is the type theory *axiomatized* by $\mathcal{H}$.

11.1.10 Definition

(i) A type inequality is called *inflationary* in a type atom α if it is of the form $\alpha \leqslant A$; it is called *deflationary* in α if it is of the form $A \leqslant \alpha$. In both cases, the type atom α is called a *subject* of the judgement. A type inequality is called *atomic* if it is of the form $\alpha \leq \beta$.

(hyp)	$\mathcal{H} \vdash A \leqslant B,$	if $(A \leqslant B) \in \mathcal{H},$
(refl)	$\mathcal{H} \vdash A \leqslant A$	
(trans)	$\dfrac{\mathcal{H} \vdash A \leqslant B \quad \mathcal{H} \vdash B \leqslant C}{\mathcal{H} \vdash A \leqslant C}$	
$(\to)$	$\dfrac{\mathcal{H} \vdash A' \leqslant A \quad \mathcal{H} \vdash B \leqslant B'}{\mathcal{H} \vdash (A \to B) \leqslant (A' \to B')}$	

Figure 11.2

(ii) A set of type inequalities $\mathcal{H}$ is called *homogeneous in* α if the subtyping judgements in $\mathcal{H}$ to which α is subject are all inflationary or they are all deflationary.

(iii) A set of type inequalities $\mathcal{H}$ is called *homogeneous* if it is homogeneous in α for every type atom α that occurs as subject.

(iv) A type theory $\mathcal{T}$ is called *homogeneous* if it is axiomatized by a homogeneous set $\mathcal{H}$ of type inequalities.

(v) The set $\mathcal{H}$ is called *inflationary* (respectively *deflationary*), if it consists only of inflationary (respectively deflationary) type inequalities.

(vi) A type theory is *inflationary* (respectively *deflationary*), if it is axiomatized by an inflationary (respectively deflationary) set of type inequalities.

11.1.11 Definition Let $\mathcal{T}$ be a type theory over $\mathbb{T} = \mathbb{T}^{\mathbb{A}}$.

(i) Write, for $A, B \in \mathbb{T}$,

$$\begin{aligned} A \leqslant_{\mathcal{T}} B &\Leftrightarrow (A \leqslant B) \in \mathcal{T}; \\ A \approx_{\mathcal{T}} B &\Leftrightarrow A \leqslant_{\mathcal{T}} B \;\&\; B \leqslant_{\mathcal{T}} A; \\ [A]_{\mathcal{T}} &= \{B \in \mathbb{T} \mid A \approx_{\mathcal{T}} B\}. \end{aligned}$$

(ii) The *type structure determined by* $\mathcal{T}$ is $\mathcal{S}_{\mathcal{T}} = \langle \mathbb{T}/{\approx_{\mathcal{T}}}, \to, \leqslant \rangle$, with

$$\begin{aligned} \mathbb{T}/{\approx_{\mathcal{T}}} &= \{[A]_{\mathcal{T}} \mid A \in \mathbb{T}\}; \\ [A]_{\mathcal{T}} \to [B]_{\mathcal{T}} &= [A \to B]_{\mathcal{T}}; \\ [A]_{\mathcal{T}} \leqslant [B]_{\mathcal{T}} &\Leftrightarrow A \leqslant_{\mathcal{T}} B. \end{aligned}$$

The last two definitions are independent of the choice of representatives.

(iii) Write $x_1{:}A_1, \ldots, x_n{:}A_n \vdash_{\mathcal{T}} M : A$, with $A_1, \ldots, A_n, A \in \mathbb{T}$, for

$$x_1{:}[A_1]_{\mathcal{T}}, \ldots, x_n{:}[A_n]_{\mathcal{T}} \vdash_{\mathcal{S}_{\mathcal{T}}} M : [A]_{\mathcal{T}}.$$

The following result relates typability and normalization of λ-terms.

11.1.12 Proposition

(i) *Let $\mathcal{T}$ be a homogeneous type theory. Then*

$$\Gamma \vdash_{\mathcal{T}} M : A \;\Rightarrow\; M \textit{ is strongly normalizing.}$$

(ii) *Let $N \in \Lambda$ be in β-nf. Then there is an inflationary type theory $\mathcal{T}$ such that N can be typed in $\vdash_{\mathcal{T}}$, i.e. for some $\mathbb{A}$, $A \in \mathbb{T}^{\mathbb{A}}$, and Γ one has*

$$\Gamma \vdash_{\mathcal{T}} N : A.$$

Proof We only sketch the proofs, leaving most of the details to the reader.

(i) The type theory $\mathcal{T}$ is axiomatized by a homogeneous set $\mathcal{H}$ of type inequalities. Interpret types as saturated sets, defined in 9.3.5, using the first two clauses of 9.3.9 and the type environment τ such that

$$\begin{array}{rcll} \tau(\alpha) & = & \bot_{\mathsf{SAT}} = \bigcap_{\mathcal{X} \in \mathsf{SAT}} \mathcal{X}, & \text{if } (\alpha \leq A) \in \mathcal{H} \text{ for some } A; \\ & = & \top_{\mathsf{SAT}} = \mathsf{SN}, & \text{if } (A \leq \alpha) \in \mathcal{H} \text{ for some } A. \\ & = & X, & \text{with } X \in \mathsf{SAT} \text{ arbitrary, otherwise.} \end{array}$$

Observe that by homogeneity of $\mathcal{H}$ the type environment τ is welldefined. By a straightforward induction on the derivations one shows, cf. Definition 9.3.14, that

$$\Gamma \vdash_{\mathcal{T}} M : A \;\Rightarrow\; \Gamma \models M : A.$$

Therefore, as in the proof of Theorem 9.3.16, every typable term belongs to a saturated set, hence is strongly normalizing.

(ii) Proceed by induction on the structure of the normal form. The only interesting case is when $N \equiv zN_1 \cdots N_k$. Then, by the induction hypothesis, there are type theories $\mathcal{T}_1, \ldots, \mathcal{T}_k$ generated by inflationary $\mathcal{H}_1, \ldots, \mathcal{H}_k$, types A_i, and bases Γ_i such that for each $i = 1, \ldots, k$ one has $\Gamma_i \vdash_{\mathcal{T}_i} N_i : A_i$. Let $\mathcal{T}$ be the type theory generated by the inflationary $\mathcal{H}_1 \cup \cdots \cup \mathcal{H}_k$. Assume that a variable x has types $X_{j(1)}, \ldots, X_{j(k)}$ in bases $\Gamma_{j(1)}, \ldots, \Gamma_{j(k)}$. For a new type variable t_x, let the statement $x : t_x$ be in Γ and let $\mathcal{H}$ be $\bigcup_i \mathcal{H}_i$ plus the type inequalities $t_x \leqslant X_{j(1)}, \ldots, t_x \leqslant X_{j(k)}$. Thus, we get a basis Γ and an inflationary $\mathcal{H}$ such that, if $\mathcal{T}$ is the type theory generated by $\mathcal{H}$, $\Gamma, z : A_1 \to \cdots \to A_k \to \beta \vdash_{\mathcal{T}} zN_1 \cdots N_k : \beta$. ∎

Subject reduction

We show now that the straightforward generalization of the notion of invertibility to subtyping yields the subject reduction property, similarly to the way it was obtained for type equality discussed in Section 9.1.

11.1.13 Definition (Invertibility for subtyping) A type structure $\mathcal{S} = \langle|\mathcal{S}|, \leqslant, \to\rangle$ is *invertible* if for all $a, b, a', b' \in |\mathcal{S}|$ one has

$$(a \to b) \leqslant (a' \to b') \;\Rightarrow\; a' \leqslant a \;\&\; b \leqslant b'.$$

The following parallels Corollary 9.1.5 of the Inversion Lemma 9.1.3 for derivations of the system $\lambda_{\leqslant}^{\mathcal{S}}$.

11.1.14 Lemma (Inversion Lemma) *Let $\mathcal{S}$ be a type structure. Then:*

(i) $\Gamma \vdash_{\lambda_{\leqslant}^{\mathcal{S}}} x : a \quad\Leftrightarrow\quad \exists a' \in \mathcal{S}.[a' \leqslant a \;\&\; (x{:}a') \in \Gamma]$.

(ii) $\Gamma \vdash_{\lambda_{\leqslant}^{\mathcal{S}}} (MN) : a \quad\Leftrightarrow\quad \exists b \in \mathcal{S}.[\Gamma \vdash_{\lambda_{\leqslant}^{\mathcal{S}}} M : (b \to a) \;\&\; \Gamma \vdash_{\lambda_{\leqslant}^{\mathcal{S}}} N : b]$.

(iii) $\Gamma \vdash_{\lambda_{\leqslant}^{\mathcal{S}}} (\lambda x.M) : a \quad\Leftrightarrow\quad \exists b, c \in \mathcal{S}.[a \geqslant (b \to c) \;\&\; \Gamma, x{:}b \vdash_{\lambda_{\leqslant}^{\mathcal{S}}} M : c]$.

Proof ($\Leftarrow$) immediate from the rules for $\vdash_{\lambda_{\leqslant}^{\mathcal{S}}}$.

($\Rightarrow$) The proof is by induction on the height of the derivation of typings $\Gamma \vdash_{\lambda_{\leqslant}^{\mathcal{S}}} M : a$. The basis is obvious, and the induction step considers the possible shapes of the term M.

(i) $M \equiv x$, then case (i) holds because the last rule must be ($\leqslant$).

(ii) $M \equiv PQ$. If the last rule applied is ($\to$E) then case (ii) holds; if the last rule applied is ($\leqslant$) then the derivation has the shape

$$\frac{\Gamma \vdash PQ : a' \quad a' \leqslant a}{\Gamma \vdash_{\lambda_{\leqslant}^{\mathcal{S}}} PQ : a}.$$

By the induction hypothesis there is a type $b \in \mathcal{S}$ such that $\Gamma \vdash_{\lambda_{\leqslant}^{\mathcal{S}}} P : b \to a'$ and $\Gamma \vdash_{\lambda_{\leqslant}^{\mathcal{S}}} Q : b$, and these judgements can be reassembled in a derivation whose last step is an application of rule ($\to$E):

$$\frac{\Gamma \vdash_{\lambda_{\leqslant}^{\mathcal{S}}} P : b \to a \quad \Gamma \vdash_{\lambda_{\leqslant}^{\mathcal{S}}} Q : b}{\Gamma \vdash_{\lambda_{\leqslant}^{\mathcal{S}}} PQ : a}.$$

(iii) $M \equiv \lambda x.N$. If the last rule is ($\to$I), then $a = b \to c$; if the last rule applied is ($\leqslant$), then $\Gamma \vdash_{\lambda_{\leqslant}^{\mathcal{S}}} \lambda x.N : a'$ and $a' \leqslant a$. By the induction hypothesis we have $\Gamma, x : b' \vdash_{\lambda_{\leqslant}^{\mathcal{S}}} N : c'$, where $b' \to c' \leqslant a'$. Then we can take $b = b'$ and $c = c'$. ∎

11.1.15 Definition Let $\Gamma = \{x_1 : a_1, \ldots, x_n : a_n\}$ and $\Gamma' = \{x_1 : a'_1, \ldots, x_n : a'_n\}$. Then we write $\Gamma' \leq \Gamma$ if $a'_i \leq a_i$ for every $1 \leq i \leq k$.

11.1.16 Lemma *Let* $\Gamma \vdash M : a$ *and* $\Gamma' \leq \Gamma$. *Then* $\Gamma' \vdash M : a$.

Proof By induction on the derivation of $\Gamma \vdash M : a$. ∎

11.1.17 Corollary (Inversion Lemma) *Let* $\mathcal{S}$ *be an invertible type structure. Then*

$$\Gamma \vdash_{\lambda^{\mathcal{S}}_{\leqslant}} \lambda x.M : (a \to b) \;\Leftrightarrow\; \Gamma, x{:}a \vdash_{\lambda^{\mathcal{S}}_{\leqslant}} M : b.$$

Proof ($\Rightarrow$) If $\Gamma \vdash_{\lambda^{\mathcal{S}}_{\leqslant}} \lambda x.M : a \to b$ then, by Lemma 11.1.14, $a \to b \geqslant a' \to b'$ and $\Gamma, x : a'c \vdash_{\lambda^{\mathcal{S}}_{\leqslant}} M : b'd$. By invertibility, $a \leqslant ca'$ and $d \leqslant bb'$. Hence $\Gamma, x : a \vdash_{\lambda^{\mathcal{S}}_{\leqslant}} M : b'$ since $\Gamma, x{:}a \leqslant \Gamma, x : a'$. Therefore $\Gamma, x : a \vdash_{\lambda^{\mathcal{S}}_{\leqslant}} M : b$, by rule ($\leqslant$).

($\Leftarrow$) By rule ($\to$I). ∎

11.1.18 Lemma *Suppose* $x \notin \mathrm{FV}(M)$, *then*

$$\Gamma, x{:}b \vdash_{\lambda^{\mathcal{S}}_{\leqslant}} M : a \;\Rightarrow\; \Gamma \vdash_{\lambda^{\mathcal{S}}_{\leqslant}} M : a.$$

Proof By induction on the derivation of $\Gamma, x{:}b \vdash_{\lambda^{\mathcal{S}}_{\leqslant}} M : a$. ∎

Now we can prove the subject reduction property for one-step $\boldsymbol{\beta\eta}$-reduction.

11.1.19 Lemma

(i) *If* $\Gamma \vdash_{\lambda^{\mathcal{S}}_{\leqslant}} \lambda x.(Mx) : a$ *and* $x \notin \mathrm{FV}(M)$, *then* $\Gamma \vdash_{\lambda^{\mathcal{S}}_{\leqslant}} M : a$.

(ii) *If moreover* $\mathcal{S}$ *is invertible, then*

$$\Gamma \vdash_{\lambda^{\mathcal{S}}_{\leqslant}} (\lambda x.M)N : a \;\Rightarrow\; \Gamma \vdash_{\lambda^{\mathcal{S}}_{\leqslant}} (M[x := N]) : a.$$

Proof Analogous to the proof of Lemma 9.1.7. ∎

11.1.20 Corollary (Subject Reduction) *Let* $\mathcal{S}$ *be an invertible type structure. Then* $\lambda^{\mathcal{S}}_{\leqslant}$ *satisfies the subject reduction property for* $\boldsymbol{\beta\eta}$.

Proof By Lemma 11.1.19. ∎

11.1.21 Example

(i) We can type self-application by an essential use of rule ($\leqslant$):

$$\{s \leqslant s \to t\} \vdash \lambda x.xx : s \to t$$

Observe that $\{s \leqslant s \to t\}$ is inflationary, hence homogeneous because it consists of one type inequality.

(ii) Type inequalities can have the same effect as simultaneous recursions:

$$\{s \leqslant s \to s, s \to s \leqslant s\} \vdash (\lambda x.xx)(\lambda x.xx) : s$$

Actually, the set $\{s \leqslant s \to s, s \to s \leqslant s\}$ lets us type every pure λ-term, and in this sense is equivalent to the simultaneous recursion $\{s = s \to s\}$. Note that this set of type inequalities is not homogeneous.

Subtyping recursive types

Now we will define μ-type theories. In the following we summarize the theory of subtyping on recursive types, seen as representations of infinite trees, introduced by Amadio and Cardelli (1993), and the equivalent formulation of Brandt and Henglein (1998).

We can turn the type algebra $\mathbb{T}^*_\mu$ into a type structure by considering approximations using the elements $\perp$ and $\top$. For most technical details, the reader is referred to the original papers, and also to the elegant exposition of Gapeyev et al. (2002).

To introduce the basic notion of subtyping in $\mathbb{T}^{\mathbb{A}}_\mu$ we add to the set $\mathbb{T}^{\mathbb{A}}_\mu$ two type constants $\perp, \top \in \mathbb{A}$ representing the minimal and maximal elements of the set of types. In this way we obtain the *partial types*. Obviously other constants and corresponding type inclusions could be added, like **nat** $\leqslant$ **int**.

First we need to define *lower approximations* $\tau \downarrow n$ and *upper approximations* $\tau \uparrow n$, for all $n \in \omega$, for an infinite tree τ.

11.1.22 Definition (Approximations of an infinite tree) Let τ be an infinite tree, its *lower* and *upper* approximations, written $\tau \downarrow n$, $\tau \uparrow n$, are the partial types defined simultaneously, for any $n \in \omega$, as follows:

$$\begin{array}{rcl@{\qquad}rcl}
\tau \downarrow 0 & = & \perp & \tau \uparrow 0 & = & \top \\
(\tau' \to \tau'') \downarrow (n+1) & = & \tau' \uparrow n \to \tau'' \downarrow n & & & \\
(\tau' \to \tau'') \uparrow (n+1) & = & \tau' \downarrow n \to \tau'' \uparrow n & & & \\
\alpha \downarrow (n+1) & = & \alpha & \alpha \uparrow (n+1) & = & \alpha,
\end{array}$$

where α is a type constant or variable.

For example we obtain the following partial types:

$$\begin{aligned}
(\alpha{\to}\beta){\to}\alpha{\downarrow}0 &= \bot \\
(\alpha{\to}\beta){\to}\alpha{\downarrow}1 &= \top{\to}\bot \\
(\alpha{\to}\beta){\to}\alpha{\downarrow}2 &= (\bot{\to}\top){\to}\alpha \\
(\alpha{\to}\beta){\to}\alpha{\downarrow}3 &= (\alpha{\to}\beta){\to}\alpha;
\end{aligned}$$

$$\begin{aligned}
(\alpha{\to}\beta){\to}\alpha{\uparrow}0 &= \top \\
(\alpha{\to}\beta){\to}\alpha{\uparrow}1 &= \bot{\to}\top \\
(\alpha{\to}\beta){\to}\alpha{\uparrow}2 &= (\top{\to}\bot){\to}\alpha \\
(\alpha{\to}\beta){\to}\alpha{\uparrow}3 &= (\alpha{\to}\beta){\to}\alpha.
\end{aligned}$$

11.1.23 Definition (Subtyping)

(i) The *subtyping relation*, written $\leqslant_{\text{fin}}$, on partial types is defined by

$$A \leqslant_{\text{fin}} A;$$

$$\bot \leqslant_{\text{fin}} A; \qquad A \leqslant_{\text{fin}} \top;$$

$$\frac{A \leqslant_{\text{fin}} B \quad B \leqslant_{\text{fin}} C}{A \leqslant_{\text{fin}} C}; \qquad \frac{A' \leqslant_{\text{fin}} A \quad B \leqslant_{\text{fin}} B'}{A \to B \leqslant_{\text{fin}} A' \to B'}.$$

(ii) For a recursive type A, let $(A)^*_\mu$ be the tree unfolding of A, as in Notation 7.5.20(ii). The subtyping relation $\leqslant^*_\mu$ is defined by

$$A \leqslant^*_\mu B \Leftrightarrow \forall n \in \omega.((A)^*_\mu) \downarrow n \leqslant_{\text{fin}} ((B)^*_\mu) \downarrow n.$$

11.1.24 Remark Let $A, B \in \mathbb{T}^{\{\bot,\top\}}_\mu$, (where $\bot$ and $\top$ are considered as distinct atomic types). Then for $=^*_\mu$ defined as in Definition 7.5.20(iii) one has

$$A =^*_\mu B \;\Leftrightarrow\; [A \leqslant^*_\mu B \;\&\; B \leqslant^*_\mu A].$$

11.1.25 Definition Define the μ-type structure $\mathcal{T}_{\text{BH}} = \langle \mathbb{T}^*_\mu, \leqslant^*_\mu, \to \rangle$, where $\leqslant^*_\mu, \to$ are lifted to equivalence classes in the obvious way.

For this notion of subtyping we can also define, following Brandt and Henglein (1998), a sound and complete co-inductive proof system.

11.1.26 Definition

(i) Let $\mathcal{H}$ be a set of subtype assumptions of the form $A \leqslant B$ for $A, B \in \mathbb{T}_\mu$. The system $(\text{BH}_\leqslant)$ is defined by the axioms and rules in Fig. 11.3. Write $\mathcal{H} \vdash_{\text{BH}_\leqslant} A \leqslant B$ if $A \leqslant B$ is derivable from assumptions in $\mathcal{H}$ in $(\text{BH}_\leqslant)$.

(hyp)	$\mathcal{H} \vdash A \leq B, \qquad \text{if } A \leq B \in \mathcal{H}$
(refl)	$\mathcal{H} \vdash A \leqslant A$
(trans)	$\dfrac{\mathcal{H} \vdash A \leqslant B \quad \mathcal{H} \vdash B \leqslant C}{\mathcal{H} \vdash A \leqslant C}$
$(\to)$	$\dfrac{\mathcal{H} \vdash A' \leqslant A \quad \mathcal{H} \vdash B \leqslant B'}{\mathcal{H} \vdash A \to B \leqslant A' \to B'}$
(unfold)	$\mathcal{H} \vdash \mu\alpha.A \leqslant A[\alpha := \mu\alpha.A]$
(fold)	$\mathcal{H} \vdash A[\alpha := \mu\alpha.A] \leqslant \mu\alpha.A$
(bot)	$\mathcal{H} \vdash \bot \leqslant A$
(top)	$\mathcal{H} \vdash A \leqslant \top$
(Arrow/Fix)	$\dfrac{\mathcal{H}, (A \to B) \leqslant (A' \to B') \vdash (A' \leqslant A) \quad \mathcal{H}, (A \to B) \leqslant (A' \to B') \vdash (B \leqslant B')}{\mathcal{H} \vdash (A \to B) \leqslant (A' \to B')}$

Figure 11.3 The system ($\mathrm{BH}_{\leqslant}$).

The above system can be proved sound and complete for $\leqslant^*_\mu$. The proof is a straightforward extension of the corresponding proof for the system (for type equality) treated in Section 8.2.

11.1.27 Theorem

$$\vdash_{\mathrm{BH}_{\leqslant}} A \leqslant B \ \Leftrightarrow\ A \leqslant^*_\mu B.$$

Proof By Brandt and Henglein (1998), Theorems 2.2. and 2.6. ∎

The formalization of Brandt and Henglein (1998) is based on a co-inductive analysis of recursive types, ultimately going back to the theory of infinite trees outlined at the end of Section 7.6, from 7.6.10. The following notion is the key to the extension of this conceptual framework to subtyping.

11.1.28 Definition (Simulations on recursive types) A *simulation* on recursive types is a binary relation $\mathcal{R}$ that satisfies:

(i) $(A \to B)\, \mathcal{R}\, (A' \to B') \ \Rightarrow\ A'\, \mathcal{R}\, A \;\&\; B\, \mathcal{R}\, B'$;
(ii) $\mu t.A\, \mathcal{R}\, B \ \Rightarrow\ A[t := \mu t.A]\, \mathcal{R}\, B$;
(iii) $A\, \mathcal{R}\, \mu t.B \ \Rightarrow\ A\, \mathcal{R}\, B[t := \mu t.B]$;

(iv) $A \mathcal{R} B \Rightarrow \mathcal{L}(A) \leqslant \mathcal{L}(B)$,

where $\mathcal{L}(A)$ is the root symbol of the tree $(A)^*_\mu$ – that is, $\mathcal{L}(A) = (A)^*_\mu(\varepsilon)$, see Definition 7.5.1 – and the ordering on symbols makes it a 'flat domain' with $\bot \leqslant s \leqslant \top$, for $s \in \{\to\} \cup \{\alpha \mid \alpha \text{ atom}\}$.

11.1.29 Proposition *Let $\mathcal{R}$ be a simulation. Then for $A, B \in \mathbb{T}_\mu$*

$$A \mathcal{R} B \Rightarrow A \leqslant^*_\mu B.$$

Proof (Brandt and Henglein (1998), Lemma 2.2.) Show by induction on $n \in \omega$ that

$$A\mathcal{R}B \Rightarrow (A)^*_\mu{\downarrow}n \leq_{\text{fin}} (B)^*_\mu{\downarrow}n.$$

See the original paper for the details. ∎

Models of subtyping

In what follows $\mathcal{D}$ will be a reflexive structure, Definition 10.1.12, with a notion of approximation, Definition 10.2.1. We will give natural truth conditions for type inequalities so that the system $\mathrm{BH}_\leqslant$ is sound when types are interpreted – like in Section 10.2 – as complete uniform subsets of a reflexive structure with a notion of approximation $\mathcal{D}$. This soundness result also motivates the contravariance of arrow types in the first argument with respect to the subtyping relation, which is hard to justify using only the definition of $\leqslant^*_\mu$. Finally, we shall point out under what conditions the reverse implication holds.

Soundness

The following lemma characterizes inclusion for elements of $\mathcal{CU}$, Definition 10.2.5(iii).

11.1.30 Lemma *Let $X, Y \in \mathcal{CU}$. Then*

$$X \subseteq Y \Leftrightarrow \forall n \in \omega . X_n \subseteq Y_n.$$

Proof If $X \subseteq Y$ and $d \in X_n$, then $d \in X$, so $d \in Y$ and $d \in Y_n$, because $d = d_n$. Conversely, if $d \in X$, then, for all $n \in \omega$, $d_n \in X_n \subseteq Y_n \subseteq Y$, so $d = \bigsqcup_{n \in \omega} d_n \in Y$. ∎

We now define a notion of satisfaction for type inequalities exploiting the stratification of $\mathcal{D}$ into levels, following the development of Section 10.2.

11.1.31 Definition Fix a reflexive structure with a notion of approximation $\mathcal{D}$. Let $A, B \in \mathbb{T}_\mu$, $k \in \omega$, and η be a type environment relative to $\mathcal{D}$.

(i) $\eta \models_k A \leqslant B \quad \Leftrightarrow \quad \mathcal{I}^k[\![A]\!]_\eta \subseteq \mathcal{I}^k[\![B]\!]_\eta$;
(ii) $\eta \models_k \mathcal{H} \quad \Leftrightarrow \quad \eta \models_k A \leqslant B$, for all judgements $A \leqslant B$ in $\mathcal{H}$;
(iii) $\mathcal{H} \models_k A \leqslant B \quad \Leftrightarrow \quad [\eta \models_k \mathcal{H} \Rightarrow \eta \models_k A \leqslant B]$, for all η.
(iv) $\mathcal{H} \models A \leqslant B \quad \Leftrightarrow \quad \mathcal{H} \models_k A \leqslant B$, for all $k \in \omega$.

11.1.32 Lemma $\mathcal{H} \vdash_{\mathrm{BH}_\leqslant} A \leqslant B \Rightarrow \mathcal{H} \models A \leqslant B$.

Proof By induction on the length of the proof that $\mathcal{H} \vdash_{\mathrm{BH}_\leqslant} A \leqslant B$. This is clear if the last rule applied is (top), (bot), (hyp), (fold) or (unfold): for the last two cases use Theorem 10.2.19(iii). Assume that the last rule applied is (Arrox/Fix). Then the last inference has the following shape:

$$\frac{\mathcal{H}, A \to B \leqslant A' \to B' \vdash A' \leqslant A \quad \mathcal{H}, A \to B \leqslant A' \to B' \vdash B \leqslant B'}{\mathcal{H} \vdash A \to B \leqslant A' \to B'}.$$

By the induction hypothesis

$$\mathcal{H}, A \to B \leqslant A' \to B' \models A' \leqslant A$$
$$\mathcal{H}, A \to B \leqslant A' \to B' \models B \leqslant B',$$

and we have to show that, for all $k \in \omega$,

$$\mathcal{H} \models_k A \to B \leqslant A' \to B'.$$

We do this by induction on k. The base case is obvious, as both sides are interpreted as $\{\perp_D\}$. For the induction step, assume that $\mathcal{H} \models_k A \to B \leqslant A' \to B'$ and assume also that $\eta \models_{k+1} \mathcal{H}$. We have to show that $\mathcal{I}^{k+1}[\![A \to B]\!]_\eta \subseteq \mathcal{I}^{k+1}[\![A' \to B']\!]_\eta$. By the induction hypothesis (on k) we have $\eta \models_k \mathcal{H}, A \to B \leqslant A' \to B'$, because if $\eta \models_{k+1} \mathcal{H}$, then also $\eta \models_k \mathcal{H}$, by the properties of the approximate interpretation of types and Lemma 11.1.30. Hence, by the hypothesis of the main induction, $\mathcal{I}^k[\![A']\!]_\eta \subseteq \mathcal{I}^k[\![A]\!]_\eta$ and $\mathcal{I}^k[\![B]\!]_\eta \subseteq \mathcal{I}^k[\![B']\!]_\eta$. Now if $d \in \mathcal{I}^{k+1}[\![A \to B]\!]_\eta = \mathcal{I}^k[\![A]\!]_\eta \to^{n+1} \mathcal{I}^k[\![B]\!]_\eta$ and $a \in \mathcal{I}^k[\![A']\!]_\eta$, then $a \in \mathcal{I}^k[\![A]\!]_\eta$. Hence $d \cdot a \in \mathcal{I}^k[\![B]\!]_\eta \subseteq \mathcal{I}^k[\![B']\!]_\eta$. Therefore $d \in \mathcal{I}^{k+1}[\![A' \to B']\!]_\eta$. ∎

11.1.33 Corollary (Soundness) *Suppose $A \leqslant^*_\mu B$. Then*

$$\forall \eta \left(\mathcal{I}[\![A]\!]_\eta \subseteq \mathcal{I}[\![B]\!]_\eta \right).$$

Completeness

The formal system $\mathrm{BH}_\leqslant$ is too weak to achieve completeness, i.e. the reverse of Corollary 11.1.33. For example (see Amadio and Cardelli (1993)), for any reflexive $\mathcal{D}$ with a notion of approximation

$$\mathcal{I}[\![\alpha \to \beta]\!]_\eta \subseteq \mathcal{I}[\![\gamma \to \top]\!]_\eta,$$

for all type variables α, β, γ, and all type environemts η relative to $\mathcal{D}$, but this fact cannot be proved in $\mathrm{BH}_{\leqslant}$. One strategy for remedying this is to add all type inequalities of this form, as do Amadio and Cardelli (1993), who interpret types under what is called the *F-semantics.* Another strategy consists in replacing the atoms $\top, \bot$, which are only needed in order to avoid the triviality of the subtype relation, by atoms $\mathbf{nat}, \mathbf{int}$ with $\mathbf{nat} \leqslant \mathbf{int}$. We use a domain $\mathcal{D}$ satisfying the equation

$$\mathcal{D} \cong \mathbb{Z}_\bot + [\mathcal{D} \to \mathcal{D}]_\bot + \{\mathtt{err}\}_\bot,$$

where $\mathbb{Z}_\bot$ is the flat CPO of integers, $E_\bot$ denotes the lifting of E, and $D+E$ is the coalesced sum of CPOs, i.e. the disjoint union of the non-bottom elements of D and E, with a new bottom element. The element $\mathtt{err}$ is the semantic value of terms that lead to run-time errors, notably those arising from type-incorrect applications. Such a $\mathcal{D}$ can be obtained as an inverse limit, endowed with a notion of approximation, having the same properties as in Definition 10.2.1. Then, types are interpreted as subsets of $\mathcal{D}$ that do not contain $\mathtt{err}$ and are closed in the Scott topology.

11.1.34 Definition Let $X \subseteq \mathcal{D}$. Then X is called an *ideal* if it is non-empty and closed in the Scott topology.

These variations on the standard setting are exploited to show that the natural relation of semantic subtyping induces a simulation on recursive types.

11.1.35 Lemma *If A, B, A', B' are ideals of $\mathcal{D}$, then*

$$(A \Rightarrow B) \subseteq (A' \Rightarrow B') \;\Rightarrow\; A' \subseteq A \,\&\, B \subseteq B'.$$

Proof Let $A \Rightarrow B \subseteq A' \Rightarrow B'$ and suppose, towards a contradiction, that $A' \not\subseteq A$. Then there is a compact element d of D such that $d \in A' \backslash A$. The step function $d \mapsto \mathtt{err}$ belongs to $A \Rightarrow B$ because no $x \sqsupseteq d$ is in A; but it does not belong to $A' \Rightarrow B'$, a contradiction.

Similarly suppose $B \not\subseteq B'$. Take $e \in B \backslash B'$. Therefore it follows that $(\bot_D \mapsto e) \in (A \Rightarrow B) \backslash (A' \Rightarrow B')$, again a contradiction. ■

11.1.36 Corollary *Define on $\mathbb{T}_\mu$ the binary relation*

$$A \,\mathcal{R}\, B \;\Leftrightarrow\; \forall \eta.\, \mathcal{I}[\![A]\!]_\eta \subseteq \mathcal{I}[\![B]\!]_\eta,$$

where η ranges over type environments that interpret type variables as ideals of $\mathcal{D}$. Then $\mathcal{R}$ is a simulation.

Proof This follows easily from Lemma 11.1.35 and Theorem 10.2.19. ■

11.1.37 Theorem (Completeness) *Let $A, B \in \mathbb{T}_\mu$. Then*

$$\forall\eta \left(\mathcal{I}[\![A]\!]_\eta \subseteq \mathcal{I}[\![B]\!]_\eta\right) \Rightarrow A \leqslant^*_\mu B.$$

Proof Define

$$C\,\mathcal{R}\,D \Leftrightarrow \forall\eta.\left(\mathcal{I}[\![C]\!]_\eta \subseteq \mathcal{I}[\![D]\!]_\eta\right).$$

Then $\mathcal{R}$ is a simulation, by Corollary 11.1.36. Now Proposition 11.1.29 applies. ∎

11.1.38 Corollary *Let $A, B \in \mathbb{T}_\mu$. Then the following are equivalent:*

(i) $A \leqslant^*_\mu B$;
(ii) $\forall\eta \left(\mathcal{I}[\![A]\!]_\eta \subseteq \mathcal{I}[\![B]\!]_\eta\right)$;
(iii) $\vdash_{\mathrm{BH}_\leqslant} A \leqslant B$.

Proof (i) ⇔ (ii). By Theorem 11.1.37 and Corollary 11.1.33.
(i) ⇔ (iii). By Theorem 11.1.27. ∎

As a corollary we obtain Theorem 10.2.33, using Remark 11.1.24.

11.1.39 Theorem $A =^*_\mu B \Leftrightarrow \forall\eta.[\mathcal{I}[\![A]\!]_\eta = \mathcal{I}[\![B]\!]_\eta]$.

11.2 The principal type structures

The notions and results in this section come from Polonsky (2011), but we simplify the proofs.

Remember that a *type structure* is of the form $\mathcal{S} = \langle|\mathcal{S}|, \leq_\mathcal{S}, \rightarrow_\mathcal{S}\rangle$; a type theory $\mathcal{T}$ is a set of type inequalities closed under derivations of the system given in Definition 11.1.9. Let $\mathcal{S}$ range over type structures and $\mathcal{T}$ over type theories. For a type theory $\mathcal{T}$, remember that $\mathcal{S}_\mathcal{T}$ is the *type structure generated by* $\mathcal{T}$, as defined in Definition 11.1.11.

Similar to Definition 9.2.2 to each untyped $M \in \Lambda$ we assign a triple $\langle\Gamma_M, \mathcal{S}_M, a_M\rangle$, where now $\mathcal{S}_M$ is a type structure, $a_M \in \mathcal{S}_M$, and Γ_M is a basis over $\mathcal{S}_M$. This $\mathcal{S}_M$ will be the *principal type structure* of M.

11.2.1 Definition Let $M \in \Lambda$. We are going to define a set of type constants $\mathbf{c}_M$, a type α_M, a basis Γ_M, a set of type inequalities $\mathcal{T}_M$, and a type algebra $\mathcal{A}_M$ with element a_M as follows. We first define, for each subterm-occurrence $L \subseteq M$ for L not a variable, a distinct type atom α_L. For variables x occuring in M we choose a fixed type α_x (for different occurrences the same α_x). Then we define the type theory $\mathcal{T}_L$ for each subterm-occurrence $L \subseteq M$, obtaining this notion also for M as we have $M \subseteq M$; see Fig. 11.4. Define

L	$\mathcal{T}_L$
x	$\emptyset$
PQ	$\mathcal{T}_P \cup \mathcal{T}_Q \cup \{\alpha_P \leq (\alpha_Q \to \alpha_{PQ})\}$
$\lambda x.P$	$\mathcal{T}_P \cup \{\alpha_x \to \alpha_P \leq (\alpha_{\lambda x.P})\}$

Figure 11.4

$\mathbf{c}_M$ as the set of all atomic types α_L and α_x occurring in M. Finally we define

$$\begin{aligned} \Gamma_M &= \{x{:}[\alpha_x] \mid x \in \mathrm{FV}(M)\}, \\ \mathcal{S}_M &= \langle \mathbb{T}^{\mathbf{c}_M}/\mathcal{T}_M, \leq, \to \rangle, \\ a_M &= [\alpha_M]_{\mathcal{T}_M}. \end{aligned}$$

The type a_M is called the *principal recursive type* of M and $\mathcal{S}_M$ its *principal type structure*. We say Γ_M is the *principal recursive basis* of M, which is empty if M is closed. The triple $\langle \Gamma_M, \mathcal{S}_M, a_M \rangle$ is called *principal recursive triple*. We call $\mathcal{T}_M$ the *principal type theory* of M. If we consider the α_L as indeterminates then $\mathcal{T}_M$ is like a simultaneous recursion, but with inequalities.

Below we will derive a principal type theorem for type structures very similar to Theorem 9.2.3 for type algebras. We need an extra definition and another lemma.

11.2.2 Definition Let D be derivation of $\Gamma \vdash_{\mathcal{S}} M : a$. For each subterm L of M let D_L be the smallest subderivation of D whose last judgement has L as subject. Then b_L denotes the type that is assigned to L in the conclusion of D_L.

11.2.3 Example Let $M \equiv \mathsf{I}y$ and D be the derivation

$$\dfrac{\dfrac{\dfrac{x{:}a \vdash x : a}{\vdash \mathsf{I} : (a {\to} a)} \quad a \leq c}{\vdash \mathsf{I} : (a \to c)} \quad y{:}a \vdash y : a}{\vdash M : c}.$$

Then $b_x = a, b_{\mathsf{I}} = a \to a, b_M = c$.

11.2.4 Lemma

(i) *Let* $\Gamma \vdash PQ : b_{PQ}$ *have subderivations* $\Gamma \vdash P : b_P$, $\Gamma \vdash Q : b_Q$. *Then*

$$b_P \leq b_Q \to b_{PQ}.$$

(ii) *Let* $\Gamma \vdash \lambda x.P : b_{\lambda x.P}$ *have as subderivation* $\Gamma, x{:}b_x \vdash P : b_P$. *Then*

$$b_x \to b_P \leq b_{\lambda x.P}.$$

Proof (i) Let the derivation D of $\Gamma \vdash PQ : b_{PQ}$ be

$$\dfrac{\dfrac{\dfrac{\vdots}{\Gamma \vdash P : b_P}\quad \vdots\ \leq}{\Gamma \vdash P : c \to b_{PQ}} \qquad \dfrac{\dfrac{\vdots}{\Gamma \vdash Q : b_Q}\quad \vdots\ \leq}{\Gamma \vdash Q : c}}{\Gamma \vdash PQ : b_{PQ}}.$$

Now $b_P \leq c \to b_{PQ}$ and $b_Q \leq c$. Hence $b_P \leq c \to b_{PQ} \leq b_Q \to b_{PQ}$.

(ii) Similarly. ■

11.2.5 Theorem *For* $M \in \Lambda$, *the principal recursive triple* $\langle \Gamma_M, \mathcal{S}_M, a_M \rangle$ *satisfies:*

(i) $\Gamma_M \vdash_{\mathcal{S}_M} M : a_M$;

(ii) $\Gamma \vdash_{\mathcal{S}} M : a \Leftrightarrow$ *there is a morphism* $h : \mathcal{S}_M \to \mathcal{S}$ *such that* $h(\Gamma_M) \subseteq \Gamma$ *and* $h(a_M) \leq a$;

(iii) *for closed* $M \in \Lambda^{\emptyset}$ *this simplifies to*

$$\vdash_{\mathcal{S}} M : a \Leftrightarrow \exists h : \mathcal{S}_M \to \mathcal{S}.h(a_M) \leq a.$$

Proof (i) As for Theorem 9.2.3(i) with $=_{\mathcal{E}}$ replaced by $\leq_{\mathcal{T}}$.

(ii) ($\Leftarrow$) By (i) and Lemmas 11.1.7, 11.1.4.

($\Rightarrow$) Similar to ($\Rightarrow$) in Theorem 9.2.3, where b_L is given by Definition 11.2.2, using Lemma 11.2.4.

(iii) By (ii). ■

11.2.6 Remark Using the principal type theories of $M, N \in \Lambda$ we can define

$$M \preccurlyeq N \Leftrightarrow \text{there exists a morphism } h : \mathcal{T}_M \to \mathcal{T}_N$$

This introduces a new non-trivial partial order relation on Λ. This relation remains non-trivial over the set of unsolvable terms, and therefore provides a new avenue for studying relations between unsolvable terms. It is not immediately obvious whether the order described above is decidable, because type theories may be equivalent without being the same (for example, $\{\alpha \leq \beta\}$ is equivalent to $\{\alpha \leq \alpha\}$). We conjecture that the $\preccurlyeq$-order on Λ is decidable.

11.3 Recursive types in programming languages

The results proved in the present Part II are relative to (slight variants of) the λ-calculus, while the languages of interest for programming practice normally include predefined term and type constants and other type constructors, as mentioned in the notion of enriched type algebra in Section 7.2. Nevertheless, the theoretical work on recursive types expounded in Chapters 7–10 constitutes a framework for modeling the uses of recursive types in programming. Many results (for example the principal type-scheme property) carry over literally to the extended languages used in practice in a straightforward manner.

In this section we review very sketchily the various forms that recursive types take in programming languages. Recursive types are orthogonal to different programming paradigms, so we shall see how they show up in imperative, functional and object-oriented languages. We shall do this mostly by means of examples, referring to the literature for more details.

Extending the language

If we consider an extension of the λ-calculus where term constants c are present, then we must assume for each of these a type $\tau(c)$ belonging to the corresponding enriched type algebra. In this case it is standard to assume that some of the atoms of $\mathbb{A}$ represent data types (for instance an atom **int** representing the set of integers). Each term constant c is then given a type $\tau(c)$ belonging to $\mathbb{T}_\mu(\mathbb{A})$ (for instance $\tau(n) = \mathbf{int}$ for all numerals n, $\tau(+) = \mathbf{int} \to \mathbf{int} \to \mathbf{int}$, etc.).

11.3.1 Remark Observe also that a rich system of types, including an encoding of the types for integer and Boolean values, can be based on recursive types, provided a minimal set of basic type constructors is available. As an example, assume that the set of basic recursive types is extended as described by the following simplified syntax (Cosmadakis, 1989):

$$\boxed{\mathbb{T}_\mu \quad ::= \quad \mathbb{A} \mid \mathbb{T}_\mu \to \mathbb{T}_\mu \mid \mu\mathbb{A}.\mathbb{T}_\mu \mid \mathbb{T}_\mu \times \mathbb{T}_\mu \mid \mathbb{T}_\mu \oplus \mathbb{T}_\mu \mid (\mathbb{T}_\mu)_\perp.}$$

Then we can define the following data types:

$$\begin{aligned}
triv &= \mu t.t;\\
\mathsf{O} &= triv_\perp;\\
bool &= \mathsf{O} \oplus \mathsf{O};\\
nat &= \mu t.\mathsf{O} \oplus t;\\
obliq &= \mu t.\mathsf{O} \oplus t_\perp;\\
vert &= \mu t.t_\perp;\\
lamb &= \mu t.\mathsf{O} \oplus (t \to t);\\
lazy &= \mu t.(t \to t)_\perp.
\end{aligned}$$

Within a formulation with explicit typing as in Section 7.1.24, we can add typed constants representing constructors and destructors for each of the new types. So, for example, we may have constants $up : A \to A_\perp$ and $down : A_\perp \to A$ (possibly with a new conversion rule $down(up(x)) = x$). This set of types is useful in formalizing a meta-language for denotational semantics. Type *lamb* then encodes the terms of the pure untyped λ-calculus, whereas *lazy* does the same for terms of the lazy λ-calculus (Abramsky, 1990). The types *nat*, *vert* and *obliq* represent three versions of a type of natural numbers.

Finding and checking types

Assume for the moment that only constants and constant types, and no other type constructors are added. Adding these is straightforward. In this case usually one must take into account that constant types are atoms. We do not allow, for instance, that **int** be equivalent to an arrow type. Moreover invertibility is always assumed. We say that the (principal) set, $\mathcal{E}_M$, of type equations is *consistent* if it does not imply any equation $\kappa = A$, where k is a constant type and A is either a different constant type or a non-atomic type expression. Then, as a consequence of Theorem 9.2.3 a term M can be typed (with respect to some invertible type algebra) iff $\mathcal{E}_M$ is consistent. We can take into account constants in the construction of $\mathcal{E}_M$ in Definition 9.2.2 by taking $\mathcal{E}_c = \{\alpha_c = \tau(c)\}$ for each constant c occurring in M. We then can easily prove that $\mathcal{E}_M$ is consistent if and only if for each atomic constant type κ the simultaneous recursion $\mathcal{E}_M$ does not contain equations $\kappa = a_1, a_1 = a_2, \ldots, a_n = C$ where $n \geq 0$ and C is either a constant type different from κ or a non-atomic type expression. This also yields an algorithm for deciding whether or not $\mathcal{E}_M$ is consistent.

All the results given in Section 9.2 still hold if we consider terms with additional constants. In some cases this makes these results more interesting. For instance, the problem of deciding whether a given term has a type with respect to some simultaneous recursion is no longer trivial since there are terms, such as $(3\,3)$, that have no type at all with respect to any simultaneous recursion. By the subject reduction theorem, however, we can still prove that a typable term in any system with recursive types cannot produce incorrect applications during its evaluation. This is one of the motivations for the practical uses of these systems in programming languages (Milner (1978)).

In real programming languages recursive types are usually defined via recursive type equations, so the decidability of the corresponding problem of typability is useful. However, in this application, the declaration of constructors, which is usually included in recursive type definitions makes, type inference much easier.

Imperative programming

Historically, a form of recursive type definition was introduced in ALGOL 68, see van Wijngaarden (1981), as "recursive mode declarations", for example

$$\textbf{mode cell} = \textbf{struct}(\textbf{ref cell}\ \texttt{next}, \textbf{int}\ \texttt{item})$$

Also in this case, equality of recursive modes can be checked by reducing the problem to that of deciding the equality of infinite regular trees (see Koster (1969)) or, equivalently, of components of solutions of finite systems of equations (Pair (1970)); the latter method is thoroughly investigated in Courcelle et al. (1974), whereas the first anticipates some ideas on which the co-inductive proof system of Brandt and Henglein (1998) is based: in particular, this algorithm also allows a certain amount of circularity because in order to decide whether two modes are equivalent, one assumes that they are and shows that this assumption does not lead to a contradiction.

Functional programming

Typed functional programming languages in the tradition of ML (Gordon et al., 1979), including Miranda and Haskell, allow recursive definitions of data-types which the type-checker exploits in inferring type schemes for programs, see Peyton-Jones (1987), Chapters 8–9. A typical example of such definitions, say, in Haskell (Peyton Jones et al., 1999), is

```
data Tree a = Leaf a | Branch (Tree a) (Tree a)
```

which declares `Tree` as a (polymorphic) *type* constructor, and `Leaf` and `Branch` as *data* constructors, so that `Leaf` has type `a -> Tree a` and `Branch` has type `Tree a -> (Tree a -> Tree a)`; an element of this type is a binary tree with leaves of type `a`, where `a` is any type, see Hudak et al. (1999), §2.2.1.

Another example from SML (a call-by-value language) is a type declaration intended to define the notion of polymorphic streams:

```
type 'a stream  = Scons of (unit -> 'a * 'a stream)
```

where `unit` is the one-element type. The function type is introduced to suspend evaluation.

Definitions like the example above also declare *constructors* (like `Leaf` and `Branch` in the example) that can be used in defining functions over elements of the data-type by pattern-matching (Burstall, 1969). In type inference, however, constructors are treated like constants and have then a specific type (or type scheme) assigned a priori.

Object-Oriented programming

Recursive types have been widely used in the theoretical study of object-oriented programming, especially in the many proposals that have been made for encoding objects as terms of λ-calculi extended with record structures. There is an extensive literature on this subject, originating with the work of Cardelli (1988); we just discuss a few motivating examples.

11.3.2 Definition We extend $\mathbb{T}_\mu = \mathbb{T}_\mu^{\mathbb{A}}$ with *record types* having the form $\{\ell_1 : A_1, \ldots, \ell_n : A_n\}$ as indicated by the following simplified syntax:

$$\boxed{\mathbb{T}_\mu \quad ::= \quad \mathbb{A} \mid \mathbb{T}_\mu \to \mathbb{T}_\mu \mid \mu\mathbb{A}.\mathbb{T}_\mu \mid \{\ell_1 : \mathbb{T}_\mu, \ldots, \ell_n : \mathbb{T}_\mu\}}$$

where $\ell_1, \ldots, \ell_n$ are distinct *labels*, and we assume that $\mathbb{A}$ contains atoms for base types such as **nat**, **int**, **bool**, ...

Correspondingly, the set of terms is extended with *record structures* of the form $\{\ell_1 = M_1, \ldots, \ell_n = M_n\}$ where $M_1, \ldots, M_n$ are terms, and constants of base types.

Record structures are tuples whose components are accessed via their labels: for $\{\ell_1 = M_1, \ldots, \ell_n = M_n\}$ the expression structure, then $\{\ell_1 = M_1, \ldots, \ell_n = M_n\}.\ell_i$ has the same meaning as M_i.

Recursive record structures can be interpreted as *objects*. For example, we may have a point object in the plane defined as:

$$\mathsf{Y}\left(\lambda s.\{\mathsf{x} = 1.0, \mathsf{y} = 2.0, \mathsf{dist} = \lambda p.\sqrt{(p.\mathsf{x} - s.\mathsf{x})^2 + (p.\mathsf{y} - s.\mathsf{y})^2}\}\right)$$

with coordinates of type **real**, x and y, and a method dist for calculating its distance from another point. Observe that this point has type

$$\mathsf{Point} \equiv \{\mathsf{x} : \mathbf{real}, \mathsf{y} : \mathbf{real}, \mathsf{dist} : \mathsf{Point} \to \mathbf{real}\},$$

a recursive record type. Further developments of this approach to modeling objects are thoroughly described in Abadi and Cardelli (1996).

11.4 Further reading

This section collects further references containing additional information on the topics discussed in Part II.

Historical

The origins of recursive types can be traced back to the beginning of theoretical computer science. On the semantical side, they appeared in the special form of recursive definitions of sets. For example, McCarthy (1963),

§2.6, discussed the set S of *sequences* of elements of a set A, defined recursively by $S = 1 + (A \times S)$ (to be read: "a sequence is either empty, or is a pair whose first component is an A and whose second component is a sequence"). McCarthy also observed that this definition implies that $1 = S - (A \times S) = S \times (1 - A)$, hence $S = 1/(1 - A)$, whose expansion gives $S = 1 + A + A^2 + A^3 + \cdots$, which describes a sequence as either empty, or consisting of two elements of A, or consisting of three elements of $A, \ldots$ (The justification of such calculations requires a good understanding of algebraic facts; see Fiore (2004) for an introduction to these, with applications to deciding if recursive types are isomorphic). Of course, the various categories of domains and the methods of solving recursive domain equations over them (Smyth and Plotkin (1982), Gunter and Scott (1990), Freyd (1990), Freyd (1991), Freyd (1992)) can be regarded as comprehensive semantical universes for recursive types, as may be seen already from the straightforward use of the category of CPOs in Section 10.3.

On the syntactical side, recursive types appeared in J.H. Morris' thesis (Morris, 1968), who considered the possibility of allowing, in the simply typed λ-calculus, 'circular' type expressions like the solution of the equation $X = X \to \alpha$. At about the same time, a form of recursive type definition was introduced in ALGOL 68 as 'recursive mode declarations' mentioned in Section 11.3.

Type inference and equivalence of recursive types

Recursive type equations such as those considered by Morris arise naturally when omitting the 'occur check' in the unification procedure invoked by the type inference algorithms of Curry (1969), Hindley (1969) and Milner (1978), as remarked already in Wand (1987). See Aho et al. (1986) for a textbook treatment. The solutions of these equations are best regarded as infinite expressions or, equivalently, infinite trees. They are always *regular* (or *rational*) trees (Courcelle, 1983, §4) a fact that yields efficient decision procedures for checking their equality. This view of recursive types as infinite trees has then been exploited systematically in Cardone and Coppo (1991), and is at the basis of this whole Part.

More recent approaches exploit the fact that the set of finite and infinite trees (over a first-order signature) is the final coalgebra of a polynomial endofunctor over the category of sets canonically associated with the signature, as in Proposition 7.6.16. This is a well-known fact in the theory of coalgebras (Rutten, 2000), and has been exploited in Fiore (1996).

Models

The first model for the implicitly typed systems was given in MacQueen et al. (1986), where types are interpreted as ideals over a Scott domain of the form $D \cong V + [D \to D]$. Recursive types are restricted to those of the form $\mu t.A$ where A is contractive in t, and are then interpreted as the unique fixed point of the induced contractive mapping over the complete metric spaces of ideals, whose existence is guaranteed by the Banach fixed point theorem, see Theorem 7.6.5.

The interpretation of recursive types described in Section 10.2, exploiting the approximation structure of domains like D, as inverse limits, (Scott, 1972), stems from Coppo (1985), where also the completeness theorems 10.2.32 and 10.2.33 were first proved. It has been shown that the technique described in that paper can be extended to model constructions for various extensions of simple recursive types: on the one hand it applies to many first-order type constructors without the contractiveness requirement, as in Cardone et al. (1994). On the other, it can be extended to second-order polymorphic types (Abadi and Plotkin (1990)) and even to bounded polymorphic types with subtyping (Cardone (1991)). Orthogonally to these applications, it is possible to adapt the constructions to explicitly typed systems, by performing them over (complete and uniform) partial equivalence relations. Concerning the interpretation of the explicitly typed systems, the domain models presented in Section 10.3 belong to the folklore of the subject.

Subtyping

The theory of subtyping recursive types has received much attention in recent years, mainly for the interest it has for the design of object-oriented programming languages. The seminal paper on this topic is Amadio and Cardelli (1993). The presentation of subtyping was refined in Brandt and Henglein (1998), who exploited in an elegant way the co-inductive basis of reasoning on infinite objects, such as the infinite trees of which recursive types can be regarded as finite notations. See also Grabmayer (2005, 2007) for work in this direction. Algorithms for deciding subtyping between μ-types, with the relative complexity issues, were studied first in Kozen et al. (1995). We have not dealt at all with this topic; pointers to the relevant literature can be found in Gapeyev et al. (2002). In Jay (2009) the (typed) lambda calculus is extended with patterns as a first-class citizen.

11.5 Exercises

Exercise 11.1 Call type theories T, T' *equivalent* if $T \preccurlyeq T'$ and $T' \preccurlyeq T$. In each of the following, show equivalence of the listed theories.

(1) $\emptyset, \{\alpha \leq \alpha\}, \{\alpha \leq \beta\}, \{\alpha \leq A\}$, where A is a type with $\alpha \notin A$.
(2) $\{\alpha \leq (\alpha \to \beta)\}, \{(\alpha \to \beta) \leq \alpha\}, \{\gamma \leq \alpha, \gamma \leq (\alpha \to \beta)\}$.

Exercise 11.2 Let **TA** be the category of type algebras and **TS** be the category of type structures. Let U : **TS** $\to$ **TA** be the forgetful map that sends $(S, \to, \leq)$ to $(S, \to)$.

(1) Show that there is a functor F : **TA** $\to$ **TS** such that U is right adjoint to F.
(2) Show that F also has a left adjoint C.
(3) Let M be a lambda term, T the principal type theory of M. Show that $C(S_T)$ is isomorphic to the principal type algebra of M in the sense of Definition 9B.2.
(4) Show that $\mathbb{T}$ is a weakly initial object in **TS** and that $S_{\{\alpha \leq (\alpha \to \alpha), (\alpha \to \alpha) \leq \alpha\}}$ is a weakly terminal object.
(5) Show that **TS** has arbitrary coproducts.
(6) Does **TS** have arbitrary finite products?
(7) *Show that the Subtyping Order refines the Typability Order induced by* $\preccurlyeq$ *between type algebras.*
(8) Show that there exist two unsolvables whose principal type theories are $\preccurlyeq$-incompatible.

PART III

INTERSECTION TYPES $\boldsymbol{\lambda}_{\cap}^{\mathcal{S}}$

In a nutshell the intersection type systems considered in this Part form a class of type assignment systems for untyped λ-calculus, extending Curry's *basic functionality*, in the context of identifications and subtyping, with a new type constructor, *intersection*. This simple move makes it possible to express naturally and in a finitary way many *operational* and *denotational* properties of terms.

Intersection types were originally introduced as a language for describing and capturing properties of λ-terms that had escaped all previous typing disciplines. For instance, they were used in order to give the first type-theoretic characterization of *strongly normalizing* terms, and later of *normalizing terms*.

It was realized early on that intersection types also had a distinctive semantical flavor: they express at a syntactical level the fact that a term belongs to suitable compact open sets in a Scott domain. Building on this intuition, intersection types were used in Barendregt et al. (1983) to introduce filter models and give a proof of the completeness of the natural semantics of simple type assignment systems in applicative structures suggested in Scott (1972).

Since then, intersection types have been used as a powerful tool both for the analysis and the synthesis of λ-models. On the one hand, intersection type disciplines provide finitary inductive definitions of interpretation of λ-terms in models. On the other hand, they are suggestive for the shape the domain model has to have in order to exhibit certain properties.

Intersection types can be viewed also as a restriction of the domain theory in logical form, see Abramsky (1991), to the special case of modeling pure lambda calculus by means of ω-algebraic complete lattices. Many properties of these models can be proved using this paradigm, which goes back to Stone duality.

Type assignment using intersection types can be parametrized by intersection type theories or intersection type structures. The various type theories (and corresponding type structures) are introduced together in order to give reasonably uniform proofs of their properties as well of those of the corresponding type assignment systems and filter models.

In the present Part III of this book both these syntactic and semantic aspects will be explored.

The interested reader can find a continuously updated bibliography maintained by Joe Wells on intersection types at <www.macs.hw.ac.uk/~jbw/itrs/bibliography.html>. Introductions to intersection types are in Cardone and Coppo (1990) and Hindley (1992).

12

An Example System

There are several systems that assign intersection types to untyped lambda terms. These will be collectively denoted by $\boldsymbol{\lambda}_\cap$. In this chapter we consider one particular system of this family, $\boldsymbol{\lambda}_\cap^{\mathrm{BCD}}$, in order to outline the concepts and related properties. Definitions and the statements of theorems will be given, but no proofs. These can be found in the subsequent chapters.

One motivation for the system presented comes from trying to modify the system $\boldsymbol{\lambda}_\rightarrow$ in such a way that not only subject reduction, but also subject expansion holds. The problem of subject expansion is the following. Suppose $\vdash_{\boldsymbol{\lambda}_\rightarrow} M : A$ and that $M' \twoheadrightarrow_{\boldsymbol{\beta}} M$. Does one have $\vdash_{\boldsymbol{\lambda}_\rightarrow} M' : A$? Let us focus on one $\boldsymbol{\beta}$-step. So let $M' \equiv (\lambda x.P)Q$ be a redex and suppose

$$\vdash_{\boldsymbol{\lambda}_\rightarrow} P[x := Q] : A. \tag{12.1}$$

Do we have $\vdash_{\boldsymbol{\lambda}_\rightarrow} (\lambda x.P)Q : A$? It is tempting to reason as follows. By assumption (12.1) Q must also have a type, say B. Then $(\lambda x.P)$ has a type $B \rightarrow A$ and therefore $\vdash_{\boldsymbol{\lambda}_\rightarrow} (\lambda x.P)Q : A$. The mistake is that in (12.1) there may be several occurrences of Q, say $Q_1 \equiv Q_2 \equiv \cdots \equiv Q_n$, having as types respectively $B_1, \ldots, B_n$. It may be impossible to find a single type for all the occurrences of Q and this prevents us from finding a type for the redex. For example

$$\vdash_{\boldsymbol{\lambda}_\rightarrow^{\mathbb{A}}} (\lambda x.\mathsf{I}(\mathsf{K}x)(\mathsf{I}x)) : A \rightarrow A,$$
$$\nvdash_{\boldsymbol{\lambda}_\rightarrow^{\mathbb{A}}} (\lambda xy.x(\mathsf{K}y)(xy))\mathsf{I} : A \rightarrow A.$$

The system introduced in this chapter with intersection types assigned to untyped lambda terms remedies the situation. The idea is that if the several occurrences of Q have to have different types $B_1, \ldots, B_n$, we give them all of these types, as follows:

$$\vdash Q : B_1 \cap \cdots \cap B_n,$$

implying that for all i one has $Q : B_i$. Then we will get

$$\begin{array}{rcll} \vdash (\lambda x.P) & : & B_1 \cap \cdots \cap B_n \rightarrow A & \text{and} \\ \vdash ((\lambda x.P)Q) & : & A. & \end{array}$$

There is, however, a second problem. In the λK-calculus, with its terms $\lambda x.P$ such that $x \notin \mathrm{FV}(P)$ there is the complication that Q may not be typable at all, as it may not occur in $P[x := Q]$! This is remedied by allowing $B_1 \cap \cdots \cap B_n$ for $n = 0$ also and writing this type as $\mathtt{U}$, to be considered as the universal type, i.e. assigned to all terms. Then in the case $x \notin \mathrm{FV}(P)$ one has

$$\begin{array}{rcll} \vdash Q & : & \mathtt{U} & \\ \vdash (\lambda x.P) & : & \mathtt{U} \to A & \text{and} \\ \vdash ((\lambda x.P)Q) & : & A. & \end{array}$$

This is the motivation for introducing a $\leq$ relation on types with largest element $\mathtt{U}$ and intersections satisfying $A \cap B \leq A, A \cap B \leq B$ and the extension of the type assignment by the sub-sumption rule

$$\Gamma \vdash M : A,\ A \leq B \ \Rightarrow\ \Gamma \vdash M : B.$$

It has as a consequence that terms such as $\lambda x.xx$ get as type $((A \to B) \cap A) \to B$, while $(\lambda x.xx)(\lambda x.xx)$ only gets $\mathtt{U}$ as type. Moreover we have subject conversion

$$\Gamma \vdash M : A \;\&\; M =_{\beta} N \ \Rightarrow\ \Gamma \vdash N : A.$$

This implies that one can create a so-called filter lambda model in which the meaning of a closed term consists of the collection of types it gets. In this way new lambda models will be obtained and new ways for studying classical models as well. It is worth noting that these models are models of the untyped lambda calculus.

The type assignment system $\boldsymbol{\lambda}_{\cap}^{\mathrm{BCD}}$ will be introduced in Section 12.1 and the corresponding filter model in Section 12.2. Results are often stated without proof, as these will appear in later chapters.

12.1 The type assignment system $\boldsymbol{\lambda}_{\cap}^{\mathrm{BCD}}$

A typical member of the family of intersection type assignment systems is $\boldsymbol{\lambda}_{\cap}^{\mathrm{BCD}}$. This system was introduced in Barendregt et al. (1983) as an extension of the initial system in Coppo and Dezani-Ciancaglini (1980).

12.1.1 Definition

(i) Define the following sets of type atoms:

$$\begin{array}{rcl} \mathbb{A}_{\infty} & = & \{\mathsf{c}_0, \mathsf{c}_1, \mathsf{c}_2, \ldots\}; \\ \mathbb{A}_{\infty}^{\mathtt{U}} & = & \mathbb{A}_{\infty} \cup \{\mathtt{U}\}; \\ \mathbb{A}^{\mathrm{BCD}} & = & \mathbb{A}_{\infty}^{\mathtt{U}}, \end{array}$$

where the type atom $\mathtt{U} \notin \mathbb{A}_\infty$ is a special symbol called the universe or universal top.

(ii) The intersection type language over $\mathbb{A}^{\mathrm{BCD}}$, denoted by $\mathbb{T} = \mathbb{T}_\cap^{\mathrm{BCD}}$, is defined by the following simplified syntax:

$$\boxed{\mathbb{T} ::= \mathbb{A}^{\mathrm{BCD}} \mid \mathbb{T}{\to}\mathbb{T} \mid \mathbb{T} \cap \mathbb{T}}$$

12.1.2 Notation

(i) Greek letters $\alpha, \beta, \ldots$ will denote arbitrary atoms in $\mathbb{A}^{\mathrm{BCD}}$.
(ii) A, B, C, D, E range over arbitrary types in $\mathbb{T}$.
(iii) In Barendregt et al. (1983) the universe was denoted by ω.
(iv) When writing intersection types we shall use the following convention: the constructor $\cap$ takes precedence over the constructor $\to$ and the constructor $\to$ associates to the right. For example

$$(A{\to}B{\to}C) \cap A{\to}B{\to}C \equiv ((A{\to}(B{\to}C)) \cap A){\to}(B{\to}C).$$

12.1.3 Remark Throughout Part III other sets $\mathbb{T}$ of types will be formed by replacing the set $\mathbb{A}^{\mathrm{BCD}}$ of type atoms by an arbitrary set $\mathbb{A}$ (finite or countably infinite). In this chapter, however, we take $\mathbb{A} = \mathbb{A}^{\mathrm{BCD}} = \mathbb{A}_\infty^{\mathtt{U}}$.

The following deductive system is intended to introduce an appropriate pre-order on $\mathbb{T}$, compatible with the operator $\to$, such that $A \cap B$ is a greatest lower bound of A and B, for each A, B.

12.1.4 Definition (Intersection type preorder) The intersection type theory BCD is the set of all statements $A \leq B$ (to be read as "A sub B"), derivable from the axioms and rules in Fig. 12.1, where $A, B, C, \ldots \in \mathbb{T}$.

12.1.5 Notation

(i) For $(A \leq B) \in \mathrm{BCD}$ we write $A \leq_{\mathrm{BCD}} B$ or $\vdash_{\mathrm{BCD}} A \leq B$ (or often just $A \leq B$ if there is little danger of confusion).
(ii) Write $A =_{\mathrm{BCD}} B$ (or $A = B$) for $A \leq_{\mathrm{BCD}} B$ & $B \leq_{\mathrm{BCD}} A$.
(iii) We write $[\mathbb{T}]$ for the set $\mathbb{T}$ modulo $=_{\mathrm{BCD}}$. For types in BCD we usually work with $[\mathbb{T}]$.
(iv) We write $A \equiv B$ for syntactic identity. For example $A \cap B \equiv A \cap B$, but $A \cap B \not\equiv B \cap A$.

12.1.6 Remark All systems in Part III have the first five axioms and rules of Fig. 12.1. They differ in the extra axioms and rules and the set of atoms.

12.1.7 Proposition *The equivalence relation* $=$ *in Notation* 12.1.5(ii) *is a congruence, i.e.* $=$ *is compatible with* $\cap$ *and* $\to$.

(refl)	$A \leq A$
(incl$_L$)	$A \cap B \leq A$
(incl$_R$)	$A \cap B \leq B$
(glb)	$\dfrac{C \leq A \quad C \leq B}{C \leq A \cap B}$
(trans)	$\dfrac{A \leq B \quad B \leq C}{A \leq C}$
($\mathtt{U}_{\text{top}}$)	$A \leq \mathtt{U}$
($\mathtt{U}\to$)	$\mathtt{U} \leq A \to \mathtt{U}$
($\to\cap$)	$(A \to B) \cap (A \to C) \leq A \to (B \cap C)$
($\to$)	$\dfrac{A' \leq A \quad B \leq B'}{(A \to B) \leq (A' \to B')}$

Figure 12.1

Proof By rules (trans), (incl$_L$), (incl$_R$) and (glb) one has

$$A = A' \ \&\ B = B' \ \Rightarrow\ (A \cap B) = (A' \cap B').$$

By rule ($\to$) one has

$$A = A' \ \&\ B = B' \ \Rightarrow\ (A \to B) = (A' \to B'). \ \blacksquare$$

12.1.8 Remark The theory BCD can be seen as a structure with a pre-order

$$\text{BCD} = \langle \mathbb{T}, \leq, \cap, \to, \mathtt{U} \rangle.$$

This means that $\leq$ is reflexive and transitive, but not anti-symmetric:

$$A \leq B \ \&\ B \leq A \ \not\Rightarrow\ A \equiv B.$$

One can go over to equivalence classes and define a partial order $\leq$ on $[\mathbb{T}]$ that satisfies antisymmetry:

$$[A] \leq [B] \ \Leftrightarrow\ A \leq B.$$

By Proposition 12.1.7, the operators $\cap$ and $\to$ can be defined on $[\mathbb{T}]$ by

$$[A] \cap [B] = [A \cap B];$$
$$[A] \to [B] = [A \to B].$$

(Ax)	$\Gamma \vdash x{:}A$	if $(x{:}A) \in \Gamma$
$(\to \mathrm{I})$	$\dfrac{\Gamma,\ x{:}A \vdash M : B}{\Gamma \vdash (\lambda x.M) : (A{\to}B)}$	
$(\to \mathrm{E})$	$\dfrac{\Gamma \vdash M : (A \to B) \quad \Gamma \vdash N : A}{\Gamma \vdash (MN) : B}$	
$(\cap \mathrm{I})$	$\dfrac{\Gamma \vdash M : A \quad \Gamma \vdash M : B}{\Gamma \vdash M : (A \cap B)}$	
$(\leq)$	$\dfrac{\Gamma \vdash M : A}{\Gamma \vdash M : B}$	if $A \leq_{\text{BCD}} B$
$(\mathtt{U}_{\text{top}})$	$\Gamma \vdash M : \mathtt{U}$	

Figure 12.2

We obtain a type structure

$$[\text{BCD}] \;=\; \langle [\mathbb{T}], \leq, \cap, \to, [\mathtt{U}] \rangle.$$

In this structure, $[\mathtt{U}]$ is the largest element (also called *top*) and $[A] \cap [B]$ is the *greatest lower bound* of $[A]$ and $[B]$.

12.1.9 Definition

(i) A *basis* is a finite set of statements of the shape $x{:}B$, where $B \in \mathbb{T}$, with all variables distinct.

(ii) The *type assignment* system $\boldsymbol{\lambda}_\cap^{\text{BCD}}$ for deriving statements of the form $\Gamma \vdash M : A$ with Γ a basis, $M \in \Lambda$ (the set of untyped lambda terms) and $A \in \mathbb{T}$ is defined by the axioms and rules in Fig. 12.2.

(iii) We say that a term M is *typable* from a given basis Γ if there is a type $A \in \mathbb{T}$ such that the judgement $\Gamma \vdash M : A$ is derivable in $\boldsymbol{\lambda}_\cap^{\text{BCD}}$. In this case we write $\Gamma \vdash_\cap^{\text{BCD}} M : A$ or just $\Gamma \vdash M : A$, if there is little danger of confusion.

12.1.10 Remark All systems of type assignment in Part III have the first five axioms and rules of Fig. 12.2

In Proposition 12.1.12 we need the notions of admissible and derived rule.

12.1.11 Definition Consider an unspecified rule of the form (possibly with several assumptions or side-conditions)

$$\frac{\Gamma \vdash M : A}{\Gamma' \vdash M' : A'} \qquad \text{if } \mathsf{p}(\Gamma, M, A) \quad (R)$$

where $\mathsf{p}(\Gamma, M, A)$ is a predicate on Γ, M, A.

(i) We say R is *admissible* if one has

$$\Gamma \vdash M : A \text{ and } \mathsf{p}(\Gamma, M, A) \ \Rightarrow \ \Gamma' \vdash M' : A'.$$

(ii) We say R is *derived* if $\mathsf{p}(\Gamma, M, A)$ is always true and there is a derivation starting from $\Gamma \vdash M : A$ that ends in $\Gamma' \vdash M' : A'$.

A derived rule is always admissible but the converse does not hold. If

$$\frac{\Gamma \vdash M : A}{\Gamma' \vdash M' : A'}$$

is a derived rule, then for all Δ one has that

$$\frac{\Gamma \cup \Delta \vdash M : A}{\Gamma' \cup \Delta \vdash M' : A'}$$

is also derived. Hence derived rules are closed under theory extension. We will only be concerned with admissible and derived rules for theories of type assignment. The statements of the next proposition are easy to prove. For instance, derivability of the rules ($\cap$E) follows immediately from rule ($\leq$). The other proofs are left to the reader.

12.1.12 Proposition

(i) *The rules* ($\cap$E)

$$\frac{\Gamma \vdash M : (A \cap B)}{\Gamma \vdash M : A} \qquad \frac{\Gamma \vdash M : (A \cap B)}{\Gamma \vdash M : B}$$

are derived in $\boldsymbol{\lambda}_\cap^{\mathrm{BCD}}$.

(ii) *The rules in Fig.* 12.3 *are admissible in the type assignment system* $\boldsymbol{\lambda}_\cap^{\mathrm{BCD}}$.

The next theorem is a particular case of an 'Inversion Lemma' (see Theorem 14.1.1).

12.1.13 Theorem *In* (i) below assume $A \neq \mathtt{U}$. Then

(i) $\Gamma \vdash x : A \ \Leftrightarrow \ \exists B \in \mathbb{T}.[(x{:}B \in \Gamma \ \&\ B \leq A]$.

(ii) $\Gamma \vdash (MN) : A \ \Leftrightarrow \ \exists B \in \mathbb{T}.[\Gamma \vdash M : (B \to A) \ \&\ \Gamma \vdash N : B]$.

(*weakening*)	$\dfrac{\Gamma \vdash M : A \quad x \notin \Gamma}{\Gamma, x{:}B \vdash M : A}$
(*strengthening*)	$\dfrac{\Gamma, x{:}B \vdash M : A \quad x \notin FV(M)}{\Gamma \vdash M : A}$
(*cut*)	$\dfrac{\Gamma, x{:}B \vdash M : A \quad \Gamma \vdash N : B}{\Gamma \vdash (M[x := N]) : A}$
($\leq$-L)	$\dfrac{\Gamma, x{:}B \vdash M : A \quad C \leq B}{\Gamma, x{:}C \vdash M : A}$
($\to$L)	$\dfrac{\Gamma, y{:}B \vdash M : A \quad \Gamma \vdash N : C \quad x \notin \Gamma}{\Gamma, x{:}(C \to B) \vdash (M[y := xN]) : A}$
($\cap$L)	$\dfrac{\Gamma, x{:}A \vdash M : B}{\Gamma, x{:}(A \cap C) \vdash M : B}$

Figure 12.3

(iii) $\Gamma \vdash \lambda x.M : A \Leftrightarrow \exists n{>}0 \exists B_1,\ldots,B_n, C_1,\ldots,C_n \in \mathbb{T}$
$\forall i \in \{1,\ldots,n\}.[\Gamma, x{:}B_i \vdash M : C_i \;\&$
$(B_1 \to C_1) \cap \cdots \cap (B_n \to C_n) \leq A]$.

(iv) $\Gamma \vdash \lambda x.M : B \to C \Leftrightarrow \Gamma, x{:}B \vdash M : C$.

12.1.14 Definition Let R be a notion of reduction. Then the rules *R-red* and *R-exp* are defined as follows:

(*R-red*)	$\dfrac{\Gamma \vdash M : A \quad M \to_R N}{\Gamma \vdash N : A}$
(*R-exp*)	$\dfrac{\Gamma \vdash M : A \quad M \leftarrow_R N}{\Gamma \vdash N : A}$

General results from Section 14.2 imply the next proposition. In more detail, Corollary 14.2.7(ii) implies admissibility of (β-*red*) and (β-*exp*), and Corollary 14.2.9 the admissibility of (η-*red*). The negative result for (η-*exp*) follows from Theorem 14.2.11(ii) together with Proposition 14.2.12(ii).

12.1.15 Proposition *The rules (β-red), (β-exp) and (η-red) are admissible in $\boldsymbol{\lambda}_\cap^{\mathrm{BCD}}$. The rule ($\eta$-exp) is not.*

The following result characterizes notions related to normalization in terms of type assignment in the system $\boldsymbol{\lambda}_\cap^{\mathrm{BCD}}$. The notation $\mathtt{U} \notin A$ means that $\mathtt{U}$ does not occur in A. The result follows from Theorem 17.2.15(i), (ii).

12.1.16 Theorem *Let $M \in \Lambda^{\emptyset}$.*

(i) *M has a head normal form* $\Leftrightarrow$ $\exists A \in \mathbb{T}.[A \neq_{\mathrm{BCD}} \mathtt{U}\ \&\ \vdash M : A]$.
(ii) *M has a normal form* $\Leftrightarrow$ $\exists A \in \mathbb{T}.[\mathtt{U} \notin A\ \&\ \vdash M : A]$.

Let M be a lambda term. For the notion 'approximant of M', see B[1984], Section 14.3. The *approximants of a term M* are obtained from the Böhm tree $\mathrm{BT}(M)$ of M by removing some branches and replacing these by a new symbol $\bot$. The set of approximants of M is denoted by $\mathcal{A}(M)$. For example, we have for the fixed-point combinator Y,

$$\mathcal{A}(\mathsf{Y}) = \{\bot\} \cup \{\lambda f.f^n\bot \mid n{>}0\}.$$

Approximants can be typed (for details see Section 17.3) by extending the typing rules from terms to approximants. For example it will be shown that

$$\begin{array}{l} \vdash \bot : \mathtt{U} \\ \vdash \lambda f.f\bot : (\mathtt{U}{\to}A_1){\to}A_1 \\ \vdash \lambda f.f(f\bot) : (\mathtt{U}{\to}A_1) \cap (A_1{\to}A_2){\to}A_2 \\ \vdots \\ \vdash \lambda f.f^n\bot : (\mathtt{U}{\to}A_1) \cap (A_1{\to}A_2) \cap \cdots \cap (A_{n-1}{\to}A_n){\to}A_n \\ \vdots \end{array}$$

The set of types of a term M will be shown to coincide with the union of the sets of types of its approximants $P \in \mathcal{A}(M)$. This will give an Approximation Theorem for the filter model of the next section. Theorem 17.3.17 is the following theorem in a more general context.

12.1.17 Theorem $\Gamma \vdash M : A \Leftrightarrow \exists P \in \mathcal{A}(M).\Gamma \vdash P : A$.

For example since $\lambda f.f^n\bot$ is, for all n, an approximant of Y, all types of the shape $(\mathtt{U}{\to}A_1) \cap \cdots \cap (A_{n-1}{\to}A_n){\to}A_n$ can be derived for Y.

Finally the question of whether or not an intersection type is inhabited is undecidable: the proof is the content of Section 17.5, see Corollary 17.5.32.

12.1.18 Theorem *The set $\{A \in \mathbb{T} \mid \exists M \in \Lambda^{\emptyset} \vdash M : A\}$ is undecidable.*

12.2 The filter model $\mathcal{F}^{\mathrm{BCD}}$

12.2.1 Definition

(i) A *complete lattice* $(\mathcal{D}, \sqsubseteq)$ is a partial order which has arbitrary least upper bounds (sups), and hence has arbitrary greatest lower bounds (infs).

(ii) A subset $Z \subseteq \mathcal{D}$ is *directed* if $Z \neq \emptyset$ and

$$\forall x, y \in Z \exists z \in Z. x, y \sqsubseteq z.$$

(iii) An element $c \in \mathcal{D}$ is *compact* (in the literature also called *finite*) if for each directed $Z \subseteq \mathcal{D}$ one has

$$c \sqsubseteq \bigsqcup Z \;\Rightarrow\; \exists z \in Z. c \sqsubseteq z.$$

Let $\mathcal{K}(\mathcal{D})$ denote the set of compact elements of $\mathcal{D}$.

(iv) A complete lattice is *ω-algebraic* if $\mathcal{K}(\mathcal{D})$ is countable, and for each $d \in \mathcal{D}$, the set $\mathcal{K}(d) = \{c \in \mathcal{K}(\mathcal{D}) \mid c \sqsubseteq d\}$ is directed and $d = \bigsqcup \mathcal{K}(d)$.

(v) If $\mathcal{D}, \mathcal{E}$ are ω-algebraic complete lattices and $f : \mathcal{D} \to \mathcal{E}$, we say that f is *Scott continuous* (or simply continuous), if for any directed $Z \subseteq \mathcal{D}$, we have

$$f(\bigsqcup Z) = \bigsqcup f(Z).$$

The set $[\mathcal{D} \to \mathcal{E}]$ consists of the continuous maps from $\mathcal{D}$ to $\mathcal{E}$ and can be ordered pointwise

$$f \sqsubseteq g \;\Leftrightarrow\; \forall d \in \mathcal{D}. f(d) \sqsubseteq g(d).$$

Then $\langle [\mathcal{D} \to \mathcal{E}], \sqsubseteq \rangle$ is again an ω-algebraic lattice.

(vi) The category **ALG** is the category whose objects are the ω-algebraic complete lattices and whose morphisms are the (Scott) continuous functions.

12.2.2 Definition

(i) A *filter* over $\mathbb{T} = \mathbb{T}^{\mathrm{BCD}}$ is a non-empty set $X \subseteq \mathbb{T}$ such that

(1) $A \in X \;\&\; A \leq_{\mathrm{BCD}} B \;\Rightarrow\; B \in X$;

(2) $A, B \in X \;\Rightarrow\; (A \cap B) \in X$.

(ii) We denote by $\mathcal{F}^{\mathrm{BCD}}$, or simply $\mathcal{F}$, the set of filters over $\mathbb{T}$.

12.2.3 Definition

(i) If $X \subseteq \mathbb{T}$ is non-empty, then the filter *generated by* X, written $\uparrow X$, is the smallest filter containing X.

(ii) A *principal* filter is of the form $\uparrow\{A\}$ for some $A \in \mathbb{T}$. We shall denote this simply by $\uparrow A$. Note that $\uparrow A = \{B \mid A \leq B\}$.

The following proposition follows easily from the above definitions.

12.2.4 Proposition

(i) *The set $\mathcal{F} = \langle \mathcal{F}, \subseteq \rangle$ is an ω-algebraic complete lattice.*

(ii) *The bottom element of* $\mathcal{F}$ *is* $\uparrow\mathtt{U}$ *and its top element is* $\mathbb{T}$.
(iii) *The compact elements of* $\mathcal{F}$ *are exactly the principal filters.*

12.2.5 Definition Let $\mathcal{D}$ be an ω-algebraic lattice and let

$$F : \mathcal{D}\to[\mathcal{D}\to\mathcal{D}]$$
$$G : [\mathcal{D}\to\mathcal{D}]\to\mathcal{D}$$

be Scott continuous. We say $\mathcal{D}$ is *reflexive* via F, G if $F \circ G = \mathtt{id}_{[\mathcal{D}\to\mathcal{D}]}$.

A reflexive element of **ALG** is also a model of the *untyped* λ-calculus in which the term interpretation is naturally defined as follows, see B[1984], Section 5.4.

12.2.6 Definition (Interpretation of terms) Let $\mathcal{D}$ be reflexive via F, G.

(i) A *term environment* in $\mathcal{D}$ is a map $\rho : \mathtt{Var}\to\mathcal{D}$. We denote by $\mathsf{Env}_{\mathcal{D}}$ the set of term environments.
(ii) If ρ is a term environment and $d \in \mathcal{D}$, then $\rho[x := d]$ is the term environment ρ' defined by

$$\begin{array}{rcll} \rho'(y) & = & \rho(y) & \text{if } y \not\equiv x; \\ \rho'(x) & = & d. & \end{array}$$

(iii) Given such ρ, the interpretation $[\![\ ]\!]^{D}_{\rho} : \Lambda\to\mathcal{D}$ is defined as

$$[\![x]\!]^{\mathcal{D}}_{\rho} = \rho(x);$$
$$[\![MN]\!]^{\mathcal{D}}_{\rho} = F([\![M]\!]^{\mathcal{D}}_{\rho})([\![N]\!]^{\mathcal{D}}_{\rho});$$
$$[\![\lambda x.M]\!]^{\mathcal{D}}_{\rho} = G(\boldsymbol{\lambda} d \in \mathcal{D}.[\![M]\!]^{\mathcal{D}}_{\rho(x:=d)}).$$

(iv) Let $M, N \in \Lambda$. Then $M = N$, is *true in* $\mathcal{D}$, written $\mathcal{D} \models M = N$, if

$$\forall\rho \in \mathsf{Env}_{\mathcal{D}}.[\![M]\!]^{\mathcal{D}}_{\rho} = [\![N]\!]^{\mathcal{D}}_{\rho}.$$

Remember the notion of a λ-model $\mathcal{D}$ given in Definition 3.1.10.

12.2.7 Theorem *Let* $\mathcal{D}$ *be reflexive via* F, G. *Then* $\mathcal{D}$ *is a* λ*-model, in particular for all* $M, N \in \Lambda$

$$\mathcal{D} \models (\lambda x.M)N = M[x: = N].$$

We state now properties of $\mathcal{F}$ which are implied by more general results proved in later sections. More precisely Proposition 12.2.8 follows from Corollary 16.2.10(i), Theorem 12.2.9 from Theorem 16.2.7, and Theorem 12.2.10 from Theorem 16.2.19.

12.2.8 Proposition *Define maps* $F : \mathcal{F}{\to}[\mathcal{F}{\to}\mathcal{F}]$ *and* $G : [\mathcal{F}{\to}\mathcal{F}]{\to}\mathcal{F}$ *by*

$$F(X)(Y) = \uparrow\{B \mid \exists A \in Y.(A{\to}B) \in X\}$$
$$G(f) = \uparrow\{A{\to}B \mid B \in f(\uparrow A)\}.$$

Then $\mathcal{F}$ *is reflexive via* F, G. *Therefore* $\mathcal{F}$ *is a model of the untyped* λ-*calculus.*

An important property of the λ-model $\mathcal{F}$ is that the meaning of a term is the set of types which can be assigned to it.

12.2.9 Theorem *For all* λ-*terms* M *one has*

$$[\![M]\!]^{\mathcal{F}}_{\rho} = \{A \mid \exists\Gamma \models \rho.\Gamma \vdash M : A\},$$

where $\Gamma \models \rho$ *iff for all* $(x{:}B) \in \Gamma$ *one has* $B \in \rho(x)$.

Lastly we note that all continuous functions are representable.

12.2.10 Theorem

$$[\mathcal{F}{\to}\mathcal{F}] = \{f : \mathcal{F}{\to}\mathcal{F} \mid f \text{ is representable}\},$$

where $f \in \mathcal{F}{\to}\mathcal{F}$ *is called* representable *if for some* $X \in \mathcal{F}$ *one has*

$$\forall Y \in \mathcal{F}.f(Y) = F(X)(Y).$$

12.3 Completeness of type assignment

12.3.1 Definition (Interpretation of types) Let $\mathcal{D}$ be reflexive via F, G and hence a λ-model. For $F(d)(e)$ we also write (as usual) $d \cdot e$.

(i) A *type environment* in $\mathcal{D}$ is a map $\xi : \mathbb{A}_\infty{\to}\mathcal{P}(\mathcal{D})$.
(ii) For $X, Y \in \mathcal{P}(\mathcal{D})$ define

$$X \Rightarrow Y = \{d \in \mathcal{D} \mid d \cdot X \subseteq Y\} = \{d \in \mathcal{D} \mid \forall x \in X.d \cdot x \in Y\}.$$

(iii) Given a type environment ξ, the interpretation $[\![\ ]\!]_\xi : \mathbb{T}{\to}\mathcal{P}(\mathcal{D})$ is defined as:

$$\begin{aligned}
[\![\mathtt{U}]\!]^{\mathcal{D}}_\xi &= \mathcal{D};\\
[\![\alpha]\!]^{\mathcal{D}}_\xi &= \xi(\alpha), && \text{for } \alpha \in \mathbb{A}_\infty;\\
[\![A{\to}B]\!]^{\mathcal{D}}_\xi &= [\![A]\!]^{\mathcal{D}}_\xi{\to}[\![B]\!]^{\mathcal{D}}_\xi;\\
[\![A \cap B]\!]^{\mathcal{D}}_\xi &= [\![A]\!]^{\mathcal{D}}_\xi \cap [\![B]\!]^{\mathcal{D}}_\xi.
\end{aligned}$$

12.3.2 Definition (Satisfaction)

(i) Given a λ-model $\mathcal{D}$, a term environment ρ, and a type environment ξ one defines the notion of satisfaction, written $\models$, as follows:

$$\mathcal{D}, \rho, \xi \models M : A \iff^{\Delta} [\![M]\!]^{\mathcal{D}}_{\rho} \in [\![A]\!]^{\mathcal{D}}_{\xi};$$

$$\mathcal{D}, \rho, \xi \models \Gamma \iff^{\Delta} \mathcal{D}, \rho, \xi \models x : B, \qquad \text{for all } (x{:}B) \in \Gamma.$$

(ii) $\Gamma \models M : A \iff^{\Delta} \forall \mathcal{D}, \rho, \xi.[\mathcal{D}, \rho, \xi \models \Gamma \Rightarrow \rho, \xi \models M : A]$.

12.3.3 Theorem (Soundness)

$$\Gamma \vdash M : A \Rightarrow \Gamma \models M : A.$$

12.3.4 Theorem (Completeness)

$$\Gamma \models M : A \Rightarrow \Gamma \vdash M : A.$$

The completeness proof is an application of the λ-model $\mathcal{F}$, see Barendregt et al. (1983) and Section 17.1, where also soundness is proved.

13

Type Assignment Systems

This chapter defines a family of systems $\boldsymbol{\lambda}_\cap^{\mathcal{T}}$ that assign intersection types to untyped lambda terms. These systems have a common set of typing rules parametric in an *intersection type theory* $\mathcal{T}$. They are obtained as a generalization of the example system $\boldsymbol{\lambda}_\cap^{\mathrm{BCD}}$ presented in Chapter 12.

In Section 13.1, we start by defining a set $\mathbb{T}^{\mathbb{A}}$ of intersection types similar to $\mathbb{T}^{\mathrm{BCD}}$, where the set of type atoms is now an arbitrary one denoted by $\mathbb{A}$. Then, we define the *intersection type theory* $\mathcal{T}$ as the set of statements of the form $A \leq_{\mathcal{T}} B$ (or just $A \leq B$) with $A, B \in \mathbb{T}^{\mathbb{A}}$ satisfying some logical rules that ensure that $\leq_{\mathcal{T}}$ is a pre-order on $\mathbb{T}^{\mathbb{A}}$. In particular, the logical rules for the intersection will ensure that $A \cap B$ is a greatest lower bound for the types A and B. Since all type theories in Part III will use the intersection operator, we keep this implicit and often simply speak about *type theories* without the word 'intersection'.

For some type theories a particular atom, denoted by $\mathtt{U}$, is selected to act as *universal type*, intended as the type of all lambda terms. The rules of type assignment are such that if $\mathtt{U} \leq A$, then A is also a universal element. So it is natural (but not strictly necessary) to require that $\mathtt{U}$ is the top element. The class of intersection type theories with a universal and top element is denoted by $\mathrm{TT}^{\mathtt{U}}$ and the one without by $\mathrm{TT}^{\text{-}\mathtt{U}}$. For the (disjoint) union we write $\mathrm{TT} = \mathrm{TT}^{\mathtt{U}} \cup \mathrm{TT}^{\text{-}\mathtt{U}}$.

Fig. 13.1 shows the thirteen specific examples of type theories that will be considered: BCD is amongst them. These theories are denoted by names (or sometimes initials) of the author(s) who first considered the λ-model induced by such a theory. In this list the given order is logical, rather than historical; some of the references define the theories directly, others deal with the corresponding filter models. In some cases the type theory was modeled after an existing domain model in order to study the image of terms and

$\mathcal{T}$	$\boldsymbol{\lambda}_\cap^{\mathcal{T}}$	$\mathcal{F}^{\mathcal{T}}$	Reference
Scott	$\boldsymbol{\lambda}_\cap^{\text{Scott}}$	$\mathcal{F}^{\text{Scott}}$	Scott (1972)
Park	$\boldsymbol{\lambda}_\cap^{\text{Park}}$	$\mathcal{F}^{\text{Park}}$	Park (1976)
CDZ	$\boldsymbol{\lambda}_\cap^{\text{CDZ}}$	$\mathcal{F}^{\text{CDZ}}$	Coppo et al. (1987)
HR	$\boldsymbol{\lambda}_\cap^{\text{HR}}$	$\mathcal{F}^{\text{HR}}$	Honsell and Ronchi Della Rocca (1992)
DHM	$\boldsymbol{\lambda}_\cap^{\text{DHM}}$	$\mathcal{F}^{\text{DHM}}$	Dezani-Ciancaglini et al. (2005)
BCD	$\boldsymbol{\lambda}_\cap^{\text{BCD}}$	$\mathcal{F}^{\text{BCD}}$	Barendregt et al. (1983)
AO	$\boldsymbol{\lambda}_\cap^{\text{AO}}$	$\mathcal{F}^{\text{AO}}$	Abramsky and Ong (1993)
Plotkin	$\boldsymbol{\lambda}_\cap^{\text{Plotkin}}$	$\mathcal{F}^{\text{Plotkin}}$	Plotkin (1993)
Engeler	$\boldsymbol{\lambda}_\cap^{\text{Engeler}}$	$\mathcal{F}^{\text{Engeler}}$	Engeler (1981)
CDS	$\boldsymbol{\lambda}_\cap^{\text{CDS}}$	$\mathcal{F}^{\text{CDS}}$	Coppo et al. (1979)
HL	$\boldsymbol{\lambda}_\cap^{\text{HL}}$	$\mathcal{F}^{\text{HL}}$	Honsell and Lenisa (1999)
CDV	$\boldsymbol{\lambda}_\cap^{\text{CDV}}$	$\mathcal{F}^{\text{CDV}}$	Coppo et al. (1981)
CD	$\boldsymbol{\lambda}_\cap^{\text{CD}}$	$\mathcal{F}^{\text{CD}}$	Coppo and Dezani-Ciancaglini (1980)

Figure 13.1 Specific type theories, type assignment systems and filter models

hence equality of terms in that model; in other cases the type theory came first and created a domain model with certain properties.

The first ten type theories of Fig. 13.1 have the universal type $\mathtt{U}$ and the remaining three do not. Explicitly,

$$\begin{array}{rl} \text{Scott, Park, CDZ, HR, DHM, BCD, AO, Plotkin, Engeler, CDS} & \in \mathrm{TT}^{\mathtt{U}} \\ \text{HL, CDV, CD} & \in \mathrm{TT}^{\text{-}\mathtt{U}}. \end{array}$$

Some of these type theories have other type atoms such as 0 and 1. We will end this Section 13.1 by proving some basic lemmas for these specific type theories. In particular, we will prove that $0 < 1 < \mathtt{U}$.

In Section 13.2 we will assign types in $\mathbb{T}^{\mathbb{A}}$ to lambda terms in Λ. Given a type theory $\mathcal{T}$, we will derive assertions of the form $\Gamma \vdash_\cap^{\mathcal{T}} M : A$ where $M \in \Lambda$, $A \in \mathbb{T}^{\mathbb{A}}$ and Γ is a set of type declarations for the variables in M. For this, we define a set of typing rules parametric in $\mathcal{T}$ denoted by $\boldsymbol{\lambda}_\cap^{\mathcal{T}}$ and called *type assignment system over* $\mathcal{T}$. This can be seen as a mapping

$$\mathcal{T} \in \mathrm{TT} \mapsto \text{set } \boldsymbol{\lambda}_\cap^{\mathcal{T}} \text{ of typing rules (and axioms) parametric in } \mathcal{T}.$$

The parameter $\mathcal{T}$ appears in rule $(\leq)$ and axiom $(\mathtt{U}_{\text{top}})$. The rule $(\leq)$ states that a lambda term has type B if it has type A and $A \leq_{\mathcal{T}} B$. The axiom $(\mathtt{U}_{\text{top}})$ states that all lambda terms have type $\mathtt{U}$ in the case $\mathcal{T} \in \mathrm{TT}^{\mathtt{U}}$. More specifically, the type assignments of the first ten type theories of Fig. 13.1 contain $(\mathtt{U}_{\text{top}})$ and the remaining three do not.

The systems $\boldsymbol{\lambda}_\cap^\mathcal{T}$ also share a set of non-parametric rules for assigning lambda terms to the types $(A \cap B)$ and $(A \to B)$. The particular use of intersection is that if a lambda term has both type A *and* type B, then it also has type $(A \cap B)$. The type $(A \to B)$ plays the same role that it does in the simply typed lambda calculus, i.e. to cover the abstraction terms.

We have an infinite collection $\{\boldsymbol{\lambda}_\cap^\mathcal{T} \mid \mathcal{T} \in \mathrm{TT}\}$ of type assignment systems which are defined by giving *only one* set of typing rules parametric in $\mathcal{T}$. Now we mention some of the advantages of having this general and common framework for all these systems:

(1) We can capture most of the intersection type assignment systems that appear in the literature as shown in Fig. 13.1.
(2) We can study general properties of $\boldsymbol{\lambda}_\cap^\mathcal{T}$ that hold for all $\mathcal{T} \in \mathrm{TT}$; this will be done in Chapter 14.

In Section 13.3 we define the notion of intersection type structure

$$\mathcal{S} = \langle |\mathcal{S}|, \leq, \cap, \to \rangle$$

where $\leq$ is now a partial order (a pre-order that is anti-symmetric) and the greatest lower bound $\cap$ is unique. The collection of type structures is denoted by TS. Given a type theory $\mathcal{T}$, one usually requires that the equivalence relation $=_\mathcal{T}$ is a congruence with respect to $\to$. Then we speak of a *compatible* type theory, having a corresponding *type structure*

$$[\mathcal{T}] = \langle [\mathbb{T}], \leq, \cap, \to \rangle.$$

Each type structure can be seen as coming from a compatible type theory; therefore compatible type theories and type structures are basically the same. In this section, we also introduce specific categories of lattices and type structures to accommodate them.

Finally in Section 13.4 we introduce the notion of *filter* over $\mathcal{T}$ as a set of types closed under intersection $\cap$ and pre-order $\leq$. If $\mathcal{T} \in \mathrm{TT}^{\mathtt{U}}$, then filters are non-empty and the smallest filter (ordered by subset inclusion) is the filter generated by $\{\mathtt{U}\}$. If $\mathcal{T} \in \mathrm{TT}^{\text{-}\mathtt{U}}$, then the empty set is considered to be a filter which in this case is the smallest one. Just as with type assignment systems, we also have the mapping

$$\mathcal{T} \in \mathrm{TT} \mapsto \text{ set } \mathcal{F}^\mathcal{T} \text{ of filters over } \mathcal{T}.$$

We define the notion of filter structure over $\mathcal{T}$ as a triple $\langle \mathcal{F}^\mathcal{T}, F^\mathcal{T}, G^\mathcal{T} \rangle$, where $F^\mathcal{T}, G^\mathcal{T}$ are operations for interpreting application and abstraction, respectively. We also have a mapping

$$\mathcal{T} \in \mathrm{TT} \mapsto \mathcal{F}^\mathcal{T} = \langle \mathcal{F}^\mathcal{T}, F^\mathcal{T}, G^\mathcal{T} \rangle.$$

In Chapter 15 a proper categorical setting is provided in order to study some interesting properties of these mappings as functors. This will be used to establish equivalences of categories of specific type structures and algebraic lattices.

In Chapter 16 we will study general conditions on $\mathcal{T}$ for ensuring that the filter structures $\mathcal{F}^{\mathcal{T}}$ are models of the untyped lambda calculus, i.e. *filter models.* This will cover the thirteen specific cases of filter models of Fig. 13.1 that appear in the literature. The first ten are models of the λ-calculus (when $\mathcal{T} \in \text{TT}^{\mathtt{U}}$): the remainder are models of the $\lambda\mathsf{I}$-calculus (when $\mathcal{T} \in \text{TT}^{\text{-}\mathtt{U}}$).

13.1 Type theories

13.1.1 Definition Let $\mathbb{A}$ be a (usually countable, i.e. finite or countably infinite) set of symbols, called *(type) atoms*. The set of *intersection types* over $\mathbb{A}$, written $\mathbb{T}_{\cap}^{\mathbb{A}}$ (if there is little danger of confusion we also denote it by $\mathbb{T}_{\cap}$, $\mathbb{T}^{\mathbb{A}}$ or just $\mathbb{T}$), is defined by the following simplified syntax:

$$\boxed{\mathbb{T} ::= \mathbb{A} \mid \mathbb{T}\to\mathbb{T} \mid \mathbb{T} \cap \mathbb{T}.}$$

13.1.2 Remark

(i) The set $\mathbb{A}$ varies in the applications of intersection types.
(ii) In Chapter 12 the set of intersection types $\mathbb{T}$ was defined over the set of atoms $\mathbb{A}_{\infty}^{\mathtt{U}} = \{\mathtt{c}_0, \mathtt{c}_1, \dots\} \cup \{\mathtt{U}\}$. We write this as

$$\mathbb{A}^{\text{BCD}} = \mathbb{A}_{\infty}^{\mathtt{U}} \text{ and } \mathbb{T}^{\text{BCD}} = \mathbb{T}^{\mathbb{A}_{\infty}^{\mathtt{U}}}.$$

(iii) The following are some of the type atoms that will be used in different contexts:

$\mathtt{c}_0, \mathtt{c}_1, \mathtt{c}_2, \dots$	indiscernible atoms
$0, 1, \mathtt{U}$	atoms with special properties.

(iv) The atom $\mathtt{U}$ is called the *universe.* Its purpose is to host all lambda terms. That of the special atom 1 varies: sometimes it hosts the strongly normalizing terms, sometimes the terms which reduce to $\lambda\mathsf{I}$-terms. Similarly 0 can host the terms that reduce to closed terms or other sets of terms. This will be determined by the properties of the type theory in which 1 and 0 occur. See Fig. 17.1 on page 729.

13.1.3 Notation

(i) Greek letters $\alpha, \beta, \dots$ range over arbitrary atoms in $\mathbb{A}$.
(ii) The letters $A, B, C, \dots$ range over types in $\mathbb{T}^{\mathbb{A}}$.

(iii) Some papers on intersection types, for example Coppo et al. (1979), use the Greek letter ω to denote the universal type, while they use the atoms 0 and 1 with the same meaning as here. Other papers, for example Dezani-Ciancaglini et al. (2003), use Ω for the universal type and ω and φ for 0 and 1.

13.1.4 Definition

(i) An *intersection type theory without (explicit) universe* over a set of type atoms $\mathbb{A}$ is the pair $\langle \mathbb{T}_{\cap}^{\mathbb{A}}, \mathcal{T} \rangle$, where $\mathbb{T}_{\cap}^{\mathbb{A}}$ is the set of intersection types over the atoms $\mathbb{A}$, and $\mathcal{T}$ is a set of sentences of the form $A \leq B$ (to be read: A sub B), with $A, B \in \mathbb{T}_{\cap}^{\mathbb{A}}$, satisfying at least the axioms and rules in Fig. 13.2. This means that, for example, $(A \leq A) \in \mathcal{T}$ and $(A \leq B), (B \leq C) \in \mathcal{T} \;\Rightarrow\; (A \leq C) \in \mathcal{T}$, for all $A, B, C \in \mathbb{T}^{\mathbb{A}}$.

(ii) An *intersection type theory with universe* $\mathtt{U}$ is a pair $\langle \mathbb{T}_{\cap}^{\mathbb{A}^{\mathtt{U}}}, \mathcal{T} \rangle$, where $\mathbb{T}_{\cap}^{\mathbb{A}^{\mathtt{U}}}$ is the set of intersection types over the atoms $\mathbb{A}^{\mathtt{U}} \triangleq \mathbb{A} \cup \{\mathtt{U}\}$, and $\mathcal{T}$ is a set of sentences of the form $A \leq B$ (to be read: A sub B), with $A, B \in \mathbb{T}_{\cap}^{\mathbb{A}^{\mathtt{U}}}$, satisfying in addition, for all $A \in \mathbb{T}^{\mathbb{A}^{\mathtt{U}}}$,

$$\boxed{(\mathtt{U}_{\mathrm{top}}) \quad A \leq \mathtt{U}.}$$

13.1.5 Notation

(i) The set of type theories *without universe* is denoted by $\mathrm{TT}^{-\mathtt{U}}$.

(ii) The set of type theories *with universe* is denoted by $\mathrm{TT}^{\mathtt{U}}$

(iii) The set of intersection type theories, written TT, is

$$\mathrm{TT} \triangleq \mathrm{TT}^{\mathtt{U}} \cup \mathrm{TT}^{-\mathtt{U}}.$$

(iv) The notion 'intersection type theory' will usually be abbreviated as 'type theory', as the 'intersection' part is default.

13.1.6 Remark

(i) An intersection type theory $\mathcal{T} \in \mathrm{TT}$ can fail to be in $\mathrm{TT}^{\mathtt{U}}$ as follows:

(a) there is a $\mathtt{U}$, but it is not declared as $\mathtt{U}$ (and hence not used in the type assignment system as universal element);

(b) there is no top.

In both cases $\mathcal{T} \in \mathrm{TT}^{-\mathtt{U}}$.

(ii) The intuition behind the special atom $\mathtt{U}$, and its name, will become clear in Section 13.2 on type assignment. The types in a TT will be assigned to untyped lambda terms. Arrow types, such as $A \to B$, will

(refl)	$A \leq A$
(incl$_L$)	$A \cap B \leq A$
(incl$_R$)	$A \cap B \leq B$
(glb)	$\dfrac{C \leq A \quad C \leq B}{C \leq A \cap B}$
(trans)	$\dfrac{A \leq B \quad B \leq C}{A \leq C}$

Figure 13.2

be assigned to abstraction terms of the form $\lambda x.M$. For $\mathcal{T} \in \mathrm{TT}^{\mathtt{U}}$ we will postulate the assignment

$$\Gamma \vdash M : \mathtt{U},$$

for all Γ and all $M \in \Lambda$. So $\mathtt{U}$ will be *a universal type*, i.e. a type assigned to all terms. Another rule of type assignment will be

$$\frac{\Gamma \vdash M : A \quad A \leq B}{\Gamma \vdash M : B}.$$

Hence if $\mathtt{U} \leq A$, then A will also be a universal type. Therefore, for type theories in which we wish to have both a universe $\mathtt{U}$ and a top $\top$, we have chosen in 13.1.4(ii) that $\mathtt{U} = \top$, i.e. we will have for all types A

$$A \leq \mathtt{U}.$$

There will be $\mathcal{T} \in \mathrm{TT}$ with a top, but without a universe. In principle the converse, a TT with a universe, but without a top, could also be considered, but we will not do so in the main theory. The system $\boldsymbol{\lambda}_\cap^{\mathrm{Krivine}^{\mathtt{U}}}$ of Krivine, see Exercise 13.10, is such an example.

13.1.7 Remark

(i) In this Part of the book $\mathcal{T}$ ranges over elements of TT.
(ii) If $\mathcal{T} \in \mathrm{TT}$ over $\mathbb{A}$, then we also write $\mathbb{T}^{\mathcal{T}} = \mathbb{T}_\cap^{\mathbb{A}}$ and $\mathbb{A}^{\mathcal{T}} = \mathbb{A}$.

13.1.8 Remark Most $\mathcal{T} \in \mathrm{TT}$ have some extra axioms or rules, the above set in Definition 13.1.4 being the minimum requirement. For example the theory BCD over $\mathbb{A} = \mathbb{A}_\infty^{\mathtt{U}}$, introduced in Chapter 12, is a $\mathrm{TT}^{\mathtt{U}}$ and has the extra axioms $(\mathtt{U}{\rightarrow})$ and $({\rightarrow}\cap)$ and rule $(\rightarrow)$.

13.1.9 Notation Let $\mathcal{T} \in \mathrm{TT}$. We write the following.

(i) $A \leq_{\mathcal{T}} B$ for $(A \leq B) \in \mathcal{T}$.
(ii) $A =_{\mathcal{T}} B$ for $A \leq_{\mathcal{T}} B \leq_{\mathcal{T}} A$.
(iii) $A <_{\mathcal{T}} B$ for $A \leq B$ & $A \neq_{\mathcal{T}} B$.
(iv) If there is little danger of confusion and $\mathcal{T}$ is clear from the context, then we will write $\leq, =, <$ for $\leq_{\mathcal{T}}, =_{\mathcal{T}}, <_{\mathcal{T}}$ respectively.
(v) We write $A \equiv B$ for syntactic identity. For example $A \cap B \equiv A \cap B$, but $A \cap B \not\equiv B \cap A$. Note that $A \cap B =_{\mathcal{T}} B \cap A$ is always true in a type theory.
(vi) We write $[\mathbb{T}]$ for $\mathbb{T}$ modulo $=_{\mathcal{T}}$.
(vii) We will use the informal notation $A_1 \cap \cdots \cap A_n$ to denote the intersection of a sequence of n types A_i for $i \in \{1, \ldots, n\}$ associating to the right. If $n = 3$ then $A_1 \cap \cdots \cap A_n$ denotes $(A_1 \cap (A_2 \cap A_3))$. In the case $n = 0$ and $\mathcal{T} \in \mathrm{TT}^{\mathtt{U}}$, we mean that the sequence is empty and $A_1 \cap \cdots \cap A_n$ denotes $\mathtt{U}$.
(viii) For a finite $I = \{1, \ldots, n\}$ we may also use the notation $\bigcap_{i \in I} A_i$ to denote $A_1 \cap \cdots \cap A_n$.

13.1.10 Proposition *The following rule is derived:*

$$(\text{mon}) \quad \frac{A \leq A' \quad B \leq B'}{A \cap B \leq A' \cap B'}.$$

Proof By (trans), (incl$_L$), (incl$_R$) and (glb). ■

The above proposition implies that for any $\mathcal{T}$, the relation $=_{\mathcal{T}}$ is compatible with the operator $\cap$. In the case that $=_{\mathcal{T}}$ is compatible with $\rightarrow$ we define the following.

13.1.11 Definition The type theory $\mathcal{T}$ is called *compatible* if the following rule is admissible:

$$(\rightarrow^{=}) \quad \frac{A = A' \quad B = B'}{(A \rightarrow B) = (A' \rightarrow B')}.$$

This implies $A =_{\mathcal{T}} A'$ & $B =_{\mathcal{T}} B' \Rightarrow (A \rightarrow B) =_{\mathcal{T}} (A' \rightarrow B')$.

13.1.12 Lemma

(i) *For any $\mathcal{T}$ one has $A \cap B =_{\mathcal{T}} B \cap A$.*
(ii) *If $\mathcal{T}$ is compatible, then $(A \cap B) \rightarrow C =_{\mathcal{T}} (B \cap A) \rightarrow C$.*

Proof (i) By (incl$_L$), (incl$_R$) and (glb).
(ii) By (i). ■

13.1.13 Remark As in Remark 12.1.8, any $\mathcal{T} \in \mathrm{TT}$ can be seen as a structure with a pre-order $\mathcal{T} = \langle \mathbb{T}, \leq, \cap, \to \rangle$. This means that $\leq$ is reflexive and transitive, but not necessarily anti-symmetric:

$$A \leq B \;\&\; B \leq A \text{ does not always imply } A \equiv B.$$

One can go over to equivalence classes and define a partial order on $[\mathbb{T}]$ by

$$[A] \leq [B] \Leftrightarrow A \leq B.$$

By Proposition 13.1.10, $\cap$ is always well defined on $[\mathbb{T}]$ by $[A] \cap [B] = [A \cap B]$. To ensure that $\to$ is well defined by $[A] \to [B] = [A \to B]$, we need to require that $=$ is compatible with $\to$. This is the case if $\mathcal{T}$ is compatible and one obtains a type structure

$$[\mathcal{T}] = \langle [\mathbb{T}], \leq, \cap, \to \rangle.$$

This structure is a meet semi-lattice, i.e. a poset with $[A] \cap [B]$ as the greatest lower bound of $[A]$ and $[B]$. If moreover $\mathcal{T} \in \mathrm{TT}^{\mathtt{U}}$, then $[\mathcal{T}]$ can be enriched to a type structure with universe $[\mathtt{U}]$ of the form

$$[\mathcal{T}] = \langle [\mathbb{T}], \leq, \cap, \to, [\mathtt{U}] \rangle.$$

This will be done in Section 13.3.

Specific intersection type theories

Now we will construct several, in total thirteen, type theories that will play an important role in later chapters, by introducing the following axiom schemes, rule schemes and axioms (see Fig. 13.3). Only two of them are non-compatible, so we obtain eleven type structures.

13.1.14 Remark

(i) The axiom scheme $(\mathtt{U}_{\mathrm{top}})$ states that the universe $\mathtt{U}$ is a top element.
(ii) In the presence of $(\to)$ the axiom-scheme $(\mathtt{U}\!\to)$ is equivalent with the axiom $\mathtt{U} \leq (\mathtt{U} \to \mathtt{U})$. Also in that case the axiom-scheme $(\to\cap)$ is equivalent with $(\to\cap^{=})$. See Exercise 13.1.

13.1.15 Definition In Fig. 13.4 a collection of elements of TT is defined. For each name $\mathcal{T}$ a set $\mathbb{A}^{\mathcal{T}}$ of atoms and a set of rules and axiom(scheme)s are given. The type theory $\mathcal{T}$ is the smallest intersection type theory over $\mathbb{A}^{\mathcal{T}}$ (see Definition 13.1.4) that satisfies the rules and axioms of $\mathcal{T}$ shown in Fig. 13.4.

13.1.16 Remark

(i) Note that CDS and CD are non-compatible, while the other eleven type theories are compatible.

Axioms	
(0_{Scott})	$(\mathtt{U}{\to}0) = 0$
(0_{Park})	$(0{\to}0) = 0$
(01)	$0 \leq 1$
$(1{\to}0)$	$(1{\to}0) = 0$
$(0{\to}1)$	$(0{\to}1) = 1$
(I)	$(1{\to}1) \cap (0{\to}0) = 1$

Axiom schemes	
$(\mathtt{U}_{\text{top}})$	$A \leq \mathtt{U}$
$(\mathtt{U}{\to})$	$\mathtt{U} \leq (A{\to}\mathtt{U})$
$(\mathtt{U}_{\text{lazy}})$	$(A{\to}B) \leq (\mathtt{U}{\to}\mathtt{U})$
$({\to}\cap)$	$(A{\to}B) \cap (A{\to}C) \leq A{\to}B \cap C$
$({\to}\cap^{=})$	$(A{\to}B) \cap (A{\to}C) = A{\to}B \cap C$

Rule schemes	
$({\to})$	$\dfrac{A' \leq A \quad B \leq B'}{(A{\to}B) \leq (A'{\to}B')}$
$({\to}^{=})$	$\dfrac{A' = A \quad B = B'}{(A{\to}B) = (A'{\to}B')}$

Figure 13.3 Possible Axioms and Rules concerning $\leq$.

(ii) The first ten type theories of Fig. 13.4 belong to $\mathrm{TT}^{\mathtt{U}}$ and the last three to $\mathrm{TT}^{\text{-}\mathtt{U}}$. In Lemma 13.1.23((i)) we will see that HL has 1 as top.

(iii) The type theories CDV and CD do not have a top at all, as shown in Lemma 13.1.23(ii) and (iii).

13.1.17 Remark The expressive power of intersection types is remarkable. This will become apparent when we will use them as a tool for characterizing properties of λ-terms (see Sections 17.2 and 17.4), and for describing different λ-models (see Section 16.3).

Much of this expressive power comes from the fact that they are endowed

$\mathcal{T}$	$\mathbb{A}^{\mathcal{T}}$	Rules	Axiom Schemes	Axioms
Scott	$\{\mathtt{U},0\}$	$(\to)$	$(\to\cap), (\mathtt{U}_{\rm top}), (\mathtt{U}\to)$	$(0_{\rm Scott})$
Park	$\{\mathtt{U},0\}$	$(\to)$	$(\to\cap), (\mathtt{U}_{\rm top}), (\mathtt{U}\to)$	$(0_{\rm Park})$
CDZ	$\{\mathtt{U},1,0\}$	$(\to)$	$(\to\cap), (\mathtt{U}_{\rm top}), (\mathtt{U}\to)$	$(01), (1\to0), (0\to1)$
HR	$\{\mathtt{U},1,0\}$	$(\to)$	$(\to\cap), (\mathtt{U}_{\rm top}), (\mathtt{U}\to)$	$(01), (1\to0), (\mathsf{I})$
DHM	$\{\mathtt{U},1,0\}$	$(\to)$	$(\to\cap), (\mathtt{U}_{\rm top}), (\mathtt{U}\to)$	$(01), (0\to1), (0_{\rm Scott})$
BCD	$\mathbb{A}^{\mathtt{U}}_{\infty}$	$(\to)$	$(\to\cap), (\mathtt{U}_{\rm top}), (\mathtt{U}\to)$	
AO	$\{\mathtt{U}\}$	$(\to)$	$(\to\cap), (\mathtt{U}_{\rm top}), (\mathtt{U}_{\rm lazy})$	
Plotkin	$\{\mathtt{U},0\}$	$(\to^{=})$	$(\mathtt{U}_{\rm top})$	–
Engeler	$\mathbb{A}^{\mathtt{U}}_{\infty}$	$(\to^{=})$	$(\to\cap^{=}), (\mathtt{U}_{\rm top}), (\mathtt{U}\to)$	–
CDS	$\mathbb{A}^{\mathtt{U}}_{\infty}$	–	$(\mathtt{U}_{\rm top})$	–
HL	$\{1,0\}$	$(\to)$	$(\to\cap)$	$(01), (0\to1), (1\to0)$
CDV	$\mathbb{A}_{\infty}$	$(\to)$	$(\to\cap)$	–
CD	$\mathbb{A}_{\infty}$	–	–	–

Figure 13.4 Various type theories

with a *pre-order relation*, $\leq$, which induces, on the set of types modulo $=$, the structure of a meet semi-lattice with respect to $\cap$. This appears natural when we think of types as subsets of a domain of discourse $\mathcal{D}$ (the interpretation of the universe $\mathtt{U}$), which is endowed with a (partial) application $\cdot : \mathcal{D}\times\mathcal{D}\to\mathcal{D}$, and when we interpret $\cap$ as set-theoretic intersection, $\leq$ as set inclusion, and give $\to$ the *realizability interpretation* $[\![A]\!]_\xi \subseteq \mathcal{D}$ for each type A. One starts by interpreting types in $\mathbb{T}_{\to}$:

$$[\![\alpha]\!]_\xi = \xi(\alpha), \ \text{ for } \alpha \not\equiv \mathtt{U};$$
$$[\![A\to B]\!]_\xi = [\![A]\!]_\xi \to [\![B]\!]_\xi = \{d\in\mathcal{D} \mid d\cdot[\![A]\!]_\xi \subseteq [\![B]\!]_\xi\}.$$

This semantics, due to Scott, is extended to intersection types by setting

$$[\![\mathtt{U}]\!]_\xi = \mathcal{D};$$
$$[\![A\cap B]\!]_\xi = [\![A]\!]_\xi \cap [\![B]\!]_\xi.$$

Given the correct TT and right domain the following holds:

$$A \leq B \text{ iff for all } \xi \text{ one has } [\![A]\!]_\xi \subseteq [\![B]\!]_\xi.$$

This type of semantics will be studied in Section 17.1.

The type $\mathtt{U}\to\mathtt{U}$ is the set of functions which, when applied to an arbitrary element, return again an arbitrary element. In that case, axiom-scheme $(\mathtt{U}\to)$ expresses the fact that all the objects in our domain of discourse are total

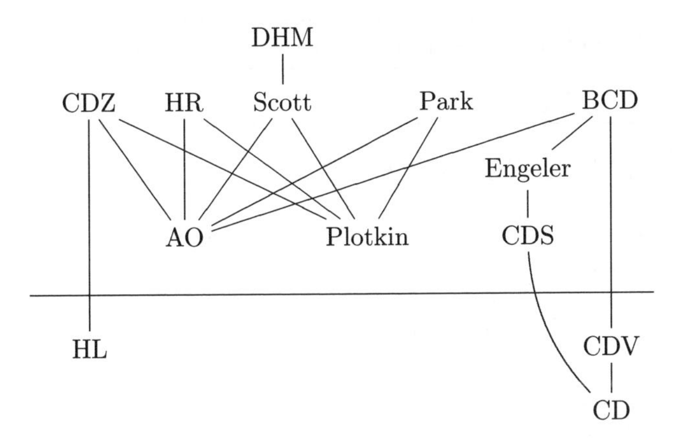

Figure 13.5 Inclusion among some intersection type theories.

functions, i.e. that $\mathtt{U}$ is equal to $A{\to}\mathtt{U}$, hence $A{\to}\mathtt{U} = B{\to}\mathtt{U}$ for all A, B (Barendregt et al. (1983)). If now we want to capture only those terms which truly represent functions, as we do for example in the lazy λ-calculus, we cannot assume axiom $(\mathtt{U}{\to})$. One may still postulate the weaker property $(\mathtt{U}_{\rm lazy})$ to make all functions total (Abramsky and Ong (1993)); this simply says that an element which is a function, because it maps A into B, maps also the whole universe into itself.

The intended interpretation of arrow types also motivates axiom $({\to}\cap)$, which implies that if a function maps A into B, and the same function also maps A into C, then actually it maps the whole A into the intersection of B and C (i.e. into $B \cap C$), see Barendregt et al. (1983).

Rule $({\to})$ is again very natural in view of the set-theoretic interpretation. It implies that the arrow constructor is contravariant in the first argument and covariant in the second. It is clear that if a function maps A into B, and we take a subset A' of A and a superset B' of B, then this function will map also A' into B', see Barendregt et al. (1983).

The rule $({\to}\cap^{=})$ is similar to the rule $({\to}\cap)$. It captures properties of the graph models for the untyped lambda calculus, see Plotkin (1975) and Engeler (1981), as we shall discuss in Section 16.3.

For Scott, Park, CDZ, HR, DHM, the axioms express peculiar properties of D_∞-like inverse limit models (see Section 16.3). For Park, CDZ, HR, DHM, and also for HL, the axioms also express properties of subsets of λ-terms (see Fig. 17.1 on page 729 and Proposition 17.2.13).

13.1.18 Remark In Fig. 13.5 we have connected $\mathcal{T}_1$ with an edge towards the higher positioned $\mathcal{T}_2$ in the case $\mathcal{T}_1 \subset \mathcal{T}_2$. In Exercise 13.15 it is shown

Class	Defining axiom(-scheme)(s) or rule(-scheme)(s)
graph	$(\to^=), (\mathtt{U}_{\rm top})$
lazy	$(\to), (\to\cap), (\mathtt{U}_{\rm top}), (\mathtt{U}_{\rm lazy})$
natural	$(\to), (\to\cap), (\mathtt{U}_{\rm top}), (\mathtt{U}\to)$
proper	$(\to), (\to\cap)$

Figure 13.6

non compatible	CD, CDS
graph	Plotkin, Engeler
lazy	AO
natural	Scott, Park, CDZ, HR, DHM, BCD
proper	HL, CDV

Figure 13.7

that the inclusions are strict. Above the horizontal line we find the elements of $\mathrm{TT}^{\mathtt{U}}$; below, those of $\mathrm{TT}^{\text{-}\mathtt{U}}$.

Some classes of type theories

Now we will consider some classes of type theories. In order to do this, we list the relevant defining properties.

13.1.19 Definition We define special subclasses of TT in Fig. 13.6.

13.1.20 Notation The sets of graph, lazy, natural and proper type theories are denoted by $\mathrm{GTT}^{\mathtt{U}}$, $\mathrm{LTT}^{\mathtt{U}}$, $\mathrm{NTT}^{\mathtt{U}}$ and PTT respectively. The following subset inclusions are easily deduced from their definition:

$$\mathrm{NTT}^{\mathtt{U}} \subset \mathrm{LTT}^{\mathtt{U}} \subset \mathrm{GTT}^{\mathtt{U}} \subset \mathrm{TT}^{\mathtt{U}} \subset \mathrm{TT};$$
$$\mathrm{PTT} \subseteq \mathrm{TT}^{\text{-}\mathtt{U}}.$$

13.1.21 Remark The type theories of Fig. 13.4 are classified in Fig. 13.7. This table indicates the typical classification for these type theories: for example, CDZ is typically natural because $\mathrm{NTT}^{\mathtt{U}}$ is the smallest set that contains CDZ.

Some properties about specific TTs

Results about proper type theories

All type theories, except CD, CDS, Plotkin, Engeler of Fig. 13.4, are proper.

13.1.22 Proposition *Let $\mathcal{T}$ be proper. Then*

(i) $(A{\to}B)\cap(A'{\to}B') \leq (A\cap A'){\to}(B\cap B')$;

(ii) $(A_1{\to}B_1)\cap\cdots\cap(A_n{\to}B_n) \leq (A_1\cap\cdots\cap A_n){\to}(B_1\cap\cdots\cap B_n)$;

(iii) $(A{\to}B_1)\cap\cdots\cap(A{\to}B_n) = A{\to}(B_1\cap\cdots\cap B_n)$.

Proof (i)
$$\begin{array}{rcl}(A{\to}B)\cap(A'{\to}B') & \leq & ((A\cap A'){\to}B)\cap((A\cap A'){\to}B') \\ & \leq & (A\cap A'){\to}(B\cap B'),\end{array}$$
by $(\to)$ and $(\to\cap)$ respectively.

(ii) Similarly (i.e. by induction on $n{>}1$, using (i) for the induction step).

(iii) By (ii) one has $(A{\to}B_1)\cap\cdots\cap(A{\to}B_n) \leq A{\to}(B_1\cap\cdots B_n)$. For $\geq$ use $(\to)$ to show that $A{\to}(B_1\cap\cdots\cap B_n) \leq (A{\to}B_i)$, for all i. ∎

It follows that the abovementioned equality and inequalities hold for Scott, Park, CDZ, HR, DHM, BCD, AO, HL, and CDV.

Results about the type theories of Fig. 13.4

13.1.23 Lemma

(i) 1 *is the top and* 0 *the bottom element in* HL.

(ii) CDV *has no top element.*

(iii) CDS, CD *have no top element.*

Proof (i) By induction on the generation of $\mathbb{T}^{\mathrm{HL}}$ one shows that $0 \leq A \leq 1$ for all $A\in\mathbb{T}^{\mathrm{HL}}$.

(ii) If α is a fixed atom and

$$\mathcal{B}_\alpha := \alpha \mid \mathcal{B}_\alpha \cap \mathcal{B}_\alpha$$

and $A\in\mathcal{B}_\alpha$, then one can show by induction on the generation of $\leq_{\mathrm{CDV}}$ that $A \leq_{\mathrm{CDV}} B \;\Rightarrow\; B\in\mathcal{B}_\alpha$. Hence if $\alpha \leq_{\mathrm{CDV}} B$, then $B\in\mathcal{B}_\alpha$. Since $\mathcal{B}_{\alpha_1}$ and $\mathcal{B}_{\alpha_2}$ are disjoint when α_1 and α_2 are two different atoms, we conclude that CDV has no top element.

(iii) Similarly to (ii), but easier. ∎

13.1.24 Remark By the above lemma, the atom 1 turns out to be the top element in HL. But 1 is not declared as a universe and hence HL is not in $\mathrm{TT}^{\mathtt{U}}$.

In the following Lemmas 13.1.25–13.1.29 we study the positions of the atoms 0, and 1 in the TTs introduced in Fig. 13.4. The principal result is that $0 < 1$ in HL and, as far as applicable, in the theories Scott, Park, CDZ, HR, DHM, and Plotkin,

$$0 < 1 < \mathtt{U}.$$

13.1.25 Lemma *Let* $\mathcal{T} \in \{\text{Scott}, \text{Park}, \text{CDZ}, \text{HR}, \text{DHM}, \text{BCD}, \text{Engeler}\}$. *Define recursively the following collection of types:*

$$\boxed{\mathcal{B} = \mathtt{U} \mid \mathbb{T}^{\mathcal{T}} \to \mathcal{B} \mid \mathcal{B} \cap \mathcal{B}.}$$

Then we have $\mathcal{B} = \{A \in \mathbb{T}^{\mathcal{T}} \mid A =_{\mathcal{T}} \mathtt{U}\}$.

Proof By induction on the generation of $A \leq_{\mathcal{T}} B$ one proves that $\mathcal{B}$ is closed upwards. This gives $\mathtt{U} \leq_{\mathcal{T}} A \Rightarrow A \in \mathcal{B}$.

By induction on the definition of $\mathcal{B}$ one shows, using $(\to)$ or $(\to^{=})$, $(\mathtt{U}_{\text{top}})$ and $(\mathtt{U}\to)$, that $A \in \mathcal{B} \Rightarrow A =_{\mathcal{T}} \mathtt{U}$.

Therefore

$$A =_{\mathcal{T}} \mathtt{U} \Leftrightarrow A \in \mathcal{B}. \blacksquare$$

13.1.26 Lemma *For* $\mathcal{T} \in \{\text{AO}, \text{Plotkin}\}$ *define recursively*

$$\boxed{\mathcal{B} = \mathtt{U} \mid \mathcal{B} \cap \mathcal{B}.}$$

Then $\mathcal{B} = \{A \in \mathbb{T}^{\mathcal{T}} \mid A =_{\mathcal{T}} \mathtt{U}\}$, *hence* $\mathtt{U}\to\mathtt{U} \neq_{\mathcal{T}} \mathtt{U}$.

Proof Similar to the proof of 13.1.23, but easier. ■

13.1.27 Lemma *For* $\mathcal{T} \in \{\text{CDZ}, \text{HR}, \text{DHM}\}$ *define by mutual induction*

$$\boxed{\begin{array}{lcl} \mathcal{B} & = & 1 \mid \mathtt{U} \mid \mathbb{T}^{\mathcal{T}} \to \mathcal{B} \mid \mathcal{H} \to \mathbb{T}^{\mathcal{T}} \mid \mathcal{B} \cap \mathcal{B} \\ \mathcal{H} & = & 0 \mid \mathcal{B} \to \mathcal{H} \mid \mathcal{H} \cap \mathbb{T}^{\mathcal{T}} \mid \mathbb{T}^{\mathcal{T}} \cap \mathcal{H}. \end{array}}$$

Then

$$\begin{array}{l} 1 \leq B \Rightarrow B \in \mathcal{B}, \\ A \leq 0 \Rightarrow A \in \mathcal{H}. \end{array}$$

Proof By induction on $\leq_{\mathcal{T}}$ one shows

$$A \leq B \Rightarrow (A \in \mathcal{B} \Rightarrow B \in \mathcal{B}) \Rightarrow (B \in \mathcal{H} \Rightarrow A \in \mathcal{H}).$$

From this the assertion follows immediately. ■

13.1.28 Lemma *We work with the theory* HL.

(i) *Define by mutual induction*

$$\boxed{\begin{array}{lcl} \mathcal{B} & = & 1 \mid \mathcal{H}{\to}\mathcal{B} \mid \mathcal{B}\cap\mathcal{B} \\ \mathcal{H} & = & 0 \mid \mathcal{B}{\to}\mathcal{H} \mid \mathcal{H}\cap\mathbb{T}^{\mathrm{HL}} \mid \mathbb{T}^{\mathrm{HL}}\cap\mathcal{H}. \end{array}}$$

Then

$$\begin{array}{lcl} \mathcal{B} & = & \{A\in\mathbb{T}^{\mathrm{HL}} \mid A =_{\mathrm{HL}} 1\}; \\ \mathcal{H} & = & \{A\in\mathbb{T}^{\mathrm{HL}} \mid A =_{\mathrm{HL}} 0.\} \end{array}$$

(ii) $0 \neq_{\mathrm{HL}} 1$ *and hence* $0 <_{\mathrm{HL}} 1$.

Proof (i) By induction on $\leq_{\mathcal{T}}$ one shows

$$A\leq B \Rightarrow (A\in\mathcal{B} \Rightarrow B\in\mathcal{B})\ \&\ (B\in\mathcal{H} \Rightarrow A\in\mathcal{H}).$$

This gives

$$(1\leq B \Rightarrow B\in\mathcal{B})\ \&\ (A\leq 0 \Rightarrow A\in\mathcal{H}).$$

By simultaneous induction on the generation of $\mathcal{B}$ and $\mathcal{H}$ one shows, by Lemma 13.1.23((i)), using the fact that 0 is the bottom element and 1 is the top element of HL, that

$$(B\in\mathcal{B} \Rightarrow B=1)\ \&\ (A\in\mathcal{H} \Rightarrow A=0).$$

Now the assertion follows immediately.

(ii) By (i). ∎

13.1.29 Proposition *In* HL *we have* $0 < 1$. *For the other members of Fig.* 13.4, *as far as applicable,*

$$0 < 1 < \mathtt{U}.$$

More precisely, in HL *one has*

(i) $0 < 1$.

In CDZ, HR, DHM *one has*

(ii) $0 < 1 < \mathtt{U}$.

Proof (i) By rule (01) and Lemma 13.1.28.

(ii) By rules (01), ($\mathtt{U}_{\mathrm{top}}$)and Lemmas 13.1.25–13.1.27. ∎

(Ax)	$\Gamma \vdash x{:}A$	if $(x{:}A) \in \Gamma$
$(\to\mathrm{I})$	$\dfrac{\Gamma, x{:}A \vdash M : B}{\Gamma \vdash \lambda x.M : A \to B}$	
$(\to\mathrm{E})$	$\dfrac{\Gamma \vdash M : A \to B \quad \Gamma \vdash N : A}{\Gamma \vdash MN : B}$	
$(\cap\mathrm{I})$	$\dfrac{\Gamma \vdash M : A \quad \Gamma \vdash M : B}{\Gamma \vdash M : A \cap B}$	
$(\leq)$	$\dfrac{\Gamma \vdash M : A \quad A \leq_{\mathcal{T}} B}{\Gamma \vdash M : B}$	
$(\mathtt{U}_{\mathrm{top}})$	$\Gamma \vdash M : \mathtt{U}$	if $\mathcal{T} \in \mathrm{TT}^{\mathtt{U}}$

Figure 13.8

13.2 Type assignment

Assignment of types from type theories

In this subsection we define an infinite collection $\{\boldsymbol{\lambda}_\cap^{\mathcal{T}} \mid \mathcal{T} \in \mathrm{TT}\}$ of type assignment systems by giving a *uniform* set of typing rules parametric in $\mathcal{T}$.

13.2.1 Definition Let $\mathcal{T} \in \mathrm{TT}$.

(i) A $\mathcal{T}$*-statement* is of the form $M : A$, with $M \in \Lambda$ and $A \in \mathbb{T}^{\mathcal{T}}$.

(ii) A $\mathcal{T}$*-declaration* is a $\mathcal{T}$-statement of the form $x : A$.

(iii) A $\mathcal{T}$*-basis* Γ is a finite set of $\mathcal{T}$-declarations, with all term variables distinct.

(iv) A $\mathcal{T}$*-assertion* is of the form

$$\Gamma \vdash M : A,$$

where $M : A$ is a $\mathcal{T}$-statement and Γ is a $\mathcal{T}$-basis.

13.2.2 Remark Let $M : A$ be a $\mathcal{T}$-statement.

(i) The term M is called the *subject* of this statement.

(ii) The type A is called its *predicate*.

13.2.3 Definition Let $\mathcal{T} \in \mathrm{TT}$. The *type assignment* system $\boldsymbol{\lambda}_\cap^{\mathcal{T}}$ derives $\mathcal{T}$-assertions by the axioms and rules in Fig. 13.8.

Note that the parameter $\mathcal{T}$ appears only in the last two rules.

13.2.4 Notation

(i) We write $\Gamma \vdash_\cap^{\mathcal{T}} M : A$ if $\Gamma \vdash M : A$ is derivable in $\boldsymbol{\lambda}_\cap^{\mathcal{T}}$.

(ii) The assertion $\vdash^{\mathcal{T}}_{\cap}$ may also be written as $\vdash^{\mathcal{T}}_{\cap}$, as $\vdash_{\cap}$, or simply as $\vdash$ if there is little danger of confusion.
(iii) We may also denote $\boldsymbol{\lambda}^{\mathcal{T}}_{\cap}$ simply by $\boldsymbol{\lambda}_{\cap}$.

13.2.5 Remark Given a type theory $\mathcal{T}$, the following two options are mutually exclusive: either the type assignment system $\boldsymbol{\lambda}^{\mathcal{T}}_{\cap}$ contains the axiom ($\mathtt{U}_{\text{top}}$); or it does not. For the specific type theories in Fig. 13.4, the situation is as follows:

(1) For the first ten type theories, i.e. Scott, Park, CDZ, HR, DHM, BCD, AO, Plotkin, Engeler, and CDS, we only get the type assignment system with the axiom ($\mathtt{U}_{\text{top}}$), since they all belong to $\text{TT}^{\mathtt{U}}$.
(2) For the remaining three type theories, i.e. HL, CDV and CD, we only get the one without this axiom, since they all belong to $\text{TT}^{\text{-}\mathtt{U}}$.

13.2.6 Remark As suggested in the introduction to Chapter 12, the type assignment systems with the axiom ($\mathtt{U}_{\text{top}}$) are closed under β-expansions and can be used to construct models of the λK-calculus. On the other hand the systems without this axiom are closed under βI-expansions (not necessarily under β) and can be used to construct models specifically for the λI-calculus.

13.2.7 Example The statements in this example will be proved in Exercises 14.7 and 14.8. Define $\boldsymbol{\omega} \equiv \lambda x.xx$, $\boldsymbol{\Omega} = \boldsymbol{\omega}\boldsymbol{\omega}$, $\mathsf{V} \equiv \lambda yz.\mathsf{K}z(yz)$ and $\mathsf{K}_* \equiv \lambda yz.z$. Then $\mathsf{V} \twoheadrightarrow_{\beta} \mathsf{K}_*$. We have the following for arbitrary $A, B \in \mathbb{T}$.

(i) For all $\mathcal{T} \in \text{TT}$:

$$\vdash^{\mathcal{T}}_{\cap} \boldsymbol{\omega} : A \cap (A \to B) \to B;$$
$$\vdash^{\mathcal{T}}_{\cap} \mathsf{V} : (B \to A) \to B \to B;$$
$$\vdash^{\mathcal{T}}_{\cap} \mathsf{K}_* : A {\to} B \to B.$$

(ii) For some $\mathcal{T} \in \text{TT}^{\text{-}\mathtt{U}}$, for example for CDV and CD, one has

$$\nvdash^{\mathcal{T}}_{\cap} \mathsf{V} : \alpha {\to} \beta \to \beta.$$

We conclude that

$$M \twoheadrightarrow_{\beta} N \;\&\; \Gamma \vdash^{\mathcal{T}}_{\cap} N : A \not\Rightarrow \Gamma \vdash^{\mathcal{T}}_{\cap} M : A,$$

i.e. in the absence of $\mathtt{U}$ subject expansion fails, even if the expanded term is typable, as observed in van Bakel (1993); this phenomenon also occurs in $\boldsymbol{\lambda}_{\to}$.
(iii) For some $\mathcal{T} \in \text{TT}^{\text{-}\mathtt{U}}$, for example for HL, CDV and CD, one has, for all A:

$$\nvdash^{\mathcal{T}}_{\cap} \mathsf{KI}\boldsymbol{\Omega} : A;$$
$$\nvdash^{\mathcal{T}}_{\cap} \boldsymbol{\Omega} : A.$$

(*weakening*)	$\dfrac{\Gamma \vdash M : A \quad x \notin \Gamma}{\Gamma, x{:}B \vdash M : A}$;
(*strengthening*)	$\dfrac{\Gamma, x{:}B \vdash M : A \quad x \notin FV(M)}{\Gamma \vdash M : A}$;
(*cut*)	$\dfrac{\Gamma, x{:}B \vdash M : A \quad \Gamma \vdash N : B}{\Gamma \vdash (M[x := N]) : A}$;
($\leq$-L)	$\dfrac{\Gamma, x{:}B \vdash M : A \quad C \leq_{\mathcal{T}} B}{\Gamma, x{:}C \vdash M : A}$;
($\to$L)	$\dfrac{\Gamma, y{:}B \vdash M : A \quad \Gamma \vdash N : C \quad x \notin \Gamma}{\Gamma, x{:}(C \to B) \vdash (M[y := xN]) : A}$;
($\cap$L)	$\dfrac{\Gamma, x{:}A \vdash M : B}{\Gamma, x{:}(A \cap C) \vdash M : B}$.

Figure 13.9 Various admissible rules.

(iv) $\not\vdash^{\mathrm{CD}} \mathsf{I} : ((\alpha \cap \beta) \to \gamma) \to ((\beta \cap \alpha) \to \gamma)$.
(v) Let $\mathcal{T} \in \mathrm{TT}^{\mathtt{U}}$. Then

$$\vdash^{\mathcal{T}}_{\cap} \boldsymbol{\Omega} : \mathtt{U};$$
$$\vdash^{\mathcal{T}}_{\cap} \mathsf{KI}\boldsymbol{\Omega} : (A \to A);$$
$$\vdash^{\mathcal{T}}_{\cap} \mathsf{V} : A \to B \to B.$$

13.2.8 Definition Denote by ($\cap$E) the rules

$$\frac{\Gamma \vdash M : (A \cap B)}{\Gamma \vdash M : A} \qquad \frac{\Gamma \vdash M : (A \cap B)}{\Gamma \vdash M : B}.$$

Notice that these rules are derived in $\boldsymbol{\lambda}^{\mathcal{T}}_{\cap}$ for all $\mathcal{T}$.

13.2.9 Lemma *For $\mathcal{T} \in \mathrm{TT}^{\text{-}\mathtt{U}}$ one has:*

(i) $\Gamma \vdash^{\mathcal{T}}_{\cap} M : A \Rightarrow \mathrm{FV}(M) \subseteq \mathrm{dom}(\Gamma)$;

(ii) $\Gamma \vdash^{\mathcal{T}}_{\cap} M : A \Rightarrow (\Gamma \upharpoonright \mathrm{FV}(M)) \vdash M : A$.

Proof (i), (ii) By straightforward induction on the derivation. ∎

Notice that $\Gamma \vdash M : A \Rightarrow \mathrm{FV}(M) \subseteq \mathrm{dom}(\Gamma)$ does not hold for $\mathcal{T} \in \mathrm{TT}^{\mathtt{U}}$, since by axiom ($\mathtt{U}_{\mathrm{top}}$) we have $\vdash^{\mathcal{T}}_{\cap} M : \mathtt{U}$ for all M.

Admissible rules

13.2.10 Proposition *The rules in Fig.* 13.9 *are admissible in* $\boldsymbol{\lambda}^{\mathcal{T}}_{\cap}$.

Proof By straightforward induction on the structure of derivations. ∎

(*multiple weakening*)	$\dfrac{\Gamma_1 \vdash M : A}{\Gamma_1 \uplus \Gamma_2 \vdash M : A}$
(*relevant* $\to$E)	$\dfrac{\Gamma_1 \vdash M : A \to B \quad \Gamma_2 \vdash N : A}{\Gamma_1 \uplus \Gamma_2 \vdash MN : B}$
(*relevant* $\cap$I)	$\dfrac{\Gamma_1 \vdash M : A \quad \Gamma_2 \vdash M : B}{\Gamma_1 \uplus \Gamma_2 \vdash M : A \cap B}$

Figure 13.10

In proofs later on in Part III we will freely use the rules of the above proposition.

As we remarked earlier, there are various equivalent alternative presentations of intersection type assignment systems. We have chosen a natural deduction presentation, where $\mathcal{T}$-bases are additive. We could just as well have taken a sequent style presentation and replace rule ($\to$E) with the three rules ($\to$L), ($\cap$L) and (cut) occuring in Proposition 13.2.10: see Barbanera et al. (1995), Barendregt and Ghilezan (2000). We could also have formulated the rules so that $\mathcal{T}$-bases 'multiply'. Note that because of the presence of the type constructor $\cap$, a special notion of *multiplication of $\mathcal{T}$-bases* is useful.

13.2.11 Definition (Multiplication of $\mathcal{T}$-bases) Denote

$$\begin{aligned}\Gamma \uplus \Gamma' = {} & \{x{:}\, A \cap B \mid x{:}\, A \in \Gamma \text{ and } x{:}\, B \in \Gamma'\}\\ & \cup\{x{:}\, A \mid x{:}\, A \in \Gamma \text{ and } x \notin \Gamma'\}\\ & \cup\{x{:}\, B \mid x{:}\, B \in \Gamma' \text{ and } x \notin \Gamma\}.\end{aligned}$$

Accordingly we denote:

$$\Gamma \Subset \Gamma' \;\Leftrightarrow\; \exists \Gamma''.\, \Gamma \uplus \Gamma'' = \Gamma'.$$

For example, $\{x{:}A, y{:}B\} \uplus \{x{:}C, z{:}D\} = \{x{:}A \cap C, y{:}B, z{:}D\}$.

13.2.12 Proposition *The rules in Fig.* 13.10 *are admissible in all* $\boldsymbol{\lambda}_\cap^{\mathcal{T}}$.

Proof By straightforward induction on derivations. ∎

In Exercise 14.23, it will be shown that we can replace rule ($\leq$) with other more transparent rules. This will be possible after we have proved appropriate 'inversion' lemmas for $\boldsymbol{\lambda}_\cap^{\mathcal{T}}$. For some very special theories, one can even omit rule ($\leq$) altogether, provided the remaining rules are reformulated 'multiplicatively' with respect to $\mathcal{T}$-bases, see for example Di Gianantonio and Honsell (1993). We shall not however follow up this line of investigation.

In $\boldsymbol{\lambda}_{\cap}^{\mathcal{T}}$, assumptions are allowed to appear in the basis without any restriction. Alternatively, we could introduce a *relevant* intersection type assignment system, where only 'minimal-base' judgements are derivable, (see Honsell and Ronchi Della Rocca (1992)). Rules such as (*relevant* $\to$E) and (*relevant* $\cap$ I), which exploit the above notion of multiplication of bases, are essential for this purpose. Relevant systems are necessary, for example, for giving finitary logical descriptions of qualitative domains as defined in Girard et al. (1989). We will not follow up this line of research either; see Honsell and Ronchi Della Rocca (1992).

Special type assignment for call-by-value λ-calculus

13.2.13 Definition The type theory EHR is defined with $\mathbb{A}^{\mathrm{EHR}} = \{\mathtt{V}\}$ and the extra rule ($\to$) and axioms ($\to\cap$) and

$$A{\to}B \leq \mathtt{V}.$$

The type assignment system $\boldsymbol{\lambda}_{\cap\mathtt{V}}^{\mathrm{EHR}}$ is defined by the axiom and rules of Definition 13.2.3 with the extra axiom

$$\boxed{(\mathtt{V}) \qquad \Gamma \vdash (\lambda x.M) : \mathtt{V}.}$$

The type theory EHR has a top, namely $\mathtt{V}$, but it is not an element of $\mathrm{TT}^{\mathtt{U}}$. Hence $\boldsymbol{\lambda}_{\cap\mathtt{V}}^{\mathrm{EHR}}$ does not contain the axiom ($\mathtt{U}_{\mathrm{top}}$). Note also that the axiom ($\mathtt{V}$) is different from ($\mathtt{U}_{\mathrm{top}}$). This type assignment system is suitable for modeling the call-by-value λ-calculus: it has been extensively studied in Ronchi Della Rocca and Paolini (2004).

13.3 Type structures

Intersection type structures

Recall that a type algebra $\mathcal{A}$, see Definition 7.1.1, has the form $\mathcal{A} = \langle |\mathcal{A}|, \to \rangle$, i.e. is just an arbitrary set $|\mathcal{A}|$ with a binary operation $\to$ on it.

13.3.1 Definition

(i) A *meet semi-lattice* (without universe) is a structure

$$\mathcal{M} = \langle |\mathcal{M}|, \leq, \cap \rangle,$$

such that $|\mathcal{M}|$ is a countable set, $\leq$ is a partial order, and, for all

$A, B \in |\mathcal{M}|$, the element $A \cap B$ (meet) is the greatest lower bound of A and B. Denote by $\mathrm{MSL}^{\text{-}\mathtt{U}}$ the set of meet semi-lattices:

(ii) A *meet semi-lattice with universe* is a similar structure

$$\mathcal{M} = \langle |\mathcal{M}|, \leq, \cap, \mathtt{U} \rangle,$$

with $\mathtt{U}$ the (unique) top element of $\mathcal{M}$. Denote by $\mathrm{MSL}^{\mathtt{U}}$ the set of meet semi-lattices with universe.

(iii) Denote by $\mathrm{MSL} = \mathrm{MSL}^{\mathtt{U}} \cup \mathrm{MSL}^{\text{-}\mathtt{U}}$ the set of meet semi-lattices with or without universe.

(iv) We have $\mathrm{MSL}^{\mathtt{U}} \cap \mathrm{MSL}^{\text{-}\mathtt{U}} = \emptyset$, as the signatures are different.

13.3.2 Definition

(i) An *(intersection) type structure* (without universe) is a type algebra with the additional structure of a meet semi-lattice:

$$\mathcal{S} = \langle |\mathcal{S}|, \leq, \cap, \rightarrow \rangle.$$

Denote by $\mathrm{TS}^{\text{-}\mathtt{U}}$ the set of type structures without universe. The relation $\leq$ and the operation $\rightarrow$ have a priori no relation with each other, but in special structures they may have.

(ii) A *type structure with universe* $\mathtt{U}$ is a type algebra that is also a meet semi-lattice with universe:

$$\mathcal{S} = \langle |\mathcal{S}|, \leq, \cap, \rightarrow, \mathtt{U} \rangle.$$

Denote by $\mathrm{TS}^{\mathtt{U}}$ the set of type structures with universe $\mathtt{U}$.

(iii) Denote by $\mathrm{TS} = \mathrm{TS}^{\mathtt{U}} \cup \mathrm{TS}^{\text{-}\mathtt{U}}$ the set of type structures with or without universe. As before $\mathrm{TS}^{\text{-}\mathtt{U}} \cap \mathrm{TS}^{\mathtt{U}} = \emptyset$.

13.3.3 Notation

(i) As the term 'intersection' is used throughout Part III, we will omit it and only speak about a *type structure*.

(ii) By abuse of language we also use $A, B, C, \ldots$ to denote arbitrary elements of type structures and we write $A \in \mathcal{S}$ for $A \in |\mathcal{S}|$.

If $\mathcal{T}$ is a type theory that is not compatible, such as CD and CDS, then $\rightarrow$ cannot be defined on the equivalence classes. But if $\mathcal{T}$ is compatible, then one can work on the equivalence classes and obtain a type structure in which $\leq$ is a partial order.

13.3.4 Proposition *Let $\mathcal{T} \in \mathrm{TT}$ be compatible. Then $\mathcal{T}$ induces a type structure $[\mathcal{T}]$:*

$$[\mathcal{T}] = \langle [\mathbb{T}], \leq, \cap, \rightarrow \rangle,$$

by defining on the $=_{\mathcal{T}}$-equivalence classes:

$$\begin{aligned} [A] \leq [B] &\Leftrightarrow A \leq B; \\ [A] \cap [B] &= [A \cap B]; \\ [A] {\to} [B] &= [A {\to} B] \end{aligned}$$

where A, B, C range over $\mathbb{T}$. If moreover $\mathcal{T}$ has universe $\mathtt{U}$, then $[\mathcal{T}]$ is a type structure with $[\mathtt{U}]$ as universe.

Proof See Remark 13.1.13. ∎

Let $\mathcal{T} \in \mathrm{TT}$. Then the type structure $[\mathcal{T}]$ is called a *syntactical type structure*.

13.3.5 Proposition *Every type structure is isomorphic to a syntactical one.*

Proof For a type structure $\mathcal{S}$, define a type theory $\mathrm{Th}(\mathcal{S})$ as follows. Take $\mathbb{A} = \mathbb{A}^{\mathrm{Th}(\mathcal{S})} = \{\underline{c} \mid c \in \mathcal{S}\}$. Then define $g : \mathbb{T}^{\mathbb{A}} {\to} \mathcal{S}$ by

$$\begin{aligned} g(\underline{c}) &= c; \\ g(A {\to} B) &= g(A) {\to} g(B); \\ g(A \cap B) &= g(A) \cap g(B). \end{aligned}$$

Define $A \leq_{\mathrm{Th}(\mathcal{S})} B \Leftrightarrow g(A) \leq_{\mathcal{S}} g(B)$. Then g induces a bijective morphism $\bar{g} : [\mathrm{Th}(\mathcal{S})] \to \mathcal{S}$ where the inverse f is defined by $f(a) = [\underline{a}]$. Moreover, if $\mathcal{S}$ has a universe $\mathtt{U}$, then $[\mathtt{U}]$ is the universe of $[\mathrm{Th}(\mathcal{S})]$ and $\bar{g}([\underline{\mathtt{U}}]) = \mathtt{U}$. ∎

13.3.6 Remark

(i) Each of the eleven compatible type theories $\mathcal{T}$ in Fig. 13.4 may be considered as the intersection type structure $[\mathcal{T}]$. For example Scott can be a name, a type theory or a type structure.

(ii) Although essentially equivalent, type structures and type theories differ in the following. In the theories the types are freely generated from a fixed set of atoms and inequality can be partially controlled by choosing the right axioms and rules. This will be explored in Chapters 14, 16 and 17. In type structures one has the antisymmetric law $A \leq B \leq A \Rightarrow A = B$, which is in line with the common theory of partial orders. This will be explored in Chapter 15.

(iii) Note that in a type theory there is up to $=_{\mathcal{T}}$ at most one universe $\mathtt{U}$, as we require it to be the top.

Next, the notion of type assignment will be extended to type structures.

These structures arise naturally from the algebraic lattices that are used towards obtaining a semantics for untyped lambda calculus.

13.3.7 Definition Let $\mathcal{S} \in \mathrm{TS}$.

(i) The notions of an $\mathcal{S}$-*statement* $M : A$, an $\mathcal{S}$-*declaration* $x{:}A$, an $\mathcal{S}$-*basis* and an $\mathcal{S}$-*assertion* $\Gamma \vdash M : A$ are as in Definition 13.2.1, only now $A \in \mathcal{S}$ is an element of the type structure $\mathcal{S}$.
(ii) The notion $\Gamma \vdash_\cap^{\mathcal{S}} M : A$ is defined by the same set of axioms and rules of $\boldsymbol{\lambda}_\cap^{\mathcal{T}}$ in Definition 13.2.3, where now $\leq_{\mathcal{S}}$ is the inequality of the structure $\mathcal{S}$.

The following result shows that for syntactic type structures type assignment is essentially the same as that coming from the corresponding lambda theory.

13.3.8 Proposition *Let $\mathcal{T} \in \mathrm{TT}$ be compatible and write*

$$[\mathcal{T}] = \langle [\mathbb{T}], \leq, \cap, \rightarrow (, [\mathtt{U}]) \rangle$$

for its corresponding type structure possibly with universe. For a type $A \in \mathcal{T}$ write its equivalence class as $[A] \in [\mathcal{T}]$. For $\Gamma = \{x_1 : B_1, \ldots, x_n : B_n\}$ a $\mathcal{T}$-basis write $[\Gamma] = \{x_1 : [B_1], \ldots, x_n : [B_n]\}$, a $[\mathcal{T}]$-basis. Then

$$\Gamma \vdash_\cap^{\mathcal{T}} M : A \;\Leftrightarrow\; [\Gamma] \vdash_\cap^{[\mathcal{T}]} M : [A].$$

Proof ($\Rightarrow$) By induction on the derivation of $\Gamma \vdash^{\mathcal{T}} M : A$.

($\Leftarrow$) Show by induction on the derivation of $[\Gamma] \vdash^{[\mathcal{T}]} M : [A]$ that for all $A' \in [A]$ and $\Gamma' = \{x_1 : B'_1, \ldots, x_n : B'_n\}$, with $B'_i \in [B_i]$ for all $1 \leq i \leq n$, one has

$$\Gamma' \vdash^{\mathcal{T}} M : A'. \blacksquare$$

Using this result we could have defined type assignment first for type structures and then for compatible type theories via translation to the type assignment for its corresponding syntactical type structure, essentially by turning the previous result into a definition.

Categories of meet semi-lattices and type structures

We will introduce some categories related to given classes of type structures for later use in Chapter 15.

13.3.9 Definition

(i) The category $\mathbf{MSL}^{\text{-}\mathtt{U}}$ has as objects the elements of $\mathrm{MSL}^{\text{-}\mathtt{U}}$ and as morphisms those maps $f : \mathcal{M} \rightarrow \mathcal{M}'$ that preserve $\leq$, $\cap$:

$$A \leq B \Rightarrow f(A) \leq' f(B);$$
$$f(A \cap B) = f(A) \cap' f(B).$$

(ii) The category $\mathbf{MSL}^{\mathtt{U}}$ has as objects the elements of $\mathrm{MSL}^{\mathtt{U}}$ and as morphisms the maps that preserve not only $\leq$, $\cap$ but also $\mathtt{U}$, i.e. $f(\mathtt{U}) = \mathtt{U}'$.

There is no natural category corresponding to MSL, as it is a hybrid set consisting of structures with and without universe.

13.3.10 Definition

(i) The category $\mathbf{TS}^{\text{-}\mathtt{U}}$ has as objects type structures and as morphisms the maps $f : \mathcal{M} \to \mathcal{M}'$, which satisfy restrictions based on inequalities:

$(m1)$ $A \leq B \Rightarrow f(A) \leq' f(B)$ (monotonicity);
$(m2)$ $f(A) \cap' f(B) \leq' f(A \cap B)$;
$(m3)$ $f(A) \to' f(B) \leq' f(A \to B)$.

By the monotonicity of f, $(m2)$ implies $f(A \cap B) = f(A) \cap' f(B)$.

(ii) The category $\mathbf{TS}^{\mathtt{U}}$ is as $\mathbf{TS}^{\text{-}\mathtt{U}}$, but based on type structures with universe. For morphisms we require

$(m4)$ $\mathtt{U}' \leq' f(\mathtt{U})$ or, equivalently, $\mathtt{U}' = f(\mathtt{U})$.

Again there is no category corresponding to TS.

The definitions of graph, lazy, natural and proper type theory translate immediately to type structures.

13.3.11 Definition We define three full subcategories of $\mathbf{TS}^{\mathtt{U}}$ and one full subcategory of $\mathbf{TS}^{\text{-}\mathtt{U}}$ by specifying in each case the objects:

(i) $\mathbf{GTS}^{\mathtt{U}}$ whose objects are the graph type structures in $\mathrm{TT}^{\mathtt{U}}$;
(ii) $\mathbf{LTS}^{\mathtt{U}}$ whose objects are the lazy type structures in $\mathrm{TT}^{\mathtt{U}}$;
(iii) $\mathbf{NTS}^{\mathtt{U}}$ whose objects are the natural type structures in $\mathrm{TT}^{\mathtt{U}}$;
(iv) $\mathbf{PTS}^{\text{-}\mathtt{U}}$ whose objects are the proper type structures in $\mathrm{TT}^{-\mathtt{U}}$.

13.4 Filters

In this section we define the notions of filter and filter structure.

13.4.1 Definition

(i) Let $\mathcal{T} \in \mathrm{TT}^{\mathtt{U}}$. Then $X \subseteq \mathbb{T}^{\mathcal{T}}$ is a *filter* over $\mathcal{T}$ if the following hold.

(1) $A \in X$ & $A \leq B \Rightarrow B \in X$;
(2) $A, B \in X \Rightarrow A \cap B \in X$;
(3) X is non-empty.

(ii) Let $\mathcal{T} \in \mathrm{TT}^{\text{-}\mathtt{U}}$. Then $X \subseteq \mathbb{T}^{\mathcal{T}}$ is a *filter* over $\mathcal{T}$ if the following hold.

(1) $A \in X$ & $A \leq B \Rightarrow B \in X$;

(2) $A, B \in X \Rightarrow A \cap B \in X$.

(iii) Write $\mathcal{F}^{\mathcal{T}} = \{X \subseteq \mathbb{T}^{\mathcal{T}} \mid X$ is a filter over $\mathcal{T}\}$.

If $\mathcal{T} \in \mathrm{TT}^{\mathtt{U}}$, then filters are sets of types containing $\mathtt{U}$ and closed under $\leq$ and $\cap$. If $\mathcal{T} \in \mathrm{TT}^{\text{-}\mathtt{U}}$, then filters may be empty.

13.4.2 Definition Let $\mathcal{T} \in \mathrm{TT}$.

(i) For $A \in \mathbb{T}^{\mathcal{T}}$ write $\uparrow A = \{B \in \mathbb{T}^{\mathcal{T}} \mid A \leq B\}$.
(ii) For $X \subseteq \mathbb{T}^{\mathcal{T}}$ define $\uparrow X$ to be the smallest filter over $\mathcal{T}$ containing X.

13.4.3 Remark

(i) If X is non-empty, then

$$\uparrow X = \{B \in \mathbb{T}^{\mathcal{T}} \mid \exists n \geq 1\, \exists A_1, \ldots, A_n \in X. A_1 \cap \cdots \cap A_n \leq B\}.$$

(ii) For $X = \emptyset$, we have that

$$\uparrow\emptyset = \begin{cases} \{A \mid A = \mathtt{U}\} = [\mathtt{U}] & \text{if } \mathcal{T} \in \mathrm{TT}^{\mathtt{U}} \\ \emptyset & \text{if } \mathcal{T} \in \mathrm{TT}^{\text{-}\mathtt{U}}. \end{cases}$$

(iii) $C \in \uparrow \{B_i \mid i \in \mathcal{I} \neq \emptyset\} \Leftrightarrow \exists J \subseteq_{\text{fin}} \mathcal{I}.[J \neq \emptyset \;\&\; \bigcap_{j \in J} B_j \leq C]$.

Complete lattices and the category **ALG** were introduced in Definition 12.2.1.

13.4.4 Proposition *Let* $\mathcal{T} \in \mathrm{TT}$.

(i) $\mathcal{F}^{\mathcal{T}} = \langle \mathcal{F}^{\mathcal{T}}, \subseteq \rangle$ *is a complete lattice, and for* $\mathcal{X} \subseteq \mathcal{F}^{\mathcal{T}}$, *the sup is*

$$\bigsqcup \mathcal{X} = \uparrow(\bigcup \mathcal{X}).$$

(ii) *For* $A \in \mathbb{T}^{\mathcal{T}}$ *one has* $\uparrow A = \uparrow\{A\}$ *and* $\uparrow A \in \mathcal{F}^{\mathcal{T}}$.
(iii) *For* $A, B \in \mathbb{T}^{\mathcal{T}}$ *one has* $\uparrow A \sqcup \uparrow B = \uparrow(A \cap B)$.
(iv) *For* $A_i \in \mathbb{T}^{\mathcal{T}}$ $(i \in \mathcal{I})$ *one has* $\bigsqcup\{\uparrow A_i \mid i \in \mathcal{I}\} = \uparrow \{A_i \mid i \in \mathcal{I}\}$.
(v) *For* $X \in \mathcal{F}^{\mathcal{T}}$ *one has*

$$\begin{aligned} X &= \bigsqcup\{\uparrow A \mid A \in X\} &= \bigsqcup\{\uparrow A \mid \uparrow A \subseteq X\} \\ &= \bigcup\{\uparrow A \mid A \in X\} &= \bigcup\{\uparrow A \mid \uparrow A \subseteq X\}. \end{aligned}$$

(vi) *The set* $\mathcal{K}(\mathcal{F}^{\mathcal{T}})$ *of finite (i.e. compact) elements of* $\mathcal{F}^{\mathcal{T}}$ *is given by*

$$\mathcal{K}(\mathcal{F}^{\mathcal{T}}) = \begin{cases} \{\uparrow A \mid A \in \mathbb{T}^{\mathcal{T}}\} & \textit{if } \mathcal{T} \in \mathrm{TT}^{\mathtt{U}} \\ \{\uparrow A \mid A \in \mathbb{T}^{\mathcal{T}}\} \cup \emptyset & \textit{if } \mathcal{T} \in \mathrm{TT}^{\text{-}\mathtt{U}}. \end{cases}$$

(vii) $\mathcal{F}^{\mathcal{T}} \in \mathbf{ALG}$.

Proof Easy. ■

Now we introduce the fundamental notion of filter structure which will be used throughout Part III of this book. Since the seminal paper of Barendregt et al. (1983), this notion has played a major role in the study of the mathematical semantics of lambda calculus.

13.4.5 Definition Let $\mathcal{T} \in \mathrm{TT}$. Define

$$\begin{array}{lll} F^{\mathcal{T}} & \in & [\mathcal{F}^{\mathcal{T}} \to [\mathcal{F}^{\mathcal{T}} \to \mathcal{F}^{\mathcal{T}}]], \qquad \text{and} \\ G^{\mathcal{T}} & \in & [[\mathcal{F}^{\mathcal{T}} \to \mathcal{F}^{\mathcal{T}}] \to \mathcal{F}^{\mathcal{T}}] \end{array}$$

as follows:

$$F^{\mathcal{T}}(X)(Y) = \uparrow\{B \in \mathbb{T}^{\mathcal{T}} \mid \exists A \in Y.(A \to B) \in X\};$$
$$G^{\mathcal{T}}(f) = \uparrow\{A \to B \mid B \in f(\uparrow A)\}.$$

Then, $\mathcal{F}^{\mathcal{T}} = \langle \mathcal{F}^{\mathcal{T}}, F^{\mathcal{T}}, G^{\mathcal{T}} \rangle$ is called the *filter structure* over $\mathcal{T}$.

It is easy to show that

$$F^{\mathcal{T}} \in [\mathcal{F}^{\mathcal{T}} \to [\mathcal{F}^{\mathcal{T}} \to \mathcal{F}^{\mathcal{T}}]] \ \& \ G^{\mathcal{T}} \in [[\mathcal{F}^{\mathcal{T}} \to \mathcal{F}^{\mathcal{T}}] \to \mathcal{F}^{\mathcal{T}}].$$

13.4.6 Remark The items 13.1.19–13.2.12 and 13.4.1–13.4.5 are about type theories, but can be translated immediately to structures and, if no $\to$ are involved, to meet semi-lattices. For example, Proposition 13.1.22 also holds for a proper type structure, hence it holds for Scott, Park, CDZ, HR, DHM, BCD, AO, HL, and CDV considered as type structures. Also 13.1.23–13.1.29 immediately yield corresponding valid statements for the corresponding type structures, though the proof for the type theories cannot be translated to proofs for the type structures because they are by induction on the syntactic generation of $\mathbb{T}$ or $\leq$. In addition, 13.2.7–13.2.12 hold for type structures, as follows immediately from Propositions 13.3.5 and 13.3.8. Finally 13.4.1–13.4.5 can be translated immediately to type structures and meet semi-lattices. In Chapter 15 we work directly with meet semi-lattices and type structures and not with type theories, because a proper partial order is needed there.

An example of the use of this remark is the following easy lemma.

13.4.7 Lemma *Let $\mathcal{T}$ be a compatible type theory. Then $\mathcal{F}^{\mathcal{T}} \cong \mathcal{F}^{[\mathcal{T}]}$ in the category* **ALG**.

Proof A filter X over $\mathcal{T}$ is mapped to

$$[X] = \{[A] \mid A \in X\}.$$

This map is 1–1, onto and preserves $\subseteq$. ∎

13.5 Exercises

Exercise 13.1 Let $\mathcal{T}$ be a type theory that satisfies $(\to)$.

(i) Show that if $(\mathtt{U}_{\rm top})$ holds, then the scheme axiom $(\mathtt{U}{\to})$ is equivalent to the following axiom:

$$\mathtt{U} \leq \mathtt{U} \to \mathtt{U}.$$

(ii) Show that $(\to\cap)$ is equivalent with $(\to\cap^{=})$.

Exercise 13.2 Let $\mathcal{T} \in \{\text{Scott}, \text{DHM}\}$. Prove that 0 is a bottom element in $\mathcal{T}$, i.e. $0 \leq_{\mathcal{T}} A$ for all $A \in \mathbb{T}^{\mathcal{T}}$.

Exercise 13.3 Let $\mathcal{T} \in \{\text{CDZ}, \text{HR}\}$. Show that $\mathtt{U} \to 0 < 0$. Conclude that 0 is not a bottom element in $\mathcal{T}$.

Exercise 13.4 Prove that for all types $A \in \mathbb{T}^{\text{AO}}$ there is an n such that

$$\mathtt{U}^n{\to}\mathtt{U} \leq_{\text{AO}} A.$$

Exercise 13.5 Show that $\Gamma, x{:}\mathtt{U} \vdash_\cap^{\mathcal{T}} M : A \;\Rightarrow\; \Gamma \vdash_\cap^{\mathcal{T}} M : A$.

Exercise 13.6 For $\mathcal{T}$ a type theory, $M, N \in \Lambda$ and $x \notin \text{dom}(\Gamma)$ show

$$\Gamma \vdash_\cap^{\mathcal{T}} M : A \quad\Rightarrow\quad \Gamma \vdash_\cap^{\mathcal{T}} M[x{:}= N] : A.$$

Exercise 13.7 Prove that if $(01), (1 \to 0)$ and $(0 \to 1)$ are axioms in $\mathcal{T}$, then, for all M in normal form, $\{x_1 : 1, \ldots, x_n : 1\} \vdash^{\mathcal{T}} M : 0$, where $\{x_1, \ldots, x_n\} \supseteq \text{FV}(M)$.

Exercise 13.8 Show that

$$M \text{ is a closed term } \Rightarrow \vdash_\cap^{\text{Park}} M : 0.$$

Later we will show the converse (Theorem 17.4.3).

Exercise 13.9 Suppose in $\mathcal{T}$ there is a type A such that $A = A{\to}A$. Then

$$\text{FV}(M) \subseteq \text{dom}(\Gamma) \;\Rightarrow\; \Gamma \vdash_\cap^{\mathcal{T}} M : A.$$

Exercise 13.10 The type theory Krivine and the type assignment system $\boldsymbol{\lambda}_\cap^{\text{Krivine}}$ of Krivine (1990) are CD and $\boldsymbol{\lambda}_\cap^{\text{CD}}$, but with rule $(\leq)$ replaced by

$$(\cap\text{E}) \quad \frac{\Gamma \vdash M : A \cap B}{\Gamma \vdash M : A} \quad \frac{\Gamma \vdash M : A \cap B}{\Gamma \vdash M : B}.$$

Similarly $\text{Krivine}^{\mathtt{U}}$ and $\boldsymbol{\lambda}_\cap^{\text{Krivine}^{\mathtt{U}}}$ are CDS and $\boldsymbol{\lambda}_\cap^{\text{CDS}}$, with $(\leq)$ replaced by $(\cap\text{E})$. Show that

(i) $\Gamma \vdash^{\text{Krivine}} M : A \quad \Leftrightarrow \quad \Gamma \vdash_\cap^{\text{CD}} M : A.$

(ii) $\Gamma \vdash^{\text{Krivine}^{\mathtt{U}}} M : A \quad \Leftrightarrow \quad \Gamma \vdash_\cap^{\text{CDS}} M : A.$

Exercise 13.11

(i) Show that $\lambda x.xxx$ and $(\lambda x.xx)\mathsf{I}$ are typable in $\boldsymbol{\lambda}_\cap^{\text{Krivine}}$.

(ii) Show that all closed terms in normal form are typable in $\boldsymbol{\lambda}_\cap^{\text{Krivine}}$.

Exercise 13.12 Show the following:

(i) $\vdash^{\text{Krivine}} \lambda z.\mathsf{KI}(zz) : \ (A{\to}B) \cap A{\to}C{\to}C.$

(ii) $\vdash^{\text{Krivine}^{\mathtt{U}}} \lambda z.\mathsf{KI}(zz) : \ \mathtt{U}{\to}C{\to}C.$

(iii) $\vdash_\cap^{\text{BCD}} \lambda z.\mathsf{KI}(zz) : \ \mathtt{U}{\to}(A{\to}B \cap C){\to}A{\to}B.$

Exercise 13.13 Let $\text{HL} \in \text{TS}^{\mathtt{U}}$ be $\langle [\mathbb{T}^{\text{HL}}], \leq_{\text{HL}}, \cap, \to, [1] \rangle$, where $[1]$ is taken as the universe. Show that $\text{HL} \notin \mathbf{LTS}^{\mathtt{U}}$ and $\text{HL} \notin \mathbf{NTS}^{\mathtt{U}}$ [Hint. Show that $1 \to 1 <_{\text{HL}} 0 \to 1$].

Exercise 13.14 Let $\mathcal{D} = \langle \mathcal{D}, \cdot \rangle$ be an applicative structure, i.e. a set with an arbitrary binary operation on it. For $X, Y \subset \mathcal{D}$ define

$$X \to Y = \{d \in \mathcal{D} \mid \forall e \in X. d \cdot e \in Y\}.$$

Consider $(\mathcal{P}(\mathcal{D}), \to, \subseteq, \cap, \mathcal{D})$, where $\mathcal{P}(\mathcal{D})$ is the power set of $\mathcal{D}$, $\subseteq$ and $\cap$ are the usual set-theoretic notions and $\mathcal{D}$ is the top of $\mathcal{P}(\mathcal{D})$. Show

(i) $(\mathcal{P}(\mathcal{D}), \to, \subseteq, \cap)$ is a proper type structure.

(ii) $\mathcal{D} = \mathcal{D} \to \mathcal{D}$.

(iii) $(\mathcal{P}(\mathcal{D}), \to, \subseteq, \cap, \mathcal{D})$ is a natural type structure.

Exercise 13.15 Show that the inclusions suggested in Fig. 13.5 are strict.

14

Basic Properties of Intersection Type Assignment

This chapter studies meta-theoretical properties of the type assignment systems. They will be crucial in Chapter 16 for the development of filter lambda models induced by these systems. The most important question here is whether the type assignment system satisfies $\boldsymbol{\beta}$-, $\boldsymbol{\eta}$-reduction or $\boldsymbol{\beta}$-, $\boldsymbol{\eta}$-expansion. Due to the intrinsic syntactic nature of the filter models, any property on the type assignment system is transferred into the filter structure. This is revealed by the Type Semantics Theorem (Theorem 16.2.7) which says that the interpretation of a term is the set of its types. The consequence of this is that the filter structure preserves the meaning of λ-terms under $\boldsymbol{\beta}$- or $\boldsymbol{\eta}$- reduction or expansion, exactly when the type assignment does.

Inversion Lemmas are proved in Section 14.1 and they essentially state when, depending on the form of M, an assertion $\Gamma \vdash^{\mathcal{T}} M : A$ holds. They are a convenient technical tool for doing proofs by induction (or by cases) on M when the typing is intricate. Instead of proving by induction on the derivation, an Inversion Lemma provides the opportunity for induction on the structure of the term.

The table in Fig. 14.1 summarizes the characterizations of the admissibility of $\boldsymbol{\beta}$ and $\boldsymbol{\eta}$ in the type assignment systems. This topic is developed in Section 14.2. On the left hand side of the table we find sufficient conditions (sometimes necessary as well) on the type theory in order to preserve the typing after reducing or expanding.

The notions of natural and proper type theory were introduced in Definition 13.1.19 because they characterize $\boldsymbol{\eta}$-reduction. The idea is to deduce which conditions are necessary to preserve the typing after $\boldsymbol{\eta}$-reducing terms. What we find turns out to be a sufficient set of conditions as well. Let $\Gamma = \{x : (A \to B)\}$. If $A \geq A'$ and $B \leq B'$, then $\Gamma \vdash \lambda y.xy : A' \to B'$ and $\lambda y.xy$ $\boldsymbol{\eta}$-reduces to x. It is easy then to see that, to deduce $\Gamma \vdash x : A' \to B'$,

Property of $\mathcal{T} \in \mathrm{TT}$	versus	property of $\boldsymbol{\lambda}_\cap^\mathcal{T}$
$\boldsymbol{\beta}$-sound	$\Rightarrow$	$\boldsymbol{\beta}$-red
$\mathcal{T} \in \mathrm{TT}$	$\Rightarrow$	$\boldsymbol{\beta}\mathsf{I}$-exp
$\mathcal{T} \in \mathrm{TT}^{\mathtt{U}}$	$\Rightarrow$	$\boldsymbol{\beta}$-exp
$\mathcal{T} \in \mathrm{TT}^{\text{-}\mathtt{U}}$ & proper	$\Leftrightarrow$	$\boldsymbol{\eta}$-red
$\mathcal{T} \in \mathrm{TT}^{\mathtt{U}}$ & natural	$\Leftrightarrow$	$\boldsymbol{\eta}$-red
$\mathcal{T} \in \mathrm{TT}^{\text{-}\mathtt{U}}$ & $\boldsymbol{\eta}$-sound	$\Leftrightarrow$	$\boldsymbol{\eta}$-exp
$\mathcal{T} \in \mathrm{TT}^{\mathtt{U}}$ & $\boldsymbol{\eta}^{\mathtt{U}}$-sound	$\Leftrightarrow$	$\boldsymbol{\eta}$-exp

Figure 14.1 Characterizing reduction and expansion

we need the axiom $(\rightarrow)$. A similar argument can be used to show that the axiom $(\rightarrow\cap)$ is also necessary.

New classes of type theories will be given for completing the picture. The notions of $\boldsymbol{\eta}$-sound and $\boldsymbol{\eta}^{\mathtt{U}}$-sound are introduced in Definition 14.2.10 to characterize $\boldsymbol{\eta}$-expansion. Roughly speaking, the condition of $\boldsymbol{\eta}$-soundness states that any type A should be equal to an intersection of arrow types. Let $\Gamma = \{x : A\}$. The term $\lambda y.xy$ will have type A in Γ if we impose the condition that there exist B, C, D, E such that:

$$\begin{array}{l}(B \rightarrow C) \leq A;\\ A \leq (D \rightarrow E);\\ B \leq D \;\&\; A \leq C.\end{array}$$

It is easy to see that if the type theory is also natural, then $A = B \rightarrow C$. The above condition is a simplification of the actual condition of $\boldsymbol{\eta}$-soundness concerning intersections.

For $\boldsymbol{\beta}$-expansion, there is no need to impose any condition, except the presence of the universal type. As explained in the introduction to Chapter 12, this was the reason for introducing intersection types. Without $(\mathtt{U}_{\mathrm{top}})$, the type assignment system only preserves $\beta\mathsf{I}$-expansion.

For $\boldsymbol{\beta}$-reduction, the condition of $\boldsymbol{\beta}$-soundness is introduced in Definition 14.1.4. To illustrate the idea, we consider a simplification of $\boldsymbol{\beta}$-soundness:

$$(A \rightarrow B) \cap (C \rightarrow D) \leq E \rightarrow F \;\Rightarrow\; E \leq A \cap C \;\&\; B \cap D \leq F.$$

If we know that $(\lambda x.M)N$ is typable, then we can apply the Inversion Lemmas and reach a point where the condition of $\boldsymbol{\beta}$-soundness appears naturally, enabling us to continue with the proof and deduce that $M[x := N]$ is typable

$\mathcal{T}$	$\boldsymbol{\beta}$-red	$\boldsymbol{\beta}$-exp	$\boldsymbol{\beta}$I-exp	$\boldsymbol{\eta}$-red	$\boldsymbol{\eta}$-exp
Scott	√	√	√	√	√
Park	√	√	√	√	√
CDZ	√	√	√	√	√
HR	√	√	√	√	√
DHM	√	√	√	√	√
BCD	√	√	√	√	×
AO	√	√	√	×	√
Plotkin	√	√	√	×	×
Engeler	√	√	√	×	×
CDS	√	√	√	×	×
HL	√	×	√	√	√
CDV	√	×	√	√	×
CD	√	×	√	×	×

Figure 14.2 Reduction and expansion in the type assignment systems

too. Suppose that

$$\begin{array}{ll} \vdash & (\lambda x.M) : E \to F \\ \vdash & N : E. \end{array}$$

This does not necessarily mean that $x : E \vdash M : F$. We could have the situation where

$$\begin{array}{l} (A \to B) \cap (C \to D) \leq E \to F \\ x : A \vdash M : B \\ x : C \vdash M : D. \end{array}$$

Our goal is to conclude that $x : E \vdash M : F$ and apply (cut) to prove eventually that $\vdash M[x := N] : F$. Currently, we see that it is enough to have the simplified condition of $\boldsymbol{\beta}$-soundness to achieve our aim.

The condition of $\boldsymbol{\beta}$-soundness is sufficient for preserving $\boldsymbol{\beta}$-red but it is not necessary. In Definition 16.2.20, an example will be given of a type theory that preserves $\boldsymbol{\beta}$-reduction but is not $\boldsymbol{\beta}$-sound.

Fig. 14.2 summarizes the results on $\boldsymbol{\beta}$ and $\boldsymbol{\eta}$ for the specific type theories of Fig. 13.4. The proofs can be found in Sections 14.2 and 14.3. The symbol '√' stands for "holds" and '×' for "fails".

Since all type theories of Fig. 14.2 are $\boldsymbol{\beta}$-sound, they all preserve $\boldsymbol{\beta}$-reduction and, obviously, they also preserve $\boldsymbol{\beta}$I-reduction. Since the type assignment systems induced by the first ten type theories contain $(\mathtt{U}_{\mathrm{top}})$, they are all closed under β-expansions. The type assignment systems for the last three are closed under βI-expansions.

It is easy to see that BCD, Engeler, CDS, CDV, CD cannot be closed under η-expansion because a constant in $\mathbb{A}_\infty$ cannot be decomposed into an intersection of arrow types. A similar argument applies to Plotkin where 0 does not have any axiom associated to it.

14.1 Inversion lemmas

In the style of Coppo et al. (1984), Alessi et al. (2003), and Alessi et al. (2006), we shall isolate special properties that will allow us to 'reverse' some of the rules of the type assignment system $\vdash_\cap^{\mathcal{T}}$, thereby achieving some form of 'generation' and '*inversion*' properties. These state necessary and sufficient conditions for an assertion $\Gamma \vdash^{\mathcal{T}} M : A$ to hold, depending on the form of M and A; see Theorems 14.1.1 and 14.1.9.

14.1.1 Theorem (Inversion Lemma for $\boldsymbol{\lambda}_\cap^{\mathcal{T}}$) *Let $\mathcal{T} \in \mathrm{TT}$ and let $\vdash$ denote $\vdash_\cap^{\mathcal{T}}$. If $\mathcal{T} \in \mathrm{TT}^{\text{-}\mathtt{U}}$, then the following statements hold unconditionally; if $\mathcal{T} \in \mathrm{TT}^{\mathtt{U}}$, then they hold under the assumption that $A \neq \mathtt{U}$ in* (i) *and* (ii) *below:*

(i) $\Gamma \vdash x : A \;\Leftrightarrow\; \Gamma(x) \leq A$;

(ii) $\Gamma \vdash MN : A \;\Leftrightarrow\; \exists k \geq 1\, \exists B_1,\dots,B_k, C_1,\dots,C_k\; [C_1 \cap \cdots \cap C_k \leq A \;\&\; \forall i \in \{1,\dots,k\}\; \Gamma \vdash M : B_i{\to}C_i \;\&\; \Gamma \vdash N : B_i]$.

(iii) $\Gamma \vdash \lambda x.M : A \;\Leftrightarrow\; \exists k \geq 1\, \exists B_1,\dots,B_k, C_1,\dots,C_k\; [(B_1{\to}C_1) \cap \cdots \cap (B_k{\to}C_k) \leq A \;\&\; \forall i \in \{1,\dots,k\}.\Gamma, x{:}B_i \vdash M : C_i]$.

Proof We only prove ($\Rightarrow$), as ($\Leftarrow$) is trivial. First consider $\mathcal{T} \in \mathrm{TT}^{\text{-}\mathtt{U}}$.

(i) By induction on derivations. We reason according which axiom or rule has been used in the final step. Only axiom (Ax), and rules ($\cap$I), ($\leq$) could have been applied. In the first case one has $\Gamma(x) \equiv A$. In the other two cases the induction hypothesis applies.

(ii) By induction on derivations. By assumption on A and the shape of the term, the final applied step has to be rule ($\to$E), ($\leq$) or ($\cap$I). In the first case the final applied rule is

$$(\to\mathrm{E})\;\frac{\Gamma \vdash M : D{\to}A \quad \Gamma \vdash N : D}{\Gamma \vdash MN : A}.$$

We can take $k = 1$ and $C_1 \equiv A$ and $B_1 \equiv D$. In the second case the final rule applied is

$$(\leq)\;\frac{\Gamma \vdash MN : B \quad B \leq A}{\Gamma \vdash MN : A}$$

and the induction hypothesis applies. Finally, in the case $A \equiv A_1 \cap A_2$ and the final applied rule is

$$(\cap\mathrm{I}) \quad \frac{\Gamma \vdash MN : A_1 \quad \Gamma \vdash MN : A_2}{\Gamma \vdash MN : A_1 \cap A_2}.$$

By the induction hypothesis there exist B_i, C_i, D_j, E_j, with $1 \leq i \leq k$, $1 \leq j \leq k'$, such that

$$\begin{array}{ll} \Gamma \vdash M : B_i{\to}C_i, & \Gamma \vdash N : B_i, \\ \Gamma \vdash M : D_j{\to}E_j, & \Gamma \vdash N : D_j, \\ C_1 \cap \cdots \cap C_k \leq A_1, & E_1 \cap \cdots \cap E_{k'} \leq A_2. \end{array}$$

Hence we are done, as $C_1 \cap \cdots \cap C_k \cap E_1 \cap \cdots \cap E_{k'} \leq A$.

(iii) Again, we use induction on derivations. We only treat the case $A \equiv A_1 \cap A_2$ and the final applied rule is $(\cap\mathrm{I})$:

$$(\cap\mathrm{I}) \quad \frac{\Gamma \vdash \lambda x.M : A_1 \quad \Gamma \vdash \lambda x.M : A_2}{\Gamma \vdash \lambda x.M : A_1 \cap A_2}.$$

By the induction hypothesis there exist B_i, C_i, D_j, E_j with $1 \leq i \leq k$, $1 \leq j \leq k'$ such that

$$\begin{array}{ll} \Gamma, x{:}B_i \vdash M : C_i, & (B_1{\to}C_1) \cap \cdots \cap (B_k{\to}C_k) \leq A_1, \\ \Gamma, x{:}D_j \vdash M : E_j, & (D_1{\to}E_1) \cap \cdots \cap (D_{k'}{\to}E_{k'}) \leq A_2. \end{array}$$

We are done, as $(B_1{\to}C_1)\cap\cdots\cap(B_k{\to}C_k)\cap(D_1{\to}E_1)\cap\cdots\cap(D_{k'}{\to}E_{k'}) \leq A$.

Now we prove $(\Rightarrow)$ in (i)–(iii) for $\mathcal{T} \in \mathrm{TT}^{\mathtt{U}}$.

(i), (ii) The condition $A \neq \mathtt{U}$ implies that axiom $(\mathtt{U}_{\mathrm{top}})$ cannot have been used in the final step. Hence the reasoning above suffices.

(iii) If $A = \mathtt{U}$, then $(\Rightarrow)$ in (iii) holds as $\mathtt{U} \to \mathtt{U} \leq \mathtt{U}$ and $\Gamma, x{:}\mathtt{U} \vdash M : \mathtt{U}$. So we may assume that $A \neq \mathtt{U}$. Then the only interesting rule is $(\cap\mathrm{I})$. Condition $A \neq \mathtt{U}$ implies that we cannot have $A_1 = A_2 = \mathtt{U}$. In the case $A_1 \neq \mathtt{U}$ and $A_2 \neq \mathtt{U}$ the result follows as above. The other cases are easier. ∎

Under some conditions (that will hold for many type theories, in particular the ones introduced in Section 13.1), the Inversion Lemma can be restated in a more memorable form. This will be done in Theorem 14.1.9.

14.1.2 Corollary (Subformula property) *Let $\mathcal{T} \in \mathrm{TT}$. Assume*

$$\Gamma \vdash_{\cap}^{\mathcal{T}} M : A \textit{ and } N \textit{ is a subterm of } M.$$

Then N is typable in an extension $\Gamma^+ = \Gamma, x_1{:}B_1, \ldots, x_n{:}B_n$ in which the variables $\{x_1, \ldots, x_n\} = \mathrm{FV}(N) - \mathrm{FV}(M)$ also get a type assigned.

Proof We can write $M \equiv C[N]$. If $\mathcal{T} \in \mathrm{TT}^{\mathtt{U}}$, then the statement is trivial as $\vdash N : \mathtt{U}$. Otherwise the statement is proved by induction on the structure of $C[\ ]$, using Theorem 14.1.1. ■

14.1.3 Proposition *Let* $\mathcal{T} \in \mathrm{TT}$. *Writing* $\vdash$ *for* $\vdash_\cap^\mathcal{T}$ *we have, for* $y \notin \mathrm{dom}(\Gamma)$,

$$\exists B\,[\Gamma \vdash N : B \;\&\; \Gamma \vdash M[x{:}= N] : A] \;\Rightarrow$$
$$\exists C\,[\Gamma \vdash N : C \;\&\; \Gamma, y{:}C \vdash M[x{:}= y] : A].$$

Proof By induction on the structure of M. ■

In the following definition, the notion of $\boldsymbol{\beta}$-soundness is introduced to prove invertibility of the rule $(\to\mathrm{I})$ and preservation of $\boldsymbol{\beta}$-reduction.

14.1.4 Definition We say $\mathcal{T}$ is $\boldsymbol{\beta}$-*sound* if

$$\begin{aligned}
&\forall k{\geq}1 \forall A_1,\ldots,A_k, B_1,\ldots,B_k, C, D.\\
&\qquad (A_1{\to}B_1) \cap \cdots \cap (A_k{\to}B_k) \leq (C{\to}D) \;\&\; D \neq \mathtt{U} \;\Rightarrow\\
&\qquad C \leq A_{i_1} \cap \cdots \cap A_{i_p} \;\&\; B_{i_1} \cap \cdots \cap B_{i_p} \leq D,\\
&\qquad \text{for some } p \geq 1 \text{ and } 1 \leq i_1,\ldots,i_p \leq k.
\end{aligned}$$

This definition translates immediately to type structures. The notion of β-soundness is important for proving invertibility of the rule $(\to\mathrm{I})$, which is crucial for the next section.

When $B' = \mathtt{U}$ and $A{\to}B \leq A'{\to}B'$, the condition of β-soundness does not imply that $A' \leq A$ and $B \leq B'$.

It will be shown that all type theories in Fig. 13.4 are $\boldsymbol{\beta}$-sound. The proof occupies 14.1.5–14.1.7.

14.1.5 Remark Note that in a TT every type A can be written uniquely, modulo the order, as

$$A \equiv \alpha_1 \cap \cdots \cap \alpha_n \cap (B_1{\to}C_1) \cap \cdots \cap (B_k{\to}C_k) \tag{14.1}$$

i.e. an intersection of atoms ($\alpha_i \in \mathbb{A}$) and arrow types.

For some of our $\mathcal{T}$ the shape (14.1) in Remark 14.1.5 can be simplified.

14.1.6 Definition For the type theories $\mathcal{T}$ of Fig. 13.4 we define for each $A \in \mathbb{T}^\mathcal{T}$ its *canonical form*, written $\mathrm{cf}(A)$, as follows.

(i) If $\mathcal{T} \in \{\mathrm{BCD}, \mathrm{AO}, \mathrm{Plotkin}, \mathrm{Engeler}, \mathrm{CDV}, \mathrm{CDS}, \mathrm{CD}\}$, then

$$\mathrm{cf}(A) \equiv A.$$

System $\mathcal{T}$	A	$\mathrm{cf}(A)$
Scott	0	$\mathtt{U}\to 0$
Park	0	$0\to 0$
CDZ, HL	0	$1\to 0$
	1	$0\to 1$
HR	0	$1\to 0$
	1	$(0\to 0)\cap(1\to 1)$
DHM	0	$\mathtt{U}\to 0$
	1	$0\to 1$
All systems	$\mathtt{U}$	$\mathtt{U}$
All systems	$B\to C$	$B\to C$
All systems	$B\cap C$	$\mathrm{cf}(B)\cap\mathrm{cf}(C)$

Figure 14.3

(ii) If $\mathcal{T}\in\{\mathrm{Scott},\mathrm{Park},\mathrm{CDZ},\mathrm{HR},\mathrm{DHM},\mathrm{HL}\}$ then the definition is by induction on A. For an atom α the canonical form $\mathrm{cf}(\alpha)$ depends on the type theory in question; moreover the mapping cf preserves $\to,\cap$ and $\mathtt{U}$. See Fig. 14.3.

14.1.7 Theorem *All type theories of Fig.* 13.4 *are* $\boldsymbol{\beta}$*-sound.*

Proof We prove the following stronger statement (induction loading). Let

$$\begin{aligned} A &\le A',\\ \mathrm{cf}(A) &= \alpha_1\cap\cdots\cap\alpha_n\cap(B_1\to C_1)\cap\cdots\cap(B_k\to C_k),\\ \mathrm{cf}(A') &= \alpha'_1\cap\cdots\cap\alpha'_{n'}\cap(B'_1\to C'_1)\cap\cdots\cap(B'_{k'}\to C'_{k'}) \end{aligned}$$

where $n,n'\ge 0, k,k'\ge 1$. Then

$$\begin{aligned} &\forall j\in\{1,\dots,k'\}.[C'_j\neq\mathtt{U}\Rightarrow\\ &\exists p{\ge}1\exists i_1,\dots,i_p\in\{1,\dots,k\}.[B'_j\le B_{i_1}\cap\cdots\cap B_{i_p}\ \&\ C_{i_1}\cap\cdots\cap C_{i_p}\le C'_j]]. \end{aligned}$$

The proof of the statement is by induction on the generation of $A\le A'$. From it $\boldsymbol{\beta}$-soundness follows easily. ∎

14.1.8 Remark From Theorem 14.1.7 it follows immediately that, for the compatible theories of Fig. 13.4, the corresponding type structures are $\boldsymbol{\beta}$-sound.

14.1.9 Theorem (Inversion Lemma II) *Let* $\mathcal{T}\in\mathrm{TT}$. *Of the following properties,* (i) *holds in general;* (ii) *provided that* $\mathcal{T}\in\mathrm{PTT}$, *and also, if* $\mathcal{T}\in\mathrm{TT}^{\mathtt{U}}$, *that* $A\neq\mathtt{U}$*; and* (iii) *provided that* $\mathcal{T}$ *is* $\boldsymbol{\beta}$*-sound.*

$$\begin{array}{llcl} \text{(i)} & \Gamma, x{:}A \vdash x : B & \Leftrightarrow & A \leq B. \\ \text{(ii)} & \Gamma \vdash (MN) : A & \Leftrightarrow & \exists B\, [\Gamma \vdash M : (B \to A)\ \&\ \Gamma \vdash N : B]. \\ \text{(iii)} & \Gamma \vdash (\lambda x.M) : (B \to C) & \Leftrightarrow & \Gamma, x{:}B \vdash M : C. \end{array}$$

Proof The proof of each ($\Leftarrow$) is easy. So we only treat ($\Rightarrow$).

(i) If $B \neq \mathtt{U}$, then the conclusion follows from Theorem 14.1.1(i). If $B = \mathtt{U}$, then the conclusion holds trivially.

(ii) Suppose $\Gamma \vdash MN : A$. Then by Theorem 14.1.1(ii) there are $B_1, \ldots, B_k$, $C_1, \ldots, C_k$, with $k \geq 1$, such that $C_1 \cap \cdots \cap C_k \leq A$, $\Gamma \vdash M : B_i \to C_i$ and $\Gamma \vdash N : B_i$ for $1 \leq i \leq k$. Hence $\Gamma \vdash N : B_1 \cap \cdots \cap B_k$ and

$$\begin{array}{rl} \Gamma \vdash M : & (B_1 \to C_1) \cap \cdots \cap (B_k \to C_k) \\ & \leq (B_1 \cap \cdots \cap B_k) \to (C_1 \cap \cdots \cap C_k) \\ & \leq (B_1 \cap \cdots \cap B_k) \to A, \end{array}$$

by Lemma 13.1.22. So we can take $B \equiv (B_1 \cap \cdots \cap B_k)$.

(iii) Suppose $\Gamma \vdash (\lambda x.M) : (B \to C)$. Then Theorem 14.1.1(iii) applies and we have for some $k \geq 1$ and $B_1, \ldots, B_k, C_1, \ldots, C_k$

$$\begin{array}{r} (B_1 \to C_1) \cap \cdots \cap (B_k \to C_k) \leq B \to C, \\ \Gamma, x{:}B_i \vdash M : C_i \text{ for all } i. \end{array}$$

If $C = \mathtt{U}$, then the assertion holds trivially, so let $C \neq \mathtt{U}$. Then by $\boldsymbol{\beta}$-soundness there exist $1 \leq i_1, \ldots, i_p \leq k$, $p \geq 1$ such that

$$\begin{array}{c} B \leq B_{i_1} \cap \cdots \cap B_{i_p}, \\ C_{i_1} \cap \cdots \cap C_{i_p} \leq C. \end{array}$$

Applying ($\leq$-L) we get

$$\begin{array}{r} \Gamma, x{:}B \vdash M : C_{i_j},\ 1 \leq j \leq p, \\ \Gamma, x{:}B \vdash M : C_{i_1} \cap \cdots \cap C_{i_p} \leq C. \ \blacksquare \end{array}$$

We give a simple example which shows that, in general, rule ($\to$E) cannot be reversed, i.e. that if $\Gamma \vdash MN : B$, then it is not always true that there exists A such that $\Gamma \vdash M : A \to B$ and $\Gamma \vdash N : A$.

14.1.10 Example Let $\mathcal{T} =$ Engeler, one of the intersection type theories of Fig. 13.4. Let $\Gamma = \{x{:}(\mathsf{c}_0 \to \mathsf{c}_1) \cap (\mathsf{c}_2 \to \mathsf{c}_3), y{:}(\mathsf{c}_0 \cap \mathsf{c}_2)\}$. Then one has

$$\Gamma \vdash_{\cap}^{\mathcal{T}} xy : \mathsf{c}_1 \cap \mathsf{c}_3.$$

But for no type B is it true that

$$\Gamma \vdash_{\cap}^{\mathcal{T}} x : B \to (\mathsf{c}_1 \cap \mathsf{c}_3) \text{ and } \Gamma \vdash_{\cap}^{\mathcal{T}} y : B.$$

14.1.11 Remark Note that in general

$$\Gamma \vdash_{\cap}^{\mathcal{T}} (\lambda x.M) : A \not\Rightarrow \exists B, C.A = (B{\to}C) \ \&\ \Gamma, x{:}B \vdash_{\cap}^{\mathcal{T}} M : C.$$

Consider $\vdash_{\cap}^{\mathrm{BCD}} \mathsf{I} : (\mathsf{c}_1{\to}\mathsf{c}_1) \cap (\mathsf{c}_2{\to}\mathsf{c}_2)$, with $\mathsf{c}_1, \mathsf{c}_2$ different type atoms.

14.1.12 Proposition *For all $\mathcal{T}$ except* AO *in Fig.* 13.4 *properties* (i), (ii) *and* (iii) *of Theorem* 14.1.9 *hold unconditionally. For $\mathcal{T}$ =* AO *they hold under the condition that $A \neq \mathtt{U}$ in* (ii).

Proof Since these $\mathcal{T}$ are proper and $\boldsymbol{\beta}$-sound, by Theorem 14.1.7, we can apply Theorem 14.1.9. Moreover, by axiom $(\to\mathtt{U})$ we have $\Gamma \vdash_{\cap}^{\mathcal{T}} M : \mathtt{U} \to \mathtt{U}$ for all Γ, M, hence we do not need to assume $A \neq \mathtt{U}$ for $\mathcal{T} \in$ {Scott, Park, CDZ, HR, DHM, BCD}. ∎

14.2 Subject reduction and expansion

Various subject reduction and expansion properties are proved for the classical $\boldsymbol{\beta}$, $\boldsymbol{\beta}\mathsf{I}$ and $\boldsymbol{\eta}$ notions of reduction. Other results can be found in Alessi et al. (2003), and Alessi et al. (2006). We consider the following rules.

$$(R\text{-}red) \quad \frac{M \to_R N \quad \Gamma \vdash M : A}{\Gamma \vdash N : A}$$

$$(R\text{-}exp) \quad \frac{M\,{}_R{\leftarrow}\, N \quad \Gamma \vdash M : A}{\Gamma \vdash N : A},$$

where R is a notion of reduction, in particular $\boldsymbol{\beta}$-, $\boldsymbol{\beta}\mathsf{I}$, or $\boldsymbol{\eta}$-reduction. If one of these rules holds in $\boldsymbol{\lambda}_{\cap}^{\mathcal{T}}$, we write $\boldsymbol{\lambda}_{\cap}^{\mathcal{T}} \models (R\text{-}\{exp, red\})$, respectively. If both hold we write $\boldsymbol{\lambda}_{\cap}^{\mathcal{T}} \models (R\text{-}cnv)$. These properties will be crucial in Section 16.2, where we will discuss (untyped) λ-models induced by these systems.

Recall that $(\lambda x.M)N$ is a $\boldsymbol{\beta}\mathsf{I}$-*redex* if $x \in \mathrm{FV}(M)$, Curry and Feys (1958).

$\boldsymbol{\beta}$-conversion

We first investigate when $\boldsymbol{\lambda}_{\cap}^{\mathcal{T}} \models (\boldsymbol{\beta}(\mathsf{I})\text{-}red)$.

14.2.1 Proposition *Let $\mathcal{T} \in$ TT. Then we have:*

(i) $\boldsymbol{\lambda}_{\cap}^{\mathcal{T}} \models (\boldsymbol{\beta}\mathsf{I}\text{-}red) \Leftrightarrow$

$$[\Gamma \vdash^{\mathcal{T}} (\lambda x.M) : (B{\to}A) \ \&\ x \in \mathrm{FV}(M) \ \Rightarrow\ \Gamma, x{:}B \vdash^{\mathcal{T}} M : A];$$

(ii) $\boldsymbol{\lambda}_{\cap}^{\mathcal{T}} \models (\boldsymbol{\beta}\text{-}red) \Leftrightarrow$

$$[\Gamma \vdash^{\mathcal{T}} (\lambda x.M) : (B{\to}A) \ \Rightarrow\ \Gamma, x{:}B \vdash^{\mathcal{T}} M : A].$$

Proof (i) ($\Rightarrow$) Assume $\Gamma \vdash \lambda x.M : B{\to}A$ & $x \in \mathrm{FV}(M)$, which implies $\Gamma, y{:}B \vdash (\lambda x.M)y : A$, by weakening and rule $(\to\mathrm{E})$ for a fresh y. Now rule ($\boldsymbol{\beta}$I-*red*) gives us $\Gamma, y{:}B \vdash M[x{:=}y] : A$. Hence $\Gamma, x{:}B \vdash M : A$.

($\Leftarrow$) Suppose $\Gamma \vdash (\lambda x.M)N : A$ & $x \in \mathrm{FV}(M)$, in order to show that $\Gamma \vdash M[x{:=}N] : A$. We may assume $A \neq \mathtt{U}$. Then Theorem 14.1.1(ii) implies $\Gamma \vdash \lambda x.M : B_i{\to}C_i$, $\Gamma \vdash N : B_i$ and $C_1 \cap \cdots \cap C_k \leq A$, for some $B_1,\ldots,B_k,C_1,\ldots,C_k$. By assumption we have $\Gamma, x{:}B_i \vdash M : C_i$. Hence by rule (*cut*), Proposition 13.2.10, one has $\Gamma \vdash M[x{:=}N] : C_i$. Therefore $\Gamma \vdash M[x{:=}N] : A$, using rules $(\cap\mathrm{I})$ and $(\leq)$.

(ii) Similarly. ■

14.2.2 Corollary *Let $\mathcal{T} \in \mathrm{TT}$ be $\boldsymbol{\beta}$-sound. Then $\boldsymbol{\lambda}_\cap^{\mathcal{T}} \models$ ($\boldsymbol{\beta}$-red).*

Proof Use Theorem 14.1.9(iii). ■

The converse of Corollary 14.2.2 does not hold. In Definition 16.2.20 we will introduce a type theory that is not $\boldsymbol{\beta}$-sound, but nevertheless induces a type assignment system satisfying ($\boldsymbol{\beta}$-*red*).

14.2.3 Corollary *Let $\mathcal{T}$ be one of the* TT *in Fig. 13.4. Then*

$$\boldsymbol{\lambda}_\cap^{\mathcal{T}} \models (\boldsymbol{\beta}\text{-}red).$$

Proof By Corollary 14.2.2 and Theorem 14.1.7. ■

Next we investigate when $\boldsymbol{\lambda}_\cap^{\mathcal{T}} \models$ ($\boldsymbol{\beta}$-*exp*). As a warm-up, suppose that $\Gamma \vdash M[x{:=}N] : A$. We would like to be able to conclude that N has a type, as it seems to be a subformula, and therefore $\Gamma \vdash (\lambda x.M)N : A$. There are two problems: N may occur several times in $M[x{:=}N]$, so that it has (should have) in fact several types. In the system $\boldsymbol{\lambda}_\to$ this problem causes the failure of rule ($\boldsymbol{\beta}$-*exp*). But in the intersection type theories one has $N : B_1 \cap \cdots \cap B_k$ if $N : B_1, \ldots, N : B_k$. Therefore $(\lambda x.M)N$ has a type if $M[x{:=}N]$ has one. The second problem arises if N does not occur at all in $M[x{:=}N]$, i.e. if the redex is a λK-redex. We would like to assign to N, as type, the intersection over an empty sequence, i.e. the universe $\mathtt{U}$. This makes ($\boldsymbol{\beta}$-*exp*) valid for $\mathcal{T} \in \mathrm{TT}^{\mathtt{U}}$, but invalid for $\mathcal{T} \in \mathrm{TT}^{\text{-}\mathtt{U}}$

14.2.4 Proposition *Let $\mathcal{T} \in \mathrm{TT}$. Then the following hold.*

(i) *Suppose $\Gamma \vdash^{\mathcal{T}} M[x{:=}N] : A$. Then*

$$\Gamma \vdash^{\mathcal{T}} (\lambda x.M)N : A \Leftrightarrow N \textit{ is typable in context } \Gamma.$$

(ii) $\boldsymbol{\lambda}_\cap^{\mathcal{T}} \models$ ($\boldsymbol{\beta}$I-*exp*) $\Leftrightarrow \forall \Gamma, M, N, A$ *with* $x \in \mathrm{FV}(M)$

$$[\Gamma \vdash^{\mathcal{T}} M[x{:=}N] : A \Rightarrow N \textit{ is typable in context } \Gamma].$$

(iii) $\boldsymbol{\lambda}_\cap^{\mathcal{T}} \models (\boldsymbol{\beta}\text{-}exp) \Leftrightarrow \forall \Gamma, M, N, A$

$$[\Gamma \vdash^{\mathcal{T}} M[x{:=}N] : A \Rightarrow N \textit{ is typable in context } \Gamma].$$

Proof (i) ($\Rightarrow$) By Theorem 14.1.1(ii).

($\Leftarrow$) Let $\Gamma \vdash M[x{:=}N] : A$ and suppose N is typable in context Γ. By Proposition 14.1.3 for some B and a fresh y one has $\Gamma \vdash N : B \;\&\; \Gamma, y{:}B \vdash M[x{:}=y] : A$. Then $\Gamma \vdash \lambda x.M : (B{\to}A)$ and hence $\Gamma \vdash (\lambda x.M)N : A$.

(ii) Similar to but simpler than (iii) below.

(iii) ($\Rightarrow$) Assume $\Gamma \vdash M[x{:=}N] : A$. Then $\Gamma \vdash (\lambda x.M)N : A$, by ($\boldsymbol{\beta}$-*exp*), hence by (i) we are done.

($\Leftarrow$) Assume $\Gamma \vdash L' : A$, with $L \to_{\boldsymbol{\beta}} L'$. By induction on the generation of $L \to_{\boldsymbol{\beta}} L'$ we get $\Gamma \vdash L : A$ from (i) and Theorem 14.1.1. ∎

14.2.5 Corollary

(i) *Let* $\mathcal{T} \in \mathrm{TT}$. *Then* $\boldsymbol{\lambda}_\cap^{\mathcal{T}} \models (\boldsymbol{\beta}\mathsf{I}\text{-}exp)$.
(ii) *Let* $\mathcal{T} \in \mathrm{TT}^{\mathtt{U}}$. *Then* $\boldsymbol{\lambda}_\cap^{\mathcal{T}} \models (\boldsymbol{\beta}\text{-}exp)$.

Proof (i) By the subformula property (Corollary 14.1.2).

(ii) Trivial, since every term has type $\mathtt{U}$. ∎

Now we can harvest results towards closure under $\boldsymbol{\beta}$-conversion.

14.2.6 Theorem

(i) *Let* $\mathcal{T} \in \mathrm{TT}$. *Then*

$$\mathcal{T} \textit{ is } \boldsymbol{\beta}\textit{-sound} \Rightarrow \boldsymbol{\lambda}_\cap^{\mathcal{T}} \models (\boldsymbol{\beta}\mathsf{I}\text{-}cnv).$$

(ii) *Let* $\mathcal{T} \in \mathrm{TT}^{\mathtt{U}}$. *Then*

$$\mathcal{T} \textit{ is } \boldsymbol{\beta}\textit{-sound} \Rightarrow \boldsymbol{\lambda}_\cap^{\mathcal{T}} \models (\boldsymbol{\beta}\text{-}cnv).$$

Proof (i) By Corollaries 14.2.2 and 14.2.5(i).

(ii) By Corollaries 14.2.2 and 14.2.5(ii). ∎

14.2.7 Corollary

(i) *Let* $\mathcal{T}$ *be a* TT *of* Fig. 13.4. *Then*

$$\boldsymbol{\lambda}_\cap^{\mathcal{T}} \models (\boldsymbol{\beta}\mathsf{I}\text{-}cnv).$$

(ii) *Let* $\mathcal{T} \in \{$Scott, Park, CDZ, HR, DHM, BCD, AO, Plotkin, Engeler, CDS$\}$. *Then* $\boldsymbol{\lambda}_\cap^{\mathcal{T}} \models (\boldsymbol{\beta}\text{-}cnv)$.

Proof (i) By Theorems 14.1.7 and 14.2.6(i).

(ii) By Theorems 14.1.7 and 14.2.6(ii). ∎

$\boldsymbol{\eta}$-conversion

First we give necessary and sufficient conditions for a system $\boldsymbol{\lambda}_\cap^\mathcal{T}$ to satisfy the rule ($\boldsymbol{\eta}$-*red*).

14.2.8 Theorem (Characterization of $\boldsymbol{\eta}$-*red*)

(i) *Let* $\mathcal{T} \in \mathrm{TT}^{-\mathtt{U}}$. *Then*

$$\boldsymbol{\lambda}_\cap^\mathcal{T} \models (\boldsymbol{\eta}\text{-}red) \Leftrightarrow \mathcal{T} \text{ is proper.}$$

(ii) *Let* $\mathcal{T} \in \mathrm{TT}^{\mathtt{U}}$. *Then*

$$\boldsymbol{\lambda}_\cap^\mathcal{T} \models (\boldsymbol{\eta}\text{-}red) \Leftrightarrow \mathcal{T} \text{ is natural.}$$

Proof (i) Similar to, but simpler than, (ii).

(ii) ($\Rightarrow$) Assume $\boldsymbol{\lambda}_\cap^\mathcal{T} \models (\boldsymbol{\eta}\text{-}red)$ towards $(\to\cap)$, $(\to)$ and $(\mathtt{U}\to)$.

For $(\to\cap)$, one has

$$x{:}(A\to B) \cap (A\to C), y{:}A \vdash xy : B \cap C,$$

hence by $(\to\mathrm{I})$ it follows that $x{:}(A\to B) \cap (A\to C) \vdash \lambda y.xy : A\to(B \cap C)$. Therefore $x{:}(A\to B) \cap (A\to C) \vdash x : A\to(B \cap C)$, by ($\boldsymbol{\eta}$-*red*). By Theorem 14.1.9(i) one can conclude $(A\to B) \cap (A\to C) \leq A\to(B \cap C)$.

For $(\to)$, suppose that $A \leq B$ and $C \leq D$, in order to show $B\to C \leq A\to D$. One has $x{:}B\to C, y{:}A \vdash xy : C \leq D$, so $x{:}B\to C \vdash \lambda y.xy : A\to D$. Therefore by ($\boldsymbol{\eta}$-*red*) it follows that $x{:}B\to C \vdash x : A\to D$ and we are done as before.

For $\mathtt{U} \leq \mathtt{U}\to\mathtt{U}$, notice that $x{:}\mathtt{U}, y{:}\mathtt{U} \vdash xy : \mathtt{U}$, so we have $x{:}\mathtt{U} \vdash \lambda y.xy : \mathtt{U}\to\mathtt{U}$. Therefore $x{:}\mathtt{U} \vdash x : \mathtt{U}\to\mathtt{U}$ and again we are done.

($\Leftarrow$) Let $\mathcal{T}$ be natural. Assume that $\Gamma \vdash \lambda x.Mx : A$, with $x \notin \mathrm{FV}(M)$, in order to show $\Gamma \vdash M : A$. If $A = \mathtt{U}$, we are done. Otherwise, by Theorem 14.1.1(iii),

$$\begin{aligned}\Gamma \vdash \lambda x.Mx : A \Rightarrow\ & \Gamma, x{:}B_i \vdash Mx : C_i, 1 \leq i \leq k, \ \& \\ & (B_1\to C_1) \cap \cdots \cap (B_k\to C_k) \leq A, \\ & \text{for some } B_1, \ldots, B_k, C_1, \ldots, C_k.\end{aligned}$$

We can suppose that $C_i \neq \mathtt{U}$ for all i. If there exists i such that $C_i = \mathtt{U}$ then, by $(\mathtt{U})$ and $(\mathtt{U}\to)$, we have $(B_i \to C_i) \cap D = \mathtt{U} \cap D = D$, for any type D. On the other hand, there is at least one $C_i \neq \mathtt{U}$, since otherwise $A \geq (B_1\to\mathtt{U}) \cap \cdots \cap (B_k\to\mathtt{U}) = \mathtt{U}$, and we would have $A = \mathtt{U}$. Hence by

Theorem 14.1.9(ii)

$$\begin{aligned}
&\Rightarrow \Gamma, x{:}B_i \vdash M : D_i{\to}C_i \text{ and}\\
&\quad\ \Gamma, x{:}B_i \vdash x : D_i, \text{ for some } D_1,\dots,D_k,\\
&\Rightarrow B_i \leq D_i, \text{ by Theorem 14.1.9(i)},\\
&\Rightarrow \Gamma \vdash M : (B_i{\to}C_i), \text{ by } ({\leq}\text{-L}) \text{ and } ({\to}),\\
&\Rightarrow \Gamma \vdash M : ((B_1{\to}C_1)\cap\cdots\cap(B_k{\to}C_k)) \leq A. \blacksquare
\end{aligned}$$

14.2.9 Corollary *Let* $\mathcal{T} \in \{\text{Scott}, \text{Park}, \text{CDZ}, \text{HR}, \text{DHM}, \text{BCD}, \text{HL}, \text{CDV}\}$. *Then* $\boldsymbol{\lambda}_\cap^{\mathcal{T}} \models (\boldsymbol{\eta}\text{-}red)$.

In order to characterize the admissibility of rule $(\boldsymbol{\eta}\text{-}exp)$, we need to introduce a further condition on type theories. This condition is necessary and sufficient to derive from the basis $x{:}A$ the same type A for $\lambda y.xy$, as we will show in the proof of Theorem 14.2.11.

14.2.10 Definition Let $\mathcal{T} \in \text{TT}$.

(i) We say $\mathcal{T}$ is $\boldsymbol{\eta}$-*sound* if, for all A, there are $k \geq 1$, $m_1,\dots,m_k \geq 1$, and $B_1,\dots,B_k$, $C_1,\dots,C_k$,

$$\begin{pmatrix} D_{11}\dots D_{1m_1}\\ \vdots \\ D_{k1}\cdots D_{km_k}\end{pmatrix}, \text{ and } \begin{pmatrix} E_{11}\cdots E_{1m_1}\\ \cdots \\ E_{k1}\cdots E_{km_k}\end{pmatrix}$$

such that

$$\begin{array}{lll}
(B_1{\to}C_1)\cap\cdots\cap(B_k{\to}C_k) & \leq & A\\
\& \ A & \leq & (D_{11}{\to}E_{11})\cap\cdots\cap(D_{1m_1}{\to}E_{1m_1})\cap\\
 & & \quad\vdots\\
 & & (D_{k1}{\to}E_{k1})\cap\cdots\cap(D_{km_k}{\to}E_{km_k})\\
\& \ B_i \leq D_{i1}\cap\cdots\cap D_{im_i} \ \& \ E_{i1}\cap\cdots\cap E_{im_i} \leq C_i, & & \\
\quad \text{for } 1\leq i\leq k. & &
\end{array}$$

(ii) Let $\mathcal{T} \in \text{TT}^{\mathtt{U}}$. Then $\mathcal{T}$ is called $\boldsymbol{\eta}^{\mathtt{U}}$-*sound* if for all $A \neq \mathtt{U}$ at least one of the following two conditions holds.

(1) There are types $B_1,\dots,B_n$ with $(B_1{\to}\mathtt{U})\cap\cdots\cap(B_n{\to}\mathtt{U}) \leq A$.

(2) There are $n \geq k \geq 1$, $m_1,\dots,m_k \geq 1$, and $B_1,\dots,B_k$, $C_1,\dots,C_k$,

$$\begin{pmatrix} D_{11}\cdots D_{1m_1}\\ \vdots \\ D_{k1}\cdots D_{km_k}\end{pmatrix}, \text{ and } \begin{pmatrix} E_{11}\cdots E_{1m_1}\\ \vdots \\ E_{k1}\cdots E_{km_k}\end{pmatrix}$$

such that

$$\begin{array}{l} (B_1{\to}C_1)\cap\cdots\cap(B_k{\to}C_k)\cap \\ \cap\,(B_{k+1}{\to}\mathtt{U})\cap\cdots\cap(B_n{\to}\mathtt{U}) \ \le\ A \\ \&\ A\ \le\ (D_{11}{\to}E_{11})\cap\cdots\cap(D_{1m_1}{\to}E_{1m_1})\cap \\ \qquad\qquad \vdots \\ \qquad\qquad (D_{k1}{\to}E_{k1})\cap\cdots\cap(D_{km_k}{\to}E_{km_k}) \\ \&\ \ B_i\ \le\ D_{i1}\cap\cdots\cap D_{im_i}\ \&\ E_{i1}\cap\cdots\cap E_{im_i}\ \le\ C_i, \\ \qquad \text{for } 1\le i\le k. \end{array}$$

The validity of η-expansion can be characterized as follows.

14.2.11 Theorem (Characterization of *η-exp*)

(i) *Let $\mathcal{T}\in\mathrm{TT}^{\text{-}\mathtt{U}}$. Then*

$$\lambda_\cap^{\mathcal{T}} \models (\eta\text{-}exp) \Leftrightarrow \mathcal{T} \text{ is } \eta\text{-sound.}$$

(ii) *Let $\mathcal{T}\in\mathrm{TT}^{\mathtt{U}}$. Then*

$$\lambda_\cap^{\mathcal{T}} \models (\eta\text{-}exp) \Leftrightarrow \mathcal{T} \text{ is } \eta^{\mathtt{U}}\text{-sound.}$$

Proof (i) ($\Rightarrow$) Assume $\lambda_\cap^{\mathcal{T}} \models (\eta\text{-exp})$. As $x{:}A \vdash x : A$, by assumption we have $x{:}A \vdash \lambda y.xy : A$. From Theorem 14.1.1(iii) it follows that $x{:}A, y{:}B_i \vdash xy : C_i$ and $(B_1{\to}C_1)\cap\cdots\cap(B_k{\to}C_k) \le A$ for some B_i, C_i. By Theorem 14.1.1(ii), for each i there exist D_{ij}, E_{ij} such that for each j one has $x{:}A, y{:}B_i \vdash x : (D_{ij}{\to}E_{ij})$, and $x{:}A, y{:}B_i \vdash y : D_{ij}$ and $E_{i1}\cap\cdots\cap E_{im_i} \le C_i$. Hence by Theorem 14.1.1(i) we have $A \le (D_{ij}{\to}E_{ij})$ and $B_i \le D_{ij}$ for all i and j. Therefore we obtain the condition of Definition 14.2.10(i).

($\Leftarrow$) Suppose that $\Gamma \vdash M : A$ in order to show $\Gamma \vdash \lambda x.Mx : A$, with x fresh. By assumption A satisfies the condition of Definition 14.2.10(i). By rule ($\le$) for all i, j we have $\Gamma \vdash M : D_{ij}{\to}E_{ij}$ and so $\Gamma, x{:}D_{ij} \vdash Mx : E_{ij}$ by rule ($\to$E). From ($\le$-L), ($\cap$I) and ($\le$) we get $\Gamma, x{:}B_i \vdash Mx : C_i$ and this implies $\Gamma \vdash \lambda x.Mx : B_i{\to}C_i$, using rule ($\to$I). So we can conclude by ($\cap$I) and ($\le$) that $\Gamma \vdash \lambda x.Mx : A$.

(ii) The proof is nearly the same as for (i).

($\Rightarrow$) Again we get $x{:}A, y{:}B_i \vdash xy : C_i$ and $(B_1{\to}C_1)\cap\cdots\cap(B_k{\to}C_k) \le A$ for some B_i, C_i. If all $C_i = \mathtt{U}$, then A satisfies the first condition of Definition 14.2.10(ii). Otherwise, consider the i such that $C_i \ne \mathtt{U}$ and reason as in the proof of ($\Rightarrow$) for (i).

($\Leftarrow$) Suppose that $\Gamma \vdash M : A$ in order to show $\Gamma \vdash \lambda x.Mx : A$, with x fresh. If A satisfies the first condition of Definition 14.2.10(ii) – that is, $(B_1{\to}\mathtt{U})\cap\cdots\cap(B_n{\to}\mathtt{U}) \le A$ – then by ($\mathtt{U}_{\text{top}}$) it follows that $\Gamma, x{:}B_i \vdash Mx : \mathtt{U}$, hence $\Gamma \vdash \lambda x.Mx : (B_1 \to \mathtt{U}) \cap \cdots \cap (B_n \to \mathtt{U}) \leq A$. Now let A satisfy the second condition. Then we reason as for ($\Leftarrow$) in (i). ■

For most intersection type theories of interest the condition of $\boldsymbol{\eta}^{(\mathtt{U})}$-soundness is deduced from the following proposition.

14.2.12 Proposition *Let $\mathcal{T} \in \mathrm{TT}$ be proper, with set $\mathbb{A}$ of atoms.*

(i) $\mathcal{T}$ *is* $\boldsymbol{\eta}$*-sound* $\Leftrightarrow$ $\forall \alpha \in \mathbb{A}\, \exists k \geq 1\, \exists B_1, \ldots, B_k, C_1, \ldots, C_k$ $\alpha = (B_1 \to C_1) \cap \cdots \cap (B_k \to C_k)$.

(ii) *Let $\mathcal{T} \in \mathrm{TT}^{\mathtt{U}}$. Then*

$$\mathcal{T} \text{ is } \boldsymbol{\eta}^{\mathtt{U}}\text{-sound} \Leftrightarrow \forall \alpha \in \mathbb{A}[\mathtt{U} \to \mathtt{U} \leq \alpha \vee \exists k \geq 1 \exists B_1, \ldots, B_k, C_1, \ldots, C_k \\ [(B_1 \to C_1) \cap \cdots \cap (B_k \to C_k) \cap (\mathtt{U} \to \mathtt{U}) \leq \alpha \\ \& \ \alpha \leq (B_1 \to C_1) \cap \cdots \cap (B_k \to C_k)]].$$

(iii) *Let $\mathcal{T} \in \mathrm{NTT}^{\mathtt{U}}$. Then*

$$\mathcal{T} \text{ is } \boldsymbol{\eta}^{\mathtt{U}}\text{-sound} \Leftrightarrow \mathcal{T} \text{ is } \boldsymbol{\eta}\text{-sound.}$$

Proof (i) ($\Rightarrow$) Suppose $\mathcal{T}$ is $\boldsymbol{\eta}$-sound. Let $\alpha \in \mathbb{A}$. Then α satisfies the condition of Definition 14.2.10(i), for some $B_1, \ldots, B_k$, $C_1, \ldots, C_k$, $D_{11}, \ldots, D_{1m_1}$, $\ldots, D_{k1}, \ldots, D_{km_1}$, $E_{11}, \ldots, E_{1m_1}, \ldots, E_{k1}, \ldots, E_{km_k}$. By ($\to\cap$) and ($\to$), using Proposition 13.1.22, it follows that

$$\begin{aligned} \alpha &\leq (D_{11} \cap \cdots \cap D_{1m_1} \to E_{11} \cap \cdots \cap E_{1m_1}) \cap \cdots \cap \\ &\quad\ (D_{k1} \cap \cdots \cap D_{km_k} \to E_{k1} \cap \cdots \cap E_{km_k}) \\ &\leq (B_1 \to C_1) \cap \cdots \cap (B_k \to C_k), \end{aligned}$$

hence $\alpha =_{\mathcal{T}} (B_1 \to C_1) \cap \cdots \cap (B_k \to C_k)$.

($\Leftarrow$) By induction on the generation of A one can show that A satisfies the condition of $\boldsymbol{\eta}$-soundness. The case $A_1 \to A_2$ is trivial and the case $A \equiv A_1 \cap A_2$ follows by the induction hypothesis and Rule (mon).

(ii) Similarly. Note that $(\mathtt{U} \to \mathtt{U}) \leq (B \to \mathtt{U})$ for all B.

(iii) Immediately by (ii) using rule ($\mathtt{U}\to$). ■

14.2.13 Corollary

(i) *Let $\mathcal{T} \in \{\mathrm{Scott}, \mathrm{Park}, \mathrm{CDZ}, \mathrm{HR}, \mathrm{DHM}, \mathrm{AO}\}$. Then $\mathcal{T}$ is $\boldsymbol{\eta}^{\mathtt{U}}$-sound.*

(ii) HL *is* $\boldsymbol{\eta}$*-sound.*

Proof Easy. For AO in (i) one applies Proposition 14.2.12(ii). ■

14.2.14 Corollary *Let $\mathcal{T} \in \{\mathrm{Scott}, \mathrm{Park}, \mathrm{CDZ}, \mathrm{HR}, \mathrm{DHM}, \mathrm{AO}, \mathrm{HL}\}$. Then*

$$\boldsymbol{\lambda}_{\cap}^{\mathcal{T}} \models (\boldsymbol{\eta}\text{-}exp).$$

Proof By the previous corollary and Theorem 14.2.11. ■

Exercise 14.21 shows that the remaining systems of Fig. 13.4 do not satisfy ($\boldsymbol{\eta}$-*exp*).

Now we can harvest results towards closure under $\boldsymbol{\eta}$-conversion.

14.2.15 Theorem

(i) *Let* $\mathcal{T} \in \mathrm{TT}^{\text{-U}}$. *Then*

$$\boldsymbol{\lambda}_\cap^{\mathcal{T}} \models (\boldsymbol{\eta}\text{-}cnv) \Leftrightarrow \mathcal{T} \text{ is proper and } \boldsymbol{\eta}\text{-sound.}$$

(ii) *Let* $\mathcal{T} \in \mathrm{TT}^{\mathtt{U}}$. *Then*

$$\boldsymbol{\lambda}_\cap^{\mathcal{T}} \models (\boldsymbol{\eta}\text{-}cnv) \Leftrightarrow \mathcal{T} \text{ is natural and } \boldsymbol{\eta}^{\mathtt{U}}\text{-sound.}$$

Proof (i) By Theorems 14.2.8(i) and 14.2.11(i).
(ii) By Theorems 14.2.8(ii) and 14.2.11(ii). ■

14.2.16 Theorem *For* $\mathcal{T} \in \{\mathrm{Scott}, \mathrm{Park}, \mathrm{CDZ}, \mathrm{HR}, \mathrm{DHM}, \mathrm{HL}\}$ *one has*

$$\boldsymbol{\lambda}_\cap^{\mathcal{T}} \models (\boldsymbol{\eta}\text{-}cnv).$$

Proof By Corollaries 14.2.9 and 14.2.14. ■

14.3 Exercises

Exercise 14.1 Let $\subset$ be the inclusion relation from Remark 13.1.18. Prove that $\mathcal{T}_1 \subset \mathcal{T}_2 \Leftrightarrow \forall \Gamma, M, A[\Gamma \vdash^{\mathcal{T}_1} M : A \Rightarrow \Gamma \vdash^{\mathcal{T}_2} M : A]$.

Exercise 14.2 Show that for each number $n \in \omega$ there is a type $A_n \in \mathbb{T}^{\mathrm{CD}}$ such that for the Church numerals $\mathbf{c}_n$ one has $\Gamma \vdash_\cap^{\mathrm{CD}} \mathbf{c}_{n+1} : A_n$, but $\Gamma \nvdash_\cap^{\mathrm{CD}} \mathbf{c}_n : A_n$.

Exercise 14.3 Show that $\mathsf{S}(\mathsf{KI})(\mathsf{II})$ and $(\lambda x.xxx)\mathsf{S}$ are typable in $\vdash_\cap^{\mathrm{CD}}$.

Exercise 14.4 Derive

$$\vdash_\cap^{\mathrm{CDZ}} (\lambda x.xxx)\mathsf{S} : 1 \text{ and } y{:}0, z{:}0 \vdash_\cap^{\mathrm{CDZ}} (\lambda x.xxx)(\mathsf{S}yz) : 0.$$

Exercise 14.5 Using the Inversion Lemmas show the following.

(i) $\nvdash_\cap^{\mathrm{CD}} \mathbf{c}_1 : \alpha \to \alpha$, where α is any constant.

(ii) $\nvdash_\cap^{\mathrm{HL}} \mathsf{K} : 0$.

(iii) $\nvdash_\cap^{\mathrm{Scott}} \mathsf{I} : 0$.

(iv) $\nvdash_\cap^{\mathrm{Plotkin}} \mathsf{I} : 0$.

Exercise 14.6 In this exercise $\vdash$ is $\vdash^{\mathrm{CD}}$ and $\leq$ is $\leq_{\mathrm{CD}}$.

(i) Let

$$A \equiv \alpha_1 \cap \cdots \cap \alpha_n \cap (B_1 \to C_1) \cap \cdots \cap (B_m \to C_m),$$
$$A' \equiv \alpha'_1 \cap \cdots \cap \alpha'_{n'} \cap (B'_1 \to C'_1) \cap \cdots \cap (B_{m'} \to C'_{m'}).$$

Suppose $A \leq A'$. Show that every α'_i is one of $\alpha_1, \ldots, \alpha_n$ and every $B'_i \to C'_i$ is one of $(B_1 \to C_1), \ldots, (B_m \to C_m)$.

(ii) Let $k \geq 1$. Show that

$$(B_1 \to C_1) \cap \cdots \cap (B_k \to C_k) \leq (B \to C) \Rightarrow$$
$$(B \to C) \equiv (B_j \to C_j), \text{ for some } 1 \leq j \leq k.$$

(iii) Show $\Gamma \vdash \lambda x.M : A \to B \Rightarrow \Gamma, x{:}A \vdash M : B$.

(iv) Show $\Gamma \vdash MN : A \Rightarrow \exists B, C\,[\Gamma \vdash M : B \to C \;\&\; \Gamma \vdash N : B]$.

Exercise 14.7 Let $\boldsymbol{\omega} = \lambda x.xx$ and $\boldsymbol{\Omega} = \boldsymbol{\omega\omega}$. Writing $\vdash$ for $\vdash^{\mathrm{CD}}$ show

$$\not\vdash \boldsymbol{\Omega} : A, \quad \text{for all types } A;$$
$$\not\vdash \mathsf{K}\mathsf{I}\boldsymbol{\Omega} : A, \quad \text{for all types } A.$$

Prove this using the following steps.

(i) $\vdash \boldsymbol{\Omega} : A \Rightarrow \exists B, C \vdash \boldsymbol{\omega} : (B \to C) \cap B$.

(ii) $\vdash \boldsymbol{\omega} : (B{\to}C) \cap B \Rightarrow \exists B', C'. \vdash \boldsymbol{\omega} : (B' \to C') \cap B'$ & B' is a proper subtype of B.

Exercise 14.8 Let $M = \lambda xy.\mathsf{K}x(xy)$. Writing $\vdash$ for $\vdash^{\mathrm{CD}}$, show one has:

$$\not\vdash M : \alpha \to \beta \to \alpha;$$
$$\not\vdash \mathsf{I} : ((\alpha \cap \beta){\to}\gamma){\to}((\beta \cap \alpha){\to}\gamma).$$

[Hint. Use 14.6.]

Exercise 14.9 We say that M and M' have the same types in Γ, written $M \sim_\Gamma M'$, if

$$\forall A\,[\Gamma \vdash M : A \Leftrightarrow \Gamma \vdash M' : A].$$

Prove that $M \sim_\Gamma M' \Rightarrow M\vec{N} \sim_\Gamma M'\vec{N}$ for all $\vec{N}$.

Exercise 14.10 Let $\mathcal{T}$ be a $\boldsymbol{\beta}$-sound type theory that satisfies $(\mathtt{U})$ and$(\mathtt{U}{\to})$. Prove that for all A, B we have that $A{\to}B = \mathtt{U} \Leftrightarrow B = \mathtt{U}$.

Exercise 14.11 Using $\boldsymbol{\beta}$-soundness, find out whether the following types are related or not with respect to $\leq_{\text{CDZ}}$:

$$\begin{aligned} (0\to(1\to1)\to0) \cap ((1\to1)\to1), \\ 0\to0\to0, \\ (0\to0)\to0, \\ 1\to(0\to0)\to1. \end{aligned}$$

Exercise 14.12 Let $\mathcal{T} \in \{\text{CDZ}, \text{HR}\}$. Consider the sequence of types defined by $A_0 = 0$ and $A_{n+1} = \mathtt{U} \to A_n$. Using Exercise 13.3 and $\boldsymbol{\beta}$-soundness, prove that $\mathcal{T}$ does not have a bottom element at all.

Exercise 14.13 Show directly that Plotkin is $\boldsymbol{\beta}$-sound by checking that it satisfies the following stronger condition:

$$\begin{aligned} &(A_1\to B_1) \cap \cdots \cap (A_n\to B_n) \leq C\to D \Rightarrow \\ &\exists k \neq 0 \, \exists i_1, \ldots, i_k . 1 \leq i_j \leq n \;\&\; C = A_{i_j} \;\&\; B_{i_1} \cap \cdots \cap B_{i_k} = D. \end{aligned}$$

Exercise 14.14 Show that Engeler is $\boldsymbol{\beta}$-sound by checking that it satisfies the following stronger condition:

$$\begin{aligned} &(A_1\to B_1) \cap \cdots \cap (A_n\to B_n) \leq C\to D \& D \neq \mathtt{U} \Rightarrow \\ &\exists k \neq 0 \, \exists i_1, \ldots, i_k . 1 \leq i_j \leq n \;\&\; C = A_{i_j} \;\&\; B_{i_1} \cap \cdots \cap B_{i_k} = D. \end{aligned}$$

Exercise 14.15 Let $\mathbb{A}^{\mathcal{T}} = \{\mathtt{U}, 0\}$ and $\mathcal{T}$ be defined by the axioms and rules of the theories Scott and Park taken together. Show that $\mathcal{T}$ is not $\boldsymbol{\beta}$-sound. [Hint: show that $\mathtt{U} \neq 0$.]

Exercise 14.16 Prove that Theorem 14.1.9(ii) still holds if the condition of properness is replaced by the following two conditions:

$$A \leq_{\mathcal{T}} B \Rightarrow C\to A \leq_{\mathcal{T}} C\to B$$

$$(A\to B) \cap (C\to D) \leq_{\mathcal{T}} A \cap C\to B \cap D.$$

Exercise 14.17 Show that for $\mathcal{T} \in \text{TT}^{\mathtt{U}}$ the condition

$$A \to B =_{\mathcal{T}} \mathtt{U} \to \mathtt{U} \Rightarrow B =_{\mathcal{T}} \mathtt{U}$$

is necessary for the admissibility of rule ($\boldsymbol{\beta}$-*red*) in $\boldsymbol{\lambda}_{\cap}^{\mathcal{T}}$. [Hint: Use Proposition 14.2.1(ii).]

Exercise 14.18 Recall the systems $\boldsymbol{\lambda}_{\cap}^{\text{Krivine}}$ and $\boldsymbol{\lambda}_{\cap}^{\text{Krivine}^{\mathtt{U}}}$ in Exercise 13.10.

(i) Show that rules ($\boldsymbol{\beta}$-*red*) and ($\boldsymbol{\beta}$I-*exp*) are admissible in $\boldsymbol{\lambda}_{\cap}^{\text{Krivine}}$, while ($\boldsymbol{\beta}$-*exp*) is not admissible.

(ii) Show that rules ($\boldsymbol{\beta}$-*red*) and ($\boldsymbol{\beta}$-*exp*) are admissible in $\boldsymbol{\lambda}_\cap^{\mathrm{Krivine}^{\mathtt{U}}}$.

Exercise 14.19 Show that for $\mathcal{T} \in \{\mathrm{AO}, \mathrm{Plotkin}, \mathrm{Engeler}, \mathrm{CDS}, \mathrm{CD}\}$ one has

$$\boldsymbol{\lambda}_\cap^{\mathcal{T}} \not\models (\boldsymbol{\eta}\text{-}red).$$

Exercise 14.20 Verify the following.

(i) Let $\mathcal{T} \in \{\mathrm{BCD}, \mathrm{Plotkin}, \mathrm{Engeler}, \mathrm{CDS}\}$. Then $\mathcal{T}$ is not $\boldsymbol{\eta}^{\mathtt{U}}$-sound.
(ii) Let $\mathcal{T} \in \{\mathrm{CDV}, \mathrm{CD}\}$. Then $\mathcal{T}$ is not $\boldsymbol{\eta}$-sound.

Exercise 14.21 Show that for $\mathcal{T} \in \{\mathrm{BCD}, \mathrm{Plotkin}, \mathrm{Engeler}, \mathrm{CDS}, \mathrm{CDV}, \mathrm{CD}\}$ one has

$$\boldsymbol{\lambda}_\cap^{\mathcal{T}} \not\models (\boldsymbol{\eta}\text{-}exp).$$

Exercise 14.22 Show that rules ($\boldsymbol{\eta}$-*red*) and ($\boldsymbol{\eta}$-*exp*) are not admissible in the systems $\boldsymbol{\lambda}_\cap^{\mathrm{Krivine}}$ and $\boldsymbol{\lambda}_\cap^{\mathrm{Krivine}^{\mathtt{U}}}$ as defined in Exercise 13.10.

Exercise 14.23 Let $\vdash$ denote derivability in the system obtained from the system $\boldsymbol{\lambda}_\cap^{\mathrm{CDV}}$ by replacing rule $(\leq)$ by the rules $(\cap\mathrm{E})$, see Definition 13.2.8, and adding the rule

$$(R\eta) \quad \frac{\Gamma \vdash \lambda x.Mx : A}{\Gamma \vdash M : A} \quad \text{if } x \notin \mathrm{FV}(M).$$

Show that $\Gamma \vdash_\cap^{\mathrm{CDV}} M : A \Leftrightarrow \Gamma \vdash M : A$.

Exercise 14.24 (Barendregt et al. (1983)) Let $\vdash$ denote derivability in the system obtained from $\boldsymbol{\lambda}_\cap^{\mathrm{BCD}}$ by replacing rule $(\leq)$ by the rules $(\cap\mathrm{E})$ and adding $(R\eta)$ as defined in Exercise 14.23. Verify that

$$\Gamma \vdash_\cap^{\mathrm{BCD}} M : A \Leftrightarrow \Gamma \vdash M : A.$$

Exercise 14.25 Let Δ be a basis that is allowed to be infinite. We define $\Delta \vdash M : A$ if there exists a finite basis $\Gamma \subseteq \Delta$ such that $\Gamma \vdash M : A$.

(i) Show that all the typability rules remain derivable for infinite bases except possibly $(\to\mathrm{I})$.
(ii) Suppose $\mathrm{dom}(\Delta)$ is the set of all the variables. Show that the rule $(\to\mathrm{I})$ is derivable if it is reformulated as

$$\Delta_x, x{:}A \vdash M : B \Rightarrow \Delta \vdash (\lambda x.M) : (A \to B),$$

with Δ_x the result of removing any $x{:}C$ from Δ.
(iii) Reformulate and prove Propositions 13.2.10, 13.2.12, Theorems 14.1.1, and 14.1.9 for infinite bases.

Exercise 14.26 A *multi-basis* Γ is a set of declarations in which the requirement that

$$x{:}A, y{:}B \in \Gamma \ \Rightarrow\ x \equiv y \ \Rightarrow\ A \equiv B$$

is dropped. Let Δ be a (possibly infinite) multi-basis. We define $\Delta \vdash M : A$ if there exists a singled (only one declaration per variable) basis $\Gamma \subseteq \Delta$ such that $\Gamma \vdash M : A$.

(i) Show that $x : \alpha_1, x : \alpha_2 \not\vdash^{\mathrm{CD}} x : \alpha_1 \cap \alpha_2$.
(ii) Show that $x : \alpha_1 \to \alpha_2, x : \alpha_1 \not\vdash^{\mathrm{CD}} xx : \alpha_2$.
(iii) Consider $\Delta = \{x : \alpha_1 \cap \alpha_2, x : \alpha_1\}$;
$A = \alpha_2$;
$B = (\alpha_1 \to \alpha_2 \to \alpha_3) \to \alpha_3$;
$M = \lambda y.yxx$.
Show that $\Delta, x : A \vdash^{\mathrm{CD}} M : B$, but $\Delta \not\vdash^{\mathrm{CD}} (\lambda x.M) : (A \to B)$.
(iv) We say that a multi-basis is closed under $\cap$ if for all $x \in \mathrm{dom}(\Delta)$ the set $\mathcal{X} = \Delta(x)$ is closed under $\cap$, i.e. $A, B \in \mathcal{X} \ \Rightarrow\ A \cap B \in \mathcal{X}$, up to equality of types in the TT under consideration.
Show that all the typability rules of Definition 13.2.3, except for ($\to$I), are derivable for (possibly infinite) multi-bases that are closed under $\cap$.
(v) Let Δ be closed under $\cap$. We define

$$\Delta[x := X] = \{y : \Delta(y) \mid y \neq x\} \cup \{x : A \mid A \in X\}.$$

Prove that the following reformulation of ($\to$I) using principal filters is derivable:

$$\frac{\Delta[x :=\uparrow B] \vdash N : C}{\Delta \vdash \lambda x.N : B \to C}.$$

(vi) Prove Propositions 13.2.10, 13.2.12, Theorems 14.1.1 and 14.1.9 for (possible infinite) multi-bases reformulating the statements whenever it is necessary.
(vii) Prove that if $\Delta(x)$ is a filter then $\{A \mid \Delta \vdash x : A\} = \Delta(x)$.

Exercise 14.27

(i) Prove that $F^{\mathcal{T}} \circ G^{\mathcal{T}} \supseteq \mathsf{Id}_{\mathcal{F}^{\mathcal{T}}}$ for all $\mathcal{T} \in \mathrm{TT}$.
(ii) Prove that $F^{\mathcal{T}} \circ G^{\mathcal{T}} \subseteq \mathsf{Id}_{\mathcal{F}^{\mathcal{T}}}$ iff $\mathcal{T}$ is β-sound.
(iii) Construct a natural type theory $\mathcal{T}$ such that $F^{\mathcal{T}} \circ G^{\mathcal{T}} \neq \mathsf{Id}_{\mathcal{F}^{\mathcal{T}}}$.

Exercise 14.28 Let $\mathcal{T} \in \mathrm{TT}^{\text{-U}}$.

(i) Prove that $G^{\mathcal{T}} \circ F^{\mathcal{T}} \subseteq \mathsf{Id}_{\mathcal{F}^{\mathcal{T}}}$ iff $\mathcal{T}$ is proper.
(ii) Prove that $G^{\mathcal{T}} \circ F^{\mathcal{T}} \supseteq \mathsf{Id}_{\mathcal{F}^{\mathcal{T}}}$ iff $\mathcal{T}$ is η-sound.

Exercise 14.29 Let $\mathcal{T} \in \mathrm{TT}^{\mathtt{U}}$.

(i) Prove that $G^{\mathcal{T}} \circ F^{\mathcal{T}} \subseteq \mathsf{Id}_{\mathcal{F}\mathcal{T}}$ iff $\mathcal{T}$ is natural.
(ii) Prove that $G^{\mathcal{T}} \circ F^{\mathcal{T}} \supseteq \mathsf{Id}_{\mathcal{F}\mathcal{T}}$ iff $\mathcal{T}$ is $\boldsymbol{\eta}^{\mathtt{U}}$-sound.

15

Type and Lambda Structures

This chapter makes connections between convenient categories for models of the untyped lambda calculus and those of certain type structures. The main result that is needed later in Section 16.3 is the equivalence between the categories of natural type structures on the one hand and natural lambda structures on the other. This can be found in Section 15.2.

As a warm-up, a prototype result in this direction is Corollary 15.1.23, stating the equivalence between the category $\mathbf{MSL}^{\mathtt{U}}$ of meet semi-lattices with universe, and the category $\mathbf{ALG}_a$ of algebraic lattices with additive morphisms that preserve compactness. The idea is that, by definition, in an algebraic lattice $\mathcal{D}$ an element d is fully determined, see Definition 12.2.1(iii), by its compact approximations:

$$d = \bigsqcup\{a \mid a \sqsubseteq d \;\&\; a \text{ compact}\}.$$

Writing $\mathcal{K}(\mathcal{D}) = \{a \in \mathcal{D} \mid a \text{ compact}\}$ and $\uparrow a = \{d \in \mathcal{D} \mid a \sqsubseteq d\}$ one has

$$a \sqsubseteq b \;\Leftrightarrow\; \uparrow b \subseteq \uparrow a.$$

Therefore it is natural to write for $a, b \in \mathcal{K}(\mathcal{D})$

$$b \leq a \;\Leftrightarrow\; a \sqsubseteq b.$$

The complete lattice $\mathcal{D}$ can be reconstructed from the meet semi-lattice $\mathcal{S}$ by a general construction that assigns to $\mathcal{S}$ the collection $\mathcal{F}^{\mathcal{S}}$ of *filters* (defined in Definition13.4.1 for type theories, see also Remark 13.4.6 for the translation to type structures), that form an algebraic lattice. We will prove the relations

$$\mathcal{S} \cong \mathcal{K}(\mathcal{F}^{\mathcal{S}});$$
$$\mathcal{D} \cong \mathcal{F}^{\mathcal{K}(\mathcal{D})}.$$

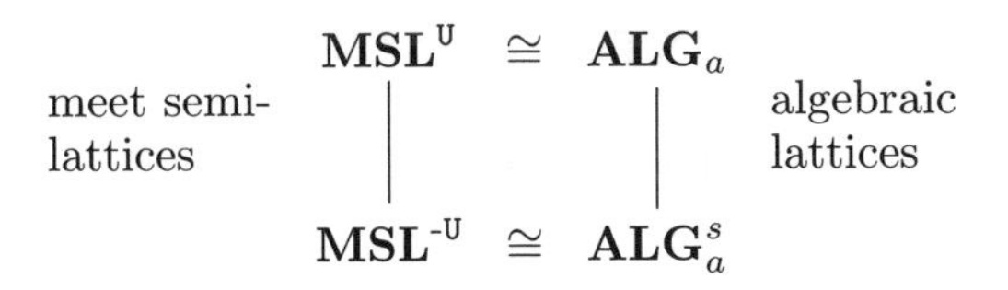

Figure 15.1 Equivalences proved in Section 15.1

In fact these isomorphisms are functional and constitute an equivalence between the categories $\mathbf{MSL}^{\mathtt{U}}$ and $\mathbf{ALG}_a$. Similarly, for the strict case, the equivalence between $\mathbf{MSL}^{\text{-}\mathtt{U}}$ and $\mathbf{ALG}_a^s$ is proved. See Fig. 15.1. For the injectivity of the map establishing $\mathcal{S} \cong \mathcal{K}(\mathcal{F}^{\mathcal{S}})$ in Proposition 15.1.22 one needs that $\leq$ is a partial order. Therefore the results of this section do not work for type theories in general.

The objects $\mathcal{D}$ in $\mathbf{ALG}$ are inadequate for interpreting untyped λ-terms. We need extra structure, $\mathcal{D}$ enriched to $\langle \mathcal{D}, F, G\rangle$, where $F \in [\mathcal{D} \to [\mathcal{D} \to \mathcal{D}]]$ and $G \in [[\mathcal{D} \to \mathcal{D}] \to \mathcal{D}]$. These are called lambda structures. There is a canonical way of defining such a pair F, G given that $\mathcal{D}$ is a filter structure obtained from a type theory; see Definition 13.4.5 and Exercise 14.27. Then we always have

$$F \circ G \sqsupseteq \mathsf{Id}_{\mathcal{D}}.$$

This does not yet imply the validity of the $\boldsymbol{\beta}$-axiom

$$(\boldsymbol{\beta})\ \ (\lambda x.M)N = M[x{:}= N].$$

For $\boldsymbol{\beta}$ to hold it suffices to require

$$(\text{fun-}\boldsymbol{\beta})\ \ F \circ G = \mathsf{Id}_{\mathcal{D}}.$$

When seeking the 'correct' category of type structures corresponding to lambda structures there are other things to consider. The axioms $(\to)$, $(\to\cap)$ and $(\mathtt{U}\to)$ are natural candidates, knowing the intended semantics of $\boldsymbol{\lambda}_{\to}$ given by Scott (1975a) generalized to $\boldsymbol{\lambda}_{\cap}$. The 'natural type structures', satisfying $(\to)$, $(\to\cap)$ and $(\mathtt{U}\to)$, exactly correspond to 'natural lambda structures' satisfying

$$F \circ G \sqsupseteq \mathsf{Id}_{[\mathcal{D} \to \mathcal{D}]} \tag{15.1}$$

$$G \circ F \sqsubseteq \mathsf{Id}_{\mathcal{D}} \tag{15.2}$$

as suggested in Exercise 14.29. Moreover, these axioms suffice to make the connection between type structures and lambda structures categorical, see Section 15.2. The equivalence in Fig. 15.2 does not, however, automatically

$$\mathbf{NTS^U} \;\cong\; \mathbf{NLS}$$

Figure 15.2 Equivalence proved in Section 15.2

give rise to λ-models. But the lambda structures induced by type structures (arising from the natural type theories) presented in Fig. 13.4 all satisfy (fun-$\boldsymbol{\beta}$) and hence are λ-models. See Exercise 14.27 for a natural lambda structure that is not a λ-model.

Weakening one of the axioms for natural type structures, taking $(\mathtt{U}_{\text{lazy}})$ instead of $(\mathtt{U}{\to})$, we obtain the 'lazy type structures'. This category is equivalent to that of 'lazy lambda structures' in which there is a weaker version of (15.2). These give rise to models of the lazy λ-calculus in which the order of a term (essentially the maximum number of consecutive λs it can have on its head) is preserved under equality, see Section 16.4.

After this success of the type structures, one might hope to find appropriate ones whose filter models are isomorphic to the well-known graph λ-models $P\omega$ by Scott and $\mathcal{D}_A$ by Engeler (see respectively Section 18.1 and 18.4.2 of B[1984]). This can be done, but we cannot rely on $P\omega$ or $\mathcal{D}_A$ as lambda structures for the following reason. The essence of the graph models is that there are maps

$$i : \mathcal{K}(\mathcal{D}) \times \text{Prime}(\mathcal{D}) \to \text{Prime}(\mathcal{D}),$$

where

$$\text{Prime}(\mathcal{D}) = \{d \in \mathcal{D} \mid \forall A \subseteq \mathcal{D}.d \sqsubseteq \bigsqcup A \;\Rightarrow\; \exists a \in A.d \sqsubseteq a\}.$$

In the case of $P\omega$ and $\mathcal{D}_A$, the subset $\text{Prime}(\mathcal{D})$ consists of the singletons. Although we can define appropriate F_i, G_i from these maps i, the reverse process does not hold. Therefore we want to incorporate i into the definition of a relevant category of structures. In order to do this, we generalize i to a coding of pairs of compact elements

$$Z : \mathcal{K}(\mathcal{D}) \times \mathcal{K}(\mathcal{D}) \to \mathcal{K}(\mathcal{D})$$

in order to be able to relate it in a simple way to type structures: the coding $Z(a, b)$ roughly corresponds to $(a \to b)$. In this way we obtain the so-called zip structures $\langle \mathcal{D}, Z\rangle$. For these structures one can define a corresponding lambda structure $\langle \mathcal{D}, F_Z, G_Z\rangle$ by

$$F_Z(x)(y) = \bigsqcup\{b \mid \exists a \sqsubseteq y.Z(a, b) \sqsubseteq x\},$$

$$G_Z(f) = \bigsqcup\{Z(a, b) \mid b \sqsubseteq f(a)\}.$$

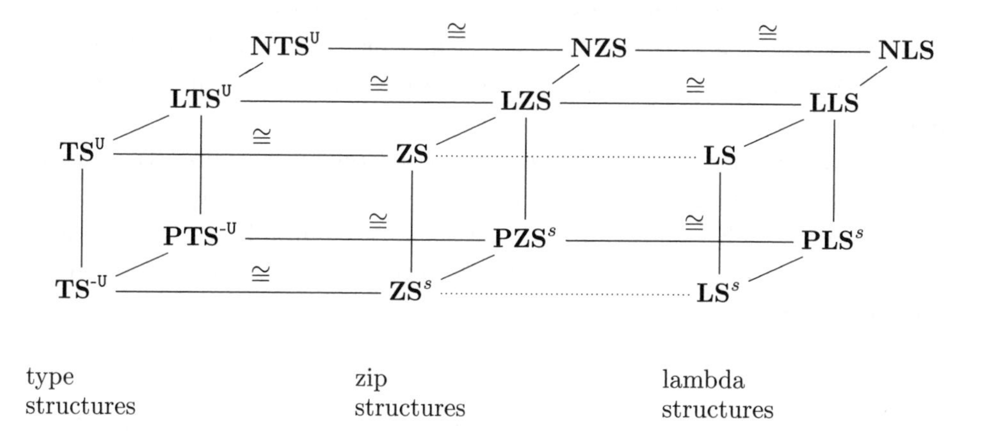

Figure 15.3 Equivalences proved in Sections 15.3 and 15.4

Finally the type structures whose filter models are $P\omega$ and $\mathcal{D}_A$, respectively, can be defined by weakening the conditions $(\rightarrow)$, $(\rightarrow\cap)$.

Even if we can establish an equivalence between the categories of type and zip structures, we lose the equivalence between the general categories of type structures and lambda structures. This situation is studied in detail in Sections 15.3 and 15.4. The equivalence between all variants of type structures and zip structures is complete, but as stated before not between those of zip structures and lambda structures.

Summarizing, there are three groups of categories of structures: the type, zip and lambda structures. The type structures are expansions of meet semi-lattices with the extra operator $\rightarrow$; the zip and lambda structures are expansions of algebraic lattices $\mathcal{D}$ with respectively a map Z merging ('zipping') two compact elements into one or the extra structure of a pair $\langle F, G\rangle$ with $F : \mathcal{D}\rightarrow[\mathcal{D}\rightarrow\mathcal{D}]$ and $G : [\mathcal{D}\rightarrow\mathcal{D}]\rightarrow\mathcal{D}$. Each time we must also consider the so-called strict case, preserving the top element, where morphisms are strict functions. Within each group there are five relevant categories, explained above. The categorical equivalences are displayed in Figs. 15.1, 15.2 and 15.3.

15.1 Meet semi-lattices and algebraic lattices

The results of this section are either known or relatively simple modifications of known ones, see Gierz et al. (1980).

Categories of meet semi-lattices

Remember the following notions, see Definitions 13.3.9–13.3.11. The category **MSL** has as objects at most countable meet semi-lattices and as morphisms maps preserving $\leq$ and $\cap$.

The category $\mathbf{MSL}^{\mathtt{U}}$ is as **MSL**, but based on meet semi-lattices with universe. So now morphisms also should preserve the universe.

The category $\mathbf{TS}^{\text{-}\mathtt{U}}$ has as objects the, at most countable, type structures and, as morphisms, the maps $f : \mathcal{S} {\to} \mathcal{S}'$, preserving $\leq$ and $\cap$, and which moreover satisfy:

$$(\text{mon-}{\to}) \quad f(s){\to}f(t) \leq f(s{\to}t).$$

The category $\mathbf{TS}^{\mathtt{U}}$ is as $\mathbf{TS}^{\text{-}\mathtt{U}}$, but based on type structures with universe. Now also morphisms should preserve the top.

In Definition 13.3.11 we defined three full subcategories of $\mathbf{TS}^{\mathtt{U}}$ and one of $\mathbf{TS}^{\text{-}\mathtt{U}}$, by specifying in each case the objects: $\mathbf{GTS}^{\mathtt{U}}$ whose objects are the graph type structures with universe; $\mathbf{LTS}^{\mathtt{U}}$ with, as objects, the lazy type structures with universe; $\mathbf{NTS}^{\mathtt{U}}$ whose objects are the natural type structures with universe; $\mathbf{PTS}^{\text{-}\mathtt{U}}$ with the proper type structures without universe as objects.

Categories of algebraic lattices

The following has already been given in Definition 12.2.1, but now we treat it in greater detail.

15.1.1 Definition

(i) A *complete lattice* is a poset $\mathcal{D} = (\mathcal{D}, \sqsubseteq)$ such that for arbitrary $X \subseteq \mathcal{D}$ the supremum $\bigsqcup X \in \mathcal{D}$ exists. Then one has also a *top* element $\top_{\mathcal{D}} = \bigsqcup \mathcal{D}$, a *bottom* element $\bot_{\mathcal{D}} = \bigsqcup \emptyset$, arbitrary *infima*

$$\bigsqcap X = \bigsqcup \{z \mid \forall x \in X. z \sqsubseteq x\}$$

and the *sup* and *inf* of two elements

$$x \sqcup y = \bigsqcup \{x, y\}, \quad x \sqcap y = \bigsqcap \{x, y\}.$$

(ii) A subset $Z \subseteq \mathcal{D}$ is called *directed* if Z is non-empty and

$$\forall x, y \in Z \ \exists z \in Z. x \sqsubseteq z \ \& \ y \sqsubseteq z.$$

(iii) An element $d \in \mathcal{D}$ is called *compact* (also called *finite*) if for every directed $Z \subseteq \mathcal{D}$ one has

$$d \sqsubseteq \bigsqcup Z \ \Rightarrow \ \exists z \in Z. d \sqsubseteq z.$$

Note that $\bot_\mathcal{D}$ is always compact and if d, e are compact, then so[1] is $d \sqcup e$.

(iv) $\mathcal{K}(\mathcal{D}) = \{d \in \mathcal{D} \mid d \text{ is compact}\}$.

(v) $\mathcal{K}^s(\mathcal{D}) = \mathcal{K}(\mathcal{D}) - \{\bot_\mathcal{D}\}$.

(vi) We call $\mathcal{D}$ an *algebraic lattice* if

$$\forall x \in \mathcal{D}. x = \bigsqcup\{e \in \mathcal{K}(\mathcal{D}) \mid e \sqsubseteq x\}.$$

We say $\mathcal{D}$ is an *ω-algebraic lattice* if moreover $\mathcal{K}(\mathcal{D})$ is countable.

When useful we will decorate $\sqsubseteq, \bigsqcup, \bigsqcap, \bot, \top, \sqcup$ and $\sqcap$ with $\mathcal{D}$, for example $\sqsubseteq_\mathcal{D}$ etc. In this chapter, $a, b, c, d, \ldots$ always denote compact elements in lattices. Generic elements are denoted by $x, y, z, \ldots$

15.1.2 Definition Let $\mathcal{D}$ be an ω-algebraic lattice.

(i) On $\mathcal{K}(\mathcal{D})$ define

$$d \leq e \Leftrightarrow e \sqsubseteq d.$$

(ii) For $x \in \mathcal{D}$, write

$$\mathcal{K}(x) = \{d \in \mathcal{K}(\mathcal{D}) \mid d \sqsubseteq x\}.$$

The following connects the notion of compact element to the notion of compact subset of a topological space.

15.1.3 Lemma *Let $\mathcal{D}$ be a complete lattice.*

(i) *Then $d \in \mathcal{D}$ is compact iff*

$$\forall Z \subseteq \mathcal{D}.[d \sqsubseteq \bigsqcup Z \Rightarrow \exists Z_0 \subseteq Z.[Z_0 \text{ is finite } \& \, d \sqsubseteq \bigsqcup Z_0]].$$

(ii) *If a, b are compact, then $a \sqcup b$ is compact.*

(iii) *For $a, b \in \mathcal{K}(\mathcal{D})$ one has $a \sqcap_{\mathcal{K}(\mathcal{D})} b = a \sqcup_\mathcal{D} b$.*

(iv) *$(\mathcal{K}(\mathcal{D}), \leq)$ is a meet semi-lattice with top.*

Proof (i) ($\Rightarrow$) Suppose $d \in \mathcal{D}$ is compact. Given $Z \subseteq \mathcal{D}$, let

$$Z^+ = \{\bigsqcup Z_0 \mid Z_0 \subseteq Z \,\&\, Z_0 \text{ finite}\}.$$

Then $Z \subseteq Z^+, \bigsqcup Z = \bigsqcup Z^+$ and Z^+ is directed. Hence

$$\begin{aligned} d \sqsubseteq \bigsqcup Z &\Rightarrow d \sqsubseteq \bigsqcup Z^+ \\ &\Rightarrow \exists z^+ \in Z^+. d \sqsubseteq z^+ \\ &\Rightarrow \exists Z_0 \subseteq Z. d \sqsubseteq \bigsqcup Z_0 \,\&\, Z_0 \text{ is finite.} \end{aligned}$$

[1] In general it is not true that if $d \sqsubseteq e \in \mathcal{K}(\mathcal{D})$, then $d \in \mathcal{K}(\mathcal{D})$; take for example $\omega + 1$ in the ordinal $\omega + \omega = \{0, 1, 2, \ldots, \omega, \omega + 1, \omega + 2, \ldots\}$. It is compact, but ω $(\sqsubseteq \omega + 1)$ is not.

($\Leftarrow$) Suppose $d \sqsubseteq \bigsqcup Z$ with $Z \subseteq \mathcal{D}$ directed. By the condition $d \sqsubseteq \bigsqcup Z_0$ for some finite $Z_0 \subseteq Z$. If Z_0 is non-empty, then by the directedness of Z there exists a $z \in Z$ such that $z \sqsupseteq \bigsqcup Z_0 \sqsupseteq d$. If Z_0 is empty, then $d = \bot_\mathcal{D}$ and we can take an arbitrary element z in the non-empty Z satisfying $d \sqsubseteq z$.

(ii) If $a \sqcup b$ is 'covered' (in the sense of $\sqsubseteq$) by the union of a family Z, then each of a, b is covered by a finite subset of Z by (i). Therefore also $a \sqcup b$ is covered by a finite subset of Z.

(iii) By (ii) $(a \sqcup_\mathcal{D} b) \in \mathcal{K}(\mathcal{D})$; now turn things around.

(iv) Immediate from (iii), noting that $\bot_\mathcal{D} \in \mathcal{K}(\mathcal{D})$ is the top. ∎

Instead of $\sqcap_\leq$ we often write $\cap_\leq$ or simply $\cap$.

15.1.4 Definition Let $\mathcal{D}, \mathcal{D}'$ be complete lattices and $f : \mathcal{D} \to \mathcal{D}'$.

(i) We say f is *(Scott) continuous* iff for all directed $X \subseteq \mathcal{D}$ one has

$$f(\bigsqcup X) = \bigsqcup f(X) \ (= \bigsqcup \{f(x) \mid x \in X\}).$$

(ii) $[\mathcal{D} \to \mathcal{D}'] = \{f : \mathcal{D} \to \mathcal{D}' \mid f \text{ is Scott continuous function}\}$.

(iii) We call f *strict* iff $f(\bot_\mathcal{D}) = \bot_{\mathcal{D}'}$.

(iv) Denote by $[\mathcal{D} \to_\mathrm{s} \mathcal{D}']$ the collection of continuous strict maps.

15.1.5 Proposition *Let $\mathcal{D}, \mathcal{D}'$ be algebraic lattices.*

(i) *Let $f \in [\mathcal{D} \to \mathcal{D}']$. Then for $x \in \mathcal{D}$*

$$f(x) = \bigsqcup \{f(a) \mid a \in \mathcal{K}(x)\}.$$

(ii) *Let $f, g \in [\mathcal{D} \to \mathcal{D}']$. Suppose $f \restriction \mathcal{K}(\mathcal{D}) = g \restriction \mathcal{K}(\mathcal{D})$. Then $f = g$.*

Proof (i) Use the fact that $x = \bigsqcup \{a \mid a \in \mathcal{K}(x)\}$ is a directed sup and that f is continuous.

(ii) By (i). ∎

15.1.6 Definition The category **ALG** has the ω-algebraic complete lattices as objects and the continuous maps as morphisms.

15.1.7 Definition

(i) $[\mathcal{D} \to \mathcal{D}']$ is partially ordered pointwise as follows:

$$f \sqsubseteq g \Leftrightarrow \forall x \in \mathcal{D}. f(x) \sqsubseteq g(x).$$

(ii) If $e \in \mathcal{D}$, $e' \in \mathcal{D}'$, then $e \mapsto e'$ is the *step function* defined by

$$(e \mapsto e')(d) = \begin{cases} e' & \text{if } e \sqsubseteq d\,, \\ \bot_{\mathcal{D}'} & \text{otherwise.} \end{cases}$$

15.1.8 Lemma *$[\mathcal{D}{\to}\mathcal{D}']$ is a complete lattice with*

$$\Big(\bigsqcup_{f\in X} f\Big)(d) = \bigsqcup_{f\in X} f(d).$$

15.1.9 Lemma *For $d, e\in\mathcal{D}$, $d', e'\in\mathcal{D}'$ and $f\in[\mathcal{D}{\to}\mathcal{D}']$ the following hold.*

(i) *d compact $\Rightarrow$ $d \mapsto d'$ is continuous.*
(ii) *$d \mapsto d'$ is continuous and $d' \neq \bot$ $\Rightarrow$ d is compact.*
(iii) *d' compact $\Leftrightarrow$ $d \mapsto d'$ compact.*
(iv) $d' \sqsubseteq f(d) \Leftrightarrow (d \mapsto d') \sqsubseteq f$.
(v) $e \sqsubseteq d \;\&\; d' \sqsubseteq e' \Rightarrow (d \mapsto d') \sqsubseteq (e \mapsto e')$.
(vi) $(d \mapsto d') \sqcup (e \mapsto e') \sqsubseteq (d \sqcap e) \mapsto (d' \sqcup e')$.

Proof Easy. ∎

15.1.10 Lemma *For all $e, d_1, \dots, d_n\in\mathcal{D}$, $e', d'_1, \dots, d'_n\in\mathcal{D}'$*

$$(e \mapsto e') \sqsubseteq (d_1 \mapsto d'_1) \sqcup \cdots \sqcup (d_n \mapsto d'_n) \Leftrightarrow$$
$$\Leftrightarrow \exists I\subseteq\{1,\dots,n\}\,[\,\textstyle\bigsqcup_{i\in I} d_i \sqsubseteq e \;\&\; e' \sqsubseteq \bigsqcup_{i\in I} d'_i\,].$$

Clearly in ($\Rightarrow$) we have $I \neq \emptyset$ if $e' \neq \bot_{\mathcal{D}'}$.

Proof Easy. ∎

15.1.11 Proposition *Let $\mathcal{D}, \mathcal{D}'\in$ **ALG**.*

(i) *For $f\in[\mathcal{D}{\to}\mathcal{D}']$ one has*

$$f = \bigsqcup\{a \mapsto a' \mid a' \sqsubseteq f(a),\; a\in\mathcal{K}(\mathcal{D}), a'\in\mathcal{K}(\mathcal{D}')\}.$$

(ii) *Let $\mathcal{D}\in$ **ALG** and let $f\in[\mathcal{D}{\to}\mathcal{D}']$ be compact. Then*

$$f = (a_1 \mapsto a'_1) \sqcup \cdots \sqcup (a_n \mapsto a'_n),$$

for some $a_1, \dots, a_n\in\mathcal{K}(\mathcal{D}), a'_1, \dots, a'_n\in\mathcal{K}(\mathcal{D}')$.
(iii) *$[\mathcal{D}{\to}\mathcal{D}']\in$ **ALG**.*

Proof (i) It suffices to show that the right and left hand sides are equal when applied to an arbitrary element $d\in\mathcal{D}$. With that in mind

$$\begin{aligned}
f(d) &= f(\bigsqcup\{a \mid a \sqsubseteq d \;\&\; a\in\mathcal{K}(\mathcal{D})\})\\
&= \bigsqcup\{f(a) \mid a \sqsubseteq d \;\&\; a\in\mathcal{K}(\mathcal{D})\}\\
&= \bigsqcup\{a' \mid a' \sqsubseteq f(a) \;\&\; a \sqsubseteq d \;\&\; a\in\mathcal{K}(\mathcal{D}), a'\in\mathcal{K}(\mathcal{D}')\}\\
&= \bigsqcup\{(a \mapsto a')(d) \mid a' \sqsubseteq f(a) \;\&\; a \sqsubseteq d \;\&\; a\in\mathcal{K}(\mathcal{D}), a'\in\mathcal{K}(\mathcal{D}')\}\\
&= \bigsqcup\{(a \mapsto a')(d) \mid a' \sqsubseteq f(a) \;\&\; a\in\mathcal{K}(\mathcal{D}), a'\in\mathcal{K}(\mathcal{D}')\}\\
&= \bigsqcup\{(a \mapsto a') \mid a' \sqsubseteq f(a) \;\&\; a\in\mathcal{K}(\mathcal{D}), a'\in\mathcal{K}(\mathcal{D}')\}(d).
\end{aligned}$$

(ii) For f compact one has, by (i),

$$f = \bigsqcup\{a \mapsto a' \mid a' \sqsubseteq f(a) \ \& \ a \in \mathcal{K}(\mathcal{D}), a' \in \mathcal{K}(\mathcal{D}')\}.$$

Hence by Lemma 15.1.3(i), for some $a_1, \ldots, a_n \in \mathcal{K}(\mathcal{D})$, $a'_1, \ldots, a'_n \in \mathcal{K}(\mathcal{D}')$,

$$f = (a_1 \mapsto a'_1) \sqcup \cdots \sqcup (a_n \mapsto a'_n). \tag{15.3}$$

(iii) It remains to show that there are only countably many compact elements in $[\mathcal{D} \to \mathcal{D}]$. Since $\mathcal{K}(\mathcal{D})$ is countable, there are only countably many expressions in the right hand side of (15.3). (The cardinality is $\leq \Sigma_n n.\aleph_0^2 = \aleph_0$.) Therefore there are countably many compact $f \in [\mathcal{D} \to \mathcal{D}]$. (There may be more expressions on the right hand side for one f, but this results in fewer compact elements.) ∎

15.1.12 Definition

(i) The category $\mathbf{ALG}_a$ has the same objects as $\mathbf{ALG}$ and, as morphisms, $\mathbf{ALG}_a(\mathcal{D}, \mathcal{D}')$ the maps $f \in [\mathcal{D} \to \mathcal{D}']$ that satisfy the 'compactness preserving' and 'additive' properties:

$$\begin{array}{ll} \text{(cmp-pres)} & \forall a \in \mathcal{K}(\mathcal{D}).f(a) \in \mathcal{K}(\mathcal{D}'); \\ \text{(add)} & \forall X \subseteq \mathcal{D}.f(\bigsqcup X) = \bigsqcup f(X). \end{array}$$

(ii) The category $\mathbf{ALG}_a^s$ has the same objects as $\mathbf{ALG}_a$ and, as morphisms $\mathbf{ALG}_a^s(\mathcal{D}, \mathcal{D}')$ the maps $f \in [\mathcal{D} \to \mathcal{D}']$ satisfying (cmp-pres), (add) and

$$\text{(s)} \quad \forall d \in \mathcal{D}.[f(d) = \bot_{\mathcal{D}'} \Rightarrow d = \bot_{\mathcal{D}}].$$

15.1.13 Remark

(i) Note that the requirement (add) implies that a morphism f is continuous (preserving sups for directed subsets X) and strict ($f(\bot_{\mathcal{D}}) = \bot_{\mathcal{D}'}$).

(ii) Recall that $\mathcal{K}^s(\mathcal{D}) = \mathcal{K}(\mathcal{D}) - \{\bot_{\mathcal{D}}\}$. Note that $\mathbf{ALG}_a^s(\mathcal{D}, \mathcal{D}')$ consists of maps satisfying

$$\begin{array}{ll} (\text{cmp-pres}^{\text{s}}) & \forall a \in \mathcal{K}^s(\mathcal{D}).f(a) \in \mathcal{K}^s(\mathcal{D}'); \\ \text{(add)} & \forall X \subseteq \mathcal{D}.f(\bigsqcup X) = \bigsqcup f(X). \end{array}$$

(iii) In contrast with Proposition 15.1.11(iii), $\mathbf{ALG}_a(\mathcal{D}, \mathcal{D}') \notin \mathbf{ALG}_a$, because from (i) of that proposition, it follows that the set of compactness-preserving functions is not closed under the operation of taking the supremum.

We need some lemmas.

15.1.14 Lemma

(i) *Let* $f : \mathcal{D} \to \mathcal{D}'$ *be a continuous function with* $\mathcal{D}, \mathcal{D}' \in \mathbf{ALG}$. *Then, for any* $X \subseteq \mathcal{D}$, $b' \in \mathcal{K}(\mathcal{D}')$,

$$b' \sqsubseteq f(\bigsqcup X) \Leftrightarrow \exists Z \subseteq_{\text{fin}} X \cap \mathcal{K}(\mathcal{D}).b' \sqsubseteq f(\bigsqcup Z).$$

(ii) *A map* $f : \mathcal{D} \to \mathcal{D}'$ *satisfies* (add) *iff* f *is Scott continuous and*

$$\forall X \subseteq_{\text{fin}} \mathcal{K}(\mathcal{D}).f(\bigsqcup X) = \bigsqcup f(X). \qquad (15.4)$$

Proof (i) Note that $\Xi = \{\bigsqcup Z \mid Z \subseteq_{\text{fin}} X \cap \mathcal{K}(\mathcal{D})\}$ is a directed set and $\bigsqcup \Xi = \bigsqcup X$. Moreover, by monotonicity of f, the set $\{f(\bigsqcup Z) \mid Z \subseteq_{\text{fin}} X \cap \mathcal{K}(\mathcal{D})\}$ is also directed. Therefore

$$\begin{aligned} b' \sqsubseteq f(\bigsqcup X) &\Leftrightarrow b' \sqsubseteq f(\bigsqcup \Xi) \\ &\Leftrightarrow b' \sqsubseteq \bigsqcup f(\Xi), && \text{since } f \text{ is continuous,} \\ &\Leftrightarrow b' \sqsubseteq \bigsqcup \{f(\bigsqcup Z) \mid Z \subseteq_{\text{fin}} X \cap \mathcal{K}(\mathcal{D})\}, && \text{by definition of } \Xi, \\ &\Leftrightarrow \exists Z \subseteq_{\text{fin}} X \cap \mathcal{K}(\mathcal{D}).b' \sqsubseteq f(\bigsqcup Z), && \text{since } b' \text{ is compact.} \end{aligned}$$

(ii) The non-trivial direction is to show, assuming f is Scott continuous and satisfies (15.4), that f is additive. By monotonicity of f we only need to show that for all $X \subseteq \mathcal{D}$,

$$f(\bigsqcup X) \sqsubseteq \bigsqcup f(X).$$

As $\mathcal{D}'$ is algebraic, it suffices to assume $b' \sqsubseteq f(\bigsqcup X)$ and conclude $b' \sqsubseteq \bigsqcup f(X)$. By (i) $\exists Z \subseteq_{\text{fin}} X \cap \mathcal{K}(\mathcal{D}).b' \sqsubseteq f(\bigsqcup Z) = \bigsqcup f(Z)$, so $b' \sqsubseteq \bigsqcup f(X)$. ∎

15.1.15 Lemma *Let* $\mathcal{S} \in \mathbf{MSL}^{\mathrm{U}}$ *be a meet semi-lattice with universe,* $I \neq \emptyset$ *and* $s, t, s_i \in \mathcal{S}$. *Then*

(i) *In* $\mathcal{F}^{\mathcal{S}}$ *we have*

$$\bigsqcup_{i \in I} \uparrow s_i = \uparrow \bigcap_{i \in I} s_i.$$

(ii) *In* $\mathcal{K}(\mathcal{F}^{\mathcal{S}})$ *we have*

$$\bigcap_{i \in I} \uparrow s_i = \uparrow \bigcap_{i \in I} s_i,$$

where the $\bigcap$ *denote the infima in* $\mathcal{S}$ *and* $\mathcal{K}(\mathcal{S})$ *respectively.*

(iii) $\uparrow s \leq_{\mathcal{K}(\mathcal{F}^{\mathcal{S}})} \uparrow t \Leftrightarrow s \leq_{\mathcal{S}} t$.

Proof From the definitions. ∎

15.1.16 Lemma

(i) *Let* $\mathcal{X} \subseteq \mathcal{F}^{\mathcal{K}(\mathcal{D})}$*. Then taking sups in* $\mathcal{D}$ *one has*

$$\bigsqcup\left(\bigcup \mathcal{X}\right) = \bigsqcup_{X \in \mathcal{X}} \left(\bigsqcup X\right).$$

(ii) *Let* $\theta \subseteq \mathcal{K}(\mathcal{D})$ *be non-empty. Then taking the sups in* $\mathcal{D}$ *one has*

$$\bigsqcup(\uparrow\theta) = \bigsqcup \theta.$$

Proof (i) Realizing that a sup (in $\mathcal{D}$) of a union of $\{Y_i\}_{i \in I} \subseteq \mathcal{K}(\mathcal{D})$ is the sup of the sups $\bigsqcup Y_i$, one has

$$\bigsqcup\left(\bigcup_{i \in I} Y_i\right) = \bigsqcup_{i \in I} \left(\bigsqcup Y_i\right).$$

The result follows by making an α-conversion $[i := X]$ and taking $I = \mathcal{X}$ and $Y_X = X$.

(ii) The filter $\uparrow\theta$ is obtained from θ by taking extensions and intersections in $\mathcal{K}(\mathcal{D})$. Now the order in $\mathcal{K}(\mathcal{D})$ is the reverse of the one induced by $\mathcal{D}$, therefore $\uparrow\theta$ is obtained by taking smaller elements and unions (in $\mathcal{D}$). But then taking the big union we get the same result. ∎

We now will establish the following equivalences of categories.

$$\mathbf{MSL}^{\mathtt{U}} \cong \mathbf{ALG}_a;$$
$$\mathbf{MSL}^{\text{-}\mathtt{U}} \cong \mathbf{ALG}_a^s.$$

Equivalence between $\mathbf{MSL}^{\mathtt{U}}$ and $\mathbf{ALG}_a$

We now define the functors establishing the equivalences between categories of meet semi-lattices and complete algebraic lattices. Remember that 13.4.1–13.4.4 can be translated immediately to meet semi-lattices.

15.1.17 Definition We define a map $\mathsf{Flt} : \mathbf{MSL}^{\mathtt{U}} \to \mathbf{ALG}_a$, that will turn out to be a functor, as follows.

(i) On objects $\mathcal{S} \in \mathbf{MSL}^{\mathtt{U}}$ one defines

$$\mathsf{Flt}(\mathcal{S}) = \langle \mathcal{F}^{\mathcal{S}}, \subseteq \rangle.$$

(ii) On morphisms $f : \mathcal{S} \to \mathcal{S}'$ one defines $\mathsf{Flt}(f) : \mathcal{F}^{\mathcal{S}} \to \mathcal{F}^{\mathcal{S}'}$ by

$$\mathsf{Flt}(f)(X) = \{s' \mid \exists s \in X . f(s) \leq s'\}.$$

15.1.18 Lemma *Let* $f \in \mathbf{MSL}^{\mathtt{U}}(\mathcal{S}, \mathcal{S}')$.

(i) *For* $X \subseteq \mathcal{S}$*, one has* $\mathsf{Flt}(f)(\uparrow X) = \uparrow \{f(s) \mid s \in X\}$.
(ii) *For* $s \in \mathcal{S}$ *one has* $\mathsf{Flt}(f)(\uparrow s) = \uparrow f(s)$.

Proof We only prove (ii). Accordingly,

$$\begin{aligned}
\mathsf{Flt}(f)(\uparrow s) &= \{s' \mid \exists t \in \uparrow s. f(t) \leq s'\}, \\
&= \{s' \mid \exists t \geq s. f(t) \leq s'\}, \\
&= \{s' \mid f(s) \leq s'\}, && \text{since } f \text{ is monotone,} \\
&= \uparrow f(s). \blacksquare
\end{aligned}$$

15.1.19 Proposition *The map* Flt *is a functor from* $\mathbf{MSL}^{\mathtt{U}}$ *to* $\mathbf{ALG}_a$.

Proof We must prove that Flt transforms a morphism in $\mathbf{MSL}^{\mathtt{U}}$ into a morphism in $\mathbf{ALG}_a$. Let $f \in \mathbf{MSL}^{\mathtt{U}}(\mathcal{S}, \mathcal{S}')$, $\uparrow s \in K(\mathcal{F}^{\mathcal{S}})$. By Lemma 15.1.18(ii) $\mathsf{Flt}(f)(\uparrow s) = \uparrow f(s)$, which is compact in $\mathcal{F}^{\mathcal{S}'}$, hence $\mathsf{Flt}(f)$ satisfies (cmp-pres).

$\mathsf{Flt}(f)$ satisfies (add). Indeed, by Lemma 15.1.14(ii) and the fact that $\mathsf{Flt}(f)$ is trivially Scott continuous, it is enough to prove that it commutes with finite joins of compact elements. Let I be non-empty. We have

$$\begin{aligned}
\mathsf{Flt}(f)(\bigsqcup_{i \in I} \uparrow s_i) &= \mathsf{Flt}(f)(\uparrow \bigcap_{i \in I} s_i), && \text{by Lemma 15.1.15(ii),} \\
&= \uparrow f(\bigcap_{i \in I} s_i), && \text{by Lemma 15.1.18(ii),} \\
&= \uparrow \bigcap_{i \in I} f(s_i), && \text{since } f \text{ commutes with } \cap, \\
&= \bigsqcup_{i \in I} \uparrow f(s_i), && \text{by Lemma 15.1.15(ii),} \\
&= \bigsqcup_{i \in I} \mathsf{Flt}(f)(\uparrow s_i) && \text{by Lemma 15.1.18(ii).}
\end{aligned}$$

If I is empty, and $\mathtt{U}, \mathtt{U}'$ are the universal tops of $\mathcal{S}, \mathcal{S}'$, respectively, then

$$\begin{aligned}
\mathsf{Flt}(f)(\bigsqcup_{\emptyset} \uparrow s_i) &= \mathsf{Flt}(f)(\uparrow \mathtt{U}), \\
&= \uparrow f(\mathtt{U}), && \text{by Lemma 15.1.18(ii),} \\
&= \uparrow \mathtt{U}', && \text{since } f \text{ preserves tops,} \\
&= \bigsqcup_{\emptyset} \mathsf{Flt}(f)(\uparrow s_i).
\end{aligned}$$

So $\mathsf{Flt}(f)$ satisfies (add). ■

It is possible to leave out conditions (cmp-pres) and (add), obtaining the category **ALG**. Then one needs to consider *approximable maps* as morphisms in the category of meet semi-lattices. See Abramsky (1991).

15.1.20 Definition We define a map $\mathsf{Cmp} : \mathbf{ALG}_a \to \mathbf{MSL}^{\mathtt{U}}$, that will turn out to be a functor, as follows.

(i) On objects $\mathcal{D} \in \mathbf{ALG}_a$ one defines Cmp by

$$\mathsf{Cmp}(\mathcal{D}) = (\mathcal{K}(\mathcal{D}), \leq).$$

(ii) On morphisms $f \in \mathbf{ALG}_a(\mathcal{D}, \mathcal{D}')$ one defines $\mathsf{Cmp}(f)$ by

$$\mathsf{Cmp}(f)(d) = f(d).$$

15.1.21 Lemma *The map* Cmp *is a functor.*

Proof Let $f \in \mathbf{ALG}_a(\mathcal{D}, \mathcal{D}')$. Note that $\mathsf{Cmp}(f) = f \upharpoonright \mathcal{K}(\mathcal{D}) : \mathcal{K}(\mathcal{D}) \to \mathcal{K}(\mathcal{D}')$, by (cmp-pres). By the fact that f is additive one has $f(\bot_{\mathcal{D}}) = \bot_{\mathcal{D}'}$, which is $f(\mathtt{U}_{\mathcal{K}(\mathcal{D})}) = \mathtt{U}_{\mathcal{K}(\mathcal{D}')}$, and

$$f(a \cap_{\mathcal{K}(\mathcal{D})} b) = f(a \sqcup_{\mathcal{D}} b) = f(a) \sqcup_{\mathcal{D}'} f(b) = f(a) \cap_{\mathcal{K}(\mathcal{D})} f(b).$$

Also f is monotonic since it is continuous. ■

In the remainder of the current section we write $\leq$ instead of $\leq_{\mathcal{K}(\mathcal{F}^{\mathcal{S}})}$.

Next we will show that the functors Flt and Cmp establish an equivalence between the categories $\mathbf{MSL}^{\mathtt{U}}$ and $\mathbf{ALG}_a$.

15.1.22 Proposition

(i) *Let $\mathcal{S} \in \mathbf{MSL}^{\mathtt{U}}$. Then $\sigma = \sigma_{\mathcal{S}} : \mathcal{S} \to \mathcal{K}(\mathcal{F}^{\mathcal{S}})$ defined by $\sigma_{\mathcal{S}}(s) = \uparrow s$ is an $\mathbf{MSL}^{\mathtt{U}}$ isomorphism. Diagrammatically,*

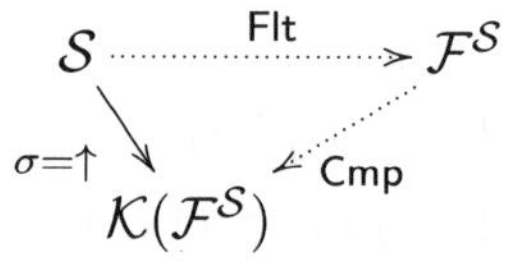

$$\begin{array}{ccc} \mathcal{S} & \xrightarrow{\mathsf{Flt}} & \mathcal{F}^{\mathcal{S}} \\ & {\scriptstyle \sigma=\uparrow}\searrow \quad \swarrow{\scriptstyle \mathsf{Cmp}} & \\ & \mathcal{K}(\mathcal{F}^{\mathcal{S}}) & \end{array}$$

Therefore $\mathcal{S} \cong \mathcal{K}(\mathcal{F}^{\mathcal{S}})$.

(ii) *Let $\mathcal{D} \in \mathbf{ALG}_a$. Then $\tau = \tau_{\mathcal{D}} : \mathcal{F}^{\mathcal{K}(\mathcal{D})} \to \mathcal{D}$ defined by $\tau(X) = \bigsqcup X$, where $\bigsqcup$ is taken in $\mathcal{D}$, is an $\mathbf{ALG}_a$ isomorphism with inverse $\tau^{-1} : \mathcal{D} \to \mathcal{F}^{\mathcal{K}(\mathcal{D})}$ defined by $\tau^{-1}(x) = \{c \in \mathcal{K}(\mathcal{D}) \mid c \sqsubseteq x\}$. Diagrammatically,*

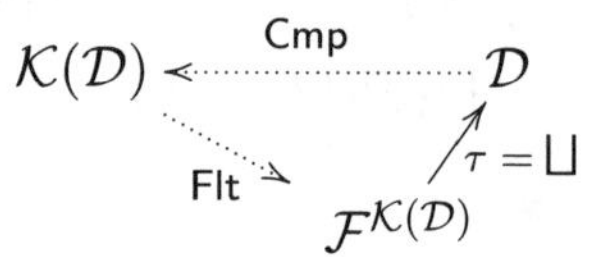

$$\begin{array}{ccc} \mathcal{K}(\mathcal{D}) & \xleftarrow{\mathsf{Cmp}} & \mathcal{D} \\ & {\scriptstyle \mathsf{Flt}}\searrow \quad \nearrow{\scriptstyle \tau=\bigsqcup} & \\ & \mathcal{F}^{\mathcal{K}(\mathcal{D})} & \end{array}$$

Therefore $\mathcal{D} \cong \mathcal{F}^{\mathcal{K}(\mathcal{D})}$.

Proof (i) By Proposition 13.4.4(v) σ is a surjection. It is also 1–1, since $\uparrow s = \uparrow t \Rightarrow s \leq t \leq s \Rightarrow s = t$. (This is the place where we need that $\leq$ is a partial order, not just a pre-order.) Moreover, σ preserves $\leq$:

$$\begin{aligned} s \leq t \quad &\Leftrightarrow \quad \uparrow t \subseteq \uparrow s \\ &\Leftrightarrow \quad \uparrow s \leq \uparrow t, \qquad \text{by definition of } \leq \text{ on } \mathcal{K}(\mathcal{F}^{\mathcal{S}}). \end{aligned}$$

Furthermore, σ preserves $\sqcap$:

$$\begin{array}{rcll}\sigma(s \sqcap t) & = & \uparrow(s \sqcap t) & \\ & = & \uparrow s \sqcup \uparrow t & \text{in } \mathcal{F}^{\mathcal{S}}\text{, by 13.4.4(iii),} \\ & = & \uparrow s \sqcap \uparrow t & \text{in } \mathcal{K}(\mathcal{F}^{\mathcal{S}})\text{, by definition of } \leq \text{ on } \mathcal{K}(\mathcal{F}^{\mathcal{S}}), \\ & = & \sigma(s) \sqcap \sigma(t). & \end{array}$$

Then $\sigma(\mathtt{U}_{\mathcal{S}}) = \uparrow\mathtt{U}_{\mathcal{S}} = \mathtt{U}_{\mathcal{K}(\mathcal{F}^{\mathcal{S}})}$, since $\{\mathtt{U}_{\mathcal{S}}\}$ is the $\subseteq$-smallest filter; hence σ preserves tops.

Finally σ^{-1} preserves $\leq, \sqcap_{\mathcal{K}(\mathcal{F}^{\mathcal{S}})}$ and $\mathtt{U}_{\mathcal{K}(\mathcal{F}^{\mathcal{S}})}$, as by Lemma 13.4.4(v), an element $c \in \mathcal{K}(\mathcal{F}^{\mathcal{S}})$ is of the form $c = \uparrow s$, with $s \in \mathcal{S}$ and $\sigma^{-1}(\uparrow s) = s$.

(ii) We have $\tau \circ \tau^{-1} = \mathbf{1}_{\mathcal{D}}$ and $\tau^{-1} \circ \tau = \mathbf{1}_{\mathcal{F}^{\mathcal{K}(\mathcal{D})}}$:

$$\begin{array}{rcll}\tau(\tau^{-1}(x)) & = & \bigsqcup\{c \in \mathcal{K}(\mathcal{D}) \mid c \sqsubseteq x\} & \\ & = & x, & \text{since } \mathcal{D} \in \mathbf{ALG}_a. \\ \tau^{-1}(\tau(X)) & = & \{c \mid c \sqsubseteq \bigsqcup X\} & \\ & = & \{c \mid c \in X\}, & \text{since one has} \\ c \sqsubseteq_{\mathcal{D}} \bigsqcup X & \Leftrightarrow & \exists x \in X \; c \sqsubseteq_{\mathcal{D}} x, & \text{as } c \text{ is compact and } X \subseteq \mathcal{K}(\mathcal{D}) \subseteq \mathcal{D} \\ & & & \text{is a filter with respect to } \leq, \\ & & & \text{so directed with respect to } \sqsubseteq \\ & \Leftrightarrow & \exists x \in X \; x \leq_{\mathcal{K}(\mathcal{D})} c & \\ & \Leftrightarrow & c \in X, & \text{as } X \text{ is a filter on } \mathcal{K}(\mathcal{D}). \end{array}$$

We still have to show that τ and τ^{-1} are morphisms. One easily sees that τ satisfies (cmp-pres). The map τ is also additive, i.e. $\bigsqcup \tau(\mathcal{X}) = \tau(\bigsqcup \mathcal{X})$ for arbitrary $\mathcal{X} \subseteq \mathcal{F}^{\mathcal{K}(\mathcal{D})}$. Indeed,

$$\begin{array}{rcll}\bigsqcup \tau(\mathcal{X}) & = & \bigsqcup_{X \in \mathcal{X}} (\bigsqcup X), & \text{by definition of } \tau, \\ & = & \bigsqcup \left(\bigcup \mathcal{X}\right), & \text{by Lemma 15.1.16(i),} \\ & = & \bigsqcup \left(\uparrow(\bigcup \mathcal{X})\right), & \text{by Lemma 15.1.16(ii),} \\ & = & \tau\left(\uparrow(\bigcup \mathcal{X})\right), & \text{by definition of } \tau, \\ & = & \tau(\bigsqcup \mathcal{X}), & \text{by Proposition 13.4.4(i).} \end{array}$$

Now we have to prove that τ^{-1} satisfies (cmp-pres) and (add). For (cmp-pres), assume that $b \in \mathcal{K}(\mathcal{D})$ and $\tau^{-1}(b) \sqsubseteq \bigsqcup X$, with X directed. Then $b \sqsubseteq \bigsqcup \tau(X)$, since τ satisfies (add). Since b is compact, there exists $x \in X$ such that $b \sqsubseteq \tau(x)$, hence $\tau^{-1}(b) \sqsubseteq x$ and we are done. For (add), let $X \subseteq \mathcal{D}$.

Then

$$\begin{aligned}\tau^{-1}(\bigsqcup X) &= \tau^{-1}(\bigsqcup\{\tau(\tau^{-1}(x)) \mid x \in X\}),\\ &= \tau^{-1}(\tau(\bigsqcup\{\tau^{-1}(x) \mid x \in X\})), \quad \text{since } \tau \text{ satisfies (add)},\\ &= \bigsqcup\{\tau^{-1}(x) \mid x \in X\}. \blacksquare\end{aligned}$$

15.1.23 Corollary *The categories* $\mathbf{MSL}^{\mathtt{U}}$ *and* $\mathbf{ALG}_a$ *are equivalent; in fact the isomorphisms in Proposition* 15.1.22 *form natural isomorphisms proving*

$$\mathsf{Cmp} \circ \mathsf{Flt} \cong \mathsf{Id}_{\mathbf{MSL}^{\mathtt{U}}} \ \&\ \mathsf{Flt} \circ \mathsf{Cmp} \cong \mathsf{Id}_{\mathbf{ALG}_a}.$$

Proof First one has to show that in $\mathbf{MSL}^{\mathtt{U}}$ the following diagram commutes:

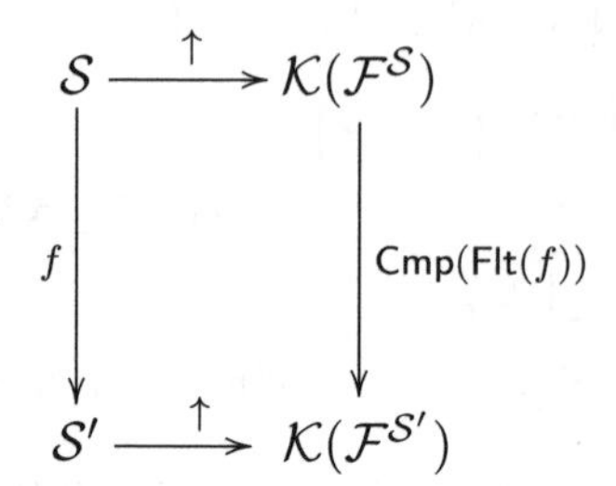

We have to prove that $\mathsf{Cmp}(\mathsf{Flt}(f))(\uparrow s) = \uparrow(f(s))$. This follows from Lemma 15.1.18(ii) and the definition of Cmp.

Second, one has to show that in $\mathbf{ALG}_a$ the following diagram commutes:

$$\begin{array}{ccc}\mathcal{F}^{\mathcal{K}(\mathcal{D})} & \xrightarrow{\ \bigsqcup\ } & \mathcal{D}\\ {\scriptstyle \mathsf{Flt}(\mathsf{Cmp}(f))}\downarrow & & \downarrow{\scriptstyle f}\\ \mathcal{F}^{\mathcal{K}(\mathcal{D}')} & \xrightarrow{\ \bigsqcup\ } & \mathcal{D}'\end{array}$$

Now for $X \in \mathcal{F}^{\mathcal{K}(\mathcal{D})}$ one has

$$\begin{aligned}\mathsf{Flt}(\mathsf{Cmp}(f))(X) &= \{d' \in \mathcal{K}(\mathcal{D}') \mid \exists d \in X.\, f(d) \leq d'\}\\ &= \{d' \in \mathcal{K}(\mathcal{D}') \mid \exists d \in X.\, d' \sqsubseteq f(d)\}.\end{aligned}$$

Hence, using also the continuity of f and the fact that X, as subset of $\langle \mathcal{D}, \sqsubseteq \rangle$, is directed, we have

$$\bigsqcup(\mathsf{Flt}(\mathsf{Cmp}(f))(X)) = \bigsqcup_{d \in X} f(d) = f(\bigsqcup X),$$

and the diagram commutes. As before we have $\mathsf{Flt} \circ \mathsf{Cmp} \cong \mathsf{Id}_{\mathbf{ALG}_a}$. $\blacksquare$

This result is a special case of Stone duality, cf. Johnstone (1986) (II, 3.3).

Equivalence between $\mathbf{MSL}^{\text{-U}}$ and $\mathbf{ALG}^s_a$

We prove that $\mathbf{MSL}^{\text{-U}} \cong \mathbf{ALG}^s_a$. The proof uses the functors Flt and a variant of Cmp, denoted by Cmp_s. The functor Flt is given in Definition 15.1.17, where now $\mathcal{F}^{\mathcal{S}}$ is taken with $\mathcal{S} \in \mathbf{MSL}^{\text{-U}}$. But now $\emptyset \in \mathcal{K}(\mathcal{F}^{\mathcal{S}})$ by Definition 13.4.1 and hence, $\mathcal{S} \not\cong \mathcal{K}(\mathcal{F}^{\mathcal{S}})$. To obtain an isomorphism, the functor Cmp_s is defined by considering the set $\mathcal{K}^s(\mathcal{D})$ of compact elements of $\mathcal{D}$ without $\bot$.

15.1.24 Definition Let $\mathcal{D} \in \mathbf{ALG}^s_a$.

(i) The functor $\mathsf{Cmp}_s : \mathbf{ALG}^s_a \to \mathbf{MSL}^{\text{-U}}$ is defined as

$$\mathsf{Cmp}_s(\mathcal{D}) = (\mathcal{K}^s(\mathcal{D}), \leq).$$

For a morphism $f \in \mathbf{ALG}^s_a(\mathcal{D}, \mathcal{D}')$ define $\mathsf{Cmp}_s(f) : \mathcal{K}^s(\mathcal{D}) \to \mathcal{K}^s(\mathcal{D}')$ by

$$\mathsf{Cmp}_s(f)(d) = f(d).$$

15.1.25 Proposition

(i) *Let $\mathcal{S} \in \mathbf{MSL}^{\text{-U}}$. Then $\sigma : \mathcal{S} \to \mathcal{K}^s(\mathcal{F}^{\mathcal{S}})$ defined by $\sigma(s) = \uparrow s$ is an $\mathbf{MSL}^{\text{-U}}$ isomorphism. Diagrammatically,*

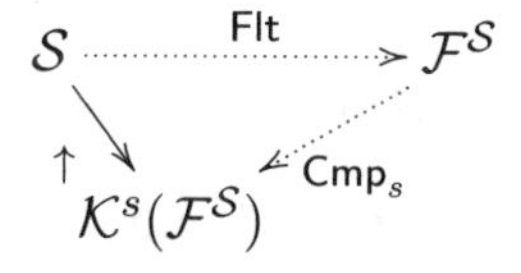

$$\begin{array}{ccc} \mathcal{S} & \overset{\mathsf{Flt}}{\cdots\cdots\rightarrow} & \mathcal{F}^{\mathcal{S}} \\ {\scriptstyle\uparrow}\searrow & & \swarrow{\scriptstyle\mathsf{Cmp}_s} \\ & \mathcal{K}^s(\mathcal{F}^{\mathcal{S}}) & \end{array}$$

Therefore $\mathcal{S} \cong \mathcal{K}^s(\mathcal{F}^{\mathcal{S}})$.

(ii) *Let $\mathcal{D} \in \mathbf{ALG}^s_a$. Then $\tau : \mathcal{F}^{\mathcal{K}^s(\mathcal{D})} \to \mathcal{D}$ defined by $\tau(X) = \bigsqcup X$, is an $\mathbf{ALG}^s_a$ isomorphism with inverse $\tau^{-1} : \mathcal{D} \to \mathcal{F}^{\mathcal{K}^s(\mathcal{D})}$ defined by*

$$\tau^{-1}(d) = \{c \in \mathcal{K}^s(\mathcal{D}) \mid c \sqsubseteq d\}.$$

Diagrammatically,

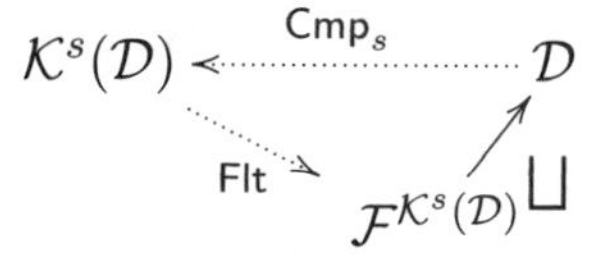

$$\begin{array}{ccc} \mathcal{K}^s(\mathcal{D}) & \overset{\mathsf{Cmp}_s}{\leftarrow\cdots\cdots} & \mathcal{D} \\ & {\scriptstyle\mathsf{Flt}}\searrow & \nearrow \\ & & \mathcal{F}^{\mathcal{K}^s(\mathcal{D})} \bigsqcup \end{array}$$

Therefore $\mathcal{D} \cong \mathcal{F}^{\mathcal{K}_s(\mathcal{D})}$.

Proof Similar to the proof of Proposition 15.1.22. ∎

15.1.26 Corollary *The categories $\mathbf{MSL}^{\text{-U}}$ and $\mathbf{ALG}^s_a$ are equivalent.*

15.1.27 Remark The map $\rho : \mathcal{F}^{\mathcal{K}^s(\mathcal{D})} \to \mathcal{F}^{\mathcal{K}(\mathcal{D})}$, given by $\rho(X) = X \cup \{\bot_{\mathcal{D}}\}$ is an isomorphism in the category $\mathbf{ALG}^s_a$, hence also in $\mathbf{ALG}_a$.

15.2 Natural type structures and lambda structures

In this section we prove for the natural type and lambda structures that

$$\begin{array}{rcl}\mathcal{S} & \simeq & \mathcal{K}(\mathcal{F}^{\mathcal{S}}),\\ \mathcal{D} & \simeq & \mathcal{F}^{\mathcal{K}(\mathcal{D})},\end{array}$$

i.e. that there is a congruence between the categories $\mathbf{NTS}^{\mathrm{U}} \cong \mathbf{NLS}$. The results of this section will be generalized using zip structures in Sections 15.3 and 15.4. Even though the results in this section follow from the said generalization, we have decided to keep the proofs here as a warm-up. Moreover, the proofs of the results in this section are more direct than those obtained as corollaries.

15.2.1 Definition Let $\mathcal{D}, \mathcal{D}' \in \mathbf{ALG}$. A *Galois connection* between $\mathcal{D}$ and $\mathcal{D}'$, written $\langle \mathsf{m}, \mathsf{n}\rangle : \mathcal{D} \to \mathcal{D}'$, is a pair of continuous functions $\langle \mathsf{m}, \mathsf{n}\rangle$ with $\mathsf{m} : \mathcal{D} \to \mathcal{D}'$, $\mathsf{n} : \mathcal{D}' \to \mathcal{D}$ such that

$$\begin{array}{ll}\text{(Galois-1)} & \mathsf{n} \circ \mathsf{m} \sqsupseteq \mathsf{Id}_{\mathcal{D}},\\ \text{(Galois-2)} & \mathsf{m} \circ \mathsf{n} \sqsubseteq \mathsf{Id}_{\mathcal{D}'};\end{array}$$

we say that m is the *left adjoint* of the Galois connection, and n the *right adjoint.*

A statement equivalent to (Galois-1, Galois-2) is

$$\text{(Galois)} \quad \forall x \in \mathcal{D}, x' \in \mathcal{D}'.\mathsf{m}(x) \sqsubseteq x' \;\Leftrightarrow\; x \sqsubseteq \mathsf{n}(x') :$$

see Exercise 15.7.

Each adjoint in a Galois connection determines the other: see Exercise 15.8. For this reason one often writes m^R for n, and n^L for m. From now on $\underline{\mathsf{m}}$ is short for $\langle \mathsf{m}, \mathsf{m}^R\rangle$.

The next lemma provides some properties of Galois connections.

15.2.2 Lemma *Let* $\underline{\mathsf{m}} : \mathcal{D} \to \mathcal{D}'$ *be a Galois connection.*

(i) m *is additive:*

$$\forall X \subseteq \mathcal{D},\ \mathsf{m}(\bigsqcup X) = \bigsqcup \mathsf{m}(X).$$

In particular m *is strict,* $\mathsf{m}(\bot) = \bot$, *since* $\bot = \bigsqcup \emptyset$.

(ii) $\mathsf{m}^R(\top) = \top$.

(iii) $d \in \mathcal{K}(\mathcal{D}) \;\Rightarrow\; \mathsf{m}(d) \in \mathcal{K}(\mathcal{D}')$.

(iv) *Let* $d, e \in \mathcal{K}(\mathcal{D})$. *Then*

$$\mathsf{m} \circ (d \mapsto e) \circ \mathsf{m}^R = (\mathsf{m}(d) \mapsto \mathsf{m}(e)).$$

Proof (i) As m is monotone we have $\bigsqcup \mathsf{m}(X) \sqsubseteq \mathsf{m}(\bigsqcup X)$. On the other hand

$$\begin{array}{rcll} \mathsf{m}(\bigsqcup X) & \sqsubseteq & \mathsf{m}(\bigsqcup \mathsf{m}^R \circ \mathsf{m}(X)), & \text{by (Galois-1)}, \\ & \sqsubseteq & \mathsf{m}(\mathsf{m}^R(\bigsqcup \mathsf{m}(X))), & \text{since } \mathsf{m}^R \text{ is monotone}, \\ & \sqsubseteq & \bigsqcup \mathsf{m}(X), & \text{by (Galois-2)}. \end{array}$$

(ii) By (Galois) applied to $x' = \top$.

(iii) Let $d \in \mathcal{K}(\mathcal{D})$ and let $\mathsf{m}(d) \sqsubseteq \bigsqcup Z$, where $Z \subseteq \mathcal{D}'$ is any directed set. Then, by (Galois), $d \sqsubseteq \mathsf{m}^R(\bigsqcup Z) = \bigsqcup \mathsf{m}^R(Z)$. Since d is compact, there exists $z \in Z$ such that $d \sqsubseteq \mathsf{m}^R(z)$, hence, by (Galois), $\mathsf{m}(d) \sqsubseteq z$. This proves that $\mathsf{m}(d)$ is compact.

(iv) Let $y \in \mathcal{D}'$. We have

$$(\mathsf{m} \circ (d \mapsto e) \circ \mathsf{m}^R)(y) = \begin{cases} \mathsf{m}(e), & \text{if } \mathsf{m}^R(y) \sqsupseteq d; \\ \mathsf{m}(\bot), & \text{otherwise.} \end{cases}$$

Note that by (Galois), $d \sqsubseteq \mathsf{m}^R(y)$ is equivalent to $\mathsf{m}(d) \sqsubseteq y$. Moreover, as previously noted, $\mathsf{m}(\bot) = \bot$. So we have

$$(\mathsf{m} \circ (d \mapsto e) \circ \mathsf{m}^R)(y) = \begin{cases} \mathsf{m}(e), & \text{if } \mathsf{m}(d) \sqsubseteq y; \\ \bot, & \text{otherwise;} \end{cases}$$

the right hand side is the definition of the step function $(\mathsf{m}(d) \mapsto \mathsf{m}(e))$. ∎

15.2.3 Definition

(i) A triple $\mathcal{D} = \langle \mathcal{D}, F, G \rangle$ is called a *lambda structure*, written LS, if $\mathcal{D} \in \mathbf{ALG}$ and $F : \mathcal{D} \to [\mathcal{D} \to \mathcal{D}]$ and $G : [\mathcal{D} \to \mathcal{D}] \to \mathcal{D}$ are continuous.

(ii) The lambda structure $\mathcal{D} = \langle \mathcal{D}, F, G \rangle$ is *natural* if $\langle G, F \rangle$ is a Galois connection, with $F = G^R$, i.e. $F \circ G \sqsupseteq \mathsf{Id}_{\mathcal{D}}$ and $G \circ F \sqsubseteq \mathsf{Id}_{\mathcal{D}}$.

Note that the notation $\langle \mathcal{D}, F, G \rangle$ is consistent with what is customary for a lambda structure, but this is in contrast with the notation for Galois connection, since the left adjoint G is put on the right.

Given a natural lambda structure $\langle \mathcal{D}, F, G \rangle$, then (Galois) implies

$$\text{(func-Galois)} \quad \forall d, e \in \mathcal{K}(\mathcal{D}), x \in \mathcal{D}.\ e \sqsubseteq F(x)(d) \ \Leftrightarrow\ G(d \mapsto e) \sqsubseteq x.$$

The next lemma shows how to build Galois connections in **ALG** from morphisms between NTS^{U}'s.

15.2.4 Lemma *Let $f \in \mathbf{NTS}^{\mathsf{U}}(\mathcal{S}, \mathcal{S}')$. Let $\overline{f}$ stand for $\mathsf{Flt}(f)$. Define $\overline{f}^R$: $\mathbf{ALG}(\mathcal{F}^{\mathcal{S}'}, \mathcal{F}^{\mathcal{S}})$ by*

$$\overline{f}^R(d') \;=\; \{s \mid f(s) \in d'\}.$$

Then $\langle \overline{f}, \overline{f}^R \rangle : \mathcal{F}^{\mathcal{S}} \to \mathcal{F}^{\mathcal{S}'}$ *is a Galois connection (so the notation* $\overline{f}^R$ *is appropriate).*

Proof We leave the proof that $\overline{f}$ is well defined and continuous as an exercise. Note that

$$[\forall s \in \mathcal{S}.\ s \in d \Rightarrow f(s) \in d'] \Leftrightarrow [\forall s' \in \mathcal{S}'.\ (\exists s \in d. f(s) \leq s') \Rightarrow s' \in d']. \tag{15.5}$$

($\Rightarrow$) Suppose that $s' \in \mathcal{S}'$ and there exists $s \in d$ such that $f(s) \leq s'$. Then $f(s) \in d'$ by the left hand side and hence $s' \in d'$ since d is a filter.

($\Leftarrow$) Let $s \in d$. Choosing $s' = f(s)$ in the right hand side, we get $f(s) \in d'$.

Now we prove that (Galois) holds for $\langle \overline{f}, \overline{f}^R \rangle$.

$$\begin{aligned}
\overline{f}(x) \subseteq x' &\Leftrightarrow \{s' \mid \exists s \in x. f(s) \leq s'\} \subseteq x' \\
&\Leftrightarrow \forall s' \in \mathcal{S}'.\ (\exists s \in x. f(s) \leq s') \Rightarrow s' \in x' \\
&\Leftrightarrow \forall s \in \mathcal{S}.\ s \in x \Rightarrow f(s) \in x', \qquad \text{by (15.5)}, \\
&\Leftrightarrow x \subseteq \{s \mid f(s) \in x'\} \\
&\Leftrightarrow x \subseteq \overline{f}^R(x'). \ \blacksquare
\end{aligned}$$

From now on $\underline{f}$ denotes the Galois connection $\langle \overline{f}, \overline{f}^R \rangle$ for $f \in \mathbf{NTS}^{\mathsf{U}}(\mathcal{S}, \mathcal{S}')$.

The next definition is necessary for introducing morphisms between lambda structures. We will explain this choice in Section 15.4.

15.2.5 Definition Let $\mathcal{D} = \langle \mathcal{D}, F, G \rangle$, $\mathcal{D}' = \langle \mathcal{D}', F', G' \rangle$ be lambda structures. A *lambda Galois connection* is a Galois connection $\underline{\mathsf{m}} = \langle \mathsf{m}, \mathsf{m}^R \rangle : \mathcal{D} \to \mathcal{D}'$ such that

$$\begin{aligned}
&\text{(lambda-gc1)} \quad \forall f \in [\mathcal{D} \to \mathcal{D}].\mathsf{m}(G(f)) \sqsubseteq G'(\mathsf{m} \circ f \circ \mathsf{m}^R); \\
&\text{(lambda-gc2)} \quad \forall x' \in \mathcal{D}', x \in \mathcal{D}. F(\mathsf{m}^R(x'))(x) \sqsubseteq \mathsf{m}^R(F'(x')(\mathsf{m}(x))).
\end{aligned}$$

If we write $f^{\mathsf{m}} = \mathsf{m} \circ f \circ \mathsf{m}^R$, then we can reformulate these conditions as

$$\begin{aligned}
&\text{(lambda-gc1)} \quad \forall f \in [\mathcal{D} \to \mathcal{D}].\mathsf{m}(G(f)) \sqsubseteq G'(f^{\mathsf{m}}); \\
&\text{(lambda-gc2)} \quad \forall x' \in \mathcal{D}', x \in \mathcal{D}.\mathsf{m}^R(x') \cdot x \sqsubseteq \mathsf{m}^R(x' \cdot (\mathsf{m}(x))).
\end{aligned}$$

See also Lemma 16.1.12.

15.2.6 Definition

(i) The category **LS** consists of lambda structures as objects and lambda Galois connections as morphisms. The composition between morphisms $\langle \mathsf{m}, \mathsf{m}^R \rangle : \mathcal{D} \to \mathcal{D}'$ and $\langle \mathsf{n}, \mathsf{n}^R \rangle : \mathcal{D}' \to \mathcal{D}''$ is given by $\langle \mathsf{n} \circ \mathsf{m}, \mathsf{m}^R \circ \mathsf{n}^R \rangle$.

(ii) The category of *natural lambda structures*, written **NLS**, is the full subcategory of **LS** which has natural lambda structures as objects.

For $\mathcal{S} \in \mathrm{TS}^{\mathtt{U}}$ the maps $F^{\mathcal{S}} : \mathcal{F}^{\mathcal{S}} \to [\mathcal{F}^{\mathcal{S}} \to \mathcal{F}^{\mathcal{S}}]$ and $G^{\mathcal{S}} : [\mathcal{F}^{\mathcal{S}} \to \mathcal{F}^{\mathcal{S}}] \to \mathcal{F}^{\mathcal{S}}$ are two continuous functions, defined in Definition 13.4.5 as

$$\begin{array}{l}\forall x \in \mathcal{F}^{\mathcal{S}}.\ F^{\mathcal{S}}(x) = \lambda\!\!\lambda y \in \mathcal{S}. \uparrow \{t \mid \exists s \in y. s \to t \in x\};\\ \forall f \in [\mathcal{F}^{\mathcal{S}} \to \mathcal{F}^{\mathcal{S}}].\ G^{\mathcal{S}}(f) = \uparrow \{s \to t \mid t \in f(\uparrow s)\}.\end{array}$$

Also remember that $x \cdot y = F^{\mathcal{S}}(x)(y)$, for any $x, y \in \mathcal{F}^{\mathcal{S}}$.

15.2.7 Lemma *Let $\mathcal{S} \in \mathbf{NTS}^{\mathtt{U}}$, $s, t \in \mathcal{S}$, $x, y \in \mathcal{F}^{\mathcal{S}}$, and $f : \mathcal{F}^{\mathcal{S}} \to \mathcal{F}^{\mathcal{S}}$. Then:*

(i) $x \cdot y = \{t \mid \exists s \in y. s \to t \in x\}$;
(ii) $t \in x \cdot \uparrow s \ \Leftrightarrow\ s \to t \in x$;
(iii) $t \in f(\uparrow s) \ \Rightarrow\ (s \to t) \in G^{\mathcal{S}}(f)$;
(iv) $G^{\mathcal{S}}(\uparrow s \mapsto \uparrow t) = \uparrow (s \to t)$;

Proof (i) Let $\{x \cdot y\} = \{t \mid \exists s \in y. s \to t \in x\}$. Then $\{x \cdot y\} \subseteq x \cdot y$ by definition. We prove that $\{x \cdot y\}$ is a filter, hence it coincides with $x \cdot y$. Now $\mathtt{U} \in \{x \cdot y\}$ by $(\mathtt{U} \to)$.

Moreover, $\{x \cdot y\}$ is upward closed. In fact, let $t \in \{x \cdot y\}$ and $t \leq t'$. Then there exists $s \in y$ such that $(s \to t) \in x$. But $(s \to t) \leq (s \to t')$ by $(\to)$, so this last type is also in the filter x. Therefore $t' \in \{x \cdot y\}$.

Finally, let $t, t' \in \{x \cdot y\}$ and $(s \to t), (s' \to t') \in x$ for some $s, s' \in y$. Then $(s \to t) \cap (s' \to t') \in x$ and $s \cap s' \in y$, as x, y are filters. By Proposition 13.1.22(ii) one has $(s \to t) \cap (s' \to t') \leq (s \cap s') \to (t \cap t')$, hence this last type is in the filter x. Therefore, by definition of application, $t \cap t' \in \{x \cdot y\}$.

(ii) By (i), $t \in x \cdot \uparrow s$ if and only if there exists $s' \in \uparrow s$ such that $(s' \to t) \in x$. By $(\to)$, this is equivalent to $(s \to t) \in x$.

(iii) Easy.

(iv) We have

$$\begin{array}{rcl} G^{\mathcal{S}}(\uparrow s \mapsto \uparrow t) & = & \uparrow \{s' \to t' \mid t' \in (\uparrow s \mapsto \uparrow t)(\uparrow s')\} \\ & = & \uparrow \{s' \to t' \mid [\uparrow s' \supseteq \uparrow s \ \&\ t' \in \uparrow t] \text{ or } t' = \mathtt{U}\} \\ & = & \uparrow (s \to t). \end{array}$$

As to the last equality, $\supseteq$ holds trivially, and $\subseteq$ by $(\to)$, $(\mathtt{U}\to)$. ■

15.2.8 Lemma *Let $\mathcal{S}$ be in $\mathbf{NTS}^{\mathtt{U}}$. Write $\mathsf{Flt}_{\mathbf{NTS}^{\mathtt{U}}}(\mathcal{S}) = \langle \mathcal{F}^{\mathcal{S}}, F^{\mathcal{S}}, G^{\mathcal{S}} \rangle$. Then*

(i) $\mathsf{Flt}_{\mathbf{NTS}^{\mathtt{U}}}(\mathcal{S}) \in \mathbf{LS}$;
(ii) $\mathsf{Flt}_{\mathbf{NTS}^{\mathtt{U}}}(\mathcal{S}) \in \mathbf{NLS}$.

Proof (i) Easy.

(ii) We claim that $\langle F^{\mathcal{S}}, G^{\mathcal{S}}\rangle$ is a Galois connection. To show $F^{\mathcal{S}}(G^{\mathcal{S}}(f)) \sqsupseteq f$, it suffices to prove this on compact elements $\uparrow s \in \mathcal{K}(\mathcal{S})$. Let $f \in [\mathcal{S} \to \mathcal{S}]$, $s \in \mathcal{S}$. We have

$$\begin{array}{rcll} F^{\mathcal{S}}(G^{\mathcal{S}}(f))(\uparrow s) & = & \{t \mid s \to t \in G^{\mathcal{S}}(f)\}, & \text{by Lemma 15.2.7((ii))}, \\ & \supseteq & \{t \mid t \in f(\uparrow s)\}, & \text{by Lemma 15.2.7(iii)}, \\ & = & f(\uparrow s). & \end{array}$$

On the other hand, let $x \in \mathcal{F}^{\mathcal{S}}$. We have

$$\begin{array}{rcll} G^{\mathcal{S}}(F^{\mathcal{S}}(x)) & = & \uparrow \{s \to t \mid t \in x \cdot \uparrow s\} & \\ & = & \uparrow \{s \to t \mid s \to t \in x\}, & \text{by Lemma 15.2.7((ii))}, \\ & \subseteq & x. \blacksquare & \end{array}$$

15.2.9 Theorem *We define the action of* $\mathsf{Flt}_{\mathbf{NTS}^{\mathtt{U}}}$ *on morphisms* $f \in \mathbf{NTS}^{\mathtt{U}}(\mathcal{S}, \mathcal{S}')$ *by*

$$\mathsf{Flt}_{\mathbf{NTS}^{\mathtt{U}}}(f) = \langle \mathsf{Flt}(f), \mathsf{Flt}(f)^R\rangle.$$

Then $\mathsf{Flt}_{\mathbf{NTS}^{\mathtt{U}}} : \mathbf{NTS}^{\mathtt{U}} \to \mathbf{NLS}$ *is a functor.*

Proof The proof will emerge from the results of Sections 15.3 and 15.4 (see in particular Proposition 15.4.27).

15.2.10 Definition

(i) Given a natural lambda structure $\mathcal{D}$, we define

$$\mathsf{Flt}_{\mathbf{NLS}}(\mathcal{D}) = \langle \mathcal{K}(\mathcal{D}), \leq, \cap, \to_{\mathcal{D}}, \mathtt{U}\rangle$$

with $a \leq b \;\Leftrightarrow\; b \sqsubseteq_{\mathcal{K}(\mathcal{D})} a$, $a \cap b \equiv a \sqcup_{\mathcal{D}} b$, $\mathtt{U} = \bot_{\mathcal{D}}$, and $a \to_{\mathcal{K}(\mathcal{D})} b \equiv G(a \mapsto b)$.

(ii) Given $\underline{\mathsf{m}} \in \mathbf{NLS}(\mathcal{D}, \mathcal{D}')$, we define

$$\mathsf{Flt}_{\mathbf{NLS}}(\underline{\mathsf{m}}) = \mathsf{m} \restriction \mathcal{K}(\mathcal{D}) : \mathcal{K}(\mathcal{D}) \to \mathcal{K}(\mathcal{D}').$$

15.2.11 Lemma $\mathsf{Flt}_{\mathbf{NLS}}$ *is a functor from* $\mathbf{NLS}$ *to* $\mathbf{NTS}^{\mathtt{U}}$.

Proof The proof will follow from the results of Sections 15.3 and 15.4. $\blacksquare$

Now we will prove that $\mathbf{NTS}^{\mathtt{U}}$ and $\mathbf{NLS}$ are equivalent categories.

15.2.12 Proposition *Let* $\mathcal{S} \in \mathbf{NTS}^{\mathtt{U}}$. *Define* $\sigma_{\mathcal{S}} : S \to \mathcal{K}(\mathcal{F}^{\mathcal{S}})$ *by*

$$\sigma_{\mathcal{S}}(s) = \uparrow s.$$

Then $\sigma_{\mathcal{S}}$ *is an isomorphism in* $\mathbf{NTS}^{\mathtt{U}}$. *Hence* $\mathcal{S} \cong \mathcal{K}(\mathcal{F}^{\mathcal{S}})$.

Proof For $\sigma_{\mathcal{S}}$ we simply write σ. We know from Proposition 15.1.22 that σ is an isomorphism of meet semi-lattices with universe, so it preserves intersections and top elements, and so does its inverse σ^{-1}. We will prove that $\sigma(s{\to}t) = \sigma(s) \to_{\mathcal{K}(\mathcal{F}^{\mathcal{S}})} \sigma(t)$, for any $s, t \in \mathcal{S}$. We have

$$\begin{aligned}
\sigma(s) \to_{\mathcal{K}(\mathcal{F}^{\mathcal{S}})} \sigma(t) &= \uparrow s \to_{\mathcal{K}(\mathcal{F}^{\mathcal{S}})} \uparrow t \\
&= G^{\mathcal{S}}(\uparrow s \mapsto \uparrow t), && \text{by Definition 15.2.10,} \\
&= \uparrow (s \to t), && \text{by Lemma 15.2.8(i),} \\
&= \sigma(s{\to}t).
\end{aligned}$$

Similarly, $\sigma^{-1}(\uparrow s \to_{\mathcal{K}(\mathcal{F}^{\mathcal{S}})} \uparrow t) = s{\to}t$. ∎

15.2.13 Proposition *Let $\mathcal{D} \in$ **NLS**. Define $\tau_{\mathcal{D}} : \mathcal{F}^{\mathcal{K}(\mathcal{D})} {\to} \mathcal{D}$ by*

$$\tau_{\mathcal{D}}(x) = \bigsqcup x.$$

Then $\tau_{\mathcal{D}}$ is an isomorphism in **NLS**. *Hence $\mathcal{D} \cong \mathcal{F}^{\mathcal{K}(\mathcal{D})}$.*

Proof Write $\tau = \tau_{\mathcal{D}}$. By Proposition 15.1.22 τ is an isomorphism of lattices. The pair $\langle \tau, \tau^{-1} \rangle$ is a Galois connection. We show that

$$\langle \tau, \tau^{-1} \rangle : \langle \mathcal{F}^{\mathcal{K}(\mathcal{D})}, F^{\mathcal{K}(\mathcal{D})}, G^{\mathcal{K}(\mathcal{D})} \rangle \to \langle \mathcal{D}, F, G \rangle$$

is lambda. Let $f : \mathcal{F}^{\mathcal{K}(\mathcal{D})} \to \mathcal{F}^{\mathcal{K}(\mathcal{D})}$ be a continuous function. We have:

$$\begin{aligned}
\tau(G^{\mathcal{K}(\mathcal{D})}(f)) &= \tau(\uparrow \{a \to_{\mathcal{K}(\mathcal{D})} b \mid b \in f(\uparrow a)\}) \\
&= \bigsqcup\{a \to_{\mathcal{K}(\mathcal{D})} b \mid b \in f(\uparrow a)\}, \\
&\quad \text{since } \bigsqcup \uparrow x = \bigsqcup x, \text{ for any } x \subseteq \mathcal{K}(\mathcal{D}), \\
&= \bigsqcup\{G(a \mapsto b \mid b \in f(\uparrow a)\}), \text{ by definition of } \to_{\mathcal{K}(\mathcal{D})}, \\
&= \bigsqcup\{G(a \mapsto b) \mid b \sqsubseteq \bigsqcup(f(\uparrow a))\} \\
&= \bigsqcup\{G(a \mapsto b) \mid b \sqsubseteq \tau \circ f \circ \tau^{-1}(a)\} \\
&= G(\bigsqcup\{a \mapsto b \mid b \sqsubseteq \tau \circ f \circ \tau^{-1}(a)\}), \\
&\quad \text{since } G \text{ is additive by Lemma 15.2.2((i)),} \\
&= G(\tau \circ f \circ \tau^{-1}).
\end{aligned}$$

The above shows in particular that $\langle \tau, \tau^{-1} \rangle$ satisfies (lambda-gc1) in Definition 15.2.5. We now prove (lambda-gc2). Of course it is sufficient to reason on compact elements, so let $a, b \in \mathsf{K}(\mathcal{D})$. Taking $x' = a$ and $x = \uparrow b$ into (lambda-gc2), we have to prove $F^{\mathcal{K}(\mathcal{D})}(\tau^{-1}(a))(\uparrow b) \sqsubseteq \tau^{-1}(F(a)(\tau(\uparrow b))$; that is,

$$F^{\mathcal{K}(\mathcal{D})}(\uparrow a)(\uparrow b) \sqsubseteq \tau^{-1}(F(a)(b)).$$

We have

$$
\begin{array}{rcll}
F^{\mathcal{K}(\mathcal{D})}(\uparrow a)(\uparrow b) & = & \{t \in \mathcal{K}(\mathcal{D}) \mid b \rightarrow_{\mathcal{K}(\mathcal{F}^{\mathcal{S}})} t \in \uparrow a\}, & \text{by Lemma 15.2.7((ii))}, \\
& = & \{t \in \mathcal{K}(\mathcal{D}) \mid G(b \mapsto t) \sqsubseteq a\}, & \text{by Definition 15.2.10}, \\
& = & \{t \in \mathcal{K}(\mathcal{D}) \mid b \mapsto t \sqsubseteq F(a)\}, & \text{since } \langle F, G\rangle \text{ is a Galois connection}, \\
& = & \{t \in \mathcal{K}(\mathcal{D}) \mid t \sqsubseteq F(a)(b)\} & \\
& = & \tau^{-1}(F(a)(b)). &
\end{array}
$$

As a consequence of the above, we have that $\langle \tau, \tau^{-1}\rangle$ satisfies (lambda-gc2). Similarly, $\langle \tau^{-1}, \tau\rangle$ is also a lambda Galois connection from $\mathcal{D}$ to $\mathcal{F}^{\mathcal{K}(\mathcal{D})}$, and it is of course the inverse of $\langle \tau, \tau^{-1}\rangle$. ∎

15.2.14 Theorem *The categories* $\mathbf{NTS^U}$ *and* **NLS** *are equivalent.*

Proof This follows from Propositions 15.2.12 and 15.2.13 almost in the same way as Corollary 15.1.23 follows from Proposition 15.1.22. There is one extra case. If $\langle \mathsf{m}, \mathsf{m}^R\rangle : \mathcal{D} \rightarrow \mathcal{D}'$, then we must show the commutativity of the following diagram

$$
\begin{array}{ccc}
\mathcal{D} & \xleftarrow{\;\bigsqcup\;} & \mathcal{F}^{\mathcal{K}(\mathcal{D})} \\
\uparrow {\scriptstyle \mathsf{m}^R} & & \uparrow {\scriptstyle \overline{\mathsf{m}\restriction\mathcal{K}(\mathcal{D})}^R} \\
\mathcal{D}' & \xleftarrow{\;\bigsqcup\;} & \mathcal{F}^{\mathcal{K}(\mathcal{D}')}
\end{array}
$$

This is done in Exercise 15.11. ∎

15.3 Type and zip structures

The aim of this section and the next one is to compare type and lambda structures, using an intermediate kind of structure, namely *zip structures.* A zip structure is a pair $\langle \mathcal{D}, Z\rangle$, where D is an object in **ALG** and $Z : \mathcal{K}(\mathcal{D}) \times \mathcal{K}(\mathcal{D}) \rightarrow \mathcal{K}(\mathcal{D})$ is the semantic counterpart of the arrow constructor in type structures; it is a set-theoretic function that 'zips' the information of two compact elements, not necessarily in such a way that the constituents can be retrieved. The various categories of zip structures are easily proven to be equivalent to the corresponding ones of type structures. So we can think of zip structures as an alternative way of describing type structures. We do not fully succeed in showing $\mathrm{TS}^U \cong \mathrm{LS}$, see Figure 15.3, because $\mathrm{ZS} \ncong \mathrm{LS}$, but we get a good 'approximation'.

15.3.1 Definition

(i) A *zip structure* is a pair $\langle \mathcal{D}, Z \rangle$ with $\mathcal{D} \in \mathbf{ALG}$ and

$$Z : \mathcal{K}(\mathcal{D}) \times \mathcal{K}(\mathcal{D}) \to \mathcal{K}(\mathcal{D})$$

an arbitrary map.

(ii) The category **ZS** has zip structures as objects. Its morphisms, $\mathbf{ZS}(\langle \mathcal{D}, Z \rangle, \langle \mathcal{D}', Z' \rangle)$, are maps $f : \mathcal{D} \to \mathcal{D}'$ such that

$$\begin{array}{ll} \text{(cmp-pres)} & \forall a. f(a) \in \mathcal{K}(\mathcal{D}'); \\ \text{(add)} & \forall X \subseteq \mathcal{D}. f(\bigsqcup X) = \bigsqcup f(X); \\ (Z\text{-mon}) & \forall a, b. f(Z(a,b)) \sqsubseteq Z'(f(a), f(b)); \end{array}$$

here $a, b, c, \ldots$ range over $\mathcal{K}(\mathcal{D})$. The second requirement implies that a morphism f is continuous (only required to preserve sups for directed sets X) and strict, i.e. $f(\bot_{\mathcal{D}}) = \bot_{\mathcal{D}'}$.

We now restrict this general framework to special zip structures.

15.3.2 Definition Let $\langle \mathcal{D}, Z \rangle$ be a zip structure.

(i) Then $\langle \mathcal{D}, Z \rangle$ is a *lazy zip structure* if the following hold.

(1) (Z-contr) $a \sqsubseteq a'$ & $b' \sqsubseteq b \Rightarrow Z(a', b') \sqsubseteq Z(a, b)$;
(2) (Z-add) $Z(a, b_1 \sqcup b_2) = Z(a, b_1) \sqcup Z(a, b_2)$;
(3) (Z-lazy) $Z(\bot_{\mathcal{D}}, \bot_{\mathcal{D}}) \sqsubseteq Z(a, b)$.

(ii) **LZS** is the full subcategory of **ZS** consisting of lazy zip structures.

15.3.3 Definition Let $\langle \mathcal{D}, Z \rangle \in \mathbf{ZS}$.

(i) Then $\langle \mathcal{D}, Z \rangle$ is a *natural zip structure* if $\langle \mathcal{D}, Z \rangle \in \mathbf{LZS}$ and moreover

$$(Z\text{-bot}) \quad Z(\bot_{\mathcal{D}}, \bot_{\mathcal{D}}) = \bot_{\mathcal{D}}.$$

(ii) **NZS** is the full subcategory of **LZS** consisting of natural zip structures.

15.3.4 Remark Since condition (Z-bot) is stronger than (Z-lazy), $\mathcal{D}$ is natural if it satisfies (Z-contr), (Z-add) and (Z-bot). In fact, (Z-bot) corresponds to $(\mathtt{U}{\rightarrow})$, and ($Z$-lazy) to the weaker $(\mathtt{U}_{\text{lazy}})$.

Strict zip structures

15.3.5 Definition Let $\mathcal{D} \in \mathbf{ALG}$. *Recall* that $\mathcal{K}^s(\mathcal{D}) = \mathcal{K}(\mathcal{D}) - \{\bot_{\mathcal{D}}\}$.

(i) A *strict zip structure* is of the form $\langle \mathcal{D}, Z \rangle$, with

$$Z : (\mathcal{K}^s(\mathcal{D}) \times \mathcal{K}^s(\mathcal{D})) \to \mathcal{K}^s(\mathcal{D}).$$

(ii) If we write $Z(a, b)$, then it is always understood that $(a, b) \in \text{dom}(Z)$.

(iii) The category $\mathbf{ZS}^s$ consists of strict zip structures as objects. Its morphisms are maps f satisfying

$$\begin{array}{ll} (\text{cmp-pres}^{\text{s}}) & \forall a \in \mathcal{K}^s(\mathcal{D}).f(a) \in \mathcal{K}^{\text{s}}(\mathcal{D}'); \\ (\text{add}) & \forall X \subseteq \mathcal{D}.f(\bigsqcup X) = \bigsqcup f(X); \\ (Z\text{-mon}) & \forall a,b.f(Z(a,b)) \sqsubseteq Z'(f(a),f(b)). \end{array}$$

15.3.6 Definition

(i) Let $\langle \mathcal{D}, Z\rangle \in \mathbf{ZS}^s$. Then $\langle \mathcal{D}, Z\rangle$ is called a *proper strict zip structure*, if it satisfies:

$$\begin{array}{ll} (Z\text{-contr}) & a \sqsubseteq a' \ \& \ b' \sqsubseteq b \ \Rightarrow \ Z(a',b') \sqsubseteq Z(a,b); \\ (Z\text{-add}) & Z(a, b_1 \sqcup b_2) = Z(a,b_1) \sqcup Z(a,b_2). \end{array}$$

(ii) $\mathbf{PZS}^s$ is the full subcategory of $\mathbf{ZS}^s$ consisting of proper strict zip structures.

15.3.7 Remark In Section 13.3 we introduced the names $\text{MSL}^{\text{U}}, \text{MSL}^{\text{-U}}$, TS^{U}, $\text{TS}^{\text{-U}}$ for collections of meet semi-lattices and type structures. The superscript U in these names denoted the fact that in the corresponding structures the top elements are distinguished points.

In Definitions 15.1.12, 15.3.1 and 15.3.5, we introduced $\mathbf{ALG}_a$, $\mathbf{ALG}_a^s$, $\mathbf{ZS}$, $\mathbf{ZS}^s$. The superscript s, to be read as 'strict', concerns the top element of $\mathcal{K}(\mathcal{D})_{\leq}$ (see Definition 15.1.2).

Equivalences between type and zip structures

Now we will extend the equivalences of Section 15.1 to the various categories of type and zip structures. We will show the following equivalences of categories:

$$\begin{aligned} \mathbf{TS}^{\text{U}} &\cong \mathbf{ZS}; \\ \mathbf{LTS}^{\text{U}} &\cong \mathbf{LZS}; \\ \mathbf{NTS}^{\text{U}} &\cong \mathbf{NZS}; \\ \mathbf{TS}^{\text{-U}} &\cong \mathbf{ZS}^s; \\ \mathbf{PTS}^{\text{-U}} &\cong \mathbf{PZS}^s. \end{aligned}$$

Note that under this correspondence there is a sort of adjunction in the superscripts of the names due to the fact that in the left hand side of this list the presence of a top is given explicitly, whereas in the right hand side the name indicates when we are *not* interested in the top (of $\mathcal{K}(\mathcal{D})_{\leq}$). In particular, note that $\mathbf{TS}$ does not correspond with $\mathbf{ZS}$.

Since the proofs are standard, we will give the full details only for the equivalence between $\mathbf{TS}^{\mathtt{U}}$ and $\mathbf{ZS}$, whilst the other cases are left to Exercises 15.15 and 15.16.

Equivalence between $\mathbf{TS}^{\mathtt{U}}$ and $\mathbf{ZS}$

First we see how the functors Flt and Cmp between $\mathbf{MSL}^{\mathtt{U}}$ and $\mathbf{ALG}_a$ induce new functors between the richer categories $\mathbf{TS}^{\mathtt{U}}$ and $\mathbf{ZS}$.

15.3.8 Definition

(i) For $\mathcal{S} \in \mathbf{TS}^{\mathtt{U}}$, define $\mathcal{Q}(\mathcal{S}) \in \mathbf{ZS}$ by

$$\mathcal{Q}(\mathcal{S}) = \langle \mathcal{F}^{\mathcal{S}}, Z^{\mathcal{S}} \rangle,$$

with $Z^{\mathcal{S}} : \mathcal{K}(\mathcal{F}^{\mathcal{S}}) \times \mathcal{K}(\mathcal{F}^{\mathcal{S}}) \to \mathcal{K}(\mathcal{F}^{\mathcal{S}})$ defined by

$$Z^{\mathcal{S}}(\uparrow s, \uparrow t) \ = \uparrow (s \to t).$$

(ii) For $\langle \mathcal{D}, Z \rangle \in \mathbf{ZS}$, define $\mathcal{M}(\langle \mathcal{D}, Z \rangle) \in \mathbf{TS}^{\mathtt{U}}$ by

$$\mathcal{M}(\langle \mathcal{D}, Z \rangle) = \langle \mathcal{K}(\mathcal{D}), \leq, \cap, \to_Z, \mathtt{U} \rangle,$$

with $a \leq b \ \Leftrightarrow \ b \sqsubseteq_{\mathcal{D}} a$, $a \cap b \equiv a \sqcup_{\mathcal{D}} b$, and $\mathtt{U} \equiv \bot_{\mathcal{D}}$, as in Definition 15.1.20, and

$$a {\to_Z} b \equiv Z(a, b).$$

(iii) The actions of the maps $\mathcal{M} : \mathbf{ZS} {\to} \mathbf{TS}^{\mathtt{U}}$ and $\mathcal{Q} : \mathbf{TS}^{\mathtt{U}} {\to} \mathbf{ZS}$ on morphisms are defined as follows. Given $f \in \mathbf{TS}^{\mathtt{U}}(\mathcal{S}, \mathcal{S}')$, $X \in \mathcal{F}^{\mathcal{S}}$ and $g \in \mathbf{ZS}(\langle \mathcal{D}, Z \rangle, \langle \mathcal{D}', Z' \rangle)$ define

$$\begin{aligned} \mathcal{Q}(f) &= \lambda X.\{t \mid \exists s \in X. f(s) \leq t\}; \\ \mathcal{M}(g) &= g \restriction \mathcal{K}(\mathcal{D}). \end{aligned}$$

15.3.9 Lemma *If $\mathcal{S} \in \mathbf{TS}^{\mathtt{U}}$ and $Z^{\mathcal{S}}$ are defined as in Definition* 15.3.8(i), *then $\langle \mathcal{K}(\mathcal{F}^{\mathcal{S}}), Z^{\mathcal{S}} \rangle \in \mathbf{ZS}$. Moreover if $\to_{Z^{\mathcal{S}}}$ is defined for $\langle \mathcal{K}(\mathcal{F}^{\mathcal{S}}), Z^{\mathcal{S}} \rangle \in \mathbf{ZS}$ as in Definition* 15.3.8(ii)*, then*

$$\uparrow s {\to_{Z^{\mathcal{S}}}} \uparrow t = \uparrow (s {\to} t).$$

Proof Immediate from Definition 15.3.8. ∎

15.3.10 Proposition *$\mathcal{M}$ is a functor from* $\mathbf{ZS}$ *to* $\mathbf{TS}^{\mathtt{U}}$.

Proof We just have to prove that $\mathcal{M}$ transforms a morphism in $\mathbf{ZS}$ into a morphism in $\mathbf{TS}^{\mathtt{U}}$. Let $\mathsf{m} : \langle \mathcal{D}, Z \rangle \to \langle \mathcal{D}', Z' \rangle$ be a morphism. By Lemma

15.1.21 we only need to show that $\mathcal{M}(\mathsf{m})$ satisfies condition $(m3)$ of Definition 13.3.10. We have

$$\begin{array}{rcll}
\mathcal{M}(\mathsf{m})(a \rightarrow_Z b) & = & \mathsf{m}(a \rightarrow_Z b), & \text{by definition of } \mathcal{M}(\mathsf{m}), \\
 & = & \mathsf{m}(Z(a,b)), & \text{by definition of } \rightarrow_Z, \\
 & \geq & Z'(\mathsf{m}(a),\mathsf{m}(b)), & \text{since } \mathsf{m} \text{ satisfies } (Z\text{-mon}), \\
 & = & \mathsf{m}(a) \rightarrow_{Z'} \mathsf{m}(b)), & \\
 & = & \mathcal{M}(\mathsf{m}(a)) \rightarrow_{Z'} \mathcal{M}(\mathsf{m}(b)), & \\
 & = & \mathcal{M}(\mathsf{m})(a) \rightarrow_{Z'} \mathcal{M}(\mathsf{m})(b). \blacksquare &
\end{array}$$

15.3.11 Proposition *$\mathcal{Q}$ is a functor from* $\mathbf{TS}^{\mathtt{U}}$ *to* $\mathbf{ZS}$.

Proof We have to prove that $\mathcal{Q}$ transforms a morphism in $\mathbf{TS}^{\mathtt{U}}$ into a morphism in $\mathbf{ZS}$. Let $f \in \mathbf{TS}^{\mathtt{U}}(\mathcal{S},\mathcal{S}')$ with arrows $\rightarrow_{Z^{\mathcal{S}}}$ and $\rightarrow_{Z^{\mathcal{S}'}}$ corresponding to $Z^{\mathcal{S}}$ and $Z^{\mathcal{S}'}$, respectively. By Proposition 15.1.19 we need only show that $\mathcal{Q}(f)$ satisfies (Z-mon). Indeed,

$$\begin{array}{rcll}
\mathcal{Q}(f)(Z^{\mathcal{S}}(\uparrow s, \uparrow t)) & = & \mathcal{Q}(f)(\uparrow (s \rightarrow_{Z^{\mathcal{S}}} t), & \text{by definition of } Z^{\mathcal{S}}, \\
 & = & \uparrow f(s \rightarrow_{Z^{\mathcal{S}}} t), & \text{by Lemma 15.1.18}, \\
 & \subseteq & \uparrow (f(s) \rightarrow_{Z^{\mathcal{S}'}} f(t)), & \text{since } f \in \mathbf{TS}^{\mathtt{U}}(\mathcal{S},\mathcal{S}'), \\
 & = & Z^{\mathcal{S}'}(\uparrow f(s), \uparrow f(t)), & \text{by definition of } Z^{\mathcal{S}'}, \\
 & = & Z^{\mathcal{S}'}(\mathcal{Q}(f)(\uparrow s), \mathcal{Q}(f)(\uparrow t)), & \text{by Lemma 15.1.18. } \blacksquare
\end{array}$$

Next we will prove that $\mathbf{ZS}$ and $\mathbf{TS}^{\mathtt{U}}$ are equivalent. To this end we show that natural isomorphisms $\mathsf{Id}_{\mathbf{TS}^{\mathtt{U}}} \simeq \mathcal{M} \circ \mathcal{Q}$ and $\mathcal{Q} \circ \mathcal{M} \simeq \mathsf{Id}_{\mathbf{ZS}}$ are given by σ and τ respectively, exactly as in the case of the equivalence between the categories $\mathbf{MSL}^{\mathtt{U}}$ and $\mathbf{ALG}_a$.

15.3.12 Proposition *Let $\mathcal{S}$ be a top type structure. Let $\sigma : \mathcal{S} \to \mathcal{K}(\mathcal{F}^{\mathcal{S}})$ be the map such that $\sigma(s) = \uparrow s$. Then σ is an isomorphism in* $\mathbf{TS}^{\mathtt{U}}$.

Proof By Proposition 15.1.22(i) we need only show that σ and σ^{-1} commute with arrows. Now σ is a bijection, hence the following suffices:

$$\begin{array}{rcll}
\sigma(s \rightarrow t) & = & \uparrow (s \rightarrow t), & \\
 & = & \uparrow s \rightarrow_{Z^{\mathcal{S}}} \uparrow t, & \text{by Lemma 15.3.9}, \\
 & = & \sigma(s) \rightarrow_{Z^{\mathcal{S}}} \sigma(t). \blacksquare &
\end{array}$$

15.3.13 Proposition *Let $\langle \mathcal{D}, Z \rangle \in$* $\mathbf{ZS}$. *Define $\tau : \mathcal{F}^{\mathcal{K}(\mathcal{D})} \to \mathcal{D}$ by*

$$\tau(x) = \bigsqcup x, \quad \textit{where the sup is taken in } \mathcal{D}.$$

Then τ is an isomorphism in $\mathbf{ZS}$.

Proof By Proposition 15.1.22(ii) we need only show that τ and τ^{-1} satisfy (Z-comm). For τ, we have

$$\begin{array}{rcll} \tau(Z^{\mathcal{K}(\mathcal{D})}(\uparrow a, \uparrow b)) & = & \tau(\uparrow (a{\rightarrow}_Z b)), & \text{by definition of } Z^{\mathcal{K}(\mathcal{D})}, \\ & = & \tau(\uparrow Z(a,b)), & \text{by definition of } \rightarrow_Z, \\ & = & \bigsqcup(\uparrow Z(a,b)) & \\ & = & Z(a,b) & \\ & = & Z(\bigsqcup \uparrow a, \bigsqcup \uparrow b) & \\ & = & Z(\tau(\uparrow a), \tau(\uparrow b)). & \end{array}$$

As to τ^{-1}, we must show

$$\tau^{-1}(Z(a,b)) = Z^{\mathcal{K}(\mathcal{D})}(\tau^{-1}(a), \tau^{-1}(b)).$$

This follows by applying τ^{-1} to both sides of the equality above, using the fact that it is 1–1. ∎

15.3.14 Theorem *The categories* $\mathbf{TS}^{\mathrm{U}}$ *and* **ZS** *are equivalent.*

Proof As in Corollary 15.1.23, using Propositions 15.3.10-15.3.13. ∎

15.4 Zip and lambda structures

In this section we do not proceed to a direct comparison between type and lambda structures, but put zip structures to work and compare them with lambda structures. We will see that there is no categorical equivalence between zip and lambda structures in general but how the correspondence is perfect in the lazy, natural, and proper cases. Then, the relation between type and lambda structures will be a consequence of combining the results of this section and the previous one. We gain clarity when comparing lambda structures with the intermediate zip structures since both have some components in common such as the ω-algebraic lattices and their compact elements. We also avoid the confusion which arises with the reversed order and the use of filters when passing from type to lambda structures or vice versa.

Justifying morphisms in LS

This subsection justifies the choice of morphisms between lambda structures. We anticipated their definitions in the previous section, and we now make that choice concrete.

15.4.1 Definition Let $\langle \mathcal{D}, Z\rangle \in \mathbf{ZS}$.

(i) Define $\Theta_{\mathcal{D}_Z}(x,y) = \{b \mid \exists a \sqsubseteq y. Z(a,b) \sqsubseteq x\}$, for $x, y \in \mathcal{D}$.

(ii) $\mathsf{F}_{\mathbf{ZS}}(\langle \mathcal{D}, Z\rangle)$ is defined as the lambda structure $\langle \mathcal{D}, F_Z, G_Z\rangle$ with the two continuous functions $F_Z : \mathcal{D} \to [\mathcal{D} \to \mathcal{D}]$ and $G_Z : [\mathcal{D} \to \mathcal{D}] \to \mathcal{D}$ defined by

$$\begin{aligned} F_Z(x) &= \lambda y.\bigsqcup \Theta_{\mathcal{D}_Z}(x, y); \\ G_Z(f) &= \bigsqcup\{Z(a, b) \mid b \sqsubseteq f(a)\}. \end{aligned}$$

(iii) Moreover define $\cdot_Z : \mathcal{D}^2 \to \mathcal{D}$, by

$$x \cdot_Z y = F_Z(x)(y).$$

15.4.2 Proposition *Let* $\mathsf{m} \in \mathbf{ZS}(\langle \mathcal{D}, Z\rangle, \langle \mathcal{D}', Z'\rangle)$. *Define* m^R, *the right adjoint of* m *by* $\mathsf{m}^R(x') = \bigsqcup\{x \mid \mathsf{m}(x) \sqsubseteq x'\}$. *Then, for all* $f : \mathcal{D} \to \mathcal{D}$, *and* $x \in \mathcal{D}$, $x' \in \mathcal{D}'$, *the pair* $\underline{\mathsf{m}} = \langle \mathsf{m}, \mathsf{m}^R\rangle$ *satisfies the following properties:*

$$\begin{aligned} \mathsf{m}(G_Z(f)) &\sqsubseteq G'_{Z'}(\mathsf{m} \circ f \circ \mathsf{m}^R) \\ F_Z(\mathsf{m}^R(x'))(x) &\sqsubseteq \mathsf{m}^R(F'_{Z'}(x')(\mathsf{m}(x))). \end{aligned}$$

Proof In order to simplify notation, we omit all the subscripts. We have

$$\begin{aligned} \mathsf{m}(G(f)) &= \mathsf{m}(\bigsqcup\{Z(a,b) \mid b \sqsubseteq f(a)\} \\ &= \bigsqcup\{\mathsf{m}(Z(a,b)) \mid b \sqsubseteq f(a)\}, && \text{by 15.2.2((i))}, \\ &\sqsubseteq \bigsqcup\{Z'(\mathsf{m}(a), \mathsf{m}(b)) \mid b \sqsubseteq f(a)\}, && \text{by } (Z\text{-mon}), \\ &\sqsubseteq \bigsqcup\{Z'(a', b') \mid \exists a, b.\mathsf{m}(a) \sqsubseteq a' \ \&\ b' \sqsubseteq \mathsf{m}(b) \ \&\ b \sqsubseteq f(a)\} \\ &= \bigsqcup\{Z'(a', b') \mid \exists a.\mathsf{m}(a) \sqsubseteq a' \ \&\ b' \sqsubseteq \mathsf{m}(f(a))\}, && \text{since } f \text{ is continuous}, \\ &= \bigsqcup\{Z'(a', b') \mid b' \sqsubseteq \mathsf{m}(f(\mathsf{m}^R(a')))\}, && \text{by (Galois)}, \\ &= G'(\mathsf{m} \circ f \circ \mathsf{m}^R). \end{aligned}$$

$$\begin{aligned} F(\mathsf{m}^R(x'))(x) &= \bigsqcup\{b \mid \exists a \sqsubseteq x.Z(a,b) \sqsubseteq \mathsf{m}^R(x')\} \\ &\sqsubseteq \bigsqcup\{b \mid \exists a \sqsubseteq x.\mathsf{m}(Z(a,b)) \sqsubseteq x'\}, && \text{by (Galois-2)}, \\ &\sqsubseteq \bigsqcup\{b \mid \exists a \sqsubseteq x.Z'(\mathsf{m}(a), \mathsf{m}(b)) \sqsubseteq x'\}, && \text{by } (Z\text{-mon}), \\ &\sqsubseteq \bigsqcup\{b \mid \exists a' \sqsubseteq \mathsf{m}(x).Z'(a', \mathsf{m}(b)) \sqsubseteq x'\} \\ &\sqsubseteq \bigsqcup\{b \mid \mathsf{m}(b) \sqsubseteq \bigsqcup\{b' \mid \exists a' \sqsubseteq \mathsf{m}(x).Z'(a', b') \sqsubseteq x'\}\} \\ &= \mathsf{m}^R(\bigsqcup\{b' \mid \exists a' \sqsubseteq \mathsf{m}(x).Z'(a', b') \sqsubseteq x'\}), && \text{see Exercise 15.8}, \\ &= \mathsf{m}^R(F'(x')(\mathsf{m}(x))). \ \blacksquare \end{aligned}$$

Recall that the category **LS** was defined in Definition 15.2.6, using Definition 15.2.5. Note that the conditions (lambda-gc1) and (lambda-gc2) of Definition 15.2.5 appear in the statement of the previous Proposition 15.4.2. As a consequence, it is immediate that $\mathsf{F}_{\mathbf{ZS}}$ can be extended to a functor.

15.4.3 Proposition *Given* $\mathsf{m} \in \mathbf{ZS}(\langle \mathcal{D}, Z \rangle, \langle \mathcal{D}', Z' \rangle)$, *define* $\mathsf{F_{ZS}}(\mathsf{m}) = \underline{\mathsf{m}}$. *Then* $\mathsf{F_{ZS}} : \mathbf{ZS} \to \mathbf{LS}$ *is a functor.*

Proof We need only show that $\mathsf{F_{ZS}}(\mathsf{m}) = \underline{\mathsf{m}}$ is in $\mathbf{LS}(\mathsf{F_{ZS}}(\mathcal{D}), \mathsf{F_{ZS}}(\mathcal{D}'))$, when $\mathsf{m} \in \mathbf{ZS}(\mathcal{D}, \mathcal{D}')$. This follows from Proposition 15.4.2. ∎

As a consequence of this proposition, the fundamental operation $\mathcal{A} = \mathsf{F_{ZS}} \circ \mathcal{Q}$ is actually a functor from $\mathbf{TS}^{\mathrm{U}}$ to $\mathbf{LS}$ (see Theorem 15.4.27). Our notion of morphism between lambda structures guarantees the functoriality of $\mathcal{A}$. In Plotkin (1993) a different definition of morphism between lambda structures is given: a morphism $\mathsf{m} : \langle \mathcal{D}, F, G \rangle \to \langle \mathcal{D}', F', G' \rangle$ between lambda structures consists of a pair of continuous functions, $\mathsf{m} : \mathcal{D} \to \mathcal{D}'$ and $\mathsf{n} : \mathcal{D}' \to \mathcal{D}$ such that

$$\begin{array}{lcl} \mathsf{m} \circ G & = & G' \circ (\mathsf{n} \to \mathsf{m}) \\ (\mathsf{m} \to \mathsf{n}) \circ F' & = & F \circ \mathsf{n} \end{array} \tag{15.6}$$

where $\mathsf{n} \to \mathsf{m} : [\mathcal{D} \to \mathcal{D}] \to [\mathcal{D}' \to \mathcal{D}']$ is defined, for every $f : \mathcal{D} \to \mathcal{D}$, by $(\mathsf{n} \to \mathsf{m})(f) = \mathsf{m} \circ f \circ \mathsf{n} : \mathcal{D}' \to \mathcal{D}'$ (and $\mathsf{n} \circ \mathsf{m} : [\mathcal{D}' \to \mathcal{D}'] \to [\mathcal{D} \to \mathcal{D}]$ is defined similarly). This choice lets us look at lambda structures as *dialgebras*, but does not guarantee functorial behavior of $\mathcal{A}$ in the general case.

The conditions (lambda-gc1) and (lambda-gc2) defining our notion of morphism between lambda structures arise from Proposition 15.4.2 and they are obviously weaker than (15.6).

Equivalence between ZS and LS?

We do not know whether the categories **ZS** and **LS** are equivalent, nor do we have a counterexample. By restricting both categories we can obtain equivalent categories, as we will see below.

Various categories of lambda structures

15.4.4 Definition Let $\langle \mathcal{D}, F, G \rangle$ be a lambda structure. Write

$$\delta := G(\bot \mapsto \bot).$$

We say that $\langle \mathcal{D}, F, G \rangle$ is a *lazy* lambda structure if the following hold:

(i) (δ-comp) $\delta \in \mathcal{K}(\mathcal{D})$;
(ii) (adj1) $\forall f \in [\mathcal{D} \to \mathcal{D}].F(G(f)) \sqsupseteq f$;
(iii) (adj2) $\forall x \in \mathcal{D}.\delta \sqsubseteq x \Rightarrow G(F(x)) \sqsubseteq x$;
(iv) ($\delta\bot$) $\forall x \in \mathcal{D}.\delta \not\sqsubseteq x \Rightarrow F(x) = \bot \mapsto \bot$.

15.4.5 Definition A *strict lambda structure*, written LS^{s}, is a triple, $\langle \mathcal{D}, F, G \rangle$, where $\mathcal{D} \in \mathbf{ALG}$, and $F : \mathcal{D} \to_{\mathrm{s}} [\mathcal{D} \to_{\mathrm{s}} \mathcal{D}]$ and $G : [\mathcal{D} \to_{\mathrm{s}} \mathcal{D}] \to_{\mathrm{s}} \mathcal{D}$ are continuous.

We now give the definition of various categories of lambda structures. This definition is an expansion of Definition 15.2.6. For the sake of completeness, we repeat the definition of the categories of lambda structures and natural lambda structures.

15.4.6 Definition

(i) The category **LS** consists of lambda structures as objects and lambda Galois connections as morphisms. The composition between morphisms $\langle \mathsf{m}, \mathsf{m}^R \rangle : \mathcal{D} \to \mathcal{D}'$ and $\langle \mathsf{n}, \mathsf{n}^R \rangle : \mathcal{D}' \to \mathcal{D}''$ is given by $\langle \mathsf{n} \circ \mathsf{m}, \mathsf{m}^R \circ \mathsf{n}^R \rangle$.

(ii) The category **LLS** is the full subcategory of **LS** which has lazy lambda structures as objects.

(iii) The category of *natural lambda structures*, written **NLS**, is the full subcategory of **LS** which has natural lambda structures as objects.

(iv) The category of *strict lambda structures*, written $\mathbf{LS}^s$, has strict lambda structures as objects and, as morphisms¡ special Galois connections $\underline{\mathsf{m}}$ such that m and m^R are strict.

(v) A *proper* lambda structure, written PLS^s is a strict lambda structure $\mathcal{D}$ such that $\langle G, F \rangle$ is a Galois connection.

(vi) $\mathbf{PLS}^s$ is the full subcategory of $\mathbf{LS}^s$ having proper lambda structures as objects.

We will establish the following equivalences of categories:

$$\mathbf{LZS} \cong \mathbf{LLS};$$
$$\mathbf{NZS} \cong \mathbf{NLS};$$
$$\mathbf{PZS}^s \cong \mathbf{PLS}^s.$$

The three equivalences are actually isomorphisms. The first two will be proved in Theorems 15.4.18 and 15.4.24, the third in Exercise 15.17.

Isomorphism between LZS and LLS

In this subsection we see how the correspondence between zip structures and lambda structures becomes very smooth (an isomorphism of categories) in the case of lazy structures. We start with some technical preliminary results.

15.4.7 Lemma *Let $\langle \mathcal{D}, F, G \rangle$ be a lazy lambda structure. Then*

$$\forall f \in [\mathcal{D} \to \mathcal{D}'] \forall x \in \mathcal{D}. f \neq (\bot \mapsto \bot) \Rightarrow [G(f) \sqsubseteq x \Leftrightarrow f \sqsubseteq F(x)].$$

Proof Do Exercise 15.12. ■

Recall $\Theta_{\mathcal{D}_Z}$ from Definition 15.4.1.

15.4.8 Remark Note that by (Z-contr) and (Z-add) one has, in **LZS**:

(i) $\forall a \in \mathcal{K}(\mathcal{D}).Z(a, \bot) = Z(\bot, \bot)$;

(ii) $\Theta_{\mathcal{D}_Z}(x, y) \neq \emptyset \Leftrightarrow \Theta_{\mathcal{D}_Z}(x, y)$ is directed.

15.4.9 Lemma *Let $\langle \mathcal{D}, Z \rangle \in \mathbf{LZS}$ and let $a, b \in \mathcal{K}(\mathcal{D})$, $x \in \mathcal{D}$, with $b \neq \bot$. Then*

$$b \sqsubseteq x \cdot a \Leftrightarrow Z(a, b) \sqsubseteq x.$$

Proof ($\Leftarrow$) follows immediately from the definition of application.

($\Rightarrow$). We have

$$\begin{array}{rcl} b \sqsubseteq x \cdot a & \Leftrightarrow & b \sqsubseteq \bigsqcup \Theta_{\mathcal{D}_Z}(x, a), \\ & \Rightarrow & \exists a_1, b_1. b \sqsubseteq b_1 \ \& \ a_1 \sqsubseteq a \ \& \ Z(a_1, b_1) \sqsubseteq x, \text{by Remark 15.4.7}, \\ & \Rightarrow & Z(a, b) \sqsubseteq x. \blacksquare \end{array}$$

15.4.10 Proposition *Let $\langle \mathcal{D}, Z \rangle$ be a lazy zip structure. Then*

(i) $G_Z(a \mapsto b) = Z(a, b)$.

(ii) $\mathsf{F}_{\mathbf{ZS}}(\langle \mathcal{D}, \mathsf{Z} \rangle) = \langle \mathcal{D}, \mathsf{F}_\mathsf{Z}, \mathsf{G}_\mathsf{Z} \rangle$ *is a lazy lambda structure.*

Proof (i) We have

$$\begin{array}{rcll} G_Z(a \mapsto b) & = & \bigsqcup \{Z(a', b') \mid b' \sqsubseteq (a \mapsto b)a'\} & \\ & = & \bigsqcup \{Z(a', b') \mid a \sqsubseteq a' \ \& \ b' \sqsubseteq b\}, & \text{by Remark 15.4.8(i)}, \\ & = & Z(a, b), & \text{by } (Z\text{-contr}). \end{array}$$

(ii) We prove that $\mathsf{F}_{\mathbf{ZS}}(\langle \mathcal{D}, \mathsf{Z} \rangle)$ satisfies the four points of Definition 15.4.4. We have $G(\bot \mapsto \bot) = Z(\bot, \bot)$, by (i). Therefore ($\delta$-comp) holds, since $Z(\bot, \bot)$ is compact.

For (adj1), it is sufficient to reason about compact elements, and prove that for every a, b,

$$b \sqsubseteq f(a) \Rightarrow b \sqsubseteq F(G(f))(a).$$

Notice that if $b \sqsubseteq f(a)$, then $b \in \Theta_{\mathcal{D}_Z}(G(f), a)$, so

$$\begin{array}{rcl} b & \sqsubseteq & \bigsqcup \Theta_{\mathcal{D}_Z}(G(f), a), \\ & = & G(f) \cdot a, \\ & = & F(G(f))(a). \end{array}$$

Now we prove (adj2). Suppose $\delta \sqsubseteq x$; that is, $Z(a, \bot) \sqsubseteq x$ for every a. Since $G(F(x)) = \bigsqcup \{Z(a, b) \mid b \sqsubseteq x \cdot a\}$, it is enough to prove that $Z(a, b) \sqsubseteq x$ whenever $b \sqsubseteq x \cdot a$. There are two cases. If $b = \bot$, then the result follows from the hypothesis. If $b \neq \bot$, then it follows from Lemma 15.4.9.

Finally we prove ($\delta\bot$). Let $\delta \not\sqsubseteq x$. Then by (Z-lazy) it follows that $Z(a, b) \not\sqsubseteq x$, for all a, b. So $F(x)(y) = \bot$, for every y. $\blacksquare$

From Propositions 15.4.10 and 15.4.3 we get the following.

15.4.11 Theorem F_{ZS} *is restricted to a functor from* **LZS** *to* **LLS**.

Going in the other direction, from every lazy lambda structure one can define a lazy zip structure. Before showing that, we need to extend Lemma 15.2.2(i), (ii) to lazy lambda structures.

15.4.12 Lemma *Let* $\langle \mathcal{D}, F, G\rangle$ *be a lazy lambda structure. Then:*

(i) G *is additive;*
(ii) $\forall f \in \mathcal{K}([\mathcal{D} \to \mathcal{D}]).G(f) \in \mathcal{K}(\mathcal{D})$.

Proof (i) Similar to the proof of Lemma 15.2.2(i).

(ii) If $f = (\bot \mapsto \bot)$ then $G(f) \in \mathcal{K}(\mathcal{D})$ by Definition 15.4.4(i). If on the other hand $f \neq (\bot \mapsto \bot)$, then the proof is similar to that of Lemma 15.2.2(ii), using Lemma 15.4.7. ∎

15.4.13 Definition Let $\mathcal{D} = \langle \mathcal{D}, F, G\rangle$ be a lazy lambda structure. Then we can define, for every $a, b \in \mathcal{K}(\mathcal{D})$,

$$Z_{F,G}(a, b) = G(a \mapsto b)$$

and

$$\mathcal{R}(\mathcal{D}) = \langle \mathcal{D}, Z_{F,G}\rangle.$$

Because of Lemma 15.4.12((ii)), $\mathcal{R}(\mathcal{D})$ is a zip structure.

15.4.14 Proposition *Let* $\mathcal{D} = \langle \mathcal{D}, F, G\rangle$ *be a lazy lambda structure. Then* $\mathcal{R}(\mathcal{D})$ *is a lazy zip structure.*

Proof First of all notice that Z is well defined since by Lemma 15.4.12((ii)), $Z(a, b)$ is a compact element.

We prove (Z-contr). Let $a \sqsubseteq a'$, $b' \sqsubseteq b$. Then, in $[\mathcal{D} \to \mathcal{D}]$, we have $a' \mapsto b' \sqsubseteq a \mapsto b$, hence $G(a' \mapsto b') \sqsubseteq G(a \mapsto b)$. By definition of Z, this implies $Z(a', b') \sqsubseteq Z(a, b)$ as desired.

We prove (Z-add). We have

$$\begin{array}{rcll} Z(a, b_1 \sqcup b_2) & = & G(a \mapsto (b_1 \sqcup b_2)), & \text{by definition,} \\ & = & G(a \mapsto b_1) \sqcup G(a \mapsto b_2), & \text{by Lemma 15.4.12((i)).} \end{array}$$

Finally, (Z-lazy) is immediate by monotonicity of G and the fact that $(\bot \mapsto \bot) \sqsubseteq (a \mapsto b)$, for all $a, b \in \mathcal{K}(\mathcal{D})$. ∎

15.4.15 Lemma *Let* $\underline{\mathsf{m}} = \langle m, m^R\rangle \in \mathbf{LLS}(\mathcal{D}, \mathcal{D}')$, *where* $\mathcal{D} = \langle \mathcal{D}, F, G\rangle$ *and* $\mathcal{D}' = \langle \mathcal{D}', F', G'\rangle$. *Define* $\mathcal{R}(\underline{\mathsf{m}}) = m$. *Then* $\mathcal{R}(\underline{\mathsf{m}}) \in \mathbf{LZS}(\mathcal{R}(\mathcal{D}), \mathcal{R}(\mathcal{D}'))$.

Proof First, $\mathcal{R}(\underline{\mathsf{m}}) = \mathsf{m}$ satisfies (cmp-pres) and (add) by Lemma 15.4.12. So we are left with proving that m satisfies (Z-mon); that is, for every $a, b \in K(D)$,

$$\mathsf{m}(Z(a,b)) \sqsubseteq Z'(\mathsf{m}(a), \mathsf{m}(b))$$

where $Z = Z_{F,G}$, $Z' = Z_{F',G'}$. We have

$$\begin{array}{rcll} \mathsf{m}(Z(a,b)) & = & \mathsf{m}(G(a \mapsto b)), & \text{by definition of } Z, \\ & \sqsubseteq & G'(\mathsf{m} \circ (a \mapsto b) \circ \mathsf{m}^R), & \text{by Definition 15.2.5(i)}, \\ & = & G'(\mathsf{m}(a) \mapsto \mathsf{m}(b)), & \text{by Lemma 15.2.2(iii)}, \\ & = & Z'(\mathsf{m}(a), \mathsf{m}(b)) \blacksquare. & \end{array}$$

From Proposition 15.4.14 and Lemma 15.4.15 we obtain the following.

15.4.16 Theorem *The map* $\mathcal{R} : \mathbf{LLS} \to \mathbf{LZS}$ *is a functor.*

Actually, $\mathcal{R}$ and $\mathsf{F}_{\mathbf{ZS}}$ set up an isomorphism between **LLS** and **LZS**. So the correspondence between **LLS** and **LZS** is perfect.

15.4.17 Theorem

(i) $\mathsf{F}_{\mathbf{ZS}} \circ \mathcal{R} = \mathsf{Id}_{\mathbf{LLS}}$.
(ii) $\mathcal{R} \circ \mathsf{F}_{\mathbf{ZS}} = \mathsf{Id}_{\mathbf{LZS}}$.

Proof (i) For every lazy lambda structure, $\mathcal{D} = \langle \mathcal{D}, F, G \rangle$, we have to prove that $\mathsf{F}_{\mathbf{ZS}}(\mathcal{R}(\mathcal{D})) = \mathcal{D}$. This is equivalent to proving that

$$F_{Z_{F,G}} = F, \quad G_{Z_{F,G}} = G.$$

The proof is left to the reader.

(ii) For every lazy zip structure $\langle \mathcal{D}, Z \rangle$, we have that $\mathcal{R}(\mathcal{A}(\langle \mathcal{D}, Z \rangle)) = \langle \mathcal{D}, Z \rangle$. So the result follows if we prove that $Z_{G_Z} = Z$. But

$$\begin{array}{rcll} Z_{G_Z}(a,b) & = & G_Z(a \mapsto b) & \\ & = & Z(a,b), & \text{by Proposition 15.4.10(i)}, \end{array}$$

hence we are done. ∎

15.4.18 Theorem *The categories* **LZS** *and* **LLS** *are isomorphic.*

Isomorphism between NZS and NLS

In this short subsection we apply the results of the previous subsection to natural zip structures.

As expected, natural lambda structures are special cases of the lazy ones: in a lambda structure $\mathcal{D} = \langle \mathcal{D}, F, G \rangle$, the maps F and G set up a Galois connection if and only if $\mathcal{D}$ is a lazy lambda structure with $\delta = \bot$.

15.4.19 Lemma *Let $\mathcal{D} = \langle \mathcal{D}, F, G \rangle$ be a lambda structure. Then $\mathcal{D}$ is natural (that is $\langle G, F \rangle$ is a Galois connection) if and only if $\mathcal{D}$ is a lazy lambda structure with $\delta = \bot$.*

15.4.20 Proposition *Let $\langle \mathcal{D}, Z \rangle$ be a natural zip structure. Then*

$$\mathsf{F}_{\mathbf{ZS}}(\langle \mathcal{D}, \mathsf{Z} \rangle) = \langle \mathcal{D}, \mathsf{F}_{\mathsf{Z}}, \mathsf{G}_{\mathsf{Z}} \rangle$$

is a natural lambda structure.

Proof The result follows immediately from Lemma 15.4.19, since $Z(\bot, \bot) = \bot$, hence $G_Z(\bot \mapsto \bot) = \bot$. ∎

So we get the following.

15.4.21 Theorem $\mathsf{F}_{\mathbf{ZS}}$ *is restricted to a functor from* **NZS** *to* **NLS**.

We now prove the other direction.

15.4.22 Proposition *Let $\mathcal{D} = \langle \mathcal{D}, F, G \rangle$ be a natural lambda structure. Then $\mathcal{R}(\mathcal{D}) = \langle \mathcal{D}, Z_{F,G} \rangle$ is a natural zip structure.*

Proof By Proposition 15.4.14, we just have to prove that $Z_{F,G}(\bot, \bot) = \bot$. By Lemma 15.4.19 we know that $G(\bot \mapsto \bot) = \bot$, so we are done by the definition of $Z_{F,G}$. ∎

15.4.23 Corollary $\mathcal{R}$ *is restricted to a functor from* **NLS** *to* **NZS**.

15.4.24 Theorem *The categories* **NLS** *and* **NZS** *are isomorphic.*

Proof From Theorem 15.4.18 and the fact that **NLS** and **NZS** are full subcategories of **LLS** and **LZS** respectively. ∎

Equivalences between type and lambda structures

In this subsection we establish the equivalences between categories of type and lambda structures. The notion of filter structure $\langle \mathcal{F}^{\mathcal{S}}, F^{\mathcal{S}}, G^{\mathcal{S}} \rangle$ over a type structure $\mathcal{S}$ given in Definition 13.4.5 induces a functor $\mathsf{Flt}_{\mathbf{TS}^{\mathtt{U}}}$ from $\mathbf{TS}^{\mathtt{U}}$ to **LS**. Since many classical models of λ-calculus are (or could be) defined as $\mathcal{A}(\mathcal{S})$ for suitable type structures $\mathcal{S}$, one fundamental question is whether it is possible to describe every lambda structure as the filter space of a suitable type structure. In the general case, lambda structures are not captured by type structures, and no categorical equivalence result seems possible. But as long as we restrict attention to the lazy, natural, or proper case, then the correspondence is perfect, and assumes the shape of categorical equivalences.

15.4.25 Definition Let $\mathcal{S} \in \mathbf{TS}^{\mathtt{U}}$. We define the functor $\mathsf{Flt}_{\mathbf{TS}^{\mathtt{U}}}$ analogously to that in Lemma 15.2.8 and Theorem 15.2.9.

(1) For $\mathcal{S} \in \mathbf{TS}^{\mathtt{U}}$, define $\mathsf{Flt}_{\mathbf{TS}^{\mathtt{U}}}(\mathcal{S}) = \langle \mathcal{F}^{\mathcal{S}}, F^{\mathcal{S}}, G^{\mathcal{S}} \rangle$.
(2) For $f \in \mathbf{TS}^{\mathtt{U}}(\mathcal{S}, \mathcal{S}')$, define $\mathsf{Flt}_{\mathbf{TS}^{\mathtt{U}}}(f) = \langle \mathsf{Flt}(f), \mathsf{Flt}(f)^R \rangle$

15.4.26 Lemma *The functor* $\mathsf{Flt}_{\mathbf{TS}^{\mathtt{U}}}$ *is the composition of* $\mathcal{Q}$ *and* $\mathsf{F}_{\mathbf{ZS}}$ *defined in* 15.3.8, 15.4.1, *and* 15.4.3.

Proof Given $f \in \mathbf{TS}^{\mathtt{U}}(\mathcal{S}, \mathcal{S}')$, it is easy to see that $\mathsf{F}_{\mathbf{ZS}}(\mathcal{Q}(\mathsf{f})) = \mathsf{Flt}_{\mathbf{TS}^{\mathtt{U}}}(\mathsf{f})$. Let $\mathcal{S} \in \mathbf{TS}^{\mathtt{U}}$. We will prove that $\mathsf{F}_{\mathbf{ZS}}(\mathcal{Q}(\mathcal{S})) = \mathsf{Flt}_{\mathbf{TS}^{\mathtt{U}}}(\mathcal{S})$. By Definitions 15.3.8 and 15.4.1, we have to prove that $F^{\mathcal{S}} = F_{\mathcal{ZS}}$ and $G^{\mathcal{S}} = G_{\mathcal{ZS}}$. Taking the suprema in $\mathcal{F}^{\mathcal{S}}$ one has

$$\begin{aligned} F^{\mathcal{S}}(X)(Y) &= \uparrow\{\uparrow A \mid \exists B \in Y.(B \to A) \in X\} \\ &= \bigsqcup\{\uparrow A \mid \exists B \in Y.\uparrow(B \to A) \subseteq X\} \\ &= \bigsqcup\{\uparrow A \mid \exists \uparrow B \subseteq Y.Z^{\mathcal{S}}(\uparrow B, \uparrow A) \subseteq X\} \\ &= F_{\mathcal{ZS}}(X)(Y). \end{aligned}$$

Moreover,

$$\begin{aligned} G^{\mathcal{S}}(f) &= \uparrow\{B \to A \mid A \in f(\uparrow B)\} \\ &= \bigsqcup\{\uparrow(B \to A) \mid A \in f(\uparrow B)\} \\ &= \bigsqcup\{Z(\uparrow B, \uparrow A) \mid \uparrow A \subseteq f(\uparrow B)\} \\ &= G_{\mathcal{ZS}}(f). \blacksquare \end{aligned}$$

15.4.27 Theorem

(i) $\mathsf{Flt}_{\mathbf{TS}^{\mathtt{U}}} : \mathbf{TS}^{\mathtt{U}} \to \mathbf{LS}$ *is a functor.*
(ii) $\mathsf{Flt}_{\mathbf{TS}^{\mathtt{U}}}$ *is restricted to a functor from* $\mathbf{LTS}^{\mathtt{U}}$ *to* $\mathbf{LLS}$.
(iii) $\mathsf{Flt}_{\mathbf{TS}^{\mathtt{U}}}$ *is restricted to a functor from* $\mathbf{NTS}^{\mathtt{U}}$ *to* $\mathbf{NLS}$.

Proof (i) By Lemma 15.4.26, $\mathsf{Flt}_{\mathbf{TS}^{\mathtt{U}}}$ is a functor since it is the composition of two functors.

(ii), (iii) By Exercise 15.15 along with Theorem 15.4.11 for the lazy case, and Theorem 15.4.21 for the natural case. ■

15.4.28 Theorem *Define* $\mathsf{Flt}_{\mathbf{NLS}} : \mathbf{LLS} \to \mathbf{LTS}^{\mathtt{U}}$ *as in Definition* 15.2.10.

(i) *Given a lazy lambda structure* $\mathcal{D}$, *we define*

$$\mathsf{Flt}_{\mathbf{NLS}}(\mathcal{D}) = \langle \mathcal{K}(\mathcal{D}), \leq, \cap, \to_{\mathcal{D}}, \mathtt{U} \rangle$$

(ii) *Given* $\underline{\mathsf{m}} \in \mathbf{LLS}(\mathcal{D}, \mathcal{D}')$, *we define*

$$\mathsf{Flt}_{\mathbf{NLS}}(\underline{\mathsf{m}}) = \mathsf{m} \restriction \mathcal{K}(\mathcal{D}) : \mathcal{K}(\mathcal{D}) \to \mathcal{K}(\mathcal{D}').$$

Then $\mathsf{Flt}_{\mathbf{NLS}}$ *is a functor. Moreover* $\mathsf{Flt}_{\mathbf{NLS}}$ *is restricted to a functor from* **NLS** *to* $\mathbf{NTS}^{\mathtt{U}}$.

Proof This follows from the fact that $\mathsf{Flt}_{\mathbf{NLS}}$ is a composition of $\mathcal{R}$ with $\mathcal{M}$ – which are functors as proved in Theorem 15.4.16 – and Corollary 15.4.23 and Exercise 15.15, both for the lazy and natural case. ■

15.4.29 Theorem *The categories* $\mathbf{LTS}^{\mathtt{U}}$ *and* **LLS** *are equivalent.*

Proof By Theorem 15.4.18 and Exercise 15.15 ■

15.4.30 Theorem *The categories* $\mathbf{NTS}^{\mathtt{U}}$ *and* **NLS** *are equivalent.*

Proof By Theorem 15.4.24 and Exercise 15.15. ■

15.4.31 Theorem *The categories* $\mathbf{PTS}^{\text{-}\mathtt{U}}$ *and* $\mathbf{PLS}^{s}$ *are equivalent.*

Proof By Exercise 15.17 and Exercise 15.16. ■

15.5 Exercises

Exercise 15.1 Let $\mathcal{S} \in \mathrm{TS}^{\text{-}\mathtt{U}}$. Show that

$$\begin{array}{ll} \emptyset \cdot X = \emptyset & \text{for all } X \in \mathcal{F}^{\mathcal{S}}; \\ X \cdot \emptyset = \emptyset & \text{for all } X \in \mathcal{F}^{\mathcal{S}}. \end{array}$$

Exercise 15.2 Let $\mathcal{S}$ be a natural and β-sound type structure. Show that

$$\uparrow \textstyle\bigcap_{i \in I}(s_i \to t_i) \cdot \uparrow s' \;=\; \textstyle\bigcap_{j \in J} t_j,$$

where $J = \{i \in I \mid s' \leq s_i\}$.

Exercise 15.3 Let $\mathcal{S}$ be a type structure with universe. Show that

$$F^{\mathcal{S}}(G^{\mathcal{S}}(\bot \mapsto \bot)) = (\bot \mapsto \bot) \Leftrightarrow$$

$$(s_1 \to \mathtt{U}) \cap \cdots \cap (s_n \to \mathtt{U}) \leq (s' \to t') \;\Rightarrow\; t' = \mathtt{U}.$$

Exercise 15.4 Let $\mathcal{S}$ be an arbitrary type structure. Show that

$$G^{\mathcal{S}}(\textstyle\bigsqcup_{i \in I}(\uparrow s_i \mapsto \uparrow t_i)) \;\supseteq\; \uparrow \textstyle\bigcap_{i \in I}(s_i \to t_i).$$

Exercise 15.5 Let $\mathcal{S}$ be a proper type structure. Show that

$$G^{\mathcal{S}}(\textstyle\bigsqcup_{i \in I}(\uparrow s_i \mapsto \uparrow t_i)) \;=\; \uparrow \textstyle\bigcap_{i \in I}(s_i \to t_i).$$

Exercise 15.6 Let $\mathcal{S}$ be a natural type structure. Show that $G^{\mathcal{S}}(\uparrow \mathtt{U} \mapsto \uparrow \mathtt{U}) = \uparrow \mathtt{U}$.

Exercise 15.7 Show that (Galois-1) and (Galois-2) in Definition 15.2.1 are equivalent to (Galois).

Exercise 15.8 Let $\langle \mathsf{m}, \mathsf{m}^R \rangle : \mathcal{D} \to \mathcal{D}'$ be a Galois connection. Show that

(i) $\mathsf{m}(x) = \bigsqcap\{x' \mid x \sqsubseteq \mathsf{m}^R(x')\}$;
(ii) $\mathsf{m}^R(x') = \bigsqcup\{x \mid \mathsf{m}(x) \sqsubseteq x'\}$.

Exercise 15.9 We consider the following conditions dual to Definition 15.2.5.

$$\begin{array}{ll} (\text{lambda-gc1}^*) & \forall f' \in [\mathcal{D}' \to \mathcal{D}'].\mathsf{m}^R(G'(f')) \sqsupseteq G(\mathsf{m}^R \circ f' \circ \mathsf{m}); \\ (\text{lambda-gc2}^*) & \forall x \in \mathcal{D}, x' \in \mathcal{D}'.F'(\mathsf{m}(x))(x') \sqsupseteq \mathsf{m}(F(x)(\mathsf{m}^R(x'))). \end{array}$$

Prove the following equivalences.

(i) (lambda-gc1) $\Leftrightarrow$ (lambda-gc1*).
(ii) (lambda-gc2) $\Leftrightarrow$ (lambda-gc2*).

Exercise 15.10 Let $f \in \mathbf{NTS}^{\text{U}}(\mathcal{S}, \mathcal{S}')$ and $f' \in \mathbf{NTS}^{\text{U}}(\mathcal{S}', \mathcal{S}'')$. Show that

$$\begin{array}{rcll} \overline{f \circ f'} & = & \overline{f} \circ \overline{f'}; & \\ \overline{f \circ f'}^R & = & \overline{f'}^R \circ \overline{f}^R; & \\ \langle \overline{\mathsf{Id}}, \overline{\mathsf{Id}}^R \rangle & = & \langle \mathsf{Id}, \mathsf{Id} \rangle, & \text{for } \mathsf{Id} = \mathsf{Id}_{\mathcal{S}} : \mathcal{S} \to \mathcal{S}. \end{array}$$

Exercise 15.11 Let $\langle \mathsf{m}, \mathsf{m}^R \rangle : \mathcal{D} \to \mathcal{D}'$ be a morphism in **LS**. Show that the following diagram is commutative:

$$\begin{array}{ccc} \mathcal{D} & \xleftarrow{\;\bigsqcup\;} & \mathcal{F}^{\mathcal{K}(\mathcal{D})} \\ {\scriptstyle \mathsf{m}^R}\uparrow & & \uparrow{\scriptstyle \overline{\mathsf{m} \restriction \mathcal{K}(\mathcal{D})}^R} \\ \mathcal{D}' & \xleftarrow{\;\bigsqcup\;} & \mathcal{F}^{\mathcal{K}(\mathcal{D}')} \end{array}$$

Note that a filter X' on $(\mathcal{K}(\mathcal{D}'), \leq)$ is a directed subset of $(\mathcal{D}', \sqsubseteq)$.

Exercise 15.12 Let $\langle \mathcal{D}, F, G \rangle$ be a lazy lambda structure, $f \in [\mathcal{D} \to \mathcal{D}]$ and $x \in \mathcal{D}$. Assume $f \neq \bot \mapsto \bot$. Show that

$$G(f) \sqsubseteq x \Leftrightarrow f \sqsubseteq F(x).$$

Exercise 15.13 Let $\langle \mathcal{D}, F, G \rangle$ and $\langle \mathcal{D}', F', G' \rangle$ be natural lambda structures and let $\underline{m} = \langle m, m^R \rangle : \mathcal{D} \to \mathcal{D}'$ be a Galois connection. Show that

$$[\forall x, y \in \mathcal{D}.m(Fxy) \sqsubseteq F'(mx)(my)] \Rightarrow$$
$$[\forall f \in [\mathcal{D} \to \mathcal{D}].m(Gf) \sqsupseteq G'(m \circ f \circ m^R)].$$

Exercise 15.14 Let $\langle D, F, G\rangle$ and $\langle D', F', G'\rangle$ be lambda structures. Assume $\mathsf{m} : \mathcal{D} \to \mathcal{D}'$ is a bijection such that m and m^{-1} are continuous and the following conditions hold for m:

$$\begin{array}{rcl}\mathsf{m}(G(f)) & = & G'(\mathsf{m} \circ f \circ \mathsf{m}^{-1});\\ \mathsf{m}(F(d)(e)) & = & F'(\mathsf{m}(d))(\mathsf{m}(e)).\end{array}$$

Prove that m induces an isomorphism of lambda structures $\underline{\mathsf{m}} = \langle \mathsf{m}, \mathsf{m}^{-1}\rangle$: $\langle \mathcal{D}, F, G\rangle \to \langle \mathcal{D}', F', G'\rangle$.

Exercise 15.15 Using restrictions of the functors $\mathcal{Q}$ and $\mathcal{M}$, prove the following equivalences of categories:

$$\mathbf{LTS}^{\mathtt{U}} \cong \mathbf{LZS};$$
$$\mathbf{NTS}^{\mathtt{U}} \cong \mathbf{NZS}.$$

Exercise 15.16 Prove the following equivalences of categories:

$$\mathbf{TS}^{\text{-}\mathtt{U}} \cong \mathbf{ZS}^{s}$$
$$\mathbf{PTS}^{\text{-}\mathtt{U}} \cong \mathbf{PZS}^{s}$$

using the functors $\mathcal{Q}_s$ and $\mathcal{M}_s$ defined as follows.

(1) For $\mathcal{S} \in \mathbf{TS}^{\text{-}\mathtt{U}}$, define $\mathcal{Q}_s(\mathcal{S}) \in \mathbf{ZS}^s$ by $\mathcal{Q}_s(\mathcal{S}) = (\mathcal{F}^{\mathcal{S}}, Z^{\mathcal{S}})$, with $Z^{\mathcal{S}}$ as in Definition 15.3.8. Note that $Z^{\mathcal{S}} : \mathcal{K}^s(\mathcal{F}^{\mathcal{S}}) \times \mathcal{K}^s(\mathcal{F}^{\mathcal{S}}) \to \mathcal{K}^s(\mathcal{F}^{\mathcal{S}})$.
   For $(\mathcal{D}, Z) \in \mathbf{ZS}^s$, define $\mathcal{M}_s(\langle \mathcal{D}, Z\rangle) \in \mathbf{TS}^{\text{-}\mathtt{U}}$ by

$$\mathcal{M}_s(\langle \mathcal{D}, Z\rangle) = \langle K^s(\mathcal{D}), \leq, \cap, \to_Z\rangle,$$

   with $\leq, \cap, \to_Z$ as in Definition 15.3.8.
(2) Given $f \in \mathbf{TS}^{\text{-}\mathtt{U}}(\mathcal{S}, \mathcal{S}')$, $X \in \mathcal{F}^{\mathcal{S}}$, and $g \in \mathbf{ZS}^s(\langle \mathcal{D}, Z\rangle, \langle \mathcal{D}', Z'\rangle)$, define

$$\begin{array}{rcl}\mathcal{Q}_s(f)(X) & = & \left\{\begin{array}{ll}\{t \mid \exists s \in X. f(s) \leq t\}, & \text{if } X \neq \bot\ (= \emptyset),\\ \bot, & \text{otherwise;}\end{array}\right.\\ \mathcal{M}_s(g) & = & g \restriction \mathcal{K}(\mathcal{D}).\end{array}$$

Exercise 15.17

(i) Let $\langle \mathcal{D}, Z\rangle \in \mathbf{ZS}^s$. Prove that the mappings F_Z and G_Z, given in Definition 15.4.1, are strict and that $F_Z : \mathcal{D} \to_{\mathrm{s}} [\mathcal{D} \to_{\mathrm{s}} \mathcal{D}]$ and $G_Z : [\mathcal{D} \to_{\mathrm{s}} \mathcal{D}] \to_{\mathrm{s}} \mathcal{D}$.
(ii) Let $\langle \mathcal{D}, F, G\rangle \in \mathbf{PLS}^s$, see Definition 15.4.6(vi). Show that if $f \neq \bot \mapsto \bot$ then $G(f) \neq \bot$ and conclude that $Z_{F,G}$, introduced in Definition 15.4.13, is a mapping from $\mathcal{K}^s(\mathcal{D}) \times \mathcal{K}^s(\mathcal{D})$ to $\mathcal{K}^s(\mathcal{D})$.
(iii) Prove that F_{ZS} is a functor from $\mathbf{PZS}^s$ to $\mathbf{PLS}^s$ (see Definition 15.4.1 and Proposition 15.4.3).

(iv) Prove that $\mathcal{R}$ is a functor from $\mathbf{PLS}^s$ to $\mathbf{PZS}^s$ (see Definition 15.4.13 and Lemma 15.4.15).

(v) Prove that the categories $\mathbf{PZS}^s$ and $\mathbf{PLS}^s$ are isomorphic.

16
Filter Models

Filter models are models of the untyped lambda calculus where terms are interpreted as sets of types. The domain of a filter model will be the set $\mathcal{F}^{\mathcal{T}}$ of filters as defined in Section 13.4, i.e.

$$[\![M]\!]^{\mathcal{F}^{\mathcal{T}}} \in \mathcal{F}^{\mathcal{T}}.$$

The application and the abstraction will be interpreted using the functions $F^{\mathcal{T}}$ and $G^{\mathcal{T}}$ of a filter structure, see Definition 13.4.5. Variations on $\mathcal{T}$ induce different interpretations $[\![\ ]\!]^{\mathcal{F}^{\mathcal{T}}}$ on λ-terms which may or may not satisfy β-conversion. This leads us to define the different classes of filter models given in Fig. 16.1. Moreover, filter models could be *extensional* or *non-extensional* depending on whether or not $[\![\ ]\!]^{\mathcal{F}^{\mathcal{T}}}$ satisfies η-conversion.

The first important property that will be proved in Section 16.2 is the so-called Type-semantics Theorem. This theorem states that the interpretation of a closed term is the set of its types. More formally, if M is closed,

$$[\![M]\!]^{\mathcal{F}^{\mathcal{T}}} = \{A \mid\ \vdash_{\cap}^{\mathcal{T}} M : A\}.$$

A first consequence of this theorem is that the interpretation in a filter structure satisfies $\beta(\mathsf{I})(\eta)$-conversion exactly when the type assignment system does. Hence, the content of Figs. 16.2 and 16.3 can be easily deduced from that of Figs. 14.1 and 14.2, respectively.

Section 16.2 also studies representability of continuous functions (Coppo

Classes of $\mathcal{F}^{\mathcal{T}}$	Rule satisfied by $[\![\ ]\!]^{\mathcal{F}^{\mathcal{T}}}$
Filter λ-model	β-conversion
Filter $\lambda\mathsf{I}$-model	$\beta\mathsf{I}$-conversion

Figure 16.1

Property of $\mathcal{T} \in \mathrm{TT}$	versus	property of $\mathcal{F}^{\mathcal{T}}$
$\mathcal{T} \in \mathrm{TT}^{\mathtt{U}}$, $\boldsymbol{\beta}$-sound	$\Rightarrow$	filter λ-model
$\mathcal{T} \in \mathrm{TT}^{\text{-}\mathtt{U}}$, $\boldsymbol{\beta}$-sound	$\Rightarrow$	filter $\lambda\mathsf{I}$-model
$\mathcal{T} \in \mathrm{TT}^{\mathtt{U}}$, $\boldsymbol{\beta}$-sound, natural, $\boldsymbol{\eta}^{\mathtt{U}}$-sound	$\Rightarrow$	extensional filter λ-model
$\mathcal{T} \in \mathrm{TT}^{\text{-}\mathtt{U}}$, $\boldsymbol{\beta}$-sound, proper, $\boldsymbol{\eta}$-sound	$\Rightarrow$	extensional filter $\lambda\mathsf{I}$-model

Figure 16.2 Conditions on type theories for inducing filter models

Class of filter model $\mathcal{F}^{\mathcal{T}}$	Type theory $\mathcal{T}$
Extensional filter λ-model	Scott, Park, CDZ, HR, DHM
Extensional filter $\lambda\mathsf{I}$-model	HL
Non-extensional filter λ-model	BCD, AO, Plotkin, Engeler, CDS
Non-extensional filter $\lambda\mathsf{I}$-model	CDV, CD

Figure 16.3 Classification of filter models

et al. (1984), Alessi et al. (2004b)). We prove that $\mathcal{T}$ is β-sound iff all continuous functions are representable in $\mathcal{F}^{\mathcal{T}}$. At the end of the section, we show an example of a non β-sound type theory called ABD which induces a filter lambda model. As a consequence, not all continuous functions are representable in $\mathcal{F}^{\mathrm{ABD}}$. This example also shows that the condition of $\boldsymbol{\beta}$-soundness is inadequate when it comes to characterizing the type theories whose type assignment is closed under β-reduction; or equivalently, whose filter structure is a λ-model.

Section 16.3 shows the relation between Scott's $\mathcal{D}_\infty$-models (see Section 18.2 of B[1984] for definition and properties of $\mathcal{D}_\infty$) and filter models. It is well known that Scott's $\mathcal{D}_\infty$-models are models of the lambda calculus that satisfy the recursive domain equation $\mathcal{D} = \mathcal{D} \to \mathcal{D}$. The construction of $\mathcal{D}_\infty$ depends not only on the initial $\mathcal{D}_0$, but also on the embedding-projection pair[1] $\langle \mathsf{i}_0, \mathsf{j}_0 \rangle$ which gives the start of the $\mathcal{D}_\infty$ construction:

$$\mathsf{i}_0 : \mathcal{D}_0 {\to} \mathcal{D}_1, \ \mathsf{j}_0 : \mathcal{D}_1 {\to} \mathcal{D}_0,$$

where $\mathcal{D}_1 = [\mathcal{D}_0 {\to} \mathcal{D}_0]$. Given the triple $t = (\mathcal{D}_0, \mathsf{i}_0, \mathsf{j}_0)$, we write $\mathcal{D}_\infty = \mathcal{D}^t_\infty$ to emphasize the dependency on t. Different t define different $\mathcal{D}^t_\infty$-models. Some instances of $t = (\mathcal{D}_0, \mathsf{i}_0, \mathsf{j}_0)$ have given rise to some specific filter models such as $\mathcal{F}^{\mathrm{Scott}}$ and $\mathcal{F}^{\mathrm{Park}}$, see Scott (1972), Park (1976), Barendregt et al.

[1] In B[1984], Definition 18.2.1, the maps i_0 and j_0 are called φ_0 and ψ_0 respectively.

(1983). Actually we will show that any $\mathcal{D}_\infty^t$ model in the category **ALG** of ω-algebraic complete lattices can be described as a filter model $\mathcal{F}^{\mathrm{CDHL}(t)}$ by considering the compact elements of $\mathcal{D}_0$ as atomic types and defining a type theory CDHL(t) that contains both the order of $\mathcal{D}_0$ and i_0, see Barendregt et al. (1983), Coppo et al. (1984), Coppo et al. (1987), Alessi et al. (2004a). We will first prove that

$$\mathcal{K}(\mathcal{D}_\infty^t) \cong [\mathrm{CDHL}(t)].$$

Then, by Proposition 15.2.13, we get our result that $\mathcal{D}_\infty^t$ can be described as $\mathcal{F}^{\mathrm{CDHL}(t)}$, i.e.

$$\mathcal{D}_\infty^t \cong \mathcal{F}^{[\mathrm{CDHL}(t)]} \cong \mathcal{F}^{\mathrm{CDHL}(t)}.$$

The converse is obviously not true. Filter models are in a sense weaker structures than $\mathcal{D}_\infty$-models. Not all of them satisfy the recursive domain equation $\mathcal{D} = \mathcal{D} \to \mathcal{D}$ for they can be non-extensional. If we restrict our attention to the extensional filter models of Fig. 16.3, then all of them, i.e. $\mathcal{F}^{\mathrm{Scott}}$, $\mathcal{F}^{\mathrm{Park}}$, $\mathcal{F}^{\mathrm{CDZ}}$, $\mathcal{F}^{\mathrm{HR}}$, and $\mathcal{F}^{\mathrm{DHM}}$, can be described as $\mathcal{D}_\infty$-models by choosing an appropriate t, see Coppo et al. (1983), Coppo et al. (1984), Alessi (1991), Dezani-Ciancaglini et al. (2004). One obtains the following versions of $\mathcal{D}_\infty$,

$$\mathcal{D}_\infty^{\mathrm{Scott}}, \mathcal{D}_\infty^{\mathrm{Park}}, \mathcal{D}_\infty^{\mathrm{CDZ}}, \mathcal{D}_\infty^{\mathrm{DHM}} \text{ and } \mathcal{D}_\infty^{\mathrm{HR}}.$$

Given $\mathcal{T} \in \{\mathrm{Scott}, \mathrm{Park}, \mathrm{CDZ}, \mathrm{DHM}, \mathrm{HR}\}$, we will associate a triple $\mathsf{init}(\mathcal{T}) = (\mathcal{D}_0, \mathsf{i}_0, \mathsf{j}_0)$ such that $\mathrm{CDHL}(\mathsf{init}(\mathcal{T})) = \mathcal{T}$. Then

$$\mathcal{D}_\infty^{\mathsf{init}(\mathcal{T})} \cong \mathcal{F}^{\mathrm{CDHL}(\mathsf{init}(\mathcal{T}))} = \mathcal{F}^{\mathcal{T}}.$$

We will write $\mathcal{D}_\infty^{\mathcal{T}} := \mathcal{D}_\infty^{\mathsf{init}(\mathcal{T})}$. Then the equation becomes

$$\mathcal{D}_\infty^{\mathcal{T}} \cong \mathcal{F}^{\mathrm{CDHL}(\mathsf{init}(\mathcal{T}))} = \mathcal{F}^{\mathcal{T}}.$$

The pleasant fact is that $\mathcal{T}$ and the triple $t = (\mathcal{D}_0, \mathsf{i}_0, \mathsf{j}_0)$ correspond to each other in a canonical way. For each of the theories $\mathcal{T} \in \{\mathrm{Scott}, \mathrm{Park}\}$, the model $\mathcal{D}_\infty^{\mathcal{T}}$ was constructed first and the natural type theory $\mathcal{T}$ came later. For $\mathcal{T} \in \{\mathrm{CDZ}, \mathrm{DHM}, \mathrm{HR}\}$ one constructed this natural type theory in order to obtain the model $\mathcal{D}_\infty^{\mathcal{T}}$ satisfying certain properties.

In spite of the fact that the non-extensional filter models do not satisfy $\mathcal{D} = \mathcal{D} \to \mathcal{D}$, in some cases it is possible to find other recursive domain equations for them; see Alessi (1991). For instance, the non-extensional filter model $\mathcal{F}^{\mathrm{AO}}$ satisfies the equation $\mathcal{D} = [\mathcal{D} \to \mathcal{D}]_\perp$ and $\mathcal{F}^{\mathrm{BCD}}$ satisfies the equation $\mathcal{D} = [\mathcal{D} \to \mathcal{D}] \times \mathsf{P}(\mathbb{A}_\infty)$, where P is the powerset operation.

Section 16.4 studies other filter models. The first model considered is AO which is shown to be computationally adequate for the lazy operational semantics; see Abramsky and Ong (1993). The second one is used as an application of intersection types to prove consistency of certain equations in the λ-calculus. Following Alessi et al. (2001) we show that $\boldsymbol{\Omega} := (\lambda x.xx)(\lambda x.xx)$ is an *easy* lambda term in the sense of Jacopini and Venturini-Zilli: $\boldsymbol{\lambda\beta} \cup \{\boldsymbol{\Omega} = M\}$, is consistent for all $M \in \Lambda$. This has been demonstrated in various ways, see for example, Jacopini (1975), Baeten and Boerboom (1979) or Mitschke's proof in B[1984], Proposition 15.3.9. Given any λ-term M, we recursively build natural intersection type theories $\mathrm{ADH}_n(M)$ in such a way that the union of these theories, called $\mathrm{ADH}(M)$, forces the interpretation of M to coincide with the interpretation of $\boldsymbol{\Omega}$, i.e.

$$\mathcal{F}^{\mathrm{ADH}(M)} \models M = \boldsymbol{\Omega}.$$

Further applications of intersection types consist of necessary conditions for filter λ-models to be *sensible* or *semi-sensible*. We will not consider this issue but see Zylberajch (1991). At the end of the section, we describe some graph models as filter models.

16.1 Lambda models

In this section we generalize the basic notions and properties of λ-models. That definition was given in Section 3.1. We now introduce also *quasi λ-models*. This makes it possible to differentiate between models for the λ-calculus and for the $\lambda\mathsf{I}$-calculus.

16.1.1 Definition

(i) Let $\mathcal{D}$ be a set and V the set of variables of the untyped lambda calculus. An *environment in $\mathcal{D}$* is a total map

$$\rho : \mathsf{V} \to \mathcal{D}.$$

The set of environments in $\mathcal{D}$ is denoted by $\mathsf{Env}_{\mathcal{D}}$.

(ii) If $\rho \in \mathsf{Env}_{\mathcal{D}}$ and $d \in \mathcal{D}$, then $\rho[x := d]$ is the $\rho' \in \mathsf{Env}_{\mathcal{D}}$ defined by

$$\rho'(y) = \begin{cases} d & \text{if } y = x, \\ \rho(y) & \text{otherwise.} \end{cases}$$

Recall that an applicative structure is a pair $\langle \mathcal{D}, \cdot \rangle$ consisting of a set $\mathcal{D}$ together with a binary operation $\cdot : \mathcal{D} \times \mathcal{D} \to \mathcal{D}$ on it.

16.1.2 Definition

(i) A *quasi λ-model* is of the form

$$\mathcal{D} = \langle \mathcal{D}, \cdot, [\![\ ]\!]^{\mathcal{D}} \rangle,$$

where $\langle \mathcal{D}, \cdot \rangle$ is an applicative structure and $[\![\ ]\!]^{\mathcal{D}} : \Lambda \times \mathsf{Env}_{\mathcal{D}} \to \mathcal{D}$ satisfies the following.

$$\begin{array}{rcll}
[\![x]\!]^{\mathcal{D}}_{\rho} & = & \rho(x); & \\
[\![MN]\!]^{\mathcal{D}}_{\rho} & = & [\![M]\!]^{\mathcal{D}}_{\rho} \cdot [\![N]\!]^{\mathcal{D}}_{\rho}; & \\
[\![\lambda x.M]\!]^{\mathcal{D}}_{\rho} & = & [\![\lambda y.M[x := y]]\!]^{\mathcal{D}}_{\rho}, & (\alpha) \\
 & & \text{provided } y \notin \mathrm{FV}(M); & \\
\forall d \in \mathcal{D}.[\![M]\!]^{\mathcal{D}}_{\rho[x:=d]} = [\![N]\!]^{\mathcal{D}}_{\rho[x:=d]} & \Rightarrow & [\![\lambda x.M]\!]^{\mathcal{D}}_{\rho} = [\![\lambda x.N]\!]^{\mathcal{D}}_{\rho}; & (\xi) \\
\rho \restriction \mathrm{FV}(M) = \rho' \restriction \mathrm{FV}(M) & \Rightarrow & [\![M]\!]^{\mathcal{D}}_{\rho} = [\![M]\!]^{\mathcal{D}}_{\rho'}. &
\end{array}$$

(ii) A λ-*model* is a quasi λ-model which satisfies

$$[\![\lambda x.M]\!]^{\mathcal{D}}_{\rho} \cdot d \quad = \quad [\![M]\!]^{\mathcal{D}}_{\rho[x:=d]} \qquad (\boldsymbol{\beta})$$

This is consistent with Definition 3.1.31.

(iii) A $\lambda\mathsf{I}$-*model* is a quasi λ-model which satisfies

$$x \in \mathrm{FV}(M) \quad \Rightarrow \quad [\![\lambda x.M]\!]^{\mathcal{D}}_{\rho} \cdot d = [\![M]\!]^{\mathcal{D}}_{\rho[x:=d]} \qquad (\boldsymbol{\beta}\mathsf{I})$$

16.1.3 Remark As noticed by R. Hindley the present definition of $\lambda\mathsf{I}$-model requires that every $\lambda\mathsf{K}$ term has an interpretation, even though the variable 'x' in '$\lambda.xM$' plays no 'active' role. But K is not in the $\lambda\mathsf{I}$ language (as defined by most people, e.g. Barendregt (1984) p.185). On the other hand, this approach agrees with Church's original λ-calculus (Church (1932, 1933)), in which K was allowed in the language but not permitted to be active in redexes.

We will simply write $[\![\ ]\!]_{\rho}$ instead of $[\![\ ]\!]^{\mathcal{D}}_{\rho}$ when there is no danger of confusion.

We have the following implications.

$$\mathcal{D}\ \lambda\text{-model} \quad \Longrightarrow \quad \mathcal{D}\ \lambda\mathsf{I}\text{-model} \quad \Longrightarrow \quad \mathcal{D}\ \text{quasi}\ \lambda\text{-model}.$$

The first class of applicative structures satisfies (β) in general, the second only for $\lambda\mathsf{I}$-redexes and the third class does not need to satisfy (β) at all, but there is an interpretation for λ-terms.

16.1.4 Definition Let $\mathcal{D} = \langle \mathcal{D}, \cdot, [\![\ ]\!] \rangle$ be a quasi λ-model.

(i) The statement $M = N$, for M, N untyped lambda terms, is *true in* $\mathcal{D}$, written $\mathcal{D} \models M = N$ if

$$\forall \rho \in \mathsf{Env}_{\mathcal{D}}.[\![M]\!]_{\rho} = [\![N]\!]_{\rho}.$$

(ii) As usual one defines $\mathcal{D} \models \chi$, where χ is any statement built up using first-order predicate logic from equations between untyped lambda terms.

(iii) A $\lambda(\mathsf{I})$-model $\mathcal{D}$ is called *extensional* iff

$$\mathcal{D} \models (\forall x.Mx = Nx) \Rightarrow M = N.$$

(iv) A $\lambda(\mathsf{I})$-model $\mathcal{D}$ is called an $\boldsymbol{\eta}$*-model* iff

$$\mathcal{D} \models \lambda x.Mx = M, \text{ for } x \notin \mathrm{FV}(M). \qquad (\boldsymbol{\eta})$$

We will see now how the notions of $\lambda(\mathsf{I})$-model and (strict) lambda structure are related, see Definition 15.2.3 and Definition 15.4.5.

16.1.5 Definition

(i) Let $\langle \mathcal{D}, F, G \rangle$ be a lambda structure, see Definition 15.2.3. We define, with the intention of constructing a quasi λ-model, the triple $\langle \mathcal{D}, \cdot_F, [[\,]]^{F,G} \rangle$ as follows,

- First we obtain an applicative structure by setting for $d, e \in \mathcal{D}$

$$d \cdot_F e = F(d)(e).$$

- Then the map $[\![\,]\!]^{F,G} : \Lambda \times \mathsf{Env}_{\mathcal{D}} \to \mathcal{D}$ is defined as:

$$\begin{aligned} [\![x]\!]^{F,G}_\rho &= \rho(x); \\ [\![MN]\!]^{F,G}_\rho &= F([\![M]\!]^{F,G}_\rho)([\![N]\!]^{F,G}_\rho); \\ [\![\lambda x.M]\!]^{F,G}_\rho &= G(\boldsymbol{\lambda} d \in \mathcal{D}.[\![M]\!]^{F,G}_{\rho[x:=d]}), \end{aligned}$$

noting that the map $\boldsymbol{\lambda} d \in \mathcal{D}.[\![M]\!]_{\rho[x:=d]}$ used for $[\![\lambda x.M]\!]_\rho$ is continuous.

(ii) Finally, let $\langle \mathcal{D}, F, G \rangle$ be a *strict* lambda structure. We define the triple $\langle \mathcal{D}, \cdot_F, [\![\,]\!]^{F,G} \rangle$ as above, changing the clause for $[\![\lambda x.M]\!]^{F,G}_\rho$ into

$$[\![\lambda x.M]\!]^{F,G}_\rho = G(\boldsymbol{\lambda} d \in \mathcal{D}. \text{ if } d = \bot_{\mathcal{D}} \text{ then } \bot_{\mathcal{D}} \text{ otherwise } [\![M]\!]^{F,G}_{\rho[x:=d]}).$$

16.1.6 Proposition *Let $\langle \mathcal{D}, F, G \rangle$ be a (strict) lambda structure. Then $\langle \mathcal{D}, \cdot_F, [\![\,]\!]^{F,G} \rangle$ is a quasi λ-model.*

Proof Easy. ∎

16.1.7 Definition Let $\langle \mathcal{D}, F, G \rangle$ be a (strict) lambda structure. Then

$$\mathcal{D} = \langle \mathcal{D}, \cdot_F, [\![\,]\!]^{F,G} \rangle$$

is called the *quasi λ-model* induced by $\langle \mathcal{D}, F, G \rangle$. We will sometimes omit the subscript from $\cdot_F$ when there is no danger of confusion.

The only requirement that a (strict) lambda structure misses to be a $\lambda(\mathsf{I})$-model is the axiom ($\boldsymbol{\beta}(\mathsf{I})$).

16.1.8 Proposition

(i) *Let* $\mathcal{D} = \langle \mathcal{D}, \cdot_F, [\![\ ]\!]^{F,G}\rangle$ *be the quasi* λ*-model induced by the lambda structure* $\langle \mathcal{D}, F, G\rangle$. *Then the following statements are equivalent.*

(1) $\mathcal{D} \models (\lambda x.M)N = M[x{:}= N]$, *for all* $M, N \in \Lambda$.

(2) $[\![\lambda x.M]\!]^{F,G}_{\rho} \cdot d = [\![M]\!]^{F,G}_{\rho(x:=d)}$, *for all* $M \in \Lambda$ *and* $d \in \mathcal{D}$.

(3) $\mathcal{D}$ *is a* λ*-model.*

(4) $\mathcal{D} \models \{M = N \mid \lambda\boldsymbol{\beta} \vdash M = N\}$.

(ii) *Let* $\mathcal{D} = \langle \mathcal{D}, \cdot_F, [\![\ ]\!]^{F,G}\rangle$ *be the quasi* λ*-model induced by the strict lambda structure* $\langle \mathcal{D}, F, G\rangle$. *Then the following statements are equivalent.*

(5) $\mathcal{D} \models (\lambda x.M)N = M[x{:}= N]$, *for all* $M, N \in \Lambda$ *such that* $x \in \mathrm{FV}(M)$.

(6) $[\![\lambda x.M]\!]^{F,G}_{\rho} \cdot d = [\![M]\!]^{F,G}_{\rho(x:=d)}$, *for all* $M \in \Lambda$ *with* $x \in \mathrm{FV}(M)$, *and* $d \in \mathcal{D}$.

(7) $\mathcal{D}$ *is a* $\lambda\mathsf{I}$*-model.*

(8) $\mathcal{D} \models \{M = N \mid \lambda\boldsymbol{\beta}\mathsf{I} \vdash M = N\}$.

Proof (i) (1)⇒(2). By (1) one has $[\![(\lambda x.M)N]\!]^{F,G}_{\rho} = [\![M[x{:}= N]]\!]^{F,G}_{\rho}$. Taking $N \equiv x$ and $\rho' = \rho(x{:}= d)$ one obtains

$$[\![(\lambda x.M)x]\!]^{F,G}_{\rho'} = [\![M]\!]^{F,G}_{\rho'},$$

hence

$$[\![\lambda x.M]\!]^{F,G}_{\rho} \cdot d = [\![M]\!]^{F,G}_{\rho'},$$

as $\rho \restriction \mathrm{FV}(\lambda x.M) = \rho' \restriction \mathrm{FV}(\lambda x.M)$.

(2)⇒(3). By Definition 16.1.5(i) and Proposition 16.1.6 all conditions for being a λ-model are fulfilled; see Definition 16.1.2.

(3)⇒(4). By Theorem 5.3.4 in B[1984].

(4)⇒(1). Trivial.

(ii) Similarly. ∎

In Definition 10.1.13, we required $F \circ G = \mathsf{Id}_{[\mathcal{D}\to\mathcal{D}]}$ for a lambda structure to be a lambda model. This condition implies representability of all continuous functions, see Lemma 16.2.15. The theory ABD defined in Definition 16.2.20 gives rise to a lambda model, where not all continuous functions are representable.

16.1.9 Corollary *Let* $\mathcal{D}$ *denote the* $\lambda(\mathsf{I})$*-model induced by the (strict) lambda structure* $\langle \mathcal{D}, F, G\rangle$. *Then*

$$\mathcal{D} \text{ is a } \lambda(\mathsf{I})\boldsymbol{\eta}\text{-model} \Leftrightarrow \mathcal{D} \text{ is an extensional } \lambda(\mathsf{I})\text{-model.}$$

Proof (⇒) Suppose that for some ρ one has for all $d \in \mathcal{D}$

$$[\![Mx]\!]^{F,G}_{\rho[x:=d]} = [\![Nx]\!]^{F,G}_{\rho[x:=d]}.$$

Then by (η) and Proposition 16.1.6(ii) one has

$$[\![M]\!]^{F,G}_{\rho} = [\![\lambda x.Mx]\!]^{F,G}_{\rho} = [\![\lambda x.Nx]\!]^{F,G}_{\rho} = [\![N]\!]^{F,G}_{\rho}.$$

(⇐) Note that by $(\beta(\mathsf{I}))$ one has $\mathcal{D} \models (\lambda x.Mx)y = My$, where x is fresh. Hence by extensionality one has $\mathcal{D} \models \lambda x.Mx = M$. ∎

Isomorphisms of λ-models

This subsection relates isomorphisms between lambda structures and lambda models.

16.1.10 Definition We say that $\mathcal{D}$ and $\mathcal{D}'$ are isomorphic λ-models (via m), written $(\mathsf{m} :)\ D \cong D'$, if m is a bijection and for all λ-terms M and environments ρ:

$$\mathsf{m}([\![M]\!]^{\mathcal{D}}_{\rho}) = [\![M]\!]^{\mathcal{D}'}_{\mathsf{m}\circ\rho}$$

16.1.11 Lemma *If two λ-models $\mathcal{D}$ and $\mathcal{D}'$ are isomorphic, then they equate the same terms, i.e. $\mathcal{D} \models M = N \Leftrightarrow \mathcal{D}' \models M = N$.*

Proof Easy. ∎

The next lemma is used to prove that an isomorphism of lambda structures is also an isomorphism of λ-models. For the converse of this lemma, see Exercise 15.14.

16.1.12 Lemma *Let $\underline{\mathsf{m}} = \langle \mathsf{m}, \mathsf{m}^R \rangle : \langle \mathcal{D}, F, G \rangle \to \langle \mathcal{D}', F', G' \rangle$ be an isomorphism between lambda structures. Then $\mathsf{m} : \mathcal{D} \to \mathcal{D}'$ is a bijective continuous map such that*

$$\begin{array}{llcl} \text{(iso-ls1)} & \mathsf{m}(G(f)) & = & G'(\mathsf{m} \circ f \circ \mathsf{m}^R); \\ \text{(iso-ls2)} & \mathsf{m}(F(d)(e)) & = & F'(\mathsf{m}(d))(\mathsf{m}(e)). \end{array}$$

If we write $f^{\mathsf{m}} = \mathsf{m} \circ f \circ \mathsf{m}^{-1}$ then we can reformulate these conditions as

$$\mathsf{m}(G(f)) = G'(f^{\mathsf{m}});$$

$$\mathsf{m}(d \cdot_F e) = \mathsf{m}(d) \cdot_{F'} \mathsf{m}(e).$$

Proof By Definition 15.2.5 we get

$$\begin{array}{ll} \text{(lambda-gc1)} & \forall f \in [\mathcal{D} \to \mathcal{D}].\mathsf{m}(G(f)) \sqsubseteq G'(\mathsf{m} \circ f \circ \mathsf{m}^R); \\ \text{(lambda-gc2)} & \forall x' \in \mathcal{D}', x \in \mathcal{D}.F(m^R(x'))(x) \sqsubseteq m^R(F'(x')(m(x))). \end{array}$$

Proof of (iso-ls1). Since $\underline{\mathsf{m}}$ is an isomorphism, we have that, besides the lambda Galois connection

$$\langle \mathsf{m}, \mathsf{m}^R \rangle : \mathcal{D} \to \mathcal{D}',$$

there is another lambda Galois connection, $\underline{\mathsf{m}}^{-1}$, which we call $\underline{\mathsf{n}} = \langle \mathsf{n}, \mathsf{n}^R \rangle : \mathcal{D}' \to \mathcal{D}$ such that

$$\begin{aligned} \underline{\mathsf{n}} \circ \underline{\mathsf{m}} &= \langle \mathsf{Id}_{\mathcal{D}}, \mathsf{Id}_{\mathcal{D}} \rangle, \\ \underline{\mathsf{m}} \circ \underline{\mathsf{n}} &= \langle \mathsf{Id}_{\mathcal{D}'}, \mathsf{Id}_{\mathcal{D}'} \rangle. \end{aligned}$$

Using composition between Galois connections, this amounts to saying

$$\left. \begin{aligned} \mathsf{n} \circ \mathsf{m} &= \mathsf{Id}_{\mathcal{D}}, \\ \mathsf{m}^R \circ \mathsf{n}^R &= \mathsf{Id}_{\mathcal{D}}, \\ \mathsf{m} \circ \mathsf{n} &= \mathsf{Id}_{\mathcal{D}'}, \\ \mathsf{n}^R \circ \mathsf{m}^R &= \mathsf{Id}_{\mathcal{D}'}. \end{aligned} \right\} \tag{16.1}$$

Looking at (16.1), we see that compositions of m and n give the identities. So $\mathsf{n} = \mathsf{m}^{-1}$. This implies that n is a right adjoint of m. But the right adjoint is unique, so $\mathsf{m}^{-1} = \mathsf{n} = \mathsf{m}^R$. For the same reason $\mathsf{n}^R = \mathsf{m}$. Therefore we have proved that $(\underline{\mathsf{m}}^{-1} = \underline{\mathsf{n}} =)\langle \mathsf{n}, \mathsf{n}^R \rangle = \langle \mathsf{m}^R, \mathsf{m} \rangle$. Note that, as $\underline{\mathsf{n}}$ is a lambda Galois connection, we have that

$$\left. \begin{array}{l} \mathsf{m}^R \text{ is (also) the left adjoint of } \mathsf{m}, \text{ and} \\ \mathsf{m} \text{ is (also) the right adjoint of } \mathsf{m}^R. \end{array} \right\} \tag{16.2}$$

We are now in the position to prove (iso-ls1). The inequality

$$\mathsf{m}(G(f)) \sqsubseteq G'(\mathsf{m} \circ f \circ \mathsf{m}^R)$$

is (lambda-gc1). As to the other direction, first of all note that, exploiting (16.2), we have that conditions (lambda-gc1) and (lambda-gc2) induce, for any $f' : \mathcal{D}' \to \mathcal{D}'$, and $x \in \mathcal{D}, x' \in \mathcal{D}'$,

$$\left. \begin{aligned} \mathsf{m}^R(G'(f')) &\sqsubseteq G(\mathsf{m}^R \circ f' \circ \mathsf{m}), \\ F'(\mathsf{m}(x))(x') &\sqsubseteq \mathsf{m}(F(x)(\mathsf{m}^R(x'))). \end{aligned} \right\} \tag{16.3}$$

So we have

$$\begin{aligned} G'(\mathsf{m} \circ f \circ \mathsf{m}^R) &= \mathsf{m} \circ \mathsf{m}^R \circ G'(\mathsf{m} \circ f \circ \mathsf{m}^R), && \text{by (16.1)}, \\ &\sqsubseteq \mathsf{m} \circ G(\mathsf{m}^R \circ \mathsf{m} \circ f \circ \mathsf{m}^R \circ \mathsf{m}), && f' = \mathsf{m} \circ f \circ \mathsf{m}^R \text{ in } (b), \\ &= \mathsf{m} \circ G(f), && \text{since } \mathsf{m}^R = \mathsf{m}^{-1}. \end{aligned}$$

Therefore we have proved $\mathsf{m}(G(f)) = G'(\mathsf{m} \circ f \circ \mathsf{m}^{-1})$.

Proof of (iso-ls2). Notice that $\mathsf{m}(F(d))(e) \sqsubseteq F'(\mathsf{m}(d))(\mathsf{m}(e))$, since

$$\begin{array}{rcll}\mathsf{m}(F(d))(e) & = & \mathsf{m}(F(\mathsf{m}^R(\mathsf{m}(d)))(e)), & \text{by (16.1)},\\ & \sqsubseteq & \mathsf{m}(\mathsf{m}^R(F'(\mathsf{m}(d))(\mathsf{m}(e)))), & \text{by (lambda-gc2)},\\ & = & F'(\mathsf{m}(d))(\mathsf{m}(e)), & \text{by (16.1)}.\end{array}$$

On the other hand, we have also $F'(\mathsf{m}(d))(\mathsf{m}(e)) \sqsubseteq \mathsf{m}(F(d)(e))$. In fact

$$\begin{array}{rcll}F'(\mathsf{m}(d))(\mathsf{m}(e)) & \sqsubseteq & \mathsf{m}(F(d)(\mathsf{m}^R(\mathsf{m}(e)))), & \text{by (16.3)},\\ & = & \mathsf{m}(F(d)(e)), & \text{by (16.1)}. \blacksquare\end{array}$$

The following proposition will be used in Corollary 16.3.32 to prove that the models $\mathcal{D}_\infty$ and $\mathcal{F}^{\text{Scott}}$ equate the same terms.

16.1.13 Proposition *Let $\mathcal{D}$ and $\mathcal{D}'$ be isomorphic as lambda structures. Then:*

(i) *$\mathcal{D}$ and $\mathcal{D}'$ are isomorphic as λ-models;*
(ii) *they equate the same terms, i.e. $\mathcal{D} \models M = N \Leftrightarrow \mathcal{D}' \models M = N$.*

Proof (i) By induction on M using Lemma 16.1.12.
(ii) Using (i) and Lemma 16.1.11. ■

16.2 Filter models

In this section, we define the notion of filter model and prove that the interpretation of a term is the set of its types (Type-semantics Theorem). Using this theorem and the results in Chapter 14, we study which conditions have to be imposed on a type theory in order to induce a filter model. At the end of this section we also study representability of continuous functions.

16.2.1 Definition Let $\mathcal{T} \in \text{TT}$. The *filter quasi λ-model* of $\mathcal{T}$ is a quasi λ-model, denoted by $\mathcal{F}^\mathcal{T}$, where $[\![\ ]\!]^{\mathcal{F}^\mathcal{T}} : \Lambda \times \mathsf{Env}_{\mathcal{F}^\mathcal{T}} \to \mathcal{F}^\mathcal{T}$ is defined by

$$\begin{array}{ll}[\![x]\!]^{\mathcal{F}^\mathcal{T}}_\rho & = \rho(x);\\ [\![MN]\!]^{\mathcal{F}^\mathcal{T}}_\rho & = \uparrow\{B \in \mathbb{T}^\mathcal{T} \mid \exists A \in [\![N]\!]^{\mathcal{F}^\mathcal{T}}_\rho .(A \to B) \in [\![M]\!]^{\mathcal{F}^\mathcal{T}}_\rho\};\\ [\![\lambda x.M]\!]^{\mathcal{F}^\mathcal{T}}_\rho & = \uparrow\{A \to B \mid B \in [\![M]\!]^{\mathcal{F}^\mathcal{T}}_{\rho[x:=\uparrow A]}\}.\end{array}$$

The notion of filter structure $\langle \mathcal{F}^\mathcal{T}, F^\mathcal{T}, G^\mathcal{T}\rangle$ given in Definition 13.4.5 contains two operations $F^\mathcal{T}$ and $G^\mathcal{T}$ that can be used for interpreting application and abstraction. These operations coincide with the way application and abstraction are interpreted in Definition 16.2.1. This leads to the following:

16.2.2 Lemma *Let $\mathcal{T} \in \text{TT}$. The filter structure $\langle \mathcal{F}^\mathcal{T}, F^\mathcal{T}, G^\mathcal{T}\rangle$ induces a quasi λ-model which coincides with the notion of filter quasi λ-model given in Definition* 16.2.1.

Proof Note first that the filter structure $\langle \mathcal{F}^{\mathcal{T}}, F^{\mathcal{T}}, G^{\mathcal{T}} \rangle$ is a lambda structure by Definition 15.2.3 and the comment just before Remark 13.4.6. In the case $\mathcal{T} \in \mathrm{TT}^{\text{-U}}$, we have that $\langle \mathcal{F}^{\mathcal{T}}, F^{\mathcal{T}}, G^{\mathcal{T}} \rangle$ is also a *strict* lambda structure, i.e.

$$F^{\mathcal{T}} \in [\mathcal{F}^{\mathcal{T}} \to_{\mathrm{s}} [\mathcal{F}^{\mathcal{T}} \to_{\mathrm{s}} \mathcal{F}^{\mathcal{T}}]] \text{ and } G^{\mathcal{T}} \in [[\mathcal{F}^{\mathcal{T}} \to_{\mathrm{s}} \mathcal{F}^{\mathcal{T}}] \to_{\mathrm{s}} \mathcal{F}^{\mathcal{T}}],$$

see Definition 15.4.5. Hence $\langle \mathcal{F}^{\mathcal{T}}, F^{\mathcal{T}}, G^{\mathcal{T}} \rangle$ induces a quasi λ-model, by Proposition 16.1.6. It is easy to see that $[\![\ ]\!]^{F^{\mathcal{T}}, G^{\mathcal{T}}} = [\![\ ]\!]^{\mathcal{F}^{\mathcal{T}}}$. ∎

We now define two classes of filter models: filter λ-models and filter λI-models.

16.2.3 Definition

(i) Let $\mathcal{T} \in \mathrm{TT}^{\mathrm{U}}$. We say that $\mathcal{F}^{\mathcal{T}}$ is a *filter model* if the filter quasi λ-model $\mathcal{F}^{\mathcal{T}} = \langle \mathcal{F}^{\mathcal{T}}, \cdot, [\![\ ]\!]^{\mathcal{F}^{\mathcal{T}}} \rangle$ is a λ-model.

(ii) Let $\mathcal{T} \in \mathrm{TT}^{\text{-U}}$. We say that $\mathcal{F}^{\mathcal{T}}$ is a *filter model* if the filter quasi λ-model $\mathcal{F}^{\mathcal{T}} = \langle \mathcal{F}^{\mathcal{T}}, \cdot, [\![\ ]\!]^{\mathcal{F}^{\mathcal{T}}} \rangle$ is a λI-model.

16.2.4 Proposition

(i) *Let $\mathcal{T} \in \mathrm{TT}^{\mathrm{U}}$. Then $\mathcal{F}^{\mathcal{T}}$ is a filter λ-model iff for all $M, N \in \Lambda$*

$$[\![(\lambda x.M)N]\!]^{\mathcal{F}^{\mathcal{T}}}_{\rho} = [\![M[x := N]]\!]^{\mathcal{F}^{\mathcal{T}}}_{\rho}.$$

(ii) *Let $\mathcal{T} \in \mathrm{TT}^{\text{-U}}$. Then $\mathcal{F}^{\mathcal{T}}$ is a filter λI-model iff for all $M, N \in \Lambda$*

$$x \in \mathrm{FV}(M) \Rightarrow [\![(\lambda x.M)N]\!]^{\mathcal{F}^{\mathcal{T}}}_{\rho} = [\![M[x := N]]\!]^{\mathcal{F}^{\mathcal{T}}}_{\rho}.$$

Proof Both equivalences follow from Proposition 16.1.8. ∎

The type-semantics result, Theorem 16.2.7, is important. It has as consequence that for a closed untyped lambda term M and a $\mathcal{T} \in \mathrm{TT}$ one has

$$[\![M]\!]^{\mathcal{F}^{\mathcal{T}}} = \{A \mid \vdash^{\mathcal{T}}_{\cap} M : A\},$$

i.e. the semantical meaning of M is the collection of its types.

16.2.5 Definition Let Γ be a context and $\rho \in \mathsf{Env}_{\mathcal{F}^{\mathcal{T}}}$. Then Γ *agrees with* ρ, written $\Gamma \models \rho$, if

$$(x : A) \in \Gamma \Rightarrow A \in \rho(x).$$

16.2.6 Proposition

(i) $\Gamma \models \rho \;\&\; \Gamma' \models \rho \Rightarrow \Gamma \uplus \Gamma' \models \rho.$

(ii) $\Gamma \models \rho[x :=\uparrow A] \Rightarrow \Gamma \backslash x \models \rho.$

Proof Immediate. ■

16.2.7 Theorem (Type-semantics Theorem) *Let* $\mathcal{T} \in \mathrm{TT}$ *and let* $\langle \mathcal{F}^{\mathcal{T}}, \cdot, [\![\ ]\!]^{\mathcal{F}^{\mathcal{T}}} \rangle$ *be its corresponding filter quasi λ-model. Then, for any* $M \in \Lambda$ *and* $\rho \in \mathsf{Env}_{\mathcal{F}^{\mathcal{T}}}$,

$$[\![M]\!]_\rho^{\mathcal{F}^{\mathcal{T}}} = \{A \mid \Gamma \vdash_\cap^{\mathcal{T}} M : A \textit{ for some } \Gamma \models \rho\}.$$

Proof We have two cases.

(i) Let $\mathcal{T} \in \mathrm{TT}^{\mathtt{U}}$. We proceed by induction on the structure of M.

Case $M \equiv x$. Then

$$\begin{array}{rcl} [\![M]\!]_\rho^{\mathcal{F}^{\mathcal{T}}} & = & \rho(x) \\ & = & \{A \mid A \in \rho(x)\} \\ & = & \{A \mid A \in \rho(x) \;\&\; x : A \vdash_\cap^{\mathcal{T}} x : A\} \\ & = & \{A \mid \Gamma \vdash_\cap^{\mathcal{T}} x : A \text{ for some } \Gamma \models \rho\}, \end{array}$$

by Definition 16.2.5 and the Inversion Lemma 14.1.1(i).

Case $M \equiv NL$. Then

$$\begin{array}{rcl} [\![M]\!]_\rho^{\mathcal{F}^{\mathcal{T}}} & = & [\![N]\!]_\rho^{\mathcal{F}^{\mathcal{T}}} \cdot [\![L]\!]_\rho^{\mathcal{F}^{\mathcal{T}}} \\ & = & \uparrow\{A \mid \exists B \in [\![L]\!]_\rho^{\mathcal{F}^{\mathcal{T}}}.(B \to A) \in [\![N]\!]_\rho^{\mathcal{F}^{\mathcal{T}}}\} \\ & = & \{A \mid \exists k{>}0 \exists B_1,\dots,B_k, C_1,\dots,C_k. \\ & & [\,(B_i{\to}C_i) \in [\![N]\!]_\rho^{\mathcal{F}^{\mathcal{T}}} \;\&\; B_i \in [\![L]\!]_\rho^{\mathcal{F}^{\mathcal{T}}} \;\&\; (\bigcap_{1 \le i \le k} C_i) \le A]\} \cup \uparrow\{\mathtt{U}\}, \\ & & \text{by definition of } \uparrow, \\ & = & \{A \mid \exists k{>}0 \exists B_1,\dots,B_k, C_1,\dots,C_k, \Gamma_1,\dots,\Gamma_k, \Delta_1,\dots,\Delta_k \\ & & [\Gamma_i, \Delta_i \models \rho \;\&\; \Gamma_i \vdash_\cap^{\mathcal{T}} N : (B_i{\to}C_i) \\ & & \;\&\; \Delta_i \vdash_\cap^{\mathcal{T}} L : B_i \;\&\; C_1 \cap \cdots \cap C_k \le A]\} \cup \uparrow\{\mathtt{U}\}, \\ & & \text{by the induction hypothesis,} \\ & = & \{A \mid \Gamma \vdash_\cap^{\mathcal{T}} NL : A \text{ for some } \Gamma \models \rho\}, \\ & & \text{taking } \Gamma = \Gamma_1 \uplus \cdots \uplus \Gamma_k \uplus \cdots \uplus \Delta_1 \uplus \cdots \uplus \Delta_k, \\ & & \text{by Theorem 14.1.1(ii) and Proposition 16.2.6(i).} \end{array}$$

Case $M \equiv \lambda x.N$. Then

$$
\begin{aligned}
&[\![\lambda x.N]\!]^{\mathcal{F}^{\mathcal{T}}}_{\rho} \\
&\quad = G^{\mathcal{T}}(\boldsymbol{\lambda} X \in \mathcal{F}^{\mathcal{T}}.[\![N]\!]^{\mathcal{F}^{\mathcal{T}}}_{\rho[x:=X]} \\
&\quad =\uparrow \{(B \to C) \mid C \in [\![N]\!]^{\mathcal{F}^{\mathcal{T}}}_{\rho[x:=\uparrow B]}\} \\
&\quad = \{A \mid \exists k>0 \exists B_1, \dots, B_k, C_1, \dots, C_k, \Gamma_1, \dots, \Gamma_k[\Gamma_i \models \rho[x: = \uparrow B_i] \\
&\quad\ \& \ \Gamma_i, x{:}B_i \vdash^{\mathcal{T}}_{\cap} N : C_i \ \& \ (B_1 \to C_1) \cap \cdots \cap (B_k \to C_k) \leq A]\},
\end{aligned}
$$

by the induction hypothesis,

$$= \{A \mid \Gamma \vdash^{\mathcal{T}}_{\cap} \lambda x.N : A \text{ for some } \Gamma \models \rho\},$$

taking $\Gamma = (\Gamma_1 \uplus \cdots \uplus \Gamma_k)\backslash x$, by Theorem 14.1.1(iii), rule $(\leq)$ and Proposition 16.2.6(ii).

(ii) Let $\mathcal{T} \in \mathrm{TT}^{\text{-}\mathtt{U}}$. Similarly, but note that in the case $M = NL$ we drop '$\cup\uparrow\{\mathtt{U}\}$' both times. ■

16.2.8 Corollary

(i) *Let* $\mathcal{T} \in \mathrm{TT}^{\mathtt{U}}$. *Then*

$$\mathcal{F}^{\mathcal{T}} \textit{ is a filter } \lambda\textit{-model} \Leftrightarrow [\Gamma \vdash^{\mathcal{T}}_{\cap} (\lambda x.M) : (B \to A) \Rightarrow \Gamma, x{:}B \vdash^{\mathcal{T}}_{\cap} M : A].$$

(ii) *Let* $\mathcal{T} \in \mathrm{TT}^{\text{-}\mathtt{U}}$. *Then*

$$\mathcal{F}^{\mathcal{T}} \textit{ is a filter } \lambda\mathsf{I}\textit{-model} \Leftrightarrow$$

$$[\Gamma \vdash^{\mathcal{T}}_{\cap} (\lambda x.M) : (B \to A) \ \& \ x \in \mathrm{FV}(M) \Rightarrow \Gamma, x{:}B \vdash^{\mathcal{T}}_{\cap} M : A].$$

Proof (i) By Propositions 16.2.4(i), 14.2.1(i) and Corollary 14.2.5(i).

(ii) By Propositions 16.2.4(ii), 14.2.1(ii) and Corollary 14.2.5(ii). ■

16.2.9 Corollary

(i) *Let* $\mathcal{T} \in \mathrm{TT}^{\mathtt{U}}$. *Then*

$$\mathcal{T} \textit{ is } \boldsymbol{\beta}\textit{-sound} \Rightarrow \mathcal{F}^{\mathcal{T}} \textit{ is a filter } \lambda\textit{-model.}$$

(ii) *Let* $\mathcal{T} \in \mathrm{TT}^{\text{-}\mathtt{U}}$. *Then*

$$\mathcal{T} \textit{ is } \boldsymbol{\beta}\textit{-sound} \Rightarrow \mathcal{F}^{\mathcal{T}} \textit{ is a filter } \lambda\mathsf{I}\textit{-model.}$$

Proof By the corollary above and Theorem 14.1.9(iii). ■

16.2.10 Corollary

(i) *Let*

$$\mathcal{T} \in \{\text{Scott}, \text{Park}, \text{CDZ}, \text{HR}, \text{DHM}, \text{BCD}, \text{AO}, \text{Plotkin}, \text{Engeler}, \text{CDS}\}.$$

Then

$$\mathcal{F}^{\mathcal{T}} \textit{ is a filter } \lambda\textit{-model.}$$

(ii) *Let* $\mathcal{T} \in \{\text{HL}, \text{CDV}, \text{CD}\}$. *Then*

$$\mathcal{F}^{\mathcal{T}} \textit{ is a filter } \lambda\mathsf{I}\textit{-model.}$$

Proof (i) By (i) of the previous corollary and Theorem 14.1.7.
(ii) By (ii) of the previous corollary, using Theorem 14.1.7. ∎

16.2.11 Proposition

(i) *Let* $\mathcal{T} \in \text{TT}^{\mathtt{U}}$. *Then*

$\mathcal{T}$ *is natural and* $\boldsymbol{\beta}$*- and* $\boldsymbol{\eta}^{\mathtt{U}}$*-sound* $\Rightarrow$ $\mathcal{F}^{\mathcal{T}}$ *is an extensional filter* λ*-model.*

(ii) *Let* $\mathcal{T} \in \text{TT}^{\text{-}\mathtt{U}}$. *Then*

$\mathcal{T}$ *is proper and* $\boldsymbol{\beta}$*- and* $\boldsymbol{\eta}$*-sound* $\Rightarrow$ $\mathcal{F}^{\mathcal{T}}$ *is an extensional filter* $\lambda\mathsf{I}$*-model.*

Proof (i) and (ii). By Corollary 16.2.9(i) and (ii), $\mathcal{F}^{\mathcal{T}}$ is a $\lambda(\mathsf{I})$-model. As to extensionality it suffices by Corollary 16.1.9 to verify, for $x \notin \text{FV}(M)$, that

$$[\![\lambda x.Mx]\!]_{\rho} = [\![M]\!]_{\rho}. \qquad (\boldsymbol{\eta})$$

This follows from Theorems 16.2.7, and 14.2.15. ∎

16.2.12 Corollary

(i) *Let* $\mathcal{T} \in \{\text{Scott}, \text{Park}, \text{CDZ}, \text{HR}, \text{DHM}\}$. *Then*

$$\mathcal{F}^{\mathcal{T}} \textit{ is an extensional filter } \lambda\textit{-model.}$$

(ii) *Let* $\mathcal{T} = \text{HL}$. *Then*

$$\mathcal{F}^{\mathcal{T}} \textit{ is an extensional filter } \lambda\mathsf{I}\textit{-model.}$$

Proof (i) and (ii) follow from Corollary 14.2.13. ∎

As shown in Meyer (1982), see also B[1984] Ch.4, a lambda structure $\mathcal{D}$ is a λ-model provided that there are elements K, $\mathsf{S}, \varepsilon \in \mathcal{D}$, satisfying certain properties. Thus, a condition for being a filter λ-model can be obtained by simply requiring the existence of such elements. This yields a characterization of the natural type theories which induce λ-models. See Alessi (1991) for the rather technical proof.

16.2.13 Theorem *Let* $\mathcal{T} \in \mathrm{NTT}^{\mathrm{U}}$.

(i) *The filter structure* $\mathcal{F}^{\mathcal{T}}$ *is a filter* λ*-model if and only if the following three conditions are fulfilled in* $\mathcal{T}$.

(K) *For all* C, E *one has*

$$\begin{array}{rl} C \leq E \Leftrightarrow & \forall D\, \exists k \geq 1, A_1, \ldots, A_k, B_1, \ldots, B_k. \\ & (A_1 \to B_1 \to A_1) \cap \cdots \cap (A_k \to B_k \to A_k) \leq C \to D \to E. \end{array}$$

(S) *For all* D, E, F, G *one has*

$$\exists H.[E \leq F \to H \;\&\; D \leq F \to H \to G] \Leftrightarrow$$

$$\left[\begin{array}{l} \exists k \geq 1, A_1, \ldots, A_k, B_1, \ldots, B_k, C_1, \ldots, C_k. \\ \;[((A_1 \to B_1 \to C_1) \to (A_1 \to B_1) \to A_1 \to C_1)\cap \\ \quad\vdots \\ \;\cap((A_k \to B_k \to C_k) \to (A_k \to B_k) \to A_k \to C_k)] \;\leq\; D \to E \to F \to G \end{array}\right].$$

(ε) *For all* C, D *one has*

$$\left[\begin{array}{l} \exists k \geq 1, A_1, \ldots, A_k, B_1, \ldots, B_k. \\ ((A_1 \to B_1) \to A_1 \to B_1) \cap \cdots \cap ((A_k \to B_k) \to A_k \to B_k) \leq (C \to D) \end{array}\right] \Leftrightarrow$$

$$\exists m \geq 1, E_1, \ldots, E_m, F_1, \ldots, F_m. C \leq (E_1 \to F_1) \cap \cdots \cap (E_m \to F_m) \leq D.$$

(ii) *The structure* $\mathcal{F}^{\mathcal{T}}$ *is an extensional filter* λ*-model iff the condition* (ε) *above is replaced by the following two.*

(ε_1) $\forall A\, \exists k \geq 1, A_1, \ldots, A_k, B_1, \ldots, B_k. A = (A_1 \to B_1) \cap \cdots \cap (A_k \to B_k)$;

(ε_2) $\forall A, B\, \exists k \geq 1, A_1, \ldots, A_k.[(A_1 \to A_1) \cap \cdots \cap (A_k \to A_k) \leq (A \to B)$
$\Leftrightarrow A \leq B]$.

Representability of continuous functions

In this subsection, following Alessi et al. (2004b), we will isolate a number of conditions on a $\mathcal{T} \in \mathrm{NTT}^{\mathrm{U}}$ to characterize properties of the set of *representable functions* in $\langle \mathcal{F}^{\mathcal{T}}, F^{\mathcal{T}}, G^{\mathcal{T}} \rangle$, i.e. the set of functions in the image of $F^{\mathcal{T}}$.

16.2.14 Definition A function $f : \mathcal{D} \to \mathcal{D}$ is called *representable* in the lambda structure $\langle \mathcal{D}, F, G \rangle$ if $f = F(d)$ for some $d \in \mathcal{D}$.

Note that since $F : \mathcal{D} \to [\mathcal{D} \to \mathcal{D}]$, all representable functions are continuous.

16.2.15 Lemma *Let $\mathcal{T} \in \mathrm{NTT}^{\mathtt{U}}$ and let $f \in [\mathcal{F}^{\mathcal{T}} \to \mathcal{F}^{\mathcal{T}}]$. Then*

$$f \textit{ is representable in } \langle \mathcal{F}^{\mathcal{T}}, F^{\mathcal{T}}, G^{\mathcal{T}} \rangle \Leftrightarrow F^{\mathcal{T}} \circ G^{\mathcal{T}}(f) = f.$$

Proof ($\Leftarrow$) Trivial.

($\Rightarrow$) Suppose $f = F^{\mathcal{T}}(X)$. We claim $F^{\mathcal{T}}(G^{\mathcal{T}}(F^{\mathcal{T}}(X))) = F^{\mathcal{T}}(X)$. We have $G^{\mathcal{T}}(F^{\mathcal{T}}(X)) = \uparrow\{A \to B \mid A \to B \in X\}$. Hence

$$A \to B \in G^{\mathcal{T}}(F^{\mathcal{T}}(X)) \Leftrightarrow A \to B \in X.$$

So $\forall Y. F^{\mathcal{T}}(G^{\mathcal{T}}(F^{\mathcal{T}}(X)))(Y) = F^{\mathcal{T}}(X)(Y)$, hence $F^{\mathcal{T}}(G^{\mathcal{T}}(f)) = f$. ∎

16.2.16 Lemma *Let $\mathcal{T} \in \mathrm{NTT}^{\mathtt{U}}$. Let $A, B \in \mathbb{T}^{\mathcal{T}}$. Then*

$$G^{\mathcal{T}}(\uparrow A \Rightarrow \uparrow B) = \uparrow(A \to B).$$

Proof

$$\begin{aligned} G^{\mathcal{T}}(\uparrow A \Rightarrow \uparrow B) &= \uparrow\{(C \to D) \mid D \in (\uparrow A \Rightarrow \uparrow B)(\uparrow C)\} \\ &= \uparrow(A \to B). \end{aligned}$$

In the last step the inclusion $\supseteq$ is trivial: $(A \to B)$ is one of the $C \to D$. Now suppose $C \to D$ is such that $D \in (\uparrow A \Rightarrow \uparrow B)(\uparrow C)$. Then there are two cases.

Case $\uparrow A \subseteq \uparrow C$. This means $C \leq A$, so $(\uparrow A \Rightarrow \uparrow B)(\uparrow C) = \uparrow B$, so $D \geq B$. Hence in this case $A \to B \leq C \to D$ by rule $(\to)$.

Case $\uparrow A \not\subseteq \uparrow C$. Then $D = \mathtt{U}$, hence $C \to D = \mathtt{U}$, by rules $(\to)$, $(\mathtt{U}\to)$, and again $A \to B \leq C \to D$. Therefore also $\subseteq$ holds in the last equation. ∎

16.2.17 Lemma *Let $\mathcal{T} \in \mathrm{NTT}^{\mathtt{U}}$, $A_1, \ldots, A_n, B_1, \ldots, B_n \in \mathbb{T}^{\mathcal{T}}$ and let the function $h : \mathcal{F}^{\mathcal{T}} \to \mathcal{F}^{\mathcal{T}}$ be given by*

$$h = (\uparrow A_1 \Rightarrow \uparrow B_1) \sqcup \cdots \sqcup (\uparrow A_n \Rightarrow \uparrow B_n).$$

Then, for all $C \in \mathbb{T}^{\mathcal{T}}$ we have that

(i) $h(\uparrow C) = \{D \mid \exists k \geq 1 \exists i_1, \ldots, i_k. [B_{i_1} \cap \cdots \cap B_{i_k} \leq D \;\&\; C \leq A_{i_1} \cap \cdots \cap A_{i_k}]\} \cup \uparrow\{\mathtt{U}\}$.

(ii) $(F^{\mathcal{T}} \circ G^{\mathcal{T}})(h)(\uparrow C) = \{D \mid (A_1 \to B_1) \cap \cdots \cap (A_n \to B_n) \leq (C \to D)\}$.

Proof (i)
$$\begin{aligned} h(\uparrow C) &= \bigsqcup\{\uparrow B_i \mid \uparrow A_i \subseteq \uparrow C \ \&\ 1 \le i \le n\} \\ &= \uparrow B_{i_1} \sqcup \cdots \sqcup \uparrow B_{i_k} \\ &\quad \text{for } \{i_1,\ldots,i_k\} = \{i \mid C \le A_i \& 1 \le i \le n\} \\ &= \uparrow(B_{i_1} \cap \cdots \cap B_{i_k}) \cup \uparrow\mathtt{U}, \text{ by Proposition 13.4.4(iii)}, \\ &= \{D \mid \exists k \ge 1 \exists i_1,\ldots,i_k. \\ &\quad B_{i_1} \cap \cdots \cap B_{i_k} \le D \ \&\ C \le A_{i_1} \cap \cdots \cap A_{i_k}\} \cup \uparrow\{\mathtt{U}\}. \end{aligned}$$

(ii) $(F^{\mathcal{T}} \circ G^{\mathcal{T}})(h)(\uparrow C) =$
$$\begin{aligned} &= F^{\mathcal{T}}(G^{\mathcal{T}}(\uparrow A_1 \Rightarrow \uparrow B_1) \sqcup \cdots \sqcup G^{\mathcal{T}}(\uparrow A_n \Rightarrow \uparrow B_n))(\uparrow C), \\ &\quad \text{by Lemma 15.2.2(i)}, \\ &= F^{\mathcal{T}}(\uparrow(A_1 \to B_1) \sqcup \cdots \sqcup \uparrow(A_n \to B_n))(\uparrow C), \text{ by Lemma 16.2.16}, \\ &= F^{\mathcal{T}}(\uparrow((A_1 \to B_1) \cap \cdots \cap (A_n \to B_n)))(\uparrow C), \\ &\quad \text{by Proposition 13.4.4(iii)}, \\ &= \{D \mid \exists E \in \uparrow C.(E \to D) \in \uparrow((A_1 \to B_1) \cap \cdots \cap (A_n \to B_n))\}, \\ &\quad \text{see Definition 13.4.5}, \\ &= \{D \mid \exists E \ge C.(A_1 \to B_1) \cap \cdots \cap (A_n \to B_n) \le (E \to D)\} \\ &= \{D \mid (A_1 \to B_1) \cap \cdots \cap (A_n \to B_n) \le (C \to D)\}, \text{ by } (\to). \ \blacksquare \end{aligned}$$

16.2.18 Notation We define the function $\mathbb{K}^{\mathcal{T}} \in \mathcal{F}^{\mathcal{T}} \to \mathcal{F}^{\mathcal{T}} \to \mathcal{F}^{\mathcal{T}}$ as

$$\mathbb{K}^{\mathcal{T}} = \lambda\!\!\lambda X \in \mathcal{F}^{\mathcal{T}}.\lambda\!\!\lambda Y \in \mathcal{F}^{\mathcal{T}}.X.$$

For each $X \in \mathcal{F}^{\mathcal{T}}$, $\mathbb{K}^{\mathcal{T}}(X)$ is a constant function in $[\mathcal{F}^{\mathcal{T}} \to \mathcal{F}^{\mathcal{T}}]$.

16.2.19 Theorem *Let $\mathcal{T} \in \mathrm{NTT}^{\mathtt{U}}$. Let **R** be the set of representable functions in $\langle \mathcal{F}^{\mathcal{T}}, F^{\mathcal{T}}, G^{\mathcal{T}} \rangle$. Then we have:*

(i) ***R** contains the bottom function $\mathbb{K}^{\mathcal{T}}(\bot_{\mathcal{F}^{\mathcal{T}}})$ iff for all C, D*

$$\mathtt{U} \le C \to D \ \Rightarrow\ \mathtt{U} \le D;$$

(ii) ***R** contains all constant functions iff for all B, C, D*

$$\mathtt{U} \to B \le C \to D \ \Rightarrow\ B \le D;$$

(iii) ***R** contains all continuous step functions iff for all A, B, C, D*

$$A \to B \le C \to D \ \&\ D \ne \mathtt{U} \ \Rightarrow\ C \le A \ \&\ B \le D;$$

(iv) ***R** contains (i.e. is the set of) all continuous functions iff $\mathcal{T}$ is β-sound.*

Proof (i) Assume that $\mathbb{K}^{\mathcal{T}}(\bot_{\mathcal{F}\mathcal{T}}) \in \boldsymbol{R}$. Then, by Lemma 16.2.15, we have $F^{\mathcal{T}}(G^{\mathcal{T}}(\mathbb{K}^{\mathcal{T}}(\bot_{\mathcal{F}\mathcal{T}}))) = \mathbb{K}^{\mathcal{T}}(\bot_{\mathcal{F}\mathcal{T}})$. Observe that

$$\begin{array}{rcll} G^{\mathcal{T}}(\mathbb{K}^{\mathcal{T}}(\bot_{\mathcal{F}\mathcal{T}})) & = & \uparrow\{A \to \mathtt{U} \mid A \in \mathbb{T}\}, & \text{by Definition 13.4.5,} \\ & = & \uparrow\mathtt{U}, & \text{since } \mathcal{T} \text{ is natural.} \end{array}$$

Hence $F^{\mathcal{T}}(\bot_{\mathcal{F}\mathcal{T}}) = \mathbb{K}^{\mathcal{T}}(\bot_{\mathcal{F}\mathcal{T}})$, so in particular $F^{\mathcal{T}}(\bot_{\mathcal{F}\mathcal{T}})(\uparrow C) = \bot_{\mathcal{F}\mathcal{T}}$, and hence

$$\begin{array}{rcll} \{D \mid \mathtt{U} \leq D\} & = & \uparrow\mathtt{U} = \bot_{\mathcal{F}\mathcal{T}} & \\ & = & F^{\mathcal{T}}(\bot_{\mathcal{F}\mathcal{T}})(\uparrow C) & \\ & = & F^{\mathcal{T}}(\uparrow\mathtt{U})(\uparrow C) & \\ & = & \{D \mid \mathtt{U} \leq (C \to D)\}, & \text{by Definition 13.4.5(i).} \end{array}$$

But then $\mathtt{U} \leq C \to D \;\Rightarrow\; \mathtt{U} \leq D$.

($\Leftarrow$) Suppose $\mathtt{U} \leq C \to D \;\Rightarrow\; \mathtt{U} \leq D$. We show that $F^{\mathcal{T}}(\bot_{\mathcal{F}\mathcal{T}}) = \mathbb{K}^{\mathcal{T}}(\bot_{\mathcal{F}\mathcal{T}})$. Indeed,

$$\begin{array}{rcll} F^{\mathcal{T}}(\bot_{\mathcal{F}\mathcal{T}})(X) & = & \{B \mid \exists A \in X.(A \to B) \in \bot_{\mathcal{F}\mathcal{T}}\} & \\ & = & \{B \mid \exists A \in X.(A \to B) \in \uparrow\mathtt{U}\} & \\ & = & \{B \mid \exists A \in X.\mathtt{U} \leq (A \to B)\} & \\ & = & \{B \mid \mathtt{U} \leq B\}, & \text{by the assumption,} \\ & = & \uparrow\mathtt{U} = \bot_{\mathcal{F}\mathcal{T}}. & \end{array}$$

(ii) Suppose that $\mathtt{U} \to B \leq C \to D \;\Rightarrow\; B \leq D$. We first show that each compact constant function $\mathbb{K}^{\mathcal{T}}(\uparrow B)$ is represented by $\uparrow(\mathtt{U} \to B)$. Indeed,

$$\begin{array}{rcll} D \in \uparrow(\mathtt{U} \to B) \cdot \uparrow C & \Leftrightarrow & C \to D \in \uparrow(\mathtt{U} \to B), & \text{by } (\to), \\ & \Leftrightarrow & \mathtt{U} \to B \leq C \to D & \\ & \Leftrightarrow & B \leq D, & \text{using the assumption,} \\ & \Leftrightarrow & D \in \uparrow B = \mathbb{K}^{\mathcal{T}}(\uparrow B)(\uparrow C). & \end{array}$$

Now we show that an arbitrary constant function $\mathbb{K}^{\mathcal{T}}(X)$ is representable. Then $X = \bigcup\{\uparrow B \mid B \in X\}$, where $\{\uparrow B \mid B \in X\}$ is directed. Notice that $\mathbb{K}^{\mathcal{T}}(X) = \bigsqcup_{B \in X} \mathbb{K}^{\mathcal{T}}(\uparrow B)$. Therefore by the representability of $\mathbb{K}^{\mathcal{T}}(\uparrow B)$ just proved, Lemma 16.2.15 and the continuity of $F^{\mathcal{T}} \circ G^{\mathcal{T}}$ we get

$$\begin{aligned}
\mathbb{K}^{\mathcal{T}}(X) &= \bigsqcup_{B\in X} \mathbb{K}^{\mathcal{T}}(\uparrow B) \\
&= \bigsqcup_{B\in X} (F^{\mathcal{T}} \circ G^{\mathcal{T}})(\mathbb{K}^{\mathcal{T}}(\uparrow B)) \\
&= (F^{\mathcal{T}} \circ G^{\mathcal{T}})(\bigsqcup_{B\in X} \mathbb{K}^{\mathcal{T}}(\uparrow B)) \\
&= (F^{\mathcal{T}} \circ G^{\mathcal{T}})(\mathbb{K}^{\mathcal{T}}(X)),
\end{aligned}$$

hence again by Lemma 16.2.15 the constant map $\mathbb{K}^{\mathcal{T}}(X)$ is representable.

Conversely, suppose that all constant functions are representable. Then $F^{\mathcal{T}} \circ G^{\mathcal{T}}(\mathbb{K}^{\mathcal{T}}(\uparrow B)) = \mathbb{K}^{\mathcal{T}}(\uparrow B)$, by Lemma 16.2.15. Therefore

$$\begin{aligned}
\mathtt{U}\to B \leq C\to D &\Rightarrow (C\to D) \in \uparrow(\mathtt{U}\to B) \\
&\Rightarrow D \in \uparrow(\mathtt{U}\to B)\cdot\uparrow C \\
&\Rightarrow D \in ((F^{\mathcal{T}} \circ G^{\mathcal{T}})(\mathbb{K}^{\mathcal{T}}(\uparrow B)))(\uparrow C), \\
&\qquad \text{since } \uparrow(\mathtt{U}\to B) \subseteq \mathbb{K}^{\mathcal{T}}(\uparrow B) \text{ by } (\to) \text{ and } (\mathtt{U}), \\
&\Rightarrow D \in (\mathbb{K}^{\mathcal{T}}(\uparrow B))(\uparrow C) = \uparrow B \\
&\Rightarrow B \leq D.
\end{aligned}$$

(iii) ($\Rightarrow$) Suppose all continuous step functions are representable. Suppose $A\to B \leq C\to D$, $D \neq \mathtt{U}$. Take $h = \uparrow A \Rightarrow \uparrow B$. By Lemma 16.2.17(ii) we have

$$\begin{aligned}
(F^{\mathcal{T}} \circ G^{\mathcal{T}})(h)(\uparrow C) &= \{E \mid A\to B \leq C\to E\} \\
h(\uparrow C) &= \{E \mid B \leq E \;\&\; C \leq A\} \cup \uparrow\mathtt{U}.
\end{aligned}$$

By the first assumption these two sets are equal. By the second assumption it follows that $C \leq A \;\&\; B \leq D$.

($\Leftarrow$) Let $h = X \Rightarrow Y$ be continuous. We have to show that

$$(F^{\mathcal{T}} \circ G^{\mathcal{T}})(h) = h. \tag{16.4}$$

By Proposition 15.1.5(ii) it suffices to show this for compact h. If $Y \neq \perp_{\mathcal{F}^{\mathcal{T}}}$, then by 15.1.9 both X, Y are compact, so $h = \uparrow A\to\uparrow B$. Then (16.4) holds by Lemma 16.2.17 and the assumption. If $Y = \perp_{\mathcal{F}^{\mathcal{T}}}$, then h is the bottom function and hence representable (the assumption in (iii) implies the assumption in (i)).

(iv) Let $\mathcal{T} \in \mathrm{NTT}^{\mathtt{U}}$. Let $h \in \mathcal{K}([\mathcal{F}^{\mathcal{T}} \to \mathcal{F}^{\mathcal{T}}])$. By Proposition 15.1.11(ii) it follows that for some $A_1, \ldots, A_n, B_1, \ldots, B_n \in \mathcal{T}$

$$h = (\uparrow A_1 \Rightarrow \uparrow B_1) \sqcup \cdots \sqcup (\uparrow A_n \Rightarrow \uparrow B_n),$$

as a finite element of $\mathcal{F}^{\mathcal{T}}$ is of the form $\uparrow A$.

($\Rightarrow$) Suppose all continuous functions are representable. Then since h above is continuous and by Lemma 16.2.15, one has

$$(F^{\mathcal{T}} \circ G^{\mathcal{T}})(h)(\uparrow C) = h(\uparrow C).$$

It follows from Lemma 16.2.17(i),(ii) that $\mathcal{T}$ is β-sound.

($\Leftarrow$) Suppose $\mathcal{T}$ is β-sound. Again by Lemma 16.2.17, for a compact continuous function h one has that $(F^{\mathcal{T}} \circ G^{\mathcal{T}})(h)$ and h coincide on the compact elements $\uparrow C$. Therefore by Proposition 15.1.5 they coincide everywhere. But then it follows again that $f = (F^{\mathcal{T}} \circ G^{\mathcal{T}})(f)$ for every continuous $f : \mathcal{F}^{\mathcal{T}} \to \mathcal{F}^{\mathcal{T}}$. Hence Lemma 16.2.15 applies. ∎

To induce a filter model $\boldsymbol{\beta}$-soundness is not necessary

The intersection type theories $\mathcal{T} \in \{\text{Scott}, \text{Park}, \text{BCD}, \text{CDZ}, \text{HR}, \text{AO}, \text{DHM}\}$ all induce filter λ-models, by Corollary 16.2.10(i). These type theories are all natural and $\boldsymbol{\beta}$-sound. Therefore by Theorem 16.2.19(iv) all continuous functions are representable in these $\mathcal{F}^{\mathcal{T}}$. In Sections 16.3 and 16.4 we will give many more filter λ-models arising from domain models. It is therefore interesting to ask whether there exist filter λ-models where *not all* continuous functions are representable. We answer affirmatively, and end this section by giving an example of a natural type theory ABD that is not $\boldsymbol{\beta}$-sound but nevertheless induces a filter λ-model $\mathcal{F}^{\text{ABD}}$. Therefore by the same theorem not all continuous functions are representable in $\mathcal{F}^{\text{ABD}}$. The model builds on an idea in Coppo et al. (1984). In Exercise 16.4 another such model, due to Alessi (1993), is constructed. See also Alessi et al. (2004b).

The theory ABD

16.2.20 Definition Let $\mathbb{A}^{\text{ABD}} = \{\mathtt{U}, \diamondsuit, \heartsuit\}$. We define ABD as the smallest natural type theory[2] that contains the axiom ($\diamondsuit$) where

$$(\diamondsuit)\ A \leq_{\text{ABD}} A[\diamondsuit := \heartsuit].$$

16.2.21 Lemma

(i) $A \leq_{\text{ABD}} B \ \Rightarrow\ A[\diamondsuit := \heartsuit] \leq_{\text{ABD}} B[\diamondsuit := \heartsuit]$.

(ii) $\Gamma \vdash^{\text{ABD}} M : A \ \Rightarrow\ \Gamma[\diamondsuit := \heartsuit] \vdash^{\text{ABD}} M : A[\diamondsuit := \heartsuit]$.

(iii) $\Gamma, \Gamma' \vdash^{\text{ABD}} M : A \ \Rightarrow\ \Gamma, \Gamma'[\diamondsuit := \heartsuit] \vdash^{\text{ABD}} M : A[\diamondsuit := \heartsuit]$.

(iv) $\Gamma, x{:}A_i \vdash^{\text{ABD}} M : B_i$ *for* $1 \leq i \leq n$ &
$(A_1 \to B_1) \cap \cdots \cap (A_n \to B_n) \leq_{\text{ABD}} C \to D \ \Rightarrow\ \Gamma, x{:}C \vdash^{\text{ABD}} M : D$.

Proof (i) By induction on the definition of $\leq_{\text{ABD}}$.

(ii) By induction on derivations using (i) for rule ($\leq_{\text{ABD}}$).

[2] ABD contains the axioms and rules of Definitions 13.1.4 and 13.1.19.

(iii) From (ii) and rule ($\leq_{\text{ABD}}$-L), taking into account that if $(x{:}B) \in \Gamma$, then $(x{:}B[\diamondsuit := \heartsuit]) \in \Gamma[\diamondsuit := \heartsuit]$ and $B \leq_{\text{ABD}} B[\diamondsuit := \heartsuit]$.

(iv) Let $\alpha_1, \dots, \alpha_n, \alpha'_1, \dots, \alpha'_{n'} \in \mathbb{A}^{\text{ABD}}$. We show by induction on the definition of $\leq_{\text{ABD}}$ that if the following statements hold

$$\begin{aligned}
&A \leq_{\text{ABD}} A',\\
&A = \alpha_1 \cap \dots \cap \alpha_n \cap (B_1 \to C_1) \cap \dots \cap (B_k \to C_k),\\
&A' = \alpha'_1 \cap \dots \cap \alpha'_{n'} \cap (B'_1 \to C'_1) \cap \dots \cap (B'_{k'} \to C'_{k'}),\\
&\Gamma, x{:}B_i \vdash^{\text{ABD}} M : C_i \quad \forall i \in \{1, \dots, k\},
\end{aligned}$$

then

$$\Gamma, x{:}B'_j \vdash^{\text{ABD}} M : C'_j \quad \forall j \in \{1, \dots, k'\}.$$

The only interesting case is when the applied rule is ($\diamondsuit$), i.e. we have

$$\begin{aligned}
&A \leq_{\text{ABD}} A[\diamondsuit := \heartsuit];\\
&A = \alpha_1 \cap \dots \cap \alpha_n \cap (B_1 \to C_1) \cap \dots \cap (B_k \to C_k);\\
&A' = A[\diamondsuit := \heartsuit].
\end{aligned}$$

By hypothesis, $\Gamma, x{:}B_i \vdash^{\text{ABD}} M : C_i$ for all $i \in \{1, \dots, k\}$, so we are done by (iii). ■

16.2.22 Theorem

(i) ABD *is a* TT *that is not* $\boldsymbol{\beta}$*-sound.*

(ii) *Nevertheless* $\mathcal{F}^{\text{ABD}}$ *is a filter* λ*-model.*

Proof (i) By definition ABD is in TT^{U}. We have $\diamondsuit \to \diamondsuit \leq_{\text{ABD}} \heartsuit \to \heartsuit$, but $\heartsuit \not\leq_{\text{ABD}} \diamondsuit$, so it is not $\boldsymbol{\beta}$-sound.

(ii) To show that $\mathcal{F}^{\text{ABD}}$ is a λ-model, it suffices, by Proposition 16.2.8, to verify that $\Gamma \vdash^{\text{ABD}} \lambda x.M : A \to B \;\Rightarrow\; \Gamma, x{:}A \vdash^{\text{ABD}} M : B$. Suppose that $\Gamma \vdash^{\text{ABD}} \lambda x.M : A \to B$. By Lemma 14.1.1(iii), there are $C_1, \dots, C_n, D_1, \dots, D_n$ such that

$$\begin{aligned}
&(C_1 \to D_1) \cap \dots \cap (C_n \to D_n) \leq_{\text{ABD}} A \to B\\
&\forall i \in \{1, \dots, n\} \Gamma, x{:}C_i \vdash^{\text{ABD}} M : D_i.
\end{aligned}$$

So, we are done by Lemma 16.2.21(iv). ■

For example the step function $\uparrow\diamondsuit \Rightarrow \uparrow\diamondsuit$ is not representable in $\mathcal{F}^{\text{ABD}}$.

16.3 $\mathcal{D}_\infty$ models as filter models

This section shows the connection between filter models and $\mathcal{D}_\infty$ models, see Scott (1972) or B[1984]. We will work in the category **ALG** of ω-algebraic complete lattices and Scott-continuous maps. In other categories, such as those of Scott domains or stable sets, filter models do not capture the $\mathcal{D}_\infty$-models in their full generality.

D_∞ models

This subsection recalls some basic concepts of the standard $\mathcal{D}_\infty$ construction and fixes some notation, see Scott (1972), B[1984], and Gierz et al. (1980). See also Definition 10.3.1, where the construction is given in a more categorical setting.

In the rest of this subsection we will recall main definitions and results presented in Section 18.2 of B[1984].

16.3.1 Definition

(i) Let $\mathcal{D}_0$ be an ω-algebraic complete lattice and

$$\langle \mathsf{i}_0, \mathsf{j}_0 \rangle$$

be an *embedding–projection* pair between $\mathcal{D}_0$ and $[\mathcal{D}_0 \to \mathcal{D}_0]$, i.e.

$$\begin{aligned} \mathsf{i}_0 &: \mathcal{D}_0 \to [\mathcal{D}_0 \to \mathcal{D}_0] \\ \mathsf{j}_0 &: [\mathcal{D}_0 \to \mathcal{D}_0] \to \mathcal{D}_0 \end{aligned}$$

are Scott continuous maps satisfying

$$\begin{aligned} \mathsf{i}_0 \circ \mathsf{j}_0 &\sqsubseteq \mathsf{Id}_{[\mathcal{D}_0 \to \mathcal{D}_0]} \\ \mathsf{j}_0 \circ \mathsf{i}_0 &= \mathsf{Id}_{\mathcal{D}_0}. \end{aligned}$$

(ii) Define a *tower* $\langle \mathsf{i}_n, \mathsf{j}_n \rangle : \mathcal{D}_n \to \mathcal{D}_{n+1}$ in the following way:

$\mathcal{D}_{n+1} = [\mathcal{D}_n \to \mathcal{D}_n]$;
$\mathsf{i}_n(f) = \mathsf{i}_{n-1} \circ f \circ \mathsf{j}_{n-1}$ for any $f \in \mathcal{D}_n$;
$\mathsf{j}_n(g) = \mathsf{j}_{n-1} \circ g \circ \mathsf{i}_{n-1}$ for any $g \in \mathcal{D}_{n+1}$.

(iii) For $d \in \Pi_{n\in\omega}\mathcal{D}_n$ write $d_n = d(n)$. The set $\mathcal{D}_\infty$ is defined by

$$\mathcal{D}_\infty = \{d \in \Pi_{n\in\omega}\mathcal{D}_n \mid \forall n \in \omega.\ d_n \in \mathcal{D}_n \ \&\ \mathsf{j}_n(d_{n+1}) = d_n\},$$

16.3.2 Notation We denote by d_n the projection on $\mathcal{D}_n$, and by d^n an element of $\mathcal{D}_n$.

16.3.3 Definition

(i) The ordering on $\mathcal{D}_\infty$ is given by

$$d \sqsubseteq e \quad \Leftrightarrow \quad \forall k \in \omega.\ d_k \sqsubseteq e_k.$$

(ii) Let $\langle \Phi_{m\infty}, \Phi_{\infty m} \rangle$ denote the standard embedding–projection pair between $\mathcal{D}_m$ and $\mathcal{D}_\infty$ defined as follows. For $d^m \in \mathcal{D}_m$, $d \in \mathcal{D}_\infty$ write

$$\Phi_{mn}(d^m) \quad = \quad \begin{cases} \mathsf{j}_n(\dots(\mathsf{j}_{m-1}(d^m))), & \text{if } m > n; \\ d^m & \text{if } m = n; \\ \mathsf{i}_{n-1}(\dots(\mathsf{i}_m(d^m))), & \text{if } m < n; \end{cases}$$

and take

$$\begin{aligned} \Phi_{m\infty}(d^m) \quad &= \quad \langle \Phi_{m0}(d^m), \Phi_{m1}(d^m), \dots, \Phi_{mn}(d^m), \dots \rangle; \\ \Phi_{\infty m}(d) \quad &= \quad d_m \ = \ d(m). \end{aligned}$$

16.3.4 Lemma $\bigsqcup X$ *exists for all* $X \subseteq \mathcal{D}_\infty$.

Proof Let $d^n \ = \ \bigsqcup\{x_n | x \in X\}$;
$e^n \ = \ \bigsqcup\{\Phi_{mn}(d^m) | m \in \omega\}$

The set $\{\Phi_{mn}(d^m) | m \in \omega\}$ is a directed set, by the monotonicity of $\mathsf{i}_m, \mathsf{j}_m$, and the fact that $\mathsf{i}_m \circ \mathsf{j}_m \sqsubseteq \mathsf{Id}_{\mathcal{D}_{m+1}}$ and $\mathsf{j}_m \circ \mathsf{i}_m = \mathsf{Id}_{\mathcal{D}_m}$. Define

$$\bigsqcup X = \lambda n \in \omega. e^n.$$

Then, by continuity of j_n, we have that $\bigsqcup X \in \mathcal{D}_\infty$, if $X \neq \emptyset$. If $X = \emptyset$, then the continuity cannot be applied, but, using $i_n(\bot_{\mathcal{D}_n}) = \bot_{\mathcal{D}_{n+1}}$ so that $j_{n+1}(\bot_{\mathcal{D}_{n+1}}) = \bot_{\mathcal{D}_n}$, we have $\bigsqcup \emptyset = \lambda n \in \omega. \bot_{\mathcal{D}_n}$. ■

16.3.5 Lemma

(i) $\mathsf{i}_n \circ \mathsf{j}_n \sqsubseteq \mathsf{Id}_{[\mathcal{D}_n \to \mathcal{D}_n]}$, $\mathsf{j}_n \circ \mathsf{i}_n = \mathsf{Id}_{\mathcal{D}_n}$.

(ii) $\forall p, q \in \mathcal{D}_n\ [\mathsf{i}_{n+1}(p \mapsto q) = (\mathsf{i}_n(p) \mapsto \mathsf{i}_n(q))\ \&$
$\mathsf{j}_{n+1}(\mathsf{i}_n(p) \mapsto \mathsf{i}_n(q)) = (p \mapsto q)]$.

(iii) $\Phi_{m\infty} \circ \Phi_{\infty m} \sqsubseteq \mathsf{Id}_\infty$ *and* $\Phi_{\infty m} \circ \Phi_{m\infty} = \mathsf{Id}_{\mathcal{D}_m}$.

(iv) $\forall e \in \mathcal{K}(\mathcal{D}_n)\ [\mathsf{i}_n(e) \in \mathcal{K}(\mathcal{D}_{n+1})]$.

(v) $\forall e \in \mathcal{K}(\mathcal{D}_n)\ [m \geq n \Rightarrow \Phi_{nm}(e) \in \mathcal{K}(\mathcal{D}_m)]$.

(vi) $\forall e \in \mathcal{K}(\mathcal{D}_n)\ [\Phi_{n\infty}(e) \in \mathcal{K}(\mathcal{D}_\infty)]$.

(vii) *If* $n \leq k \leq m$ *and* $d \in \mathcal{D}_n$, $e \in \mathcal{D}_k$, *then*

$$\Phi_{nk}(d) \sqsubseteq e \Leftrightarrow \Phi_{nm}(d) \sqsubseteq \Phi_{km}(e) \Leftrightarrow \Phi_{n\infty}(d) \sqsubseteq \Phi_{k\infty}(e).$$

(viii) $\Phi_{mn} = \Phi_{\infty n} \circ \Phi_{m\infty}$.

(ix) $\forall a, b \in \mathcal{D}_n\ [(\Phi_{n\infty}(a) \mapsto \Phi_{n\infty}(b)) = \Phi_{n\infty} \circ (a \mapsto b) \circ \Phi_{\infty n}]$.

(x) *For any* $k, n > 0$ *and* $f \in D_k$ *we have*

$$\Phi_{kn}(f) = \Phi_{(k-1)(n-1)} \circ f \circ \Phi(n-1)(k-1).$$

Proof (i) and (ii): By induction on n.

(iii) follows from (i).

(iv) and (v) and (vi): By Lemma 15.2.2(ii), observing that the following pairs are all Galois connections:

- $\langle i_n, j_n \rangle$;
- $\langle \Phi_{nm}, \Phi_{mn} \rangle$ for $n \leq m$;
- $\langle \Phi_{n\infty}, \Phi_{\infty n} \rangle$.

Parts (vii) and (viii) are left as exercises for the reader.

(ix) follows from 15.2.2(iii).

(x) Left as an exercise. ∎

16.3.6 Lemma $\bigsqcup_{n \in \omega} \Phi_{n\infty} \circ \Phi_{\infty n} = \mathsf{Id}_{\mathcal{D}_\infty}$.

Proof Since $\langle \Phi_{n\infty}, \Phi_{\infty n} \rangle$ is an embedding–projection pair, we have for all $n \in \omega$ $\Phi_{n\infty} \circ \Phi_{\infty n} \sqsubseteq \mathsf{Id}_{\mathcal{D}_\infty}$, hence for all $d \in \mathcal{D}_\infty$

$$\bigsqcup_{n \in \omega} \Phi_{n\infty} \circ \Phi_{\infty n}(d) \sqsubseteq d.$$

On the other hand, for all $k \in \omega$, we have

$$\begin{array}{lll} (\bigsqcup_{n \in N} \Phi_{n\infty} \circ \Phi_{\infty n}(d))_k & \sqsupseteq (\Phi_{k\infty} \circ \Phi_{\infty k}(d))_k, & \text{because } (-)_k \text{ is monotone,} \\ & = \Phi_{\infty k}(d), & \text{as } \forall x \in \mathcal{D}_k.(\Phi_{k\infty}(x))_k = x, \\ & = d_k. & \end{array}$$

Therefore, in addition

$$\bigsqcup_{n \in \omega} \Phi_{n\infty} \circ \Phi_{\infty n}(d) \sqsupseteq d,$$

and we are done. ∎

The next lemma characterizes the compact elements of $\mathcal{D}_\infty$ and $[\mathcal{D}_\infty \to \mathcal{D}_\infty]$.

16.3.7 Lemma

(i) $d \in \mathcal{K}(\mathcal{D}_\infty) \Leftrightarrow \exists k, e \in \mathcal{K}(\mathcal{D}_k).\Phi_{k\infty}(e) = d$.

(ii) $f \in \mathcal{K}([\mathcal{D}_\infty \to \mathcal{D}_\infty]) \Leftrightarrow \exists k, g \in \mathcal{K}(D_{k+1}).f = \Phi_{k\infty} \circ g \circ \Phi_{\infty k}$.

Proof (i) ($\Rightarrow$) Let $d \in \mathcal{K}(\mathcal{D}_\infty)$. Then $d = \bigsqcup_{n \in \omega} \Phi_{n\infty}(d_n)$, by Lemma 16.3.6.

Since d is compact, there exists $k \in \omega$ such that $d = \Phi_{k\infty}(d_k)$. Now we prove that $d_k \in \mathcal{K}(\mathcal{D}_k)$. Let $X \subseteq \mathcal{D}_k$ be directed. Then

$$\begin{array}{llll} d_k \sqsubseteq \bigsqcup X & \Rightarrow & d \sqsubseteq \Phi_{k\infty}(\bigsqcup X) & \\ & \Rightarrow & d \sqsubseteq \bigsqcup \Phi_{k\infty}(X), & \text{since } \Phi_{k\infty} \text{ is continuous,} \\ & \Rightarrow & \exists x \in X. d \sqsubseteq \Phi_{k\infty}(x), & \text{for some } k \text{ since } d \text{ is compact,} \\ & \Rightarrow & \Phi_{\infty k}(d) \sqsubseteq \Phi_{\infty k} \circ \Phi_{k\infty}(x) & \\ & \Rightarrow & d_k \sqsubseteq x. & \end{array}$$

This proves that $d_k \in \mathcal{K}(\mathcal{D}_k)$. ($\Leftarrow$) By Lemma 16.3.5(vi).

(ii) ($\Rightarrow$) By Lemma 16.3.6, we have

$$f = \bigsqcup_{n \in \omega} \Phi_{n\infty} \circ (\Phi_{\infty n} \circ f \circ \Phi_{n\infty}) \circ \Phi_{\infty n}.$$

Using similar arguments to those in the proof of ((i)), we obtain:

$$\begin{array}{l} \exists k \in \omega. f = \Phi_{k\infty} \circ (\Phi_{\infty k} \circ f \circ \Phi_{k\infty}) \circ \Phi_{\infty k}; \\ (\Phi_{\infty k} \circ f \circ \Phi_{k\infty}) \in \mathcal{K}(\mathcal{D}_{k+1}). \end{array}$$

Put $g = (\Phi_{\infty k} \circ f \circ \Phi_{k\infty})$.

($\Leftarrow$) Easy. ∎

16.3.8 Lemma

(i) $\forall x \in \mathcal{D}_\infty . x = \bigsqcup \{e \in \mathcal{K}(\mathcal{D}_\infty) \mid e \sqsubseteq x\}$.
(ii) $\mathcal{K}(\mathcal{D}_\infty)$ *is countable.*
(iii) $\mathcal{D}_\infty \in$ **ALG**.

Proof (i) Let $x \in \mathcal{D}_\infty$ and $U_x = \{e \in \mathcal{K}(\mathcal{D}_\infty) \mid e \sqsubseteq x\}$. Clearly, $\bigsqcup U_x \sqsubseteq x$. Now let $f = \bigsqcup U_x$ in order to show $x \sqsubseteq f$. By definition of sup in $\mathcal{D}_\infty$ we have

$$f_n = \bigsqcup V(n,x), \quad \text{where } V(n,x) = \{e_n \in \mathcal{D}_n \mid e \in \mathcal{K}(\mathcal{D}_\infty) \ \& \ e \sqsubseteq x\}.$$

Since $\mathcal{D}_n$ is algebraic, we have that

$$x_n = \bigsqcup W(n,x), \quad \text{where } W(n,x) = \{d \in \mathcal{K}(\mathcal{D}_n) \mid d \sqsubseteq x_n\}.$$

We will prove $W(n,x) \subseteq V(n,x)$. Suppose $d \in W(n,x)$. Then $d \in \mathcal{K}(\mathcal{D}_n)$ and $d \sqsubseteq x_n$. Let $e = \Phi_{n\infty}(d)$. Then,

- $d = \Phi_{\infty n} \circ \Phi_{n\infty}(d) = e_n$.
- $e = \Phi_{n\infty}(d) \in \mathcal{K}(\mathcal{D}_\infty)$, by Lemma 16.3.5(vi).
- $e = \Phi_{n\infty}(d) \sqsubseteq \Phi_{n\infty}(x_n) \sqsubseteq x$, by monotonicity of $\Phi_{n\infty}$ and Lemma 16.3.5(iii).

Hence $d \in V(n,x)$. Now indeed $x \sqsubseteq f$, as clearly $x_n \sqsubseteq f_n$.

(ii) By Proposition 15.1.11 one has $\mathcal{D}_n \in \mathbf{ALG}$ for each n. Hence $\mathcal{K}(\mathcal{D}_n)$ is countable for each n. But then also $\mathcal{K}(\mathcal{D}_\infty)$ is countable, by Lemma 16.3.7(i).

(iii) By (i), and (ii). ∎

16.3.9 Definition Define

$$F_\infty : \mathcal{D}_\infty \to [\mathcal{D}_\infty \to \mathcal{D}_\infty]$$
$$G_\infty : [\mathcal{D}_\infty \to \mathcal{D}_\infty] \to \mathcal{D}_\infty$$

as

$$F_\infty(d) = \bigsqcup_{n\in\omega} (\Phi_{n\infty} \circ d_{n+1} \circ \Phi_{\infty n});$$
$$G_\infty(f) = \bigsqcup_{n\in\omega} \Phi_{(n+1)\infty}(\Phi_{\infty n} \circ f \circ \Phi_{n\infty}).$$

16.3.10 Theorem (Scott (1972)) *Let $\mathcal{D}_\infty$ be constructed from $\mathcal{D}_0 \in \mathbf{ALG}$ and a projection pair $\mathsf{i}_0, \mathsf{j}_0$. Then $\mathcal{D}_\infty \in \mathbf{ALG}$ and $\mathcal{D}_\infty$ with F_∞, G_∞ is reflexive. Moreover,*

$$F_\infty \circ G_\infty = \mathsf{Id}_{[\mathcal{D}_\infty \to \mathcal{D}_\infty]} \;\;\&\;\; G_\infty \circ F_\infty = \mathsf{Id}_{\mathcal{D}_\infty}.$$

It follows that $\mathcal{D}_\infty$ is an extensional λ-model.

Proof For the proof that F_∞ and G_∞ are inverse of each other for Scott's model $\mathcal{D}_\infty$, with embedding–projection pair $i_0(d) = \lambda e.d$ and $j_0(f) = f(\bot)$, see B[1984], Theorem 18.2.16. For a proof in the general case see Plotkin (1982). By Proposition 16.1.9 it follows that $\mathcal{D}_\infty$ is an extensional λ-model. ∎

16.3.11 Corollary *Let $\mathcal{D}_\infty$ be constructed from $\mathcal{D}_0 \in \mathbf{ALG}$ and a projection pair $\mathsf{i}_0, \mathsf{j}_0$. Then $\langle \mathcal{D}_\infty, F_\infty, G_\infty \rangle$ is in* **NLS**.

Proof Immediate from the theorem. ∎

$\mathcal{D}_\infty$ as a filter λ-model

In this subsection we follow Alessi (1991), Alessi et al. (2004a). Let $\mathcal{D}_\infty$ be constructed from the triple $t = (\mathcal{D}_0, \mathsf{i}_0, \mathsf{j}_0)$. To emphasize the dependence on t we write $\mathcal{D}_\infty = \mathcal{D}^t_\infty$. From the previous corollary and Proposition 15.2.13 it follows that $\mathcal{D}^t_\infty \cong \mathcal{F}^{\mathcal{K}(\mathcal{D}^t_\infty)}$. In this subsection we associate with $t = (\mathcal{D}_0, \mathsf{i}_0, \mathsf{j}_0)$ a family of intersection type theories CDHL(t). These type theories are compatible and they can be considered as type structures, also denoted by CDHL(t). We will show that

$$\mathcal{K}(\mathcal{D}^t_\infty) \cong \mathrm{CDHL}(t),$$

hence

$$\mathcal{D}^t_\infty \cong \mathcal{F}^{\text{CDHL}(t)}.$$

The name of the family of type theories CDHL(t) comes from Coppo et al. (1984) where this construction was first discussed. Other relevant references are Coppo et al. (1987), which presents the filter λ-model induced by the type theory CDZ, Honsell and Ronchi Della Rocca (1992), where the filter λ-models induced by the type theories Park, HR and other models are considered, and Alessi (1991), Di Gianantonio and Honsell (1993), Plotkin (1993), where the relation between applicative structures and type theories is studied.

16.3.12 Definition Let $t = (\mathcal{D}_0, \mathsf{i}_0, \mathsf{j}_0)$ be given. We define the type theory CDHL(t) as follows:

(i) the partial order $\leq_0$ on $\mathcal{K}(\mathcal{D}_0)$ is defined by

$$d \leq_0 e \Leftrightarrow d \sqsupseteq e \quad d, e \in \mathcal{K}(\mathcal{D}_0);$$

(ii) $\mathbb{T}^{\text{CDHL}(t)} = \mathcal{K}(\mathcal{D}_0) \mid \mathbb{T}^{\text{CDHL}(t)} \to \mathbb{T}^{\text{CDHL}(t)} \mid \mathbb{T}^{\text{CDHL}(t)} \cap \mathbb{T}^{\text{CDHL}(t)}$.

(iii) write $\mathtt{U}$ for $\bot_{\mathcal{D}_0}$. Let CDHL(t) be the smallest natural type theory[3] on $\mathbb{T}^{\text{CDHL}(t)}$ that contains the following extra axiom and rules.

$$(\sqcup) \qquad c \cap d =_{\text{CDHL}(t)} c \sqcup d,$$

$$(\leq_0) \qquad \frac{c \leq_0 d}{c \leq_{\text{CDHL}(t)} d}$$

$$(\mathsf{i}_0) \qquad \frac{\mathsf{i}_0(e) = (c_1 \mapsto d_1) \sqcup \cdots \sqcup (c_n \mapsto d_n)}{e =_{\text{CDHL}(t)} (c_1 \to d_1) \cap \cdots \cap (c_n \to d_n)},$$

where $c, d, e, c_1, d_1, \ldots, c_n, d_n \in \mathcal{K}(\mathcal{D}_0)$ and $\sqcup$ is the least upper bound for the ordering $\sqsubseteq$ on $\mathcal{K}(\mathcal{D}_0)$. Note that for $c, d, e \in \mathcal{K}(\mathcal{D}_0)$ one has $(c \mapsto d), \mathsf{i}_0(e) \in \mathcal{D}_1$.

(iv) Now CDHL(t) is compatible by rule ($\to$), i.e. the relation $=_{\text{CDHL}(t)}$ is a congruence with respect to $\to$, so it can be considered as a type structure; we denote it by [CDHL(t)]. Note that $[\text{CDHL}(t)] \in \mathbf{NTS}^{\mathtt{U}}$.

The proof of the next lemma follows easily from Definition 16.3.12.

16.3.13 Lemma $c_1 \cap \cdots \cap c_n =_{\text{CDHL}(t)} c_1 \sqcup \cdots \sqcup c_n$.

The proof that $\mathcal{K}(\mathcal{D}^t_\infty) \cong \text{CDHL}(t)$ occupies 16.3.14–16.3.19. First we classify the types in $\mathbb{T}^{\text{CDHL}(t)}$ according to the maximal number of nested arrow occurrences they may contain.

[3] CDHL(t) contains the axiom and rules of Definitions 13.1.4 and 13.1.19.

16.3.14 Definition

(i) We define the map depth : $\mathbb{T}^{\mathrm{CDHL}(t)} \to \omega$ by:

$$\begin{array}{rcll} \mathrm{depth}(c) & = & 0, & \text{for } c \in \mathcal{K}(\mathcal{D}_0); \\ \mathrm{depth}(A \to B) & = & \max\{\mathrm{depth}(A), \mathrm{depth}(B)\} + 1; & \\ \mathrm{depth}(A \cap B) & = & \max\{\mathrm{depth}(A), \mathrm{depth}(B)\}. & \end{array}$$

(ii) Denote $\mathbb{T}_n^{\mathrm{CDHL}(t)} = \{A \in \mathbb{T}^{\mathrm{CDHL}(t)} \mid \mathrm{depth}(A) \leq n\}$.

Note that depth differs from the map dpt in Definition 1.1.22.

We can associate to each type in $\mathbb{T}_n^{\mathrm{CDHL}(t)}$ an element in $\mathcal{D}_n$: this will be crucial for defining the required isomorphism (see Definition 16.3.21).

16.3.15 Definition We define for each $n \in \omega$ a map $w_n : \mathbb{T}_n^{\mathrm{CDHL}(t)} \to \mathcal{K}(\mathcal{D}_n^t)$ by a double recursion on n and on the construction of types in $\mathbb{T}^{\mathrm{CDHL}(t)}$:

$$\begin{array}{lcl} w_n(c) & = & \Phi_{0n}(c); \\ w_n(A \cap B) & = & w_n(A) \sqcup w_n(B); \\ w_n(A \to B) & = & (w_{n-1}(A) \mapsto w_{n-1}(B)). \end{array}$$

16.3.16 Remark From Lemma 15.1.9 we get

$$\forall A \in \mathbb{T}_n^{\mathrm{CDHL}(t)}.w_n(A) \in \mathcal{K}(\mathcal{D}_n^t).$$

16.3.17 Lemma *Let $n \leq m$ and $A \in \mathbb{T}_n^{\mathrm{CDHL}(t)}$. Then $\Phi_{m\infty}(w_m(A)) = \Phi_{n\infty}(w_n(A))$.*

Proof We show by induction on the definition of w_n that $w_{n+1}(A) = \mathsf{i}_n(w_n(A))$. Then the desired equality follows from the definition of the function Φ. The only interesting case is when $A \equiv B \to C$. We get

$$\begin{array}{rcll} w_{n+1}(B \to C) & = & w_n(B) \mapsto w_n(C), & \text{by definition,} \\ & = & \mathsf{i}_{n-1}(w_{n-1}(B)) \mapsto \mathsf{i}_{n-1}(w_{n-1}(C)), & \text{by induction,} \\ & = & \mathsf{i}_n(w_{n-1}(B) \mapsto w_{n-1}(C)), & \\ & & \text{by Lemma 16.3.5(ii),} & \\ & = & \mathsf{i}_n(w_n(B \to C)), & \\ & & \text{by Definition 16.3.15.} \blacksquare & \end{array}$$

The maps w_n reverse the order between types.

16.3.18 Lemma *Let* $\mathrm{depth}(A \cap B) \leq n$. *Then*

$$A \leq_{\mathrm{CDHL}(t)} B \Rightarrow w_n(B) \sqsubseteq w_n(A).$$

Proof The proof is by induction on the definition of $\leq_{\mathrm{CDHL}(t)}$. We consider only two cases.

Case ($\to$). That $A \leq_{\mathrm{CDHL}(t)} B$ follows from $A \equiv C{\to}D$, $B \equiv E{\to}F$, $E \leq_{\mathrm{CDHL}(t)} C$ and $D \leq_{\mathrm{CDHL}(t)} F$. Then

$$\begin{array}{rl} E \leq_{\mathrm{CDHL}(t)} C \;\&\; D \leq_{\mathrm{CDHL}(t)} F \Rightarrow & \\ \Rightarrow & w_{n-1}(C) \sqsubseteq w_{n-1}(E) \;\&\; w_{n-1}(F) \sqsubseteq w_{n-1}(D), \\ & \text{by the induction hypothesis,} \\ \Rightarrow & w_{n-1}(E) \mapsto w_{n-1}(F) \sqsubseteq w_{n-1}(C) \mapsto w_{n-1}(D) \\ \Rightarrow & w_n(B) \sqsubseteq w_n(A). \end{array}$$

Case $e =_{\mathrm{CDHL}(t)} (c_1{\to}d_1) \cap \cdots \cap (c_k{\to}d_k)$ follows from $i_0(e) = (c_1 \mapsto d_1) \sqcup \cdots \sqcup (c_k \mapsto d_k)$. We show by induction on $n \geq 1$ that

$$w_n(e) = (w_{n-1}(c_1) \mapsto w_{n-1}(d_1)) \sqcup \cdots \sqcup (w_{n-1}(c_k) \mapsto w_{n-1}(d_k)).$$

It trivially holds for $n = 1$, so let $n > 1$.

$$\begin{array}{rl} w_n(e) = \mathsf{i}_{n-1}(w_{n-1}(e)) & \\ = & \mathsf{i}_{n-1}((w_{n-2}(c_1) \mapsto w_{n-2}(d_1)) \sqcup \cdots \sqcup (w_{n-2}(c_k) \mapsto w_{n-2}(d_k))) \\ = & \mathsf{i}_{n-1}(w_{n-2}(c_1) \mapsto w_{n-2}(d_1)) \sqcup \cdots \sqcup \mathsf{i}_{n-1}(w_{n-2}(c_k) \mapsto w_{n-2}(d_k)) \\ = & (\mathsf{i}_{n-2}(w_{n-2}(c_1)) \mapsto \mathsf{i}_{n-2}(w_{n-2}(d_1))) \sqcup \cdots \\ & \cdots \sqcup (\mathsf{i}_{n-2}(w_{n-2}(c_k)) \mapsto \mathsf{i}_{n-2}(w_{n-2}(d_k)) \\ = & (w_{n-1}(c_1) \mapsto w_{n-1}(d_1)) \sqcup \cdots \sqcup (w_{n-1}(c_k) \mapsto w_{n-1}(d_k)). \blacksquare \end{array}$$

Also the reverse implication of Lemma 16.3.18 holds.

16.3.19 Lemma *Let* $\mathrm{depth}(A \cap B) \leq n$. *Then*

$$w_n(B) \sqsubseteq w_n(A) \Rightarrow A \leq_{\mathrm{CDHL}(t)} B.$$

Proof By induction on $\mathrm{depth}(A \cap B)$.
If $\mathrm{depth}(A \cap B) = 0$ we have $A \equiv \bigcap_{i\in I} c_i$, $B = \bigcap_{j\in J} d_j$. Then

$$\begin{array}{rl} w_n(B) \sqsubseteq w_n(A) & \Rightarrow \bigsqcup_{j\in J} \Phi_{0n}(d_j) \sqsubseteq \bigsqcup_{i\in I} \Phi_{0n}(c_i) \\ & \Rightarrow \Phi_{n0}(\bigsqcup_{j\in J} \Phi_{0n}(d_j)) \sqsubseteq \Phi_{n0}(\bigsqcup_{i\in I} \Phi_{0n}(c_i)) \\ & \Rightarrow \bigsqcup_{j\in J} (\Phi_{n0} \circ \Phi_{0n})(d_j) \sqsubseteq \bigsqcup_{i\in I} (\Phi_{n0} \circ \Phi_{0n})(c_i) \\ & \Rightarrow \bigsqcup_{j\in J} d_j \sqsubseteq \bigsqcup_{i\in I} c_i \\ & \Rightarrow A \leq_{\mathrm{CDHL}(t)} B. \end{array}$$

Otherwise, let

$$A \equiv (\bigcap_{i\in I} c_i) \cap (\bigcap_{l\in L} (C_l \to D_l)),$$
$$B \equiv (\bigcap_{h\in H} d_h) \cap (\bigcap_{m\in M} (E_m \to F_m)).$$

By rule (i_0), we have that

$$c_i =_{\mathrm{CDHL}(t)} \bigcap_{j\in J_i} (a_j \to b_j), \quad d_h =_{\mathrm{CDHL}(t)} \bigcap_{k\in K_h} (e_k \to f_k),$$

where $a_j, b_j, e_k, f_k \in \mathcal{K}(\mathcal{D}_0)$. Now for all $n \geq 1$

$$w_n(c_i) = \bigsqcup_{j\in J_i} (w_{n-1}(a_j) \mapsto w_{n-1}(b_j))),$$
$$w_n(d_h) = (\bigsqcup_{k\in K_h} (w_{n-1}(e_k) \mapsto w_{n-1}(f_k))),$$

since by Lemma 16.3.18 the function w_n identifies elements in the equivalence classes of $=_{\mathrm{CDHL}(t)}$. Therefore

$$\bigsqcup_{h\in H} (\bigsqcup_{k\in K_h} w_{n-1}(e_k) \mapsto w_{n-1}(f_k)) \sqcup (\bigsqcup_{m\in M} w_{n-1}(E_m) \mapsto w_{n-1}(F_m)) \sqsubseteq$$
$$\bigsqcup_{i\in I} (\bigsqcup_{j\in J_i} w_{n-1}(a_j) \mapsto w_{n-1}(b_j)) \sqcup (\bigsqcup_{l\in L} w_{n-1}(C_l) \mapsto w_{n-1}(D_l)).$$

Hence for each $h \in H$, $k \in K_h$ we have

$$(w_{n-1}(e_k) \mapsto w_{n-1}(f_k)) \sqsubseteq \bigsqcup_{i\in I} (\bigsqcup_{j\in J_i} w_{n-1}(a_j) \mapsto w_{n-1}(b_j)) \sqcup$$
$$(\bigsqcup_{l\in L} w_{n-1}(C_l) \mapsto w_{n-1}(D_l)).$$

Suppose $w_{n-1}(f_k) \neq \bot_{\mathcal{D}_n}$. By Lemma 15.1.10 there exist $I' \subseteq I$, $J_i' \subseteq J_i$, $L' \subseteq L$ such that

$$\bigsqcup_{i\in I'} (\bigsqcup_{j\in J_i'} w_{n-1}(a_j)) \sqcup (\bigsqcup_{l\in L'} w_{n-1}(C_l)) \sqsubseteq w_{n-1}(e_k),$$
$$\bigsqcup_{i\in I'} (\bigsqcup_{j\in J_i'} w_{n-1}(b_j)) \sqcup (\bigsqcup_{l\in L'} w_{n-1}(D_l)) \sqsupseteq w_{n-1}(f_k).$$

Notice that all types involved in the two above judgments have depths strictly less than depth$(A \cap B)$:

(i) the depth of a_j, b_j, e_k, f_k is 0, since they are all atoms in $\mathcal{K}(\mathcal{D}_0)$;
(ii) the depth of $C_l, \mathcal{D}_l$ is strictly smaller than the one of $A \cap B$, since they are subterms of an arrow in A.

Then by induction and by Lemma 16.3.13 we obtain

$$e_k \leq_{\mathrm{CDHL}(t)} \bigcap_{i\in I'} \Big(\bigcap_{j\in J'_i} a_j\Big) \cap \bigcap_{l\in L'} C_l,$$

$$f_k \geq_{\mathrm{CDHL}(t)} \bigcap_{i\in I'} \Big(\bigcap_{j\in J'_i} b_j\Big) \cap \bigcap_{l\in L'} D_l.$$

Therefore, by $(\rightarrow)$ and Proposition 13.1.22, we have $A \leq_{\mathrm{CDHL}(t)} e_k{\rightarrow}f_k$. If $w_{n-1}(f_k) = \bot_{\mathcal{D}_n}$, then $w_{n-1}(f_k) = \Phi_{0n}(f_k)$ since $f_k \in \mathcal{K}(\mathcal{D}_0)$. This gives $f_k = \Phi_{n0} \circ \Phi_{0n}(f_k) = \Phi_{n0}(\bot_{\mathcal{D}_n}) = \bot_{\mathcal{D}_0}$ because $j_n(\bot_{\mathcal{D}_{n+1}}) = \bot_{\mathcal{D}_n}$. Since $f_k = \bot_{\mathcal{D}_0}$ implies $A \leq_{\mathrm{CDHL}(t)} e_k \rightarrow f_k$ we are done.

In a similar way we can prove that $A \leq_{\mathrm{CDHL}(t)} E_m{\rightarrow}F_m$, for any $m \in M$. Putting together these results we get $A \leq_{\mathrm{CDHL}(t)} B$. ■

16.3.20 Proposition *$\langle \mathcal{K}(\mathcal{D}^t_\infty), \leq_\infty, \cap, \rightarrow_\infty, \mathtt{U}\rangle$ is a natural type structure, where $\leq_\infty$ is the reverse order on $\mathcal{K}(\mathcal{D}^t_\infty)$, $a \rightarrow_\infty b = G_\infty(a \mapsto b)$, $\cap$ is the least upper bound of $\mathcal{D}^t_\infty$ and $\mathtt{U} = \bot_{\mathcal{D}^t_\infty}$.*

Proof By Lemma 15.2.11. ■

We can now prove the isomorphism in $\mathbf{NTS}^{\mathtt{U}}$ between $\mathcal{K}(\mathcal{D}^t_\infty)$ and CDHL(t) seen as type structure, i.e. [CDHL(t)].

16.3.21 Definition For $A \in \mathbb{T}^{\mathrm{CDHL}(t)}$ write

$$f([A]) = \Phi_{r\infty}(w_r(A)),$$

where $r \geq \mathrm{depth}(A)$.

16.3.22 Theorem *In $\mathbf{NTS}^{\mathtt{U}}$ one has $[\mathrm{CDHL}(t)] \cong \mathcal{K}(\mathcal{D}^t_\infty)$ via f.*

Proof First of all notice that f is well defined, in the sense that it does not depend on either the type chosen in $[A]$ or the depth r. In fact let $B, B' \in [A]$, and let $p \geq \mathrm{depth}(B)$, $p' \geq \mathrm{depth}(B')$. Fix any $q \geq p, p'$. Then we have

$$\begin{aligned}
\Phi_{p\infty}(w_p(B)) &= \Phi_{q\infty}(w_q(B)), && \text{by Lemma 16.3.17,}\\
&= \Phi_{q\infty}(w_q(B')), && \text{by Lemma 16.3.18,}\\
&= \Phi_{p'\infty}(w_{p'}(B')), && \text{by Lemma 16.3.17.}
\end{aligned}$$

Write $f(A)$ for $f([A])$. Now f is injective by Lemma 16.3.19 and monotone by Lemma 16.3.18. From Lemma 15.1.11(ii) we get immediately

$$\mathcal{K}(\mathcal{D}_{n+1}) = \{c_1 \mapsto d_1 \sqcup \cdots \sqcup c_m \mapsto d_m | c_i, d_i \in \mathcal{K}(\mathcal{D}_n)\}.$$

It is easily proved by induction on n that w_n is surjective on $\mathcal{K}(\mathcal{D}_n)$, hence f is surjective by Lemma 16.3.7(i). The function f^{-1} is monotone by Lemma

16.3.19. Taking into account that the order $\leq_\infty$ on $\mathcal{K}(\mathcal{D}^t_\infty)$ is the reversed of $\sqsubseteq$ of $\mathcal{D}^t_\infty$ and that $d \to_\infty e = G_\infty(d \mapsto e)$, we need to show

$$A \leq_{\mathrm{CDHL}(t)} B \to C \;\Leftrightarrow\; f(A) \sqsupseteq G_\infty(f(B) \mapsto f(C)) \tag{16.5}$$

In order to prove (16.5), let $r \geq \max\{\mathrm{depth}(A), \mathrm{depth}(B \to C)\}$ (in particular it follows $\mathrm{depth}(B), \mathrm{depth}(C) \leq r-1$). We have

$$\begin{aligned}
G_\infty(f(B) \mapsto f(C)) &= G_\infty(\Phi_{(r-1)\infty}(w_{r-1}(B)) \mapsto \Phi_{(r-1)\infty}(w_{r-1}(C))) \\
&= G_\infty(\Phi_{(r-1)\infty} \circ (w_{r-1}(B) \mapsto w_{r-1}(C)) \circ \Phi_{\infty(r-1)}), \\
&\quad \text{by Lemma 16.3.5(viii)}, \\
&= \Phi_{r\infty}(w_{r-1}(B) \mapsto w_{r-1}(C)), \quad \text{by the definition} \\
&\quad \text{of } G \text{ and Lemma 16.3.5(viii) and (x)}, \\
&= \Phi_{r\infty}(w_r(B \to C)), \quad \text{by the definition of } w_r.
\end{aligned}$$

Finally we have

$$\begin{aligned}
A \leq_{\mathrm{CDHL}(t)} B \to C &\Leftrightarrow w_r(A) \sqsupseteq w_r(B \to C), \\
&\quad \text{by Lemmas 16.3.18 and 16.3.19}, \\
&\Leftrightarrow \Phi_{r\infty}(w_r(A)) \sqsupseteq \Phi_{r\infty}(w_r(B \to C)), \\
&\quad \text{since } \Phi_{r\infty} \text{ is an embedding}, \\
&\Leftrightarrow \Phi_{r\infty}(w_r(A)) \sqsupseteq G_\infty(f(B) \mapsto f(C)), \quad \text{as above}, \\
&\Leftrightarrow f(A) \sqsupseteq G_\infty(f(B) \mapsto f(C)). \\
&\Leftrightarrow f(A) \leq_\infty f(B) \to_\infty f(C)
\end{aligned}$$

So we have proved (16.5) and the proof is complete. ∎

16.3.23 Theorem *$\mathcal{F}^{[\mathrm{CDHL}(t)]} \cong \mathcal{D}^t_\infty$ in* **NLS**, *via the map*

$$\hat{f}(X) = \bigsqcup\{f(B) \mid B \in X\},$$

satisfying $\hat{f}(\uparrow A) = f(A)$.

Proof Let $f : [\mathrm{CDHL}(t)] \to \mathcal{K}(\mathcal{D}^t_\infty)$ be the isomorphism in $\mathbf{NTS}^{\mathrm{U}}$. By Theorem 15.2.9 we know that $\mathsf{Flt}_{\mathbf{NTS}^{\mathrm{U}}}$ is a functor from $\mathbf{NTS}^{\mathrm{U}}$ to **NLS**. Then

$$\mathsf{Flt}_{\mathbf{NTS}^{\mathrm{U}}}(f) : \mathcal{F}^{[\mathrm{CDHL}(t)]} \to \mathcal{F}^{\mathcal{K}(\mathcal{D}^t_\infty)}$$

is an isomorphism in **NLS** where $\mathcal{A}(f)(X) = \{B \mid \exists A \in X. f(A) \sqsubseteq B\}$.

By Proposition 15.2.13 we have that

$$\tau : \mathcal{F}^{\mathcal{K}(\mathcal{D}^t_\infty)} \to \mathcal{D}^t_\infty$$

is an isomorphism in **NLS** where $\tau(X) = \bigsqcup X$.

(0_{Scott})	$(\mathtt{U}\to 0) = 0$
(0_{Park})	$(0\to 0) = 0$
(01)	$0 \le 1$
$(1\to 0)$	$(1\to 0) = 0$
$(0\to 1)$	$(0\to 1) = 1$
(I)	$(1\to 1) \cap (0\to 0) = 1$

Figure 16.4

The composition of τ and $\mathsf{Flt}_{\mathbf{NTS}^{\mathtt{U}}}(f)$ is an isomorphism from $\mathcal{F}^{[\mathrm{CDHL}(t)]}$ to $\mathcal{D}_\infty^t$ explicitly given by

$$\begin{aligned} \tau \circ \mathsf{Flt}_{\mathbf{NTS}^{\mathtt{U}}}(f)(X) &= \bigsqcup\{B \mid \exists A \in X. f(A) \sqsubseteq B\} \\ &= \bigsqcup\{f(A) \mid A \in X\} = \hat{f}(X). \blacksquare \end{aligned}$$

16.3.24 Corollary $\mathcal{F}^{\mathrm{CDHL}(t)} \cong \mathcal{D}_\infty^t$ *in* **NLS**.

Proof By Lemma 13.4.7. ■

Specific models $\mathcal{D}_\infty$ as filter models

In this subsection we apply Theorem 16.3.23 to $\mathcal{D}_\infty^t$ models constructed from specific triples $t = (\mathcal{D}_0, \mathsf{i}_0, \mathsf{j}_0)$, five in total, each satisfying a specific model-theoretic property. For each $\mathcal{T} \in \{\mathrm{Scott}, \mathrm{Park}, \mathrm{CDZ}, \mathrm{DHM}, \mathrm{HR}\}$, we associate a triple $\mathsf{init}(\mathcal{T}) = (\mathcal{D}_0, \mathsf{i}_0, \mathsf{j}_0)$ such that $\mathrm{CDHL}(\mathsf{init}(\mathcal{T})) = \mathcal{T}$. By Corollary 16.3.24,

$$\mathcal{D}_\infty^{\mathsf{init}(\mathcal{T})} \cong \mathcal{F}^{\mathrm{CDHL}(\mathsf{init}(\mathcal{T}))} = \mathcal{F}^{\mathcal{T}}.$$

We will write $\mathcal{D}_\infty^{\mathcal{T}} := \mathcal{D}_\infty^{\mathsf{init}(\mathcal{T})}$, so that this reads smoothly as $\mathcal{D}_\infty^{\mathcal{T}} \cong \mathcal{F}^{\mathcal{T}}$.

16.3.25 Remark

(i) Recall the type-theoretic axioms in Fig. 16.4.
(ii) In Definition 13.1.15, each $\mathcal{T} \in \{\mathrm{Scott}, \mathrm{Park}, \mathrm{CDZ}, \mathrm{DHM}, \mathrm{HR}\}$ was defined by specifying its set of atoms $\mathbb{A}^{\mathcal{T}}$ and some extras (as well as the axioms $(\to\cap)$, $(\mathtt{U}_{\mathrm{top}})$ and $(\mathtt{U}\to)$ and the rule $(\to)$); see Fig. 16.5.

16.3.26 Definition For each $\mathcal{T} \in \{\mathrm{Scott}, \mathrm{Park}, \mathrm{CDZ}, \mathrm{DHM}, \mathrm{HR}\}$, we associate a triple $\mathsf{init}(\mathcal{T}) = (\mathcal{D}_0^{\mathcal{T}}, \mathsf{i}_0^{\mathcal{T}}, \mathsf{j}_0^{\mathcal{T}})$ defined as follows.

$\mathcal{T}$	Atoms $\mathbb{A}^\mathcal{T}$	Axioms of $\mathcal{T}$
Scott	$\{\mathtt{U}, 0\}$	(0_{Scott})
Park	$\{\mathtt{U}, 0\}$	(0_{Park})
CDZ	$\{\mathtt{U}, 0, 1\}$	$(01), (1\to 0), (0\to 1)$
DHM	$\{\mathtt{U}, 0, 1\}$	$(01), (0\to 1), (0_{Scott})$
HR	$\{\mathtt{U}, 0, 1\}$	$(01), (1\to 0), (\mathsf{I})$

Figure 16.5

(1) We define $\mathcal{D}_0^\mathcal{T}$ as either a two point chain or a three point chain depending on $\mathcal{T}$ as follows:

$\mathcal{T}$	$\mathcal{D}_0^\mathcal{T}$
Scott, Park	$\mathtt{U} \sqsubseteq 0$
CDZ, DHM, HR	$\mathtt{U} \sqsubseteq 1 \sqsubseteq 0$

(2) We define $\mathsf{i}_0^\mathcal{T} : \mathcal{D}_0^\mathcal{T} \to [\mathcal{D}_0^\mathcal{T} \to \mathcal{D}_0^\mathcal{T}]$ as

$$\begin{aligned}
\mathsf{i}_0^\mathcal{T}(\mathtt{U}) &= \mathtt{U} \mapsto \mathtt{U} \ \text{ for any } \mathcal{T} \in \{\text{Scott, Park, CDZ, DHM, HR}\} \\
\mathsf{i}_0^\mathcal{T}(1) &= \begin{cases} 0 \mapsto 1 & \text{if } \mathcal{T} \in \{\text{CDZ, DHM}\} \\ (1 \mapsto 1) \sqcup (0 \mapsto 0) & \text{if } \mathcal{T} \in \{\text{HR}\} \end{cases} \\
\mathsf{i}_0^\mathcal{T}(0) &= \begin{cases} \mathtt{U} \mapsto 0 & \text{if } \mathcal{T} \in \{\text{Scott, DHM}\} \\ 0 \mapsto 0 & \text{if } \mathcal{T} \in \{\text{Park}\} \\ 1 \mapsto 0 & \text{if } \mathcal{T} \in \{\text{CDZ, HR}\} \end{cases}
\end{aligned}$$

(3) Then we define $\mathsf{j}_0^\mathcal{T} : [\mathcal{D}_0^\mathcal{T} \to \mathcal{D}_0^\mathcal{T}] \to \mathcal{D}_0^\mathcal{T}$ as

$$\mathsf{j}_0^\mathcal{T}(f) = \bigsqcup\{d \in \mathcal{D}_0^\mathcal{T} \mid \mathsf{i}_0^\mathcal{T}(d) \sqsubseteq f\} \qquad \text{for } f \in [\mathcal{D}_0^\mathcal{T} \to \mathcal{D}_0^\mathcal{T}].$$

It is easy to prove that $\langle \mathsf{i}_0^\mathcal{T}, \mathsf{j}_0^\mathcal{T} \rangle$ is an embedding–projection pair from $\mathcal{D}_0^\mathcal{T}$ to $[\mathcal{D}_0^\mathcal{T} \to \mathcal{D}_0^\mathcal{T}]$, so we can build $\mathcal{D}_\infty^{\mathsf{init}(\mathcal{T})}$ following the steps outlined in Definition 16.3.1.

16.3.27 Lemma *Let $\mathcal{T} \in \{\text{Scott, Park, CDZ, DHM, HR}\}$ and $c_1, \dots, c_n$, $d_1, \dots, d_n$, $e_1, \dots, e_k$, $f_1, \dots, f_k \in \mathcal{D}_0^\mathcal{T}$. Then*

$$\begin{aligned}
&(e_1 \to f_1) \cap \cdots \cap (e_k \to f_k) =_\mathcal{T} (c_1 \to d_1) \cap \cdots \cap (c_n \to d_n) \Leftrightarrow \\
&(c_1 \mapsto d_1) \sqcup \cdots \sqcup (c_n \mapsto d_n) = (e_1 \mapsto f_1) \sqcup \cdots \sqcup (e_k \mapsto f_k).
\end{aligned}$$

Proof It suffices to prove

$$(c \mapsto d) \sqsubseteq (e_1 \mapsto f_1) \sqcup \cdots \sqcup (e_k \mapsto f_k) \Leftrightarrow (e_1 \to f_1) \cap \cdots \cap (e_k \to f_k) \leq_\mathcal{T} (c \to d).$$

Now, $(c \mapsto d) \sqsubseteq (e_1 \mapsto f_1) \sqcup \cdots \sqcup (e_k \mapsto f_k)$

$\Leftrightarrow \exists I \subseteq \{1, \ldots, k\}\, [\sqcup_{i \in I} e_i \sqsubseteq c \;\&\; d \sqsubseteq \sqcup_{i \in I} f_i]$, by Lemma 15.1.10,

$\Leftrightarrow \exists i_1, \ldots, i_p \in \{1, \ldots, k\}\, [c \leq_{\mathcal{T}} e_{i_1} \cap \cdots \cap e_{i_p} \;\&\; f_{i_1} \cap \cdots \cap f_{i_p} \leq_{\mathcal{T}} d]$,

$\Leftrightarrow (e_1 \to f_1) \cap \cdots \cap (e_k \to f_k) \leq_{\mathcal{T}} (c \to d)$, by β-soundness, $(\to)$ and $(\to\cap)$. ■

16.3.28 Corollary *The definition of* $\mathsf{i}_0^{\mathcal{T}}$ *is canonical. By this we mean that we could have given equivalently the following definition.*

$$\mathsf{i}_o^{\mathcal{T}}(e) = (c_1 \mapsto d_1) \sqcup \cdots \sqcup (c_n \mapsto d_n) \;\Leftrightarrow\; e =_{\mathcal{T}} (c_1 \to d_1) \cap \cdots \cap (c_n \to d_n).$$

Proof Immediate, by the definition of $\mathsf{i}_0^{\mathcal{T}}$, the axioms $(\mathtt{U}_{\mathrm{top}})$ and $(\mathtt{U}\to)$, the special axioms (0_{Scott}), (0_{Park}), $(1\to 0)$, $(0\to 1)$, and (I) respectively, and the previous lemma. ■

16.3.29 Proposition *For* $\mathcal{T} \in \{\mathrm{Scott}, \mathrm{Park}, \mathrm{CDZ}, \mathrm{DHM}, \mathrm{HR}\}$ *one has*

$$\mathcal{T} = \mathrm{CDHL}(\mathsf{init}(\mathcal{T})).$$

Proof First of all, we have that $\mathbb{T}^{\mathcal{T}} = \mathbb{T}^{\mathrm{CDHL}(\mathsf{init}(\mathcal{T}))}$ because

$$\begin{aligned} \mathbb{A}^{\mathcal{T}} &= \mathcal{K}(\mathcal{D}_0^{\mathcal{T}}) && \text{by Definition 16.3.26} \\ &= \mathbb{A}^{\mathrm{CDHL}(\mathsf{init}(\mathcal{T}))} && \text{by Definition 16.3.12} \end{aligned}$$

since each $\mathcal{D}_0^{\mathcal{T}}$ only contains compact elements. It remains to show that

$$A \leq_{\mathrm{CDHL}(\mathsf{init}(\mathcal{T}))} B \;\Leftrightarrow\; A \leq_{\mathcal{T}} B.$$

$(\Rightarrow)$. This follows by induction on the generation of $\leq_{\mathrm{CDHL}(t)}$ where $t = \mathsf{init}(\mathcal{T})$. Now $\mathcal{T}$ satisfies the axioms $(\to\cap), (\mathtt{U}_{\mathrm{top}})$ and is closed under rule $(\to)$, since it is a natural type theory. It remains to show that the extra axiom and rules are valid in this theory.

Axiom $(\sqcup)$. Then, $c \cap d =_{\mathrm{CDHL}(t)} c \sqcup d$, with $c, d \in \mathcal{K}(\mathcal{D}_0^{\mathcal{T}}) = \mathcal{D}_0^{\mathcal{T}}$, we have, say, $c \sqsubseteq d$. Then $c \sqcup d = d$. Again we have $d \leq_{\mathcal{T}} c$. Therefore $d \leq_{\mathcal{T}} c \cap d \leq_{\mathcal{T}} d$, and hence $c \cap d =_{\mathcal{T}} d = c \sqcup d$.

Rule $(\leq_0)$. Then, $c \leq_{\mathrm{CDHL}(t)} d$ because $d \sqsubseteq c$, we have $d = \mathtt{U}, c = 0$ or $d = c$. Then in all cases $c \leq_{\mathcal{T}} d$, by the axioms $(\mathtt{U}_{\mathrm{top}})$ and (01).

Rule (i_0) where $\mathsf{i}_0 = \mathsf{i}_0^{\mathcal{T}}$. Suppose $e =_{\mathrm{CDHL}(t)} (c_1 \to d_1) \cap \cdots \cap (c_n \to d_n)$, because

$$\mathsf{i}_0^{\mathcal{T}}(e) = (c_1 \mapsto d_1) \sqcup \cdots \sqcup (c_n \mapsto d_n).$$

Then $e =_{\mathcal{T}} (c_1 \to d_1) \cap \cdots \cap (c_n \to d_n)$, by Corollary 16.3.28.

$(\Leftarrow)$. The axioms and rules $(\mathtt{U}_{\mathrm{top}}), (\mathtt{U} \to), (\to\cap)$ and $(\to)$ hold for $\mathcal{T}$ by definition. Moreover, all axioms extra of $\mathcal{T}$ hold in $\mathrm{CDHL}(\mathsf{init}(\mathcal{T}))$, by Definition 16.3.26 and the rules $(\leq_0)$, (i_0) of Definition 16.3.12. ■

Now we can obtain the following result.

16.3.30 Corollary *Let $\mathcal{T} \in \{\text{Scott}, \text{Park}, \text{CDZ}, \text{DHM}, \text{HR}\}$. Then in the category* **NLS** *we have*

$$\mathcal{F}^{\mathcal{T}} \cong \mathcal{D}_\infty^{\mathcal{T}}.$$

Proof
$$\begin{array}{rcll} \mathcal{F}^{\mathcal{T}} & = & \mathcal{F}^{\text{CDHL}(\mathsf{init}(\mathcal{T}))}, & \text{by Proposition 16.3.29,} \\ & \cong & \mathcal{D}_\infty^{\mathcal{T}}, & \text{by Corollary 16.3.24.} \ \blacksquare \end{array}$$

We will end this subsection by explaining why we are interested in the various models $\mathcal{D}_\infty^t$. In B[1984], Theorem 19.2.9, the following result is proved.

16.3.31 Theorem (Hyland and Wadsworth) *Let $t = (\mathcal{D}_0, \mathsf{i}_0, \mathsf{j}_0)$, where $\mathcal{D}_0$ is a CPO (or object of* **ALG***) with at least two elements and*

$$\begin{array}{rcll} \mathsf{i}_0(d) & = & \lambda\!\!\lambda e \in \mathcal{D}_0.d, & \textit{for } d \in \mathcal{D}_0, \\ \mathsf{j}_0(f) & = & f(\bot_{\mathcal{D}_0}), & \textit{for } f \in [\mathcal{D}_0 \to \mathcal{D}_0]. \end{array}$$

Then for $M, N \in \Lambda$ (untyped lambda terms) and $C[\,]$ ranging over contexts

$$\mathcal{D}_\infty^t \models M = N \ \Leftrightarrow\ \forall C[\,].(C[M] \textit{ is solvable} \ \Leftrightarrow\ C[N] \textit{ is solvable}).$$

In particular, the local structure of $\mathcal{D}_\infty^t$, i.e. $\{M = N \mid \mathcal{D}_\infty^t \models M = N\}$, is independent of the initial $\mathcal{D}_0$.

16.3.32 Corollary *For t as in the theorem one has for closed terms M, N*

$$\mathcal{D}_\infty^t \models M = N \ \Leftrightarrow\ \forall A \in \mathbb{T}^{\text{Scott}}\, [\vdash_\cap^{\text{Scott}} M : A \ \Leftrightarrow\ \vdash_\cap^{\text{Scott}} N : A].$$

Proof Let $M, N \in \Lambda^\emptyset$. Then

$$\begin{array}{rcl} \mathcal{D}_\infty^t \models M = N & \Leftrightarrow & \mathcal{D}_\infty^{\text{Scott}} \models M = N, \ \text{by Theorem 16.3.31,} \\ & \Leftrightarrow & \mathcal{F}^{\text{Scott}} \models M = N, \\ & & \quad \text{by Corollary 16.3.30 and Proposition 16.1.13,} \\ & \Leftrightarrow & \forall A \in \mathbb{T}^{\text{Scott}}\, [\vdash_\cap^{\text{Scott}} M : A \ \Leftrightarrow\ \vdash_\cap^{\text{Scott}} N : A], \\ & & \text{by Theorem 16.2.7.} \ \blacksquare \end{array}$$

The model $\mathcal{D}_\infty^{\text{Park}}$ was introduced to provide contrast with the following result, see B[1984], 19.3.6.

16.3.33 Theorem (Park) *Let t be as in* 16.3.31*. Then for the untyped λ-term*

$$\mathsf{Y}_{\text{Curry}} \equiv \lambda f.(\lambda x.f(xx))(\lambda x.f(xx))$$

one has

$$[\![\mathsf{Y}_{\text{Curry}}]\!]^{\mathcal{D}_\infty^t} = \mathsf{Y}_{\text{Tarski}},$$

where $\mathsf{Y}_{\text{Tarski}}$ *is the least fixed-point combinator on* $\mathcal{D}^t_\infty$.

The model $\mathcal{D}^{\text{Park}}_\infty$ has been constructed to give $\mathsf{Y}_{\text{Curry}}$ a meaning different from $\mathsf{Y}_{\text{Tarski}}$.

16.3.34 Theorem (Park) $[\![\mathsf{Y}_{\text{Curry}}]\!]^{\mathcal{D}^{\text{Park}}_\infty} \neq \mathsf{Y}_{\text{Tarski}}$.

Now this model can be obtained as a simple filter model $\mathcal{D}^{\text{Park}}_\infty \cong \mathcal{F}^{\text{Park}}$ and therefore, by Corollary 16.3.30, one has

$$[\![\mathsf{Y}_{\text{Curry}}]\!]^{\mathcal{F}^{\text{Park}}} \neq \mathsf{Y}_{\text{Tarski}}.$$

Other domain equations

Results similar to Theorem 16.3.22 can be given also for other, non-extensional, inverse limit λ-models. These are obtained as solutions of domain equations involving different functors. For instance one can solve the equations

$$\begin{aligned}
\mathcal{D} &= [\mathcal{D} \to \mathcal{D}] \times A \\
\mathcal{D} &= [\mathcal{D} \to \mathcal{D}] + A \\
\mathcal{D} &= [\mathcal{D} \to_\perp \mathcal{D}] \times A \\
\mathcal{D} &= [\mathcal{D} \to_\perp \mathcal{D}] + A,
\end{aligned}$$

useful for the analysis of models for restricted λ-calculi. In all such cases one gets concise type-theoretic descriptions of the λ-models obtained as fixed points of such functors corresponding to suitable choices of the mapping G, see Coppo et al. (1983). Solutions of these equations will be discussed below. At least the following result is worthwhile mentioning in this respect, see Coppo et al. (1984) for a proof.

16.3.35 Proposition *The filter* λ*-model* $\mathcal{F}^{\text{BCD}}$ *is isomorphic to* $\langle \mathcal{D}, F, G\rangle$, *where* $\mathcal{D}$ *is the initial solution of the domain equation*

$$\mathcal{D} = [\mathcal{D} \to \mathcal{D}] \times \mathsf{P}(\mathbb{A}_\infty),$$

and the pair $\langle F, G\rangle$ *is the Galois connection induced by the embedding that sends each function* f *to the minimal element in the extensionality classes of* f.

16.4 Other filter models

Lazy λ-calculus

Intersection types are flexible enough to allow for the description of λ-models which are computationally adequate for the lazy operational semantics (Abramsky and Ong (1993)). Following Berline (2000) we define the notion of lazy λ-model.

16.4.1 Definition

(i) The *order* of an untyped lambda term is

$$\text{order}(M) = \sup\{n \mid \exists N.M \twoheadrightarrow_{\beta} \lambda x_1 \cdots x_n.N\},$$

i.e. the upper bound of the number of its initial abstractions modulo β-conversion. So order$(M) \in \omega \cup \{\infty\}$.

(ii) A λ-model $\mathcal{D}$ is *lazy* if

$$[\mathcal{D} \models M = N \ \& \ \text{order}(M) = k] \Rightarrow \text{order}(N) = k.$$

For example order$(\mathbf{\Omega}) = 0$, order$(\mathsf{K}) = 2$ and order$(\mathsf{YK}) = \infty$.

16.4.2 Proposition *Let $\mathcal{F}^{\mathcal{T}}$ be a filter λ-model. Then $\mathcal{F}^{\mathcal{T}}$ is lazy iff*

$$\forall \Gamma, A.[\Gamma \vdash_{\cap}^{\mathcal{T}} M : A \Leftrightarrow \Gamma \vdash_{\cap}^{\mathcal{T}} N : A] \Rightarrow \text{order}(M) = \text{order}(N),$$

i.e. if M and N have the same types, then they have the same order.

Proof By 16.2.7(i). ■

A very simple type theory is AO, see Fig. 13.4. This gives a lazy λ-model, which is discussed in Abramsky and Ong (1993). In that paper the following result is proved, where for a $\mathcal{D} \in \mathbf{ALG}$ its lifting $\mathcal{D}_{\perp}$ is defined as the domain obtained by adding a new bottom element.

16.4.3 Theorem (Abramsky and Ong (1993)) *Let $\mathcal{D}_{\infty}^{\text{lazy}}$ be the initial solution of the domain equation $\mathcal{D} \cong [\mathcal{D} \to \mathcal{D}]_{\perp}$ in* **ALG**. *Then $\mathcal{D}_{\infty}^{\text{lazy}} \cong \mathcal{F}^{\text{AO}}$.*

The theory AO can also be used to prove the completeness of the so-called *F-semantics*, see Dezani-Ciancaglini and Margaria (1986).

The λI-calculus

Models of the λI-calculus are considered in Honsell and Lenisa (1993, 1999). Similarly to Theorem 16.3.22, one has the following.

16.4.4 Theorem (Honsell and Lenisa (1999)) *Let $\mathcal{D}_{\infty}^{\mathsf{I}}$ be the inverse limit solution of the domain equation $[\mathcal{D} \to_{\perp} \mathcal{D}] \cong \mathcal{D}$. Then $\mathcal{D}_{\infty}^{\mathsf{I}} \cong \mathcal{F}^{\mathcal{T}}$, for some proper type theory $\mathcal{T}$ with $\mathbb{A}^{\mathcal{T}} = \mathcal{K}(\mathcal{D}_0)$.*

Honsell and Lenisa (1999) discusses a filter structure which gives a *computationally adequate* model for the *perpetual* operational semantics and a mathematical model for the maximal sensible λI-theory.

A filter model equating an arbitrary closed term to $\mathbf{\Omega}$

In Jacopini (1975) it was proved by an analysis of conversion that the lambda term $\mathbf{\Omega}$ is easy, i.e. for any closed lambda term M the equation $\mathbf{\Omega} = M$ is consistent. This fact was proved by a Church–Rosser argument by Mitschke (see Mitschke (1976) or B[1984] Proposition 15.3.9). A model-theoretical proof was given by Baeten and Boerboom (1979), where it was shown that for any closed M one has

$$\mathcal{P}(\omega) \models \mathbf{\Omega} = M,$$

for a particular way of coding pairs on the set of natural numbers ω. We will now present the proof of this fact based on Alessi et al. (2001), using intersection types. For an arbitrary closed λ-term M we will build a filter model $\mathcal{F}^{\mathrm{ADH}(M)}$ such that

$$\mathcal{F}^{\mathrm{ADH}(M)} \models \mathbf{\Omega} = M.$$

We first examine which types can be assigned to $\boldsymbol{\omega} := \lambda x.xx$ and $\mathbf{\Omega} := \boldsymbol{\omega}\boldsymbol{\omega}$.

16.4.5 Lemma *Let $\mathcal{T}$ be a natural type theory that is $\boldsymbol{\beta}$-sound. Then*

(i) $\vdash_\cap^\mathcal{T} \boldsymbol{\omega} : A \to B \;\Leftrightarrow\; A \leq_\mathcal{T} A \to B;$

(ii) $\vdash_\cap^\mathcal{T} \mathbf{\Omega} : B \;\Leftrightarrow\; \exists A \in \mathbb{T}^\mathcal{T}. \vdash_\cap^\mathcal{T} \boldsymbol{\omega} : A \leq_\mathcal{T} (A \to B).$

Proof (i) ($\Leftarrow$) Suppose $A \leq_\mathcal{T} (A{\to}B)$. Then

$$\begin{aligned} x{:}A &\vdash_\cap^\mathcal{T} x : (A{\to}B) \\ x{:}A &\vdash_\cap^\mathcal{T} xx : B \\ &\vdash_\cap^\mathcal{T} \lambda x.xx : (A{\to}B). \end{aligned}$$

($\Rightarrow$) Suppose $\vdash_\cap^\mathcal{T} \boldsymbol{\omega} : (A \to B)$. If $B =_\mathcal{T} \mathtt{U}$, then $A \leq_\mathcal{T} \mathtt{U} =_\mathcal{T} (A \to B)$, by axiom $(\mathtt{U}\to)$. Otherwise, by Theorem 14.1.9,

$$\begin{aligned} \vdash_\cap^\mathcal{T} \lambda x.xx : (A{\to}B) \quad &\Rightarrow \quad x{:}A \vdash_\cap^\mathcal{T} xx : B, \\ &\Rightarrow \quad x{:}A \vdash_\cap^\mathcal{T} x : C, \quad x{:}A \vdash_\cap^\mathcal{T} x : (C{\to}B), \text{ for some } C, \\ &\Rightarrow \quad A \leq_\mathcal{T} (C{\to}B) \leq_\mathcal{T} (A{\to}B), \qquad \text{by } (\to). \end{aligned}$$

(ii) ($\Leftarrow$) Immediate.

($\Rightarrow$) If $B =_\mathcal{T} \mathtt{U}$, then $\vdash_\cap^\mathcal{T} \boldsymbol{\omega} : \mathtt{U} \leq_\mathcal{T} \mathtt{U}{\to}B$. If $B \neq_\mathcal{T} \mathtt{U}$, then by Theorem 14.1.9(ii) one has $\vdash_\cap^\mathcal{T} \boldsymbol{\omega} : (A{\to}B)$, $\vdash_\cap^\mathcal{T} \boldsymbol{\omega} : A$, for some A. By (i) one has $A \leq_\mathcal{T} A{\to}B$. ∎

We associate to each type the maximum number of nested arrows in the left-most path.

16.4.6 Definition Let $\mathcal{T}$ be a type theory. For $A \in \mathbb{T}^{\mathcal{T}}$ its *type nesting*, written $\#(A)$, is defined recursively on types as follows:

$$\begin{array}{lcll} \#(A) & = & 0 & \text{if } A \in \mathbb{A}^{\mathcal{T}}; \\ \#(A \to B) & = & \#(A) + 1; & \\ \#(A \cap B) & = & \max\{\#(A), \#(B)\}. & \end{array}$$

For $\boldsymbol{\eta}^{\mathtt{U}}$-sound and natural type theories, Lemma 16.4.5(ii) can be strengthened using type nesting. First we need the following lemma which shows that any type A with $\#(A) \geq 1$ is equivalent to an intersection of arrows with the same type nesting.

16.4.7 Lemma *Let $\mathcal{T}$ be a $\boldsymbol{\eta}^{\mathtt{U}}$-sound and natural type theory. Then for all $A \in \mathbb{T}^{\mathcal{T}}$ with $\#(A) \geq 1$, there exist $C_1, \ldots, C_m, D_1, \ldots, D_m$ such that*

$$\begin{array}{rcl} A & =_{\mathcal{T}} & (C_1 \to D_1) \cap \cdots \cap (C_m \to D_m); \\ \#(A) & = & \#((C_1 \to D_1) \cap \cdots \cap (C_m \to D_m)). \end{array}$$

Proof Every type A is an intersection of arrow types and atoms. Since $\mathcal{T}$ is $\boldsymbol{\eta}^{\mathtt{U}}$-sound and natural, the atoms can be replaced by an intersection of arrows between atoms. As $\#(A) \geq 1$ this does not increase the type nesting. ∎

16.4.8 Lemma *Let $\mathcal{T}$ be a natural type theory which is $\boldsymbol{\beta}$ and $\boldsymbol{\eta}^{\mathtt{U}}$-sound. Then*

$$\vdash_{\cap}^{\mathcal{T}} \boldsymbol{\Omega} : B \;\Rightarrow\; \exists A \in \mathbb{T}^{\mathcal{T}} [\vdash_{\cap}^{\mathcal{T}} \boldsymbol{\omega} : A \leq_{\mathcal{T}} A \to B \;\&\; \#(A) = 0].$$

Proof Let $\vdash_{\cap}^{\mathcal{T}} \boldsymbol{\Omega} : B$. If $B =_{\mathcal{T}} \mathtt{U}$ take $A \equiv \mathtt{U}$. Otherwise, by Lemma 16.4.5((ii)), there exists $A \in \mathbb{T}^{\mathcal{T}}$ such that $\vdash_{\cap}^{\mathcal{T}} \boldsymbol{\omega} : A$ and $A \leq_{\mathcal{T}} A \to B$. We show using value induction on $n = \#(A)$ that we can take an alternative A' with $\#(A') = 0$. If $n = 0$ we are done, so suppose $n \geq 1$. By Lemma 16.4.7, we may assume that A is of the form $A \equiv (C_1 \to D_1) \cap \cdots \cap (C_m \to D_m)$. Now $A \leq_{\mathcal{T}} A \to B$, hence $A \leq_{\mathcal{T}} C_{i_1} \cap \cdots \cap C_{i_p}$ and $D_{i_1} \cap \cdots \cap D_{i_p} \leq_{\mathcal{T}} B$, with $p > 0$ and $1 \leq i_1, \ldots, i_p \leq m$, since $\mathcal{T}$ is $\boldsymbol{\beta}$-sound. Hence,

$$\begin{array}{rcl} \vdash_{\cap}^{\mathcal{T}} \boldsymbol{\omega} : A & \Rightarrow & \vdash_{\cap}^{\mathcal{T}} \boldsymbol{\omega} : (C_{i_k} \to D_{i_k}), 1 \leq k \leq p, \\ & \Rightarrow & C_{i_k} \leq_{\mathcal{T}} (C_{i_k} \to D_{i_k}), \text{ by 16.4.5((i))}, \\ & \Rightarrow & C_{i_1} \cap \cdots \cap C_{i_p} \leq_{\mathcal{T}} (C_{i_1} \to D_{i_1}) \cap \cdots \cap (C_{i_p} \to D_{i_p}) \\ & & \leq_{\mathcal{T}} C_{i_1} \cap \cdots \cap C_{i_p} \to D_{i_1} \cap \cdots \cap D_{i_p}, \text{ since } \mathcal{T} \text{ is natural}, \\ & & \leq_{\mathcal{T}} (C_{i_1} \cap \cdots \cap C_{i_p} \to B), \text{ as } D_{i_1} \cap \cdots \cap D_{i_p} \leq_{\mathcal{T}} B. \end{array}$$

Now take $A' \equiv C_{i_1} \cap \cdots \cap C_{i_p}$. Then $\#(A') < n$ and we are done by the induction hypothesis. ∎

Now let $M \in \Lambda^{\emptyset}$. We will build the desired model satisfying $\models \boldsymbol{\Omega} = M$ by taking the union of a countable sequence of type theories $\mathrm{ADH}_n(M)$ defined in a suitable way to force the final interpretation of M to coincide with the interpretation of $\boldsymbol{\Omega}$. In the following, $\langle\cdot,\cdot\rangle$ denotes any bijection between $\omega\times\omega$ and ω.

16.4.9 Definition

(i) Define the following increasing sequence of intersection type theories $\mathrm{ADH}_n(M)$ by recursion on $n\in\omega$, specifying the atoms, axioms and rules:

$\mathbb{A}^{\,\mathrm{ADH}_0(M)} = \{\mathtt{U},0\}$;

$\mathrm{ADH}_0(M) =$ Scott, the smallest natural type theory that contains (0_{Scott});

$\mathbb{A}^{\,\mathrm{ADH}_{n+1}(M)} = \mathbb{A}^{\,\mathrm{ADH}_n(M)} \cup \{\xi_{\langle n,m\rangle} \mid m\in\omega\}$, with $\xi_{\langle n,m\rangle}$ fresh constraints;

$\mathrm{ADH}_{n+1}(M)$ is the smallest natural type theory that contains $\mathrm{ADH}_n(M)$ and the following infinite set of axioms

$$\xi_{\langle n,m\rangle} = (\xi_{\langle n,m\rangle} \to W_{\langle n,m\rangle}) \text{ for } m\in\omega,$$

where $\langle W_{\langle n,m\rangle}\rangle_{m\in\omega}$ is any enumeration of the countably infinite set

$$\{A \mid \vdash_{\cap}^{\mathrm{ADH}_n(M)} M : A\}.$$

(ii) We define $\mathrm{ADH}(M)$ as follows:

$$\mathbb{A}^{\mathrm{ADH}(M)} = \bigcup_{n\in\omega} \mathbb{A}^{\mathrm{ADH}_n(M)}; \quad \mathrm{ADH}(M) = \bigcup_{n\in\omega} \mathrm{ADH}_n(M).$$

16.4.10 Proposition *$\mathrm{ADH}(M)$ is a $\boldsymbol{\beta},\boldsymbol{\eta}^{\mathtt{U}}$-sound natural type theory.*

Proof We can immediately see that $\mathrm{ADH}(M)$ is $\boldsymbol{\beta}$ and $\boldsymbol{\eta}^{\mathtt{U}}$-sound: they hold for $\mathrm{ADH}_0(M)$, and not adding $\le$ between arrows means $\boldsymbol{\beta}$ is preserved, while equating each fresh constant to an arrow means η is assured. By construction all the $\mathrm{ADH}_n(M)$ are natural type theories. The validity of rule $(\to)$ in $\mathrm{ADH}(M)$ follows by a 'compactness' argument: if $A' \le_{\mathrm{ADH}(M)} A$ and $B \le_{\mathrm{ADH}(M)} B'$, then $A' \le_{\mathrm{ADH}_n(M)} A$ and $B \le_{\mathrm{ADH}_m(M)} B'$; but then we have $(A\to B) \le_{\mathrm{ADH}_{\max\{n,m\}}(M)} (A'\to B')$ and hence $(A\to B) \le_{\mathrm{ADH}(M)} (A'\to B')$. Similarly one verifies that $(\to\cap), (\mathtt{U}_{\mathrm{top}})$, and $(\mathtt{U}\to)$ all hold in $\mathrm{ADH}(M)$. Therefore the type theory $\mathrm{ADH}(M)$ is natural. ∎

16.4.11 Theorem *$\mathcal{F}^{\mathrm{ADH}(M)}$ is an extensional filter λ-model.*

Proof By Propositions 16.4.10 and 16.2.11. ∎

We now need to show that some types cannot be deduced for $\boldsymbol{\omega}$.

16.4.12 Lemma $\nvdash_{\cap}^{\mathrm{ADH}(M)} \boldsymbol{\omega} : 0$ *and* $\nvdash_{\cap}^{\mathrm{ADH}(M)} \boldsymbol{\omega} : (0 \to 0) \to 0 \to 0.$

Proof Define the set $\mathcal{E}_{\mathtt{U}} \subseteq \mathbb{T}(\mathbb{A}^{\mathrm{ADH}(M)})$ as the minimal set such that:

$$\begin{array}{lcl} \mathtt{U} \in \mathcal{E}_{\mathtt{U}}; & & \\ A \in \mathbb{T}^{\mathrm{ADH}(M)}, B \in \mathcal{E}_{\mathtt{U}} & \Rightarrow & (A \to B) \in \mathcal{E}_{\mathtt{U}}; \\ A, B \in \mathcal{E}_{\mathtt{U}} & \Rightarrow & (A \cap B) \in \mathcal{E}_{\mathtt{U}}; \\ W_i \in \mathcal{E}_{\mathtt{U}} & \Rightarrow & \xi_i \in \mathcal{E}_{\mathtt{U}}. \end{array}$$

Claim: $A \in \mathcal{E}_{\mathtt{U}} \Leftrightarrow A =_{\mathrm{ADH}(M)} \mathtt{U}$.

($\Rightarrow$) By induction on the definition of $\mathcal{E}_{\mathtt{U}}$, using axiom ($\mathtt{U} \to$).

($\Leftarrow$) By induction on $\leq_{\mathrm{ADH}(M)}$ it follows that

$$\mathcal{E}_{\mathtt{U}} \ni B \leq_{\mathrm{ADH}(M)} A \Rightarrow A \in \mathcal{E}_{\mathtt{U}}.$$

Hence if $A =_{\mathrm{ADH}(M)} \mathtt{U}$, one has $\mathcal{E}_{\mathtt{U}} \ni \mathtt{U} \leq_{\mathrm{ADH}(M)} A$ and thus $A \in \mathcal{E}_{\mathtt{U}}$.

As $0 \notin \mathcal{E}_{\mathtt{U}}$, it follows by the claim that

$$0 \neq_{\mathrm{ADH}(M)} \mathtt{U}. \tag{16.6}$$

Similarly one has $0 \to 0 \notin \mathcal{E}_U$, hence $0 \to 0 \neq_{\mathrm{ADH}(M)} U$.

Suppose towards a contradiction that $\vdash_{\cap}^{\mathrm{ADH}(M)} \boldsymbol{\omega} : 0$. Then $\vdash^{\mathrm{ADH}(M)} \boldsymbol{\omega} : \mathtt{U} \to 0$, by ($0_{\mathrm{Scott}}$). By Lemma 16.4.5((i)) we get $\mathtt{U} \leq_{\mathrm{ADH}(M)} (\mathtt{U} \to 0) =_{\mathrm{ADH}(M)} 0 \leq_{\mathrm{ADH}(M)} \mathtt{U}$, i.e. $\mathtt{U} =_{\mathrm{ADH}(M)} 0$, contradicting (16.6).

Similarly, from $\vdash_{\cap}^{\mathrm{ADH}(M)} \boldsymbol{\omega} : (0 \to 0) \to 0 \to 0$, by Lemma 16.4.5((i)) we get $0 \to 0 \leq_{\mathrm{ADH}(M)} (0 \to 0) \to (0 \to 0)$, which implies $0 \to 0 \leq_{\mathrm{ADH}(M)} 0 \leq_{\mathrm{ADH}(M)} \mathtt{U} \to 0$, by $\boldsymbol{\beta}$-soundness and (0_{Scott}). Therefore $\mathtt{U} =_{\mathrm{ADH}(M)} 0$, contradicting (16.6). ∎

We finally are able to prove the main theorem.

16.4.13 Theorem *Let $M \in \Lambda^{\emptyset}$. Then $\mathcal{F}^{\mathrm{ADH}(M)}$ is a non-trivial extensional λ-model such that $\mathcal{F}^{\mathrm{ADH}(M)} \models M = \boldsymbol{\Omega}$.*

Proof The model is non-trivial since clearly $\vdash_{\cap}^{\mathrm{ADH}(M)} \mathsf{I} : (0 \to 0) \to 0 \to 0$ and, by Lemma 16.4.12, we have $\nvdash_{\cap}^{\mathrm{ADH}(M)} \boldsymbol{\omega} : (0 \to 0) \to 0 \to 0$, hence $\mathcal{F}^{\mathrm{ADH}(M)} \not\models \mathsf{I} = \boldsymbol{\omega}$.

We must show that $[\![M]\!] = [\![\boldsymbol{\Omega}]\!]$. Suppose that $W \in [\![M]\!]$. Then

$$\begin{array}{rcll} \vdash_{\cap}^{\mathrm{ADH}(M)} M : W & \Rightarrow & \vdash_{\cap}^{\mathrm{ADH}_n(M)} M : W, & \text{for some } n, \\ & \Rightarrow & \xi_i =_{\mathrm{ADH}_{n+1}(M)} (\xi_i \to W), & \text{for some } i, \\ & \Rightarrow & \vdash_{\cap}^{\mathrm{ADH}(M)} \boldsymbol{\Omega} : W, & \\ & \Rightarrow & W \in [\![\boldsymbol{\Omega}]\!]. & \end{array}$$

This proves $[\![M]\!] \subseteq [\![\boldsymbol{\Omega}]\!]$.

Now suppose $B \in [\![\boldsymbol{\Omega}]\!]$, i.e. $\vdash_{\cap}^{\text{ADH}(M)} \boldsymbol{\Omega} : B$. Then by Lemma 16.4.8 there exists A such that $\#(A) = 0$ and $\vdash_{\cap}^{\text{ADH}(M)} \boldsymbol{\omega} : A \leq_{\text{ADH}(M)} A \to B$. Let $A \equiv \bigcap_{i \in I} \psi_i$, with $\psi_i \in \mathbb{A} = \{\mathtt{U}, 0, \xi_0, \dots\}$. By Lemma 16.4.12 we have $\psi_i \neq_{\text{ADH}(M)} 0$. Hence it follows that $A =_{\text{ADH}(M)} \mathtt{U}$ or $A =_{\text{ADH}(M)} \bigcap_{j \in J}(\xi_j)$, for some finite $J \subseteq \omega$. Since $\mathtt{U} =_{\text{ADH}(M)} (\mathtt{U} \to \mathtt{U})$ and $\xi_j =_{\text{ADH}(M)} (\xi_j \to W)$ we get $A =_{\text{ADH}(M)} (\mathtt{U} \to \mathtt{U})$ or $A =_{\text{ADH}(M)} \bigcap_{j \in J}(\xi_j \to W_j)$. Since $A \leq_{\text{ADH}(M)} A \to B$ it follows by $\boldsymbol{\beta}$-soundness that in the first case $\mathtt{U} \leq_{\text{ADH}(M)} B$ or in the second case $\bigcap_{j \in L} W_j \leq_{\text{ADH}(M)} B$, for some $L \subseteq J$. Since each W_j is in $[\![M]\!]$, we have in both cases $B \in [\![M]\!]$. This shows $[\![\boldsymbol{\Omega}]\!] \subseteq [\![M]\!]$ and we are done. ∎

Graph models as filter models

For a set X we use $\mathsf{P}(X)$ to denote the power-set of X and $\mathsf{P}_{\text{fin}}(X)$ to denote the set of finite subsets of X.

Engeler's Model

16.4.14 Definition (Engeler (1981)) Let $\mathbb{A}_\infty$ be a countable set of atoms.

(i) Define Em as the smallest set satisfying $\mathsf{Em} = \mathbb{A}_\infty \cup (\mathsf{P}_{\text{fin}}(\mathsf{Em}) \times \mathsf{Em}))$.

(ii) Define
$$F_{\mathsf{Em}} : \mathsf{P}(\mathsf{Em}) \to [\mathsf{P}(\mathsf{Em}) \to \mathsf{P}(\mathsf{Em})]$$
$$G_{\mathsf{Em}} : [\mathsf{P}(\mathsf{Em}) \to \mathsf{P}(\mathsf{Em})] \to \mathsf{P}(\mathsf{Em})$$
by
$$F_{\mathsf{Em}}(X) = \bigsqcup\{u \mapsto e \mid \langle u, e\rangle \in X\}$$
$$G_{\mathsf{Em}}(f) = \{\langle u, e\rangle \mid e \in f(u)\}.$$

16.4.15 Theorem *$F_{\mathsf{Em}}, G_{\mathsf{Em}}$ satisfy $F_{\mathsf{Em}} \circ G_{\mathsf{Em}} = \mathsf{Id}$, making $\mathsf{P}(\mathsf{Em})$ a λ-model.*

Proof See Engeler (1981). ∎

16.4.16 Theorem *Let the type theory* Engeler *be as defined in Fig.* 13.4. *Then $\mathsf{P}(\mathsf{Em}) \cong \mathcal{F}^{\text{Engeler}}$ are isomorphic as λ-structures (λ-models).*

Proof See Plotkin (1993). ∎

Scott's $\mathsf{P}(\omega)$ model

Following the original notation by Scott, we use ω to denote the set of natural numbers.

16.4.17 Notation

(i) Let $\lambda\!\!\lambda nm.\langle n,m\rangle : \omega \times \omega \to \omega$ be the polynomially defined bijection

$$\langle n,m\rangle = \frac{1}{2}(n+m)(n+m+1)+m.$$

(ii) Let $\lambda\!\!\lambda n.e_n : \omega \to \mathsf{P}_{\mathrm{fin}}(\omega)$ be a bijection, for example, that defined by

$$e_n = \{k_0,\ldots,k_{m-1}\} \text{ with } k_0 < k_1 < \cdots < k_{m-1} \Leftrightarrow n = \Sigma_{i<m}2^{k_i}.$$

16.4.18 Definition [Scott (1972)] Let $\gamma : \mathsf{P}_{\mathrm{fin}}(\omega) \times \omega \to \omega$ be the bijection defined by

$$\gamma(e_n,m) = \langle n,m\rangle.$$

(i) Define $F_\omega : \mathsf{P}\omega \to [\mathsf{P}\omega \to \mathsf{P}\omega]$ by

$$F_\omega(X) = \bigsqcup\{u \mapsto i \mid \gamma(u,i) \in X\}.$$

(ii) $G_\omega : [\mathsf{P}\omega \to \mathsf{P}\omega] \to \mathsf{P}\omega$ by

$$G_\omega(f) = \{\gamma(u,i) \mid i \in f(u)\}$$

for all $f \in [\mathsf{P}\omega \to \mathsf{P}\omega]$.

16.4.19 Proposition *Define for $X, Y \in \mathsf{P}\omega$ the application*

$$X \cdot_{\mathsf{P}\omega} Y = \{m \mid \exists e_n \subseteq Y\ \langle n,m\rangle \in X\}.$$

Then $F_\omega(X)(Y) = X \cdot_{\mathsf{P}\omega} Y$ is a (more common) equivalent definition for F_ω.

Proof Do Exercise 16.5. ■

16.4.20 Theorem *$\mathsf{P}\omega$ is a λ-model via F_ω, G_ω.*

Proof See Scott (1972). ■

16.4.21 Theorem *Define*

$$\begin{array}{rcl} \mathbb{A}^{\text{Scott-}\omega} & = & \omega \\ \text{Scott-}\omega & = & \text{Engeler} \cup \{\bigcap_{k \in e}(k \to n) = \gamma(e,n) \mid e \in \mathsf{P}_{\mathrm{fin}}(\omega), n \in \omega\}. \end{array}$$

Then $\mathsf{P}\omega \cong \mathcal{F}^{\text{Scott-}\omega}$ are isomorphic as natural λ-structures (λ-models).

Proof See Alessi (1991). ■

Plotkin's Model

16.4.22 Definition (Plotkin (1993)) Let 0 be an atom.

(i) Define Pm as the smallest set such that

$$\mathsf{Pm} = \{0\} \cup (\mathsf{P}_{\text{fin}}(\mathsf{Pm}) \times \mathsf{P}_{\text{fin}}(\mathsf{Pm})).$$

(ii) Define
$$\begin{array}{rcl} F_{\mathsf{Pm}} & : & \mathsf{P}(\mathsf{Pm}) \to [\mathsf{P}(\mathsf{Pm}) \to \mathsf{P}(\mathsf{Pm})] \\ G_{\mathsf{Pm}} & : & [\mathsf{P}(\mathsf{Pm}) \to \mathsf{P}(\mathsf{Pm})] \to \mathsf{P}(\mathsf{Pm}) \end{array}$$
by

$$F_{\mathsf{Pm}}(X) = \bigsqcup\{u \mapsto v \mid \langle u, v\rangle \in X\}$$
$$G_{\mathsf{Pm}}(f) = \{\langle u, v\rangle \mid v \subseteq f(u)\}.$$

16.4.23 Theorem *$F_{\mathsf{Pm}}, G_{\mathsf{Pm}}$ satisfy $F_{\mathsf{Pm}} \circ G_{\mathsf{Pm}} = \mathsf{Id}$, turning $\mathsf{P}(\mathsf{Pm})$ into a λ-model.*

Proof See Plotkin (1993). ∎

16.4.24 Theorem *Let the type theory* Plotkin *be as defined in Fig. 13.4. Then $\mathsf{P}(\mathsf{Pm}) \cong \mathcal{F}^{\text{Plotkin}}$ are isomorphic as natural λ-structures (λ-models).*

Proof See Plotkin (1993). ∎

16.5 Exercises

Exercise 16.1 Check the following equalities:

$$\begin{array}{rcl} [\![\lambda x.y]\!]^{\mathcal{F}^{\text{CDV}}}_{\rho_0} & = & \emptyset; \\ [\![\lambda x.y]\!]^{\mathcal{F}^{\text{HL}}}_{\rho_1} & = & \uparrow 0; \\ [\![\lambda x.y]\!]^{\mathcal{F}^{\text{AO}}}_{\rho_0} & = & \uparrow (\mathtt{U} \to \mathtt{U}); \\ [\![\lambda x.y]\!]^{\mathcal{F}^{\text{EHR}}}_{\rho_0} & = & \uparrow \mathtt{V}, \end{array}$$

where $\rho_0(y) = \uparrow \emptyset$ and $\rho_1(y) = \uparrow 0$.

Exercise 16.2

(i) Define $\mathsf{K}^\infty \equiv \mathsf{YK}$. This term is called the ogre. Find a type for it in the system $\lambda\cap^{\text{AO}}$.
(ii) Show that ogre inhabits all types in $\lambda\cap^{\text{AO}}$. [Hint. Use (i) and Exercise 13.4.]

Exercise 16.3 Prove using the results of Exercise 16.2 that $[\![\mathsf{K}^\infty]\!]^{\mathcal{F}^{\text{AO}}}_{\rho} = \mathcal{F}^{\text{AO}}$.

Exercise 16.4 Define $\mathsf{t} : \mathbb{T}(\{0,1,\mathtt{U}\}) \to \mathbb{T}(\{0,1,\mathtt{U}\})$ recursively:

$$\begin{array}{rcll} \mathsf{t}(\alpha) & = & \alpha, & \text{where } \alpha \in \{\mathtt{U}, 0\}; \\ \mathsf{t}(1) & = & \mathtt{U}; & \\ \mathsf{t}(A \to B) & = & A \to \mathsf{t}(B); & \\ \mathsf{t}(A \cap B) & = & \mathsf{t}(A) \cap \mathsf{t}(B). & \end{array}$$

The intersection type theory Alessi is axiomatized by rule $(\to)$ and axioms $(\to\cap)$, $(\mathtt{U}_{\text{top}})$, $(\mathtt{U}\to)$, (01), $(1\to 0)$, $(0\to 1)$, see Fig. 13.4, and (t), $(\mathsf{t}\to)$, where

$$\begin{array}{ll} (\mathsf{t}) & A \leq \mathsf{t}(A); \\ (\mathsf{t}\to) & A \to B \leq \mathsf{t}(A) \to \mathsf{t}(B). \end{array}$$

If $\Gamma = \{x_1{:}A_1, \ldots, x_n{:}A_n\}$, then write $\mathsf{t}(\Gamma) = \{x_1{:}\mathsf{t}(A_1), \ldots, x_n{:}\mathsf{t}(A_n)\}$. Show:

(i) The map t is idempotent, i.e. $\mathsf{t}(\mathsf{t}(A)) = \mathsf{t}(A)$.
(ii) $A \to \mathsf{t}(B) =_{\text{Alessi}} \mathsf{t}(A) \to \mathsf{t}(B)$.
(iii) $A \leq_{\text{Alessi}} B \Rightarrow \mathsf{t}(A) \leq_{\text{Alessi}} \mathsf{t}(B)$.
(iv) $\Gamma \vdash^{\text{Alessi}} M : A \Rightarrow \mathsf{t}(\Gamma) \vdash^{\text{Alessi}} M : \mathsf{t}(A)$.
(v) $\Gamma, \Gamma' \vdash^{\text{Alessi}} M : A \Rightarrow \Gamma, \mathsf{t}(\Gamma') \vdash^{\text{Alessi}} M : \mathsf{t}(A)$.
(vi) $\forall i \in I.\ \Gamma, x{:}A_i \vdash^{\text{Alessi}} M : B_i\ \&\ \bigcap_{i \in I}(A_i \to B_i) \leq_{\text{Alessi}} C \to D \Rightarrow \Gamma, x{:}C \vdash^{\text{Alessi}} M : D$.
(vii) Alessi is not $\boldsymbol{\beta}$-sound. [Hint. $1 \to 0 \leq_{\text{Alessi}} \mathtt{U} \to 0$.]
(viii) $\mathcal{F}^{\text{Alessi}}$ is a filter λ-model. [Hint. Modify Theorem 16.2.22 and Lemma 16.2.21(iv).]
(ix) The step function $\uparrow 1 \Rightarrow \uparrow 0$ is not representable in $\mathcal{F}^{\text{Alessi}}$.

Actually, $\mathcal{F}^{\text{Alessi}}$ is the inverse limit solution of the domain equation $\mathcal{D} \simeq [\mathcal{D} \to \mathcal{D}]$ taken in the category of t-*lattices*, whose objects are ω-algebraic lattices $\mathcal{D}$ endowed with a finitary additive projection $\delta : \mathcal{D} \to \mathcal{D}$ and whose morphisms $f : (\mathcal{D}, \delta) \to (\mathcal{D}', \delta')$ are continuous functions such that $\delta' \circ f \sqsubseteq f \circ \delta$. See Alessi (1993), Alessi et al. (2004b) for details.

Exercise 16.5 Prove Proposition 16.4.19.

Exercise 16.6 Prove Theorem 16.4.16 using the mapping

$$f : \mathcal{P}_{\text{fin}}(\mathsf{Em}) \to \mathbb{T}^{\text{Engeler}}$$

defined by

$$\begin{array}{rcl} f(\emptyset) & = & \mathtt{U}, \\ f(\{\mathsf{a}\}) & = & \mathsf{a}, \\ f(u \cup \{e\}) & = & f(u) \cap f(\{e\}) \\ f(\{\langle u, e \rangle\}) & = & f(u) \to f(\{e\}), \end{array}$$

where $u \in \mathcal{P}_{\text{fin}}(\mathsf{Em}), e \in \mathsf{Em}, \mathsf{a} \in \mathbb{A}_\infty$.

Exercise 16.7 Prove Theorem 16.4.21 using the mapping

$$f : \mathsf{P}\omega \to \mathbb{T}^{\text{Alessi}}$$

defined by

$$\begin{array}{lcl} f(\emptyset) & = & \mathtt{U}, \\ f(\{i\}) & = & i, \\ f(u \cup \{i\}) & = & f(u) \cap i. \end{array}$$

where $u \in \mathcal{P}_{\text{fin}}(\omega), i \in \omega$.

Exercise 16.8 Prove Theorem 16.4.24 using the mapping

$$f : \mathcal{P}_{\text{fin}}(\mathsf{Pm}) \to \mathbb{T}^{\text{Plotkin}}$$

defined by

$$\begin{array}{lcl} f(\emptyset) & = & \mathtt{U}, \\ f(\{\omega\}) & = & \omega, \\ f(u \cup \{a\}) & = & f(u) \cap f(\{a\}), \\ f(\{\langle u, v\rangle\}) & = & f(u) \to f(v). \end{array}$$

where $u, v \in \mathcal{P}_{\text{fin}}(\mathsf{Pm})$, $a \in \mathsf{Pm}$.

Exercise 16.9 Let $\mathcal{T} \in \mathrm{TT}^{\text{-}\mathtt{U}}$. Prove the following statements.

(1) The filter quasi λ-model $\mathcal{F}^{\mathcal{T}} = \langle \mathcal{F}^{\mathcal{T}}, \cdot, [\![\ ]\!]^{\mathcal{F}^{\mathcal{T}}} \rangle$ is not a λ-model [Hint. Consider the constant function $\lambda x.y$.]
(2) The filter quasi λ-model $\mathcal{F}^{\mathcal{T}} = \langle \mathcal{F}^{\mathcal{T}}, \cdot, [\![\ ]\!]^{\mathcal{F}^{\mathcal{T}}} \rangle$ does not satisfy $\boldsymbol{\beta}$-conversion.
(3) The filter structure $\mathcal{F}^{\mathcal{T}} = \langle \mathcal{F}^{\mathcal{T}}, F^{\mathcal{T}}, G^{\mathcal{T}} \rangle$ is not reflexive.
(4) Not all continuous functions are representable in $\mathcal{F}^{\mathcal{T}} = \langle \mathcal{F}^{\mathcal{T}}, F^{\mathcal{T}}, G^{\mathcal{T}} \rangle$.

Exercise 16.10 Let $\mathcal{F} = \mathcal{F}^{\text{BCD}}$ be the model described in Section 16.2. To every intersection type $A \in \mathbb{T}^{\text{BCD}}$ associate a *height*, written $|\,A\,|$, as follows:

(i) $|\,\mathtt{U}\,| = 0$;
(ii) $|\,\mathsf{c}_i\,| = 1$, for all $\mathsf{c}_i \in \mathbb{A}^{\text{BCD}}$ such that $\mathsf{c}_i \neq \mathtt{U}$;
(iii) $|\,A \cap B\,| = \max\{|\,A\,|, |\,B\,|\}$;
(iv) $|\,A \to B\,| = 1 + \max\{|\,A\,|, |\,B\,|\}$.

Given a filter $d \in \mathcal{F}$, define

$$d[n] =_{\text{def}} \{A \in d \mid |A| \leq n\}$$

and set $d_n =_{\text{def}} \uparrow d[n]$, where $\uparrow X$ is the filter generated by the set X (this is the intersection of all filters containing X). Prove that the mappings $d \mapsto d_n : \mathcal{F} \to \mathcal{F}$ define a notion of approximation over $\mathcal{F}$. [Hint. First, show that, for a filter d, the set $d[n]$ is closed under finite intersections. As a consequence, $B \in \uparrow d[n]$ if and only if $B \geq A$ for some $A \in d[n]$. Second, prove that $d_0 = \uparrow\{\mathtt{U}\}$. Third, in order to prove some of the equations for a notion of approximation, you may need the following properties of the preorder on intersection types.

(a) If $B \neq \mathtt{U}$ and $A' \to B' \leq A \to B$, then $A \leq A'$ and $B' \leq B$.
(b) If $C \leq A \to B$ and $|C| \leq n + 1$, then there exist A', B' such that $|A'|, |B'| \leq n$ and $C \leq A' \to B' \leq A \to B$.

For the proof of (b), define the relation $\lessdot$ on types by the same axioms and rules as $\leq$, with the exception of reflexivity and transitivity, then show that $A \leq B$ if and only if $A(\lessdot)^*B$, where $(\lessdot)^*$ is the reflexive transitive closure of $(\lessdot)$.]

17
Advanced Properties and Applications

This chapter proves some properties of intersection types in relation to terms and models.

Section 17.1 defines a realizability interpretation of types (Barendregt et al. (1983)). Types are interpreted as subsets of a domain of discourse $\mathcal{D}$. Assuming a (partial) application $\cdot : \mathcal{D}\times\mathcal{D}\to\mathcal{D}$, and a type environment ξ, i.e. a mapping from type atoms to subsets of $\mathcal{D}$, we can define an interpretation $[\![A]\!]_\xi = [\![A]\!]^{\mathcal{T}}_\xi \subseteq \mathcal{D}$ for each type A in $\mathbb{T}_\to$ by giving $\to$ the *realizability interpretation*, i.e.

$$\begin{aligned} [\![\alpha]\!]_\xi &= \xi(\alpha); \\ [\![A\to B]\!]_\xi &= [\![A]\!]_\xi \to [\![B]\!]_\xi = \{d\in\mathcal{D} \mid d\cdot[\![A]\!]_\xi \subseteq [\![B]\!]_\xi\}. \end{aligned}$$

This semantics, due to Scott (1975a), can be extended to intersection types by interpreting $\mathtt{U}$ as the domain of discourse and the intersection $\cap$ on types as set-theoretic intersection, i.e.

$$\begin{aligned} [\![\mathtt{U}]\!]_\xi &= \mathcal{D}; \\ [\![A\cap B]\!]_\xi &= [\![A]\!]_\xi \cap [\![B]\!]_\xi. \end{aligned}$$

The first requirement will be met by considering only ξ with $\xi(\mathtt{U}) = \mathcal{D}$. Then, $\leq$ is interpreted as inclusion between sets. For interpreting λ-terms, it is enough to require that $\mathcal{D}$ is a quasi λ-model, depending on a valuation ρ of the term variables in $\mathcal{D}$. One says that $\mathcal{D}$ *satisfies* Γ under ρ,ξ, written $\mathcal{D},\rho,\xi \models \Gamma$, if

$$\mathcal{D},\rho,\xi \models x : A, \quad \text{for all } (x{:}A)\in\Gamma.$$

Finally, Γ *satisfies* $M : A$, which we write as $\Gamma \models M : A$, is defined by

$$\Gamma \models M : A \Leftrightarrow \forall\mathcal{D},\rho,\xi.[\mathcal{D},\rho,\xi \models \Gamma \Rightarrow \mathcal{D},\rho,\xi \models M : A].$$

$\mathcal{T}$	$\mathsf{Ctx}^{\mathcal{T}}$	$\mathsf{Set}^{\mathcal{T}}(0)$	$\mathsf{Set}^{\mathcal{T}}(1)$	$\mathsf{Set}^{\mathcal{T}}(\mathtt{U})$
Park	$\emptyset$	$\Lambda^{\emptyset}\!\uparrow\!\boldsymbol{\beta}$	$-$	Λ
CDZ	Γ_0	PN	N	Λ
HR	Γ_1	$\emptyset$	$\Lambda^{\mathsf{I}}\!\uparrow\!\boldsymbol{\beta}$	Λ
DHM	Γ_0	PHN	HN	Λ
HL	Γ_0	PSN	SN	$-$

Figure 17.1 Context and Set associated to 0, 1 and U.

First soundness is proved, i.e.

$$\Gamma \vdash^{\mathcal{T}}_{\cap} M : A \;\Rightarrow\; \Gamma \models^{\mathcal{T}}_{\cap} M : A.$$

Completeness is the converse implication. Not all type theories satisfy completeness. We will prove that the natural type theories are exactly the ones that satisfy completeness:

$$[\Gamma \models^{\mathcal{T}}_{\cap} M : A \;\Rightarrow\; \Gamma \vdash^{\mathcal{T}}_{\cap} M : A] \text{ iff } \mathcal{T} \in \mathrm{NTT}^{\mathtt{U}}.$$

The proof of completeness for a natural type theory $\mathcal{T}$ follows by taking $\mathcal{D} = \mathcal{F}^{\mathcal{T}}$ where $\mathcal{F}^{\mathcal{T}}$ is the filter quasi λ-model over $\mathcal{T}$.

In Barendregt et al. (1983) the completeness of the type theory BCD was used to show the completeness of simple types: for all $M \in \Lambda^{\emptyset}$ and $A \in \mathbb{T}^{\mathbb{A}}_{\rightarrow}$

$$\vdash_{\boldsymbol{\lambda}_{\rightarrow}} M : A \;\Leftrightarrow\; \models M : A.$$

In Sections 17.2 and 17.4 intersection types will be used to characterize some properties of λ-terms. We consider the following properties (subsets) of λ-terms: strong normalization, normalization, head normalization and the persistent variants of these. A term has *persistently* property Q if, for all appropriate (to be defined in Section 17.2) arguments $\vec{N}$, the term $M\vec{N}$ has property Q. The sets of untyped lambda terms having these properties are denoted by $\mathsf{SN}, \mathsf{N}, \mathsf{HN}, \mathsf{PSN}, \mathsf{PN}, \mathsf{PHN}$. For a set of terms $X \subseteq \Lambda$ write

$$X\!\uparrow\!\boldsymbol{\beta} = \{M \mid \exists N \in X. M \twoheadrightarrow_{\boldsymbol{\beta}} N\}.$$

We denote by Γ_α the set of type declarations which associate α to all variables. For $\mathcal{T} \in \{\mathrm{Park}, \mathrm{CDZ}, \mathrm{HR}, \mathrm{DHM}, \mathrm{HL}\}$, Fig. 17.1 defines a context $\mathsf{Ctx}^{\mathcal{T}}$ and a subset $\mathsf{Set}^{\mathcal{T}}(c)$ of lambda terms for each type atom $c \in \{0, 1, \mathtt{U}\}$.
We will show the following characterizations

$$M \in \mathsf{Set}^{\mathcal{T}}(c) \text{ iff } \mathsf{Ctx}^{\mathcal{T}} \vdash^{\mathcal{T}} M : c.$$

The characterizations for $\mathtt{U}$ are immediate; those for 1 are given in Theorems 17.2.15 (namely, CDZ, DHM, HL) and 17.4.9 (HR); those for 0 in Theorem 17.4.3 (Park), Lemma 17.4.6 (HR); the remainder can be found in Dezani-Ciancaglini et al. (2005) (CDZ, DHM), and Tatsuta and Dezani-Ciancaglini (2006) (HL). The characterizations are located in separate sections because two different methods are used to prove them:

(1) In Section 17.2 we use the type interpretation defined in Section 17.1 and the standard technique of type stable sets;
(2) In Section 17.4 we use the Approximation Theorem presented in Section 17.3.

In Section 17.3, we introduce appropriate notions of approximants for almost all the type theories of Fig. 13.4. Intuitively, an approximant is a partial term in the computation that does not contain redexes. In some cases, the approximants are obtained by replacing redexes by $\perp$ and in other cases by just freezing them. In the case of Scott, CDZ, DHM and BCD, the whole context containing a redex in a head position is replaced by $\perp$. In the case of AO, the notion of approximant is relaxed and abstractions are not replaced by $\perp$. In the case of Park and HR, the redexes are frozen by inserting a constant before the abstraction. We will show that a type can be derived for a term *if and only if* it can be derived for an approximant of that term (Approximation Theorem). A common and uniform proof is given of this theorem for all the type theories mentioned above (Dezani-Ciancaglini et al. (2001)). The proof technique used is a variant of stable sets over a Kripke applicative structure. In Section 17.4 some applications of the Approximation Theorem are given. Amongst these, the characterizations for Park and HL mentioned in Fig. 17.1.

Finally in Section 17.5 it will be shown that given Γ, A one cannot decide inhabitation for the type theory CDV, i.e. the existence of an M such that $\Gamma \vdash^{\text{CDV}} M : A$. Also for BCD the inhabitation is undecidable, see Urzyczyn (1999). On the other hand, in the type theory AO all types are inhabited, see Exercise 16.2, hence inhabitation is decidable. The question of decidability of inhabitation for several other type theories remains open.

The contents of the first four sections of this chapter can be regarded as applications of intersection types.

17.1 Realizability interpretation of types

The natural set-theoretic semantics for type assignment in $\boldsymbol{\lambda}_{\rightarrow}$ based on untyped λ-models was given in Scott (1975a) where it was shown that

$$\Gamma \vdash_{\boldsymbol{\lambda}_{\rightarrow}} M : A \Rightarrow \Gamma \models M : A.$$

Scott asked whether the converse (completeness) holds. In Barendregt et al. (1983) the notion of semantics was extended to intersection types and completeness was proved for $\boldsymbol{\lambda}_{\cap}^{\mathrm{BCD}}$ via the corresponding filter model. Completeness for $\boldsymbol{\lambda}_{\rightarrow}$ then follows by a conservativity result. In Hindley (1983) an alternative direct proof of completeness for $\boldsymbol{\lambda}_{\rightarrow}$ was given using a term model. Variations of the semantics are presented in Dezani-Ciancaglini et al. (2003).

Recall that quasi λ-models are defined in Definition 16.1.2 and are λ-models without the requirement that ($\boldsymbol{\beta}$) holds. Using this notion one can distinguish between models for the λI-calculus and the full λ-calculus.

17.1.1 Definition (Type Interpretation) Let $\mathcal{D} = \langle \mathcal{D}, \cdot, [\![\;]\!]^{\mathcal{D}} \rangle$ be a quasi λ-model and let $\mathcal{T}$ be an intersection type theory over the atoms $\mathbb{A}^{\mathcal{T}}$.

(i) Recall that for $X, Y \subseteq \mathcal{D}$ we defined

$$X \Rightarrow Y = \{d \in \mathcal{D} \mid \forall e \in X. d \cdot e \in Y\}.$$

(ii) The *type interpretation* induced by the type environment $\xi : \mathbb{A}^{\mathcal{T}} \to \mathsf{P}(\mathcal{D})$, with $\xi(\mathtt{U}) = \mathcal{D}$ for $\mathcal{T} \in \mathrm{TT}^{\mathtt{U}}$, is the map $[\![\;]\!]^{\mathcal{D}}_{\xi} : \mathbb{T}^{\mathcal{T}} \to \mathsf{P}(\mathcal{D})$ defined as:

$$\begin{aligned} [\![\alpha]\!]^{\mathcal{D}}_{\xi} &= \xi(\alpha), \\ [\![A \to B]\!]^{\mathcal{D}}_{\xi} &= [\![A]\!]^{\mathcal{D}}_{\xi} \Rightarrow [\![B]\!]^{\mathcal{D}}_{\xi}, \\ [\![A \cap B]\!]^{\mathcal{D}}_{\xi} &= [\![A]\!]^{\mathcal{D}}_{\xi} \cap [\![B]\!]^{\mathcal{D}}_{\xi}. \end{aligned}$$

The above definition is the extension to intersection types of the *simple semantics* for simple types of Scott (1975a), generalized by allowing $\mathcal{D}$ to be just a quasi λ-model instead of a λ-model.

It is easy to verify that $[\![\mathtt{U} \to \mathtt{U}]\!]^{\mathcal{D}}_{\xi} = \mathcal{D}$ for all $\mathcal{D}, \xi$.

In order to prove soundness, we have to check that the interpretation preserves the typability rules. We already know that the interpretation preserves the typing rules for application and intersection. This is because $\cap$ is interpreted as intersection on sets and $\to$ is interpreted as the arrow induced by the application $\cdot$ on $\mathcal{D}$. The following definition is necessary before we can establish that the interpretation preserves the remaining two typability rules: abstraction and subtyping.

17.1.2 Definition Let $\mathcal{D} = \langle \mathcal{D}, \cdot, [\![\;\;]\!]_{\mathcal{D}} \rangle$ be a quasi λ-model and $\xi : \mathbb{A}^{\mathcal{T}} \to \mathsf{P}(\mathcal{D})$ be a type environment.

(i) The pair $(\mathcal{D},\xi)$ is $\to$-*good* if, for all $A,B\in\mathbb{T}^{\mathcal{T}}$, and for all environments ρ, terms M and variables x,

$$[\forall d\in [\![A]\!]^{\mathcal{D}}_{\xi}.\ [\![M]\!]^{\mathcal{D}}_{\rho[x:=d]}\in [\![B]\!]^{\mathcal{D}}_{\xi}] \Rightarrow [\![\lambda x.M]\!]^{\mathcal{D}}_{\rho}\in [\![A]\!]^{\mathcal{D}}_{\xi}\to[\![B]\!]^{\mathcal{D}}_{\xi}.$$

(ii) The pair $(\mathcal{D},\xi)$ *preserves* $\leq_{\mathcal{T}}$ iff, for all $A,B\in\mathbb{T}^{\mathcal{T}}$, one has

$$A\leq_{\mathcal{T}} B \Rightarrow [\![A]\!]^{\mathcal{D}}_{\xi}\subseteq [\![B]\!]^{\mathcal{D}}_{\xi}.$$

We now introduce the semantics of type assignment.

17.1.3 Definition (Semantic Satisfiability) Let $\mathcal{T}\in\mathrm{TT}^{\mathtt{U}}$.

(i) Let $\mathcal{D}=\langle\mathcal{D},\cdot,[\![\ ]\!]^{\mathcal{D}}\rangle$ be a quasi λ-model. Define

$$\begin{array}{lcl}\mathcal{D},\rho,\xi\models M:A & \Leftrightarrow & [\![M]\!]^{\mathcal{D}}_{\rho}\in [\![A]\!]^{\mathcal{D}}_{\xi};\\ \mathcal{D},\rho,\xi\models\Gamma & \Leftrightarrow & \mathcal{D},\rho,\xi\models x:B, \text{for all } (x{:}B)\in\Gamma.\end{array}$$

(ii) We say that Γ *satisfies* $M:A$, written $\Gamma\models^{\mathcal{T}}_{\cap} M:A$, iff

$$\mathcal{D},\rho,\xi\models\Gamma \Rightarrow \mathcal{D},\rho,\xi\models M:A,$$

for all $\mathcal{D},\xi,\rho$ such that $(\mathcal{D},\xi)$ is $\to$-good and preserves $\leq_{\mathcal{T}}$.

Derivability in the type system implies semantic satisfiability, as shown in the next theorem.

17.1.4 Theorem (Soundness) *For all* $\mathcal{T}\in\mathrm{TT}^{\mathtt{U}}$ *one has*

$$\Gamma\vdash^{\mathcal{T}}_{\cap} M:A \Rightarrow \Gamma\models^{\mathcal{T}}_{\cap} M:A.$$

Proof By induction on the derivation of $\Gamma\vdash^{\mathcal{T}}_{\cap} M:A$. Rules ($\to$E), ($\cap$I) and ($\mathtt{U}_{\mathrm{top}}$) are sound by the definition of type interpretation (Definition 17.1.1).

As to the soundness of rule ($\to$I), assume $\Gamma,x{:}A\vdash^{\mathcal{T}}_{\cap} M:B$ in order to show $\Gamma\models^{\mathcal{T}}(\lambda x.M):(A\to B)$. Assuming $\mathcal{D},\rho,\xi\models\Gamma$ we have to show

$$[\![\lambda x.M]\!]^{\mathcal{D}}_{\rho}\in [\![A]\!]^{\mathcal{D}}_{\xi}\to[\![B]\!]^{\mathcal{D}}_{\xi}.$$

Let $d\in [\![A]\!]^{\mathcal{D}}_{\xi}$. We are done if we can show

$$[\![M]\!]^{\mathcal{D}}_{\rho[x:=d]}\in [\![B]\!]^{\mathcal{D}}_{\xi},$$

because $(\mathcal{D},\xi)$ are $\to$-good. Now $\mathcal{D},\rho[x:=d],\xi\models\Gamma,x{:}A$, hence, by the induction hypothesis for $\Gamma,x{:}A\vdash^{\mathcal{T}}_{\cap} M:B$, we have $[\![M]\!]^{\mathcal{D}}_{\rho[x:=d]}\in [\![B]\!]^{\mathcal{D}}_{\xi}$.

Rule ($\leq$) is sound, as we consider only $(\mathcal{D},\xi)$ that preserve $\leq_{\mathcal{T}}$. ∎

Completeness

Now we characterize the complete theories.

17.1.5 Notation Let $\mathcal{T} \in \mathrm{NTT}^{\mathtt{U}}$ and $\mathcal{F}^{\mathcal{T}} = \langle \mathcal{F}^{\mathcal{T}}, \cdot, [\![\ ]\!]^{\mathcal{F}^{\mathcal{T}}} \rangle$ its corresponding filter quasi λ-model, see Definition 16.2.1.

(i) Denote by $\xi^{\mathcal{T}} : \mathbb{A}^{\mathcal{T}} \to \mathsf{P}(\mathcal{F}^{\mathcal{T}})$ the type environment defined by

$$\xi^{\mathcal{T}}(\alpha) = \{X \in \mathcal{F}^{\mathcal{T}} \mid \alpha \in X\}.$$

(ii) Denote by $[\![\ ]\!]^{\mathcal{T}} : \mathbb{T}^{\mathcal{T}} \to \mathsf{P}(\mathcal{F}^{\mathcal{T}})$ the mapping $[\![\ ]\!]^{\mathcal{F}^{\mathcal{T}}}_{\xi^{\mathcal{T}}}$.

The mapping $[\![\ ]\!]^{\mathcal{T}} : \mathbb{T}^{\mathcal{T}} \to \mathsf{P}(\mathcal{F}^{\mathcal{T}})$ turns out to have the property of associating to each type A the set of filters which contain A (thus preserving the property which defines $\xi^{\mathcal{T}}$ in the basic case of type atoms).

17.1.6 Proposition *Let $\mathcal{T} \in \mathrm{NTT}^{\mathtt{U}}$. Then we have*

$$[\![A]\!]^{\mathcal{T}} \quad = \quad \{X \in \mathcal{F}^{\mathcal{T}} \mid A \in X\}.$$

Proof By induction on A. The only interesting case is when A is an arrow type. If $A \equiv B \to C$ we have

$$\begin{aligned}
[\![B\to C]\!]^{\mathcal{T}} &= \{X \in \mathcal{F}^{\mathcal{T}} \mid \forall Y \in [\![B]\!]^{\mathcal{T}}.\, X \cdot Y \in [\![C]\!]^{\mathcal{T}}\}, && \text{by definition,}\\
&= \{X \in \mathcal{F}^{\mathcal{T}} \mid \forall Y.\, B \in Y \Rightarrow C \in X \cdot Y\}, && \text{by induction,}\\
&= \{X \in \mathcal{F}^{\mathcal{T}} \mid C \in X \cdot \uparrow B\}, && \text{by monotonicity,}\\
&= \{X \in \mathcal{F}^{\mathcal{T}} \mid C \in \uparrow\{C' \mid \exists B' \in \uparrow B. B' \to C' \in X\}\}, && \text{by the definition of}\\
& && \text{filter application,}\\
&= \{X \in \mathcal{F}^{\mathcal{T}} \mid B \to C \in X\}, && \text{by } (\to) \text{ and } (\mathtt{U}\to). \blacksquare
\end{aligned}$$

17.1.7 Lemma *Let $\mathcal{T} \in \mathrm{NTT}^{\mathtt{U}}$. Then $(\mathcal{F}^{\mathcal{T}}, \xi^{\mathcal{T}})$ is $\to$-good and preserves $\le_{\mathcal{T}}$.*

Proof In order to show $[\![\lambda x.M]\!]^{\mathcal{T}}_{\rho} \cdot X \in [\![B]\!]^{\mathcal{T}}$, which establishes that $(\mathcal{F}^{\mathcal{T}}, \xi^{\mathcal{T}})$ is $\to$-good, suppose that $X \in [\![A]\!]^{\mathcal{T}}$ is such that

$$[\![M]\!]^{\mathcal{T}}_{\rho[x:=X]} \in [\![B]\!]^{\mathcal{T}}.$$

By Proposition 17.1.6 we have $B \in [\![M]\!]^{\mathcal{T}}_{\rho[x:=X]}$, hence $B \in f(X)$, where we have put $f = \lambda\!\!\lambda d.[\![M]\!]^{\mathcal{T}}_{\rho[x:=d]}$. Since by Lemma 15.2.8(ii) one has $f \sqsubseteq F^{\mathcal{T}}(G^{\mathcal{T}}(f))$, it follows that $B \in F^{\mathcal{T}}(G^{\mathcal{T}}(f))(X)$. Hence $[\![\lambda x.M]\!]^{\mathcal{T}}_{\rho} \cdot X = F^{\mathcal{T}}(G^{\mathcal{T}}(f))(X) \in [\![B]\!]^{\mathcal{T}}$, by Definition 16.1.5(i) and Proposition 17.1.6.

As an immediate consequence of Proposition 17.1.6 we get

$$A \le_{\mathcal{T}} B \Leftrightarrow \forall X \in \mathcal{F}^{\mathcal{T}}.[A \in X \Rightarrow B \in X] \Leftrightarrow [\![A]\!]^{\mathcal{T}} \subseteq [\![B]\!]^{\mathcal{T}},$$

and therefore $(\mathcal{F}^{\mathcal{T}}, \xi^{\mathcal{T}})$ preserves $\le_{\mathcal{T}}$. $\blacksquare$

Now we can prove the desired completeness result.

17.1.8 Theorem (Completeness) *Let $\mathcal{T} \in \mathrm{TT}^{\mathtt{U}}$.*

(i) *Then* $\forall \Gamma, M, A$ *we have* $[\Gamma \models_\cap^\mathcal{T} M : A \Rightarrow \Gamma \vdash_\cap^\mathcal{T} M : A]$ *iff* $\mathcal{T} \in \mathrm{NTT}^{\mathtt{U}}$.

(ii) *Let* $\mathcal{T} \in \mathrm{NTT}^{\mathtt{U}}$. *Then*

$$\Gamma \models_\cap^\mathcal{T} M : A \Leftrightarrow \Gamma \vdash_\cap^\mathcal{T} M : A.$$

Proof (i) ($\Rightarrow$) It is easy to verify that all type interpretations validate rule ($\to$) and the axiom ($\mathtt{U}_{\mathrm{top}}$). As to axiom ($\to\cap$), consider the $\mathcal{T}$-basis $\Gamma = \{x{:}(A \to B) \cap (A \to C)\}$. From Definition 17.1.1 we get

$$\Gamma \models_\cap^\mathcal{T} x : A{\to}(B \cap C).$$

Hence, by hypothesis, we have $\Gamma \vdash_\cap^\mathcal{T} x : A{\to}(B{\cap}C)$. Using Theorem 14.1.9(i) it follows that $(A{\to}B){\cap}(A{\to}C) \leq_\mathcal{T} A{\to}B{\cap}C$. Therefore axiom ($\to\cap$) holds.

As to axiom ($\mathtt{U}\to$)

$$\begin{array}{lll} \models_\cap^\mathcal{T} x : \mathtt{U} \to \mathtt{U} & \Rightarrow & \vdash_\cap^\mathcal{T} x : (\mathtt{U}{\to}\mathtt{U}) \\ & \Rightarrow & x{:}\mathtt{U} \vdash_\cap^\mathcal{T} x : (\mathtt{U}{\to}\mathtt{U}) \\ & \Rightarrow & \mathtt{U} \leq_\mathcal{T} (\mathtt{U}{\to}\mathtt{U}), \qquad \text{by Theorem 14.1.9(i).} \end{array}$$

This proves ($\Rightarrow$).

($\Leftarrow$) Now suppose $\Gamma \models^\mathcal{T} M : A$ towards proving $\Gamma \vdash_\cap^\mathcal{T} M : A$. We use the filter quasi λ-model $\langle \mathcal{F}^\mathcal{T}, \cdot, [\![\ ]\!]^\mathcal{T}\rangle$. Now $\Gamma \models_\cap^\mathcal{T} M : A$ implies $[\![M]\!]_{\rho_\Gamma}^\mathcal{T} \in [\![A]\!]^\mathcal{T}$, by Lemma 17.1.7, where

$$\rho_\Gamma(x) = \begin{cases} \uparrow A & \text{if } x{:}A \in \Gamma, \\ \uparrow \mathtt{U} & \text{otherwise.} \end{cases}$$

We conclude $\Gamma \vdash_\cap^\mathcal{T} M : A$, using Proposition 17.1.6 and Theorem 16.2.7(i).

(ii) By Proposition 17.1.4 and (i). ∎

17.1.9 Corollary *For* $\mathcal{T} \in \{\mathrm{Scott}, \mathrm{Park}, \mathrm{CDZ}, \mathrm{HR}, \mathrm{DHM}, \mathrm{BCD}\}$ *one has, for all* $M \in \Lambda^\emptyset$,

$$\models_\cap^\mathcal{T} M : A \Rightarrow \vdash_\cap^\mathcal{T} M : A.$$

17.1.10 Remark In Barendregt et al. (1983) the completeness of the type theory BCD was used to show the completeness of simple types via the following conservativity result

$$\forall A \in \mathbb{T}_\to^\mathbb{A} \forall M \in \Lambda [\vdash_\cap^{\mathrm{BCD}} M : A \Rightarrow \vdash_{\lambda_\to} M : A\,].$$

This solved an open problem of Scott (1975a).

17.1.11 Corollary *For $M \in \Lambda^{\emptyset}$ and $A \in \mathbb{T}^{\mathbb{A}}_{\rightarrow}$ one has*

$$\vdash_{\boldsymbol{\lambda}_{\rightarrow}} M : A \Leftrightarrow \models M : A.$$

Proof ($\Rightarrow$) This is soundness, Proposition 3.1.37, part ($\Rightarrow$).

$$\begin{array}{lcll} (\Leftarrow)\ \models_{\boldsymbol{\lambda}_{\rightarrow}} M : A & \Rightarrow & \models^{\mathrm{BCD}}_{\cap} M : A & \\ & \Rightarrow & \vdash^{\mathrm{BCD}}_{\cap} M : A, & \text{by Corollary 17.1.9,} \\ & \Rightarrow & \vdash_{\boldsymbol{\lambda}_{\rightarrow}} M : A, & \text{by Remark 17.1.10.} \ \blacksquare \end{array}$$

Similar results for $\mathcal{T} \in \mathrm{PTT}$ can be found in Dezani-Ciancaglini et al. (2003); see also Exercises 17.3 and 17.4.

17.2 Characterizing syntactic properties

In this section we will see the intersection type systems at work in the characterization of several properties of λ-terms. Since types are preserved by reduction, it is only possible to characterize properties which induce equivalences that are preserved by reduction. In particular we will consider some normalization properties of λ-terms, i.e. the standard properties of having a head normal form or a normal form, and of being strongly normalizable. First we recall some basic definitions.

17.2.1 Definition

(i) A lambda term M is called *$\boldsymbol{\beta}$-strongly normalizing* (*SN*) if there is no infinite $\boldsymbol{\beta}$-reduction starting with M. This is equivalent to being $\boldsymbol{\beta\eta}$-SN.

(ii) Write $\mathsf{SN} = \{M \mid M \text{ is strongly normalizing}\}$.

For example $\mathsf{SK} \in \mathsf{SN}$, but for $\boldsymbol{\Omega} \equiv (\lambda x.xx)(\lambda x.xx)$ one has $\mathsf{SK}\boldsymbol{\Omega} \notin \mathsf{SN}$, even if the last term has a normal form.

17.2.2 Lemma (van Raamsdonk et al. (1999)) *The set* SN *is the smallest set of terms closed under the following rules.*

$$\frac{M_1 \in \mathsf{SN}, \ldots, M_n \in \mathsf{SN}}{xM_1 \cdots M_n \in \mathsf{SN}}\ n \geq 0$$

$$\frac{M \in \mathsf{SN}}{\lambda x.M \in \mathsf{SN}}$$

$$\frac{M[x := N]M_1 \cdots M_n \in \mathsf{SN} \qquad N \in \mathsf{SN}}{(\lambda x M)NM_1 \cdots M_n \in \mathsf{SN}}\ n \geq 0$$

Proof Let $\mathcal{S}N$ be the set defined by these rules. We show

$$M \in \mathcal{S}N \Leftrightarrow M \in \mathsf{SN}.$$

($\Rightarrow$) By induction on the generation of $\mathcal{S}N$.

($\Leftarrow$) Suppose that M is strongly normalizing. Let $||M||$, the *norm* of M, be the length of the longest reduction path starting with M. We prove that $M \in \mathcal{S}N$ by induction on the pair $(||M||, M)$, lexicographically ordered by the usual ordering on natural numbers and the subterm ordering. If M is a normal form, then $M \in \mathcal{S}N$. In the case $||M|| = n > 0$, we have three cases, namely $x\vec{M}$, $\lambda x.N$ and $(\lambda x.P)N\vec{M}$. In the first two cases, the result follows by the induction hypothesis applied to subterms, where the norm is the same or has decreased. In the last case, the induction hypothesis is applied to $P[x := N]M_1 \cdots M_n$ and N, where the norm strictly decreases. ■

17.2.3 Definition

(i) A term M is *persistently head normalizing* if $M\vec{N}$ has a head normal form for all terms $\vec{N}$.
(ii) A term M is *persistently normalizing* if $M\vec{N}$ has a normal form for all normalizable terms $\vec{N}$.
(iii) A term M is *persistently strongly normalizing* if $M\vec{N}$ is strongly normalizing for all strongly normalizing terms $\vec{N}$.

The notion of persistently normalizing terms was introduced in Böhm and Dezani-Ciancaglini (1975).

17.2.4 Notation Several classes of lambda terms are denoted by initials:

$$\begin{aligned}
\mathsf{HN} &= \{M \mid M \text{ has a head normal form}\};\\
\mathsf{PHN} &= \{M \mid M \text{ is persistently head normalizing}\};\\
\mathsf{N} &= \{M \mid M \text{ has a normal form}\};\\
\mathsf{PN} &= \{M \mid M \text{ is persistently normalizing}\};\\
\mathsf{SN} &= \{M \mid M \text{ is strongly normalizing}\};\\
\mathsf{PSN} &= \{M \mid M \text{ is persistently strongly normalizing}\}.
\end{aligned}$$

The inclusions below follow immediately by definition, except those of the last line, namely $\mathsf{PSN} \subseteq \mathsf{PN} \subseteq \mathsf{PHN}$, which are proved in Exercise 17.6:

$$\begin{array}{ccccc}
\mathsf{SN} & \subseteq & \mathsf{N} & \subseteq & \mathsf{HN}\\
\cup\!\shortmid & & \cup\!\shortmid & & \cup\!\shortmid\\
\mathsf{PSN} & \subseteq & \mathsf{PN} & \subseteq & \mathsf{PHN}
\end{array}$$

The inclusion $\mathsf{PSN} \subseteq \mathsf{PN}$ can also be obtained by comparing Figs. 13.5 and 17.1. It is easy to find examples to show that all these inclusions are strict.

17.2.5 Example

(i) $\mathsf{K}x\boldsymbol{\Omega} \in \mathsf{PN}$ but not in SN, hence not in PSN.
(ii) $(\lambda x.\mathsf{K}x\boldsymbol{\Omega}) \in \mathsf{N}$ but not in SN nor in PHN, hence not in PN.
(iii) $x\boldsymbol{\Omega} \in \mathsf{PHN}$ but not in N, hence not in PN.
(iv) $x \in \mathsf{PSN}$.
(v) $\mathsf{II} \in \mathsf{SN}$, but not in PN, hence not in PSN.

Stable sets

We will use the standard proof technique of type stable sets (Krivine (1990)) to characterize some of the syntactic classes.

17.2.6 Definition The open term model of the lambda calculus consists of arbitrary λ-terms modulo $\boldsymbol{\beta}$-conversion. That is, $\mathcal{M}_{\Lambda(\boldsymbol{\beta})} = \langle \Lambda, \cdot, [\![\ ]\!]^{\Lambda} \rangle$ and we have

$$\begin{aligned} [\![M]\!]^{\Lambda}_{\rho} &= \{N \mid N =_{\boldsymbol{\beta}} M[\vec{x} := \rho(\vec{x})]\}, \qquad \text{where } \vec{x} = FV(M), \\ [\![M]\!]^{\Lambda}_{\rho} \cdot [\![N]\!]^{\Lambda}_{\rho} &= [\![MN]\!]^{\Lambda}_{\rho}. \end{aligned}$$

The substitution $M[x := \rho(x)]$ is to be interpreted as follows. If $\rho(x) = [P]_{=_{\boldsymbol{\beta}}}$, then $M[x := \rho(x)] = M[x := P]$; this is independent of the choice of the representative P.

17.2.7 Remark In $\mathcal{M}_{\Lambda(\boldsymbol{\beta})} = \langle \Lambda, \cdot, [\![\ ]\!]^{\Lambda} \rangle$ one has, for $X, Y \subseteq \mathcal{M}_{\Lambda(\boldsymbol{\beta})}$,

$$X \Rightarrow Y = \{M \in \Lambda \mid \forall N \in X\ MN \in Y\}.$$

17.2.8 Definition Let $X \subseteq \Lambda$.

(i) We say X is *closed under head expansion of redexes*, written h$\uparrow$-closed, if

$$P[x := Q]\vec{R} \in X \text{ implies } (\lambda x.P)Q\vec{R} \in X.$$

The term Q is called the *argument* of the head expansion.
(ii) We say X is HN-*stable* if $X \subseteq \mathsf{HN}$, it contains $x\vec{M}$ for all $\vec{M} \in \Lambda$, and is h$\uparrow$-closed.
(iii) X is N-*stable* if $X \subseteq \mathsf{N}$, it contains $x\vec{M}$ for all $\vec{M} \in \mathsf{N}$, and is h$\uparrow$-closed.
(iv) A set X is SN-*stable* if $X \subseteq \mathsf{SN}$, it contains $x\vec{M}$ for all $\vec{M} \in \mathsf{SN}$ and is closed under head expansion of redexes, whose arguments are in SN.

From the above definition and Lemma 17.2.2 we easily get the following.

17.2.9 Proposition *Let* $\mathsf{S} \in \{\mathsf{HN}, \mathsf{N}, \mathsf{SN}\}$ *and* $X, Y \subseteq \Lambda$. *Then:*

(i) *The class* S *is* S*-stable.*
(ii) PHN *is* HN*-stable and* PN *is* N*-stable.*
(iii) *If* X, Y *are* S*-stable, then* $(X \to Y)$ *and* $(X \cap Y)$ *are* S*-stable.*
(iv) *If* Y *is* HN*-stable and* $X \neq \emptyset$*, then* $(X \to Y)$ *is* HN*-stable.*

Proof Simple. ∎

17.2.10 Definition (Type environments)

(i) The type environment
$\xi = \xi^1_{\mathrm{BCD}} : \mathbb{A}_\infty \to \mathcal{P}(\Lambda)$ in the open term model $\mathcal{M}_{\Lambda(\boldsymbol{\beta})}$ is defined as

$$\xi(\alpha) = \mathsf{HN}, \quad \text{for } \alpha \in \mathbb{A}_\infty.$$

(ii) The type environment $\xi = \xi^2_{\mathrm{BCD}}$ in $\mathcal{M}_{\Lambda(\boldsymbol{\beta})}$ is defined as

$$\xi(\alpha) = \mathsf{N}, \quad \text{for } \alpha \in \mathbb{A}_\infty.$$

(iii) The type environment $\xi = \xi_{\mathrm{DHM}}$ in $\mathcal{M}_{\Lambda(\boldsymbol{\beta})}$ is defined as:

$$\begin{aligned} \xi(0) &= \mathsf{PHN}; \\ \xi(1) &= \mathsf{HN}. \end{aligned}$$

(iv) The type environment $\xi = \xi_{\mathrm{CDZ}}$ in $\mathcal{M}_{\Lambda(\boldsymbol{\beta})}$ is defined as:

$$\begin{aligned} \xi(0) &= \mathsf{PN}; \\ \xi(1) &= \mathsf{N}. \end{aligned}$$

(v) The type environment $\xi = \xi_{\mathrm{CDV}}$ in $\mathcal{M}_{\Lambda(\boldsymbol{\beta})}$ is defined as

$$\xi(\alpha) = \mathsf{SN}, \quad \text{for } \alpha \in \mathbb{A}_\infty.$$

(vi) The type environment $\xi = \xi_{\mathrm{HL}}$ in $\mathcal{M}_{\Lambda(\boldsymbol{\beta})}$ is defined as:

$$\begin{aligned} \xi(0) &= \mathsf{PSN}; \\ \xi(1) &= \mathsf{SN}. \end{aligned}$$

17.2.11 Lemma

(i) $[\![A]\!]_{\xi^1_{\mathrm{BCD}}}$ *and* $[\![A]\!]_{\xi_{\mathrm{DHM}}}$ *are* HN*-stable.*
(ii) $[\![A]\!]_{\xi^2_{\mathrm{BCD}}}$ *and* $[\![A]\!]_{\xi_{\mathrm{CDZ}}}$ *are* N*-stable.*
(iii) $[\![A]\!]_{\xi_{\mathrm{CDV}}}$ *and* $[\![A]\!]_{\xi_{\mathrm{HL}}}$ *are* SN*-stable.*

Proof All items follow easily from Proposition 17.2.9. ∎

We shall show that, for each type environment $\xi_{\mathcal{T}}$ of Definition 17.2.10, $(\mathcal{M}_{\Lambda(\boldsymbol{\beta})}, \xi_{\mathcal{T}})$ are $\to$-good – see Definition 17.1.2(i) – and preserve $\leq_{\mathcal{T}}$. The proof occupies 17.2.12–17.2.14.

17.2.12 Lemma

$$
\begin{array}{rrcl}
\text{(i)} & M \in \mathsf{SN}, N \in \mathsf{PSN} & \Rightarrow & M[x := N] \in \mathsf{SN}. \\
\text{(ii)} & M \in \mathsf{SN}, N \in \mathsf{PSN} & \Rightarrow & MN \in \mathsf{SN}. \\
\text{(iii)} & M \in \mathsf{N}, N \in \mathsf{PN} & \Rightarrow & M[x := N] \in \mathsf{N}. \\
\text{(iv)} & M \in \mathsf{N}, N \in \mathsf{PN} & \Rightarrow & MN \in \mathsf{N}. \\
\text{(v)} & M \in \mathsf{HN}, N \in \mathsf{PHN} & \Rightarrow & M[x := N] \in \mathsf{HN}. \\
\text{(vi)} & M \in \mathsf{HN}, N \in \mathsf{PHN} & \Rightarrow & MN \in \mathsf{HN}.
\end{array}
$$

Proof The first two statements follow using the inductive definition of SN given in 17.2.2. The rest follow by an easy induction on the (head) normal form of M. ∎

17.2.13 Proposition

$$
\begin{array}{rrcl}
\text{(i)} & \mathsf{PSN} & = & (\mathsf{N} \Rightarrow \mathsf{PSN}). \\
\text{(ii)} & \mathsf{SN} & = & (\mathsf{PSN} \Rightarrow \mathsf{SN}). \\
\text{(iii)} & \mathsf{PN} & = & (\mathsf{N} \Rightarrow \mathsf{PN}). \\
\text{(iv)} & \mathsf{N} & = & (\mathsf{PN} \Rightarrow \mathsf{N}). \\
\text{(v)} & \mathsf{PHN} & = & (\mathsf{HN} \Rightarrow \mathsf{PHN}). \\
\text{(vi)} & \mathsf{HN} & = & (\mathsf{PHN} \Rightarrow \mathsf{HN}).
\end{array}
$$

Proof All cases are immediate except the inclusions $\mathsf{SN} \subseteq (\mathsf{PSN} \Rightarrow \mathsf{SN})$, $\mathsf{N} \subseteq (\mathsf{PN} \Rightarrow \mathsf{N})$ and $\mathsf{HN} \subseteq (\mathsf{PHN} \Rightarrow \mathsf{HN})$. These follow easily from Lemma 17.2.12(ii), (iv) and (vi). ∎

17.2.14 Lemma *For all $\xi_{\mathcal{T}}$ of Definition 17.2.10 we have the following:*

(i) $\forall N \in [\![B]\!]_{\xi_{\mathcal{T}}}, [M[x := N] \in [\![A]\!]_{\xi_{\mathcal{T}}} \Rightarrow (\lambda x.M) \in [\![B \to A]\!]_{\xi_{\mathcal{T}}}]$.

(ii) $A \leq_{\mathcal{T}} B \Rightarrow [\![A]\!]_{\xi_{\mathcal{T}}} \subseteq [\![B]\!]_{\xi_{\mathcal{T}}}$.

That is, for all $\xi_{\mathcal{T}}$ of Definition 17.2.10, the pair $(\mathcal{M}_{\Lambda(\beta)}, \xi_{\mathcal{T}})$ is $\to$-good and preserve $\leq_{\mathcal{T}}$.

Proof (i) If either $\mathcal{T} \neq$ CDV or $\mathcal{T} =$ CDV and $N \in \mathsf{SN}$ one easily shows, by induction on A using Proposition 17.2.9, that $M[x := N] \in [\![A]\!]_{\xi_{\mathcal{T}}}$ implies $(\lambda x.M)N \in [\![A]\!]_{\xi_{\mathcal{T}}}$. The conclusion follows from the definition of $\to$.

(ii) By induction on the generation of $\leq_{\mathcal{T}}$, using Proposition 17.2.13. ∎

In the following result several important syntactic properties of lambda terms are characterized by typability with respect to some intersection type theory. We define $\Gamma_0^M = \{x_1{:}0, \ldots, x_n{:}0\}$, where $\{x_1, \ldots, x_n\} = \mathrm{FV}(M)$.

17.2.15 Theorem (Characterization Theorems)

(i)
$$M \in \mathsf{N} \Leftrightarrow \forall \mathcal{T} \in \mathrm{TT}^{\mathtt{U}}\, \exists \Gamma, A.\Gamma \vdash_\cap^{\mathcal{T}} M : A \;\&\; \mathtt{U} \notin \Gamma, A$$
$$\Leftrightarrow \exists \Gamma, A.\Gamma \vdash_\cap^{\mathrm{BCD}} M : A \;\&\; \mathtt{U} \notin \Gamma, A$$
$$\Leftrightarrow \Gamma_0^M \vdash_\cap^{\mathrm{CDZ}} M : 1.$$

(ii)
$$M \in \mathsf{HN} \Leftrightarrow \forall \mathcal{T} \in \mathrm{TT}^{\mathtt{U}}\, \exists \Gamma\, \exists n, m \in \omega.\Gamma \vdash_\cap^{\mathcal{T}} M : (\mathtt{U}^m \to A)^n \to A$$
$$\Leftrightarrow \exists \Gamma, A.\Gamma \vdash_\cap^{\mathrm{BCD}} M : A \;\&\; A \neq_{\mathrm{BCD}} \mathtt{U}$$
$$\Leftrightarrow \Gamma_0^M \vdash_\cap^{\mathrm{DHM}} M : 1.$$

(iii)
$$M \in \mathsf{SN} \Leftrightarrow \forall \mathcal{T} \in \mathrm{TT}\, \exists \Gamma, A.\Gamma \vdash_\cap^{\mathcal{T}} M : A$$
$$\Leftrightarrow \exists \Gamma, A.\Gamma \vdash_\cap^{\mathrm{CDV}} M : A$$
$$\Leftrightarrow G_0^M \vdash_\cap^{\mathrm{HL}} M : 1.$$

Proof For each of the items (i), (ii), and (iii) the statement is

$$A \Leftrightarrow B \Leftrightarrow C \Leftrightarrow D.$$

We first show for each of the three items that we have $(A \Rightarrow B)$, hence trivially also $(A \Rightarrow C)$, and that $(A \Rightarrow D)$.

(i) By Corollary 14.2.5(ii) it suffices to consider M in normal form. The proof is by induction on M. The only interesting case is $M \equiv x\vec{M}$, where $\vec{M} \equiv M_1 \cdots M_m$. By the induction hypothesis we have $\Gamma_j \vdash^{\mathcal{T}} M_j : A_j$, for some Γ_j, A_j not containing $\mathtt{U}$ and for $j \leq m$. This implies that

$$\uplus_{j \leq m} \Gamma_j \uplus \{x{:}A_1 \to \cdots \to A_m \to A\} \vdash_\cap^{\mathcal{T}} x\vec{M} : A.$$

Therefore, $\forall \mathcal{T} \in \mathrm{TT}^{\mathtt{U}}\, \exists \Gamma, A.\Gamma \vdash_\cap^{\mathcal{T}} M : A \;\&\; \mathtt{U} \notin \Gamma, A$ in particular for $\mathcal{T} =$ BCD.

For $\lambda_\cap^{\mathrm{CDZ}}$ we also show by induction on M in normal form that $\Gamma_0^M \vdash M : 1$. If $M \equiv x\vec{M}$ then $\Gamma_0^M \vdash_\cap^{\mathrm{CDZ}} M_j : 1$ by the induction hypothesis and weakening. As $0 = 1 \to 0$ in CDZ, this implies $\Gamma_0^M \vdash_\cap^{\mathrm{CDZ}} x\vec{M} : 0$. By rule $(\leq_{\mathrm{CDZ}})$ we conclude $\Gamma_0^M \vdash_\cap^{\mathrm{CDZ}} M : 1$.

If $M \equiv \lambda y.N$ then by the induction hypothesis we have $\Gamma_0^M, y : 0 \vdash_\cap^{\mathrm{CDZ}} N : 1$ and this implies $\Gamma_0^M \vdash_\cap^{\mathrm{CDZ}} M : 0 \to 1$. Hence $\Gamma_0^M \vdash_\cap^{\mathrm{CDZ}} M : 1$ by rule $(\leq_{\mathrm{CDZ}})$.

(ii) Again assume $M \equiv \lambda x_1 \cdots x_n.xM_1 \cdots M_m$ is in head normal form. Then

$$x{:}(\mathtt{U}^m \to A) \vdash_\cap^{\mathcal{T}} xM_1 \cdots M_m : A, \qquad \text{by } (\to\mathrm{E}), \text{ hence}$$

$$x{:}(\mathtt{U}^m \to A) \vdash_\cap^{\mathcal{T}} M : (\mathtt{U}^m \to A)^n \to A, \qquad \text{by } (\textit{weakening}) \text{ and } (\to\mathrm{I}).$$

As A is arbitrary we can get the type $\neq \mathtt{U}$ in $\mathcal{T} =$ BCD. We get

$$x{:}(\mathtt{U}^m \to 0) \vdash_\cap^{\mathrm{DHM}} M : (\mathtt{U}^m \to 0)^n \to 0,$$

taking $\mathcal{T} = \mathrm{DHM}$ and $A \equiv 0$. This implies

$$x{:}0 \vdash_{\cap}^{\mathrm{DHM}} M : 1,$$

using $(\mathtt{U}^m \to 0) =_{\mathrm{DHM}} 0$, as $(\mathtt{U} \to 0) =_{\mathrm{DHM}} 0$, and $((\mathtt{U}^m \to 0)^n \to 0) \leq_{\mathrm{DHM}} 1$, as $0 \leq_{\mathrm{DHM}} 1$ and $1 =_{\mathrm{DHM}} 0 \to 1$.

(iii) By induction on the structure of strongly normalizing terms following Definition 17.2.2. We only consider the case $M \equiv (\lambda x.R)N\vec{M}$ with $\vec{M} \equiv M_1 \cdots M_n$ and with both $R[x := N]\vec{M}$ and N being strongly normalizing. By the induction hypothesis there are Γ, A, Γ', B such that $\Gamma \vdash_{\cap}^{\mathcal{T}} R[x := N]\vec{M} : A$ and $\Gamma' \vdash_{\cap}^{\mathcal{T}} N : B$. We get $\Gamma \uplus \Gamma' \vdash_{\cap}^{\mathcal{T}} R[x := N]\vec{M} : A$ and $\Gamma \uplus \Gamma' \vdash_{\cap}^{\mathcal{T}} N : B$, so if $n = 0$ we are done by Theorem 14.2.4(i). If $n > 0$, then by iterated applications of Theorem 14.1.1(ii) to $\Gamma \vdash_{\cap}^{\mathcal{T}} R[x := N]\vec{M} : A$ we obtain

$$\Gamma \vdash_{\cap}^{\mathcal{T}} R[x := N] : B_1^{(i)} \to \cdots \to B_n^{(i)} \to B^{(i)} \quad \Gamma \vdash_{\cap}^{\mathcal{T}} M_j : B_j^{(i)}, \ (j \leq n)$$

and $\bigcap_{i \in I} B^{(i)} \leq_{\mathcal{T}} A$ for some $I, B_j^{(i)} (j \leq n), B^{(i)} \in \mathbb{T}^{\mathcal{T}}$. As in the case $n = 0$ we obtain $\Gamma \uplus \Gamma' \vdash_{\cap}^{\mathcal{T}} (\lambda x.R)N : B_1^{(i)} \to \cdots \to B_m^{(i)} \to B^{(i)}$. So we can conclude $\Gamma \uplus \Gamma' \vdash_{\cap}^{\mathcal{T}} (\lambda x.R)N\vec{M} : A$. Finally, $\Gamma_0^M \vdash_{\cap}^{\mathrm{HL}} M : 1$ follows from the observation that 1 is the top and 0 the bottom element in HL, see Lemma 13.1.23((i)).

($\Leftarrow$) Now we show for the three items that we have $(C \Rightarrow A)$, hence trivially also $(B \Rightarrow A)$, and that $(D \Rightarrow A)$. Write $\rho_0(x) = x$ for all $x \in \mathsf{V}$.

(i) Suppose $\Gamma \vdash_{\cap}^{\mathrm{BCD}} M : A$ and $\mathtt{U} \notin A, \Gamma$. By soundness (Theorem 17.1.4) it follows that $\Gamma \models_{\cap}^{\mathrm{BCD}} M : A$. By Lemmas 17.2.14 and 17.2.11(ii) one has $\Lambda(\boldsymbol{\beta}), \rho_0, \xi_{\mathrm{BCD}}^2 \models \Gamma$. Hence, $M \in [\![A]\!]_{\xi_{\mathrm{BCD}}^2} \subseteq \mathsf{N}$, again by Lemma 17.2.11(ii).

Suppose $\Gamma_0^M \vdash_{\cap}^{\mathrm{CDZ}} M : 1$. By soundness it follows that $\Gamma_0^M \models_{\cap}^{\mathrm{CDZ}} M : 1$. By Lemmas 17.2.14 and 17.2.11(ii) one has $\Lambda(\boldsymbol{\beta}), \rho_0, \xi_{\mathrm{CDZ}} \models \Gamma$. Hence, $M \in [\![1]\!]_{\xi_{\mathrm{CDZ}}} = \mathsf{N}$, by Definition 17.2.10(iv).

(ii) Suppose $\Gamma \vdash_{\cap}^{\mathrm{BCD}} M : A \neq \mathtt{U}$. Then $\Gamma \models_{\cap}^{\mathrm{BCD}} M : A$ by soundness. By Lemmas 17.2.14 and 17.2.11(i) one has $\Lambda(\boldsymbol{\beta}), \rho_0, \xi_{\mathrm{BCD}}^1 \models \Gamma$. Therefore it follows that $M \in [\![A]\!]_{\xi_{\mathrm{BCD}}^1} \subseteq \mathsf{HN}$, again by Lemma 17.2.11(ii).

Suppose $\Gamma_0^M \vdash_{\cap}^{\mathrm{DHM}} M : 1$. Again by soundness one has $\Gamma_0^M \models_{\cap}^{\mathrm{DHM}} M : 1$. By Lemmas 17.2.14 and 17.2.11(i) we obtain $\Lambda(\boldsymbol{\beta}), \rho_0, \xi_{\mathrm{DHM}} \models \Gamma$. Hence, by Definition 17.2.10(iii) one has $M \in [\![1]\!]_{\xi_{\mathrm{DHM}}} = \mathsf{HN}$.

(iii) Suppose $\Gamma \vdash_{\cap}^{\mathrm{CDV}} M : A$. Again by soundness $\Gamma \models_{\cap}^{\mathrm{CDV}} M : A$. By Lemmas 17.2.14 and 17.2.11(iii) one has $\Lambda(\boldsymbol{\beta}), \rho_0, \xi_{\mathrm{CDV}} \models \Gamma$. Hence $M \in [\![A]\!]_{\xi_{\mathrm{CDV}}} \subseteq \mathsf{SN}$, by Lemma 17.2.11(iii).

The implication $\Gamma_0^M \vdash_\cap^{\mathrm{HL}} M : 1 \Rightarrow M \in \mathsf{SN}$ is proved similarly. ■

17.2.16 Remark

(i) For $\mathcal{T} \in \mathrm{TT}^{\mathtt{U}}$ one has

$$\exists A, \Gamma . \Gamma \vdash_\cap^{\mathcal{T}} M : A \;\&\; \mathtt{U} \neq_{\mathcal{T}} A \not\Rightarrow M \in \mathsf{HN}.$$

Take for example $\mathcal{T} = \mathrm{Park}$, then $\vdash_\cap^{\mathrm{Park}} \mathbf{\Omega} : 0 \neq_{\mathrm{Park}} \mathtt{U}$, by Theorem 17.4.3, but this term is unsolvable, hence without head normal form, see B[1984].

(ii) There are many proofs of Theorem 17.2.15(iii) in the literature, for example Pottinger (1980), Leivant (1986), van Bakel (1992), Krivine (1990), Ghilezan (1996), Amadio and Curien (1998). As observed in Venneri (1996) all but Amadio and Curien (1998) contain some bugs, which in Krivine (1990) can be easily remedied with a suitable non-standard notion of length of reduction path.

(iii) In Coppo et al. (1987) persistently normalizing normal forms were given a similar characterization using the notion of replaceable variable. Other classes of terms were characterized in Dezani-Ciancaglini et al. (2005) and Tatsuta and Dezani-Ciancaglini (2006).

17.3 Approximation theorems

Crucial results for the study of the equational theory of ω-algebraic λ-models are the *approximation theorems*, see for example Hyland (1975/76), Wadsworth (1976), B[1984], Longo (1988), Ronchi Della Rocca (1988), Honsell and Ronchi Della Rocca (1992). An approximation theorem expresses the interpretation of any λ-term, even a non-terminating one, as the supremum of the interpretations of suitable *normal forms*, called the *approximants* of the term, in an appropriate *extended language*. Approximation theorems are very useful in proving, for instance, *Computational Adequacy* of models with respect to *operational semantics*, see for example B[1984], Honsell and Ronchi Della Rocca (1992). There are other possible methods for showing computational adequacy, both semantical and syntactical, for example Hyland (1975/76), Wadsworth (1976), Honsell and Ronchi Della Rocca (1992), Abramsky and Ong (1993), but the method based on approximation theorems is usually the most straightforward. However, proving an Approximation Theorem for a given model theory is usually rather difficult. Most of the proofs in the literature are based on the technique of *indexed reduction*, see Wadsworth (1976), Honsell and Ronchi Della Rocca (1992), Abramsky and Ong (1993). However, when the model in question is a filter model, by applying duality, the approximation theorem can be rephrased as follows: the types of a given term are all and only the types of its approximants. This change in perspective opens the way to proving approximation theorems using the syntactical machinery of proof theory, such as *logical predicates* and *computability* techniques.

The aim of the present section is to show in a uniform way that the Approximation Theorem is satisfied by all the type assignment systems that induce filter models isomorphic to the models in Scott (1972), Park (1976), Coppo et al. (1987), Honsell and Ronchi Della Rocca (1992), Dezani-Ciancaglini et al. (2005), Barendregt et al. (1983), Abramsky and Ong (1993). To this end following Dezani-Ciancaglini et al. (2001) we use a technique that can be constructed as a version of stable sets over a Kripke applicative structure. In Ronchi Della Rocca and Paolini (2004) the Approximation Theorem is given also for the type assignment system $\boldsymbol{\lambda}_{\cap \mathsf{V}}^{\mathrm{EHR}}$ defined in Definition 13.2.13.

For almost all the type theories of Fig. 13.4 which induce λ-models we introduce appropriate notions of *approximants* which agree with the λ-theories of different models and therefore also with the type theories describing these models. Then we will prove that all types of an approximant of a given term (with respect to the appropriate notion of approximants) are also types of the given term. Finally we show the converse, namely that the types which can be assigned to a term can also be assigned to at least one approximant of that term. Hence a type can be derived for a term *if and only if* it can be derived for an approximant of that term. We end the section showing some applications of the Approximation Theorem.

Approximate normal forms

In this section we consider two extensions of λ-calculus, both obtained by adding a constant. The first one is the well-known language $\lambda\bot$, see B[1984]. The other extension is obtained by adding the constant Φ and is discussed in Honsell and Ronchi Della Rocca (1992).

17.3.1 Definition

(i) The set $\Lambda\bot$ of $\lambda\bot$-*terms* is obtained by adding the constant *bottom*, written $\bot$, to the formation rules of terms.

(ii) The set $\Lambda\Phi$ of $\lambda\Phi$-*terms* is obtained by adding the constant Φ to the formation rules of terms.

We consider two mappings ($\Box_\bot$ and $\Box_L$) from λ-terms to $\lambda\bot$-terms and one mapping ($\Box_\Phi$) from λ-terms to $\lambda\Phi$-terms. These mappings differ in the translation of $\boldsymbol{\beta}$-redexes. Clearly the values of these mappings are $\boldsymbol{\beta}$-irreducible terms, i.e. normal forms for an extended language. We call such a term an *approximate normal form* or an *anf* for short.

17.3.2 Definition The mappings $\Box_\perp : \Lambda \to \Lambda\perp$, $\Box_L : \Lambda \to \Lambda\perp$, $\Box_\Phi : \Lambda \to \Lambda\Phi$ are recursively defined as follows:

$$\begin{array}{rcl} \Box(\lambda \vec{x}.y\vec{M}) & = & \lambda \vec{x}.y\Box(M_1)\cdots\Box(M_m); \\ \Box_\perp(\lambda \vec{x}.(\lambda y.R)N\vec{M}) & = & \perp; \\ \Box_L(\lambda \vec{x}.(\lambda y.R)N\vec{M}) & = & \lambda \vec{x}.\perp; \\ \Box_\Phi(\lambda \vec{x}.(\lambda y.R)N\vec{M}) & = & \lambda \vec{x}.\Phi\Box_\Phi(\lambda y.R)\Box_\Phi(N)\Box_\Phi(M_1)\cdots\Box_\Phi(M_m), \end{array}$$

where $\Box \in \{\Box_\perp, \Box_L, \Box_\Phi\}$, $\vec{M} \equiv M_1 \cdots M_m$ and $m \geq 0$.

The mapping $\Box_\perp$ is related to the Böhm-tree of untyped lambda terms, and $\Box_L$ to the Lévy–Longo trees; see van Bakel et al. (2002), where these trees are related to intersection types.

In order to give the appropriate approximation theorem we will use the mapping $\Box_\perp$ for the type assignment systems $\boldsymbol{\lambda}_\cap^{\text{Scott}}$, $\boldsymbol{\lambda}_\cap^{\text{CDZ}}$, $\boldsymbol{\lambda}_\cap^{\text{DHM}}$, $\boldsymbol{\lambda}_\cap^{\text{BCD}}$, the mapping $\Box_L$ for the type assignment system $\boldsymbol{\lambda}_\cap^{\text{AO}}$, and the mapping $\Box_\Phi$ for the type assignment systems $\boldsymbol{\lambda}_\cap^{\text{Park}}$, $\boldsymbol{\lambda}_\cap^{\text{HR}}$. Every one of the above mappings associates a set of approximants to each λ-term in the standard way.

17.3.3 Definition Let $\mathcal{T} \in \{\text{Scott}, \text{Park}, \text{CDZ}, \text{HR}, \text{DHM}, \text{BCD}, \text{AO}\}$. The set $\mathcal{A}_\mathcal{T}(M)$ of $\mathcal{T}$*-approximants* of M is defined by

$$\mathcal{A}_\mathcal{T}(M) = \{P \mid \exists M'.\ M \twoheadrightarrow_{\boldsymbol{\beta}} M' \text{ and } P \equiv \Box(M')\},$$

where

$$\begin{array}{rcll} \Box & = & \Box_\perp, & \text{for } \mathcal{T} \in \{\text{Scott}, \text{CDZ}, \text{DHM}, \text{BCD}\}, \\ \Box & = & \Box_L, & \text{for } \mathcal{T} \in \{\text{AO}\}, \\ \Box & = & \Box_\Phi, & \text{for } \mathcal{T} \in \{\text{Park}, \text{HR}\}. \end{array}$$

We extend the typing to $\lambda\perp$-terms and to $\lambda\Phi$-terms by adding two different axioms for Φ and nothing for $\perp$.

17.3.4 Definition

(i) Let $\mathcal{T} \in \{\text{Scott}, \text{CDZ}, \text{DHM}, \text{BCD}, \text{AO}\}$. We extend the definition of type assignment $\Gamma \vdash_\cap^\mathcal{T} M : A$ to $\lambda\perp$-terms by letting M, N in Definition 13.2.3 range over $\Lambda\perp$.

(ii) We extend the type assignment $\boldsymbol{\lambda}_\cap^{\text{Park}}$ to $\lambda\Phi$-terms by adding the axiom

$$(\text{Ax-}\Phi\text{-Park})\ \Gamma \vdash_\cap^{\text{Park}} \Phi : 0.$$

(iii) We extend the type assignment $\boldsymbol{\lambda}_\cap^{\mathrm{HR}}$ to $\lambda\Phi$-terms by adding the axiom

$$(\text{Ax-}\Phi\text{-HR})\ \Gamma \vdash_\cap^{\mathrm{HR}} \Phi : 1.$$

We do not introduce different notations for these extended type assignment systems concerning terms in $\Lambda\bot\Phi$. It is easy to verify that the Inversion Lemmas (Theorems 14.1.1 and 14.1.9) remain valid. In addition to these the following result is relevant.

17.3.5 Proposition

(i) *Let* $\mathcal{T} \in \{\text{Scott}, \text{CDZ}, \text{DHM}, \text{BCD}, \text{AO}\}$. *Then*

$$\Gamma \vdash_\cap^{\mathcal{T}} \bot : A \Leftrightarrow A =_{\mathcal{T}} \mathtt{U}.$$

(ii) $\Gamma \vdash_\cap^{\mathrm{Park}} \Phi : A \Leftrightarrow 0 \leq_{\mathrm{Park}} A$.
(iii) $\Gamma \vdash_\cap^{\mathrm{HR}} \Phi : A \Leftrightarrow 1 \leq_{\mathrm{HR}} A$.

17.3.6 Lemma *Let* $\mathcal{T} \in \{\text{Scott}, \text{Park}, \text{CDZ}, \text{HR}, \text{DHM}, \text{BCD}, \text{AO}\}$.

(i) $M_1 \twoheadrightarrow_\beta M_2\ \&\ \Gamma \vdash_\cap^{\mathcal{T}} \Box(M_1) : A \Rightarrow \Gamma \vdash_\cap^{\mathcal{T}} \Box(M_2) : A$.
(ii) *If* $P, P' \in \mathcal{A}_{\mathcal{T}}(M)$, $\Gamma \vdash_\cap^{\mathcal{T}} P : A$ *and* $\Gamma \vdash_\cap^{\mathcal{T}} P' : B$, *then*

$$\exists P'' \in \mathcal{A}_{\mathcal{T}}(M).\Gamma \vdash_\cap^{\mathcal{T}} P'' : A \cap B.$$

Proof (i) For $\mathcal{T} \in \{\text{Scott}, \text{CDZ}, \text{DHM}, \text{BCD}, \text{AO}\}$ the proof follows by induction on the structure of the term distinguishing cases (being or not in head normal form) and using the Inversion Lemmas.

For $\mathcal{T} \in \{\text{Park}, \text{HR}\}$ it suffices to consider the case $M_1 \equiv (\lambda x.M)N$ and $M_2 \equiv M[x := N]$. Note that $\Box_\Phi(M[x := N])$ is $\Box_\Phi(M)$, where the occurrences of x have been replaced by $\Phi\Box_\Phi(N)$ if they are functional and N is an abstraction, and by $\Box_\Phi(N)$ otherwise. More formally, define the mapping $\overline{\Box_\Phi} : \Lambda \to \Lambda\Phi$ by

$$\overline{\Box_\Phi}(M) = \begin{cases} \Phi\Box_\Phi(M) & \text{if } M \equiv \lambda x.M' \\ \Box_\Phi(M) & \text{otherwise} \end{cases}$$

and the mapping $\{\ \}_y^x : \Lambda \to \Lambda$ by

$$\begin{aligned} \{z\}_y^x &= z \\ \{M_1 M_2\}_y^x &= \begin{cases} y\{M_2\}_y^x & \text{if } M_1 \equiv x \\ \{M_1\}_y^x\{M_2\}_y^x & \text{otherwise} \end{cases} \\ \{\lambda z.M\}_y^x &= \lambda z.\{M\}_y^x. \end{aligned}$$

Then $\Box_\Phi(M_1 M_2) = \overline{\Box_\Phi}(M_1)\Box_\Phi(M_2)$ and one can check, by induction on M, that $\Box_\Phi(M[x := N]) \equiv \Box_\Phi(\{M\}_y^x)[x := \Box_\Phi(N)][y := \overline{\Box_\Phi}(N)]$ for y fresh.

We may assume $A \neq_{\mathcal{T}} \mathtt{U}$. Then from $\Gamma \vdash_{\cap}^{\mathcal{T}} \Phi(\lambda x.\Box_\Phi(M))\Box_\Phi(N) : A$ we get

$$\Gamma \vdash_{\cap}^{\mathcal{T}} \Phi(\lambda x.\Box_\Phi(M)) : C \to A, \quad \Gamma \vdash_{\cap}^{\mathcal{T}} \Box_\Phi(N) : C$$

for some C, by Theorem 14.1.9(ii). By Lemma 13.1.25 we have $C \to A \neq_{\mathcal{T}} \mathtt{U}$, so again by Theorem 14.1.9(ii) we get, for some B,

$$\Gamma \vdash_{\cap}^{\mathcal{T}} \Phi : B \to C \to A, \quad \Gamma \vdash_{\cap}^{\mathcal{T}} \lambda x.\Box_\Phi(M) : B.$$

For $\mathcal{T} = \text{Park}$ we get $0 \leq_{\text{Park}} B \to C \to A$ by Proposition 17.3.5(ii). This implies $B \leq_{\text{Park}} 0$, $C \leq_{\text{Park}} 0$, and $0 \leq_{\text{Park}} A$, since $0 =_{\mathcal{T}} 0 \to 0$, since Park is $\boldsymbol{\beta}$-sound by Theorem 14.1.7, $(C{\to}A) \neq_{\mathcal{T}} \mathtt{U}$ and $A \neq_{\mathcal{T}} \mathtt{U}$. We obtain by rule $(\leq)$ $\Gamma \vdash_{\cap}^{\text{Park}} \lambda x.\Box_\Phi(M) : 0$ and $\Gamma \vdash_{\cap}^{\text{Park}} \Box_\Phi(N) : 0$. We get $\Gamma, x{:}0 \vdash_{\cap}^{\text{Park}} \Box_\Phi(M) : 0$ (by Theorem 14.1.9(iii)) and $\Gamma \vdash_{\cap}^{\text{Park}} \Phi\Box_\Phi(N) : 0$ since $0 =_{\text{Park}} 0 \to 0$. Now $\Box_\Phi(\{M\}^x_y)$ equals $\Box_\Phi(M)$ with some occurrences of x replaced by the fresh variable y. Hence $\Gamma, y{:}0, x{:}0 \vdash_{\cap}^{\text{Park}} \Box_\Phi(\{M\}^x_y){:}0$. So we conclude $\Gamma \vdash_{\cap}^{\text{Park}} \Box_\Phi(\{M\}^x_y)[x := \Box_\Phi(N)][y := \overline{\Box_\Phi}(N)] : A$ by rules (cut) and $(\leq)$.

For $\mathcal{T} = \text{HR}$ we get $1 \leq_{\text{HR}} B \to C \to A$ from $\Gamma \vdash_{\cap}^{\text{HR}} \Phi : B \to C \to A$ by Theorem 17.3.5(iii). This implies either ($B \leq_{\text{HR}} 1$ and $1 \leq_{\text{HR}} C \to A$) or ($B \leq_{\text{HR}} 0$ and $0 \leq_{\text{HR}} C \to A$) since $1 =_{\text{HR}} (1 \to 1) \cap (0 \to 0)$ and HR is $\boldsymbol{\beta}$-sound by Theorem 14.1.7, $C{\to}A \neq_{\text{HR}} \mathtt{U}$, and $A \neq_{\text{HR}} \mathtt{U}$ (note that $1 \cap 0 = 0$). Similarly in the first case from $1 \leq_{\text{HR}} C \to A$ we get either $C \leq_{\text{HR}} 1$ and $1 \leq_{\text{HR}} A$ or $C \leq_{\text{HR}} 0$ and $0 \leq_{\text{HR}} A$. In the second case from $0 \leq_{\text{HR}} C \to A$ we get $C \leq_{\text{HR}} 1$ and $0 \leq_{\text{HR}} A$, since $0 =_{\text{HR}} 1 \to 0$.

To sum up, using rule $(\leq)$ we have the following alternative cases.

- $\Gamma \vdash_{\cap}^{\text{HR}} \lambda x.\Box_\Phi(M) : 1$, $\Gamma \vdash_{\cap}^{\text{HR}} \Box_\Phi(N) : 1$, and $1 \leq_{\text{HR}} A$;
- $\Gamma \vdash_{\cap}^{\text{HR}} \lambda x.\Box_\Phi(M) : 1$, $\Gamma \vdash_{\cap}^{\text{HR}} \Box_\Phi(N) : 0$, and $0 \leq_{\text{HR}} A$;
- $\Gamma \vdash_{\cap}^{\text{HR}} \lambda x.\Box_\Phi(M) : 0$, $\Gamma \vdash_{\cap}^{\text{HR}} \Box_\Phi(N) : 1$, and $0 \leq_{\text{HR}} A$.

From Theorem 14.1.9 (iii) these alternative cases become

- $\Gamma, x{:}1 \vdash_{\cap}^{\text{HR}} \Box_\Phi(M) : 1$, and $\Gamma \vdash_{\cap}^{\text{HR}} \Phi\Box_\Phi(N) : 1$;
- $\Gamma, x{:}0 \vdash_{\cap}^{\text{HR}} \Box_\Phi(M) : 0$, and $\Gamma \vdash_{\cap}^{\text{HR}} \Phi\Box_\Phi(N) : 0$;
- $\Gamma, x{:}1 \vdash_{\cap}^{\text{HR}} \Box_\Phi(M) : 0$, and $\Gamma \vdash_{\cap}^{\text{HR}} \Phi\Box_\Phi(N) : 1$.

so we can conclude as in $\mathcal{T} = \text{Park}$.

(ii) By hypothesis there are M_1, M_2 such that $M \twoheadrightarrow_{\boldsymbol{\beta}} M_1$, $M \twoheadrightarrow_{\boldsymbol{\beta}} M_2$ and $P \equiv \Box(M_1)$, $P' \equiv \Box(M_2)$. By the Church–Rosser property of $\to_{\boldsymbol{\beta}}$ we can find M_3 such that $M_1 \twoheadrightarrow_{\boldsymbol{\beta}} M_3$ and $M_2 \twoheadrightarrow_{\boldsymbol{\beta}} M_3$. By (i) we can choose $P'' \equiv \Box(M_3)$. ∎

Approximation theorem – Part 1

It is useful to introduce the following definition.

17.3.7 Definition Let $\mathcal{T} \in \{\text{Scott}, \text{Park}, \text{CDZ}, \text{HR}, \text{DHM}, \text{BCD}, \text{AO}\}$. Write

$$[A]^{\mathcal{T}}_{\Gamma} = \{M \mid \exists P \in \mathcal{A}_{\mathcal{T}}(M).\ \Gamma \vdash^{\mathcal{T}}_{\cap} P : A\}.$$

By definition we get that $M \in [A]^{\mathcal{T}}_{\Gamma}$ and $N \twoheadrightarrow_{\beta} M$ imply $N \in [A]^{\mathcal{T}}_{\Gamma}$. Moreover $\Gamma \Subset \Gamma'$ implies $[A]^{\mathcal{T}}_{\Gamma} \subseteq [A]^{\mathcal{T}}_{\Gamma'}$ for all types $A \in \mathbb{T}^{\mathcal{T}}$.

In this subsection we prove that, if $M \in [A]^{\mathcal{T}}_{\Gamma}$, then there exists a derivation of $\Gamma \vdash^{\mathcal{T}}_{\cap} M : A$.

17.3.8 Proposition *Let* $\mathcal{T} \in \{\text{Scott}, \text{Park}, \text{CDZ}, \text{HR}, \text{DHM}, \text{BCD}, \text{AO}\}$. *Then*

$$M \in [A]^{\mathcal{T}}_{\Gamma} \Rightarrow \Gamma \vdash^{\mathcal{T}}_{\cap} M : A.$$

Proof Write $P \equiv \Box(M)$ with $\Box = \Box_{\perp}$ for $\mathcal{T} \in \{\text{Scott}, \text{CDZ}, \text{DHM}, \text{BCD}\}$,
$\Box = \Box_{L}$ for $\mathcal{T} = \text{AO}$,
$\Box = \Box_{\Phi}$ for $\mathcal{T} \in \{\text{Park}, \text{HR}\}$.

By Corollary 14.2.5(ii) it is sufficient to show that for each of the mentioned $\mathcal{T}$ one has

$$\Gamma \vdash^{\mathcal{T}}_{\cap} P : A \Rightarrow \Gamma \vdash^{\mathcal{T}}_{\cap} M : A. \tag{17.1}$$

In this proof we just write $\vdash$ for $\vdash^{\mathcal{T}}_{\cap}$.

For $\mathcal{T} \in \{\text{Scott}, \text{CDZ}, \text{DHM}, \text{BCD}, \text{AO}\}$ the implication (17.1) follows from Proposition 17.3.5((i)) and the definition of the mappings $\Box_{\perp}$ and $\Box_{L}$.

For $\mathcal{T} \in \{\text{Park}, \text{HR}\}$ we prove (17.1) by induction on M, assuming $A \neq_{\mathcal{T}} \mathtt{U}$.

Case $M \equiv x$. Trivial.

Case $M \equiv \lambda x.M'$. Then $P \equiv \lambda x.P'$ where $P' \equiv \Box_{\Phi}(M')$. From $\Gamma \vdash P : A$ by Theorem 14.1.1(iii) we get

$$\Gamma, x{:}B_i \vdash P' : C_i \text{ and } \textstyle\bigcap_{i \in I}(B_i \to C_i) \leq A$$

for some I, B_i, C_i. We find by induction $\Gamma, x{:}B_i \vdash M' : C_i$ and so we conclude $\Gamma \vdash M : A$ using rules ($\to$I), ($\cap$I) and ($\leq$).

Case $M \equiv M_1M_2$, *with* M_1 *not an abstraction.* Then $P \equiv P_1P_2$ where $P_1 \equiv \Box_{\Phi}(M_1)$ and $P_2 \equiv \Box_{\Phi}(M_2)$. By Theorem 14.1.9(ii) from $\Gamma \vdash P : A$ we get $\Gamma \vdash P_1 : B \to A$, $\Gamma \vdash P_2 : B$ for some B. By induction this implies $\Gamma \vdash M_1 : B \to A$ and $\Gamma \vdash M_2 : B$, hence $\Gamma \vdash M \equiv M_1M_2 : A$.

Case $M \equiv M_1M_2$, *where* M_1 *is an abstraction.* Then $P \equiv \Phi P_1 P_2$ where $P_1 \equiv \Box_{\Phi}(M_1)$ and $P_2 \equiv \Box_{\Phi}(M_2)$. As in the proof of Lemma

17.3.6(i) from $\Gamma \vdash P : A$, where $A \neq_{\mathcal{T}} \mathtt{U}$, we get $\Gamma \vdash \Phi : B \to C \to A$, $\Gamma \vdash P_1 : B$, $\Gamma \vdash P_2 : C$ for some B, C. By induction this implies $\Gamma \vdash M_1 : B$ and $\Gamma \vdash M_2 : C$.

For $\mathcal{T} = \text{Park}$, as in the proof of Lemma 17.3.6(i), we get $\Gamma \vdash M_1 : 0$ and $\Gamma \vdash M_2 : 0$. We can conclude $\Gamma \vdash M : A$ using rules $(\leq_{\text{Park}})$ and $(\to\text{E})$ since $0 =_{\text{Park}} 0 \to 0$.

For $\mathcal{T} = \text{HR}$ as in the proof of Lemma 17.3.6((i)) we have the following alternative cases:

- $\Gamma \vdash^{\text{HR}}_{\cap} M_1 : 1$, $\Gamma \vdash^{\text{HR}}_{\cap} M_2 : 1$, and $1 \leq_{\text{HR}} A$;
- $\Gamma \vdash^{\text{HR}}_{\cap} M_1 : 1$, $\Gamma \vdash^{\text{HR}}_{\cap} M_2 : 0$, and $0 \leq_{\text{HR}} A$;
- $\Gamma \vdash^{\text{HR}}_{\cap} M_1 : 0$, $\Gamma \vdash^{\text{HR}}_{\cap} M_2 : 1$, and $0 \leq_{\text{HR}} A$.

It is easy to verify that in all cases we can derive $\Gamma \vdash M : A$ from (I) and $(1{\to}0)$ using rules $(\leq_{\text{HR}})$ and $(\to\text{E})$. ∎

Approximation theorem – Part 2

In order to prove the converse of Proposition 17.3.8 we will use a Kripke-like version of stable sets (Mitchell (1996)). First we need a technical result.

17.3.9 Lemma *Let* $\mathcal{T} \in \{\text{Scott}, \text{Park}, \text{CDZ}, \text{HR}, \text{DHM}, \text{BCD}, \text{AO}\}$. *Let* $A, B \in \mathbb{T}$, $M \in \Lambda$, $z \notin \text{FV}(M)$, *and* G *a context. Write* $\Gamma' = \Gamma, z : B$. *Assume* $A \neq_{\mathcal{T}} \mathtt{U}$ *for* $\mathcal{T} = \text{AO}$. *Then*

$$Mz \in [A]^{\mathcal{T}}_{\Gamma'} \Rightarrow M \in [B \to A]^{\mathcal{T}}_{\Gamma}.$$

Proof Let $P \in \mathcal{A}_{\mathcal{T}}(Mz)$ and $\Gamma' \vdash P : A$. We show by cases on P and M that there is $\hat{P} \in \mathcal{A}_{\mathcal{T}}(M)$ such that $\Gamma \vdash \hat{P} : B \to A$.

There are two possibilities.

(a) $Mz \twoheadrightarrow_{\beta} M'z$ and $P \equiv \Box(M'z)$;
(b) $Mz \twoheadrightarrow_{\beta} (\lambda x.M')z \to_{\beta} M'[x := z]$ and $P \in \mathcal{A}_{\mathcal{T}}(M'[x := z])$.

In case (a) again there are two possibilities.

(a′) $M' \equiv yM_1 \cdots M_m, \quad m \geq 0$;
(a″) $M' \equiv (\lambda y.M_0)M_1 \cdots M_m, \quad m \geq 0$.

In total there are four cases:

(a′)] $P \equiv P'z$ and $P' \equiv y\Box(M_1) \cdots \Box(M_m) \in \mathcal{A}_{\mathcal{T}}(M)$;
(a″) $P \equiv \bot$ and $\mathcal{T} \in \{\text{Scott}, \text{CDZ}, \text{DHM}, \text{BCD}, \text{AO}\}$;
(a‴) $P \equiv \Phi P'z$, $P' \equiv \Box(\lambda y.M_0)\Box(M_1) \cdots \Box(M_m) \in \mathcal{A}_{\mathcal{T}}(M)$ for $\mathcal{T} \in \{\text{Park}, \text{HR}\}$;
(b) $M \twoheadrightarrow_{\beta} \lambda x.M'$ and $P \in \mathcal{A}_{\mathcal{T}}(M'[x := z])$.

Case $P \equiv P'z$, *where* $P' \in \mathcal{A}_\mathcal{T}(M)$. Then we can choose $\hat{P} \equiv P'$. This is clear if $A =_\mathcal{T} \mathtt{U}$ because by assumption $\mathcal{T} \neq \mathrm{AO}$, hence we have $(\mathtt{U}{\to})$. Now let $A \neq_\mathcal{T} \mathtt{U}$. Then, by Theorem 14.1.9(ii), from $\Gamma' \vdash P : A$ we get $\Gamma' \vdash P' : C \to A$, $\Gamma' \vdash z : C$ for some C. By Theorem 14.1.9(i) $B \leq C$ and we conclude using $(\leq)$ and (*strengthening*) that $\Gamma \vdash P' : B \to A$.

Case $P \equiv \bot$. By Proposition 17.3.5(ii) we have $A =_\mathcal{T} \mathtt{U}$. By assumption, $\mathcal{T} \neq \mathrm{AO}$. Hence, we have rule $(\mathtt{U}{\to})$.

Case $P \equiv \Phi P'z$, *where* $P' \in \mathcal{A}_\mathcal{T}(M)$ *and* $\mathcal{T} \in \{\mathrm{Park}, \mathrm{HR}\}$. Now we show that we can choose $\hat{P} \equiv P'$. Again let $A \neq_\mathcal{T} \mathtt{U}$. Then from $\Gamma' \vdash P : A$ we see by Theorem 14.1.9(ii) and (i) that $\Gamma' \vdash \Phi : C \to D \to A$, $\Gamma' \vdash P' : C$, and $\Gamma' \vdash z : D$, for some C, D with $B \leq D$.

For $\mathcal{T} = \mathrm{Park}$, using Proposition 17.3.5(ii) as in the proof of Lemma 17.3.6(i), we get $C \leq_{\mathrm{Park}} 0$, $D \leq_{\mathrm{Park}} 0$, and $0 \leq_{\mathrm{Park}} A$ (remember that $0 =_{\mathrm{Park}} 0{\to}0$).

Similarly for $\mathcal{T} = \mathrm{HR}$, using Proposition 17.3.5(iii), we get either $C \leq_{\mathrm{HR}} 1$, $D \leq_{\mathrm{HR}} 1$, and $1 \leq_{\mathrm{HR}} A$; or $C \leq_{\mathrm{HR}} 1$, $D \leq_{\mathrm{HR}} 0$, and $0 \leq_{\mathrm{HR}} A$; or $C \leq_{\mathrm{HR}} 0$, $D \leq_{\mathrm{HR}} 1$, and $0 \leq_{\mathrm{Park}} A$ – remember that $1 =_{\mathrm{HR}} (1{\to}1) \cap (0{\to}0)$ and $0 =_{\mathrm{HR}} 1{\to}0$. In all cases we can conclude that $C \leq D \to A \leq B \to A$ and therefore by $(\leq)$ and (*strengthening*) that $\Gamma \vdash P' : B \to A$.

Case $M \twoheadrightarrow_\beta \lambda x.M'$ *and* $P \in \mathcal{A}_\mathcal{T}(M'[x := z])$. If $\Box = \Box_\bot$ and $P \equiv \bot$, then we choose $\hat{P} \equiv P$, otherwise $\hat{P} \equiv \lambda z.P$. ■

The following crucial definition is somewhat involved. It amounts essentially to the definition of the natural set-theoretic semantics of intersection types over a suitable Kripke applicative structure, where bases play the role of worlds.[1] In order to keep the treatment elementary we will not develop the full theory of the natural semantics of intersection types in Kripke applicative structures. The definition below is rather long, since we have different cases for the type 0 and for arrow types according to the different type theories under consideration.

17.3.10 Definition (Kripke type interpretation) Let $\mathcal{T} \in \{\mathrm{Scott}, \mathrm{Park},$

[1] As already observed in Berline (2000) we cannot use stable sets here as we did in Section 17.2, since we need to take into account not only the λ-terms and their types, but also the $\mathcal{T}$-bases.

CDZ, HR, DHM, BCD, AO}. Define $[\![A]\!]^{\mathcal{T}}_{\Gamma}$ for $A \in \mathbb{T}^{\mathcal{T}}$ as follows:

$$\begin{array}{lcl}
[\![\alpha]\!]^{\mathcal{T}}_{\Gamma} & = & [\alpha]^{\mathcal{T}}_{\Gamma}, \qquad \text{for } \alpha \in \mathbb{A}_\infty \cup \{\mathtt{U}, 1\}; \\
[\![0]\!]^{\mathcal{T}}_{\Gamma} & = & \begin{cases} \{M \mid \forall \vec{N}.M\vec{N} \in [0]^{\mathcal{T}}_{\Gamma}\}, & \text{for } \mathcal{T} \in \{\text{Scott}, \text{DHM}\}; \\ \{M \mid \forall \Gamma' \supseteq \Gamma \, \forall \vec{N} \in [1]^{\mathcal{T}}_{\Gamma'}.M\vec{N} \in [0]^{\mathcal{T}}_{\Gamma'}\}, & \text{for } \mathcal{T} \in \{\text{CDZ}, \text{HR}\}; \\ [0]^{\mathcal{T}}_{\Gamma}, & \text{for } \mathcal{T} = \text{Park}; \end{cases} \\
[\![A \to B]\!]^{\mathcal{T}}_{\Gamma} & = & \begin{cases} \{M \mid \forall \Gamma' \supseteq \Gamma \, \forall N \in [\![A]\!]^{\mathcal{T}}_{\Gamma'}.MN \in [\![B]\!]^{\mathcal{T}}_{\Gamma'}\}, & \text{if } \mathcal{T} \neq \text{AO or } B \neq_{\text{AO}} \mathtt{U}; \\ [A \to B]^{\mathcal{T}}_{\Gamma}, & \text{if } \mathcal{T} = \text{AO} \ \& \ B =_{\text{AO}} \mathtt{U}; \end{cases} \\
[\![A \cap B]\!]^{\mathcal{T}}_{\Gamma} & = & [\![A]\!]^{\mathcal{T}}_{\Gamma} \cap [\![B]\!]^{\mathcal{T}}_{\Gamma}.
\end{array}$$

The reason that the definition of $[\![A \to B]\!]^{\mathcal{T}}_{\Gamma}$ is somewhat involved is to make Lemma 17.3.12(ii) also valid for $\mathcal{T} = \text{AO}$.

17.3.11 Proposition

(i) $M \in [\![A]\!]^{\mathcal{T}}_{\Gamma}$ *and* $N \twoheadrightarrow_\beta M$ *imply* $N \in [\![A]\!]^{\mathcal{T}}_{\Gamma}$.

(ii) $\Gamma \Subset \Gamma'$ *implies* $[\![A]\!]^{\mathcal{T}}_{\Gamma} \subseteq [\![A]\!]^{\mathcal{T}}_{\Gamma'}$ *for all types* $A \in \mathbb{T}^{\mathcal{T}}$.

Proof Easy. ∎

Lemmas 17.3.12 and 17.3.15, and Theorem 17.3.17 below are standard.

17.3.12 Lemma *Let* $\mathcal{T} \in \{\text{Scott}, \text{Park}, \text{CDZ}, \text{HR}, \text{DHM}, \text{BCD}, \text{AO}\}$. *Then*

(i) $x\vec{M} \in [A]^{\mathcal{T}}_{\Gamma} \Rightarrow x\vec{M} \in [\![A]\!]^{\mathcal{T}}_{\Gamma}$.

(ii) $[\![A]\!]^{\mathcal{T}}_{\Gamma} \subseteq [A]^{\mathcal{T}}_{\Gamma}$.

Proof Parts (i) and (ii) are proved simultaneously by induction on A. We consider only some interesting cases.

(i) *Case* $A \equiv 0$ *and* $\mathcal{T} = \text{CDZ}$. Let $\Gamma' \supseteq \Gamma$ and $\vec{N} \in [1]^{\text{CDZ}}_{\Gamma'}$. Clearly

$$P \in \mathcal{A}_{\text{CDZ}}(x\vec{M}) \text{ and } \vec{Q} \in \mathcal{A}_{\text{CDZ}}(\vec{N}) \Rightarrow P\vec{Q} \in \mathcal{A}_{\text{CDZ}}(x\vec{M}\vec{N}).$$

Hence

$$\begin{array}{rll}
x\vec{M} \in [0]^{\text{CDZ}}_{\Gamma} & \Rightarrow \ x\vec{M}\vec{N} \in [0]^{\text{CDZ}}_{\Gamma'} & \text{by rules } (\leq_{\text{CDZ}}) \text{ and } (\to\text{E}) \\
 & & \text{since } 0 =_{\text{CDZ}} 1 \to 0, \\
 & \Rightarrow \ x\vec{M} \in [\![0]\!]^{\text{CDZ}}_{\Gamma} & \text{by Definition 17.3.10.}
\end{array}$$

Case $A \equiv B \to C$. Let $\Gamma' \supseteq \Gamma$ and $\mathcal{T} \neq \text{AO}$ or $C \neq_{\text{AO}} \mathtt{U}$ and let $N \in [\![B]\!]^{\mathcal{T}}_{\Gamma'}$.

Then $[\![B]\!]^{\mathcal{T}}_{\Gamma'} \subseteq [B]^{\mathcal{T}}_{\Gamma'}$ by induction on (ii). Hence

$$\begin{array}{rcll} x\vec{M} \in [A]^{\mathcal{T}}_{\Gamma} & \Rightarrow & x\vec{M}N \in [C]^{\mathcal{T}}_{\Gamma'} & \text{by rule } (\to\text{E}), \\ & \Rightarrow & x\vec{M}N \in [\![C]\!]^{\mathcal{T}}_{\Gamma'} & \text{by induction on (i)} \\ & \Rightarrow & x\vec{M} \in [\![B \to C]\!]^{\mathcal{T}}_{\Gamma} & \text{by Definition 17.3.10.} \end{array}$$

(ii) *Case* $A \equiv B{\to}C$ *and* $\mathcal{T} \neq \text{AO}$ *or* $C{\neq_{\text{AO}}}\texttt{U}$. Let $\Gamma' = \Gamma, z{:}\,B$ with z fresh, and suppose $M \in [\![B \to C]\!]^{\mathcal{T}}_{\Gamma}$; since $z \in [\![B]\!]^{\mathcal{T}}_{\Gamma,z:B}$ by induction on (i), we have

$$\begin{array}{rcll} M \in [\![B \to C]\!]^{\mathcal{T}}_{\Gamma} \text{ and } z \in [\![B]\!]^{\mathcal{T}}_{\Gamma,z:B} & \Rightarrow & Mz \in [\![C]\!]^{\mathcal{T}}_{\Gamma'} & \text{by Definition 17.3.10,} \\ & \Rightarrow & Mz \in [C]^{\mathcal{T}}_{\Gamma'} & \text{by induction on (ii,} \\ & \Rightarrow & M \in [B{\to}C]^{\mathcal{T}}_{\Gamma} & \text{by Lemma 17.3.9.} \end{array}$$

Case $A \equiv B \cap C$. This follows from $[\![B \cap C]\!]^{\mathcal{T}}_{\Gamma} = [\![B]\!]^{\mathcal{T}}_{\Gamma} \cap [\![C]\!]^{\mathcal{T}}_{\Gamma}$ and the induction hypothesis using Lemma 17.3.6(ii). ∎

The following lemma essentially states that the Kripke type interpretations agree with the corresponding type theories.

17.3.13 Lemma

(i) *Let* $\mathcal{T} \in \{\text{CDZ}, \text{DHM}\}$. *Then*

$$M \in [A]^{\mathcal{T}}_{\Gamma,z:0} \;\&\; N \in [\![0]\!]^{\mathcal{T}}_{\Gamma} \;\Rightarrow\; M[z := N] \in [A]^{\mathcal{T}}_{\Gamma}.$$

(ii) *Let* $\mathcal{T} \in \{\text{Scott}, \text{Park}, \text{CDZ}, \text{HR}, \text{DHM}, \text{BCD}, \text{AO}\}$. *Then*

$$\forall A, B \in \mathbb{T}^{\mathcal{T}} [A \leq_{\mathcal{T}} B \;\Rightarrow\; [\![A]\!]^{\mathcal{T}}_{\Gamma} \subseteq [\![B]\!]^{\mathcal{T}}_{\Gamma}].$$

Proof (i) We may assume $A \neq_{\mathcal{T}} \texttt{U}$.

Consider first $\mathcal{T} = \text{CDZ}$. If $M \in [A]^{\text{CDZ}}_{\Gamma,z:0}$, then there is a $P \in \mathcal{A}_{\text{CDZ}}(M)$ such that $\Gamma, z{:}0 \vdash^{\text{CDZ}}_{\cap} P : A$. The assertion is proved by induction on P.

Case $P \equiv \bot$. Trivial.

Case $P \equiv \lambda x.P'$. Then $M \twoheadrightarrow_{\beta} \lambda x.M'$ and $P' \in \mathcal{A}_{\text{CDZ}}(M')$. Moreover from $\Gamma, z{:}0 \vdash^{\text{CDZ}}_{\cap} P : A$ we get $\Gamma, z{:}0, x{:}B_i \vdash^{\text{CDZ}}_{\cap} P' : C_i$ and by Theorem 14.1.1(iii) $\bigcap_{i \in I}(B_i{\to}C_i) \leq_{\text{CDZ}} A$ for some I and $B_i, C_i \in \mathbb{T}^{\text{CDZ}}$. By induction, for each $i \in I$ there is a $P_i \in \mathcal{A}_{\text{CDZ}}(M'[z := N])$ such that $\Gamma, x{:}B_i \vdash^{\text{CDZ}}_{\cap} P_i : C_i$. Let $P_i = \Box(M_i)$, where $M'[z := N] \twoheadrightarrow_{\beta} M_i$ and let M'' be a common reduct of the M_i and $P'' \equiv \Box(M'')$. Then $P'' \in \mathcal{A}_{\text{CDZ}}(M'[z := N])$ and, by Lemma 17.3.6(i), $\Gamma, x{:}B_i \vdash^{\text{CDZ}}_{\cap} P'' : C_i$, for all $i \in I$. Clearly $\lambda x.P'' \in \mathcal{A}_{\text{CDZ}}(M[z := N])$ and by construction $\Gamma \vdash^{\text{CDZ}}_{\cap} \lambda x.P'' : A$.

Case $P \equiv x\vec{P}$. Then $M \twoheadrightarrow_{\beta} x\vec{M}$ and $\vec{P} \in \mathcal{A}_{\text{CDZ}}(\vec{M})$. From $\Gamma, z{:}0 \vdash^{\text{CDZ}}_{\cap} P :$

A we get $\Gamma, z{:}0 \vdash_\cap^{\text{CDZ}} x : \vec{B} \to A$ and, by Theorem 14.1.9(ii) and Lemma 13.1.25, $\Gamma, z{:}0 \vdash_\cap^{\text{CDZ}} \vec{P} : \vec{B}$. By induction there are $\vec{P'} \in \mathcal{A}_{\text{CDZ}}(\overrightarrow{M}[z := N])$ such that $\Gamma \vdash_\cap^{\text{CDZ}} \vec{P'} : \vec{B}$. If $x \neq z$ we are done since $x\vec{P'} \in \mathcal{A}_{\text{CDZ}}(M[z := N])$ and we can derive $\Gamma \vdash_\cap^{\text{CDZ}} x\vec{P'} : A$ using ($\to$E). Otherwise, by Theorem 14.1.9(i), $\Gamma, z{:}0 \vdash_\cap^{\text{CDZ}} z : \vec{B} \to A$ implies $0 \leq_{\text{CDZ}} \vec{B} \to A$. Since $0 =_{\text{CDZ}} \vec{1} \to 0$ in CDZ, which is $\boldsymbol{\beta}$-sound by Theorem 14.1.7, this implies $\vec{B} \leq_{\text{CDZ}} \vec{1}$ and $0 \leq_{\text{CDZ}} A$ by Lemma 13.1.25. So we get $\Gamma \vdash_\cap^{\text{CDZ}} \vec{P'} : \vec{1}$, i.e. one has $\overrightarrow{M}[z := N] \in \overrightarrow{[1]_\Gamma^{\text{CDZ}}}$. Now, by by Definition 17.3.10,

$$N \in [\![0]\!]_\Gamma^{\text{CDZ}} \ \&\ \overrightarrow{M}[z := N] \in \overrightarrow{[1]_\Gamma^{\text{CDZ}}} \ \Rightarrow\ M[z := N] \in [0]_\Gamma^{\text{CDZ}}.$$

Since $0 \leq_{\text{CDZ}} A$ we get $M[z := N] \in [A]_\Gamma^{\text{CDZ}}$.

Now consider $\mathcal{T} = \text{DHM}$. Then the proof is similar but easier. In the case $P \equiv z\vec{P}$ the implication $N \in [\![0]\!]_\Gamma^{\text{DHM}} \ \Rightarrow\ M[z := N] \in [A]_\Gamma^{\text{DHM}}$ follows from Definition 17.3.10.

(ii) We treat the cases related to $A{\to}B{\leq}\mathtt{U}{\to}\mathtt{U}$ in AO, $(0{\to}1){=}1$, $0{=}(1{\to}0)$ in CDZ, $(1{\to}1) \cap (0{\to}0){=}1$ in HR, and $(0{\to}0){=}0$ in Park.

Proof of $[\![A{\to}B]\!]_\Gamma^{\text{AO}} \subseteq [\![\mathtt{U}{\to}\mathtt{U}]\!]_\Gamma^{\text{AO}}$. If $B =_{\text{AO}} \mathtt{U}$, then

$$[\![A{\to}B]\!]_\Gamma^{\text{AO}} = [A{\to}B]_\Gamma^{\text{AO}} \subseteq [\mathtt{U}{\to}\mathtt{U}]_\Gamma^{\text{AO}} = [\![\mathtt{U}{\to}\mathtt{U}]\!]_\Gamma^{\text{AO}}.$$

If, on the other hand, $B \neq_{\text{AO}} \mathtt{U}$, then $M \in [\![A{\to}B]\!]_\Gamma^{\text{AO}}$. Write $\Gamma' = \Gamma, z{:}A$. Then $z \in [A]_{\Gamma'}^{\text{AO}}$, hence by Lemma 17.3.12(i) we have $z \in [\![A]\!]_{\Gamma'}^{\text{AO}}$. Therefore $Mz \in [\![B]\!]_{\Gamma'}^{\text{AO}} \subseteq [B]_{\Gamma'}^{\text{AO}}$, and so Lemma 17.3.12(ii) implies

$$M \in [A{\to}B]_\Gamma^{\text{AO}} \subseteq [\mathtt{U}{\to}\mathtt{U}]_\Gamma^{\text{AO}} = [\![\mathtt{U}{\to}\mathtt{U}]\!]_\Gamma^{\text{AO}}.$$

Proof of $[\![0{\to}1]\!]_\Gamma^{\text{CDZ}} \subseteq [\![1]\!]_\Gamma^{\text{CDZ}}$. We have

$$\begin{array}{rcll} [\![0{\to}1]\!]_\Gamma^{\text{CDZ}} & \subseteq & [0{\to}1]_\Gamma^{\text{CDZ}}, & \text{by Lemma 17.3.12((ii))},\\ & = & [1]_\Gamma^{\text{CDZ}}, & \text{since } 0{\to}1 =_{\text{CDZ}} 1,\\ & = & [\![1]\!]_\Gamma^{\text{CDZ}}, & \text{by Definition 17.3.10.} \end{array}$$

Proof of $[\![1]\!]_\Gamma^{\text{CDZ}} \subseteq [\![0{\to}1]\!]_\Gamma^{\text{CDZ}}$. Suppose that $\Gamma' \supseteq \Gamma$, $M \in [\![1]\!]_\Gamma^{\text{CDZ}}$ and $N \in [\![0]\!]_{\Gamma'}^{\text{CDZ}}$, in order to show that $MN \in [\![1]\!]_{\Gamma'}^{\text{CDZ}}$. By Definition 17.3.10, we have $[\![1]\!]_\Gamma^{\text{CDZ}} = [1]_\Gamma^{\text{CDZ}}$. If $M \in [1]_\Gamma^{\text{CDZ}}$, then there is $P \in \mathcal{A}_{\text{CDZ}}(M)$ such that $\Gamma \vdash_\cap^{\text{CDZ}} P : 1$. We will show $MN \in [\![1]\!]_{\Gamma'}^{\text{CDZ}}$ by distinguishing cases of P.

Case $P \equiv \bot$. By Proposition 17.3.5(i) one has $1 =_{\text{CDZ}} \mathtt{U}$, contradicting Proposition 13.1.29. So this case is impossible.

Case $P \equiv \lambda z.P'$. Then $M \twoheadrightarrow_{\beta} \lambda z.M'$ and $P' \in \mathcal{A}_{\mathrm{CDZ}}(M')$. As $\Gamma \vdash_{\cap}^{\mathrm{CDZ}} P : 1$ we get $\Gamma, z{:}0 \vdash_{\cap}^{\mathrm{CDZ}} P' : 1$ by Theorem 14.1.9(iii), since $1 =_{\mathrm{CDZ}} 0{\to}1$. This implies $M' \in [1]_{\Gamma,z:0}^{\mathrm{CDZ}}$. We may assume that $z \notin \mathrm{dom}(\Gamma')$. Then also $M' \in [1]_{\Gamma',z:0}^{\mathrm{CDZ}}$. Therefore

$$\begin{array}{rcll} MN & \twoheadrightarrow_{\beta} & (\lambda z.M')N & \\ & \to_{\beta} & M'[z := N] & \\ & \in & [1]_{\Gamma'}^{\mathrm{CDZ}}, & \text{by (i)}, \\ & = & [\![1]\!]_{\Gamma'}^{\mathrm{CDZ}}. & \end{array}$$

Case $P \equiv x\vec{P}$. Note that $\Gamma \vdash_{\cap}^{\mathrm{CDZ}} P : 1$ implies $\Gamma \vdash_{\cap}^{\mathrm{CDZ}} P : 0{\to}1$, since $1 =_{\mathrm{CDZ}} 0{\to}1$. By Lemma 17.3.12(ii) we have $[\![0]\!]_{\Gamma'}^{\mathrm{CDZ}} \subseteq [0]_{\Gamma'}^{\mathrm{CDZ}}$, hence there is $P' \in \mathcal{A}_{\mathrm{CDZ}}(N)$ such that $\Gamma' \vdash_{\cap}^{\mathrm{CDZ}} P' : 0$. We get $\Gamma' \vdash_{\cap}^{\mathrm{CDZ}} PP' : 1$. As $PP' \in \mathcal{A}_{\mathrm{CDZ}}(MN)$ we conclude that $MN \in [\![1]\!]_{\Gamma'}^{\mathrm{CDZ}}$.

Proof of $[\![1{\to}0]\!]_{\Gamma}^{\mathrm{CDZ}} \subseteq [\![0]\!]_{\Gamma}^{\mathrm{CDZ}}$. We have $[\![1{\to}0]\!]_{\Gamma}^{\mathrm{CDZ}} \subseteq [1{\to}0]_{\Gamma}^{\mathrm{CDZ}}$, by Lemma 17.3.12((ii)) and, from Definition 17.3.7, $[1{\to}0]_{\Gamma}^{\mathrm{CDZ}} = [0]_{\Gamma}^{\mathrm{CDZ}}$, as $1{\to}0 =_{\mathrm{CDZ}} 0$. Moreover, using Definition 17.3.10, it follows that

$$\begin{aligned} [\![1{\to}0]\!]_{\Gamma}^{\mathrm{CDZ}} &= \{M \mid \forall \Gamma' \supseteq \Gamma, \forall N \in [\![1]\!]_{\Gamma'}^{\mathrm{CDZ}}.MN \in [\![0]\!]_{\Gamma'}^{\mathrm{CDZ}}\} \\ &= \{M \mid \forall \Gamma' \supseteq \Gamma, \forall N \in [1]_{\Gamma'}^{\mathrm{CDZ}}.MN \in [\![0]\!]_{\Gamma'}^{\mathrm{CDZ}}\} \\ &\subseteq \{M \mid \forall \Gamma' \supseteq \Gamma, \forall N, \vec{N} \in [1]_{\Gamma'}^{\mathrm{CDZ}}.MN\vec{N} \in [0]_{\Gamma'}^{\mathrm{CDZ}}\}. \end{aligned}$$

From $[\![1{\to}0]\!]_{\Gamma}^{\mathrm{CDZ}} \subseteq [0]_{\Gamma}^{\mathrm{CDZ}}$ and

$$[\![1{\to}0]\!]_{\Gamma}^{\mathrm{CDZ}} \subseteq \{M \mid \forall \Gamma' \supseteq \Gamma, \forall N, \vec{N} \in [1]_{\Gamma'}^{\mathrm{CDZ}}.MN\vec{N} \in [0]_{\Gamma'}^{\mathrm{CDZ}}\}$$

we can conclude

$$[\![1{\to}0]\!]_{\Gamma}^{\mathrm{CDZ}} \subseteq \{M \mid \forall \Gamma' \supseteq \Gamma, \vec{N} \in [1]_{\Gamma'}^{\mathrm{CDZ}}.M\vec{N} \in [0]_{\Gamma'}^{\mathrm{CDZ}}\} = [\![0]\!]_{\Gamma}^{\mathrm{CDZ}}.$$

Proof of $[\![0]\!]_{\Gamma}^{\mathrm{CDZ}} \subseteq [\![1{\to}0]\!]_{\Gamma}^{\mathrm{CDZ}}$. Again using Definition 17.3.10 one has

$$\begin{array}{rcl} M \in [\![0]\!]_{\Gamma}^{\mathrm{CDZ}} & \Rightarrow & \forall \Gamma' \supseteq \Gamma, \forall N, \vec{N} \in [1]_{\Gamma'}^{\mathrm{CDZ}}.MN\vec{N} \in [0]_{\Gamma'}^{\mathrm{CDZ}} \\ & \Rightarrow & \forall \Gamma' \supseteq \Gamma, \forall N \in [1]_{\Gamma'}^{\mathrm{CDZ}}.MN \in [\![0]\!]_{\Gamma'}^{\mathrm{CDZ}} \\ & \Rightarrow & M \in [\![1{\to}0]\!]_{\Gamma}^{\mathrm{CDZ}}. \end{array}$$

Proof of $[\![(1{\to}1){\cap}(0{\to}0)]\!]^{\rm HR}_{\Gamma} \subseteq [\![1]\!]^{\rm HR}_{\Gamma}$. By Lemma 17.3.12(ii) one has

$$\begin{aligned}[\![(1{\to}1){\cap}(0{\to}0)]\!]^{\rm HR}_{\Gamma} &\subseteq [(1{\to}1){\cap}(0{\to}0)]^{\rm HR}_{\Gamma}\\ &= [1]^{\rm HR}_{\Gamma}\\ &= [\![1]\!]^{\rm HR}_{\Gamma},\end{aligned}$$

using Definition 17.3.7, $(1{\to}1){\cap}(0{\to}0) = 1$ and Definition 17.3.10.

Proof of $[\![1]\!]^{\rm HR}_{\Gamma} \subseteq [\![(1{\to}1){\cap}(0{\to}0)]\!]^{\rm HR}_{\Gamma}$. Let $\Gamma' \supseteq \Gamma$.

$$\begin{aligned}M \in [\![1]\!]^{\rm HR}_{\Gamma} &\Rightarrow M \in [1]^{\rm HR}_{\Gamma}\\ &\Rightarrow \exists P \in \mathcal{A}_{\rm HR}(M)\ \Gamma \vdash^{\rm HR}_{\cap} P : 1, \quad \text{by Definition 17.3.7.}\end{aligned} \tag{17.2}$$

$$\begin{aligned}N \in [\![1]\!]^{\rm HR}_{\Gamma'} &\Rightarrow N \in [1]^{\rm HR}_{\Gamma'}\\ &\Rightarrow \exists P' \in \mathcal{A}_{\rm HR}(N)\ \Gamma' \vdash^{\rm HR}_{\cap} P' : 1, \text{ by Definition 17.3.7.}\end{aligned} \tag{17.3}$$

Let $\hat{P} \equiv \Phi P P'$ if P is a lambda-abstraction and $\hat{P} \equiv P P'$ otherwise. Then

$$\begin{aligned}(17.2) \text{ and } (17.3) &\Rightarrow \Gamma' \vdash^{\rm HR} \hat{P} : 1, && \text{by (Ax-}\Phi\text{-HR)}, (\leq_{\mathcal{T}}), ({\to}\text{E}),\\ &\Rightarrow MN \in [1]^{\rm HR}_{\Gamma'}, && \text{since } \hat{P} \in \mathcal{A}_{\rm HR}(MN),\\ &\Rightarrow MN \in [\![1]\!]^{\rm HR}_{\Gamma'}\\ &\Rightarrow M \in [\![1 \to 1]\!]^{\rm HR}_{\Gamma}.\end{aligned}$$

$$\begin{aligned}N \in [\![0]\!]^{\rm HR}_{\Gamma'} &\Rightarrow N \in [0]^{\rm HR}_{\Gamma'}\\ &\Rightarrow \exists P' \in \mathcal{A}_{\rm HR}(N)\ \Gamma' \vdash^{\rm HR}_{\cap} P' : 0, \text{ by Definition 17.3.7.}\end{aligned} \tag{17.4}$$

$$\begin{aligned}\vec{N} \in [\![1]\!]^{\rm HR}_{\Gamma'} &\Rightarrow \vec{N} \in [1]^{\rm HR}_{\Gamma'}\\ &\Rightarrow \exists \vec{P} \in \mathcal{A}_{\rm HR}(\vec{N})\ \Gamma' \vdash^{\rm HR}_{\cap} \vec{P} : \vec{1}, \quad \text{by Definition 17.3.7.}\end{aligned} \tag{17.5}$$

Let $\hat{P} \equiv \Phi P P' \vec{P}$ if P is a lambda-abstraction and $\hat{P} \equiv P P' \vec{P}$ otherwise. Then

$$\begin{aligned}(17.2), (17.4) \text{ and } (17.5) &\Rightarrow \Gamma' \vdash^{\rm HR} \hat{P} : 0, && \text{by (Ax-}\Phi\text{-HR)}, (\leq_{\mathcal{A}}), ({\to}\text{E})\\ &\Rightarrow MN\vec{N} \in [0]^{\rm HR}_{\Gamma'} && \text{since } \hat{P} \in \mathcal{A}_{\rm HR}(MN\vec{N})\\ &\Rightarrow MN \in [\![0]\!]^{\rm HR}_{\Gamma'}\\ &\Rightarrow M \in [\![0 \to 0]\!]^{\rm HR}_{\Gamma}.\end{aligned}$$

Proof of $[\![0{\to}0]\!]^{\rm Park}_{\Gamma} \subseteq [\![0]\!]^{\rm Park}_{\Gamma}$. Let $M \in [\![0{\to}0]\!]^{\rm Park}_{\Gamma}$ and $\Gamma' = \Gamma, z : 0$, where

$z \notin \mathrm{FV}(M)$.

$$\begin{aligned} z \in [0]^{\mathrm{Park}}_{\{z:0\}} &\Rightarrow z \in [\![0]\!]^{\mathrm{Park}}_{\{z:0\}} \\ &\Rightarrow Mz \in [\![0]\!]^{\mathrm{Park}}_{\Gamma'} \\ &\Rightarrow Mz \in [0]^{\mathrm{Park}}_{\Gamma'} \\ &\Rightarrow M \in [0]^{\mathrm{Park}}_{\Gamma}, \quad \text{by Lemma 17.3.9 and } (0 \to 0) \leq_{\mathrm{Park}} 0, \\ &\Rightarrow M \in [\![0]\!]^{\mathrm{Park}}_{\Gamma}. \end{aligned}$$

Proof of $[\![0]\!]^{\mathrm{Park}}_{\Gamma} \subseteq [\![0 \to 0]\!]^{\mathrm{Park}}_{\Gamma}$. Let $\Gamma' \sqsupseteq \Gamma$. Then we have

$$\begin{aligned} M \in [\![0]\!]^{\mathrm{Park}}_{\Gamma} &\Rightarrow M \in [0]^{\mathrm{Park}}_{\Gamma} \\ &\Rightarrow \exists P \in \mathcal{A}_{\mathrm{Park}}(M)\ \Gamma \vdash^{\mathrm{Park}}_{\cap} P : 0, \quad \text{by Definition 17.3.7.} \end{aligned} \tag{17.6}$$

$$\begin{aligned} N \in [\![0]\!]^{\mathrm{Park}}_{\Gamma'} &\Rightarrow N \in [0]^{\mathrm{Park}}_{\Gamma'} \\ &\Rightarrow \exists P' \in \mathcal{A}_{\mathrm{Park}}(N)\ \Gamma' \vdash^{\mathrm{Park}}_{\cap} P' : 0, \quad \text{by Definition 17.3.7.} \end{aligned} \tag{17.7}$$

Let $\hat{P} \equiv \Phi P P'$ if P is a lambda-abstraction and $\hat{P} \equiv PP'$ otherwise. Then

$$\begin{aligned} (17.6) \text{ and } (17.7) &\Rightarrow \Gamma' \vdash^{\mathrm{Park}}_{\cap} \hat{P} : 0, && \text{by (Ax-}\Phi\text{-Park)}, (\leq_{\mathrm{Park}}), (\to\mathrm{E}) \\ &\Rightarrow MN \in [0]^{\mathrm{Park}}_{\Gamma'}, && \text{since } \hat{P} \in \mathcal{A}_{\mathrm{Park}}(MN) \\ &\Rightarrow MN \in [\![0]\!]^{\mathrm{Park}}_{\Gamma'} \\ &\Rightarrow M \in [\![0 \to 0]\!]^{\mathrm{Park}}_{\Gamma}. \ \blacksquare \end{aligned}$$

17.3.14 Definition (Semantic Satisfiability) Let ρ be a mapping from term variables to terms and write $[\![M]\!]_\rho = M[\vec{x} := \rho(\vec{x})]$, where $\vec{x} = \mathrm{FV}(M)$. We write

(i) $\mathcal{T}, \rho, \Gamma \models M : A \Leftrightarrow [\![M]\!]_\rho \in [\![A]\!]^{\mathcal{T}}_{\Gamma}$.
(ii) $\mathcal{T}, \rho, \Gamma' \models \Gamma \Leftrightarrow \mathcal{T}, \rho, \Gamma' \models x : B$, for all $(x{:}B) \in \Gamma$;
(iii) $\Gamma \models^{\mathcal{T}}_{\cap} M : A \Leftrightarrow \mathcal{T}, \rho, \Gamma' \models \Gamma \Rightarrow [\mathcal{T}, \rho, \Gamma' \models M : A$, for all $\rho, \Gamma']$.

In line with the previous remarks, the following result can be constructed also as the soundness of the natural semantics of intersection types over a particular Kripke applicative structure, where bases play the role of worlds.

17.3.15 Lemma *Let* $\mathcal{T} \in \{\mathrm{Scott}, \mathrm{Park}, \mathrm{CDZ}, \mathrm{HR}, \mathrm{DHM}, \mathrm{BCD}, \mathrm{AO}\}$. *Then*

$$\Gamma \vdash^{\mathcal{T}}_{\cap} M : A \Rightarrow \Gamma \models^{\mathcal{T}}_{\cap} M : A.$$

Proof The proof is by induction on the derivation of $\Gamma \vdash^{\mathcal{T}}_{\cap} M : A$.

Cases (Ax), $(Ax\text{-}\mathtt{U})$. Immediate.

Cases $(\to\mathrm{E})$, $(\cap\mathrm{I})$. By induction.

Case ($\leq$). By Lemma 17.3.13((ii)).

Case ($\to$I). Suppose $M \equiv \lambda y.R$, $A \equiv B \to C$ and $\Gamma, y : B \vdash_\cap^{\mathcal{T}} R : C$.

Subcase $\mathcal{T} \neq \mathrm{AO}$ *or* $C \neq_{\mathrm{AO}} \mathtt{U}$. Suppose $\mathcal{T}, \rho, \Gamma' \models \Gamma$ in order to show $[\![\lambda y.R]\!]_\rho \in [\![B \to C]\!]^{\mathcal{T}}_{\Gamma'}$. Let $\Gamma'' \supseteq \Gamma'$ and $T \in [\![B]\!]^{\mathcal{T}}_{\Gamma''}$. Then by the induction hypothesis $[\![R]\!]_{\rho[y:=T]} \in [\![C]\!]^{\mathcal{T}}_{\Gamma''}$. We may assume $y \notin \rho(x)$ for all $x \in \mathrm{dom}(\Gamma)$. Then one has $[\![\lambda y.R]\!]_\rho T \to_\beta [\![R]\!]_{\rho[y:=T]}$ and hence $[\![\lambda y.R]\!]_\rho T \in [\![C]\!]^{\mathcal{T}}_{\Gamma''}$, by Proposition 17.3.11. Therefore $[\![\lambda y.R]\!]_\rho \in [\![B \to C]\!]^{\mathcal{T}}_{\Gamma'}$.

Subcase $\mathcal{T} = \mathrm{AO}$ *and* $C =_{\mathrm{AO}} \mathtt{U}$. The result follows easily from

$$\lambda x.\bot \in \mathcal{A}_{\mathrm{AO}}([\![\lambda y.R]\!]_\rho) \quad \text{for all } \rho,$$

and we can derive $\vdash_\cap^{\mathrm{AO}} \lambda x.\bot : B \to C$, using (Ax-$\mathtt{U}$), ($\to$I) and ($\leq_{\mathrm{AO}}$). Therefore $[\![M]\!]_\rho \in [A]^{\mathcal{T}}_\Gamma$, implying $[\![M]\!]_\rho \in [\![A]\!]^{\mathcal{T}}_\Gamma$ by Definition 17.3.10. ■

Now we can prove the converse of Proposition 17.3.8.

17.3.16 Proposition *Let* $\mathcal{T} \in \{\mathrm{Scott}, \mathrm{Park}, \mathrm{CDZ}, \mathrm{HR}, \mathrm{DHM}, \mathrm{BCD}, \mathrm{AO}\}$. *Then*

$$\Gamma \vdash_\cap^{\mathcal{T}} M : A \Rightarrow M \in [A]^{\mathcal{T}}_\Gamma.$$

Proof Let $\rho_0(x) = x$. By Lemma 17.3.12(i) $\mathcal{T}, \rho_0, \Gamma \models \Gamma$. Then $\Gamma \vdash_\cap^{\mathcal{T}} M : A$ implies $M = [\![M]\!]_{\rho_0} \in [\![A]\!]^{\mathcal{T}}_\Gamma$ by Lemma 17.3.15. So we conclude $M \in [A]^{\mathcal{T}}_\Gamma$ by Lemma 17.3.12(ii). ■

17.3.17 Theorem (Approximation Theorem) *Let* $\mathcal{T} \in \{\mathrm{Scott}, \mathrm{Park}, \mathrm{CDZ}, \mathrm{HR}, \mathrm{DHM}, \mathrm{BCD}, \mathrm{AO}\}$. *Then*

$$\Gamma \vdash_\cap^{\mathcal{T}} M : A \;\Leftrightarrow\; \exists P \in \mathcal{A}_{\mathcal{T}}(M).\Gamma \vdash_\cap^{\mathcal{T}} P : A.$$

Proof By Propositions 17.3.8 and 17.3.16. ■

17.3.18 Corollary *Let* $\mathcal{T} \in \{\mathrm{Scott}, \mathrm{Park}, \mathrm{CDZ}, \mathrm{HR}, \mathrm{DHM}, \mathrm{BCD}, \mathrm{AO}\}$. *Let M be an untyped lambda term. Then*

$$[\![M]\!]^{\mathcal{F}^{\mathcal{T}}}_\rho = \{A \mid \Gamma \vdash_\cap^{\mathcal{T}} P : A \textit{ for some } P \in \mathcal{A}_{\mathcal{T}}(M) \textit{ and some } \Gamma \models \rho\}.$$

Proof By Theorem 16.2.7 and the Approximation Theorem. ■

Another way of writing this is

$$\begin{aligned} [\![M]\!]^{\mathcal{F}^{\mathcal{T}}}_\rho &= \bigcup_{P \in \mathcal{A}_{\mathcal{T}}(M)} [\![P]\!]^{\mathcal{F}^{\mathcal{T}}}_\rho \\ &= \bigcup_{P \in \mathcal{A}_{\mathcal{T}}(M)} \{A \mid \Gamma \vdash_\cap^{\mathcal{T}} P : A \text{ for some } \Gamma \models \rho\}. \end{aligned}$$

This gives the motivation for the name 'Approximation Theorem'. Theorem 17.3.17 was first proved for $\mathcal{T} = \text{BCD}$ in Barendregt et al. (1983), for $\mathcal{T} = \text{CDZ}$ in Coppo et al. (1987), for $\mathcal{T} = \text{Scott}$ in Ronchi Della Rocca (1988), for $\mathcal{T} = \text{Park}$ and $\mathcal{T} = \text{HR}$ in Honsell and Ronchi Della Rocca (1992), and for $\mathcal{T} = \text{AO}$ in Abramsky and Ong (1993).

17.4 Applications of the approximation theorem

As discussed in Section 16.2, type theories give rise in a natural way to *filter λ-models*. Properties of $\mathcal{F}^{\mathcal{T}}$ with $\mathcal{T} \in \{\text{Scott}, \text{CDZ}, \text{AO}, \text{BCD}, \text{Park}, \text{HR}\}$ can be easily derived using Theorem 17.3.17. For instance, one can check the following:

- The models $\mathcal{F}^{\text{Scott}}, \mathcal{F}^{\text{CDZ}}, \mathcal{F}^{\text{DHM}}, \mathcal{F}^{\text{BCD}}$ are sensible, i.e. equate unsolvables.
- The top element in $\mathcal{F}^{\text{AO}}$ is the interpretation of the terms of order ∞.
- The model $\mathcal{F}^{\text{Park}}$ characterizes the terms reducible to closed terms.
- The model $\mathcal{F}^{\text{HR}}$ characterizes the terms reducible to $\lambda\mathsf{I}$-terms.

The rest of this section is devoted to the proof of these properties. Other uses of the Approximation Theorem can be found in the corresponding relevant papers, i.e. Barendregt et al. (1983), Coppo et al. (1987), Ronchi Della Rocca (1988), Honsell and Ronchi Della Rocca (1992), Abramsky and Ong (1993).

17.4.1 Theorem *The models $\mathcal{F}^{\mathcal{T}}$ with $\mathcal{T} \in \{\text{Scott}, \text{CDZ}, \text{DHM}, \text{BCD}\}$ are sensible, i.e. for all unsolvable terms M, N one has*

$$[\![M]\!]_{\rho}^{\mathcal{F}^{\mathcal{T}}} = [\![N]\!]_{\rho}^{\mathcal{F}^{\mathcal{T}}}.$$

Proof From Corollary 17.3.18 of the Approximation Theorem, it follows immediately that $\perp$ is the only approximant of an unsolvable term for the mapping $\square_{\perp}$, and hence the assertion follows from Proposition 17.3.5(i). ∎

Recall that according to Definition 16.4.1(i) an untyped lambda term M is of order ∞ if

$$\forall n \exists N_n \in \Lambda M \twoheadrightarrow_{\beta} \lambda x_1 \cdots \lambda x_n . N_n.$$

17.4.2 Theorem *Let M be an untyped lambda term. Then the following are equivalent.*

(i) *M is of order ∞.*
(ii) $\vdash_{\cap}^{\text{AO}} M : A$ *for all types* $A \in \mathbb{T}^{\text{AO}}$.
(iii) $[\![M]\!]_{\rho}^{\mathcal{F}^{\text{AO}}} = \mathbb{T}^{\text{AO}} \in \mathcal{F}^{\text{AO}}$ *for all valuations ρ.*

Proof Write $\vdash, \leq$ for $\vdash^{\mathrm{AO}}_{\cap}, \leq_{\mathrm{AO}}$, respectively.

(i)⇔(ii). It is easy to check by structural induction on types (see Exercise 13.4) that

$$\forall A \in \mathbb{T}^{\mathrm{AO}} \exists n \in \omega.(\mathtt{U}^n \to \mathtt{U}) \leq A.$$

So by the Approximation Theorem it suffices to show that, if $P \in \Lambda\bot$ is an approximate normal form, then we have

$$\vdash P : (\mathtt{U}^n \to \mathtt{U}) \;\Leftrightarrow\; P \equiv \lambda x_1 \cdots \lambda x_n.P' \text{ for some } P'.$$

(⇐) By axiom ($\mathtt{U}_{\mathrm{top}}$) and rule (→I).

(⇒) Assume towards a contradiction that $P \equiv \lambda x_1 \cdots \lambda x_m.P'$ for $m < n$ and that P' is of the form $\bot$ or $x\vec{P}$. Then by Theorem 14.1.9(iii) we have

$$\vdash P : (\mathtt{U}^n \to \mathtt{U}) \;\Rightarrow\; \{x_1{:}\mathtt{U}, \ldots, x_m{:}\mathtt{U}\} \vdash P' : (\mathtt{U}^{n-m} \to \mathtt{U}).$$

But this latter judgement can neither be derived if $P' \equiv \bot$, by Proposition 17.3.5(i) and Lemma 13.1.26, nor if $P' \equiv x\vec{P}$, by Theorem 14.1.1(i) and (ii) and Lemma 13.1.26.

(ii)⇔(iii). Suppose $\vdash M : A$ for all A. Then by Theorem 16.2.7

$$[\![M]\!]^{\mathcal{F}^{\mathrm{AO}}}_{\rho} = \{A \mid \Gamma \vdash M : A \text{ for some } \Gamma \models \rho\} = \mathbb{T}^{\mathrm{AO}},$$

for all ρ. This is the top element in $\mathcal{F}^{\mathrm{AO}}$. Conversely, if $[\![M]\!]^{\mathcal{F}^{\mathrm{AO}}}_{\rho} = \top = \mathbb{T}^{\mathrm{AO}}$, for all ρ, then take $\rho_0(x) = \uparrow\mathtt{U}$. Then $\Gamma_{\rho_0} = \{x{:}\mathtt{U} \mid x \in \mathtt{Var}\}$. Hence

$$\begin{aligned} \mathbb{T}^{\mathrm{AO}} &= [\![M]\!]^{\mathcal{F}^{\mathrm{AO}}}_{\rho_0} \\ &= \{A \mid \Gamma \vdash M : A \text{ for some } \Gamma \models \rho_0\}, && \text{by Theorem 16.2.7,} \\ &= \{A \mid \vdash M : A\}, && \text{by Exercise 13.5.} \end{aligned}$$

Therefore $\vdash M : A$ for all $A \in \mathbb{T}^{\mathrm{AO}}$. ■

We denote by $\Lambda_{\downarrow\Lambda^{\emptyset}}$ the set of terms which reduce to a closed term.

17.4.3 Theorem *Let $M \in \Lambda$. Then*

$$M \in \Lambda_{\downarrow\Lambda^{\emptyset}} \Leftrightarrow \vdash^{\mathrm{Park}}_{\cap} M : 0.$$

Proof By the Approximation Theorem it suffices to check that if $P \in \Lambda\Phi$ is an approximate normal form and $\mathcal{V}$ is a finite set of term variables:

$$\{x{:}0 \mid x \in \mathcal{V}\} \vdash^{\mathrm{Park}}_{\cap} P : 0 \;\Leftrightarrow\; FV(P) \subseteq \mathcal{V}.$$

(⇐) By an easy induction on P, using the fact that $0 = 0{\to}0$.

(⇒) By induction on P. Lemma 13.1.25 shows $\vec{B}{\to}0 \neq \mathtt{U}$.

Case $P \equiv \lambda y.Q$. By Theorem 14.1.9(iii) and the induction hypothesis for Q.

Case $P \equiv y\vec{P}$. By Theorem 14.1.9(ii) we have $\Gamma \vdash_{\cap}^{\text{Park}} y : \vec{B}\to 0$ and $\Gamma \vdash_{\cap}^{\text{Park}} \vec{P} : \vec{B}$. Hence by Theorem 14.1.9(i) one has $y \in \mathcal{V}$ and $0 \leq \vec{B}\to 0$. By $\boldsymbol{\beta}$-soundness and $0 = 0\to 0$ we get $B_i \leq 0$. Thus $\Gamma \vdash_{\cap}^{\text{Park}} P_i : 0$ and hence $\text{FV}(P_i) \subseteq \mathcal{V}$, by the induction hypothesis.

Case $P \equiv \Phi\vec{P}$. Similar to the previous case. ∎

Lastly we work out the characterization of terms reducible to λI-terms.

17.4.4 Lemma *Let $m > 0$. Then*

$$1 \leq_{\text{HR}} A_1 \to \cdots \to A_m \to 0 \;\Rightarrow\; [A_i = 0, \text{ for some } i].$$

Proof By induction on m.

Case $m = 1$. Then $1 = (1 \to 1) \cap (0 \to 0) \leq A_1 \to 0$. By $\boldsymbol{\beta}$-soundness, the only possible case is $A_1 \leq 0$. Since 0 is the least element, we have that $A_1 = 0$.

Case $m > 1$. Similarly, using the induction hypothesis. ∎

17.4.5 Notation Write $\Gamma_1^P = \{x{:}1 \mid x \in FV(P)\}$.

17.4.6 Lemma *There is no approximate normal form P with $\Gamma_1^P \vdash_{\cap}^{\text{HR}} P : 0$.*

Proof Suppose towards a contradiction there exists such a P. By induction on P, using the Inversion Lemmas, the result will be shown.

Case $P \equiv \lambda z.P'$. Then $\Gamma_1^P \vdash P : 0$ implies $\Gamma_1^P, z{:}1 \vdash P' : 0$ and the induction hypothesis applies.

Case $P \equiv P_0 P_1 \cdots P_m$, *where P_0 is either a variable z or Φ.* Then

$$\Gamma_1^P \vdash P_0 : A_1 \to \cdots \to A_m \to 0 \tag{17.8}$$

and $\Gamma_1^{P_i} \vdash P_i : A_i$ for some A_i, by the Inversion Lemma 14.1.9(ii) and (*strengthening*). Then $1 \leq A_1 \to \cdots \to A_m \to 0$, by (17.8). By Lemma 17.4.4, there exists an i such that $A_i = 0$. Then $\Gamma \vdash P_i : 0$. By the induction hypothesis, this is impossible. ∎

17.4.7 Lemma *Let $P \in \Lambda\Phi$ be an approximate normal form. Suppose that $\Gamma_1^P \vdash^{\text{HR}} P : 1$. Then P is a λI-term.*

Proof By induction on P.

Case $P \equiv \lambda z.P'$. By the Inversion Lemma 14.1.9(iii), $\Gamma_1^{P'} \vdash P' : 1$ and $\Gamma_1^{P'} \uplus z : 0 \vdash P' : 0$. By the induction hypothesis, we have that P' is a λI-term. It remains to prove that $z \in FV(P')$. If not, then we could remove it from the context and get $\Gamma_1^{P'} \vdash P' : 0$, contradicting Lemma 17.4.6.

Case $P \equiv P_0 P_1 \cdots P_n$, *where* $P_0 = x$ *or* $P_0 = \Phi$. By the Inversion Lemma 14.1.9(ii) we have $\Gamma_1^P \vdash P_0 : A_1 \to \cdots \to A_n \to 1$ and $\Gamma_1^P \vdash P_i : A_i$ for all $i > 0$. By the Inversion Lemma 14.1.9(i), for $P_0 = x$ or Proposition 17.3.5(iii), for $P_0 = \Phi$, we get $1 \leq A_1 \to \cdots \to A_n \to 1$. Since $0 \leq 1$ and $0 = 1 \to 0$, we see that $1 \to \cdots \to 1 \to 0 \leq A_1 \to \cdots \to A_n \to 1$. So $A_i \leq 1$, by $\boldsymbol{\beta}$-soundness. Hence, by rules (*strengthening*) and ($\leq$), we have $\Gamma_1^{P_i} \vdash P_i : 1$ for all $i > 0$. Therefore, by the induction hypothesis, all P_i are λI-terms. ■

17.4.8 Lemma *Let* $P \in \Lambda\Phi$ *be an approximate normal form.*

(i) *If* P *is a* λI*-term, then* $\Gamma_1^P \vdash^{\mathrm{HR}} P : 1$.
(ii) *If* P *is a* λI*-term and* $x \in FV(P)$*, then* $\Gamma_1^P \uplus \{x : 0\} \vdash^{\mathrm{HR}} P : 0$.

Proof By simultaneous induction on P.

(i) *Case* $P \equiv y$. Trivial.

Case $P \equiv \lambda z.P'$. If P is a λI-term, so is P' and $z \in FV(P')$. By the induction hypothesis (i), we have that $\Gamma_1^{P'} \vdash P' : 1$. By the induction hypothesis (ii), we have that $\Gamma_1^{P'} \uplus \{z : 0\} \vdash P' : 0$. Since $1 \cap 0 = 0$ and $1 = (1 \to 1) \cap (0 \to 0)$, we get $\Gamma_1^P \vdash P : 1$ by rules ($\to$I),($\cap$I), ($\leq$).

Case $P = P'P''$. Then $\Gamma_1^{P'} \vdash P' : 1 \leq 1 \to 1$ and $\Gamma_1^{P''} \vdash P'' : 1$, by the induction hypothesis (i). Hence $\Gamma_1^P \vdash P : 1$ by rules (*weakening*), ($\leq$) and ($\to$E).

(ii) *Case* $P \equiv x$. Trivial.

Case $P \equiv \lambda z.P'$. Since $x \in FV(P)$, it is also true that $x \in FV(P')$. Then

$$\Gamma_1^{P'} \uplus \{x : 0\} \vdash P' : 0,$$

by the induction hypothesis (ii). By rules ($\to$I), ($\leq$) with $1 \to 0 = 0$ we get $\Gamma_1^P \uplus \{x : 0\} \vdash P : 0$.

Case $P \equiv P'P''$. We have two subcases.

Subcase $x \in FV(P')$. By the induction hypothesis (ii) it follows that

$$\Gamma_1^{P'} \uplus \{x : 0\} \vdash P' : 0 = 1 \to 0.$$

By the induction hypothesis (i) $\Gamma_1^{P''} \vdash P'' : 1$. Hence by (*weakening*) and ($\to$E) we conclude $\Gamma_1^P \uplus \{x : 0\} \vdash P : 0$.

Subcase $x \in FV(P'')$. Again $\Gamma_1^{P''} \uplus \{x : 0\} \vdash P'' : 0$. By the induction hypothesis (i) we have $\Gamma_1^{P'} \vdash P' : 1 \leq 0 \to 0$. Hence by (*weakening*) and ($\to$E) we conclude $\Gamma_1^P \uplus \{x : 0\} \vdash P : 0$. ■

Now we can characterize the set $\Lambda_{\downarrow\Lambda^{\mathsf{I}}}$ of terms that reduce to λI-terms in a type-theoretic way.

17.4.9 Theorem *Let M be a lambda term. Then*

$$M \in \Lambda_{\downarrow\Lambda^{\mathsf{I}}} \Leftrightarrow \Gamma_1^M \vdash_\cap^{\mathrm{HR}} M : 1.$$

Proof By Theorem 17.3.17 the types of M are all and only the types of its approximants. By Lemmas 17.4.7 and 17.4.8(i) an approximate normal form is a $\lambda\mathsf{I}$-term iff it has the type 1 in HR from the basis which gives type 1 to all its free variables. We conclude observing that if $\square_\Phi(M)$ is a $\lambda\mathsf{I}$-term, then M is a $\lambda\mathsf{I}$-term as well. ∎

Theorem 17.4.9 was first proved in Honsell and Ronchi Della Rocca (1992) by purely semantic means.

The following results are translations of the results in Theorems 17.4.3, 17.4.9 and 17.2.15 to filter models.

17.4.10 Proposition *Let $M \in \Lambda^{\emptyset}$ be a closed lambda term. Then*

(i) *M has a normal form* $\Leftrightarrow [\![M]\!]^{\mathcal{D}_\infty^{\mathrm{CDZ}}} \sqsupseteq 1$.
(ii) *M is solvable* $\Leftrightarrow [\![M]\!]^{\mathcal{D}_\infty^{\mathrm{DHM}}} \sqsupseteq 1$.
(iii) *M reduces to a $\lambda\mathsf{I}$-term* $\Leftrightarrow [\![M]\!]^{\mathcal{D}_\infty^{\mathrm{HR}}} \sqsupseteq 1$.
(iv) *M is strongly normalizing* $\Leftrightarrow [\![M]\!]^{\mathcal{F}^{\mathrm{CDV}}} \neq \emptyset$.

Let $M \in \Lambda$ be an (open) lambda term. Then

(v) *M reduces to a closed term* $\Leftrightarrow [\![M]\!]_\rho^{\mathcal{D}_\infty^{\mathrm{Park}}} \sqsupseteq 0$, *for all ρ.*

Proof (i)–(ii) We explain the situation for (i), the other case being similar.

$$\begin{aligned}
M \text{ has a normal form} &\Leftrightarrow \vdash_\cap^{\mathrm{CDZ}} M : 1, \quad \text{by Theorem 17.2.15(i)},\\
&\Leftrightarrow [\![M]\!]^{\mathcal{F}^{\mathrm{CDZ}}} \ni 1, \quad \text{by 16.2.7},\\
&\Leftrightarrow [\![M]\!]^{\mathcal{F}^{\mathrm{CDZ}}} \sqsupseteq \uparrow 1, \quad \text{by the definition of filters},\\
&\Leftrightarrow [\![M]\!]^{\mathcal{D}_\infty^{\mathrm{CDZ}}} \sqsupseteq \hat{f}(\uparrow 1) = f(1), \text{ by Theorem 16.3.23},\\
&\Leftrightarrow [\![M]\!]^{\mathcal{D}_\infty^{\mathrm{CDZ}}} \sqsupseteq \Phi_{0,\infty}(1), \quad \text{since } 1 \in \mathcal{D}_0,\\
&\Leftrightarrow [\![M]\!]^{\mathcal{D}_\infty^{\mathrm{CDZ}}} \sqsupseteq 1,
\end{aligned}$$

by the identification of $\mathcal{D}_0$ as a subset of $\mathcal{D}_\infty$.

(iii) Similar, using Theorem 17.4.9.

(iv) As (i), but simpler as the step towards $\mathcal{D}_\infty$ is not made. Note that $\mathcal{F}^{\mathrm{CDV}}$ gives rise to a $\lambda\mathsf{I}$-model, not to a λ-model.

(v) Similar, using Theorem 17.4.3. ∎

We do not have a characterization such as (i) in the proposition for $\mathcal{D}_\infty^{\mathrm{Scott}}$, since it was shown in Wadsworth (1976) that there is a closed term J without a normal form such that $\mathcal{D}_\infty^{\mathrm{Scott}} \models \mathsf{I} = \mathsf{J}$.

17.5 Undecidability of inhabitation

In this section we consider type theories with infinitely many type atoms, as described in Section 13.1. To fix ideas, we are concerned here with the theory $\mathcal{T}$ = CDV. Since we do not consider other type theories, in this section the symbols $\vdash$ and $\leq$ stand for $\vdash_{\cap}^{\text{CDV}}$ and $\leq_{\text{CDV}}$, respectively. Moreover $\mathbb{T} = \mathbb{T}^{\text{CDV}}$.

We investigate the *inhabitation problem* for this type theory, which is to determine, for a given type A, if there exists a closed term of type A (an inhabitant). In symbols, the problem can be presented as follows:

$$\vdash ? : A$$

A slightly more general variant of the problem is the inhabitation problem *relativized* to a given context Γ:

$$\Gamma \vdash ? : A$$

It is, however, not difficult to show that these two problems are equivalent.

17.5.1 Lemma *Let $\Gamma = \{x_1{:}A_1, \ldots, x_n{:}A_n\}$. Then the following are equivalent.*

(1) *There exists a term $M \in \Lambda$ such that $\Gamma \vdash M : A$.*
(2) *There exists a term $N \in \Lambda$ such that $\vdash N : A_1 \to \cdots \to A_n \to A$.*

Proof (1) $\Rightarrow$ (2) Define $N \equiv \lambda x_1 \cdots x_n . M$. Apply the rule ($\to$I) n times.

(2) $\Rightarrow$ (1) Take $M \equiv N x_1 \cdots x_n$ and apply the rules ($\to$E) and (*weakening*) n times. ■

The main result (Theorem 17.5.31) of the present section is that type inhabitation is undecidable for $\mathcal{T}$ = CDV. Compare this to Remark 2.4.16 and Statman (1979a), stating that for simple types the problem is decidable in polynomial space.

By Theorem 17.2.15 and Corollary 14.2.3 we need to consider only inhabitants in normal form. The main idea of the undecidability proof is based on the following observation. The process of solving an instance of the inhabitation problem can be seen as a certain (solitary) game of building trees. In this way, one can obtain a combinatorial representation of the computational contents of the inhabitation problem (for a restricted class of types). We call this model a 'tree game'. In order to win a tree game, the player may be forced to execute a computation of a particular automaton ('typewriter automaton'). Thus, the global strategy of the proof is as follows. We make the following abbreviations.

17.5.2 Definition We introduce the following decision problems.

EQA	:	Emptiness Problem for Queue Automata;
ETW	:	Emptiness Problem for Typewriter Automata;
WTG	:	Problem of determining whether one can Win a Tree Game;
IHP	:	Inhabitation Problem in $\lambda_\cap^{\text{CDV}}$.

These are problems in the following sense. In each case there is a set involved: the set of natural number codes for the description of the automata or types involved; the problem consists in determining whether a candidate element of the set actually belongs to it. It is well known that EQA is undecidable; see, for example, Kozen (1997). Let $P_1, P_2 \subseteq \omega$, seen as problems on (types coded as) natural numbers. The notion $P_1 \leq_m P_2$ denotes many–one reducibility of problem P_1 to problem P_2. It is defined by

$$P_1 \leq_m P_2 \stackrel{\Delta}{\Longleftrightarrow} \forall n \in \omega.[P_1(n) \Leftrightarrow P_2(f(n))],$$
$$\text{for some computable function } f : \omega \to \omega.$$

Suppose $P_1 \leq_m P_2$. This implies that if P_2 is decidable, then so is P_1. Or taking the contrapositive, if P_1 is undecidable, then so is P_2. The following inequalities will show that IHP is undecidable.

EQA	$\leq_m$	ETW	(Lemma 17.5.26);
ETW	$\leq_m$	WTG	(Proposition 17.5.30);
WTG	$\leq_m$	IHP	(Corollary 17.5.24).

Basic properties

We begin with some basic observations concerning the relation $\leq$.

17.5.3 Lemma *Let $n > 0$.*

(i) *Let $\alpha \in \mathbb{A}_\infty$, and let none of the $A_1, \dots, A_n \in \mathbb{T}$ be an intersection. Then*

$$A_1 \cap \cdots \cap A_n \leq \alpha \Rightarrow \exists i.\alpha \equiv A_i.$$

(ii) *Let $\alpha_1, \dots, \alpha_n \in \mathbb{A}_\infty$ and $A \in \mathbb{T}$ not be an intersection. Then*

$$\alpha_1 \cap \cdots \cap \alpha_n \leq A \Rightarrow \exists i.A \equiv \alpha_i.$$

(iii) *Let $\alpha_1, \dots, \alpha_n \in \mathbb{A}_\infty$ and $A \in \mathbb{T}$. Then*

$$\alpha_1 \cap \cdots \cap \alpha_n \leq A \Rightarrow A \equiv \alpha_{i_1} \cap \cdots \cap \alpha_{i_k},$$

for some $k \geq 0$ and $1 \leq i_1 < \cdots < i_k \leq n$.

Proof (i), (ii) See Exercise 17.18.

(iii) Let $A \equiv B_1 \cap \cdots \cap B_k$ with $k > 0$ and let the B_j not be intersections. Then for each j one has $\alpha_1 \cap \cdots \cap \alpha_n \leq B_j$ and can apply (ii) to show that $B_j \equiv \alpha_{i_j}$. ∎

17.5.4 Lemma *Let $A_1, \ldots, A_n, B \in \mathbb{T}$ and $\alpha_1, \ldots, \alpha_n, \beta \in \mathbb{A}_\infty$, with $n > 0$. Then*

$$(A_1 \to \alpha_1) \cap \cdots \cap (A_n \to \alpha_n) \leq (B \to \beta) \Rightarrow \exists i\, [\beta \equiv \alpha_i \;\&\; B \leq A_i].$$

Proof By Theorem 14.1.7 the type theory CDV is $\boldsymbol{\beta}$-sound. Hence by the assumption it follows that $B \leq (A_{i_1} \cap \cdots \cap A_{i_k})$ and $(\alpha_{i_1} \cap \cdots \cap \alpha_{i_k}) \leq \beta$. By Lemma 17.5.3(ii) one has $\beta \equiv \alpha_{i_p}$, for some $1 \leq p \leq k$, and the conclusion follows. ∎

17.5.5 Lemma *If $\Gamma \vdash \lambda x.M : A$ then $A \notin \mathbb{A}_\infty$.*

Proof Suppose $\Gamma \vdash \lambda x.M : \alpha$. By Lemma 14.1.1(iii) it follows that there are $n > 0$ and $B_1, \ldots, B_n, C_1, \ldots, C_n$ such that $\Gamma, x : B_i \vdash M : C_i$, for $1 \leq i \leq n$, and $(B_1 \to C_1) \cap \cdots \cap (B_n \to C_n) \leq \alpha$. This is impossible by Lemma 17.5.3(i). ∎

Game contexts

In order to prove that a general decision problem is undecidable, it is enough to identify a 'sufficiently difficult' fragment of the problem and prove undecidability of that fragment. Such an approach is often useful. This is because restricting the consideration to specific instances may simplify the analysis of the problem. Of course the choice should be done in such a way that the 'core' of the problem remains within the selected special case. This is the strategy we are applying for our inhabitation problem. Namely, we restrict our analysis to the following special case of relativized inhabitation:

$$\Gamma \vdash ? : \alpha,$$

where α is a type atom, and Γ is a 'game context', the notion of game context being defined as follows.

17.5.6 Definition

(i) If $\mathcal{X}, \mathcal{Y} \subseteq \mathbb{T}$ are sets of types, then

$$\mathcal{X} \to \mathcal{Y} = \{X \to Y \mid X \in \mathcal{X},\ Y \in \mathcal{Y}\}.$$
$$\mathcal{X} \cap \mathcal{Y} = \{X \cap Y \mid X \in \mathcal{X},\ Y \in \mathcal{Y}\}.$$
$$\mathcal{X}^\cap = \{A_1 \cap \cdots \cap A_n \mid n \geq 1 \;\&\; A_1, \ldots, A_n \in \mathcal{X}\}.$$

If $A \equiv A_1 \cap \cdots \cap A_n$, and each A_i is not an intersection, then the A_i are called the *components* of A.

(ii) We consider the following sets of types:

(1) $\mathcal{A} = \mathbb{A}_\infty^\cap$;
(2) $\mathcal{B} = (\mathbb{A}_\infty \to \mathbb{A}_\infty)^\cap$;
(3) $\mathcal{C} = (\mathcal{D} \to \mathbb{A}_\infty)^\cap$, where
(4) $\mathcal{D} = (\mathcal{B} \to \mathbb{A}_\infty) \sqcap (\mathcal{B} \to \mathbb{A}_\infty)$.

(iii) Types in $\mathcal{A} \cup \mathcal{B} \cup \mathcal{C}$ are called *game types*.
(iv) A type context Γ is a *game context* if all types in Γ are game types.

We show some properties of type judgements involving game types.

17.5.7 Lemma *For a game context* Γ *the following three properties hold:*

(i) $\Gamma \vdash xM : A \;\Rightarrow\; A \in \mathcal{A} \;\&\; \exists n > 0.[\Gamma(x) \equiv (E_1 \to \alpha_1) \cap \cdots \cap (E_n \to \alpha_n) \;\&$
$\exists k > 0\, \exists i_1, \ldots, i_k.[1 \leq i_1 < \cdots < i_k \leq n \;\&$
$A \equiv \alpha_{i_1} \cap \cdots \cap \alpha_{i_k} \;\&\; \Gamma \vdash M : (E_{i_1} \cap \cdots \cap E_{i_k})]]$.

(ii) $\Gamma \vdash xM : \alpha \;\Rightarrow\; \exists n > 0.[\Gamma(x) \equiv (E_1 \to \alpha_1) \cap \cdots \cap (E_n \to \alpha_n) \;\&$
$\exists i.[1 \leq i \leq n \;\&\; \alpha \equiv \alpha_i \;\&\; \Gamma \vdash M : E_i]]$.

(iii) $\Gamma \not\vdash xMN : A$.

Proof (i) Suppose $\Gamma \vdash xM : A$. By Lemma 14.1.9(ii) we have for some type B that $\Gamma \vdash x : (B \to A)$ and $\Gamma \vdash M : B$. Then $\Gamma(x) \leq B \to A$, using Lemma 14.1.1(i). By Lemma 17.5.3(ii), this cannot happen if $\Gamma(x)$ is in $\mathcal{A}$. Thus, since $\Gamma(x)$ is a game type, it has the form $(E_1 \to \alpha_1) \cap \cdots \cap (E_n \to \alpha_n)$. Then,

$$(E_1 \to \alpha_1) \cap \cdots \cap (E_n \to \alpha_n) \equiv \Gamma(x) \leq (B \to A),$$

Since CDV is $\boldsymbol{\beta}$-sound, using Lemma 17.5.3(iii), we have $B \leq E_{i_1} \cap \cdots \cap E_{i_k}$ and $\alpha_{i_1} \cap \cdots \cap \alpha_{i_k} \equiv A$, for some $k > 0$ and i_j such that $1 \leq i_j \leq n$.

(ii) By (i) and Lemma 17.5.3(ii).

(iii) By (i), using the fact that $B \to A \neq \alpha_{i_1} \cap \cdots \cap \alpha_{i_k}$, by Lemma 17.5.3(ii). ∎

17.5.8 Lemma *If* A *is a game type and* $D \in \mathcal{D}$, *then* $A \not\leq D$.

Proof Suppose $A \leq D \leq (B \to \alpha)$, with $B \in \mathcal{B}$.

The case $A \in \mathcal{A}$ is impossible by Lemma 17.5.3(ii).

If $A \in \mathcal{B}$, then $(\alpha_1 \to \beta_1) \cap \cdots \cap (\alpha_n \to \beta_n) \leq B \to \alpha$ and hence $B \leq \alpha_i$ for some i, by Lemma 17.5.4. Using Lemma 17.5.3(i) this is also impossible. If $A \in \mathcal{C}$, then $(D_1 \to \beta_1) \cap \cdots \cap (D_n \to \beta_n) \leq B \to \alpha$ and hence $B \leq D_i \in \mathcal{D}$ for some i, by Lemma 17.5.4. We have already shown that this is impossible. ∎

For game contexts the Inversion Lemma 14.1.9 can be extended as follows.

17.5.9 Lemma *Let* Γ *be a game context, and let* M *be in normal form.*

(i) *If* $\Gamma \vdash M : (B_1 \to \alpha_1) \cap (B_2 \to \alpha_2) \in \mathcal{D}$, *with* $B_i \in \mathcal{B}$, *then* $M \equiv \lambda y.N$, *and* $\Gamma, y{:}B_i \vdash N : \alpha_i$ *for* $i = 1, 2$.

(ii) *If* $\Gamma \vdash M : \alpha$, *with* $\alpha \in \mathbb{A}_\infty$, *then there are two mutually exclusive possible subcases.*

- *M is a variable z and $\Gamma(z)$ is in $\mathcal{A}$, where α is one of the components.*
- *$M \equiv xN$, where $\Gamma(x) \equiv (E_1 \to \beta_1) \cap \cdots \cap (E_n \to \beta_n)$, with $\alpha \equiv \beta_i$ and $\Gamma \vdash N : E_i$, for some $1 \leq i \leq n$.*

Proof (i) Note first that by Lemma 17.5.7 the term M cannot be an application. If it is a variable x, then $\Gamma(x) \leq (B_1 \to \alpha_1) \cap (B_2 \to \alpha_2)$, by Lemma 14.1.1(i). This contradicts Lemma 17.5.8, because $\Gamma(x)$ is a game type. It follows that $M = \lambda y.N$ and $\Gamma \vdash \lambda y.N : (B_i \to \alpha_i)$. Then for $i = 1, 2$ one has $\Gamma, y{:}B_i \vdash N_i : \alpha_i$, by Lemma 14.1.9(iii).

(ii) Observe that M is not an abstraction, by Lemma 17.5.5. If $M \equiv z$, then $\Gamma(z) \leq \alpha$ and $\Gamma(z)$ is a game type, i.e. in $\mathcal{A} \cup \mathcal{B} \cup \mathcal{C}$. By Lemma 17.5.3(i) one has $\Gamma(z) \in \mathcal{A}$, with α as one of the components.

If M is an application, i.e., $M \equiv zM_1 \cdots M_m$, with $m > 0$, write

$$\Gamma(z) = (E_1 \to \beta_1) \cap \cdots \cap (E_n \to \beta_n).$$

As it is a game type we have $\Gamma(z) \notin \mathcal{A}$, by Theorem 14.1.9(ii). Then $M \equiv zN$, by Lemma 17.5.7(iii). By (ii) of the same lemma one has $\alpha \equiv \beta_i$ and $\Gamma \vdash N : E_i$ for some i. ∎

Tree games

In order to show the undecidability of inhabitation for CDV we will introduce a certain class of *tree games*. Within a game one can have a particular '*play*'. (Chess is a game, but the match between Kasparov and Karpov consisted of several plays). These form an intermediate step in our construction. The idea of a tree game is to represent, in an abstract way, the crucial combinatorial behavior of proof search in CDV. We focus on establishing WTG $\leq_m$ IHP.

17.5.10 Definition Let Σ be a finite alphabet; its elements are called *labels*.

(1) A *local move* (over Σ) is a finite non-empty set B of pairs of labels.

(2) A *global move* (over Σ) is a finite non-empty set C of triples of the form

$$\langle \langle X, b \rangle, \langle Y, c \rangle, d \rangle,$$

where $b, c, d \in \Sigma$ and X, Y are local moves.

(3) A *tree game* (over Σ) is a triple of the form

$$G = \langle a, A, \{C_1, \ldots, C_n\} \rangle,$$

where $a \in \Sigma$, $A \subseteq \Sigma$ and $C_1, \ldots, C_n$ are global moves. We call a the *initial label* and A the set of *final labels.*

Before we explain the rules of the game, we give an interpretation of the constituents of the tree games in terms of types.

17.5.11 Definition Let Σ be a finite subset of $\mathbb{A}_\infty$, the infinite set of type atoms, and let G be a tree game over Σ. Moves of G, and the set of final labels, can be interpreted as types of CDV as follows.

(1) If $A = \{a_1, \ldots, a_n\}$, then $\tilde{A} = a_1 \cap \cdots \cap a_n$.
(2) If $B = \{\langle a_1, b_1 \rangle, \ldots, \langle a_n, b_n \rangle\}$, then $\tilde{B} = (a_1 \to b_1) \cap \cdots \cap (a_n \to b_n)$.
(3) If $C = \{\langle \langle B_1, b_1 \rangle, \langle B_1', b_1' \rangle, c_1 \rangle, \ldots, \langle \langle B_n, b_n \rangle, \langle B_n', b_n' \rangle, c_n \rangle\}$, then

$$\tilde{C} = (((\tilde{B}_1 \to b_1)\cap(\tilde{B}_1' \to b_1')) \to c_1)\cap\cdots\cap(((\tilde{B}_n \to b_n)\cap(\tilde{B}_n' \to b_n')) \to c_n).$$

Note that $\tilde{A} \in \mathcal{A}, \tilde{B} \in \mathcal{B}$ and $\tilde{C} \in \mathcal{C}$.

A tree game is a *solitary game*, i.e. there is only one player. Starting from an initial position, the player can non-deterministically choose a sequence of moves, and wins if (s)he can manage to reach a final position. Every position (configuration) of the game is a finite labelled tree, and at every step the depth of the tree is increasing.

17.5.12 Definition Let $G = \langle a, A, \{C_1, \ldots, C_n\}\rangle$ be a tree game over Σ. A *position* T of G is a finite labelled tree, see Fig. 17.2, satisfying the following conditions:

- the root is labelled by the initial symbol a;
- every node has at most two children;
- nodes at the same level (the same distance from the root) have the same number of children (in particular all leaves are at the same level);
- all nodes are labeled by elements of Σ;
- in addition, if a node v has two children v' and v'', then the branches $\langle v, v' \rangle$ and $\langle v, v'' \rangle$ are labeled by local moves.

17.5.13 Definition Let $G = \langle a, A, \{C_1, \ldots, C_n\}\rangle$ be a tree game.

(1) The *initial position* of G is the tree with a unique node labeled a.
(2) A position T is *winning* if all labels of the leaves of T are in A.

17.5.14 Definition Let T be a position in a game $G = \langle a, A, \{C_1, \ldots, C_n\}\rangle$, and let v be a leaf of T. Let k be such that all nodes in T at level $k-1$ have two children as shown in Fig. 17.3. There is a node u at level $k-1$ which is an ancestor of v; one of the children of u, say u', is also an ancestor of

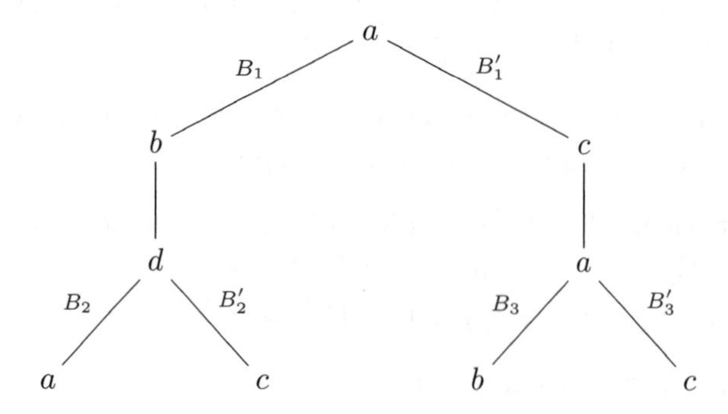

Figure 17.2 An example position

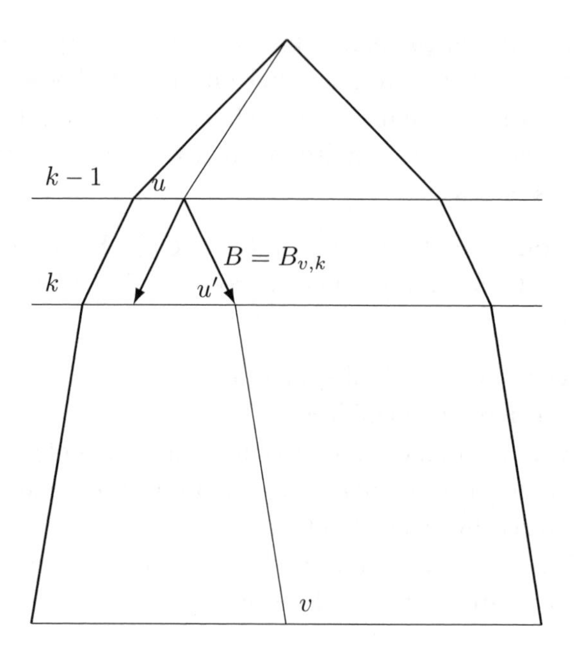

Figure 17.3 Local move X associated to node v

v (possibly improper, i.e., it may happen that $u' = v$). Assume that B is the label of the branch $\langle u, u' \rangle$. Then we say that B is the *kth local move associated to* v, and we write $B = B_{v,k}$.

Now we can finally describe the *rules of the game*.

17.5.15 Definition

(i) Let $G = \langle a, A, \{C_1, \dots, C_n\} \rangle$ and let T be a (current) position in G. There are two ways of obtaining a *next position*.

(1) The player can perform a 'global' step, by first selecting one of the global moves C_i and then performing the following actions for each leaf v of T:

(a) choose a triple $\langle \langle B, b \rangle, \langle B', b' \rangle, c \rangle \in C_i$ such that c is the label of v;

(b) create two children of v, say v' and v'', labelled b and b', respectively;

(c) label the branch $\langle v, v' \rangle$ by B and the branch $\langle v, v'' \rangle$ by B'.

The step is only allowed if the appropriate actions can be performed at every leaf: otherwise the resulting tree is not a position.

(2) The player can also perform a 'local' step. This begins with a choice of a level $k > 0$ of T such that each node at level $k - 1$ has two children. Then, for each leaf v of T, the player executes the following actions:

(a) choose a pair $\langle a, b \rangle \in B_{v,k}$ such that b is the label of v;

(b) create a single child of v, labeled a.

Again the step is only allowed if the appropriate actions can be performed at every leaf: otherwise the resulting tree is not a position.

(ii) If a position T' is reachable from T with help of one global step C_i, we write

$$T \Rightarrow^{C_i} T'.$$

If T' is obtained from T by a local step defined via level k, we write

$$T \Rightarrow^{k} T'.$$

If T' is reachable from T in one step (global or local), then we write $T \Rightarrow T'$.

(iii) A position T in G is called *favorable* if there is a position T' with

$$T \Rightarrow^* T' \text{ and } T' \text{ is winning,}$$

where $\Rightarrow^*$ is the reflexive transitive closure of $\Rightarrow$.

(iv) The game G is *solvable* (i.e. the player can win) if the initial position is favorable.

The following example gives an idea of the tree games and, moreover, is important for our principal construction.

17.5.16 Example Consider the *tree game* $G_0 = \langle 1, \{c\}, \{C_1, C_2\} \rangle$, over the alphabet $\Sigma = \{1, 2, a, b, c\}$, where

$$C_1 = \{\langle \langle \{\langle a, a\rangle\}, 1\rangle, \langle \{\langle b, a\rangle\}, 2\rangle, 1\rangle, \langle \langle \{\langle a, b\rangle\}, 1\rangle, \langle \{\langle b, b\rangle\}, 2\rangle, 2\rangle\};$$
$$C_2 = \{\langle \langle \{\langle c, a\rangle\}, a\rangle, \langle \{\langle c, a\rangle\}, a\rangle, 1\rangle, \langle \langle \{\langle c, b\rangle\}, a\rangle, \langle \{\langle c, b\rangle\}, a\rangle, 2\rangle\}.$$

Fig. 17.4 demonstrates a possible winning position T of the game. Note that this position can actually be reached from the initial one in six steps, so that the player can win G_0. These six steps are:

$$T_0 \Rightarrow^{C_1} T_1 \Rightarrow^{C_1} T_2 \Rightarrow^{C_2} T_3 \Rightarrow^{1} T_4 \Rightarrow^{2} T_5 \Rightarrow^{3} T_6 = T,$$

where each T_i is of depth i. The reader should observe that every sequence of steps leading from T_0 to a winning position must obey a similar pattern:

$$T_0 \Rightarrow^{C_1} T_1 \Rightarrow^{C_1} \cdots \Rightarrow^{C_1} T_{n-1} \Rightarrow^{C_2} T_n \Rightarrow^{1} T_{n+1} \Rightarrow^{2} \cdots \Rightarrow^{n} T_{2n},$$

where T_{2n} is winning. Thus, the game must consist of two phases: first a number of applications of C_1 followed by a single application of C_2; and then a sequence of steps using only local moves. What is important in our example is that the order of local steps is fully determined. Indeed, at the position T_{n+k-1} the only action possible is '$\Rightarrow^k$'. That is, one must apply the kth local moves associated to the leaves (the moves labeling branches at depth k). This is forced by the distribution of symbols a, b at depth $n+k-1$.

Let us emphasize a few properties of our games. First, a game is a non-deterministic process, and there are various sequences of steps possible. We can have winning sequences (reaching a winning position), and infinitely long sequences, but also 'deadlocks' when no rule is applicable. Note that there are various levels of non-determinism here: we can choose between the C_i and ks and then between various elements of the chosen set C_i (respectively $B_{v,k}$). It is an important property of the game that the actions performed at various leaves during a local step may be different, as different moves $B_{v,k}$ were 'declared' before at the corresponding branches of the tree.

We now explain the relationship between tree games and term search in CDV. Since we deal with intersection types, it is not unexpected that we need sometimes to require one term to have many types, possibly within different contexts. This leads to the following definition.

17.5.17 Definition Let $k, n > 0$. Let $\vec{A}_1, \ldots, \vec{A}_n$ be n sequences of k types:

$$\vec{A}_i = A_{i1}, \ldots, A_{ik}.$$

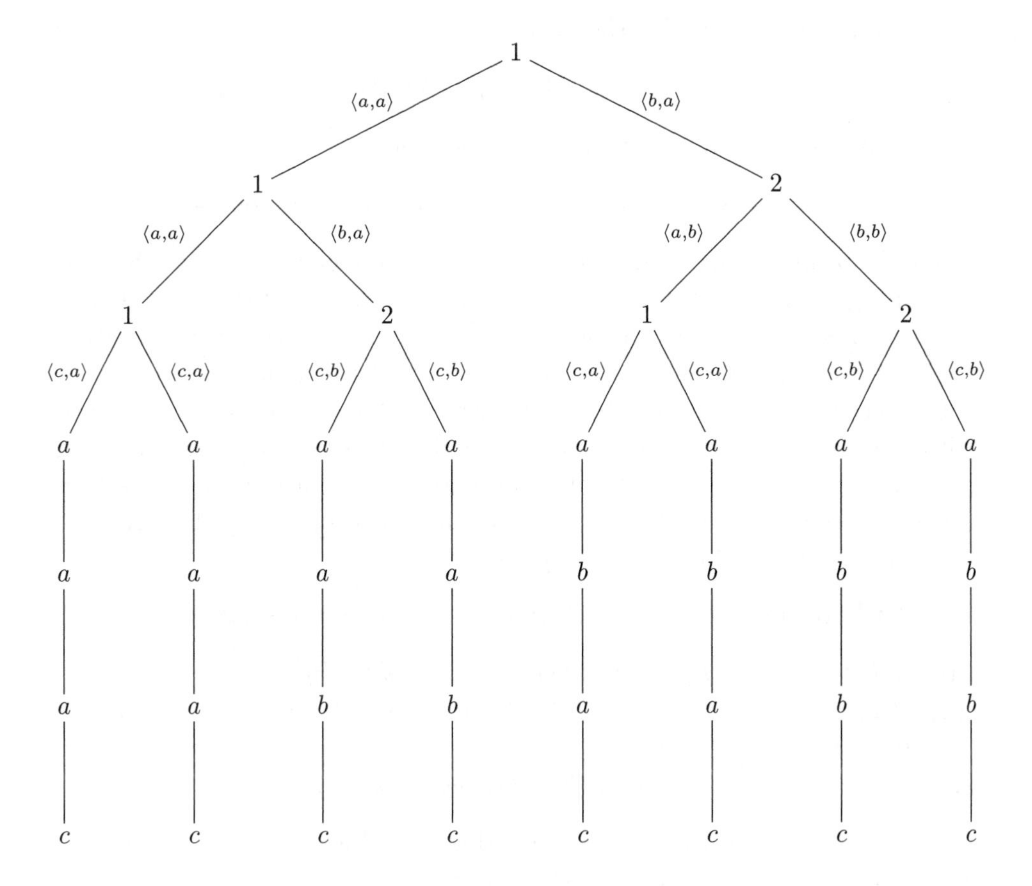

Figure 17.4 A winning position in G_0 of Example 17.5.16.

Let $\Gamma_i = \{x_1{:}A_{i1}, \ldots, x_k{:}A_{ik}\}$ and let $\alpha_i \in \mathbb{A}_\infty$, for $1 \leq i \leq n$. An instance of a *generalized inhabitation problem* (*gip*) is a finite set of pairs $P = \{\langle\, \Gamma_i, \alpha_i \,\rangle \mid i = 1, \ldots, n\}$, where each Γ_i and α_i are as above. A *solution* of P is a term M such that $\Gamma_i \vdash M : \alpha_i$ holds for each i. Then we say that M *solves* P. This is equivalent to requiring that

$$\vdash \lambda\vec{x}.M : (\vec{A}_1 \to \alpha_1) \cap \cdots \cap (\vec{A}_n \to \alpha_n).$$

17.5.18 Definition Let $G = \langle\, a, A, \{C_1, \ldots, C_n\} \,\rangle$ be a tree game and let T be a position in G.

(i) Define $\Gamma_G = \{x_0{:}\tilde{A}, x_1{:}\tilde{C}_1, \ldots, x_n{:}\tilde{C}_n\}$.

(ii) Let J be the set of all numbers k such that every node in T at level $k-1$ has two children. We associate a new variable y_k to each $k \in J$.

Define for a leaf v of T the basis

$$\Gamma_v = \{y_k{:}\tilde{B}_{v,k} \mid k \in J\}.$$

(iii) Define the generalized inhabitation problem

$$P_T^G = \{\langle \Gamma_G \cup \Gamma_v, a_v\rangle \mid v \text{ is a leaf of } T \text{ with label } a_v\}.$$

The following lemma states the exact correspondence between inhabitation and games. Let us first make one comment that perhaps may help to avoid confusion. We deal here with quite a restricted form of inhabitation problem (only game contexts), which implies that the lambda terms we are constructing have a 'linear' shape (Exercise 17.20). Thus we have to deal with trees not because of the shape of lambda terms (as often happens with proof search algorithms) but solely because of the nature of intersection types and the need to solve various inhabitation problems uniformly (i.e., to solve a *generalized inhabitation problem*).

17.5.19 Lemma *Let $G = \langle a, A, \{C_1, \dots, C_n\}\rangle$ be a tree game.*

(i) *If T is a winning position of G, then x_0 solves P_T^G.*
(ii) *If $T_1 \Rightarrow^{C_i} T_2$ and N solves $P_{T_2}^G$, then $x_i(\lambda y_k.N)$ solves $P_{T_1}^G$, where k is the depth of T_2 and $i > 0$.*
(iii) *If $T_1 \Rightarrow^k T_2$ and N solves $P_{T_2}^G$, then $y_k N$ solves $P_{T_1}^G$.*

Proof (i) Since T is winning, we know that $a_v \in A$ for each leaf v. It follows $x_0{:}\tilde{A} \vdash x_0 : a_v$ and therefore $\Gamma_G \cup \Gamma_v \vdash x_0 : a_v$.

(ii) Since $T_1 \Rightarrow^{C_i} T_2$, it follows that each leaf v of T_1 has two children v' and v'' in T_2 with the branches $\langle v, v'\rangle$ and $\langle v, v''\rangle$ labeled by B' and B'' respectively, such that $\langle\langle B', a_{v'}\rangle, \langle B'', a_{v''}\rangle, a_v\rangle \in C_i$. Let k =level(v). As N solves $P_{T_2}^G$ one has

$$\begin{array}{rcl}\Gamma_G \cup \Gamma_v, y_k{:}\tilde{B}' & \vdash & N : a_{v'};\\ \Gamma_G \cup \Gamma_v, y_k{:}\tilde{B}'' & \vdash & N : a_{v''}.\end{array}$$

So $\Gamma_G \cup \Gamma_v \vdash \lambda y_k.N : (\tilde{B}' \to a_{v'}) \cap (\tilde{B}'' \to a_{v''})$. Hence $\Gamma_G \cup \Gamma_v \vdash x_i(\lambda y_k.N) : a_v$.

(iii) Since $T_1 \Rightarrow^k T_2$, we know each leaf v of T_1 has a child v' in T_2 such that $\langle a_{v'}, a_v\rangle \in B_{v,k}$. Then $y_k{:}\tilde{B}_{v,k} \vdash y_k : a'_v \to a_v$ and $\Gamma_G \cup \Gamma_v \vdash N : a_{v'}$, by assumption. Therefore

$$\Gamma_G \cup \Gamma_v \vdash y_k N : a_v. \blacksquare$$

17.5.20 Corollary *Let G be a tree-game. For positions T one has*

$$T \textit{ is favorable} \Rightarrow P_T^G \textit{ has a solution.}$$

Proof By induction on the number of steps needed to reach a winning position and the lemma. ∎

For the converse we need the following result.

17.5.21 Lemma *Let T_1 be a position in a tree game G and let M be a solution in $\boldsymbol{\beta}$-nf of $P^G_{T_1}$. Then we have one of the following cases.*

(1) $M \equiv x_0$ and T_1 is winning.
(2) $M \equiv y_k N$ and N is the solution of $P^G_{T_2}$, for some T_2 with $T_1 \Rightarrow^k T_2$.
(3) $M \equiv x_i(\lambda y_k.N)$ and N is the solution of $P^G_{T_2}$, for some T_2 with $T_1 \Rightarrow^{C_i} T_2$.

Proof *Case $M \equiv z$.* As M is a solution of $P^G_{T_1}$, one has

$$\Delta_v = \Gamma_G \cup \Gamma_v \vdash z : a_v,$$

for all leaves v of T_1. Then by Lemma 14.1.1(i) one has $\Delta_v(z) \leq a_v$. Hence by Lemma 17.5.3(i) $a_v \in A$ and $z = x_0$. Therefore T_1 is winning.

Case M is an application. Then for all leaves v of T_1

$$\Delta_v = \Gamma_G \cup \Gamma_v \vdash M : a_v.$$

Lemma 17.5.9(ii) implies $M \equiv zN$ and $\Delta_v(z) \equiv (E_1 \to \beta_1)\cap\cdots\cap(E_n \to \beta_n)$, with $\Delta_v \vdash N : E_j$ and $a_v = \beta_j$, for some j. Now choose a leaf v of T_1. As Δ_v is a game context there are only two possibilities:

$$E_j \in \mathbb{A}_\infty \text{ or } E_j \in \mathcal{D}.$$

Subcase $E_j \in \mathbb{A}_\infty$. Let $E_j \equiv \alpha_j$. Now $z \notin \mathrm{dom}(\Gamma_G)$, hence $z \in \mathrm{dom}(\Gamma_v)$ and so

$$z{:}(\alpha_1{\to}\beta_1) \cap \cdots \cap (\alpha_n{\to}\beta_n)$$

is $y_k : \tilde{B}_{v,k}$, for some k. Also for each leaf w of T_1 one has $z \in \mathrm{dom}(\Gamma_w)$ and $z{:}\Gamma_w(z)$ is $y_k{:}\tilde{B}_{w,k}$. Define $T_1 \Rightarrow^k T_2$ by giving each leaf v of T_1 with label β_j a child v' with label α_j. We have $\Delta_v \vdash N : \alpha_j$ and $y_k{:}\tilde{B}_{v,k} \vdash y_k : \alpha_j{\to}\beta_j$. Hence $\Delta_{v'} \vdash M : a_{v'}$.

Subcase $E_j \in \mathcal{D}$. Then $\Delta_v(z) \equiv (E_1{\to}\beta_1)\cap\cdots\cap(E_n{\to}\beta_n)$. Hence $z : \Delta_v(z)$ is $x_i{:}\tilde{C}_i$ for some i. So $z \in \mathrm{dom}(\Gamma_G)$ and therefore $\Delta_w(z) = \Delta_v(z)$ for all leaves w of T_1. Let $C_i = \{\ldots, \langle\langle B_j, \alpha_j\rangle, \langle B'_j \alpha'_j\rangle, \beta_j\rangle, \ldots\}$ and $E_j = (\tilde{B}_j{\to}\alpha_j)\cap(\tilde{B}'_j{\to}\alpha'_j)$. Define the move $T_1 \Rightarrow^{C_i} T_2$ as follows. Give each leaf v with label β_j two children v' and v'' with labels α_j and α'_j, respectively. Label the branches with B_j and B'_j, respectively. Let $k = \mathrm{depth}(T_1)$. One has $\Delta_v \vdash N : E_j$, hence by Lemma 17.5.9(i) one has $N \equiv \lambda y_k.N'$, with

$$\Delta_v, y_k{:}\tilde{B}_j \vdash N'{:}\alpha_j \ \&\ \Delta_v, y_k{:}\tilde{B}'_j \vdash N' : \alpha'_j.$$

Therefore $\Delta_{v'} \vdash N' : a_{v'}$ and $\Delta_{v''} \vdash N' : a_{v''}$.

Case M is an abstraction. This is impossible, by Lemma 17.5.5. ∎

17.5.22 Corollary *Let G be a tree game. For positions T one has*

$$T \text{ is favorable} \Leftrightarrow P_T^G \text{ has a solution.}$$

Proof (⇒) This was Corollary 17.5.20.

(⇐) Let M be a solution of P_T^G. By Theorem 17.2.15(ii) one may assume that M is in normal form. The conclusion follows from the previous lemma by induction on the size of M. ∎

17.5.23 Theorem *Let G be a tree game with initial position* $\{a\}$*. Then*

$$G \text{ is solvable} \Leftrightarrow P_{\{a\}}^G \text{ has a solution.}$$

Proof Immediate from the previous corollary. ∎

17.5.24 Corollary WTG $\leq_m$ IHP, *i.e. winning a tree-game is many–one reducible to the inhabitation problem.*

Proof By the theorem, since $P_{T_0}^G = P_a^G$ is an inhabitation problem. ∎

Typewriters

In order to simplify our construction we introduce the auxiliary notion of a typewriter automaton. Informally, a typewriter automaton is just a reusable finite-state transducer. At each step, it reads a symbol, replaces it with a new one and changes the internal state. But at the end of the word, our automaton moves its reading and printing head back to the beginning of the tape and continues. This goes on until a final state is reached. That is, a typewriter automaton is a special kind of a linear-bounded automaton, see Kozen (1997).

17.5.25 Definition

(i) A (deterministic) *typewriter automaton* $\mathcal{A}$ is a tuple of the form

$$\mathcal{A} = \langle \Sigma, Q, q_0, F, \varrho \rangle,$$

where Σ is a finite alphabet, Q is a finite set of states, $q_0 \in Q$ is an initial state and $F \subseteq Q$ is a set of final states. The last component is a transition function

$$\varrho : (Q - F) \times (\Sigma \cup \{\epsilon\}) \to Q \times (\Sigma \cup \{\epsilon\}),$$

which must satisfy the following condition: whenever $\varrho(q, a) = (p, b)$, then either $a, b \in \Sigma$ or $a = b = \epsilon$.

(ii) A *configuration* (*instantaneous description*, of $\mathcal{A}$ is represented by a triple $\langle w, q, v \rangle$, where (as usual) $wv \in \Sigma^*$ is the tape contents, $q \in Q$ is the current state, and the machine head points at the first symbol of v.

(iii) The *next instantaneous description function* $\overline{\varrho}$ is defined as follows:

- $\overline{\varrho}(\langle w, q, av \rangle) = \langle wb, p, v \rangle$, if $a \neq \epsilon$ and $\varrho(q, a) = (p, b)$;
- $\overline{\varrho}(\langle w, q, \epsilon \rangle) = \langle \epsilon, p, w \rangle$, if $\varrho(q, \epsilon) = (p, \epsilon)$.

(iv) The language $L^{\mathcal{A}}$ accepted by $\mathcal{A}$ is the set of all $w \in \Sigma^*$, such that $\overline{\varrho}^k(\langle \epsilon, q_0, w \rangle) = \langle u, q, v \rangle$, for some k and $q \in F$, and $uv \in \Sigma^*$.

Recall from Definition 17.5.2 that ETW is the emptiness problem for typewriter automata and EQA is the emptiness problem for queue automata. The latter is undecidable, see Kozen (1997). We need the following.

17.5.26 Lemma EQA $\leq_m$ ETW.

Proof Exercise 17.23. ∎

It follows that also ETW is undecidable.

Our goal is now to represent typewriters as games, in order to establish ETW $\leq_m$ WTG. We begin with a refinement of Example 17.5.16. In what follows, triples of the form $\langle \langle B_1, b \rangle, \langle B_2, c \rangle, d \rangle$ will be represented graphically as

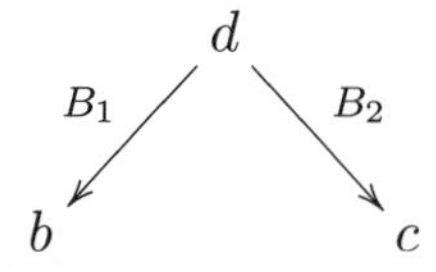

in order to enhance readability.

17.5.27 Definition The alphabet Σ_1 is the following Cartesian product.

$$\Sigma_1 = \{\bot, a, b\} \times \{\bot, 1, 2\} \times \{\text{loc}, \text{glo}\}.$$

We define a *tree game* $G_1 = \langle \langle \bot, 1, \text{glo} \rangle, \Sigma_1, \{C_1, C_2, C_3\} \rangle$. The set of accepting labels is Σ_1, because we are interested in all possible 'plays' (i.e. instances) of the game. The moves are defined as

(i) *global move* C_1 consists of the following two triples:

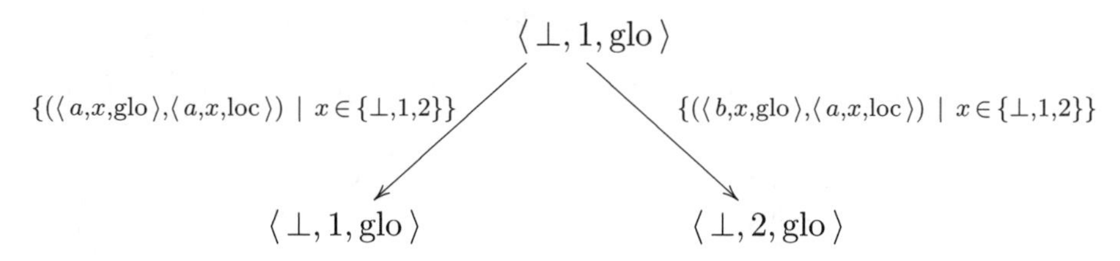

and

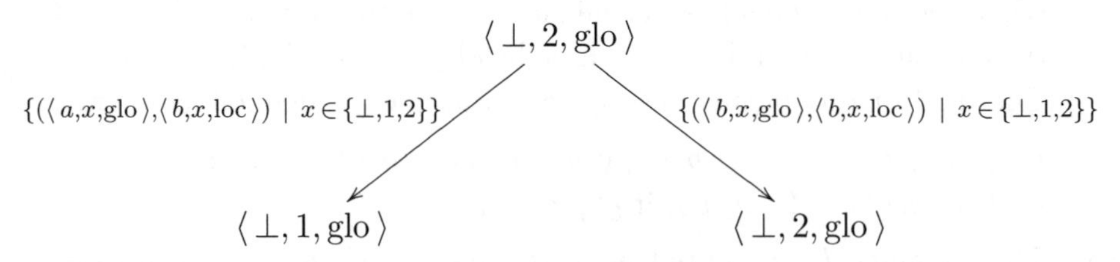

Before we define other global moves, let us point out that C_1 is very similar to rule C_1 of the game G_0 in Example 17.5.16 (observe the a and b in the first component and 1 and 2 in the second one).

(ii) *global move* C_2 consists again of two triples:

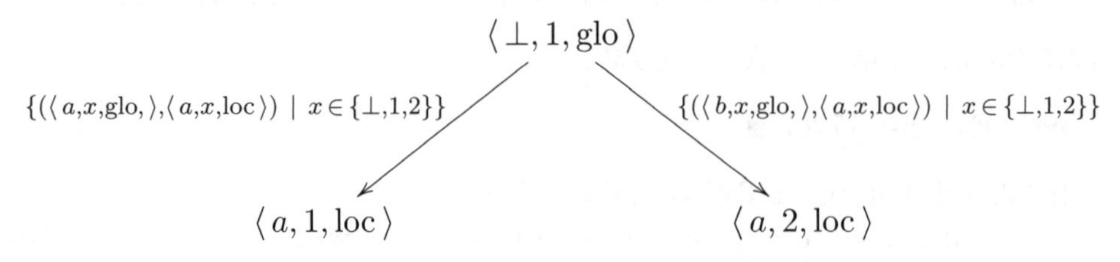

and

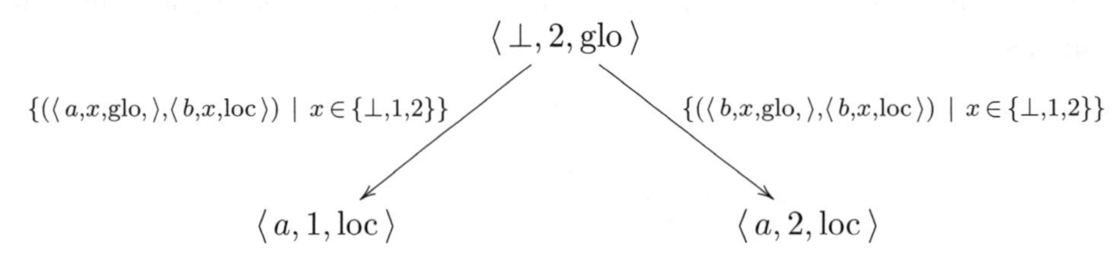

(iii) *global move* C_3 consists of the triples (where $z \in \{a, b\}$, i.e., $z \neq \perp$)

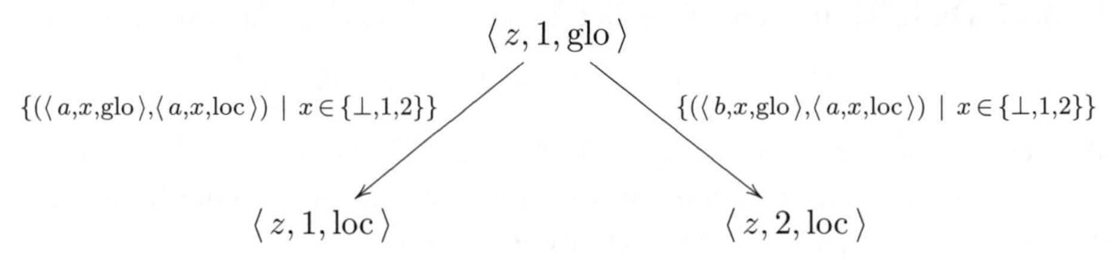

and

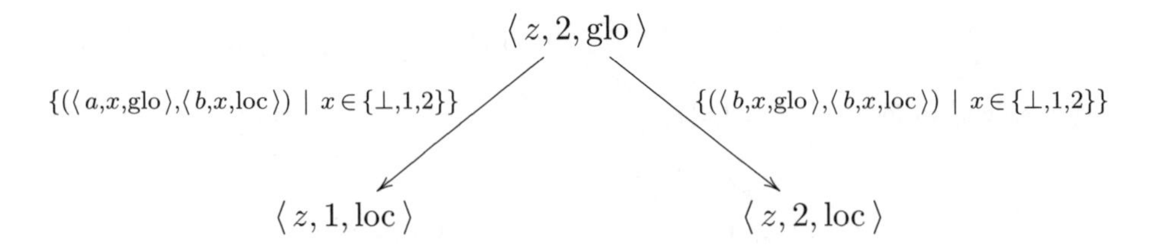

17.5.28 Lemma *Every play of G_1 must have the following sequence of moves:*

$$\begin{array}{cccccccccccc}
 & C_1 & & C_1 & & C_1 & & C_1 & \ldots & & C_1 & & C_2 \\
1 & C_3 & 2 & C_3 & 3 & C_3 & 4 & C_3 & \ldots & m & C_3 & m+1 & C_3 \\
m+2 & C_3 & m+3 & C_3 & m+4 & C_3 & .. & .. & \ldots & .. & .. & & ..
\end{array}$$

That is, the game starts with m times a C_1 move (possibly $m = 0$ or $m = \infty$), followed by a single C_2 move. Then the local and global C_3 moves alternate. It is convenient to see a play as divided into phases. Each phase consists of $m{+}1$ 'rounds'; in the first phase a 'round' is a single global move, in the other phases it consists of two steps: one local and one global. The local move declared at any global step is executed $m{+}1$ phases later, after exactly $2m{+}1$ steps.

Proof Exercise 17.22. ■

17.5.29 Definition Let G_1 be as above and let $\mathcal{A} = \langle \Sigma^{\mathcal{A}}, Q, q_0, F^{\mathcal{A}}, \varrho \rangle$ be a typewriter automaton. We define a game $G^{\mathcal{A}}$ as follows.

(i) The alphabet of $G^{\mathcal{A}}$ is the Cartesian product $\Sigma_1 \times Q \times (\Sigma^{\mathcal{A}} \cup \{\bot, \epsilon\})$.

(ii) For each local move B of G_1 and each $\beta \in \Sigma^{\mathcal{A}} \cup \{\epsilon\}$, we define a local move $B_\beta = \{(\langle a, q, \beta \rangle, \langle b, q, \bot \rangle) \mid q \in Q \text{ and } \langle a, b \rangle \in B\}$.

(iii) If $\Delta = \langle \langle B, b \rangle, \langle B', b' \rangle, c \rangle \in C_1 \cup C_2$, then we define Δ_β as the triple

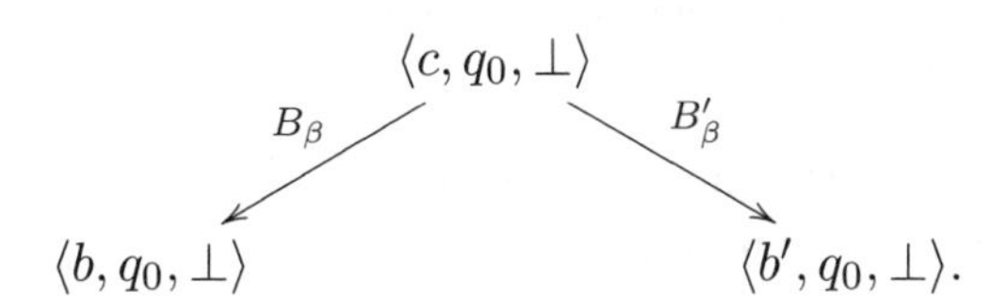

(iv) For each triple $\Delta = \langle \langle B, b \rangle, \langle B', b' \rangle, c \rangle \in C_3$ we define a set $C^{\mathcal{A}}_\Delta$ consisting of all triples of the form

$$\langle c, q, \alpha \rangle \quad \xrightarrow{B_\beta} \langle b, p, \bot \rangle, \qquad \xrightarrow{B'_\beta} \langle b', p, \bot \rangle,$$

where $\varrho(q, \alpha) = (p, \beta)$.

(v) Define

$$\begin{array}{rcll}
C^{\mathcal{A}}_1(\beta) & = & \{\Delta_\beta \mid \Delta \in C_1\}, & \text{for } \beta \in \Sigma^{\mathcal{A}}; \\
C^{\mathcal{A}}_2 & = & \{\Delta_\epsilon \mid \Delta \in C_2\}; & \\
C^{\mathcal{A}}_3 & = & \bigcup\{C^{\mathcal{A}}_\Delta \mid \Delta \in C_3\}. &
\end{array}$$

(vi) The initial symbol of $G^{\mathcal{A}}$ is $a = \langle a_1, q_0, \bot \rangle$, where $a_1 = \langle \bot, 1, G \rangle$, the initial symbol of G_1.

(vii) The set of final symbols is $A = \Sigma_1 \times (\Sigma^{\mathcal{A}} \cup \{\bot, \epsilon\}) \times F^{\mathcal{A}}$.

(viii) Finally, we take $G^{\mathcal{A}} = \langle a, A, \{C_1^{\mathcal{A}}(\beta) \mid \beta \in \Sigma^{\mathcal{A}}\} \cup \{C_2^{\mathcal{A}}, C_3^{\mathcal{A}}\} \rangle$.

17.5.30 Proposition *Let $\mathcal{A}$ be a typewriter that accepts the language $L_{\mathcal{A}}$. Then*

$$L_{\mathcal{A}} \neq \emptyset \Leftrightarrow G^{\mathcal{A}} \textit{ is solvable.}$$

Hence ETW $\leq_m$ WTG.

Proof Our game $G^{\mathcal{A}}$ behaves as a 'Cartesian product' of G_1 and $\mathcal{A}$. Informally speaking, there is no communication between the first component and the other two. In particular we have the following.

(1) Lemma 17.5.28 remains true with G_1 replaced by $G^{\mathcal{A}}$ and C_1, C_2, C_3 replaced respectively by $C_1^{\mathcal{A}}(\beta)$, $C_2^{\mathcal{A}}$, $C_3^{\mathcal{A}}$. That is, a legitimate play of $G^{\mathcal{A}}$ must look like

$$\begin{array}{cccccccccc} & C_1^{\mathcal{A}}(\beta_1) & & C_1^{\mathcal{A}}(\beta_2) & \ldots & & C_1^{\mathcal{A}}(\beta_m) & & C_2^{\mathcal{A}} \\ 1 & C_3^{\mathcal{A}} & 2 & C_3^{\mathcal{A}} & \ldots & m & C_3^{\mathcal{A}} & m+1 & C_3^{\mathcal{A}} \\ m+2 & C_3^{\mathcal{A}} & m+3 & C_3^{\mathcal{A}} & \ldots & & & & \end{array}$$

(2) If a position T of $G^{\mathcal{A}}$ can be reached from the initial position, then the second and third components of labels are always the same for all nodes at every fixed level of T.

Consider a play of the game $G^{\mathcal{A}}$ as in (1) above. Note that this sequence is fully determined by the choice of m and $\beta_1, \ldots, \beta_m$. Also observe that β_i, for $i = 1, \ldots, m$, are the third components of the labels of all leaves of T_{m+2i}. Let w denote the word $\beta_1\beta_2\cdots\beta_m$ and $O^{\mathcal{A}}(w)$ the 'opening' $C_1^{\mathcal{A}}(\beta_1), \ldots, C_1^{\mathcal{A}}(\beta_m)$ in the game $G^{\mathcal{A}}$.

Claim. *Typewriter $\mathcal{A}$ accepts w iff $O^{\mathcal{A}}(w)$ leads to a winning position in $G^{\mathcal{A}}$.*

We shall now prove the claim. Let β_j^k denote the symbol contained in the jth cell of the tape of $\mathcal{A}$, after the machine has completed $k-1$ full phases (the tape has been fully scanned $k-1$ times). That is, β_j^k is the symbol to be read during the kth phase. Of course β_j^1 is β_j. For uniformity write $\beta_{m+1}^k = \epsilon$. Further, let q_j^k be the internal state of the machine, just before it reads the jth cell for the kth time (i.e., after $k-1$ full phases). The reader will easily show that for all k and all $j = 1, \ldots, m+1$ the following statements hold:

(i) the third component of labels of all leaves of $T_{(2k-1)(m+1)+2j-1}$ is β_j^k;
(ii) the second component of labels of all leaves of $T_{(2k-1)(m+1)+2j-2}$ is q_j^k.

In particular, the second and third component of these labels are solely determined by the depth of the appropriate node. That is, the computation-related information is the same in every node at any given level; the tree shape is only needed to ensure the pattern of alternating global and local moves.

To have a closer look at the simulation, assume that $\varrho(\beta, q) = (\alpha, p)$. Then a global move changes labels of the form $(\cdot, \beta, q)$ (where we ignore the first coordinate) into $(\cdot, \bot, p)$ and creates a local move containing pairs of the form $(\cdot, \alpha, q') \to (\cdot, \bot, q')$. The role of the local move is to pass the information about α (the symbol to be written in place of β) to the next phase of the game.

The next local move brings the information from the previous phase about the symbol in the next tape cell, and then we have again a global move, and so on. Consider for example an initial word $\beta_1\beta_2\beta_3$ and let the automaton take on the following sequence of configurations:

$$
\begin{aligned}
&(\varepsilon, q_0, \beta_1\beta_2\beta_3) \to (\gamma_1, q_1, \beta_2\beta_3) \to (\gamma_1\gamma_2, q_2, \beta_3) \to (\gamma_1\gamma_2\gamma_3, q_3, \varepsilon) \to \\
&\quad \to (\varepsilon, p_0, \gamma_1\gamma_2\gamma_3) \to (\delta_1, p_1, \gamma_2\gamma_3) \to (\delta_1\delta_2, p_2, \gamma_3) \to (\delta_1\delta_2\delta_3, p_3, \varepsilon) \to \\
&\qquad \to (\varepsilon, r_0, \delta_1\delta_2\delta_3) \to \cdots
\end{aligned}
$$

This corresponds to a play with these labels on the consecutive levels of the tree: $(\cdot, \bot, q_0) \xrightarrow{\beta_1} (\cdot, \bot, q_0) \xrightarrow{\beta_2} (\cdot, \bot, q_0) \xrightarrow{\beta_3} (\cdot, \bot, q_0) \xrightarrow{\varepsilon}$ $(\cdot, \beta_1, q_0) \xrightarrow{\gamma_1} (\cdot, \bot, q_1) \to (\cdot, \beta_2, q_1) \xrightarrow{\gamma_2} (\cdot, \bot, q_2) \to (\cdot, \beta_3, q_2) \xrightarrow{\gamma_3} (\cdot, \bot, q_3) \to$ $(\cdot, \varepsilon, q_3) \xrightarrow{\varepsilon} (\cdot, \bot, p_0) \to (\cdot, \gamma_1, p_0) \xrightarrow{\delta_1} (\cdot, \bot, p_1) \to (\cdot, \gamma_2, p_1) \xrightarrow{\delta_2} (\cdot, \bot, p_2) \to$ $(\cdot, \gamma_3, p_2) \xrightarrow{\delta_3} (\cdot, \bot, p_3) \to (\cdot, \varepsilon, p_3) \xrightarrow{\varepsilon} (\cdot, \bot, r_0) \to (\cdot, \delta_1, r_0) \longrightarrow \cdots$.

Fig. 17.5 illustrates the induction hypothesis by showing labels of a node at depth $(2k-1)(m+1)+2j-2$ together with her daughter and grandchildren. The claim follows when (ii) is applied to the final states.

It follows immediately from 1 and the claim that $L_\mathcal{A} \neq \emptyset$ iff there is a strategy for winning $G^\mathcal{A}$. Hence the emptiness problem for typewriter automata can be reduced to the problem of winning tree games. ∎

17.5.31 Theorem *The inhabitation problem for* $\lambda_\cap^{\mathrm{CDV}}$ *is undecidable.*

Proof By Corollary 17.5.24, Lemma 17.5.26 and Proposition 17.5.30. ∎

17.5.32 Corollary *The inhabitation problem for* $\lambda_\cap^{\mathrm{BCD}}$ *is undecidable.*

Proof Do Exercise 17.26. ∎

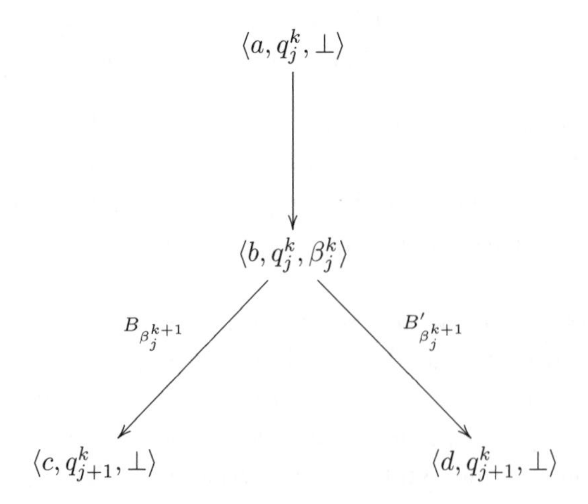

Figure 17.5 Simulation of a single machine step

Remarks

The proof of undecidability of the inhabitation problem presented in this section is a modified version of the original proof in Urzyczyn (1999).

The following notion of rank has been given for intersection types by Leivant (1983a); we denote it by *i-rank* in order to distinguish it from the rank in Definition 1.1.22.

17.5.33 Definition

$$
\begin{array}{rcll}
\text{i-rank}(A) & = & 0, & \text{for simple types } A; \\
\text{i-rank}(A \cap B) & = & \max(1, \text{i-rank}(A), \text{i-rank}(B)); & \\
\text{i-rank}(A \to B) & = & \max(1 + \text{i-rank}(A), \text{i-rank}(B)), & \text{if } \cap \text{ occurs in } A \to B.
\end{array}
$$

It should be observed that all types in game contexts are of i-rank at most 3. Thus, the relativized inhabitation problem is undecidable for contexts of i-rank 3, and the inhabitation problem (with empty contexts) is undecidable for types of i-rank 4. It is decidable for i-rank 3, see Urzyczyn (2009). Decidability for i-rank 2 is done in Exercise 17.24. Some other decidable cases are discussed in Kurata and Takahashi (1995).

From the point of view of the *formulas-as-types* principle, (the 'Curry–Howard isomorphism'), the inhabitation problem should correspond to a provability problem for a certain logic. It is however not at all obvious what should be the logic corresponding to intersection types. We deal here with a 'proof–functional' rather than 'truth–functional' connective $\cap$, which is called sometimes 'strong conjunction': a proof of $A \cap B$ must *be* a single proof of both A and B, rather than merely *contain* two separate proofs of A and B. See Lopez-Escobar (1983), Mints (1989), and Alessi and Barbanera (1991) for the discussion of strong conjunction logics.

Various authors defined Church-style calculi for intersection types, which is another way leading to understanding the logic of intersection types. We refer the reader to Venneri (1994), Dezani-Ciancaglini et al. (1997), Wells et al. (1997), Capitani et al. (2001), Ronchi Della Rocca (2002), and Pimentel et al. (2012), Liquori and Ronchi Della Rocca (2007), Bono et al. (2008) for this approach.

Undecidability of inhabitation in $\boldsymbol{\lambda}_\cap^{\mathrm{CDV}}$ vs. $\boldsymbol{\lambda}$-definability in $\mathcal{M}_X$ During the completion of this book, Salvati (2009) showed that the undecidability of inhabitation follows directly from the undecidability of λ-definability in full type structures over a finite set, Theorem 4.1.25. The main idea is a simple embedding of the elements of $\mathcal{M}_X$ into the intersection types in $\mathbb{T}_\cap^X$.

17.5.34 Definition Let X be a finite set. Consider the full type structure $\mathcal{M}_X$ over X; see Definition 2.4.18 Elements of $\mathcal{M}_X = \bigcup_{A\in\mathbb{T}^0_\to} X(A)$ can be encoded as elements of $\mathbb{T}_\cap^X$. To avoid confusion between simple and intersection types, we write $A, B, \ldots \in \mathbb{T}^0_\to$ and $\sigma, \xi, \ldots \in \mathbb{T}_\cap^X$. This helps to disambiguate $\vdash$ as being either $\vdash^{\mathrm{Cu}}_{\boldsymbol{\lambda}^0_\to}$ or $\vdash^{\mathrm{CDV}}_{\boldsymbol{\lambda}_\cap}$. For $d\in X(A)$ define $\xi_d \in \mathbb{T}_\cap^X$ by induction on the structure of $A\in\mathbb{T}^0_\to$.

$$\begin{array}{rcll} \xi_d &=& d, & \text{if } A = 0; \\ \xi_d &=& \bigcap\limits_{e\in X(B)} (\xi_e \to \xi_{de}), & \text{if } A = B{\to}C. \end{array}$$

17.5.35 Lemma *Let $d\in X(A{\to}B), e\in X(A)$. Then we have:*

(i) $\Gamma \vdash M : \xi_d \;\&\; \Gamma \vdash N : \xi_e \;\Rightarrow\; \Gamma \vdash (MN) : \xi_{de}$;
(ii) $\Gamma \vdash (\lambda x.N) : \xi_d \;\Leftrightarrow\; \forall e\in X(A)\,[\Gamma, x : \xi_e \vdash N : \xi_{de}]$.

Proof (i) Since $\xi_d \leq \xi_e \to \xi_{de}$.

(ii) ($\Rightarrow$) By (i) and the subject reduction property for $\vdash$, since it is $\vdash_\cap^{\mathrm{CDV}}$. ($\Leftarrow$) By the rules ($\to$I) and ($\cap$I). ∎

17.5.36 Lemma

(i) *Let $\alpha\in X$ be a type atom and $\sigma\in\mathbb{T}_\cap^X$. Then*

$$\alpha \leq_{\mathrm{CDV}} \sigma \text{ or } \sigma \leq_{\mathrm{CDV}} \alpha \;\Rightarrow\; \sigma \equiv \alpha.$$

(ii) *Let* $d, e \in \mathcal{M}_X$. *Then*

$$\xi_d \leq_{\text{CDV}} \xi_e \Rightarrow d = e.$$

Proof (i) By induction on the derivation of say $\alpha \leq \sigma$ in $\mathcal{T}^{\text{CDV}}$.

(ii) By induction on the joint size of ξ_d and ξ_e. If either of the two types is an atom, the result follows by (i). Otherwise, both types are intersections of arrows and we have $d \in X(A \to B)$ and $e \in X(C \to D)$, for some $A, B, C, D \in \mathbb{T}^0$. Thus for all $c \in X(C)$

$$\xi_d = (\xi_{a_1} \to \xi_{da_1}) \cap \cdots \cap (\xi_{a_k} \to \xi_{da_k}) \leq \xi_c \to \xi_{ec},$$

where $X(A) = \{a_1, \ldots, a_k\}$. By Theorem 14.1.7 ($\boldsymbol{\beta}$-soundness) it follows that

$$\xi_c \leq \xi_{a_{i_1}} \cap \cdots \cap \xi_{a_{i_p}} \quad (\leq a_{i_j}) \tag{17.9}$$

$$\xi_{da_{i_1}} \cap \cdots \cap \xi_{da_{i_p}} \leq \xi_{ec}, \tag{17.10}$$

for a non-empty subset $\{a_{i_1}, \ldots, a_{i_p}\} \subseteq X(A)$. By the induction hypothesis, we have $c = a_{i_1} = \cdots = a_{i_p} = a$, say, and $A = C$. Hence (17.10) boils down to $\xi_{dc} \leq \xi_{ec}$. Again by the induction hypothesis one has $dc = ec$ and $B = D$. As c was arbitrary, it follows that the functions d and e are equal. ∎

17.5.37 Definition An X-*basis* is of the form

$$\Gamma = \{x_1 : \xi_{d_1}, \ldots, x_n : \xi_{d_n}\},$$

with $d_1, \ldots, d_n \in \mathcal{M}_X$.

17.5.38 Lemma *Define the sets* $\mathcal{N}, \mathcal{V} \subseteq \Lambda$ *as the smallest one such that*

$$N_1, \ldots, N_n \in \mathcal{N},\ n \geq 0 \Rightarrow xN_1 \ldots N_n \in \mathcal{V}$$
$$M \in \mathcal{V} \Rightarrow M \in \mathcal{N}$$
$$M \in \mathcal{N} \Rightarrow (\lambda x.M) \in \mathcal{N}.$$

Then

$$\mathcal{N} = \{M \in \Lambda \mid M \textit{ in } \boldsymbol{\beta}\textit{-nf}\}$$
$$\mathcal{V} = \{M \in \mathcal{N} \mid M \equiv x\vec{N}, \textit{ for some } \vec{N} = N_1 \ldots N_n\}.$$

17.5.39 Lemma *Consider two predicates* P, Q *on* Λ*:*

$P(M)$:= *for all* X*-bases* Γ *and* $d_1, d_2 \in X(A)$ *one has*

$$\Gamma \vdash M : \xi_d \ \&\ \Gamma \vdash M : \xi_{d'} \Rightarrow d = d'.$$

$Q(M)$:= M *is of the form* $M \equiv xN_1 \cdots N_n$, $n \geq 0$, *and for all* X*-bases*

Γ and $\sigma \in \mathbb{T}_\cap^X$ one has

$$\Gamma \vdash M : \sigma \;\Rightarrow\; \exists e_1, \dots, e_n \in \mathcal{M}_X.[\Gamma \vdash N_i : \xi_{e_i}, 1 \leq i \leq n,$$
$$\& \; \Gamma \vdash M : \xi_{ce_1 \dots e_n} \leq \sigma], \text{ where } \Gamma(x) = \xi_c.$$

Then for all $M \in \Lambda$ one has

$$M \in \mathcal{N} \;\Rightarrow\; P(M);$$
$$M \in \mathcal{V} \;\Rightarrow\; Q(M).$$

Proof By 'induction on the simultaneous inductive definition' of $\mathcal{V}$ and $\mathcal{N}$ in Lemma 17.5.38. It suffices to show

$$P(N_1), \dots, P(N_n) \;\Rightarrow\; Q(xN_1 \dots N_n) \tag{17.11}$$
$$P(N_1), \dots, P(N_n) \;\Rightarrow\; P(xN_1 \dots N_n) \tag{17.12}$$
$$P(M) \;\Rightarrow\; P(\lambda x.M). \tag{17.13}$$

To prove (17.11), suppose $P(N_1), \dots, P(N_n)$ towards $Q(xN_1 \dots N_n)$. We do this by induction on n. For $n = 0$ the assumption is $\Gamma \vdash x : \sigma$. By Theorem 14.1.1(i) one has $\Gamma(x) \leq \sigma$, i.e. $\Gamma \vdash x : \xi_c \leq \sigma$, with $\Gamma(x) = \xi_c$. Therefore $Q(x)$. Now suppose $P(N_1), \dots, P(N_{n+1})$ towards $Q(xN_1 \dots N_{n+1})$. Assume $\Gamma \vdash xN_1 \dots N_{n+1} : \sigma$. Then by Theorem 14.1.9(ii) one has

$$\exists \nu \in \mathbb{T}_\cap^X.[\Gamma \vdash xN_1 \cdots N_n : \nu \to \sigma \;\&\; \Gamma \vdash N_{n+1} : \nu].$$

Then $Q(xN_1 \cdots N_n)$ by the induction hypothesis, so for some $e_1, \dots, e_n \in \mathcal{M}_X$

$$\Gamma \vdash N_i : \xi_{e_i} (1 \leq i \leq n) \;\&\; \Gamma \vdash xN_1 \cdots N_n : \xi_{ce_1 \cdots e_n} \leq \nu \to \sigma.$$

Write $b = ce_1 \cdots e_n$. By Lemma 17.5.35(i) the intersection type ξ_b cannot be a type atom (in X), hence writing $X(\nu) = \{a_1, \dots, a_k\}$ we have

$$\xi_b = (\xi_{a_1} \to \xi_{ba_1}) \cap \cdots \cap (\xi_{a_k} \to \xi_{ba_k}) \leq \nu \to \sigma.$$

By Theorem 14.1.7 one has, for some non-empty $\{i_1, \dots, i_p\} \subseteq \{1, \dots, k\}$,

$$\begin{aligned} \nu \;&\leq\; \xi_{a_{i_1}} \cap \cdots \cap \xi_{a_{i_p}} \\ & \quad\; \xi_{ba_{i_1}} \cap \cdots \cap \xi_{ba_{i_p}} \;\leq \sigma. \end{aligned}$$

Since $P(N_{n+1})$ and $\Gamma \vdash N_{n+1} : \nu \leq \xi_{a_i}$ we have $a_{i_1} = \cdots = a_{i_k} = a$, say. Noting that

$$\Gamma \vdash xN_1 \cdots N_n : \xi_b \leq \xi_a \to \xi_{ba}$$
$$\Gamma \vdash N_{k+1} : \nu \leq \xi_a,$$

taking $e_{n+1} = a$ it follows that

$$\Gamma \vdash xN_1 \cdots N_{n+1} : \xi_{ba} = \xi_{ce_1 \cdots e_{n+1}} \leq \sigma.$$

Therefore indeed $Q(xN_1 \cdots N_{n+1})$.

As to (17.12), writing $M \equiv xN_1 \cdots N_n$ suppose $P(N_i)$ towards $P(M)$. Assume

$$\Gamma \vdash M : \xi_d \;\&\; \Gamma \vdash M : \xi_{d'}$$

towards $d = d'$. As $Q(M)$, by (17.11), we have

$$\exists e_1, \ldots, e_n.\Gamma \vdash M : \xi_{ce_1\cdots e_n} \leq \xi_d \;\&\; \Gamma \vdash N_i : \xi_{e_i},$$
$$\exists e'_1, \ldots, e'_n.\Gamma \vdash M : \xi_{ce'_1\cdots e'_n} \leq \xi_{d'} \;\&\; \Gamma \vdash N_i : \xi_{e'_i}.$$

Since $P(N_i)$ we have $e_i = e'_i$. Therefore by Lemma 17.5.35(ii) we indeed have $d = c\vec{e} = c\vec{e'} = d'$.

As to (17.13), suppose $P(M)$ towards $P(\lambda x.M)$. Assume $\Gamma \vdash \lambda x.M : \xi_d$ and $\Gamma \vdash \lambda x.M : \xi_e$, in order to show $d = e$. By Theorem 14.1.1(iii) and Lemma 17.5.36(ii) it follows that $d \in X(A \to B)$. So $\xi_d \leq (\xi_c \to \xi_{dc})$ for all $c \in X(A)$. Then, for all $c \in X(A)$,

$$\begin{aligned} \Gamma &\vdash \lambda x.M : \xi_c \to \xi_{dc}, \\ \Gamma, x : \xi_c &\vdash M : \xi_{dc}, \qquad \text{by Theorem 14.1.9(iii)}. \end{aligned}$$

Similarly $\Gamma, x : \xi_c \vdash M : \xi_{ec}$. By $P(M)$ we have $\forall c \in X(A).dc = ec$. Therefore $d = e$. ■

17.5.40 Proposition *Let $A \in \mathbb{T}^0$, $M \in \Lambda_{\to}(A)$, and $d \in X(A)$. For a valuation ρ in $\mathcal{M}_X$, write $\Gamma_\rho(x) = \xi_{\rho(x)}$. Then:*

(i) $[\![M]\!]_\rho = d \Leftrightarrow \Gamma_\rho \restriction \mathrm{FV}(M) \vdash |M| : \xi_d$;

(ii) *If $M \in \Lambda^{\emptyset}_{\to}(A)$, then*

$$[\![M]\!] = d \Leftrightarrow \vdash |M| : \xi_d.$$

Proof (i) ($\Rightarrow$) By induction on M one shows

$$\Gamma_\rho \restriction \mathrm{FV}(M) \vdash |M| : \xi_{[\![M]\!]_\rho}.$$

We write Γ_ρ for $\Gamma_\rho \restriction \mathrm{FV}(M)$.

Case $M \equiv x^A$. Then $\Gamma_\rho \vdash x : \xi_{\rho(X)}$, as $\Gamma(x) = \xi_{\rho(x)}$.

Case $M \equiv NL$. By the induction hypothesis one has

$$\Gamma_\rho \vdash |N| : \xi_{[\![N]\!]_\rho} \;\&\; \Gamma_\rho \vdash |L| : \xi_{[\![L]\!]_\rho}.$$

Therefore, by Lemma 17.5.35(i),

$$\Gamma_\rho \vdash |NL| : \xi_{[\![N]\!]_\rho [\![L]\!]_\rho} = \xi_{[\![NL]\!]_\rho}.$$

Case $M \equiv \lambda x^B.N$. By the induction hypothesis one has for all $b \in X(B)$

$$\Gamma_{\rho[x:=b]} \vdash |N| : \xi_{[\![N]\!]_{\rho[x:=b]}}.$$

Hence

$$\begin{aligned} \Gamma_\rho, x : \xi_b \vdash |N| : \xi_{[\![N]\!]_{\rho[x:=b]}} &\Rightarrow \Gamma_\rho \vdash \lambda x.|N| : \xi_b \to \xi_{[\![N]\!]_{\rho[x:=b]}} \\ &\Rightarrow \Gamma_\rho \vdash |\lambda x^B.N| : \bigcap_{b \in X(B)} .\xi_b \to \xi_{[\![N]\!]_{\rho[x:=b]}} = \xi_{[\![\lambda x^B.N]\!]_\rho}. \end{aligned}$$

($\Leftarrow$) Assume $\Gamma_\rho \vdash |M| : \xi_d$. Then, by normalization and subject reduction,

$$\Gamma_\rho \vdash |M^{\rm nf}| : \xi_d,$$

where $M^{\rm nf}$ is the $\boldsymbol{\beta}$-nf of M. By ($\Rightarrow$) applied to $M^{\rm nf}$ and $d = [\![M^{\rm nf}]\!]_\rho$ one has

$$\Gamma_\rho \vdash |M^{\rm nf}| : \xi_{[\![M^{\rm nf}]\!]_\rho}.$$

By Lemmas 17.5.38 and 17.5.39 one has $P(M^{\rm nf})$. Therefore

$$d = [\![M^{\rm nf}]\!]_\rho = [\![M]\!]_\rho.$$

(ii) By (i). ∎

17.5.41 Lemma *For $d \in X(A)$ with $A \in \mathbb{T}^0$, define $\xi_d^\sim = A$. For an X-basis $\Gamma = \{x_1{:}\xi_{d_1}, \dots, x_n{:}\xi_{d_n}\}$, define $\Gamma^\sim = \{x_1{:}\xi_{d_1}^\sim, \dots, x_n{:}\xi_{d_n}^\sim\}$. Then for $M \in \Lambda$ in $\boldsymbol{\beta}$-nf one has*

$$\Gamma \vdash_{\boldsymbol{\lambda}_\cap} M : \xi_d \Rightarrow \Gamma^\sim \vdash^{\rm Cu}_{\boldsymbol{\lambda}^0_\to} M : \xi_d^\sim.$$

Proof By induction on the generation of M.

Case $M \equiv x$. Then $\Gamma \vdash_{\boldsymbol{\lambda}_\cap} x : \xi_d$ implies $\Gamma(x) \leq \xi_d$, so $\Gamma(x) = \xi_d$, by Lemma 17.5.36(ii). Then $\Gamma^\sim \vdash^{\rm Cu}_{\boldsymbol{\lambda}^0_\to} x : \xi_d^\sim$.

Case $M \equiv \lambda x.N$. Then $\Gamma \vdash_{\boldsymbol{\lambda}_\cap} \lambda x.N : \xi_e = \bigcap_{a \in X(A)}(\xi_a \to \xi_{ea})$, with $e \in X(A \to B)$ implies $\Gamma, x{:}\xi_a \vdash_{\boldsymbol{\lambda}_\cap} N : \xi_{ea}$, for all $a \in X(A)$. Therefore by Lemma 17.5.35(ii) one has

$$\Gamma^\sim \vdash^{\rm Cu}_{\boldsymbol{\lambda}^0_\to} (\lambda x.N) : (A \to B).$$

Case $M \equiv xN_1 \cdots N_n$. Suppose $\Gamma \vdash_{\boldsymbol{\lambda}_\cap} xN_1 \cdots N_n : \xi_d$. Then by Lemmas 17.5.39 and 17.5.36(ii) one has $d = ca_1 \cdots a_n$ with $\Gamma(x) = \xi_c$, and $\Gamma \vdash_{\boldsymbol{\lambda}_\cap} N_i : \xi_{a_i}$, with $a_i \in X(A_i)$, and $c \in X(A_1 \to \cdots \to A_n \to B)$. By the induction hypothesis one has

$$\Gamma^\sim \vdash^{\rm Cu}_{\boldsymbol{\lambda}^0_\to} N_i : A_i \;\&\; \Gamma^\sim \vdash x : (A_1 \to \cdots \to A_n \to B).$$

Therefore

$$\Gamma^{\sim} \vdash^{\mathrm{Cu}}_{\boldsymbol{\lambda}^0_{\to}} xN_1 \cdots N_n : B = \xi^{\sim}_d,$$

as $d = ca_1 \cdots a_n \in X(B)$ by Lemma 17.5.35(i). ∎

17.5.42 Theorem (Salvati) *The undecidability of inhabitation in $\mathcal{T}^{\mathrm{CDV}}$ follows by a reduction from the undecidability of λ-definability, Theorem 4.1.25.*

Proof Given $A \in \mathbb{T}^0$ and $d \in X(A)$ we claim that

$$\exists M \in \Lambda^{\emptyset}_{\to}(A)\ [\![M]\!] = d \Leftrightarrow \exists M \in \Lambda^{\emptyset} \vdash_{\boldsymbol{\lambda}_\cap} M : \xi_d.$$

($\Rightarrow$) By Proposition 17.5.40(ii).

$$\begin{array}{llll}
(\Leftarrow)\ \vdash_{\boldsymbol{\lambda}_\cap} M \in \xi_d & \Rightarrow & \vdash^{\mathrm{Cu}}_{\boldsymbol{\lambda}^0_{\to}} M : A, & \text{by 17.5.41,} \\
& \Rightarrow & M^+ \in \Lambda^{\emptyset}_{\to}(A)\ \&\ |M^+| = M, & \text{by 1.2.19(ii),} \\
& \Rightarrow & [\![M^+]\!] = d, & \text{by 17.5.40(ii).}
\end{array}$$

Therefore d is definable iff ξ_d is inhabited. This yields a reduction of the definability problem to the inhabitation problem. ∎

In Salvati et al. (2012) the equivalence between undecidability of λ-definability of elements in models of $\boldsymbol{\lambda}_{\to}$ over a finite set and that of inhabitation of a type in $\boldsymbol{\lambda}^{\mathrm{CDV}}_\cap$ has been proved.

17.6 Exercises

Exercise 17.1 Show by means of examples that the type theories Plotkin and Engeler are not complete, i.e. we do not have $\Gamma \models^{\mathcal{T}} M : A \Leftrightarrow \Gamma \vdash^{\mathcal{T}}_\cap M : A$, for $\mathcal{T} = $ Plotkin or $\mathcal{T} = $ Engeler.

Exercise 17.2 Show that $\Sigma(\{0\}, \mathrm{CDV})$ is the smallest complete type theory (with respect to the order of Fig. 13.5).

Exercise 17.3 Show that for all $\mathcal{T} \in \mathrm{PTT}$ one has

$$\Gamma \vdash^{\mathcal{T}}_\cap M : A \Rightarrow \Gamma \models^{\mathcal{T}}_\cap M : A.$$

[Hint. Adapt Definition 17.1.1 to proper intersection type theories in the obvious way.]

Exercise 17.4 Let $\mathcal{T} \in \mathrm{TT}^{\text{-}\mathrm{U}}$. Take $\mathcal{F}^{\mathcal{T}} = \langle \mathcal{F}^{\mathcal{T}}, \cdot, [\![\ ]\!]^{\mathcal{F}^{\mathcal{T}}} \rangle$ as in 16.2.1; define the type environment $\xi^{\mathcal{T}} : \mathbb{A}^{\mathcal{T}} \to \mathsf{P}(\mathcal{F}^{\mathcal{T}})$ by

$$\xi^{\mathcal{T}}(\alpha) = \{X \in \mathcal{F}^{\mathcal{T}} \mid \alpha \in X\};$$

and let $[\![\ ]\!]^{\mathcal{T}} : \mathbb{T}^{\mathcal{T}} \to \mathsf{P}(\mathcal{F}^{\mathcal{T}})$ be the map $[\![\ ]\!]^{\mathcal{F}^{\mathcal{T}}}_{\xi^{\mathcal{T}}}$ defined by deleting the first clause from Definition 17.1.1. Show:

(i) $[\![A]\!]^{\mathcal{T}} = \{X \in \mathcal{F}^{\mathcal{T}} \mid A \in X\}$;
(ii) $\mathcal{F}^{\mathcal{T}}$, $\xi^{\mathcal{T}}$ are $\to$-good and preserve $\leq_{\mathcal{T}}$;
(iii) $[\Gamma \models^{\mathcal{T}}_{\cap} M : A \Rightarrow \Gamma \vdash^{\mathcal{T}}_{\cap} M : A] \Leftrightarrow \mathcal{T} \in \mathrm{PTT}$;
(iv) $\Gamma \models^{\mathcal{T}}_{\cap} M : A \Leftrightarrow \Gamma \vdash^{\mathcal{T}}_{\cap} M : A$;

Exercise 17.5 Show that all quasi λ-models and all type environments preserve $\leq_{\mathrm{BCD}}$.

Exercise 17.6 Dezani-Ciancaglini et al. (2005), Tatsuta and Dezani-Ciancaglini (2006). Show the following.

(i) The terms that are in PHN reduce to terms of the form $\lambda\vec{x}.y\vec{M}$ where $y \notin \{\vec{x}\}$. Are these all of them? Is this enough to characterize them?
(ii) The terms that are in PN reduce to terms of the form $\lambda\vec{x}.y\vec{M}$ where $y \notin \{\vec{x}\}$ and $\vec{M} \in \mathsf{N}$. Are these all of them? Is this enough to characterize them?
(iii) The terms that are in PSN strongly reduce to terms of the form $\lambda\vec{x}.y\vec{M}$ where $y \notin \{\vec{x}\}$ and $\vec{M} \in \mathsf{N}$. Are these all of them? Is this enough to characterize them?
(iv) Conclude that $\mathsf{PSN} \subset \mathsf{PN} \subset \mathsf{PHN}$.

Exercise 17.7 Show that PHN is HN-stable and PN is N-stable.

Exercise 17.8 Let $\mathcal{M}_{\Lambda(\boldsymbol{\beta})} = \langle \Lambda, \cdot, [\![\ ]\!]^{\Lambda} \rangle$ be the term model of $\boldsymbol{\beta}$-equality and $[M]$ the equivalence class of M under $\boldsymbol{\beta}$-equality. Let a term M be persistently head normalizing if $M\vec{N}$ has a head normal form for all terms $\vec{N}$ (see Definition 17.2.3). Define the type environment

$$\xi_{\mathrm{Scott}}(0) = \{[M] \mid M \text{ is persistently head normalizing}\}.$$

Prove that $(\mathcal{M}_{\Lambda(\boldsymbol{\beta})}, \xi_{\mathrm{Scott}})$ preserves $\leq_{\mathrm{Scott}}$.

Exercise 17.9 Let $\mathcal{M}_{\Lambda(\boldsymbol{\beta})}$ and $[M]$ be as in Exercise 17.8. Define the type environment

$$\xi_{\mathrm{Park}}(0) = \{[M] \mid M \text{ reduces to a closed term}\}.$$

Prove that $(\mathcal{M}_{\Lambda(\boldsymbol{\beta})}, \xi_{\mathrm{Park}})$ preserves $\leq_{\mathrm{Park}}$.

Exercise 17.10 A term $(\lambda x.M)N$ is a $\beta\mathsf{N}$-*redex* if $x \notin FV(M)$ or [N is either a variable or a closed strongly normalizing term]; see Honsell and Lenisa (1999). We denote by $\twoheadrightarrow_{\beta\mathsf{N}}$ the induced reduction. Show that if Γ

assigns types to all free variables in N, i.e. $x \in \mathrm{dom}(\Gamma)$ for all $x \in \mathrm{FV}(N)$, then

$$\Gamma \vdash_{\cap}^{\mathrm{HL}} M : A \,\&\, N \twoheadrightarrow_{\beta\mathsf{N}} M \;\Rightarrow\; \Gamma \vdash_{\cap}^{\mathrm{HL}} N : A.$$

[Hint: use Theorem 17.2.15(iii).]

Exercise 17.11 Show that in the assignment system $\lambda_{\cap}^{\mathrm{AO}}$ the terms typable with type $\mathtt{U} \to \mathtt{U}$ for a suitable context are precisely the lazy normalizing ones, i.e. the terms which reduce either to an abstraction or to a (λ-free) head normal form.

Exercise 17.12 Show that in $\lambda_{\cap}^{\mathrm{EHR}}$, as defined in Definition 13.2.13, the terms typable with type $\mathtt{V}$ in the context all whose predicates are $\mathtt{V}$ are precisely the terms which reduce either to an abstraction or to a variable using the call-by-value β-reduction rule.

Exercise 17.13 Show that all type interpretation domains and all type environments agree with AO.

Exercise 17.14 Using the Approximation Theorem, show that:

- there is no type deducible for $\mathbf{\Omega}$ in the system $\vdash_{\cap}^{\mathrm{CDV}}$;
- $\vdash_{\cap}^{\mathrm{BCD}} \mathbf{\Omega} : A$ iff $A =_{\mathrm{BCD}} \mathtt{U}$;
- $\vdash_{\cap}^{\mathrm{Park}} \mathbf{\Omega} : A$ iff $0 \leq_{\mathrm{Park}} A$.

Exercise 17.15 Using the Approximation Theorem, show that in the system $\lambda_{\cap}^{\mathrm{AO}}$ the set of types deducible for $\boldsymbol{\omega\omega}$ is strictly included in the set of types deducible for $\mathsf{K}(\mathbf{\Omega})$.

Exercise 17.16 Using the Approximation Theorem, show that in the system $\lambda_{\cap}^{\mathrm{Scott}}$ the set of types deducible for J and I coincide.

Exercise 17.17 Prove that typability in $\boldsymbol{\lambda}_{\cap}^{\mathrm{CDV}}$ is undecidable.

Exercise 17.18 Prove Lemma 17.5.3(i) and (ii). [Hint. Similar to the proof of Lemma 13.1.23.]

Exercise 17.19 Consider the type assignment systems $\lambda_{\cap}^{\mathrm{Krivine}}$ and $\lambda_{\cap}^{\mathrm{Krivine}^{\mathtt{U}}}$ as defined in Exercise 13.10.

(i) Prove an analog of Lemma 17.5.9.
(ii) Prove that if Γ is a game context then $\Gamma \vdash^{\mathrm{Krivine}} M : \alpha$ and $\Gamma \vdash^{\mathrm{Krivine}^{\mathtt{U}}} M : \alpha$ are equivalent to $\Gamma \vdash M : \alpha$, for all type variables α. Conclude that type inhabitation remains undecidable without $(\leq)$.
(iii) Prove that the type $\delta \cap (\alpha \to \beta) \cap (\alpha \to \gamma) \to \delta \cap (\alpha \to \beta \cap \gamma)$ is inhabited in $\lambda_{\cap}^{\mathrm{BCD}}$ but is not inhabited in $\lambda_{\cap}^{\mathrm{Krivine}^{\mathtt{U}}}$.

Exercise 17.20 Let Γ be a game context and $\alpha \in \mathbb{A}_\infty$. Prove that if $\Gamma \vdash M : \alpha$ then every node in the Böhm tree of M (see B[1984], Ch. 10) has at most one branch.

Exercise 17.21 Complete the proofs of Theorem 17.5.23 and Corollary 17.5.24.

Exercise 17.22 Prove Lemma 17.5.28. [Hint. Compare G_1 to the game G_0 of Example 17.5.16. Observe that in each sequence of positions

$$T_{(2k+1)n}, \ldots, T_{(2k+3)n},$$

the odd steps behave as an initial phase of G_0, while the even steps behave as a final phase of G_0. Writing

$$\begin{aligned} \bot_i &= \langle \bot, i, G \rangle; \\ A &= \langle a, 1, \{G, L\} \rangle \langle a, 2, \{G, L\} \rangle; \\ B &= \langle b, 1, \{G, L\} \rangle \langle b, 2, \{G, L\} \rangle, \end{aligned}$$

Fig. 17.6 illustrates the case of two initial C_1 steps (the canopy of a tree is the collection of its leaves). If one starts with m moves of C_1 ($0<m<\infty$), then the canopy of position $m+2+2k$ will be $(A^{2^{m-1}}B^{2^{m-1}})^{2^k}$. Note that $m=0$ or $m=\infty$ yield possible plays of the game.]

position #	via move	canopy of position
0		$\bot_1$
1	C_1	$\bot_1\bot_2$
2	C_1	$(\bot_1\bot_2)^2$
3	C_2	A^2A^2
4	1	A^2B^2
5	C_3	A^4B^4
6	2	$(A^2B^2)^2$
7	C_3	$(A^4B^4)^2$
8	3	$(A^2B^2)^4$
9	C_3	$(A^4B^4)^4$
10	5	$(A^2B^2)^8$
11	C_3	$(A^4B^4)^8$
12	7	$(A^2B^2)^{16}$
...	...	...
$4+2k$		$(A^2B^2)^{2^k}$

Figure 17.6

Exercise 17.23 Prove Lemma 17.5.26. [Hint. Encode a queue automaton (also called a Post machine), i.e. a deterministic finite automaton with a queue, into a typewriter, thus reducing the halting problem for queue automata to the emptiness problem for typewriters. One possible way of doing this is as follows. Represent a queue, say '011100010', as a string of the form

$$\$\$\cdots\$<011100010>\sharp\cdots\sharp\sharp,$$

with a certain number of the \$s and $\sharp$s. The initial empty queue is just '$<>\sharp\cdots\sharp\sharp$'. Now an *insert* instruction means: *replace '$>$' with a digit and replace the first '$\sharp$' with '$>$'*, and similarly for a *remove*. The number of \$s increases after each *remove*, while the suffix of $\sharp$s shrinks after each *insert*, so that the queue 'moves to the right'. If the number of the initial suffix of $\sharp$s is sufficiently large, then a typewriter automaton can verify the queue computation.]

Exercise 17.24 Kuśmierek (2007). Prove that in $\lambda_\cap^{\mathrm{CDV}}$ the inhabitation problem for types of i-rank at most 2 is decidable. See definition of i-rank after Theorem 17.5.31. Note that if the i-rank of $B{\to}C$ is at most 2 then the i-rank of B is at most 1.

Exercise 17.25 Kuśmierek (2007).

(i) Let $\iota = (\alpha \to \alpha) \cap (\beta \to \beta)$ and define for $k = 0, \ldots, n$ the type

$$A_k = \alpha \to \iota^k \to (\alpha \to \beta) \to (\beta \to \alpha)^{n-k} \to \beta.$$

Prove that the shortest inhabitant of $A_1, \ldots, A_n$ is of length exponential in n. Can you modify the example so that the shortest inhabitant is of double exponential length?

(ii) How long (in the worst case) is the shortest inhabitant (if it exists) of a given type whose i-rank (see Definition 17.5.33) is 2? [Hard.]

Exercise 17.26 Show that inhabitation in $\boldsymbol{\lambda}_\cap^{\mathrm{BCD}}$ is undecidable. [Hint. Show that $\boldsymbol{\lambda}_\cap^{\mathrm{BCD}}$ is conservative over $\boldsymbol{\lambda}_\cap^{\mathrm{CDV}}$ for $\boldsymbol{\beta}$-normal forms; that is,

$$\Gamma \vdash^{\mathrm{BCD}} M : A \Leftrightarrow \Gamma \vdash^{\mathrm{CDV}} M : A,$$

for all $A \in \mathbb{T}^{\mathrm{CDV}}$, Γ with types from $\mathbb{T}^{\mathrm{CDV}}$, and $M \in \Lambda$ in $\boldsymbol{\beta}$-normal form. Use Theorem 17.2.15(i).]

References

Abadi, M., and Cardelli, L. 1996. *A Theory of Objects*. Springer.

Abadi, M., and Fiore, M. P. 1996. Syntactic considerations on recursive types. Pages 242–252 of: *Logic in Computer Science*. IEEE Computer Society Press.

Abadi, M., and Plotkin, G. D. 1990. A PER model of polymorphism and recursive types. Pages 355–365 of: *Logic in Computer Science*. IEEE Computer Society Press.

Abelson, H., Dybvig, R. K., Haynes, C. T., Rozas, G. J., IV, N. I. Adams, Friedman, D. P., Kohlbecker, E., Jr., G. L. Steele, Bartley, D. H., Halstead, R., Oxley, D., Sussman, G. J., Brooks, G., Hanson, C., Pitman, K. M., Wand, M., Clinger, W., and Rees, J. 1991. Revised report on the algorithmic language Scheme. *ACM SIGPLAN Lisp Pointers*, **IV**(3), 1–55.

Abramsky, S. 1990. The lazy lambda calculus. Pages 65–116 of: *Research Topics in Functional Programming*, Turner, D. A. (ed). Addison-Wesley.

Abramsky, S. 1991. Domain theory in logical form. *Annals of Pure and Applied Logic*, **51**(1–2), 1–77.

Abramsky, S., and Jung, A. 1994. Domain theory. Pages 1–168 of: *Handbook for Logic in Computer Science*, vol. 3, Abramsky, S., Gabbay, D. M., and Maibaum, T. S. E. (eds). Clarendon Press.

Abramsky, S., and Ong, C.-H. L. 1993. Full abstraction in the lazy lambda calculus. *Information and Computation*, **105**(2), 159–267.

Ackermann, W. 1928. Zum Hilbertschen Aufbau der reellen Zahlen. *Mathematische Annalen*, **99**, 118–133.

Aczel, P. 1988. *Non-Well-Founded Sets*. Center for the Study of Language and Information, Stanford.

Aho, A. V., Sethi, R., and Ullman, J. D. 1986. *Compilers*. Addison-Wesley.

Alessi, F. 1991. *Strutture di Tipi, Teoria dei Domini e Modelli del Lambda Calcolo*. Ph.D. thesis, Torino University.

Alessi, F. 1993. *The p model*. Internal Report, Udine University.

Alessi, F., and Barbanera, F. 1991. Strong conjunction and intersection types. Pages 64–73 of: *Mathematical Foundations of Computer Science*. Lecture Notes in Computer Science. Springer.

Alessi, F., Dezani-Ciancaglini, M., and Honsell, F. 2001. Filter models and easy terms. Pages 17–37 of: *Italian Conference on Theoretical Computer Science.* Lecture Notes in Computer Science, vol. 2202. Springer.

Alessi, F., Barbanera, F., and Dezani-Ciancaglini, M. 2003. Intersection types and computational rules. Pages 1–15 of: *Workshop on Logic, Language, Information and Computation.* Electronic Notes in Theoretical Computer Science, **84**. Elsevier.

Alessi, F., Dezani-Ciancaglini, M., and Honsell, F. 2004a. Inverse limit models as filter models. Pages 3–25 of: *International Workshop on Higher-Order Rewriting.* RWTH Aachen.

Alessi, F., Barbanera, F., and Dezani-Ciancaglini, M. 2004b. Tailoring filter models. Pages 17–33 of: *Types.* Lecture Notes in Computer Science, vol. 3085. Springer.

Alessi, F., Barbanera, F., and Dezani-Ciancaglini, M. 2006. Intersection types and lambda models. *Theoretical Computer Science*, **355**(2), 108–126.

Allouche, J.-P., and Shallit, J. 2003. *Automatic Sequences.* Cambridge University Press.

Amadio, R. M. 1991. Recursion over realizability structures. *Information and Computation*, **91**(1), 55–85.

Amadio, R. M., and Cardelli, L. 1993. Subtyping recursive types. *ACM Transactions on Programming Languages and Systems*, **15**(4), 575–631.

Amadio, R. M., and Curien, P.-L. 1998. *Domains and Lambda-Calculi.* Cambridge University Press.

Andrews, Peter B. 2002. *An Introduction to Mathematical Logic and Type Theory: To Truth Through Proof.* Applied Logic, vol. 27. Springer.

Appel, A. W. 1992. *Compiling with Continuations.* Cambridge University Press.

Appel, A. W., and McAllester, D. 2001. An indexed model of recursive types for foundational proof-carrying code. *ACM Transactions on Programming Languages and Systems*, **23**(5), 657–683.

Ariola, Z. M., and Klop, J. W. 1994. Cyclic lambda graph rewriting. Pages 416–425 of: *Logic in Computer Science.* IEEE Computer Society Press.

Ariola, Z. M., and Klop, J. W. 1996. Equational term graph rewriting. *Fundamenta Informaticae*, **26**(3-4), 207–240.

Aspinall, D., and Compagnoni, A. 1996. Subtyping dependent types. Pages 86–97 of: *Logic in Computer Science.* IEEE Computer Society Press.

Augustson, L. 1999. Cayenne – a language with dependent types. Pages 240–267 of: *Advanced Functional Programming.* Lecture Notes in Computer Science, vol. 1608. Springer.

Avigad, J., Donnelly, K., Gray, D., and Raff, Paul. 2007. A formally verified proof of the prime number theorem. *ACM Transactions on Computational Logic*, **9**(1-2), 1–23.

Baader, F., and Nipkow, T. 1998. *Term Rewriting and All That.* Cambridge University Press.

Baader, F., and Snyder, W. 2001. Unification theory. Pages 447–533 of: *Handbook of Automated Reasoning*, vol. I, Robinson, J.A., and Voronkov, A. (eds). Elsevier.

Backus, J. W. 1978. Can programming be liberated from the von Neumann style? *Communication of the ACM*, **21**, 613–641.

Baeten, J., and Boerboom, B. 1979. Ω can be anything it shouldn't be. *Indagationes Mathematicae*, **41**, 111–120.

van Bakel, S. 1992. Complete restrictions of the intersection type discipline. *Theoretical Computer Science*, **102**(1), 135–163.

van Bakel, S. 1993. Principal type schemes for the strict type assignment system. *Journal of Logic and Computation*, **3**(6), 643–670.

van Bakel, S., Barbanera, F., Dezani-Ciancaglini, M., and de Vries, F.-J. 2002. Intersection types for lambda-trees. *Theoretical Computer Science*, **272**(1-2), 3–40.

Baldridge, J. 2002. *Lexically Specified Derivational Control in Combinatory Categorial Grammar*. Ph.D. thesis, University of Edinburgh.

Barbanera, F., Dezani-Ciancaglini, M., and de'Liguoro, U. 1995. Intersection and union types: syntax and semantics. *Information and Computation*, **119**(2), 202–230.

Barendregt, H. P. 1974. Pairing without conventional restraints. *Zeitschrift für Mathematische Logik und Grundlagen der Mathematik*, **20**, 289–306.

Barendregt, H. P. 1984. *The Lambda Calculus: its Syntax and Semantics*. Revised edition. North-Holland.

Barendregt, H. P. 1991. Self-interpretation in lambda calculus. *Journal of Functional Programming*, **1**(2), 229–233.

Barendregt, H. P. 1992. Lambda calculi with types. Pages 117–309 of: *Handbook for Logic in Computer Science*, Abramsky, S., Gabbay, D. M., and Maibaum, T. S. E. (eds). Oxford University Press.

Barendregt, H. P. 1994. Discriminating coded lambda terms. Pages 141–151 of: *From Universal Morphisms to Megabytes: A Baayen Space-Odyssey*, Apt, K.R., Schrijver, A.A., and Temme, N.M. (eds). CWI.

Barendregt, H. P. 1995. Enumerators of lambda terms are reducing constructively. *Annals of Pure and Applied Logic*, **73**, 3–9.

Barendregt, H. P. 1996. The quest for correctness. Pages 39–58 of: *Images of SMC Research*. CWI.

Barendregt, H. P., and Barendsen, E. 1997. Efficient computations in formal proofs. *Journal of Automated Reasoning*, 321–336.

Barendregt, H. P., and Ghilezan, S. 2000. Lambda terms for natural deduction, sequent calculus and cut-elimination. *Journal of Functional Programming*, **10**, 121–134.

Barendregt, H. P., and Rezus, A. 1983. Semantics for classical Automath and related systems. *Information and Control*, **59**, 127–147.

Barendregt, H., and Wiedijk, F. 2005. The challenge of computer mathematics. *Transactions A of the Royal Society*, **1835**, 2351–2375.

Barendregt, H. P., Coppo, M., and Dezani-Ciancaglini, M. 1983. A filter lambda model and the completeness of type assignment. *The Journal of Symbolic Logic*, **48**(4), 931–940.

Barendregt, H. P., Bunder, M., and Dekkers, W. 1993. Systems of illative combinatory logic complete for first order propositional and predicate calculus. *The Journal of Symbolic Logic*, **58**(3), 89–108.

Barendregt, H., Manzonetto, G., and Plasmeijer, R. 2013. The imperative and functional programming paradigm. In *Alan Turing – His Work and Impact*, Cooper, B. and van Leeuwen, J. (eds). Elsevier.

Barendsen, E., and Smetsers, J. E. W. 1993. Conventional and uniqueness typing in graph rewrite systems (extended abstract). Pages 41–51 of: *Foundations of Software Technology and Theoretical Computer Science.* Lecture Notes in Computer Science, vol. 761. Springer.

Barendsen, E., and Smetsers, J. E. W. 1996. Uniqueness typing for functional languages with graph rewriting semantics. *Mathematical Structures in Computer Science*, **6**(6), 579–612.

Bekič, H. 1984. Programming languages and their definition. *Selected Papers*, C. B. Jones (ed). Lecture Notes in Computer Science, vol. 177. Springer.

van Benthem, J. F. A. K. 1995. *Language in Action: Categories, Lambdas, and Dynamic Logic.* The MIT Press.

van Benthem, J. F. A. K., and ter Meulen, A. (eds). 1997. *Handbook of Logic and Language.* Elsevier and MIT Press.

van Benthem Jutting, L. S. 1977. *Checking Landau's "Grundlagen" in the Automath System.* Ph.D. thesis, Eindhoven University of Technology.

Berarducci, A., and Böhm, C. 1993. A self-interpreter of lambda calculus having a normal form. Pages 85–99 of: *Computer Science Logic.* Lecture Notes in Computer Science, vol. 702. Springer.

Berline, C. 2000. From computation to foundations via functions and applications: the lambda-calculus and its webbed models. *Theoretical Computer Science*, **249**, 81–161.

Bernardi, R. 2002. *Reasoning with Polarity in Categorial Type Logic.* Ph.D. thesis, Utrecht Institute of Linguistics OTS.

Bertot, Y., and Castéran, P. 2004. *Interactive Theorem Proving and Program Development.* Texts in Theoretical Computer Science. Springer.

Bezem, M. A. 1985. Strong normalization of bar recursive terms without using infinite terms. *Archiv für Mathematische Logik und Grundlagenforschung*, **25**, 175–181.

Birkedal, L., and Harper, R. 1999. Constructing interpretations of recursive types in an operational setting. *Information and Computation*, **155**, 3–63.

Böhm, C. 1966. The CUCH as a formal and description language. Pages 179–197 of: *Formal Languages Description Languages for Computer Programming.* North-Holland.

Böhm, C. (ed). 1975. *λ-calculus and Computer Science Theory.* Lecture Notes in Computer Science, vol. 37. Springer.

Böhm, C., and Berarducci, A. 1985. Automatic synthesis of typed Λ-programs on term algebras. *Theoretical Computer Science*, **39**, 135–154.

Böhm, C., and Dezani-Ciancaglini, M. 1975. λ-terms as total or partial functions on normal forms. Pages 96–121 of: *λ-Calculus and Computer Science Theory.* Lecture Notes in Computer Science, vol. 37. Springer.

Böhm, C., and Gross, W. 1966. Introduction to the CUCH. Pages 35–65 of: *Automata Theory*, Caianiello, E.R. (ed). Academic Press.

Böhm, C., Piperno, A., and Guerrini, S. 1994. Lambda-definition of function(al)s by normal forms. Pages 135–154 of: *European Symposium on Programming*, vol. 788. Springer.

Bono, V., Venneri, B., and Bettini, L. 2008. A typed lambda calculus with intersection types. *Theoretical Computer Science*, **398**(1-3), 95–113.

Bove, A., Dybjer, P., and Norell, U. 2009. A brief overview of Agda – a functional language with dependent types. Pages 73–78 of: *Theorem Proving in Higher Order Logics*. Lecture Notes in Computer Science, vol. 5674. Springer.

Brandt, M., and Henglein, F. 1998. Coinductive axiomatization of recursive type equality and subtyping. *Fundamenta Informaticæ*, **33**, 309–338.

Breazu-Tannen, V., and Meyer, A. R. 1985. Lambda calculus with constrained types. Pages 23–40 of: *Logics of Programs*. Lecture Notes in Computer Science, vol. 193. Springer.

Bruce, K. B., Meyer, A. R., and Mitchell, J. C. 1990. The semantics of second-order lambda calculus. Pages 213–272 of: *Logical Foundations of Functional Programming*. University of Texas at Austin Year of Programming Series. Addison Wesley.

de Bruijn, N. G. 1968. *AUTOMATH, a language for mathematics*. Tech. rept. 68-WSK-05. T.H.-Reports.

de Bruijn, N. G. 1970. The mathematical language AUTOMATH, its usage and some of its extensions. Pages 29–61 of: *Symposium on Automatic Demonstration*. Lecture Notes in Mathematics, no. 125. Springer.

de Bruijn, N. G. 1972. Lambda calculus notation with nameless dummies, a tool for automatic formula manipulation, with application to the Church-Rosser theorem. *Indagationes Mathematicae*, **34**, 381–392.

de Bruijn, N. G. 1994a. A survey of the project Automath. Pages 141–161 of: *Selected Papers on Automath*, Nederpelt, R. P., Geuvers, J. H., and de Vrijer, R. C. (eds). Studies in Logic and the Foundations of Mathematics, **133**. North-Holland.

de Bruijn, N. G. 1994b. Reflections on Automath. Pages 201–228 of: *Selected Papers on Automath*, Nederpelt, R. P., Geuvers, J. H., and de Vrijer, R. C. (eds). Studies in Logic and the Foundations of Mathematics, **133**. North-Holland.

Burstall, R. M. 1969. Proving properties of programs by structural induction. *Computer Journal*, **12**(1), 41–48.

Buszkowski, W., and Penn, G. 1990. Categorial grammars determined from linguistic data by unification. *Studia Logica*, **49**(4), 431–454.

Buszkowski, W., Marciszewski, W., and Benthem, J. F. A. K. van (eds). 1988. *Categorial Grammar*. Linguistic & Literary Studies in Eastern Europe **25**. John Benjamins.

Capitani, B., Loreti, M., and Venneri, B. 2001. Hyperformulae, parallel deductions and intersection types. *Electronic Notes in Theoretical Computer Science*, **50**(2), 1–18.

Capretta, V., and Valentini, S. 1998. A general method for proving the normalization theorem for first and second order typed λ-calculi. *Mathematical Structures in Computer Science*, **9**(6), 719–739.

Caprotti, O., and Oostdijk, M. 2001. Formal and efficient primality proofs by use of computer algebra oracles. *Journal of Symbolic Computation*, **32**, 55–70.

Cardelli, L. 1988. A semantics of multiple inheritance. *Information and Computation*, **76**(2-3).

Cardone, F. 1989. Relational semantics for recursive types and bounded quantification. Pages 164–178 of: *Automata, Languages and Programming*. Lecture Notes in Computer Science, vol. 372. Springer.

Cardone, F. 1991. Recursive types for fun. *Theoretical Computer Science*, **83**, 29–56.

Cardone, F. 2002. A coinductive completeness proof for the equivalence of recursive types. *Theoretical Computer Science*, **275**, 575–587.

Cardone, F., and Coppo, M. 1990. Two extensions of Curry's type inference system. Pages 19–76 of: *Logic and Computer Science*, Odifreddi, P. (ed). APIC Studies in Data Processing, vol. 31. Academic Press.

Cardone, F., and Coppo, M. 1991. Type inference with recursive types. Syntax and Semantics. *Information and Computation*, **92**(1), 48–80.

Cardone, F., and Coppo, M. 2003. Decidability properties of recursive types. Pages 242–255 of: *Italian Conference on Theoretical Computer Science*. Lecture Notes in Computer Science, vol. 2841. Springer.

Cardone, F., and Hindley, J. R. 2009. Lambda-calculus and combinators in the 20th century. Pages 723–817 of: *Handbook of the History of Logic*, vol. 5: Logic from Russell to Church, Gabbay, D. M., and Woods, J. (eds). Elsevier.

Cardone, F., Dezani-Ciancaglini, M., and de'Liguoro, U. 1994. Combining type disciplines. *Annals of Pure and Applied Logic*, **66**(3), 197–230.

Church, A. 1932. A set of postulates for the foundation of logic (1). *Annals of Mathematics*, **33**, 346–366.

Church, A. 1933. A set of postulates for the foundation of logic (2). *Annals of Mathematics*, **34**, 839–864.

Church, A. 1936. An unsolvable problem of elementary number theory. *American Journal of Mathematics*, **58**, 354–363.

Church, A. 1940. A formulation of the simple theory of types. *The Journal of Symbolic Logic*, **5**, 56–68.

Church, A. 1941. *The calculi of lambda-conversion.* Princeton University Press. Annals of Mathematics Studies, no. 6.

Church, A., and Rosser, J. B. 1936. Some properties of conversion. *Transactions of the American Mathematical Society*, **39**, 472–482.

Clinger, W., and (editors), Jonathan Rees. 1991. Revised^4 Report on the Algorithmic Language Scheme. *LISP Pointers*, **IV**(3), 1–55.

Comon, H., and Jurski, Y. 1998. Higher-order matching and tree automata. Pages 157–176 of: *Computer Science Logic*. Lecture Notes in Computer Science, vol. 1414. Springer.

Coppo, M. 1985. A completeness theorem for recursively defined types. Pages 120–129 of: *Automata, Languages and Programming*. Lecture Notes in Computer Science, vol. 194. Springer.

Coppo, M., and Dezani-Ciancaglini, M. 1980. An extension of the basic functionality theory for the λ-calculus. *Notre Dame Journal of Formal Logic*, **21**(4), 685–693.

Coppo, M., Dezani-Ciancaglini, M., and Sallé, P. 1979. Functional characterization of some semantic equalities inside lambda-calculus. Pages 133–146 of: *Automata, Languages and Programming.*

Coppo, M., Dezani-Ciancaglini, M., and Venneri, B. 1981. Functional characters of solvable terms. *Zeitschrift für Mathematische Logik und Grundlagen der Mathematik*, **27**(1), 45–58.

Coppo, M., Dezani-Ciancaglini, M., and Longo, G. 1983. Applicative information systems. Pages 35–64 of: *Colloquium on Trees in Algebra and Programming.* Lecture Notes in Computer Science, vol. 159. Springer.

Coppo, M., Dezani-Ciancaglini, M., Honsell, F., and Longo, G. 1984. Extended type structures and filter lambda models. Pages 241–262 of: *Logic Colloquium.* North-Holland.

Coppo, M., Dezani-Ciancaglini, M., and Zacchi, M. 1987. Type theories, normal forms, and D_∞-lambda-models. *Information and Computation*, **72**(2), 85–116.

Cosmadakis, S. 1989. Computing with recursive types (Extended Abstract). Pages 24–38 of: *Logic in Computer Science.* IEEE Computer Society Press.

Courcelle, B. 1983. Fundamental properties of infinite trees. *Theoretical Computer Science*, **25**, 95–169.

Courcelle, B., Kahn, G., and Vuillemin, J. 1974. Algorithmes d'équivalence et de réduction à des expressions minimales, dans une classe d'équations récursives simples. Pages 200–213 of: *Automata, Languages and Programming.* Lecture Notes in Computer Science, vol. 14. Springer.

Cousineau, G., Curien, P.-L., and Mauny, M. 1987. The categorical abstract machine. *Science of Computer Programming*, **8(2)**, 173–202.

Crossley, J. N. 1975. Reminiscences of logicians. Pages 1–62 of: *Algebra and Logic*, Crossley, J. N. (ed). Lecture Notes in Mathematics, vol. 450. Springer.

Curien, P.-L. 1993. *Categorical Combinators, Sequential Algorithms, and Functional Programming.* 2nd edition. Progress in Theoretical Computer Science. Birkhäuser.

Curry, H. B. 1934. Functionality in combinatory logic. *Proceedings of the National Academy of Science of the USA*, **20**, 584–590.

Curry, H. B. 1969. Modified basic functionality in combinatory logic. *Dialectica*, **23**, 83–92.

Curry, H. B., and Feys, R. 1958. *Combinatory Logic.* Vol. I. North-Holland.

David, R., and Nour, K. 2007. An arithmetical proof of the strong normalization for the *lambda*-calculus with recursive equations on types. Pages 84–101 of: *Typed Lambda Calculi and Applications.* Lecture Notes in Computer Science, vol. 4583. Springer.

Davis, M. 1973. Hilbert's tenth problem is unsolvable. *American Mathematical OPmonthly*, **80**, 233–269.

Davis, M., Robinson, J., and Putnam, H. 1961. The decision problem for exponential Diophantine equations. *Annals of Mathematics,* second series, **74**(3), 425–436.

Dedekind, R. 1901. *Essays on the Theory of Numbers.* Open Court Publishing Company. Translation by W.W. Beman of *Stetigkeit und irrationale Zahlen* (1872) and *Was sind und was sollen die Zahlen?* (1888), reprinted 1963 by Dover Press.

Dekkers, W. 1988. Reducibility of types in typed lambda calculus. Comment on: "On the existence of closed terms in the typed λ-calculus, I" *(Statman (1980a)). Information and Computation*, **77**(2), 131–137.

Dekkers, W., Bunder, M., and Barendregt, H. P. 1998. Completeness of the propositions-as-types interpretation of intuitionistic logic into illative combinatory logic. *The Journal of Symbolic Logic*, **63**(3), 869–890.

Dezani-Ciancaglini, M., and Margaria, I. 1986. A characterization of F-complete type assignments. *Theoretical Computer Science*, **45**(2), 121–157.

Dezani-Ciancaglini, M., Ghilezan, S., and Venneri, B. 1997. The "relevance" of intersection and union types. *Notre Dame Journal of Formal Logic*, **38**(2), 246–269.

Dezani-Ciancaglini, M., Honsell, F., and Motohama, Y. 2001. Approximation theorems for intersection type systems. *Journal of Logic and Computation*, **11**(3), 395–417.

Dezani-Ciancaglini, M., Honsell, F., and Alessi, F. 2003. A complete characterization of complete intersection-type preorders. *ACM Transactions on Computational Logic*, **4**(1), 120–147.

Dezani-Ciancaglini, M., Ghilezan, S., and Likavec, S. 2004. Behavioural inverse limit lambda-models. *Theoretical Computer Science*, **316**(1-3), 49–74.

Dezani-Ciancaglini, M., Honsell, F., and Motohama, Y. 2005. Compositional characterization of λ-terms using Intersection Types. *Theoretical Computer Science*, **340**(3), 459–495.

Di Gianantonio, P., and Honsell, F. 1993. An abstract notion of application. Pages 124–138 of: *Typed Lambda Calculi and Applications.* Lecture Notes in Computer Science, vol. 664. Springer.

Došen, K. 1992. A brief survey of frames for the Lambek calculus. *Zeitschrift für Mathematische Logik und Grundlagen der Mathematik*, **38**, 179–187.

Dowek, G. 1994. Third order matching is decidable. *Annals of Pure and Applied Logic*, **69**(2-3), 135–155.

van Draanen, J.-P. 1995. *Models for Simply Typed Lambda-Calculi with Fixed Point Combinators and Enumerators.* Ph.D. thesis, Catholic University of Nijmegen.

Dyckhoff, R., and Pinto, L. 1999. Permutability of proofs in intuitionistic sequent calculi. *Theoretical Computer Science*, **212**, 141–155.

van Eekelen, M. C. J. D., and Plasmeijer, M. J. 1993. *Functional Programming and Parallel Graph Rewriting.* Addison-Wesley.

Elbers, H. 1996. *Personal communication.*

Endrullis, J., Grabmayer, C., Klop, J. W., and van Oostrom, V. 2011. On equal μ-terms. *Theoretical Computer Science*, **412**, 3175–3202. Festschrift in Honour of Jan Bergstra.

Engeler, E. 1981. Algebras and combinators. *Algebra Universalis*, **13**(3), 389–392.

ten Eikelder, H. M. M. 1991. *Some algorithms to decide the equivalence of recursive types.* URL: <alexandria.tue.nl/extra1/wskrap/publichtml/9211264.pdf>.

Euclid of Alexandria. –300. *The Elements.* English translation in Heath (1956).

Fiore, M. 1996. A coinduction principle for recursive data types based on bisimulation. *Information and Computation*, **127**(2), 186–198.

Fiore, M. 2004. Isomorphisms of generic recursive polynomial types. *SIGPLAN Notices*, **39**(1), 77–88.

Flajolet, P., and Sedgewick, R. 1993. *The average case analysis of algorithms: counting and generating functions.* Tech. rept. 1888. INRIA.

Fortune, S., Leivant, D., and O'Donnel, M. 1983. The expressiveness of simple and second-order type structures. *Journal of the ACM*, **30**(1), 151–185.

Fox, A. 2003. Formal specification and verification of ARM6. In: Basin, D. A., and Wolff, B. (eds), *Theorem Proving in Higher Order Logics 2003.* Lecture Notes in Computer Science, vol. 2758. Springer.

Freyd, P. J. 1990. Recursive types reduced to inductive types. Pages 498–507 of: *Logic in Computer Science.* IEEE Computer Society Press.

Freyd, P. J. 1991. Algebraically complete categories. Pages 131–156 of: *Como Category Theory Conference.* Lecture Notes in Mathematics, vol. 1488. Springer.

Freyd, P. J. 1992. Remarks on algebraically compact categories. Pages 95–106 of: *Applications of Categories in Computer Science.* London Mathematical Society Lecture Notes Series, vol. 177. Cambridge University Press.

Friedman, H. M. 1975. Equality between functionals. In: *Logic Colloqium.* Lecture Notes in Mathematics, vol. 453. Springer.

Gandy, R. O. 1980. An early proof of normalization by A. M. Turing. Pages 453–457 of: *To H. B. Curry: Essays on Combinatory Logic, Lambda Calculus and Formalism* (Hindley and Seldin (1980)).

Gapeyev, V., Levin, M. Y., and Pierce, B. C. 2002. Recursive subtyping revealed. *Journal of Functional Programming*, **12**(6), 511–548.

Gentzen, G. 1936a. Die Widerspruchfreiheit der reinen Zahlentheorie. *Mathematische Annalen*, **112**, 493–565. Translated as 'The consistency of arithmetic', in Szabo (1969).

Gentzen, G. 1936b. Untersuchungen über das logischen Schliessen. *Mathematische Zeitschrift*, **39**, 405–431. Translation in: *Collected Papers of Gerhard Gentzen*, ed. M. E. Szabo, North-Holland [1969], 68-131.

Gentzen, G. 1943. Beweisbarkeit und Unbeweisbarkeit von Anfangsfällen der transfiniten Induktion in der reinen Zahlentheorie. *Mathematische Annalen*, **120**, 140–161.

Gentzen, G. 1969. Investigations into logical deduction. Pages 68–131 of: *The Collected Papers of Gerhard Gentzen*, Szabo, M. E. (ed). North-Holland.

Ghilezan, S. 1996. Strong normalization and typability with intersection types. *Notre Dame Journal of Formal Logic*, **37**(1), 44–52.

Ghilezan, S. 2007. *Terms for natural deduction, sequent calculus and cut elimination in classical logic.* URL: <http://www.cs.ru.nl/barendregt60/essays/>. Essays Dedicated to Henk Barendregt on the Occasion of his 60th Birthday.

Gierz, G. K., Hofmann, K. H., Keimel, K., Lawson, J. D., Mislove, M. W., and Scott, D. S. 1980. *A Compendium of Continuous Lattices*. Springer.

Girard, J.-Y. 1971. Une extension de l'interprétation de Gödel à l'analyse, et son application à l'élimination des coupures dans l'analyse et la théorie des types. Pages 63–92 of: *Scandinavian Logic Symposium*. Studies in Logic and the Foundations of Mathematics, vol. 63. North-Holland.

Girard, J.-Y. 1995. Linear logic: its syntax and semantics. In: *Advances in Linear Logic*, Girard, J.-Y., Lafont, Y., and Regnier, L. (eds). London Mathematical Society Lecture Note Series. Cambridge University Press.

Girard, J.-Y., Lafont, Y. G. A., and Taylor, P. 1989. *Proofs and Types*. Cambridge Tracts in Theoretical Computer Science, vol. 7. Cambridge University Press.

Gödel, K. 1931. Über formal unentscheidbare Sätze der Principia Mathematica und verwandter Systeme I. *Monatshefte für Mathematik und Physik*, **38**, 173–198. German; English translation in Heijenoort (1967), pages 592-618.

Gödel, K. 1958. Ueber eine bisher noch nicht benützte Erweiterung des finiten Standpunktes. *Dialectica*, **12**, 280–287.

Goguen, J., Thatcher, J. W., Wagner, E. G., and Wright, J. B. 1977. Initial algebra semantics and continuous algebras. *Journal of the ACM*, **24**, 68–95.

Goldfarb, W. D. 1981. The undecidability of the second-order unification problem. *Theoretical Computer Science*, **13**(2), 225–230.

Gonthier, G. 2008. Formal proof – the four-color theorem. *Notices of the American Mathematical Society*, **55**(11), 1382–1393.

Gordon, A. D. 1994. *Functional Programming and Input/Output*. Distinguished Dissertations in Computer Science. Cambridge University Press.

Gordon, M., Milner, R., and Wadsworth, C. P. 1979. *Edinburgh LCF. A Mechanical Logic of Computation*. Lecture Notes in Computer Science, vol. 78. Springer.

Gordon, M.J.C., and Melham, T.F. (eds). 1993. *Introduction to HOL: A Theorem Proving Environment for Higher Order Logic*. Cambridge University Press.

Grabmayer, C. 2005. *Relating Proof Systems for Recursive Types*. Ph.D. thesis, Vrije Universiteit Amsterdam.

Grabmayer, C. 2007. A duality between proof systems for cyclic term graphs. *Mathematical Structures in Computer Science*, **17**, 439–484.

Grégoire, B., Théry, L., and Werner, B. 2006. A computational approach to Pocklington certificates in type theory. Pages 97–113 of: *Functional and Logic Programming*. Lecture Notes in Computer Science, vol. 3945. Springer.

de Groote, P. 1995. *The Curry–Howard Isomorphism*. Cahiers du Centre de Logique, vol. 8. Academia-Bruylant.

de Groote, P., and Pogodalla, S. 2004. On the expressive power of abstract categorial grammars: Representing context-free formalisms. *Journal of Logic, Language and Information*, **13**(4), 421–438.

Grzegorczyk, A. 1964. Recursive objects in all finite types. *Fundamenta Mathematicae*, **54**, 73–93.

Gunter, C. A. 1992. *Semantics of Programming Languages: Structures and Techniques*. MIT Press.

Gunter, C. A., and Scott, D. S. 1990. Semantic domains. Pages 633–674 of: *Handbook of Theoretical Computer Science*, vol. B, Leeuwen, J. Van (ed). North-Holland, MIT-Press.

Hales, T. C. 2005. A proof of the Kepler conjecture. *Annals of Mathematics*, **162**(3), 1065–1185.

Harrington, L. A., Morley, M. D., Ščedrov, A., and Simpson, S. G. (eds). 1985. *Harvey Friedman's Research on the Foundations of Mathematics*. Studies in Logic and the Foundations of Mathematics, vol. 117. North-Holland.

Harrison, J. 2009a. Formalizing an analytic proof of the prime number theorem. *Journal of Automated Reasoning*, **43**(3), 243–261.

Harrison, J. 2009b. HOL Light: an overview. Pages 60–66 of: *Theorem Proving in Higher Order Logics*. Springer.

Harrop, R. 1958. On the existence of finite models and decision procedures for propositional calculi. *Proceedings of the Cambridge Philosophical Society*, **54**, 1–13.

Heath, T. L. 1956. *The Thirteen Books of Euclid's Elements*. Dover Publications.

Heijenoort, J. van (ed). 1967. *From Frege to Gödel: A Source Book in Mathematical Logic, 1879 -1931*. Harvard University Press.

Henderson, P. 1980. *Functional Programming: Application and Implementation*. Prentice-Hall.

Henkin, L. 1950. Completeness in the theory of types. *The Journal of Symbolic Logic*, **15**, 81–91.

Herbelin, H. 1995. A lambda calculus structure isomorphic to Gentzen-style sequent calculus structure. Pages 61–75 of: *Computer Science Logic*. Lecture Notes in Computer Science, vol. 933. Springer.

Hilbert, D., and Ackermann, W. 1928. *Grundzüge der Theoretischen Logik*. Die Grundlehren der Mathematischen Wissenschaften in Einzeldars tellungen, Band XXVII. Springer.

Hindley, J. R. 1969. The principal type-scheme of an object in combinatory logic. *Transactions of the American Mathematical Society*, **146**, 29–60.

Hindley, J. R. 1983. The completeness theorem for typing λ-terms. *Theoretical Computer Science*, **22**, 127–133.

Hindley, J. R. 1992. Types with intersection: an introduction. *Formal Aspects of Computing*, **4**(5), 470–486.

Hindley, J. R. 1997. *Basic Simple Type Theory*. Cambridge University Press.

Hindley, J. R., and Seldin, J. P. (eds). 1980. *To H. B. Curry: Essays on Combinatory Logic, Lambda Calculus and Formalism*. Academic Press.

Hinze, Ralf, Jeuring, J., and Löh, A. 2007. Comparing approaches to generic programming. In: Backhouse, R., Gibbons, J., Hinze, R., and Jeuring, J. (eds), *Datatype-Generic Programming 2006*. Lecture Notes in Computer Science, vol. 4719. Springer.

Hodges, A. 1983. *The Enigma of Intelligence*. Unwin paperbacks.

Hofmann, M. 1995. A simple model for quotient types. Pages 216–234 of: *Typed Lambda Calculi and Applications*. Lecture Notes in Computer Science, vol. 902. Springer.

Honsell, F., and Lenisa, M. 1993. Some results on the full abstraction problem for restricted lambda calculi. Pages 84–104 of: *Mathematical Foundations of Computer Science*. Springer.

Honsell, F., and Lenisa, M. 1999. Semantical analysis of perpetual strategies in λ-calculus. *Theoretical Computer Science*, **212**(1-2), 183–209.

Honsell, F., and Ronchi Della Rocca, S. 1992. An approximation theorem for topological lambda models and the topological incompleteness of lambda calculus. *Journal of Computer and System Sciences*, **45**(1), 49–75.

Howard, W. A. 1970. Assignment of ordinals to terms for primitive recursive functionals of finite type. Pages 443–458 of: *Intuitionism and Proof Theory*, Kino, A., Myhill, J., and Vesley, R. E. (eds). Studies in Logic and the Foundations of Mathematics. North-Holland.

Howard, W. A. 1980. The formulas-as-types notion of construction. Pages 479–490 of: *To H. B. Curry: Essays on Combinatory Logic, Lambda Calculus and Formalism* (Hindley and Seldin (1980)).

Hudak, P., Peyton Jones, S., Wadler, P., Boutel, B., Fairbairn, J., Fasel, J., Guzman, M. M., Hammond, K., Hughes, J., Johnsson, T., Kieburtz, D., Nikhil, R., Partain, W., and Peterson, J. 1992. Report on the programming language Haskell: a non-strict, purely functional language (Version 1.2). *ACM SIGPLAN Notices*, **27**(5), 1–164.

Hudak, P., Peterson, J., and Fasel, J. H. 1999. *A gentle introduction to Haskell 98.* Technical Report. Dept. of Computer Science, Yale University, USA.

Huet, G. P. 1975. A unification algorithm for typed lambda-calculus. *Theoretical Computer Science*, **1**, 27–57.

Hughes, R. J. M. 1984. *The Design and Implementation of Programming Languages.* Ph.D. thesis, University of Oxford.

Hughes, R. J. M. 1989. Why functional programming matters. *The Computer Journal*, **32**(2), 98–107.

Hutton, G. 2007. *Programming in Haskell.* Cambridge University Press.

Hyland, M. 1975/76. A syntactic characterization of the equality in some models for the lambda calculus. *Proceedings of the London Mathematical Society (2)*, **12**(3), 361–370.

Iverson, K. E. 1962. *A Programming Language.* Wiley.

Jacopini, G. 1975. A condition for identifying two elements of whatever model of combinatory logic. Pages 213–219 of: *λ-Calculus and Computer Science theory.* Lecture Notes in Computer Science, vol. 37. Springer.

Jay, B. 2009. *Pattern Calculus.* Springer.

Johnsson, T. 1984. Efficient compilation of lazy evaluation. *SIGPLAN Notices*, **19**(6), 58–69.

Johnstone, P. T. 1986. *Stone Spaces.* Cambridge University Press.

Joly, T. 2001a. Constant time parallel computations in lambda-calculus. *Theoretical Computer Science*, **266**(1), 975–985.

Joly, T. 2001b. The finitely generated types of the lambda-calculus. Pages 240–252 of: *Typed Lambda Calculi and Applications.* Lecture Notes in Computer Science, vol. 2044. Springer.

Joly, T. 2002. *On λ-definability II: the fixed type problem and finite generation of types.* Unpublished. Author's email: `<Thierry.Joly@pps.jussieu.fr>`.

Joly, T. 2003. Encoding of the halting problem into the monster type & applications. Pages 153–166 of: *Typed Lambda Calculi and Applications.* Lecture Notes in Computer Science, vol. 2701. Springer.

Joly, T. 2005. On lambda-definability I: the fixed model problem and generalizations of the matching problem. *Fundamenta Informaticae*, **65**(1–2), 135–151.

Jones, J. P. 1982. Universal Diophantine equation. *The Journal of Symbolic Logic*, **47**(3), 549–571.

Jones, M. P. 1993. A system of constructor classes: overloading and implicit higher-order polymorphism. Pages 52–61 of: *Functional Programming Languages and Computer Architecture.* ACM Press.

Kamareddine, F., Laan, T., and Nederpelt, R. 2004. *A Modern Perspective on Type Theory: From its Origins until Today.* Applied Logic Series, vol. 29. Kluwer Academic Publishers.

Kanazawa, M. 1998. *Learnable Classes of Categorial Grammars.* Cambridge University Press.

Kaufmann, M., Manolios, P., and Moore, J. S. 2000. *Computer-Aided Reasoning: An Approach.* Kluwer.

Kfoury, A.J., and Wells, J. 1995. New notions of reduction and non-semantic proofs of strong β-normalization in typed λ-calculi. Pages 311–321 of: *Logic in Computer Science.* IEEE Computer Society Press.

Kleene, S. C. 1936. Lambda-definability and recursiveness. *Duke Mathematical Journal*, **2**, 340–353.

Kleene, S. C. 1952. *Introduction to Metamathematics.* The University Series in Higher Mathematics. van Nostrand.

Kleene, S. C. 1959a. Countable functionals. Pages 81–100 of: *Constructivity in Mathematics*, Heyting, A. (ed). Studies in Logic and the Foundations of Mathematics. North-Holland.

Kleene, S. C. 1959b. Recursive functionals and quantifiers of finite types. I. *Transactions of the American Mathematical Society*, **91**, 1–52.

Kleene, S. C. 1975. Reminiscences of logicians. Reported by J. N. Crossley. Pages 1–62 of: *Algebra and Logic.* Lecture Notes in Mathematics, vol. 450. Springer.

Klein, G., Elphinstone, K., Heiser, G., Andronick, J., Cock, D., Derrin, P., Elkaduwe, D., Engelhardt, K., Kolanski, R., Norrish, M., Sewell, T., Tuch, H., and Winwood, S. 2009. seL4: formal verification of an OS kernel. Pages 207–220 of: *ACM Symposium on Principles of Operating Systems*, Matthews, J.N., and Anderson, Th. (eds). Big Sky.

Klop, J. W. 1980. *Combinatory Reduction Systems.* Ph.D. thesis, Utrecht University.

Klop, J. W. 1992. Term Rewriting Systems. Pages 1–116 of: *Handbook of Logic in Computer Science*, Abramsky, S., Gabbay, D. M., and Maibaum, T. S. E. (eds). Oxford University Press.

Klop, J. W., and de Vrijer, R. C. 1989. Unique normal forms for lambda calculus with surjective pairing. *Information and Computation*, **80**(2), 97–113.

Klop, J.-W., Oostrom, V. van, and Raamsdonk, F. van. 1993. Combinatory reduction systems: introduction and survey. *Theoretical Computer Science*, **121**, 279–308.

Koopman, P., and Plasmeijer, M. J. 1999. Efficient combinator parsers. Pages 120–136 of: *Implementation of Functional Languages*, Hammond, K., Davie, A. J. T., and Clack, C. (eds). Lecture Notes in Computer Science, vol. 1595. Springer.

Koopman, P., and Plasmeijer, M. J. 2006. Fully automatic testing with functions as specifications. Pages 35–61 of: *Selected Lectures of the 1st Central European Functional Programming School*. Lecture Notes in Computer Science, vol. 4164. Springer.

Koster, C. H. A. 1969. On infinite modes. *ALGOL Bulletin*, **30**, 86–89.

Koymans, C. P. J. 1982. Models of the lambda calculus. *Information and Control*, **52**(3), 306–323.

Kozen, D. 1977. Complexity of finitely presented algebras. Pages 164–177 of: *ACM Symposium on Theory of Computing*. ACM Press.

Kozen, D. 1997. *Automata and Computability*. Springer.

Kozen, D., Palsberg, J., and Schwartzbach, M. I. 1995. Efficient recursive subtyping. *Mathematical Structures in Computer Science*, **5**(1), 113–125.

Kreisel, G. 1959. Interpretation of analysis by means of constructive functionals of finite types. Pages 101–128 of: *Constructivity in Mathematics*, Heyting, A. (ed). North-Holland.

Kripke, S. A. 1965. Semantical analysis of intuitionistic logic I. Pages 92–130 of: *Formal Systems and Recursive Functions*, Crossley, J. N. and Dummett, M. (eds). North-Holland.

Krivine, J.-L. 1990. *Lambda-Calcul Types et Modèles*. Masson. English translation Krivine (1993).

Krivine, J.-L. 1993. *Lambda-Calculus, Types and Models*. Ellis Horwood. Translated from the 1990 French original by René Cori.

Kurata, T., and Takahashi, M. 1995. Decidable properties of intersection type systems. Pages 297–311 of: *Typed Lambda Calculi and Applications*. Lecture Notes in Computer Science, vol. 902. Springer.

Kurtonina, N. 1995. *Frames and Labels. A Modal Analysis of Categorial Inference*. Ph.D. thesis, OTS Utrecht, ILLC Amsterdam.

Kurtonina, N., and Moortgat, M. 1997. Structural control. Pages 75–113 of: *Specifying Syntactic Structures*, Blackburn, Patrick, and de Rijke, Maarten (eds). Center for the Study of Language and Information, Stanford.

Kuśmierek, D. 2007. The inhabitation problem for rank two intersection types. Pages 240–254 of: *Typed Lambda Calculi and Applications*. Lecture Notes in Computer Science, vol. 4583.

Lambek, J. 1958. The mathematics of sentence structure. *American Mathematical Monthly*, **65**, 154–170. Also in Buszkowski et al. (1988).

Lambek, J. 1961. On the calculus of syntactic types. Pages 166–178 of: *Structure of Language and its Mathematical Aspects*, Jacobson, R. (ed). Proceedings of the Symposia in Applied Mathematics, vol. XII. American Mathematical Society.

Lambek, J. 1980. From λ-calculus to Cartesian closed categories. Pages 375–402 of: *To H. B. Curry: Essays on Combinatory Logic, Lambda Calculus and Formalism Hindley* (Hindley and Seldin (1980)).

Lambek, J., and Scott, P. J. 1981. Intuitionist type theory and foundations. *Journal of Philosophical Logic*, **10**, 101–115.

Landau, E. 1960. *Grundlagen der Analysis*. 3rd edition. Chelsea Publishing Company.

Landin, P. J. 1964. The mechanical evaluation of expressions. *The Computer Journal*, **6**(4), 308–320.

Lax, P. D. 2002. *Functional Analysis*. Pure and Applied Mathematics. Wiley.

Lehmann, D. J., and Smyth, M. B. 1981. Algebraic specification of data types: a synthetic approach. *Mathematical Systems Theory*, **14**, 97–139.

Leivant, D. 1983a. Polymorphic Type Inference. Pages 88–98 of: *Symposium on Principles of Programming Languages*. ACM Press.

Leivant, D. 1983b. Reasoning About functional programs and complexity classes associated with type disciplines. Pages 460–469 of: *Symposium on Foundations of Computer Science*. IEEE Computer Society Press.

Leivant, D. 1986. Typing and computational properties of lambda expressions. *Theoretical Computer Science*, **44**(1), 51–68.

Leivant, D. 1990. Discrete polymorphism. Pages 288–297 of: *ACM Conference on LISP and Functional Programming*. ACM Press.

Leroy, X. 2009. A formally verified compiler back-end. *Journal of Automated Reasoning*, **43**(4), 363–446.

Lévy, J.-J. 1978. *Réductions Correctes et Optimales dans le Lambda-Calcul*. Ph.D. thesis, Université Paris VII.

Liquori, L., and Ronchi Della Rocca, S. 2007. Intersection types à la Church. *Information and Computation*, **205**(9), 1371–1386.

Loader, R. 1997. *An algorithm for the minimal model*. Unpublished manuscript, obtainable from `<homepages.ihug.co.nz/~suckfish/papers/papers.html>`.

Loader, R. 2001a. Finitary PCF is not decidable. *Theoretical Computer Science*, **266**(1–2), 341–364.

Loader, R. 2001b. The undecidability of lambda definability. Pages 331–342 of: *Church Memorial Volume: Logic, Language, and Computation* (Zeleny and Anderson (2001)).

Loader, R. 2003. Higher order β matching is undecidable. *Logic Journal of the Interest Group in Pure and Applied Logics*, **11**(1), 51–68.

Longo, G. 1988. The lambda-calculus: connections to higher type recursion theory, proof theory, category theory. *Annals of Pure and Applied Logic*, **40**, 93–133.

Lopez-Escobar, E. G. K. 1983. Proof functional connectives. Pages 208–221 of: *Methods in Mathematical Logic*. Lecture Notes in Mathematics, vol. 1130.

Luo, Z., and Pollack, R. 1992. *The LEGO Proof Development System: A User's Manual*. Tech. rept. ECS-LFCS-92-211. University of Edinburgh.

MacQueen, D., Plotkin, G. D., and Sethi, R. 1986. An ideal model for recursive polymorphic types. *Information and Control*, **71**((1/2)), 95–130.

Mairson, H. G. 1992. A simple proof of a theorem of Statman. *Theoretical Computer Science*, **103**(2), 387–394.

Makanin, G. S. 1977. The problem of solvability of equations in a free semigroup. *Mathematics of the USSR-Sbornik*, **32**(2), 129–198.

Martin-Löf, P. 1984. *Intuitionistic Type Theory.* Studies in Proof Theory. Bibliopolis.

Marz, M. 1999. An algebraic view on recursive types. *Applied Categorical Structures*, **7**(1–2), 147–157.

Matiyasevič, Y. V. 1972. Diophantine representation of recursively enumerable predicates. *Mathematical Notes*, **12**(1), 501–504.

Matiyasevič, Y. V. 1993. *Hilbert's Tenth Problem.* Foundations of Computing Series. MIT Press.

Mayr, R., and Nipkow, T. 1998. Higher-order rewrite systems and their confluence. *Theoretical Computer Science*, **192**(1), 3–29.

McCarthy, J. 1963. A basis for a mathematical theory of computation. Pages 33–70 of: *Computer Programming and Formal Systems*, Braffort, P., and Hirschberg, D. (eds). North-Holland.

McCarthy, J., Abrahams, P.W., Edwards, D. J., Hart, T. P., and Levin, M. I. 1962. *LISP 1.5 Programmer's Manual.* MIT Press.

McCracken, N. J. 1979. *An investigation of a programming language with a polymorphic type structure.* Ph.D. thesis, Syracuse University.

Melliès, P.-A. 1996. *Description Abstraite des Systèmes de Réécriture.* Ph.D. thesis, Université de Paris.

Mendler, N. P. 1987. *Inductive Definitions in Type Theory.* Ph.D. thesis, Department of Computer Science, Cornell University.

Mendler, N. P. 1991. Inductive types and type constraints in the second-order lambda calculus. *Annals of Pure and Applied Logic*, **51**, 159–172.

Meyer, A. R. 1982. What is a model of the lambda calculus? *Information and Control*, **52**(1), 87–122.

Milner, R. 1978. A theory of type polymorphism in programming. *Journal of Computer and System Sciences*, **17**, 348–375.

Milner, R., Tofte, M., Harper, R., and McQueen, D. 1997. *The Definition of Standard ML.* The MIT Press.

Mints, G. E. 1989. The completeness of provable realizability. *Notre Dame Journal Formal Logic*, **30**, 420–441.

Mints, G. E. 1996. Normal forms for sequent derivations. Pages 469–492 of: *Kreiseliana. About and Around Georg Kreisel*, Odifreddi, P. (ed). A.K. Peters.

Mitchell, J. 1996. *Foundation for Programmimg Languages.* MIT Press.

Mitschke, G. 1976. *λ-Kalkül, δ-Konversion und axiomatische Rekursionstheorie.* Tech. rept. Preprint 274. Technische Hochschule, Darmstadt.

Mogensen, T. Æ. 1992. Efficient self-interpretation in lambda calculus. *Journal of Functional Programming*, **2**(3), 345–364.

Moller, F., and Smolka, S. A. 1995. On the computational complexity of bisimulation. *ACM Computing Surveys*, **27**(2), 287–289.

Montague, R. 1973. The proper treatment of quantification in ordinary English. In: Hintikka, J., Moravcsik, J. M. E., and Suppes, P. (eds), *Approaches to Natural Language.* Dordrecht.

Moot, R. 2002. Proof nets for linguistic analysis. Pages 65–72 of: *Proceedings of TAG+9, The 9th International Workshop on Tree Adjoining Grammars and Related Formalisms, Tübingen*, Gardent, Claire and Sarkar, A. (eds).

Moot, R. 2008. *Lambek Grammars, Tree Adjoining Grammars and Hyperedge Replacement Grammars.* Ph.D. thesis, Utrecht University.

Moot, R., and Retoré, C. 2012. *The Logic of Categorial Grammars.* Lecture Notes in Computer Science, vol. 6850. Springer.

Morrill, G. 1994. *Type Logical Grammar.* Kluwer.

Morris, J. H. 1968. *Lambda-calculus Models of Programming Languages.* Ph.D. thesis, Massachusetts Institute of Technology.

Muzalewski, M. 1993. *An Outline of PC Mizar.* Brussels: Fondation Philippe le Hodey.

Nadathur, G., and Miller, D. 1988. An overview of λProlog. Pages 810–827 of: *Logic Programming Conference.* MIT Press.

Nederpelt, R. P. 1973. *Strong Normalisation in a Typed Lambda Calculus with Lambda Structured Types.* Ph.D. thesis, Eindhoven University.

Nederpelt, R. P., Geuvers, J. H., and de Vrijer, R. C. (eds). 1994. *Selected Papers on Automath.* Studies in Logic and the Foundations of Mathematics, no. 133. North-Holland.

Nerode, A., Odifreddi, P., and Platek, R. In preparation. *The Four Noble Truths of Logic.* To appear.

Nikhil, R. S. 2008. Bluespec: a general-purpose approach to high-Level synthesis based on parallel atomic transactions. Pages 129–146 of: *High-Level Synthesis*, Coussy, Philippe and Morawiec, Adam (eds). Springer.

Nipkow, T., Paulson, L. C., and Wenzel, M. 2002. *Isabelle/HOL. A Proof Assistant for Higher-Order Logic.* Lecture Notes in Computer Science, vol. 2283. Springer.

Oostdijk, M. 1996. *Proof by Calculation.* M.Phil. thesis, Nijmegen University.

van Oostrom, V. 2007. *α-free-μ.* Unpublished manuscript, `Vincent.vanOostrom@phil.uu.nl`.

Padovani, V. 1996. *Filtrage d'ordre Supérieur.* Ph.D. thesis, Université de Paris VII.

Padovani, V. 2000. Decidability of fourth order matching. *Mathematical Structures in Computer Science*, **3**(10), 361–372.

Pair, C. 1970. Concerning the syntax of ALGOL 68. *ALGOL Bulletin*, **31**, 16–27.

Parigot, M. 1992. $\lambda\mu$-calculus: an algorithmic interpretation of classical natural deduction. Pages 190–201 of: *Logic Programming and Automated Reasoning. Proceedings of the International Conference LPAR'92, St. Petersburg*, Voronkov, A. (ed). Lecture Notes in Computer Science, vol. 624. Springer.

Parikh, R. 1973. On the length of proofs. *Transactions of the American Mathematical Society*, **177**, 29–36.

Park, D. 1976. *The Y combinator in Scott's lambda calculus models.* Theory of Computation Report 13. University of Warwick, Department of Computer Science.

Pentus, M. 1993. Lambek grammars are context free. Pages 429–433 of: *Logic in Computer Science.* IEEE Computer Society Press.

Pentus, M. 2006. Lambek calculus is NP-complete. *Theoretical Computer Science*, **357**(1–3), 186–201.

Péter, R. 1967. *Recursive Functions.* 3rd revised edition. Academic Press.

Peyton-Jones, S. 1987. *The Implementation of Functional Programming Languages.* Prentice Hall.

Peyton Jones, S. (ed). 2003. *Haskell 98 Language and Libraries: the Revised Report.* Cambridge University Press.

Peyton Jones, S. L., and Wadler, P. 1993. Imperative functional programming. Pages 71–84 of: *Principles of Programming Languages.* ACM Press.

Peyton Jones, S. and Hughes, J. [Editors]; Augustsson, L., Barton, D., Boutel, B., Burton, W., Fraser, S., Fasel, J., Hammond, K., Hinze, R., Hudak, P., Johnsson, T., Jones, M., Launchbury, J., Meijer, E., Peterson, J., Reid, A., Runciman, C., and Wadler, P. 1999. *Haskell 98 — A Non-strict, Purely Functional Language.* URL <www.haskell.org/definition/>.

Peyton Jones, S., Vytiniotis, D., Weirich, S., and Washburn, G. 2006. Simple unification-based type inference for GADTs. Pages 50–61 of: *International Conference on Functional Programming*, ACM Press..

Pierce, B. C. 2002. *Types and Programming Languages.* MIT Press.

Pil, M. R. C. 1999. Dynamic types and type dependent functions. Pages 169–185 of: *Implementation of Functional Languages*, Hammond, K., Davie, A. J. T., and Clack, C. (eds). Lecture Notes in Computer Science, vol. 1595. Springer.

Pimentel, E., Ronchi Della Rocca, S., and Roversi, L. 2012. Intersection types from a proof-theoretic perspective. *Fundamenta Informaticae*, **121**(1–4), 253–274.

Plasmeijer, M. J., and van Eekelen, M. 2002. *Concurrent Clean language report (version 2.1).* <www.cs.ru.nl/~clean>.

Plasmeijer, R., Achten, P., and Koopman, P. 2007. iTasks: Executable specifications of interactive work flow systems for the Web. Pages 141–152 of: *International Conference on Functional Programming.* ACM Press.

Platek, R. A. 1966. *Foundations of Recursion Theory.* Ph.D. thesis, Stanford University.

Plotkin, G. D. 1975. Call-by-name, call-by-value and the λ-calculus. *Theoretical Computer Science*, **1**(2), 125–159.

Plotkin, G. D. 1977. LCF considered as a programming language. *Theoretical Computer Science*, **5**, 225–255.

Plotkin, G. D. 1980. Lambda-definability in the full type hierarchy. Pages 363–373 of: *To H. B. Curry: Essays on Combinatory Logic, Lambda Calculus and Formalism* (Hindley and Seldin (1980)).

Plotkin, G. D. 1982. *The category of complete partial orders: a tool for making meanings.* Postgraduate lecture notes, Edinburgh University.

Plotkin, G. D. 1985. *Lectures on Predomains and Partial Functions.* Center for the Study of Language and Information, Stanford.

Plotkin, G. D. 1993. Set-theoretical and other elementary models of the λ-calculus. *Theoretical Computer Science*, **121**(1-2), 351–409.

Poincaré, H. 1902. *La Science et l'Hypothèse.* Flammarion.

Polonsky, A. 2011. *Proofs, Types and Lambda Calculus.* Ph.D. thesis, University of Bergen, Norway.

Post, E. 1947. Recursive unsolvability of a problem of Thue. *The Journal of Symbolic Logic*, **12**(1), 1–11.

Pottinger, G. 1977. Normalization as a homomorphic image of cut-elimination. *Annals of Mathematical Logic*, **12**, 323–357.

Pottinger, G. 1980. A type assignment for the strongly normalizable λ-terms. Pages 561–77 of: *To H. B. Curry: Essays on Combinatory Logic, Lambda Calculus and Formalism* (Hindley and Seldin (1980)).

Pottinger, G. 1981. The Church–Rosser theorem for the typed λ-calculus with surjective pairing. *Notre Dame Journal of Formal Logic*, **22**(3), 264–268.

Prawitz, D. 1965. *Natural Deduction.* Almqvist & Wiksell.

Prawitz, D. 1971. Ideas and results in proof theory. Pages 235–307 of: *Scandinavian Logic Symposium*, Fenstad, J. E. (ed). North-Holland.

van Raamsdonk, F. 1996. *Confluence and Normalisation for Higher-Order Rewriting.* Ph.D. thesis, Vrije Universiteit.

van Raamsdonk, F., Severi, P., Sørensen, M.H., and Xi, H. 1999. Perpetual reductions in lambda calculus. *Information and Computation*, **149**(2), 173–225.

Reynolds, J. C. 1972. Definitional interpreters for higher-order programming languages. Pages 717–740 of: *ACM National Conference.* ACM Press.

Reynolds, J. C. 1993. The discoveries of continuations. *LISP and Symbolic Computation*, **6**(3/4), 233–247.

Robertson, N., Sanders, D., Seymour, P., and Thomas, R. 1997. The four-colour theorem. *Journal of Combinatorial Theory. Series B*, **70**(1), 2–44.

Robinson, J. A. 1965. A machine-oriented logic based on the resolution principle. *Journal of the ACM*, **12**(1), 23–41.

Rogers Jr., H. 1967. *Theory of Recursive Functions and Effective Computability.* McGraw-Hill.

Ronchi Della Rocca, S. 1988. *Lecture Notes on Semantics and Types.* Tech. rept. Torino University.

Ronchi Della Rocca, S. 2002. Intersection typed lambda-calculus. Pages 163–181 of: *Intersection Types and Related Systems.* Electronic Notes in Theoretical Computer Science, **70**, no. 1. Elsevier.

Ronchi Della Rocca, S., and Paolini, L. 2004. *The Parametric Lambda Calculus: A Metamodel for Computation.* Texts in Theoretical Computer Science, vol. XIII. Springer.

Rosser, J. B. 1984. Highlights of the history of the lambda-calculus. *Annals of the History of Computing*, **6**(4), 337–349.

Rutten, J. J. M. M. 2000. Universal coalgebra: a theory of systems. *Theoretical Computer Science*, **249**(1), 3–80.

Rutten, J. J. M. M. 2005. A coinductive calculus of streams. *Mathematical Structures in Computer Science*, **15**(1), 93–147.

Salvati, S. 2009. Recognizability in the simply typed lambda-calculus. Pages 48–60 of: *Workshop on Logic, Language, Information and Computation.* Lecture Notes in Computer Science, vol. 5514. Springer.

Salvati, S., Manzonetto, G., Gehrke, M., and Barendregt, H.P. 2012. Loader and Urzyczyn are logically related. Pages 364–376 of: *Automata, Languages, and Programming – 39th International Colloquium, ICALP 2012*, Czumaj, A.,

Mehlhorn, K., Pitts, A.M., and Wattenhofer, R. (eds). LNCS, vol. 7392. Springer.

Schmidt-Schauß, M. 1999. Decidability of behavioural equivalence in unary PCF. *Theoretical Computer Science*, **216**(1-2), 363–373.

Schrijvers, T., Jones, S. L. Peyton, Sulzmann, M., and Vytiniotis, D. 2009. Complete and decidable type inference for GADTs. Pages 341–352 of: *Proceedings of the 14th ACM SIGPLAN International Conference on Functional Programming, ICFP 2009, Edinburgh, Scotland, UK, August 31 – September 2, 2009*, Hutton, G., and Tolmach, A. P. (eds). ACM.

Schubert, A. 1998. Second-order unification and type inference for Church-style polymorphism. Pages 279–288 of: *Symposium on Principles of Programming Languages*.

Schwichtenberg, H. 1975. Elimination of higher type levels in definitions of primitive recursive functionals by means of transfinite recursion. Pages 279–303 of: *Logic Colloquium*. Studies in Logic and the Foundations of Mathematics, vol. 80. North-Holland.

Schwichtenberg, H. 1976. Definierbare Funktionen im λ-Kalkül mit Typen. *Archief für Mathematische Logik*, **25**, 113–114.

Schwichtenberg, H. 1999. Termination of permutative conversion inintuitionistic Gentzen calculi. *Theoretical Computer Science*, **212**(1-2), 247–260.

Schwichtenberg, H., and Berger, U. 1991. An inverse of the evaluation functional for typed λ-calculus. Pages 203–211 of: *Logic in Computer Science*. IEEE Computer Society Press.

Scott, D. S. 1970. Constructive validity. Pages 237–275 of: *Symposium on Automated Demonstration*. Lecture Notes in Mathematics, vol. 125. Springer.

Scott, D. S. 1972. Continuous lattices. Pages 97–136 of: *Toposes, Algebraic Geometry and Logic*. Lecture Notes in Mathematics, vol. 274. Springer.

Scott, D. S. 1975a. Open problem 4. Page 369 of: *λ-calculus and Computer Science Theory* (Böhm (1975)).

Scott, D. S. 1975b. Some philosophical issues concerning theories of combinators. Pages 346–366 of: *λ-calculus and Computer Science Theory* (Böhm (1975)).

Scott, D. S. 1976. Data types as lattices. *SIAM Journal on Computing*, **5**, 522–587.

Scott, D. S. 1980. Relating theories of the λ-calculus. Pages 403–450 of: *To H. B. Curry: Essays on Combinatory Logic, Lambda Calculus and Formalism* (Hindley and Seldin (1980)).

Severi, P., and Poll, E. 1994. Pure type systems with definitions. Pages 316–328 of: *Logical Foundations of Computer Science*. Lecture Notes in Computer Science, vol. 813. Springer.

Sheeran, M. 2005. Hardware design and functional programming: a perfect match. *Journal of Universal Computer Systems*, **11**(7), 1135–1158.

Smyth, M. B., and Plotkin, G. D. 1982. The category-theoretic solution of recursive domain equations. *SIAM Journal on Computing*, **11**(4), 761–783.

Sørensen, M. H. 1997. Strong normalization from weak normalization in typed λ-calculi. *Information and Computation*, **133**, 35–71.

Sørensen, M. H., and Urzyczyn, P. 2006. *Lectures on the Curry-Howard Isomorphism*. Elsevier.

Spector, C. 1962. Provable recursive functionals of analysis. Pages 1–27 of: *Recursive Function Theory*, Dekker, J. C. E. (ed). American Mathematical Society.

Statman, R. 1979. The typed λ-calculus is not elementary recursive. *Theoret. Comput. Sci.*, **9**(1), 73–81.

Statman, R. 1979a. Intuitionistic propositional logic is polynomial-space complete. *Theoret. Comput. Sci.*, **9**(1), 67–72.

Statman, R. 1980a. On the existence of closed terms in the typed λ-calculus. I. Pages 511–534 of: *To H. B. Curry: Essays on Combinatory Logic, Lambda Calculus and Formalism* (Hindley and Seldin (1980)).

Statman, R. 1980b. On the existence of closed terms in the typed λ-calculus. III. Dept. of Mathematics, CMU, Pittsburgh, USA.

Statman, R. 1981. On the existence of closed terms in the typed λ calculus. II. Transformations of unification problems. *Theoret. Comput. Sci.*, **15**(3), 329–338.

Statman, R. 1982. Completeness, invariance and λ-definability. *The Journal of Symbolic Logic*, **47**(1), 17–26.

Statman, R. 1985. Equality between functionals revisited. Pages 331–338 of: *Harvey Friedman's Research on the Foundations of Mathematics* (Harrington et al. (1985)). Amsterdam: North-Holland.

Statman, R. 1994. *Recursive types and the subject reduction theorem.* Technical Report 94–164. Carnegie Mellon University.

Statman, R. 2000. Church's lambda delta calculus. Pages 293–307 of: *Logic for Programming and Automated Reasoning (Reunion Island, 2000).* Lecture Notes in Comput. Sci., vol. 1955. Springer.

Statman, R. 2007. On the complexity of α-conversion. *The Journal of Symbolic Logic*, **72**(4), 1197–1203.

Steedman, M. 2000. *The Syntactic Process.* MIT Press.

Steele Jr., G. L. 1978. *RABBIT: A Compiler for SCHEME.* Tech. rept. AI-TR-474. Artificial Intelligence Laboratory, MIT.

Steele Jr., G. L. 1984. *Common Lisp: The Language.* Digital Press.

Stenlund, S. 1972. *Combinators, λ-Terms and Proof Theory.* Synthese Library. D. Reidel.

Stirling, C. 2009. Decidability of higher-order matching. *Logical Methods in Computer Science*, **5**(3), 1–52.

Støvring, K. 2006. Extending the extensional lambda calculus with surjective pairing is conservative. *Logical Methods in Computer Science*, **2**(2:1), 1–14.

Sudan, G. 1927. Sur le nombre transfini ω^ω. *Bulletin mathématique de la Société Roumaine des Sciences*, **30**, 11–30.

Szabo, M. E. (ed). 1969. *Collected Papers of Gerhard Gentzen.* North-Holland.

Tait, W. W. 1965. Infinitely long terms of transfinite type I. Pages 176–185 of: *Formal Systems and Recursive Functions*, Crossley, J., and Dummett, M. (eds). North-Holland.

Tait, W. W. 1967. Intensional interpretations of functionals of finite type. I. *The Journal of Symbolic Logic*, **32**, 198–212.

Tait, W. W. 1971. Normal form theorem for barrecursive functions of finite type. Pages 353–367 of: *Scandinavian Logic Symposium.* North-Holland.

Tatsuta, M., and Dezani-Ciancaglini, M. 2006. Normalisation is insensible to lambda-term identity or difference. Pages 327–336 of: *Logic in Computer Science.* IEEE Computer Society Press.

Terese. 2003. *Term Rewriting Systems.* Cambridge University Press.

Terlouw, J. 1982. On definition trees of ordinal recursive functionals: reduction of the recursion orders by means of type level raising. *The Journal of Symbolic Logic*, **47**(2), 395–402.

Thatcher, J. W. 1973. Tree automata: an informal survey. Chap. 4, pages 143–172 of: *Currents in the Theory of Computing*, Aho, A.V. (ed). Prentice-Hall.

Thompson, S. 1995. *Miranda, The Craft of Functional Programming.* Addison-Wesley.

Tiede, H. J. 2001. Lambek calculus proofs and tree automata. Pages 251–265 of: *Logical Aspects of Computational Linguistics.* Lecture Notes in Artificial Intelligence, vol. 2014. Springer.

Tiede, H. J. 2002. Proof tree automata. Pages 143–162 of: *Words, Proofs, and Diagrams*, Barker-Plummer, D., Beaver, D., van Benthem, J. F. A. K., and Scotto di Luzio, P. (eds). Center for the Study of Language and Information, Stanford.

Troelstra, A. S. 1973. *Metamathematical Investigation of Intuitionistic Arithmetic and Analysis.* Lecture Notes in Mathematics, no. 344. Springer.

Troelstra, A. S. 1999. Marginalia on sequent calculi. *Studia Logica*, **62**(2), 291–303.

Troelstra, A. S., and Schwichtenberg, H. 1996. *Basic Proof Theory.* Cambridge University Press.

Turner, D. A. 1976. *The SASL language manual.* `http://www.eis.mdx.ac.uk/staffpages/dat/saslman.pdf`.

Turner, D. A. 1979. A new implementation technique for applicative languages. *Software – Practice and Experience*, **9**, 31–49.

Turner, D. A. 1981. The semantic elegance of functional languages. Pages 85–92 of: *Functional Programming Languages and Computer Architecture.* ACM Press.

Turner, D. A. 1985. Miranda, a non-strict functional language with polymorphic types. Pages 1–16 of: *Functional Programming Languages and Computer Architecture.* Lecture Notes in Computer Science, vol. 201. Springer.

Urban, C., and Tasson, C. 2005. Nominal techniques in Isabelle/HOL. Pages 38–53 of: *Automated deduction – CADE-20.* Lecture Notes in Computer Science, vol. 3632. Springer.

Urban, C., Berghofer, S., and Norrish, M. 2007. Barendregt's variable convention in rule inductions. Pages 35–50 of: *Proc. of the 21st International Conference on Automated Deduction (CADE).* LNAI, vol. 4603.

Urzyczyn, P. 1999. The emptiness problem for intersection types. *The Journal of Symbolic Logic*, **64**(3), 1195–1215.

Urzyczyn, P. 2009. Inhabitation of low-rank intersection types. Pages 356–370 of: *Typed Lambda Calculi and Applications.* Lecture Notes in Computer Science, vol. 5608. Springer.

Venneri, B. 1994. Intersection types as logical formulae. *Journal of Logic and Computation*, **4**(2), 109–124.

Venneri, B. 1996. *Private Communication.* Florence University.

Vermaat, W. 2006. *The Logic of Variation. A Cross-Linguistic Account of wh-Question Formation.* Ph.D. thesis, Utrecht University.

Vogel, H. 1976. Ein starker Normalisationssatz für die barrekursiven Funktionale. *Archiv für Mathematische Logik und Grundlagenforschung*, **18**, 81–84.

Voigtländer, J. 2009. Free theorems involving type constructor classes: functional pearl. Pages 173–184 of: *ACM SIGPLAN International Conference on Functional Programming.* ACM Press.

Vouillon, J., and Melliès, P.-A. 2004. Semantic types: a fresh look at the ideal model for types. Pages 24–38 of: *Principles of Programming Languages.* ACM Press.

de Vrijer, R. C. 1987a. *Surjective Pairing and Strong Normalization: Two Themes in Lambda Calculus.* Ph.D. thesis, University of Amsterdam.

de Vrijer, R.C. 1987b. Exactly estimating functionals and strong normalization *Indagationes Mathematicae*, **49**, 479–493.

de Vrijer, R. C. 1989. Extending the lambda calculus with surjective pairing is conservative. Pages 204–215 of: *Logic in Computer Science.* IEEE Computer Society Press.

Wadsworth, C. P. 1971. *Semantics and Pragmatics of the Lambda-Calculus.* Ph.D. thesis, University of Oxford.

Wadsworth, C. P. 1976. The relation between computational and denotational properties for Scott's D_∞-models of the lambda-calculus. *SIAM Journal of Computing*, **5**(3), 488–521.

Wand, M. 1987. A simple algorithm and proof for type inference. *Fundamenta Informaticae*, **10**, 115–122.

Wansing, H. 2002. Sequent systems for modal logics. Pages 61–145 of: *Handbook of Philosophical Logic*, vol. 8, Gabbay, D., and Guenthner, F. (eds). Kluwer.

Wells, J. B. 1999. Typability and type checking in system F are equivalent and undecidable. *Annals of Pure and Applied Logic*, **98**(1-3), 111–156.

Wells, J. B., Dimock, A., Muller, R., and Turbak, F. 1997. A typed intermediate language for flow-directed compilation. Pages 757–771 of: *Theory and Practice of Software Development.* Lecture Notes in Computer Science, no. 1214.

Whitehead, A.N., and Russell, B. 1910–1913. *Principia Mathematica.* Cambridge University Press.

Wiedijk, F. 2006. *The Seventeen Provers of the World.* Lecture Notes in Computer Science, vol. 3600. Springer.

van Wijngaarden, A. 1981. Revised report of the algorithmic language Algol 68. *ALGOL Bulletin*, 1–119.

Xi, H. 1997. Weak and strong beta normalisations in typed λ-calculi. Pages 390–404 of: *Typed Lambda Calculi and Applications.* Lecture Notes in Computer Science, vol. 1210. Springer.

Zeleny, M., and Anderson, C.A. (eds). 2001. *Church Memorial Volume: Logic, Language, and Computation.* Kluwer.

Zucker, J. 1974. Cut-elimination and normalization. *Annals of Mathematical Logic*, **7**, 1–112.

Zylberajch, C. 1991. *Syntaxe et Semantique de la Facilité en Lambda-Calcul.* Ph.D. thesis, Université de Paris VII.

Indices

There is an index of terms, of citations (authors of cited references), and of symbols. The index of symbols is subdivided as follows:

Terms general, operations, classes, relations, theories;
Types/Propositions general, operations, classes, relations, theories, classes of theories;
Assignment theories, derivation;
Models general, operations on, classes, relations, interpretations in;
Miscellaneous general, operations, classes, relations;
Categories general, functors.

The meaning of the main division is as follows. There are objects and relations between these. Using the operations one constructs new objects. By collecting objects one obtains classes, by collecting valid relations one obtains theories. Categories consist of classes of objects together with operations.

Index of terms

Index of Citations

Index of symbols

QM LIBRARY
(MILE END)